Regional Climate Studies

Series editors

Hans-Jürgen Bolle†, München, Germany
Massimo Menenti, GA Delft, The Netherlands
S. Ichtiaque Rasool, Bethesda, USA

The North Sea region as seen from satellite on 16 April 2003. *Sensor* Terra-Modis. *Credit* Jacques Descloitres, MODIS Rapid Response Team, NASA/GSFC; Saharan dust signals have been removed

Markus Quante · Franciscus Colijn
Editors

North Sea Region Climate Change Assessment

Editors
Markus Quante
Institute of Coastal Research
Helmholtz-Zentrum Geesthacht
Geesthacht
Germany

Franciscus Colijn
Institute of Coastal Research
Helmholtz-Zentrum Geesthacht
Geesthacht
Germany

DOI 10.1007/978-3-319-39745-0

Library of Congress Control Number: 2016941308

Foreword

Climate change is a major threat for the 21st century and beyond as recognised by the world's governments who have funded the five assessments of the Intergovernmental Panel on Climate Change (IPCC) and numerous special reports since the 1980s. These efforts have been important in supporting global climate policy, culminating in the recent Paris Agreement on reducing future greenhouse gas emissions. In contrast, adaptation happens on smaller scales than climate mitigation and very different and more detailed information is required to support such decisions. A number of regional and local assessments have been produced with these issues in mind. As examples, in North America there have been several national assessments of the implications of climate change as well as city level studies such as for New York. In Europe there have been assessments at the EU scale such as the ACACIA and CLIMSAVE projects, the Baltic Sea region, national assessments such as the Delta Commission in the Netherlands, and city assessments such as for London and Hamburg.

The international North Sea Region Climate Change Assessment (NOSCCA) contributes substantial new insight into these efforts for the greater North Sea Region, constituting the first such assessment for this region. While North Sea societies have always faced climatic risk, the challenges are growing due to human-induced climate change mainly forced by enhanced greenhouse gas emissions, and often other significant non-climate drivers are in operation. At the same time, the available knowledge of climate change and its implications has expanded impressively over the past few decades. However, there is a challenge to synthesise and communicate this information in accessible and useful forms. The present assessment rises to these challenges to provide science-based information on climate change on the scale of adaptation decision-makers.

The independent and voluntary assessment team come from across the North Sea region. The component chapters have all been subject to extensive peer review and modification to promote wide and inclusive perspectives. Collectively, the chapters address a range of issues embracing climate science, ecosystems and socio-economics providing a unique integrated perspective which can support decision-makers and policy development. The authorship and editorship team are to be commended for their supreme efforts, establishing a platform for further assessments and updates as needed. The approach is readily transferable and might be transferred to other interested regions.

Robert J. Nicholls
Professor of Coastal Engineering
Review Editor for the North Sea Region Climate Change Assessment (NOSCCA)
University of Southampton, UK

Preface

Climate change impacts show wide regional variability; their strength, nature and evolution depending on the principal features of the area in which they are occurring. To cope responsibly with its impacts, decision-makers and authorities need sound information on the specifics of climate change in their region. The science community would also benefit from a comprehensive analysis of the state-of-knowledge on regional climate change and its effects.

The North Sea region is a precious natural and cultural environment and a major economic entity within Europe. The North Sea is one of the world's richest fishing grounds as well as being one of the busiest seas with respect to marine traffic and its related infrastructure, oil and gas extraction is also of high economic value. More recently the area has become a major site for wind energy, with many large offshore wind farms. Climate change impacts are expected to have profound effects on North Sea ecosystems and economic development. Despite its importance, until now a comprehensive analysis of climate change and its impacts for the region as a whole had not been attempted. Some nationally-focused studies with an emphasis on climate change projections have been published in recent years, such as the UK Climate Projections—Marine and Coastal Projections and the KNMI'14 Climate Scenarios for The Netherlands to name but two examples,[1] and these have all been considered in the present study.

A few years ago, inspired by our colleague Hans von Storch, we initiated an international climate change assessment of the North Sea region. We adopted a similar approach to that successfully employed for reviews of knowledge on climate change in the Baltic Sea basin, published in 2008 and 2015.[2] This activity was named the North Sea Region Climate Change Assessment—NOSCCA—and has involved around 200 climate scientists in different research areas from all countries around the North Sea, as well as a few from more distant localities. NOSCCA developed into an independent international initiative, with all scientists involved contributing their time and effort on a voluntary basis as there was no extra funding available.

Present knowledge of climate change in the North Sea region has been evaluated mainly using peer-reviewed publications on climate change in the physical systems and its effects on land and marine systems. Two types of impact studies were envisaged: those concerning specific ecosystems and those related to specific human activities causing degradation of the environment.

After an introductory chapter on the North Sea region and its characteristics in terms of geography, geology, hydrography, present-day climate and ecology, Part I describes the climate change experienced over the past 200 years, described separately in each of three chapters on the atmosphere, the North Sea and river flow. Part II examines projections of

[1]Lowe, JA, Howard TP, Pardaens A, Tinker J, Holt J, Wakelin S, Milne G, Leake J, Wol J, Horsburgh K, Reeder T, Jenkins G, Ridley J, Dye S, Bradley S. (2009) UK Climate Projections science report: Marine and coastal projections. Met Office Hadley Centre, Exeter, UK; KNMI (2015): KNMI'14 climate scenarios for the Netherlands; A guide for professionals in climate adaptation, KNMI, De Bilt, The Netherlands, 34 pp.

[2]The BACC Author Team (2008), Assessment of Climate Change for the Baltic Sea Basin. Regional Climate Studies, Springer-Verlag, 473pp; The BACC II Author Team (2015) Second Assessment of Climate Change for the Baltic Sea Basin, Regional Climate Studies, Springer, 501pp.

future climate with separate chapters on the atmosphere, the North Sea, and river flow and urban drainage. The impacts of recent and future climate change on marine, coastal, lake and terrestrial ecosystems are presented in Part III. The report concludes with a consideration of climate change impacts on socio-economic sectors, Part IV contains chapters on fisheries, agricultural systems, offshore activities related to the energy sector, urban climate, air quality, recreation, coastal protection and finally coastal management and governance. Important background information is presented in five annexes to the report. An overall summary containing key statements from the different chapters precedes the main body of the book.

Climate change and its impacts on ecosystems has received much attention for many years. However, assessing the impacts of climate change on natural systems is far from straightforward. Environmental impacts resulting from non-climate drivers often make it very difficult to clearly establish the specific effects of climate change, which are already hard to attribute due to the difficulties of discriminating between natural variability and human interventions and their potential interactions. As a result, for many of the topics addressed in this assessment, other drivers have also been discussed, especially those that may mask potential climate change signals. Strict detection and attribution has not been undertaken here, mainly due to the lack of relevant published work. This could be the subject of a follow-up activity.

This assessment is a joint effort of 35 Lead Authors and a large group of contributing authors, who were willing to share their knowledge on many different aspects of the North Sea region and to contribute to compiling the different chapters. The process has been overseen by an international Scientific Steering Committee; the members are listed in the section 'About NOSCCA'. A review phase involving a sovereign review editor and more than 60 external reviewers was crucial to establishing an independent and scientifically sound product. All authors worked without financial support for this book and were supported by their respective institutions. We are extremely grateful for their contributions. Authors and reviewers are acknowledged and listed by name on the following pages. The open access publication of this report was made possible by funds provided by various institutions, which are listed in the acknowledgements section.

We consider this assessment to be the most comprehensive study of climate change in the North Sea region to date. It is hoped that NOSCCA will be of use to decision-makers in the many countries surrounding the North Sea as well as to those who are responsible for planning and implementing climate change adaptation in the region. We hope this assessment will stimulate further monitoring and topical studies on climate change in this ecologically and economically important region of Europe and as a result will increase the effectiveness of decision-making at the local level.

Geesthacht, Germany

Markus Quante
Francisclus Colijn

Acknowledgements

Book Production, Meetings, and Talks

A comprehensive and thorough assessment of climate change in the North Sea region would not be possible without the tremendous effort of many experts analysing regional climate change and its impacts and compiling dedicated topical chapters or reviewing manuscripts. Therefore, our thanks go primarily to the lead authors and contributing authors, who through excellent teamwork have produced the most comprehensive assessment of climate change in the North Sea region to date. We also thank the many reviewers, whose work has been crucial in ensuring the high scientific standard of this assessment report. Lead authors, contributing authors and disclosed reviewers are listed by name and institution on the following pages.

The entire review process was defined and overseen by an independent review editor. We are extremely grateful to Prof. Robert J. Nicholls of the University of Southampton, UK, for taking on and so competently accomplishing this important function.

The NOSCCA initiative was advised and supported, throughout the entire process, by an international Scientific Steering Committee (SSC), whose contributions are greatly appreciated. The members of the SSC are introduced in the section 'About NOSCCA' of this front matter.

Working efficiently with a large group of experts from many different institutes and countries in Europe profits from face-to-face exchange. Therefore several meetings of NOSCCA lead authors and members of the SSC were held during the writing and revision phases. The support—including financial—of some of our colleagues is greatly appreciated. Monika Breuch-Moritz of the Federal Maritime and Hydrographic Agency (BSH), Germany, hosted the first meeting of the SSC in close proximity to Hamburg harbour. The initial gathering of the NOSCCA lead authors together with the members of the SSC took place in the Royal Netherlands Academy of Arts and Sciences in Amsterdam. This meeting was made possible by Hein J.W. de Baar from the Royal Netherlands Institute for Sea Research and the University of Groningen, The Netherlands. The second lead author meeting took place in the Carlsberg Academy in Copenhagen, Denmark, and was arranged by Eigil Kaas of the Niels Bohr Institute, Denmark. Jaap Kwadijk from Deltares, The Netherlands, arranged the third lead author meeting at the Deltares subsidiary in Delft. The final lead author meeting took place in the Chile House in Hamburg and was hosted by Daniela Jacob from the Climate Service Centre Germany of the Helmholtz-Zentrum Geesthacht.

The various topics of the envisaged climate change assessment were introduced during our meetings by invited keynote speakers. For their inspiring talks we thank Bas Amelung (Wageningen University, The Netherlands), Peter Burkill (Plymouth University, UK), Jens Hesselbjerg Christensen (Danish Meteorological Institute), Ken Drinkwater (Institute of Marine Research, Norway), Kirstin Halsnæs (Technical University of Denmark), Daniela Jacob (Climate Service Centre Germany), Albert Klein Tank (Royal Netherlands Meteorological Institute), Jaap Kwadijk (Deltares, The Netherlands), John K. Pinnegar (Centre for Environment, Fisheries and Aquaculture Science, UK), Marcus Reckermann (Baltic Earth),

Markku Rummukainen (Lund University, Sweden), Lesley Salt (Royal Netherlands Institute for Sea Research), Corinna Schrum (University of Bergen, Norway, and Helmholtz-Zentrum Geesthacht, Germany), Hendrik M. van Aken (Royal Netherlands Institute for Sea Research), and Hans von Storch (Helmholtz-Zentrum Geesthacht, Germany).

Producing such an extensive book is not possible without the technical support of particular individuals, to whom we are extremely grateful. Special thanks go to our colleague Ingeborg Nöhren; Ingeborg was deeply involved in technical editing, obtaining reproduction permissions and improving most of the many graphics. Graphical expertise was also provided by Beate Gardeike and Bianca Seth. Merja Helena Tölle, Marcus Lange, and Sabine Hartmann supported us in coordinating NOSCCA during its initial phase. Sönke Rau helped formatting the chapters. Insa Puchert conducted an actor analysis for the North Sea region. Thanks to Ina Frings for maintaining the NOSCCA homepage.

Last but by no means least, we thank Carolyn Symon (UK) for professional language editing and many useful editorial suggestions.

Reproduction Permission

As is inevitable for a review of this type, a great number of published figures and tables were reprinted here. Every effort was made to obtain permission from the copyright holders. We apologise for any inadvertent infringement of copyright that may still have occurred despite our best efforts; if such a case is brought to our attention we will certainly rectify this in any future reprint. We thank all copyright holders that granted reproduction permission free of charge.

Open Access

To foster a wider outreach and to enhance the availability of our climate change review to young researchers and students, it was recently decided to release the NOSCCA report as an open access publication. This was made possible by shared funding with contributions provided by the following institutions and programmes: Cluster of Excellence 'Integrated Climate System Analysis and Prediction' at the University of Hamburg (CLISAP; Germany), Danish Meteorological Institute (DMI; Denmark), German Meteorological Service (DWD; Germany), Met Office (UK), Royal Netherlands Meteorological Institute (KNMI; The Netherlands), Swedish Meteorological and Hydrological Institute (SMHI; Sweden), Technical University of Denmark (DTU; Denmark), University of Bergen (Norway), and the Library and Institute of Coastal Research of the Helmholtz-Zentrum Geesthacht (Germany). Thank you very much for the essential support at short notice.

We sincerely hope that we have not forgotten anyone. Thank you so much to all of you for your tremendous effort and support, which together has made this assessment possible.

Geesthacht, Germany

Markus Quante
Franciscus Colijn

Contents

Lead Authors

Chapter 1: Introduction to the Assessment—Characteristics of the Region

Markus Quante
Institute of Coastal Research, Helmholtz-Zentrum Geesthacht, Geesthacht, Germany
markus.quante@hzg.de

Franciscus Colijn
Institute of Coastal Research, Helmholtz-Zentrum Geesthacht, Geesthacht, Germany
franciscus.colijn@hzg.de

Part I—Recent Climate Change (Past 200 Years)

Chapter 2: Recent Change—Atmosphere

Martin Stendel
Department for Arctic and Climate, Danish Meteorological Institute (DMI), Copenhagen, Denmark
mas@dmi.dk

Chapter 3: Recent Change—North Sea

John Huthnance
National Oceanography Centre, Liverpool, UK
jmh@noc.ac.uk

Ralf Weisse
Institute of Coastal Research, Helmholtz-Zentrum Geesthacht, Geesthacht, Germany
ralf.weisse@hzg.de

Chapter 4: Recent Change—River Flow

Jaap Kwadijk
Deltares, Delft, The Netherlands, and Faculty of Engineering and Technology, Department of Water Engineering and Management, Twente University, Enschede, The Netherlands
jaap.kwadijk@deltares.nl

Nigel W. Arnell
Walker Institute for Climate System Research, University of Reading, UK
n.w.arnell@reading.ac.uk

Part II—Future Climate Change

Chapter 5: Projected Change—Atmosphere

Wilhelm May
Research and Development Department, Danish Meteorological Institute (DMI), Copenhagen, Denmark, and Centre for Environmental and Climate Research, Lund University, Lund, Sweden
wm@dmi.dk; wilhelm.may@cec.lu.se

Chapter 6: Projected Change—North Sea

Corinna Schrum
Geophysical Institute, University of Bergen, Norway, and Institute of Coastal Research, Helmholtz-Zentrum Geesthacht, Geesthacht, Germany
corinna.schrum@gfi.uib.no; corinna.schrum@hzg.de

Jason Lowe
Met Office Hadley Centre, Exeter, UK
jason.lowe@metoffice.gov.uk

H.E. Markus Meier
Research Department, Swedish Meteorological and Hydrological Institute, Norrköping, Sweden, and Department of Physical Oceanography and Instrumentation, Leibniz-Institute for Baltic Sea Research (IOW), Rostock, Germany
markus.meier@smhi.se; markus.meier@io-warnemuende.de

Chapter 7: Projected Change—River Flow and Urban Drainage

Patrick Willems
Department of Civil Engineering, KU Leuven, Leuven, Belgium
patrick.willems@bwk.kuleuven.be

Benjamin Lloyd-Hughes
Walker Institute for Climate System Research, University of Reading, Reading, UK
B.LloydHughes@reading.ac.uk

Part III—Impacts of Recent and Future Climate Change on Ecosystems

Chapter 8: Environmental Impacts—Marine Ecosystems

Keith M. Brander
National Institute of Aquatic Resources, Technical University of Denmark, Charlottenlund, Denmark
kbr@aqua.dtu.dk

Geir Ottersen
Institute of Marine Research (IMR), Bergen, Norway, and CEES Centre for Ecological and Evolutionary Synthesis, University of Oslo, Oslo, Norway
geir.ottersen@imr.no

Chapter 9: Environmental Impacts—Coastal Ecosystems

Jan P. Bakker
Groningen Institute for Evolutionary Life Sciences, University of Groningen, Groningen, The Netherlands
j.p.bakker@rug.nl

Chapter 10: Environmental Impacts—Lake Ecosystems

Rita Adrian
Leibniz-Institute of Freshwater Ecology and Inland Fisheries, Berlin, Germany
adrian@igb-berlin.de

Dag Olav Hessen
Department of Biosciences, University of Oslo, Oslo, Norway
d.o.hessen@ibv.uio.no

Chapter 11: Environmental Impacts—Terrestrial Ecosystems

Norbert Hölzel
Institute of Landscape Ecology, University of Münster, Münster, Germany
norbert.hoelzel@uni-muenster.de

Thomas Hickler
Senckenberg Biodiversity and Climate Research Centre (BiK-F), and Department of Physical Geography, Goethe University, Frankfurt, Germany
thomas.hickler@senckenberg.de

Lars Kutzbach
Institute of Soil Science, CEN, University of Hamburg, Hamburg, Germany
lars.kutzbach@uni-hamburg.de

Part IV—Climate Change Impacts on Socio-economic Sectors

Chapter 12: Socio-economic Impacts—Fisheries

John K. Pinnegar
Centre for Environment, Fisheries and Aquaculture Science (Cefas), Lowestoft, UK, and School of Environmental Sciences, University of East Anglia (UEA), Norwich, UK
john.pinnegar@cefas.co.uk

Chapter 13: Socio-economic Impacts—Agricultural Systems

Jørgen Eivind Olesen
Department of Agroecology—Climate and Bioenergy, Aarhus University, Aarhus, Denmark
jorgene.olesen@agrsci.dk

Chapter 14: Socio-economic Impacts—Offshore Activities/Energy

Kirsten Halsnæs
Systems Analysis, DTU Management Engineering, Lyngby, Denmark
khal@dtu.dk

Martin Drews
Systems Analysis, DTU Management Engineering, Lyngby, Denmark
mard@dtu.dk

Chapter 15: Socio-economic Impacts—Urban Climate

K. Heinke Schlünzen
Institute of Meteorology, CEN, University of Hamburg, Hamburg, Germany
heinke.schluenzen@uni-hamburg.de

Chapter 16: Socio-economic Impacts—Air Quality

Stig Bjørløw Dalsøren
CICERO Center for International Climate and Environmental Research-Oslo, Oslo, Norway
s.b.dalsoren@cicero.uio.no

Chapter 17: Socio-economic Impacts—Recreation

Edgar Kreilkamp
Institute of Corporate Development, Leuphana University of Lüneburg, Lüneburg, Germany
edgar.kreilkamp@uni.leuphana.de

Chapter 18: Socio-economic Impacts—Coastal Protection

Hanz D. Niemeyer
Independent Consultant, ret. Coastal Research Station, Norderney, Germany
hanz.niemeyer@yahoo.com

Chapter 19: Socio-economic Impacts—Coastal Management and Governance

Job Dronkers
Deltares, Delft, and Netherlands Centre for Coastal Research, Delft, The Netherlands
j.dronkers@hccnet.nl

Tim Stojanovic
Department of Geography and Sustainable Development, University of St Andrews, UK
tas21@st-andrews.ac.uk

Annexes

Annex 1: What is NAO?

Abdel Hannachi
Department of Meteorology, Stockholm University, Stockholm, Sweden
a.hannachi@misu.su.se

Annex 2: Climate Model Simulations for the North Sea Region

Diana Rechid
Climate Service Center Germany (GERICS), Helmholtz-Zentrum Geesthacht, Hamburg, Germany
diana.rechid@hzg.de

Annex 3: Uncertainties in Climate Change Projections

Markku Rummukainen
Centre for Environmental and Climate Research, Lund University, Lund, Sweden
markku.rummukainen@cec.lu.se

Annex 4: Emission Scenarios for Climate Projections

Markus Quante
Institute of Coastal Research, Helmholtz-Zentrum Geesthacht, Geesthacht, Germany
markus.quante@hzg.de

Annex 5: Facts and Maps

Ingeborg Nöhren
Institute of Coastal Research, Helmholtz-Zentrum Geesthacht, Geesthacht, Germany
ingeborg.noehren@hzg.de

Contributing Authors

Andreas C.W. Baas Department of Geography, King's College London, London, UK

Jan P. Bakker Groningen Institute for Evolutionary Life Sciences, University of Groningen, Groningen, The Netherlands

Jesper Bartholdy Department of Geosciences and Natural Resource Management, University of Copenhagen, Copenhagen, Denmark

Gé Beaufort Hydr. Engineering Cons., De Meern, The Netherlands

Gregory Beaugrand CNRS, Laboratoire d'Océanologie de Géosciences (LOG), Wimereux, France

Christian Bjørnæs CICERO Center for International Climate and Environmental Research-Oslo, Oslo, Norway

Thorsten Blenckner Stockholm Resilience Centre, Stockholm University, Stockholm, Sweden

Sylvia I. Bohnenstengel Department of Meteorology, University of Reading, Reading, UK; *Present Address:* MetOffice@Reading, University of Reading, Reading, UK

Keith M. Brander National Institute of Aquatic Resources, Technical University of Denmark, Charlottenlund, Denmark

Katharina Bülow Climate Service Center Germany (GERICS), Helmholtz-Zentrum Geesthacht, Hamburg, Germany

William W.L. Cheung Institute for the Ocean and Fisheries, University of British Columbia, Vancouver, Canada

Niels-Erik Clausen DTU Wind Energy, Roskilde, Denmark

Franciscus Colijn Institute of Coastal Research, Helmholtz-Zentrum Geesthacht, Geesthacht, Germany

Huib de Vriend Independent Consultant, Oegstgeest, The Netherlands

Mark de Weerd TAUW, Deventer, The Netherlands

Alberto Elizalde Max Planck Institute for Meteorology, Hamburg, Germany

Georg H. Engelhard Centre for Environment, Fisheries and Aquaculture Science (Cefas), Lowestoft, UK; School of Environmental Sciences, University of East Anglia (UEA), Norwich, UK

Anette Ganske Federal Maritime and Hydrographic Agency (BSH), Hamburg, Germany

Stefan Garthe Research and Technology Centre (FTZ), University of Kiel, Büsum, Germany

Anita Gilles Institute for Terrestrial and Aquatic Wildlife Research, University of Veterinary Medicine Hannover, Foundation, Büsum, Germany

Iris Grabemann Institute of Coastal Research, Helmholtz-Zentrum Geesthacht, Geesthacht, Germany

Ivan Haigh Ocean and Earth Science, University of Southampton, Southampton, UK

Abdel Hannachi Department of Meteorology, Stockholm University, Stockholm, Sweden

Werner Härdtle Institute of Ecology, Leuphana University of Lüneburg, Lüneburg, Germany

Hartmut Heinrich Federal Maritime and Hydrographic Agency (BSH), Hamburg, Germany

Helena Herr Institute for Terrestrial and Aquatic Wildlife Research, University of Veterinary Medicine Hannover, Foundation, Büsum, Germany

Roland Hiederer Institute for Environment and Sustainability, European Commission Joint Research Centre, Ispra, Italy

Helmut Hillebrand Institute for Chemistry and Biology of the Marine Environment, Carl von Ossietzky University of Oldenburg, Oldenburg, Germany

Sabine Hilt Leibniz-Institute of Freshwater Ecology and Inland Fisheries (IGB), Berlin, Germany

Solfrid Hjøllo Institute of Marine Research (IMR), Bergen, Norway

Jürgen Holfort Federal Maritime and Hydrographic Agency (BSH), Rostock, Germany

Jason Holt National Oceanography Centre, Liverpool, UK

Erik Jeppesen Department of Bioscience, Aarhus University, Aarhus, Denmark

Laurence Jones Centre for Ecology and Hydrology, Bangor, UK

Miranda C. Jones NF-UBC Nereus Program, Fisheries Centre, University of British Columbia, Vancouver, Canada

Jan Eiof Jonson Norwegian Meteorological Institute, Oslo, Norway

Hans Joosten Institute of Botany and Landscape Ecology, Ernst-Moritz-Arndt University of Greifswald, Greifswald, Germany

Andrew Kenny Centre for Environment, Fisheries and Aquaculture Science (Cefas), Lowestoft, UK; School of Environmental Sciences, University of East Anglia (UEA), Norwich, UK

Elizabeth C. Kent National Oceanography Centre, Southampton, UK

Aart Kroon Department of Geosciences and Natural Resource Management, University of Copenhagen, Copenhagen, Denmark

Wilfried Kühn Institute of Oceanography, CEN, University of Hamburg, Hamburg, Germany

Gregor C. Leckebusch School of Geography, Earth and Environmental Sciences, University of Birmingham, Birmingham, UK

Christiana Lefebvre German Meteorological Service (DWD), Hamburg, Germany

David M. Livingstone Department of Water Resources and Drinking Water, EAWAG Swiss Federal Institute of Aquatic Science and Technology, Dübendorf, Switzerland

Peter Loewe Federal Maritime and Hydrographic Agency (BSH), Hamburg, Germany

Ina Lorkowski Federal Maritime and Hydrographic Agency (BSH), Hamburg, Germany

Moritz Mathis Max Planck Institute for Meteorology, Hamburg, Germany

Claudia Mauser IFOK GmbH, Bensheim, Germany

Roberto Mayerle Research and Technology Centre Westcoast, Kiel University, Kiel, Germany

H.E. Markus Meier Research Department, Swedish Meteorological and Hydrological Institute, Norrköping, Sweden; Department of Physical Oceanography and Instrumentation, Leibniz-Institute for Baltic Sea Research (IOW), Rostock, Germany

Jaak Monbaliu Department of Civil Engineering, KU Leuven, Leuven, Belgium

Kjell Arne Mork Institute of Marine Research (IMR), Bergen, Norway

Christopher Moseley Max Planck Institute for Meteorology, Hamburg, Germany

Christoph Mudersbach Institute of Water and Environment, Bochum University of Applied Sciences, Bochum, Germany

Ingeborg Nöhren Institute of Coastal Research, Helmholtz-Zentrum Geesthacht, Geesthacht, Germany

Jørgen Eivind Olesen Department of Agroecology—Climate and Bioenergy, Aarhus University, Aarhus, Denmark

Johannes Pätsch Institute of Oceanography, CEN, University of Hamburg, Hamburg, Germany

Myron A. Peck Institute of Hydrobiology and Fisheries Science, CEN, University of Hamburg, Hamburg, Germany

Julie Pietrzak Environmental Fluid Mechanics Section, Delft University of Technology, Delft, The Netherlands

Thomas Pohlmann Institute of Oceanography, CEN, University of Hamburg, Hamburg, Germany

Markus Quante Institute of Coastal Research, Helmholtz-Zentrum Geesthacht, Geesthacht, Germany

Adriaan D. Rijnsdorp Institute for Marine Research and Ecosystem Studies (IMARES), Wageningen University, IJmuiden, The Netherlands

Burkhardt Rockel Institute of Coastal Research, Helmholtz-Zentrum Geesthacht, Geesthacht, Germany

Gerben Ruessink Department of Physical Geography, Faculty of Geosciences, Utrecht University, Utrecht, The Netherlands

Markku Rummukainen Centre for Environmental and Climate Research, Lund University, Lund, Sweden

Lesley Salt Royal Netherlands Institute for Sea Research, Texel, The Netherlands

Frederik Schenk Bolin Centre for Climate Research, University of Stockholm, Stockholm, Sweden

Natalija Schmelzer Federal Maritime and Hydrographic Agency (BSH), Rostock, Germany

Corinna Schrum Geophysical Institute, University of Bergen, Bergen, Norway; Institute of Coastal Research, Helmholtz-Zentrum Geesthacht, Geesthacht, Germany

John Siddorn Met Office, Exeter, UK

Ursula Siebert Institute for Terrestrial and Aquatic Wildlife Research, University of Veterinary Medicine Hannover, Foundation, Büsum, Germany

Hein Rune Skjoldal Institute of Marine Research (IMR), Bergen, Norway

Morten D. Skogen Institute of Marine Research (IMR), Bergen, Norway

Tim Smyth Plymouth Marine Laboratory, Plymouth, UK

Alejandro Jose Souza National Oceanography Centre, Liverpool, UK

Martin Stendel Department for Arctic and Climate, Danish Meteorological Institute (DMI), Copenhagen, Denmark

Andreas Sterl Royal Netherlands Meteorological Institute (KNMI), De Bilt, The Netherlands

Horst Sterr Department of Geography, Kiel University, Kiel, Germany

Jian Su Institute of Oceanography, CEN, University of Hamburg, Hamburg, Germany

Jürgen Sündermann Institute of Oceanography, CEN, University of Hamburg, Hamburg, Germany

Stijn Temmerman Ecosystem Management Research Group, Antwerp University, Antwerp, Belgium

Helmuth Thomas Department of Oceanography, Dalhousie University, Halifax, Canada

Birger Tinz German Meteorological Service (DWD), Hamburg, Germany

Merja Helena Tölle Department of Geography, Climatology, Climate Dynamics and Climate, Justus-Liebig-University Gießen, Gießen, Germany

Ian Townend Ocean and Earth Sciences, University of Southampton, Winchester, UK

Holger Toxvig Madsen Danish Coastal Authority, Lemvig, Denmark

Dennis Trolle Department of Bioscience—Lake Ecology, Aarhus University, Silkeborg, Denmark

Ingrid Tulp Institute for Marine Resources and Ecosystem Management (IMARES), IJmuiden, The Netherlands

Uwe Ulbrich Institute of Meteorology, Freie Universität Berlin, Berlin, Germany

Justus van Beusekom Institute of Coastal Research, Helmholtz-Zentrum Geesthacht, Geesthacht, Germany

Martijn van de Pol Department of Evolution, Ecology and Genetics, Research School of Biology, Australian National University, Canberra, Australia; Department of Animal Ecology, Netherlands Institute of Ecology (NIOO-KNAW), Wageningen, The Netherlands

Else van den Besselaar Royal Netherlands Meteorological Institute (KNMI), De Bilt, The Netherlands

Gerard van der Schrier Royal Netherlands Meteorological Institute (KNMI), De Bilt, The Netherlands

Sytze van Heteren Department of Geomodeling, TNO-Geological Survey of the Netherlands, Utrecht, The Netherlands

Jakobus van Huissteden Department of Earth Sciences, University Amsterdam, Amsterdam, The Netherlands

Nele Marisa von Bergner Institute for Tourism and Leisure ITF, HTW Chur, Chur, Switzerland

Thomas Wahl College of Marine Science, University of South Florida, St. Petersburg, USA; University of Southampton, Southampton, UK

Sarah Wakelin National Oceanography Centre, Liverpool, UK

Philip Woodworth National Oceanography Centre, Liverpool, UK

Tim Woollings Atmospheric Physics, Clarendon Laboratory, University of Oxford, Oxford, UK

Andreas Wurpts Coastal Research Station, Norderney, Germany

Reviewers

Review Editor

Robert J. Nicholls
Engineering and the Environment, University of Southampton, UK
R.J.Nicholls@soton.ac.uk

Reviewers
(Listed are in alphabetical order reviewers, who disclosed their identity)

Jürgen Alheit
Leibniz Institute for Baltic Research (IOW), Rostock, Germany
juergen.alheit@io-warnemuende.de

Rob Allan
Met Office, Exeter, UK
rob.allan@metoffice.gov.uk

Eric Audsley
Operational Research, Cranfield University, UK
EAudsley@aol.com; e.audsley@cranfield.ac.uk

Jan P. Bakker
Groningen Institute for Evolutionary Life Sciences, University of Groningen, Groningen, The Netherlands
j.p.bakker@rug.nl

Alexander Baklanov
Danish Meteorological Institute, Copenhagen, Denmark
alb@dmi.dk

Frank Berendse
Plant Ecology and Nature Conservation, Wageningen University, The Netherlands
frank.berendse@wur.nl

Barbara Berx
Marine Scotland, Marine Laboratory, Aberdeen, UK
B.Berx@marlab.ac.uk

Fred Bosveld
The Royal Netherlands Meteorological Institute (KNMI), De Bilt, The Netherlands
fred.bosveld@knmi.nl

Jørgen Brandt
Department of Environmental Science, Aarhus University, Denmark
jbr@envs.au.dk

Ida Brøker
DHI, Hørsholm, Denmark
ibh@dhigroup.com

James Bullock
Centre for Ecology & Hydrology, Wallingford, UK
jmbul@ceh.ac.uk

Hans Burchard
Leibniz Institute for Baltic Research (IOW), Rostock, Germany
hans.burchard@io-warnemuende.de

Virginia Burkett
US Geological Survey, Reston, Virginia, USA
virginia_burkett@usgs.gov

Matthew Chamberlain
CSIRO, Oceans and Atmosphere, Hobart, Australia
matthew.chamberlain@csiro.au

Mark Dickey-Collas
ICES, Copenhagen, Denmark
mark.dickey-collas@ices.dk

Markus Donat
Climate Change Research Centre, University of New South Wales, Sydney, Australia
m.donat@unsw.edu.au

Stephen Dye
Cefas, Lowestoft, UK
stephen.dye@cefas.co.uk

Henrik Eckersten
Swedish University of Agricultural Sciences, Uppsala, Sweden
henrik.eckersten@slu.se

Karen Edelvang
GEUS, Copenhagen, Denmark
kae@geus.dk

Kay-Christian Emeis
Institute of Coastal Research, Helmholtz-Zentrum Geesthacht, Germany
kay.emeis@hzg.de

Jon French
Department of Geography, University College London, UK
j.french@ucl.ac.uk

Audrey Geffen
Department of Biology, University of Bergen, Norway
audrey.geffen@bio.uib.no

Simon Gosling
School of Geography, University of Nottingham, UK
Simon.Gosling@nottingham.ac.uk

Lukas Gudmundsson
Institute for Atmospheric and Climate Science, ETH Zürich, Switzerland
lukas.gudmundsson@env.ethz.ch

Jacqueline M. Hamilton
Research Unit Sustainability and Global Change, CEN, University of Hamburg, Germany
jacqueline_m_hamilton@yahoo.co.uk

Lars Anders Hansson
Aquatic Ecology, Lund University, Sweden
lars-anders.hansson@biol.lu.se

Gareth Harrison
School of Engineering, University of Edinburgh, UK
Gareth.Harrison@ed.ac.uk

Jan Geert Hiddink
School of Ocean Sciences, Bangor University, UK
ossc06@bangor.ac.uk

Jochen Hinkel
Global Climate Forum, Berlin, Germany
hinkel@globalclimateforum.org

Ian Holman
Cranfield Water Science Institute, Cranfield University, UK
i.holman@cranfield.ac.uk

Jose A. Jiménez
Department of Civil and Environmental Engineering, Universitat Politecnica de Catalunya, Barcelona, Spain
jose.jimenez@upc.edu

Jürgen Jensen
Department of Civil Engineering, Siegen University, Germany
juergen.jensen@uni-siegen.de

Birgit Klein
Federal Maritime and Hydrographic Agency (BSH), Hamburg, Germany
birgit.klein@bsh.de

Sven Kotlarski
Federal Office of Meteorology and Climatology MeteoSwiss, Zurich, Switzerland
sven.kotlarski@meteoswiss.ch

Jürg Luterbacher
Department of Geography, Justus-Liebig-University Gießen, Germany
juerg.luterbacher@geogr.uni-giessen.de

Rob Maas
National Institute for Public Health and the Environment (RIVM), Bilthoven, The Netherlands
rob.maas@rivm.nl

Robert Marrs
School of Environmental Sciences, University of Liverpool, UK
Calluna@liverpool.ac.uk

Valéry Masson
Météo France, National Centre for Meteorological Research, Toulouse, France
valery.masson@meteo.fr

Andreas Matzarakis
University of Freiburg, Chair for Environmental Meteorology, and Research Centre Human Biometeorology, German Meteorological Service (DWD), Freiburg, Germany
andreas.matzarakis@meteo.uni-freiburg.de; andreas.matzarakis@dwd.de

Erik van Meijgaard
The Royal Netherlands Meteorological Institute (KNMI), De Bilt, The Netherlands
vanmeijg@knmi.nl

Jack Middelburg
Department of Earth Sciences, University of Utrecht, The Netherlands
J.b.m.middelburg@uu.nl

Hans Middelkoop
Department of Physical Geography, University of Utrecht, The Netherlands
H.Middelkoop@uu.nl

Birger Mo
SINTEF ENERGY, Trondheim, Norway
Birger.Mo@sintef.no

Luke Myers
Engineering and the Environment, University of Southampton, UK
L.E.MYERS@soton.ac.uk

Claas Nendel
Leibniz Centre for Agricultural Landscape Research (ZALF), Institute of Landscape Systems Analysis, Müncheberg, Germany
nendel@zalf.de

Ingeborg Nöhren
Institute of Coastal Research, Helmholtz-Zentrum Geesthacht, Geesthacht, Germany
ingeborg.noehren@hzg.de

Gualbert Oude Essink
Faculty of Geosciences, University of Utrecht, The Netherlands
gualbert.oudeessink@deltares.nl

Myron A. Peck
Institute of Hydrobiology and Fisheries Science, CEN, University of Hamburg, Hamburg, Germany
myron.peck@uni-hamburg.de

Tom Rientjes
Faculty of Geo-Information Science and Earth Observation (ITC), University of Twente, The Netherlands
rientjes@itc.nl; t.h.m.rientjes@utwente.nl

Markku Rummukainen
Centre for Environmental and Climate Research, Lund University, Lund, Sweden
markku.rummukainen@cec.lu.se

Enrique Sánchez
Faculty of Environmental Sciences and Biochemistry, University of Castilla-La Mancha (UCLM), Toledo, Spain
E.Sanchez@uclm.es

Daniel Scott
Geography and Environmental Management, University of Waterloo, Canada
dj2scott@uwaterloo.ca

Øystein Skagseth
Institute of Marine Research, Bergen, Norway
oystein.skagseth@imr.no

Thomas Spencer
Cambridge Coastal Research Unit, University of Cambridge, UK
ts111@cam.ac.uk

Horst Sterr
Department of Geography, Kiel University, Kiel, Germany
sterr@geographie.uni-kiel.de

Dietmar Straile
Limnological Institute, University of Konstanz, Germany
Dietmar.Straile@uni-konstanz.de

Justus van Beusekom
Institute of Coastal Research, Helmholtz-Zentrum Geesthacht, Geesthacht, Germany
Justus.Beusekom@hzg.de

David Viner
Mott MacDonald Consultancy, Cambridge, UK
David.Viner@mottmac.com

James Voogt
Department of Geography, University of Western Ontario, London, Canada
javoogt@uwo.ca

Craig Williamson
Department of Biology, Miami University, Oxford, USA
craig.williamson@miamiOH.edu

Kai Wirtz
Institute of Coastal Research, Helmholtz-Zentrum Geesthacht, Germany
kai.wirtz@hzg.de

Acronyms and Abbreviations

20CR	20th century reanalysis
AGCM	Atmospheric general circulation model
AH	Azores high
AMO	Atlantic multidecadal oscillation
AMOC	Atlantic meridional overturning circulation
AMSL	Absolute mean sea level
AO	Arctic oscillation
AOGCM	Atmosphere–Ocean general circulation model
AR4	Fourth assessment report (IPCC)
AR5	Fifth assessment report (IPCC)
A_T	Total alkalinity
AVHRR	Advanced very high resolution radiometer
Bft	Beaufort scale
CH_4	Methane
CMIP3	Coupled model intercomparison project phase 3
CMIP5	Coupled model intercomparison project phase 5
CO	Carbon monoxide
CO_2	Carbon dioxide
CO_2eq	Carbon dioxide equivalent
CPR	Continuous plankton recorder
CPUE	Catch-per-unit-effort
CTD	Conductivity-temperature-depth profiler
DGVM	Dynamic global vegetation model
DIC	Dissolved inorganic carbon
DOC	Dissolved organic carbon
DOM	Dissolved organic matter
DON	Dissolved organic nitrogen
ECMWF	European Centre for Medium-Range Weather Forecasts
EEA	European Environment Agency
EEZ	Exclusive economic zone
ENSO	El Niño Southern Oscillation
ENW	Equivalent neutral wind
EOF	Empirical orthogonal function
ESM	Earth system model
ETM	Estuarine turbidity maximum
EU	European Union
EUR	Euro
GCM	General circulation model/Global climate model
GEV	Generalised extreme value
GHG	Greenhouse gas
GIA	Glacial isostatic adjustment

GNP	Gross national product
GPS	Global positioning system
HAT	Highest astronomical tide
H_s	Significant wave height
ICES	International Council for the Exploration of the Sea
ICOADS	International Comprehensive Ocean-Atmosphere Data Set
ICZM	Integrated coastal zone management
IL	Icelandic low
IMO	International Maritime Organization
IPCC	Intergovernmental Panel on Climate Change
ka	Thousand years ago
LW	Longwave (radiation)
ma	Million years ago
MHT	Mean high tide
MSL	Mean sea level
MSLP	Mean sea-level pressure
N	Nitrogen
N_2O	Nitrous oxide
NAM	Northern annular mode
NAO	North atlantic oscillation
NCEP	National Centers for Environmental Prediction
netPP	Net primary production
NH_3	Ammonia
NH_4	Ammonium
NMAT	Night marine air temperature
nmVOC	Non-methane volatile organic compounds
NO	Nitrogen oxide
NO_2	Nitrogen dioxide
NO_3	Nitrate
NO_X	Nitrogen oxides
NPP	Net primary productivity
O_3	Ozone
OA	Ocean acidification
OGCM	Ocean general circulation model
P	Phosphorus
PAH	Polycyclic aromatic hydrocarbon
PAN	Peroxyacetyl nitrate
PBL	Planetary boundary layer
PCB	Polychlorinated biphenyl
$p\mathrm{CO}_2$	Partial pressure of carbon dioxide
PEA	Potential energy anomaly
$PM_{2.5}$	Particles of less than 2.5 μm in diameter
PM_{10}	Particles of less than 10 μm in diameter
PM_{coarse}	Particles between PM_{10} and $PM_{2.5}$ in diameter
POC	Particulate organic carbon
RCM	Regional climate model
RCP	Representative concentration pathway (IPCC)
RCSM	Regional climate system model
RMSL	Relative mean sea level
ROFI	Region of freshwater influence
S	Sulphur
SBT	Sea-bed temperature

SD	Standard deviation
SEC	Surface elevation change
SLP	Sea-level pressure
SLR	Sea-level rise
SO_2	Sulphur dioxide
SPM	Suspended particulate matter
SRES	Special Report on Emission Scenarios (IPCC)
SSB	Spawning stock biomass
SSC	Suspended sediment concentration
SSS	Sea-surface salinity
SST	Sea-surface temperature
Sv	Sverdrup, 10^6 m^3/sec
SW	Shortwave (radiation)
T	Annual wave period
TAC	Total allowable catch
TAR	Third assessment report (IPCC)
UHI	Urban heat island effect
UK	United Kingdom
USD	US dollar
VOC	Volatile organic compound
VOS	Voluntary observing ship
WETCHIMP	Wetland and Wetland CH_4 Intercomparison of Models Project
WHO	World Health Organization
WMO	World Meteorological Organization

About NOSCCA

Ongoing and future anthropogenic climate change is widely recognised as a major scientific and societal issue, with huge economic consequences. The North Sea and its adjacent land areas is one of the major economic regions of the world and a place for settlement and commerce for millions of people. Like many other areas, this region is already facing a changing climate and projections indicate that impacts will become even stronger in the coming decades.

Knowledge of climate change has increased massively over the past few decades, which enables a more strategic response to climate-related risk. For example, the Intergovernmental Panel on Climate Change (IPCC) has released a series of major climate change assessments; the first in 1990 and the latest in 2013/2014. But although reliable information on the characteristics and impacts of climate change at a regional scale is essential for scientists, responsible authorities and stakeholders in the regions, it is arguably still limited. Even the most recent IPCC assessment (AR5, published in 2013 and 2014) could not report the desired level of detail for many regions of interest—including the North Sea.

In 2010, the Institute of Coastal Research of the Helmholtz-Zentrum Geesthacht in Germany initiated a comprehensive climate change assessment for the Greater North Sea region and adjacent land areas, referred to as the 'North Sea Region Climate Change Assessment' (NOSCCA). The purpose of this assessment is to review and analyse the scientifically legitimised knowledge of climate change and its impacts across the entire region. The NOSCCA approach is similar in format to the IPCC approach and close to that of a climate change assessment compiled for the Baltic Sea Basin (BACC).[3]

The challenges for NOSCCA as a full assessment of climate change in the North Sea region were first to get access to the scattered information, second to render it comparable, and finally to prepare an assessment of climate change based on the entire body of material. This synthesis is based entirely on scientifically legitimate published work, with the emphasis on peer-reviewed journal articles or book chapters wherever possible. Conference proceedings and reports from scientific institutes and governmental agencies (such as meteorological services or oceanographic centres) have also been evaluated. Reports from bodies with a mainly non-scientific agenda were excluded. In cases where a clear consensus on a climate change issue could not be found in the literature this is clearly stated and if appropriate different views are reported or knowledge gaps highlighted.

The 'North Sea region' as envisaged in the NOSCCA context comprises the Greater North Sea, as defined by OSPAR and the land domains of the bounding countries, which are part of the catchment area and which have a coastline along the Greater North Sea. Thus the Skagerrak, Kattegat and English Channel belong to the area of interest.

From the start, NOSCCA has been an independent international initiative involving scientists from all countries in the region. NOSCCA authors are predominately from universities and public research institutes. There was no special or external funding for NOSCCA activities, all contributions were made on a voluntary basis and scientists relied on

[3]The BACC Author Team (2008), Assessment of Climate Change for the Baltic Sea Basin. Regional Climate Studies, Springer-Verlag, 473pp; The BACC II Author Team (2015) Second Assessment of Climate Change for the Baltic Sea Basin, Regional Climate Studies, Springer, 501pp.

their institutional resources and support. Writing teams guided by Lead Authors compiled the chapters. Lead Authors have played a crucial role in the overall process as they were responsible for the respective writing teams and are responsible for the content as well as the overall quality of their chapters. All climate change chapters were subject to independent scientific review. NOSCCA cooperates with the *International Council for the Exploration of the Sea* (ICES) and is a *Land-Ocean Interactions in the Coastal Zone* (LOICZ) affiliated project, information exchange with the OSPAR Commission was agreed upon. The entire process was coordinated by a team based at the Institute of Coastal Research at the Helmholtz-Zentrum Geesthacht.

From initialisation to the final product, the NOSCCA process was overseen by an international Scientific Steering Committee (SSC), whose members were selected to represent the North Sea countries and a wide range of expertise relevant to marine and terrestrial climate change. The role of the SSC was to formulate and determine the procedure leading to the final assessment report and to outline the topics to be addressed. Another important responsibility of the SCC was to select Lead Authors for the different chapters. The SSC was also involved in initialising the external review process. The NOSCCA SSC members are Hein J.W. de Baar (Royal Netherlands Institute for Sea Research and University of Groningen, The Netherlands), Monika Breuch-Moritz (Federal Maritime and Hydrographic Agency, Hamburg, Germany), Peter Burkill (Plymouth University, UK), Franciscus Colijn (Chair; Helmholtz-Zentrum Geesthacht, Germany), Ken Drinkwater (Institute of Marine Research, Bergen, Norway), Kevin Horsburgh (National Oceanography Centre, Liverpool, UK), Eigil Kaas (Niels Bohr Institute, Copenhagen, Denmark), Albert M.G. Klein Tank (Royal Netherlands Meteorological Institute, De Bilt, The Netherlands), Hartwig Kremer (United Nations Environment Programme, Copenhagen, Denmark), Georges Pichot (Management Unit of the North Sea Mathematical Models, Brussels, Belgium), Markus Quante (Helmholtz-Zentrum Geesthacht, Germany), Hans von Storch (Helmholtz-Zentrum Geesthacht, Germany), Göran Wallin (University of Gothenburg, Sweden) and Karen Helen Wiltshire (Alfred Wegener Institute, Bremerhaven, Germany).

To ensure an independent review process an external review editor was assigned, who is not involved in any other NOSCCA activity. The renowned climate change scientist Professor Robert J. Nicholls from the University of Southampton, Engineering and the Environment, UK, kindly agreed to take on this task. The review editor defined the overall review process and together with the SSC Chair, selected and invited the individual reviewers. The review process was overseen and undertaken with the assistance of the NOSCCA coordination team. Three independent reviewers, preferably from different countries were assigned to each climate change chapter. Only the introductory chapter and the annexes were reviewed by expert colleagues or authors of other chapters. The review editor had the final say in the case of conflicting opinions.

The NOSCCA process began in October 2010, when the SSC was formed during a meeting in Hamburg hosted by the Federal Maritime and Hydrographic Agency (BSH). The first meeting of Lead Authors together with the members of the SSC took place at the Royal Netherlands Academy of Arts and Sciences in Amsterdam in October 2011. The second Lead Author meeting took place at the Carlsberg Academy in Copenhagen in October 2012, where the Lead Authors agreed on the layout of the various chapters. The third Lead Author meeting was held in June 2013 at Deltares in Delft. The NOSCCA review phase began in spring 2014, and the external review was complete by the end of spring 2015. A final Lead Author meeting was held in June 2015 at the Climate Service Centre Germany in Hamburg. Key findings of all chapters were exchanged and discussed. All revised chapters were available by the end of 2015. All chapters were then subject to language editing before the final material was sent to the publisher in spring 2016. The final text was published as a print and open access book in summer 2016.

The NOSCCA initiative and process has been introduced at various meetings, symposia and conferences. Together with the Baltic Earth consortium a joint BACC-NOSCCA session *Climate change and its impacts in the Baltic and North Sea regions: Observations and model projections* was conducted during the European Geosciences Union General Assembly in 2015 and in 2016, where the first results were presented to the scientific public.

The assessment report comprises 19 chapters each allocated to one of four topical parts. Five annexes complement the climate change chapters with background knowledge. The assessment comprises past (the last 200 years) and current climate change, and climate change projections to the end of the century for the North Sea, the atmosphere and river flows; impacts of climate change on marine, coastal, and terrestrial ecosystems; and on socio-economic sectors, such as fisheries, agricultural systems, recreation, offshore activities, urban climate, air quality, coastal protection and coastal zone management. Long-term climate change was not an extensive theme of the present report; a few aspects are covered in the section *Geological and Climatic Evolution of the North Sea Basin* of the introductory chapter. Also *detection and attribution* and *adaptation measures* were not dealt with in depth in this first assessment but may be topics of follow-up activities. Concerning terminology, it should be noted that NOSCCA essentially follows the IPCC definition of the term "Climate change", and "anthropogenic" is explicitly added to that term when human causes are attributable. "Climate variability" is used, when referring to variations unrelated to anthropogenic influences.

The annexes cover the North Atlantic Oscillation (NAO), climate model simulations for the North Sea region, uncertainties in climate change projections, and emission scenarios for climate projections. The final annex provides facts about the Greater North Sea Region and geographical maps.

The NOSCCA report is written for a broad target readership ranging from scientists of different disciplines to authorities, agencies, decision makers and stakeholders acting in the North Sea region. It also aims to assist in the development of robust regional and local adaptation strategies.

Markus Quante

Overall Summary

The entire North Sea region is experiencing a changing climate and all available projections suggest the region will exhibit a wide range of climate change impacts over the coming decades. Among the robust results of this assessment are that the entire region is warming, and that the warming is almost certain to continue throughout this century; also that sea level is rising and will continue to rise at a rate close to the global average. Substantial natural variability in the North Sea region (from annual to multi-decadal time scales) makes it challenging to isolate regional climate change signals and impacts for some parameters. This is the case both for the observational period and for regional climate change projections and impact studies.

Projecting regional climate change and impacts for the North Sea region is currently limited by the small number of regional coupled model runs available and the lack of consistent downscaling approaches, both for marine and terrestrial impacts. The wide spread in results from multi-model ensembles indicates the present uncertainty in the amplitude and spatial pattern of the projected changes in sea level, temperature, salinity and primary production. For moderate climate change, anthropogenic drivers such as changes in land use, agricultural practice, river flow management or pollutant emissions are often more important for impacts on ecosystems than climate change.

The NOSCCA key findings that follow are provided as short statements. Quantifying the effects, changes or impacts has largely been avoided as this would require additional annotations or geographical specification. The aim here is to provide a concise summary of the major outcome of NOSCCA.

Recent Climate Change (Past 200 years)

Atmosphere

Temperature has increased everywhere in the North Sea region, especially in spring and in the north. Due to the lower heat capacity of land, land temperatures rise much faster than sea temperatures. The imbalance between the two is now nearly half a degree. Linear trends in the annual mean land temperature anomalies are about 0.17 °C per decade (for the period 1950–2010) and about 0.39 °C per decade (for the period 1980–2010). Generally, more warm and fewer cold extremes are observed.

There are indications that the persistence (duration) of circulation types has increased, with the consequence that 'atmospheric blocking' has become more frequent, thus contributing to the observation that extremes have become 'more extreme'. It is unclear how this is related to the decline in Arctic sea ice.

An observed north-eastward shift in storm tracks agrees with projections from climate models forced by increased greenhouse gas concentrations. This is a new phenomenon that has not been observed before.

While the number of deep cyclones (but not the number of all cyclones) has increased, whether storminess as a whole has increased cannot be determined: although reanalyses show an increase in storminess over time, observations do not. Variability from decade to decade is large, and clear trends cannot be identified. Furthermore, reanalyses can suffer from homogeneity issues and observations from errors made during digitization, emphasising the need for a manual quality check for the latter.

Overall, precipitation has increased in the northern North Sea region and decreased in the south, summers have become warmer and drier and winters have become wetter. Heavy precipitation events have become more extreme.

North Sea

There is strong evidence of surface warming in the North Sea especially since the 1980s. Warming is greatest in the south-east, exceeding 1 °C since the end of the 19th century.

Absolute mean sea level in the North Sea rose by about 1.6 mm/year over the past 100–120 years, comparable with the global rise. Extreme levels rose primarily because of this rise in mean sea level.

The North Sea is a sink for atmospheric carbon dioxide (CO_2); uptake declined over the last decade owing to lower pH and higher temperatures.

Short-term variations in all variables (including sea-surface temperature and sea level) exceed climate-related changes over the past two centuries. This is especially true for salinity, currents (varying with tides, winds, and seasonal density), waves, storm surges and suspended particulate matter (varying with currents, river inputs and seasonal stratification).

Coastal erosion is extensive but irregular and some coastlines are accreting. Evidence for a link to climate change has not yet been established.

River Flow

Rivers draining into the North Sea show considerable interannual and decadal variability in annual discharge. In northern areas this is closely associated with variation in the North Atlantic Oscillation, particularly in winter.

Discharge to the North Sea in winter appears to be increasing, but there is little evidence of a widespread trend in summer inflow. Higher winter temperatures appear to have led to higher winter flows, as winter precipitation increasingly falls as rain rather than snow.

To date, no significant trends in response to climate change are apparent for most of the individual rivers discharging into the North Sea.

Future Climate Change

Atmosphere

A marked mean warming of 1.7–3.2 °C is projected for the end of the 21st century (2071–2100, with respect to 1971–2000) for different scenarios (RCP4.5 and RCP8.5, respectively), with stronger warming in winter than in summer and relatively strong warming over southern Norway. The overall warming is accompanied by intensified extremes related to daily maximum temperature and reduced extremes related to daily minimum temperature, both in terms of strength and frequency.

Simulations project marked future changes in some aspects of the large-scale circulation over the Atlantic-European region, of which the North Sea region is part.

Changes in the storm track with increased cyclone density over western Europe in winter and reduced cyclone density on the southern flank of the storm track over western Europe in summer are projected to occur towards the end of the 21st century.

A general tendency for more frequent strong westerly winds and for less frequent easterly winds in the central North Sea as well as in the German Bight in the course of the 21st century was projected using SRES A1B and SRES B1 scenarios.

Projections suggest an increase in mean precipitation during the cold season and a reduction during the warm season for the period 2071–2100 relative to 1971–2000, as well as a pronounced increase in the intensity of heavy daily precipitation events, particularly in winter and a considerable increase in the intensity of extreme hourly precipitation in summer.

North Sea

Consistent results are found for projections regarding a warming of the surface water to the end of the century (about 1–3 °C; A1B scenario). Exact numbers are not given due to differences in spatial averaging and reference periods from published studies.

Coherent findings from published climate change impact studies include an overall rise in sea level, an increase in ocean acidification and a decrease in primary production.

Larger uncertainties exist for projected changes in salinity, mostly a freshening was reported, but contrasting signals were also projected. Uncertainties for projected changes in extreme sea level and waves are large.

Model studies reveal large uncertainties in future changes in net primary production with decreases ranging from 1 to 36 % (and not statistically significant across all parts of the North Sea region).

Substantial natural variability in the North Sea region from annual to multi-decadal time scales is a particular challenge for isolating and projecting regional climate change impacts. Separating natural variations and regional climate change impacts is a remaining task for the North Sea.

River Flows and Urban Drainage

Increased hydrological risks due to more intense hydrological extremes in the North Sea region such as flooding along rivers, droughts and water scarcity, are projected by climate models and are of socio-economic importance for the region. Risk is particularly enhanced in winter due to increases in the volume and intensity of precipitation.

Models project that peak flow in many rivers may be up to 30 % higher by 2100, and in some rivers even higher.

The impacts projected lead both to opportunities and challenges in water management, agricultural practices, biodiversity and aquatic ecosystems.

The exposure and vulnerability of cities in the North Sea region to changes in extreme hydrometeorological and hydrological conditions are expected to increase due to greater urban land take, rising urban population growth, a concentration of population in cities and an aging population. Business-as-usual approaches are no longer feasible for these cities.

Impacts of Recent and Future Climate Change on Ecosystems

Marine Ecosystems

The marine ecosystem of the North Sea is highly productive, intensively exploited and well-studied. The changing North Sea environment is affecting biological processes and organisation at all scales, including the physiology, reproduction, growth, survival, behaviour and transport of individuals. The distribution, dynamics and evolution of populations and trophic structure are also affected.

Long-term knowledge and exploitation of the North Sea indicates that climate affects marine biota in complex ways. Climate change influences the distribution of all taxa, but other factors (fishing, biological interactions) are also important.

The distribution and abundance of many species have changed. Warmer water species have become more abundant and species richness (biodiversity) has increased. This will have consequences for sustainable levels of harvesting and other ecosystem services in the future.

Coastal Ecosystems

Accelerated sea-level rise, changes in the wave climate and storms may result in a narrowing of dunes and salt marshes where they cannot spread inland, particularly in the case of a narrow and steep foreshore. The relative importance of accelerated sea-level rise, changes in the wave climate, storms, and local sediment availability and their interactions are poorly understood. Human impacts on geomorphology and sediment transport interact with the potential impacts of climate change.

Estuaries and most mainland marshes will survive sea-level rise. Back-barrier salt marshes with lower suspended sediment concentrations and tidal ranges may be more vulnerable. Depressions away from salt-marsh edges and creeks on back-barrier marshes may be at particular risk.

Plant and animal communities can suffer habitat loss in dunes and salt marshes through high wave energy. Natural succession, and management practices such as grazing and mowing have a strong impact. Minor storm floodings in spring negatively affect breeding birds. Invasive species may change competitive interactions.

Plant and animal communities are affected by changes in temperature and precipitation and by atmospheric deposition of nitrogen. Their interactions result in faster growth of competitive species. Increased plant production may cause losses of slow-growing and low-statured plant species.

Lake Ecosystems

The North Sea region contains a vast number of lakes. These freshwaters and the biota they contain are highly vulnerable to climate change.

Lakes in the North Sea region have experienced a range of physical, chemical and biological changes due to climatic drivers over past decades. Lake temperatures have increased, ice-cover duration has decreased and major changes have occurred in the fluxes of dissolved organic matter and key elements such as nitrogen, phosphorus, silicate, iron and calcium.

Together, all physical and chemical changes have had a profound impact on the biota from algae to fish and biodiversity, and these impacts are predicted to proceed and intensify in the future.

Terrestrial Ecosystems

There is strong empirical evidence of changes in phenology in many plant and animal taxa and northward range expansions of mobile thermophilous animals.

There is limited empirical evidence of climate-induced changes in vegetation patterns and ecosystem processes (carbon cycling) in terrestrial ecosystems. Predictions concerning vegetation patterns and ecosystem processes are almost exclusively based on modelling approaches.

Climate change projections and impact studies suggest a northward shift in vegetation zones, enhanced carbon release from soils, and increased export of dissolved organic carbon to aquatic ecosystems.

Future climate change is likely to increase net primary production in the North Sea region due to warmer conditions and longer growing seasons, as long as future climate change is moderate and summer precipitation does not decrease as strongly as projected in some of the more extreme climate scenarios. The physiological effects of increasing atmospheric CO_2 levels and increasing N-mineralisation in the soil may also play a significant role, but to an as yet uncertain extent.

Climate Change Impacts on Socio-economic Sectors

Fisheries

North Sea fisheries may be impacted by climate change in various ways. Consequences of rapid temperature rise are already being felt in terms of shifts in species distribution and variability in stock recruitment.

Although an expanding body of research exists on this topic, there are still many knowledge gaps, especially with regard to understanding how fishing fleets themselves might be impacted by underlying biological changes and what this might mean for regional economies.

It is clear that fish communities and the fisheries that target them will almost certainly be very different in 50 or 100 years from now and that management and governance will need to adapt accordingly.

Agricultural Systems

Climate change impacts on agricultural production will vary across the North Sea region, both in terms of crops grown and yields obtained. Increased productivity and wider scope of crops is expected for northern areas. Larger risks of summer drought and associated effects will be a challenge in southern parts. In general, more extreme weather events may severely disrupt crop production.

Given adequate water and nutrient supply, a doubling of atmospheric CO_2 concentration could lead to yield increases of 20–40 % for most crops grown in the North Sea region.

Increased risks of nutrient (nitrogen and phosphorus) loadings from agricultural land to aquatic systems are likely with projected climate change.

The challenge in the North Sea region will be to ensure sustainable growth in agricultural production without negatively affecting the environment and natural resources.

Offshore Activities/Energy

There is no doubt that energy systems and offshore activities in the North Sea region will be impacted by climate change.

While most studies suggest an increase in hydropower potential, climate projections are highly uncertain regarding how much the future potential of other renewable energy sources such as wind, solar, terrestrial biomass, or emerging technologies like wave, tidal or marine biomass could be affected, positively or negatively.

Both offshore and onshore activities in the North Sea region (of which offshore wind, oil and gas dominate) are highly vulnerable to extreme weather events, in terms of extreme wave heights, storms and storm surges.

Urban Climate

About 80 % of the population within the North Sea countries lives in an urban area and this percentage is projected to rise. Some larger metropolitan areas in the region are generally located in low altitude areas. This is especially true for the urban areas of the Netherlands (Amsterdam, Rotterdam, The Hague and Utrecht), as well as for Antwerp, London and Hamburg.

There are indications that climate change in the North Sea region, potentially affecting urban climate and thus the health and welfare of city dwellers, is now apparent and includes drier and warmer summers, more intense precipitation, sea-level rise and hinterland flooding.

Cities must adapt to climate change. Despite broad similarities between urban areas, in terms of mitigation and adaptation to climate change there are large location-specific differences with regard to city planning needs. As cities themselves strongly contribute to greenhouse gas emissions, there is an opportunity for them to change both simultaneously: adapting to and mitigating climate change.

Air Quality

In the North Sea region, poor air quality has serious implications for human health and the related societal costs are considerable.

The effects on air quality of emission changes since preindustrial times are stronger than the effects of climate change. Model simulations suggest this is also the case for future air quality in the region, but substantial variation between model results implies considerable uncertainty.

If the reductions in air pollutant emissions expected through increasingly stringent policy measures are not achieved, any increase in the severity or frequency of heat waves may have severe consequences for air quality.

Recreation

Sea-level rise, coastal erosion and storms can destroy coastal infrastructure and alter coastal landscapes. Rebuild costs and a decline in tourism revenue can have significant economic impacts. Nevertheless, tourism in the North Sea area is expected to profit from rising temperatures, lower summer precipitation and a longer season. Destination attractiveness is largely determined by thermal environmental assets. However, landscape changes, natural and man-made, such as reduced beach width and higher sea walls, may decrease destination appeal.

Tourists are unlikely to change travel behaviour. Coping with climate change and its effects will require changes in government policy and innovative approaches from tourism suppliers. Investment cycles should be made on a long-term basis.

Coastal Protection

All countries around the North Sea with coastal areas vulnerable to flooding due to storm surges are ready to take up the challenges expected to occur as a consequence of climate change. Scenarios of accelerating sea-level rise leading to sea levels by 2100 of up to 1 m or more above present day, in some countries accompanied by increased storm surge set-up and wave energy, have been used as a basis for evaluation and planning of the adaptation of coastal protection strategies and schemes.

Coastal protection strategies differ widely from country to country, not only in terms of distinct geographical boundary conditions but also in terms of the length of planning periods, the amount of regulations and budgeting.

All countries, except Denmark and the UK, which allow coastal retreat at some stretches of their coasts, aim at keeping the current protection line in place to protect the hinterland. Combatting coastal erosion by nourishments is currently the most effective solution used for sandy coastlines and will continue to be a major tool for balancing climate change impacts in these environments.

Coastal Management and Governance

Broadly shared assessments of the urgency of adaptation are hampered by the difficulty of identifying the climate-driven component of observed change in the coastal zone. Due to uncertainty about the extent and timing of climate-driven impacts, current adaptation plans focus on no-regret measures.

The most considered no-regret measures in the North Sea countries are spatial planning in the coastal zone (set-back lines), coastal nourishment, reinforcement of existing protection structures and wetland restoration including managed realignment schemes.

In Germany, the Netherlands and Belgium coastal adaptation is steered by national and regional programmes and plans. The UK and the Scandinavian countries pursue active public involvement by transferring adaptation responsibilities to private stakeholders and partnerships.

The NOSCCA Author Team

1 Introduction to the Assessment—Characteristics of the Region

Markus Quante, Franciscus Colijn, Jan P. Bakker, Werner Härdtle, Hartmut Heinrich, Christiana Lefebvre, Ingeborg Nöhren, Jørgen Eivind Olesen, Thomas Pohlmann, Horst Sterr, Jürgen Sündermann and Merja Helena Tölle

Abstract

This scene-setting chapter provides the basis for the climate change-related assessments presented in later chapters of this book. It opens with an overview of the geography, demography and major human activities of the North Sea and its boundary countries. This is followed by a series of sections describing the geological and climatic evolution of the North Sea basin, the topography and hydrography of the North Sea (i.e. boundary forcing; thermohaline, wind-driven and tidally-driven regimes; and transport processes), and its current atmospheric climate (focussing on circulation, wind, temperature, precipitation, radiation and cloud cover). This physical description is followed by a review of North Sea ecosystems. Marine and coastal ecosystems are addressed in terms of ecological habitats, ecological dynamics, and human-induced stresses representing a threat (i.e. eutrophication, harmful algal blooms, offshore oil and gas, renewable energy, fisheries, contaminants, tourism, ports, non-indigenous species and climate change). Terrestrial coastal range

Electronic supplementary material Supplementary material is available in the online version of this chapter at 10.1007/978-3-319-39745-0_1.

M. Quante (✉) · F. Colijn (✉) · I. Nöhren
Institute of Coastal Research, Helmholtz-Zentrum Geesthacht, Geesthacht, Germany
e-mail: Markus.Quante@hzg.de

F. Colijn
e-mail: Franciscus.Colijn@hzg.de

I. Nöhren
e-mail: ingeborg.noehren@hzg.de

J.P. Bakker
Groningen Institute for Evolutionary Life Sciences, University of Groningen, Groningen, The Netherlands
e-mail: j.p.bakker@rug.nl

W. Härdtle
Institute of Ecology, Leuphana University of Lüneburg, Lüneburg, Germany
e-mail: werner.haerdtle@uni.leuphana.de

H. Heinrich
Federal Maritime and Hydrographic Agency (BSH), Hamburg, Germany
e-mail: Hartmut.Heinrich@bsh.de

C. Lefebvre
German Meteorological Service (DWD), Hamburg, Germany
e-mail: Christiana.Lefebvre@dwd.de

J.E. Olesen
Department of Agroecology—Climate and Bioenergy, Aarhus University, Aarhus, Denmark
e-mail: jorgene.olesen@agrsci.dk

T. Pohlmann · J. Sündermann
Institute of Oceanography, CEN, University of Hamburg, Hamburg, Germany
e-mail: Thomas.Pohlmann@uni-hamburg.de

J. Sündermann
e-mail: Juergen.Suendermann@uni-hamburg.de

H. Sterr
Department of Geography, Kiel University, Kiel, Germany
e-mail: sterr@geographie.uni-kiel.de

M.H. Tölle
Department of Geography, Climatology, Climate Dynamics and Climate, Justus-Liebig-University Gießen, Gießen, Germany
e-mail: Merja.Toelle@geogr.uni-giessen.de

M. Quante and F. Colijn (eds.), *North Sea Region Climate Change Assessment*, Regional Climate Studies, DOI 10.1007/978-3-319-39745-0_1

vegetation is addressed in terms of natural vegetation (salt marshes, dunes, moors/bogs, tundra and alpine vegetation, and forests), semi-natural vegetation (heathlands and grasslands), agricultural areas and artificial surfaces.

1.1 Introduction

To provide a sound basis for the climate change related assessments presented in later chapters of this book, this introduction to the North Sea region reviews both the natural features of the region and the human-related aspects. This overview includes a comprehensive description of North Sea hydrography, the current climate of the region, and the various marine and coastal ecosystems. The depth to which these topics can be addressed in this introductory chapter is limited, so additional reference material is provided in Annex 5 to this report. To date, there are few sources that cover the North Sea region as a whole and that address natural, social, and economic issues. Notable among those that are available are the OSPAR Quality Status Report for the Greater North Sea Region (OSPAR 2000) and a characterisation of the North Sea Large Marine Ecosystem by McGlade (2002). Comprehensive reviews of North Sea physical oceanography are provided by Otto et al. (1990), Rodhe (1998) and Charnock et al. (1994, reprinted 2012), and of physical-chemical-biological interaction processes within the North Sea by Rodhe et al. (2006) and Emeis et al. (2015).

1.2 North Sea and Bounding Countries—A General Overview

Markus Quante, Franciscus Colijn, Horst Sterr

1.2.1 Geography of the Region

The North Sea region is situated just north of the boundary (~50°N) between the warm and cool temperate biogeographic regions, classified from north to south as Alpine North, Boreal, Atlantic North, Nemoral, and Atlantic Central (Metzger et al. 2005). The North Sea is a semi-enclosed marginal sea of the North Atlantic Ocean situated on the north-west European shelf. It opens widely into the Atlantic Ocean at its northern extreme and has a smaller opening to the Atlantic Ocean via the Dover Strait and English Channel in the south-west. To the east there is a connection with the Baltic Sea. The transition zone between the North Sea and Baltic Sea is located in a sea area between the Skagerrak and the Danish Straits, named the Kattegat. The continental bounding of the North Sea is well defined by the coastlines of the United Kingdom (Scotland and England) in the west, southern Norway, southern Sweden and Denmark (Jutland) in the east and Germany, the Netherlands, Belgium and France in the south (Fig. 1.1). The 'North Sea region' as defined in the NOSCCA context comprises the Greater North Sea and the land domains of the bounding countries, which form part of the catchment area. There are countries in the North Sea catchment area without a coastline, namely Liechtenstein, Luxembourg and larger parts of Switzerland and the Czech Republic. These countries are outside the scope of this assessment. A more formal definition of the boundary to the Atlantic Ocean may follow the internationally accepted setting proposed by the OSPAR Commission, where the western boundary of the so-called Greater North Sea (i.e. OSPAR Region II) is marked by the 5°W meridian and the northern boundary by the imaginary line along 62°N (OSPAR 2000). This definition encloses the

Fig. 1.1 North Sea region (map produced using Ocean Data View)

entire English Channel in the west. In the east, the Skagerrak and Kattegat are considered by OSPAR to be part of the Greater North Sea. Overall, the North Sea extends about 900 km in the north-south direction and about 500 km in the east-west direction. Including all its estuaries and fjords the Greater North Sea spans a surface area of about 750,000 km^2, the estimated water volume amounts to 94,000 km^3 (OSPAR 2000).

1.2.1.1 Catchment Area and Freshwater Supply

The total catchment area of the North Sea is about 850,000 km^2. Major rivers discharging freshwater into the North Sea are the Forth, Humber, Thames, Seine, Meuse, Scheldt, Rhine/Waal, Ems, Weser, Elbe, and Glomma. Some of these rivers form larger estuaries such as the Weser and Elbe in the German Bight and the Thames and Humber at the English east coast, while the mouth of the Rhine-Meuse-Scheldt system is more delta-like. The annual input of freshwater from all rivers is highly variable and within the range 295–355 km^3, with melt water from Norway and Sweden contributing about one third (OSPAR 2000). River runoff to the North Sea is considerably less than the freshwater input from the Baltic Sea. When runoff to the Danish Belts and Sounds is excluded, the net freshwater supply to the Baltic Sea is typically around 445 km^3 $year^{-1}$ (Bergström 2001). Overall, the salinity of Baltic Sea water is much lower than that of the North Sea. In the Kattegat, brackish water enters the North Sea in a surface flow with a counter flow of salty and oxygen-rich water to the Baltic Sea at depth (see Sect. 1.4.2). Total river runoff to the Baltic Sea is a good marker for the lower bound of freshwater delivered to the North Sea, since precipitation over the Baltic Proper is on average higher than evaporation, typically by about 10–20 % (e.g. Omstedt et al. 2004; Leppäranta and Myrberg 2009). According to Omstedt et al. (2004) the net outflow to the North Sea, excluding the Kattegat and Belt Sea water budget, is around 15,500 m^3 s^{-1} (or 488 km^3 $year^{-1}$) with an interannual variability of ± 5000 m^3 s^{-1}.

1.2.1.2 Coastal Types

The coasts of the North Sea display a large variety of landscapes, among the largest in the world. They vary from mountainous coasts interspersed by fjords and cliffs with pebble beaches to low cliffs with valleys and sandy beaches with dunes. There is a broad contrast between the northern North Sea coasts and the southern North Sea coasts; while the northern coasts are more mountainous and rocky the southern coasts are often sandy or muddy. This reflects regional differences in geology, glaciation history and vertical tectonic movements. The northern coastlines have experienced isostatic vertical uplift since the disappearance of the huge ice masses after the Weichselian glaciation (see Sect. 1.3.1).

The following description of the North Sea coastlines uses the morphological characterisation of the coastal landscapes by Sterr (2003). It should also be noted that the North Sea coast of the UK has been well documented in a set of dedicated volumes prepared by the UK Joint Nature Conservation Committee (Doody et al. 1993; Barne et al. 1995a, b, 1996a, b, c, 1997a, b, c, 1998a, b).

Starting in the northwest, the coasts of the Shetland and Orkney Islands and northern Scotland are mountainous showing a morphologically strongly structured rocky appearance. The area has a rugged and open character and predominantly comprises cliffed landscapes. The rocks of this region include some of the oldest in Great Britain.

The coasts of southern Scotland and northern England also feature cliffs of various sizes, which were shaped by several glaciation events and erosion. But in contrast to the northern part of Scotland, this region has a much gentler topography. Pebble beaches and intersections by river valleys are typical of this coastal region. The Firth of Tay and Firth of Forth are the most prominent features along the coastline; the latter being one of the major UK estuaries. The coast of northern England has fewer bays, headlands or estuaries compared to that of southern Scotland. Nevertheless it is still varied with cliffs alternating with stretches of lower relief.

The southern North Sea coast of the UK including the Humber and Thames estuaries comprises low-lying land that alternates with soft glacial rock cliffs. This makes it vulnerable to flooding and coastal erosion, and almost all of the open coast has some form of man-made defence along it (Leggett et al. 1999). Embedded in this area is a large indentation called the Wash—effectively a large bay into which four mid-sized rivers discharge. Most parts of the Wash are shallow and several large mudflats and sand flats are exposed at low tide. Much of the English Channel coast comprises cliffs, of both soft and harder rock. This results in a varied landscape with all the major coastal geomorphological structure types present. In the south-western segment the coast is punctuated by many narrow, steep-sided estuaries, with the River Exe the only typical coastal plain estuary in this region.

On the southern side of the North Sea especially between the Belgian lowland and the northern tip of Jutland an extended shallow coast has formed. The natural coastline of this region was strongly modified by storm floods during medieval times and has now changed considerably with the development of towns and harbours, land reclamation projects and coastal protection structures. The south-eastern North Sea coast of France, Belgium and the Netherlands is characterised by coastal dunes, sandy beaches and often a gentle shoreface. In the Delta area of the south-western Netherlands the straight coastline is interrupted by a complex of larger estuaries and tidal basins (Lahousse et al.

Fig. 1.2 Examples of the different types of coastline around the North Sea: cliff coast near St Andrews, Scotland (**a**), sandy beaches and dunes in Jutland, Denmark (**b**), tidal flats of the Wadden Sea, Germany (**c**), Norwegian fjord (**d**). *Photo copyright* T. Stojanovic (**a**), M. Quante (**b**, **d**), M. Stock (**c**)

1993). Some of these estuaries were closed in recent times. The subsequent Dutch coast, between Hoek van Holland and Den Helder is characterised by an uninterrupted coastal barrier with high dunes.

The Wadden Sea and barrier islands characterise the coast between the IJsselmeer in the Netherlands and the Blavandshuk peninsula *Skallingen* in Denmark. The Wadden Sea comprises a shallow body of water with tidal sand- and mudflats, which cover over two-thirds of the area, as well as salt marshes and other wetlands. The morphology of the Wadden Sea was described by Ehlers and Kunz (1993). It has the world's largest continuous belt of bare tidal flats, which are partially sheltered by a sandy barrier (Reise et al. 2010; Kabat et al. 2012). In 2009, the Wadden Sea was added to the UNESCO World Heritage List.

The West- and East Frisian barrier islands in front of the Dutch coast and the coast of lower Saxony in Germany are basically sandy dune islands and form an entity with the Wadden Sea and adjacent beaches of the mainland. Further north, the North Frisian and Danish islands have a different origin; they were originally parts of the mainland, their cores comprising glacial till from the Saalian Glaciation. Following Holocene sea-level rise, a north-south-trending barrier was formed from these islands, which fostered moorland growth and sheltered the mainland from medieval storm floods. Later, this segment suffered massive land losses during several disastrous medieval storm floods.

At the Danish west coast of Jutland large coastal dune areas have formed, some reaching several kilometres inland. There are no major estuaries along that coast and spits shelter former bays from the open sea. Overall, the graded shorelines of Jutland are similar in character to those in Belgium and the central part of the Netherlands.

In Sweden and in southern Norway up to Stavanger the typical skerry coasts are found, while the Norwegian coastline further north is mountainous and often dissected by deep fjords. The Norwegian and Swedish mainland is sheltered from the open ocean by a more or less continuous archipelago. Overall, the Swedish and Norwegian coast is strongly structured and dissected leading to a distinct interleaving of land and sea. The images in Fig. 1.2 show the diverse character of the different parts of the North Sea coast.

1.2.1.3 Demography

The North Sea plays a key role in one of the world's major economic regions, it is a place for settlement and commerce for millions of people and thus parts of its coastal area are densely populated. Establishing the overall population of the North Sea region without a strict geographical definition of the enclosed area is not straight forward. About 185 million people live within its catchment area (OSPAR 2000), which includes landlocked countries such as Liechtenstein, Luxembourg and larger parts of Switzerland and the Czech Republic. Although about 168 million people live within the catchments of countries bordering the North Sea, many probably consider they have no direct relation to the North Sea. Because a clear distinction between coastal zones and inland areas is difficult to perceive, to estimate population size for the North Sea region it may be more appropriate to include only those inhabitants whose daily living or economic activities are linked with the sea. The coastal regions as defined by the European Union (Eurostat 2011) could serve as a starting point for such a census. Here, a coastal region is a statistical region defined at NUTS level 3 (district level) that has a coastline or more than half of its population living within 50 km of the sea. According to Eurostat (2011) in 2008 about 205 million people lived in the coastal regions of the EU, of which 20.6 % or about 42 million lived in the North Sea maritime basin as defined by the EU (which excludes the Dover Strait, English Channel and Norwegian coast). Adding the number of people living in the relevant districts on both sides of the English Channel and the Dover Strait (about 18 million in 2008 according to Eurostat NUTS3) and an estimated 2.5 million Norwegians, it may be concluded that roughly 70 million people live in the North Sea region and use the coastline and marine environment in a number of ways. This estimate is reasonably close to an earlier estimate of 50 million people reported by Sündermann and Pohlmann (2011) and compiled in connection with the SYCON project (Sündermann et al. 2001a).

Population density varies widely around the North Sea, and is highest along the southern coast and lowest along the eastern coast. Heavily populated areas are found in the river basins of the Elbe, Weser, Rhine, Meuse, Scheldt, Seine, Thames, and Humber. The regions showing the highest population densities, with maxima exceeding 1000 inhabitants km^{-2} are found near the coast in the Netherlands and Belgium. In contrast, densities of less than 50 inhabitants km^{-2} are common along the coasts of Norway and Scotland. Four different types of region can be distinguished with respect to population density: the Netherlands and Belgium with very high density (>300 inhabitants km^{-2}), the UK and Germany with high density (>200 inhabitants km^{-2}), France and Denmark with medium density (>100–200 inhabitants km^{-2}), and Sweden and Norway with low density (<50 inhabitants km^{-2}). Along the North Sea coastline, the highest crude rates of population growth in recent years were in the English, Belgian and Dutch regions (Eurostat 2011).

Several large cities or agglomerations are situated in the North Sea region. In 2009, about 64 % of the population of EU coastal regions bordering the North Sea lived in predominantly urbanised areas (Eurostat 2011). Greater London with an official population of more than 8 million in 2012 is by far the most populous municipality, with the overall London metropolitan area having an estimated population of 12 to 14 million. Other large cities are Hamburg (1.77 million; metropolitan area 4.3 million), Amsterdam (0.75 million; metropolitan area 2.2 million), Oslo (0.61 million; metropolitan area 1.9 million), Rotterdam (0.58 million; metropolitan area 1.2 million), Bremen (0.55 million), Gothenburg (0.51 million), Edinburgh (0.48 million), The Hague (0.47 million), and Antwerp (0.47 million). In the Netherlands, the Randstad—a megalopolis comprising the four largest Dutch cities (Amsterdam, Rotterdam, The Hague and Utrecht) and the surrounding areas, including several midsize towns such as Haarlem, Delft, Leiden and Zoetermeer—has a population of 7.1 million and forms one of the largest conurbations in Europe. Another agglomeration has developed in northern England around the city of Newcastle; the Tyne and Wear City Region has more than 1.6 million inhabitants.

1.2.2 Major Human and Economic Activities

The North Sea region is a major economic entity within Europe and a very busy marine area with respect to human activities. The importance of the North Sea is determined by its geographic position off north-western Europe and the economic status of its surrounding countries. Its importance depends primarily on its transport function with several important harbours located on the North Sea coasts. This section provides only a few very general statements on coastal industries and agriculture in north-western Europe since detailed information on these sectors can be found in geography textbooks, such as that by Blouet (2012).

Various types of industry are established along the North Sea coasts, including metal and metal processing, chemicals, oil refineries, and shipbuilding, and these are mainly clustered in specific geographic locations. In the UK the coastal industries are found near the estuaries of the rivers Thames, Tyne, Tees and in the Firth of Forth as well as in the

Southampton area. On the French Channel coast industrial activities are concentrated in the Calais-Dunkerque region and around the Seine estuary. In Belgium the coastal industry is chiefly found in the Antwerp area near the Scheldt estuary. The estuaries of the Scheldt, Meuse and Rhine and the greater Amsterdam area are the major industrial regions in the Netherlands. German coastal industries are concentrated near the banks of the rivers Elbe, Weser, Ems and Jade. At the North Sea coast of Denmark, some industry occurs around the town of Esbjerg, more industrial activity is situated on the east coast of Jutland. In southern Sweden, major industrial activities settled around Gothenburg at the Kattegat. Along the southern and western coasts of Norway, industries developed in the innermost part of fjords, mostly connected to larger cities like Oslo, Stavanger and Bergen but also at sites where hydroelectric power is generated (OSPAR 2000). Most of these industrial activities are located either around an estuary or directly on the coast, and for efficient exchange of goods and materials several important harbours have developed.

Europe's busiest ports in cargo tonnage and container units are situated on the southern North Sea coast. Rotterdam (Netherlands) is the busiest reporting 291.1 million tonnes of bulk cargo throughput and over 12.3 million twenty-foot equivalent units (TEU) in 2014, followed by Antwerp (Belgium) with 62.8 million tonnes of bulk cargo and 9.0 million TEU and Hamburg (Germany) with 43.0 million tonnes of bulk cargo and 9.7 million TEU (Port of Rotterdam 2015). London and Felixstowe and several other British harbours together with Le Havre (France), Amsterdam (Netherlands), Bremen/Bremerhaven (Germany), and Gothenburg (Sweden) comprise the remainder. As these harbours play an important role in global trade the Straits of Dover and the North Sea proper contain some of the most heavily trafficked sea routes in the world. Several million tonnes of cargo are transported over the North Sea annually with about 280,000 ship movements a year (OSPAR 2000). At any given time, 900 to 1200 large ships are traversing the North Sea. The economy of north-western Europe depends strongly on the North Sea as a major transport corridor.

In north-western Europe there are several extended areas of intensive field-crop farming, these are concentrated in eastern England, northern Germany and large parts of the Netherlands. Areas of intensive animal production or fruit and vegetable farming are found in the coastal and southern areas of western Denmark and parts of Germany, the Netherlands, northern Belgium and northern Brittany. The main land areas in the North Sea region used for agriculture are introduced in Sect. 1.7.

The North Sea also fulfils a series of other economic, military and recreational functions. All coastal countries have declared an exclusive economic zone (EEZ); one of three area categories recognised by the United Nations Convention on the Law of the Sea (UNCLOS). An EEZ can extend from the territorial sea to 200 nautical miles from the baseline (commonly the low-water mark). Within this zone the coastal state has the sovereign right of exploitation of marine resources, including energy generation from water and wind.

All coastal states extract oil or natural gas from the North Sea with up to a thousand production platforms, depending on how they are counted; oil and gas are extracted in the northern North Sea and in the southern North Sea it is mostly gas. In 2007, production for the North Sea as a whole totalled 205 million tonnes of oil and 173 million tonnes of gas in oil equivalents (OSPAR 2010).

Although fisheries are a minor activity in terms of gross national product (GNP), thousands of fishing boats (over 5800) operate over the rich North Sea fishing grounds and are responsible for total landings of just over a million tonnes of fish and shellfish each year, primarily by British, Danish, French and Dutch fleets. In 2012, catches amounted to EUR 436 million (UK), EUR 327 million (Denmark), EUR 241 million (Netherlands), and EUR 196 million (France), together accounting for 81 % of the total value of landings in the North Sea (STECF 2014). Following a period of increasing depletion, fish stocks in the North Sea are now improving, owing to major reductions in the regional fishing industry five to ten years ago.

Exploitation of living resources other than from fin fisheries is restricted to the catch of shrimp and shellfish such as mussels and cockles in coastal seas like the Wadden Sea, although strong restrictions are in place in this area owing to its protected status as a World Heritage site. Mariculture is mainly restricted to coastal inlets, especially to fjords in Norway and Scotland.

A new economic activity in the North Sea is the establishment of large wind farms with up to 100 turbines per farm. A strong increase in the marine area covered by wind farms is expected especially in the German EEZ, but construction is occurring all along the coasts of Belgium, Denmark, the Netherlands and the UK. By 2015 there were 29 wind farms in the North Sea with a total installed capacity of 6000 MW, nine additional farms are under construction.[1]

Tourism is an important economic factor along the Belgian, Dutch, German, Danish and UK coasts. Tourists and daily visitors in the Wadden Sea region along the Dutch, German and Danish coasts are an especially important economic factor. Between 1998 and 2007 the annual number of visitors to the North Sea region increased from 50 million to 80 million (OSPAR 2010), potentially increasing pressure on the environment.

[1]Wikipedia, List of offshore wind farms in the North Sea. Accessed 15 October 2015.

Economic activities and the discharge of rivers draining extensive industrial and agricultural areas within the North Sea catchment, result in high nutrient and pollutant loads to the North Sea. Eutrophication has a long-term impact on the coastal regions of the North Sea, but some of its effects such as low oxygen levels in bottom waters in the German Bight have decreased over the past two decades. However, the North Sea coast remains on the list of OSPAR-designated eutrophication problem areas (OSPAR 2010). Riverine and direct pollution by heavy metals (cadmium, lead, mercury) also decreased substantially between 1990 and 2006 (OSPAR 2010). Polycyclic aromatic hydrocarbons (PAHs) and polychlorinated biphenyls (PCBs) are still widespread in the North Sea region (OSPAR 2010), with a large proportion of the sites monitored showing unacceptable levels. The burden of pollution by persistent organic pollutants continues albeit their character changes over time, owing to the phasing out of some compounds and the introduction of others such as flame retardants (Theobald 2011).

Nature conservation in the North Sea is based on the designation of EU protected areas. By the end of 2012, marine Natura 2000 sites covered nearly 18 % of waters in the North Sea region (EEA 2015).

The various and extensive uses of the North Sea collectively place a strong environmental pressure on the entire marine ecosystem.

1.3 Geology and Topography of the North Sea

Hartmut Heinrich

1.3.1 Geological and Climatic Evolution of the North Sea Basin

1.3.1.1 General Settings

The North Sea Basin in its present shape started to form after the end of the Variscian Orogeny during the Permian Period about 260 million years (Ma) ago. In a subsiding area between the Palaeozoic mountain chains with the Scottish and London Massifs in the west, the Brabant, Rhenohercynian and Bohemian Massifs in the south and Fennoscandia and Greenland in the north a tectonically very active basin formed which has accumulated about 10 km of sediments in certain locations (the Horn, Central and Viking Graben) that vary in character from terrestrial to fully marine. Over the same period, and as a consequence of plate tectonics Europe has drifted northwards from the equator towards its present position. In parallel, the climate of the area has successively evolved from extremely arid to subtropical, moderate, and finally over the last million years into the highly variable conditions of the glacial period (Ziegler 1990; Gatcliff et al. 1994; Balson et al. 2001; Lyngsie et al. 2006).

1.3.1.2 Permian to Holocene

During the **Permian [296–250 Ma]**[2] the North Sea region was situated close to the equator. To the North the desert-like hot and dry area was connected to the open ocean through a gap between Greenland and Norway leading to the deposition of different types of chemical sediments comprising carbonates, anhydride/gypsum and various types of salt that later formed structures that acted as traps for commercially exploitable hydrocarbons (Glennie 1984; Johnson et al. 1993). The paleogeographic setting is shown in Fig. 1.3.

At the onset of the **Triassic [250–200 Ma]** the region arrived in the zone of the northern trade winds. The marine basin began to rise culminating during the Lower Triassic in a desert-like landscape with fluvial deposits. Later, its southern part for roughly 10 million years [243–235 Ma, Muschelkalk] became part of the shallow marine German-Polish Basin where under subtropical climatic conditions mainly lime sediments were deposited.

During the **Jurassic [200–145 Ma]** the North Sea region was still located in the subtropical climate belt. This was a period of strong tectonic activity on Earth. The Tethys Ocean progressed from East Asia westward across the ancient Pangaea continent dividing it into the Gondwana continent (India-Africa-South America-Australia-Antarctica) and Eurasia. At the same time, in the south Gondwana began to split into an African and a South American part forming the Proto-Southern Atlantic. The growing mid-ocean ridges led to substantial sea-level rise.

As a result the North Sea basin developed wide connections to the Tethys Ocean and the Proto-Atlantic Ocean with far greater exchange of water than before. Within the North Sea area tectonic movements caused fluctuations in the distribution of land and sea that created a variety of depositional environments including floodplains and fluvio-deltaic to shallow marine systems. Towards the south as water deepened clays and limestone were deposited. These types of marine sediment indicate a warm productive environment with oxygen-depleted conditions at least in the bottom waters. The organic matter deposited during the mid-Jurassic is the source of most hydrocarbons in the North Sea exploited today.

The tectonic rise of the North Sea basin reached its greatest extent around the onset of the **Cretaceous [145–65 Ma]** and about half the area was dry land,

[2]All ages from Deutsche Stratigraphische Kommission (Menning and Hendrich 2012).

Fig. 1.3 Paleogeography of the European area during the Permo-Triassic (*upper*, ~250 Ma), the Upper Cretaceous (*middle*, ~75 Ma), and the Miocene (*lower*, ~13 Ma). The current position of the North Sea is marked by the *red circle* (graphical reproduction permitted by Prof. Blakey; http://cpgeosystems.com/paleomaps.html)

accompanied by a drop in global sea level (Haq et al. 1987). Continental rifting in the Atlantic Ocean began to separate North America from Europe (see Fig. 1.3). Uplift of the London-Brabant-Rhenish-Bohemian Massifs largely closed the connection to the Tethys Ocean in the south, whereas the connection via the Viking Graben to the Proto-Arctic Ocean remained open.

Later, when sea level started rising again a connection to the Proto-North Atlantic developed and, via the Paris Basin in the west and the Polish Straits in the east, water exchange with the Tethys again intensified. During the Lower Cretaceous [140–100 Ma] the fluctuating but generally rising sea level in conjunction with the warm climate generated a succession of clay and calcareous deposits. At the Lower-to-Upper Cretaceous boundary the North Sea Basin was fully flooded. In the Upper Cretaceous [100–65 Ma] continental rifting in the North Sea area accelerated leading to strong subsidence of the Central Graben.

During the period of high sea level water exchange between the North Sea area and the surrounding open oceans through the wide straits resulted in a warm and well oxygenated marine environment. Both the very warm climate and the high global sea level are linked with the development of a super plume in the Earth's mantle in the western Pacific Ocean that was accompanied by very strong volcanic activity from 120 to 80 Ma outgassing huge quantities of carbon dioxide (CO_2).

During the **Tertiary [65–2.6 Ma]** the North Sea region arrived in its present latitudinal position. The climate changed dramatically from relatively warm towards boreal conditions. Antarctica arrived at its present position at the South Pole and was climatically isolated from the rest of the globe. Huge ice sheets began to form in Antarctica as well as on Greenland and on high mountain areas leading to a global cooling and a notable drop in sea level. Connecting the Proto Atlantic Ocean with the now cold Arctic Ocean and in the North Sea the closure of the seaways to southern Europe at the Oligocene–Miocene boundary [24 Ma] enabled much cooler water to move into the North Sea Basin changing the climate in the northern half of Europe. The first ice-rafted debris appeared in the northern North Sea around the end of the Tertiary [2.4 Ma] (Rasmussen et al. 2008). Sea level was increasingly controlled by glacio-eustatic changes.

Increasing erosion in the rising land areas bordering the wider North Sea and the Polish Basin completely changed the coastal landscape. Extensive deltas developed at the mouth of river Rhine and on the present western Baltic Sea–Polish Platform areas; both deltas progressing slowly towards the centre of the North Sea basin. During the warm periods of the early Tertiary these deltas were covered with dense paralic forests and swamps that led to coal formation. At the Miocene–Pliocene boundary [1.8 Ma] the North Sea had reached a geographic and bathymetric size close to that of the present day.

In the southern and central parts of the North Sea area the deltaic regime of the Upper Tertiary proceeded into the **Quaternary [2.6–0 Ma]** with increasing rates of deposition, while the north comprised a deeper pelagic depositional environment. The faunal composition points towards water temperatures similar to those of today in the southern North Sea (Nilsson 1983).

The **Pleistocene [2.6–0.012 Ma]** in the North Sea area is a time of extremely variable climatic conditions. Beginning in the Middle Pleistocene [~1 Ma] glacial and interglacial types of sedimentary processes dominated the North Sea Basin. During glacial periods the sea level dropped several tens to hundred or more metres exposing a dry landscape, which was sometimes covered by ice sheets. These ice sheets usually originated in the Norwegian-Swedish-Scottish Mountains. They carried large amounts of eroded rock debris of varying grain size into the basin forming moraines and other glacigenic features. During the temperate interglacial periods the North Sea experienced repeated changes in sea level that led to brackish to marine deposits (Ehlers 1983; Schwarz 1991; Sejrup et al. 2000; Ehlers and Gibbard 2008; Gibbard and Cohen 2008).

As a result of erosion from the temporary ice covers and alternating transgressive and regressive sea levels the climatic record of the North Sea area is very fragmentary. The oldest traces of a glaciation in the North Sea area are reported from the Netherlands [1.8–1.2 Ma; MIS[3] 34–36] and Denmark [~0.850 Ma; MIS 22–19]. The first glaciation documented, which covered the southern North Sea as far south as 52°N was the Elsterian/Anglian glaciation [~0.48–0.42 Ma; MIS 12]. At the end of the glaciation [0.425 Ma] a catastrophic flood from a collapsed ice-damned lake in the southern North Sea area is deemed to have cut a first gorge into the Weald–Artois chalk range that is now Dover Strait. A second breach of that barrier occurred roughly 240,000 years later during the Saale/Wolstonian glaciation.

The subsequent Holstein/Hoxnian interglacial [0.424–0.374 Ma; MIS 11] in terms of water temperature and geographical extent is meant to be an analogue of the present North Sea. A very narrow Dover Strait did not allow a significant inflow of temperate water from the Bay of Biscay.

The Saale/Wolstonian Glacial complex [0.350–0.130 Ma; MIS 10–6] is a succession of cold and slightly milder periods mostly represented by glacio-marine deposits in the North Sea. There were two phases of greater ice advance, the Warthe and Drenthe stadials that formed moraines off the North Sea coasts.

[3]MIS: Marine Isotope Stage.

The most recent interglacial period was the Eemian/Ipswichian [0.130–0.115 Ma; MIS 5e]. The climate at that time is considered to have been similar to that of the present day, and possibly warmer for some period. The maximum global sea level was 4–6 m higher than today (Dutton and Lambeck 2012) such that marine sediments of Eemian age extend inland far beyond the present coastlines.

In the subsequent Weichselian/Devensian glaciation [0.115–0.017 Ma; MIS 5d − 2] the climate slowly cooled until 0.075 Ma [MIS 4]. At that time temperature dropped rapidly and initiated a first ice advance from the Scandinavian mountains towards the North Sea coasts. Within the next 50,000 years the climate switched on millennial time scales between cool (Dansgaard-Öschger events) and extremely cold (Heinrich events) until the maximum extent of inland ice was reached at about 0.020 Ma [MIS 2]. In parallel, sea level dropped discontinuously to 120 m below the present level. A pro-glacial landscape with rivers, marshes, lakes and lagoons subsequently developed in the North Sea Basin, as illustrated in Fig. 1.3. Sediments representing the coldest part of the Weichselian/Devensian in the southern North Sea are mainly of aeolian and riverine origin. There is still debate about a possible closure of the Norwegian and Scottish ice sheets. Closure of these ice sheets is supported by glacio-lacustrine sediments found in depressions on top of Dogger Tail End and in the Elbe Urstrom Valley area.

The melting of the ice sheets began at about 0.019 Ma. However, global warming was discontinuous and showed several rapid changes, such as cold excursions like Heinrich Event 1 [17.5–15.0 ka[4]] and the Antarctic Cold Reversal [14.7–14.2 ka] in the southern hemisphere. Warm excursions included the Bölling and Alleröd periods [14.7–12.7 ka]. The concomitant rate of sea-level rise changed abruptly. During the Bölling period, for example, sea level rose about 20 m within 500 years (Melt Water Pulse 1a). The Pleistocene ended with the Younger Dryas [12.7–11.7 ka], a cold spell of still unknown origin.

The **Holocene [0.012–0 Ma]** in the North Sea area is characterised by increasing warmth and a rising sea level that rapidly flooded the flat glacial landscape. The present sea level was reached about 2000 years ago, and continues to rise slowly.

The Holocene climatic warming following the Younger Dryas cold spell did not evolve continuously. During the Preboreal [11.6–10.7 ka] temperatures in the northern hemisphere rose rapidly. Temperate conditions developed in the North Sea region in the Boreal [10.7–9.3 ka]. The warmest period of the Holocene was the Atlantic [9–5 ka]. Temperatures and sea level were higher than today, as was precipitation. The composition of pollen in sediments shows the onset of extended human influence (Mesolithic) on the flora of north-western Europe. During the subsequent Subboreal [5–2.5 ka] the climate was slightly cooler than in the Atlantic period and drier. The final stage of the Holocene, the Sub Atlantic was a period of climatic oscillations with a general tendency to cooler and wetter conditions than in the Subboreal. Its warmest period was the medieval Climatic Optimum [900–1100 AD]. Temperatures then dropped towards the Little Ice Age [1300–1850 AD].

The post-glacial inundation of the North Sea north of the Dogger-Fischer Bank Ridge started at the end of the Younger Dryas [12.7–11.7 ka]. Along the northern margin of the North Sea rock debris was dumped by icebergs from the Scottish and Norwegian mountain glaciers. At about 9 ka sea water entered the southern North Sea through the gap that forms the northernmost parts of the Elbe Urstrom Valley between Dogger Tail End and Fischer Bank/Jyske Rev (Konradi 2000). At about 8.5 ka the Dogger Bank became an island owing to the flooding of the Silver Pit depression between England and the Dogger Bank. During this time the southern North Sea landscape evolved from dry land via a shallow swampy environment into a brackish lake or lagoon with lagoon-type sedimentation. Slightly later, at 8.3 ka the ingress of marine waters from the Northeast Atlantic through the English Channel established full marine conditions in the North Sea. The Dogger Bank Island with its Mesolithic settlements was finally drowned around 8.1 ka. During the short period between the first marine ingression through the Elbe Valley and the flooding of the Dogger Bank, sea level rose at an average rate of 1.25 m per century culminating in the latter half of this period at a rate of 2.1 m per century. Later the rate of rise slowed to 0.30–0.35 m per century. The rate of sea level rise finally approached present-day values at about 2.8 ka. A unique sea level curve for the North Sea is still not available due to very different and partly unresolved local effects of post-glacial rebound and other related tectonic movements (Shennan et al. 2006; Vink et al. 2007; Baeteman et al. 2011).

Although it is likely that Early Holocene hunter and gatherer communities occupied the vegetated landscape, apart from the Mesolithic human traces found on Dogger Bank evidence has not been found to support this. Settling along the coasts and in the marshlands started to increase with the onset of the Sub Atlantic (Iron Age), which included the building of simple dykes and terps as protection against storm surges and changes in sea level. Human settlements in the southern coastal area of the North Sea can be traced back to 3 ka.

[4]ka: calendar years before Present (absolute age).

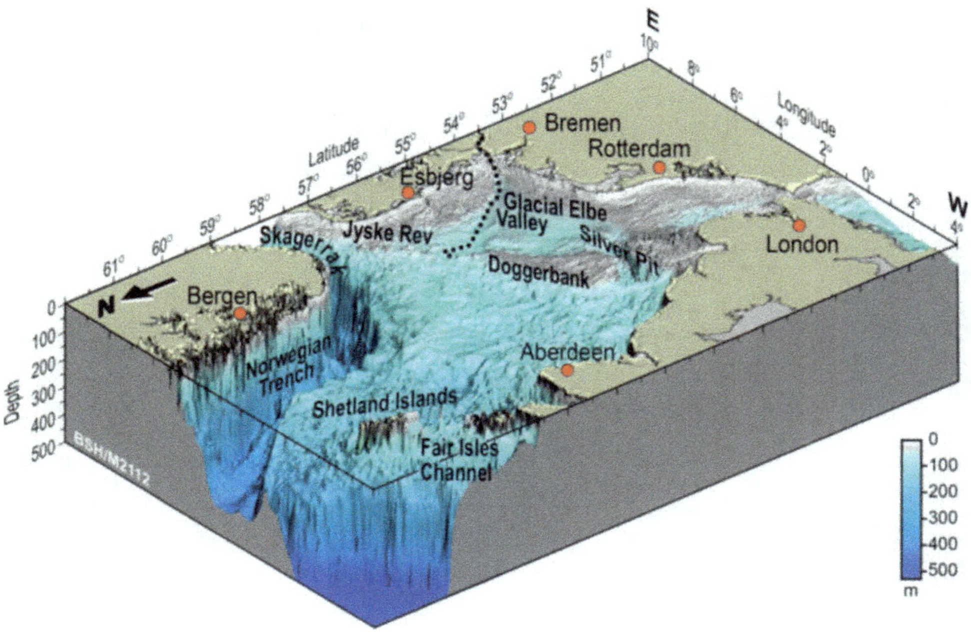

Fig. 1.4 Bathymetry of the North Sea

1.3.2 Topography of the North Sea

The North Sea is a continental shelf sea of the North Atlantic Ocean. It is bounded in the west by the British Isles, in the north by Norway, in the east by the Jutland Peninsula and North Frisia and in the south by the East and West Frisian Coast. In the north it is widely open to the deep Norwegian Sea. In its south-western corner the North Sea is connected to the Celtic Sea and subsequently to the Northeast Atlantic through the shallow Dover Strait/Pas de Calais with a minimum width of 30 km. A shallow connection to the Baltic Sea exists via the Skagerrak, Kattegat and Danish Belt Sea between the Jutland Peninsula, the Danish Islands and Scandinavia.

The North Sea dips gently from the shallow Frisian coasts in the south towards the continental margin along the deep Norwegian Sea. Average water depth in the North Sea is about 90 m. The present-day bathymetry of the North Sea is shown in Fig. 1.4. The western part of the northern basin has a trench 230 m deep, 30 km long, and 1–2 km wide, named the Devils Hole. Along the Norwegian coast into the Skagerrak spans a deep furrow, the Norwegian Trench which is up to 700 m deep.

The North Sea is mostly relatively flat. A ridge, the Dogger Bank–Jyske Rev subdivides the North Sea into a southern and a northern basin. The southern basin has a maximum depth of 50 m. North of this ridge the basin declines smoothly towards the shelf edge at about 200 m depth.

The Dogger Bank–Jyske Rev complex extends from east of Flamborough Head in England to the northern tip of the Jutland Peninsula. In the west the Silver Pit, a southeast-northwest trending gap about 100 km wide and up to 80 m deep separates the central Dogger Bank from its western part that extends to the coast of Norfolk and Suffolk. A smaller gap about 60 m deep right in the middle of the ridge, the breakthrough of the glacial Elbe Valley, separates the Dogger Bank/Dogger Tail End from the Jyske Rev. Minimum water depth on the Dogger Bank is 15 m below sea level. The Jyske Rev rises from 35 m at its western tip to the coast of northern Jutland. The two gaps enable deeper central North Sea water masses to pass from the northern basin to the southern basin.

1.4 Hydrography—Description of North Sea Physics

Jürgen Sündermann, Thomas Pohlmann

As generally in the ocean, the physical system of the North Sea is determined by the space-time distribution of seven macroscopic variables (Krauss 1973): temperature, salinity, density, pressure and the three components of the velocity vector. All additional physical quantities (such as water level elevation, thermocline depth, energy, and momentum) can then be calculated from the former. Current knowledge of all seven variables is based on field observations, remote sensing and model simulations.

The three-dimensional fields of temperature and salinity and their low-frequency variability are well-known, including statistical parameters such as error bounds, evidence, and confidence. Density can be calculated with high accuracy by means of the equation of state. Pressure can be related to sea-level elevation at the surface or within the water column,

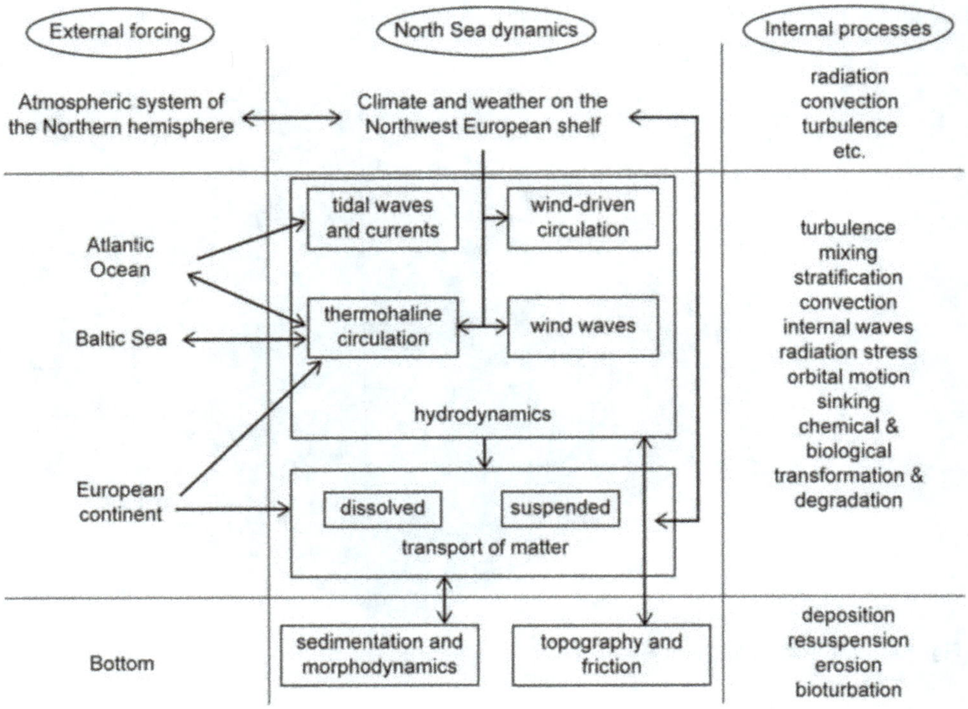

Fig. 1.5 Scheme of interactions within the physical system

which can be observed. Current velocities are only measured at certain points/sections and over defined periods. The best overall information today is given by hydrodynamic models combined with measurements assimilated into the model (Köhl and Stammer 2008).

Dissolved and suspended substances (such as nutrients, contaminants, and particulate material) may also be considered physical properties and their distributions are widely observed. Modelling requires knowledge of the sources and sinks of these substances.

To understand the physical system (and enable its simulation by models), knowledge of the interactions and processes within the current and transport regime including boundary forcing is also fundamental (Fig. 1.5).

North Sea dynamics are characterised by the regional interaction of the atmosphere, hydrosphere and lithosphere on the Northwest European shelf, and exhibit a broad spectrum of spatial and temporal scales (Fig. 1.6). Specific features of the physical system include turbulence, baroclinic eddies, internal waves, surface waves, convection, and tides.

Owing to the many interactions between them the physical, chemical and biological compartments of the North Sea system cannot be separated. Certainly, physics is essential for transport of chemical and biological substances and for the dynamics of the North Sea ecosystem as a whole. But there is also feedback depending on the scales considered, for example, in radiation flux, albedo, surface films, and surface roughness.

1.4.1 Boundary Forcing

The North Sea is part of the Northwest European shelf; its physical state is essentially determined by the adjacent Atlantic Ocean and the European continent including the Baltic Sea. Furthermore, the atmosphere of the northern globe is essential for the external forcing (Schlünzen and Krell 2004).

1.4.1.1 Atlantic Ocean

The open connection with the Atlantic Ocean allows the free exchange of matter, heat and momentum between the two seas. Planetary waves generated by astronomical and atmospheric forces in the ocean propagate over the shelf break into the North Sea generating changes in sea level, tides and water mass transport. In contrast, continental fresh water discharges influence the water characteristics of the North Atlantic (Figs. 1.7, 1.8 and 1.9).

There is an inflow of cold and salty Atlantic water into the deeper northern central basin of the North Sea and along the Norwegian Trench up to the Skagerrak. Less salty coastal waters circulate in an anti-clockwise gyre in the southern North Sea basin ultimately joining the Baltic Sea outflow (Winther and Johannessen 2006).

The total transport at the northern entrance exhibits a net outflow from the North Sea with an average of 2 Sverdrups (Sv) and annual variations of around 0.4 Sv (Fig. S1.4.1 in the Electronic (E-)Supplement, Schrum and Siegismund 2001).

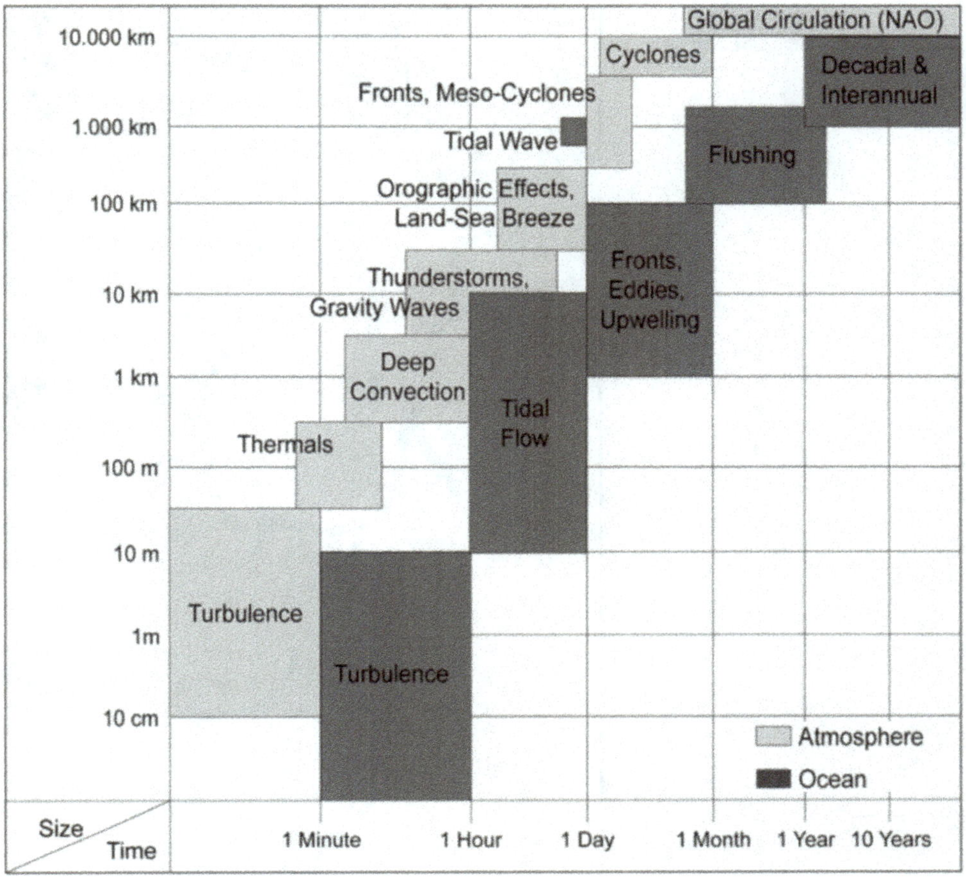

Fig. 1.6 Scale spectrum of the physical system (Sündermann et al. 2001b)

Decadal variability within the Atlantic Ocean, mainly driven by the North Atlantic Oscillation (NAO) (Hjøllo et al. 2009) and to a lesser extent by the Atlantic Multi-decadal Oscillation (AMO), is transferred to the North Sea. For heat this mainly occurs through the atmosphere, less through direct exchange of water masses, as is shown by the correlation pattern of NAO versus sea-surface temperature anomalies (Fig. S1.4.2 in the E-Supplement). High values in the central North Sea indicate this interrelation. Correlations reach their minimum at the north-western entrance and in the Southern Bight, indicating the influence of the advective heat transport from the Atlantic Ocean in these areas.

Net transport at the North Sea border is a superposition of continental outflow and Atlantic inflow. Substances moved by the respective currents enter the North Sea (salt, nutrients) or the Atlantic Ocean (fresh water, contaminants).

1.4.1.2 European Continent and Baltic Sea

The continental influence on the water characteristics of the North Sea comprises freshwater supply, and the input of dissolved and suspended matter: nutrients, contaminants, sediments.

Owing to continental freshwater discharge and the positive difference 'precipitation minus evaporation', the Baltic Sea exhibits a mass surplus resulting in a mean net outflow of 15,500 $m^3\ s^{-1}$ (Omstedt et al. 2004). More precisely, from the North Sea perspective there is a permanent inflow of freshwater, superimposed by a weak outflow of saltwater. During episodes of decadal frequency this outflow can strongly increase (salt water intrusion events) and renew and ventilate the Baltic Sea deep water.

Through its geostrophic balance the Baltic Sea outflow causes an eastward elevation in sea level of the order of centimetres within the Kattegat.

1.4.1.3 Atmosphere

Through the downward flux of momentum the atmosphere significantly controls the general circulation of the North Sea as well as the vertical structure of the water column due to turbulence. The frequency distribution of the wind direction and speed determines the major current patterns in the North Sea (see Sect. 1.4.3). The wind forcing exhibits a predominantly southwestern direction with significant decadal variability.

The wind further controls the spectrum of surface waves in the North Sea, and cyclones can lead to strong storm surges. The atmosphere influences the heat budget via vertical heat fluxes and their variability. Precipitation on the Northwest European shelf influences North Sea salinity and its seasonal variability directly or via continental discharge (see Sect. 1.5.4).

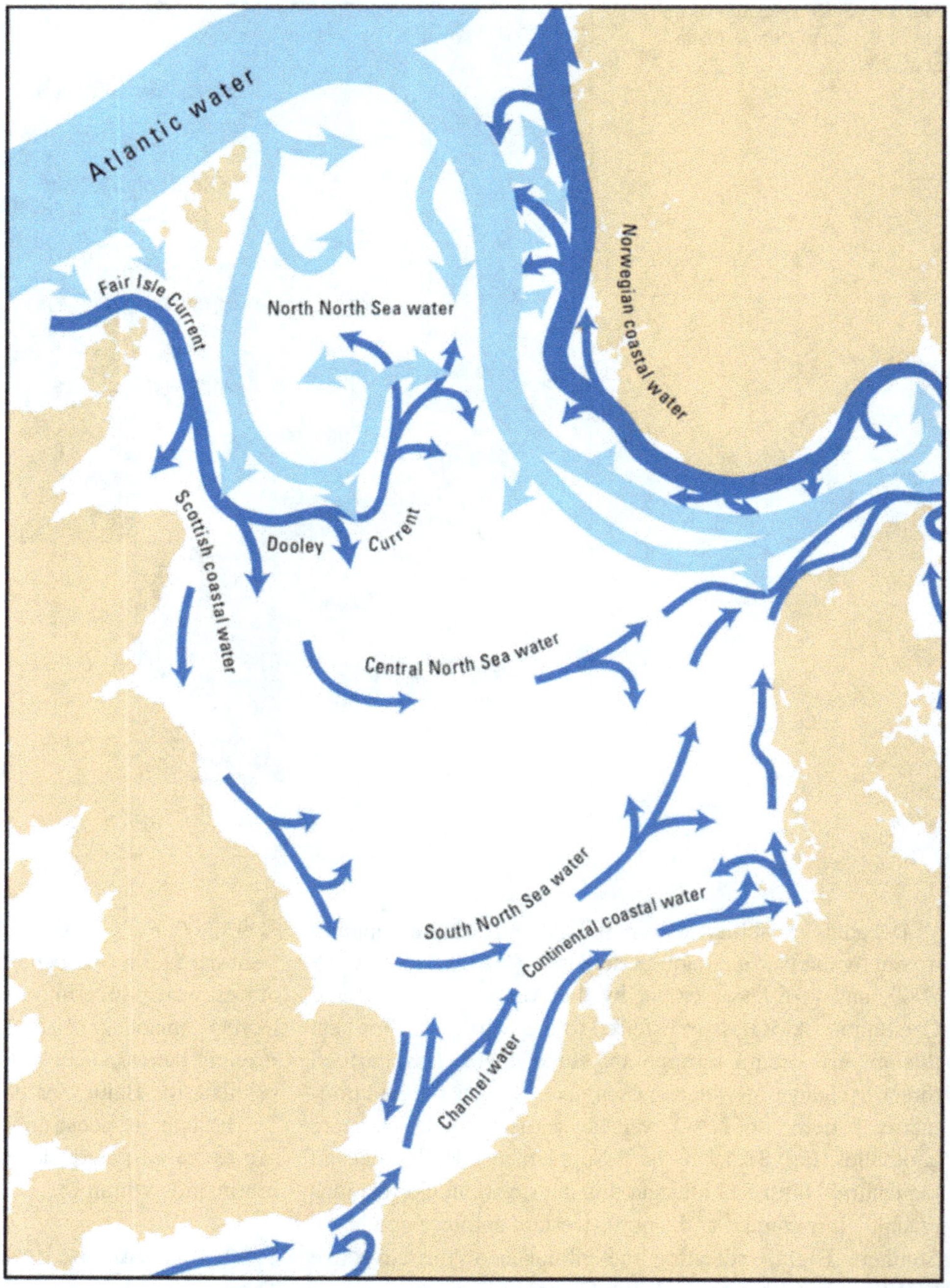

Fig. 1.7 Circulation system of the North Sea (after OSPAR 2000)

1.4.1.4 Seabed

The seabed represents a source or sink of sediment (by erosion, resuspension, or deposition) and associated substances. Related biogeochemical processes are important. Morphodynamics affect topography and as a consequence currents and waves.

1.4.2 Thermohaline Regime

The thermodynamic state of the North Sea is characterised by the 4-dimensional distributions of temperature and salinity (Figs. 1.8 and 1.9) and the related variable baroclinic circulation.

The strong seasonal variation in sea-surface temperature is the most evident low-frequency periodic feature in the North Sea. The annual mean shows a relatively homogeneous water mass with a tongue of warmer Atlantic water inflowing through the English Channel. Superimposed on this is the seasonal wave with amplitude increasing from northwest to southeast.

The annual mean salinity distribution reflects the inflow of Atlantic water through the northern entrance and the English Channel as well as the fresh water supply from the

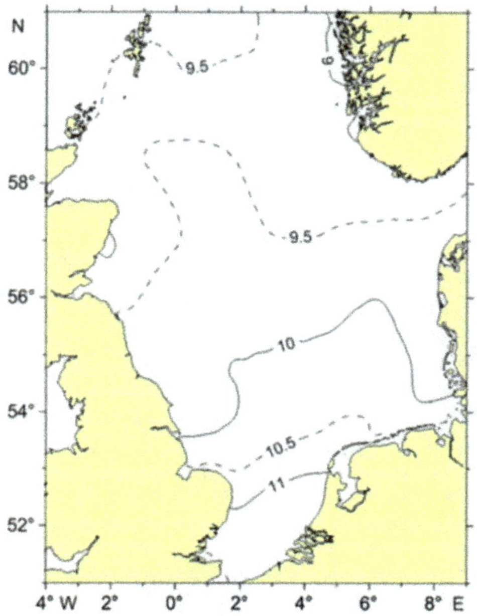

Fig. 1.8 Sea-surface temperature in the North Sea: annual mean (*left*) and seasonal variability (*right*) in °C, for the period 1900–1996 (data from Janssen et al. 1999)

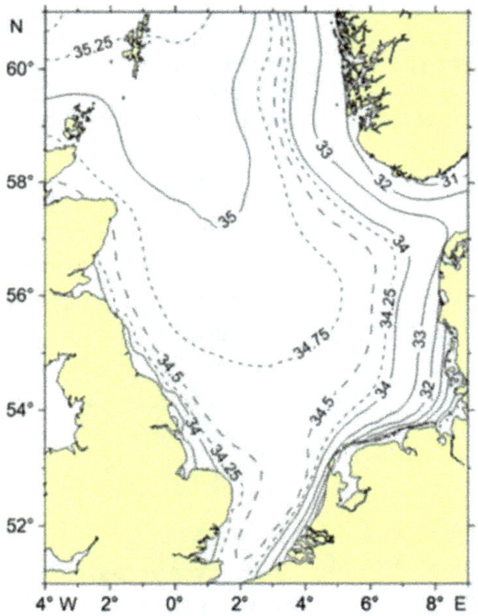

Fig. 1.9 Sea-surface salinity in the North Sea: annual mean (*left*) and seasonal variability (*right*) in psu, for the period 1900–1996 (data from Janssen et al. 1999)

European continent including the Baltic Sea outflow. Seasonal variation is most obvious in the southern and eastern coastal regions.

Figure 1.8 (right) shows the dominant seasonal temperature cycle. As explained in Sect. 1.4.1, temperature is mainly determined by heat exchange with the atmosphere. In the vertical, temperature development also shows significant differences between the southern and northern North Sea. While in the southern North Sea vertically homogenous conditions occur all year round, in the northern North Sea a summer thermocline develops in waters of around 30 to 40 m in depth (Fig. 1.10 and S1.4.3 in the E-Supplement).

Using the equation of state (Gill 1982), the baroclinic pressure gradients can be determined from the temperature and salinity fields and thus the geostrophic current system (Fig. S1.4.4 in the E-Supplement). This represents the density driven flow, which is superimposed by the wind driven current and the stronger tidal flow (see Sects. 1.4.3 and 1.4.4 and Pohlmann 2003). Tidal residual currents of the M_2 constituent are shown in Fig. S1.4.6 in the E-Supplement.

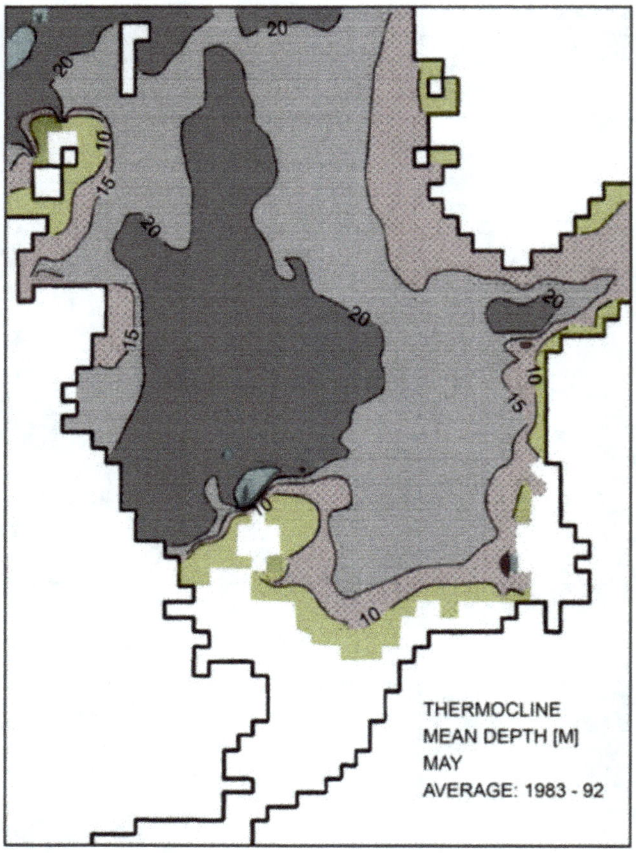

Fig. 1.10 Climatological monthly mean extension and depth of the thermocline in May (Pohlmann 1996)

For further studies on this topic see Luyten et al. (2003) and Meyer et al. (2011).

1.4.3 Wind-Driven Regime

The wind-induced circulation is clearly the dominant permanent regime, characterising the mean current system of the North Sea. Tidal currents may be stronger, but are almost periodic with relatively small net transports. Figure 1.11 shows the basic patterns of the wind-driven currents according to wind direction. Owing to the prevailing southwesterly winds on the Northwest European shelf an intense anti-clockwise circulation is dominant, which occasionally (in the case of easterly winds) reverses. For north-westerly and south-easterly winds, states of stagnation appear.

The NAO strengthens or weakens these patterns (see Fig. 1.12).

Storm surges constitute the greatest potential natural hazard for coastal communities in the North Sea region. An analysis of historical storm surges (Sündermann 1994) indicates that these extreme events fall into two classes:

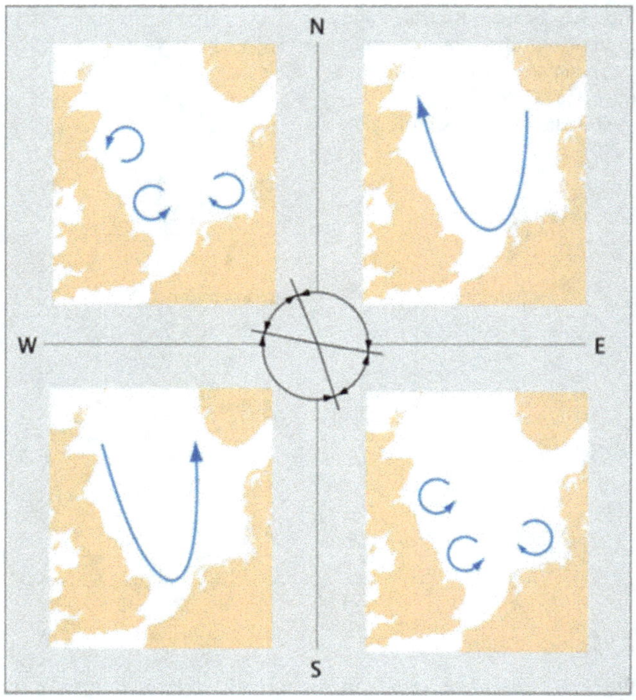

Fig. 1.11 Basic wind-driven circulation patterns in the North Sea. The four current states correspond to the four wind direction sectors in the central diagram (after Sündermann and Pohlmann 2011)

- Static type: low pressure track Iceland–northern North Sea–Scandinavia; extended, cold low; long-lasting, but not necessarily extreme winds push water into the German Bight. Example: 17 February 1962.
- Dynamic type: low pressure track Subtropical Atlantic–Great Britain–Denmark; small-scale, warm low; a short-lived, rotating extreme wind moves the North Sea water like a centrifuge raising sea level along all coasts. Example: 3 January 1976.

The storm surges mentioned are extreme, but not worst cases. Even under present-day conditions (i.e. without climate change) the full spectrum of possible surges allows for severely higher floods.

Surface waves are generated by the wind. Although their influence on large-scale circulation and transport is generally relatively small, it is significant for dynamical processes such as generation of turbulence, shear stress, radiation stress, sediment erosion, wave set-up in coastal areas, and extreme waves. The spectrum of wind waves is dependent on the fetch and duration of wind action (Fig. 1.13). Figure 1.14 depicts the first EOF pattern (Empirical Orthogonal Functions; represent the basic modes of the oscillation system) of the characteristic wave height. In combination with the time series of its principal component (not shown), it represents 82.5 % of total variance. Characteristic wave heights show the strongest variability in the open northern

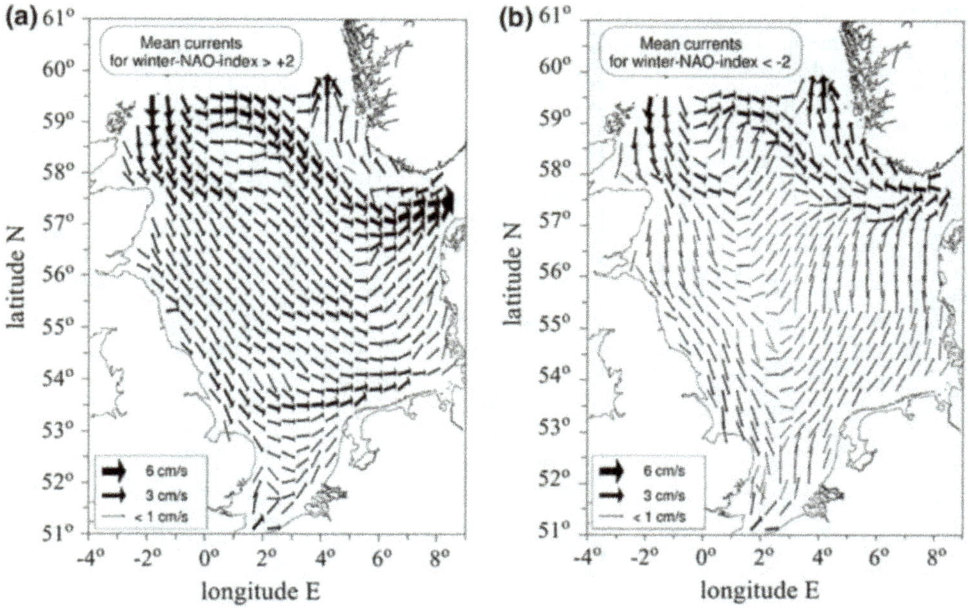

Fig. 1.12 North Sea circulation during prevailing positive winter NAO conditions (*left*) and negative winter NAO conditions (*right*) (Schrum and Siegismund 2001)

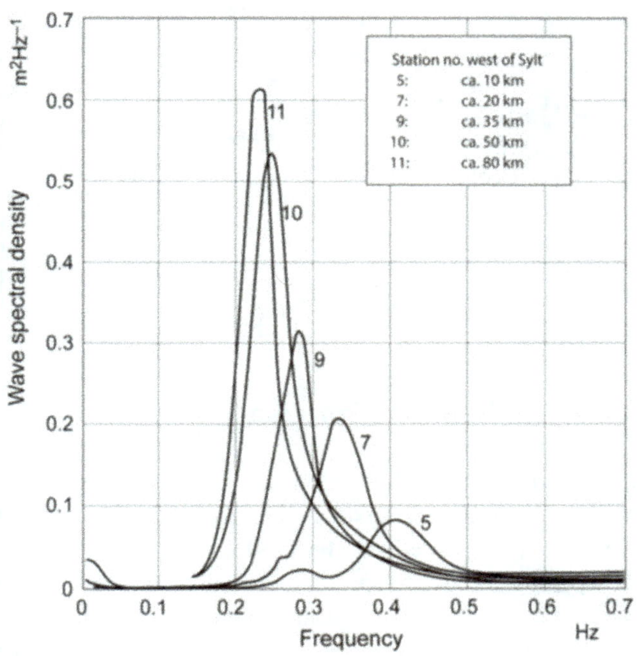

Fig. 1.13 JONSWAP spectrum (Hasselmann et al. 1973). Wave spectra of a developing sea for different fetches. Evolution with increasing distance from shore (numbers refer to stations off the island of Sylt)

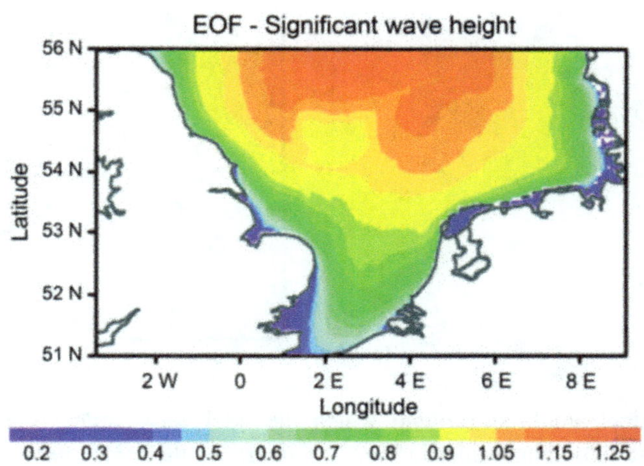

Fig. 1.14 Characteristic wave heights (m) in the North Sea (Stanev et al. 2009)

North Sea; in shallow regions and near the coast, variability is significantly reduced.

1.4.4 Tidally-Driven Regime

The North Sea dynamics are significantly influenced by astronomical tides. These are co-oscillations with the autonomous tidal waves of the Atlantic Ocean. The specific geometry of the North Sea basin implies eigen-periods and hence resonance in the semi-diurnal spectral range (Fig. 1.15). The superposition of the semidiurnal principal lunar and solar tides ($M_2 + S_2$) causes a significant spring-neap rhythm. The tidal currents may reach a speed of tens of centimetres per second and dominate any other flow, especially as they move the entire water column.

Tidal currents give rise to strong mixing of water masses, preventing thermohaline stratification in the shallow southern North Sea. In the Wadden Sea, tides cause the periodic exposure of large areas of seabed.

Tidal elevations penetrate from the Atlantic moving anti-clockwise as a Kelvin wave through the entire basin. There are three amphidromic points, i.e. points with zero

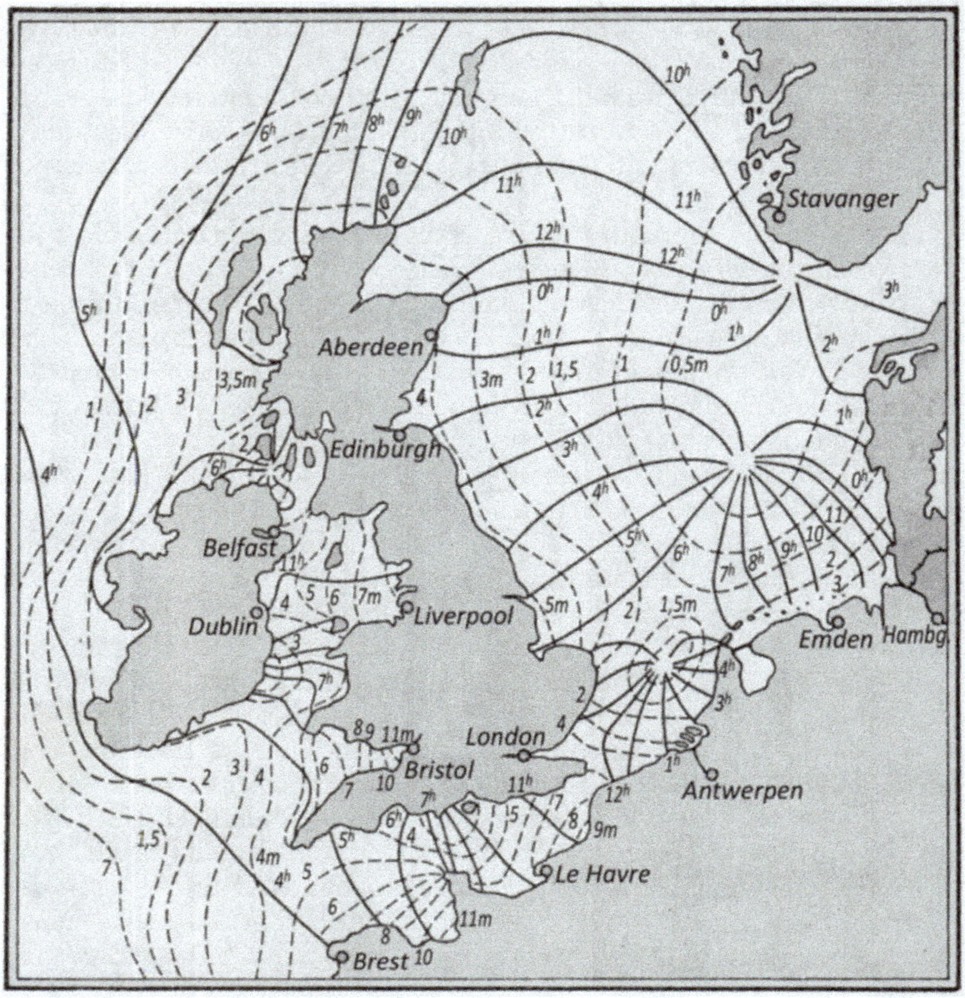

Fig. 1.15 Co-tidal lines for $M_2 + S_2$, the major tidal constituents in the North Sea (Sager 1959)

tidal amplitude around which the tidal wave rotates (see solid co-phase lines in Fig. 1.15 depicting the same phase of the tide). The co-range lines join places having the same tidal range or amplitude (dashed lines in Fig. 1.15).

Within one period the tidal currents form elliptic stream figures with positive or negative orientation (see Fig. S1.4.5 in the E-Supplement).

Nonlinear processes generate non-harmonic tidal motions and, as a consequence, non-vanishing residual currents with maximum values of the order of 10 cm (see also Fig. S1.4.6 in the E-Supplement).

Tides superimpose any other motion in the North Sea. They dissipate energy, mix water masses, and alter the coastline. Backhaus et al. (1986) showed that whenever a constant flow component is combined with a time-dependent periodic tide, there is a considerable reduction in the resulting residual flow.

Jungclaus and Wagner (1988) showed that sea-level rise leads to a shift in the central amphidromy towards the northwest and thus to higher tidal amplitudes in the southern North Sea.

1.4.5 Transport Processes

Dissolved and particulate inorganic and organic substances in the North Sea waters are transported by the current regimes as just described (Figs. S1.4.7 and S1.4.8 in the E-Supplement). Their spread and deposition control the biological and sedimentological systems of the North Sea.

The cyclonic current system of the North Sea means that as a generalisation substances entering the North Sea via river outflows are transported along the English, then Dutch, German, and Danish coasts into the Norwegian Trench, with significant scatter due to actual atmospheric conditions. Time integration yields residence times of the order of 6–12 months in the northern North Sea, to three years in the German Bight and along the north-eastern British coast (see Fig. S1.4.9 in the E-Supplement).

Besides advection in the flow field the concentrations of substances within North Sea waters are also the result of processes such as mixing, biogeochemical cycling, deposition, and resuspension. For particulate matter, sinking, flocculation, seabed processes such as bioturbation are also

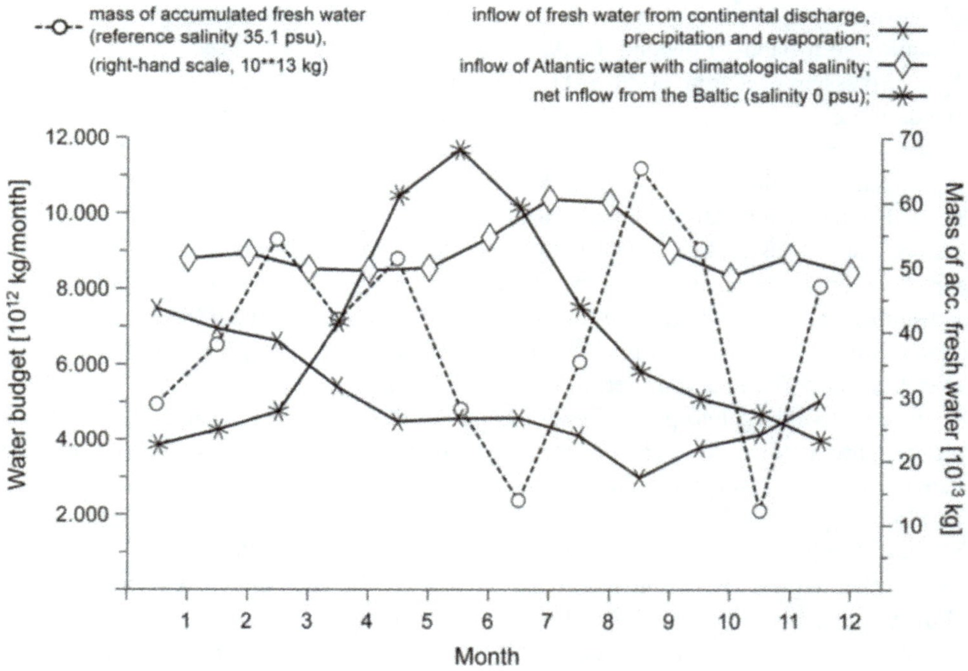

Fig. 1.16 North Sea water budget for one climatological year (Damm 1997)

important. Particulate transport drives the morphological changes occurring within the North Sea region.

Damm (1997) calculated the North Sea freshwater budget using the balance of inflows and outflows from long-term field records (Fig. 1.16).

1.5 Current Atmospheric Climate

Christiana Lefebvre

For climate descriptions, the data base should comprise a continuous period of at least 30 years in order to be representative in terms of variations and extremes. To account for climate change, the period 1971–2000 was chosen as the reference period for this study, because it seems to reflect features of the current climate better than the current World Meteorological Organization (WMO) climatological standard period 1961–1990. However, some analyses are also considered based on periods that differ slightly from 1971 to 2000. The current atmospheric climate as described in this section is based on analyses of observations for various climate parameters: air pressure, air temperature, precipitation, wind speed and direction, sunshine duration and cloud cover. In addition to analyses based on in situ data, for which coverage is poor in some sea areas, climate analyses are also based on the European Centre for Medium-Range Weather Forecasts (ECMWF) 40-Year Re-analysis (ERA-40; Uppala et al. 2005) and regional climate model (RCM) results from the ENSEMBLES project (Van der Linden and Mitchell 2009). Climatologies based on remote sensing from satellites are also used.

1.5.1 Atmospheric Circulation

The North Sea is situated in the west wind drift of the mid-latitudes between the subtropical high pressure belt in the south and the polar low pressure trough in the north. Westerly upper winds steer extratropical lows from the North Atlantic to northern Europe interrupted by relatively short anticyclonic periods. This is accompanied by frequent changes in air masses of different thermal and moisture-related properties, which drives continuous change in the synoptic-scale weather conditions over periods of days to weeks.

1.5.1.1 Sea-Level Pressure

Figure 1.17 shows the spatial distribution of the mean monthly sea-level pressure (SLP) across the North Sea region based on daily hemispheric analysis of SLP by the UK Met Office provided by the British Atmospheric Data Centre for the period 1971–2000 (Loewe 2009). Charts of the mean annual and mean monthly SLP for the period 1971–2000 based on data from the ERA-40 reanalysis and a hindcast of the ENSEMBLES RCMs forced by the ERA-40 reanalysis using a horizontal grid resolution of 25 × 25 km are reported by Bülow et al. (2013). The circulation pattern shows a distinct annual cycle that results from the interaction of the predominant air pressure centres across the North Atlantic: the Icelandic Low and the Azores High. On average, air pressure rises from northwest to southeast, while the standard deviation decreases. The strongest air pressure gradients are observed in autumn and winter. This is accompanied by the strengthening of low pressure systems due to a greater temperature difference between the polar and subtropical regions at this time of year. In the cold season, the direction of mean air flow varies between westerly in November and December and southwesterly from January to March. From March to April the intensity of mean air flow over the North Sea region decreases markedly. The Azores High starts to extend into parts of mid-Europe. In May, the air pressure gradient is weakest. In June and July, the extension of the Azores High causes on average a weak northwesterly air flow. In the English Channel and southwestern North Sea, mean air pressure is highest in July, while other regions have maximum air pressure in May (Fig. 1.18). Charts of mean SLP and standard deviation for the year as a whole, and for January and July from the ENSEMBLES RCMs and ERA-40 reanalysis for the period 1971–2000 are discussed by Bülow et al. (2013). The ERA-40 spatial air pressure distribution for January and July fits well with the corresponding analyses of UK MetOffice data in Fig. 1.17. The standard deviation derived from the ERA-40 reanalysis decreases in general from the northwest (area of the Orkney and Shetland Islands) to the southeast (southern Germany), for the annual mean (from 13–14 to 8–9 hPa), in January (from 16–17 to 10–11 hPa) and in July (from 8–9 to 4–5 hPa). The differences between the RCM SLP fields and ERA-40 reanalysis show a range of different patterns. The annual cycles of SLP from the ENSEMBLES RCMs and ERA-40 reanalysis for four North Sea subregions (northwest, northeast, southwest, southeast) show good agreement (Bülow et al. 2013), except for one RCM showing a permanent positive bias. Anders (2015) showed that the ERA-40 driven RCMs are able to reproduce different weather types, which were derived from the ERA-40 mean SLP field. Most of the RCMs reproduce the weather regime classification after Jenkinson and Collison (1977) centred over the North Sea with a coverage of 80–99 %. In general, agreement is better in winter than in summer.

1.5.1.2 North Atlantic Circulation

Circulation in the North Sea region is strongly influenced by the North Atlantic Oscillation (NAO). The NAO is described in detail in Annex 1 at the end of this report. The NAO is evident throughout the year, but it is most pronounced during winter and accounts for more than a third of the total variance in the SLP field over the North Atlantic (Hurrell and Van Loon 1997). During positive phases of the NAO, the meridional SLP gradient over the North Atlantic is enhanced. In the North Sea region, a positive NAO phase during the winter is usually accompanied by enhanced winter storm activity, above-average air temperatures and above-average precipitation in Scotland and Norway (Hurrell 1995; Hense and Glowienka-Hense 2008). Negative phases of the NAO, characterised by a weaker than normal Atlantic meridional SLP gradient are often linked with below-average air temperatures in northern Europe (Trigo et al. 2002). A prevailingly negative NAO phase influenced circulation from the mid-1950s to the 1978/1979 winter (Hurrell 1995). A period of high positive NAO in the late 1980s and early 1990s was connected with a high frequency of strong winter storms across the North Atlantic and high wind speeds across western Europe (e.g. Hurrell and Van Loon 1997). Figure 1.19 shows the frequency of extreme North Atlantic lows with core pressure of 950 hPa or less from November to March based on analysed weather charts (Franke 2009). Counts for November to March since 1956 show a sudden increase in winter 1988/1989 to 15 strong storms and a decrease since the late 1990s with the exception of the extraordinarily mild winter of 2006/2007 with 16 intense lows. The highest number occurred in winter 2013/2014 with 18 storms, causing severe damage along the western European coasts through storm surges.

1.5.2 Wind

Wind is the most important meteorological parameter for the North Sea and its coastal areas for two reasons. First, severe wind storms have the potential for high impact damage. They cause rough sea conditions, and increase risk of storm surges in coastal regions and thus damage to coastal settlements, infrastructure, agricultural land and forests, as well as increased coastal erosion in some areas. Second, wind is of increasing interest as a source of renewable energy. Widespread construction of onshore wind farms in coastal areas is now being followed by a significant increase in the development of offshore wind farms, and this requires reliable wind statistics for site identification and wind resource mapping as well as for user-specific forecasts of wind conditions for construction work and wind resource management.

Wind speed and direction across the North Sea area are driven by the large-scale pressure field and are modified at a local level by orographic effects caused by adjacent isles and coastal features. Wind speed is also affected by vertical temperature differences between the atmosphere and the sea surface, even for wind speeds of 7 Bft and above (Baas et al. 2015). The wind climate is characterised by pronounced seasonality. In general, the winds from November to March

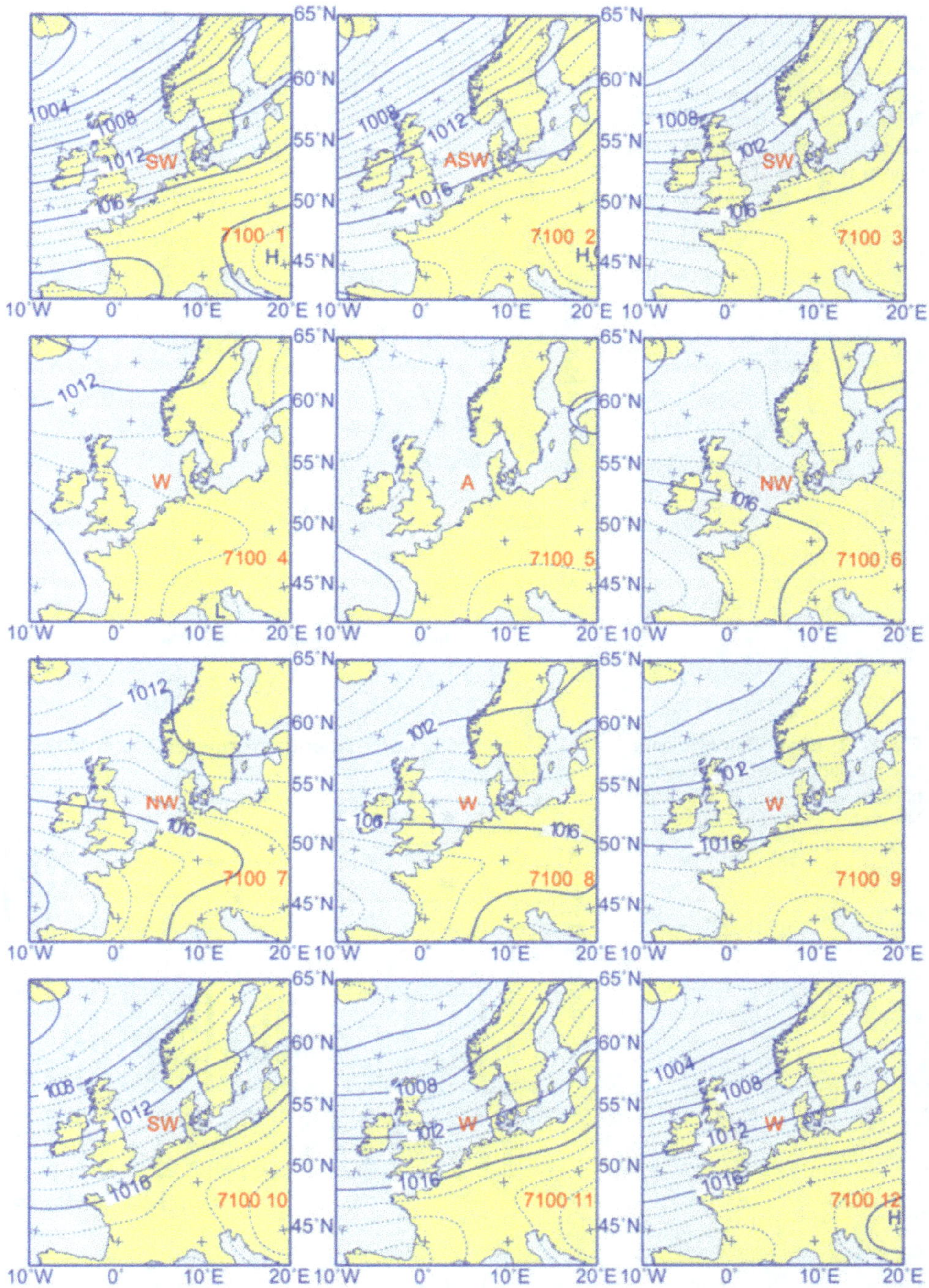

Fig. 1.17 Spatial distribution of mean monthly sea-level pressure (hPa) across the North Sea region for the period 1971–2000 (Loewe 2009). Numbers in the lower right of the subplots indicate time period and month

are stronger due to larger temperature differences between the subtropical and polar regions leading to more intense low-pressure systems.

1.5.2.1 Ship-Based Wind Speed Observations

The spatial distribution of mean monthly and annual wind speed based on ship observations was reported for the period 1981–1990 by Michaelsen et al. (2000), while Stammer et al. (2014) reported mean wind speed for June and December for the period 1981–2010. Due to uncertainties in wind data time series using ship-based observations arising from, among others, the switch from estimating wind speed from wave heights to direct measurements, no common measure height, and the effect of ship type and superstructure on measurements, it was decided to not include the results without further treatment into the KLIWAS North Sea Climatology (Stammer et al. 2014).

1.5.2.2 Remotely-Sensed Wind Speeds

In regions, where in situ observations are sparse, satellite observations with their extended spatial coverage are of great advantage. Satellite-based data sets covering long

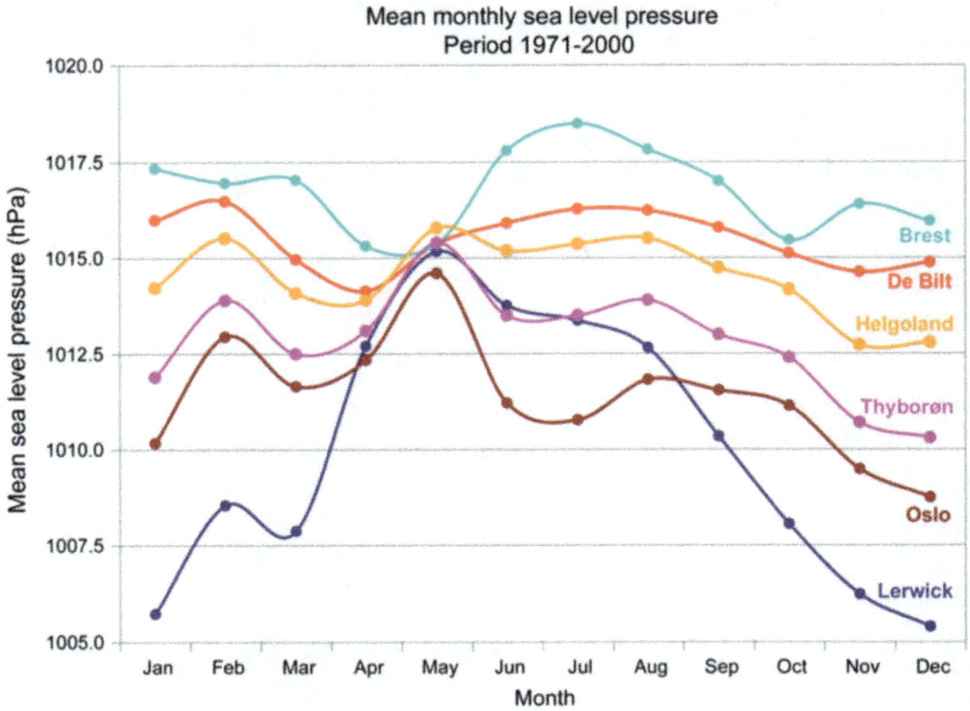

Fig. 1.18 Annual cycle of mean sea-level pressure (hPa) at selected land stations in the North Sea region for the period 1971–2000. Station positions are shown in the E-Supplement to this chapter (Fig. S1.5.1)

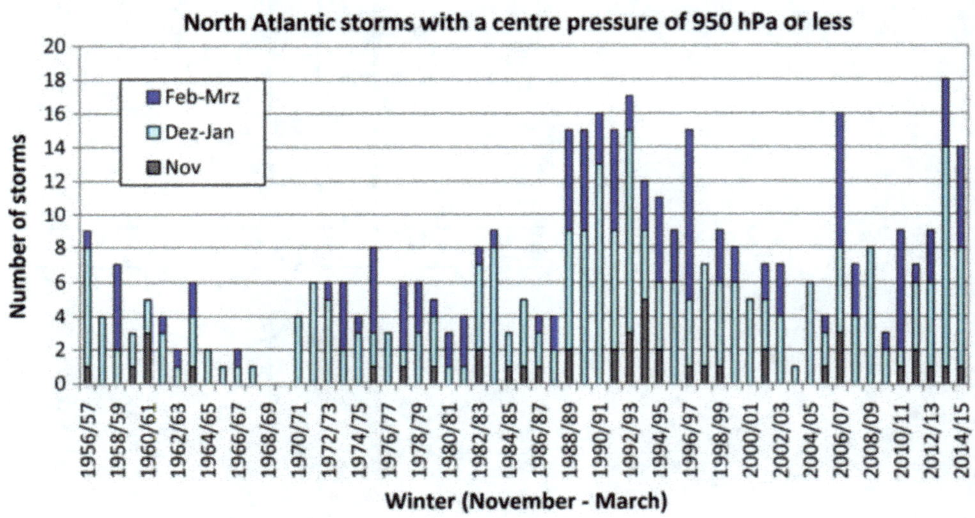

Fig. 1.19 Intense North Atlantic low-pressure systems with a core pressure of 950 hPa or less from November to March (Franke 2009, updated)

periods are also suitable for deriving climatologies. Figure 1.20 shows seasonal mean wind distributions developed from a 10-year data set of twice-daily (06:00 and 18:00 UTC) observations from NASA's QuikSCAT mission for the North Sea and Baltic Sea (Karagali et al. 2014). The data show equivalent neutral wind (ENW) at 10 m above the sea surface from rain-free observations assuming neutral atmospheric stratification. The ENWs originate from transmitted radar signals that are backscattered by small-scale waves at the sea surface, to which empirical algorithms are applied. The spatial distributions show higher wind speeds in winter and autumn as a result of increased storm activities as low-pressure systems cross the eastern North Atlantic and North Sea. In addition, the atmosphere in autumn and early winter is often colder than the ocean surface leading to prevailing unstable conditions that drive momentum transfer from higher atmospheric layers and result in higher wind speeds. Because neutral atmospheric stratification was assumed in the data processing, the satellite data may not exactly correspond to real conditions and may overestimate the true wind. In contrast, the strong lee effects in the western North Sea from the land effect of the British Isles may be less pronounced, because using ENWs leads to underestimation of true wind speed in the case of frequently stable atmospheric stratification. Besides atmospheric stability, rain affects the backscattered signal causing an

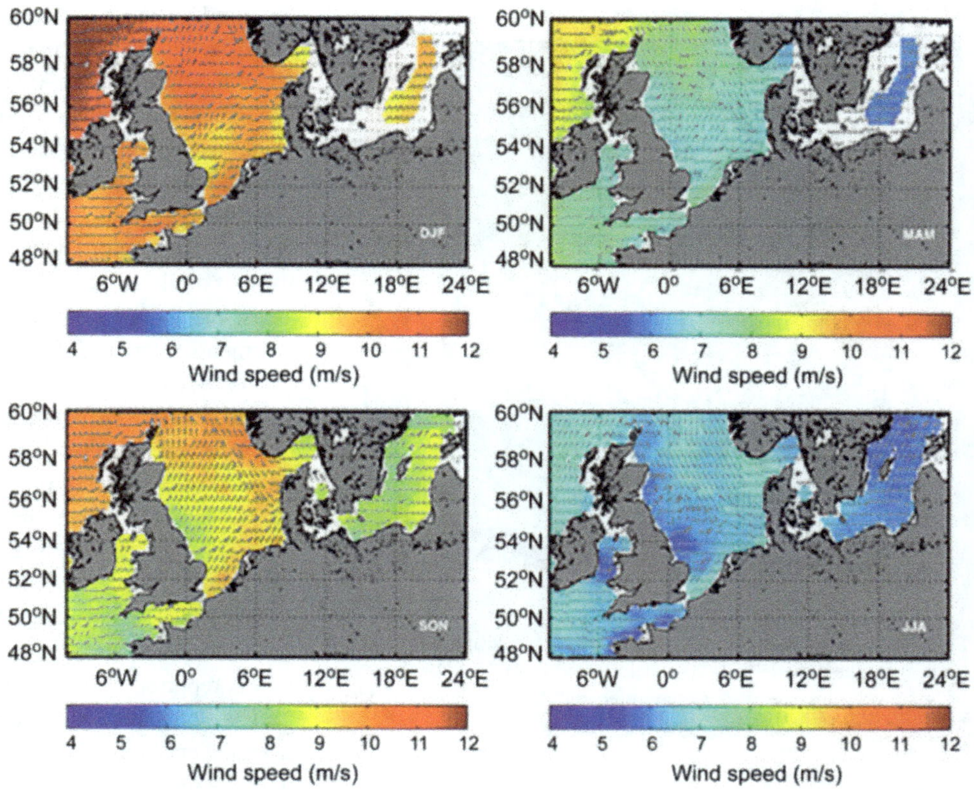

Fig. 1.20 Spatial distribution of seasonal mean wind speed ($m\ s^{-1}$) and the most frequently observed wind direction (*arrows*) derived from QuikSCAT satellite data for the period November 1999 to October 2009 (Karagali et al. 2014)

increase in retrieved wind speed. While rain-contaminated QuikSCAT winds could be excluded over sea areas but not always over coastal areas, wind speed estimates from QuikSCAT are higher than in situ observations in areas near land. Uncertainties in the QuikSCAT wind characteristics may also arise due to ice cover in the Baltic Sea, spatial differences in sea surface temperature and sea surface currents (Karagali et al. 2014).

Karagali et al. (2014) compared the QuikSCAT wind speeds with in situ measurements at three locations (platforms) in the North Sea: Greater Gabbard (off the Suffolk coast, UK), FINO 1 (north of the island Borkum, Germany) and Horns Rev 1 (off Esbjerg, Denmark). Owing to the mast's position relative to the wind farms and its proximity to land, only QuikSCAT and in situ winds from the south to north sectors (175°–13°) were considered. In this sector, wind speeds match well on average. The correlation is high (r 0.92) and the mean bias (in situ minus satellite) does not exceed $-0.23\ m\ s^{-1}$. For wind speeds above $3\ m\ s^{-1}$ the bias is close to zero with a standard deviation of $1.2\ m\ s^{-1}$. Figure 1.21 shows the change in in situ (all three stations) and QuikSCAT wind speeds over the year as well as wind speeds residuals (in situ minus satellite wind speeds). The mean wind speeds exhibit a pronounced annual cycle with a minimum ($\sim 7\ m\ s^{-1}$) in July and a maximum ($\sim 10.5\ m\ s^{-1}$) in December and January. Compared with in situ data, QuikSCAT winds are lower from April to July and higher from

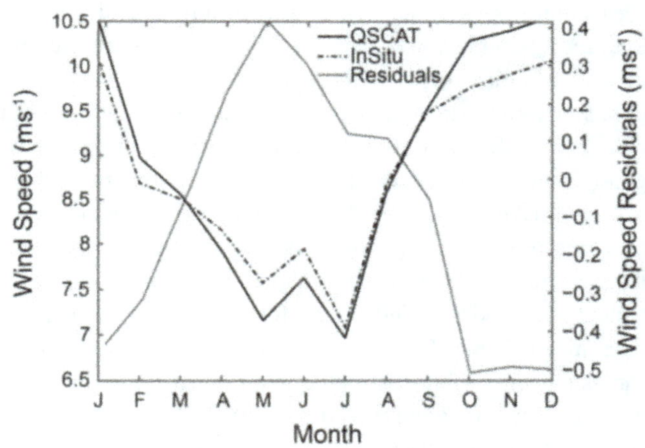

Fig. 1.21 Monthly variation of mean wind speed from QuikSCAT (*black line*), in situ observations from all stations (*dashed line*) and wind speed residuals (*grey line*) (Karagali et al. 2014)

October to March. The residuals (which do not exceed $\pm 0.5\ m\ s^{1}$) show the highest positive values in May and the highest negative values during October to December. For the North Sea area as a whole, an even stronger annual cycle but lower wind speeds results from the NCEP/NCAR reanalysis (Kalnay et al. 1996) for the earlier 40-year period 1959–1997. Siegismund and Schrum (2001) derived a mean wind speed of about $9\ m\ s^{-1}$ for October to January from this dataset, which is about 50 % higher than for April to August.

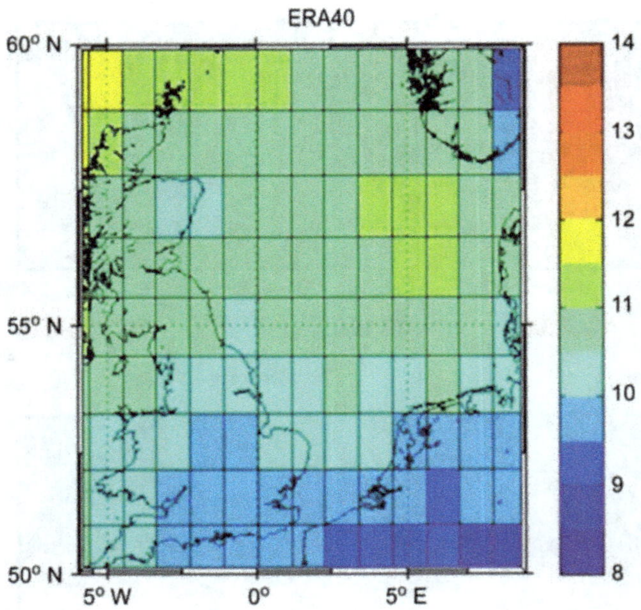

Fig. 1.22 Spatial distribution of mean annual geostrophic wind speed (m s^{-1}) for the period 1971–2000 determined from ERA-40 reanalyses of sea-level pressure data (figure prepared by A. Ganske, Federal Maritime and Hydrographic Agency, Germany)

1.5.2.3 Geostrophic Wind Speeds

Measurements of wind speed are usually insufficient for long-term wind analyses because they are easily affected by inhomogeneities, such as through changes in instrumentation, immediate surroundings, and location. To avoid such deficiencies, Schmidt and von Storch (1993) instead used geostrophic wind speeds derived from air pressure differences. This method is widely used and is often not only based on observed air pressure values but also on reanalysis data. Figure 1.22 shows the spatial distribution of mean annual geostrophic wind speed for the period 1971–2000 determined from ERA-40 reanalyses of SLP. The graphic shows a decrease in geostrophic wind speed from northwest to southeast and wind speeds of 10.5–11.5 m s^{-1} in the central North Sea. These wind speeds exceed those from ship-based observations for the shorter period 1981–1990 (Michaelsen et al. 2000) by more than 2 m s^{-1}.

A comparison of geostrophic wind fields derived from the ERA-40 reanalysis and the higher spatially resolved ERA-Interim reanalyses from the ECMWF (Berrisford et al. 2009) of SLP for the period 1981–2000 show greater spatial variation in the ERA-Interim fields, but differences relative to the ERA-40 geostrophic wind speeds of less than 1 m s^{-1} (Ganske et al. 2012). The shorter period used for this comparison shows higher wind speeds than seen in Fig. 1.22, because it is more influenced by the intensive wind phase from the late 1980s to the end of the 1990s. Additional graphics as well as a discussion of frequency distributions for wind speed in different sea areas and seasons are provided in the E-Supplement to this chapter (S1.5.3).

Evaluating the temporal and statistical characteristics of the coastal wind climate along the North Sea in the ERA-40 reanalysis and the ENSEMBLES RCMs by comparing them against wind measurements in the Netherlands and Germany, Anders (2015) found that ERA-40 reproduces the wind field very well in regions where high quality observational data were assimilated at high temporal frequency. Outside these areas, the RCMs mostly show better agreement with measurements over land. The correlation between ERA-40 and observations is between 0.75 and 0.95, and for most RCMs and observations is between 0.6 and 0.9. The models seem to overestimate mean wind speed at inland sites. For the North Sea region as a whole (but excluding coastal waters), five RCMs generate annual mean wind speeds that differ by less than ±0.5 m s^{-1} from ERA-40 wind speeds while the remaining seven RCMs differ by ±1.5 m s^{-1} at most. Six RCMs depict smaller wind fields than ERA-40, partly due to fewer wind speeds above 10 m s^{-1}. More detailed information is reported by Bülow et al. (2013).

The occurrence of high wind speeds across the North Sea from the end of the 1980s to the end of the 1990s agrees well with the increased frequency of strong North Atlantic winter storms (lows with a core pressure of 950 hPa or less—Fig. 1.19). Until 1974/1975 these extreme lows mainly occurred between December and January. Their occurrence has since increased in November as well as March (Franke 2009), which supports the results of Siegismund and Schrum (2001).

Studies estimating storminess from homogenous station pressure records show pronounced decadal variability but no robust sign of any long-term trend (WASA Group 1998; Bärring and von Storch 2004). This is not the case for results from studies using the 20CR reanalysis (Compo et al. 2011). For these, upper percentiles of daily wind speed (Donat et al. 2011) or geostrophic winds derived from air pressure (Krueger et al. 2013) indicate a significant upward trend in storminess in many parts of western and northern Europe. Figure 1.23 shows trends in the annual 95th percentile of daily maximum wind speed and trends in the days with gales during the period 1871–2008. Donat et al. (2011) assumed that the 20CR reanalysis may suffer from some inhomogeneities due to changes of station density.

1.5.2.4 Wind Direction

Wind direction across the different sea areas of the North Sea is determined by the large-scale air pressure distribution. Near islands and coasts winds are modified by orographic effects, depending on the shape and orientation of the adjacent coastlines. Figure 1.24 shows the annual frequencies of wind force as observed from ships in the period 1971–2000

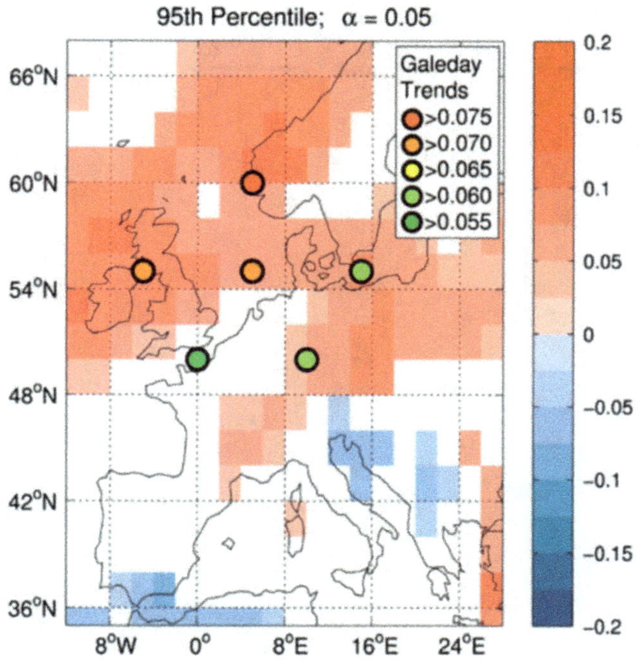

Fig. 1.23 Trends in the annual 95th percentile of daily maximum wind speed in the 20CR ensemble mean for the period 1871–2008 (*unit* standard deviation per 10 years). Only significant trends are plotted ($p \leq 0.05$, Mann-Kendall-Test). *Circles* indicate trends in gale day at that location (Donat et al. 2011)

for four categories (1–3, 4–5, 6–7 and 8–12 Beaufort scale) and eight directions. On average, winds from the southwest and west predominate. Owing to the orography, only in the sea areas between the Shetland Islands and the Norwegian coast (Viking, Utsira north) are winds most frequently from the south; west of the southern Norwegian coast (Utsira south) winds from the northwest predominate. A wind force of 8 Bft and above occurred in 6–9 % of observations for the central and eastern North Sea areas north of 56°N. While the prevailing wind directions from October to March are southerly to southwesterly, northwesterly to northerly winds predominate in the northern and central North Sea in spring and summer. Over the course of a year, easterly winds are most frequent in May, when the frequency of winds exceeding 6 Bft is lowest.

Frequency distributions of daily mean wind direction from the ERA-40 reanalysis and ENSEMBLES RCMs for the four sea areas considered for the period 1971–2000 are discussed by Bülow et al. (2013).

1.5.2.5 Sea Breeze

With calm conditions and strong insolation (i.e. high pressure situations), a local wind regime develops due to uneven heating of the land and sea surface in coastal areas. According to Steele et al. (2015), who referred to Atkinson (1981) and Simpson (1994) 'The sea breeze is defined as a circulation which is induced by a thermal contrast, between the land and sea that overcomes the strength of the background or gradient wind'. The development of sea breezes occurs most often in spring and summer, when the North Sea is relatively cold and the sun reaches its minimum zenith angle (Tijm et al. 1999). At mid-latitudes, sea breezes usually have an inland penetration of 5–50 km (Atkinson 1981). In southern England (Simpson et al. 1977) and the Netherlands (Tijm et al. 1999), favourable conditions for sea-breeze development may result in inland penetration of 100 km or more. Offshore sea breezes can extend similar distances (Arrit 1989; Finkele 1998). Thus, sea breezes are an important component of the coastal wind climate and are of growing interest due to the development of wind farms in offshore and coastal areas and their need for reliable wind forecasts. Based on data from the Weather Research and Forecasting model (WRF, version 3.3.1; Skamarock and Klemp 2008), Steele et al. (2015) derived a sea-breeze climatology for five coastlines around the southern North Sea by considering three sea-breeze types (as described by Miller et al. 2003): a pure sea breeze (where the sea breeze forms in opposition to the gradient wind) and the corkscrew and the backdoor sea-breeze, where the gradient wind has an along-shore component. Pure sea breezes induce offshore calm zones, defined as regions where the 10-m simulated wind is <1 m s^{-1}. The corkscrew sea breeze is the strongest type of sea breeze. They induce coastal jets, which are defined as local wind speed maxima within 1 km of the coast (Capon 2003). The 10-m simulated wind speeds reach about 5 m s^{-1} and cause a net increase in wind energy on a given day of up to 10 % (Steele et al. 2015). The backdoor sea breeze is the weakest type of sea breeze. In many cases an intensifying backdoor sea breeze reduces the extent of the offshore calm zones. Table 1.1 shows the annual frequency of the different sea-breeze types for each coastline in the period 1 January to 31 December 2012. There is clearly significant variability between coasts and this is attributed to coastal orientation relative to the prevailing wind direction. Season length also differs between coasts; mostly extending from May to September, along the coasts of East Norfolk, Suffolk and Essex sea breezes can occur from March to October. Variations in coastal orientation or the presence of another coastline can cause interactions between sea breezes.

The magnitude of the differential heating and the strength of the opposing large-scale wind flow are important for the offshore extent and inland penetration of the sea breeze. Studies in southern England show that most sea breezes occur in conditions of near-calm or with offshore winds of less than 2–3 m s^{-1} at the surface (Tijm et al. 1999). Using two-dimensional model simulations, Finkele (1998) found the offshore sea-breeze propagation speed (−3.4 m s^{-1}) to be about double that for inland sea-breeze propagation (1.6 m s^{-1}) under light (−2.5 m s^{-1}) offshore geostrophic

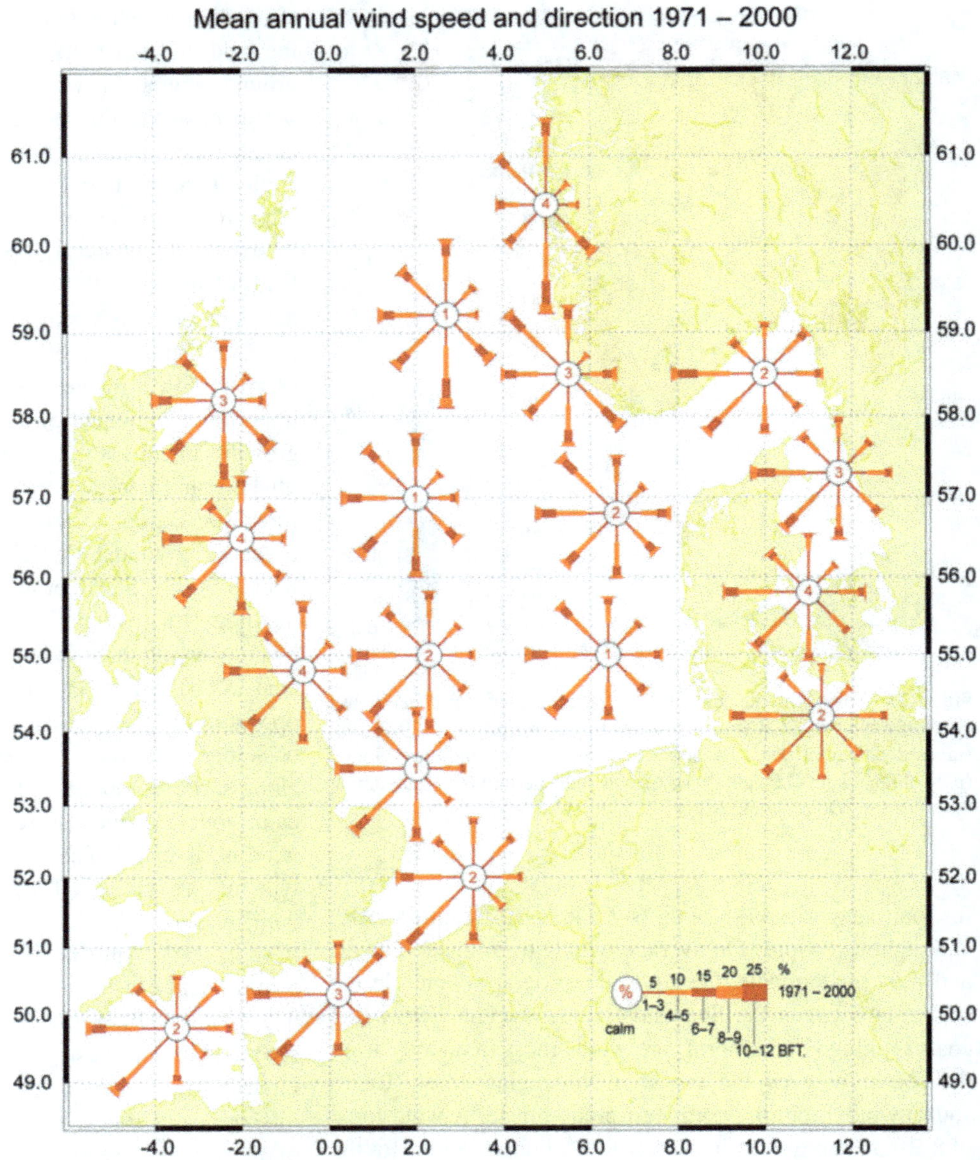

Fig. 1.24 Annual distribution of wind forces in Beaufort (Bft) derived from ship observations for different sea areas of the North Sea in the period 1971–2000. The length of each branch is proportional to the percentage frequency of the respective wind direction and wind force class. *Numbers in circles* denote the frequency of calm conditions (courtesy of the German Meteorological Service)

Table 1.1 Sea-breeze frequency (days) by sea-breeze type on southern North Sea coastlines between 1 January and 31 December 2012 (Steele et al. 2015)

Coast	Pure	Corkscrew	Backdoor	Total
North Norfolk	117	96	51	264
East Norfolk	166	169	0	335
Suffolk + Essex	46	167	13	226
Netherlands	146	76	71	293
South Kent	21	122	9	154
Total	496	630	144	1270

wind conditions (without a coast-parallel geostrophic wind component and a maximum sensible heat flux of about 380 W m^{-2}), which agreed well with measurements. The offshore sea-breeze propagation speed is defined as the speed at which the seaward extent of the sea breeze grows offshore. It is almost constant under light offshore geostrophic wind conditions and non-linear under moderate (−5.0 m s^{-1}) and strong (−7.5 m s^{-1}) geostrophic wind conditions. In the case of light offshore geostrophic wind conditions the inland sea-breeze propagation speed is almost constant until early afternoon and then increases during late afternoon and through the evening, which agrees with the findings of Tijm et al. (1999). Moderate geostrophic winds cause a non-linear inland sea-breeze propagation speed with the sea breeze slowing after having reached the coast. For strong geostrophic winds the sea-breeze circulation stays totally offshore.

The onshore sea breeze during the day alternates with the oppositely directed, but usually weaker, land breeze which develops during the night.

1.5.3 Air Temperature

1.5.3.1 Monthly and Annual Means

Offshore, the temperature of the lower atmosphere is mainly determined by sea surface temperature. For the North Sea, the two branches of the North Atlantic Current entering through the English Channel and between Scotland and the Shetland Islands respectively, have a significant influence (see Sect. 1.4). Air temperature is also influenced by the European land masses and latitude. The result is that mean air temperatures over the North Sea are above average for sea areas at similar latitudes elsewhere (Korevaar 1990). Spatial distributions of monthly mean air temperature across the North Sea region based on observations from ships and light vessels are available for the periods 1961–1980 (Korevaar 1990), 1981–1990 (Michaelsen et al. 2000) and 1971–2000 (Stammer et al. 2014). The latter are displayed in Fig. 1.25 and show only small differences relative to the climatology of Michaelsen et al. (2000), and for various periods the monthly means are highly correlated with the NCEP-RA1 (1950–2010), NCEP-RA2 (1979–2010), ERA-Interim (1979–2010), ERA-40 (1957–2002) and 20CR (1950–2010) reanalyses. The correlation coefficients of 0.98 and 0.99 exceed the 99 % level of significance (Stammer et al. 2014). Comparisons of the ERA 40 reanalysis with the ENSEMBLES RCMs for the period 1971–2000 are reported by Bülow et al. (2013). Annual air temperatures of the ERA-40 reanalysis are roughly 7.0–8.0 °C in the coastal areas of Norway and Sweden and 12–13 °C in the western English Channel. Due to differences in radiation absorption and heat capacity, the air over the North Sea is warmer than over the continent from autumn to spring and cooler from May to July. In winter (December to March), temperatures decrease from southwest to northeast, due to the declining influence of the warm North Atlantic Current and the advection of maritime air masses towards the east and the increasing influence of cold continental air masses. The lowest monthly means occur in February, ranging from below 0 °C in the Skagerrak and Kattegat area to 8–9 °C in the western English Channel. Towards summer, warming of the air over land proceeds faster and stronger than over the sea, leading to a shift in the orientation of the isotherms. The southern part of the North Sea is then warmer than the other regions. August is the warmest month of the year.

The annual cycle of mean air temperature at land stations in the North Sea region is shown in Fig. 1.26. Warming in summer is lowest in the far northwestern area (Lerwick on the Shetland Islands), where the highest mean monthly temperature is only 12 °C (August). The highest winter temperatures occur in Brest, in the western part of northwestern France, due to the influence of the North Atlantic Current.

The amplitudes of daily temperature as well as those of monthly mean temperature are more pronounced in coastal areas than in the central North Sea. Air temperatures show high variability, especially in winter, as reflected in a large standard deviation. In the period 1981–1990, the standard deviation increases from about 1 °C in the northwestern North Sea region, to about 3 °C along the western coast of Germany and Denmark and about 4 °C along the Swedish coastline in the western Baltic Sea region. The standard deviation is lowest from April to October and especially in summer with values mostly below 1.0 °C. In the inflow area between Scotland and the Shetland Islands it is about 0.4 °C in July (Michaelsen et al. 2000).

Figure 1.27 shows variation in air temperature over the course of a year from ship-based observations for the period 1950–2007 for different sea areas of the North Sea (BSH 2009). These distributions agree well with those for eight sea areas 2° × 2° in extent and four light vessels (Korevaar 1990), although the sea areas compared are not congruent. The coldest month is February with mean temperatures between 2 °C in the German Bight and 9 °C south of Scilly. The warmest month is August with mean temperatures between 13 °C in the area Shetland/Orkneys and about 15.5 °C in the German Bight and southern North Sea. Annual amplitude ranges from about 8 °C in the sea areas Shetland/Orkneys and South of Scilly and 14–15 °C in the sea areas Skagerrak and German Bight. The annual cycles for the northwestern, northeastern, southwestern and southeastern parts of the North Sea based on ERA-40 data and various ENSEMBLES RCMs show considerable deviations with differences of up to 2 °C in winter, but agree well in June and July.

1.5.3.2 Extremes

Significant deviations from the monthly mean are linked to specific air pressure patterns. The warmest European seasons since records began, autumn 2006 and winter 2006/2007 had temperature anomalies of up to 4 °C across the land areas surrounding the North Sea (WMO 2007, 2008). While the unusually warm autumn was linked to a negative NAO phase (−1.62 in September and −2.24 in October), the warm winter was associated with a positive

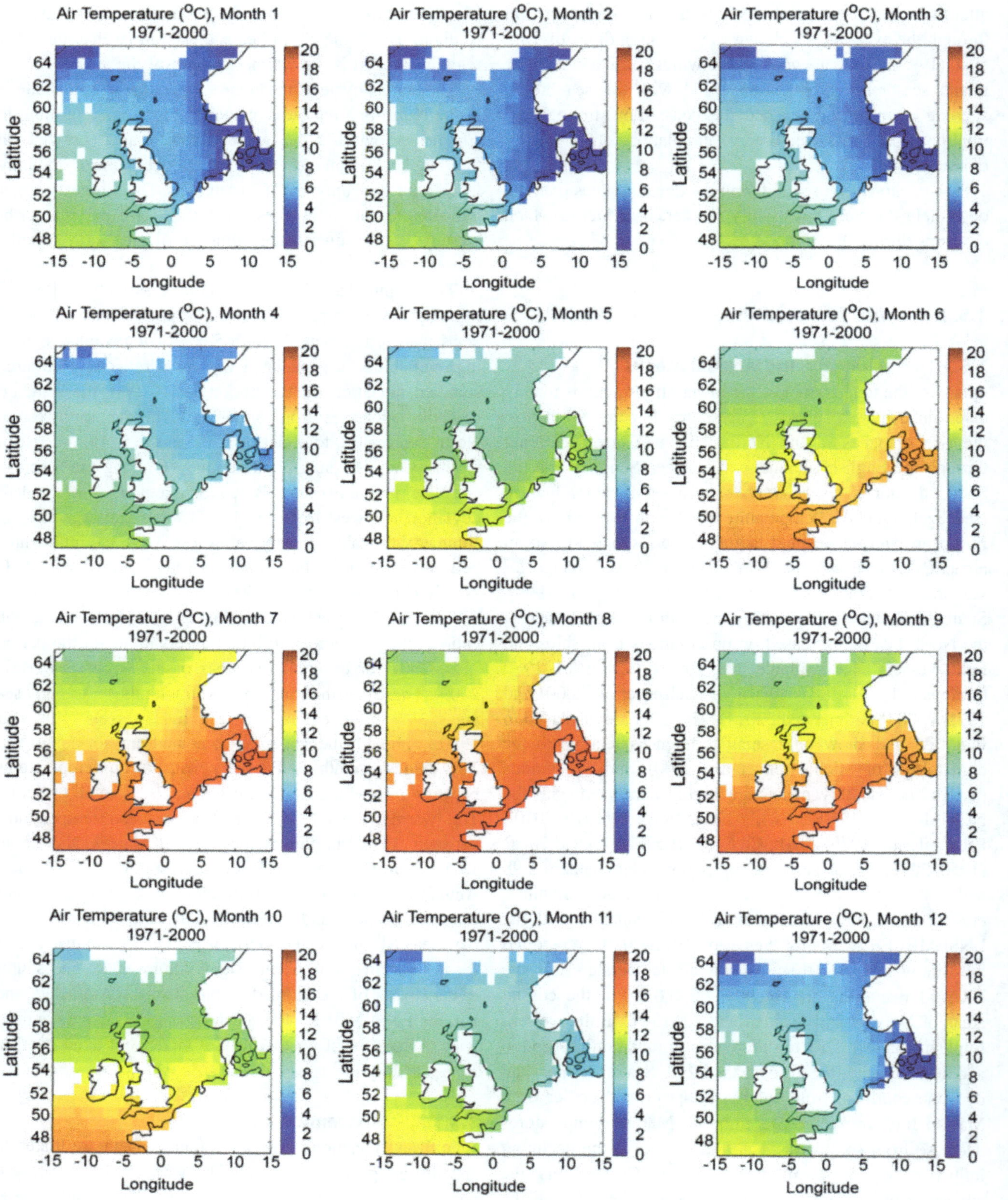

Fig. 1.25 Spatial distribution of monthly mean air temperature (°C) across the North Sea region based on in situ data for the period 1971–2000 (Stammer et al. 2014)

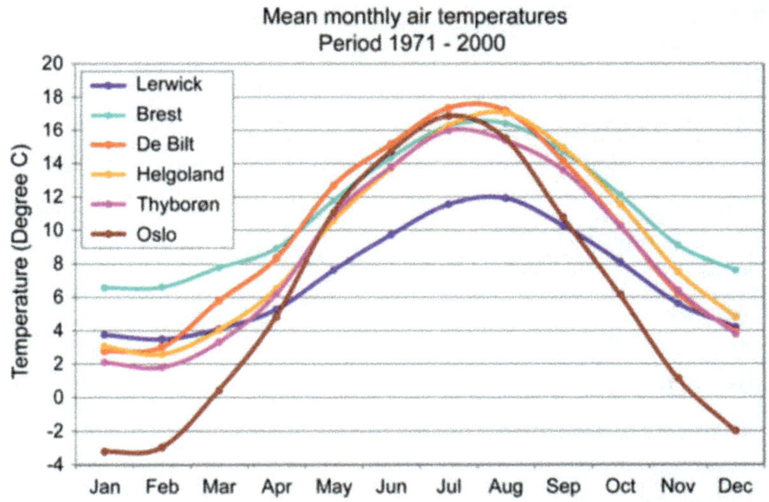

Fig. 1.26 Annual cycle in mean air temperature (°C) at land stations around the North Sea for the period 1971–2000. Station locations are shown in the E-Supplement to this chapter (Fig. S1.5.1)

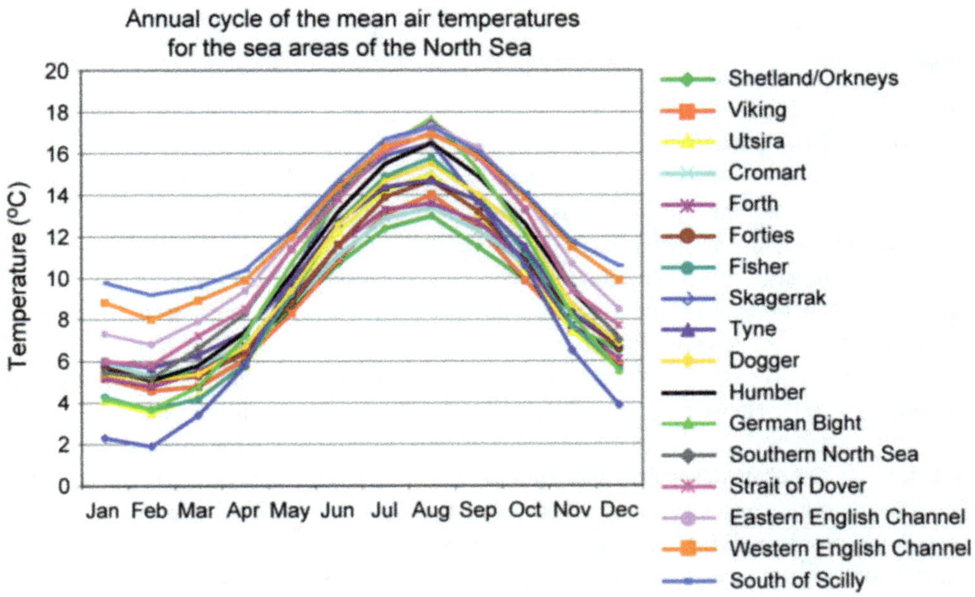

Fig. 1.27 Annual cycle in air temperature (°C) for the sea areas of the North Sea (BSH 2009). Sea area locations are shown in the E-Supplement to this chapter (Fig. S1.5.1)

NAO phase (+1.32 in December and +0.22 in January) (Luterbacher et al. 2007). The record negative NAO (−2.4) in winter 2009/2010 was linked to a cold and snowy winter influencing the area from northwestern Europe to central Asia (WMO 2010; Osborn 2011). By coupling mean SLP fields with temperature extremes (monthly occurrence of cold night and warm day temperatures) in Europe for the period 1961–2010, Andrade et al. (2012) found that in winter and spring the NAO is not only the major driver of mean temperature, but also of the occurrence of temperature extremes. Andrade et al. (2012) showed the leading modes for all four seasons. For winter, spring and autumn, there is coherent coupling for both extreme maximum and extreme minimum temperatures. The East Atlantic Oscillation defined by Barnston and Livezey (1987) also helps explain the occurrence of warm temperatures due to the advection of warm air with southwesterly winds into the region. This is the leading mode for temperature extremes in autumn. Cold air from the North and an absence of cloud cover are responsible for very cold nights on land. In summer, exceptionally long warm episodes are linked to a blocking high pressure system centred over the British Isles or over the European continent, stopping the westward propagation of North Atlantic pressure systems and leading to warm air advection and low cloudiness.

Table 1.2 shows extreme minimum and maximum air temperatures, determined from ship-based observations for the period 1950–2007 (99.9th percentile, the values were inferior to or exceeded, respectively, in 0.1 % of all measurements). Due to good data coverage, they can be considered more representative for the different North Sea areas than extreme temperatures derived from ERA records because ERA-40 temperatures are available at 6-h intervals only and are also smoothed by gridding.

Table 1.2 Extreme minimum and maximum air temperature (°C) for various sea areas from ship-based observations in the period 1950–2007 (BSH 2009)

Sea area	Extreme minimum temperature	Extreme maximum temperature
Shetland/Orkneys	−2	20
Viking	−4	21
Utsira	−8	23
Cromarty, Forth, Forties	−3	21
Fisher	−6	24
Skagerrak	−9	24
Tyne	−1	22
Dogger, Humber	−4	24
German Bight	−6	25
Southern North Sea	−5	25
Strait of Dover	−5	24
Eastern English Channel	−2	25
Western English Channel	0	25
South of Scilly	2	24

1.5.4 Precipitation

Precipitation across the North Sea is mainly supplied by Atlantic low pressure systems and their associated frontal systems. Their variation in occurrence and storm tracks causes regional and interannual variability. Strong positive phases of the NAO tend to be associated with above-average precipitation, especially over the northern North Sea area in winter (Trigo et al. 2002; Hense and Glowienka-Hense 2008).

1.5.4.1 Precipitation Frequency

In situ measurements of precipitation over the sea are extremely problematic. Many factors influence measurements, such as the movement and shape of the vessel, sea spray and wind. There are few measured precipitation data for sea areas. Those that are available are measured on light-vessels using a conical rain gauge, and on oil and gas platforms. The fleet of voluntary observing ships (VOS) does not measure precipitation amount, only information about the occurrence of precipitation. Analyses of data for different sea areas of the North Sea show precipitation frequency is mostly highest in November, sometimes in December or January (BSH 2009). Also, that precipitation frequency is higher over the northern North Sea (about 20–27 % of observations) than the English Channel, western North Sea and Skagerrak (10–16 % of observations), the latter due to lee effects of the Scandinavian mountains. The probability of precipitation is lowest in May.

1.5.4.2 Precipitation Amount

Prior to the satellite era, precipitation amounts across the North Sea were estimated in different studies by analysing and interpolating gauge measurements at coastal stations. As it was assumed that precipitation amounts for sea areas are generally less than over neighbouring land areas, different reduction factors were applied to coastal data leading to a broad range of estimates for the annual total for the North Sea area as a whole (440–800 mm; Barrett et al. 1991). Since the late 1970s, precipitation information has been remotely sensed from satellites. Their advantage is the provision of values for regions with few in situ measurements and the provision of spatially continuous data. However, satellite-based precipitation is not measured directly, but derived from radiance measurements. Validation of satellite-based data with in situ measurements from rain gauges on ships within the Baltic Sea shows precipitation is underestimated by remote sensing, because the detectability of small-scale precipitation typical of convective weather conditions is too low, although the detectability in cases of prevailing stratiform clouds fits well (Bumke et al. 2012). A first estimation for mean monthly, seasonal and annual precipitation totals was determined from the Scanning Multispectral Microwave Radiometer (SMMR) data on the Nimbus-7 satellite for the period 1978–1987 by retrieving the relationship of 37 GHz horizontal channel brightness temperatures to rain-rates (mm h^{-1}) from the calibration of the SMMR data by contemporaneous data from the Camborne radar station (Cornwall) of the UK Meteorological Office and published by Barrett et al. (1991). Barrett et al. examined the North Sea region from 47° to 63°N and 03°W to 13°E, excluding the land and mixed land/sea parts. Mean annual precipitation for the period October 1978 to August 1987 was estimated at 425 mm. Andersson et al. (2010) investigated monthly means of the quality-controlled

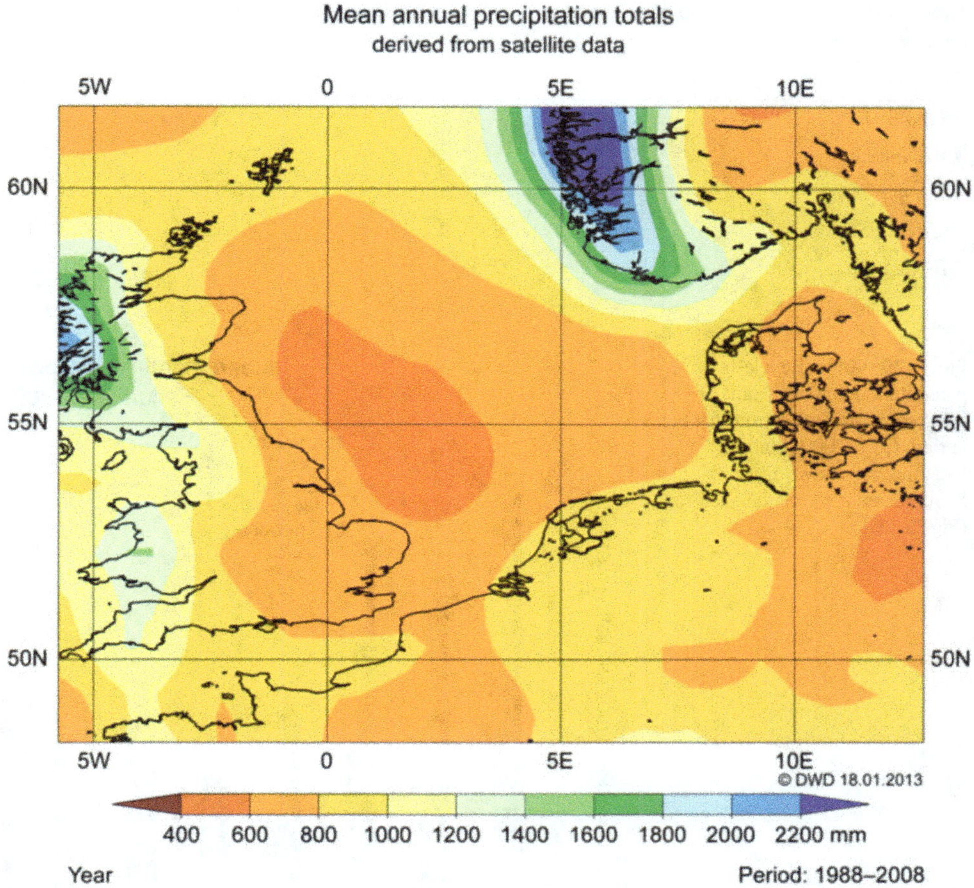

Fig. 1.28 Mean annual spatial distribution of precipitation (mm) across the North Sea region derived from the combined HOAPS-3 and GPCC data set for the period 1988–2008 (figure prepared by the German Meteorological Service, DWD)

HOAPS-3 (Hamburg Ocean Atmosphere Parameters and Fluxes from Satellite) data set with a grid resolution of 0.5° × 0.5° for the period 1988–2008. The data were derived from the Special Sensor Microwave Imager (SSM/I) operating on the polar-orbiting Defence Meteorological Satellite Program (DMSP) satellites for the sea area. For the land area this was completed with 'Full Data reanalysis Product Version 4' provided by the Global Precipitation Climatology Centre (GPCC; Rudolf and Schneider 2005), which only uses land-based rain gauge measurements. Figure 1.28 shows the annual spatial distribution of HOAPS precipitation data for the North Sea region calculated by the German Meteorological Service. Mean annual precipitation for the region from 50° to 62°N and 03°W to 13°E and excluding the land and mixed land/sea parts is 810 mm. This significant difference is partly explained by the different analysis periods (see below) and ongoing improvements in the instrumentation onboard satellites. But further research will be necessary, since comparisons of the ERA-40 reanalysis with the ENSEMBLES RCMs (Bülow et al. 2013) and those from different reanalyses with data sets of gridded observations from precipitation gauges (Lorenz and Kunstmann 2012) display considerable differences. An example for the central North Sea area (1.5–5.5°E, 54–58°N) is presented in Table 1.3 including a comparison of mean annual precipitation derived from the HOAPS satellite data and four reanalyses: ERA-Interim, ERA-40, NCEP/CFSR and MERRA.

Precipitation amount shows considerable intra-annual (see a selection of months in the E-Supplement Fig. S1.5.8) and interannual variability. For the reference period October 1978 to August 1987, February and/or April were the driest months. February 1986 was unusually dry with widespread means of less than 10 mm, even in coastal areas. Some areas were even rainless (e.g. in the region of De Bilt). Because February 1985 and 1986 received well below-average precipitation, the February mean for the period 1979–1987 was about 10 mm less than for the period 1971–2000 and about half the average for the period 1988–2008. The difference was greatest in the Shetland Islands. At Lerwick, the February mean was 62 mm for the period 1979–1987, 108 mm for 1971–2000 and 144 mm for 1988–2008. The HOAPS data show May as the driest month with average precipitation amounts of less than 25 mm in an extended area east and north of the British Islands. Spring is the driest season except in the western English Channel and the region around the Shetland Islands. Precipitation amount is mostly highest in autumn. The increase in precipitation amount in autumn is caused by the increase in low-pressure activity and convective rains due to the destabilisation of

Table 1.3 Mean annual precipitation (mm) for the central North Sea area (1.5–5.5°E, 54–58°N) derived from the satellite data set HOAPS and four reanalyses

Data set	1979–2001	1988–2008
HOAPS	–	643
ERA-Interim	812	800
ERA-40	691	–
NCEP-CFSR	966	1000
MERRA	754	772

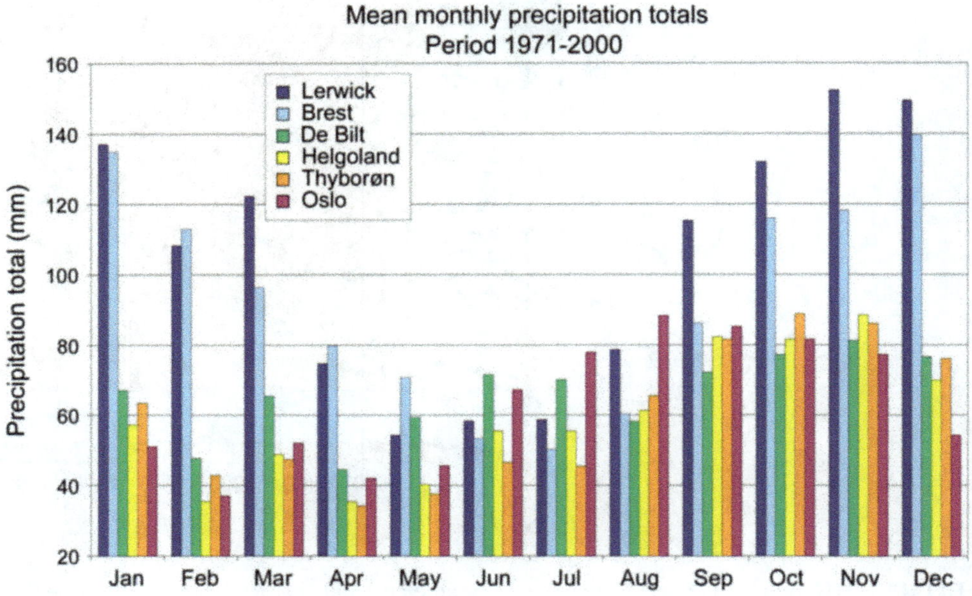

Fig. 1.29 Annual cycle in precipitation amount (mm) at selected North Sea stations for the period 1971–2000. Station locations are shown in the E-Supplement to this chapter (Fig. S1.5.1)

cold air masses moving over the warm North Sea (Lefebvre and Rosenhagen 2008).

In coastal areas, precipitation is intensified by coastal convergence and increasing friction effects. The annual cycle of the precipitation totals for stations in coastal areas of the North Sea region for the reference period 1971–2000 are shown in Fig. 1.29. There are significant differences in precipitation amount from October to April when the stations around the northern North Sea and the western part of the English Channel record about twice the amount of the other regions.

In winter, there is a pronounced correlation between precipitation amount in the northern North Sea region and the NAO. A positive NAO phase is linked to above-average precipitation, especially in Scotland and Norway and below-average precipitation in the Mediterranean region (Hense and Glowienka-Hense 2008). Studies for a box across northern Europe including the North Sea except for the southern part (10°W–20°E 55°–75°N) show the correlation between precipitation amount and NAO in winter (DJFM) seems to be stronger for the land area (correlation coefficient $r = 0.84$) than across the sea ($r = 0.38$), with all correlations statistically significant at the 99 % confidence level of the t-test (Andersson et al. 2010).

1.5.5 Radiation

1.5.5.1 Sunshine Duration

Sunshine duration depends on latitude and daytime cloud cover conditions. Measurements of sunshine duration are not included in the observational routine of the VOS fleet and statistics are not derived from measurements on light-vessels.

Annual cycles for some land stations in the North Sea region are shown in Fig. 1.30. Because there is a gradient in sunshine duration from land to sea, their representativeness for sea areas is limited. Change in daylength throughout the year shows a predictable pattern with sunshine duration in the North Sea region at a maximum in summer and a minimum in winter. On average May and July are the sunniest months with about 170 h of sunlight in Lerwick, 250 h in Helgoland and 290 h in Skagen Fyr (adjusted, not shown) for the period 1971–2000. Sunshine duration is lowest in

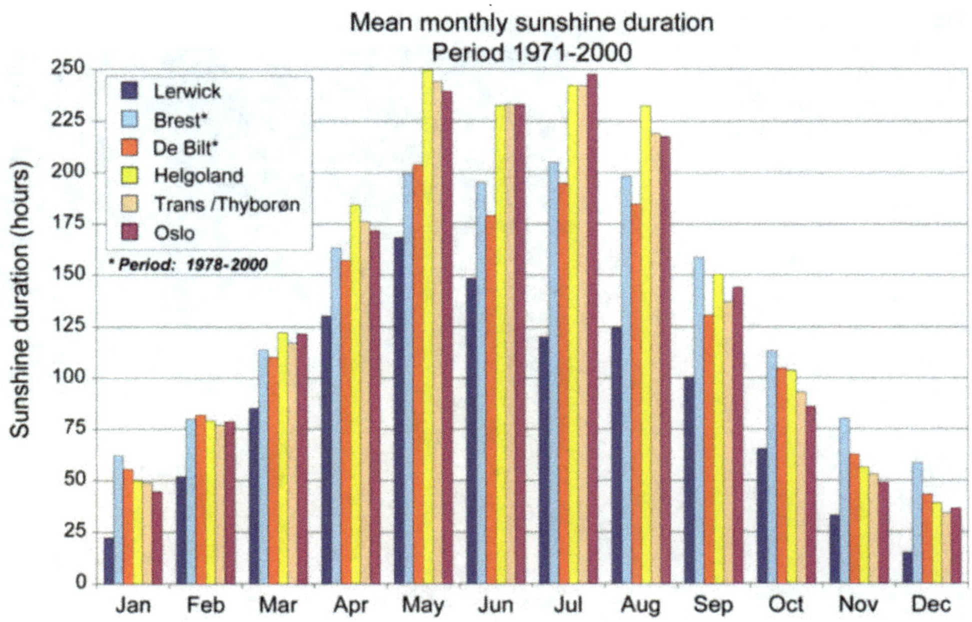

Fig. 1.30 Annual cycle of mean sunshine duration (hours/month) at selected land stations for the period 1971–2000. Station locations are shown in the E-Supplement to this chapter (Fig. S1.5.1)

December with mean values of 15 h in Lerwick to 60 h in Brest. From May to August, sunshine duration in the northern and western parts of the North Sea and the English Channel is about 20–50 h less than in the southern and eastern parts.

Recent research studies use various methods for deriving sunshine duration from satellite observations. For example, using cloud type data (Good 2010) or solar incoming direct radiation (Kothe et al. 2013), based on observations from the SEVIRI (Spinning Enhanced Visible and Infrared Imager) instrument on the Meteosat Second Generation satellite. A comparison of products generated by both methods for Europe with in situ observations showed that satellite-based sunshine duration is within ±1 h day^{-1} compared to the high-quality Baseline Surface Network or surface synoptic station measurements (Kothe et al. 2013). The procedure for deriving sunshine duration from satellite data for an area including the North Sea was recently established by the UK MetOffice, but the data have not yet been analysed from a climate perspective.

1.5.5.2 Global Radiation

There is a strong relationship between sunshine duration and global radiation. Global radiation is the sum of direct solar radiation and diffuse sky radiation received by a unit horizontal surface. It is a function of latitude and depends on atmospheric scattering and absorption in the presence of clouds and atmospheric particles. More than 90 % of solar irradiance is absorbed by the ocean having an important impact on thermal structure and density-induced motions within the ocean (Bülow et al. 2013). In the network of EUMETSAT, the Satellite Application Facility on Climate Monitoring (CM SAF) provides a satellite-based data set of surface irradiance (Posselt et al. 2012). In the North Sea region, mean annual global radiation increases from north to south with the highest annual average irradiance of 130–140 W m^{-2} in the English Channel, for results derived from observations of the MVIRI instruments on the geostationary Meteosat satellites for the period 1983–2005 (compare Fig. S1.5.11 in the E-Supplement).

Global radiation exhibits a pronounced seasonal cycle with a minimum in winter shown by the December irradiance and a maximum in summer shown by that for June (Fig. 1.31). Over the course of a year, global radiation varies from about 5 to 200 W m^{-2} in the northern North Sea and between 25–30 and 230–240 W m^{-2} in the English Channel. The highest values are recorded in the Skagerrak (250 W m^{-2}) and English Channel (260 W m^{-2}) in June. In both areas, lee effects may account for this high irradiance as well as the influence of the ridge of the Azores High which stretches towards central Europe across the English Channel in summer (compare Fig. 1.17).

1.5.6 Cloud Cover

The spatial distribution of mean annual cloud cover for the period 1982–2009 is shown in Fig. 1.32 based on measurements by the Advanced Very High Resolution Radiometer (AVHRR) on the polar-orbiting NOAA and Metop satellites (CLARA data set). Cloud cover is highest in the northwestern sea areas and decreases southward. In coastal areas, cloudiness is affected by lee and luv effects depending on the exposure of the coastline to the prevailing

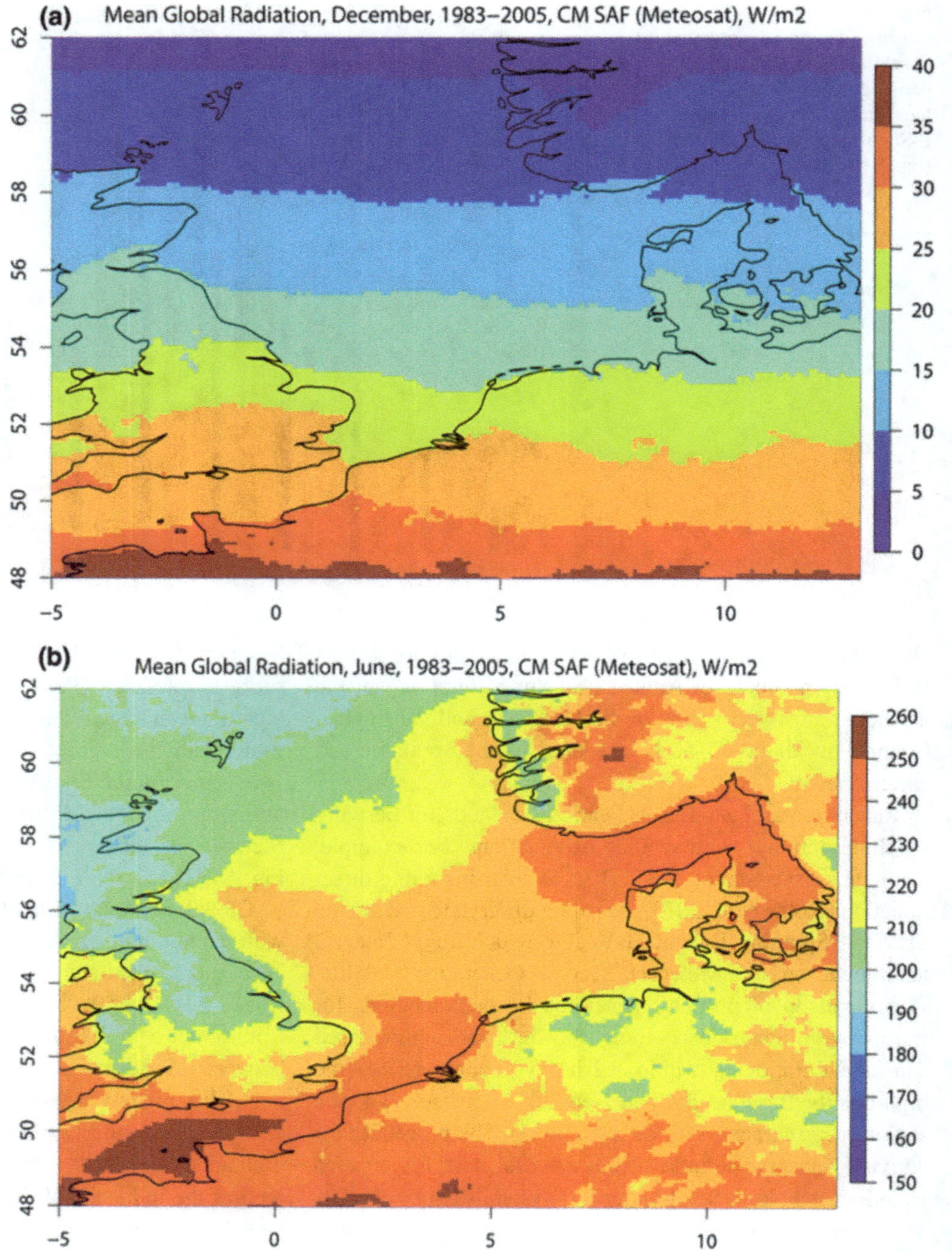

Fig. 1.31 Mean monthly global radiation (W m^{-2}) in December (*upper*) and June (*lower*) for the period 1983–2005 derived from satellite data by CM-SAF

wind direction. The Skagerrak is especially affected by a pronounced reduction in cloudiness due to the lee effect of the Scandinavian mountains. In particular, during weather situations with high reaching northerly airflow at the rear of large low pressure systems air humidity decreases significantly on their leeward side resulting in cloud dissolution as shown in Fig. 1.33. This Norwegian fohn effect is observed about 10 to 20 times a year. The mean monthly and annual cloud cover derived from ship observations (BSH 2009) shows cloud cover is generally lowest in May, the month with a high frequency of anticyclonic situations, and highest in winter, except for the sea areas along the northeastern coast of Great Britain where lee effects in the predominantly westerly air flow cause a reduction. The amplitude of the annual cycle is smallest (0.3 octa) in the sea areas Forth and Tyne and highest (1.5 octa) in the Skagerrak, where the cloud cover is lowest in summer.

A comparison of the mean annual cycle in cloud cover derived from the CLARA data set and ERA-40 over the German Bight for the period 1982–2002 shows an underestimation in the reanalysis data, which is smallest in winter and highest in summer (Bülow et al. 2014).

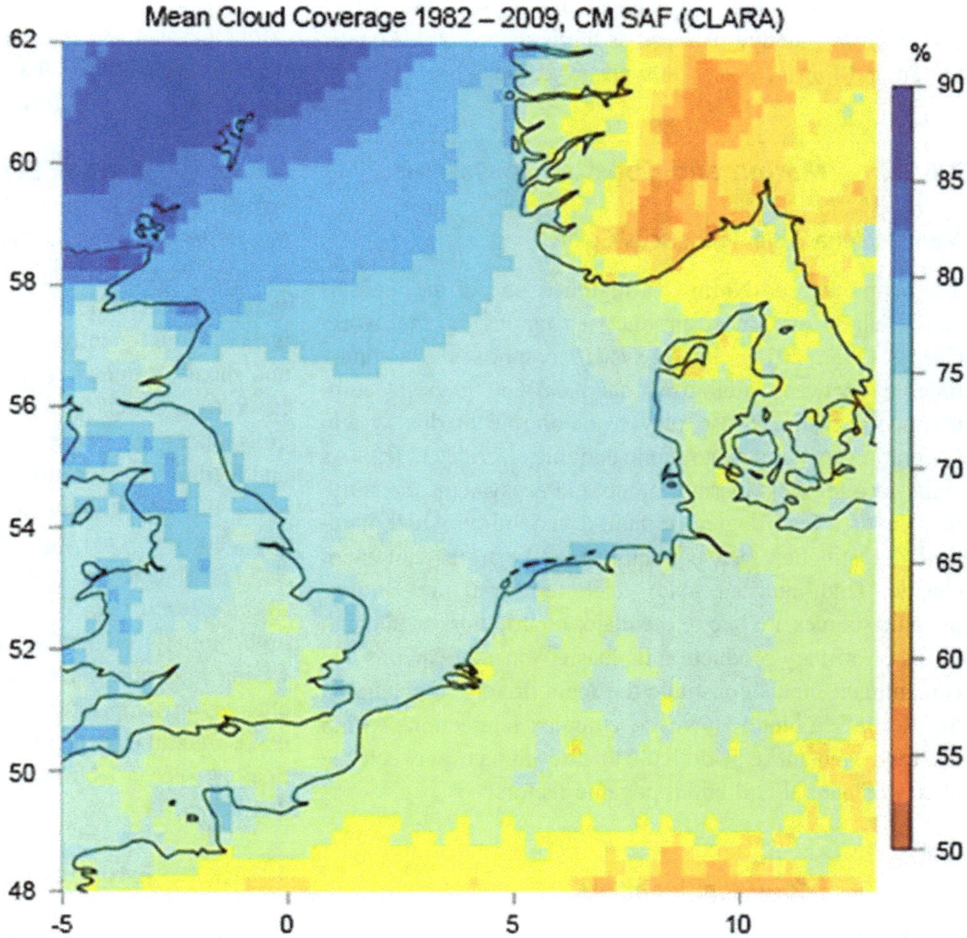

Fig. 1.32 Mean annual cloud coverage (in percentage) for the period 1982–2009 derived from satellite data by CM-SAF

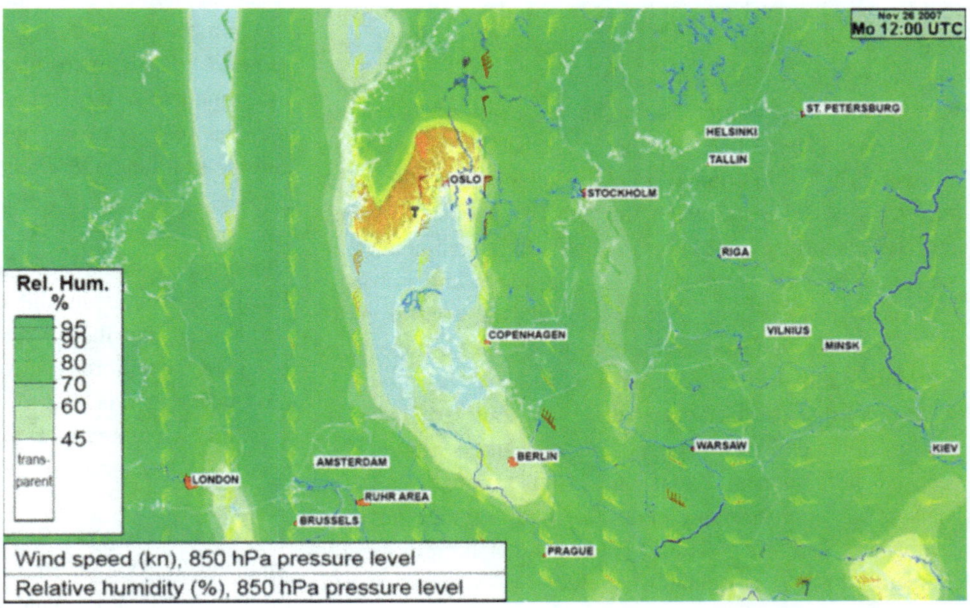

Fig. 1.33 Example of the lee effect off Scandinavia (Lefebvre and Rosenhagen 2008)

Spatial distributions of mean monthly cloud cover derived from satellite data across the eastern North Atlantic Ocean and Europe since 2009 are available.[5]

1.6 Marine and Coastal Ecosystems

Merja Helena Tölle, Franciscus Colijn

The semi-enclosed North Sea region is one of the biologically richest and most productive regions in the world (Emeis et al. 2015). Its ecosystem comprises a complex interplay between biological, chemical and physical compartments. Humans also play a major role in this system integrating social and economic activities (see Sect. 1.2). As is the case for most marine and coastal ecosystems the North Sea shows a high degree of natural variability, which hampers a distinction between human and natural causes of change. High nutrient loads from terrestrial and anthropogenic sources are one of the major contributors to the high levels of primary production in coastal waters. Fisheries and contaminant inputs constitute the main drivers of change in the North Sea biota. However, complex interactions within the food web make it difficult to discriminate between the effects of natural and anthropogenic factors.

1.6.1 Ecological Habitats

The North Sea is delimited by Dover Strait between the UK and France to the south (50°30′N 2°E), to the north by a line between Scotland and Norway north of the Fladen Ground (62°N), and by the Kattegat between Sweden and Denmark to the east (see Sect. 1.2 for more details). The North Sea can be subdivided into the shallow Southern Bight with vertically mixed water masses and a deeper, seasonally stratified northern part which is subject to significant nutrient-rich North Atlantic inflow. The southern North Sea receives warm oceanic water through the English Channel, experiences strong tidal currents, has a high sediment load, and receives large amounts of contaminants and nutrients from continental rivers discharging into the coastal zone. The whole North Sea is subject to atmospheric deposition from land- and ship-based sources.

The North Sea coastline includes a variety of habitats, such as fjords and rocky shores, estuaries and coastal deltas, beaches with dunes, banks including sandbanks, cliffs, islands, salt marshes and intertidal mudflats (see detailed review in Sect. 1.2). These coastal habitats are mainly characterised by a seabed covered by seagrasses and macroalgae or by sandy, gravelly or muddy sediments (ICES 2011). Coastal and estuarine habitats serve as spawning and nursery grounds for fish (Van Dijk 1994) and breeding grounds for coastal birds.

1.6.1.1 Wadden Sea

About 60 % of the intertidal area at the south-eastern North Sea shores occurs in the Wadden Sea (Reise et al. 2010). The Wadden Sea is a very shallow wetland area with a high sediment load in the channels due to strong tidal currents and waves. Consequently this region features sandy mud flats and shoals, seagrass meadows, and oyster and mussel beds. It serves as a rich food source and habitat for seals, waders, gulls, ducks and geese (Reise et al. 1994), provides a resting and feeding area for millions of migratory birds, and is used as a nursery ground by fish (e.g. plaice *Pleuronectes* spp. and sole *Solea* spp.). The appearance of the sandy tidal flats is shaped by the faecal mounds of the lugworm *Arenicola marina*. The Wadden Sea is characterised by a high biomass of benthic species (up to 100 g ash free dry weight m^{-2}) that can cope with tidal exposure and extreme changes in temperature, salinity and turbidity, due to tidal exposure and currents. Since 2009 the international Wadden Sea has been a UNESCO World Heritage Site (Reise et al. 2010).

1.6.1.2 Estuaries and Fjords

Estuaries of large rivers (e.g. Rhine, Meuse, Elbe, Weser, Ems, Tyne, Thames and Seine) extend along the North Sea coast and many harbour ports. All the estuaries have a characteristic turbidity maximum caused by the mixing of fresh and saline water. Due to their low and variable salinity the number of plant and animal species in the estuaries is reduced relative to sea and freshwater. There are some fully estuarine fish living almost their entire lifecycle in estuarine habitats (e.g. common goby *Potamoschistus microps* and European flounder *Platichthys flesus*).

A great variety of fjords can be found along the Norwegian coastline. These are typically deep, long and narrow, showing significant biological gradients from head to mouth. Plankton blooms in the coastal water may be passively transported into the fjords. The reefs of the upper sediment layers of these fjords (between 40 and 400 m) may comprise cold-water corals (e.g. *Lophelia pertusa* and *Gorgonocephalus caputmedusae*) and benthic communities including species such as sea stars, brittle stars and sea urchins.

1.6.1.3 Dogger Bank

The shallow Dogger Bank is a large sandbank situated east off the coast of England in the central North Sea and is influenced predominantly by Atlantic water from the north and Channel water from the south (Bo Pedersen 1994). Due

[5]http://www.dwd.de/DE/klimaumwelt/klimaueberwachung/europa/europa_node.html.

to year-round primary production this is a productive fisheries area (Kröncke and Knust 1995).

1.6.1.4 Helgoland, and the Shetland and Orkney Islands

The offshore island Helgoland in the Southern Bight and the Shetland and Orkney islands off the north-east of Scotland form rocky outcrops surrounded by soft sediments. Due to rich prey in the surrounding waters millions of seabirds use these islands for breeding. Sand dunes are inhabited partly by seals. The hard substrate around these islands serves as a holdfast for kelp (*Laminaria* spp.) (Lüning 1979).

1.6.2 Ecosystem Dynamics

1.6.2.1 Communities, Food Webs and the Seasonal Cycle

The North Sea is a temperate sea with a clear seasonal production cycle. In winter, most of the primary production is light-limited. In spring distinct phytoplankton blooms occur at the surface of the water column, mainly due to higher light levels and rising temperature. Carbon or energy is transferred from phytoplankton to herbivorous zooplankton to carnivorous zooplankton and finally to (jelly) fish, seabirds, and marine mammals. Figure 1.34 illustrates the trophic structure and energy flows of the North Sea food web. All organism groups discussed in later chapters of this report are shown in this schematic. Links between the pelagic and benthic components are indicated (i.e. settling detritus and suspension feeders). The spring phytoplankton bloom is initialised in the southern regions of the North Sea (south of Dogger Bank) in late winter/early spring, developing later in the northern part of the North Sea (Colebrook and Robinson 1965). The timing of the spring bloom depends primarily on increased light availability. The spring diatom bloom is responsible for a large proportion of the annual primary production which is succeeded by late summer and/or autumn blooms of dinoflagellates (Cushing 1959). Diatoms may bloom again in autumn but are then less numerous (Reid 1978). Grazing by secondary consumers occurs between May and September. Copepods consume about 10–20 % of the primary production of the spring bloom in the coastal zone (Baars and Fransz 1984). Roff et al. (1988) found the spring bloom is also grazed by heterotrophic protists which may provide an alternative food source for copepods. Pelagic herbivores such as copepods are preyed on by invertebrate carnivores (e.g. arrowworms and jellyfish). Meso- and macrozooplankton are also preyed on by pelagic fish, which are in turn consumed by larger fish, cetaceans and seabirds. A major proportion of the primary production, and the zooplankton and their faeces is transferred through sedimentation to benthic communities. There

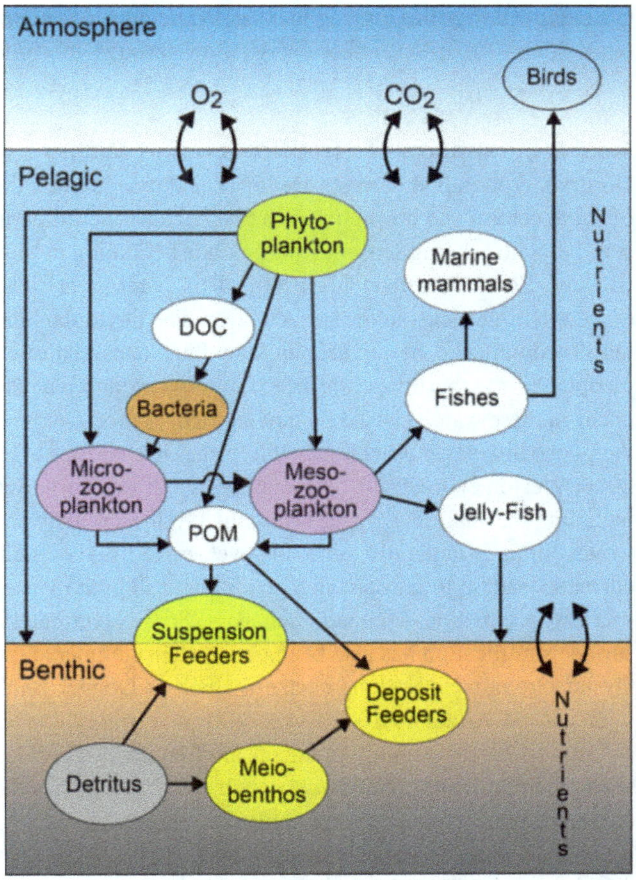

Fig. 1.34 Schematic overview of the trophic structure and flows of energy in the North Sea

they can be consumed by macrobenthos (shellfish, worms) and meiobenthos which serve as food for invertebrate carnivores (e.g. squid) and demersal fish (Dagg et al. 1982; Joris et al. 1982; Nicolajsen et al. 1983; Reid et al. 2009) or are mineralised by bacteria (Van Es and Meyer-Reil 1982). Invertebrate carnivores and demersal fish are in turn consumed by larger fish and other carnivores (such as seals) and some ultimately end up in the human diet. In the post-bloom period parts of the benthic biomass can be resuspended and remineralised by bacteria and protozoa. However, it should be noted that the actual North Sea food web is more complex than implied by this simple overview, owing to the large number of flora and fauna present in the North Sea and the many interrelations between these species and their various life stages.

Recruitment success of fish stocks and fisheries depends strongly on zooplankton production, and on their size and composition. Marine mammals and seabirds depend on fish stocks. Zooplankton peak abundance relies on primary production by phytoplankton. Complex interactions between the various components of the food web makes the North Sea ecosystem vulnerable to physical changes (such as in currents, salinity, mixing regimes or temperature; Schlüter et al. 2008, 2010) that affect energy flow through the food web.

1.6.2.2 Impacts of Stratification and Mixing

North of a rough line between Denmark and the Humber, the offshore central and northern North Sea becomes stratified in summer with a strong seasonal thermocline beginning in May (Becker 1981; Turrell et al. 1996). This leads to a rapid depletion of nutrients in the surface layer. The phytoplankton may be dominated by autotrophic nanoflagellates and other picoplankton at this time. Subsurface summer production may occur in these waters, based on new nutrients introduced into the pycnocline layer associated with fronts (Richardson et al. 2000; Weston et al. 2005). The coastal area of the North Sea including the shallow Dogger Bank, the Oyster Grounds and Dover Strait is generally vertically well-mixed due to tidal currents, leading to nutrient-rich conditions and high rates of primary production (Legendre et al. 1986). Nevertheless, surface stratification has also been observed in summer at the Oyster Grounds and on the northern slope of Dogger Bank (Greenwood et al. 2010). Fronts or frontal zones due to salinity and/or temperature gradients are frequently observed in the German Bight, off Flamborough Head, and along the coasts leading to high local phytoplankton biomass.

1.6.2.3 Carbon and Nutrient Cycles

A detailed North Sea field study initiated in 2001 revealed that bottom topography exerts fundamental controls on carbon dioxide (CO_2) fluxes and productivity, with the 50 m depth contour acting as a biogeochemical boundary between two distinct regimes. In the deeper northern areas, seasonal stratification facilitates the export of particulate organic matter (POM) from the surface mixed layer to the subsurface layer with the consequence that biologically-fixed CO_2 is replenished from the atmosphere. After entering the subsurface layer the POM is respired, releasing dissolved inorganic carbon (DIC) that is either exported to the deeper Atlantic or brought back to the surface during autumn and winter once the seasonal stratification has broken down (Thomas et al. 2004; Bozec et al. 2006; Wakelin et al. 2012), see Fig. 1.34. At the annual scale the northern parts act as a sink for atmospheric CO_2, and the high productivity driving the CO_2 fixation is largely fuelled by nutrient inputs from the Atlantic Ocean (Pätsch and Kühn 2008). In the southern part of the North Sea the shallow, well-mixed water column prevents the settling of POM from the euphotic zone, with the result that both production and respiration occur within the well-mixed water column. Except for a short period during the spring bloom, the effects of POM production and respiration cancel out and the CO_2 system appears to be temperature-controlled (Thomas et al. 2005a; Schiettecatte et al. 2006, 2007; Prowe et al. 2009). Productivity in this area relies on terrestrial nutrients to a far greater degree than in the northern North Sea, and nutrients are a limiting factor during the productive period (Pätsch and Kühn 2008; Loebl et al. 2009). Carbon cycling in the North Sea is dominated by its interaction with the North Atlantic Ocean, which serves as both a source and a sink for CO_2 transport due to the high rate of water exchange (Thomas et al. 2005b; Kühn et al. 2010). Thus, input from the Baltic Sea and river loads play a crucial role, since these constitute net imports of carbon (and other biogeochemical tracers) into the North Sea. As previously stated, nutrient concentrations are high in coastal areas relative to the offshore region and show considerable spatial variability. Concentrations decrease with distance from the shore. Riverine winter nutrient concentrations may be up to 50 times higher than offshore values (e.g. nitrogen in the river Ems; Howarth et al. 1994). In autumn and winter, nutrients accumulate due to intense mineralisation and peak in late winter. An average winter nitrate concentration in the central North Sea was estimated at about 8 μmol l^{-1} (Brockmann and Wegner 1996). In winter, nutrient concentrations in coastal waters are mostly influenced by river inputs and are only marginally influenced by biological processes. An exception is the Dogger Bank area where primary production continues all-year round and there is no winter peak in nutrient concentrations. Phosphate and silicate are usually the first nutrients to become depleted in coastal waters during spring slowing the growth of diatoms in the spring bloom. Excess nitrate is then taken up by flagellates and other plankton such as *Phaeocystis* sp. (Billen et al. 1991). In much of the North Sea, nutrients become depleted in the upper layer due to summer phytoplankton growth following stratification. An increase in surface nutrient concentrations can be seen in autumn after mineralisation has occurred in deeper water layers below the thermocline and when this nutrient-rich bottom water is brought up to the euphotic zone by stormy autumn weather. Nitrogen to phosphorus ratios show that nitrogen is a potential limiting factor in the central North Sea, whereas phosphate is a potential limiting factor in the coastal area and off England (Brockmann et al. 1990). The atmosphere, Atlantic inflow and several rivers contribute to the total nutrient load in the North Sea. Atmospheric nitrogen input to the North Sea is estimated at 300–600 kt (Richardson and Bo Pederson 1998). However, the spatial distribution of the atmospheric input is unclear. About 1411 kt N $year^{-1}$ is advected through the English Channel (Laane et al. 1993), with nutrient inputs entering the North Sea through the Atlantic Ocean estimated at 8870 kt N $year^{-1}$ and 494 kt P $year^{-1}$ (Brion et al. 2004). Total riverine inputs to the continental coastal zone are estimated to average 722 kt N $year^{-1}$ and 48 kt P $year^{-1}$ (Radach and Pätsch 2007).

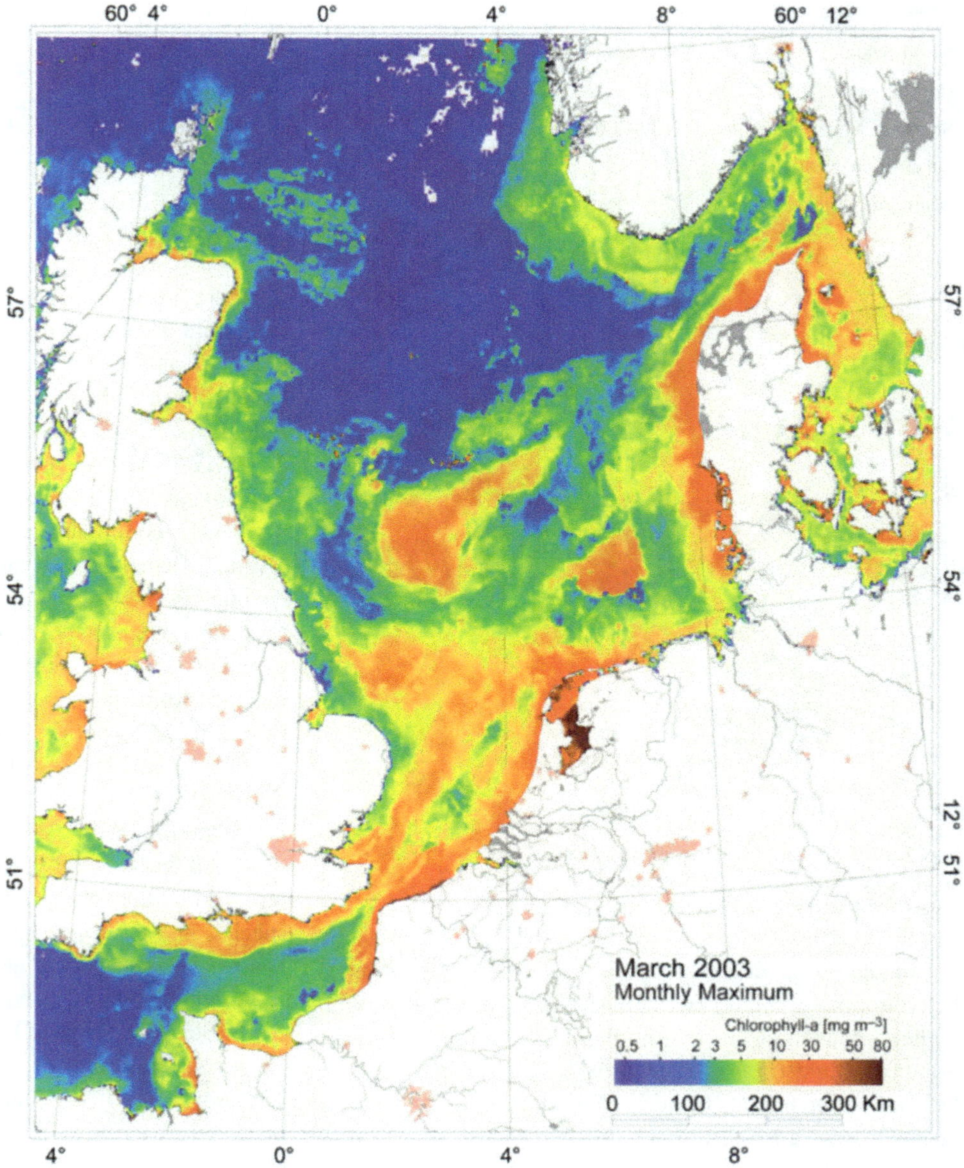

Fig. 1.35 Maximum chlorophyll-*a* concentration in the North Sea in March 2003, based on an ENVISAT/MERIS satellite image (Peters et al. 2005)

1.6.2.4 Production and Biomass

In addition to regulation by predators, phytoplankton growth and abundance are regulated by aspects of their physical and chemical environment, such as light intensity, wind, and nutrient concentration (e.g. Sverdrup 1953; Margalef 1997; Schlüter et al. 2012). Zooplankton biomass depends on the growth and quality of their food as well as on hydrodynamic and chemical factors.

Annual new production from the input of nitrogen to the North Sea is roughly 15.6 million t C (Richardson and Bo Pederson 1998). About 40 % of annual new production is estimated to be associated with the spring bloom in surface waters of the stratified regions of the North Sea, another 40 % with production in the coastal waters, and the rest with the deep chlorophyll maximum (Riegman et al. 1990).

There is a considerable variability in phytoplankton biomass across the North Sea with high values occurring inshore in the coastal regions of southern England, along the continental coast to Denmark and Norway, within tidal fronts, and at the Oyster Ground and Dogger Bank (Peters et al. 2005). The highest chlorophyll-*a* concentrations are observed along the southern coastal areas in spring (Fig. 1.35). The spatial distribution and productivity of benthic organisms in the North Sea are related to currents, and to variations in temperature and food availability (Heip and Craeymeersch 1995). High primary productivity in the eutrophic coastal zone is mostly due to terrestrial nutrient inputs from rivers and atmospheric deposition. Estimates of annual primary production in the North Sea are highest in coastal areas, and decrease northward from the southern to the northern North Sea (see Table 1.4).

Table 1.4 Estimates of primary production over an annual cycle for the North Sea and its subareas

North Sea area	Annual primary production, g C m^{-2}	Source
Coastal area	400	Cadee (1992)
Southern North Sea	150–200	Reid et al. (1990), Heip et al. (1992), Joint and Pomeroy (1993)
Offshore of Netherlands	375	Bot and Colijn (1996)
Northern North Sea	70–90	Steele (1974)

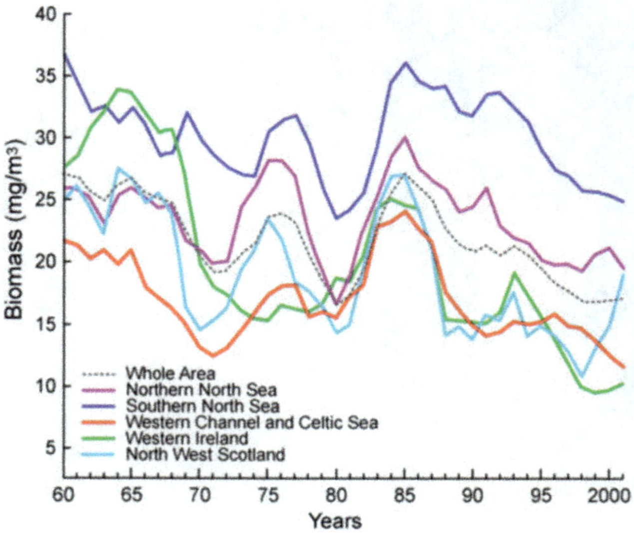

Fig. 1.36 Long-term trend in annual average biomass for 29 major copepods and cladocerans in five North Sea sub-areas and North Sea as a whole (Mackinson and Daskalov 2008)

Annual production of secondary producers (formation of heterotrophic biomass) is estimated at 2–20 g C m^{-2}, with higher production in the northern North Sea than the Southern Bight (Fransz et al. 1991). This gradient may be because large amounts of primary production in the shallow southern areas are transferred to benthic communities and so are unavailable for consumption by zooplankton. De Wilde et al. (1992) reported a significant decline in zooplankton abundance across the North Sea between 1960 and 1980. As shown in Fig. 1.36 this decline was followed by a subsequent recovery and then a new and ongoing decline. The long-term trend in zooplankton biomass for the North Sea as a whole is well documented through the variability observed in its sub-areas.

Estimates of annual secondary and higher production for the North Sea are given in Table 1.5. The benthic fauna production estimate for Oyster Ground (36 g C m^{-2} $year^{-1}$) includes 21 g C m^{-2} $year^{-1}$ for macrofauna and 6 g C m^{-2} $year^{-1}$ for meiofauna (De Wilde et al. 1984). Biomass of brown and green algae increased in the Wadden Sea and the Skagerrak-Kattegat area between 1960 and 1990 (Beukema 1989; Josefson et al. 1993). Kröncke and Knust (1995) found a decrease in total benthic biomass at the Dogger Bank between 1950 and the 1980s.

There are over 200 species of fish in the North Sea and their total annual productivity has been estimated at 1.8 g C m^{-2} (Daan et al. 1990). Cushing (1984) reported a large increase in the abundance of gadoids such as whiting *Merlangius merlangus*, haddock *Melanogrammus aeglefinus*, cod *Gadus morhua* and Norway pout *Trisopterus esmarkii* accompanied by a decrease in herring *Clupea harengus* in the 1960s, commonly referred to as the 'gadoid outburst'. Overall, demersal fish and their size composition have declined in the North Sea since the start of the 20th century (Pope and Macer 1996). The commercial fishery was heavily exploited in the 1960s, when landings peaked and then declined following a ban on herring in 1977 from a change in EU fisheries policy. The fishery reopened in 1981 and landings increased until 1988 followed by a record low in the 1990s.

Of the roughly 10 million t of fish and shellfish biomass produced each year in the North Sea, around 25 % is removed by the fishery, 50 % is consumed by predatory fish species, and the rest is consumed by birds and mammals or lost to disease (Daan et al. 1990).

About 2.3 million t of fish and shellfish were landed in the North Sea in 1999 (FAO 2003). Fish catches over the past 60 years seem strongly related to levels of primary production (Chassot et al. 2010). Pelagic species such as herring and mackerel are the most important in landings, as well as smaller species such as sardines, and anchovy. Demersal landings include cod, haddock, whiting, saithe *Pollachius virens*, plaice and sole. Detailed statistics of fish landings are available from the International Council for the Exploration of the Sea (ICES 2008; FAO Annual reports; see also Chaps. 8 and 12).

The sedimentary environment of the North Sea contains up to 5000 species of macrobenthic and meiobenthic invertebrates (Heip and Craeymeersch 1995). Total macrobenthic biomass decreases northward from 51.5°N while biodiversity and density increase (Heip et al. 1992). Rehm and Rachor (2007) also showed benthic biomass to decrease with distance from the coast. According to Heip and Craeymeersch (1995), northern species of the macrobenthic community extend southward to the northern slope of Dogger Bank and southern species extend northward to the 100 m depth contour. They also found the separation between northern and southern species to disappear at around the 70 m depth contour in the central North Sea.

Table 1.5 Estimates of annual secondary and higher production for the North Sea

Group	Annual secondary and higher production (g C m^{-2})	Source
Macrobenthos	2.4	Heip et al. (1992)
Fish	1.8	Daan et al. (1990)
Meiofauna	10	Heip and Craeymeersch (1995)
Phytobenthos	Negligible	Beukema (1989), Josefson et al. (1993)
Secondary production	2–20	Fransz et al. (1991)
Benthic fauna of Oyster Ground	36	De Wilde et al. (1984)

Several mammal species occur in the North Sea. The harbour seal *Phoca vitulina* is particularly abundant and there are estimated to be 4500–5000 in the German Bight, 6000 in the Wash, and 7000–10000 on the British east coast. For other common species such as harbour porpoise *Phocoena phocoena*, bottlenosed dolphin *Tursiops truncatus*, and less common species such as common dolphin *Delphinus delphis*, white beaked dolphin *Lagenorhynchus albirostris*, white-sided dolphin *L. acutus*, beluga *Delphinapterus leucas*, pilot whale *Globicephala melas* and minke whale *Balaenoptera acutorostrata* all populations are either stable or are declining due to pollution, fewer prey or accidental bycatch by the fishing fleet (FAO 2003).

Bellamy et al. (1973) listed 71 species of bird in the North Sea coastal and offshore areas.

1.6.3 Current Status and Threats

Human-induced stresses represent major threats to the North Sea ecosystem. These include overfishing, eutrophication, ocean acidification, climate change, recreation, offshore mining, wind farms, shipping, dumping of waste, changes in food web dynamics, and the introduction of non-indigenous species (Richardson et al. 2009). Potential threats on the North Sea ecosystem could be reduced by better management based on a concept of zones with different utilisation levels (Fock et al. 2014).

1.6.3.1 Eutrophication

Sewage effluents, leaching from agricultural land, and atmospheric nitrogen deposition are responsible for the high nutrient input to the North Sea (Druon et al. 2004). Aquaculture is another important factor. Enrichment by nitrates and phosphates may encourage blooms of particular phytoplankton species, such as *Phaeocystis* sp. or *Noctiluca* sp. in the coastal area. Under calm weather conditions massive algal blooms may settle on the sediment surface and cause hypoxia in the near-bottom water as seen in the German Bight and at the Danish Coast (De Wilde et al. 1992) resulting in mass mortality of macrobenthos. Present conditions reflect a strong decrease (70 %) in phosphate input since 1985 and a moderate reduction (about 50 %) in nitrogen for two major rivers Rhine and Elbe (van Beusekom et al. 2009).

1.6.3.2 Harmful Algal Blooms

Harmful algal blooms (HABs) have been reported in several regions of the North Sea and linked to changes in nutrients, temperature, salinity, and the North Atlantic Oscillation (ICES 2011). Their relation to anthropogenic drivers is unclear. Harmful algal blooms may impact ecosystem function and thus influence economic functions, including human health. However, several species known to cause HABs (e.g. *Phaeocystis* spp., and several dinoflagellates and raphidophytes) were already present in the North Sea at the beginning of 20th century (Peperzak 2003). Laboratory studies indicate that a rise in temperature may enhance growth, but that not all species tested responded in the same way (Peperzak 2003).

1.6.3.3 Offshore Oil and Gas

Vast oil and gas reserves formed during the North Sea's long geological history are currently being exploited. A large number of oil and gas production platforms widely distributed over the North Sea are responsible for supplying energy to European countries. Waters around the platforms are sometimes contaminated by chemicals including substances that affect benthic communities. Detailed studies showed effects of drilling activities (discharge of oil-based muds) only up to about 1 km from the platform (Daan and Mulder 1996).

1.6.3.4 Renewable Energy

As the use of renewable energy from offshore wind farms and from wave/tidal power plants increases, so the threat to seabirds and marine mammals will increase. The long-term impacts of renewable energy infrastructure on the North Sea ecosystem are not yet clear. Initial studies based on a UK wind farm show that plant and animal growth on piles may offer a substantial additional food supply for marine mammals. Studies by Garthe (Kiel University, unpubl.) on telemetered northern gannet *Morus bassanus* from Helgoland show birds can avoid wind farms to some extent.

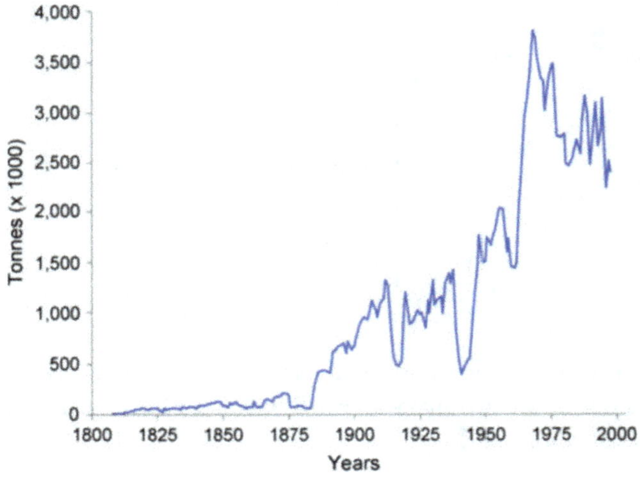

Fig. 1.37 Long-term trend in North Sea fish catches since the 1800s (Mackinson and Daskalov 2008)

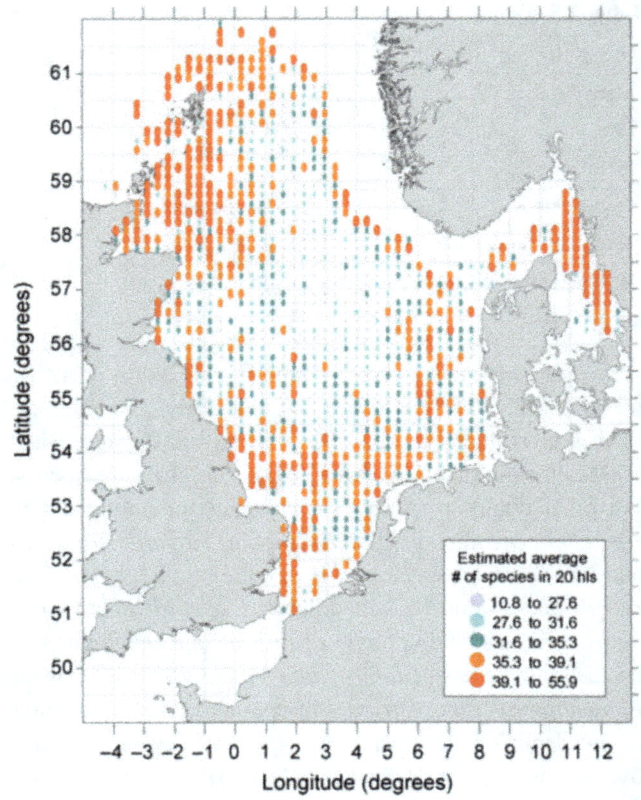

Fig. 1.38 Spatial distribution of species richness (1977–2005) for all North Sea fish species based on International Bottom Trawl Survey (IBTS) data (ICES 2008)

Long-term avoidance of wind farms by marine mammals such as harbour porpoise is still debated (Dähne et al. 2013). To date, only a small number of the planned wind farms have been realised (see also Sect. 1.2).

1.6.3.5 Fisheries

The North Sea ecosystem is overfished (Cury et al. 2000; FAO 2012). Bottom and beam trawling are a major threat for marine organisms because they damage the seabed and alter mature benthic communities. Based on long-term data, Fock et al. (2014) stated that the age, volume and size composition of commercial fish stocks has shifted since 1902. As a result, species composition has shown significant changes in recent decades. For example, slow-reproducing fish such as sharks and rays declined and may only be encountered in areas with low fishing density. With beam trawling, the benthic community structure has changed from long-lived species (e.g. sharks and rays) to short-lived opportunistic and scavenging species (e.g. crabs and shrimps). Seagrasses declined along the Dutch, German and Danish coasts as a result of bottom trawling fisheries and an epidemic of wasting disease in the 1930s. Exploitation of North Sea fish stocks has increased since 1945 as shown by trends in fishery mortality among all groups of exploited fish species (Daan et al. 1990). As a result of measures to allow fish stocks to recover, total catches have declined since the late 1960s (Fig. 1.37). Even though pollution and overfishing in the North Sea (Cabeçadas et al. 1999) have resulted in major changes in species composition and shifts in species distribution, species richness based on International Bottom Trawl Survey (IBTS) data has increased over the past 30 years (ICES 2008); see

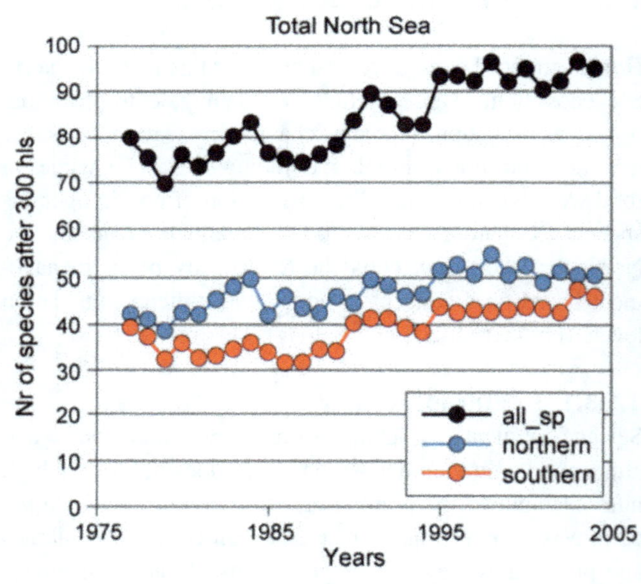

Fig. 1.39 Trends in species richness for all North Sea fish species: for the northern North Sea, the southern North Sea and the North Sea as a whole based on International Bottom Trawl Survey (IBTS) data (ICES 2008)

Fig. 1.38 for the spatial distribution of fish species richness and Fig. 1.39 for the long-term trend.

1.6.3.6 Contaminants

Contaminants enter the North Sea through various pathways and have had adverse and toxic effects on many species including cold-water corals, seabirds and marine mammals. Oil and oily wastes enter the North Sea from drilling operations, activities on platforms, shipping, harbours and ports, tanker spills and illegal discharges. Tankers and cargo vessels discharge ballast water containing non-indigenous species into the North Sea potentially with severe consequences for the ecosystem. Heavy metals and persistent organic pollutants enter the North Sea via atmospheric deposition and discharges from coastal industries and rivers (Chester et al. 1994). The quantities of most contaminants reaching the North Sea have decreased substantially since around 1985 (OSPAR 2010; Laane et al. 2013).

After entering the marine environment, contaminated particles mix with non-contaminated particles. Depending on the transport mechanism associated with seasonal and short-term variations in waves and primary productivity these particles can be widely dispersed or be deposited and so remain within the area. At the seabed older material becomes contaminated by mixing with newly supplied material. Thus, after sedimentation contaminants may be resuspended and redistributed. This causes a long-term transport of toxic persistent compounds towards the North Atlantic up to the Arctic Ocean. Apart from physical mechanisms, bioturbation by infaunal communities may also redistribute pollutants into overlying waters (Kersten 1988).

Chlorinated hydrocarbons (such as polychlorinated biphenyls; PCBs) were responsible for a dramatic reduction in the numbers of harbour seals in the Wadden Sea (Reijnders 1982), but after a ban on hunting and a strong reduction in the use of PCBs the population has since increased strongly (Wolff et al. 2010). Exposure to tributyltin (TBT) and polycyclic aromatic hydrocarbons (PAHs) also resulted in adverse effects on organisms (Laane et al. 2013). Some of the persistent organic pollutants entering the North Sea are biomagnified through the marine food web.

1.6.3.7 Tourism

Tourism is exerting high pressure on some coastal areas. Species and habitats could be affected by those activities that disturb wildlife, damage seabed habitat, and which cause noise disturbance and pollution.

1.6.3.8 Ports

Port development is expected to continue as a result of increasing ship traffic in the short-sea and containerised cargo markets. An example is the new deep water Jade-Weser-Port in the Southern Bight, which is accessible by the largest container vessels. A recurring issue with major harbours is the need for dredging and then disposal of the often contaminated sludge. Such as occurs with the deepening of the Elbe river with the port of Hamburg.

1.6.3.9 Non-indigenous Species

Many studies report the occurrence of non-indigenous species in the North Sea that may have been introduced via ballast water from cargo ships, shipping or mariculture. Some new species are also reported that were misidentified in the past (Gómez, 2008). Two examples of imported species are the American comb jelly *Mnemiopsis leidyi* first recorded in the North Sea in 2006 (Faasse and Bayha 2006) and the Pacific oyster *Crassostrea gigantea* which was introduced in the 1980s on selected plots in the Wadden Sea used for cultivation of shellfish (Drinkwaard 1998). A combination of over-exploitation of native oyster *Ostrea edulis* and import of the Pacific oyster has almost driven the former to vanish (Reise et al. 2010). Jellyfish concentrations are expected to increase across the entire North Sea in the next 100 years, owing to the projected rise in water temperature and fall in pH (Attrill et al. 2007; Blackford and Gilbert 2007). The ability of jellyfish to reproduce in a short time may affect pelagic and coastal ecosystems (Purcell 2005; Schlüter et al. 2010). An example of a non-indigenous phytoplankton species that is now well established in the North Sea is the diatom *Coscinodiscus wailesii*, which originated in the Pacific Ocean (Edwards et al. 2001; Wiltshire et al. 2010).

1.6.3.10 Climate Change

Chapters 3 and 6 provide a detailed description of climate change in the marine system. The following examples show that biological and physical effects of climate change are probably already being observed within the North Sea.

Major south-north changes in distribution and abundance related to rising temperatures have been detected for two key calanoid copepods, the cold-water species *Calanus finmarchicus* and the warm-water species *C. helgolandicus* (Planque and Fromentin 1996). Annual abundance of *C. finmarchicus*, which was usually high in the northern North Sea, decreased and the population has shifted north out of the North Sea (Beaugrand et al. 2009). Former high abundance of *C. helgolandicus* in the central and southern North Sea declined in that area and higher abundances were found further north (Helaouët and Beaugrand 2007).

During the summer period after the onset of stratification decreased dissolved oxygen concentrations in the bottom water at the North Dogger and Oyster Grounds were observed related to an increase in seasonal bottom water temperature (Greenwood et al. 2010).

1.7 Terrestrial Coastal Range Vegetation

Werner Härdtle, Jan P. Bakker, Jørgen Eivind Olesen

1.7.1 Natural Vegetation

1.7.1.1 Salt Marshes

Coastal salt marshes (also tidal marshes) are natural or semi-natural (non-grazed or grazed, respectively) halophytic ecosystems. They are habitats of the upper coastal intertidal zone of the North Sea and thus represent the transition between marine and terrestrial ecosystems (Ellenberg and Leuschner 2010).

Salt marshes establish in areas where plants that can trap sediment are inundated by water carrying suspended sediment. Salt marshes are often associated with muddy substrates and occur as 'foreland salt marshes' at the mainland coast (including sheltered estuaries and inlets), as well as in the lee of some North Sea islands (such as the Frisian Islands).

Salt marshes are highly dynamic ecosystems, and sites are characterised by daily inundation with salt water, sedimentation, and (locally) erosion processes. Species diversity is low and the vegetation comprises halophytic plants with life forms such as grasses, dwarf shrubs and herbs. Species composition and zonation are mainly determined by inundation frequency and soil salinity. Based on their topography, salt marshes are classified as low, medium or high salt marshes. A low salt marsh has more than 200 inundation events per year. Typical plant species include glassworts (*Salicornia* spp.), the dwarf shrub *Halimione portulacoides* and the grass *Puccinellia maritima*. High salt marshes, in contrast, are only flooded during extremely high tides and during storms. Typical species include black rush *Juncus gerardii* and grasses such as *Festuca rubra* agg. and *Agropyron repens* agg.

Salt marshes are highly productive ecosystems. They serve as a sink for organic matter and sediments, and play a crucial role in the littoral zone food web. In the past, most salt marshes were cut for hay and grazed by livestock, and so can be considered semi-natural. To maintain biodiversity, these semi-natural systems are often managed by livestock grazing for nature conservation purposes.

1.7.1.2 Dunes

Dunes are present on shorelines where fine sediments (mainly sand) are transported landward by a combination of wind and waves. Large dune areas occur on the North Sea coast, for example, in Denmark (to the north of Esbjerg), in the Netherlands (to the west of Den Helder) and on the Frisian Islands (Ellenberg and Leuschner 2010).

From an ecological perspective dunes are extreme habitats. Important stressors include water and nutrient shortage, unstable substrate, and salt spray. Dune formation is closely associated with plant succession and pedogenetic processes that occur with dune ageing. The natural succession starts with an embryo dune, where soils have low humus contents but are buffered due to the presence of calcium carbonate from seashells. Embryo dunes become foredunes as sand is accumulated by dune grasses such as *Ammophila arenaria* and *Elymus europaeus*. Further developmental stages are grey dunes (dominated by grasses or sedges such as *Carex arenaria*) and brown dunes (often characterised by dwarf shrubs such as *Calluna vulgaris* and *Empetrum nigrum*). Dune ageing is accompanied by soil acidification, the accumulation of humus and decreasing impact of wind and salt spray. In many places, eutrophication and dune disturbance have led to the establishment and proliferation of shrub species such as sea-buckthorn *Hippophae rhamnoides* and bramble *Rubus fruticosus*.

These successional stages often form a mosaic with duneslacks, low lying inter-dune areas protected from inundation and often separating foredunes from older phases of dune development, where a near-surface (brackish) water table favours a diverse and species-rich community, including bog-stars *Parnassia palustris*, marsh helleborine *Epipactis palustris*, marsh pennywort *Hydrocotyle vulgaris*, various other marsh orchids, rushes, sedges, and in mature stages also creeping willow *Salix repens*.

In the past, most dunes (except foredunes) were grazed by livestock and so can be considered semi-natural. To maintain biodiversity and prevent bush encroachment, dunes are often managed by livestock grazing for nature conservation purposes.

1.7.1.3 Moors (Bogs)

Moors, such as bogs (or mires) are wetlands that accumulate peat (i.e. organic matter originating from dead plant material). In bogs, peat is often formed by mosses (bryophytes), as well as grass species or sedges (Dierssen 1996). Waterlogged soils, in which the breakdown of (dead) organic material is inhibited by a lack of oxygen, are an important prerequisite for peat formation. The gradual accumulation of decayed plant material in a bog functions as a carbon sink and so provides an important ecosystem service.

Bogs are widely distributed in the temperate and boreal zone and are particularly well developed under an oceanic climate with high precipitation and low evapotranspiration (e.g. Scotland, western Scandinavia; Dierssen 1996; see Fig. 1.40).

In bogs the interstitial water of the peat body is acidic and low in nutrients (particularly nitrogen). Where the water supply is from precipitation, bogs are termed as ombrotrophic. If peat formation takes place under the influence of a

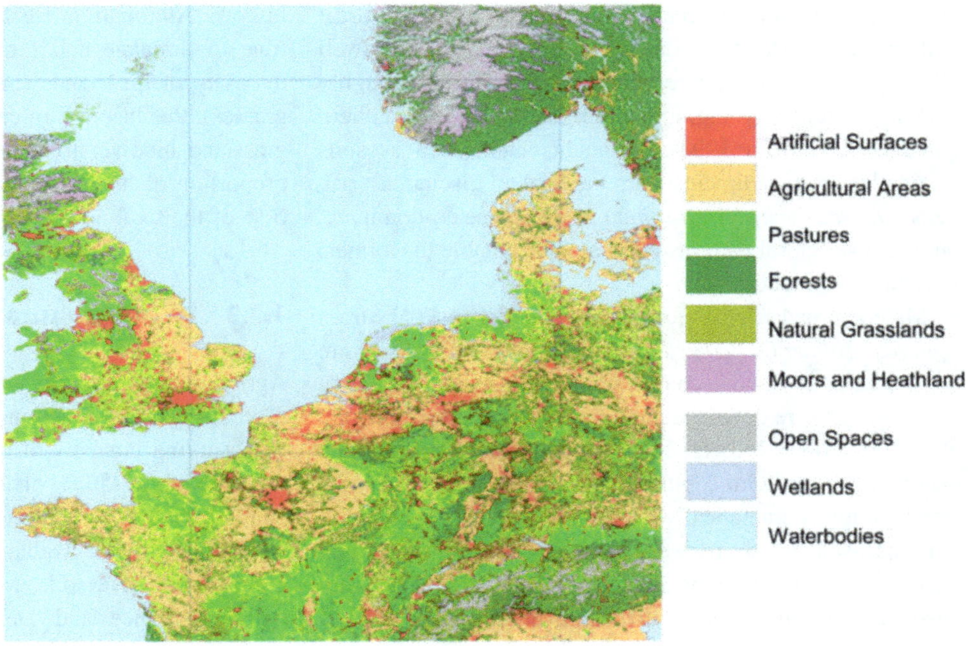

Fig. 1.40 Current (2006) land cover in the North Sea region according CORINE, a programme initiated by the EU to unify the classification of vegetation types and the description of different forms of land use (www.eea.europa.eu/data-and-maps/data/corine-land-cover-2006-raster-2)

high groundwater table, habitats are termed 'fens'. Low soil fertility and cool climates contribute to slow plant growth, as well as to low decomposition rates of organic material. In many bogs, the peat layer is 10 m thick or more.

Species typical of bogs are highly specialised. Most are capable of tolerating the combination of low nutrient levels and waterlogging. Dominant species include mosses of the genus *Sphagnum* and dwarf shrubs of the Ericaceae. Bogs are susceptible to fertilisation and drainage. Both factors may cause rapid and irreversible shifts in species composition.

1.7.1.4 Tundra and Alpine Vegetation

Tundra (including alpine vegetation types) occurs above the timber line and so is typical of the alpine zone in Scotland and Scandinavia (roughly between 1200 and 1700 m above sea level). Sites are mostly acidic and growing seasons are between 45 and 90 days long. The vegetation comprises perennial (mostly tuft forming) grasses, sedges, and dwarf shrubs of the Ericaceae. Cryptogams such as bryophytes and lichens are also very common (Dierssen 1996).

1.7.1.5 Forests

Forests are an important habitat type in the coastal range vegetation of the North Sea region. Under natural conditions forests would cover more than 90 % of this region, but currently cover only about a third (Bohn et al. 2002/2003).

Birch forests (with *Betula pubescens* agg.) are the dominant forest type in western Scandinavia (so-called western boreal and nemoral-montane birch forests) but are also developed as pioneer stands under temperate conditions in north-western parts of central Europe and in Great Britain (Dierssen 1996; Bohn et al. 2002/2003). Soils are mostly acidic (podzolic) and poor in nutrient supply (particularly for nitrogen). In the canopy, birch is the dominant trees species but pine trees (*Pinus sylvestris*) and willows (*Salix glauca*) may also be present. In contrast to the low number of vascular plants, birch forests are characterised by a highly diverse moss flora, among which species such as *Rhytidiadelphus loreus*, *Hylocomium umbratum* and *Rhodobryum roseum* are particularly typical.

Natural spruce forests are typical of central and southern parts of Scandinavia (western boreal and hemiboreal spruce forests; Dierssen 1996). In the canopy, spruce *Picea abies* is the dominant tree species, but pine trees (*Pinus sylvestris*) in the boreal zone and fir (*Abies sibirica*) and some broad-leaved trees in the hemiboreal zone may form mixed species stands. Soils are acidic, but base and nitrogen supply is better than in birch forest ecosystems. In the understorey, grasses (e.g. *Deschampsia flexuosa*) and bryophytes such as *Barbilophozia lycopodioides* and *Hylocomium umbratum* are common.

Forests dominated by European beech *Fagus sylvatica* are typical of large areas of temperate Europe (Ellenberg 2009; Ellenberg and Leuschner 2010). Beech trees cover a wide range of ecological site conditions; from strongly acidic soils (e.g. Podzols) to neutral soils (e.g. Luvisols) (Härdtle et al. 2008). The extreme shade tolerance of beech is a key factor in its competitive advantage over other European broad-leaved tree species. As a result, many natural beech stands are monospecific. In the understorey, typical herb species include *Anemone nemorosa*, *Viola reichenbachiana*, and *Galium odoratum* (Härdtle et al. 2008). Beech forests are the prevailing forest type in north-western parts of central Europe, in southern Great Britain, and in Denmark (Bohn et al. 2002/2003; Rodwell 2003). The northern range limit of beech forests roughly coincides with 58°N (e.g. in northern Denmark and southern Sweden; Diekmann 1994). At acidic sites (such as soils originating from sediments from the Saalian glaciation), oak trees (*Quercus petraea*, *Q. robur*) may become co-dominant, but these structural patterns often are attributable to (former) silvicultural practices.

Mixed broad-leaved forests dominated by oak trees (*Quercus robur*, *Q. petraea*) and ash *Fraxinus excelsior* are the prevailing natural coastal range vegetation of Great Britain and so are developed under an oceanic climate (Bohn et al. 2002/2003; Rodwell 2003). Oak and ash trees form the canopy layer, and elm *Ulmus glabra* may be present in some stands (but has dramatically decreased in recent years due to elm disease). Sites are well supplied with nutrients and water, and nutrient-demanding species such as *Sanicula europaea*, *Brachypodium sylvaticum*, and *Primula acaulis* are frequent in the understorey vegetation.

1.7.2 Semi-natural Vegetation

1.7.2.1 Heathlands

Heathlands (including heather moorland) constitute one of the oldest cultural landscapes in north-western Europe (Gimingham 1972). In the 18th century, they were widespread throughout Great Britain, Belgium, the Netherlands, northwest Germany, Denmark, and Norway, and were the product of long-lasting extensive heathland farming systems (Sutherland 2004). In many regions, heath areas declined by more than 30 % during the 20th century, mainly due to afforestation activities and the conversion of heaths to arable land. Heaths are typical of strongly acidic soils (e.g. Podzols) and their structure is characterised by a dominance of dwarf shrubs such as *Calluna vulgaris*, *Empetrum nigrum*, and *Erica tetralix*. Heaths are today considered habitats of high conservation value, since they host a large proportion of the biodiversity typical of open landscapes in north-western Europe.

1.7.2.2 Grasslands

Grasslands (low-intensity pastures and meadows) are areas where the vegetation is dominated by grasses (Poaceae) and other herbaceous (non-woody) plants (forbs, sedges, rushes). The majority of grasslands in temperate climates are 'semi-natural', since their maintenance depends on human activities such as low-intensity farming (Ellenberg and Leuschner 2010). Thus, grasslands are subject to grazing and cutting regimes, which in turn keep the landscape open and prevent the encroachment of trees or shrubs. Unfertilised grasslands are characterised by a large diversity of plants and insects. Although intensification in land-use management (due to drainage and fertilisation) has improved the productivity of grasslands (in favour of cultivated varieties of grasses), this has in turn caused a significant decline in total grassland biodiversity. In central Europe, for example, the proportion of low-intensity pastures and meadows is below 6 % of the total agricultural landscape.

1.7.3 Agricultural Areas

Agricultural landscapes comprise areas that are primarily used for food production (i.e. arable land and grasslands with varying intensity of use). In north-western Europe, for example, about 50 % of the total area is currently in agricultural use (for which the proportion of arable land ranges from 30 to 70 %). Arable land is primarily associated with highly productive soils, for example in southeast England, northern France and Denmark (see Fig. 1.40), whereas grasslands are located in areas with high rainfall and/or at sites that are less suitable for cultivation. Much of the agricultural land in north-western Europe is artificially drained, either by ditches or tile drains, to ensure that the land can be farmed.

A major proportion of the agricultural area is used for livestock production, and many different production systems are applied. This covers purely grassland-based beef or dairy farming systems in the wetter western part of the region, intensive dairy farming systems based on grass and maize in large areas along the North Sea, and intensive pig and poultry farming systems in areas where arable agriculture dominates.

The countries in north-western Europe have economically well performing agricultural sectors that have a high value addition per farm (Giannakis and Bruggeman 2015). This is due to well-trained farmers supported by a highly competitive agricultural sector that supplies inputs and processes and markets the agricultural products. The productivity of the agricultural systems increased greatly over the 20th century and a high proportion of the produce was used for food production (Gingrich et al. 2015). Current trends in agricultural land use in north-western Europe are primarily driven by the globalisation of agricultural markets and a transition from a rural to an urban society (van Vliet et al. 2015).

Use of the agricultural area for producing bioenergy is growing, either by using existing crops for biofuel (e.g. rapeseed for biodiesel or maize for biogas) or by growing dedicated biofuel crops (e.g. *Miscanthus* or willow as biomass crops). There is also increased consideration for using crop residues for bioenergy purposes, although this competes with the need for organic matter input to soils to maintain fertility.

1.7.4 Artificial Surfaces

The area covered by artificial surfaces (e.g. for residential areas, industrial and commercial sites) amounts to about 4 % and is mainly related to urban centres (Fig. 1.40).

References

Anders I (2015) Regional climate modelling: The Eastern European "summer drying" problem and the representation of coastal surface wind speed in a multi model ensemble. Dissertationsschrift der Naturwissenschaften im Fachbereich Geowissenschaften der Universität Hamburg

Andersson A, Bakan S, Grassl H (2010) Satellite derived precipitation and freshwater flux variability and its dependency on the North Atlantic Oscillation. Tellus A 62:453–468

Andrade C, Leite SM, Santos JA (2012) Temperature extremes in Europe: overview of their driving atmospheric patterns. Nat Hazards Earth Syst Sci 12:1671–1691

Arrit RW (1989) Numerical modelling of the offshore extent of sea breezes. Q J Roy Meteor Soc 115:547–570

Atkinson BW (1981) Meso-scale Atmospheric Circulation. Academic Press

Attrill M, Wright J, Edwards M (2007) Climate related increases in jellyfish frequency suggest a more gelatinous future for the North Sea. Limnol Oceanogr 52:480–485

Baars MA, Fransz HG (1984) Grazing pressure of copepods on the phytoplankton stock of the central North Sea. Neth J Sea Res 18:120–142

Baas P, Bosveld FC, Burgers G (2015) The impact of atmospheric stability on the near-surface wind over sea in storm conditions. Wind Energy. doi:10.1002/we.1825

Backhaus J, Pohlmann T, Hainbucher D (1986) Regional aspects of the circulation on the North European shelf. ICES Hydrogr Comm C38

Baeteman C, Waller M, Kiden P (2011) Reconstructing middle to late Holocene sea-level change: A methodological review with particular reference to 'A new Holocene sea-level curve for the southern North Sea' presented by K.-E. Behre. Boreas 40:557–572

Balson P, Butcher A, Holmes R, Johnson H, Lewis M, Musson R, Henni P, Jones S, Leppage P, Rayner J, Tuggey G (2001) North Sea Geology. British Geological Survey, Technical Report TR_008, Strategic Environmental Assessment – SEA 2

Barne JH, Robson CF, Kaznowska SS, Doody JP, Davidson NC (1995a) Region 5. North-east England: Berwick-upon-Tweed to Filey Bay. Joint Nature Conservation Committee, UK

Barne JH, Robson CF, Kaznowska SS, Doody JP, Davidson NC (1995b) Region 6. Eastern England: Flamborough Head to Great Yarmouth. Joint Nature Conservation Committee, UK

Barne JH, Robson CF, Kaznowska SS, Doody JP, Davidson NC (1996a) Region 3. North-east Scotland: Cape Wrath to St Cyrus. Joint Nature Conservation Committee, UK

Barne JH, Robson CF, Kaznowska SS, Doody JP, Davidson NC (1996b) Region 9. Southern England: Hayling Island to Lyme Regis. Joint Nature Conservation Committee, UK

Barne JH, Robson CF, Kaznowska SS, Doody JP, Davidson NC, Buck AL (1996c) Region 10. South-west England: Seaton to the Roseland Peninsula. Joint Nature Conservation Committee, UK

Barne JH, Robson CF, Kaznowska SS, Doody JP, Davidson NC, Buck AL (1997a) Region 1. Shetland. Joint Nature Conservation Committee, UK

Barne JH, Robson CF, Kaznowska SS, Doody JP, Davidson NC, Buck AL (1997b) Region 2. Orkney. Joint Nature Conservation Committee, UK

Barne JH, Robson CF, Kaznowska SS, Doody JP, Davidson NC, Buck AL (1997c) Region 4. South-east Scotland: Montrose to Eyemouth. Joint Nature Conservation Committee, UK

Barne JH, Robson CF, Kaznowska SS, Doody JP, Davidson NC, Buck AL (1998a) Region 7. South-east England: Lowestoft to Dungeness. Joint Nature Conservation Committee, UK

Barne JH, Robson CF, Kaznowska SS, Doody JP, Davidson NC, Buck AL (1998b) Region 8. Sussex: Rye Bay to Chichester Harbour. Joint Nature Conservation Committee, UK

Barnston AG, Livezey RE (1987) Classification, seasonality and persistence of low-frequency atmospheric circulation patterns. Mon Wea Rev 115:1083–1126

Barrett EC, Beaumont MJ, Corlyon AM (1991) A Satellite-derived Rainfall Atlas of the North Sea 1978–87. Remote Sensing Unit Department of Geography, University of Bristol

Bärring L, von Storch H (2004) Scandinavian storminess since about 1800. Geophys Res Lett 31:L20202 doi:10.1029/2004GL020441

Beaugrand G, Luczak C, Edwards M (2009) Rapid biogeographical plankton shifts in the North Atlantic Ocean. Glob Change Biol 15:1790–1803

Becker GA (1981) Beiträge zur Hydrographie und Wärmebilanz der Nordsee. Dtsch Hydro Z 5:1–262

Bellamy DJ, Edwards P, Hirons MJD, Jones DJ, Evans PR (1973) Resources of the North Sea and some interactions. In: Goldberg D (ed) North Sea Science. MIT Press, pp 383–399

Bergström S (2001) Variability and change in precipitation and runoff to the Baltic Sea. A Workshop on climate variability and change in the Baltic Sea area. KNMI De Bilt, Netherlands, 12 November 2001

Berrisford P, Dee D, Fielding K, Fuentes M, Kållberg P, Kobayashi S, Uppala S (2009) The ERA-Interim archive. ECMWF, ERA Report Series 1

Beukema JJ (1989) Long-term changes in macrozoobenthic abundance on the tidal flats of the western part of the Dutch Wadden Sea. Helgol Mar Res 43:405–415

Billen G, Lancelot C, Meybeck M (1991) N, P and Si retention along the aquatic continuum from land to ocean. In: Mantoura RFC, Martin JM, Wollast J (eds). Ocean margin processes in global change. John Wiley & Sons, pp 19–44

Blackford JC and Gilbert FJ (2007) pH variability and CO_2 induced acidification in the North Sea. J Mar Syst 64:229–241

Blouet BW (2012) The EU and Neighbors: A Geography of Europe in the Modern World. 2nd edition, John Wiley & Sons

Bo Pedersen F (1994) The oceanographic and biological cycle succession in shallow sea fronts in the North Sea and the English Channel. Estuar Coast Shelf Sci 38:249–269

Bohn U, Neuhäusl R, Gollub G, Hettwer C, Neuhäuslova Z, Schlüter H, Weber H (2002/2003) Maps of the potential natural vegetation of

Europe 1:2,500,000. Part 1: Text; Part 2: Figure legends; Part 3: Maps. Landwirtschaftsverlag, Germany

Bot PVM, Colijn F (1996) A method for estimating primary production from chlorophyll concentrations with results showing trends in the Irish Sea and the Dutch coastal zone. ICES J Mar Sci 53:945–950

Bozec Y, Thomas H, Schiettecatte L-S, Borges AV, Elkalay K, de Baar HJW (2006) Assessment of the processes controlling the seasonal variations of dissolved inorganic carbon in the North Sea. Limnol Oceanogr 51:2746–2762

Brion N, Baeyens W, De Galan S, Elskens M, Laane RWPM (2004) The North Sea: source or sink for nitrogen and phosphorus to the Atlantic Ocean? Biogeochemistry 68:277–296

Brockmann UH, Wegner G (1996) Nutrient gradients in the North Sea during winter (1984-1993). In: Proceedings of Scientific Symposium on the North Sea Quality Status Report 1993, pp 178–181. Ministry of the Environment and Energy, Denmark

Brockmann UH, Laane RWP, Postma H (1990) Cycling of nutrient elements in the North Sea. Neth J Sea Res 26:239–264

BSH (2009) Naturverhältnisse Nordsee und Englischer Kanal. Teil B zu den Seehandbüchern für die Nordsee und den Kanal [Natural Conditions North Sea and English Channel. Part B of the Marine Manual for the North Sea and English Channel, Federal Maritime and Hydrographic Agency] Bundesamt für Seeschifffahrt und Hydrographie (BSH) Report 20062

Bülow K, Ganske A, Heinrich H, Hüttl-Kabus S, Klein B, Klein H, Möller J, Rosenhagen G, Schade N, Tinz B (2013) Comparing meteorological fields of the ENSEMBLES regional climate models with ERA-40-data over the North Sea. KLIWAS-Schriftenreihe, KLIWAS-21/2013

Bülow K, Ganske A, Heinrich H, Hüttl-Kabus S, Klein B, Klein H, Löwe P, Möller J, Schade N, Tinz B, Heinrich H, Rosenhagen G (2014) Ozeanische und atmosphärische Referenzdaten und Hindcast-Analysen für den Nordseeraum [Oceanographic and atmospheric reference data and hind-cast analyses for the North Sea area]. KLIWAS-Schriftenreihe, KLIWAS-30/2014

Bumke K, Fenning K, Strehz A, Mecking R, Schröder M (2012) HOAPS precipitation validation with ship-borne rain gauge measurements over the Baltic Sea. Tellus A 64:18486, http://dx.doi.org/10.3402/tellusa.v64i0.18486

Cabeçadas G, Nogueira M, Brogueira MJ (1999) Nutrient dynamics and productivity in three European estuaries. Mar Pollut Bull 38:1092–1096

Cadee GC (1992) Trends in Marsdiep phytoplankton. In: Danker N, Smit CJ, Scholl M (eds). Present and Future Conservation of the Wadden Sea. Neth Inst Sea Res Publ Ser 20:143–149

Capon RA (2003) Wind speed-up in the Dover Straits with the Met Office new dynamics model. Meteorol Appl 10:229–237

Charnock H, Dyer KR, Huthnance JM, Liss PS, Simpson JH, Tett PB (eds) (1994) Understanding the North Sea System. Chapman & Hall (reprinted 2012 by Springer)

Chassot E, Bonhommeau S, Dulvy NK, Melin F, Watson R, Gascuel D, Le Pape O (2010) Global marine primary production constrains fisheries catches. Ecol Lett 13:495–505

Chester R, Bradshaw GF, Ottley CJ, Harrison RM, Merrett JL, Preston MR, Rendell AR, Kane MM, Jickells TD (1994) The atmospheric distributions of trace metals, trace organics and nitrogen sources over the North Sea. In: Charnock H, Dyer KR, Huthnance JM, Liss PS, Simpson JH, Tett PB (eds). Understanding the North Sea system. Chapman and Hall, pp 153–163

Colebrook JM, Robinson GA (1965) Continuous plankton records: seasonal cycles of phytoplankton and copepods in the north-eastern Atlantic and the North Sea. B Mar Ecol 6:123–139

Compo GP, Whitaker JS, Sardeshmukh PD, Matsui N, Allan RJ, Yin X, Gleason Jr BE, Vose RS, Rutledge G, Bessemoulin P, Brönnimann S, Brunet M, Crouthamel RI, Grant AN, Groisman PY, Jones PD, Kruk MC, Kruger AC, Marshall GJ, Maugeri M, Mok HY, Nordli Ø, Ross TF, Trigo RM, Wang XL, Woodruff SD, Worley SJ (2011) The Twentieth Century Reanalysis Project. Q J Roy Meteor Soc 137:1–28

Cury P, Bakun A, Crawford RJ, Jarre A, Quinones RA, Shannon LJ, Verheye HM (2000) Small pelagics in upwelling systems: patterns of interaction and structural changes in "wasp-waist" ecosystems. ICES J Mar Sci 57:603–618

Cushing DH (1959) On the nature of production in the sea. Fishery Invest Lond Ser 2, 22:1–40

Cushing DH (1984) The gadoid outburst in the North Sea. J Cons Int Explor Mer 41:159–166

Daan R, Mulder M (1996) On the short-term and long-term impact of drilling activities in the Dutch sector of the North Sea. ICES J Mar Sci 53:1036–1044

Daan N, Bromley PJ, Hislop JRG, Nielsen NA (1990) Ecology of North Sea fish. Neth J Sea Res 26:343–368

Dagg MJ, Vidal J, Whiteledge TE, Iverson RL, Goering JJ (1982) The feeding, respiration, and excretion of zooplankton in the Bering Sea during a spring bloom. Deep Sea Res 29:45–63

Dähne M, Gilles A, Lucke K, Peschko V, Adler S, Krügel K, Sundermeyer J, Siebert U (2013) Effects of pile-driving on harbour porpoises (*Phocoena phocoena*) at the first offshore wind farm in Germany. Environ Res Lett 8: doi:10.1088/1748-9326/8/2/025002

Damm P (1997) Die saisonale Salzgehalts- und Frischwasserverteilung in der Nordsee und ihre Bilanzierung. Ber Zentr Meeres- u Klimaforsch, Univ Hamburg, Reihe B, Nr 28

De Wilde PAWJ, Berghuis EM, Kok A (1984) Structure and energy demand of the benthic community of the Oyster Ground, central North Sea. Neth J Sea Res 18:143–159

De Wilde PAWJ, Jenness MI, Duineveld GCA (1992) Introduction into the ecosystem of the North Sea: hydrography, biota, and food web relationships. Neth J Aquat Ecol 26:7–18

Diekmann M (1994) Deciduous forest vegetation in Boreo-nemoral Scandinavia. Acta Phytogeographica Suecica 80:1–116

Dierssen K (1996) Vegetation Nordeuropas [Vegetation of Northern Europe]. Ulmer

Donat MG, Renggli D, Wild S, Alexander LV, Leckebusch GC, Ulbrich U (2011) Reanalysis suggests long-term upward trends in European storminess since 1871. Geophys Res Lett 38: L14703 doi:10.1029/2011GL047995

Doody JP, Johnston C, Smith B (eds) (1993) Directory of the North Sea Coastal Margin. Joint Nature Conservation Committee, UK

Drinkwaard AC (1998) Introductions and developments of oysters in the North Sea area: a review. Helgoland Mar Res 52:301–308

Druon JN, Schrimpf W, Dobricic S, Stips A (2004) Comparative assessment of largescale marine eutrophication: North Sea area and Adriatic Sea as case studies. Mar Ecol-Prog Ser 272:1–23

Dutton A, Lambeck K (2012) Ice volume and sea level during the last interglacial. Science 337:216–219

Edwards M, Reid PC, Planque B (2001) Long-term and regional variability of phytoplankton biomass in the Northeast Atlantic (1960–1995). ICES J Mar Sci 58:39–49

EEA (2015) Marine Protected Areas in Europe's Seas. European Environment Agency, Report No 3/2015

Ehlers J (ed) (1983) Glacial Deposits in North-West Europe. A.A. Balkema

Ehlers J, Gibbard P (2008) Extent and chronology of Quaternary glaciation. Episodes 31:211–218

Ehlers J, Kunz H (1993) Morphology of the Wadden Sea: natural processes and human interference. In: Hillen R, Verhagen HJ (eds) Coastlines of the Southern North Sea. Coastlines of the World, American Society of Civil Engineers, pp 65–84

Ellenberg H (2009) Vegetation Ecology of Central Europe. 4th Edn. Cambridge University Press

Ellenberg H, Leuschner C (2010) Vegetation Mitteleuropas mit den Alpen [Vegetation of Central Europe and the Alps]. 6th Edn. Ulmer

Emeis K-C, van Beusekom J, Callies U, Ebinghaus R., Kannen A, Kraus G, Kröncke I, Lenhart H, Lorkowski I, Matthias V, Möllmann C, Pätsch J, Scharfe M, Thomas H, Weisse R, Zorita E (2015) The North Sea – A Shelf Sea In The Anthropocene. J Mar Syst 141:18–33

Eurostat (2011) Eurostat Regional Yearbook 2011. Eurostat Statistical Books, European Commission

Faasse M, Bayha KM (2006) The ctenophore *Mnemiopsis leidyi* A. Agassiz 1865 in coastal waters of the Netherlands: an unrecognized invasion? Aquat Inv 1:270–277

FAO (2003) Trends in Oceanic Captures and Clustering of Large Marine Ecosystems: Two studies based on the FAO capture database. FAO Fish Tech Pap 435

FAO (2012) The State of the World Fisheries and Aquaculture: 2012. United Nations Food and Agriculture Organization

Finkele K (1998) Inland and offshore propagation speeds of a sea breeze from simulations and measurements. Boundary-Layer Meteorol 87:307–329

Fock HO, Kloppmann M, Probst WN (2014) An early footprint of fisheries: changes for a demersal fish assemblage in the German Bight from 1902-1932 to 1991-2009. J Sea Res 85:325–335

Franke R (2009) Die Nordatlantischen Orkantiefs seit 1956 [The North Atlantic storms since 1956]. Naturwissenschaftliche Rundschau 7:349–356

Fransz HG, Colebrook JM, Gamble JC, Krause M (1991) The zooplankton of the North Sea. Neth J Sea Res 28:1–52

Ganske A, Rosenhagen G, Heinrich H (2012) Validation of Geostrophic Wind Fields in ERA-40 and ERA Interim Reanalyses for the North Sea Area, Poster for 4th WCRP, International Conference on Reanalyses, Silver Spring Maryland, USA

Gatcliff RW, Richards PC, Smith K, Graham CC, McCormac M, Smith NJP, Long D, Cameron TDJ, Evans D, Stevenson AG, Bulat J, Ritchie JD (1994) United Kingdom Offshore Regional Report: The Geology of the Central North Sea. British Geological Survey

Giannakis G, Bruggeman A (2015) The highly variable economic performance of European agriculture. Land Use Policy 45:26–35

Gibbard P, Cohen KM (2008) Global chronostratigraphical correlation table for the last 2.7 million years. Episodes 31:243–247

Gill A (1982) Atmosphere-Ocean Dynamics. Academic Press

Gimingham CH (1972) Ecology of heathlands, Chapman and Hall

Gingrich S, Niedertscheide M, Kastner T, Haberl H, Cosor G, Krausmann F, Kuemmerle T, Müller D, Reith-Musel A, Jepsen MR, Vadineanu A, Erb K-H (2015) Exploring long-term trends in land use change and aboveground human appropriation of net primary production in nine European countries. Land Use Policy 47:426–438

Glennie KW (ed) (1984) Introduction to the Petroleum Geology of the North Sea. Blackwell Scientific

Gómez F (2008) Phytoplankton invasions: comments on the validity of categorizing the nonindigenous dinoflagellates and diatoms in European seas. Mar Pollut Bull 56:620–628

Good E (2010) Estimating daily sunshine duration over the UK from geostationary satellite data. Weather 65:324–328

Greenwood N, Parker ER, Fernand L, Sivyer DB, Weston K, Painting SJ, Kröger S, Forster RM, Lees HE, Mills DK, Laane RWPM (2010) Detection of low bottom water oxygen concentrations in the North Sea: implications for monitoring and assessment of ecosystem health. Biogeosciences 7:1357–1373

Haq BU, Hardenbol J, Vail PR (1987) Chronology of fluctuating sea levels since the Triassic. Science 235:1156–1167

Härdtle W, Ewald J, Hölzel N (2008) Wälder des Tieflandes und der Mittelgebirge. Ökosysteme Mitteleuropas in geobotanischer Sicht [Forest of the Central European lowlands and low mountain range. Ecosystems of Central Europe]. 2nd Edn. Ulmer

Hasselmann K, Barnett T, Bouws E, Carlson H, Cartwright D, Enke K, Ewing J, Gienapp H, Hasselmann D, Krusemann P, Meerburg A, Müller P, Olbers D, Richter K, Sell W, Walden H (1973) Measurements of wind-wave growth and swell decay during the Joint North Sea Wave Project (JONSWAP). Deutsches Hydrographisches Institut

Heip C, Craeymeersch JA (1995) Benthic community structures in the North Sea. Helgoland Mar Res 49:313–328

Heip C, Basford D, Craeymeersch JA, Dewarumez JM, Dörjes J, DeWilde P, Duineveld G, Eleftheriou A, Herman PMJ, Niermann U, Kingston P, Künitzer A, Rachor E, Rumoh, H, Soetaert K, Soltwedel T (1992) Trends in biomass, density and diversity of North Sea macrofauna. ICES J Mar Sci 49:13–22

Helaouët P, Beaugrand G (2007) Macroecology of *Calanus finmarchicus* and *C. helgolandicus* in the North Atlantic Ocean and adjacent seas. Mar Ecol-Prog Ser 345:147–165

Hense A, Glowienka-Hense R (2008) Auswirkungen der Nordatlantischen Zirkulation [Effects of the North Atlantic Circulation]. Promet 34:89–94

Hjøllo SS, Skoge MD, Svendsen E (2009) Exploring currents and heat within the North Sea using a numerical model. J Mar Syst 78:180–192

Howarth MJ, Dyer KR, Joint IR, Hydes DJ, Purdie DA, Edmunds H, Jones JE, Lowry RK, Moffat TJ, Pomeroy AJ, Proctor R (1994) Seasonal cycles and their spatial variability. In: Charnock H, Dyer KR, Huthnance JM, Liss PS, Simpson JH, Tett PB (eds). Understanding the North Sea system. Chapman and Hall pp 5–25

Hurrell JW (1995) Decadal trends in the North Atlantic Oscillation: Regional temperatures and precipitation. Science 269:676–679

Hurrell JW, Van Loon H (1997) Decadal variations in climate associated with the North Atlantic Oscillation. Climate Change 36:301–326

ICES (2008) Report of the ICES Advisory Committee 2008. Book 6: North Sea. International Council for the Exploration of the Sea (ICES)

ICES (2011) Status Report on Climate Change in the North Atlantic. ICES Coop Res Rep No. 310

Janssen F, Schrum C, Backhaus J (1999) A climatological data set for temperature and salinity in the North Sea and the Baltic Sea. Dt Hydrogr Z 51(S9):5–245

Jenkinson AF, Collison FP (1977) An initial Climatology of Gales over the North Sea. Synoptic Climatology Branch Memorandum, No. 62. UK Met Office

Johnson H, Richards PC, Long D, Graham CC (1993) United Kingdom Offshore Regional Report: The Geology of the Northern North Sea. British Geological Survey

Joint I, Pomeroy A (1993) Phytoplankton biomass and production in the North Sea. Mar Ecol-Prog Ser 99:169–182

Joris C, Billen G, Lancelot C, Daro MH, Mommaerts JP, Bertels A, Bossicart M, NIjs J, Hecq JH (1982) A budget of carbon cycling in the Belgian coastal zone: relative roles of zooplankton, bacterioplankton and benthos in the utilization of primary production. Neth J Sea Res 16:260–275

Josefson AB, Jensen JN, Ærtebjerg G (1993) The benthos community structure anomaly in the late 1970s and early 1980s – a result of a major food pulse? J Exp Mar Biol Ecol 172:31–45

Jungclaus J, Wagner V (1988) Meeresspiegelanstieg und seine Auswirkungen auf das Schwingungssystem der Nordsee. Diplomarbeit, Univ Hamburg

Kabat P, Bazelmans J, van Dijk J, Herman PMJ, van Oijen T, Pejrup M, Reise K, Speelman H, Wolff WJ (2012) The Wadden Sea Region: Towards a science for sustainable development. Ocean Coast Manage 68:4–17

Kalnay E, Kanamitsu M, Kistler R, Collins W, Deaven D, Gandin L, Iredell M, Saha S, White G, Woollen J, Zhu Y, Leetmaa A, Reynolds R, Chelliah M, Ebisuzaki W, Higgins W, Janowiak J, Mo KC, Ropelewski C, Wang J, Jenne R, Joseph D (1996) The NCEP/NCAR 40-Year Reanalysis Project. Bull Amer Meteor Soc 77:437–471

Karagali I, Peña A, Badger M, Hasager CB (2014) Wind characteristics in the North and Baltic Seas from QuikSACT satellite. Wind Energ 17:123–140

Kersten M (1988) Suspended matter and sediment transport. In: Pollution of the North Sea: An Assessment. Salomons W, Bayne BL, Duursma EK, Förstner U (eds), Springer pp 36–58

Köhl A, Stammer D (2008) Variability of meridional overturning in the North Atlantic from the 50-year GECCO state estimation. J Phys Oceanog 38:1913–1930

Konradi PB (2000) Biostratigraphy and environment of the Holocene marine transgression in the Heligoland Channel, North Sea. B Geol Soc Denmark 47:71–79

Korevaar CG (1990) North Sea Climate based on observations from ships and lightvessels. Kluwer Academic Publishers

Kothe S, Good E, Obregón A, Ahrens B, Nitsche H (2013) Satellite-based sunshine duration for Europe. Remote Sens 5:2943–2972

Krauss W (1973) Methods and Results of Theoretical Oceanography, Vol. 1: Dynamics of the Homogeneous and the Quasihomogeneous Ocean. Borntraeger

Kröncke I, Knust R (1995) The Dogger Bank: a special ecological region in the central North Sea. Helg Meeres 49:335–353

Krueger O, Schenk F, Feser F, Weisse R (2013) Inconsistencies between long-term trends in storminess derived from the 20CR reanalysis and observations. J Climate 26:868–874

Kühn W, Pätsch J, Thomas H, Borges AV, Schiettecatte L-S, Bozec Y, Prowe F (2010). Nitrogen and carbon cycling in the North Sea and exchange with the North Atlantic - a model study Part II. Carbon budget and fluxes. Cont Shelf Res 30:1701–1716

Laane RWPM, Groeneveld G, De Vries A, Van Bennekom J, Sydow S (1993). Nutrients (P, N, Si) in the Channel and the Dover Strait: seasonal and year-to-year variation and fluxes to the North Sea. Oceanologica Acta 16:607–616

Laane RWPM, Vethaak AD, Gandrass J, Vorkamp K, Kohler A, Larsen MM, Strand J (2013) Chemical contaminants in the Wadden Sea: Sources, transport, fate and effects. J Sea Res 82:10–53

Lahousse B, Clabaut P, Chamley H, van der Valk L (1993). Morphology of the southern North Sea coast from Cape Blanc-Nez (F) to Den Helder (NL). In: Hillen R, Verhagen HJ (eds) Coastlines of the Southern North Sea. Coastlines of the World, American Society of Civil Engineers, pp 85–95

Lefebvre C, Rosenhagen G (2008) The Climate of the North Sea and Baltic Sea Region. Die Küste, 74:45–59

Legendre L, Demers S, Lefaivre D (1986). Biological production at marine ergoclines. In: Nihoul JCJ (ed) Marine Interfaces Ecohydrodynamics. Elsevier pp 191–207

Leggett DJ, Lowe JP, Cooper NJ (1999) Beach evolution on the southern North Sea coast. In: Edge BL (ed.), Coastal Engineering 1998, Part IV: Coastal Processes and Sediment Transport, pp 2759–2772

Leppäranta M, Myrberg K (2009) Physical Oceanography of the Baltic Sea. Springer Praxis Books

Loebl M, Colijn F, Beusekom JEE, Baretta-Bekker JG, Lancelot C, Philippart CJM, Rousseau V (2009) Recent patterns in potential phytoplankton limitation along the NW European continental coast. J Sea Res 61:34–43

Loewe P (2009) System Nordsee, Zustand 2005 im Kontext langzeitlicher Entwicklungen [North Sea System, State 2005 in the context of long-term developments]. Berichte des BSH 44

Lorenz C, Kunstmann H (2012) The hydrological cycle in three state-of-the-art reanalyses: Intercomparison and performance analysis. J Hydrometeor 13:1397–1420

Lüning K (1979) Growth strategies of three *Laminaria* species (Phaeophyceae) inhabiting different depth zones in the sublittoral region of Helgoland (North Sea). Mar Ecol-Prog Ser 1:195–207

Luterbacher J, Liniger MA, Menzel A, Estrella N, Della-Marta PM, Pfister C, Rutishauser T, Xoplaki E (2007) Exceptional European warmth of autumn 2006 and winter 2007: Historical context, the understanding dynamics, and its phenological impacts. Geophys Res Lett 34, L12704, doi:10.1029/2007GL029951

Luyten PJ, Jones JE, Proctor R (2003) A numerical study of the long- and short-term temperature variability and thermal circulation in the North Sea. J Phys Oceanog 33:37–56

Lyngsie SB, Thybo H, Rasmussen TM (2006) Regional geological and tectonic structures of the North Sea area from potential field modelling. Tectonophysics 413:147–160

Mackinson S, Daskalov G (2008) An ecosystem model of the North Sea to support an ecosystem approach to fisheries management: description and parameterisation. Cefas, Sci Ser Tech Rep No 142

Margalef R (1997) Our biosphere. In: Kinne O (ed) Excellence in Ecology Book 10. Ecology Institute, Oldendorf/Luhe, Germany

McGlade JM (2002) The North Sea large marine ecosystem. In: Sherman K, Skjoldal HR (eds). Large Marine Ecosystems of the North Atlantic – Changing States and Sustainability. Elsevier Science BV pp 339–412.

Menning M, Hendrich A (2012) Stratigraphische Tabellen von Deutschland Kompakt 2012 (STDK 2012). Potsdam (GFZ)

Metzger MJ, Bunce RGH, Jongman RHG, Mücher CA, Watkins JW (2005) A climatic stratification of the environment of Europe. Global Ecol Biogeogr 14:549–563

Meyer EMI, Pohlmann T, Weisse R (2011) Thermodynamic variability and change in the North Sea (1948-2007) derived from a multidecadal hindcast. J Mar Syst 86:35–44

Michaelsen K, Krell U, Reinhardt V, Graßl H, Kaufeld L (2000) Climate of the North Sea. Einzelveröffentlichung No. 118. 2. korr. Auflage, Deutscher Wetterdienst Offenbach

Miller STK, Keim D, Talbot RW, Mao H (2003) Sea breeze: Structure, forecasting, and impacts. Rev Geophys 41:1011–1042

Nicolajsen H, Mohlenberg F, Kiorboe T (1983). Algal grazing rate by the planktonic copepod *Centropages hamatus* and *Pseudocalanus* sp.: diurnal and seasonal variation during the spring phytoplankton bloom in the Oresund. Ophelia 22(1):15–31

Nilsson T (1983) Pleistocene: Geology and life in the Quaternary Ice Age. EnkeVerlag

Omstedt A, Elken J, Lehmann A, Piechura J (2004) Knowledge of the Baltic Sea physics gained during the BALTEX and related programmes. Progr Oceanogr 63:1–28

Osborn TJ (2011) Winter 2009/2010 temperatures and a record-breaking North Atlantic Oscillation index. Weather 66:19–21

OSPAR (2000) Quality Status Report 2000. Region II: Greater North Sea. OSPAR Commission, London

OSPAR (2010) Quality Status Report 2010. OSPAR Commission, London

Otto L, Zimmerman JTF, Furnes GK, Mork M, Saetre R, Becker G (1990) Review of the physical oceanography of the North Sea. Neth J Sea Res 26:161–238

Pätsch J, Kühn W (2008) Nitrogen and carbon cycling in the North Sea and exchange with the North Atlantic – a model study. Part I. Nitrogen budget and fluxes. Cont Shelf Res 28:767–787

Peperzak L (2003) Climate change and harmful algal blooms in the North Sea. Acta Oecologica 24:139–144

Peters SWM, Eleveld M, Pasterkamp R, van der Woerd H, Devolder M, Jans S, Park Y, Ruddick K, Block T, Brockmann C,

Doerffer R, Krasemann H, Röttgers R, Schönfeld W, Jorgensen PV, Tilstone G, Martinez-Vicente V, Moore G, Sorensen K, Hokedal J, Johnson TM, Lomsland ER, Aas E (2005) Atlas of Chlorophyll-*a* concentration for the North Sea based on MERIS imagery of 2003. 3rd Edn. Vrije Universiteit, Netherlands

Planque B, Fromentin JM (1996) Calanus and environment in the eastern North Atlantic. 1. Spatial and temporal patterns of *C. finmarchicus* and *C. helgolandicus*. Mar Ecol-Prog Ser 134:101–109

Pohlmann T (1996) Calculating the development of the thermal vertical stratification in the North Sea with a three-dimensional baroclinic circulation model. Cont Shelf Res 16:163–194

Pohlmann T (2003) Eine Bewertung der hydro-thermodynamischen Nordseemodellierung. Ber Zentr Meeres- u Klimaforsch, Univ Hamburg, Reihe B, Nr 46

Pope JG, Macer CT (1996) An evaluation of the stock structure of North Sea cod, haddock, and whiting since 1920, together with a consideration of the impacts of fisheries and predation effects on their biomass and recruitment. ICES J Mar Sci 53:1157–1169

Port of Rotterdam (2015) Port Statistics. www.portofrotterdam.com/sites/default/files/upload/Port-Statistics/Port-Statistics/index.html#1 Accessed 15 October 2015

Posselt R, Müller RW, Stöckli R, Trentmann J (2012) Remote sensing of solar surface radiation for climate monitoring – the CM-SAF retrieval in international comparison. Remote Sens Environ 118:186–198

Prowe F, Thomas H, Pätsch J, Kühn W, Bozec Y, Schiettecatte, L-S, Borges AV, de Baar HJW (2009) Mechanisms controlling the air–sea CO_2 flux in the North Sea. Cont Shelf Res 29:1801–1808

Purcell JE (2005) Climate effects on formation of jellyfish and ctenophore blooms: a review. J Mar Biol Ass UK 85:461–476

Radach G, Pätsch J (2007) Variability of continental riverine freshwater and nutrient inputs into the North Sea for the years 1977–2000 and its consequences for the assessment of eutrophication. Estuaries Coasts 30:66–81

Rasmussen ES, Heilmann-Clausen C, Waagstein R, Eidvin T (2008) The Tertiary of Norden. Episodes 31:66–72

Rehm P, Rachor E (2007) Benthic macrofauna communities of the submersed Pleistocene Elbe valley in the southern North Sea. Helgoland Mar Res 61:127–134

Reid PC (1978) Continuous plankton records: large-scale changes in the abundance of phytoplankton in the North Sea from 1958–1973. Rapp P V Reun Cons Int Explor Mer 172:384–389

Reid PC, Lancelot C, Gieskes WWC, Hagmeier E, Weichart G (1990). Phytoplankton of the North Sea and its dynamics: a review. Neth J Sea Res 26:295–331

Reid PC, Fischer AC, Lewis-Brown E, Meredith MP, Sparrow M, Andersson AJ, Antia A, et al. (2009) Impacts of the oceans on climate change. Adv Mar Biol 56:1–150

Reijnders PJH (1982) Threads to the harbor seal population in the Wadden Sea. In: Reijnders PJH, Wolff WJ (Eds) Marine Mammals of the Wadden Sea. Balkema pp 38–47

Reise K, Herre E, Sturm M (1994) Biomass and abundance of macrofauna in intertidal sediments of Königshafen in the northern Wadden Sea. Helgoland Mar Res 48:201–215

Reise K, Baptist M, Burbridge P, Dankers N, Fischer L, Flemming B, Oost AP, Smit C (2010) The Wadden Sea – A Universally Outstanding Tidal Wetland. Wadden Sea Ecosystem No. 29. Common Wadden Sea Secretariat, Wilhelmshaven, Germany, pp 7–24

Richardson K, Bo Pedersen F (1998) Estimation of new production in the North Sea: consequences for temporal and spatial variability of phytoplankton. ICES J Mar Sci 55:574–580

Richardson K, Visser AW, Bo Pedersen F (2000) Subsurface phytoplankton blooms fuel pelagic production in the North Sea. J Plankt Res 22:1663–1671

Richardson AJ, Bakun A, Hays GC, Gibbons MJ (2009) The jellyfish joyride: causes, consequences and management responses to a more gelatinous future. Trends Ecol Evol 24:312–322

Riegman R, Malschaert H, Colijn F (1990) Primary production of phytoplankton at a frontal zone located at the northern slope of the Dogger Bank (North Sea). Mar Biol 105:329–336

Rodhe J (1998) The Baltic and North Seas: A process-oriented review of the physical oceanography. In: Robinson A, Brink K (eds). The Seas. Harvard University Press, Vol. 11, pp 699–732

Rodhe J, Tett P, Wulff F (2006) The Baltic and North seas: a regional review of some important physical-chemical-biological interaction processes. In: Robinson A, Brink K (eds). Harvard University Press, Vol. 14B, pp 1029–1071

Rodwell JS (2003) British Plant Communities. Volume 1: woodlands and scrub. Cambridge University Press

Roff JC, Middlebrook K, Evans F (1988) Long-term variability in North Sea zooplankton off the Northumberland coast: productivity of small copepods and analysis of trophic interactions. J Mar Biol Ass UK 68:143–164

Rudolf B, Schneider U (2005) Calculation of gridded precipitation data for the global land-surface using in-situ gauge observations. Proc. Second Workshop of the Int. Precipitation Working Group, Monterey, CA, pp 231–247

Sager G (1959) Gezeiten und Schiffahrt. Fachbuchverlag, Leipzig

Schiettecatte LS, Gazeau F, Van der Zee C, Brion N, Borges AV (2006) Time series of the partial pressure of carbon dioxide (2001–2004) and preliminary inorganic carbon budget in the Scheldt plume (Belgian coast waters). Geochem Geophy Geosy 7:Q06009, doi:10.1029/2005GC001161

Schiettecatte LS, Thomas H, Bozec Y, Borges AV (2007) High temporal coverage of carbon dioxide measurements in the Southern Bight of the North Sea. Mar Chem 106:161–173

Schlünzen KH, Krell U (2004) Atmospheric parameters for the North Sea: a review. Mar Bio 34:1–52

Schlüter MH, Merico A, Wiltshire KH, Greve W, von Storch H (2008) A statistical analysis of climate variability and ecosystem response in the German Bight. Ocean Dynam 58:169–186

Schlüter MH, Merico A, Reginatto M, Boersma M, Wiltshire KH (2010) Phenological shifts of three interacting zooplankton groups in relation to climate change. Glob Change Biol 16:3144–3153

Schlüter MH, Kraberg A, Wiltshire KH (2012) Long-term changes in the seasonality of selected diatoms related to grazers and environmental conditions. J Sea R 67:91–97

Schmidt H, von Storch H (1993) German Bight storms analysed. Nature 365:791–791

Schrum C, Siegismund F (2001) Modellkonfiguration des Nordsee/Ostseemodells. 40 Jahre NCEP-Integration. Ber Zentr Meeres- u Klimaforsch, Univ Hamburg, Reihe B, Nr 4

Schwarz C (1991) Quartärprofile im südlichen Nordseebecken: Sedimentologische Analyse und lithostratigraphische Konnektierung. Diss. Fachber. Erdwissenschaften Universität Hannover

Sejrup HP, Larsen E, Landvik J, King EL, Haflidason H, Nesje A (2000) Quaternary glaciations in southern Fennoscandia: Evidence from southwestern Norway and the northern North Sea region. Quaternary Sci Rev 19:667–685

Shennan I, Bradley S, Milne G, Brooks A, Bassett S, Hamilton S (2006) Relative sea-level changes, glacial isostatic modelling and ice-sheet reconstructions from the British Isles since the Last Glacial Maximum. J Quaternary Sci 21:585–599

Siegismund F, Schrum C (2001) Decadal changes in the wind forcing over the North Sea. Clim Res 18:39–45

Simpson JE (1994) Sea Breeze and Local Winds. Cambridge University Press

Simpson JE, Mansfield DA, Milford JR (1977) Inland penetration of sea-breeze fronts. Quart J R Met Soc 103:47–76

Skamarock WC, Klemp JB (2008) A time-split non-hydrostatic atmospheric model for weather research and forecasting applications. J Comput Phys 227:3465–3485

Stammer D, Bersch M, Sadikni R, Jahnke-Bornemann A, Gouretski V, Hinrichs I, Heinrich H, Klein B, Klein H, Schade N, Rosenhagen G, Tinz B, Lefebvre C (2014) Die KLIWAS Nordseeklimatologie für ozeanographische und meteorologische In-Situ Daten [The North Sea climatology of the KLIWAS project for oceanographic and meterological in-situ data]. KLIWAS-Schriftenreihe 59, 45 S.

Stanev E, Dobrynin M, Pleskachevsky A, Grayek S, Günther H (2009) Bed shear stress in the Southern North Sea as an important driver for suspended sediment dynamics. Ocean Dyn 59:183–194

STECF (2014) The 2014 Annual Economic Report on the EU Fishing Fleet. Scientific, Technical and Economic Committee for Fisheries (STECF-14-16)

Steele JH (1974) The Structure of Marine Ecosystems. Harvard University Press

Steele CJ, Dorling SR, von Glasow R, Bacon J (2015) Modelling sea-breeze climatologies and interactions on coasts in the southern North Sea: implications for offshore wind energy. Q J Roy Meteor Soc 141:1821-1835

Sterr H (2003) Geographische Charakterisierung der Nordseeregion. In: Lozán JL, Rachor E, Reise K, Sündermann JV Westernhagen H (eds). Warnsignale aus Nordsee & Wattenmeer - Eine aktuelle Umweltbilanz, GEO, Hamburg, pp 40–46

Sündermann J (1994) Extreme Sturmfluten an der deutschen Nordseeküste. Geogr Rundschau 46:10,598–599

Sündermann J, Pohlmann T (2011) A brief analysis of North Sea physics. Oceanologia 53:663–689

Sündermann J, Beddig S, Könecke I, Radach G, Schlünzen H (eds) (2001a) The changing North Sea: Knowledge, speculation and new challenges – Synthesis and new conception of North Sea Research (SYCON)

Sündermann J, Beddig S, Kröncke I, Radach G, Schlünzen H (eds) (2001b) Synthesis and new conception of North Sea research. Ber Zentr Meeres- u Klimaforsch, Univ Hamburg, Reihe Z, Nr 3

Sutherland N (2004) Conservation and management of north-west European upland moorlands. Heathguard, The Heathland Centre, Lygra, Norway

Sverdrup HU (1953) On conditions for the vernal blooming of phytoplankton. J Cons Int Explor Mer 18:287–295

Theobald N (2011) Emerging persistent organic pollutants in the marine environment. In: Quante M, Ebinghaus R, Flöser G (eds.). Persistent Pollution – Past, Present, Future. Springer Verlag

Thomas H, Bozec Y, Elkalay K, de Baar HJW (2004) Enhanced open ocean storage of CO_2 from shelf sea pumping. Science 304:1005–1008

Thomas H, Bozec Y, Elkalay K, de Baar HJW, Borges AV, Schiettecatte L-S (2005a) Controls of the surface water partial pressure of CO_2 in the North Sea. Biogeosciences 2:323–334

Thomas H, Bozec Y, de Baar HJW, Elkalay K, Frankignoulle M, Schiettecatte L-S, Kattner G, Borges AV (2005b) The carbon budget of the North Sea. Biogeosciences 2:87–96

Tijm ABC, Van Delden AJ, Holtslag AAM (1999) The inland penetration of sea breezes. Contr Atmos Phys 317–328

Trigo RM, Osborn TJ, Corte-Real J (2002) The North Atlantic Oscillation influence on Europe: climate impacts and associated physical mechanisms. Clim Res 20:9–17

Turrell WR, Slesser G, Payne R, Adams RD, Gillibrand PA (1996) Hydrography of the East Shetland Basin in relation to decadal North Sea variability. ICES J Mar Sci 53:899–916

Uppala SM, Kållberg PW, Simmons AJ, Andrae U, Da Costa Bechthold V, Fiorino M, Gibson JK, Haseler J, Hernandez A, Kelly GA, Li X, Onogi K, Saarinen S, Sokka N, Allan RP, Andersson E, Arpe K, Balmaseda MA, Beljaars ACM, Van De Berg L, Bidlot J, Bormann N, Caires S, Chevallier F, Dethof A, Dragosavac M, Fisher M, Fuentes M, Hagemann S, Hólm E, Hoskins BJ, Isaksen L, Janssen PAEM, Jenne R, Mcnally AP, Mahfouf J-F, Morcrette J-J, Rayner NA, Saunders RW, Simon P, Sterl A, Trenberth KE, Untch A, Vasiljevic D, Viterbo P, Woollen J (2005) The ERA-40 re-analysis. Q J Roy Meteor Soc 131:2961–3012

van Beusekom JEE, Bot P, Carstensen J, Goebel J, Lenhart H, Pätsch J, Petenati T, Raabe T, Reise K, Wetsteijn B (2009). Eutrophication. In: Marencic H, de Vlas J (eds), Quality Status Report 2009. Wadden Sea Ecosystem. No 25. Common Wadden Sea Secretariat

Van der Linden P, Mitchell JFB (eds) (2009) ENSEMBLES: Climate Change and its Impacts: Summary of research and results from the ENSEMBLES project. Met Office Hadley Centre, UK

Van Dijk HWJ (1994) Integrated management and conservation of Dutch coastal areas: Premises, practice and solutions. Mar Pollut Bull 29:609–616

Van Es FB, Meyer-Reil L-A (1982) Biomass and metabolic activity of heterotrophic marine bacteria. Adv Microb Ecol 6:111–170

van Vliet J, de Groot HLF, Rietveld P, Verburg PH (2015) Manifestations and underlying drivers of agricultural land use change in Europe. Landscape Urban Plan 133:24–36

Vink A, Steffen H, Reinhard L, Kaufmann G (2007) Holocene relative sea-level change, isostatic subsidence and the radial viscosity structure of the mantle of northwest Europe (Belgium, the Netherlands, Germany, southern North Sea). Quaternary Sci Rev 26:3249–3275

Wakelin SL, Holt JT, Blackford JC, Allen JI, Butenschön M, Artioli Y (2012) Modeling the carbon fluxes of the northwest European continental shelf: validation and budgets. J Geophys Res 117: C05020 doi:10.1029/2011JC007402

Wasa Group (1998) Changing waves and storms in the Northeast Atlantic? Bull Am Met Soc 79:741–760

Weston K, Fernand L, Mills DK, Delahunty R, Brown J (2005) Primary production in the deep chlorophyll maximum of the northern North Sea. J Plankton Res 27:909–922

Wiltshire KH, Kraberg A, Bartsch I, Boersma M, Franke HD, Freund J, Gebühr C, Gerdts G, Stockmann K, Wichels A (2010) Helgoland Roads, North Sea: 45 years of change. Estuaries Coasts 33:295–310

Winther NG, Johannessen JA (2006) North Sea circulation: Atlantic inflow and its destination. J Geophys Res 111:C12018, doi:10.1029/2005JC003310

WMO (2007) WMO Statement on the status of the global climate in 2006. WMO-No. 1016, World Meteorological Organization

WMO (2008) WMO Statement on the status of the global climate in 2007. WMO-No. 1031, World Meteorological Organization

WMO (2010) Assessment of the observed extreme conditions during the 2009/2010 boreal winter. WMO/TD-No 1550. World Meteorological Organization

Wolff WJ, Bakker JP, Laursen K, Reise K (2010) The Wadden Sea Quality Status Report. Synthesis Report 2010. In: Wadden Sea Ecosystem No.29. Common Wadden Sea Secretariat, Wilhelmshaven

Ziegler PA (1990) Geological Atlas of Western and Central Europe. 2nd edition, Shell International Petroleum

Part I

Recent Climate Change (Past 200 Years)

Recent Change—Atmosphere

2

Martin Stendel, Else van den Besselaar, Abdel Hannachi, Elizabeth C. Kent, Christiana Lefebvre, Frederik Schenk, Gerard van der Schrier and Tim Woollings

Abstract

This chapter examines past and present studies of variability and changes in atmospheric variables within the North Sea region over the instrumental period; roughly the past 200 years. The variables addressed are large-scale circulation, pressure and wind, surface air temperature, precipitation and radiative properties (clouds, solar radiation, and sunshine duration). Temperature has increased everywhere in the North Sea region, especially in spring and in the north. Precipitation has increased in the north and decreased in the south. There has been a north-eastward shift in storm tracks, which agrees with climate model projections. Due to large internal variability, it is not clear which aspects of the observed changes are due to anthropogenic activities and which are internally forced, and long-term trends are difficult to deduce. The number of deep cyclones seems to have increased (but not the total number of cyclones). The persistence of circulation types seems to have increased over the past century, with 'more extreme' extreme events. Changes in extreme weather events, however, are difficult to assess due to changes in instrumentation, station relocations, and problems with digitisation. Without thorough quality control digitised datasets may be useless or even counterproductive. Reanalyses are useful as long as biases introduced by inhomogeneities are properly addressed. It is unclear to what extent circulation over the North Sea region is controlled by distant factors, especially changes in Arctic sea ice.

Electronic supplementary material Supplementary material is available in the online version of this chapter at 10.1007/978-3-319-39745-0_2.

M. Stendel (✉)
Department for Arctic and Climate, Danish Meteorological Institute (DMI), Copenhagen, Denmark
e-mail: mas@dmi.dk

E. van den Besselaar · G. van der Schrier
Royal Netherlands Meteorological Institute (KNMI), De Bilt, The Netherlands
e-mail: besselaar@knmi.nl

G. van der Schrier
e-mail: schrier@knmi.nl

A. Hannachi
Department of Meteorology, Stockholm University, Stockholm, Sweden
e-mail: a.hannachi@misu.su.se

E.C. Kent
National Oceanography Centre, Southampton, UK
e-mail: eck@noc.ac.uk

C. Lefebvre
German Meteorological Service (DWD), Hamburg, Germany
e-mail: Christiana.Lefebvre@dwd.de

F. Schenk
Bolin Centre for Climate Research, University of Stockholm, Stockholm, Sweden
e-mail: frederik.schenk@geo.su.se

T. Woollings
Atmospheric Physics, Clarendon Laboratory, University of Oxford, Oxford, UK
e-mail: Tim.Woollings@physics.ox.ac.uk

M. Quante and F. Colijn (eds.), *North Sea Region Climate Change Assessment*,
Regional Climate Studies, DOI 10.1007/978-3-319-39745-0_2

2.1 Introduction

Situated in northern central Europe, the North Sea exhibits large climate variability with inflow of a wide range of air masses from arctic to subtropical. For this reason, it is difficult to differentiate between natural and externally forced variability, despite large amounts of historical data. This chapter examines past and present studies of variability and changes in atmospheric variables over the instrumental period; roughly the last 200 years. Research areas lacking consensus in the scientific community are highlighted to stimulate further research.

The main driver of atmospheric variability in the North Sea region is the North Atlantic Oscillation (NAO). Despite its apparent long-term irregularity, the NAO exhibits extended periods of positive or negative index values. No consensus exists with respect to the size of the fraction of interannual NAO variance that cannot be explained by random forcing and is therefore probably influenced by external forcing. Slowly varying natural factors with an effect on European climate, such as the Atlantic Multidecadal Oscillation (AMO), may superimpose long-term trends on atmospheric variability and so be difficult to distinguish from the anthropogenic climate change signal.

The source of atmospheric and surface data influences the results obtained, even in the comparatively data-rich North Sea region. Based on reanalysis data, several studies find positive trends in storm activity over the North Sea region and a northeast shift in storm tracks over the past few decades. However, studies based on direct or indirect historical records of long-term variations in pressure, wind or wind-related proxies, mostly do not identify robust long-term trends. This counter-intuitive result is explained by uncertainties in the long-term historical wind and atmospheric pressure observations, and additional uncertainties arising from the lack of quality control when digitising old data as well as potential biases in the reanalyses due to the fact that the underlying amount of available data is not constant in time. Nevertheless, the northeast shift in storm tracks appears to be a new phenomenon. In contrast, the increase in wind speed and storminess in the latter half of the 20th century does not seem to be unprecedented within the context of historical observations. There are indications of an increase in the number of deep cyclones (but not in the total number of cyclones). There are also indications that the persistence of circulation types has increased over the past century.

Temperatures have increased both over land and over the North Sea. There is a distinct signal in the number of frost days and the number of summer days. While there is a clear winter and spring warming signal over the Baltic Sea region, this is not as clear for the North Sea region. As expected, the variability in marine temperatures on seasonal timescales is less than for the land temperatures.

Precipitation over land and, but to a lesser extent, over sea is positively correlated with the NAO, and on longer time scales with the AMO. There are indications of an increase in precipitation in the north of the region and a decrease in the south, in agreement with the north-eastward shift in the storm tracks. There are also indications that extreme precipitation events have become more extreme and that return periods have decreased.

From the few datasets available on radiative properties, it may be concluded that there are non-negligible trends together with potential uncertainties and land-sea inhomogeneities which make it difficult to assess these quantities in detail.

Climate change in the North Sea region cannot be investigated in isolation. In particular, what the relation is between changes in the Arctic cryosphere and trends in storminess, number of cyclones, persistence of circulation anomalies and extreme events further south, is an open research question. As analyses of the latter often rely on small datasets covering relatively short time scales, it is difficult to draw statistically significant conclusions. It is therefore essential to make available the large amount of data from past decades that have not yet been digitised. However, it is essential to thoroughly quality-check the data.

2.2 Large-Scale Circulation

2.2.1 Circulation Over the North Sea Region in a Climatological Perspective

From a climatological perspective, the North Sea region is characterised by strong ocean-atmosphere interactions, especially during winter, compared to other regions at similar latitudes (Furevik and Nilsen 2005). These interactions involve transfer of momentum, moisture and various trace gases, mainly carbon dioxide (Takahashi et al. 2002). In addition to the recent warming trend (Delworth and Knutson 2000; Johannessen et al. 2004), the North Sea and nearby regions witnessed climate change during the early 20th century, which was large in comparison to similar latitudes elsewhere (von Storch and Reichardt 1997; Gönnert 2003).

Atmospheric circulation in the European/North Atlantic region plays an important role in the regional climate of the North Sea and surrounding land areas (Hurrell 1995; Slonosky et al. 2000, 2001). It is mainly described by the NAO (e.g. Hurrell et al. 2003) which is an expression of the zonality of the atmospheric flow. The North Sea region is controlled by two large-scale quasi-stationary atmospheric patterns, the Icelandic Low (IL) and the Azores High (AH) plus a thermally driven system over Eurasia with high pressure in winter and low pressure in summer. The dominant flow is therefore westerly, although any other wind direction is also frequently observed, and one of the main factors controlling air-sea interactions in the North Sea region is wind stress. Large-scale processes also constitute one of the main driving mechanisms responsible for the connection between local processes and global change. It is therefore important to pay attention to the recent changes in large-scale flow directly affecting the North Sea region.

The remainder of Sect. 2.2 reviews the status of the large-scale atmospheric variability affecting the North Sea region by focusing on the major teleconnection patterns and their effect on the jet stream. The NAO can be seen as the European expression of a larger-scale phenomenon, the Arctic Oscillation (AO). The relationship between the NAO and AO is briefly discussed in the following section, while a general description of the NAO and its properties is given in Annex 1. The NAO varies on a wide range of time scales from days to decades, reflecting interactions with surface conditions including sea-surface temperature (SST) and sea ice. These changes translate into changes in pressure and winds (Sect. 2.3), temperature (Sect. 2.4) and precipitation (Sect. 2.5) and also affect other variables, like sunshine (Sect. 2.6).

2.2.2 NAO and AO

The strength of the westerlies and the eddy-driven jet stream over the North Atlantic and western Europe are controlled by various factors including the pressure difference between the IL and AH as the main centres of action of the NAO (Wanner et al. 2001; Hurrell et al. 2003; Budikova 2009). The NAO and its changes can be understood as signals in the surface pressure field of jet stream variation (Hurrell and Deser 2009) and, as such, are often referred to as the regional expression of the AO, which describes sea-level pressure variations between the Arctic and northern hemisphere lower latitudes (Budikova 2012) or, in other words, variability in the strength of the polar vortex. The AO, also termed the Northern Annular Mode (NAM), was first identified by Lorenz (1951) and named by Thompson and Wallace (1998). Its positive phase is characterised by low surface pressure in the Arctic and a generally zonal (west to east) jet stream, thus keeping cold air in the Arctic. When the AO index is negative, there is high pressure in the Arctic and a stronger meridional (north to south or vice versa) component of the jet stream so that cold air can extend to lower latitudes. With respect to the North Sea region, the state of the AO controls the westerly flow and the storm track.

The AO is strongly correlated with the NAO, and the latter can be viewed as the signature of the former over the North Atlantic region. Therefore, the discussion below and in Annex 1 focuses on the NAO. Note, however, that there is extensive literature on the relationship between the NAO and AO, including studies on the robustness and consistency of one versus the other (Ambaum et al. 2001). Even though the AO and NAO are strongly correlated, their relationship is not linear (Kravtsov et al. 2006; de Viron et al. 2013). A fully three-dimensional picture discussing the connection between the AO and stratospheric circulation anomalies is provided by, for example, Ripesi et al. (2012).

2.2.3 Temporal and Spatial Changes in the NAO

Given the importance of the NAO in the North Atlantic and European climate, substantial efforts have been made to understand its variability in order to gain insight into its potential predictability. The NAO index varies on a wide range of time scales ranging from days to decades. As Fig. A1.1 in Annex 1 shows, the long-term behaviour of the NAO is irregular, and there is large interannual and inter-decadal variability, reflecting the interaction with the surface, including SST and sea ice. Focusing on the 20th century, a period with predominantly positive NAO index values prevailed in the 1920s, followed by mainly negative values in the 1960s. Since then, a positive trend has been observed, which means more zonal circulation with mild and wet winters and increased storminess in central and northern Europe, including the North Sea region (e.g. Hurrell et al. 2003). This was especially the case in the early 1990s, raising claims that this behaviour was 'due to anthropogenic climate change'. After the mid-1990s, however, there was a tendency towards more negative NAO index values, in other words a more meridional circulation and according to Jones et al. (1997), Slonosky et al. (2000, 2001) and Moberg et al. (2006), the strongly positive NAO phase in the early 1990s should be seen as an element of multi-decadal variation comparable to that at the start of the 20th century rather than as a trend towards more positive NAO values. It should also be noted that the winter of 2010/2011 had one of the most negative NAO indices in the record (Jung et al. 2011; Pinto and Raible 2012).

Intraseasonal variability in the NAO can be reasonably well described by a Markov process or first-order autoregressive (AR1) model (Feldstein 2000), although the observed skewness of the NAO index (Woollings et al. 2010; Hannachi et al. 2012) or the particularly enhanced persistence of the negative phase (Barnes and Hartmann 2010) are not captured very well. Nonlinear Rossby wave breaking mechanisms have been proposed to explain the intraseasonal variability in the NAO (e.g. Benedict et al. 2004; Franzke et al. 2004; Woollings et al. 2008).

Interannual variability in the winter mean NAO index is also found to be linked to the intraseasonal transitions between the positive and negative phases of the NAO pattern (Luo et al. 2012).

Several studies have attempted to quantify the fraction of interannual NAO variance that can be explained by 'climate noise', namely by random sampling of the intraseasonal variations (Feldstein 2000; Keeley et al. 2009; Franzke and Woollings 2011). As the NAO exhibits no preferred periods on the interannual and longer timescale (Hurrell and Deser 2009), the results of these studies differ widely, meaning as yet no consensus on the fraction of interannual NAO variability that cannot be explained by such noise and which is thus likely to be forced externally (Stephenson et al. 2000; Feldstein 2002; Rennert and Wallace 2009). Several external forcing mechanisms have been proposed, including bottom boundary conditions of local SST (Rodwell et al. 1999; Marshall et al. 2001) and sea ice (Strong and Magnusdottir 2011), volcanoes (Fischer et al. 2007), solar activity (Shindell et al. 2001; Spangehl et al. 2010; Ineson et al. 2011), and even stratospheric influence (Blessing et al. 2005; Scaife et al. 2005), including the quasi-biennial oscillation (Marshall and Scaife 2009) and stratospheric water vapour trends (Joshi et al. 2006). Remote SST forcing of the NAO originating from as far as the Indian Ocean was proposed by Hoerling et al. (2001) and Kucharski et al. (2006), while Cassou (2008) proposed an influence of the Madden-Julian Oscillation, but no consensus has been reached.

The positive trend in the NAO from the 1960s to the 1990s has been a particular focus of interest, and it has been a concern that many climate models have been unable to simulate the observed trends (Gillett 2005). Even though some models simulate quite large natural variability (Selten et al. 2004; Semenov et al. 2008), they still fall short of the strongest observed 30-year trend (Scaife et al. 2009). While Gillett et al. (2003) suggested that anthropogenic forcing could have contributed to the positive NAO trend there is no agreement on this, and concerns over the ability of models to represent NAO variability mean that attribution attempts should be treated with caution. The downturn in the NAO since the mid-1990s has brought its relation to climate change under further doubt (Cohen and Barlow 2005).

The NAO pattern is not entirely stationary, neither geographically nor with respect to season (Fig. 2.1). Although the amplitude is greatest and the explained variance highest in winter, the NAO is present all year round with varying strength and position. In particular, there is a westward shift of the southern centre of action in spring and an eastward displacement of both centres in autumn. Portis et al. (2001) developed a seasonally and geographically varying 'mobile' NAO, which is obtained from sea-level pressure data by taking into account the migration of the AH nodal points. Other techniques also exist, such as optimally interpolated patterns, trend empirical orthogonal functions (EOFs; Hannachi 2007a, 2008) and cluster analysis (Cheng and Wallace 1993; Hannachi 2007b, 2010).

Regression of the NAO on near-surface winds (Fig. 2.2) illustrates the well-known fact that the positive NAO phase is accompanied by stronger than average westerlies in the mid-latitudes right across the North Sea region into Scandinavia (Hurrell and Deser 2009), thus contributing to enhanced precipitation in the northern mid-latitudes.

2.2.4 Other Modes of Variability

The NAO is essentially a signal of variations in the Atlantic eddy-driven jet stream, but one pattern is not sufficient to fully describe the jet variability. The eddy-driven jet stream variability and regimes can, in fact, be described by at least two modes of variability—the NAO and the East Atlantic Pattern (EA; Wallace and Gutzler 1981; Woollings et al. 2010; Hannachi et al. 2012). The latter is defined as the second prominent mode of atmospheric low frequency variability over the North Atlantic. It appears throughout the year, but is more prominent in winter. Comprising a north-south dipole of anomalies, the EA pattern[1] resembles the NAO, but with anomaly centres displaced south-eastward and thus is sometimes interpreted as a 'southward shifted' NAO (e.g. Barnston and Livezey 1987). The positive phase of the pattern is associated with wetter-than-average conditions over northern Europe and Scandinavia. The EA pattern has particularly strong fluctuations in the low frequency component showing a consistent positive trend over recent decades. However, this is indicative of trends over the Mediterranean region rather than the North Sea region (Woollings and Blackburn 2012). Trouet et al. (2012) introduced a 'summer NAO' pattern with a blocking high over the British Isles and a low over south-eastern Europe in its positive phase which is most pronounced on interannual timescales and resembles the EA.

A third pattern that has some impact on the North Sea and Scandinavia is the Scandinavian pattern[1] (Wallace and Gutzler 1981), characterised primarily by a centre over Scandinavia and weaker centres of opposite polarity over western Europe. The positive phase of the Scandinavian pattern is frequently linked to the occurrence of atmospheric blocking over northern Europe (Barriopedro et al. 2006) and is therefore characterised by below-average precipitation in this region, as well as by large interannual and decadal variability over the past 50 years (Croci-Maspoli et al. 2007; Tyrlis and Hoskins 2008), which may be related to longer term variability of blocking across the Atlantic Ocean

[1] www.cpc.ncep.noaa.gov/data/teledoc/telecontents.shtml.

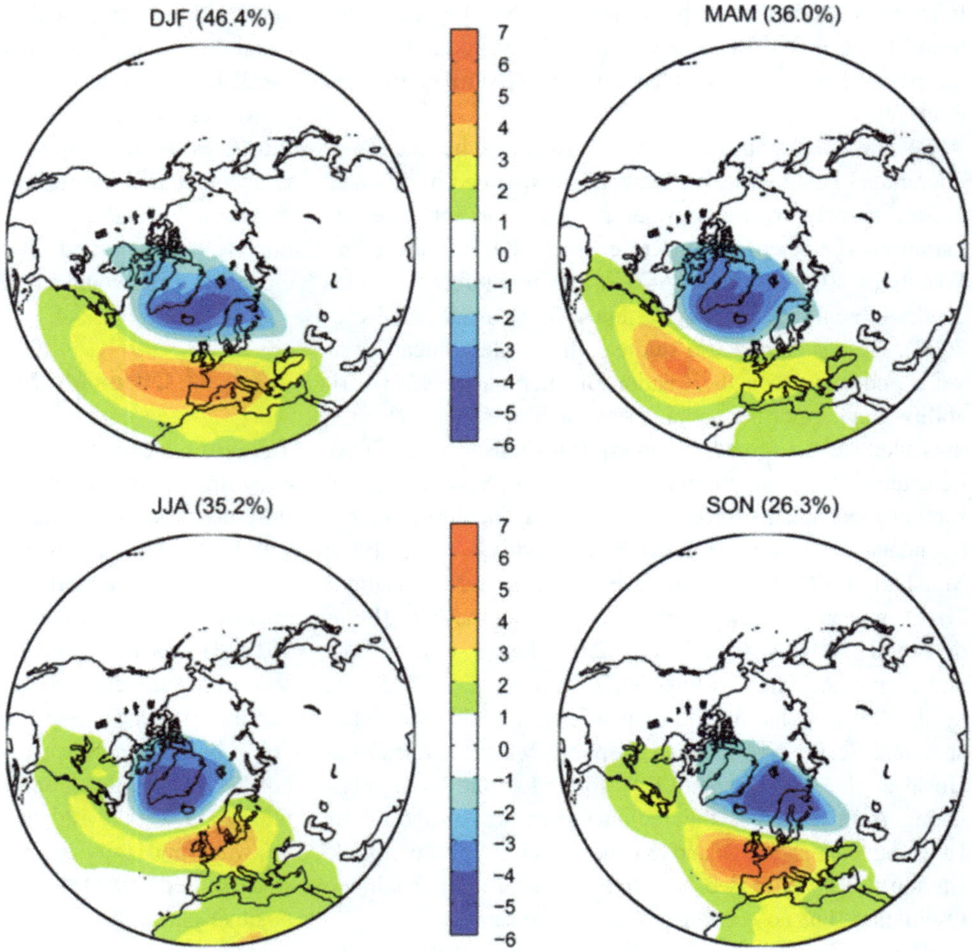

Fig. 2.1 Leading EOF of seasonal mean sea-level pressure (SLP) anomalies over the North Atlantic (20°–0°N, 90°W–40°E for the period 1948–2014. The percentage of explained variance is given above each panel

(Häkkinen et al. 2011). Before the 1980s the positive phase was dominant, this was followed by a negative phase between 1980 and 2000. The pattern amplitude has weakened over the past decade compared to the earlier part of the record.

Finally, on longer timescales, atmospheric conditions in the North Sea region are significantly influenced by the Atlantic Multidecadal Oscillation (AMO—more correctly termed Atlantic Multidecadal Variability, as there is no temporal regularity), which describes basin-wide variations in the temperature of the North Atlantic (Knight et al. 2005a, b) on the order of decades. These are particularly important in summer (Sutton and Hodson 2005) when the atmospheric response resembles a pattern termed the 'summer NAO' (Folland et al. 2009; Ionita et al. 2012b), see also Fig. 2.1. Warm periods were observed prior to 1880, between 1930 and 1965 and after 1995, and cool periods between 1900 and 1930 and between 1965 and 1995.

2.2.5 Summary

The NAO is the dominant mode of near-surface pressure variability over the North Atlantic and Europe, including the North Sea region. Amplitude and explained variance are largest in winter, but the NAO impacts the North Sea region throughout the year. Despite its apparent long-term irregularity, the NAO exhibits extended periods of positive or negative index values. It is therefore important to quantify the fraction of interannual NAO variance that cannot be explained by random forcing and so is likely to be influenced by external forcing. There is no consensus on the size of this fraction, or on the possible external forcing mechanisms.

2.3 Atmospheric Pressure and Wind

A typical characteristic of the climatology of the North Sea region is the large variability in meteorological variables on multiple time scales. The strong increase in wave height and storminess between the 1970s and the 1990s over the North Sea and North Atlantic (Carter and Draper 1988; Hogben 1994) raised public concern about a roughening wind climate and speculations about whether global warming might have an impact on storminess (Schmidt and von Storch 1993). With the availability of many more observations and gridded

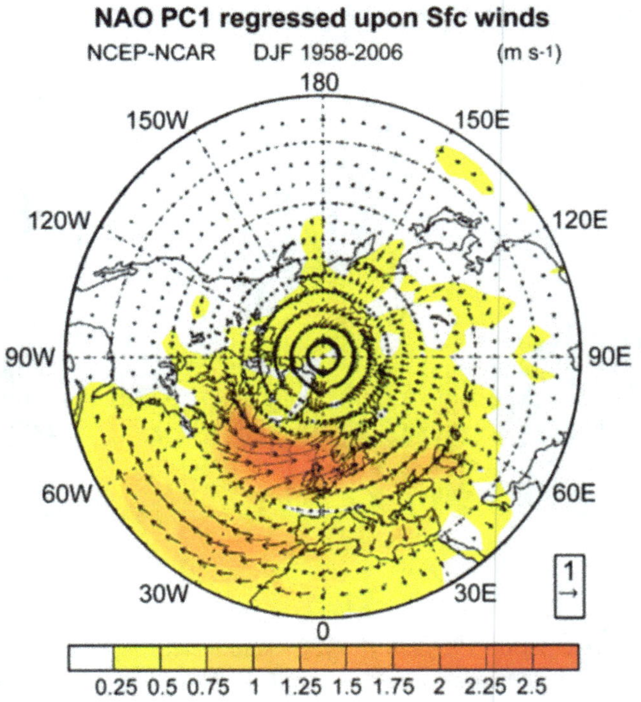

Fig. 2.2 Wind speed and direction associated with a 1 standard deviation change in the NAO index. The index is obtained from an EOF analysis of NCEP/NCAR sea-level pressure data (1958–2006) over the North Atlantic sector. Colour scale: m s^{-1}, unit vectors: 1 m s^{-1} (Hurrell and Deser 2009)

reanalysis data sets, detailed studies have significantly improved understanding of the atmospheric circulation and related winds over the North Atlantic and North Sea.

Owing to the large climate variability, results regarding changes or trends in the wind climate are strongly dependent on the period and region considered (Feser et al. 2015b). Through the strong link to large-scale atmospheric variability over the North Atlantic, conclusions about changes over the North Sea region are best understood in a wider spatial context. The following sections summarise studies on variations and trends in pressure and wind for recent decades and place these in the context of studies about changes over the last roughly 200 years.

2.3.1 Atmospheric Circulation and Wind Since Around 1950

The period since about 1950 is relatively well covered by observational data. The beginning of the satellite period in 1979 led to further substantial improvements in global data coverage, especially over the oceans and in data-sparse regions. Although in situ wind observations allow direct analysis of this variable, in particular over the sea (e.g. International Comprehensive Ocean-Atmosphere Data Set, ICOADS; Woodruff et al. 2011), the information is often predominantly local and inhomogeneities make the straightforward use of these data difficult, even for recent decades (Annex 1). Examples include an increase in roughness length over time due to growing vegetation or building activities, inhomogeneous wind data over the German Bight from 1952 onwards (Lindenberg et al. 2012) or 'atmospheric stilling' in continental surface wind speeds due to widespread changes in land use (Vautard et al. 2010).

Most studies therefore do not use direct wind observations, but instead rely on reanalysis products such as NCEP/NCAR (from 1948 onwards; Kalnay et al. 1996; Kistler et al. 2001), ERA40 (from 1958 onwards; Uppala et al. 2005) or, more recently, ERA-Interim, starting in 1979 (Dee et al. 2011) and the 20th Century Reanalysis 20CR (from 1871 onwards; Compo et al. 2011) and other reanalysis products, see Electronic (E-)Supplement Sect. S2.2. Making use of all available observations, a frozen scheme for the data assimilation of observations into state-of-the-art climate models is used to minimise inhomogeneities caused by changes in the observational record over time. However, studies indicate that these inhomogeneities cannot be fully eliminated (see E-Supplement S2). In addition, systematic differences between the underlying forecast models, such as due to their different spatial resolutions (Trigo 2006; Raible et al. 2008) and differences in detection and tracking algorithms (Xia et al. 2012) may affect cyclone statistics (for example changes in their intensity, number and position). Apart from these differences and inhomogeneities, the number of detected cyclones and their intensities show very high correlations between reanalyses (Weisse et al. 2005; Raible et al. 2008).

Three recent studies cover a continental-scale area. Franke (2009) manually counted the number of strong low pressure systems (central pressure below 950 hPa) over the North Atlantic (north of 30°N) from weather maps of the Maritime Department of the German Weather Service ('Seewetteramt'), see Fig. 2.3, which shows generally weak activity prior to 1988 and enhanced activity for the following decade, followed by a decrease to 2006.

Since then, the number of deep cyclones has again increased despite the predominantly negative NAO (Fig. A1.1 in Annex 1), and the maximum value with 18 such cyclones was observed in 2013/2014. Despite large decadal variations, there is still a positive trend in the number of deep cyclones over the last six decades, which is consistent with results based on NCEP reanalyses since 1958 over the northern North Atlantic Ocean (Lehmann et al. 2011). Using an analogue-based field reconstruction of daily pressure fields over central to northern Europe (Schenk and Zorita 2012), the increase in deep lows over the region might be unprecedented since 1850 (Schenk 2015). Barredo (2010) investigated adjusted storm losses in the period 1970–2008 on a European scale but did not find any trends despite a roughened wind climate. Based on the CoastDat2 reanalysis

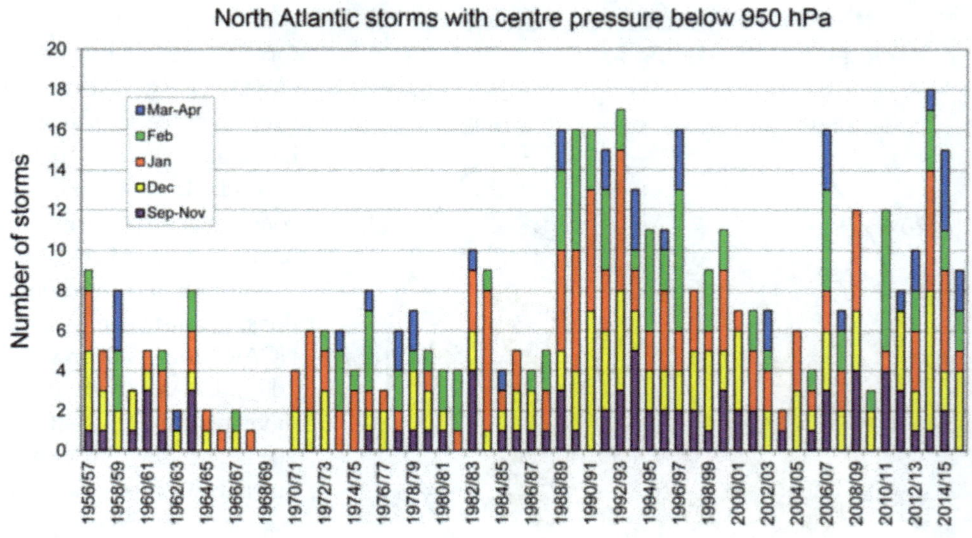

Fig. 2.3 Number of low pressure systems on the North Atlantic with a core pressure of 950 hPa or below, 1956/57–2015/16 (after Franke 2009, updated)

(Geyer 2014) and using E-OBS pressure data (van den Besselaar et al. 2011), von Storch et al. (2014) did not find robust evidence for supporting claims that the intensity of the two strong storms in late 2013 would be beyond historical occurrences and that the recent clustering of storms should be related to anthropogenic influence.

2.3.1.1 North Sea Region

Different studies based on reanalyses confirm the strong increase in wind speeds and wave heights observed over the North Sea region since the 1970s. Covering the period since about 1950, a positive trend is visible in annual storm activity in the NCEP, ERA40 and 20CR datasets, although the most recent decade shows a decrease in wind speed (e.g. Matulla et al. 2007; Donat et al. 2011) with no notable trend in mean wind speed for the period as a whole (1948–2014) over the North Sea (Fig. 2.4).

Siegismund and Schrum (2001) analysed decadal changes in wind forcing over the North Sea based on the NCEP reanalysis for the period 1958–1997 (Fig. 2.4). Over the 40 years, mean annual wind speeds increased by 10 %, mainly due to an increase in autumn and winter (ONDJ) after the 1960s and in late winter (FM) since the mid-1980s, but no trend was found for summer. Increased wind speeds are accompanied by an increase in WSW wind directions in autumn and winter (ONDJ) over the last three decades compared to the first (1958–1967). The enhanced mean winter wind speeds agree with an increase in the mean winter NAO index with a correlation of 0.69 for the year-to-year variations (Fig. 2.4). An update of the graphic for 1948–2014 shows a return to average wind conditions in the last decade with no notable trend remaining. The updated correlation with the NAO is 0.73.

Weisse et al. (2005) compared an NCEP-driven regional climate simulation (50 km resolution) with wind speeds from marine stations and found relatively good agreement. For the period 1958–2001, they found increasing storminess over most marine areas north of 45°N with a small, but significant positive trend over the North Sea and Norwegian Sea. The relative increase in storm frequency is largest over the southern North Sea across Denmark towards the Baltic Sea (1–2 % per year). The number of storms was lowest during the 1970s (with some notable exceptions, in particular the 'Capella' storm in January 1976) and peaked around 1990–1995. However, since then, a decrease in storm frequency has been observed which is confirmed by other studies (e.g. Matulla et al. 2007). Based on a high-resolution model hindcast forced by NCEP reanalyses for the storm season (November to March) 1958–2002, simulated storm-related sea-level variations confirm a significant positive trend for the Frisian and Danish coast (Weisse and Plüß 2006) while insignificant changes in mean and 90th percentile water levels are found for the UK, the Dutch coast and the German Bight. However, Weisse and Plüß also noted that positive trends in observations are higher than those in the NCEP-driven hindcast (see Chap. 3).

In contrast to the strong increase in wind speed in the NCEP reanalysis, Smits et al. (2005) found no increase in geostrophic wind speeds and even a decrease in homogenised wind observations for inland stations in the Netherlands for the period 1962–2002, while coastal stations show an increase consistent with NCEP. Smits et al. (2005) claimed that this is due to inconsistencies in the NCEP data, but it might also be that the 'atmospheric stilling' postulated by Vautard et al. (2010; see E-Supplement Sect. S2.1) can explain these differences.

2.3.1.2 Northern North Atlantic Region

As variations in atmospheric circulation and the wind climate over the North Sea show a high co-variability with

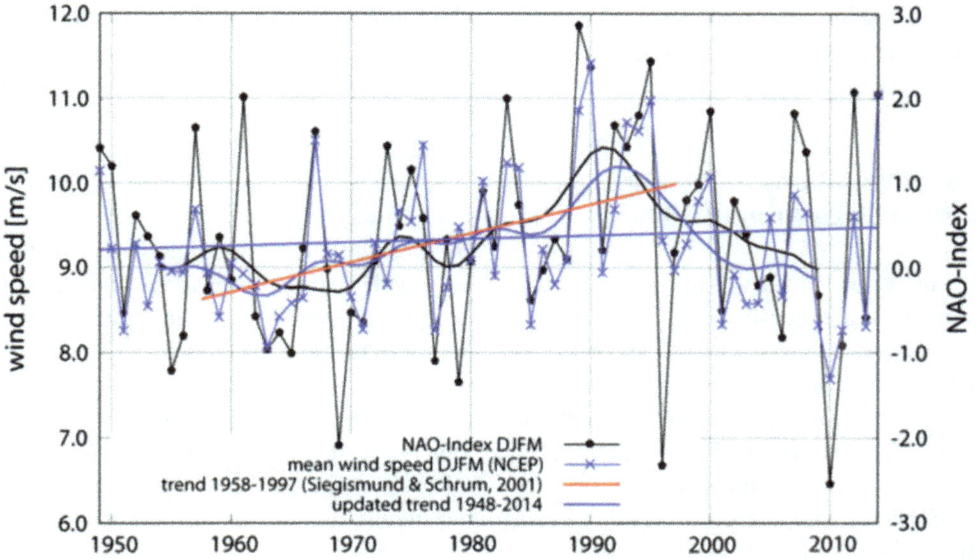

Fig. 2.4 Time series of mean seasonal wind speed derived from NCEP/NCAR reanalysis over the North Sea (*blue*) and the NAO index (*black*) for winter (DJFM) 1948–2014 recalculated and updated as in Siegismund and Schrum (2001), by F. Schenk. The positive trend (*red linear fit*) from this study has ended due to more average wind conditions in the last decade (*blue linear fit*). *Smoothed lines* are shown to highlight decadal-scale variations (11-year Hamming window)

large-scale variations and cyclonic activity over the North Atlantic, the majority of studies focus on the whole Euro-Atlantic region rather than on the North Sea alone. These studies show that the increase in wind speed or storminess is related to a general intensification of storm tracks over the North Atlantic north of about 55°N.

Based on NCEP reanalyses, Chang and Fu (2002) found a significant increase of about 30 % in decadal mean winter (DJF) storm track intensity for the period 1948–1998, with values about 30 % higher in the 1990s than during the late 1960s and early 1970s. The strengthening of storm track intensity is most pronounced over the North Atlantic albeit the trend compared to the few available conventional measurements seems to be overestimated in the NCEP reanalysis (Chang and Fu 2002; Harnik and Chang 2003; see E-Supplement Sect. S2.2). Geng and Sugi (2001) also found a significant increase in the number of North Atlantic cyclones and a significant intensifying trend for the cyclone central pressure gradient.

Similar results were found by Raible et al. (2008) based on ERA40 and NCEP for the period 1958–2001. The authors highlighted seasonal differences and reported a slight increase in number and a significant increase in intensity in winter (DJF) for the northern North Atlantic including the North Sea region (55°N–70°N, 45°W–15°E), a negative tendency in summer (JJA) and a non-significant increase in autumn (SON). The intensification in winter is stronger in NCEP than in ERA40 and also includes spring (MAM) in agreement with a similar increase in the number of deep lows (<980 hPa) in both seasons (Lehmann et al. 2011). The number of deep lows shows a minimum in the 1970s, followed by a strong increase.

The general enhancement in winter storm track intensity is accompanied by a northward shift in the storm track of 2–5° (depending on the data set) for NCEP (1948–1997; Chang and Fu 2002), ERA15 (1979–1997; Sickmoeller et al. 2000) and ERA40 (1958–2001; Wang et al. 2006), in agreement with a shift and intensification of deep lows (<980 hPa) towards the NE over the North Atlantic in the period 1948–2008 (Lehmann et al. 2011).

2.3.1.3 Southern North Atlantic Region

The general increase in the number of deep cyclones and storminess over the northern North Atlantic and North Sea is accompanied by partly opposing tendencies for the mid-latitudes south of 55–60°N (Gulev et al. 2001), suggesting a general northward shift in the cyclone tracks, consistent with findings of McCabe et al. (2001).

The north-south contrast in the sign of trends was also confirmed by Trigo (2006) who applied an objective detection and tracking algorithm to NCEP and ERA40 for winter (DJFM) 1958–2000 to produce a storm-track database for different stages in the cyclone lifecycle over the North Atlantic and Europe (20°–70°N; 85°W–70°E). On a seasonal basis, the trend is generally positive at higher latitudes (mostly due to an increased frequency of moderate and intense storms) and negative in the subtropical belt. Wang et al. (2006) and Raible et al. (2008) also drew similar conclusions. All these results consistently show a northward shift in mean storm track position since about 1950 (Feser et al. 2015a).

2.3.2 Regional Variations in Pressure and Wind Since Around 1800

Observed changes in cyclone characteristics and winds over the last 40–60 years pose the question as to whether these changes merely reflect (multi-)decadal variations or whether they reflect long-term change. This section summarises current knowledge about the historical evolution of pressure and wind in the last 200 years over the Euro-Atlantic region.

Information about long-term variations in pressure and wind rely on multiple direct and indirect observations. They provide qualitative to semi-quantitative historical descriptions (the latter quite far back in time), such as storm surge-related damage on the Dutch coast since the 15th century (de Kraker 1999) or daily weather diaries like those at the observatory of Armagh (Ireland) since 1798 (Hickey 2003). Direct measurements such as surge levels at Liverpool since 1768 (Woodworth and Blackman 2002) and Cuxhaven since 1843 (Dangendorf et al. 2014) or wind records in the Dublin region since 1715 (Sweeney 2000) also provide important information on variations in the storm climate. The difficulty with these observations is that they often represent local conditions and so often exhibit inhomogeneities to an unknown extent.

As recommended by WASA Group (1998), most studies use pressure observations (which are more homogeneous over time than wind measurements) to derive wind and storm indices (e.g. WASA Group 1998; Klein Tank et al. 2002). The usefulness of typically-used pressure-based indices has recently been re-assessed and confirmed. Even single-station pressure indices such as strong pressure changes over 6- or 24-h periods or the annual number of deep lows provide useful information about long-term variations in the wind and storm climate (Krueger and von Storch 2011). The information content to describe long-term variations in the statistics of pressure and wind is even higher for indices of geostrophic wind speeds calculated from triangles of daily pressure observations (Krueger and von Storch 2012). The correlation of geostrophic wind speeds calculated from station triplets with real model wind speeds is especially high over open terrain and sea areas (i.e. regions that often lack conventional observations). Although pressure-based indices provide only an indirect link to real wind speeds or storminess, they can be considered a valid approach for assessing long-term statistics of pressure and wind (Krueger and von Storch 2011, 2012).

As historical information on pressure and wind mostly relates to regional or local scale rather than gridded fields, the studies presented in the following sections are discussed by region. Figure 2.5 provides an overview of potential long-term trends in the wind and storm climate.

2.3.2.1 North Atlantic and Iceland

The region north of around 55–60°N is of special interest regarding changes in the intensity or position of the main storm tracks. Historical information here is limited to the Shetland, Orkney and Faroe Islands as well as Iceland. The longest pressure-based wind index, the annual mean of absolute pressure changes over 24 h, suggests a significant positive trend over Iceland in the period 1823–2006 (Hanna et al. 2008), but no robust trend exists over the Norwegian Sea (since 1833) or the North Sea (since 1874). The latter is consistent with the result of Schmith et al. (1998) who found no significant trend for absolute daily pressure tendencies for stations around the NE Atlantic for winter 1871–1997. Analysis of high annual geostrophic wind speed percentiles over the NE Atlantic also indicates no significant change since the late 19th century (WASA Group 1998; Alexandersson et al. 2000; Matulla et al. 2007; Wang et al. 2009a, 2011). For the shorter period 1923–2008, positive trends exist over the northern NE Atlantic for spring (Wang et al. 2009a) which is in agreement with the intensification and northeast shift in cyclone activity in the last 60 years.

A positive trend also exists for the annual frequency of zonal weather types in the winter half-year 1881–1992 (Schiesser et al. 1997). As this weather type (Großwetterlagen) classification (Baur 1937; Hess and Brezowsky 1952, 1977; Hoy et al. 2012) relies on historical weather maps over the North Atlantic and Europe, the trend should be viewed with caution due to an improvement over time in detecting smaller lows (E-Supplement S2).

2.3.2.2 British Isles

There are many historical wind and wind-related documents and records for Great Britain and Ireland. Although robust trend estimates have not been undertaken, available information suggests large multi-decadal variations but no overall long-term trends (e.g. Sweeney 2000 for the number of storms per decade from historical reports of the Dublin region 1715–1999). For the shorter period 1903–1999, however, adjusted wind observations do show a decrease in the decadal number of storms exceeding 50 knots (25.7 m s^{-1}). The record of storms from a daily weather diary of Armagh (Ireland) 1798–1999 (Hickey 2003) also indicates similarly large variations to those of recent decades, although observer bias reduces reliability over time. Anemometer readings at the station show no obvious trend in the number of gale days per year for the period 1883–1999.

For the Irish Sea, tide gauge records at Liverpool provide an indirect estimate of long-term variations in storms. These show a negative tendency for the annual maximum surge at

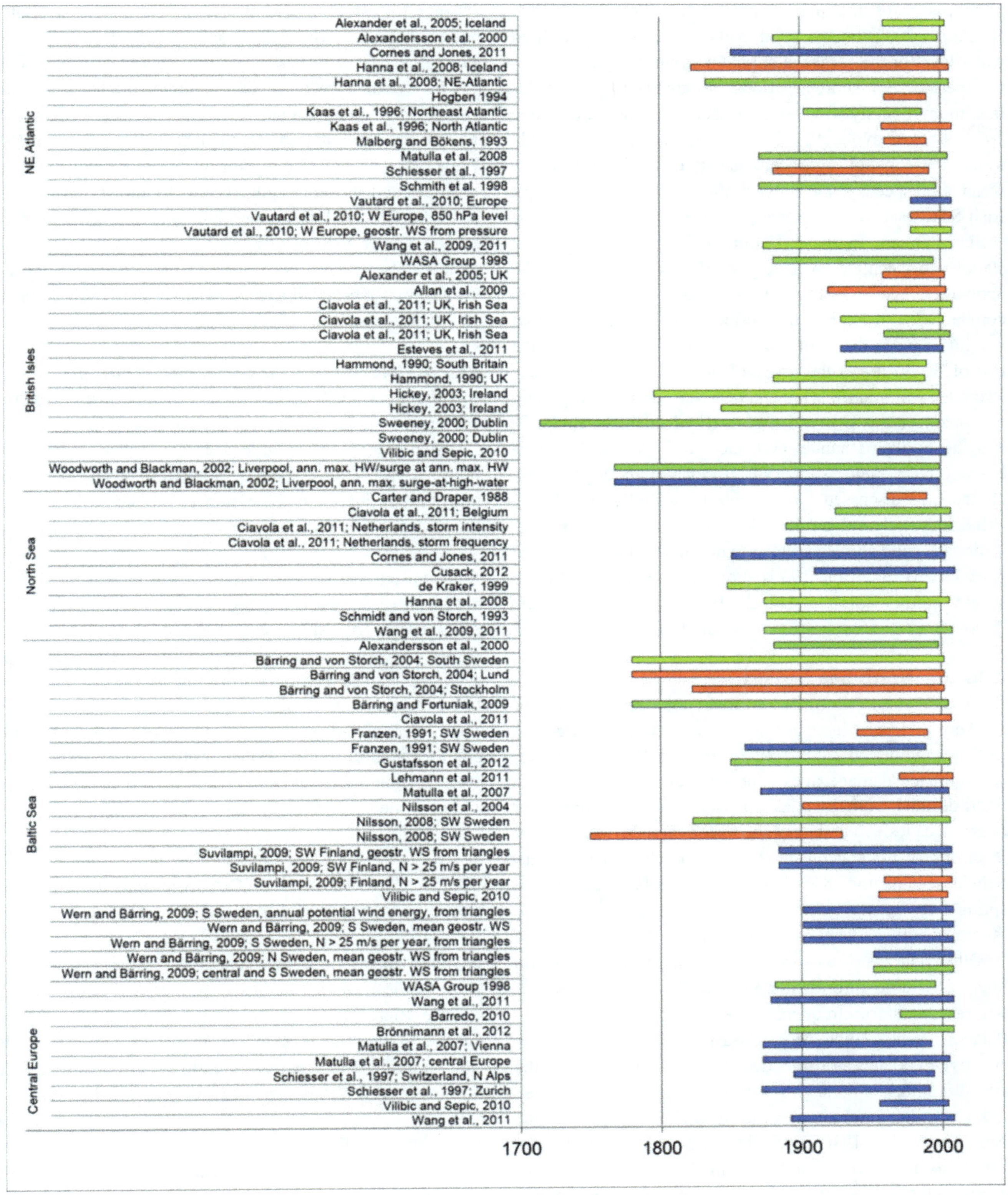

Fig. 2.5 Long-term trends for storminess over the Euro-Atlantic region based on different historical and observational sources (Feser et al. 2015a). *Red*, *blue* and *green colours* indicate positive, negative or no trend, respectively

high water for the period 1768–1999 (Woodworth and Blackman 2002) but no trend in the annual maximum high water level or the surges at annual maximum high water for this period. For shorter periods, no trends are found for maximum monthly wind speed observations for the Irish Sea 1929–2002 (Ciavola et al. 2011) while Esteves et al. (2011) found a weak but significant negative trend for monthly mean wind speeds at the Bidston observatory on the northern Irish Sea coast over the same period.

For southern England, Hammond (1990) used different stations to calculate an annual windiness index taking the annual average of monthly mean wind speeds for the Boscombe Down area. For the period 1881–1989, the annual windiness index does not show any long-term trend. For the end of the 17th and the first half of the 18th century (Late Maunder Minimum), wind indices were derived from ship logbooks for the Øresund region (Frydendahl et al. 1992) and the English Channel (Wheeler et al. 2009). Ship logbooks offer a unique source of information about past wind climates, as discussed for example by Küttel et al. (2009). Wind indices based on these logbooks suggest a generally stationary wind climate with large decadal variations. For the period 1920–2004, the 1930s show another period of more severe storms for the British Isles based on extreme three-hourly pressure changes (Allan et al. 2009).

2.3.2.3 North Sea Region

Limited historical information about dike repair costs for northern Flanders indicates no obvious visible long-term trend for the period 1488–1609 (de Kraker 1999). However, storm-related damage does appear to reflect similarly large multi-decadal variations as for storm observations over recent decades. An update of the historical index using water level observations at Flushing at the mouth of the Western Scheldt estuary for 1848–1990 shows a notable increase in spring tides in the 1990s which at least partly reflects the 30 cm rise in sea level over this period.

Other datasets do not indicate long-term trends in storm intensity. Surge information for the Netherlands shows a decrease in storm frequency over the period 1890–2008 (Ciavola et al. 2011). Also, Cusack (2012) found a weak negative tendency for the decadal running mean of the annual number of damaging storms and a related storm loss index calculated from homogenised wind observations of the Netherlands for 1910–2010, but did find large decadal variations for stormy conditions in the 1920s and 1990s. Storm intensity estimates derived from wind, wave and surge observations from Belgium for the period 1925–2007 show no trend (Hossen and Akhter 2015).

An analysis of geostrophic wind speeds for the German Bight shows no robust trends for the period 1876–1990 (Schmidt and von Storch 1993). But when the record is extended to include the period up to 2012 (Fig. 2.6) a tendency for decreasing wind speed in the upper percentiles becomes visible, corroborating direct wind, surge and wave observations from Belgium and the Netherlands, and findings by Rosenhagen et al. (2011). Analogue-based storminess shows a good correlation with the German Bight index and indicates no long-term trend since 1850 (Schenk 2015). Wang et al. (2011) found significant negative trends over the North Sea and surrounding land areas for the 99th percentile of geostrophic wind in summer, but no robust trends in other seasons. Direct wind observations from Skagen in northern Denmark also suggest decreasing overall storminess for the period 1860–2012 with extremely high storminess prior to 1875 (Clemmensen et al. 2014).

2.3.2.4 Northern Alps and Central Europe

Although not directly linked to the North Sea wind climate, observations from further south are also useful to help understand variations in large-scale atmospheric circulation and give indications about a northward displacement in storm tracks. For Vienna, the number of gale days (above 8 Bft, 17.2 m s^{-1}) show a clear decrease for the period 1872–1992 (Matulla et al. 2007), however based on non-homogenised observations. This is corroborated by significant negative trends in the number of days exceeding 7, 8 and 9 Bft (>13.9, 17.2 and 20.8 m s^{-1}, respectively) in northern Switzerland for the period 1894–1994 (Schiesser et al. 1997). The duration of strong winds (>7 Bft) also shows a negative trend for Zürich for 1871–1991 except in winter (Brönnimann et al. 2012). Negative trends are also found for central Europe (Matulla et al. 2007; Wang et al. 2011). In contrast, Stucki et al. (2014) found no clear trends, but large interdecadal variability, over Switzerland.

2.3.3 Trends in the 20th Century Reanalysis Since 1871

As shown in Fig. 2.5, the majority of studies using observational storm proxies find no robust trends, some even a negative tendency, for the wind and storm climate in historical pressure and wind observations. In contrast, 20CR (Compo et al. 2011) suggests significant upward trends for storminess over data-sparse regions like the NE Atlantic (Donat et al. 2011) while Bett et al. (2013) did not find a clear trend over Europe. Closer inspection reveals that the agreement of 20CR and wind observations over land like Zürich is reasonable (Brönnimann et al. 2012), but there are discrepancies over sea (Krueger et al. 2013; Schenk 2015). This is because 20CR, like other reanalyses, assimilates all available pressure observations at a given time step which leads to a strong increase in assimilated land pressure observations (and to a lesser extent also sea pressure observations) over time. Following Krueger et al. (2013),

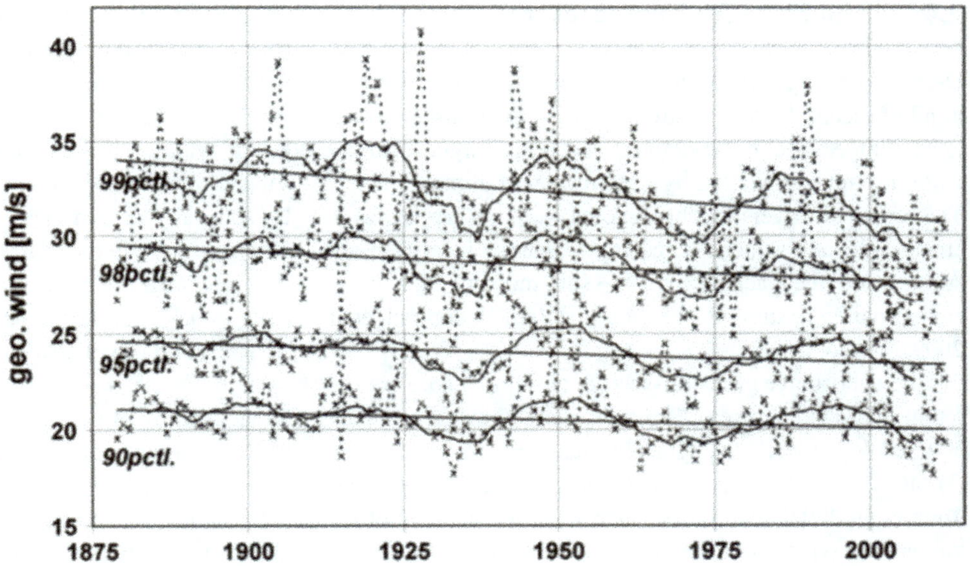

Fig. 2.6 High annual percentiles of geostrophic wind speeds over the German Bight after Schmidt and von Storch (1993) updated and reproduced for 1879–2012. Running 11-year means and linear trends are displayed to highlight long-term variations (data by G. Rosenhagen, figure by F. Schenk)

inconsistencies between 20CR and pressure-based storm indices over data-sparse regions increase back in time as the number of assimilated stations by 20CR, mainly over sea areas, decreases. Spurious pressure trends in data-sparse regions, identified in the NCEP/NCAR reanalysis (Hines et al. 2000) might also affect 20CR (E-Supplement Sect. S2.3).

2.3.4 Summary

Different studies mainly based on reanalysis data show positive trends in storm activity over the NE Atlantic and North Sea together with a northeast shift in the position of storm tracks over the last 40–60 years. This is also reflected in a roughened wind and wave climate, although a return to average conditions beginning at the end of the 20th century has clearly reduced trends from earlier publications. As summarised in Fig. 2.5 direct or indirect historical records of long-term variations in pressure, wind or wind-related proxies mostly show no robust long-term trends for the last 100 years or more. Large decadal variations seem to dominate for centuries.

While the increase in wind speeds and storminess in the latter half of the 20th century does not seem unprecedented in the context of historical observations, the northeast shift in storm tracks in this period may be a new phenomenon. The long-term decrease north of the Alps mainly results from a less stormy period during the 1990s compared to the North Sea and North Atlantic while the period at the end of the 19th century is comparably windy. The less stormy 1990s further south are consistent with the northeast shift in storm tracks and the decrease in winter cyclone activity in the mid-latitudes. This northeast shift together with the trend pattern of decreasing cyclone activity for southern mid-latitudes and increasing trends north of 55–60°N after around 1950 seems consistent with scenario simulations to 2100 under increasing greenhouse gas concentrations (e.g. Ulbrich et al. 2009; Feser et al. 2015a; see Chap. 5). This corroborates the findings by Wang et al. (2009b) that combined anthropogenic and natural forcing had a detectable influence on this pattern of atmospheric circulation, storminess and ocean wave heights during boreal winter 1955–2004 while an analysis for the first half of the 20th century is less likely to be dominated by external forcing.

Uncertainties remain not only for long historical wind and pressure observations (e.g. Lindenberg et al. 2012; Wang et al. 2014), but also for 20CR that to a large extent relies on these observations (Brönnimann et al. 2013; Krueger et al. 2013; Dangendorf et al. 2014; Schenk 2015). These are discussed in detail in E-Supplement S2. As a better understanding of long-term variations versus trends, and their link to atmospheric circulation is crucial for any regional climate change analysis, data rescue initiatives and digitisation initiatives such as data.rescue@home (www.data-rescue-at-home.org) or oldWeather (www.oldweather.org) are essential for further improvements towards the homogenisation of observations and reanalyses prior to about 1950. Such data can then be used in reanalysis projects such as ACRE (Atmospheric Circulation Reconstructions over the Earth; www.met-acre.org). However, in the light of problems apparently introduced into the WASA dataset during the digitisation step (see E-Supplement Sect. S2.3), it is also essential to thoroughly quality-check this type of data.

2.4 Surface Air Temperature

Despite the large variability in temperature, the warming trend of recent decades is strong enough to be discernible in local temperature observations, and it is larger than the warming trend simulated by state-of-the-art climate models. The principal drivers for this 'excess warming' appear to be changes in atmospheric circulation, mainly in winter and spring, and feedbacks involving soil moisture and cloud cover, mainly in summer and autumn (Van Oldenborgh et al. 2009).

The data sources for near-surface air temperature are different over land and sea. Terrestrial measurements are made at fixed locations, with typically standardised installations (WMO 2010) and at a reference height of 2 m (e.g. Klein Tank et al. 2002). In contrast, marine air temperature observations are typically made aboard moving ships (ICOADS; Woodruff et al. 2011), adjusted to a common reference height of 10 m (necessary because the typical observation height has increased by about 20 m over the period of record; Kent et al. 2013). Only a few fixed station measurements exist, such as on oil platforms. Observations of marine air temperature from ships are affected by daytime heating biases, and to avoid these problems datasets (for example from the Hadley Centre) are constructed using night-time observations only. Alternatively, both day and night observations, with adjustments for daytime heating following Berry et al. (2004), can be used. The North Sea region is relatively well sampled, but observations are sparse in the 19th Century and, more recently, during the Second World War.

2.4.1 Terrestrial Surface Air Temperature

The first decade of the 21st century was characterised by some extreme seasons. The hot summer of 2003 was probably unprecedented for at least 500 years in western Europe (Luterbacher et al. 2004), but was even surpassed in extremity by the East European summer of 2010 (Barriopedro et al. 2011). Summer and autumn 2006 and winter 2006/2007 were also exceptionally warm (Luterbacher et al. 2007; Cattiaux et al. 2009). On the other hand, winter 2010/2011 had a very negative NAO index (see Sect. 2.2), but was much warmer than comparable winters with a similarly negative NAO index (Cattiaux et al. 2010). Figure 2.7 shows time series of annually averaged land air temperature for the North Sea region, defined here as the area between 48°N and 62°N and 6°W and 10°E, for various data sets. The graphic shows 2014 to be unprecedentedly warm, even though none of the four seasons was the warmest on record (winter ranks 2nd after 2006/2007, spring 3rd after 2007 and 2011, summer 14th and autumn 2nd after 2006), and the previous maximum from 2011 was exceeded by almost 0.5°C. The datasets used are the CRUTEM4v (from UEA/CRU; Jones et al. 2012), GHCN-M version 3 (NOAA/NCDC, Peterson and Vose 1997; Jones and Moberg 2003), GISTEMP (NASA/GISS; Hansen et al. 2010) and BerkeleyEarth (http://berkeleyearth.org/), which have been subject to a homogeneity adjustment, supplemented by the E-OBS daily dataset version 10.0 (Haylock et al. 2008). To compare the different datasets, the grids of the global datasets are regridded to match that of the E-OBS grid (0.5° × 0.5°; van der Schrier et al. 2013).

The similarity between these estimates of temperature over the North Sea region is evident, with only minor differences in trend values (Table 2.1). Over the period 1980–2010, the trend in annual averaged daily mean temperature is approximately 0.38 °C decade^{-1}. Trend values are based on a linear least-square approximation to the data. Table 2.1 also gives temperature change for the whole of Europe (30°–75°N, 12°W–45°E plus Iceland, based on E-OBS), the northern hemisphere land and the global land area, both based on CRUTEM4. For 1980–2010, the warming trend in the North Sea region is smaller than that of Europe as a whole, but larger than the average over the northern hemisphere and global land areas.

In all datasets the period from the early 1990s onwards is warmest. Figure 2.8 highlights annual temperatures of the past few decades, averaged over the North Sea region and relative to the 1961–1990 climatology, based on the E-OBS dataset. The grey bars in Fig. 2.8 indicate the estimated uncertainties which take into account both errors introduced by spatial interpolation over areas without observations, by inhomogeneities in the temperature data that result from station relocations or instrument changes etc., and by urbanisation, as documented by van der Schrier et al. (2013) and Chrysanthou et al. (2014). The uncertainties indicate that although it is not possible to be 100 % certain about the ranking of individual years, the positive overall trend since the 1980s is very pronounced and 2014 stands out, even taking the uncertainties into account.

Ionita et al. (2012b) examined the connection between diurnal temperature range (DTR) and atmospheric circulation. They found that modes of interannual winter DTR variability are strongly related to the NAO and, to a lesser extent, the AMO, whereas in summer DTR variability is mainly influenced by a blocking pattern over Europe.

2.4.2 Number of Frost Days and Summer Days

According to Della-Marta et al. (2007), the length of western European heat waves has doubled since 1880 and Europe's climate has seen more warm extremes. This is illustrated in Fig. 2.9 which shows the difference in the annual number of frost days (minimum temperature <0 °C) and summer days

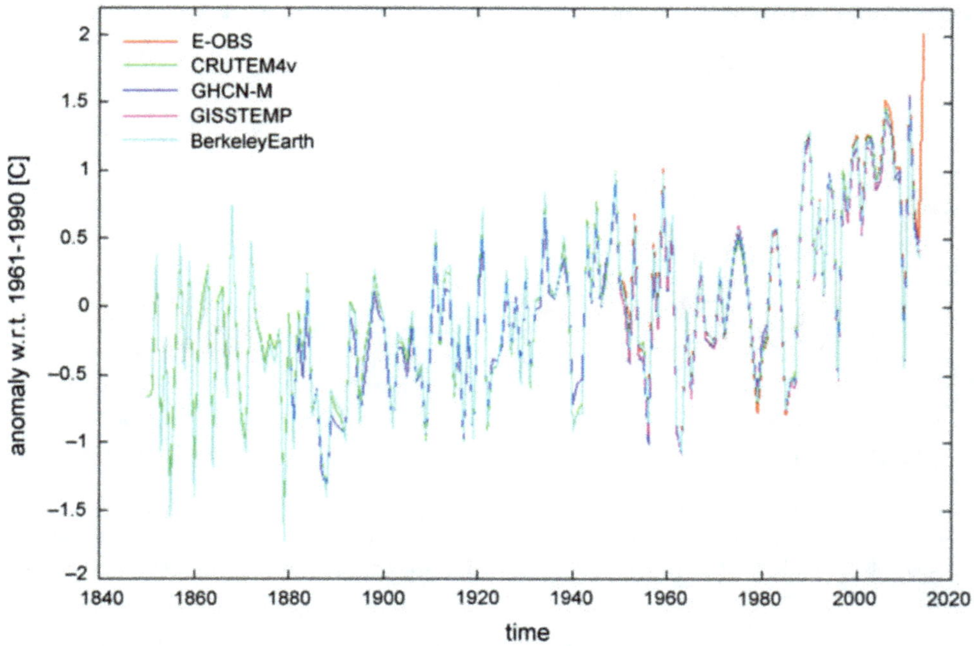

Fig. 2.7 Land-based annual mean air temperatures averaged over the North Sea region (48°–62°N, 6°W–10°E) with respect to the 1961–1990 climatology as calculated by E-OBS (*red*), CRUTEM4v (*green*), GHCN-M (*blue*), GISTEMP (*purple*) and the Berkeley Earth dataset (*light blue*)

Table 2.1 Linear temperature trends (°C decade^{-1}) over 1950–2010 and 1980–2010 for the North Sea region for CRUTEM4v, GHCN-D, GISTEMP, BerkeleyEarth and E-OBS

	CRUTEM4v	GHCN-D	GISTEMP	Berkeley Earth	E-OBS	Europe	NH land	Global land
1950–2010	0.210	**0.174**	0.228	**0.157**	0.204	**0.179**	**0.199**	**0.172**
1980–2010	**0.383**	**0.389**	**0.389**	**0.353**	**0.408**	**0.414**	**0.337**	**0.267**

The last three columns give trends for Europe (based on E-OBS), the northern hemisphere land and global land temperatures (based on CRUTEM4), respectively. Numbers in bold indicate that the trend is statistically significant at the 5 % level based on a t-test accounting for the reduced degrees of freedom due to autocorrelation (von Storch and Zwiers 1999)

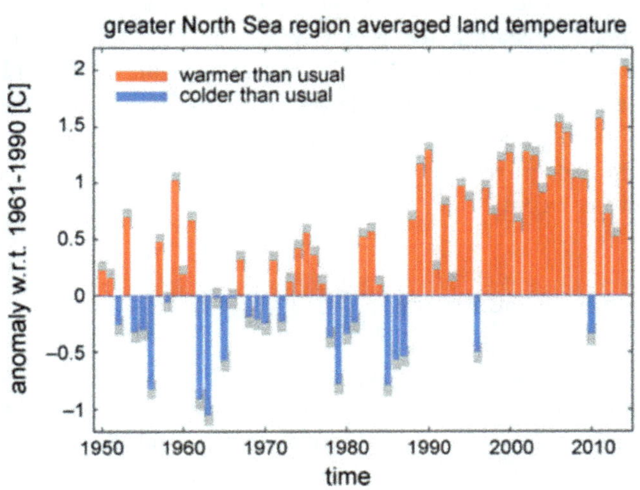

Fig. 2.8 Annual averages for land-based air temperature over the North Sea region with respect to the 1961–1990 climatology as calculated by the E-OBS dataset. The uncertainty estimate for the E-OBS data is included as *grey boxes*

(maximum temperature ≥25 °C) between 1981–2010 and 1951–1980 (based on E-OBS data). The change in these indices is not spatially consistent (in contrast to the increase in annual averaged temperature—not shown). All differences are statistically significant at the 5 % level using a one-sided Student t-test. The figure shows that the number of frost days has declined almost everywhere, with the strongest decreases found in the northern and eastern parts of the domain. The number of summer days has also increased almost everywhere, with the smallest increases in Scotland, northern England and Scandinavia and the largest in northern France.

2.4.3 Night Marine Air Temperature

As for land-based temperatures, the night marine air temperature (NMAT) also increased over the period 1856–2010 (Fig. 2.10). Two datasets were used, an uninterpolated 5° monthly mean dataset for 1880–2010 (HadNMAT2; Kent et al. 2013) and an interpolated (using a large-scale

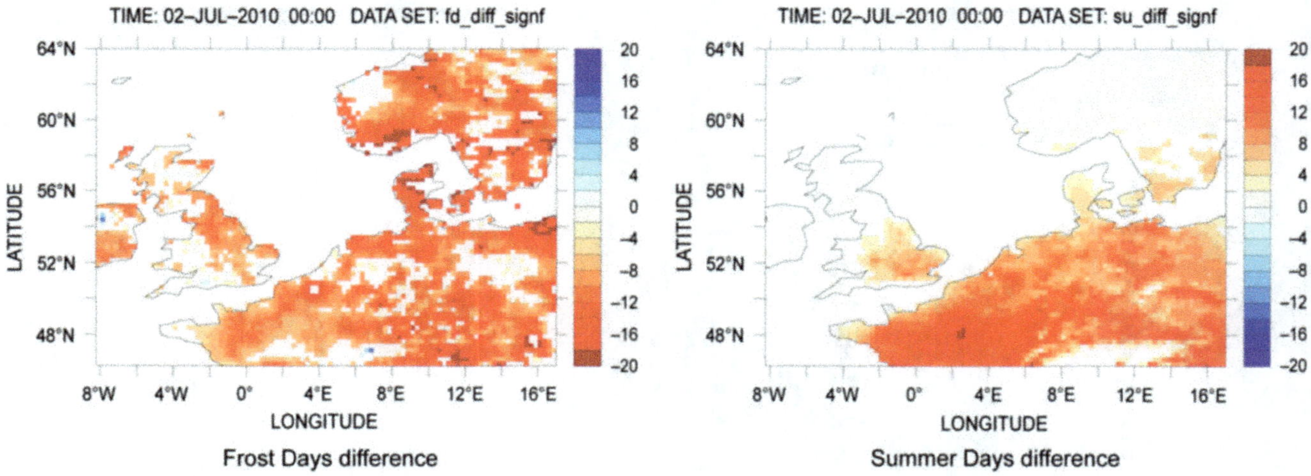

Fig. 2.9 Difference between the 1981–2010 and 1951–1980 climatological values of the annual number of frost days (*left*, daily minimum temperature <0 °C) and summer days (*right*, daily maximum temperature ≥25 °C). Grid squares with missing data or where the difference did not pass the 95 % significance level using a Student t-test, are white. Calculations based on E-OBS data

reconstruction technique; Rayner et al. 2003) 5° monthly mean dataset for 1856–2001. Differences between these datasets are larger than for land temperatures, especially around 1900 and during and just after the Second World War. In the latter period, sampling is sparse and non-standard observing practices necessitated adjustments to the observations (Kent et al. 2013). After about 1950, agreement improves. Linear trends in air temperature, adjusted for day-time heating biases (Berry and Kent 2009) show similar values.

Figure 2.10 also indicates that for marine air temperature the values in the most recent decade are likely to be the warmest on record, although uncertainty is large in the early part of the record due to sparse sampling.

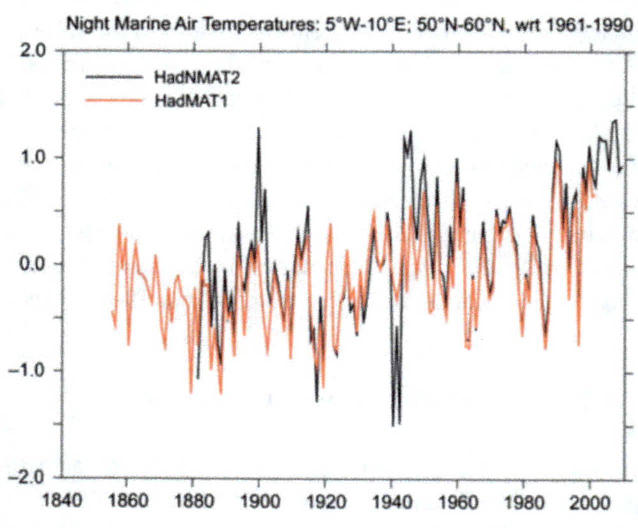

Fig. 2.10 Annual average night marine air temperature anomalies (°C) for the region 50°–60°N, 5°W–10°E from 5° monthly mean datasets: HadNMAT2 (*black*) and HadMAT1 (*red*)

Seasonal time series of the marine air temperature data sets show broadly similar variability to land-based temperatures (Fig. 2.11), but with a smaller amplitude. Very recently, the differences again increase, but this seems to be due to sparse observations and changes in the marine observing system (Kent et al. 2007, 2013).

2.4.4 Comparison of Land and Marine Air Temperatures

A comparison of land and marine temperatures (Fig. 2.12) shows general agreement. The lower plot in each panel depicts three estimates of the land-marine air temperature difference over the North Sea region based on E-OBS data for the land component and three different marine air temperature datasets: the NOCv2.0 dataset (Berry and Kent 2009, 2011), the HadNMAT2 dataset and the HadMAT1 dataset. Due to the much larger heat capacity of water, the difference series between the land and marine air temperature shows a residual positive trend over the last few decades of the record.

2.4.5 Summary

There is generally good agreement between the different temperature data sets over the oceans and over land. While temperatures have clearly increased over land, the NMAT shows there has also been an increase over the North Sea, even though the variability on seasonal timescales is smaller than for the land temperatures. Furthermore, due to the large heat capacity of water it takes much longer to warm the ocean than the land. In addition, heat is transported away

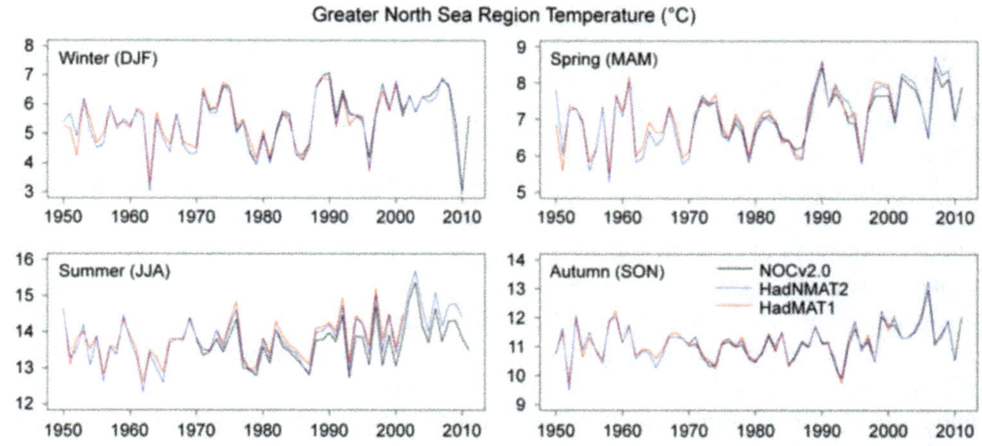

Fig. 2.11 Seasonal mean night marine air temperature (°C) from NOCv2.0 (1970–2010), HadNMAT2 (1950–2010) and HadMAT1 (1950–2001). HadNMAT2 and HadMAT1 were averaged to the same 1° grid as NOCv2.0 and masked to the NOCv2.0 land mask

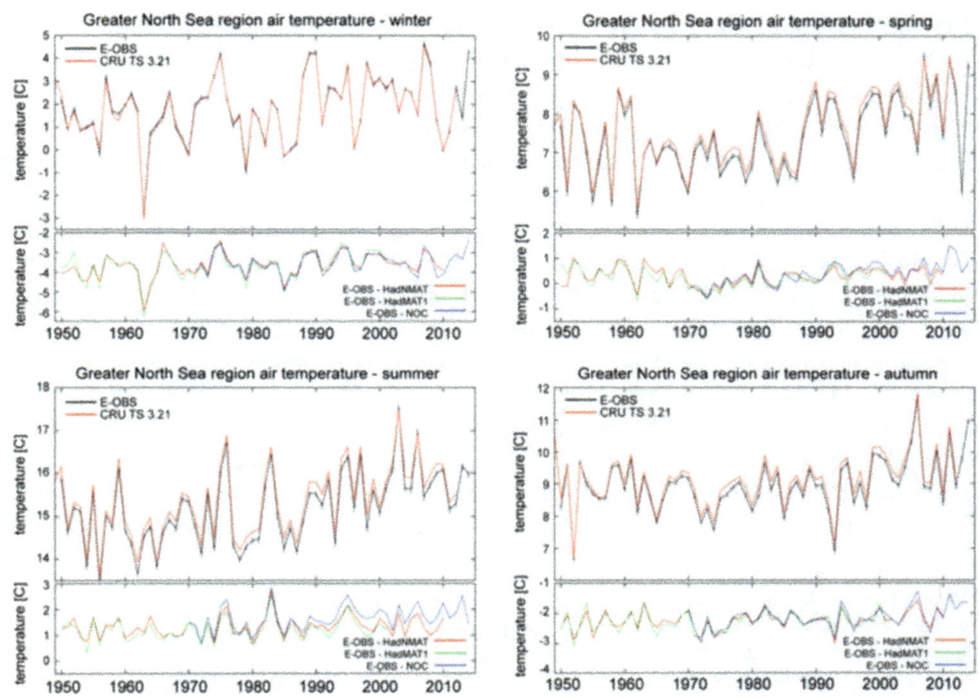

Fig. 2.12 Land-based air temperature over the North Sea region for winter (*upper left*), spring (*upper right*), summer (*lower left*) and autumn (*lower right*) based on the CRU TS 3.10 (Harris et al. 2014; *red*) and E-OBS (*black*) datasets. The uncertainty estimate for the E-OBS data is included as *grey boxes*. The *lower plots* show the difference between land-based air temperature and marine air temperature for this region, for three marine air temperature datasets

from the surface into deeper waters, where it cannot be directly measured. Thus, it would be expected that the land warms faster than the sea as long as the radiative forcing is positive. This is exactly the situation being observed, and according to the datasets, the imbalance is up to several tenths of a degree.

2.5 Precipitation

2.5.1 Precipitation Over Land in the North Sea Region

In a warmer climate, the atmospheric water vapour content is likely to rise due to the increase in saturation water vapour pressure with air temperature, as described by the Clausius-Clapeyron relation, and to result in an intensification of rainfall (Held and Soden 2006; O'Gorman and Schneider 2009). Evidence of higher amounts and more extreme precipitation has already been reported (e.g. Groisman et al. 2005; Moberg et al. 2006; Donat et al. 2013; Hartmann et al. 2013). Even though floods are a recurring event in Europe, attempts have been made to link increased flood risk to changes in the frequency of atmospheric blocking events (Lavers et al. 2012) or to anthropogenic climate change (Pall et al. 2011).

In a global study, Donat et al. (2013) showed a weak increase in the number of days exceeding 10 mm of precipitation (R10 mm) over the northern parts of Europe (but statistically significant only over eastern Europe at the 96 %

level). Over the Iberian Peninsula, a non-significant decrease in this metric is observed. A non-significant increase in the contribution of extreme precipitation events to the total precipitation amount (R95pTOT) is also observed over Great Britain and Scandinavia.

The European Climate Assessment and Dataset (ECA&D, Klein Tank et al. 2002) is a collection of daily station observations of 12 elements (of which five are gridded) and contains (as of March 2014) data from nearly 8000 stations across Europe and the Mediterranean. The station time series are updated on a regular basis using data provided by the national meteorological and hydrological services (NMHSs), universities or, before updates from these institutions are available, synoptic messages from the Global Telecommunication System (GTS). ECA&D receives time series from 61 data providers for 62 countries (as of March 2014).

Figure 2.13 compares precipitation for three stations with data since the beginning of the 20th century; Cambridge (UK), Stromsfoss Sluse (Norway) and De Bilt (Netherlands). The time series show strong interannual and decadal variability. A general upward trend is visible in the Dutch and Norwegian time series with trends of 14.52 mm decade^{-1} in the annual data over the 1901–2014 period for De Bilt and 28.81 mm decade^{-1} over the 1901–1950 period for Stromsfoss Sluse. Both trends are (just) statistically significant at the 5 % level following a t-test accounting for autocorrelation in the time series (von Storch and Zwiers 1999). A long-term trend in the UK time series is less pronounced. The Norwegian time series exhibits an enhanced trend since the mid-1990s, especially in summer. A weak drying trend since the 1990s, although not unprecedented, is visible in the UK series. It is also clear that, under certain circumstances, the entire area is influenced by high pressure for extended periods (e.g. in 1921) such that the whole North Sea area remains very dry.

Trends in annual land precipitation are positive almost everywhere over the North Sea region for the period 1951–2012 (Fig. 2.14, top panel). The greatest increase in precipitation is observed in winter (Fig. 2.14, lower left), especially along the west coast of Norway, over southern Sweden, parts of Scotland and the Netherlands and Belgium. Further inland, trends are much smaller and statistically non-significant almost everywhere. In summer, there is no evidence of increasing precipitation trends along the coast of western Norway, while the contrast between trends in coastal regions and more inland regions of the European mainland increases considerably as the latter show negative trends in summer (Fig. 2.14, lower right).

Winter and spring in northern Europe (defined as the land area north of 48°N) show an overall decreasing trend in return periods of extreme precipitation (van den Besselaar et al. 2013), which is indicative of increasing precipitation extremes. The trend is most pronounced in the 5-day precipitation amount in northern Europe during spring. The 5-day amount which is statistically a 20-year event over the 1951–1970 period becomes an approximately 8-year event in the 1991–2010 period.

For annual 5-day and 10-day precipitation amounts in the UK, Fowler and Kilsby (2003) found significant decadal-level changes in many regions. For the 10-day precipitation amount, the 50-year event during 1961–1990 became an 8-, 11- and 25-year event in eastern, southern and

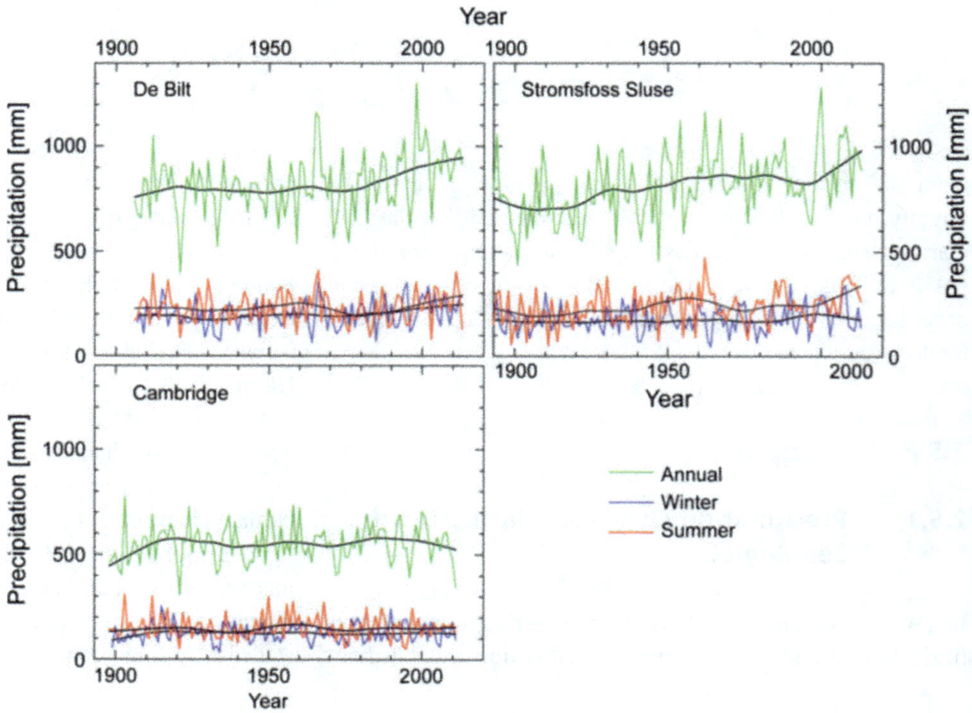

Fig. 2.13 Annual, winter and summer precipitation series for three stations from the ECA&D dataset; De Bilt (Netherlands, *top left*), Stromsfoss Sluse (Norway, *top right*) and Cambridge (UK, *lower left*). A low-pass filter is applied for the *black curves*

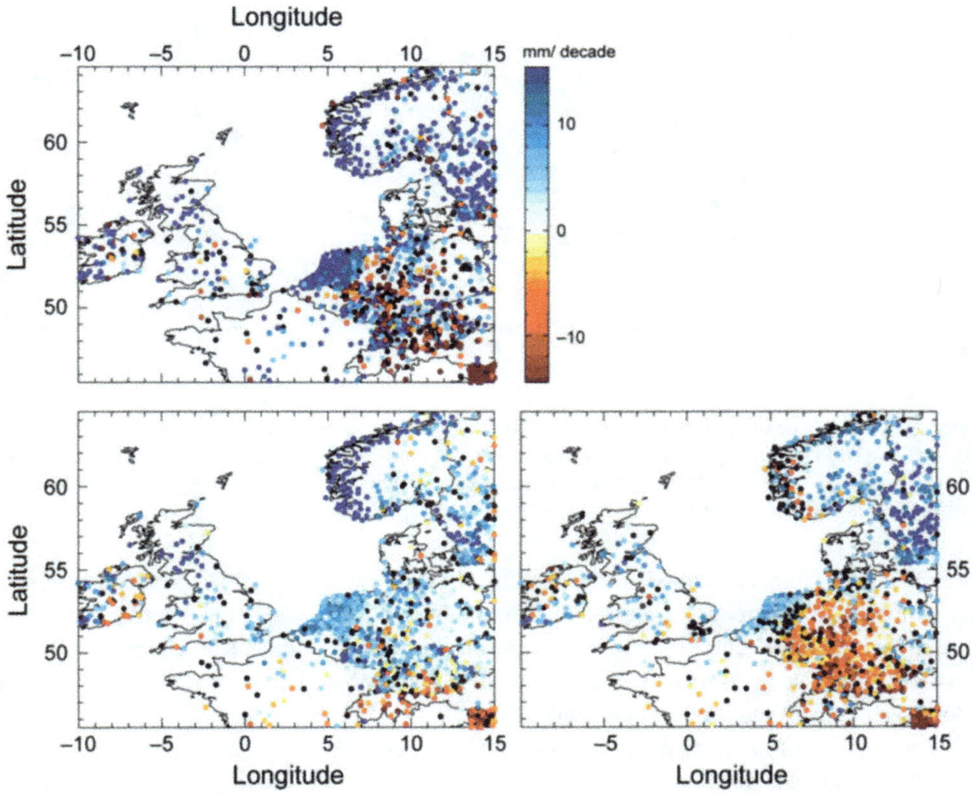

Fig. 2.14 Linear least-squares fit trends in annual (*top*), winter (*lower left*) and summer (*lower right*) precipitation over the period 1951–2012 in mm decade^{-1}. *Blue circles* denote a trend towards wetter conditions, while *orange* and *red circles* denote a trend towards drier conditions, both significant at the 5 % level ($p \leq 0.05$). *Black circles* fail to be significant at the 25 % level ($p \leq 0.25$) and are added to the figure to illustrate areas without any significant trend. *Source* ECA&D

northern Scotland, respectively, during the 1990s. In northern England the average return period has also halved.

Cortesi et al. (2012) analysed the precipitation concentration index, which is a measure of the amount of precipitation on a day with precipitation. The north-western coast of Europe shows relatively low values for this index (i.e. evenly distributed precipitation) compared to more Mediterranean climate types. No clear spatial pattern was detected in the trends in the index.

Groisman et al. (2005) studied total precipitation and frequency of intense precipitation in several regions of the world, including Fennoscandia. They found a significant increase in the annual totals and in the frequency of very heavy annual and summer precipitation events, where 'very heavy' precipitation events are defined by counting the upper 0.3 % of daily rainfall events (relating to a daily event that occurs once every 3–5 years).

There is temporal variability in trends when studying precipitation indices of extremes. For example, the trend in precipitation fraction due to very wet days, related to the 95th percentile in daily sums (R95pTOT), shows a different picture when the trends are determined over the period 1951–1978 compared to 1979–2012 (Fig. 2.15). Along the coasts of south-eastern England and the Netherlands, there is no trend apparent for the period 1951–1978, while the period 1979–2012 has an increasing trend for several stations in these areas.

Care should be taken if the precipitation fraction exceeding the 95th percentile (R95pTOT) is determined over a climatological period of several decades, since extremes may have increased disproportionally and thus the shape of the distribution may have changed. For example, an index S95pTOT, using the Weibull shape parameter instead of an explicit estimate of the 95th percentile, can be used (Leander et al. 2014). Northern Europe shows a (significant) increase in R95pTOT, but this is far less pronounced for S95pTOT. Since R95pTOT cannot distinguish between a shift in the median of the probability distribution for precipitation and a change in only the tail of the distribution, trends are generally 'more negative' for S95pTOT, especially over southern Scandinavia, the Netherlands, Germany and the UK.

Zolina et al. (2009) introduced a new index for R95pTOT, making use of a gamma distribution for wet day precipitation amounts and the associated theoretical distribution of the fractional contribution of the wettest days to the seasonal or annual total. The trend results for their new index are similar to R95pTOT.

Another way of analysing changes in precipitation is by counting the number of wet days. An example of this is the index CWD (maximum number of consecutive wet days, here defined as the number of days with precipitation ≥1 mm). Trends in the station records for the period 1951–2012 are shown in Fig. 2.16, which indicates that most of the stations in the North Sea region show a slight increasing

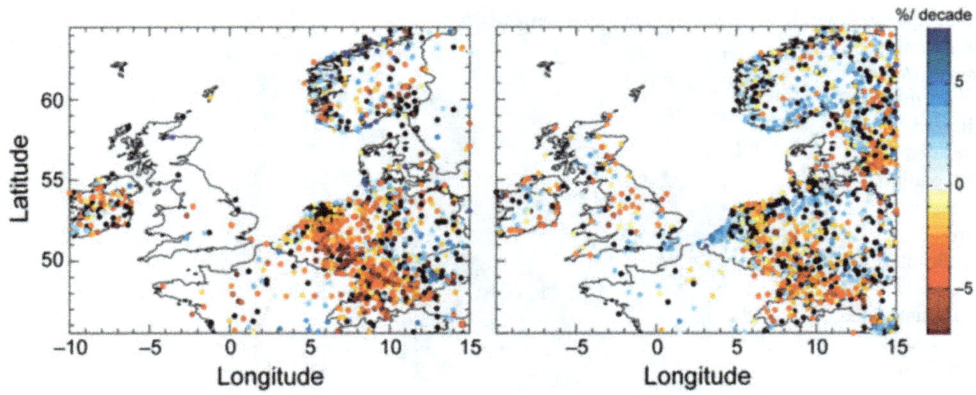

Fig. 2.15 Linear trends in the precipitation fraction due to very wet days (R95pTOT) in winter over the periods 1951–1978 (*left*) and 1979–2012 (*right*). *Blue circles* denote a trend towards wetter conditions, while *orange* and *red circles* denote a trend towards drier conditions, both significant at the 5 % level ($p \leq 0.05$). *Black circles* fail to be significant at the 25 % level ($p \leq 0.25$) and are added to the figure to illustrate areas without any significant trend. *Source* ECA&D

trend in the annual number of consecutive wet days. A similar map for trends in the maximum number of consecutive dry days (CDD) does not indicate a coherent change in the region. The majority of stations show trend values that do not meet even the 25 % significance level.

An example of interaction between North Sea waters and coastal climate was documented by Lenderink et al. (2009) for a month with extreme precipitation in the coastal region of the Netherlands (August 2006), where precipitation amounts were four times higher than the climatological average. Preceded by an extremely warm July (see Sect. 2.4) with very high sea surface temperatures in the North Sea at the end of July, favourable atmospheric flow conditions transported large amounts of moisture onto land, producing excessive rainfall in an area less than 50 km from the coastline. This phenomenon seems to be a robust finding since the positive trend in the difference between coastal and inland precipitation observed in the Netherlands is not sensitive to the period analysed.

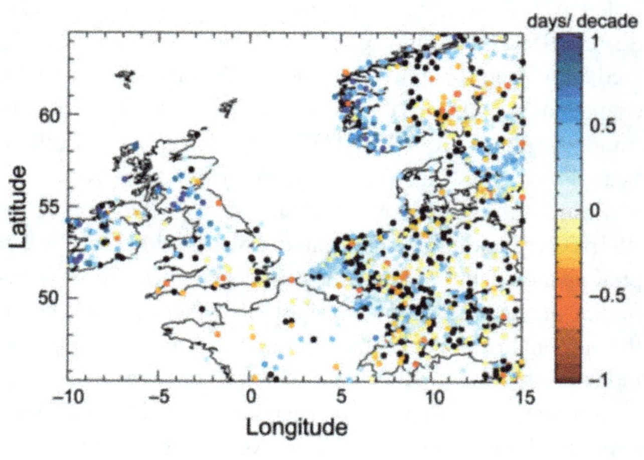

Fig. 2.16 Linear trend in annual maximum number of consecutive wet days (CWD) over the period 1951–2012. *Blue circles* denote a trend towards wetter conditions, while *orange* and *red circles* denote a trend towards drier conditions, both significant at the 5 % level ($p \leq 0.05$). *Black circles* fail to be significant at the 25 % level ($p \leq 0.25$) and are added to the figure to illustrate areas without any significant trend. *Source* ECA&D

2.5.2 Precipitation Over the North Sea

Only limited information is available for precipitation over oceans in general and the North Sea in particular. Almost no in situ measurements exist, which means it is necessary to rely on satellite observations using passive microwave detectors. HOAPS (Hamburg Ocean-Atmosphere Parameters and Fluxes) is one such dataset (Andersson et al. 2010, 2011). This covers the period 1988–2008 and is the only generally available satellite-based dataset for which fields of precipitation and evaporation over the oceans are consistently derived (Andersson et al. 2011). Over land, the dataset is gauge-based. Figure 2.17 shows the geographical distribution of annual average precipitation for the North Sea region (Fennig et al. 2012). Over most of the North Sea, the mean annual precipitation is between 600 and 800 mm, although values below 600 mm are also found off the east coast of England. Most coastal regions receive more than 800 mm, and in some mountainous regions (Scotland, Norway) more than 2000 mm are observed. While land stations in the south of the region have most rain in winter with a weak secondary summer maximum, further north and generally over the sea, there is only a maximum in winter, and May and June are the driest months.

Table 2.2 shows annual precipitation totals over the central North Sea region (54°–58°N, 1.5°–5.5°E) from the few available datasets.

There are considerable differences between the various estimates of precipitation, even in reasonably data-rich regions like the North Sea. There are also large differences

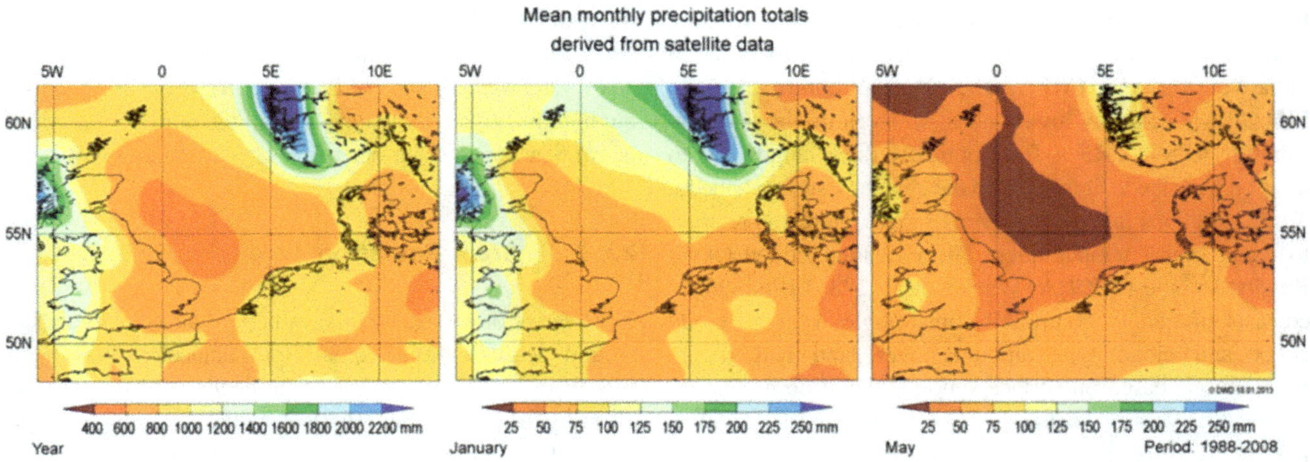

Fig. 2.17 Precipitation over the North Sea area. Ocean: HOAPS dataset (Fennig et al. 2012), land: gauge-based. Annual sum (*left*), monthly sum for January (*centre*), and monthly sum for May (*right*) for the period 1988–2008. All data in mm

Table 2.2 Estimates of annual average precipitation over the North Sea region from different reanalyses and satellite-based datasets

Dataset	1979–2001	1988–2008	Source
HOAPS	–	643	Andersson et al. (2010)
ERA-Interim	812	800	Simmons et al. (2010) and Berrisford et al. (2009)
ERA40	691		Uppala et al. (2005)
Coastdat2 (cDII.00)	853	861	Geyer (2014)
NCEP-CFSR	966	1000	Saha et al. (2010)
MERRA	754	772	Rienecker et al. (2011)

All units in mm

between periods, highlighting the problems in deriving trends in precipitation (Bengtsson et al. 2004). Precipitation in reanalyses depends on the moisture flux divergence, a rather weakly constrained quantity, which in turn does not depend on direct observations, but on the assimilation of satellite radiances (see e.g. Lorenz and Kunstmann 2012).

2.5.3 Summary

An assessment of temporal variability shows that precipitation over land and, but somewhat weaker, over sea is positively correlated with the NAO. Winters with strong positive NAO anomalies show distinct peaks in precipitation and amounts up to twice the average over the North Sea region (Andersson et al. 2010). On longer time scales, drought conditions over central Europe are also connected to the AMO (Atlantic Multidecadal Oscillation; Ionita et al. 2012a). Generally speaking, precipitation is more variable than temperature, and agreement between datasets is less. Nevertheless, there are indications of an increase in precipitation to the north of the North Sea region and a decrease to the south, in agreement with the projected north-eastward shift in the storm tracks. In many regions, there are also indications that extreme precipitation events have become more extreme and that return periods have decreased.

2.6 Radiative Properties

Meteorological observations aboard ships usually do not include measurements of sunshine duration and radiation. As a result there are few data available, and these are mainly from isolated field campaigns on research vessels. In contrast, cloud parameters are often observed routinely, although the quality of observations varies widely. The following discussion of clouds, solar radiation and sunshine duration therefore relies mainly on studies concerning a wider area, but these data should also be valid for the North Sea region.

2.6.1 Clouds

Clouds have a significant impact on the Earth's radiation budget. They affect incoming solar shortwave (SW) radiation (by reflecting this back to space) as well as outgoing thermal longwave (LW) radiation (by reducing its emission to space). The difference between the actual radiative flux and that under clear sky conditions is referred to as cloud radiative forcing (CRF). The largest contribution to LW CRF is made by high clouds, whereas the largest contribution to SW CRF is from optically thick clouds due to their higher albedo compared to the clear sky surface albedo. Thus, variations in cloudiness are of great interest in relation to rising global temperatures. The Extended Edited Cloud Report Archive (EECRA; Warren et al. 1986, 1988, 2006) consists of quality controlled climatologies of total cloud cover and cloud type amounts over land and ocean, respectively, based on surface synoptic cloud observations.

Few analyses of changes in cloud cover exist, and even less for the North Sea region. A decrease has been observed in global high cloud cover over almost all land regions since 1971 and most ocean regions since 1952. Norris (2008) analysed global mean time series from gridded surface observations of low-, mid- and upper-level clouds as well as total cloud cover and satellite cloud observations over land and ocean based on EECRA, other surface synoptic cloud reports from land since 1971, the ship-based ICOADS which includes observations since 1952, and satellite observations available from July 1983 in the International Satellite Cloud Climatology Project (ISCCP). Norris (2008) found inconsistencies for the overlapping period of in situ and satellite data except for high clouds.

Over Europe, variability in total winter cloud cover is strongly connected to the NAO. Because the NAO was undergoing a positive trend during a study by Warren et al. (2006), there is a strong positive trend in total cloud cover over Norway at this time (Fig. 2.18). No clear trend is visible further south in the North Sea region (Thompson et al. 2000; Hense and Glowienka-Hense 2008).

Warren et al. (2006) also analysed cloud types observed at European land stations in relation to the NAO/AO signal in winter for the period 1971–1996. Not surprisingly, the strongest correlation was for nimbostratus (Fig. 2.19). Across the western parts of Europe correlations are negative, but high positive correlations exist in the northern part of the North Sea region from northern Scotland to Norway.

2.6.2 Solar Radiation

Time series of measured solar radiation data at various sites around the globe show decreasing irradiances on the order of 6–9 W m^{-2} (corresponding to a decline of 4–6 % over 30 years) after the mid-1950s ('global dimming'; Gilgen et al. 1998; Stanhill and Cohen 2001; Liepert and Tegen 2002) and mainly over land, and subsequent increases since the mid-1980s ('global brightening'; Wild et al. 2005; Norris and Wild 2007) which cannot be explained by variations in solar irradiance or cloudiness alone (Wild 2009), but are largely due to marked changes in the amount of anthropogenic aerosol particles after the Second World War (Stanhill and Cohen 2001; Liepert and Tegen 2002; Streets et al. 2006; Norris and Wild 2007). Since the 1980s, air-quality regulations have led to a decline in air pollution, as can be seen from time series of optical depth (Mishchenko et al. 2007; Ruckstuhl et al. 2008). More recent studies (e.g. Granier et al. 2011; Lee et al. 2013; Myhre et al. 2013; Shindell et al. 2013) corroborate these findings.

For Europe, Norris and Wild (2007) examined changes in SW downward radiation and total cloud cover to distinguish the effects of cloud variability from long-term aerosol influences in the period 1971–2002 (Fig. 2.20). Their 'cloud cover radiative effect' (CCRE), defined as the radiative effects of changes in cloud cover, is derived from daytime synoptic surface observations and ISCCP data, subtracted from the downward radiation obtained from the Global

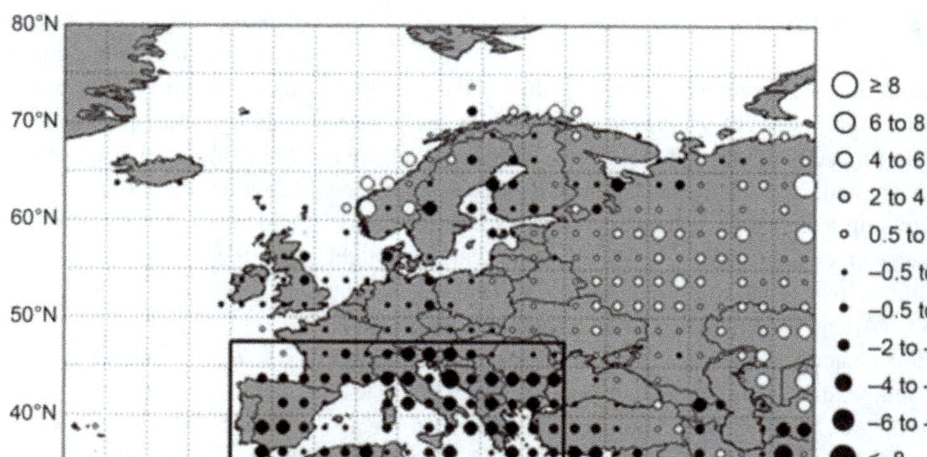

Fig. 2.18 Linear trends in total cloud cover in percent per decade for 2.5° × 2.5° boxes in Europe and North Africa in winter (DJF) for the period 1971–1996. The size of each *dot* indicates the magnitude of the trend (Warren et al. 2006)

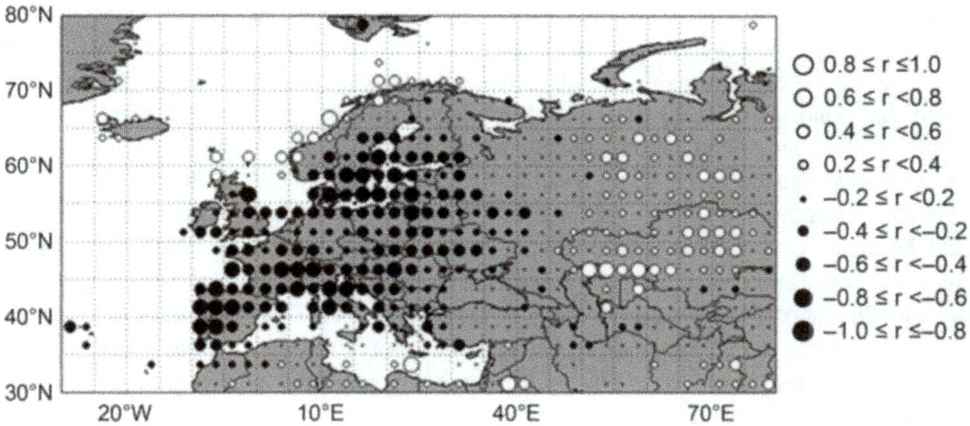

Fig. 2.19 Correlation of nimbostratus anomalies with the Arctic Oscillation index, for 2.5° × 2.5° boxes in Europe and North Africa in winter (DJF) for the period 1971–1996. The size of each *dot* indicates the magnitude of the correlation coefficient (Warren et al. 2006)

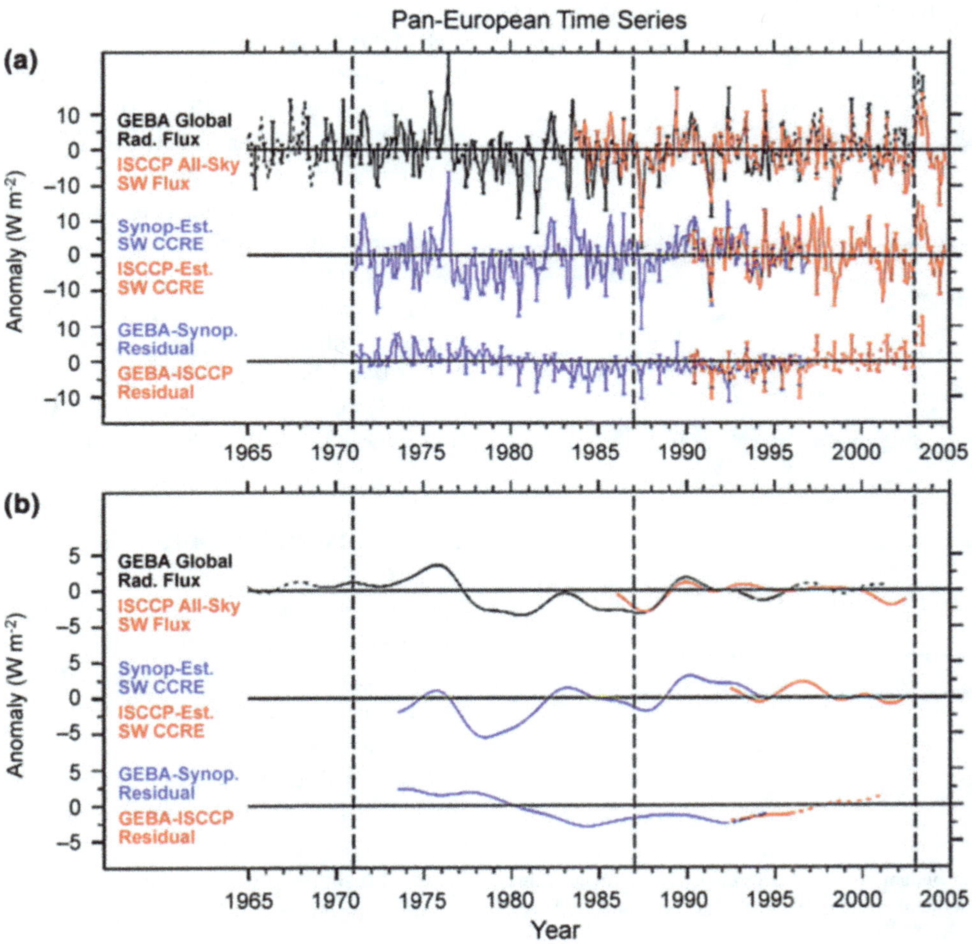

Fig. 2.20 Time series of monthly anomalies averaged over grid boxes covering most of Europe with **a** a 1-2-1 filter and **b** a 61-point 5-year Lanczos low-pass filter for GEBA global radiation flux (*upper, black*), ISCCP all-sky downward SW radiation flux (*upper, red*), SW cloud cover radiation effect (CCRE) estimated from synoptic reports of total cloud cover (*middle, blue*), SW CCRE estimated from ISCCP total cloud cover amount (*middle, red*), residual anomalies after removing synoptic-estimated SW CCRE from GEBA global radiation (*lower, blue*) and residual anomalies after removing ISCCP-estimated SW CCRE from GEBA global radiation (*lower, red*). *Dashed values* indicate where less than 75 % of the grid boxes contributed to the GEBA time series. *Small vertical bars* denote 95 % confidence intervals for June and December anomalies, and *vertical dashed lines* mark the start and end times for trend calculations (Norris and Wild 2007)

Energy Balance Archive (GEBA). The resulting time series comprises variations in clear-sky solar flux as well as radiative effects of changes in cloud albedo that are not linearly correlated to the cloud cover. They found a high correlation (r = 0.88) between global radiation anomalies and the estimated cloud cover radiative effect on monthly and sub-decadal timescales, but the time series of differences show dimming and brightening as well as low frequency trends with minima related to the volcanic eruptions of El Chichón (Mexico) and Pinatubo (Philippines). Decreasing trends for the period 1971–1986 and then increasing trends for 1987–2002 are found for coastal areas of the North Sea region (not shown), but these are mostly not statistically significant (Ruckstuhl et al. 2008, Ruckstuhl and Norris 2009).

Aerosol particles influence the radiation budget and hence air temperature in two ways. The direct aerosol radiative effect refers to clear-sky cases, when solar radiation is directly scattered (mainly by sulphate) or absorbed (mainly by black carbon), while the indirect aerosol effect enhances cloud albedo via an increase in the number of aerosol particles that act as condensation nuclei creating smaller droplets and prolonging cloud lifetime due to a decrease in droplet size and less precipitation loss.

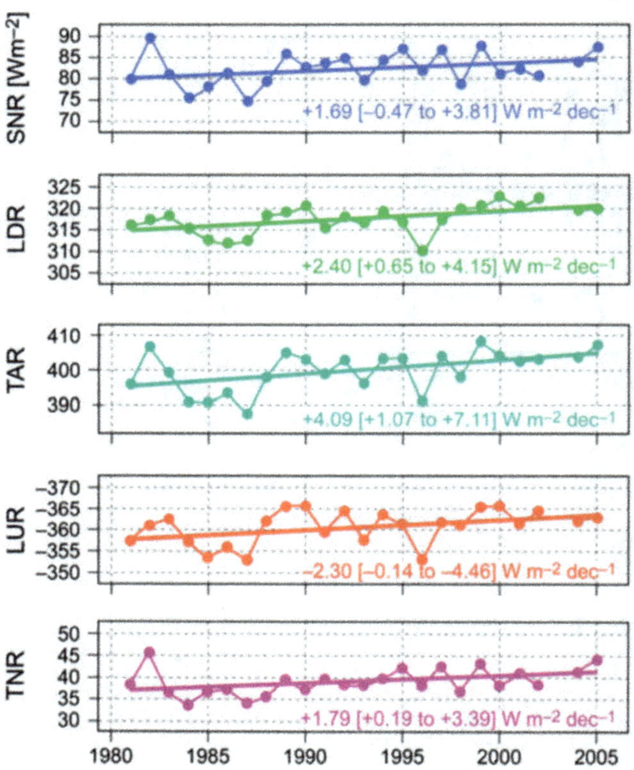

Fig. 2.21 Radiation budget and surface forcing. Annual mean values (W m^{-2}) for the individual components of the surface radiation budget for eight stations in northern Germany: SW net radiation (SNR), LW downward radiation (LDR), total absorbed radiation (TAR), LW upward radiation (LUR) and total net radiation (TNR) from 1981 to 2005 (missing 2003 data). Downward fluxes are positive and upward fluxes negative. Trends in W m^{-2} $decade^{-1}$ with the 95 % confidence interval in brackets. TNR, the balance between downward and upward fluxes at the surface representing the energy available for sensible and latent energy fluxes increases primarily due to the increase of water vapour in the atmosphere (Philipona et al. 2009)

For the period 1981–2005, Ruckstuhl et al. (2008) estimated the direct and indirect aerosol effects by determining the SW downward radiation for cloud-free and cloudy conditions from eight sites in northern Germany. Excluding the sunny year 2003, the net LW forcing under cloud-free skies is 0.84 W m^{-2} $decade^{-1}$ (range: 0.49–1.20), whereas the SW net forcing from changes in cloudiness is 0.56 W m^{-2} $decade^{-1}$ (range: −0.91 to 2.00), resulting in a total cloud forcing of 0.16 W m^{-2} $decade^{-1}$ (range: −0.26 to 0.57). Thus, the direct aerosol effect has a much larger impact on climate forcing than the indirect aerosol and other cloud effects.

Philipona et al. (2009) found an increase in LW downward radiation over Germany, based on observations for 1981–2005, due to rising temperature and humidity and to the increase in greenhouse gas concentration, but found no effect of changes in cloudiness (Fig. 2.21). The total net LW radiation (the difference between incoming and outgoing radiation at the surface) depends on temperature and absolute humidity, which itself is dependent on temperature.

For northern Germany, the LW forcing due to greenhouse gases including water vapour resulted in 0.95 W m^{-2} $decade^{-1}$ (range: 0.26–1.64), while the part due to water vapour feedback alone is 0.60 W m^{-2} $decade^{-1}$ (range: 0.16–1.04) (Philipona et al. 2009). Thus, the total SW forcing is three times larger than the LW forcing from rising atmospheric levels of anthropogenic greenhouse gases.

2.6.3 Sunshine Duration

Operational measurements of sunshine duration started at most weather stations in the 1930s or 1940s, mainly using the Campbell-Stokes heliograph. However, over recent decades, new electronic–optical equipment has increasingly been used, causing data quality issues and thus consistency problems within the time series of sunshine duration data (Augter 2013). As the existing sunshine duration database for the open sea is insufficient for climate analyses, the results presented in this chapter are based on coastal or island stations only. Sunshine duration depends on three factors: daylength (which is a function of latitude and season), amount of daytime clouds, and atmospheric opacity. Cloudiness and opacity are influenced by meteorological conditions and the latter also by aerosol concentration, which can be very different over land and sea. For Germany, Schönwiese and Janoschitz (2005) analysed changes in sunshine duration for the periods 1951–2000 and 1971–

2000 and found no obvious trends. It is not clear whether the increase in global radiation is related to an increase in sunshine duration.

2.6.4 Summary

From the few available datasets on radiative properties, it may be concluded that there are non-negligible trends together with potential uncertainties and land-sea inhomogeneities which make it difficult to assess these quantities in detail.

2.7 Summary and Open Questions

It is not obvious how atmospheric circulation has changed in the North Sea region over the last roughly 200 years. Further research is therefore necessary to understand climate change versus climate variability. One open research question is the extent to which circulation over the North Sea region is controlled by distant factors. In particular, whether there is a link between changes in the Arctic cryosphere and atmospheric circulation further south, including over the North Sea region. Overland and Wang (2010) highlighted a connection between the recent decrease in Arctic sea ice and cold winters in several areas of Europe. With the ongoing decline in sea ice in the Arctic, any such effect on circulation patterns would be important for climate in the North Sea region. Rahmstorf et al. (2015) proposed a proxy-based connection between the observed cooling in the North Atlantic south of Greenland and a weakening of the Atlantic Meridional Overturning Circulation (AMOC) partly due to increased melting of the Greenland Ice Sheet and subsequent freshening of the surface waters. Changes in the strength of the AMOC, however, are still debated and Zhang (2008) and more recently, Tett et al. (2014) have stated that its strength has actually increased.

Owing to large internal variability, it is unclear which part of the observed atmospheric changes is due to anthropogenic activities and which is internally forced. Slowly varying natural factors with an effect on European climate, such as the AMO (Petoukhov and Semenov 2010), may superimpose long-term trends and therefore be difficult to distinguish from the anthropogenic climate change signal.

There are signs of an increase in the number of deep cyclones (but not in the total number of cyclones). There are also indications that the persistence of circulation types has increased over the last century or so (Della-Marta et al. 2007). It is an open question whether this is also related to the decline in Arctic sea ice.

Another open question is whether there have been changes in extreme weather events. However, most studies rely on small datasets covering relatively short time periods, which makes it is difficult to draw statistically significant conclusions. As short time series and a lack of homogeneous data make it impossible to obtain reliable trend estimates, it is important to make available and homogenise the large number of data from past decades that have not yet been digitised. However, shown by the case of the erroneous pressure digitisations in the WASA dataset (see E-Supplement Sect. S2.3), it is essential for data to be thoroughly quality-checked. Experience from the WASA data suggests that this step requires human expertise and cannot be fully automated. On the other hand, further reanalyses, which may be considered a 'best-possible' time-space interpolator for observed data, can be useful as long as any bias that is potentially introduced through new instruments, station relocations etc. is properly addressed. The same is true for existing reanalyses, as it is unclear how homogeneous reanalyses can be that rely only on surface observations such as 20CR (Compo et al. 2011).

Temperature has increased in the North Sea region, and there is a clear signal in the annual number of frost days or summer days. While there is a clear winter and spring warming signal over the Baltic Sea region (Rutgersson et al. 2014), this is not as clear for the North Sea region. For precipitation, it is difficult to deduce long-term trends; however, there are indications of longer precipitation periods and 'more extreme' extreme events.

Other quantities, such as clouds, radiation or sunshine duration, are difficult to judge owing to a general lack of data.

References

Alexandersson H, Tuomenvirta H, Schmith T, Iden K (2000) Trends of storms in NW Europe derived from an updated pressure data set. Clim Res 14:71–73

Allan R, Tett S, Alexander L (2009) Fluctuations in autumn-winter severe storms over the British Isles: 1920 to present. Int J Climatol 29:357–371

Ambaum MHP, Hoskins BJ, Stephenson DB (2001) Arctic Oscillation or North Atlantic Oscillation? J Clim 14:3495–3507

Andersson A, Bakan S, Graßl H (2010) Satellite derived precipitation and freshwater flux variability and its dependence on the North Atlantic Oscillation. Tellus A 62:453–468

Andersson A, Klepp C, Bakan S, Graßl H, Schulz J (2011) Evaluation of HOAPS-3 ocean freshwater flux components. J Appl Meteor Climatol 50:379–398

Augter G (2013) Vergleich der Referenzmessungen des Deutschen Wetterdienstes mit automatisch gewonnenen Messwerten [Comparison of reference measurements of the German Weather Service with automatically obtained measurements]. Berichte des Deutschen Wetterdienstes 238

Barnes EA, Hartmann DL (2010) Dynamical feedbacks and the persistence of the NAO. J Atmos Sci 67:851–865

Barnston AG, Livezey RE (1987) Classification, seasonality and persistence of low frequency atmospheric circulation patterns. Mon Wea Rev 115:1083–1126

Barredo JI (2010) No upward trend in normalised windstorm losses in Europe: 1970–2008. Nat Hazard Earth Syst Sci 10:97–104

Barriopedro D, Garcìa-Herrera R, Lupo AR, Hernández E (2006) A climatology of northern hemisphere blocking. J Climate 19:1042–1063

Barriopedro D, Fischer EM, Luterbacher J, Trigo RM, Garcia-Herrera R (2011) The hot summer of 2010: Redrawing the temperature record map of Europe. Science 332:220–224

Baur F (1937) Einführung in die Großwetterforschung [Introduction to Grosswetter research; in German]. Verlag BG Teubner, Leipzig

Benedict JJ, Lee S, Feldstein SB (2004) Synoptic view of the North Atlantic Oscillation. J Atmos Sci 61:121–144

Bengtsson L, Hagemann S, Hodges KI (2004) Can climate trends be calculated from reanalysis data? J Geophys Res 109:D11111 doi:10.1029/2004JD004536

Berrisford P, Dee DK, Fuentes M, Kållberg P, Kobayashi S, Uppala S (2009). The ERA-Interim archive. ERA-40 Report Series No. 1. European Centre for Medium-Range Weather Forecasts, UK

Berry DI, Kent EC (2009) A new air-sea interaction gridded dataset from ICOADS with uncertainty estimates. Bull Am Met Soc 90:645–656

Berry DI, Kent EC (2011) Air-sea fluxes from ICOADS: The construction of a new gridded dataset with uncertainty estimates. Int J Climatol 31:987–1001

Berry DI, Kent EC, Taylor PK (2004) An analytical model of heating errors in marine air temperatures from ships. J Atmos Ocean Technol 21:1198–1215

Bett PE, Thornton HE, Clark RT (2013) European wind variability over 140 yr. Adv Sci Res 10:51–58

Blessing S, Fraedrich K, Junge M, Kunz T, Linkheit F (2005) Daily North Atlantic Oscillation (NAO) index: statistics and its stratospheric polar vortex dependence. Meteorol Z 14:763–769

Brönnimann S, Martius O, von Waldow H, Welker C, Luterbacher J, Compo G, Sardeshmukh PD, Usbeck T (2012) Extreme winds at northern mid-latitudes since 1871. Meteorol Z 21:13–27

Brönnimann S, Martius O, Franke J, Stickler A, Auchmann R (2013) Historical weather extremes in the "Twentieth Century Reanalysis". In: Brönnimann S, Martius O (eds) (2013) Weather Extremes During the Past 140 Years. Geographica Bernensia G89:7–17

Budikova D (2009) Role of Arctic sea ice in global atmospheric circulation: A review. Global Plan Change 68:149–163

Budikova D (2012) Northern Hemisphere climate variability: Character, forcing mechanisms, and significance of the North Atlantic/Arctic Oscillation. Geogr Compass 6:401–422

Carter DJ, Draper L (1988) Has the north-east Atlantic become rougher? Nature 332:494

Cassou C (2008) Intraseasonal interaction between the Madden-Julian Oscillation and the North Atlantic Oscillation. Nature 455:523–527

Cattiaux J, Vautard R, Yiou P (2009) Origins of the extremely warm European fall of 2006. Geophys Res Lett 36:L06713 doi:10.1029/2009GL037339

Cattiaux J, Vautard R, Cassou C, Yiou P, Masson-Delmotte V, Codron F (2010) Winter 2010 in Europe: A cold extreme in a warming climate. Geophys Res Lett 37:L20704 doi:10.1029/2010GL044613

Chang EK, Fu Y (2002) Interdecadal variations in Northern Hemisphere winter storm track intensity. J Climate 15:642–658

Cheng X, Wallace JM (1993) Cluster analysis of the Northern Hemisphere wintertime 500 hPa height field: Spatial patterns. J Atmos Sci 50:2674–2696

Chrysanthou A, van der Schrier G, van den Besselaar EJM, Klein Tank AMG, Brandsma T (2014) The effects of urbanization on the rise of the European temperature since 1960. Geophys Res Lett 41:7716–7722 doi:10.1002/2014GL061154

Ciavola P, Ferreira O, Haerens P, van Koningsveld M, Armaroli C, Lequeux Q (2011) Storm impacts along European coastlines. Part 1: The joint effort of the MICORE and ConHaz Projects. Environ Sci Policy 14:912–923

Clemmensen LB, Hansen KWT, Kroon A (2014) Storminess variation at Skagen, northern Denmark since AD 1860: Relations to climate change and implications for coastal dunes. Aeolian Res 15:101–112

Cohen J, Barlow M (2005) The NAO, the AO, and global warming: How closely related? J Climate 18:4498–4513

Compo GP, Whitaker JS, Sardeshmukh PD, Matsui N, Allan RJ, Yin, X, Gleason BE, Vose RS, Rutledge G, Bessemoulin P, Brönnimann S, Brunet M, Crouthamel RI, Grant AN, Groisman PY, Jones PD, Kruk M, Kruger AC, Marshall GJ, Maugeri M, Mok HY, Nordli Ø, Ross TF, Trigo RM, Wang XL, Woodruff SD, Worley SJ (2011) The twentieth century reanalysis project. Q J Roy Met Soc 137:1–28

Cortesi N, Gonzalez-Hidalgo JC, Brunetti M, Martin-Vide J (2012) Daily precipitation concentration across Europe 1971–2010. Nat Hazards Earth Syst Sci 12:2799–2810

Croci-Maspoli M, Schwierz C, Davies H (2007) Atmospheric blocking: Space-time links to the NAO and PNA. Clim Dyn 29:713–725

Cusack S (2012) A 101 year record of windstorms in the Netherlands. Clim Change 116:693–704

Dangendorf S, Müller-Navarra S, Jensen J, Schenk F, Wahl T, Weisse R (2014) North Sea storminess from a novel storm surge record since AD 1843. J Climate 27:3582–3595

de Kraker A (1999) A method to assess the impact of high tides, storms and storm surges as vital elements in climatic history - the case of stormy weather and dikes in the northern part of Flanders, 1488 to 1609. Clim Change 43:287–302

de Viron O, Dickey JO, Ghil M (2013) Global models of climate variability. Geophys Res Lett 40:1832–1837

Dee DP, Uppala SM, Simmons AJ, Berrisford P, Poli P, Kobayashi S, Andrae U, Balmaseda MA, Balsamo G, Bauer P, Bechtold P, Beljaars ACM, van de Berg L, Bidlot J, Bormann N, Delsol C, Dragani R, Fuentes M, Geer AJ, Haimberger L, Healy SB, Hersbach H, Hólm EV, Isaksen L, Kållberg P, Köhler M, Matricardi M, McNally AP, Monge-Sanz BM, Morcrette JJ, Park BK, Peubey C, de Rosnay,Tavolato C, Thépaut JN, Vitart F (2011) The ERA-Interim reanalysis: configuration and performance of the data assimilation system. Quart J Roy Met Soc 137:553–597

Della-Marta PM, Haylock MR, Luterbacher J, Wanner H (2007) Doubled length of western European summer heat waves since 1880. J Geophys Res 112:D15103 doi:10.1029/2007JD008510

Delworth TL, Knutson TR (2000) Simulation of early 20th century global warming. Science 287:2246–2250

Donat MG, Renggli D, Wild S, Alexander LV, Leckebusch GC, Ulbrich U (2011) Reanalysis suggests long-term upward trends in European storminess since 1871. Geophys Res Lett 38:L14703 doi:10.1029/2011GL047995

Donat MG, Alexander LV, Yang H, Durre I, Vose R, Dunn RJH, Willett KM, Aguilar E, Brunet M, Caesar J, Hewitson B, Jack C, Klein Tank AMG, Kruger AC, Marengo J, Peterson TC, Renom M,

Oria Rojas C, Rusticucci M, Salinger J, Elrayah AS, Sekele SS, Srivastava AK, Trewin B, Villarroel C, Vincent LA, Zhai P, Zhang X, Kitching S (2013) Updated analyses of temperature and precipitation extreme indices since the beginning of the twentieth century: The HadEX2 dataset. J Geophys Res 118:2098–2118

Esteves LS, Williams JJ, Brown JM (2011) Looking for evidence of climate change impacts in the eastern Irish Sea. Nat Hazards and Earth System Sci 11:1641–1656

Feldstein SB (2000) Teleconnection and ENSO: The timescale, power spectra, and climate noise properties. J Climate 13:4430–4440

Feldstein SB (2002) The recent trend and variance increase of the annular mode. J Climate 15:88–94

Fennig K, Andersson A, Bakan S, Klepp C, Schroeder M (2012) Hamburg Ocean Atmosphere Parameters and Fluxes from Satellite Data - HOAPS 3.2 - Monthly means / 6-hourly composites. Satellite Application Facility on Climate Monitoring. doi:10.5676/EUM_SAF_CM/HOAPS/V001

Feser F, Barcikowska M, Krueger O, Schenk F, Weisse R, Xia L (2015a) Storminess over the North Atlantic and Northwestern Europe – A Review. Q J Roy Met Soc. 141:350–382

Feser F, Barcikowska M, Haeseler S, Lefebvre C, Schubert-Frisius M, Stendel M, von Storch H, Zahn M (2015b) Hurricane Gonzalo and its extratropical transition to a strong European storm. In: Explaining Extreme Events of 2014 from a Climate Perspective. Bull Amer Met Soc 96:S51–S55

Fischer EM, Luterbacher J, Zorita E, Tett SFB, Casty C, Wanner H (2007) European climate response to tropical volcanic eruptions over the last half millennium. Geophys Res Lett 34:L05707 doi:10.1029/2006GL027992

Folland CK, Knight J, Linderholm HW, Fereday D, Ineson S, Hurrell JW (2009) The summer North Atlantic Oscillation: Past, present, and future. J Climate 22:1082–1103

Fowler H, Kilsby C (2003) A regional frequency analysis of United Kingdom extreme rainfall from 1961 to 2000. Int J Climatol 23:1313–1334

Franke R (2009) Die nordatlantischen Orkantiefs seit 1956. Naturwissenschaftliche Rundschau 62:349–356

Franzke C, Woollings T (2011) On the persistence and predictability properties of North Atlantic climate variability. J Climate 24:466–472

Franzke C, Lee S, Franzke SB (2004) Is the North Atlantic Oscillation a breaking wave? J Atmos Sci 61:145–160

Frydendahl K, Frich P, Hansen C (1992) Danish weather observations 1685–1715. Danish Met Inst Tech Rep 92-3

Furevik T, Nilsen JEØ (2005) Large-scale atmospheric circulation variability and its impacts on the Nordic Seas ocean climate - a review. In Drange H, Dokken T, Furevik T, Gerdes R, Berger W (eds.) The Nordic Seas: An Integrated Perspective. Geophys Mon Ser 158.

Geng Q, Sugi M (2001) Variability of the North Atlantic cyclone activity in winter analyzed from NCEP-NCAR reanalysis data. J Climate 14:3863–3873

Geyer B (2014) High-resolution atmospheric reconstruction for Europe 1948–2012: coastDat2. Earth Syst Sci Data 6:147–164

Gilgen H, Wild M, Ohmura A (1998) Means and trends of shortwave irradiance at the surface estimated from Global Energy Balance Archive data. J Climate 11:2042–2061

Gillett NP (2005) Climate modelling: Northern Hemisphere circulation. Nature 437:496

Gillett NP, Zwiers FW, Weaver AJ, Stott PA (2003) Detection of human influence on sea-level pressure. Nature 422:292–294

Gönnert G (2003) Sturmfluten und Windstau in der Deutschen Bucht, Charakter, Veränderungen und Maximalwerte im 20. Jahrhundert. Die Küste 67

Granier C, Bessagnet B, Bond T, d'Angiola A, van der Gon HD, Frost GJ, Heil A, Kaiser JW, Kinne S, Klimont Z, Kloster S, Lamarque J-F, Liousse C, Masui T, Meleux F, Mieville A, Ohara T, Raut J-C, Riahi K, Schultz MG, Smith SJ, Thompson A, van Aardenne J, van der Werf GR, van Vuuren DP (2011) Evolution of anthropogenic and biomass burning emissions of air pollutants at global and regional scales during the 1980–2010 period. Clim Change 109:163–190

Groisman P, Knight R, Easterling D, Karl T, Hegerl G, Razuvaev V (2005) Trends in intense precipitation in the climate record. J Climate 18:1326–1350

Gulev SK, Zolina O, Grigoriev S (2001) Extratropical cyclone variability in the Northern Hemisphere winter from the NCEP/NCAR reanalysis data. Clim Dyn 17:795–809

Häkkinen S, Rhines PB, Worthen DL (2011) Atmospheric blocking and Atlantic multidecadal ocean variability. Science 334:655–659

Hammond JM (1990). Storm in a teacup or winds of change? Weather 45:443–448

Hanna E, Cappelen J, Allan R, Jónsson T, LeBlancq F, Lillington T, Hickey K (2008) New insights into North European and North Atlantic surface pressure variability, storminess, and related climatic change since 1830. J Climate 21:6739–6766

Hannachi A (2007a): Pattern hunting in climate: a new method for trends in gridded climate data. Int J Climatol 27:1–15

Hannachi A (2007b) Tropospheric planetary wave dynamics and mixture modeling: Two preferred regimes and a regime shift. J Atmos Sci 64:3521–3541

Hannachi A (2008) A new set of orthogonal patterns in weather and climate: Optimally interpolated patterns. J Climate 21:6724–6738

Hannachi A (2010) On the origin of planetary-scale extratropical circulation regimes. J Atmos Sci 67:1382–1401

Hannachi A, Woollings T Fraedrich K (2012) The North Atlantic jet stream: a look at preferred positions, paths and transitions. Q J Roy Met Soc 138:862–877

Hansen, J, Ruedy R, Sato M, Lo K (2010) Global surface temperature change. Rev Geophys 48:RG4004 doi:10.1029/2010RG000345

Harnik N, Chang EK (2003) Storm track variations as seen in radiosonde observations and reanalysis data. J Climate 16:480–495

Harris I, Jones PD, Osborn TJ, Lister DH (2014) Updated high-resolution grids of monthly climatic observations - the CRU TS 3.10 dataset. Int J Climatol 34:623–642

Hartmann DL, Klein Tank AMG, Rusticucci M, Alexander LV, Brönnimann S, Charabi Y, Dentener FJ, Dlugokencky EJ, Easterling DR, Kaplan A, Soden BJ, Thorne PW, Wild M, Zhai PM (2013) Observations: atmosphere and surface. In: Climate Change 2013: The Physical Science Basis. Contribution of Working Group I to the Fifth Assessment Report of the Intergovernmental Panel on Climate Change. Stocker TF, Qin D, Plattner G-K, Tignor M, Allen SK, Boschung J, Nauels A, Xia Y, Bex V, Midgley PM (eds) Cambridge University Press

Haylock MR, Hofstra N, Klein Tank AMG, Klok EJ, Jones PD, New M (2008) A European daily high-resolution gridded dataset of surface temperature and precipitation for 1950–2006. J Geophys Res 113: D20119 doi:10.1029/2008JD10201

Held IM, Soden BJ (2006) Robust responses of the hydrological cycle to global warming. J Climate 19:5686–5699

Hense A, Glowienka-Hense R (2008) Auswirkungen der Nordatlantischen Oszillation. [Climate impact of the North Atlantic Oscillation; in German]. Promet 34:89–94

Hess P, Brezowsky H (1952) Katalog der Großwetterlagen Europas [Catalogue of European Grosswetterlagen; in German]. Ber Dt Wetterd in der US-Zone 33:39

Hess P, Brezowsky H (1977) Katalog der Großwetterlagen Europas 1881–1976 (3. verb. u. erg. Aufl.), Ber. Dt. Wetterd. 15(113)

Selbstverlag des Deutschen Wetterdienstes, Offenbach am Main, Germany
Hickey KR (2003) The storminess record from Armagh Observatory, Northern Ireland, 1796–1999. Weather 58:28–35
Hines KM, Bromwich DH, Marshall G J (2000) Artificial surface pressure trends in the NCEP–NCAR Reanalysis over the Southern Ocean and Antarctica. J Climate 13:3940–3952
Hoerling MP, Hurrell JW, Xu T (2001) Tropical origins for recent North Atlantic climate change. Science 292:90–92
Hogben N (1994) Increases in wave heights over the North Atlantic: A review of the evidence and some implications for the naval architect. Trans R Inst Naval Arch 5:93–101
Hossen MA, Akhter F (2015) Study of the wind speed, rainfall and storm surges for the Scheldt Estuary in Belgium. Int J Sci Tech Res 4:130–134
Hoy A, Sepp M, Matschullat J (2012) Atmospheric circulation variability in Europe and northern Asia (1901 to 2010). Theor Appl Climatol 113:105–126
Hurrell JW (1995) Decadal trends in the North Atlantic Oscillation, regional temperatures and precipitation. Science 269:676–679
Hurrell JW, Deser C (2009) North Atlantic climate variability: The role of the North Atlantic Oscillation. J Mar Sys 78:28–41
Hurrell JW, Kushnir Y, Ottersen G, Visbeck M (2003) An overview of the North Atlantic Oscillation. In: The North Atlantic Oscillation: Climatic significance and environmental impact. Geoph Monog Series 134:1–36
Ineson S, Scaife AA, Knight JR, Manners JC, Dunstone NJ, Gray LJ, Haigh JD (2011) Solar forcing of winter climate variability in the northern hemisphere. Nature Geosci 4:753–757
Ionita M, Lohmann G, Rimbu N, Chelcea S, Dima M (2012a) Interannual to decadal summer drought variability over Europe and its relationship to global sea surface temperature. Clim Dyn 38:363–377
Ionita M, Lohmann G, Rimbu N, Scholz P (2012b) Dominant modes of diurnal temperature range variability over Europe and their relationship with large-scale atmospheric circulation and sea surface temperature anomaly patterns. J Geophys Res 117:D15111, doi:10.1029/2011JD016669
Johannessen OM, Bengtsson L, Miles MW, Kuzmina SI, Semenov VA, Alekseev GV, Nagurnyi AP, Zakharov VF, Bobylev LP, Pettersson LH, Hasselmann K, Cattle HP (2004) Arctic climate change: Observed and modelled temperature and sea-ice variability. Tellus A 56:328–341
Jones PD, Moberg A (2003) Hemispheric and large-scale surface air temperature variations: An extensive revision and an update to 2001. J Climate 16:206–223
Jones PD, Jonsson T, Wheeler D (1997) Extension to the North Atlantic Oscillation using early instrumental pressure observations from Gibraltar and south-west Iceland. Int J Climatol 17:1433–1450
Jones, PD, Lister DH, Osborn TJ, Harpham C, Salmon M, Morice CP (2012) Hemispheric and large-scale land surface air temperature variations: An extensive revision and an update to 2010. J Geophys Res 117:D05127 doi:10.1029/2011JD017139
Joshi MM, Charlton AJ, Scaife AA (2006) On the influence of stratospheric water vapor changes on the tropospheric circulation. Geophys Res Lett 33:L09806. doi:10.1029/2006GL025983
Jung T, Vitart F, Ferranti L, Morcrette JJ (2011) Origin and predictability of the extreme negative NAO winter of 2009/10. Geophys Res Lett 38:L07701 doi:10.1029/2011GL046786
Kalnay E, Kanamitsu M, Kistler R, Collins W, Deaven D, Gandin L, Iredell M, Saha S, White G, Woollen J, Zhu Y, Leetmaa A, Reynolds R, Chelliah M, Ebisuzaki W, Higgins W, Janowiak J, Mo KC, Ropelewski C, Wang J, Jenne R, Joseph D (1996) The NCEP/NCAR 40-Year Reanalysis Project. Bull Am Met Soc 77:437–471
Keeley SPE, Sutton RT, Shaffrey, LC (2009) Does the North Atlantic Oscillation show unusual persistence on intraseasonal timescales? Geophys Res Lett 36:L22706. doi:10.1029/2009GL040367
Kent, EC, Woodruff SD, Berry DI (2007) Metadata from WMO Publication No. 47 and an Assessment of Voluntary Observing Ships Observation Heights in ICOADS. J Atmos Ocean Tech 24:214–234
Kent EC, Rayner NA, Berry DI, Saunby M, Moat BI, Kennedy JJ, Parker DE (2013) Global analysis of night marine air temperature and its uncertainty since 1880, the HadNMAT2 Dataset. J Geophys Res 118:1281–1298
Kistler R, Collins W, Saha S, White G, Woollen J, Kalnay E, Chelliah M, Ebisuzaki W, Kanamitsu M, Kousky V, van den Dool H, Jenne R, Fiorino M (2001) The NCEP-NCAR 50-Year Reanalysis: Monthly Means CD-ROM and Documentation. Bull Am Met Soc 82:247–267
Klein Tank AMG, Wijngaard JB, Können GP, Böhm R, Demarée G, Gocheva A, Mileta M, Pashiardis S, Hejkrlik L, Kern-Hansen C, Heino R, Bessemoulin P, Müller-Westermeier G, Tzanakou M, Szalai S, Pálsdóttir T, Fitzgerald D, Rubin S, Capaldo M, Maugeri M, Leitass A, Bukantis A, Aberfeld R, van Engelen AFV, Forland E, Mietus M, Coelho F, MaresC, Razuvaev V, Nieplova E, Cegnar T, Antonio López J, Dahlström B, Moberg A, Kirchhofer W, Ceylan A, Pachaliuk O., Alexander LV, Petrovic P (2002) Daily dataset of 20th-century surface air temperature and precipitation series for the European Climate Assessment. Int J Climatol 22:1441–1453
Knight JR, Folland CK, Scaife AA (2005a) Climate impacts of the Atlantic Multidecadal Oscillation. Geophys Res Lett 33:L17706. doi:10.1029/2006GL026242
Knight JR, Allan RJ, Folland CK, Vellinga M, Mann ME (2005b) A signature of persistent natural thermohaline circulation cycles in observed climate. Geophys Res Lett 32:L20708. doi:10.1029/2005GL024233
Kravtsov S, Robertson AW, Ghil M (2006) Multiple regimes and low frequency oscillations in the Northern Hemisphere's zonal-mean flow. J Atmos Sci 63:840–860
Krueger O, von Storch H (2011) Evaluation of an air pressure based proxy for storm activity. J Climate 24:2612–2619
Krueger O, von Storch H (2012) The informational value of pressure-based single-station proxies for storm activity. J Atmos Ocean Technol 29:569–580
Krueger O, Schenk F, Feser F, Weisse R (2013) Inconsistencies between long-term trends in storminess derived from the 20CR reanalysis and observations. J Climate 26:868–874
Kucharski F, Molteni F, Bracco A (2006) Decadal interactions between the western tropical Pacific and the North Atlantic Oscillation. Clim Dyn 26:79–91
Küttel M, Xoplaki E, Gallego D, Luterbacher J, García-Herrera R, Allan R, Barriendos M, Jones PD, Wheeler D, Wanner H (2009) The importance of ship log data: reconstructing North Atlantic, European and Mediterranean sea level pressure fields back to 1750. Clim Dyn 34:1115–1128
Lavers DA, Villarini G, Allan RP, Wood EF, Wade AJ (2012) The detection of atmospheric rivers in atmospheric reanalyses and their links to British winter floods and the large-scale climatic circulation. J Geophys Res 117:D20106. doi:10.1029/2012JD018027
Leander R, Buishand TA, Klein Tank AMG (2014) An alternative index for the contribution of precipitation on very wet days to the total precipitation. J Climate 27:1365–1378

Lee YH, Lamarque J-F, Flanner MG, Jiao C, Shindell DT, Berntsen T, Bisiaux MM, Cao J, Collins WJ, Curran M, Edwards R, Faluvegi G, Ghan S, Horowitz LW, McConnell JR, Ming J, Myhre G, Nagashima T, Naik V, Rumbold SDT, Skeie RB, Sudo K, Takemura T, Thevenon F, Xu B, Yoon J-H (2013) Evaluation of preindustrial to present-day black carbon and its albedo forcing from Atmospheric Chemistry and Climate Model Intercomparison Project (ACCMIP). Atmos Chem Phys 13:2607–2634

Lehmann A, Getzlaff K, Harlaß J (2011) Detailed assessment of climate variability in the Baltic Sea area for the period 1958 to 2009. Clim Res 46:185–196

Lenderink G, van Meijgaard E, Selten F (2009) Intense coastal rainfall in the Netherlands in response to high sea surface temperatures: analysis of the event of August 2006 from the perspective of a changing climate. Clim Dyn 32:19–33

Liepert B, Tegen I (2002) Multidecadal solar radiation trends in the United States and Germany and direct tropospheric aerosol forcing. J Geophys Res 107:L02806. doi:10.1029/2001JD000760

Lindenberg J, Mengelkamp HT, Rosenhagen G (2012) Representativity of near surface wind measurements from coastal stations at the German Bight. Meteorol Z 21:99–106

Lorenz E (1951) Seasonal and irregular variations of the Northern Hemisphere sea-level pressure profile. J. Meteor 8:52–59

Lorenz C, Kunstmann H (2012) The hydrological cycle in three state-of-the-art reanalyses: Intercomparison and performance analysis. J Hydromet 13:1397–1420

Luo D, Cha J, Feldstein SB (2012) Weather regime transitions and the interannual variability of the North Atlantic Oscillation. Part I: A likely connection. J Atmos Sci 69:2329–2346

Luterbacher J, Dietrich D, Xoplaki E, Grosjean M, Wanner H (2004) European seasonal and annual temperature variability, trends, and extremes since 1500. Science 303:1499–1503

Luterbacher J, Liniger, MA, Menzel A, Estrella N, Della-Marta PM, Pfister C, Rutishauser T, Xoplaki E (2007) Exceptional European warmth of autumn 2006 and winter 2007: Historical context, the underlying dynamics, and its phenological impacts. Geophys Res Lett 34:L12704. doi:10.1029/2007GL029951

Marshall AG, Scaife AA (2009) Impact of the QBO on surface winter climate. J Geophys Res 114:D18110. doi:10.1029/2009JD011737

Marshall J, Johnson H, Goodman J (2001) A study of the interaction of the North Atlantic Oscillation with the ocean circulation. J Climate 14:1399–1421

Matulla C, Schöner W, Alexandersson H, von Storch H, Wang X (2007) European storminess: late nineteenth century to present. Clim Dyn 31:125–130

McCabe GJ, Clark MP, Serreze MC (2001) Trends in Northern Hemisphere surface cyclone frequency and intensity. J Climate 14:2763–2768

Mishchenko MI, Geogdzhayev IV, Rossow WB, Cairns B, Carlson BE, Lacis AA, Liu L, Travis LD (2007) Long-term satellite record reveals likely recent aerosol trend. Science 315:1543

Moberg A, Jones PD, Lister D, Walther A, Brunet M, Jacobeit J, Alexander LV, Della-Marta PM, Luterbacher J, Yiou P, Chen D, Klein Tank AMG, Saladié O, Sigró J, Aguilar E, Alexandersson H, Almarza C,Auer I, Barriendos M, Begert M, Bergström H, Böhm R, Butler C. J. Caesar J, Drebs A, Founda D, Gerstengarbe F-W, Micela G, Tolasz R, Tuomenvirta H, Werner PC, Linderholm,H, Philipp A, Wanner H, Xoplaki E (2006) Indices for daily temperature and precipitation extremes in Europe analyzed for the period 1901–2000. J Geophys Res 111:2156–2202

Myhre G, Samset BH, Schulz M, Balkanski Y, Bauer S, Berntsen TK, Bian H, Bellouin N, Chin M, Diehl T, Easter RC, Feichter J, Ghan SJ, Hauglustaine D, Iversen T, Kinne S, Kirkevåg A, Lamarque J-F, Lin G, Liu X, Lund MT, Luo G, Ma X, van Noije T, Penner JE, Rasch PJ, Ruiz A, Seland Ø, Skeie RB, Stier P, Takemura T, Tsigaridis K, Wang P, Wang Z, Xu L, Yu H, Yu F, Yoon J-H, Zhang K, Zhang H, Zhou C (2013) Radiative forcing of the direct aerosol effect from AeroCom Phase II simulations. Atmos Chem Phys 13:1853–1877

Norris JR (2008) Observed interdecadal changes in cloudiness: real or spurious. In: Brönnimann S, Luterbacher J, Ewen T, Diaz HF, Stolarski RS, Neu U (eds) Climate Variability and Extremes During the Past 100 Years. Springer

Norris JR, Wild M (2007) Trends in aerosol radiative effects over Europe inferred from observed cloud cover, solar "dimming" and solar "brightening". J Geophys Res 112:D08214 doi:10.1029/2006JD007794

O'Gorman PA, Schneider T (2009) The physical basis for increases in precipitation extremes in simulations of 21st century climate change. Proc Natl Acad Sci USA 106:14773–14777

Overland JE, Wang M (2010) Large-scale atmospheric circulation changes are associated with the recent loss of Arctic sea ice. Tellus A 62:1–9

Pall P, Stone DA, Stott PA, Nozawa T, Hilberts AGJ, Lohmann D, Allen MR (2011) Anthropogenic greenhouse gas contribution to flood risk in England and Wales in autumn 2000. Nature 470:382–385

Peterson TC, Vose RS (1997) An overview of the global historical climatology network temperature data base. Bull Am Met Soc 78:2837–2849

Petoukhov V, Semenov VA (2010) A link between reduced Barents-Kara sea ice and cold winter extremes over northern continents. J Geophys Res 115:D21111 doi:10.1029/2009JD013568

Philipona R, Behrens K, Ruckstuhl C (2009) How declining aerosols and rising greenhouse gases forced rapid warming in Europe since the 1980s. Geophys Res Lett 36:L02806 doi:10.1029/2008GL036350

Pinto JG, Raible CC (2012) Past and recent changes in the North Atlantic oscillation. Clim Change 3:79–90

Portis DH, Walsh JE, El-Hamly M, Lamb PJ (2001) Seasonality of the North Atlantic Oscillation. J Climate 14:2069-2078

Rahmstorf S, Box JE, Feulner G, Mann ME, Robinson A, Rutherford S, Schaffernicht EJ (2015) Exceptional twentieth-century slowdown in Atlantic Ocean overturning circulation. Nature Clim Change 5:475–480

Raible C, Della-Marta PM, Schwierz C, Blender R (2008) Northern Hemisphere extratropical cyclones: a comparison of detection and tracking methods and different reanalyses. Mon Wea Rev 136:880–897

Rayner NA, Parker DE, Horton EB, Folland CK, Alexander LV, Rowell DP, Kent EC, Kaplan A (2003) Global analyses of sea surface temperature, sea ice, and night marine air temperature since the late nineteenth century. J Geophys Res 108:4407 doi:10.1029/2002JD002670

Rennert KJ, Wallace JM (2009) Cross-frequency coupling, skewness, and blocking in the Northern Hemisphere winter circulation. J Climate 22:5650–5666

Rienecker MR, Suarez MJ, Gelaro R, Todling R, Julio Bacmeister EL, Bosilovich MG, Schubert SD, Takacs L, Kim G-K, Bloom S, Chen J, Collins D, Conaty A, da Silva A, Gu W, Joiner J, Koster RD, Lucchesi R, Molod A, Owens T, Pawson S, Pegion P, Redder CR, Reichle R, Robertson FR, Ruddick AG, Sienkiewicz M, Woollen J (2011) MERRA: NASA's modern-era retrospective analysis for research and applications. J. Climate 24:3624–3648

Ripesi O, Ciciulla F, Maimone F, Pelino V (2012) The February 2010 Arctic Oscillation Index and its stratospheric connection. Q J Roy Met Soc 138:1961–1969

Rodwell MJ, Rowell DP, Folland CK (1999) Oceanic forcing of the wintertime North Atlantic Oscillation and European climate. Nature 398:320–323

Rosenhagen G, Schatzmann M, Schrön A (2011) Das Klima der Metropolregion auf Grundlage meteorologischer Messungen und Beobachtungen. In: v Storch H, Claussen M (eds) Klimabericht für die Metropolregion Hamburg

Ruckstuhl C, Norris JR (2009) How do aerosol histories affect solar "dimming" and "brightening" over Europe? IPCC-AR4 models versus observations. J Geophys Res 114:D00D04 doi:10.1029/2008JD011066

Ruckstuhl C, Philipona R, Behrens K, Coen MC, Dürr B, Heimo A, Mätzler C, Nyeki S, Ohmura A, Vuilleumier L, Weller M, Wehrli C, Zelenka A (2008) Aerosol and cloud effects on solar brightening and recent rapid warming. Geophys Res Lett 35:L12708 doi:10.1029/2008GL034228

Rutgersson A, Jaagus J, Schenk F, Stendel M (2014) Observed changes and variability of atmospheric parameters in the Baltic Sea region during the last 200 years. Clim Res 61:177–190

Saha S, Moorthi S, Pan H-L, Wu X, Wang J, Nadiga S, Tripp P, Kistler R, Woollen J, Behringer D, Liu H, Stokes D, Grumbine R, Gayno G, Wang J, Hou Y-T, Chuang H-Y, Juang H-MH, Sela J, Iredell M, Treadon R, Kleist D, van Delst P, Keyser D, Derber J, Ek M, Meng J, Wei H, Yang R, Lord S, van den Dool H, Kumar A, Wang W, Long C, Chelliah M, Xue Y, Huang B, Schemm J-K, Ebisuzaki W, Lin R, Xie P, Chen M, Zhou S, Higgins W, Zou C-Z, Liu Q, Chen Y, Han Y, Cucurull L, Reynolds RW, Rutledge G, Goldberg M (2010) The NCEP climate forecast system reanalysis. Bull Amer Meteor Soc 91:1015–1057

Scaife AA, Knight JR, Vallis G, Folland CK (2005) A stratospheric influence on the winter NAO and north Atlantic surface climate. Geophys Res Lett 32:L18715 doi:10.1029/2005GL023226

Scaife AA, Kucharski F, Folland CK, Kinter J, Brönnimann S, Fereday D, Fischer AM, Grainger S, Jin EK, Kang IS, Knight JR, Kusunoki S, Lau NC, Nath MJ, Nakaegawa T, Pegion P, Schubert S, Sporyshev P, Syktus J, Yoon JH, Zeng N, Zhou T (2009) The CLIVAR C20C project: selected twentieth century climate events. Clim Dyn 33:603–614

Schenk F (2015) The analog-method as statistical upscaling tool for meteorological field reconstructions over Northern Europe since 1850. Dissertation Univ Hamburg

Schenk F, Zorita E (2012) Reconstruction of high resolution atmospheric fields for Northern Europe using analog-upscaling. Clim Past 8:1681–1703

Schiesser HH, Pfister C, Bader J (1997) Winter storms in Switzerland north of the Alps 1864/1865–1993/1994. Theor Appl Climatol 58:1–19

Schmidt H, von Storch H (1993) German Bight storms analysed. Nature 365:791

Schmith T, Kaas E, Li TS (1998) Northeast Atlantic winter storminess 1875-1995 re-analysed. Clim Dyn 14:529–536

Schönwiese CD, Janoschitz R (2005) Klima-Trendatlas Deutschland, 1901-2000. Berichte Inst Atmosph Umwelt Univ Frankfurt 4

Selten FM, Branstator GW, Dijkstra HA, Kliphuis M (2004) Tropical origins for recent and future Northern Hemisphere climate change. Geophys Res Lett 31:L21205 doi:10.1029/2004GL020739

Semenov VA, Latif M, Jungclaus JH, Park W (2008) Is the observed NAO variability during the instrumental record unusual? Geophys Res Lett 35:L11701 doi:10.1029/2008GL033273

Shindell DT, Schmidt GA, Mann ME, Rind D, Waple A (2001) Solar forcing of regional climate change during the Maunder minimum. Science 294:2149–2152

Shindell DT, Lamarque J-F, Schulz M, Flanner M, Jiao C, Chin M, Young PJ, Lee Y.H, Rotstayn L, Mahowald N, Milly G, Faluvegi G, Balkanski Y, Collins WJ, Conley AJ, Dalsoren S, Easter R, Ghan S, Horowitz L, Liu X, Myhre G, Nagashima T, Naik V, Rumbold ST, Skeie R, Sudo K, Szopa S, Takemura T, Voulgarakis A, Yoon J-H, Lo F (2013) Radiative forcing in the ACCMIP historical and future climate simulations. Atmos Chem Phys 13:2939–2974

Sickmoeller M, Blender R, Fraedrich K (2000) Observed winter cyclone tracks in the Northern Hemisphere in re-analysed ECMWF data. Q J Roy Met Soc 126:591–620

Siegismund F, Schrum C (2001) Decadal changes in the wind forcing over the North Sea. Clim Res 18:39–45

Simmons AJ, Willett KM, Jones PD, Thorne PW, Dee DP (2010) Low-frequency variations in surface atmospheric humidity, temperature, and precipitation: Inferences from reanalyses and monthly gridded observational data sets. J Geophys Res 115:D01110 doi:10.1029/2009JD012442

Slonosky VC, Jones PD, Davies TD (2000) Variability of the surface atmospheric circulation over Europe, 1774–1995. Int J Climatol 20:1875–1897

Slonosky VC, Jones PD, Davies TD (2001) Atmospheric circulation and surface temperature in Europe from the 18th century to 1995. Int J Climatol 21:63–75

Smits A, Klein Tank AM, Können GP (2005) Trends in storminess over the Netherlands, 1962–2002. Int J Climatol 25:1331–1344

Spangehl T, Cubasch U, Raible CC, Schimanke S, Korper J, Hofer D (2010) Transient climate simulations from the Maunder minimum to present day: role of the stratosphere. J Geophys Res 115:D00110 doi:10.1029/2009JD012358

Stanhill G, Cohen S (2001) Global dimming: A review of the evidence for a widespread and significant reduction in global radiation with discussion of its probable causes and possible agricultural consequences. Agric for Meteorol 107:255–278

Stephenson DB, Pavan V, Bojariu R (2000) Is the North Atlantic Oscillation a random walk? Int J Climatol 20:1–18

Streets DG, Wu Y, Chin M (2006) Two-decadal aerosol trends as a likely explanation of the global dimming/brightening transition. Geophys Res Lett 33:L15806 doi:10.1029/2006GL026471

Strong C, Magnusdottir G (2011) Dependence of NAO variability on coupling with sea ice. Clim Dyn 36:1681–1689

Stucki P, Brönnimann S, Martius O, Welker C, Imhof M, von Wattenwyl N, Philipp N (2014) A catalogue of high-impact windstorms in Switzerland since 1859. Nat Haz Earth Syst Sci Discuss 2:3821–3862

Sutton RT, Hodson DLR (2005) Atlantic Ocean forcing of North American and European summer climate. Science 309:115–118

Sweeney J (2000) A three-century storm climatology for Dublin 1715–2000. Irish Geogr 33:1–14

Takahashi T, Sutherland SC, Sweeney C, Poisson A, Metzl N, Tilbrook B, Bates N, Wanninkhof R, Feely RA, Sabine C (2002) Global sea-air CO_2 flux based on climatological surface ocean pCO_2, and seasonal biological and temperature effects. Deep Sea Res II 49:1601–1622

Tett SFB, Sherwin TJ, Shravat A, Browne O (2014) How much has the North Atlantic Ocean overturning circulation changed in the last 50 years? J Climate 27:6325–6342

Thompson DWJ, Wallace JM (1998) The Arctic oscillation signature in the wintertime geopotential height and temperature fields. Geophys Res Lett 25:1297–1300

Thompson DWJ, Wallace JM, Hegerl GC (2000) Annual modes in the extratropical circulation: Part II: Trends. J Climate 13:1018–1036

Trigo IF (2006) Climatology and interannual variability of storm-tracks in the Euro-Atlantic sector: a comparison between ERA-40 and NCEP/NCAR reanalyses. Clim Dyn 26:127–143

Trouet V, Panayotov MP, Ivanova A, Frank D (2012) A pan-European summer teleconnection mode recorded by a new temperature reconstruction from the northeastern Mediterranean (ad 1768–2008). The Holocene 22:887–898

Tyrlis E, Hoskins BJ (2008) Aspects of a Northern Hemisphere atmospheric blocking climatology. J Atmos Sci 65:1638–1652

Ulbrich U, Leckebusch GC, Pinto JG (2009) Extra-tropical cyclones in the present and future climate: a review. Theor Appl Climatol 96:117–131

Uppala SM, Kållberg PW, Simmons AJ, Andrae U, Da Costa Bechtold V, Fiorino M, Gibson JK, Haseler J, Hernandez A, Kelly GA, Li X, Onogi K, Saarinen S, Sokka N, Allan RP, Andersson E, Arpe K, Beljaars ACM, Van De Berg L, Bidlot J, Bormann N, Caires S, Chevallier F, Dethof A, Dragosavac M, Fisher M, Fuentes M, Hagemann S, Hólm E, Hoskins BJ, Isaksen L, Janssen PAEM, Jenne R, McNally AP, Mahfouf J-F, Morcrette J-J, Rayner NA, Saunders RW, Simon P, Sterl A, KE Trenberth, Untch A, Vasiljevic D, Viterbo P, Woollen J (2005) The ERA-40 re-analysis. Q J ROY METEOR SOC 131, Issue 612, 2961–3012. J (2005) The ERA-40 re-analysis. Q J Roy Met Soc 131:2961–3012

van den Besselaar EJM, Haylock MR, Klein Tank AMG, van der Schrier G (2011) A European daily high-resolution observational gridded data set of sea level pressure. J Geophys Res 116:D11110 doi:10.1029/2010JD015468

van den Besselaar EJM, Klein Tank AMG, Buishand TA (2013) Trends in European precipitation extremes over 1951–2010. Int J Climatol 33:2682–2689

van der Schrier G, van den Besselaar, EJM, Klein Tank AMG,Verver G (2013) Monitoring European averaged temperature based on the E-OBS gridded dataset. J Geophys Res 118:5120–5135

van Oldenborgh GJ, Drijfhout SS, van Ulden A, Haarsma R, Sterl A, Severijns C, Hazeleger W, Dijkstra H (2009) Western Europe is warming much faster than expected. Clim Past 5:1–12

Vautard RJ, Yiou P, Thépaut J-N, Ciais P (2010) Northern Hemisphere atmospheric stilling partly attributed to an increase in surface roughness. Nat Geosci 3:756–761

von Storch H, Reichardt H (1997) A scenario of storm surge statistics for the German Bight at the expected time of doubled atmospheric carbon dioxide concentration. J Climate 10:2653–2662

von Storch H, Zwiers FW (1999) Statistical analysis in climate research. Cambridge University Press

von Storch H, Feser F, Haeseler S, Lefebvre C, Stendel M (2014) A violent mid-latitude storm in Northern Germany and Denmark, 28 October 2013. Special Supplement to the Bull Am Met Soc 95:76–78

Wallace JM, Gutzler DS (1981) Teleconnections in the geopotential height field during the Northern Hemisphere winter. Mon Wea Rev 109:784–812

Wang XL, Swail VR, Zwiers FW (2006) Climatology and changes of extratropical cyclone activity: comparison of ERA40 with NCEP-NCAR reanalysis for 1958-2001. J Climate 19:3145–3166

Wang XL, Zwiers F, Swail V, Feng Y (2009a) Trends and variability of storminess in the Northeast Atlantic region, 1874–2007. Clim Dyn 33:1179–1195

Wang XL, Swail V, Zwiers F, Zhang X, Feng Y (2009b) Detection of external influence on trends of atmospheric storminess and northern oceans wave heights. Clim Dyn 32:189–203

Wang XL, Wan H, Zwiers FW, Swail VR, Compo GP, Allan RJ, Vose RS, Jourdain S, Yin X (2011) Trends and low-frequency variability of storminess over western Europe, 1878–2007. Clim Dyn 37:2355–2371

Wang XL, Feng Y, Compo GP, Zwiers FW, Allan RJ, Swail VR, Sardeshmukh PD (2014) Is the storminess in the Twentieth Century Reanalysis really inconsistent with observations? A reply to the comment by Krueger at al. (2013b). Clim Dyn 42:1113–1125

Wanner H, Brönnimann S, Casty C, Gyalistras D, Luterbacher J, Schmutz C, Stephenson DB, Xoplaki E (2001) North Atlantic Oscillation - concepts and studies. Surv Geophys 22:321–381

Warren SG, Hahn CJ, London J, Chervin RM, Jenne RL (1986) Global distribution of total cloud cover and cloud type amounts over land. NCAR Tech. Note TN-273 + STR

Warren SG, Hahn CJ, London J, Chervin RM, Jenne RL (1988) Global distribution of total cloud cover and cloud type amounts over the ocean. NCAR Tech. Note TN-317 + STR

Warren SG, Eastman RM, Hahn CJ (2006) A survey of changes in cloud cover and cloud types over land from surface observations, 1971-96. J Climate 20:717–738

WASA Group (1998) Changing waves and storms in the Northeast Atlantic? Bull Am Met Soc 79:741–760

Weisse R, Plüß A (2006) Storm-related sea level variations along the North Sea coast as simulated by a high-resolution model 1958–2002. Ocean Dyn 56:16–25

Weisse R, von Storch H, Feser F (2005) Northeast Atlantic and North Sea storminess as simulated by a regional climate model during 1958–2001 and comparison with observations. J Climate 18:465–479

Wheeler D, Garcia-Herrera R, Wilkinson CW, Ward C (2009) Atmospheric circulation and storminess derived from Royal Navy logbooks: 1685 to 1750. Clim Change 110:257–280

Wild M (2009) Global dimming and brightening: A review. J Geophys Res 114:D00D16 doi:10.1029/2008JD011470

Wild M, Gilgen H, Roesch A, Ohmura A, Long CN, Dutton EG, Forgan B, Kallis A, Russak V, Tsvetkov A (2005) From dimming to brightening: Decadal changes in solar radiation at Earth's surface. Science 308:847–850

WMO (2010) WMO Guide to Meteorological Instruments and Methods of Observation, World Meteorological Organisation Publication No. 8, WMO, Geneva

Woodruff SD, Worley SJ, Lubker SJ, Ji Z, Freeman JE, Berry DI, Brohan P, Kent EC, Reynolds RW, Smith SR, Wilkinson C (2011) ICOADS Release 2.5 and Data Characteristics. Int J Climatol 31:951–967

Woodworth PL, Blackman DL (2002) Changes in extreme high waters at Liverpool since 1768. Int J Climatol 22:697–714

Woollings TJ, Blackburn M (2012) The North Atlantic jet stream under climate change and its relation to the NAO and EA patterns. J Climate 25:886–902

Woollings TJ, Hoskins BJ, Blackburn M, Berrisford P (2008) A New Rossby wave-breaking interpretation of the North Atlantic oscillation. J Atmos Sci 65:609–626

Woollings TJ, Hannachi A, Hoskins B, Turner A (2010) A regime view of the North Atlantic Oscillation and its response anthropogenic forcing. J Climate 23:1291–1307

Xia L, Zahn M, Hodges KI, Feser F (2012) A comparison of two identification and tracking methods for polar lows. Tellus A 64:17196

Zhang R (2008) Coherent surface-subsurface fingerprint of the Atlantic meridional overturning circulation. Geophys Res Lett 35:L20705 doi:10.1029/2008GL035463

Zolina O, Simmer C, Belyaev K, Kapala A, Gulev S (2009) Improving estimates of heavy and extreme precipitation using daily records from European rain gauges. J Hydromet 10:701–716

Recent Change—North Sea

3

John Huthnance, Ralf Weisse, Thomas Wahl, Helmuth Thomas, Julie Pietrzak, Alejandro Jose Souza, Sytze van Heteren, Natalija Schmelzer, Justus van Beusekom, Franciscus Colijn, Ivan Haigh, Solfrid Hjøllo, Jürgen Holfort, Elizabeth C. Kent, Wilfried Kühn, Peter Loewe, Ina Lorkowski, Kjell Arne Mork, Johannes Pätsch, Markus Quante, Lesley Salt, John Siddorn, Tim Smyth, Andreas Sterl and Philip Woodworth

Electronic supplementary material Supplementary material is available in the online version of this chapter at 10.1007/978-3-319-39745-0_3.

J. Huthnance (✉) · A.J. Souza · P. Woodworth
National Oceanography Centre, Liverpool, UK
e-mail: jmh@noc.ac.uk

A.J. Souza
e-mail: ajso@noc.ac.uk

P. Woodworth
e-mail: plw@noc.ac.uk

R. Weisse (✉) · J. van Beusekom · F. Colijn · M. Quante
Institute of Coastal Research, Helmholtz-Zentrum Geesthacht, Geesthacht, Germany
e-mail: ralf.weisse@hzg.de

J. van Beusekom
e-mail: Justus.Beusekom@hzg.de

F. Colijn
e-mail: franciscus.colijn@hzg.de

M. Quante
e-mail: markus.quante@hzg.de

T. Wahl
College of Marine Science, University of South Florida, St. Petersburg, USA

T. Wahl
University of Southampton, Southampton, UK
e-mail: t.wahl@soton.ac.uk

H. Thomas
Department of Oceanography, Dalhousie University, Halifax, Canada
e-mail: helmuth.thomas@posteo.org

J. Pietrzak
Environmental Fluid Mechanics Section, Delft University of Technology, Delft, The Netherlands
e-mail: J.D.Pietrzak@tudelft.nl

S. van Heteren
Department of Geomodeling, TNO-Geological Survey of the Netherlands, Utrecht, The Netherlands
e-mail: sytze.vanheteren@tno.nl

N. Schmelzer · J. Holfort
Federal Maritime and Hydrographic Agency (BSH), Rostock, Germany
e-mail: natalija.schmelzer@bsh.de

J. Holfort
e-mail: juergen.holfort@bsh.de

I. Haigh
Ocean and Earth Science, University of Southampton, Southampton, UK
e-mail: I.D.Haigh@soton.ac.uk

S. Hjøllo · K.A. Mork
Institute of Marine Research (IMR), Bergen, Norway
e-mail: solfrid.hjollo@imr.no

K.A. Mork
e-mail: kjell.arne.mork@imr.no

E.C. Kent
National Oceanography Centre, Southampton, UK
e-mail: eck@noc.ac.uk

W. Kühn · J. Pätsch
Institute of Oceanography, CEN, University of Hamburg, Hamburg, Germany
e-mail: wilfried.kuehn@uni-hamburg.de

J. Pätsch
e-mail: johannes.paetsch@uni-hamburg.de

P. Loewe · I. Lorkowski
Federal Maritime and Hydrographic Agency (BSH), Hamburg, Germany
e-mail: peter.loewe@bsh.de

I. Lorkowski
e-mail: ina.lorkowski@bsh.de

L. Salt
Royal Netherlands Institute for Sea Research, Texel, The Netherlands
e-mail: lesleyannesalt@gmail.com

M. Quante and F. Colijn (eds.), *North Sea Region Climate Change Assessment*, Regional Climate Studies, DOI 10.1007/978-3-319-39745-0_3

Abstract

This chapter discusses past and ongoing change in the following physical variables within the North Sea: temperature, salinity and stratification; currents and circulation; mean sea level; and extreme sea levels. Also considered are carbon dioxide; pH and nutrients; oxygen; suspended particulate matter and turbidity; coastal erosion, sedimentation and morphology; and sea ice. The distinctive character of the Wadden Sea is addressed, with a particular focus on nutrients and sediments. This chapter covers the past 200 years and focuses on the historical development of evidence (measurements, process understanding and models), the form, duration and accuracy of the evidence available, and what the evidence shows in terms of the state and trends in the respective variables. Much work has focused on detecting long-term change in the North Sea region, either from measurements or with models. Attempts to attribute such changes to, for example, anthropogenic forcing are still missing for the North Sea. Studies are urgently needed to assess consistency between observed changes and current expectations, in order to increase the level of confidence in projections of expected future conditions.

3.1 Introduction

John Huthnance, Ralf Weisse

Physical variables, most obviously sea temperature, relate closely to climate change and strongly affect other properties and life in the sea. This chapter discusses past and ongoing change in the following physical variables within the North Sea: temperature, salinity and stratification (Sect. 3.2), currents and circulation (Sect. 3.3), mean sea level (Sect. 3.4) and extreme sea levels, i.e. contributions from wind-generated waves and storm surges (Sect. 3.5). Also considered are carbon dioxide (CO_2), pH, and nutrients (Sect. 3.6), oxygen (Sect. 3.7), suspended particulate matter and turbidity (Sect. 3.8), coastal erosion, sedimentation and morphology (Sect. 3.9) and sea ice (Sect. 3.10). The distinctive character of the Wadden Sea is addressed in Sect. 3.11, with a particular focus on sediments and nutrients. The chapter covers the past 200 years. Chapter 1 described the North Sea context and physical process understanding, so the focus of the present chapter is on the historical development of evidence (measurements, process understanding and models), the form, duration and accuracy of the evidence available (further detailed in Electronic (E-) Supplement S3) and what the evidence shows in terms of the state and trends in the respective variables.

J. Siddorn
Met Office, Exeter, UK
e-mail: john.siddorn@metoffice.gov.uk

T. Smyth
Plymouth Marine Laboratory, Plymouth, UK
e-mail: tjsm@pml.ac.uk

A. Sterl
Royal Netherlands Meteorological Institute (KNMI), De Bilt, The Netherlands
e-mail: sterl@knmi.nl

3.2 Temperature, Salinity and Stratification

John Huthnance, Elizabeth C. Kent, Tim Smyth, Kjell Arne Mork, Solfrid Hjøllo, Peter Loewe

3.2.1 Historical Perspective

Observations of sea-surface temperature (SST) have been made in the North Sea since 1823, but were sparse initially. The typical number of observations per month (from ships, and moored and drifting buoys) increased from a few hundred in the 19th century to more than 10,000 in recent decades, despite the Voluntary Observing Ship (VOS) fleet declining from a peak of about 7700 ships worldwide in 1984/85 to about 4000 in 2009 (www.vos.noaa.gov/vos_scheme.shtml). Early SST observations used buckets (Kent et al. 2010); adjustments of up to ~0.3 °C in the annual mean, and 0.6 °C in winter, may be needed for these early data owing to sample heat loss or gain (Folland and Parker 1995; Smith and Reynolds 2002; Kennedy et al. 2011a, b). The adjustments depend on large-scale forcing and assumptions about measurement methods—local variations add uncertainty. Cooling water intake temperatures have been measured on ships since the 1920s but data quality is variable, sometimes poor (Kent et al. 1993). Temperature sensors on ships' hulls became more numerous in recent decades (Kent et al. 2010). About 70 % of in situ observations in 2006 came from moored and drifting buoys

(Kennedy et al. 2011b). Other modern shipboard methods include radiation thermometers, expendable bathythermographs (XBTs) and towed thermistors (Woodruff et al. 2011). Satellite estimates of SST are regularly available using Advanced Very High Resolution Radiometers (AVHRR; from 1981) and passive microwave radiometers (with little cloud attenuation; from 1997).

Below the sea surface, temperature was measured by reversing (mercury) thermometers until the 1960s. Since then, electronic instruments lowered from ships (conductivity-temperature-depth profilers; CTDs) enable near-continuous measurements. Since about 2005, multi-decadal model runs have become increasingly available and now provide useful information on temperature distribution to complement the observational evidence (see E-Supplement Sect. S3.1).

Early salinity estimates used titration-based chemical analysis of recovered water samples (from buckets and water intakes) and from lowered sample bottles. Titration estimates usually depended on assuming a constant relation between chlorinity and total dissolved salts (a subject of discussion since 1900), with typical error O(0.01 ‰). Since the 1960s–1970s lowered CTD conductivity cells enable near-continuous measurements, calibrated by comparing the conductivity of water samples against standardised sea water; typical error O(0.001 ‰). Consistent definition of salinity has continued to be a research topic (Pawlowicz et al. 2012).

Thermistors and conductivity cells as on CTDs now record temperature and salinity of (near-surface) intake water on ships. Since the late 1990s, CTDs on profiling 'Argo' floats have greatly increased available temperature and salinity data for the upper 2000 m of the open ocean (www.argo.ucsd.edu). Although not available for the North Sea, these data greatly improve estimates of open-ocean temperature and salinity and thereby North Sea model estimates by better specifying open-ocean boundary conditions.

The history of stratification estimates, based on profiles of temperature and salinity (or at least near-surface and near-bottom values), corresponds with that of subsurface temperature and salinity.

Detail on time-series evidence for coastal and offshore temperature and salinity variations is given in E-Supplement S3.1 and S3.2.

3.2.2 Temperature Variability and Trends

3.2.2.1 Northeast Atlantic

Most water entering the North Sea comes from the adjacent North Atlantic via Rockall Trough and around Scotland. The North Atlantic has had relatively cool periods (1900–1925, 1970–1990) and warm periods (1930–1960, since 1990; Holliday et al. 2011; Dye et al. 2013a; Ivchenko et al. 2010 using 1999–2008 Argo float data). Adjacent to the north-west European shelf, however, different Atlantic water sources make varying contributions (Holliday 2003). For Rockall Trough surface waters, the period 1948–1965 was about 0.8 °C warmer on average than the period 1876–1915 (Ellett and Martin 1973). Subsequently, temperatures of upper water (0–800 m) in Rockall Trough and Atlantic water on the West Shetland slope (Fig. 3.1) oscillated with little trend until around 1994. Temperatures then rose,

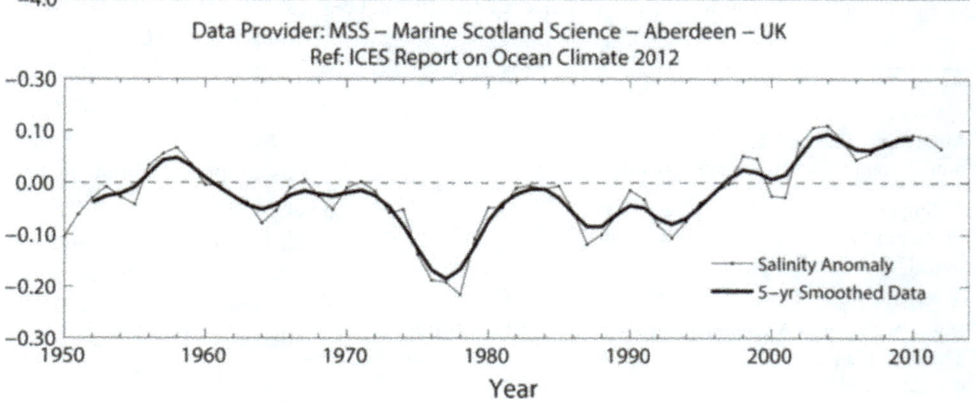

Fig. 3.1 Atlantic Water in the Faroe–Shetland Channel slope current. Temperature (*upper*) and salinity (*lower*) anomalies relative to the 1981–2010 average (Beszczynska-Möller and Dye 2013)

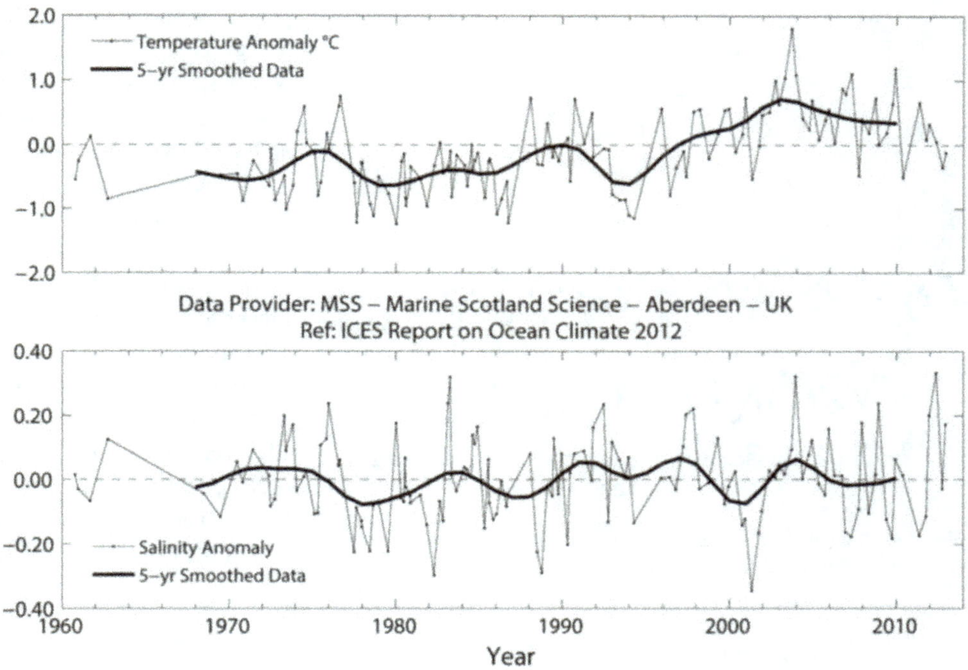

Fig. 3.2 Fair Isle Current entering the northern North Sea from the west and north of Scotland. Annual upper water temperature (*upper*) and salinity (*lower*) anomalies relative to the 1981–2010 average (Beszczynska-Möller and Dye 2013)

peaked in 2006, and subsequently cooled to early 2000s values (Berx et al. 2013; Beszczynska-Möller and Dye 2013; Holliday and Cunningham 2013).

West and north of Britain, the HadISST data set shows an SST trend of 0.2–0.3 °C decade^{-1} over the period 1983–2012, which is higher than the global average (Rayner et al. 2003; see Dye et al. 2013a among several references). Thus positive temperature anomalies exceeding one standard deviation (based on the period 1981–2010) were widespread in adjacent Atlantic Water and the northern North Sea during 2003–2012 (Beszczynska-Möller and Dye 2013). In fact, several authors suggest an inverse relation between Subpolar Gyre strength and the extent of warm saline water (e.g. Hátún et al. 2005; Johnson and Gruber 2007; Haekkinen et al. 2011).

3.2.2.2 North Sea

In Atlantic Water inflow to the North Sea at the western side of the Norwegian Trench (Utsira section, 59.3°N), 'core' temperature has risen by about 0.8 °C since the 1970s and about 1 °C near the seabed in the north-western part of the section (estimated from Holliday et al. 2009). Figure 3.2 shows long-term temperature variability in the Fair Isle Current flowing into the North Sea on the shelf.

For the North Sea as a whole, annual average SST derived from six gridded data sets (Fig. 3.3) shows relatively cool SST from 1870, especially in the early 1900s, 'plateaux' in the periods 1932–1939 and 1943–1950, and then overall decline to a minimum around 1988 (anomaly about −0.8 °C). This was followed by a rise to a peak in 2008 (anomaly about 1 °C) and subsequent fall. SST trends

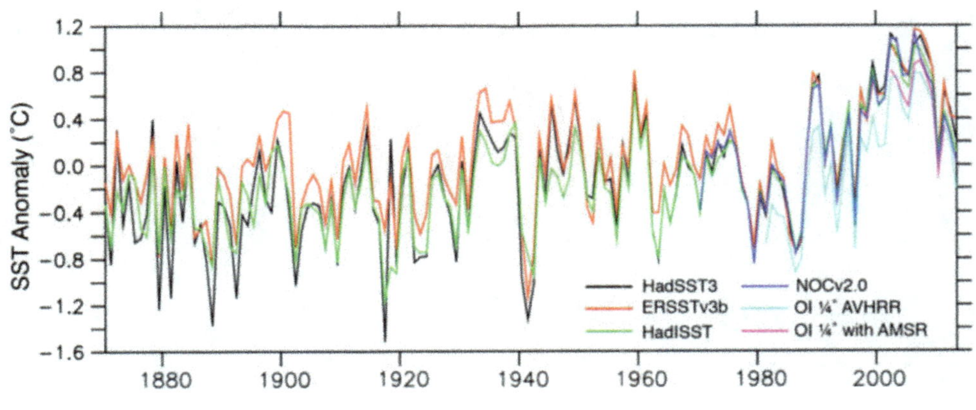

Fig. 3.3 North Sea region annual sea-surface temperature (SST) anomalies relative to the 1971–2000 average, for the datasets in E-Supplement Table S3.1 (figure by Elizabeth Kent, UK National Oceanography Centre)

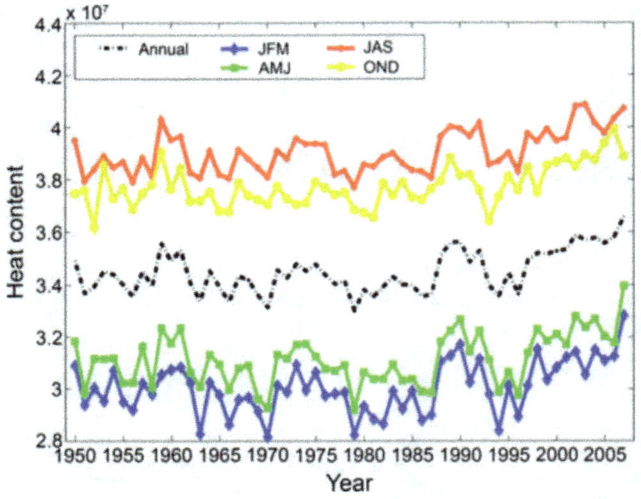

Fig. 3.4 Annual and seasonal mean North Sea heat content (10^7 J m^{-3}) (reprinted from Meyer et al. 2011)

generally show also in heat content (Hjøllo et al. 2009; Meyer et al. 2011) and in all seasons (Fig. 3.4), despite winter-spring variability exceeding summer-autumn variability. The increase in North Sea heat content between 1985 and 2007 was about 0.8×10^{20} J, much less than the seasonal range (about 5×10^{20} J) and comparable with inter-annual variability (Hjøllo et al. 2009).

Despite an inherent anomaly adjustment time-scale of just a few months (Fig. 3.5 and Meyer et al. 2011), the longer-term decline in SST from the 1940s to 1980s and subsequent marked rise to the early 2000s are widely reported. The basis is in observations, for example those shown by McQuatters-Gollop et al. (2007 using HADISST v1.1; see Fig. 3.6 and E-Supplement Table S3.1), Kirby et al. (2007), Holt et al. (2012, including satellite SST data, Fig. 3.7) and multi-decadal hindcasts, such as those of Meyer et al. (2011) and Holt et al. (2012). Particular features noted

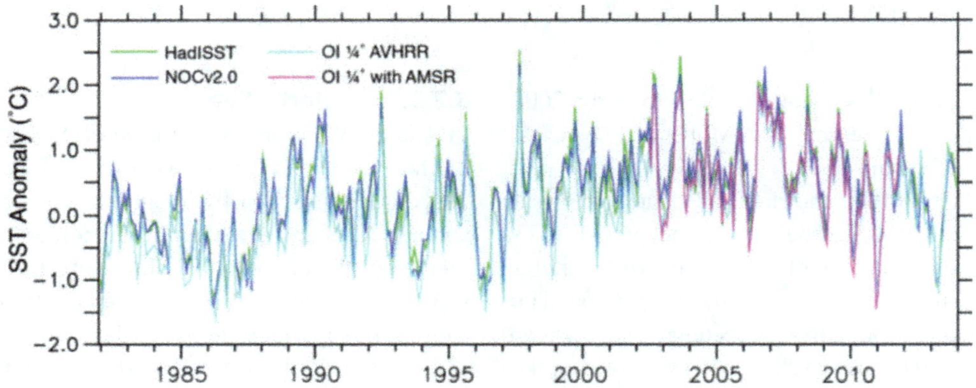

Fig. 3.5 North Sea region monthly sea-surface temperature (SST) anomalies relative to 1971–2000 monthly averages, for the gridded datasets in E-Supplement Table S3.1 with resolution of 1° or finer. Sharp month-to-month variability indicates an inherent anomaly 'adjustment' time of just a few months (figure by Elizabeth Kent, UK National Oceanography Centre)

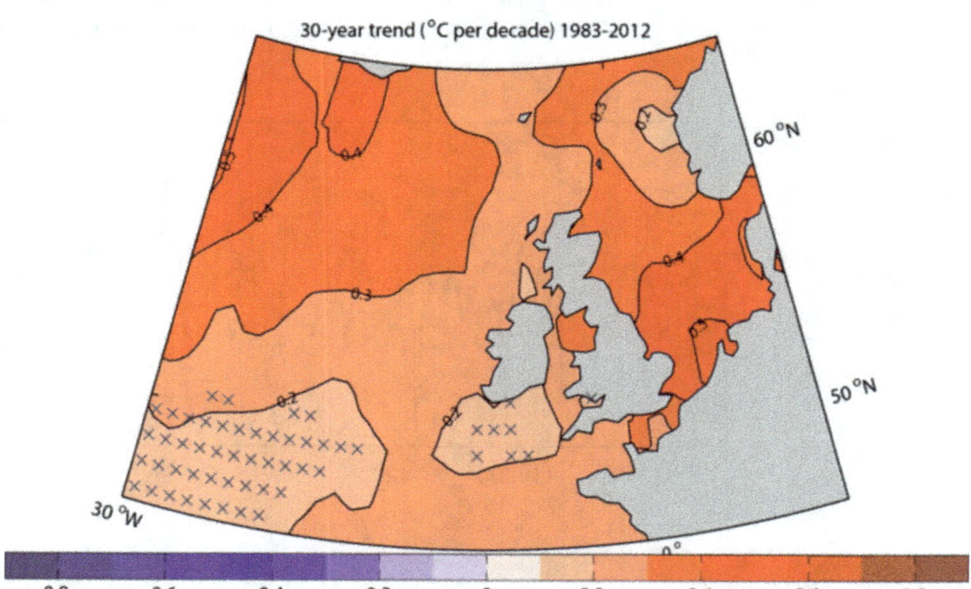

Fig. 3.6 Linear sea-surface temperature trends (°C decade^{-1}) in annual values for the period 1983–2012. From the HadISST1 dataset (Rayner et al. 2003). Hatched areas: trend not significantly different from zero at 95 % confidence level (Dye et al. 2013a, see Acknowledgement)

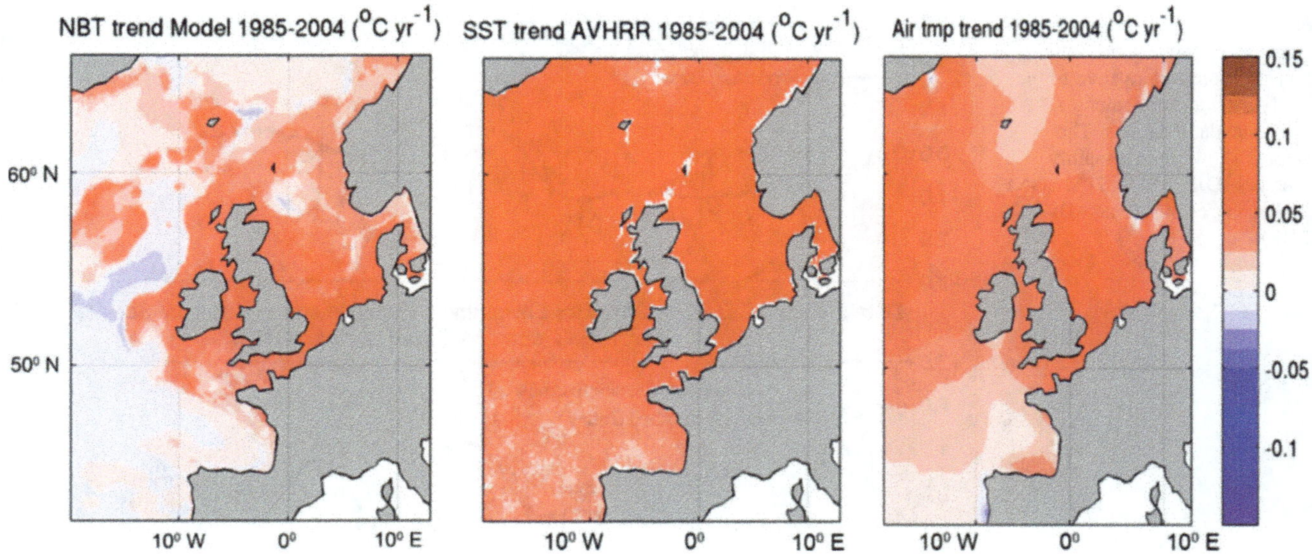

Fig. 3.7 Linear trends for the period 1985–2004 in model near-bed temperature (*left*), satellite sea-surface temperature (SST; *middle*) and 2-m ERA40 air temperature (*right*) (Holt et al. 2012)

are rapid cooling in the period 1960–1963, rapid warming in the late 1980s, followed by cooling again in the early 1990s and then resumed warming to about 2006. The warming trends of the 1980s to 2000s are widely reported to be significant (e.g. Holt et al. 2012) and are mainly but not entirely accounted for by trends in air temperature (see hindcasts of Meyer et al. 2011; Holt et al. 2012). Observed North Sea winter bottom temperature between 1983 and 2012 shows a typical trend of 0.2–0.5 °C decade^{-1} (Dye et al. 2013a) superimposed on by considerable interannual variability.

3.2.2.3 Regional Variations

The rise in North Sea SST since the 1980s increased from north (trend <0.2 °C decade^{-1}) to south (trend 0.8 °C decade^{-1}; Fig. 3.6; McQuatters-Gollop et al. 2007). Based on HadISST1 for the period 1987–2011, the EEA (2012) showed warming of 0.3 °C decade^{-1} in the Channel, 0.4 °C decade^{-1} off the Dutch coast, and less than 0.2 °C decade^{-1} at 60°N off Norway.

The German Bight shows the largest warming trend in recent decades (Fig. 3.6) with a rapid SST rise in the late 1980s (Wiltshire et al. 2008; Meyer et al. 2011). Variability is also large, between years O(1 °C) and longer term (Wiltshire et al. 2008; Meyer et al. 2011; Holt et al. 2012). At Helgoland Roads Station (54° 11′N, 7° 54′E) decadal SST trends since 1873 show the warming after the early 1980s was the strongest.

For southern North Sea SST, the 1971–2010 ferry data (Fig. 3.8) show a rise of O(2 °C) from 1985/6 to 1989; the five-year smoothing emphasises a late 1980s rise of about 1.5 °C followed by 5- to 10-year fluctuations superimposed on a slow decline from the early 1990s to about 1 °C above the 1971–1986 average (smoothed values). Model hindcast spatial averages between Dover Strait and 54.5°N (water column mostly well-mixed; Alheit et al. 2012 based on Meyer et al. 2011) also show cold winters for 1985 to 1987 but the 1990 winter as the warmest since 1948 (and winter 2007 as warmer again). Anomalies (observations and model results) became mainly positive from the late 1980s apart from a dip in the early 1990s. This all illustrates the late 1980s temperature rise.

The Dutch coastal zone shows a trend of rising SST since 1982 (van Aken 2010), despite a very cold winter in 1996 (January–March; about 4 °C below the 1969–2008 average; van Hal et al. 2010). Factors contributing to this rise are thermal inertia (seasonally), winds and cloudiness or bright sunshine (van Aken 2010). The 1956–2003 Marsdiep winter temperature (Tsimplis et al. 2006) and Wadden Sea winter and spring temperature (van Aken 2008) were significantly correlated with the winter North Atlantic Oscillation (NAO) index (see Annex 1). However, decadal to centennial temperature variations (a cooling of about 1.5 °C over the period 1860–1890 and a similar warming in the last 25 years) were not related to long-term changes in the NAO.

The western English Channel (50.03°N, 4.37°W) warmed in the 1920s and 1930s (Southward 1960); after a dip it warmed again in the 1950s, cooled in the 1960s and warmed over the full water column from the mid-1980s to the early 2000s (0.6 °C decade^{-1}, Smyth et al. 2010; see E-Supplement Fig. S3.2). The greatest (1990s) temperature rise coincided with a decrease in median wind speed (from 3.5 to 2.75 m s^{-1}) and an increase in surface solar irradiation (of about 20 %), both correlated with changes in the NAO (Smyth et al. 2010).

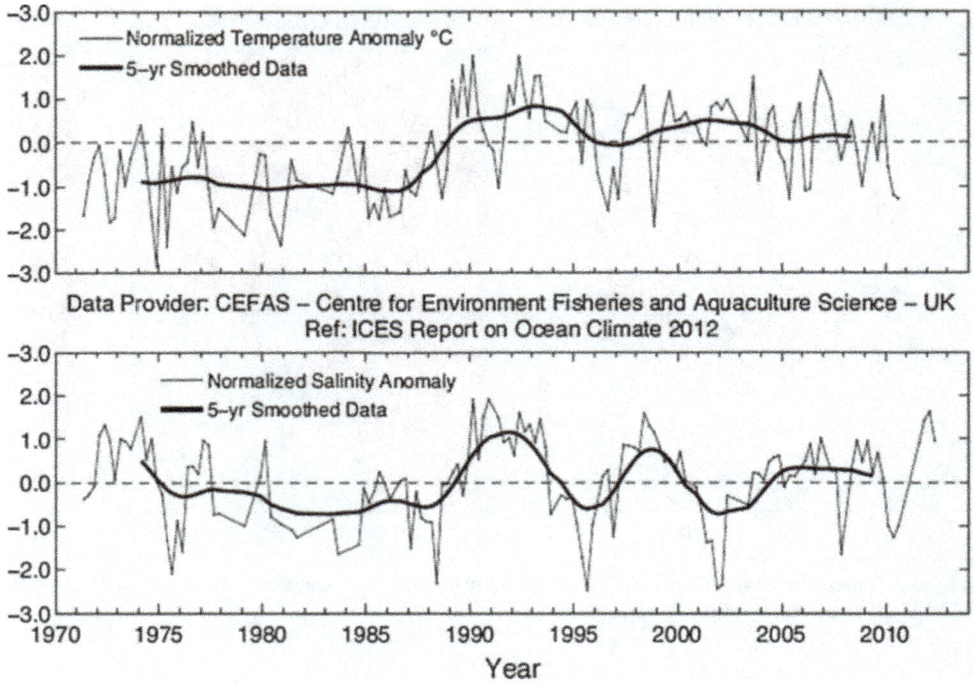

Fig. 3.8 Ferry-based sea-surface temperature (*upper*) and salinity (*lower*) anomalies relative to the 1981–2010 average, along 52°N at six standard stations. The graphic shows three-monthly averages (DJF, MAM, JJA, SON) (Beszczynska-Möller and Dye 2013)

Off northern Denmark and Norway, coastal waters in winter (JFM) were 0.8–1.3 °C warmer in the period 2000–2009 than the period 1961–1990 (Albretsen et al. 2012); the corresponding rise at 200 m depth was 0.55–0.8 °C. Winter–spring observed SST in the Kattegat and Danish Straits rose by about 1 °C between 1897–1901 and the 1980s, and again by about 1 °C to the 1990s–2006 period (Henriksen 2009). Summer–autumn trends were not as clear.

3.2.3 Salinity Variability and Trends

3.2.3.1 Northeast Atlantic

North Atlantic surface salinity shows pronounced interannual and multi-decadal variability. In the Subpolar Gyre salinity variations are correlated with SST such that high salinities usually coincide with anomalously warm water and vice versa (such as in Rockall Trough; Beszczynska-Möller and Dye 2013). On decadal time scales, upper-layer salinity is also positively correlated with the winter NAO, especially in the eastern part of the gyre (Holliday et al. 2011). Shelf-sea and oceanic surface waters to the north and west of the UK had a salinity maximum in the early 1960s and a relatively fresh period in the 1970s, associated with the so-called Great Salinity Anomaly (Dickson et al. 1988). In Rockall Trough the minimum occurred about 1975 (Dickson et al. 1988) and was followed by increasing salinities, interrupted by a mid-1990s minimum (Holliday et al. 2010; Hughes et al. 2012; Sherwin et al. 2012).

Correspondingly, the Fair Isle—Munken section (~2°W 59.5°N to 6°W 61°N across the Faroe-Shetland Channel) at 50–100 m depth showed an upward salinity trend of 0.075 decade^{-1} during the period 1994–2011 (Fig. 3.1; Berx et al. 2013). Likewise, the salinity of Atlantic water inflow to the Nordic Seas through Svinøy section (to the north-west off Norway through ~4°E 63°N) has increased by about 0.15 since the 1970s (Holliday et al. 2008; Beszczynska-Möller and Dye 2013), for example by 0.08 from 1992 to 2009 (Mork and Skagseth 2010).

3.2.3.2 North Sea

Salinity has shown a long-term (1958–2003) increase around northern Scotland (Leterme et al. 2008) and (1971–2012) in the northern North Sea (Fig. 3.9). This is confirmed by Hughes et al. (2012) who charted pentadal-mean upper-ocean salinity showing positive anomalies (relative to the 1971–2000 mean) since 1995 in the northern North Sea most influenced by the Atlantic. Linkage to more saline Atlantic inflow has been suggested (Corten and van de Kamp 1996).

On the western side of the Norwegian Trench and in the central northern North Sea (Utsira section, 59.3°N), influenced by Atlantic water, salinity has increased by about 0.05 since the late 1970s (when values were relatively stable after the Great Salinity Anomaly; Beszczynska-Möller and Dye 2013). On the other hand, salinity in the Fair Isle Current shows interannual variability and no clear long-term trend (Fig. 3.2), being influenced by the fresher waters of the Scottish Coastal Current from west of Scotland.

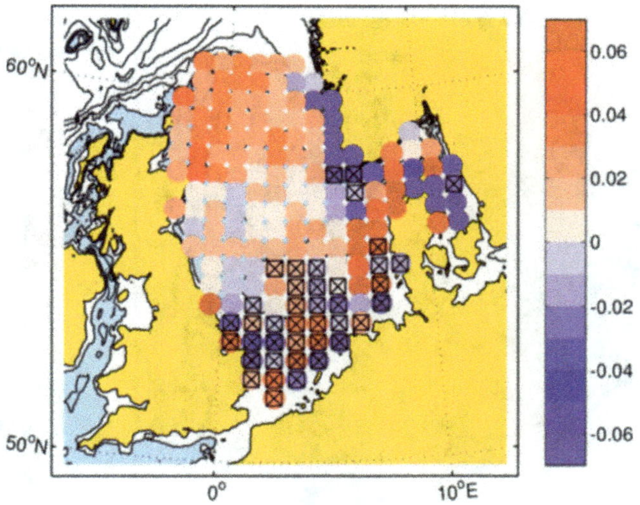

Fig. 3.9 Linear trend per decade in winter bottom salinity, from International Bottom Trawl Survey (IBTS) Quarter 1 data, 1971–2012. Values are calculated for ICES rectangles with more than 30 years of data (hatched areas: trend not significantly different from zero at 95 % confidence level, Dye et al. 2013b; see Acknowledgement, updated from UKMMAS 2010, courtesy of S. Hughes, Marine Scotland Science)

Coastal regions of the southern North Sea, notably the German Bight, are influenced by fluvial inputs (primarily from the rivers Rhine and Elbe) as well as Atlantic inflows (Heyen and Dippner 1998; Janssen 2002). Away from coastal waters, the influence of Atlantic inflow dominates. For the German Bight, Heyen and Dippner (1998) reported no substantial trends in sea-surface salinity (SSS) for the period 1908–1995, a result confirmed by earlier analysis of Helgoland Roads SSS for the period 1873–1993 (Becker et al. 1997) and the analyses of Janssen (2002). German Bight studies (e.g. Fig. 3.10) agree on a temporal minimum around 1982 and a maximum during the early 1990s with a difference of about 0.7 between the two. 1971–2010 ferry data (Fig. 3.8) show pentadal fluctuations with a temporal minimum and maximum also around 1982 and the early 1990s respectively.

The western English Channel (50.03°N, 4.37°W), away from the coast, is influenced by North Atlantic water, showing a similar increase in salinity in recent years (Holliday et al. 2010). Local weather effects (mixed vertically by tidal currents) add to interannual salinity variability which is much greater than in the open ocean. For example, station L4 off Plymouth experiences pulses of surface freshening after intense summer rain increases riverine input (Smyth et al. 2010). However, there is no clear trend over a century of measurements (see also E-Supplement Fig. S3.3, E-Supplement Sect. S3.2).

In the Kattegat and Skagerrak, salinities are affected by low-salinity Baltic Sea outflow. Skagerrak coastal waters in winter (January–March) were up to 0.5 more saline in the period 2000–2009 than the period 1961–1990, but further west and north around Norway their salinity decreased slightly (Albretsen et al. 2012). Shorter-term variability is larger. Salinity variability in the Kattegat and Skagerrak exceeds that in Atlantic water, owing to varying Baltic outflow (see Sect. 3.3) and net precipitation minus evaporation in catchments.

Salinity variability on all time scales to multi-decadal exceeds and obscures any potential long-term trend. For

Fig. 3.10 Winter bottom salinity from the ICES International Bottom Trawl Survey (IBTS) dataset at Viking Bank, Dogger Bank and German Bight, together with annual mean salinity from Helgoland Roads (Holliday et al. 2010; see Acknowledgement)

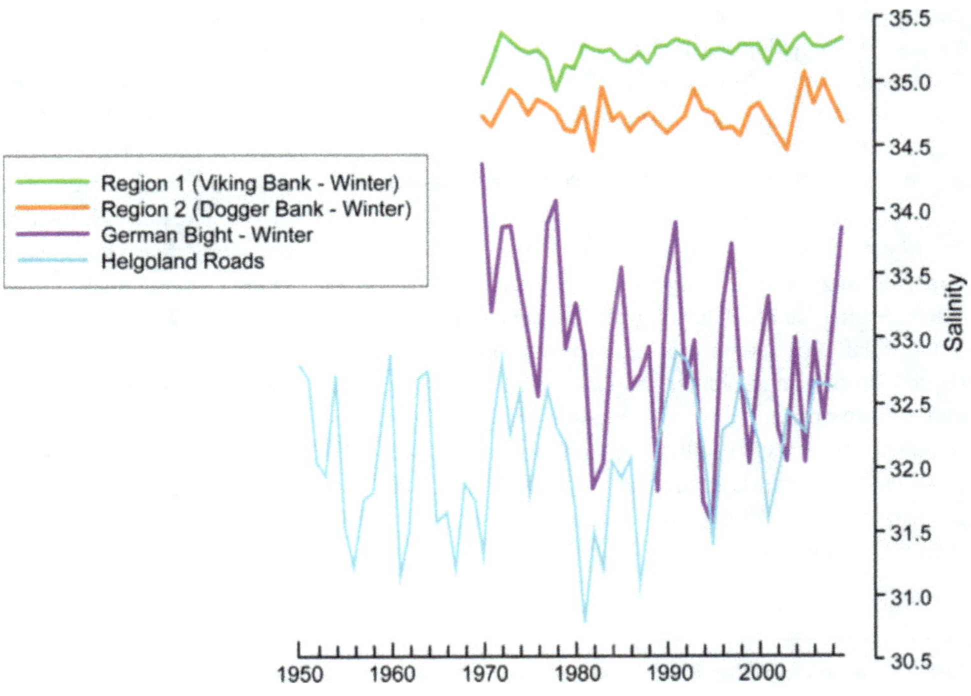

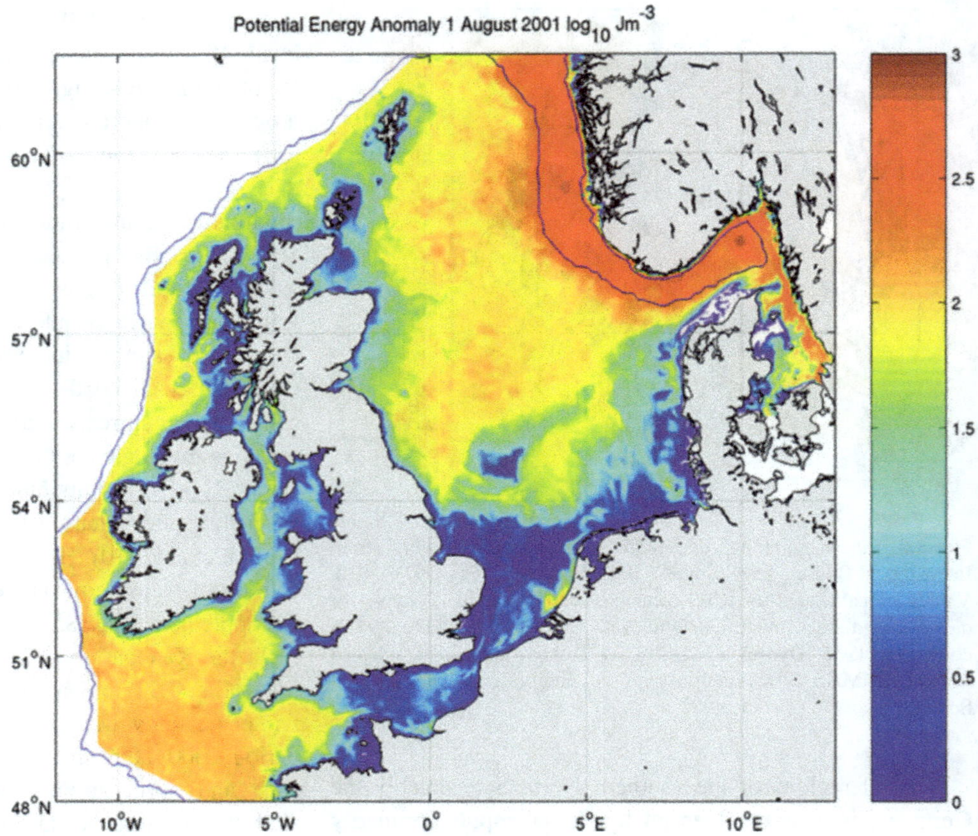

Fig. 3.11 Distribution of potential energy anomaly (energy required to completely mix the water column; log scale, 1 August 2001) (Holt and Proctor 2008)

example, in winter 2005, a series of storms drove much high-salinity Atlantic water across the north-west boundary into the North Sea as far south as Dogger Bank and bottom-water salinity exceeded 35 in 63 % of the North Sea area (Loewe 2009). Adjacent Atlantic waters in the period 2002–2010 (Hughes et al. 2011) show positive salinity anomalies of more than two (one) standard deviation in Rockall Trough (Faroe-Shetland Channel) while the North Sea has no comparably clear signal.

3.2.4 Stratification Variability and Trends

Stratification is a key control on shelf-sea marine ecosystems. Strong stratification inhibits vertical exchange of water. Spring–summer heating reduces near-surface density where tidal currents are too weak to mix through the water depth (Simpson and Hunter 1974), typically where depth is about 50 m or more. The configuration of summer-stratified regions controls much of the average flow in shelf seas (Hill et al. 2008). Mixed-layer data are available albeit only on a 2° grid.[1] The distribution of summer stratification (mainly thermal) is illustrated in Figs. 3.11 and 3.12.

Annual time series of ECOHAM4 simulated thermocline characteristics averaged over the North Sea were reported by Lorkowski et al. (2012). The maximum depth of the thermocline[2] is much more variable interannually than its mean depth. Thermocline intensity shows no trend and only moderate variability. The annual number of days with a mean thermocline greater than 0.2 °C m^{-1} ranged from 31 to 101. The warmest summer in the period simulated (2003) hardly shows in any thermocline characteristics (Lorkowski et al. 2012). In the north-western North Sea, the strength of thermal stratification varies interannually (with no clear trend but periodicity of about 7–8 years; Sharples et al. 2010). The multi-decadal hindcast by Meyer et al. (2011) for the North Sea confirmed that variability in stratification is mainly interannual. In seasonally stratified regions, Holt et al. (2012) modelling showed 1985–2004 warming trends to be greater at the surface than at depth (reflecting an increase in stratification), especially in the central North Sea, at frontal areas of Dogger Bank, in an area north-east of Scotland and in inflow to the Skagerrak. They also found this pattern in annual trends of ICES (International Council for the Exploration of the Sea) data, albeit limited by a lack of seasonal resolution.

[1]www.ifremer.fr/cerweb/deboyer/mld/home.php.

[2]Defined here as (existence of) the uppermost vertical temperature gradient $\Delta T/\Delta z \geq 0.1$; T(°C) is temperature, z (m) is depth.

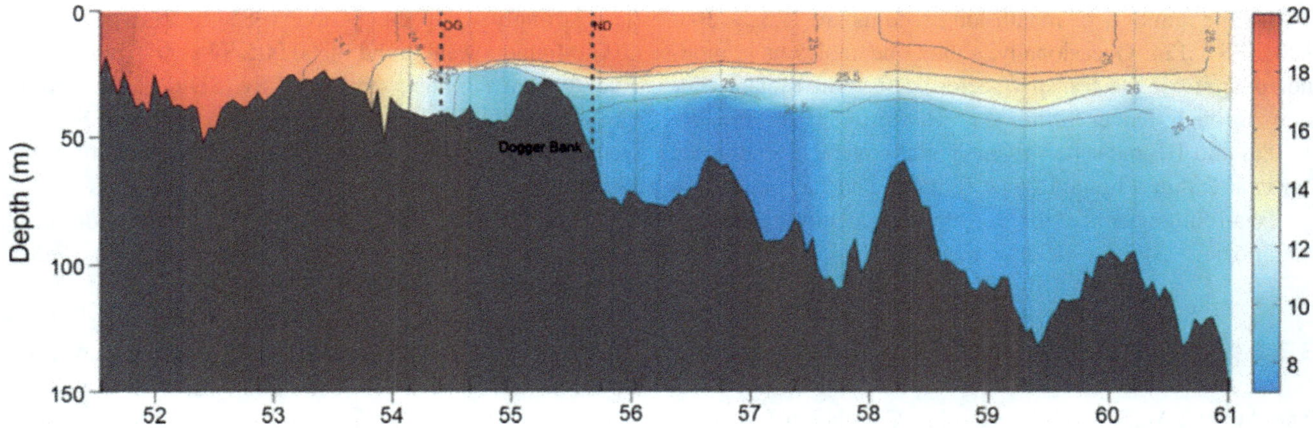

Fig. 3.12 South-north section of potential temperature (°C) near 2.5°E (but further east around Dogger Bank), August 2010 (Queste et al. 2013)

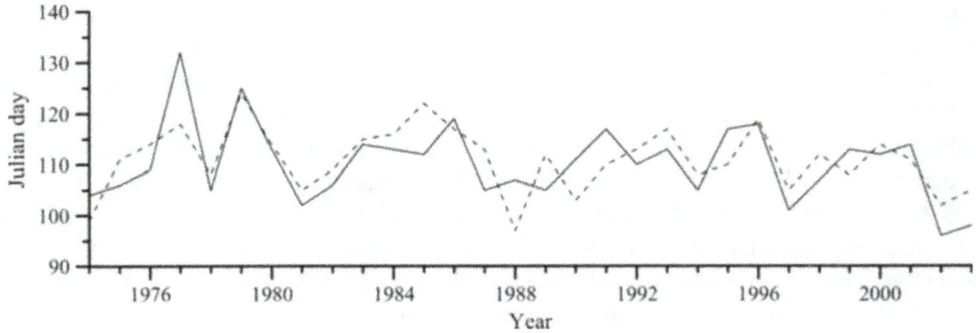

Fig. 3.13 Modelled timing (Julian day) of spring stratification (when the surface-bottom temperature difference first exceeds 0.5 °C for at least three days; *solid line*) and spring bloom (*dashed line*) between 1974 and 2003 in 60 m water depth near 1.4°W 56.2°N (reprinted from Fig. 5a of Sharples et al. 2006)

Lorkowski et al. (2012) found the time of initial thermocline development to vary between Julian days 54 and 107, with relatively large values (i.e. a late start) from 1970 to 1977. Other evidence also suggests a recent trend to earlier thermal stratification (Young and Holt 2007, albeit for the Irish Sea). The timing of spring stratification in the north-western North Sea was modelled for the period 1974–2003 and compared with observed variability by Sharples et al. (2006; Fig. 3.13). Persistent stratification typically begins (on 21 April ± three weeks range) as tidal currents decrease from springs to neaps. The main meteorological control is air temperature; since the mid-1990s its rise seems to have caused stratification to be an average of one day earlier per year with wind stress (linked to the NAO) having had some influence before the 1990s. Holt et al. (2012), modelling 1985–2004, found an extension to the stratified season in the central North Sea and north-east of Scotland.

In estuarine outflow regions, strong short-term and interannual variability in precipitation (hence fluvial inputs) and tidal mixing mask any longer-term trends in stratification (timing or strength).

3.3 Currents and Circulation

John Huthnance, John Siddorn, Ralf Weisse

3.3.1 Historical Perspective

The earliest evidence for circulation comes from hydrographic sections, for time scales longer than a day, and from drifters, observed by chance or deliberately deployed. Prior to satellite tracking (of floats or drogued buoys), typically only drifters' start and end points would be known; temporal and spatial resolution were lacking. Moored current meters record time series at one location; their use was rare until the 1960s. Within the area (5°W–13°E, 48°N–62°N) the international current meter inventory at the British Oceanographic Data Centre[3] records just 27 year-long records and 3025 month-long records to 2008; by decade from the

[3]https://www.bodc.ac.uk/data/information_and_inventories/current_meters/search.

1950s, the numbers of month-long records are 1, 32, 1306, 1201, 381, 124. Occasionally, submarine cables have monitored approximate transport across a section (notably for flow through Dover Strait; e.g. Robinson 1976; Prandle 1978a) and HF radar has given spatial coverage for surface currents within a limited range (Prandle and Player 1993).

Detail on evidence for currents, circulation and their variations is given in E-Supplement Sect. S3.3.

3.3.2 Circulation: Variability and Trends

The Atlantic Meridional Overturning Circulation (AMOC), and its warm north-eastern limb in the Subpolar Gyre, influence the flow and properties of Atlantic Water bordering and partly flowing onto the north-west European shelf and into the North Sea. The AMOC has much seasonal and some interannual variability: mean 18.5 Sv (SD ~ 3 Sv) for April 2004 to March 2009 (Sv is Sverdrup, 10^6 m^3 s^{-1}) (McCarthy et al. 2012). The AMOC probably also varies on decadal time scales (e.g. Latif et al. 2006). Longer-term trends are not yet determined (Cunningham et al. 2010) even though Smeed et al. (2014) found the mean for April 2008 to March 2012 to be significantly less than for the previous four years. The Subpolar Gyre extent correlates with the NAO (Lozier and Stewart 2008). It strengthened overall from the 1960s to the mid-1990s, then decreased (Hátún et al. 2005). While the Subpolar Gyre was relatively weak in the period 2000–2009, more warm, salty Mediterranean and Eastern North Atlantic waters flowed poleward around Britain (Lozier and Stewart 2008; Hughes et al. 2012). Negative NAO also correlates with more warm water in the Faroe-Shetland Channel (Chafik 2012). However, observations show no significant longer-term trend in Atlantic Water transport to the north-east past Scotland and Norway (Orvik and Skagseth 2005; Mork and Skagseth 2010; Berx et al. 2013).

Inflow of oceanic waters to the North Sea from the Atlantic Ocean, primarily in the north driven by prevailing south-westerly winds, has been modelled by Hjøllo et al. (2009; 1985–2007), Holt et al. (2009) and using NORWECOM/POM (3-D hydrodynamic model; Iversen et al. 2002; Leterme et al. 2008, for 1958–2003; Albretsen et al. 2012). Relative to the long-term mean, results show weaker northern inflow between 1958 and 1988; within this period, there were increases in the 1960s and early 1970s, a decrease from 1976 to 1980 and an increase in the early and mid-1980s. The northern inflow was greater than the long-term mean in 1988 to 1995 with a maximum in 1989 (McQuatters-Gollop et al. 2007) but smaller again in 1996 to 2003. This inflow is correlated positively with salinity, SST (less strongly) and the NAO (especially in winter), and negatively with discharges from the rivers Elbe and Rhine (less strongly). For the period 1985–2007, Hjøllo et al. (2009) found a weak trend of −0.005 Sv $year^{-1}$ in modelled Atlantic Water inflows (mean 1.7 Sv, SD 0.41 Sv, correlation with NAO ~0.9). Strong flows into the North Sea (and Nordic Seas) frequently correspond to high-salinity events (Sundby and Drinkwater 2007).

Dover Strait inflow, of the order 0.1 Sv (Prandle et al. 1996), was smaller than the long-term mean from 1958 to 1981 and then greater until 2003 (Leterme et al. 2008). Baltic Sea outflow variations (modelled freshwater relative to salinity 35.0) correlate with winds, resulting sea-surface elevation and NAO index; correlation coefficients with the NAO were 0.57 during the period 1962–2004 and 0.74 during 1980–2004 (Hordoir and Meier 2010; Hordoir et al. 2013). Days-to-months variability O(0.1 Sv) in North Sea—Baltic Sea exchange far exceeds the mean Baltic Sea outflow of the order 0.01 Sv or any trend therein.

North Sea outflows and inflows (plus net precipitation minus evaporation) have to balance on a time scale of just a few days. Off-shelf flow is persistent in the Norwegian Trench and in a bottom layer below the poleward along-slope flow (Holt et al. 2009; Huthnance et al. 2009). A modelled time series for 1958–1997 (Schrum and Sigismund 2001) shows an average outflow of about 2 Sv, little clear trend but consistency with the above interannual variations in inflow.

A MyOcean (project) reanalysis of the region 40°–65°N by 20°W–13°E for the period 1984–2012 was undertaken with the NEMO model version 3.4 (Madec 2008; for details on this application see MyOcean 2014). Transports normal to transects were calculated following NOOS (2010): averaging flow over 24.8 h to give a tidal mean at each model point across the transect; then area-weighting for transports, separating the mean negative and mean positive flows. For the Norway–Shetland transect, flow in the west is dominantly into the North Sea and makes a significant contribution to exchange with the wider Atlantic; circulation is partially density-driven during summer and confined to the coastal waters east of Shetland. Mean inflow is 0.56 Sv with significant seasonality and interannual variability but no obvious trend. In the east sector of the Norway–Shetland transect, flow is both into and out of the North Sea, strongly steered by the Norwegian Trench and includes the Norwegian Coastal Current, resulting in a larger outflow than inflow. Mean net flow is 1.3 Sv (SD 0.97 Sv) representing large seasonal and interannual variability, especially in the outflow.

Net circulation within the North Sea is shown schematically in Fig. 1.7. Tidal currents are important, primarily semi-diurnal with longer-period modulation (Sect. 1.4.4); locally values exceed 1.2 m s^{-1} in the Pentland Firth, off East Anglia and in Dover Strait. Other important current contributions are due to winds (Sect. 1.4.3 shows representative flow patterns) and to differences in density (Sect. 1.4.2) including estuarine outflows (e.g. van Alphen et al. 1988), varying on time scales from hours to seasons (e.g. Turrell et al. 1992) to decades. Hence flows can be very variable in time; they also vary strongly with location.

Wind forcing is the most variable factor; water transports in one storm (typically in winter; time-scale hours to a day) can be significant relative to a year's total. 50-year return values for currents in storm surges have been estimated at 0.4–0.6 m s^{-1} in general, but exceed 1 m s^{-1} locally off Scottish promontories, in Dover Strait, west of Denmark and over Dogger Bank (Flather 1987). These extreme currents are directed anti-clockwise around the North Sea near coasts, and into the Skagerrak.

In summer-stratified areas (Sect. 3.2.4) cold bottom water is nearly static (velocity tends to zero at the sea bed due to friction). Between stratified and mixed areas, relatively strong density gradients are expected to drive near-surface flows anti-clockwise around the dense bottom water (Hill et al. 2008). These flows, of the order 0.3 m s^{-1} but sometimes >1 m s^{-1} in the Norwegian Coastal Current, are liable to baroclinic instability developing meanders, scale 5–10 km (e.g. Badin et al. 2009; their model shows eddy variability increasing in late summer with increased stratification). Such meanders are prominent north of Scotland over the continental slope and off Norway where the fresher surface layer increases stratification.

When a region of freshwater influence (ROFI) is stratified, cross-shore tidal currents may develop; for example, according to de Boer et al. (2009) surface currents rotate clockwise and bottom currents anti-clockwise in the Rhine ROFI when stratified. These authors also found cyclical upwelling there due to tidal currents going offshore at the surface and onshore below.

The winter mean circulation of the North Sea is organised in one anti-clockwise gyre with typical mean velocities of about 10 cm s^{-1} (Kauker and von Storch 2000). On shorter time scales the circulation is highly variable. Kauker and von Storch (2000) identified four regimes. Two are characterised by a basin-wide gyre with clockwise (15 % of the time) or anti-clockwise (30 % of the time) orientation. The other two regimes are characterised by the opposite regimes of a bipolar pattern with maxima in the southern and northern parts of the North Sea (45 % of the time). For 10 % of the time the circulation nearly ceased. Kauker and von Storch (2000) found that only 40 % of the one-gyre regimes persist for longer than five days while the duration of the bipolar circulation patterns rarely exceeded five days. Accordingly, short-term variability typically dominates transports; tidal flows dominate instantaneous transports (positive and negative volume fluxes across sections) and meteorological phenomena dominate residual (net) transports.

Mean residual transports are generally smaller than their variability. Many transects show strong seasonality as meteorological conditions drive surges, river runoff and ice melt. No trend in transports has been seen in these data: limited duration of available data and large variability in the transports on time scales of days, seasons and interannually makes discerning trends difficult.

In the German Bight, anti-clockwise circulation is about twice as frequent as clockwise, and prevails during south-westerly winds typical of winter storms, giving rapid transports through the German Bight (Thiel et al. 2011, on the basis of Pohlmann 2006). Loewe (2009) associated clockwise flow with high-pressure and north-westerly weather types, anti-clockwise flow with south-westerly weather types, and flow towards the north or north-west with south-easterly weather types. However, Port et al. (2011) found that the wind-current relation changes away from the coast owing to dependence on density effects, the coastline and topography.

On longer time scales the variability of the North Sea circulation and thus transports is linked to variations in the large-scale atmospheric circulation. Emeis et al. (2015) reported results of an EOF analysis (see von Storch and Zwiers 1999) of monthly mean fields of vertically integrated volume transports derived from a multi-decadal model hindcast (Fig. 3.14; see also Mathis et al. 2015). Regions of particularly high variability include the inflow areas of Atlantic waters via the northern boundary of the North Sea and the English Channel, respectively. The time coefficient associated with the dominant EOF mode overlaid with the NAO index illustrates the relation with variability of the large-scale atmospheric circulation (Fig. 3.14). Positive EOF coefficient (intensified inflow of Atlantic waters) corresponds with a positive NAO, i.e. enhanced westerly winds, which in turn result in an intensified anti-clockwise North Sea circulation (Emeis et al. 2015); opposites also hold.

In summary, multiple forcings cause currents to vary on a range of time and space scales, including short scales relative to which measurements are sparse. Hence trends are of lesser significance and hard to discern. Moreover, causes of trends in flows are difficult to diagnose; improvements are needed in observational data (quantity and quality). Reliance is placed on models, which need improvement (in formulation, forcing) for currents other than tides and storm surges.

3.4 Mean Sea Level

Thomas Wahl, Philip Woodworth, Ivan Haigh, Ralf Weisse

Changes in mean sea level (MSL) result from different aspects of climate change (e.g. the melting of land-based ice, thermal expansion of sea water) and climate variability (e.g.

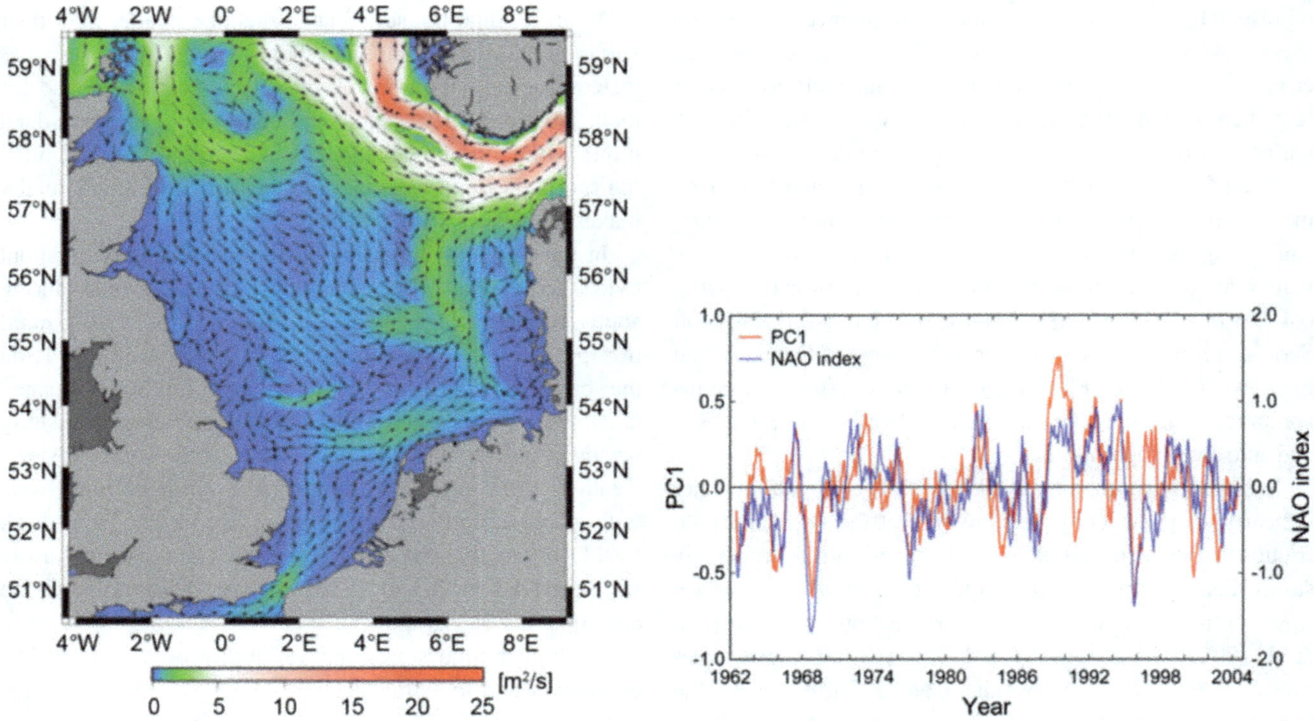

Fig. 3.14 Dominant EOF of monthly mean vertically integrated volume transports obtained from a 3D baroclinic simulation (1962–2004) explaining 75.8 % of the variability. Vectors indicate directions of transport anomalies while colours indicate magnitudes (*left* Emeis et al. 2015); Corresponding coefficient time series (*red*) and NAO index (*blue*) (*right* Hurrell et al. 2013). Shown are moving annual averages based on monthly values

changes in wind forcing related to the NAO or El Niño–Southern Oscillation) and occur over all temporal and spatial scales. MSL is sea level averaged into monthly or annual mean values, which are the parameters of most interest to climate researchers (Woodworth et al. 2011). The focus in this chapter is on the last 200 years, when direct 'modern' measurements of sea level are available from tide gauges and high precision satellite radar altimeter observations. MSL can be inferred indirectly over this period (and thousands of years earlier) using proxy records from salt-marsh sediments and the fossils within them (Gehrels and Woodworth 2013) or archaeology (e.g. fish tanks built by the Romans), and over much longer time scales (thousands to millions of years) using other paleo-data (e.g. geological records, from corals or isotopic methods).

The North Sea coastline has one of the world's most densely populated tide gauge networks, with many (>15) records spanning 100 years or longer and a few going back almost continuously to the early 19th century. The tide gauges of Brest and Amsterdam also provide some data for parts of the 18th century and are among the longest sea level records in the world. Since 1992, satellite altimetry has provided near-global coverage of MSL. The advantage of altimetry is that it records geocentric sea level (i.e. measurements relative to the centre of the Earth). By contrast, tide gauges measure the relative changes between the ocean surface and the land itself; hence, the term 'relative mean sea level' (RMSL), and it is this that is of most relevance to coastal managers, engineers and planners. Calculation from tide gauge records of changes in 'geocentric mean sea level' (sometimes referred to as 'absolute mean sea level'; AMSL) requires the removal of non-climate contributions to sea level change, which arise both from natural processes (e.g. tectonics, glacial isostatic adjustment GIA) and from anthropogenic processes (e.g. subsidence caused by ground water abstraction). Tide gauge records can be corrected using estimates of vertical land motion from (i) models which predict the main geological aspect of vertical motion, namely GIA (e.g. Peltier 2004); (ii) geological information near tide gauge sites (e.g. Shennan et al. 2012); and (iii) direct measurements made at or near tide gauge locations using continuous global positioning system (GPS) or absolute gravity (e.g. Bouin and Wöppelmann 2010). Rates of vertical land movement have also been estimated by comparing trends derived from altimetry data and tide gauge records (e.g. Nerem and Mitchum 2002; Garcia et al. 2007; Wöppelmann and Marcos 2012).

Paleo sea level data from coastal sediments, the few long (pre-1900) tide gauge records and reconstructions of MSL, made by combining tide gauge records with altimetry measurements (e.g. Church and White 2006, 2011; Jevrejeva et al. 2006, 2008; Merrifield et al. 2009), indicate that there was an increase in the rate of global MSL rise during the late 19th and early 20th centuries (e.g. Church et al. 2010; Woodworth et al. 2011; Gehrels and Woodworth 2013). Over the last 2000 to 3000 years, global MSL has been near present-day levels with fluctuations not larger than about ±0.25 m on time scales of a few hundred years (Church et al. 2013) whereas the global average rate of rise estimated for the 20th century was 1.7 mm $year^{-1}$ (Bindoff et al. 2007). Measurements from altimetry suggest that the rate of MSL rise has almost doubled over the last two decades; Church and White (2011) estimated a global trend of 3.2 ± 0.4 mm $year^{-1}$ for the period 1993–2009. Milne et al. (2009) assessed the spatial variability of MSL trends derived from altimetry data and found that local trends vary by as much as −10 to +10 mm $year^{-1}$ from the global average value for the period since 1993, due to regional effects influencing MSL changes and variability (e.g. non-uniform contributions of melting glaciers and ice sheets, density anomalies, atmospheric forcing, ocean circulation, terrestrial water storage). This highlights the importance of regional assessments. Examining whether past MSL has risen faster or slower in certain areas compared to the global average will help to provide more reliable region-specific MSL rise projections for coastal engineering, management and planning.

There have been very few region-wide studies of MSL changes in the North Sea. The first detailed study was by Shennan and Woodworth (1992), who used geological and tide gauge data from sites around the North Sea to infer secular trends in MSL in the late Holocene and 20th century (up until the late 1980s). They concluded that a systematic offset of 1.0 ± 0.15 mm $year^{-1}$ in the tide gauge trends, compared to those derived from the geological data, could be interpreted as the regional average rate of geocentric MSL change over the 20th century; this is significantly less than global rates over this period. They also showed that part of the interannual MSL variability of the region was coherent, and they represented this as an index, created by averaging the de-trended MSL time series. Like Woodworth (1990), they found no evidence for a statistically significant acceleration in the rates of MSL rise for the 20th century.

Since then many other investigations of MSL changes have been undertaken for specific stretches of the North Sea coastline, mostly on a country-by-country basis, as for example by Araújo (2005), Araújo and Pugh (2008), Wöppelmann et al. (2006, 2008) and Haigh et al. (2009) for the English Channel; by van Cauwenberghe (1995, 1999) and Verwaest et al. (2005) for the Belgian coastline; Jensen et al. (1993) and Dillingh et al. (2010) for the Dutch coastline; Jensen et al. (1993), Albrecht et al. (2011), Albrecht and Weisse (2012) and Wahl et al. (2010, 2011) for the German coastline; Madsen (2009) for the Danish coastline; Richter et al. (2012) for the Norwegian coastline; and by Woodworth (1987) and Woodworth et al. (1999, 2009a) for the United Kingdom (UK). The most detailed analysis of 20th century geocentric MSL changes was undertaken by Woodworth et al. (2009a). They estimated that geocentric MSL around the UK rose by 1.4 ± 0.2 mm $year^{-1}$ over the 20th century; faster (but not significantly faster at 95 % confidence) than the earlier estimate by Shennan and Woodworth (1992) for the whole North Sea and slower (but not significantly slower at 95 % confidence level) than the global 20th century rate.

A recent investigation undertaken by Wahl et al. (2013) aimed at updating the results of the Shennan and Woodworth (1992) study, using tide-gauge records that are now 20 years longer across a larger network of sites, altimetry measurements made since 1992, and more precise estimates of vertical land movement made since then with the development of advanced geodetic techniques. They analysed MSL records from 30 tide gauges covering the entire North Sea coastline (Fig. 3.15). Trends in RMSL were found to vary significantly across the North Sea region due to the influence of vertical land movement (i.e. land uplift in northern Scotland, Norway and Denmark, and land subsidence elsewhere). The accuracy of the estimated trends was also influenced by considerable interannual variability present in many of the MSL time series. The interannual variability was found to be much greater along the coastlines of the Netherlands, Germany and Denmark, compared to Norway, the UK east coast and the English Channel (Fig. 3.16).

However, using correlation analyses, Wahl et al. (2013) showed that part of the variability was coherent throughout the region, with some differences between the Inner North Sea (number 4 anti-clockwise to 26 in Fig. 3.15) and the English Channel. Following Shennan and Woodworth (1992), they represented this coherent part of the variability by means of MSL indices (Fig. 3.16). Geocentric MSL trends of 1.59 ± 0.16 and 1.18 ± 0.16 mm $year^{-1}$ were obtained for the Inner North Sea and English Channel indices, respectively, for the period 1900–2009 (data sets were corrected for GIA to remove the influence of vertical land movement). For the North Sea region as a whole, the geocentric MSL trend was 1.53 ± 0.16 mm $year^{-1}$. These results are consistent with those presented by Woodworth et al. (2009a) for the UK (i.e. an AMSL trend of 1.4 ± 0.2 mm $year^{-1}$ for the 20th century), but were significantly different from those presented by Shennan and Woodworth (1992) for the North Sea region (i.e. a geocentric MSL trend of 1.0 ± 0.15 mm $year^{-1}$ for the period from 1901 to the late 1980s). For the 'satellite period' (i.e. 1993 to 2009) the geocentric MSL trend was estimated to be 4.00 ± 1.53 mm $year^{-1}$ from the North Sea tide gauge records. This trend is faster but not significantly different from the global geocentric MSL trend for the same period (i.e. 3.20 ± 0.40 mm $year^{-1}$ from satellite altimetry and

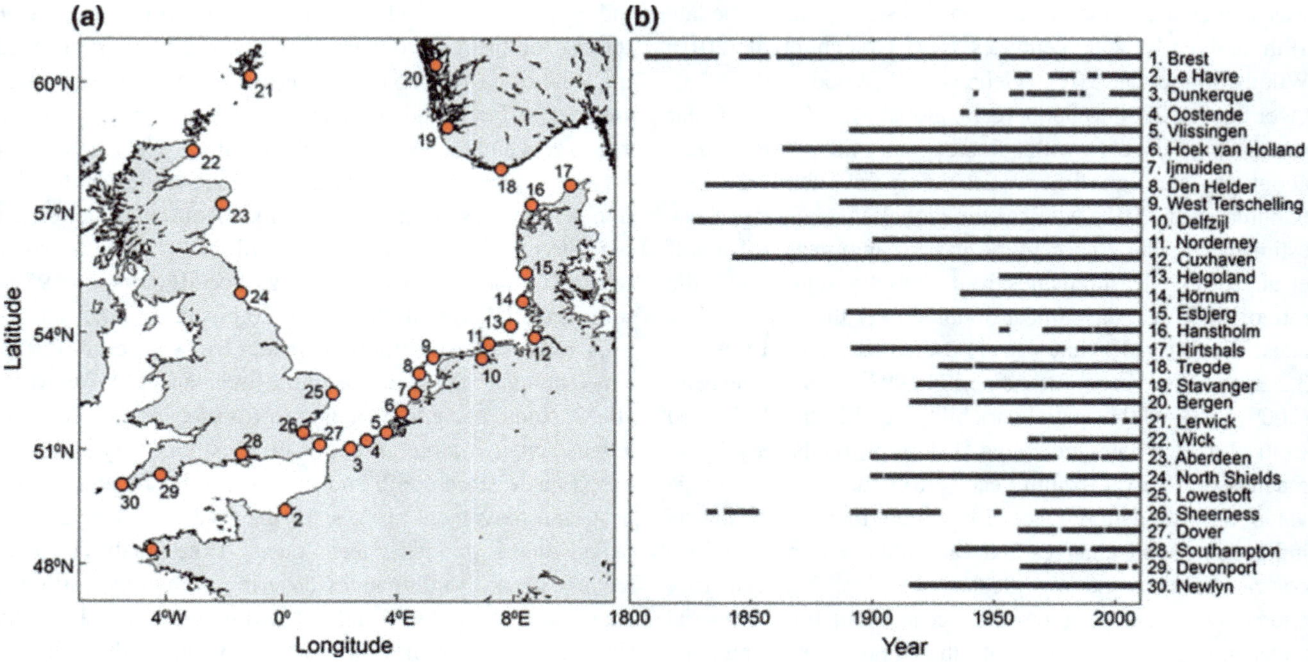

Fig. 3.15 Study area, tide gauge locations and length of individual mean sea level data sets (Wahl et al. 2013)

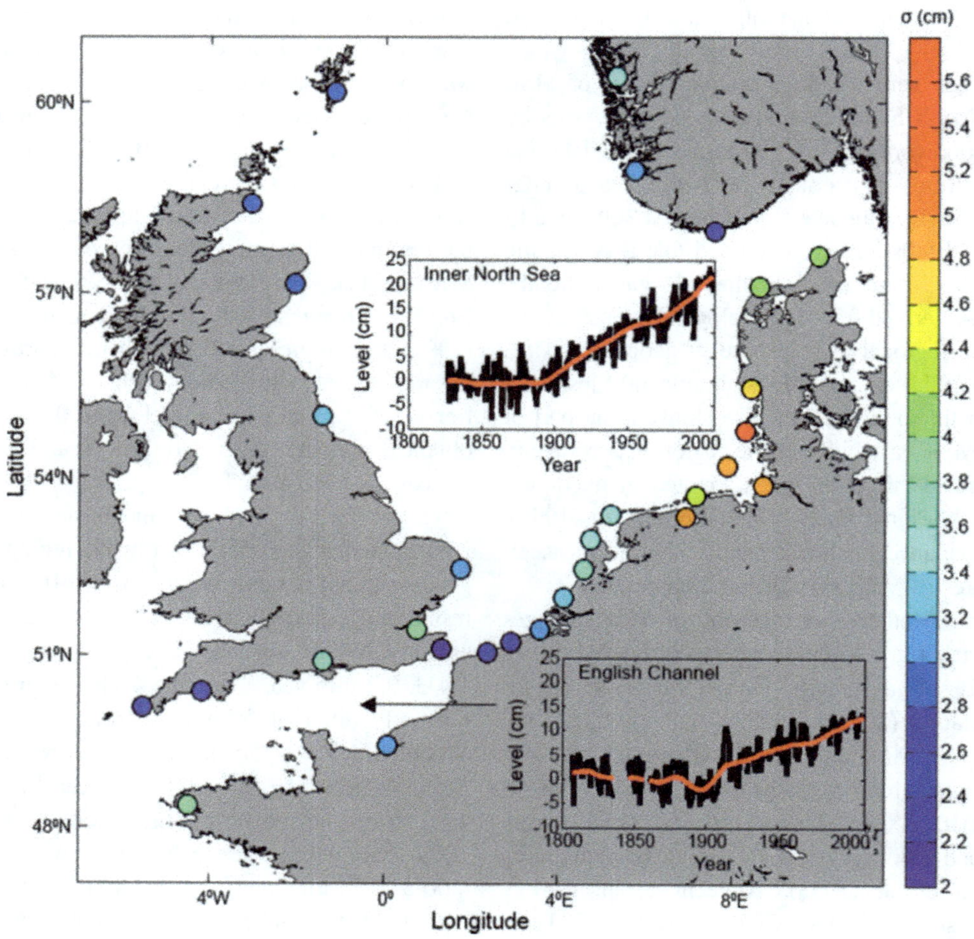

Fig. 3.16 Standard deviation from de-trended annual mean sea level (MSL) time series from 30 tide gauge sites around the North Sea; *upper inset* MSL index for the Inner North Sea (*black*) together with the non-linear sea-surface anomaly (SSA) smoothed time series (*red*); *lower inset* MSL index for the English Channel (*black*) together with the non-linear SSA smoothed time series (*red*) (after Wahl et al. 2013)

2.80 ± 0.80 mm year^{-1} from tide gauge data; Church and White 2011). In summary, the observed long-term changes in sea-level rise (SLR) in the North Sea do not differ significantly from global rates over the same period.

In recent years there has also been considerable focus on the issue of 'acceleration in rates of MSL rise'. Several methods have been applied to examine non-linear changes in long MSL time series from individual tide gauge sites and global or regional reconstructions (see Woodworth et al. 2009b, 2011 for a synthesis of these studies). Wahl et al. (2013) used singular system analysis (SSA) with an embedding dimension of 15 years for smoothing the MSL indices for the Inner North Sea and English Channel (Fig. 3.16). Periods of SLR acceleration were detected at the end of the 19th century and in the 1970s; a period of deceleration occurred in the 1950s. Several authors (e.g. Miller and Douglas 2007; Woodworth et al. 2010; Sturges and Douglas 2011; Calafat et al. 2012) suggested that these periods of acceleration/deceleration are associated with decadal MSL fluctuations arising from large-scale atmospheric changes. The recent rates of MSL rise were found to be faster than on average, with the fastest rates occurring at the end of the 20th century. These rates are, however, still comparable to those observed during the 19th and 20th centuries.

3.5 Extreme Sea Levels

Ralf Weisse, Andreas Sterl

Extreme sea levels pose significant threats (such as flooding and/or erosion) to many of the low-lying coastal areas along the North Sea coast. Two of the more recent examples are the events of 31 January/1 February 1953 and 16/17 February 1962 that caused extreme sea levels along much of the North Sea coastline and that were associated with a widespread failure of coastal protection, mostly in the UK, the Netherlands and Germany (e.g. Baxter 2005; Gerritsen 2005). Since then, coastal defences have been substantially enhanced along much of the North Sea coastline.

Extreme sea levels usually arise from a combination of factors extending over a wide range of spatial and temporal scales comprising high astronomical tides, storm surges (also referred to as meteorological residuals caused by high wind speeds and inverse barometric pressure effects) and extreme sea states (wind-generated waves at the ocean surface) (Weisse et al. 2012). On longer time scales, rising MSL may increase the risk associated with extreme sea levels as it modifies the baseline upon which extreme sea levels act; that is, it tends to shift the entire frequency distribution towards higher values.

The large-scale picture may be modified by local conditions. For example, for given wind speed and direction the magnitude of a storm surge may depend on local bathymetry or the shape of the coastline. Extreme sea states may become depth-limited in very shallow water and effects such as wave set-up (Longuet-Higgins and Stewart 1962) may further raise extreme sea levels. Moreover, there is considerable interaction among the different factors contributing to extreme sea levels, especially in shallow water. For example, for the UK coastline Horsburgh and Wilson (2007) reported a tendency for storm surge maxima to occur most frequently on the rising tide arising primarily from tide-surge interaction. Mean SLR may modify tidal patterns and several authors report changes in tidal range associated with MSL changes. For M2 tidal ranges, estimates vary from a few centimetres increase in the German Bight for a 1-m SLR (e.g. Kauker 1999) to 35 cm in the same area for a 2-m SLR (Pickering et al. 2011). So far, reasons for these differences are not elaborated on in the peer-reviewed literature.

Large sectors of the North Sea coastline are significantly affected by storm surges. A typical measure to assess the weather-related contributions relative to the overall variability is the standard deviation of the meteorological residuals (Pugh 2004). Typically, this measure varies from a few centimetres for open ocean islands hardly affected by storm surges to tens of centimetres for shallow water subject to frequent meteorological extremes (Pugh 2004). For the German Bight, values are in the order of approximately 30–40 cm indicating that storm surges provide a substantial contribution to the total sea level variability (Weisse and von Storch 2009). There is also pronounced seasonal variability with the most severe surges generally occurring within the winter season from November to February reflecting the corresponding cycle in severe weather conditions (Weisse and von Storch 2009).

Extreme sea level variability and change for Cuxhaven, Germany is illustrated in Fig. 3.17. Here a statistical approach was used to separate effects due to changes in MSL and to storm surges (von Storch and Reichardt 1997). The approach is based on the assumption that changes in MSL will be visible both in mean and in extreme sea levels as these changes tend to shift the entire frequency distribution towards higher values. Changes in the statistics of storm surges, on the other hand, will not be visible in the mean but only in the extremes. Following this idea, variations in the extremes may be analysed for example by subtracting trends in annual means from higher annual percentiles while variations and changes in the mean may be obtained by analysing the means themselves. Figure 3.17 shows the result of such an analysis for Cuxhaven, Germany. It can be inferred that the meteorological part (i.e. storm surges) shows pronounced decadal and interannual variability but no substantial long-term trend. The decadal variations are broadly consistent with observed variations in storm activity in the area (e.g. Rosenhagen and Schatzmann 2011; Weisse et al.

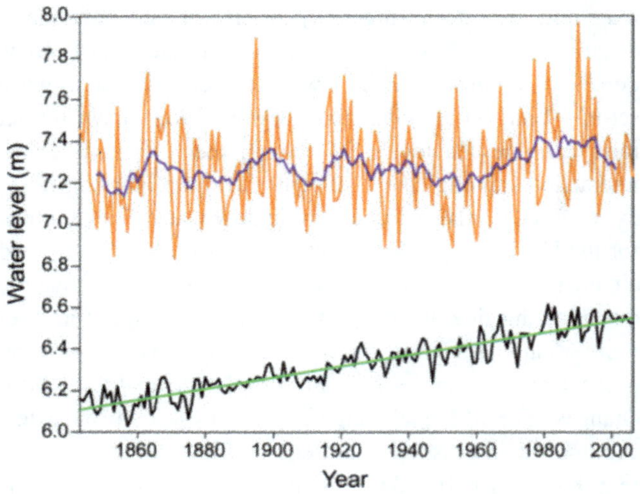

Fig. 3.17 Annual mean high water and linear trend (in m) for the period 1843–2012 at Cuxhaven, Germany (*lower*) and annual 99th percentile of the approximately twice-daily high-tide water levels at Cuxhaven after subtraction of the linear trend in the annual mean levels (*upper*); an 11-year running mean is also shown in the upper panel (redrawn and updated after von Storch and Reichardt 1997)

2012). Figure 3.17 also reveals that extreme sea levels substantially increased over the study period but that changes are primarily a consequence of corresponding changes in MSL and not of storm activity.

An alternative approach to analyse changes in extreme sea levels caused by changing meteorological conditions is by using numerical tide-surge models for hindcasting extended periods over past decades. Such hindcasts are usually set up using present-day bathymetry and are driven by observed (reanalysed) atmospheric wind and pressure fields. In such a design any observed changes in extreme sea levels result solely from meteorological changes while contributions from all other effects such as changes in MSL or local construction works are explicitly removed. Generally, and consistent with the results obtained from observations, such studies do not show any long-term trend but pronounced decadal and interannual variability consistent with observed changes in storm activity (e.g. Langenberg et al. 1999; Weisse and Pluess 2006).

In the analysis of von Storch and Reichardt (1997) annual mean high water is used as a proxy to describe changes in the mean. Climatically induced changes in annual mean high water statistics result principally from two different contributions: (i) corresponding changes in MSL and/or (ii) changes in tidal dynamics. Separating both contributions, Mudersbach et al. (2013) found for Cuxhaven from 1953 onwards that, apart from changes in MSL, extreme sea levels have also increased as a result of changing tidal dynamics. Reasons for the observed changes in tidal variation remain unclear. While increasing MSL represents a potential driver discussed by some authors (e.g. Mudersbach et al. 2013) the magnitude of the observed changes is too large compared to expectations from modelling studies (e.g. Kauker 1999; Pickering et al. 2011) and other contributions (such as those caused by local construction works) could not be ruled out (e.g. Hollebrandse 2005). Other potential reasons for changes in tidal constituents are referred to by Woodworth (2010) and Müller (2012) but have not been explored for the North Sea.

Systematic measurements of sea state parameters exist only for periods much shorter than those from tide gauges. In the late 1980s and early 1990s a series of studies analysed changes in mean and extreme wave heights in the North Atlantic and the North Sea (e.g. Neu 1984; Carter and Draper 1988; Bacon and Carter 1991; Hogben 1994). These were typically based on time series of 15 to at most 25 years and, while reporting a tendency towards more extreme sea states, all authors concluded that the time series were too short for definitive statements on longer-term changes. As for storm surges, numerical models are therefore frequently used to make inferences about past long-term changes in wave climate. Such models are either used globally (e.g. Cox and Swail 2001; Sterl and Caires 2005) or regionally for the North Sea and adjacent sea areas (e.g. WASA-Group 1998; Weisse and Günther 2007). For the North Sea, the latter found considerable interannual and decadal variability in the hindcast wave data consistent with existing knowledge on variations in storm activity.

Results from numerical studies should be complemented with those from statistical approaches. While numerical studies may represent variability and changes with fine spatial and temporal detail, the period for which such studies are possible is presently limited to a few decades. Statistical approaches may bridge the gap by providing information for longer time spans, but are usually limited in spatial and/or temporal detail. Such approaches were used by Kushnir et al. (1997), WASA-Group (1998), Woolf et al. (2002) and Vikebø et al. (2003), exploiting different statistical models between sea-state parameters and large-scale atmospheric conditions. Generally these approaches illustrate the substantial interannual and decadal variability inherent in the North Sea and North Atlantic wave climate. While longer periods are covered, the authors described periods of decreases and increases in extreme wave conditions. For example, Vikebø et al. (2003) described an increase in severe wave heights emerging around 1960 and lasting until about 1999 and concluded that this increase is not unusual when longer periods are considered. This indicates that changes extending over several decades, i.e. typical periods covered by numerical or observational based studies, should be viewed in the light of decadal variability obtained by analysing longer time series.

3.6 Carbon Dioxide, pH, and Nutrients

Helmuth Thomas, Johannes Pätsch, Ina Lorkowski, Lesley Salt, Wilfried Kühn, John Huthnance

Drivers and consequences of climate change are usually discussed from the perspective of physical processes. As such, Sects. 3.2 and 3.3 focus on aspects of physical water column properties (sea temperature, salinity and stratification) and physical interaction with adjacent water bodies (circulation and currents), and climate-change-driven alterations of these. While biogeochemical properties clearly respond to changes in physical conditions, changes can also be modulated by anthropogenic changes in the chemical conditions. These include increasing atmospheric CO_2 levels, ocean acidification as a consequence, and eutrophication/oligotrophication. Relevant time scales can co-vary with those of climate change processes, however they may also be distinctly different (e.g. Borges and Gypens 2010). Furthermore, effects of direct anthropogenic changes (such as nutrient inputs) and feedbacks between anthropogenic and climate changes (atmospheric CO_2 and warming, for example) can be synergistic (amplify each other) or antagonistic (diminish each other). Eutrophication and oligotrophication, feedbacks to changes in physical properties and their effects on productivity in the North Sea have been investigated using models (e.g. Lenhart et al. 2010; Lancelot et al. 2011). Results have been used by international bodies and regulations such as OSPAR, the European Water Framework Directive (EC 2000) and the Marine Strategy Framework Directive. A summary was recently given by Emeis et al. (2015).

The main focus of this section is on the carbonate and pH system of the North Sea and its vulnerability to climate and anthropogenic change. To address these issues, large systematic observational studies were initiated in the early 2000s by an international consortium led by the Royal Netherlands Institute of Sea Research (e.g. Thomas et al. 2005b; Bozec et al. 2006). Observational studies have been supplemented by modelling studies (e.g. Blackford and Gilbert 2007; Gypens et al. 2009; Prowe et al. 2009; Borges and Gypens 2010; Kühn et al. 2010; Liu et al. 2010; Omar et al. 2010; Artioli et al. 2012, 2014; Lorkowski et al. 2012; Wakelin et al. 2012; Daewel and Schrum 2013).

The North Sea is one of the best studied and most understood marginal seas in the world and so offers a unique opportunity to identify biogeochemical responses to climate variability and change. To better understand the sensitivity of the North Sea biogeochemistry to climate and anthropogenic change, this section first discusses some of the main responses to variability in the dominant regional climate mode—the NAO—based on observational data for 2001, 2005 and 2008. The effects of long-term perturbations on the major processes regulating biogeochemical conditions in the North Sea are then discussed based on results from multi-decadal ecosystem model runs. Observations on longer time scales exist locally off the Netherlands, Helgoland and elsewhere but are all from sites close to the coast where strong offshore gradients in nutrients and primary productivity (e.g. Baretta-Bekker et al. 2009; Artioli et al. 2014) affect CO_2.

3.6.1 Observed Responses to Variable External Forcing

In deeper areas of the North Sea, beyond the 50 m depth contour, primary production and CO_2 fixation are supported by seasonal stratification and by nutrients, which are a limiting factor and largely originate from the Atlantic Ocean (Pätsch and Kühn 2008; Loebl et al. 2009). Sinking particulate organic matter facilitates the replenishment of biologically-fixed CO_2 by atmospheric CO_2. Respiration of particulate organic matter below the surface layer releases metabolic dissolved inorganic carbon (DIC) which is either exported to the deeper Atlantic or mixed back to the surface in autumn and winter (Thomas et al. 2004, 2005b; Bozec et al. 2006; Wakelin et al. 2012). These northern areas of the North Sea act as a net annual sink for atmospheric CO_2.

By contrast, in the south (depth <50 m), the absence of stratification causes respiration and primary production to occur within the well-mixed water column. Except during the spring bloom, the effects of particulate organic carbon (POC) production and respiration cancel out and the CO_2 system is largely temperature-controlled (Thomas et al. 2005a; Schiettecatte et al. 2006, 2007; Prowe et al. 2009). Total production in this area is high in global terms; terrestrial nutrients contribute, especially in the German Bight, but in the shallow south, primary production is based largely on recycled nutrients with little net fixation of CO_2.

Beyond the biologically-mediated CO_2 controls, North Atlantic waters, flushing through the North Sea, dominate the carbonate system (Thomas et al. 2005b; Kühn et al. 2010) but may have only small net budgetary effects. The Baltic Sea outflow and river loads constitute net imports of carbon to the North Sea and modify the background conditions set by North Atlantic waters.

Basin-wide observations of DIC, pH, and surface temperature during the summers of 2001, 2005 and 2008 (Salt et al. 2013) reveal the dominant physical mechanisms regulating the North Sea pH and CO_2 system. pH and CO_2 system responses to interannual variability in climate and weather conditions (NAO, local heat budgets, wind and fluxes to or from the Atlantic, the Baltic Sea and rivers, see also Sects. 3.2 and 3.3) are also considered to be the responses that climate change will trigger. Interannual variability appears generally more pronounced than long-term trends (e.g. Thomas et al. 2008).

The NAO index (Hurrell 1995; Hurrell et al. 2013) is commonly established for the winter months (DJF), although its impacts have been identified at various time scales. Many processes in the North Sea are reported to be correlated with the winter NAO, even if they occur in later seasons. Two aspects may explain an apparent delay between the trigger (i.e. winter NAO) and the response (the timing of the actual process): preconditioning and hysteresis (Salt et al. 2013).

An example of pre-conditioning is the water mass exchange between the North Atlantic Ocean and the North Sea. This exchange is enhanced during years of positive NAO (Winther and Johannessen 2006) and leads to an increased nutrient inventory in the North Sea and to higher annual productivity in spring and summer (Pätsch and Kühn 2008). Hysteresis can be characteristic of the North Sea's response to the NAO. Stronger westerly winds in winter, correlated with the winter NAO, push North Sea water into the Baltic Sea, a process that in turn leads to an enhanced outflow from the Baltic Sea into the North Sea in subsequent seasons (Hordoir and Meier 2010).

The bottom topographic divide of the North Sea, at about 40–50 m depth, is reflected in DIC, pH and temperature distributions (Figs. 3.18, 3.19 and 3.20) with higher DIC and temperature, and lower pH observed in the south, which is under stronger influence of terrestrial waters. In summer 2001, the year with the most negative NAO, the lowest DIC values and highest pH values were observed across the entire basin, whereas 2005 and 2008 were both characterised by higher DIC and lower pH, with some variability in these patterns across the North Sea. Summer 2005 had the coolest surface waters.

For winter NAO values, 2001 was the most negative (−1.9), 2005 was effectively neutral (0.12) and 2008 was positive (2.1). Weaker winds and circulation in the North Sea are associated with negative NAO (see Sects. 1.4.3 and 3.3.2) and reduce the upward mixing of cold winter water (Salt et al. 2013). Hence, metabolic DIC accumulated in deeper waters during the preceding autumn and winter

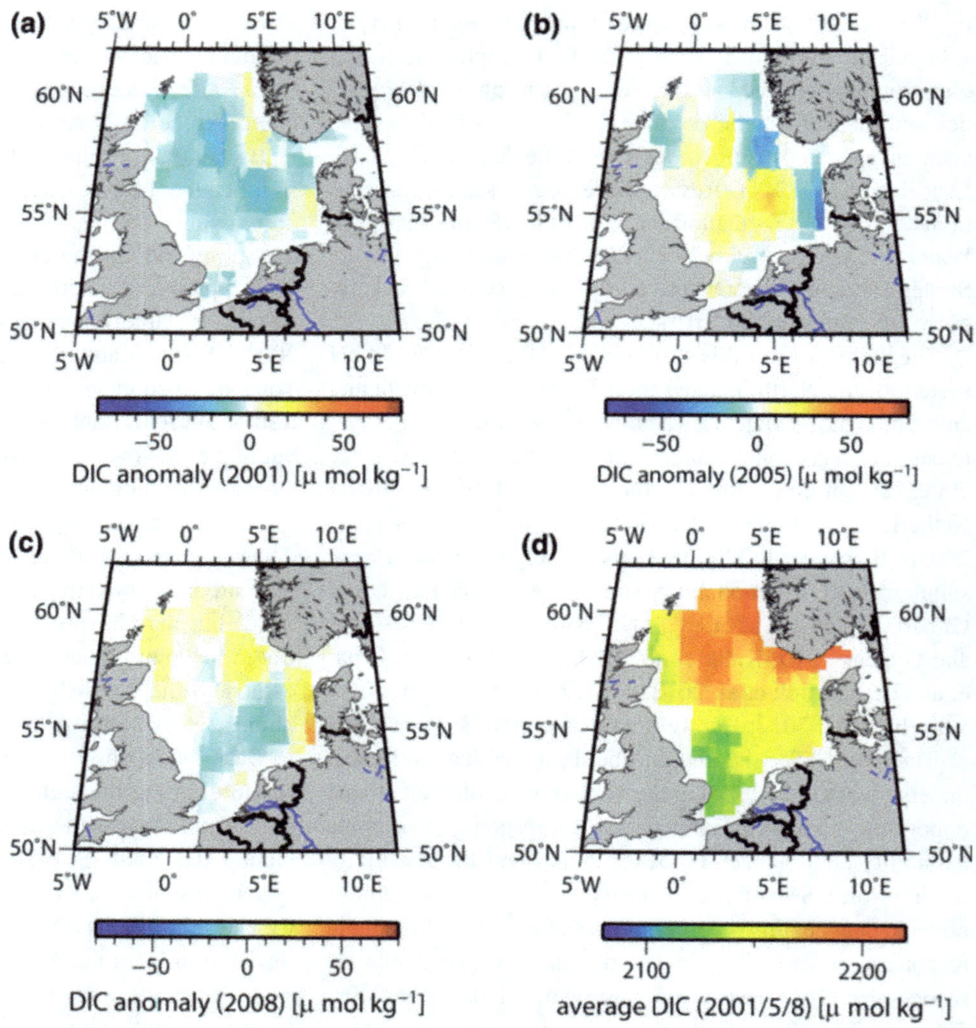

Fig. 3.18 Observed variability in surface water dissolved inorganic carbon (DIC) concentrations. All observations were made in summer (August/September) of the years 2001, 2005 and 2008 (Salt et al. 2013). Anomalies are shown relative to the average observed values for these years (also shown; figure by Helmuth Thomas, Dalhousie University, Canada)

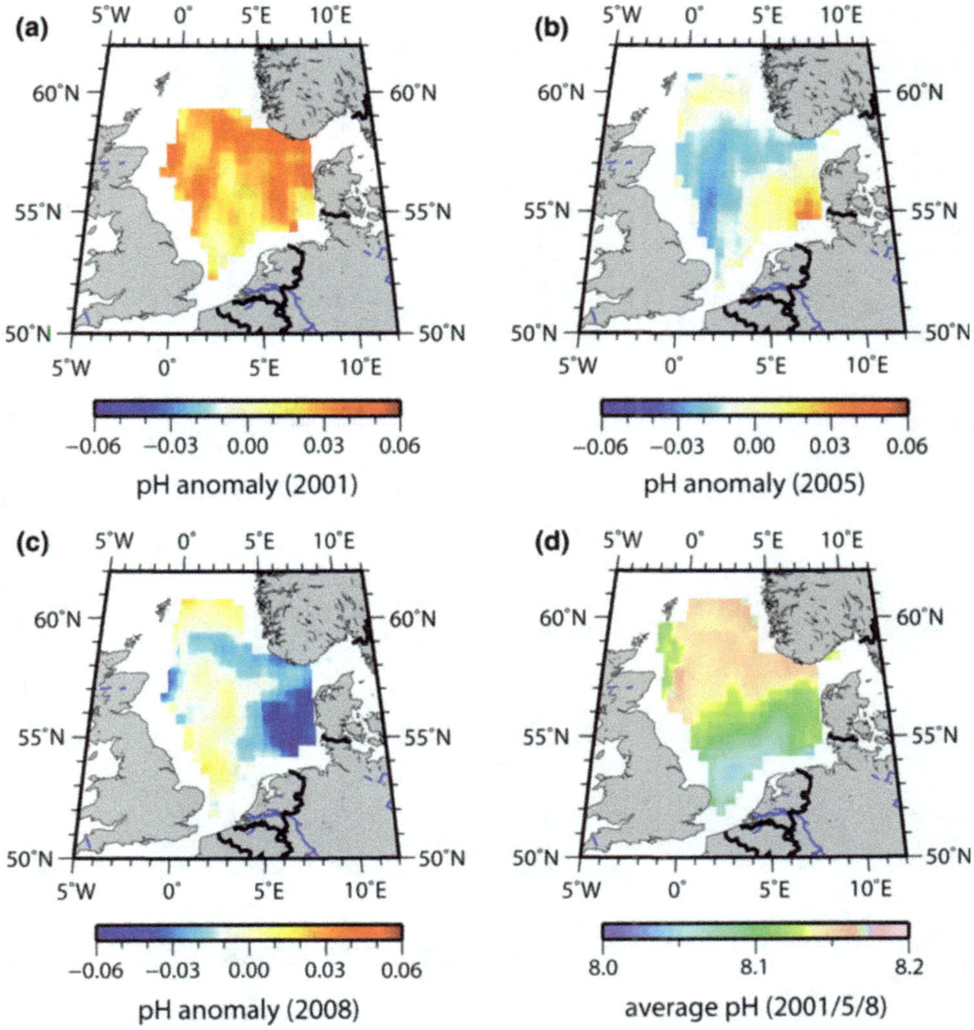

Fig. 3.19 Observed variability in surface water pH. All observations were made in summer (August/September) of the years 2001, 2005 and 2008 (Salt et al. 2013). Anomalies are shown relative to the average observed values for these years (figure by Helmuth Thomas, Dalhousie University, Canada)

(Thomas et al. 2004) was mixed into surface waters to a lesser extent in 2001 than in 2005 or 2008 when wind or circulation-driven mixing was stronger (see also Salt et al. 2013), which explained the elevated surface DIC and lower pH in 2005 and 2008 relative to 2001 (Figs. 3.18 and 3.19).

The striking difference between 2001 and 2005 in the northern North Sea (Thomas et al. 2007) was reinforced by the warmer summer with a shallower mixed layer in 2001 (Salt et al. 2013: their Fig. 5). Comparable biological activity caused the shallower mixed layer of 2001 to experience stronger biological DIC drawdown on a concentration basis, resulting in higher pH, than in 2005 (Figs. 3.18 and 3.19).

Interaction with the North Atlantic Ocean also causes variability in the CO_2 system, partly explained by NAO-dependent circulation changes (Thomas et al. 2008; Watson et al. 2009). Figure 3.21 shows the net flow of water in the first half of the three respective years. 2008 (positive NAO) has the strongest north-western inflow of DIC-enriched North Atlantic waters to the North Sea, via the Fair Isle Current and Pentland Firth, although 2001 had strong inflow from the north which recirculated out of the North Sea quickly off Norway (Lorkowski et al. 2012).

Such an influence of North Atlantic inflow is supported by strong correlations between changes in the inventories of salinity and corrected DIC (i.e. accounting for biological effects) during the periods 2001–2005 and 2005–2008 (Salt et al. 2013). Mean values of partial pressure of CO_2 (pCO_2) in the water (331.6 ppm in 2001, 352.5 ppm in 2005, 364.0 ppm in 2008) reflect the large change between 2001 and 2005 and the moderate change between 2005 and 2008. Also, strong NAO-driven anti-clockwise circulation in the North Sea in 2008 intensified the distinct characteristics of the southern and northern North Sea and sharpened the transition between them (e.g. high to low pH, see Salt et al. 2013: their Fig. 2).

Modelling results (Lorkowski et al. 2012) agree with several of these findings: a mixed layer shallower in 2001 and 2008 than in 2005, which had the coolest summer surface waters; central North Sea DIC concentrations about 10 μmol/kg less than average in 2001.

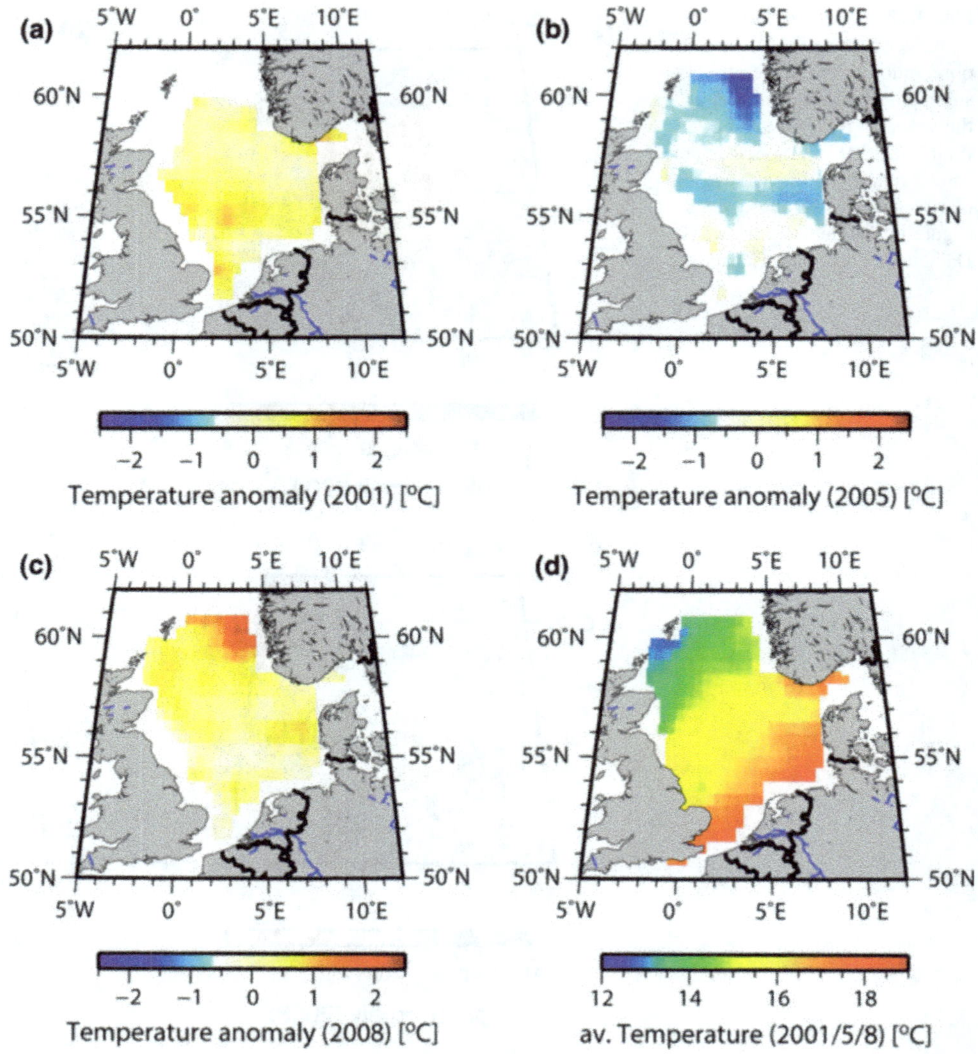

Fig. 3.20 Observed variability in sea surface temperature. All observations were made in summer (August/September) of the years 2001, 2005 and 2008 (Salt et al. 2013). Anomalies are shown relative to the average observed values for these years (figure by Helmuth Thomas, Dalhousie University, Canada)

In summary, three factors regulate the North Sea's CO_2 system and thus reveal points of vulnerability to climate change and more direct anthropogenic influences: local weather conditions (including water temperature in the shallower southern North Sea), circulation patterns, and end-member properties of relevant water masses (Atlantic Ocean, German Bight and Baltic Sea). Thus a positive NAO increases Atlantic Ocean and Baltic Sea inflow, the anti-clockwise circulation, carbon export out of the Norwegian Trench below the surface (limiting out-gassing) and hence the effectiveness of the shelf-sea CO_2 'pump' (Salt et al. 2013). If the NAO is positive together with higher SST, a shallower mixed layer favours lower surface pCO_2 and higher pH in the northern North Sea. These factors can be considered key to regulation of the North Sea's response to climate change and more direct anthropogenic influences.

3.6.2 Model-Based Interannual Variations in Nitrogen Fluxes

The North Sea is a net nitrogen sink for the Atlantic Ocean, due to efficient flushing by North Atlantic water with strong nitrogen concentrations and to large rates of benthic denitrification in the southern North Sea (Pätsch and Kühn 2008). This is the case despite large nitrogen inputs from the rivers and atmosphere. There is net production of inorganic nitrogen from organic compounds.

Pätsch and Kühn (2008) investigated nitrogen fluxes in 1995 and 1996 as the NAO shifted from very strong positive conditions in winter 1994/1995 to extreme negative conditions in winter 1995/1996. Due to enhanced ocean circulation on the Northwest European Shelf, the influx of total nitrogen from the North Atlantic was much stronger in 1995

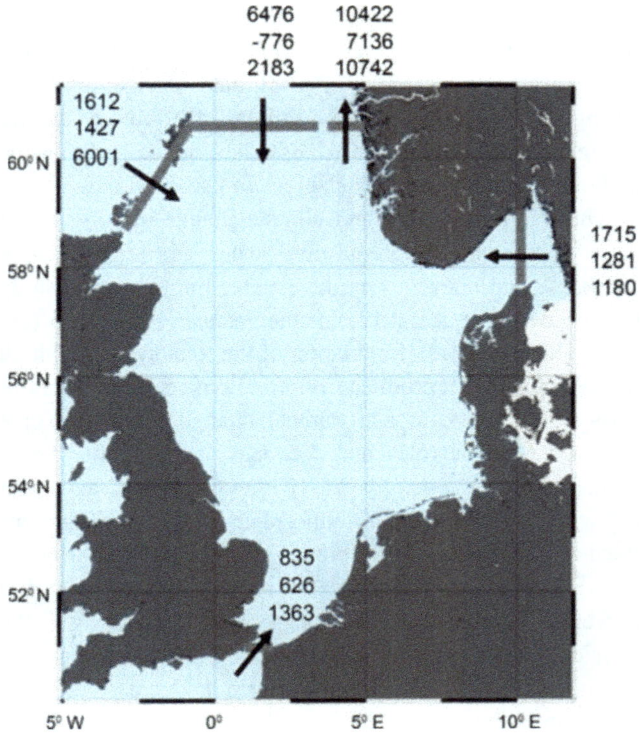

Fig. 3.21 Simulated cumulative net flux of water from 1 January to 30 June (km^3 per half year) for the years 2001 (*upper values*), 2005 (*middle values*), 2008 (*lower values*) (Lorkowski et al. 2012)

(NAO positive) than in 1996. River input of nitrogen was also larger in 1995 than 1996. While the import of organic nitrogen was similar for both years, the import of inorganic nitrogen was larger in 1995 than in 1996. The ecosystem response was stronger dominance of remineralisation over production of organic nitrogen in 1996 with negative NAO conditions.

According to this simulation, in 1996 (with extreme negative winter NAO) the net-heterotrophic state of the North Sea was stronger than in 1995. As a result, the biologically-driven air-to-sea flux of CO_2 was larger in 1995 than in 1996 (Kühn et al. 2010). In other words, in positive NAO years stronger fixation of inorganic nitrogen and inorganic carbon facilitates stronger biological CO_2 uptake. This carbon is exported into the adjacent North Atlantic in positive NAO years, as reported above. The balance between respiration and production in regulating DIC and pCO_2 conditions thus acts in synergy with the processes discussed in Sect. 3.6.1. At regional and sub-regional scales, modelling studies have investigated the concurrent impacts of eutrophication, increases in atmospheric CO_2 and climate change on the Southern Bight of the North Sea (Gypens et al. 2009; Borges and Gypens 2010; Artioli et al. 2014). The studies clearly highlight the complex effects of the individual drivers, as well as the different time scales of impact. Eutrophication, oligotrophication and temperature variability affect the CO_2 system at interannual to decadal time scales. Long-term trends of increases in atmospheric CO_2 and rising temperature have begun to cause tangible effects (e.g. Artioli et al. 2014) although, to date, these have been much less pronounced than effects at shorter time scales.

3.6.3 Ocean Acidification and Eutrophication

The interplay of the different anthropogenic and climate change processes, as well as their different, obviously overlapping time scales, can be exemplified with respect to the long-term effects of ocean acidification and the shorter-term effects of eutrophication/oligotrophication. Effects of eutrophication are closely related to the trend of ocean acidification, since both affect DIC concentrations and the DIC/A_T ratio (A_T: total alkalinity) in coastal waters, and thus CO_2 uptake capacity. Increased nutrient loads may lead to enhanced respiration of organic matter, which releases DIC and thus lowers pH. On shorter time scales, enhanced respiration overrides ocean acidification, which acts at centennial time scales (e.g. Borges and Gypens 2010; Artioli et al. 2014). (Surface-ocean pH has declined by 0.1 over the industrial era, in the North Sea as well as globally, and a hundred times faster in recent decades than during the previous 55 million years; EEA 2012).

If eutrophication-enhanced respiration of organic matter exhausts available oxygen, respiration then takes place through anaerobic pathways. Denitrification is crucial here; the biogeochemical consequences of depleted oxygen are many. Under eutrophic conditions, release of nitrate (NO_3) by enhanced respiration is controlled by the amount of available oxygen. If oxygen is depleted, NO_3 is converted to nitrogen gas (N_2). Any further input of NO_3 stimulates denitrification. The lost NO_3 is not available for biological production, thus the system is losing reactive nitrogen (Pätsch and Kühn 2008) as with eutrophication in the Baltic Sea (Vichi et al. 2004). A transition from aerobic to anaerobic processes has consequences for CO_2 uptake capacity and pH regulation: denitrification driven by allochthonous NO_3 releases alkalinity in parallel with the metabolic DIC, with a DIC/A_T ratio of 1:1.

Compared with aerobic respiration, which gives a DIC/A_T ratio of −6.6, the release of alkalinity in denitrification increases the CO_2 and pH buffer capacity of the waters, in turn buffering ocean acidification. Since denitrification is irreversible, the increased CO_2 and pH buffer capacity will persist on time scales relevant for climate change. In other words, if eutrophication yields anaerobic metabolic pathways, this constitutes a negative feedback to climate change, since more CO_2 can be absorbed from the atmosphere, which in turn dampens the CO_2 greenhouse gas effect.

Other anaerobic pathways such as sulphate or iron reduction give even lower DIC/A_T release ratios (Chen and Wang 1999; Thomas et al. 2009); those may be reversible, however. Reduced (nitrogen-) nutrient input (i.e. oligotrophication) thus comes with a negative feedback with regard to ocean acidification: a desirable reduction in NO_3 release enhances vulnerability of the coastal ecosystem to ocean acidification, since most organic matter respiration is on or in shallow surface sediments (Thomas et al. 2009; Burt et al. 2013, 2014).

3.6.4 Variability on Longer Time Scales

Climate, CO_2 and more direct anthropogenic drivers also determine the variability of carbon fluxes in the North Sea. They can all be indicated as negative or positive feedback mechanisms for CO_2 exchange with the atmosphere and thus as feedbacks on climate change. The main direct anthropogenic impact on the carbon cycle, mostly for the southern North Sea, is the input of bio-reactive tracers, namely nutrients, via the atmosphere and rivers. Indirect anthropogenic drivers include acidification due to the ongoing increase in atmospheric pCO_2. Climate change processes (rising SST and changes in salinity distribution due to changes in circulation and winds) also induce shifts in the carbonate system and thus changes in carbon fluxes.

These anthropogenic and climate-change drivers, which act at interannual to decadal time scales, and their potential feedbacks and impacts were investigated in the model study by Lorkowski et al. (2012) for the years 1970 to 2006 (extended here to 2009). Simulation of the total system with all drivers included reproduced observations. Scenarios, mimicking anthropogenic and climate change processes, give insight into their roles and feedback mechanisms. These scenarios were generally run without biology, and with either fixed temperature or atmospheric CO_2 concentrations fixed at 1970 values. Both 'biotic' and 'abiotic' scenarios are shown here (Figs. 3.22 and 3.23, respectively), the latter to prevent biological feedbacks overshadowing the physically-driven and biogeochemically-driven responses.

The 'standard' simulation showed a decrease in CO_2 uptake from the atmosphere in the last decade (Fig. 3.22), an increase in SST by 0.027 °C year^{-1} and a decrease in winter

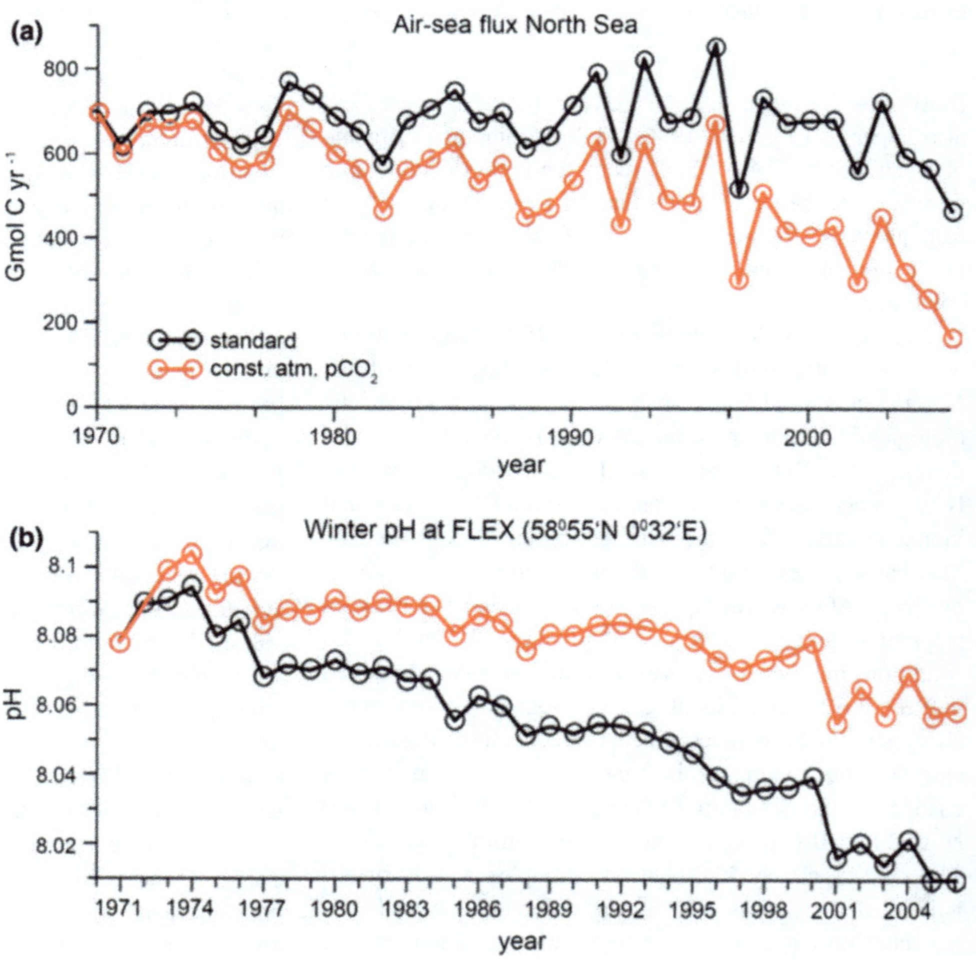

Fig. 3.22 Carbon dioxide (CO_2) air-sea fluxes for the total North Sea (*upper*, black curve reprinted from Fig. 5a in Lorkowski et al. 2012) and winter pH at one station in the northern North Sea (*lower*). Standard simulation (*black*); repeated annual cycle of atmospheric CO_2 (*red*) (figure by Helmuth Thomas, Dalhousie University, Canada)

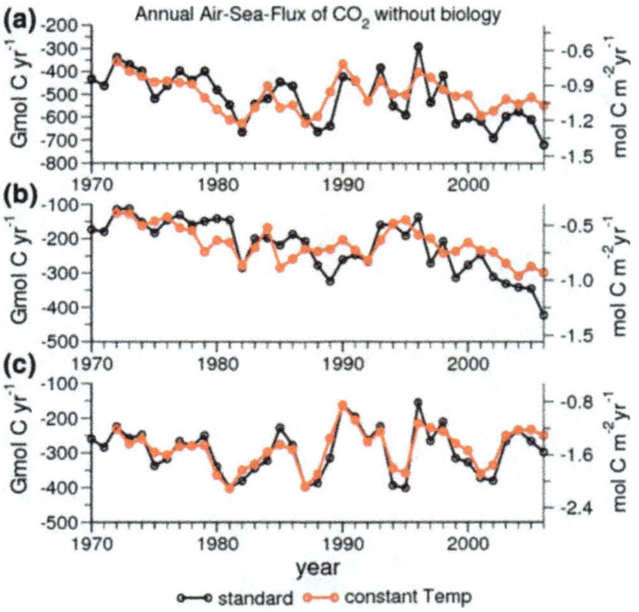

Fig. 3.23 Annual air-sea carbon dioxide (CO_2) flux for 'abiotic' simulations: total North Sea (*upper*), northern North Sea (*middle*), southern North Sea (*lower*). *Black* Results for standard conditions (Fig. 8 in Lorkowski et al. 2012); *red* results from the simulation with a repeated annual cycle of 1972 temperature. NB. Scales differ between the plots (figure by Helmuth Thomas, Dalhousie University, Canada)

pH by 0.002 $year^{-1}$ (Lorkowski et al. 2012). Thus climate change alone (i.e. rising sea temperature) thermodynamically raises the pCO_2 and reduces CO_2 uptake in the North Sea. Furthermore, warming waters cause a lower pH, thus increased surface water acidity (Fig. 3.22).

Increasing atmospheric pCO_2 during the 'standard' simulation increases the gradient between seawater and atmospheric pCO_2 and increases the (net-) CO_2 uptake. To investigate this, the standard simulation is compared with a simulation using a repeated 1970 annual cycle of atmospheric pCO_2 (Fig. 3.22). 1970 pCO_2 (with rising temperature in common) leads to a smaller air-sea flux and less CO_2 uptake. pH decreases less than in the standard simulation (Fig. 3.22). Thus the simulations show enhanced CO_2 uptake in the North Sea as a consequence of rising atmospheric pCO_2, in turn increasing North Sea acidification as a 'local' process. This experiment also shows that for today's carbonate-system-status the increase in atmospheric CO_2 has a stronger impact on air-sea flux of CO_2 than the reduction in the buffer capacity by the ongoing acidification. This trend in acidification might be overlain on shorter time scales by advective processes (Thomas et al. 2008; Salt et al. 2013) as discussed in Sect. 3.6.1, by eutrophication (Gypens et al. 2009; Borges and Gypens 2010; Artioli et al. 2014) or by variability in biological activity.

Climate change enhances the hydrologic cycle, which means enhanced precipitation and river runoff, which drive changes in surface water salinity. Salinity decrease generally represents a dilution of DIC and A_T, with the DIC-effect dominating the A_T-effect on pCO_2 and pH (e.g. Thomas et al. 2008). Changes in salinity also alter the equilibrium conditions of the carbonate system (a minor effect): on addition of freshwater, pCO_2 decreases and pH increases. In coastal areas, precipitation-evaporation effects are confounded by changes in the mixing ratios of the dominant water masses, i.e., runoff and the oceanic end-member; higher salinity can mean a larger proportion of oceanic water relative to river runoff and vice versa. A sensitivity study, with salinity reduced by 1 (compared with the standard setup) and no biological processes, showed 10 % less outgassing, slightly counteracting the effect of rising temperature. In summary, rising temperature reduces uptake of atmospheric CO_2; increasing atmospheric pCO_2 or reduced salinity increases net uptake of atmospheric CO_2.

3.7 Oxygen

John Huthnance, Franciscus Colijn, Markus Quante

Oxygen is of concern because depletion (hypoxia) adversely affects ecosystem functioning and can lead to fish mortality. Air-sea exchange and photosynthesis tend to keep upper waters oxygenated; oxygen concentrations can be strongest in the thermocline associated with a sub-surface chlorophyll maximum (Queste et al. 2013). However, oxygen concentration near the sea bed can be reduced by organic matter respiration below stable stratification, breakdown of detrital organic matter in the sediment and lack of oxygen supply (by advection or vertical mixing). Temperature is also a factor; warmer waters can contain less oxygen but increase metabolic rates. Extra nutrients from rivers and estuaries can increase the amount of respiring organic matter. In the North Sea, most areas are well-oxygenated but some areas are prone to low oxygen concentrations near the bottom—the Oyster Grounds (central North Sea), off the Danish coast (Karlson et al. 2002) and locally near some estuaries, as in the German Bight. Climate change may influence oxygen concentrations through changes in absolute water temperature as well as through changes in temperature gradient, storm intensity and frequency, and related changes in mixing.

Data are available from the International Council for the Exploration of the Sea (ICES) for the past 100 years or so, research cruises (notably August 2010; Queste et al. 2013) and models (e.g. Meire et al. 2013; Emeis et al. 2015). The deep oxygen distribution and its relation to stratification is illustrated in Fig. 3.24.

There is strong interannual variability in the oxygen concentration of the bottom water in late summer. Published

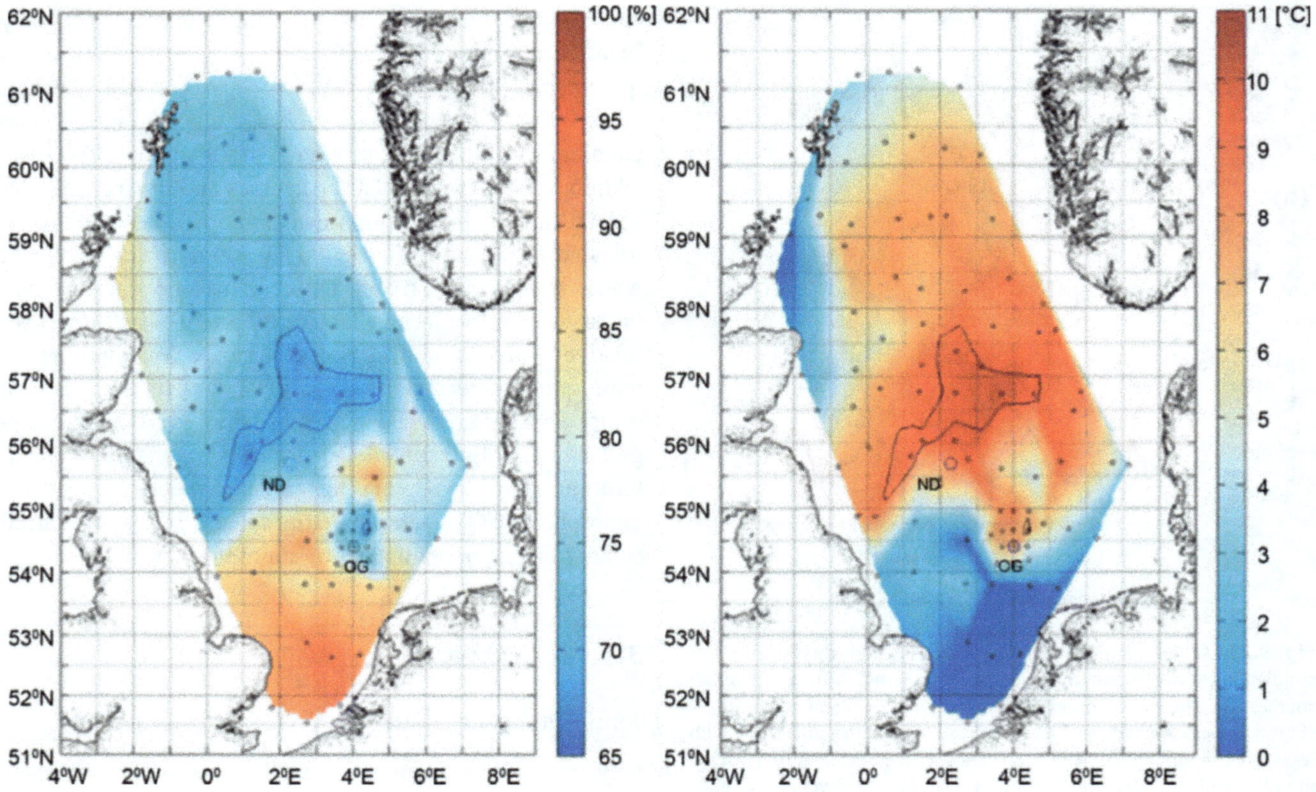

Fig. 3.24 August 2010 CTD casts for the shallowest 3 m and deepest 3 m of the water column. Deepest oxygen saturation (%, *left*) and temperature difference between surface and bottom mixed layer (°C, *right*). The 70 % saturation contour highlights similarity in the distributions (Queste et al. 2013)

oxygen minimum values vary from 65 to 220 µM (Meire et al. 2013). A value of 65 µM in the Oyster Grounds indicates that this area could be on the brink of hypoxia in exceptional years. Models suggest that it is the effect of warming on stratification, rather than on decreased oxygen solubility or enhanced respiration rates, that is the main physical factor affecting bottom oxygen concentrations (Meire et al. 2013; Emeis et al. 2015). Comparing the periods 1970–1979 and 2000–2009, Emeis et al. (2015) found the more recent period to show a longer period of stratification in the middle of the North Sea with increased apparent oxygen utilisation and a lower September oxygen minimum around 6°E 56°N. They also found decreased September oxygen concentrations in the south-eastern German Bight around 54°S. However, storms can promote sediment resuspension and subsequent oxygen consumption, such as in the Oyster Grounds (Greenwood et al. 2010). Observed changes in summer bottom temperature and oxygen concentration show a strong relation (Queste et al. 2013 and Fig. 3.25).

Climate change, for example raised water temperatures, is expected to have a negative impact on oxygen concentrations in surface waters, and deepening of the thermocline will reduce the bottom mixed layer and may cause further depletion of oxygen concentrations in deeper layers. However, quantifying temperature effects is difficult, owing to climate-related effects on nutrient inputs to the North Sea as well as on local mixing characteristics and the duration of reduced oxygen concentrations.

3.8 Suspended Particulate Matter and Turbidity

Julie Pietrzak, Alejandro Jose Souza, John Huthnance

Suspended particulate matter (SPM) is a significant agent of change in morphology, it also transports pollutants, redistributes nutrients and modifies the light climate (Capuzzo et al. 2013), hence its role in modulating primary production. Suspended particulate matter includes plankton.

3.8.1 Sources in the North Sea

The seabed is an important source of SPM in the North Sea. Rivers and cliffs are also important sources in certain areas. Cliff sources are very variable interannually.

Fig. 3.25 Five-yearly values of an 11-year running mean of summer bottom-mixed-layer temperature (°C, *upper*), oxygen concentration (μmol dm^{-3}, *middle*) and oxygen saturation (%, *lower*) in the stratified central North Sea. The data are for June to September below 30 m, in regions deeper than 45 m and in grid 'squares' (1° longitude × 0.5° latitude) north of 56°N with more than five data points (total of 16,250 measurements). *Error bar* length shows two standard errors. The *horizontal line* represents the overall mean of the time series. The number of grid squares retained and the standard deviation are indicated below each data point (Queste et al. 2013)

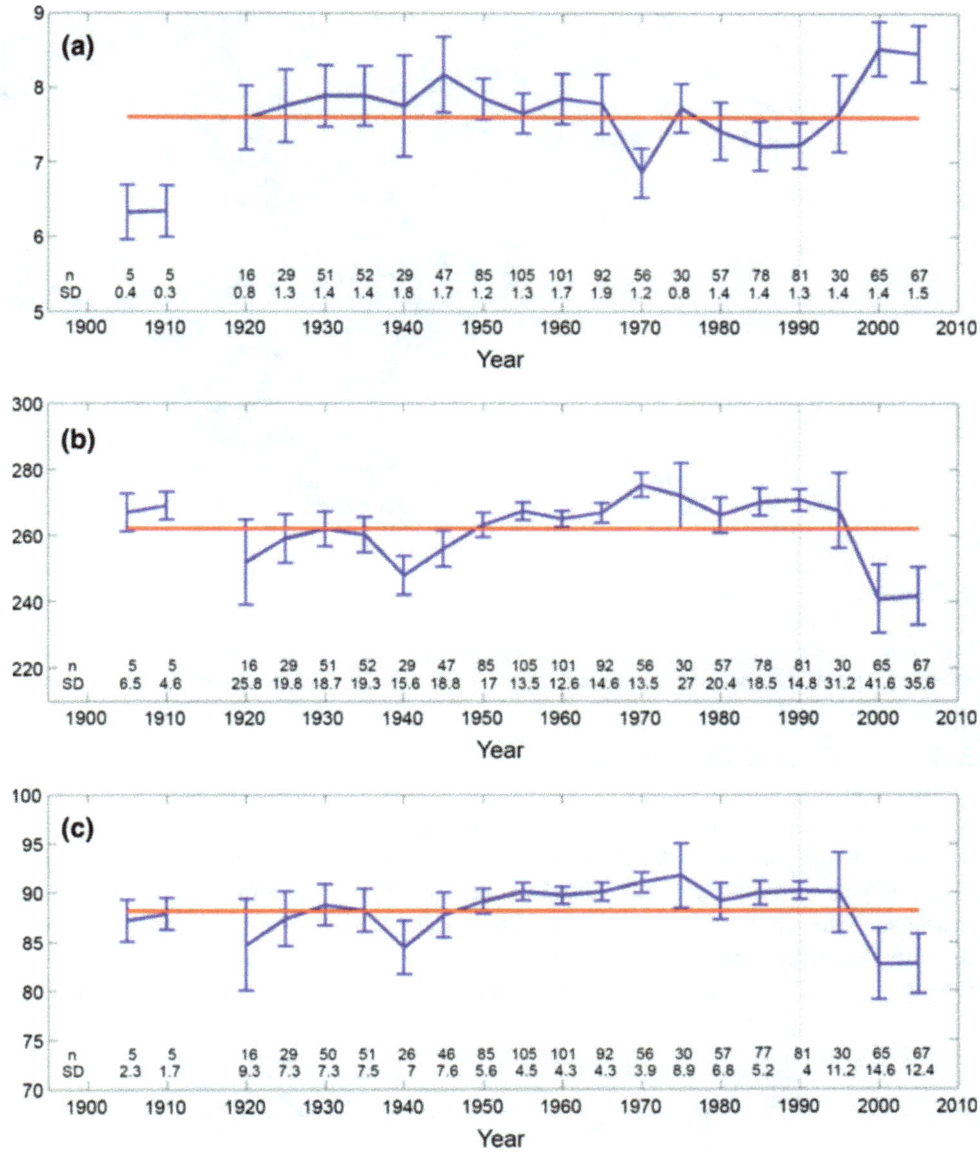

Bottom sediment distributions in the North Sea have been reported by many authors (such as Eisma and Kalf 1987 and Nedwell et al. 1993, using data from the UK NERC North Sea Project). More recently, Dobrynin et al. (2010; Fig. 3.26) compiled a seabed fine-sediment distribution that combines in situ data (Puls et al. 1997) and satellite data (Gayer et al. 2006).

The main river sources are the Elbe, Weser, Ems, Rhine, Thames, Welland, Humber, Tees, and Tyne, as well as the Forth, Ijssel and the Nordzeekanaal. River discharge and consequently SPM load (Fig. 3.27) show strong interannual variability. For example, freshwater discharge and SPM loads for the major continental rivers (i.e. Rhine, Elbe and Weser) were much less in 2003 than in 2002, owing to the low precipitation and very high temperatures in central and western Europe in 2003 (Gayer et al. 2006).

Long-term measurements of annual mean amounts of SPM eroded from English cliffs imply the following average rates: Suffolk, 50 kg s^{-1}; Norfolk, 45 kg s^{-1}; and Holderness, 58 kg s^{-1}. SPM loads eroded from cliffs are dependent on whether storm or calm conditions occur (Fig. 3.27). According to Gayer et al. (2006) cliff erosion appears to start when significant wave heights near the coast exceed 2 m. Sediment composition suggests that SPM for alongshore transport off the Belgian and Dutch coasts is largely supplied by sediment transported through Dover Strait from the erosion of the French and British cliff coasts (Irion and Zöllmer 1999; Fettweis and van den Eynde 2003). The transport through Dover Strait largely exceeds the fluvial input by rivers such as the Rhine-Meuse estuary (de Nijs 2012). The annual sediment influx shows large interannual variations which appear to reflect differences in number and duration of storms (van Alphen 1990).

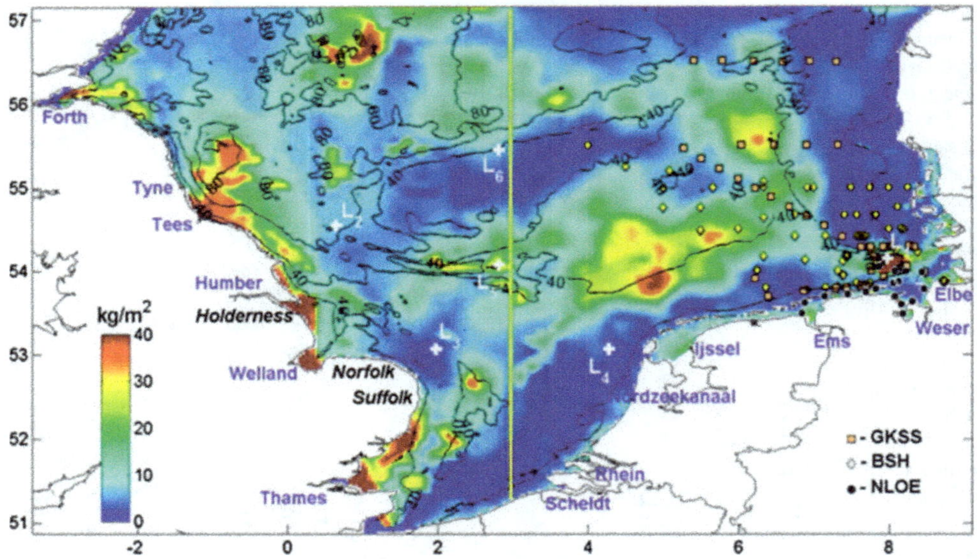

Fig. 3.26 Sea-bed fine sediment (<20 μm; settling velocities 0.01–0.1 mm s^{-1}). Distribution in kg m^{-2} shown by *colour shading*; bathymetry (m) shown by *black contour lines*. The distribution is the result of combining sediment grain analysis, satellite data and numerical model results. The map (reprinted from Dobrynin et al. 2010) also shows the most important rivers and cliffs. (Labels $L_{1\text{-}6}$ and the *vertical yellow line* are not relevant here). *Marked points* show sites of data used by Dobrynin et al. (2010)

3.8.2 Overall Distribution

Sediment is transported either as bed load (typically for coarse material) or suspended load (SPM). Advances in observational techniques, from water samples to in situ instruments (transmissometers, optical backscatter and Laser In Situ Scattering and Transmissometery—LISST) and reliable use of optical remote sensing (e.g. AVHRR, SeaWiFS, recent MODIS and MERIS) have increased understanding of SPM distribution. Remote sensing techniques provide a synoptic view of the sea surface at fine temporal (daily) and spatial (kilometre) resolution, providing information on variability in SPM distribution. The ability to estimate near-surface SPM loads, relatively continuously and at synoptic scale, has allowed study of surface SPM over seasonal cycles: for example it has been observed that high

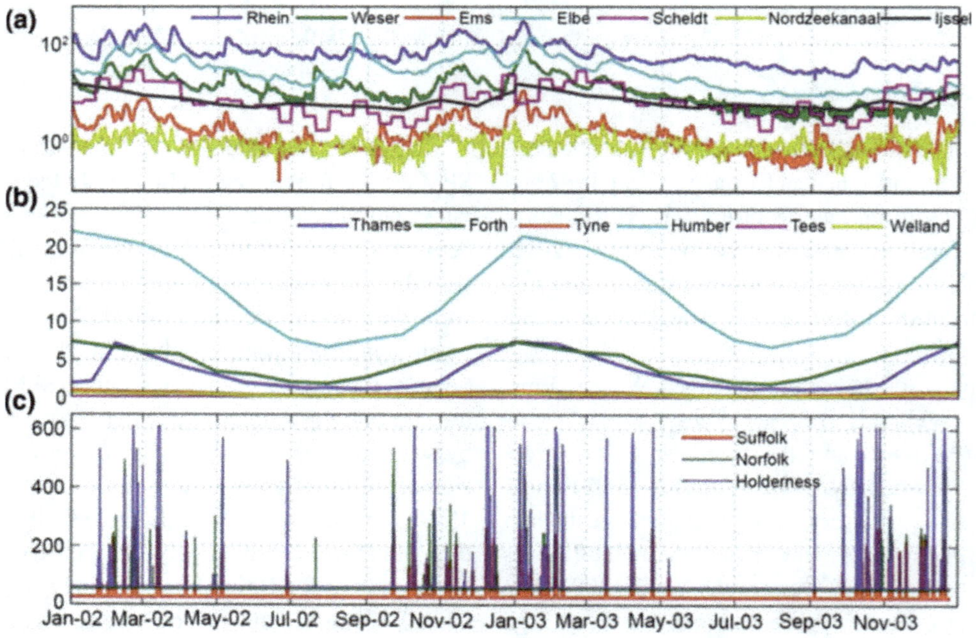

Fig. 3.27 Sources of SPM (kg s^{-1}) to the North Sea from continental Europe (*upper*), UK rivers (*middle*) and UK cliffs (*lower*) during 2002 and 2003. Each river's SPM load is the product of freshwater discharge (m^3 s^{-1}) and annual mean SPM concentration (kg m^{-3}; Gayer et al. 2004). River discharges are a combination of measured data and climatology (reprinted from Dobrynin et al. 2010)

SPM concentrations evolve during winter, with much lower values in summer (Eleveld et al. 2004, 2006, 2008). However, it should be noted that satellites provide information concerning the sea surface only. When the water column is well mixed SPM can be remotely observed at the sea surface; when it is stratified (as in the Rhine ROFI) SPM can only be observed remotely for high discharge events and close to the mouth (of the Rotterdam Waterway), see Sect. 3.8.5.

Suspension of particles off the bed needs stronger currents (including waves and turbulence) than the limit for the same particles to settle. If the currents allow settling, there is still transport until the particles reach the bed. These biases, tidal straining and current asymmetries cause net transport of SPM. Moored instruments have allowed better understanding of tidal, spring-neap and vertical variability (Jones et al. 1996).

Eleveld et al. (2008) found that annual, winter and summer remote sensing SPM observations highlighted the dominant North Sea water types as characterised by Lee (1980) in terms of salinity (see also Otto et al. 1990). Pietrzak et al. (2011; Fig. 3.28) used remote sensing images to characterise the water types in terms of SPM, relating the SPM to stratification in the southern North Sea. They found pronounced seasonal and spring-neap variability, and highlighted the significant role played by tides and winds in controlling stratification, and hence the distribution of SPM at the sea surface. Prevailing circulations (river plumes and overall anti-clockwise circulation) also influence SPM distribution. (The *southern* North Sea is emphasised fhaving

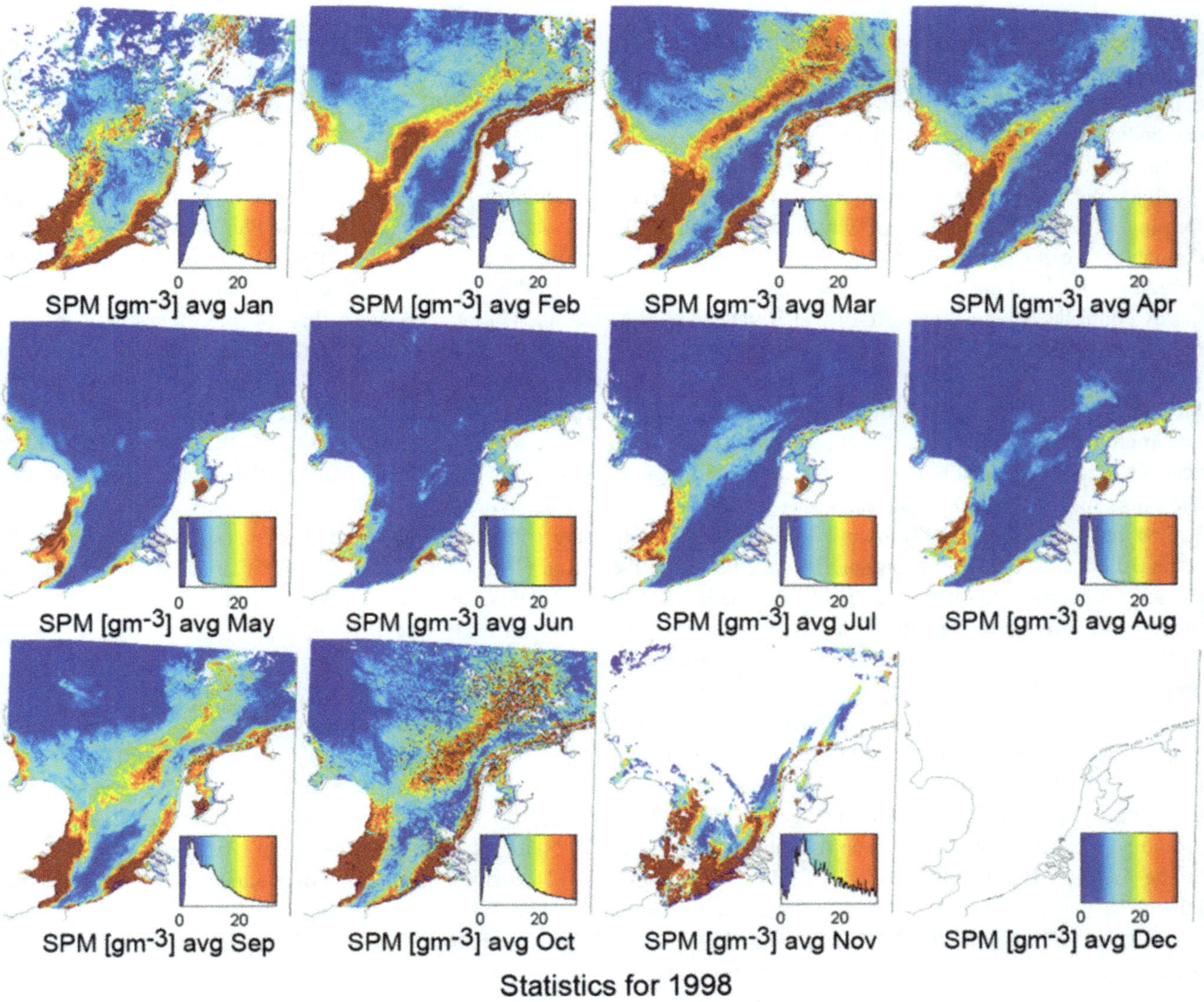

Fig. 3.28 Monthly average surface suspended particulate matter (SPM) based on SeaWiFs data from 1998 (reprinted from Pietrzak et al. 2011). The *insets* are the *colour bars* and the *white areas* the histograms of the pixel values, where the x-axis corresponds to the SPM pixel value and the y-axis to the relative frequency distribution of the pixel values

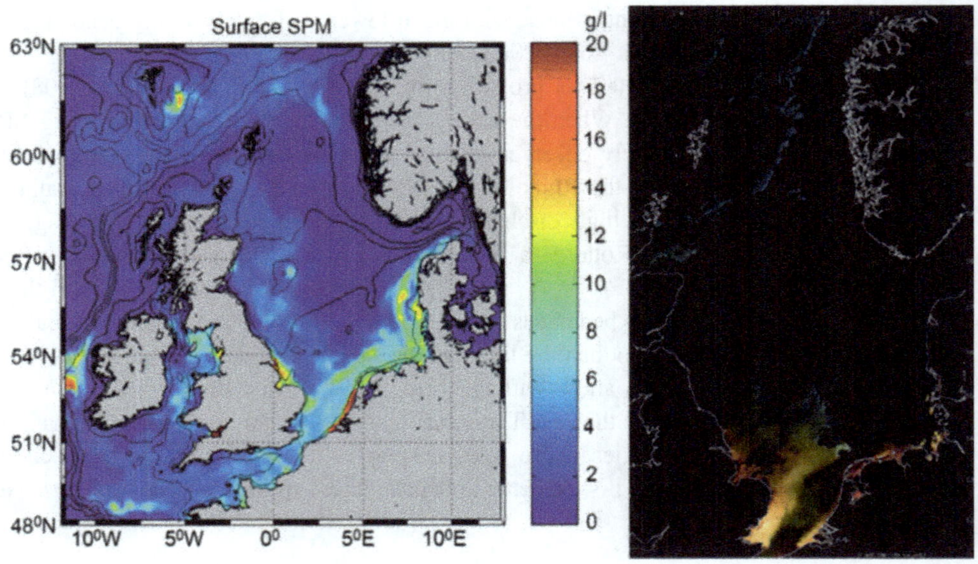

Fig. 3.29 Surface suspended particulate matter (SPM) distribution (g l^{-1}) for March from a numerical model (*left*) and satellite false colour (*right*). Areas of strong concentration qualitatively agree with areas of strong reflectance, showing the East Anglia and Rhine plumes (reprinted from Souza et al. 2007)

greater SPM concentrations than the northern North Sea due to differences in the sources, stronger currents and shallower depth, i.e. a larger Stokes number (Souza 2013), and correspondingly more studies and available data).

The waters exiting the rivers Rhine, Tees, Humber and Wash are deflected to the right under the Coriolis influence, forming classic river plumes of which the Rhine ROFI dominates the southernmost North Sea. A zone of high turbidity extends along the Belgian and Dutch coasts, primarily controlled by the Rhine ROFI which transports SPM northwards along the Dutch coast (Dronkers et al. 1990; Visser et al. 1991; de Kok 1992; de Ruijter et al. 1992; McCandliss et al. 2002). The Flemish Banks turbidity maximum off Belgium (Fig. 3.28) is present throughout the year but is much reduced in summer; studies have disagreed about its cause. Off the Dutch coast, Visser et al. (1991) and Suijlen and Duin (2001, 2002) found a local SPM minimum about 30 km offshore. In the German Bight, SPM from the Rhine ROFI appears to merge with SPM from the Weser and Elbe, before arriving in the Skagerrak (Simpson et al. 1993).

Within the North Sea, an overall anti-clockwise circulation extends south to around East Anglia, before turning into strong eastward flow at about 53°N across the southern North Sea, as a residual tidal and meteorologically-forced flow (e.g. Dyer and Moffat 1998; Holt and James 1999). Off East Anglia, this eastward flow has high turbidity (Fig. 3.29) extending from the Norfolk banks across the southern Bight to the German Bight (McCave 1987). This 'East-Anglia Plume' (Dyer and Moffat 1998) comprises southward-flowing waters from the Tees, Humber and Wash and a southern source of fresh water from the Thames (Thames waters deflect to the left, unlike those of other rivers entering the North Sea). The plume location is evident in SPM images (Fig. 3.29) and can be linked to the frontal boundary that separates well-mixed water in the southern Bight from seasonally-stratified central North Sea waters (Eisma and Kalf 1987; Hill et al. 1993), especially the Flamborough Head and Frisian Fronts.

Plume currents can be enhanced by the thermohaline circulation (jets associated with tidal mixing fronts; Hill et al. 1993). The East-Anglia Plume eventually joins the northward flow from Dover Strait and the Rhine ROFI (Prandle 1978b; Prandle et al. 1993). The plume carries an estimated annual SPM flux of 6.6 million tonnes, from English rivers and cliffs (Sündermann 1993), eastwards across the southern North Sea (Howarth et al. 1993). Both the East-Anglia Plume and the seasonal thermocline have a large impact on the transport of SPM across the southern North Sea. Holt and James (1999) found that deposition typically occurs along the 40 m depth contour. Their results are consistent with those of Eisma (1981) and Eisma and Kalf (1987), who found that the main areas of fine sediment accumulation are the Oyster Grounds and the strip along the German Bight (see also Fig. 3.26).

3.8.3 Tidal Influence

Figure 3.30 shows the spring-neap SPM cycle and highlights the important role played by tides in the surface distribution of SPM in the southern North Sea.

Deposition and erosion of sediment are related to critical values of the bed shear stress. Fine sediments are deposited when the bed shear stress is less than critical (0.1–0.2 N m^{-2}) and are typically eroded if the stress exceeds 0.4–0.5 N m^{-2} (Puls and Sündermann 1990; Holt and James 1999; Souza et al. 2007). In principle the tidal bed stress can be used to characterise regions where the tides can resuspend

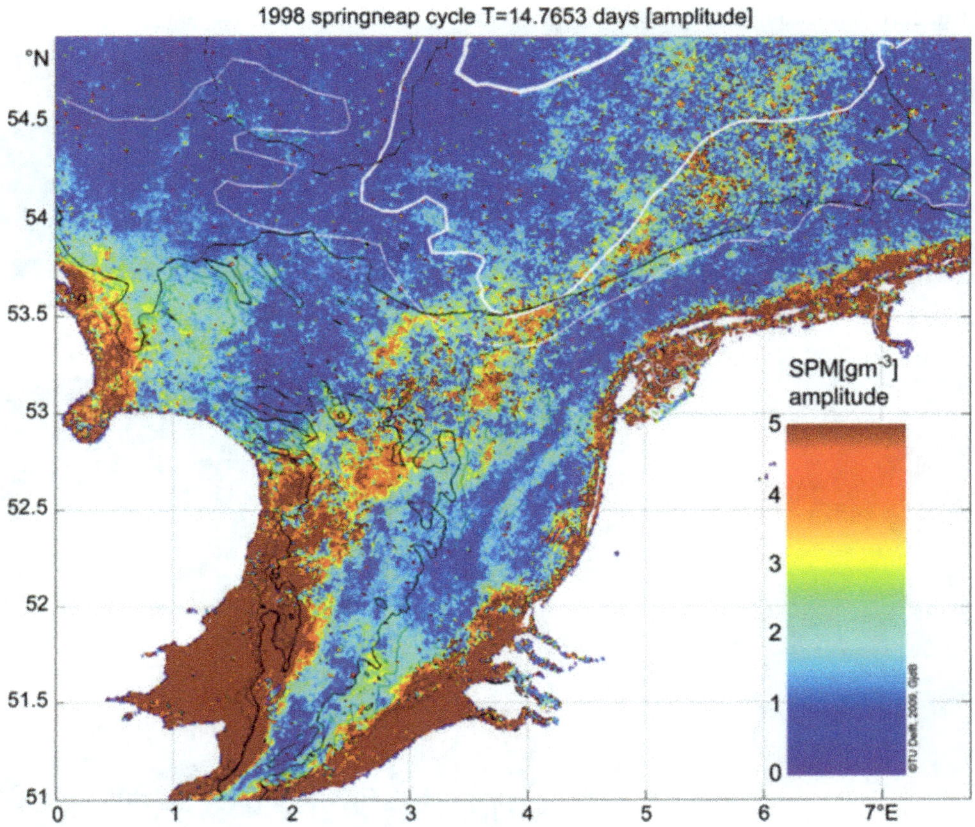

Fig. 3.30 The spring-neap harmonic using SPM images from SeaWiFS in 1998 (reprinted from Pietrzak et al. 2011). *Black lines* show bathymetry contours: 25 m (thinnest), 30 m (medium) and 35 m (thickest). *White lines* show values of the Simpson and Hunter (1974) criterion (Sect. 3.2.4): S = 1 (*thinnest line*), S = 1.5 (*medium*) and S = 2 (*thickest line*). Areas with S < 1 are well mixed

bed material. For example, in the Belgium coastal zone, the water column is always well mixed by tidal currents which also cause the SPM maxima in this zone (Lacroix et al. 2004).

Jago et al. (1993) showed typical behaviour of SPM: quarter-diurnal maxima due to tidal resuspension and semi-diurnal maxima due to tidal advection of ambient SPM concentration gradients. The result of these processes is that SPM time series show two maxima per tidal cycle, one larger than the other. SPM also responds to the spring-neap tidal cycle as shown by numerical simulations (Souza et al. 2007; see Fig. 3.31); monthly variability is also clear in the deposition of material and is present in SPM and q^2 (where q is the turbulent velocity). The model has been extensively validated using North Sea Project data including that for station CS (e.g. Holt and James 1999; Holt et al. 2005; Allen et al. 2007).

3.8.4 Wave Effects

Data from the northern Rhine ROFI after storms show a sudden increase in SPM over the entire coastal zone, suggesting local resuspension of sediment (Suijlen and Duin 2001, 2002) and an important effect of waves. Waves are strongly seasonal in the North Sea. Significant wave heights (*Hs*) in the winter half year are usually much higher than in the summer half year over the entire North Sea. For example, Dobrynin et al. (2010) showed that during 2002–2003 seasonally-averaged *Hs* was up to 1.5 m higher for the winter season.

The largest values of *Hs* are usually found in the open North Sea (e.g. Weisse and Günther 2007; Dobrynin et al. 2010), owing to a combination of topography effects with predominant wind direction and storms coming from the North Atlantic. In shallow regions, effects such as bottom boundary dissipation of waves can be significant even during the (calmer) summer, although they are usually more important during winter. For example, in winter over Dogger Bank, Dobrynin et al. (2010) found enhanced combined wave-current stress resulting in very high SPM concentrations (>50 mg l^{-1}).

3.8.5 Impact of Stratification

Stratification (Sect. 3.2) has an impact on the vertical distribution of SPM through reduced turbulence at the pycnocline. The effect of *thermal* stratification is particularly noticeable in the East-Anglia Plume around the stratified areas close to the Frisian Front. Moreover, at the 55° 30′N station shown in Fig. 3.31, as thermal stratification develops the SPM

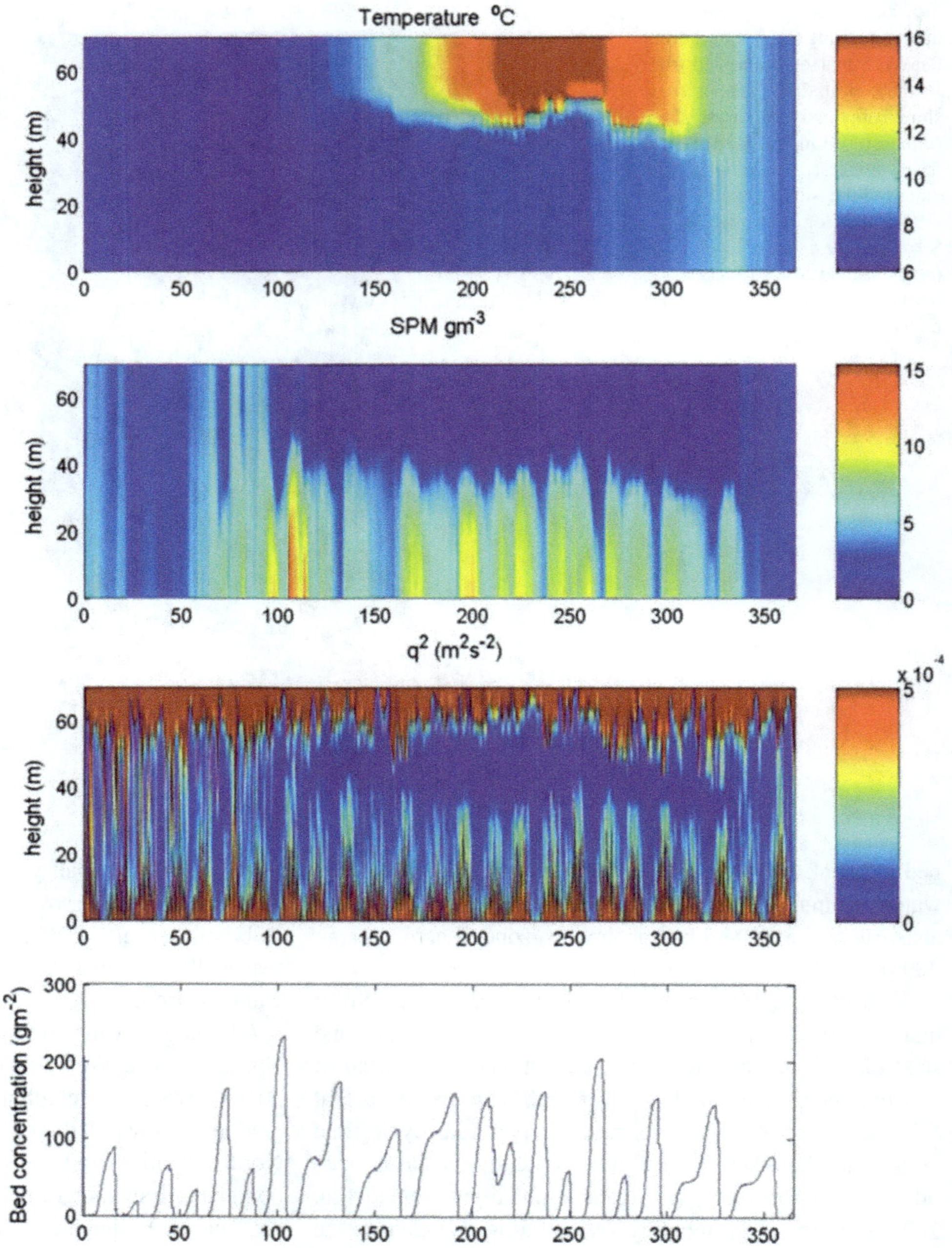

Fig. 3.31 Modelled time series at 55° 30′N 0° 50′E (North Sea Project station CS); time in days. Temperature (°C, *upper*), SPM (g m^{-3}, *upper middle*), q^2 (turbulence kinetic energy × 2, m^2 s^{-2}, *lower middle*), bed concentration (g m^{-2}, *lower*) (Souza et al. 2007)

concentrates in the lower 40 m of the water column, with values up to 15 g m^{-3}, but decreases to zero at the surface. *Haline* stratification appears to be significant especially within the Rhine ROFI and other adjacent ROFI areas such as the German Bight. The Rhine ROFI switches between well-mixed and stratified within a tidal cycle and through the spring-neap cycle (de Boer et al. 2006); haline stratification can develop at neap tides throughout the year and thermal stratification can be important in summer. As found in field studies of the Rotterdam Waterway, with stratification and reduced turbulence, any SPM advected over the salt wedge settles (de Nijs et al. 2010). Observations (e.g. Joordens et al. 2001) and numerical simulations (Fig. 3.31) show how SPM is trapped beneath the pycnocline when stratified. When the water column is mixed, turbulence and SPM can reach the surface; when a pycnocline develops, it inhibits turbulence, preventing upward flux of SPM and turbulence to the surface layer. Although de Boer et al. (2006) found differences to Heaps' model for the Rhine ROFI, cross-shore flow associated with estuarine-type circulation (Heaps 1972) tends to be offshore near the surface and onshore near the bed, giving a bias to onshore transport of settling SPM.

3.8.6 Seasonal Variability

Pleskachevsky et al. (2005) found that most SPM transport occurs in winter when cliff erosion along the English coasts is greatest. Numerical simulations of Holt and James (1999) highlighted seasonal variability of SPM. They found that SPM is only measured in time series of the water column in a series of discrete events associated with stronger winds, which resuspend bed material and mix the water column. Likewise, Souza et al. (2007) found their modelled water column to be well-mixed with very similar surface and bed SPM distributions in the East-Anglia Plume in February. In contrast, during summer the water column is stratified, almost all SPM settles out and an increase in net deposition correlates with decreased wind stress; moreover, less SPM is supplied by erosion of Holderness cliffs.

Satellite images of surface SPM show significant annual and seasonal variability (Eleveld et al. 2004, 2006, 2008; Pietrzak et al. 2011; see Fig. 3.28). Large values of surface SPM were observed in winter in Southern Bight coastal waters, especially in the Rhine ROFI, the East-Anglia Plume and Frisian Front. Summer minima occur throughout the southern North Sea, with low SPM values from April to August; in August almost all surface SPM in the Dutch coastal zone has gone. Pietrzak et al. (2011) showed the influence of the Rhine ROFI, East-Anglia Plume and fronts on the intra-annual distribution of SPM.

De Nijs (2012) found siltation rates in the Dutch waters to vary over the year and correlate with variations in sediment supply; the availability of sediment at Dover Strait and the river boundaries is typically greater between late autumn and spring. Verlaan and Spanhoff (2000) found massive siltation events, caused by storms, to occur near the mouth of the Rotterdam Waterway; a few such events determine annual siltation rates.

3.8.7 Interannual and Long-Term Variability

Strong year-to-year variation in sediment supply and required dredging near the mouth of the Rotterdam Waterway (de Nijs 2012) highlight the importance of interannual meteorological forcing.

Most long-term records of SPM are from satellites or sea-surface water samples collected along the Dutch coast (Suijlen and Duin 2001, 2002). However, near surface values vary in relation to stratification, making it difficult to use these data to infer trends or study impacts of climate change. Nevertheless, recent work continuing that of Pietrzak et al. (2011) indicates that interannual variability in wind stress, river discharge and heating due to variations in the NAO may have a pronounced impact on SPM distribution in the North Sea. Fettweis et al. (2012) classified surface SPM distributions according to 11 weather types, emphasising dependence in different locations on different hydrodynamic and wave conditions. Thus Southern Bight SPM is strongly influenced by advection, while exposure to waves favours resuspension in the central North Sea and German Bight and there is some overall positive correlation with the NAO.

Many coasts around the southern North Sea, notably the Dutch coast, are highly engineered, making it difficult to assess climate change impacts on sediment dynamics. However, supply of SPM for transport along the Belgian and Dutch coast, from French and English cliff coasts (Sect. 3.8.1), is determined by prevailing meteorological conditions and associated periods of large waves. This explains (seasonal and) interannual variations in the transport of SPM into the southern North Sea through Dover Strait. Climate-change effects on river discharge, storm tracks and associated winds and waves (intensity) are likely to affect the supply and distribution of SPM in the coastal zone and thus sediment distribution within the North Sea.

3.9 Coastal Erosion, Sedimentation and Morphology

Sytze van Heteren, John Huthnance

3.9.1 Historical Perspective

Coastal erosion is a key element of coastal behaviour and a useful—though imperfect—indicator of climate change. It is the active removal of sediment from various environments that marks the transition from sea to land, generally forcing the coastline landward. For cliffs and bluff coasts, erosion is irreversible. For sandy or muddy coasts, accretion and erosion may alternate. Coastal erosion may result in loss of land, destruction of sea defences and flooding. It has been measured for many centuries.

Coastal erosion takes place on different time scales. Long-term changes are commonly driven by relative sea-level rise and the associated creation of 'accommodation space' available for potential sediment accumulation. Shorter-term changes show regular patterns and irregular, partly short-lived effects of waves, tides, storm surges, slope processes and local or regional sediment dynamics. Long-term measurements enable distinction between the effects of these short- and long-term drivers, both natural and human-induced.

Systematic observation of coastal erosion, by periodically mapping or measuring the North Sea water lines and frontal dunes, started in earnest during the 17th century.

Annual coastline monitoring started in 1843, in the western Netherlands. Between 1840 and 1857, 124 oak poles were driven about 3 m into the beach sand, initiating a network of beach poles with 1000-m spacing that spans the entire Dutch coast and still operates today. For each pole, cross-shore distances to the low- and high-water lines and dune foot have been logged annually ever since. Starting in 1965, some 1450 transects at 250-m spacing have been profiled near-annually from at least the frontal dune to about 1000 m seaward of the dune foot. Below low water, data come primarily from single-beam echo sounding. Above low-water, photogrammetry was used before 1996 and laser altimetry since.

Monitoring records for other countries bordering the North Sea are shorter, commonly limited to local or regional studies, and used mainly to assess the need for coastal maintenance and protection measures. In Denmark, soundings and levelling for coastal monitoring started in 1874 in the Thyborøn area along lines spaced 600–1000 m apart (Thyme 1990). Since 1957, the entire west coast of Jutland has been monitored for cross-shore coastline migration at least several times per decade, at similar spacing (Kystdirektoratet 2008). In Belgium, coastline changes have been monitored nationally since the late 1970s, using echo-sounding, topographic surveying, photogrammetry and laser altimetry (since the late 1990s). In the United Kingdom, one of the longest records concerns cliff behaviour at Holderness, Yorkshire, from 1951 to the present day (Brown 2008) using erosion posts, on average 500 m apart along the cliff edge. More extensive regional surveys have been conducted since 1992, when the Environment Agency started annual monitoring of winter and summer cliff and beach profile change at 1-km intervals between the Humber and Thames estuaries.

Where systematic monitoring networks to quantify behaviour of the entire coastline do not exist, useful information on local or regional coastal erosion is provided by analyses of historical to recent maps, aerial photographs (extensively used in mapping since the Second World War) and satellite images. In Germany, series of maps were analysed by Mroczek (1980) to deduce rates of erosion or accretion for the North Sea coast over more than one hundred years. In Britain, maps from the late 16th century provide the oldest former positions of coastal bluffs and cliffs.

Detail on evidence for (possibly varying) coastal erosion rates is given in E-Supplement Sect. S3.4.

3.9.2 Understanding Coastal Erosion: State, Variability and Trends

Erosion is widespread along the central and southern parts of the North Sea coastline, with about 25 % of the Danish-to-Scottish coastline eroding. Farther north, erosion is rare. Erosion percentages for the North Sea part of the coastline of countries bordering the North Sea vary widely. On a country-by-country basis, the percentage eroding is as follows: UK (22 %), France (76 %), Belgium (40 %), Netherlands (30 %), Germany (14 %), Denmark, north to Skagen (57 %), Sweden, south to Marstrand (0 %), and Norway (0 %). The data were calculated from EUrosion/EMODnet data (http://onegeology-europe.brgm.fr/geoportal/viewer.jsp) and the values are approximate. They exclude areas with no data (e.g. back-barrier shorelines are excluded for some countries like the Netherlands) and the Skagerrak is included.

In Norway, the coastline is mostly rocky. Combined with limited and only recent relative sea-level rise, there is no significant coastline change. In Sweden, Rydell et al. (2004) showed that erosion is limited to very few small areas with pocket beaches and bluffs in Västra Götaland County.

In Denmark, coastal erosion affects most of the North Sea shoreline, as summarised by Sørensen (2013). Much of the northern Jutland headland coast has eroded about 2–4 m year^{-1} over the last 20 years, with maximum erosion in central bays and maximum deposition on the north-western side of eroding headlands (Christiansen and Bowman 1990). The central west coast, which is dominated by barrier beaches and bounded by glacial bluffs, has been the most vulnerable, with natural erosion rates of 2–8 m year^{-1}. Only the southernmost barrier beaches around Vejers, between Nymindegab and Blaavands Huk, are accreting naturally (Aagaard 2011). Horns Rev, which marks the southern end of the barrier-beach coast, acts as a natural groyne, protecting the island coast farther south from wave attack (Meesenburg 1996). This coast is in overall sediment balance, with parts of the barrier islands growing seaward, supplied by sediment from north and south (Sørensen 2013). Here, the southern ends of the islands are the most vulnerable to erosion.

Storm surges induce the most prominent changes along the Danish North Sea coast (Fruergaard et al. 2013). They lead to ephemeral and permanent barrier breaching, and drive the episodic transfer of large volumes of sediment from dunes and beaches to the nearshore and shoreface. During subsequent healing phases, which may last several decades, much of this sediment returns to the coast. South of Blavands Huk, persistent onshore migration of nearshore bars, governed by high-energy dissipative conditions, explains long-term coastline behaviour (Aagaard et al. 2004). Bar welding widens the beach and temporarily increases the amount of sand available for aeolian transport. Two trends can be observed on the Danish coast. First, most natural changes in coastal erosion are overprinted by increasing beach and shoreface nourishment (Fig. 3.32). Second, natural changes seem to indicate a switch to or acceleration of coastal erosion.

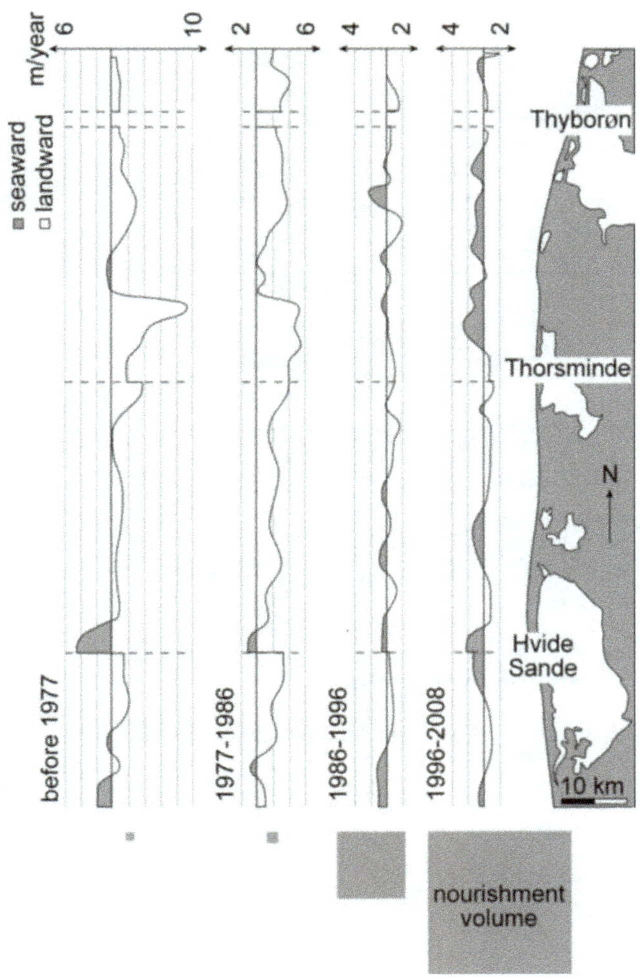

Fig. 3.32 West Jutland coastline advance/retreat rates for four periods and corresponding nourishment volumes (2.5 million m^3 $year^{-1}$ is represented by the *large box*). The 'y'-axis is located along the coast as in the sketch map; *horizontal lines* are at lagoon entrances (Sørensen 2013)

Along the German North Sea coast, erosion has had the most impact on the barrier-island coasts in the north and west, away from the central estuaries. At present, erosion occurs only along the 100 km of the coastal length that is not protected by dikes. Rates vary between <1 and 8 m $year^{-1}$. Most erosion has resulted from storm surges (Kelletat 1992) which have also reshaped the overall barrier and back-barrier morphology. Eroded sediments are partially regained during subsequent fair-weather periods, after temporary storage in the nearshore zone, and may be redistributed on neighbouring coasts. Through spit progradation, Sylt (one of the North Frisian Islands) has been growing northwards and southwards.

Acceleration of erosion along the German North Sea coast over the last 120 years is evident. Dette and Gärtner (1987) found average loss on the west coast of Sylt to have increased from 0.9 m $year^{-1}$ in the period 1870–1952 to 1.5 m $year^{-1}$ in the period 1952–1984, in spite of increasing efforts to protect the coast (Fig. 3.33). It is tempting to link this increase to acceleration of sea-level rise, but an increase in extreme events, with warmer, stormier winters, must also be considered (Salman et al. 2004).

The Dutch long-term monitoring record, starting in 1843, shows that the central (Holland) coastline receded relatively moderately (about 1 m $year^{-1}$) throughout the 19th and 20th centuries (van der Meulen et al. 2013). During this period, the coastline became increasingly defended by groynes and seawalls, which stabilised the coastline but resulted in steepening of the nearshore zone and the shoreface. Monitoring data also show that major storms have caused interruptions of long-term trends. Short-lived and rapid recession is commonly followed by extended post-storm periods of little erosion or even recovery, after which erosion resumes at the pre-storm rate (Fig. 3.34). Along the southwestern estuarine coast, erosion is linked to nearshore channel activity alongside island heads. Figure 3.35 shows long-term coastal erosion in North-Holland reversed by large-scale nourishment.

The west-Frisian islands fringing the Wadden Sea show large spatial and temporal differences in coastal erosion and accretion, similar to those observed along the German coast; local processes clearly overprint the long-term regional trend in coastal behaviour (Oost et al. 2012). Island growth and dune development reflect sand supply from the ebb deltas and shoreface, with episodes of rapid accretion linked to bars merging with the islands. Coastal erosion and dune scarping are associated with exposure to high-energy waves, as ebb-tidal deltas shift, or to strong currents, as marginal flood channels are forced toward the coast. The adjacent barrier islands of Ameland and Schiermonnikoog showed opposite behaviour during the last 150 years: Ameland receded and extended only slightly in a longshore direction, whereas Schiermonnikoog shifted seawards and accreted eastwards by more than 6 km (Oost 1995).

In Belgium, beaches and dunes showed long-term erosion until human intervention intensified (Charlier 2013). Currently, the dune foot is growing seaward in most places, aided by sand nourishments and hard coastal-protection structures (De Wolf 2002). Beaches behave more heterogeneously. West of Oostende, most are stable or show slight accretion. East of Oostende, erosive, stable and accretionary stretches alternate (De Wolf 2002).

In France, beach accretion is most usual southwest of Dunkerque, beach erosion between Dunkerque and the French-Belgian border (Bryche et al. 1993). Swell action is responsible for coastline recession of about 0.75 m $year^{-1}$ (Clabaut et al. 2000). Detailed topographic surveys of the upper beaches and dune fronts fringing the French North Sea indicate that periods of greater storminess do not necessarily result in more rapid retreat or more general coastal erosion. Here, coastline behaviour at a decade-to-century time scale

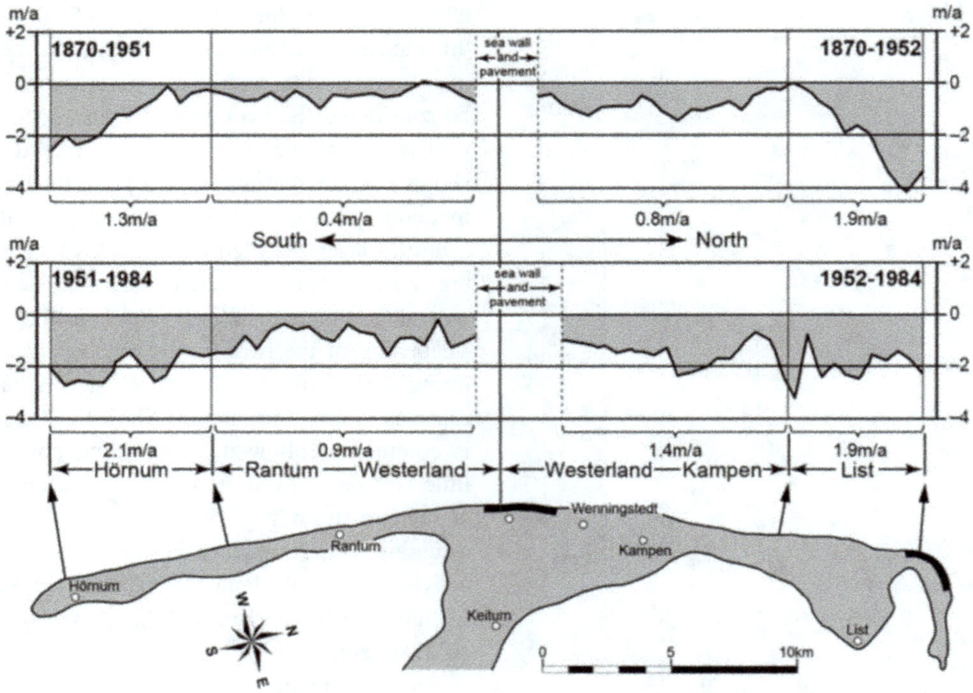

Fig. 3.33 Annual rates of coastal change on the west coast of Sylt island from 1870 to 1951/52, and from 1951/52 to 1984 (Besch 1987)

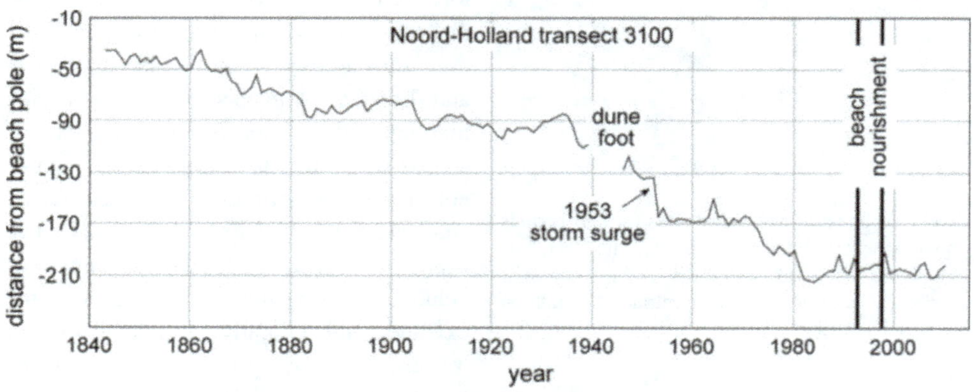

Fig. 3.34 Long-term erosion of the western Dutch coast just north of Bergen aan Zee, from annual monitoring data. The major effect of the 1953 record storm surge is clearly visible. It was followed by 20 years of relative stability before the pre-storm trend of coastal erosion continued (figure by Sytze van Heteren, Geological Survey of the Netherlands)

also depends on occurrences of high water levels (Vasseur and Héquette 2000), on local sediment budgets and on nearshore bathymetry (Ruz and Meur-Férec 2004; Chaverot et al. 2005). Although net loss of sediment from coastal systems prevails, see Fig. 3.36 for example, some coastline recession is counterbalanced by sand accumulation on top of the dunes and on their land-facing slopes (Clabaut et al. 2000).

The North Sea coasts of England and Scotland are dominated by cliffs and bluffs, with shorter stretches of pebble beaches, several estuaries, dwindling tidal flats and few dunes. The English areas at risk of erosion are mostly those where the North Sea is fringed by beaches or bluffs (Blott et al. 2013). Coastal erosion percentages are estimated as 27 % for north-eastern England, 56 % for Yorkshire and Humber, 9 % for Lincolnshire, 13 % for Norfolk to Essex and 31 % for south-eastern England (EUROSION 2004). Intertidal zones of saltmarsh and mudflat have been disappearing at a typical loss rate of 1–1.5 % per year over a period of more than 50 years. This loss is linked, though not solely attributable, to coastal squeeze. The lateral recession rates of coastal cliffs and bluffs vary with rock type (French 2001). In Scotland, most cliffs consist of resistant rock; significant erosion is limited to beaches.

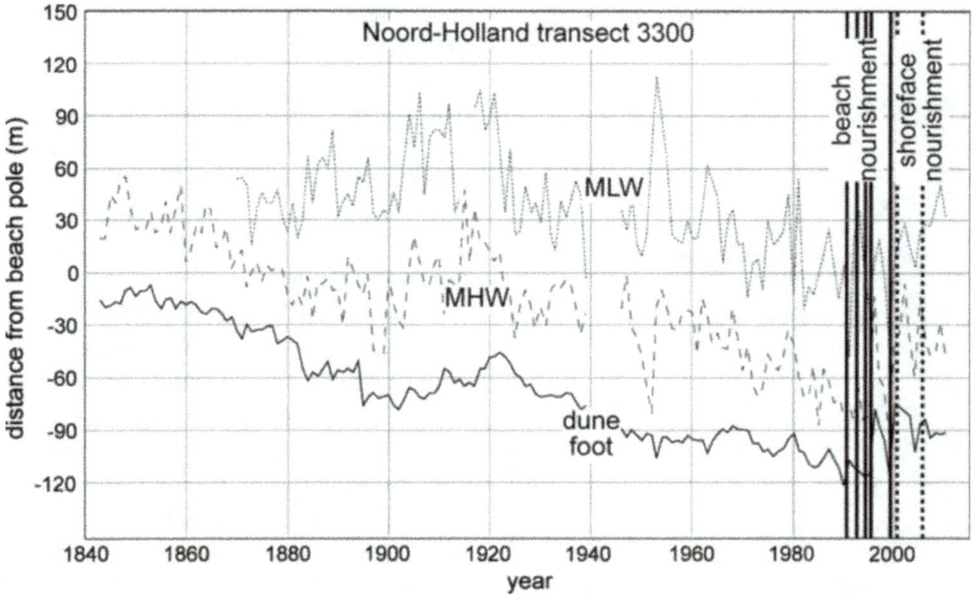

Fig. 3.35 Long-term coastal erosion at the Bergen aan Zee monitoring station in North-Holland, reversed with the advent of large-scale nourishment (*vertical lines on the right*) in the 1990s. The dune foot (*solid line*) receded more than 100 m between 1850 and 1990, as indicated by its distance to the beach pole. Similar patterns, although more diffuse, are shown by the mean-high-water (*dashed line*) and mean-low-water (*dotted*) lines (data from Deltares, figure by Sytze van Heteren, Geological Survey of the Netherlands)

Frontal dunes along beaches show average erosion rates of 1 m year^{-1} (Pye et al. 2007). Bluffs are particularly common in east Yorkshire, Humberside and East Anglia; they show very variable spatial and temporal erosion patterns, a function of complex glacial geology. Changes are episodic with no discernible trend. At Dunwich, for example, phases of accelerated coastal retreat (e.g. 1863–1880, 2.57 m year^{-1}; 1903–1919, 3.53 m year^{-1}) have alternated with periods of relative stability (e.g. 1826–1863, 0.06 m year^{-1}; 1882/3–1903, 0.08 m year^{-1}; Carr 1979). Landslides, protection by shingle beaches, cliff or bluff material, pore-water pressure and hydrodynamics all play a role (Brooks and Spencer 2010). These factors interact with longer-term drivers, including shifts in dominant weather patterns and changes in

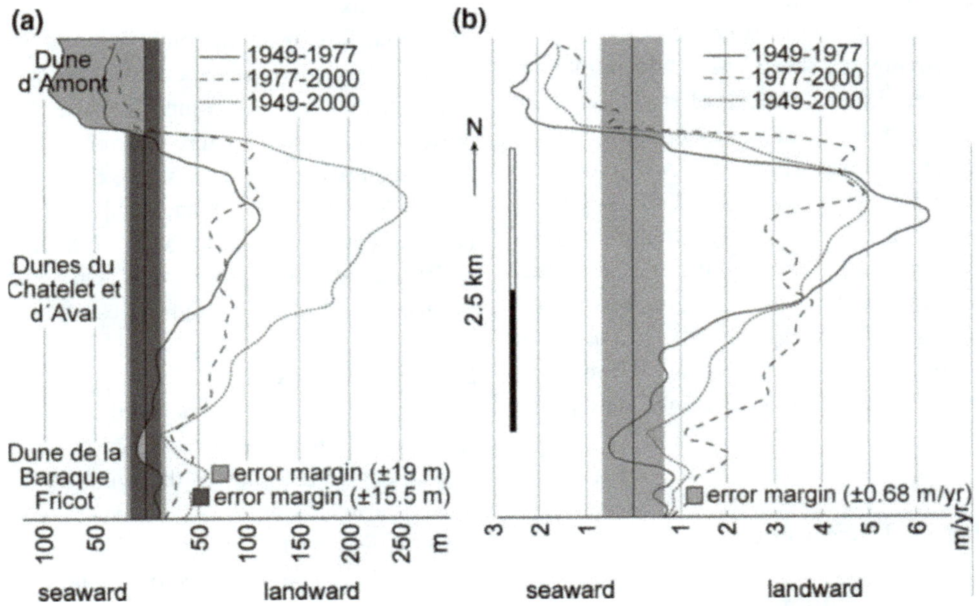

Fig. 3.36 Coastline evolution in the Bay of Wissant from 1949 to 2000 (*left*) and mean rates of coastline evolution for 1949–1977, 1977–2000 and 1949–2000 (*right*) (Aernouts and Héquette 2006)

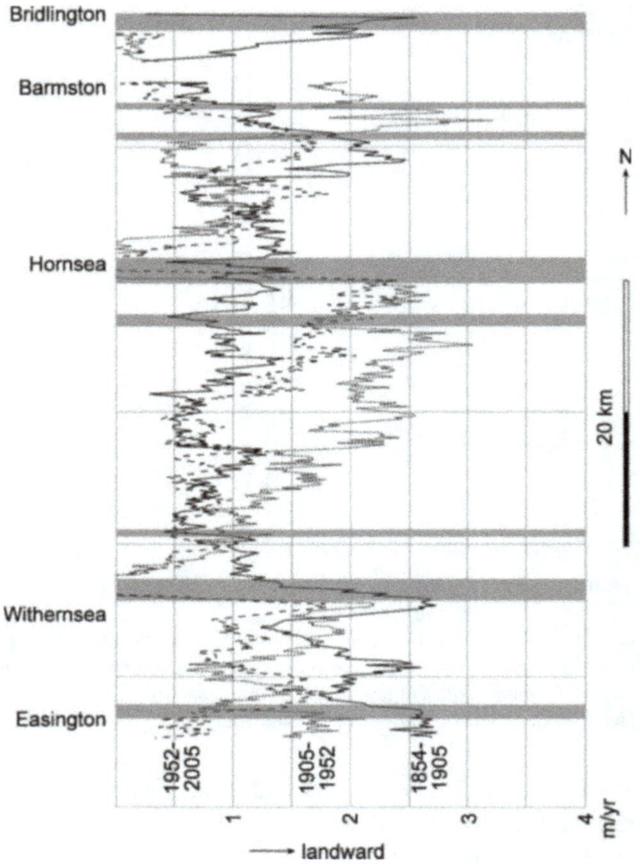

Fig. 3.37 Cliff-top retreat rates at Holderness calculated in three approximate 50-year periods. Areas defended in 2005 are in *grey*. Retreat rates vary within each zone (Brown 2008)

the rate of relative sea-level rise. As receding bluffs and cliffs expose new material, they may steepen, flatten or even disappear (Brooks and Spencer 2010). Feedbacks also operate. Large landslides provide the steep coast with temporary protection from the sea. Longshore drift of eroded sand and gravel leads to net accumulation at nearby beaches.

The Holderness bluff coast (Fig. 3.37) was eroded by more than 300 m in 150 years (Valentin 1954). At Happisburgh rates of erosion increased to average about 8 m year^{-1} in 1992–2004 (Poulton et al. 2006), significantly faster than the long-term averages of 0.9 m year^{-1} for North Norfolk in 1880–1967 (Cambers 1976; Thomalla and Vincent 2003), 2.3–3.5 m year^{-1} for Benacre–Southwold and 0.9 m year^{-1} for Dunwich–Minsmere in 1883–2008 (Brooks and Spencer 2010).

3.9.3 Offshore Morphology

3.9.3.1 History and Evidence

Long bathymetric time series (many decades to centuries) are rare and generally localised in the context of port approaches and dredging (for navigation) or aggregate extraction. 200-year records of an ebb-tidal delta in the Deben estuary, eastern England, and a 180-year record in the outer Thames estuary, have been analysed (Burningham and French 2006, 2011). Horrillo-Caraballo and Reeve (2008) interpreted sandbank configurations off Great Yarmouth using historic charts over a 150-year period. There are no field studies relating such series to climate-dependent factors (e.g. wave heights).

Van der Molen et al. (2004) numerically modelled millennial-scale morphodynamics of an idealised, semi-enclosed, energetic tidal shelf sea with dimensions and tidal characteristics resembling the Southern Bight of the North Sea. Several local process studies have related morphological change to wider-scale sediment transport and to tides, wind-driven flows and waves, for example eight years of current profile monitoring at Marsdiep inlet, Netherlands (Buijsman and Ridderinkhof 2008a, b) and a 1977–2003 sidescan sonar and multibeam backscatter record in the German Bight (Diesing et al. 2006).

3.9.3.2 Features

The English Channel is shallow and tidally dominated with waves mainly from the west-southwest during storms; it has limited sediment sources and is marked by extensive reworking of a relatively thin sediment cover (Paphitis et al. 2010). The central Channel seabed is covered by coarse-grained material. Wide areas are occupied by sand-sized sediments with various bedforms: ripples, sand-waves, longitudinal bedforms and sandbanks. Fine-grained sediments are confined to coastal embayments, rias, estuaries and open-coast intertidal flats (Paphitis et al. 2010).

Sandbanks occur widely in the North Sea (Knaapen 2009); they dominate the southern North Sea except near the mouth of the Rhine. In some areas, bank crests comprise fine sand whereas troughs comprise material too coarse to be moved by local tidal currents alone: such as shoreface-connected ridges off the East Frisian barrier-island coast (Son et al. 2012) and banks off Thorsminde (west Denmark; Anthony and Leth 2002). Sand-waves also occur widely, most commonly around sand banks. There is an extensive field of large sand-waves off the Netherlands.

3.9.3.3 Dynamics

Van der Molen (2002) modelled the influence of tides, wind and waves on net sand transport in the present southern North Sea. Results showed that wind-driven flow and waves only contribute significantly to net sand transport by tides when acting together where tidal currents are small. However, various combinations of forcing dominate net sand transport in different regions of the southern North Sea. Tides dominate in the southern, middle and north-western parts of the Southern Bight and in the region of The Wash.

Tides, wind-driven flow and waves are all important in the north-eastern part of the Southern Bight. Wind-driven flow and waves dominate north of the Frisian Islands, in the German Bight and on Dogger Bank. In the Channel, Reynaud et al. (2003) inferred predominant control by tidal dynamics, with mobile sediments in the central and eastern Channel, and longer-term influence of sea-level rise.

Contrasts in sand-wave character are related to differences in the relative importance of suspended load transport and of tidal currents and waves near the bed (Van Dijk and Kleinhans 2005; stronger tidal residual currents tend to cause faster sand-wave migration, waves tend to flatten sand-wave crests). In Marsdiep inlet, sand-waves typically migrate in the flood direction and not necessarily with predicted bed-load and suspended load transport; Buijsman and Ridderinkhof (2008b) hypothesised that advection of suspended sand and lag effects may govern sand-wave migration.

Tidal sandbank height (60–90 % of water depth) and shape are controlled by the mode of sediment transport and hydrodynamic conditions (Roos et al. 2004). Bedload transport under symmetrical tidal conditions leads to high spiky banks. Profiles are lowered and smoothed by relaxation of suspended sediment, wind-wave stirring and tidal asymmetry; this last factor also causes profiles to be asymmetric. Thus strong tidal currents and their residuals, with enhancement of bed stress by waves, are the main hydrodynamic agents for (long-term changes in) sandbank morphology. Examples are off Great Yarmouth (Horrillo-Caraballo and Reeve 2008), Westhinder sandbank (Deleu et al. 2004) and Kwinte Bank (Giardino et al. 2010) off Belgium, the outer troughs of shoreface-connected ridges off the East Frisian barrier-island coast (Son et al. 2012). Near the mouth of the river Rhine, freshwater outflow affects the direction of tidal ellipses and residual flow, suppressing the formation of open ridges (Knaapen 2009). Other studies of local system dynamics (E-Supplement Sect. S3.4.1) emphasise the role of tidal asymmetries with wave-enhanced transport, and show that shoals evolve, in some cases cyclically.

3.9.3.4 Evolution in Relation to Climate

Idealised simulations (Van der Molen et al. 2004) suggest that a basin resembling the Southern Bight may be expected to export sediment, deepen and expand by accumulation of eroded sediment in the deeper waters to the north, owing to asymmetry in the amphidromic tidal velocities. Sea-level changes affect tidal wavelength, hence this sediment distribution, and deepening reduces evolution rate. However, changes in sea level and tides are small relative to average water depths on time scales of decades to a century (Sect. 3.4).

In practice, repeat surveys tend to show stability of larger (kilometre-scale) features or patterns over periods of nine months (off Thorsminde on the Danish west coast; Anthony and Leth 2002), some years (e.g. Son et al. 2012) or even decades (Diesing et al. 2006). There is no clear evidence of any regionally coherent response to large-scale historical forcing such as sea-level rise. However, harbour or dike works and large-scale dredging induce wider change, such as to Texel inlet (Elias et al. 2006) and Kwinte Bank (Degrendele et al. 2010). Detecting abiotically-induced climate-related changes in morphological evolution could only be expected where the influence of storms and waves is significant in net sediment transport, namely in the north-east of the Southern Bight (notably the German Bight) and on Dogger Bank (van der Molen 2002).

Climate-related change is more likely where benthic organisms play an important role in the development and dynamics of bedforms, as so-called 'ecosystem engineers'. Although no field studies exist that link changes in benthic habitats and indicator species to bedform development and mobility, an increasing number of modelling studies suggest that such a link exists. Borsje et al. (2009) showed that including the abundance of three dominant eco-engineers (*Lanice conchilega*, *Tellina fabula*, *Echinocardium cordatum*) gives a more accurate prediction of sand-wave occurrence on the Dutch continental shelf than modelling without biology. Very little work has been done on eco-engineers, climate change and seabed morphology; this is an important subject for future research.

3.10 Sea Ice

Natalija Schmelzer, Jürgen Holfort, Ralf Weisse

In the North Sea, ice does not form in every winter season. Ice formation generally depends on the weather regimes prevailing over Europe and their temporal stability, and on the morphological characteristics of the North Sea.

Typically, in autumn and winter, prevailing westerly weather regimes bring relatively mild air masses from the Atlantic Ocean into the North Sea area. These weather regimes also cause inflows of relatively warm, high-salinity Atlantic water into the North Sea, preventing or delaying the seasonal cooling of North Sea water and hence ice formation. By contrast, easterly weather regimes cause rapid cooling of the water.

Large stationary high-pressure zones over northern Scandinavia and the European polar seas, and stable anti-cyclonic regimes over eastern Europe, are particularly effective in this respect. The extent and duration of ice cover in the North Sea are governed by the number, intensity, and length of freezing periods and by the timing of their occurrence. The North Sea comprises open sea areas, Wadden Sea areas, and tributaries; these play an important role in the

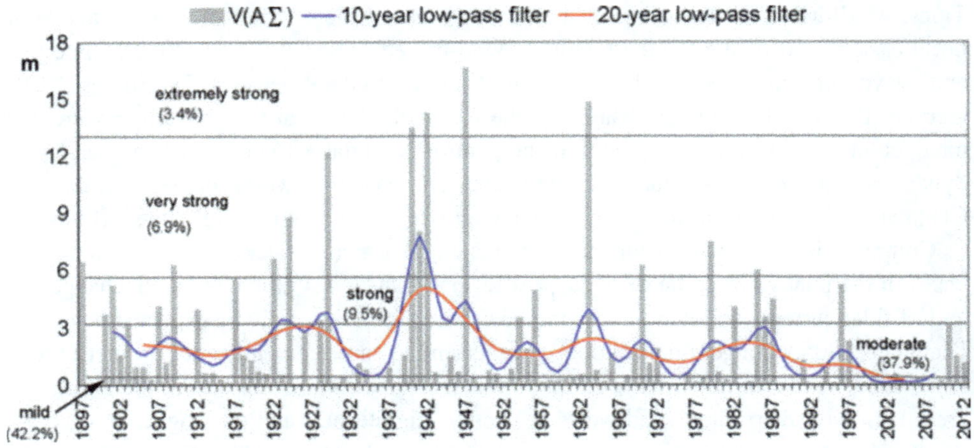

Fig. 3.38 Annual accumulated areal ice volume $V_{A\Sigma}$ on the German North Sea coast for 1897 to 2012 (update from http://www.bsh.de/de/Meeresdaten/Beobachtungen/Eis/Eiswinter2010_11.pdf)

development of ice conditions. Peer-reviewed literature on long-term changes in sea-ice conditions in the North Sea is sparse.

For the German North Sea coast, ice winters are classified on the basis of the accumulated areal ice volume $V_{A\Sigma}$ (in metres)

$$V_{A\Sigma} = \frac{1}{n}\sum_{j}\sum_{k}(NH)_{jk} \quad (3.1)$$

where n denotes the number of observations (stations); N is the fractional ice concentration varying between 0 and 1; and H represents ice thickness in metres at station $k = 1...n$ on day j when there is ice (Koslowski 1989). Available data allow the classification of different types of ice winter on the German North Sea coast for the period 1897–2012 (Fig. 3.38).

In the past 52 years (1961–2012), 26 (50 %) winter seasons on the North Sea coast were very weak or weak, 18 (35 %) were moderate and eight (15 %) were strong, very strong or extremely strong. In comparison with the 116-year time span shown in Fig. 3.38, the occurrence of extremely strong and very strong ice winters has decreased while there has been a simultaneous increase in winters with low ice-cover conditions.

Ice formation on the tidal flats of the Wadden Sea normally begins in mid- to late January (BSH 2008). Ice-cover duration varies widely, in space and time. In moderate ice winters, ice occurs on 10–20 days in the sheltered inner coastal waters of the North Frisian Wadden Sea, and on up to 10 days in open navigation channels, which is comparable to ice formation in the East Frisian Wadden Sea.

In strong and very strong ice winters, ice cover in the sheltered navigation channels of the North Frisian Wadden Sea lasts from 55 to 75 days on average, and in open navigation channels from 45 to 55 days, similar to the East Frisian Wadden Sea. In near-shore tidal flats, the most common type of ice is fast ice or rafted/ridged ice; in outer tidal flats, ice floes and slash or shuga predominate, kept in motion by wind and tidal forces.

In the open part of the German Bight, the remaining heat content of North Sea water in early winter is so large that ice rarely forms. Shuga and ice floes occurred in offshore waters to the west and northwest of Helgoland in about 8 % of all winter seasons, being last observed in late January 1970 and before that in February and March 1963; they do not originate near Helgoland but are carried there from the coastal area by tidal currents and by easterly winds persisting for long periods. In the offshore waters of the German Bight off the North and East Frisian islands, ice forms only in very strong or extremely strong ice winters.

Level ice thicknesses reach 10–15 cm in most winters, 15–30 cm in strong ice winters, and as much as 30–50 cm in very strong ice winters. The higher ice thickness categories are most likely to occur in February and early March. Ice thickness data refer to level ice, which occurs primarily in the form of ice cakes and small- to medium-sized floes. A typical phenomenon caused by tidal influences in areas of the Wadden Sea is rafting and ridging of initially level ice, which may cause ice walls several metres high. In such areas, floebits of about 1–3 m thickness comprising frozen ice cakes and small to medium-sized floes of coarse compacted ice are also likely to occur, especially in winters with long freezing periods.

Thawing causes rapid retreat of ice in south-eastern North Sea coastal areas. Westerly winds push warmer, high-salinity North Sea water toward the coast, accelerating the melting process started by meteorological influences. Ice in south-eastern North Sea coastal waters normally melts completely by the end of February. In very strong to extremely strong ice winters during which maximum ice formation is not reached until mid-February, the last remnants of ice may melt as late as the end of March.

The frequency of ice occurrence has decreased since 1961, analogous to the development of ice conditions in the western Baltic Sea (Schmelzer and Holfort 2012). Although there is presently a trend toward milder ice winters, strong to extremely strong ice winters in the area of the western Baltic Sea and North Sea are still likely in the future (Vavrus et al. 2006; Kodra et al. 2011).

3.11 Wadden Sea

Justus van Beusekom

3.11.1 Background

The Wadden Sea, fringing the coast of the south-eastern North Sea, is the largest coherent tidal flat system of the temperate world. Its outstanding geological and biological importance makes it a world nature heritage site (Reise et al. 2010). The Wadden Sea is the result of the intricate interplay of tidal forces, sediment supply and moderate sea-level rise (Reise 2013). Since its beginning some 8000 years ago, the Wadden Sea has been increasingly affected by humans (Lotze et al. 2005). Around 1000 years ago, coastal people started to transform the coast, ultimately embanking about half of the original Wadden Sea (Reise 2005).

Owing to this coastal squeeze, hydrodynamic forces increased leading to a loss of fine material (Flemming and Nyandwi 1994). Accelerated sea-level rise may also have contributed to increased hydrodynamics and a loss of mudflats (Dolch and Hass 2008). Also, the estuaries intersecting the Wadden Sea have dramatically changed due to diking and dredging (Reise 2005).

The Wadden Sea receives large amounts of riverine fresh water, either directly as (branches of) rivers debouch into it (e.g. IJsselmeer, Ems, Weser, Elbe, Eider, Varde A) or indirectly (especially from the Rhine and Maas). River water from the Rhine and Maas is transported to the Wadden Sea along with residual North Sea currents flowing predominantly anti-clockwise around the North Sea. Thus, nutrient and contaminant loads can be imported into the Wadden Sea and large amounts may be present. Given this background, three interacting aspects of environmental quality are addressed in this section: SPM, nutrients and contaminants.

3.11.2 Suspended Matter Dynamics

The Wadden Sea is characterised by relatively high SPM concentrations compared to the North Sea. To maintain these gradients, some type of accumulation process must be active (Postma 1954). Earlier explanations focusing on Wadden Sea processes included the biases described at the end of Sect. 3.8 (see also Postma 1954; van Straaten and Kuenen 1957; Burchard et al. 2008). Biological interactions such as filter-feeding or microphytobenthic microfilms (Staats et al. 2001; Andersen et al. 2010) increase the retention efficiency of the Wadden Sea. Thus the Wadden Sea proper can be regarded as an effective 'keeper' of SPM (van Beusekom et al. 2012), at least on a seasonal scale.

The North Sea, on the other hand, can be seen as the 'pusher' (van Beusekom et al. 2012) of suspended organic matter into the Wadden Sea via estuarine-type circulation (see Sect. 3.8.5; Postma 1981). The Wadden Sea is heterotrophic: locally-produced and imported organic matter is remineralised (Postma 1984) as supported by carbon budgets (van Beusekom et al. 1999). The ecological importance of organic matter import from the North Sea for the productivity of the Wadden Sea was discussed by Verwey (1952).

Changes have occurred in SPM dynamics in the Wadden Sea and its intersecting estuaries: De Jonge et al. (2014) showed that dredging of the Ems estuary caused a regime shift in hydrodynamics leading to hyperturbid conditions. Dredging of the Rhine estuary and subsequent dumping along the Dutch coast ultimately led to increased SPM concentrations in the western Dutch Wadden Sea (de Jonge and de Jong 2002). SPM dynamics were related to riverine runoff and de Jonge and de Jong (2002) suggested that increased runoff due to global climate change will enhance SPM dynamics in the Dutch Wadden Sea.

3.11.3 Changes in Nutrient and Organic Matter Dynamics

Riverine nutrient discharges are the main drivers of Wadden Sea eutrophication. Historic nutrient concentrations for the Rhine, compiled by van Bennekom and Wetsteijn (1990), show a clear increase in concentrations after the Second World War and a decrease since the mid-1980s due to the implementation of wastewater treatment and better agricultural practice (e.g. de Jong 2000) (see also Chaps. 11 and 13).

Increased nutrient availability has led to increased primary production and increased turnover of organic matter and nutrients. Changes in primary production are demonstrated by the Marsdiep time series (western Dutch Wadden Sea) initiated by Cadée (Cadée and Hegeman 2002), showing an increase in primary production until the 1990s and then a decrease. Present levels are about 120–200 g C m^{-2} $year^{-1}$ in the western and northern Wadden Sea (Loebl et al. 2007; Philippart et al. 2007). Most monitoring programmes started in the 1980s or later, thus only documenting changes after the maximum eutrophication had occurred. Summer chlorophyll levels are a good proxy for Wadden-Sea-wide eutrophication, correlating with riverine

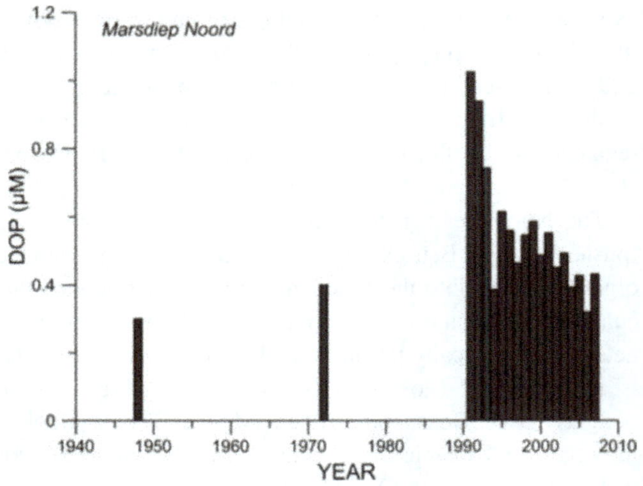

Fig. 3.39 Long-term changes in summer concentration of dissolved organic phosphorus in the Marsdiep area (western Dutch Wadden Sea) (figure by Justus van Beusekom after van Beusekom and de Jonge 2012)

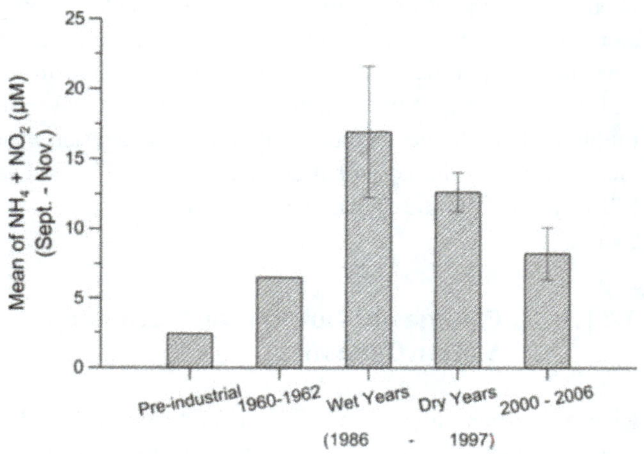

Fig. 3.40 Mean autumn ammonium and nitrite (µM) during pre-industrial times, in 1960–1962 (Postma 1966), in the five wettest (1986–1988, 1994, 1995) and five driest (1991–1993, 1996, 1997) years between 1986 and 1997, and in 2000–2006 in the western Dutch Wadden Sea (figure by Justus van Beusekom after van Beusekom and de Jonge 2002, updated with data from van Beusekom et al. 2009b). For the latter three groups, the standard deviation is shown

nutrient discharges (van Beusekom et al. 2009a) and demonstrating a gradual decrease in eutrophication.

Changes in organic matter and nutrient turnover covering both the increase and decrease in eutrophication are again best demonstrated for the western Dutch Wadden Sea. De Jonge and Postma (1974) documented a two- to three-fold increase in phosphate and particulate and dissolved organic phosphorus concentrations between the 1950s and 1972 (Fig. 3.39). Between 1961 (Postma 1966) and the 1990s, autumn levels of ammonium and nitrite (indicating organic matter turnover; van Beusekom and de Jonge 2002) also increased about two- to five-fold depending on whether wet years with high riverine loads or dry years with low riverine loads are compared (Fig. 3.40). From the 1990s, dissolved organic phosphorus levels and autumn ammonium and nitrite levels have decreased (van Beusekom et al. 2009a; van Beusekom and de Jonge 2012) as chlorophyll levels and riverine nutrient loads generally decreased.

Since the mid-1980s, several monitoring programmes have covered the entire Wadden Sea. These data document large regional differences in eutrophication. In general, levels are higher in the southern Wadden Sea than in the northern Wadden Sea, but some overlap exists (van Beusekom et al. 2009a). These differences are also captured by summer chlorophyll levels (average May–September) and autumn ammonium and nitrite levels (average September–November) suggesting that both parameters are useful proxies for describing Wadden Sea eutrophication. At a few stations only, dissolved nutrients are monitored and show the same spatial differences: high values in the Dutch Wadden Sea and low values in the northern Wadden Sea (Sylt; van Beusekom and de Jonge 2012).

The factors responsible for the regional differences are not yet known. Van Beusekom et al. (2012) suggested two possibilities: differences in the amount of imported organic matter or the size of the receiving tidal basins. Evidence was presented to show that the size of the receiving tidal basin, in particular the distance between barrier islands and the mainland, may lead to a dilution of the import signal, whereas in narrow basins with a small distance between the islands the imported organic matter is concentrated.

The general decrease in eutrophication has resulted in lower chlorophyll levels and, in the northern Wadden Sea, to a decrease in green macroalgae (van Beusekom et al. 2009a) and has possibly contributed to an increase in seagrass (Reise and Kohlus 2008).

3.11.4 Contaminants

Within the framework of the Wadden Sea Quality Status Reports, Bakker et al. (2009) presented an overview of hazardous substances. Dissolved heavy metal concentrations in the Wadden Sea (mercury, cadmium, copper, zinc and lead) are not monitored and the focus is on sediment contamination. The main reduction in riverine heavy metal loads—the principal source for the Wadden Sea—was during the 1980s and 1990s. Notable decreases were observed in the river Elbe after the end of the German Democratic Republic. Since then heavy metal concentrations have remained similar or decreased slightly. Sediment concentrations of mercury and lead still pose a risk in a majority of Wadden Sea sub-regions.

Xenobiotic compounds in the Wadden Sea are a major concern because most are persistent, bio-accumulative and toxic (Bakker et al. 2009). In general, riverine inputs and environmental concentrations have decreased. For instance, a ban on tributyltin (TBT) has proved very successful, but effects can still be observed, for example on snails. Polychlorinated biphenyls (PCBs) are still widespread but concentrations are decreasing. Levels of lindane and DDT are also decreasing, but occasional erosion of old deposits leads to fluctuating concentrations in the Wadden Sea. Of particular future concern (as possible hormone disruptors) are newly developed xenobiotics, which include flame retardants, perfluorinated sulfonates and phthalates. These substances are not regularly monitored and little is known about their ecological effects.

3.11.5 Relevance of Climate Change

Climate change is expected to affect the sediment composition of the Wadden Sea: directly by sea-level rise (e.g. Dolch and Hass 2008) and indirectly by altered wind regimes (de Jonge and van Beusekom 1995). It is an open question whether sediment import into the Wadden Sea can compensate for increased sea-level rise or whether the Wadden Sea will ultimately drown (CPSL 2010). Nevertheless, sea-level rise will necessitate new strategies for coastal protection with yet unknown consequences for dikes, hydrodynamics and morphology (Reise 2013).

Climate change will drive complex and interacting effects on Wadden Sea water quality. One aspect of climate change is weather extremes. Extreme high river flows may transport large amounts of toxic substances to the Wadden Sea affecting filter feeders and fish (Einsporn et al. 2005). Likewise, increased nutrient fluxes during high river flows or extreme wet years lead to greater organic matter turnover (Fig. 3.40). De Jonge and de Jong (2002) suggested that increased river runoff will increase SPM levels in the Wadden Sea.

Higher temperatures will have complex and interacting effects on organic matter turnover in the Wadden Sea (Reise and van Beusekom 2008). For instance, higher temperatures enhance zooplankton dynamics (Martens and van Beusekom 2008) but suppress spring phytoplankton blooms through enhanced grazing (van Beusekom et al. 2009a). In general, higher temperatures are expected to enhance organic matter turnover with as yet unknown effects on the Wadden Sea food web.

3.12 Summary

In general, temperature variability on all time scales to multi-decadal tends to obscure longer-term trends. This variability is probably a greater source of uncertainty than lack of surface temperature data. Nevertheless, evidence of exceptional warming, especially since the 1980s, is very strong. In adjacent Atlantic waters (Faroe-Shetland Channel) and the northern North Sea, there has been a positive temperature anomaly of more than one standard deviation in most years since the mid-1990s; more than two standard deviations in a majority of years between 2002 and 2010 (Hughes et al. 2011). The temperature rise is not uniform in space, with largest rises (exceeding 1 °C since the end of the 19th century) in the south-east. Models provide some evidence of increasing duration of summer stratification away from estuarine outflow regions.

Shorter-term variations in salinity exceed any climate-related changes.

The Atlantic Meridional Overturning Circulation is very variable with no clear trend to date. However, changes in northern inflow to the North Sea correlate with changes in the NAO. Otherwise currents are highly variable on various time scales (tides, winds, seasonal density), and one storm can be significant compared to a year's integrated transport. The evidence is strong, coming from models as much as from measurements; however, observations are sparse at any one time and brief relative to climate-change time scales.

Over the past 100 to 120 years, absolute mean sea level in the North Sea rose by about 1.6 mm year^{-1}, comparable to the rates of global mean sea-level rise. Extreme sea levels have increased over the past 100–150 years in the North Sea, mainly due to a rise in mean sea level. Evidence for changes in sea level is very strong. Waves and storm surges (resulting from the weather) show pronounced variation on time scales of years and decades but no substantial long-term trend.

Carbon dioxide, pH and nutrients basin-wide are influenced by the circulation pattern, especially inflow from the Atlantic, local weather conditions (correlating with the NAO) and properties of component water masses. However, measurements on long time scales relating to climate change are only local, made close to the coast and affected by strong offshore gradients. There is net CO_2 uptake from the atmosphere, attributable to areas stratified in summer. The North Sea is a net nitrogen sink for the Atlantic. Model results suggest a long-term decrease in pH with relatively large variability due to shorter-term changes of circulation.

Higher temperatures tend to reduce oxygen concentrations near the surface; at depth, concentrations depend on vertical mixing for which variable weather is an important driver where depletion is a concern.

Suspended matter and turbidity are very variable, influenced by river inputs, seasons, tidal resuspension and advection (spring-neap modulation), waves and stratification. The North Sea is generally turbid in the unstratified south with transport to the north-east.

Coastal erosion is extensive but irregular; where it occurs, long-term rates are often 1 m year^{-1} or more. Some sectors accrete. Climate-related change in evolving morphology is not yet known.

Ice occurrence is restricted to shallow waters of the southern and eastern North Sea and has decreased over the last 50 years. Nevertheless, some severe ice winters are still expected in future.

In the Wadden Sea, higher temperatures are expected to enhance organic matter turnover.

Much work has focused on detecting long-term change in the North Sea region, either from measurements or model results. In other regions, there have been attempts to attribute such changes to, for example, anthropogenic forcing (e.g. Barkhordarian et al. 2012). However, comparable studies are still missing for the North Sea. Such studies are urgently needed to assess consistency between observed changes and current expectations, in order to increase the level of confidence in projections of expected future conditions.

Acknowledgments Figures 3.6, 3.9, 3.10 are reproduced under the Open Government Licence v3.0: www.nationalarchives.gov.uk/doc/open-government-licence/version/3.

References

Aagaard T (2011) Sediment transfer from beach to shoreface: The sediment budget of an accreting beach on the Danish North Sea Coast. Geomorphology 135:143–157

Aagaard T, Davidson-Arnott R, Greenwood B, Nielsen J (2004) Sediment supply from shoreface to dunes: Linking sediment transport measurements and long-term morphological evolution. Geomorphology 60:205–224

Aernouts D, Héquette A (2006) L'évolution du rivage et des petits-fonds en baie de Wissant pendant le XXe siècle (Pas-de-Calais, France). Géomorphologie: relief, processus, environnement [online]: http://geomorphologie.revues.org/477; doi:10.4000/geomorphologie.477

Albrecht F, Weisse R (2012) Pressure effects on past regional sea level trends and variability in the German Bight. Ocean Dyn 62:1169–1186

Albrecht F, Wahl T, Jensen J, Weisse R (2011) Determining sea level change in the German Bight. Ocean Dyn 61:2037–2050

Albretsen J, Aure J, Sætre R, Danielssen DS (2012) Climatic variability in the Skagerrak and coastal waters of Norway. ICES J Mar Sci 69:758–763

Alheit J, Pohlmann T, Casini M, Greve W, Hinrichs R, mathis M, O'driscoll K, Vorberg R, Wagner C (2012) Climate variability drives anchovies and sardines into the North and Baltic Seas. Prog Oceanogr 96:128–139

Allen JI, Holt JT, Blackford J, Proctor R (2007) Error quantification of a high-resolution coupled hydrodynamic-ecosystem coastal-oceanmodel: Part 2. Chlorophyll-a, nutrients and SPM. J Mar Sys 68:381–404

Andersen TJ, Lanuru M, van Bernem C, Pejrup M, Riethmueller R (2010) Erodibility of a mixed mudflat dominated by microphytobenthos and *Cerastoderma edule*, East Frisian Wadden Sea, Germany. Estuar, Coastal Shelf Sci 87:197–206.

Anthony D, Leth JO (2002) Large-scale bedforms, sediment distribution and sand mobility in the eastern North Sea off the Danish west coast. Mar Geol 182:247–263

Araújo I (2005) Sea level variability: Examples from the Atlantic coast of Europe. Ph.D. Thesis, University of Southampton, United Kingdom

Araújo I, Pugh DT (2008) Sea levels at Newlyn 1915–2005: Analysis of trends for future flooding risks. J Coast Res 24(sp3):203–212

Artioli Y, Blackford JC, Butenschön M, Holt JT, Wakelin SL, Thomas H, Borges AV, Allen JI (2012) The carbonate system in the North Sea: sensitivity and model validation. J Mar Sys 102–104:1–13

Artioli Y, Blackford JC, Nondal G, Bellerby RGJ, Wakelin SL, Holt JT, Butenschön M, Allen JI (2014) Heterogeneity of impacts of high CO_2 on the North Western European Shelf. Biogeosciences 11:601–612

Bacon S, Carter D (1991). Wave climate changes in the North Atlantic and the North Sea. Int J Climatol 11:545–558

Badin G, Williams RG, Holt JT, Fernand LJ (2009) Are mesoscale eddies in shelf seas formed by baroclinic instability of tidal fronts? J Geophys Res 114:C10021, doi:10.1029/2009JC005340

Bakker J, Lüerßen G, Marencic H, Jung K (2009) Hazardous substances. In: Thematic Report 5.1 In: Marencic H, de Vlas J (eds) Quality Status Report 2009, Wadden Sea Ecosystem. No 25. Common Wadden Sea Secretariat, Trilateral Monitoring and Assessment Group, Wilhelmshaven, Germany

Baretta-Bekker JG, Baretta JW, Latuhihin MJ, Desmit X, Prins TC (2009) Description of the long-term (1991–2005) temporal and spatial distribution of phytoplankton carbon biomass in the Dutch North Sea. J Sea Res 61:50–59

Barkhordarian A, Bhend J, von Storch H (2012) Consistency of observed near surface temperature trends with climate change projections over the Mediterranean region. Clim Dyn 38:1695–1702

Baxter P (2005) The East Coast Great Flood, 31 January–1 February 1953: A summary of the human disaster. Phil Trans R Soc A 363:1293–1312

Becker GA, Frohse A, Damm P (1997) The Northwest European shelf temperature and salinity variability. German J. Hydr. (Ocean. Dynamics) 49:135–151

Berx B, Hansen B, Østerhus S, Larsen KM, Sherwin T, Jochumsen K (2013) Combining in-situ measurements and altimetry to estimate volume, heat and salt transport variability through the Faroe Shetland Channel. Ocean Sci 9:639–654

Besch H-W (1987) Sylt-Naturräumliche Gliederung und Umwandlung durch Meer und Mensch. In: Hofmeister B, Voss F (eds), Beiträge zur Geographie der Küsten und Meere - Sylt 1986 und Berlin 1987. Berliner Geogr. Studien 25

Beszczynska-Möller A, Dye SR (eds) (2013) ICES Report on Ocean Climate 2012. ICES Coop Res Rep 321

Bindoff NL, Willebrand J, Artale V, Cazenave A, Gregory J, Gulev S, Hanawa K, Le Quéré C, Levitus S, Nojiri Y, Shum CK, Talley LD, Unnikrishnan A (2007) Observations: Oceanic climate change and sea level. In: Solomon, S., Qin, D., Manning, M., Chen, Z.,

Marquis, M., Averyt, K.B., Tignor M., Miller H.L (eds) Climate Change 2007: The Physical Science Basis. Contribution of Working Group I to the Fourth Assessment Report of the Intergovernmental Panel on Climate Change. Cambridge University Press

Blackford JC, Gilbert FJ (2007) pH variability and CO_2 induced acidification in the North Sea. J. Mar Sys 64:229–241

Blott SJ, Duck RW, Phillips MR, Pontee NI, Pye K, Williams A (2013) Great Britain. In: Pranzini E, Williams A (eds) Coastal Erosion and Protection in Europe. Routledge, Abingdon, pp. 173–208

Borges AV, Gypens N (2010) Carbonate chemistry in the coastal zone responds more strongly to eutrophication than to ocean acidification. Limnol Oceanogr 55:346–353

Borsje BW, Hulscher SJMH, Herman PMJ, De Vries MB (2009) On the parameterization of biological influences on offshore sand wave dynamics. Ocean Dyn 59:659–670

Bouin MN, Wöppelmann G (2010) Land motion estimates from GPS at tide gauges: A geophysical evaluation. Geophys J Int 180:193–209

Bozec Y, Thomas H, Schiettecatte L-S, Borges AV, Elkalay K, de Baar HJW (2006) Assessment of the processes controlling the seasonal variations of dissolved inorganic carbon in the North Sea. Limnol Oceanogr 51:2746–2762

Brooks SM, Spencer T (2010) Temporal and spatial variations in recession rates and sediment release from soft rock cliffs, Suffolk coast, UK. Geomorphology 124:26–41

Brown S (2008) Soft cliff retreat adjacent to coastal defences, with particular reference to Holderness and Christchurch Bay, UK. PhD Thesis University of Southampton

Bryche A, de Putter B, De Wolf P (1993) French and Belgian Coast from Dunkirk to De Panne: a case study of transborder cooperation in the framework of the Interreg initiative of the European community. In: Coastal Zone: Proceedings of the Symposium on Coastal and Ocean Management

BSH (2008) Umweltbericht zum Raumordnungsplan für die deutsche ausschließliche Wirtschaftszone (AWZ) Teil Nordsee. Bundesamt für Seeschiffahrt und Hydrographie (BSH)

Buijsman MC, Ridderinkhof H (2008a) Long-term evolution of sand waves in the Marsdiep inlet. I: High-resolution observations. Cont Shelf Res 28:1190–1201

Buijsman MC, Ridderinkhof H (2008b) Long-term evolution of sand waves in the Marsdiep inlet. II: Relation to hydrodynamics. Cont Shelf Res 28:1202–1215

Burchard H, Flöser G, Staneva J, Badewien TH, Riethmüller R (2008) Impact of density gradients on net sediment transport into the Wadden Sea. J Phys Oceanogr 38:566–587

Burningham H, French J (2006) Morphodynamic behaviour of a mixed sand-gravel ebb-tidal delta: Deben estuary, Suffolk, UK. Mar Geol 225:23–44

Burningham H, French J (2011) Seabed dynamics in a large coastal embayment: 180 years of morphological change in the outer Thames estuary. Hydrobiologia 672:105–119

Burt WJ, Thomas H, Fennel K, Horne E (2013) Sediment-water column fluxes of carbon, oxygen and nutrients in Bedford Basin, Nova Scotia, inferred from 224Ra measurements. Biogeosci 10:53–66

Burt WJ, Thomas H, Pätsch J, Omar AM, Schrum C, Daewel U, Brenner H, de Baar HJW (2014) Radium isotopes as a tracer of sediment-water column exchange in the North Sea. Glob Biogeochem Cyc 28, doi:10.1002/2014GB004825

Cadée GC, Hegeman J (2002) Phytoplankton in the Marsdiep at the end of the 20th century; 30 years monitoring biomass, primary production, and Phaeocystis blooms. J Sea Res 48:97–110

Calafat FM, Chambers DP, Tsimplis MN (2012) Mechanisms of decadal sea level variability in the Eastern North Atlantic and the Mediterranean Sea. J Geophys Res 117 C09022, doi:10.1029/2012JC008285

Cambers G (1976) Temporal scales in coastal erosion systems. Trans Inst British Geographers 1:246–256

Capuzzo E, Painting SJ, Forster RM, Greenwood N, Stephens DT, Mikkelsen OA (2013) Variability in the sub-surface light climate at ecohydrodynamically distinct sites in the North Sea. Biogeochemistry 113:85–103

Carr AP (1979) Sizewell-Dunwich Banks Field Study. Topic Report: 2. Long-term changes in the coastline and offshore banks. Report 89, Institute of Oceanographic Sciences, Taunton

Carter D, Draper L (1988). Has the north-east Atlantic become rougher? Nature 332:494

Chafik L (2012) The response of the circulation in the Faroe-Shetland Channel to the North Atlantic Oscillation. Tellus A 64:18423, doi:10.3402/tellusa.v64i0.18423

Charlier RH (2013) Belgium. In: Pranzini E, Williams A (eds) Coastal Erosion and Protection in Europe. Routledge, Abingdon, pp. 158–172

Chaverot S, Héquette A, Cohen O (2005) Evolution of climatic forcings and potentially eroding events on the coast of Northern France. Proceedings 5th International Conference on Coastal Dynamics, Barcelona, Spain, April 2005, Cd-Rom, 11 p.

Chen C-TA, Wang S-L (1999) Carbon, alkalinity and nutrient budgets on the East China Sea continental shelf. J Geophys Res 104:20675–20686

Christiansen C, Bowman D (1990) Long-term beach and shoreface changes, NW Jutland, Denmark: effects of a change in wind direction. In: Beukema JJ et al. (eds) Expected Effects of Climatic Change on Marine Coastal Ecosystems, pp. 113–122

Church JA, White NJ (2006) A 20th century acceleration in global sea-level rise. Geophys Res Lett 33: L01602, doi:10.1029/2005GL024826

Church JA, White NJ (2011) Sea-level rise from the late 19th to the early 21st Century. Surveys in Geophysics 32:585–602

Church J, Aarup T, Woodworth P, Wilson W, Nicholls R, Rayner R, Lambeck K, Mitchum G, Steffen K, Cazenave A, Blewitt G, Mitrovica J, Lowe J (2010). Sea-level rise and variability: synthesis and outlook for the future. In: Church J, Woodworth P, Aarup T, Wilson S (eds) Understanding Sea-level Rise and Variability. Wiley & Sons

Church JA, Clark PU, Cazenave A, Gregory JM, Jevrejeva S, Levermann A, Merrifield MA, Milne GA, Nerem RS, Nunn PD, Payne AJ, Pfeffer WT, Stammer D, Unnikrishnan AS (2013) Sea level change. In: Climate Change 2013: The Physical Science Basis. Contribution of Working Group I to the Fifth Assessment Report of the Intergovernmental Panel on Climate Change. Stocker TF, Qin D, Plattner G-K, Tignor M, Allen SK, Boschung J, Nauels A, Xia Y, Bex V, Midgley PM (eds.). Cambridge University Press

Clabaut P, Chamley H, Marteel H (2000) Évolution récente des dunes littorales à l'est de Dunkerque (Nord de la France) [Recent coastal dunes evolution, East of Dunkirk, Northern France]. Géomorphologie: relief, processus, environnement 6:125–136

Corten A, van de Kamp G (1996) Variation in the abundance of southern fish species in the southern North Sea in relation to hydrography and climate. ICES J Mar Sci 53:1113–1119

Cox AT, Swail VR (2001) A global wave hindcast over the period 1958–1997: Validation and climate assessment. J Geophys Res 106C:2313–2329

CPSL (2010) Third Report. The role of spatial planning and sediment in coastal risk management. Wadden Sea Ecosystem No. 28. Common Wadden Sea Secretariat, Trilateral Working Group on Coastal Protection and Sea Level Rise (CPSL), Wilhelmshaven, Germany: 1-51

Cunningham S, Marsh R, Wood R, Wallace C, Kuhlbrodt T, Dye S (2010) Atlantic Heat Conveyor (Atlantic Meridional Overturning

Circulation). In MCCIP Annual Report Card 2010-11, MCCIP Science Review, 14 pp, www.mccip.org.uk/arc

Daewel U, Schrum C (2013) Simulating long-term dynamics of the coupled North Sea and Baltic Sea ecosystem with ECOSMO II: Model description and validation. J Mar Sys 119–120:30–49

de Boer GJ, Pietrzak JD, Winterwerp JC (2006) On the vertical structure of the Rhine region of freshwater influence. Ocean Dyn 56:198–216

de Boer GJ, Pietrzak JD, Winterwerp JC (2009) SST observations of upwelling induced by tidal straining in the Rhine ROFI. Cont Shelf Res 29:263–277

de Jong F (2000) Marine Eutrophication in Perspective. On the Relevance of Ecology for Environmental Policy. Springer

de Jonge VN, de Jong DJ (2002) 'Global change' Impact of inter-annual variation in water discharge as a driving factor to dredging and spoil disposal in the river Rhine system and of turbidity in the Wadden Sea. Est Coast Shelf Sci 55:969–991

de Jonge VN, Postma H (1974) Phosphorus compounds in the Dutch Wadden Sea. Netherland J Sea Res 8:139–153

de Jonge VN, van Beusekom JEE (1995) Wind-induced and tide-induced resuspension of sediment and microphytobenthos from tidal flats in the Ems estuary. Limnol Oceanogr 40:766–778

de Jonge VN, Schuttelaars HM, van Beusekom JEE, Talke SA, de Swart HE (2014) The influence of channel deepening on estuarine turbidity levels and dynamics, as exemplified by the Ems estuary. Est Coast Shelf Sci 139:46–59

de Kok JM (1992) A three-dimensional finite difference model for computation of near- and far-field transport of suspended matter near a river mouth. Cont Shelf Res 12:625–642

de Nijs MAJ (2012) On sedimentation processes in a stratified estuarine system. PhD Thesis, Delft University of Technology

de Nijs MAJ, Winterwerp JC, Pietrzak JD (2010) The effects of internal flow structure on sediment entrapment in the Rotterdam Waterway. J Phys Oceanogr 40:2357–2380

de Ruijter WPM, Van der Giessen A, Groenendijk FC (1992) Current and density structure in the Netherlands coastal zone. In: Prandle D (ed) Coast Estuar Studies 40: Dynamics and exchanges in estuaries and the coastal zone. AGU, Washington, USA

De Wolf P (2002) Kusterosie en verzanding van het Zwin. In: Van Lancker V et al. (eds) Colloquium Kustzonebeheer vanuit geo-ecologische en economische invalshoek. Oostende (B), 16–17 May 2002. Genootschap van Gentse Geologen (GGG)- Vlaams Instituut van de Zee (VLIZ). VLIZ Special Publication 10: Oostende, Belgium

Degrendele K, Roche M, Schotte P, Van Lancker V, Bellec V, Bonne W (2010) Morphological evolution of the Kwinte Bank central depression before and after the cessation of aggregate extraction. J Coast Res SI 51:77–86

Deleu S, Van Lancker V, Van den Eynde D, Moerkerke G (2004) Morphodynamic evolution of the kink of an offshore tidal sandbank: the Westhinder Bank (Southern North Sea). Cont Shelf Res 24:1587–1610

Dette HH, Gärtner J (1987) Erfahrungen mit der Versuchssand-vorspülung vor Hörnum im Jahre 1983. Die Küste 45:209–258

Dickson RR, Meincke J, Malmberg SA, Lee AJ (1988) The "Great Salinity Anomaly" in the northern North Atlantic 1968–1982. Prog Oceanogr 20:103–151

Diesing M, Kubicki A, Winter C, Schwarzer K (2006) Decadal scale stability of sorted bedforms, German Bight, southeastern North Sea. Cont Shelf Res 26:902–916

Dillingh D, Fedor B, de Ronde J (2010) Definitiezeespiegelstijgingvoorbepalingsuppletiebehoefte. Deltares Report 1201993-002, Delft

Dobrynin M, Gayer G, Pleskachevsky A, Günther H (2010) Effect of waves and currents on the dynamics and seasonal variations of suspended particulate matter in the North Sea. J Mar Sys 82:1–20

Dolch T, Hass HC (2008) Long-term changes of intertidal and subtidal sediment composition in a tidal basin in the northern Wadden Sea (SE North Sea). Helgol Mar Res 62:3–11

Dronkers J, Van Alphen JSLJ, Borst JC (1990) Suspended sediment transport processes in the southern North Sea. In: Cheng RT (ed) Residual Currents and Long Term Transport. Springer

Dye SR, Hughes SL, Tinker J, Berry DI, Holliday NP, Kent EC, Kennington K, Inall M, Smyth T, Nolan G, Lyons K, Andres O, Beszczynska-Möller A (2013a) Impacts of climate change on temperature (air and sea). MCCIP Science Review 2013:1–12. doi:10.14465/2013.arc01.001-012

Dye SR, Holliday NP, Hughes SL, Inall M, Kennington K, Smyth T, Tinker J, Andres O, Beszczynska-Möller A (2013b) Climate change impacts on the waters around the UK and Ireland: Salinity. MCCIP Science Review 2013:60–66

Dyer KR, Moffat TJ (1998) Fluxes of suspended matter in the East Anglian plume Southern North Sea. Cont Shelf Res 18:1311–1331

EC (2000) Directive of the European Parliament and of the council 2000/60/EC establishing a framework for community action in the field of water policy, European Union. The European Parliament, The Council, PE-CONS 3639/1/00 REV 1 EN

EEA (2012) Climate Change, Impacts and Vulnerability in Europe 2012. European Environment Agency, Copenhagen

Einsporn S, Broeg K, Koehler A (2005) The Elbe flood 2002 – toxic effects of transported contaminants in flatfish and mussels of the Wadden Sea. Mar Poll Bull 50:423–429

Eisma D (1981) Supply and deposition of suspended matter in the North Sea. In: Nio S-D, Shüttenhelm RTE, Van Weering TJCE (eds), Holocene marine sedimentation in the North Sea basin. Special Publication 5 of the International Association of Sedimentologists. Blackwell Scientific Publishers

Eisma D, Kalf J (1987) Dispersal, concentration and deposition of suspended matter in the North Sea. J Geol Soc London 144:161–178

Eleveld MA, Pasterkamp R, Van der Woerd HJ (2004) A survey of total suspended matter in the southern North Sea based on the 2001 SeaWiFS data. EARSeL eProceedings 3:166–178, www.eproceedings.org

Eleveld MA, Pasterkamp R, van der Woerd HJ, Pietrzak J (2006) Suspended particulate matter from SeaWiFS data: statistical validation, and verification against the physical oceanography of the southern North Sea. Ocean Optics OOXVIII, (9–14 Oct 2006, Montreal)

Eleveld MA, Pasterkamp R, van der Woerd HJ, Pietrzak JD (2008) Remotely sensed seasonality in the spatial distribution of sea-surface suspended particulate matter in the southern North Sea. Estuarine Coast Shelf Sci 80:103–113

Elias EPL, Cleveringa J, Buijsman MC, Roelvink JA, Stive MJF (2006) Field and model data analysis of sand transport patterns in Texel Tidal inlet (the Netherlands). Coast Eng 53:505–529

Ellett DJ, Martin JHA (1973) The physical and chemical oceanography of the Rockall Channel. Deep-Sea Res. 20:585–625

Emeis K-C, van Beusekom J, Callies U, Ebinghaus R, Kannen A, Kraus G, Kröncke I, Lenhart H, Lorkowski I, Matthias V, Möllmann C, Pätsch J, Scharfe M, Thomas H, Weisse R, Zorita E (2015) The North Sea – a shelf sea in the Anthropocene. J Mar Sys 141:18-33

EUROSION (2004) Living with Coastal Erosion in Europe: Sediment and Space for Sustainability. Part II – Maps and Statistics. Online at: www.eurosion.org/reports-online/part2.pdf

Fettweis M, van den Eynde D (2003) The mud deposits and the high turbidity in the Belgian-Dutch coastal zone, southern bight of the North Sea. Cont Shelf Res 23:669–691

Fettweis M, Monbaliu J, Baeye M, Nechad B, van den Eynde D (2012) Weather and climate induced spatial variability of surface suspended particulate matter concentration in the North Sea and the English Channel. Method Oceanogr 3–4:25–39

Flather RA (1987) Estimates of extreme conditions of tide and surge using a numerical model of the north-west European continental shelf. Estuarine, Coast Shelf Sci 24:69–93

Flemming B, Nyandwi N (1994) Land reclamation as a cause of fine-grained sediment depletion in backbarrier tidal flats (southern North Sea). Netherlands J Aquat Ecol 28:299–307

Folland CK, Parker DE (1995) Correction of instrumental biases in historical sea surface temperature data. Quart J Roy Met Soc 121:319–367

French PW (2001) Coastal Defences - Processes, Problems and Solutions. Routledge

Fruergaard M, Andersen TJ, Johannessen PN, Nielsen LH, Pejrup M (2013) Major coastal impact induced by a 1000-year storm event. Scientific Reports 3: no. 1051, doi:10.1038/srep01051

Garcia D, Vigo I, Chao BF, Martinez MC (2007) Vertical crustal motion along the Mediterranean and Black Sea coast derived from ocean altimetry and tide gauge data. Pure Appl Geophys 164:851–863

Gayer G, Dick S, Pleskachevsky A, Rosental W (2004) Modellierung von Schwebstofftransporten in Nordsee und Ostsee. Berichte des BSH 36. BSH, Hamburg

Gayer G, Dick S, Pleskachevsky A, Rosenthal W (2006) Numerical modelling of suspended matter transport in the North Sea Ocean Dyn 56:62–77

Gehrels WR, Woodworth PL (2013) When did modern rates of sea-level rise start? Global and Plan Change 100:263–277

Gerritsen H (2005) What happened in 1953? The big flood in the Netherlands in retrospect. Phil Trans R Soc A 363:1271–1291

Giardino A, Van den Eynde D, Monbaliu J (2010) Wave effects on the morphodynamic evolution of an offshore sand bank. J Coast Res 51:127–140

Greenwood N, Parker ER, Fernand L, Sivyer DB, Weston K, Painting SJ, Kröger S, Forster RM, Lees HE, Mills DK, Laane RWPM (2010) Detection of low bottom water oxygen concentrations in the North Sea; implications for monitoring and assessment of ecosystem health. Biogeosciences 7:1357–1373

Gypens N, Borges AV, Lancelot C (2009) Effect of eutrophication on air-sea CO_2 fluxes in the coastal Southern North Sea: a model study of the past 50 years. Glob Change Biol 15:1040–1056

Haekkinen S, Rhines PB, Worthen DL (2011). Warm and saline events embedded in the meridional circulation of the northern North Atlantic. J Geophys Res 116: C03006, doi:10.1029/2010JC006275

Haigh ID, Nicholls RJ, Wells NC (2009) Mean sea-level trends around the English Channel over the 20th century and their wider context. Cont Shelf Res 29:2083–2098

Hátún H, Sando A, Drange H, Bentsen M (2005) Seasonal to decadal temperature variations in the Faroe–Shetland inflow waters. In: Drange H, Dokken T, Furevik T, Gerdes R, Berger W (eds) The Nordic Seas: an Integrated Perspective: Oceanography, Climatology, Biogeochemistry, and Modeling. American Geophysical Union, pp. 239–250

Heaps NS (1972) Estimation of density currents in Liverpool Bay. Geophys J Roy Astron Soc 30:415–432

Henriksen P (2009) Long-term changes in phytoplankton in the Kattegat, the Belt Sea, the Sound and the western Baltic Sea. J Sea Res 61:114–123

Heyen H, Dippner JW (1998) Salinity variability in the German Bight in relation to climate variability. Tellus 50A:545–556

Hill AE, James ID, Linden PF, Matthews JP, Prandle D, Simpson JH, Gmitrowicz EM, Smeed DA, Lwiza KMM, Durazo R, Fox AD, Bowers DG (1993) Dynamics of tidal mixing fronts in the North Sea. Phil Trans Roy Soc London A 343:431–446

Hill AE, Brown J, Fernand L, Holt J, Horsburgh KJ, Proctor R, Raine R, Turrell WR (2008) Thermohaline circulation of shallow tidal seas. Geophys Res Lett 35:L11605, doi:10.1029/2008GL033459

Hjøllo SS, Skogen MD, Svendsen E (2009) Exploring currents and heat within the North Sea using a numerical model. J Mar Sys 78:180–192

Hogben N (1994) Increases in wave heights over the North Atlantic: a review of the evidence and some implications for the naval architect. Trans Roy Inst Naval Arch W5:93–101.

Hollebrandse FAP (2005) Temporal development of the tidal range in the Southern North Sea. Ph.D. Thesis, Delft University of Technology, Netherlands

Holliday NP (2003) Air-sea interaction and circulation changes in the northeast Atlantic. J Geophys Res 108:C3259, doi:10.1029/2002JC001344

Holliday NP, Cunningham SA (2013) The Extended Ellett Line: Discoveries from 65 years of marine observations west of the UK. Oceanography 26:156–163

Holliday NP, Hughes SL, Bacon S, Beszczynska-Möller A, Hansen B, Lavín A, Loeng H, Mork KA, Østerhus S, Sherwin T, Walczowski W (2008) Reversal of the 1960s – 1990s freshening trend in the North Atlantic and Nordic Seas. Geophys Res Lett 35: L03614, doi:10.1029/2007GL032675

Holliday NP, Hughes SL, Beszczynska-Möller A (eds) (2009) ICES Report on Ocean Climate 2008. ICES Coop Res Rep 298

Holliday NP, Hughes SL, Dye S, Inall M, Read J, Shammon T, Sherwin T, Smyth T (2010) Salinity. In: MCCIP Annual Report Card 2010-11, MCCIP Science Review, www.mccip.org.uk/arc

Holliday NP, Hughes SL, Borenäs K, Feistel R, Gaillard F, Lavìn A, Loeng H, Mork KA, Nolan G, Quante M, Somavilla R (2011) Long-term physical variability in the North Atlantic Ocean. In: Reid PC, Valdés L (eds) ICES Status Report on Climate Change in the North Atlantic. ICES Coop Res Rep 310: 21–46.

Holt JT, James ID (1999) A simulation of the southern North Sea in comparison with measurements from the North Sea Project Part 2, Suspended Particulate Matter. Cont Shelf Res 19:1617–1642

Holt J, Proctor R (2008) The seasonal circulation and volume transport on the northwest European continental shelf: a fine-resolution model study. J Geophys Res 113:C06021, doi:10.1029/2006JC004034

Holt JT, Allen JI, Proctor R, Gilbert F (2005) Error quantification of a high resolution coupled hydrodynamic–ecosystem coastal–ocean model: part 1 model overview and assessment of the hydrodynamics. J Mar Sys 57:167–188

Holt J, Wakelin SL, Huthnance JM (2009) The down-welling circulation of the northwest European continental shelf: a driving mechanism for the continental shelf carbon pump. Geophys Res Lett 36:L14602, doi:10.1029/2009GL038997

Holt J, Hughes S, Hopkins J, Wakelin SL, Holliday NP, Dye S, González-Pola C, Hjøllo SS, Mork KA, Nolan G, Proctor R, Read J, Shammon T, Sherwin T, Smyth T, Tattersall G, Ward B, Wiltshire K (2012) Multi-decadal variability and trends in the temperature of the northwest European continental shelf: A model-data synthesis. Prog Oceanogr 106:96–117

Hordoir R, Meier HEM (2010) Freshwater fluxes in the Baltic Sea: A model study. J Geophys Res 115:C08028, doi:10.1029/2009JC005604

Hordoir R, Dieterich C, Basu C, Dietze H, Meier HEM (2013) Freshwater outflow of the Baltic Sea and transport in the Norwegian Current: A statistical correlation analysis based on a numerical experiment. Cont Shelf Res 64:1–9

Horrillo-Caraballo JM, Reeve DE (2008) Morphodynamic behaviour of a nearshore sandbank system: The Great Yarmouth Sandbanks, UK. Mar Geol 254:91–106

Horsburgh K, Wilson C (2007) Tide-surge interaction and its role in the distribution of surge residuals in the North Sea. J Geophys Res 112: C08003, doi:10.1029/2006JC004033

Howarth MJ, Dyer KR, Joint IR, Hydes DJ, Purdie DA, Edmunds H, Jones JE, Lowry RK, Moffat TJ, Pomroy AJ, Proctor R (1993) Seasonal cycles and their spatial variability. Phil Trans Roy Soc Lond A 343:383–403

Hughes SL, Holliday NP, Beszczynska-Möller A (eds) (2011) ICES Report on Ocean Climate 2010. ICES Coop Res Rep 309

Hughes SL, Holliday NP, Gaillard F, ICES Working Group on Oceanic Hydrography (2012) Variability in the ICES/NAFO region between 1950 and 2009: observations from the ICES Report on Ocean Climate. ICES J Mar Sci 69:706–719

Hurrell JW (1995) Decadal trends in the North Atlantic Oscillation: Regional temperatures and precipitation. Science 269:676–679

Hurrell JW, National Center for Atmospheric Research Staff (eds) (2013) The Climate Data Guide: Hurrell North Atlantic Oscillation (NAO) Index (station-based). https://climatedataguide.ucar.edu/climate-data/hurrell-north-atlantic-oscillation-nao-index-station-based

Huthnance JM, Holt JT, Wakelin SL (2009) Deep ocean exchange with west-European shelf seas. Ocean Sci 5:621–634

Irion G, Zöllmer V (1999) Clay mineral associations in fine-grained surface sediments of the North Sea. J Sea Res 41:119–128

Ivchenko V, Wells N, Aleynik D, Shaw A (2010) Variability of heat and salinity content in the North Atlantic in the last decade. Ocean Sci 6:719–735

Iversen SA, Skogen MD, Svendsen E (2002) Availability of horse mackerel (*Trachurus trachurus*) in the north-eastern North Sea, predicted by the transport of Atlantic water. Fish Oceanogr 11:245–250

Jago CF, Bale AJ, Green MO, Howarth MJ, Jones SE, McCave IN, Millward GE, Morris AW, Rowden AA, Williams JJ, Hydes D, Turner A, Huntley D, Van Leussen W (1993) Resuspension processes and seston dynamics, southern North Sea [and discussion]. Phil Trans Roy Soc London A 343:475–491

Janssen F (2002) Statistische Analyse mehrjähriger Variabilität der Hydrographie in Nord- und Ostsee. Dissertation Fachbereich Geowissenschaften, Universität Hamburg, Germany

Jensen J, Hofstede J, Kunz H, De Ronde J, Heinen P, Siefert W (1993) Long-term water level observations and variations. Coastal Zone '93, Coastlines of the Southern North Sea

Jevrejeva S, Grinsted A, Moore JC, Holgate S (2006) Nonlinear trends and multiyear cycles in sea level records. J Geophys Res 111: C09012, doi.org/10.1029/2005JC003229

Jevrejeva S, Moore JC, Grinsted A, Woodworth PL (2008) Recent global sea level acceleration started over 200 years ago? Geophys Res Lett 35:L08715, doi:10.1029/2008GL033611

Johnson GC, Gruber N (2007) Decadal water mass variations along 20 degrees W in the Northeastern Atlantic Ocean. Prog Oceanogr 73:277–295

Jones SE, Jago CF, Simpson JH (1996) Modelling suspended sediment dynamics in tidally stirred and periodically stratified waters: Progress and pitfalls. In: Pattiaratchi C (ed), Mixing in Estuaries and Coastal Seas, AGU, Coast Estuar Studies 50

Joordens JCA, Souza AJ, Visser AW (2001) The Influence of tidal straining and wind on suspended matter and phytoplankton dynamics in the Rhine outflow region. Cont Shelf Res 21:301–325

Karlson K, Rosenberg R, Bonsdorff E (2002) Temporal and spatial large-scale effects of eutrophication and oxygen deficiency on benthic fauna in Scandinavian and Baltic waters – a review. Oceanogr Mar Biol 40:427–489

Kauker F (1999) Regionalization of climate model results for the North Sea. Ph.D. Thesis, Universität Hamburg, Germany

Kauker F, von Storch H (2000) Statistics of "synoptic circulation weather" in the North Sea as derived from a multiannual OGCM simulation. J Phys Oceanogr 30:3039–3049

Kelletat D (1992) Coastal erosion and protection measures at the German North Sea coast. J Coast Res 8:699–711

Kennedy JJ, Rayner NA, Smith RO, Saunby M, Parker DE (2011a) Reassessing biases and other uncertainties in sea-surface temperature observations since 1850 part 1: measurement and sampling errors. J Geophys Res 116:D14103, doi:10.1029/2010JD015218

Kennedy JJ, Rayner NA, Smith RO, Saunby M, Parker DE (2011b) Reassessing biases and other uncertainties in sea-surface temperature observations since 1850, part 2: biases and homogenisation. J Geophys Res 116:D14104, doi:10.1029/2010JD015220

Kent EC, Taylor PK, Truscott BS, Hopkins JS (1993) The accuracy of voluntary observing ship's meteorological observations – Results of the VSOP-NA. J Atmos Oceanic Tech 10:591–608

Kent EC, Kennedy JJ, Berry DI, Smith RO (2010) Effects of instrumentation changes on ocean surface temperature measured in situ. Climate Change 1:718–728

Kirby RR, Beaugrand G, Lindley JA, Richardson AJ, Edwards M, Reid PC (2007) Climate effects and benthic-pelagic coupling in the North Sea. Mar Ecol-Prog Ser 330:31–38

Knaapen MAF (2009) Sandbank occurrence on the Dutch continental shelf in the North Sea. Geo-Mar Lett 29:17–24

Kodra E, Steinhaeuser K, Ganguly AR (2011) Persisting cold extremes under 21st-century warming scenarios. Geophys Res Lett 38: L08705, doi:10.1029/2011GL047103

Koslowski G (1989) Die flächenbezogene Eisvolumensumme, eine neue Maßzahl für die Bewertung des Eiswinters an der Ostseeküste Schleswig-Holsteins und ihr Zusammenhang mit dem meteorologischen Charakter des meteorologischen Winters. Dt Hydrogr Z 42:61–80

Kühn W, Pätsch J, Thomas H, Borges AV, Schiettecatte L-S, Bozec Y, Prowe F (2010) Nitrogen and carbon cycling in the North Sea and exchange with the North Atlantic - a model study Part II. Carbon budget and fluxes. Cont Shelf Res 30:1701–1716

Kushnir Y, Cardone VJ, Greenwood JG, Cane MA (1997) The recent increase in North Atlantic wave heights. J. Climate 10:2107–2113

Kystdirektoratet (2008) Vestkysten 2008. Kystdirektoratet, Lemvig

Lacroix G, Ruddick K, Ozer J, Lancelot C (2004) Modelling the impact of the Scheldt and Rhine/Meuse plumes on the salinity distribution in Belgian waters (southern North Sea). J Sea Res 52:149–163

Lancelot C, Thieu V, Polard A., Garnier J, Billen G, Hecq W, Gypens N (2011) Ecological and economic effectiveness of nutrient reduction policies on coastal *Phaeocystis* colony blooms in the Southern North Sea: an integrated modeling approach. Sci Tot Env 409:2179–2191

Langenberg H, Pfizenmayer A, von Storch H, Sündermann J (1999) Storm-related sea level variations along the North Sea coast: natural variability and anthropogenic change. Cont Shelf Res 19:821–842

Latif M, Böning C, Willebrand J, Biastoch A, Dengg J, Keenlyside N, Schweckendiek U, Madec G (2006) Is the thermohaline circulation changing? J Climate 19:4631–4637

Lee AJ (1980) North Sea: Physical oceanography. In: Banner FT, Collins MB, Massie KS (eds) The North-West European Shelf Seas: The seabed and the sea in motion. 2. Physical and chemical oceanography, and physical resources. Elsevier Oceanography Series, 24B, pp. 467–493

Lenhart H-J, Mills DK, Baretta-Bekker H, van Leeuwen SM, van der Molen J, Baretta JW, Blaas M, Desmit X, Kühn W, Lacroix G, Los HJ, Ménesguen A, neves R, Proctor R, Ruardij P, Skogen MD, Vanhoutte-Brunier A, Villars MT, Wakelin S (2010) Predicting the consequences of nutrient reduction on the eutrophication status of the North Sea. J Mar Sys 81:148–170

Leterme SC, Pingree RD, Skogen MD, Seuront L, Reid PC, Attrill MJ (2008) Decadal fluctuations in North Atlantic water inflow in the North Sea 1958–2003: impacts on temperature and phytoplankton populations. Oceanologia 50:59–72

Liu KK, Atkinson L, Quinones R, Talaue-McManus L (eds) (2010) Carbon and Nutrient Fluxes in Continental Margins. Springer

Loebl, M., Dolch T, van Beusekom JEE (2007) Annual dynamics of pelagic primary production and respiration in a shallow coastal basin. J Sea Res 58:269–282

Loebl M, Colijn F, Beusekom JEE, Baretta-Bekker JG, Lancelot C, Philippart CJM, Rousseau V (2009) Recent patterns in potential phytoplankton limitation along the NW European continental coast. J Sea Res 61:34–43

Loewe P (ed) (2009) System Nordsee – Zustand 2005 im Kontext langzeitlicher Entwicklungen. Report 44, Bundesamt für Seeschifffahrt und Hydrographie, Hamburg and Rostock

Longuet-Higgins M, Stewart R (1962) Radiation stress and mass transport in gravity waves, with application to 'surf beats'. J Fluid Mech 13:481–503

Lorkowski I, Pätsch J, Moll A, Kühn W (2012) Interannual variability of carbon fluxes in the North Sea from 1970 to 2006 – Competing effects of abiotic and biotic drivers on the gas-exchange of CO_2. Estuar Coast Shelf Sci 100:38–57

Lotze HK, Reise K, Worm B, van Beusekom JEE, Busch M, Ehlers A, Heinrich D, Hoffmann RC, Holm P, Jensen C, Knottnerus OS, Langjanki N, Prummel W, Vollmer M, Wolff WJ (2005) Human transformation of the Wadden Sea ecosystem through time: a synthesis. Helgol Mar Res 59:84–95

Lozier MS, Stewart NM (2008) On the temporally varying northward penetration of Mediterranean Overflow Water and eastward penetration of Labrador Sea Water. J Phys Oceanogr 38:2097–2103

Madec G (2008) NEMO reference manual, ocean dynamics component, Institut Pierre-Simon Laplace, technical report

Madsen KS (2009) Recent and future climatic changes in temperature, salinity, and sea level of the North Sea and the Baltic Sea. PhD Thesis, Niels Bohr Institute, University of Copenhagen

Martens P, van Beusekom JEE (2008) Zooplankton response to a warmer northern Wadden Sea. Helg Mar Res 62:67–75

Mathis M, Elizalde A, Mikolajewicz U, Pohlmann T (2015) Variability patterns of the general circulation and sea water temperature in the North Sea. Prog Oceanogr 135:91–112

McCandliss RR, Jones SE, Hearn M, Latter R, Jago CF (2002) Dynamics of suspended particles in coastal waters (southern North Sea) during a spring bloom. J Sea Res 47:285–302

McCarthy G, Frajka-Williams E, Johns WE, Baringer MO, Meinen CS, Bryden HL, Rayner D, Duchez A, Roberts C, Cunningham SA (2012) Observed interannual variability of the Atlantic meridional overturning circulation at 26.5 degrees N. Geophys Res Lett 39: L19609, doi:10.1029/2012GL052933

McCave IN (1987) Fine sediment sources and sinks around the East Anglian Coast. J Geol Soc London 144:149–152

McQuatters-Gollop A, Raitsos DE, Edwards M, Pradhan Y, Mee LD, Lavender SJ, Attrill MJ (2007) A long-term chlorophyll data set reveals regime shift in North Sea phytoplankton biomass unconnected to nutrient trends. Limnol Oceanogr 52:635–648

Meesenburg H (1996) Man's role in changing the coastal landscapes in Denmark. GeoJournal 39:143–151

Meire L, Soetaert KER, Meysman FJR (2013) Impact of global change on coastal oxygen dynamics and risk of hypoxia. Biogeosciences 10:2633–2653

Merrifield M, Merrifield S, Mitchum G (2009) An anomalous recent acceleration of global sea level rise. J Clim 22:5772–5781

Meyer EMI, Pohlmann T, Weisse R (2011) Thermodynamic variability and change in the North Sea (1948–2007) derived from a multidecadal hindcast. J Mar Sys 86:35–44

Miller L, Douglas BC (2007) Gyre-scale atmospheric pressure variations and their relation to 19th and 20th century sea level rise. Geophys Res Lett 34:L16602, http://dx.doi.org/10.1029/2007GL030862

Milne GA, Gehrels WR, Hughes CW, Tamisiea ME (2009) Identifying the causes of sea-level change. Nat Geosci 2:471–478

Mork KA, Skagseth Ø (2010) A quantitative description of the Norwegian Atlantic Current by combining altimetry and hydrography. Ocean Sci 6:901–911

Mroczek P (1980) Zu einer Karte der Veränderungen der Uferlinie der deutschen Nordseeküste in den letzten 100 Jahren im Maßstab 1:200.000

Mudersbach C, Wahl T, Haigh I, Jensen J (2013) Trends in high sea levels of German North Sea gauges compared to regional mean sea level changes. Cont Shelf Res, 65:111–120

Müller M (2012) The influence of changing stratification conditions on barotropic tidal transport and its implications for seasonal and secular changes of tides. Cont Shelf Res 47:107–118

MyOcean (2014) Atlantic-European North West Shelf – Ocean Physics Reanalysis from Met Office (1985-2012). http://www.myocean.eu/web/69-myocean-interactive-catalogue.php?option=com_csw&view=details&product_id=NORTHWESTSHELF_REANALYSIS_PHYS_004_009

Nedwell DB, Parkes RJ, Upton AC, Assinder DJ (1993) Seasonal fluxes across the sediment-water interface, and processes within sediments. Phil Trans Roy Soc London A 343:519–529

Nerem RS, Mitchum GT (2002) Estimates of vertical crustal motion derived from differences of TOPEX/POSEIDON and tide gauge sea level measurements. Geophys Res Lett 29: http://dx.doi.org/10.1029/2002GL015037

Neu H (1984) Interannual variations and longer-term changes in the sea state of the North Atlantic. J Geophys Res 89:6397–6402

NOOS (2010) NOOS Activity: Exchange of computed water, salt and heat transport across selected transects. North West European Shelf Operational Oceanographic System (NOOS). www.noos.cc/fileadmin/user_upload/exch_transports_NOOS-BOOS_2010-11-11.pdf

Omar AM, Olsen A, Johannessen T, Hoppema M, Thomas H, Borges AV (2010) Spatio-temporal variations of fCO_2 in the North Sea. Ocean Sci 6:77–89

Oost AP (1995) Dynamics and sedimentary development of the Dutch Wadden Sea with emphasis on the Frisian Inlet; a study of the barrier islands, ebb-tidal deltas and drainage basins. Thesis, Utrecht, Geologica Ultraiectina, 126

Oost AP, Hoekstra P, Wiersma A, Flemming B, Lammerts EJ, Pejrup M, Hofstede J, van der Valk B, Kiden P, Bartholdy J, van der Berg MW, Vos PC, de Vries S, Wang ZB (2012) Barrier island management: Lessons from the past and directions for the future. Ocean Coast Manage 68:18–38

Orvik KA, Skagseth Ø (2005) Heat flux variations in the eastern Norwegian Atlantic Current toward the Arctic from moored instruments, 1995–2005. Geophys Res Lett 32:L14610, doi:10.1029/2005GL023487

Otto L, Zimmerman JTF, Furnes GK, Mork M, Saetre R, Becker G (1990) Review of the physical oceanography of the North Sea. Netherlands J Sea Res 26:161–238

Paphitis D, Bastos AC, Evans G, Collins MB (2010) The English Channel (La Manche): evolution, oceanography and sediment dynamics – a synthesis. In: Whittaker JE, Hart MB (eds) Micropalaeontology, Sedimentary Environments and Stratigraphy: a tribute to Dennis Curry (1921–2001). pp. 99–132

Pätsch J, Kühn W (2008) Nitrogen and carbon cycling in the North Sea and exchange with the North Atlantic – a model study. Part I. Nitrogen budget and fluxes. Cont Shelf Res 28:767–787

Pawlowicz R, McDougall T, Feistel R, Tailleux R (2012) Preface: An historical perspective on the development of the Thermodynamic Equation of Seawater – 2010. Ocean Sci 8:161–174

Peltier WR (2004) Global glacial isostasy and the surface of the ice-age earth: The ICE-5G(VM2) model and GRACE. Ann Rev Earth Planet Sci 32:111–149

Philippart CJM, Beukema JJ, Cadée GC, Dekker R, Goedhart PW, van Iperen JM, Leopold MF, Herman PNJ (2007) Impact of nutrient reduction on coastal communities. Ecosystems 10:96–119

Pickering M, Wells N, Horsburgh K, Green J (2011) The impact of future sea-level rise on the European Shelf tides. Cont Shelf Res 35:1–15

Pietrzak JD, de Boer GJ, Eleveld M (2011) Mechanisms controlling the intra-annual mesoscale variability of SST and SPM in the southern North Sea. Cont Shelf Res 31:594–610

Pleskachevsky A, Gayer G, Horstmann J, Rosenthal W (2005) Synergy of satellite remote sensing and numerical modeling for monitoring of suspended particulate matter. Ocean Dyn 55:2–9

Pohlmann T (2006) A meso-scale model of the central and southern North Sea: consequences of an improved resolution. Cont Shelf Res 26:2367–2385

Port A, Gurgel K-W, Stanev J, Schulz-Stellenfleth J, Stanev EV (2011) Tidal and wind-driven surface currents in the German Bight: HFR observations versus model simulations. Ocean Dyn 61:1567–1585

Postma H (1954) Hydrography of the Dutch Wadden Sea. Archives néerlandaises de Zoologie 10:405–511

Postma H (1966) The cycle of nitrogen in the Wadden Sea and adjacent areas. Netherlands J Sea Res 3:186–221

Postma H (1981) Exchange of materials between the North Sea and the Wadden Sea. Mar Geol 40:199–215

Postma H (1984) Introduction to the symposium on organic matter in the Wadden Sea. Neth Inst S 10:15–22

Poulton CVL, Lee JR, Jones LD, Hobbs PRN, Hall M (2006) Preliminary investigation into monitoring coastal erosion using terrestrial laser scanning: case study at Happisburgh, Norfolk, UK. Bull Geol Soc Norfolk 56:45–65

Prandle D (1978a) Monthly-mean residual flows through Dover Strait, 1949–1972. J Mar Biol Assoc UK 58:965–973

Prandle D (1978b) Residual flows and elevations in the Southern North Sea. Proc Roy Soc Lond A 259:189–228

Prandle D, Player R (1993) Residual current through the Dover Strait measured by HF radar. Estuar Coast Shelf Sci 37:635–653

Prandle D, Loch SG, Player RJ (1993) Tidal flow through the Straits of Dover. J Phys Oceanogr 23:23–37

Prandle D, Ballard G, Flatt D, Harrison AJ, Jones SE, Knight PJ, Loch S, McManus J, Player R, Tappin A (1996) Combining modelling and monitoring to determine fluxes of water, dissolved and particulate metals through the Dover Strait. Cont Shelf Res 16:237–257

Prowe F, Thomas H, Pätsch J, Kühn W, Bozec Y, Schiettecatte L-S, Borges AV, de Baar HJW (2009) Mechanisms controlling the air–sea CO_2 flux in the North Sea. Cont Shelf Res 29:1801–1808

Pugh D (2004) Changing Sea Levels: Effects of Tides, Weather and Climate. Cambridge University Press

Puls W, Sündermann J (1990) Simulation of suspended sediment dispersion in the North Sea. Coast Estuar Studies 38:356–372

Puls W, Pohlmann T, Sündermann J (1997) Suspended particulate matter in the southern North Sea: application of a numerical model to extend NERC North Sea project data interpretation. Deutsche Hydrografische Zeitschrift 49:307–327

Pye K, Saye SE, Blott SJ (2007) Sand Dune Processes and Management for Flood and Coastal Defence. Part 2. Sand Dune Processes and Morphology. Joint Defra/Environment Agency Flood and Coastal Erosion Risk Management R & D Programme, R & D Technical report FD1302/TR

Queste BY, Fernand L, Jickells TD, Heywood KJ (2013) Spatial extent and historical context of North Sea oxygen depletion in August 2010. Biogeochemistry 113:53–68

Rayner NA, Parker DE, Horton EB, Folland CK, Alexander LV, Rowell DP, Kent EC, Kaplan A (2003) Global analyses of sea surface temperature, sea ice, and night marine air temperature since the late nineteenth century. J Geophys Res 108(D14):4407, doi:10.1029/2002JD002670

Reise K (2005) Coast of change: habitat loss and transformations in the Wadden Sea. Helgoland Mar Res 59:9–21

Reise K (2013) A Natural History of the Wadden Sea. Riddled by Contingencies. Wadden Academy, Leeuwarden, NL

Reise K, Kohlus J (2008) Seagrass recovery in the Northern Wadden Sea? Helgoland Mar Res 62:77–84

Reise K, van Beusekom JEE (2008) Interactive effects of global and regional change on a coastal ecosystem. Helgoland Mar Res 62: 85–91

Reise K, Baptist M. Burbridge P, Dankers NMJA, Fischer L, Flemming BW, Oost AP, Smit C (2010) The Wadden Sea – A Universally Outstanding Tidal Wetland. Wadden Sea Ecosystem 29:7–24

Reynaud JY, Tessier B, Auffret JP, Berné S, de Baptist M, Marsset T, Walker P (2003) The offshore Quaternary sediment bodies of the English Channel and its Western Approaches. J Quat Sci 18:361–371

Richter K, Nilsen JEØ, Drange H (2012) Contributions to sea level variability along the Norwegian coast for 1960–2010. J Geophys Res 117:C05038, http://dx.doi.org/10.1029/2011JC007826

Robinson IS (1976) Theoretical analysis of use of submarine cables as electromagnetic oceanographic flowmeters. Phil Trans Roy Soc Lond A 280:355–396

Roos PC, Hulscher SJMH, Knaapen MAF, Van Damme RM (2004) The cross-sectional shape of tidal sandbanks: Modeling and observations. J Geophys Res 109:F02003, doi:10.1029/2003JF000070

Rosenhagen G, Schatzmann M (2011) Das Klima der Metropolregion auf Grundlage meteorologischer Messungen und Beobachtungen. In: von Storch H, Claussen M (eds) Klimabericht für die Metropolregion Hamburg. Springer

Ruz MH, Meur-Férec C (2004) Influence of high water levels on aeolian sand transport: upper-beach/dune evolution on a macrotidal coast, Wissant Bay, Northern France. Geomorphology 60:73–87

Rydell B, Angerud P, Hågeryd A-C (2004) Omfattning av stranderosion i Sverige Översiktlig kartläggning av erosionsförhållanden: Metodik och redovisning. Varia 543, Statens Geotekniska Institut, Linköping

Salman A, Lombardo S, Doody P (2004) Living with coastal erosion in Europe: Sediment and Space for Sustainability. EUROSION.

Salt L, Thomas H, Prowe AEF, Borges AV, De Baar HJW (2013) Variability of North Sea pH and CO_2 in response to North Atlantic Oscillation forcing. J Geophys Res (Biogeosciences) 118:1584–1592

Schiettecatte L-S, Gazeau F, Van der Zee C, Brion N, Borges AV (2006) Time series of the partial pressure of carbon dioxide (2001–2004) and preliminary inorganic carbon budget in the Scheldt plume (Belgian coast waters). Geochem Geophy Geosy 7:Q06009. doi:10.1029/2005GC001161

Schiettecatte L-S, Thomas H, Bozec Y, Borges AV (2007) High temporal coverage of carbon dioxide measurements in the Southern Bight of the North Sea. Mar Chem 106:161–173

Schmelzer N, Holfort J (eds) (2012) Climatological Ice Atlas for the western and southern Baltic Sea (1961–2010). BSH Hamburg und Rostock

Schrum C, Sigismund F (2001) Modellkonfiguration des Nordsee/Ostsee-Modells. 40-Jahres NCEP Integration. Ber Zent Meeres- Klimaforsch Univ Hamburg B(4)

Sharples J, Ross ON, Scott BE, Greenstreet SPR, Fraser H (2006) Inter-annual variability in the timing of stratification and the spring bloom in the North-western North Sea. Cont Shelf Res 26:733–751

Sharples J, Holt J, Dye S (2010) Shelf sea stratification. In: MCCIP Annual Report Card 2010-11, MCCIP Science Review, www.mccip.org.uk/arc

Shennan I, Woodworth PL (1992) A comparison of late Holocene and twentieth century sea level trends from the UK and North Sea region. Geophys J Int 109:96–105

Shennan I, Milne G, Bradley S (2012) Late Holocene vertical land motion and relative sea-level changes: lessons from the British Isles. J Quart Sci 27:64–70

Sherwin TJ, Read JF, Holliday NP, Johnson C (2012) The impact of changes in North Atlantic Gyre distribution on water mass characteristics in the Rockall Trough. ICES J Mar Sci 69:751–757

Simpson JH, Hunter JR (1974) Fronts in the Irish Sea. Nature 250: 404–406

Simpson JH, Bos WG, Schirmer F, Souza AJ, Rippeth TP, Jones SE, Hydes D (1993) Periodic stratification in the Rhine ROFI in the North Sea. Oceanol Acta 16:23–32

Smeed DA, McCarthy GD, Cunningham SA, Frajka-Williams E, Rayner D, Johns WE, Meinen CS, Baringer MO, Moat BI, Duchez A, Bryden HL (2014) Observed decline of the Atlantic meridional overturning circulation 2004–2012. Ocean Sci 10:29–38

Smith TM, Reynolds RW (2002) Bias corrections for historical sea surface temperatures based on marine air temperatures. J Climate 15:73–87

Smyth TJ, Fishwick JR, Al-Moosawi L, Cummings DG, Harris C, Kitidis V, rees A, Martinez-Vicente V, Woodward EMS (2010) A broad spatio-temporal view of the Western English Channel observatory. J Plankton Res 32:585–601

Son CS, Flemming BV, Bartholomae A, Chun SS (2012) Long-term changes in morphology and textural sediment characteristics in response to energy variation on shoreface-connected ridges off the East Frisian barrier-island coast, southern North Sea. J Sed Res 82:385–399

Sørensen P (2013) Denmark. In: Pranzini E, Williams A (eds) Coastal erosion and protection in Europe. Routledge

Southward AJ (1960) On changes of sea temperature in the English Channel. J Mar Biol Assoc UK 42:275–375

Souza AJ (2013) On the use of the Stokes number to explain frictional tidal dynamics and water column structure in shelf seas. Ocean Sci 9:391–398

Souza A, Holt J, Proctor R (2007) Modelling SPM on the NW European Shelf Seas. In: Balson PS, Collins MB (eds) Coastal and Shelf Sediment Transport. Geol Soc London, Spec Publ 274:147–158

Staats N, de Deckere EMGT, de Winder B, Stal LJ (2001) Spatial patterns of benthic diatoms, carbohydrates and mud on a tidal flat in the Ems-Dollard estuary. Hydrobiologia 448:107–115

Sterl A, Caires S (2005) Climatology, variability and extrema of ocean waves. The web-based KNMI/ERA-40 Wave Atlas. Int J Climatol 25:963–977

KNMI/ERA-40 Wave Atlas. Int J Climatol 25:963–977

Sturges W, Douglas BC (2011) Wind effects on estimates of sea level rise. J Geophys Res 116:C06008, http://dx.doi.org/10.1029/2010JC006492

Suijlen JM, Duin RNM (2001) Variability of near-surface total suspended-matter concentrations in the Dutch coastal zone of the North Sea: Climatological study on the suspended matter concentrations in the North Sea. Report RIKZ/OS/2001.150X. RIKZ, The Hague, Netherlands

Suijlen JM, Duin RNM (2002) Atlas of near-surface total suspended matter concentrations in the Dutch coastal zone of the North Sea. Report RIKZ/2002.059. RIKZ, The Hague, Netherlands

Sundby S, Drinkwater K (2007). On the mechanisms behind salinity anomaly signals of the northern North Atlantic. Progr Oceanogr 73:190–202

Sündermann J (1993) Suspended particulate matter in the North Sea: Field observations and model simulations. Phil Trans Roy Soc Lond A 343:423–430

Thiel M, Hinojosa IA, Joschko T, Gutow L (2011) Spatio-temporal distribution of floating objects in the German Bight (North Sea). J Sea Res 65:368–379

Thomalla F, Vincent CE (2003) Beach response to shore-parallel breakwaters at Sea Palling, Norfolk, UK. Estuar, Coast Shelf Sci 56:203–212

Thomas H, Bozec Y, Elkalay K, deBaar HJW (2004) Enhanced open ocean storage of CO_2 from shelf sea pumping. Science 304:1005–1008

Thomas H, Bozec Y, Elkalay K, deBaar HJW, Borges AV, Schiettecatte L-S (2005a) Controls of the surface water partial pressure of CO_2 in the North Sea. Biogeosciences 2:323–334

Thomas H, Bozec Y, de Baar HJW, Elkalay K, Frankignoulle M, Schiettecatte L-S, Kattner G, Borges AV (2005b) The carbon budget of the North Sea. Biogeosciences 2:87–96

Thomas H, Prowe F, van Heuven S, Bozec Y, deBaar HJW, Schiettecatte L-S, Suykens K, Koné K, Borges AV, Lima ID, Doney SC (2007) Rapid decline of the CO_2 buffering capacity in the North Sea and implications for the North Atlantic Ocean. Global Biogeochem Cy 21:GB4001, doi:10.1029/2006GB002825

Thomas H, Prowe F, Lima ID, Doney SC, Wanninkhof R, Greatbatch RJ, Corbière A, Schuster U (2008) Changes in the North Atlantic Oscillation influence CO_2 uptake in the North Atlantic over the past 2 decades. Global Biogeochem Cy 22: doi:10.1029/2007GB003167

Thomas H, Schiettecatte L-S, Suykens K, Koné YJM, Shadwick EH, Prowe F, Bozec Y, de Baar HJW, Borges AV (2009) Enhanced ocean carbon storage from anaerobic alkalinity generation in coastal sediments. Biogeosciences 6:267–274

Thyme F (1990) Beach nourishment on the west coast of Jutland. J Coast Res 6:201–210

Tsimplis MN, Shaw AGP, Flather RA, Woolf DK (2006) The influence of the North Atlantic Oscillation on the sea-level around the northern European coasts reconsidered: the thermosteric effects. Phil Trans Roy Soc London A 364:845–856

Turrell WR, Henderson EW, Slesser G, Payne R, Adams RD (1992) Seasonal changes in the circulation of the northern North Sea. Cont Shelf Res 12:257–286

UKMMAS (2010) Charting Progress 2 Feeder Report: Ocean Processes. Huthnance J (ed.). Department for Environment Food and Rural Affairs on behalf of UKMMAS (UK Marine Monitoring and Assessment Strategy)

Valentin H (1954) Land loss at Holderness. Reprinted in 1971: Steers JA (ed) Applied coastal geomorphology. Macmillan, London

van Aken HM (2008) Variability of the water temperature in the western Wadden Sea on tidal to centennial time scales. J Sea Res 60:227–234

van Aken HM (2010) Meteorological forcing of long-term temperature variations of the Dutch coastal waters. J Sea Res 63:143–151

van Alphen JSLJ (1990) A mud balance for Belgian-Dutch coastal waters between 1969 and 1986. Netherlands J Sea Res 25:19–30

van Alphen JSLJ, de Ruijter WPM, Borst JC (1988) Outflow and three-dimensional spreading of Rhine river water in the Netherlands coastal zone. In: Dronkers J, van Leussen W (eds) Physical Processes in Estuaries. Springer van Bennekom AJ, Wetsteijn FJ (1990) The winter distribution of nutrients in the southern bight of the North Sea (1961–1978) and in the estuaries of the Scheldt and the Rhine/Meuse. Netherlands J Sea Res 25:75–87

van Beusekom JEE, de Jonge VN (2002) Long-term changes in Wadden Sea nutrient cycles: importance of organic matter import from the North Sea. Hydrobiologia 475/476:185–194

van Beusekom JEE, de Jonge VN (2012) Dissolved organic phosphorus: An indicator of organic matter turnover? Estuar Coast Shelf Sci 108:29–36

van Beusekom JEE, Brockmann UH, Hesse KJ, Hickel W, Poremba K, Tillmann U (1999) The importance of sediments in the transformation and turnover of nutrients and organic matter in the Wadden Sea and German Bight. German J Hydrogr 51:245–266

van Beusekom JEE, Bot P, Carstensen J, Goebel J, Lenhart H, Pätsch J, Petenati T, Raabe T, Reise K, Wetsteijn B (2009a) Eutrophication. Thematic Report 6 in: Marencic H, de Vlas J (eds) Quality Status Report 2009, Wadden Sea Ecosystem. No 25. Common Wadden Sea Secretariat, Trilateral Monitoring and Assessment Group, Wilhelmshaven, Germany

van Beusekom JEE, Loebl M, Martens P (2009b) Distant riverine nutrient supply and local temperature drive the long-term phytoplankton development in a temperate coastal basin. J Sea Res 61:26–33

van Beusekom JEE, Buschbaum C, Reise K (2012) Wadden Sea tidal basins and the mediating role of the North Sea in ecological processes: scaling up of management? Ocean Coast Manage 68:69–78

van Cauwenberghe C (1995) Relative sea level rise: further analyses and conclusions with respect to the high water, the mean sea and the low water levels along the Belgian coast. Report 37, HydrografischeDienst der Kust

van Cauwenberghe C (1999) Relative sea level rise along the Belgian coast: analyses and conclusions with respect to the high water, the mean sea and the low water levels. Report 46, HydrografischeDienst der Kust

van der Meulen F, Van der Valk B, Arens B (2013) The Netherlands. In Pranzini E, Williams A (eds) Coastal Erosion and Protection in Europe. Routledge, Abingdon

van der Molen J (2002) The influence of tides, wind and waves on the net sand transport in the North Sea. Cont Shelf Res 22:2739–2762

van der Molen J, Gerrits J, de Swart HE (2004) Modelling the morphodynamics of a tidal shelf sea. Cont Shelf Res 24:483–507

van Dijk TAGP, Kleinhans MG (2005) Processes controlling the dynamics of compound sand waves in the North Sea, Netherlands. J Geophys Res 110:F04S10, doi:10.1029/2004JF000173

van Hal R, Smits K, Rijnsdorp AD (2010) How climate warming impacts the distribution and abundance of two small flatfish species in the North Sea. J Sea Res 64:76–84

van Straaten LMJU, Kuenen PH (1957) Accumulation of fine-grained sediments in the Dutch Wadden Sea. Geol Mijnb 19:319–354

Vasseur B, Héquette A (2000) Storm surges and erosion of coastal dunes between 1957 and 1988 near Dunkerque (France), southwestern North Sea. Geol Soc Lond, Spec Publ 175:99–107

Vavrus S, Walsh JE, Chapman WL, Portis D (2006) The behavior of extreme cold air outbreaks under greenhouse warming. Int J Climatol 26:1133–1147

Verlaan PAJ, Spanhoff R (2000) Massive sedimentation events at the mouth of the Rotterdam Waterway. J Coast Res 16:458–469

Verwaest T, Viaene P, Verstraeten J, Mostaert F (2005) De zeespiegelstijgingmeten, begrijpen en afblokken. De Grote Rede 15:15–25

Verwey J (1952) On the ecology of cockle and mussel in the Dutch Wadden Sea. Archives néerlandaises de Zoologie 10:171–239

Vichi M, Ruardij P, Baretta JW (2004) Link or sink: a modelling interpretation of the open Baltic biogeochemistry. Biogeosciences 1:79–100

Vikebø F, Furevik T, Furnes G, Kvamstø NG, Reistad M (2003) Wave height variations in the North Sea and on the Norwegian Continental Shelf, 1881–1999. Cont Shelf Res 23:251–263

Visser M, de Ruijter WPM, Postma L (1991) The distribution of suspended matter in the Dutch coastal zone. Netherlands J Sea Res 27:127–143

von Storch H, Reichardt H (1997) A scenario of storm surge statistics for the German Bight at the expected time of doubled atmospheric carbon dioxide concentration. J Clim 10:2653–2662

von Storch H, Zwiers F (1999) Statistical Analysis in Climate Research. Cambridge University Press

Wahl T, Jensen J, Frank T (2010) On analysing sea level rise in the German Bight since 1844. Nat Hazards Earth Syst Sci 10:171–179

Wahl T, Jensen J, Frank T, Haigh ID (2011) Improved estimates of mean sea level changes in the German Bight over the last 166 years. Ocean Dyn 61:701–715

Wahl T, Haigh I, Woodworth P, Albrecht F, Dillingh D, Jensen J, Nicholls RJ, Weisse R, Wöppelmann G (2013) Observed mean sea level changes around the North Sea coastline from 1800 to present. Earth-Sci Rev 124:51–67

Wakelin SL, Holt JT, Blackford JC, Allen JI, Butenschön M, Artioli Y (2012) Modeling the carbon fluxes of the northwest European continental shelf: validation and budgets. J Geophys Res 117: C05020, doi:10.1029/2011JC007402

WASA-Group (1998) Changing waves and storms in the Northeast Atlantic? Bull Am Met Soc 79:741–760

Watson AJ, Schuster U, Bakker DCE, Bates N, Corbiere A, Gonzalez-Davila M, Friedrich T, Hauck J, Heinze C, Johannessen T, Kortzinger A, Metz N, Olaffson J, Olsen A, Oschlies A, Pfeil B, Santano-Casiano JM, Steinhoff T, Telszewski M, Rios A, Wallace DWR, Wanninkhof R (2009) Tracking the variable North Atlantic sink for atmospheric CO_2. Science 326:1391–1393

Weisse R, Günther H (2007) Wave climate and long-term changes for the Southern North Sea obtained from a high-resolution hindcast 1958–2002. Ocean Dyn 57:161–172

Weisse R, Pluess A (2006) Storm-related sea level variations along the North Sea coast as simulated by a high-resolution model 1958–2002. Ocean Dyn 56:16–25

Weisse R, von Storch H (2009) Marine Climate and Climate Change: Storms, Wind Waves and Storm Surges. Springer

Weisse R, von Storch H, Niemeyer H, Knaack H (2012) Changing North Sea storm surge climate: An increasing hazard? Ocean Coast Manag 68:58–68

Wiltshire KH, Malzahn AM, Wirtz K, Greve W, Janisch S, Mangelsdorf P, Manly BFJ, Boersma M (2008) Resilience of North Sea phytoplankton spring bloom dynamics: an analysis of long-term data at Helgoland Roads. Limnol Oceanogr 53:1294–1302

Winther NG, Johannessen JA (2006) North Sea circulation: Atlantic inflow and its destination. J Geophys Res 111:C12018, doi:10.1029/2005JC003310

Woodruff SD, Worley SJ, Lubker SJ, Ji Z, Freeman JE, Berry DI, Brohan P, Kent EC, Reynolds RW, Smith SR, Wilkinson C (2011) ICOADS Release 2.5 and Data Characteristics. Int J Climatol 31:951–967

Woodworth PL (1987) Trends in U.K. mean sea level. Mar Geod 11:57–87

Woodworth PL (1990) A search for accelerations in records of European mean sea level. Int J Climatol 10:129–143

Woodworth PL (2010) A survey of recent changes in the main components of the ocean tide. Cont Shelf Res 30:1680–1691

Woodworth PL, Tsimplis MN, Flather RA, Shennan I (1999) A review of the trends observed in British Isles mean sea level data measured by tide gauges. Geophys J Int 136:651–670

Woodworth PL, Teferle FN, Bingley RM, Shennan I, Williams SDP (2009a) Trends in UK mean sea level revisited. Geophys J Int 176:19–30

Woodworth PL, White NJ, Jevrejeva S, Holgate SJ, Gehrels WR (2009b) Evidence for the accelerations of sea level on multi-decade and century timescales. Int J Climatol 29:777–789

Woodworth PL, Pouvreau N, Wöppelmann G (2010) The gyre-scale circulation of the North Atlantic and sea level at Brest. Ocean Sci 6:185–190

Woodworth PL, Menéndez M, Gehrels WR (2011) Evidence for century-timescale acceleration in mean sea levels and for recent changes in extreme sea levels. Surv in Geophys 32:603–618

Woolf DK, Challenor PG, Cotton PD (2002) Variability and predictability of the North Atlantic wave climate. J Geophys Res 107: C3145, doi:10.1029/2001JC001124

Wöppelmann G, Marcos M (2012) Coastal sea level rise in southern Europe and the non-climate contribution of vertical land motion. J Geophys Res 117:C01007, doi:10.1029/2011JC007469

Wöppelmann G, Pouvreau N, Simon B (2006). Brest sea level record: A time series construction back to the early eighteenth century. Ocean Dyn 56:487–497

Wöppelmann G, Pouvreau N, Coulomb A, Simon B, Woodworth PL (2008) Tide gauge datum continuity at Brest since 1711: France's longest sea-level record. Geophys Res Lett 35:L22605, doi:10.1029/2008GL035783

Young EF, Holt JT (2007) Prediction and analysis of long-term variability of temperature and salinity in the Irish Sea. J Geophys Res 112:C01008, doi:10.1029/2005JC003386

Recent Change—River Flow

4

Jaap Kwadijk, Nigel W. Arnell, Christoph Mudersbach, Mark de Weerd, Aart Kroon and Markus Quante

Abstract

This chapter reviews recent trends and variability in river flows to the North Sea. The main contributors are the River Elbe and the River Rhine. In addition to these large rivers many smaller rivers also discharge into the North Sea. However, by far the biggest contributor is the Baltic Sea outflow. Observation records for the major rivers draining into the North Sea are relatively long, while records for the smaller rivers are typically much shorter. Variability in flow is dependent on variations in weather—mainly precipitation and temperature—from year to year, but also on a wide range of direct and indirect human interventions in the North Sea basin. Rivers draining into the North Sea show considerable interannual and decadal variability in annual discharge. In northern areas this is closely associated with variation in the North Atlantic Oscillation, particularly in winter. Discharge to the North Sea in winter appears to be increasing, but there is little evidence of a widespread trend in summer inflow. Higher winter temperatures appear to have led to higher winter flows, as winter precipitation increasingly falls as rain rather than snow. To date, no significant trends in response to climate change are apparent for most of the individual rivers discharging into the North Sea.

4.1 Introduction

The waters flowing into the North Sea from the surrounding land masses add 296–354 km^3 of fresh water to the North Sea each year. The main contributors are the River Elbe and the River Rhine. In addition to these large rivers a great number of smaller rivers also discharge into the North Sea (Table 4.1). However, by far the biggest contributor to the North Sea is the Baltic Sea. Discharging about 470 km^3 year^{-1} this fresh water inflow exceeds the total contribution from the entire North Sea catchment area.

J. Kwadijk (✉)
Deltares, Delft, The Netherlands
e-mail: jaap.kwadijk@deltares.nl

J. Kwadijk
Faculty of Engineering and Technology, Department of Water Engineering and Management, University of Twente, Enschede, The Netherlands

N.W. Arnell (✉)
Walker Institute for Climate System Research, University of Reading, Reading, UK
e-mail: n.w.arnell@reading.ac.uk

C. Mudersbach
Institute of Water and Environment, Bochum University of Applied Sciences, Bochum, Germany
e-mail: christoph.mudersbach@hs-bochum.de

M. de Weerd
TAUW, Deventer, The Netherlands
e-mail: markde.weerd@gmail.com

A. Kroon
Department of Geosciences and Natural Resource Management, University of Copenhagen, Copenhagen, Denmark
e-mail: ak@ign.ku.dk

M. Quante
Institute of Coastal Research, Helmholtz-Zentrum Geesthacht, Geesthacht, Germany
e-mail: markus.quante@hzg.de

M. Quante and F. Colijn (eds.), *North Sea Region Climate Change Assessment*, Regional Climate Studies, DOI 10.1007/978-3-319-39745-0_4

Table 4.1 Main sources of fresh water inflow to the North Sea (adapted from OSPAR 2000)

	Discharge (km^3 $year^{-1}$)	Catchment area (km^2)
Norwegian North Sea coast	58–70	45,500
Skagerrak and Kattegat coasts	58–70	102,200
Danish and German coasts (Elbe)	32	219,900
Dutch and Belgian coasts (Rhine, Meuse and Scheldt)	91–97	221,400
English and French Channel coasts (including Seine)	9–37	137,000
English east coast (including Tyne, Tees, Humber, Thames)	32	74,500
Scottish coast (including Forth)	16	41,000
Total North Sea region (excluding the Baltic Sea)	296–354	841,500
Baltic Sea	470	1,650,000

Climate change is expected to affect the hydrological cycle and so will alter fresh water inflow to the North Sea (see Chaps. 5 and 7). In northern Europe, a trend has been observed towards more intense winter precipitation (EEA 2008, 2012). More inflow to the North Sea in winter is therefore expected.

Change in annual discharge is the usual variable assessed in relation to freshwater flow, but because rivers also transport vast quantities of sediment, nutrients and contaminants, climate change impacts in coastal areas will also result from changes in river regime and in the magnitude and timing of high and low flows. Short periods of high river flow can transport large sediment, nutrient and contaminant loads to the North Sea. This may affect coastal water quality and thus the functioning of coastal ecosystems (see Chap. 7). Although the total loads carried into the North Sea during periods of low flow may be small, nutrient and contaminant concentrations can be high. Meanwhile saline water can penetrate estuaries much further upstream as the river flow is less able to restrict its ingress. As a result, this chapter addresses observed changes in average annual flow as well as in high and low flows. As the rivers Rhine and Elbe and the Baltic Sea outflow contribute most of the fresh water, the focus is on observed changes in these inputs.

4.2 Detecting and Attributing Trends

Detecting and attributing temporal and spatial trends in river flow needs long records. Flow measurements based on water level gauges started in the 18th century in Europe, while more extensive monitoring networks of the European rivers were established only towards the end of the 19th century (Brazdil et al. 2006). Historical documentary sources such as newspapers, diaries, economic records, and log books may be used to reconstruct change occurring before the start of the instrumental record as these often report the occurrence and date of extreme events (Buisman 1996, 1998, 2000, 2006; Pfister 1999). Change occurring prior to the historical period must be reconstructed from the geological record.

According to Glaser and Stangl (2003) the frequency of floods has changed considerably over the past 700 years in central Europe. They concluded that flood frequency has higher natural variability than would be expected from present-day observations.

The records of discharge measurements for many of the major rivers draining into the North Sea are long—spanning several decades—while river flow records in smaller catchments are relatively short, and the longest records are clustered in specific regions (Hannaford et al. 2013). As changes over time in flow regimes are often a complex function of different trends and patterns of variability in the many parts of a catchment it is therefore difficult to characterise and understand reasons for interannual variability in river flow from observations alone. It is possible to 'fill in the gaps' using river flow simulated by a global or regional hydrological model driven by gridded climate input data (e.g. Jones et al. 2006). Stahl et al. (2012) used an ensemble of gridded hydrological models to simulate past variations in river flow across Europe. They demonstrated that although there were differences between models, they did reproduce the observed trends in the areas where there were data, and therefore that the broad spatial patterns of variability simulated by the models were robust. Stahl et al. (2012) noted, however, that the models tended to perform less well in those parts of Europe affected by snowfall and snowmelt. Also, that the models did not take account of human intervention within the catchment areas, which has substantially affected flow regimes in many of the rivers draining into the North Sea. As a result, Stahl et al. (2012) concluded that models should only be used to simulate 'natural' patterns of variability over time.

The large interannual variations in average, low and high river flow make it difficult to establish trends in hydrological data series (e.g. Wilby 2006; Conway 2013; Hannaford et al. 2013). Even in the well-instrumented basins of the rivers Elbe and Rhine, where records are available for 100 years or more, trends are easily obscured by the variability (Diermanse et al. 2010; Bormann et al. 2011), and so the length of the observation record can significantly affect the ability to identify trends over time. Attributing observed trends to changes in climate or other factors is even more difficult

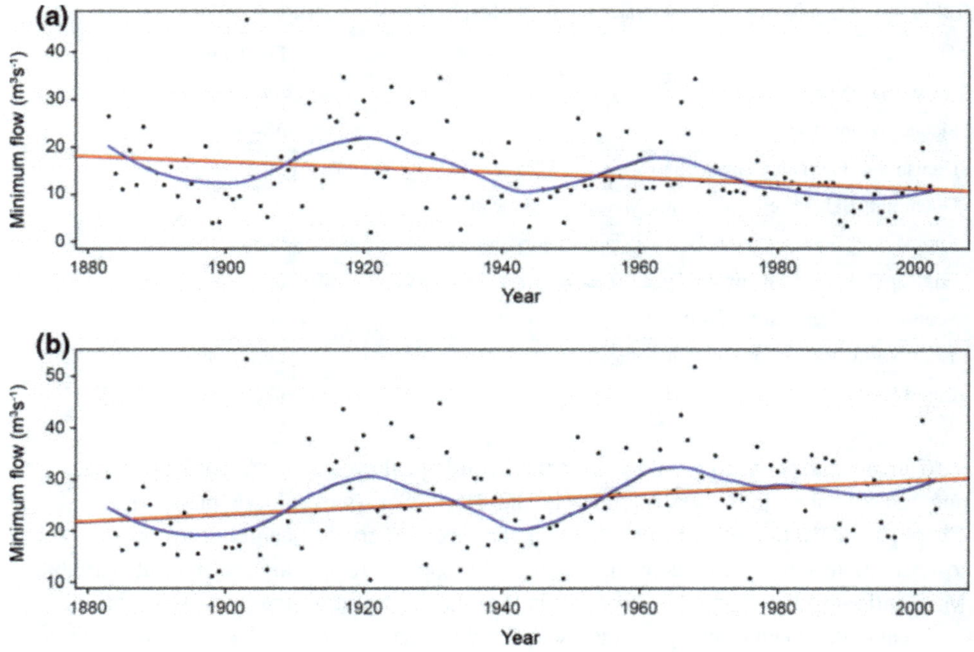

Fig. 4.1 Linear trends and the locally-weighted regression (Loess) smoothing curve for the 30-day minimum-flow time series on the River Thames at Kingston in southwest London, contrasting gauged (*upper*) and naturalised (*lower*) flow records (from Hannaford and Marsh 2006, modified)

because climate change effects are complex and changes in catchment conditions including land use and human interventions occur simultaneously. Some opposing effects of climate change—such as increasing precipitation and increasing evaporation—may cancel each other out, while other effects may have major consequences for river discharge. For example, in the case of the River Rhine where increasing precipitation may have led to higher peak flows, and where canalisation and a reduction in floodplain areas have steepened the peak flows (e.g. Engel 1997, 1999; CHR/KHR 1993; Ebel and Engel 1994). The effect of changes in land use on peak flow magnitude and frequency is another issue of fierce debate (e.g. O'Connell et al. 2007). According to Pfister et al. (2004), changes in land use, particularly urbanisation, can have significant local effects in small river basins (headwaters) with respect to flooding, especially during heavy local rainstorms. In the larger basins of the rivers Rhine and Meuse, however, there is no evidence that land-use change has had significant effects on peak flow in these rivers (Pfister et al. 2004). Water abstraction can also affect natural trends in flow variability and may even reverse them, as it is indicated in Fig. 4.1 which shows trends in minimum flow for the lower River Thames since the late 19th century (Hannaford and Marsh 2006). The gauged flow indicates a decreasing trend, with the decline in low flow primarily attributable to the seven-fold increase in abstraction upstream of the gauging station, to meet much of London's water needs. Minimum flow corrected for abstraction (here termed 'naturalised' flow) shows the opposite trend, and suggests an almost 40 % increase in low flow since the late 19th century.

The sensitivity of river basins to climate change differs from basin to basin. Lawrence and Hisdal (2011) found climate impact effects on peak flows in Norway to depend on the size of the catchment and the river regime: sensitivity to climate change reduces as catchment size increases and the contribution of snowmelt and melt from glaciers (relative to rainfall) to flow peaks increases. Coastal catchments in south-western Norway are particularly sensitive to changes in climate.

4.3 Observed Changes in Annual and Seasonal Flow

4.3.1 River Rhine

Compared to the extensive literature on future projections of River Rhine discharge, there are relatively few studies on changes in discharge during the instrumental record. The most recent international study was conducted by the International Commission on the Hydrology of the River Rhine (Belz et al. 2007; Belz 2010). This investigated discharge from the various regions using data from 38 main gauges. Each gauge measured daily mean discharge in a sub-catchment of the River Rhine. The longest time series covered the period 1901–2007 and the shortest started in 1951. On the basis of data from the Lobith gauging station near the German-Dutch border (Rhine-km 862.2), Belz et al. (2007) concluded that the observations show a clear increase in average discharge in winter (December to February) over the last century. At the Lobith gauging station over the period 1901–2000 this increase in winter amounts about 12 %, from an average of 2300–2600 $m^3\ s^{-1}$. In fact, the same trend of increasing winter discharge appears in nearly

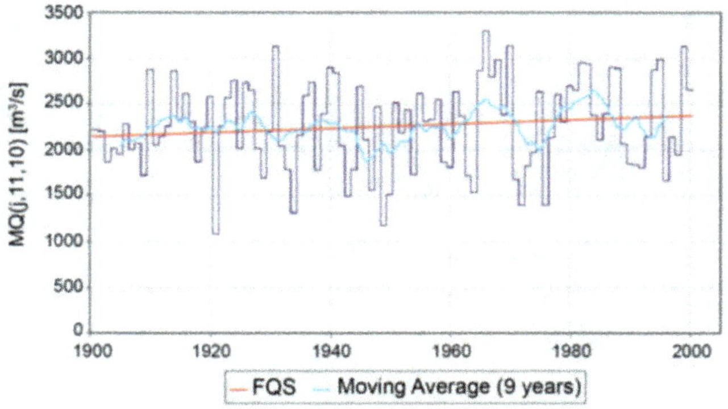

Fig. 4.2 Annual mean river discharges over the 20th century, nine-year moving average and trend (using the sum of errors method—FQS—significant at a 80 %-level Mann-Kendall-test) for the lower Rhine at the Rees gauge (adapted from Belz 2010)

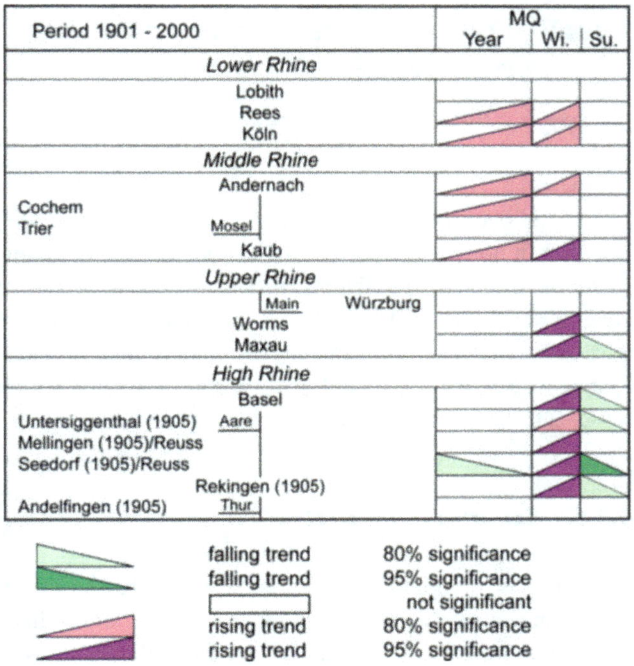

Fig. 4.3 Changes in average annual discharge for the Rhine basin through the 20th century (translated from Belz et al. 2007)

all sections of the River Rhine (Belz et al. 2007). There appears to be a slight increasing trend in annual discharge for the Rees gauge in the lower Rhine which is shown in Fig. 4.2. This trend is significant at the 80 % level for a Mann-Kendall test (Belz et al. 2007). In summer a few gauges in the upper Rhine section show significant (at the 80 % level for a Mann-Kendall test) declining trends (Belz et al. 2007). Belz et al. (2007) attributed the changes in winter discharge to an increase in the amount of winter precipitation. An increase in winter precipitation in the Rhine catchment has been reported by many studies (Caspary and Bardossy 1995; Blochliger and Neidhofer 1998; Konnen 1999; Uhlenbrook et al. 2001). Higher temperatures

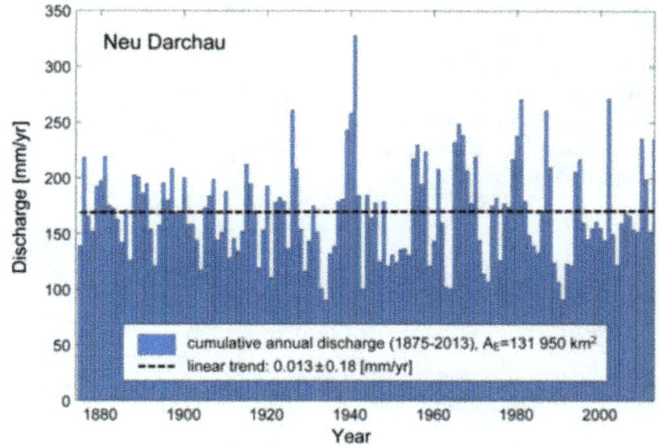

Fig. 4.4 Total annual discharge at the Neu Darchau gauging station on the River Elbe (Elbe-km 536) for the period 1875–2013 (Mudersbach et al. 2013)

leading to more precipitation falling as rain rather than snow have also led to an increase in winter discharge. The findings of Belz et al. (2007) are summarised schematically in Fig. 4.3.

4.3.2 River Elbe

Mudersbach et al. (2013) calculated the cumulative annual discharge at the Neu Darchau gauging station on the River Elbe (Elbe-km 536), about 75 km southeast of Hamburg for the period 1875–2013 (Fig. 4.4). They found no significant long-term trend, but did find strong decadal variability. This long-term variability was also reported by Ionita et al. (2011), who analysed annual mean discharge at the Neu Darchau station based on mean monthly discharge data for the period 1902–2002. They found strong decadal variability with a dominant period of 20 years. Analysing different River Elbe discharge data series from the Dresden gauge downstream to the Neu Darchau gauge, Markovic and Koch (2006) identified statistically significant low frequency oscillations with periods of 7.1 years and 10–14 years occurring in addition to the seasonal cycle, indicating the occurrence of extended dry and wet periods.

4.3.3 River Meuse

Min (2006) analysed the reconstructed discharge series at the Monsin gauging station on the River Meuse, Belgium, for the period 1912–2002, investigating both a long-term (1912–2002) and shorter term period (1950–2002). Average discharge over the long-term was 270 $m^3 s^{-1}$ (de Wit et al. 2001). For the average annual discharge (see Fig. 4.5), neither a change-point nor a trend was found in either record

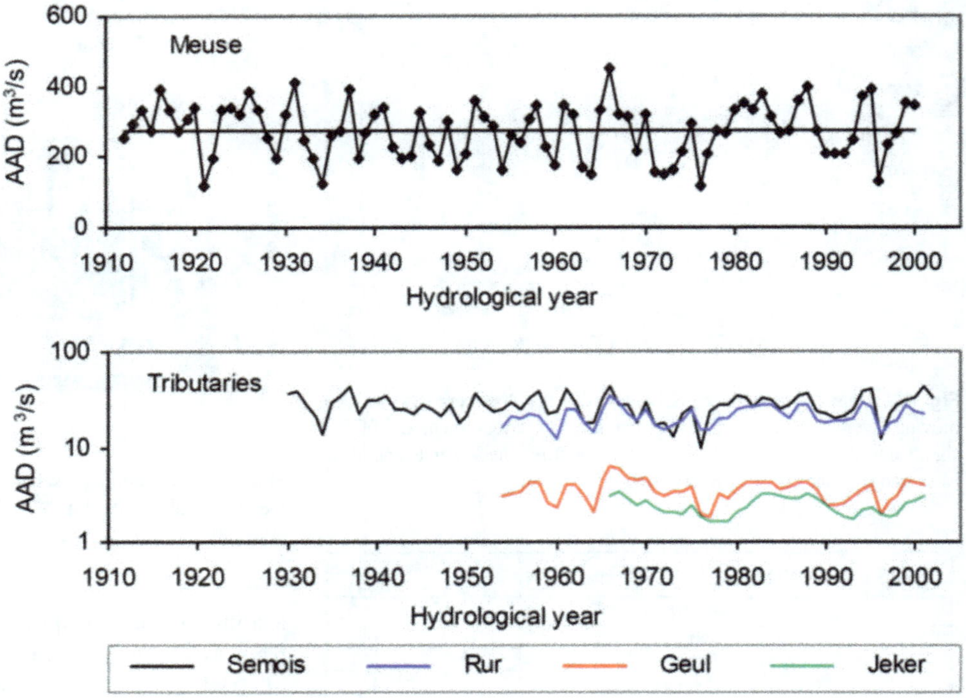

Fig. 4.5 Annual average discharge in the River Meuse near Monsin (Meuse-km 586) for 1912–2000 and selected tributaries. *Note* a logarithmic scale is used for the tributaries (adapted from Min 2006)

(de Wit et al. 2001; Min 2006). However, a slight increase in average spring discharge was seen between 1978 and 1989 (mainly March), and a decrease in average autumn discharge from 1933 onwards. The increase in average spring discharge could be attributed to an increase in precipitation, and thus to climate variability rather than to land-use change or climate change (Min 2006).

Ward (2009) suggested that discharge in the Meuse basin was higher during the 20th century than in previous centuries, identifying a 2.5 % difference between average annual discharge in the 19th and 20th centuries and an even greater difference relative to earlier centuries. Ward (2009) attributed the difference in annual average discharge between the 19th and 20th centuries to climate change, since there was little change in land use over this period.

4.3.4 UK Rivers

Hannaford and Marsh (2006) and Hannaford and Buys (2012) applied trend tests to time series of discharge and measures of low flow in UK rivers. They reported little variability in discharge and low flow since the early 1960s. However, an increasing trend in annual discharge was apparent for some catchments in Scotland.

Figure 4.6 shows seasonal discharge (summer, winter) for six 'large' UK rivers discharging into the North Sea. Although there are no clear long-term trends, considerable decadal variability is apparent. Table 4.2 shows the coefficient of variation in seasonal discharge from the six rivers (calculated over the common period 1970–2011). Variability is lowest in northern rivers and highest in the rivers draining southern England. In the northern rivers variability is greatest in summer, while in the southern rivers variability is greatest in autumn. Interannual variability in winter discharge in Scotland (River Tay) is strongly related to the NAO, but the correlations are less strong further south.

4.3.5 Scandinavian Rivers

In Scandinavia, both the seasonal river discharge and the ice regime are strongly influenced by large-scale atmospheric circulation processes over the North Atlantic that are closely correlated with the NAO index (HELCOM 2013). Reported changes in average annual flow from the Norwegian and Kattegat coasts are small. A slight increase in discharge was reported by Hellström and Lindström (2008) which they attributed to a slight increase in annual precipitation. However, higher temperatures lead to less snowfall, earlier snow melt and thus to shifts in river regime where flow in winter tends to increase.

4.3.6 Inflow from the Baltic Sea

Long-term data series on the inflow of water from the Baltic Sea to the North Sea are lacking as measurements of this inflow are not made. Overall, it may be assumed that the total river discharge into the Baltic Sea is a good

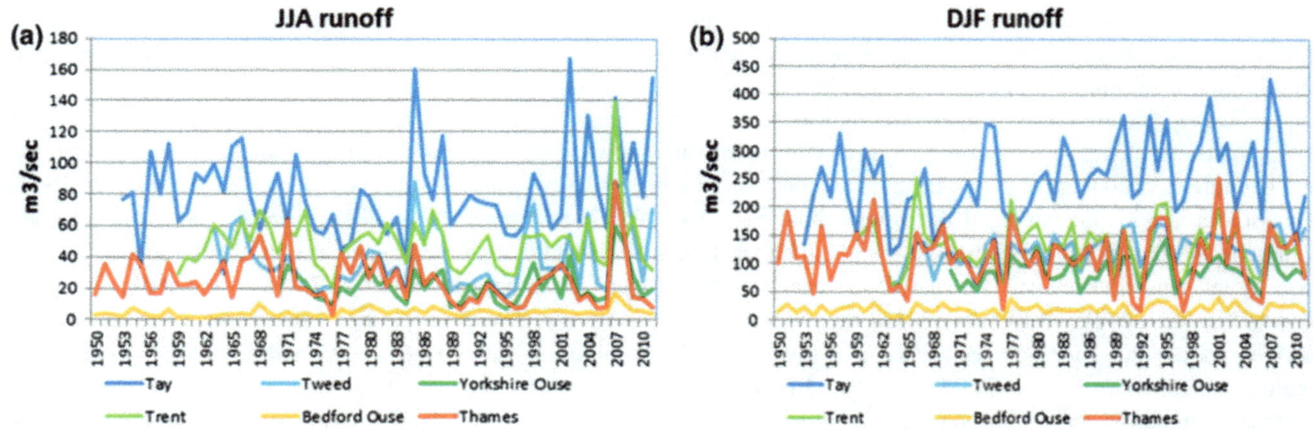

Fig. 4.6 Mean discharge in summer (*left*) and winter (*right*) for selected UK rivers draining into the North Sea. Data from the UK National River Flows Archive (www.ceh.ac.uk/data/nrfa)

Table 4.2 Coefficient of variation in seasonal flows for six UK rivers that discharge into the North Sea (1970–2011) and the correlation between winter flow and the strength of the winter North Atlantic Oscillation (NAO) index

River (north to south)	Winter (DJF)	Spring (MAM)	Summer (JJA)	Autumn (SON)	Annual	Correlation between winter flow and the NAO index
Tay at Ballathie	0.26	0.27	0.39	0.30	0.16	0.71
Tweed at Norham	0.23	0.29	0.56	0.41	0.21	0.20
Yorkshire Ouse at Skelton	0.28	0.39	0.52	0.46	0.22	0.24
Trent at Colwick	0.30	0.35	0.39	0.46	0.22	0.11
Bedford Ouse at Bedford	0.44	0.51	0.54	0.79	0.33	−0.02
Thames at Kingston	0.48	0.47	0.72	0.77	0.36	0.01

Data from the UK National River Flows Archive (www.ceh.ac.uk/data/nrfa)

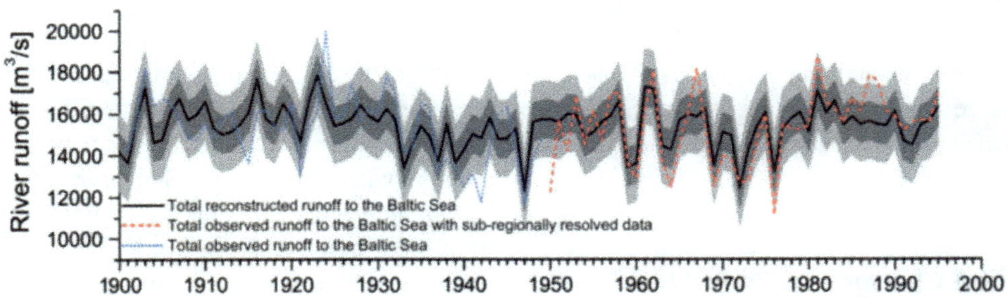

Fig. 4.7 Observed (*blue and red lines*) and reconstructed (*black line*) total river discharge to the Baltic Sea over the 20th century (adapted from Hansson et al. 2011)

approximation for the lower bound of water inflow into the North Sea since precipitation over the Baltic Proper is on average higher than evaporation (typically between 10 and 25 %; see study summaries by Omstedt et al. 2000, 2004; Hennemuth et al. 2003; Leppäranta and Myrberg 2009). Mean annual river discharge into the Baltic Sea is reported by different observational studies as between 470 and 485 km^3 $year^{-1}$ for the past century (Cyberski and Wróblewski 2000; Hansson et al. 2011). According to Omstedt et al. (2004) net outflow to the North Sea, excluding the Kattegat and Belt Sea water budget, is around 15,500 m^3 s^{-1} (corresponding to 488 km^3 $year^{-1}$) with an interannual variability of ±5000 m^3 s^{-1}.

Figure 4.7 shows observed and reconstructed annual total river discharge to the Baltic Sea during the 20th century. Decadal and regional variability is large, but no *significant* long-term change has been detected in total river discharge to the Baltic Sea during the last 500 years (Hansson et al. 2011).

Owing to climate change in the Baltic Sea basin, Störmer (2011) expected an increase in water temperature, a decrease in salinity and a decrease in summer river discharge. Analysis of discharge sensitivity to temperature indicates that the southern Baltic Sea basin may become drier with rising air temperature (Omstedt et al. 2014). Based on reconstructions, Hansson et al. (2011) concluded that total annual river discharge to the Baltic Sea has decreased by 3 % for each 1 °C rise in air temperature over the past 500 years (see also Omstedt et al. 2014). HELCOM (2013) reported that this increase is not due to the large decadal and regional variability in discharge. HELCOM (2013) also reported increasing trends in annual, winter and spring stream flow, but found no trend for autumn. The shift in discharge towards the winter period was attributed to rising temperature.

Changes (trends and variability) in inflow from the Baltic Sea largely depend on changes in river discharge. Wilson et al. (2010) analysed a dataset of 151 streamflow records from the Nordic countries and found an increase in streamflow over the periods 1920–2005, 1941–2005 and 1961–2000 in the annual data and the winter and spring data. Trends identified in summer flows differed between the three periods, whereas no trend was found for autumn. In all three periods, a signal towards earlier snowmelt-driven high flows was clear, as was the tendency towards more severe low flows in summer in southern and eastern Norway. These trends in streamflow result from changes in both temperature and precipitation, although the temperature-related signal is stronger than the influence of precipitation. Changes in the observed annual and seasonal discharge from rivers in the Baltic Sea catchment were discussed by Käyhkö et al. (2015). They concluded that statistically significant increasing trends were apparent in annual river discharge and winter discharge due to the rise in air temperature and subsequent snowmelt, while spring discharge had decreased due to less snow available.

4.4 Observed Changes in Peak Flow and Low Flow

Owing to the increased attention to flood risk more studies are now investigating trends in the frequency and magnitude of peak flows than in changes in average or low flows.

4.4.1 River Rhine

Toonen (2013) examined the lower Rhine flooding regime between 1350 and 2011 and found no permanent change over this period, but did mention the non-stationarity of the series. There appeared to be more frequent minor floods during the Little Ice Age (AD 1550–1850). Devastating floods in this period were associated with ice jams, but events of extreme discharge were not recorded (Toonen 2013).

Several authors have reported a noticeable increase in flow peak frequency in the Rhine basin over the past 100 years (Pfister et al. 2004; Pinter et al. 2006; Belz et al. 2007; Diermanse et al. 2010). Pinter et al. (2006) found a statistically significant increase in the frequency of flow peaks ranging from >5000 to 7500 $m^3\ s^{-1}$ at nine gauges in Germany. These gauges have records of 50 years of nearly continuous daily water level and discharge observations. Increase in frequency for the very large peak flows (>8000 $m^3\ s^{-1}$) is difficult to detect with high (>90 %) confidence because of the relatively small number of these extreme events during the last century. Diermanse et al. (2010) found the increase in annual maximum discharge for the Lobith gauge (running from 1901 to 2003) to be 8 $m^3\ s^{-1}$ per year over the observation period (or 13 % of the average annual maximum), but this trend was not statistically significant for four different tests (Pearson t-test, Spearman's rank correlation test, Mann-Kendall test and Wilcoxon-Mann-Whitney test). The significance levels of the tests were between 15 and 32 %.

This recent increase in peak flow is attributed to the increase in winter precipitation and increased snow melt in winter (Pfister et al. 2004; Belz et al. 2007) and to an increase in westerly atmospheric circulation types (Pfister et al. 2004).

4.4.2 River Elbe

Mudelsee et al. (2003) investigated discharge data sets for the past 80–150 years in the rivers Elbe and Odra basin. For the River Elbe (Dresden gauge), they found a decreasing trend in winter peak flows which they attributed to fewer ice damming events. Summer peak flows do not show any significant trend. Mudelsee et al. (2003) analysed floods in the rivers Elbe and Odra over the past 500 years focusing on relations with large-scale atmospheric circulation over Europe. They found significantly decreasing trends in winter peak flows in both rivers but no significant trends in summer peak flows during the 20th century.

In a study based on data from 78 gauges across Germany, Bormann et al. (2011) found no significant trends in different flood parameters. They found trends of different sign along the Elbe and emphasised the strong dependence of the trend on the underlying record length.

Analysing discharge for the period 1875–2011 at the Neu Darchau gauge, Mudersbach et al. (2013) found some evidence of an increase in the number of years with high annual flows over the past few decades. This study included trend analysis of six different flood indicators: annual maximum floods, annual winter and summer maximum flows, the two-largest flow peaks per year and two peak-over-threshold time series. They found a downward trend in frequency of winter peak discharges and no trend in frequency of summer peak discharges. However, with the exception of one peak-over-threshold time series, none of the trends were significant at the 95 % confidence level.

Changes in low flows have received less attention than changes in peak flows. Belz et al. (2007) reported a redistribution of summer discharge to winter in the Alpine region of the River Rhine but that this disappeared further downstream. For the River Elbe, IKSE (2012) conducted a trend analysis over the period 1961–2005 using different gauging stations. They found different trends at different stations and the trends could not be attributed to the slightly increasing trend in precipitation. They concluded that other human influences such as the operation of large storage reservoirs, especially those of the Moldau cascade were probably responsible for the changes in river flow.

4.4.3 River Meuse

Peak discharges have also increased in magnitude and frequency over the last century in the River Meuse (de Wit et al. 2001; Min 2006; Ward 2009; Diermanse et al. 2010). Diermanse et al. (2010) found an increasing trend in annual maximum discharge for the Borgharen gauge (running from 1911 to 2003) of 3.4 $m^3\ s^{-1}$ per year over the observation period (or 23 %). The significance levels of the tests were between 10 and 22 %. Min (2006) and Ward (2009) studied the period after 1984, analysing maximum winter discharges using a peak-over-threshold method, and found a statistically significant increase for this period.

4.4.4 UK Rivers

Hannaford and Marsh (2006) and Hannaford and Buys (2012) found no significant trends for low flow in UK rivers. They did find significant increasing trends for low flows over the period 1973–2002 but these were influenced by a sequence of notably dry years at the start of the study period and were not observed over the most recent 40-year period analysed. Jones et al. (2006) concluded that for the period 1885–2002, there were relatively few low flow periods in UK rivers between 1980 and 2002. There are some indications of an increasing frequency of peak flows in many catchments (Hannaford and Buys 2012), especially in upland areas, but there is considerable year-to-year variability and the strength of trends is very dependent on the length of record used.

4.4.5 Scandinavian Rivers

With respect to the Scandinavian region, peak flow frequency and intensity differ over time (Lindström and Bergström 2004; Thodsen 2007). Some studies suggest a slight increase in flood discharge but according to overview studies covering many rivers there is no statistically significant trend over the last century (Forland et al. 2000 for Norway; Hyvarinen 2003 for Finland; Lindström and Bergström 2004 for Sweden; Bering Ovesen et al. 2000; Thodsen 2007 for Denmark).

4.5 Conclusions

This chapter reviews the main trends and variations in river flows to the North Sea based on observations from the many gauges throughout the catchments draining into the North Sea. Most of the inflows are from a small number of large river basins, and the overall trends and variability in discharge from these large rivers are the combined effects of much variability in the different parts of the basin. Variability in river flow over time is dependent on variations in the weather—mainly precipitation and temperature—from year to year, as well as on a wide range of direct and indirect human interventions within the basin. Direct interventions include the construction of reservoirs along the river network and the abstraction and return of water for domestic, industrial and, to a much lesser extent, agricultural purposes. Indirect interventions include changes in land use. Although data records for the major rivers draining into the North Sea are relatively long, records for the smaller rivers are typically much shorter.

Analyses of observed flow records lead to three broad conclusions. First, there is considerable interannual and decadal variability in river flow in all areas draining into the North Sea. In northern areas this is closely associated with variation in the strength and direction of the NAO index, particularly in winter. Second, there are some indications of increasing discharge to the North Sea in winter, but little evidence of a widespread trend in summer; the magnitude of the trend, however, appears to depend on the length of record used and the technique used to estimate trends. Third, there is evidence that higher winter temperatures have led to increased winter river flow particularly to the Baltic Sea, as winter precipitation increasingly falls as rain rather than snow.

To date, no significant climate-related trends have been shown for most of the rivers discharging into the North Sea.

References

Belz JU (2010) Das Abflussregime des Rheins und seiner Nebenflüsse im 20. Jahrhundert – Analyse, Veränderungen, Trends [The flow regime of the River Rhine and its tributaries in the 20th century – analysis, changes, trends]. Hydrol Wasserbew 54:4–17

Belz JU, Brahmer G, Buiteveld H, Engel H, Grabher R, Hodel H, Krahe P, Lammersen R., Larina M, Mendel HG, Meuser A, Müller G, Plonka B, Pfister L, van Vuuren W (2007) Das Abflussregime des Rheins und seiner Nebenflüsse im 20. Jahrhundert. Schriftenreihe der KHR, Bd. 1–22. Konlenz und Lelystad

Bering Ovesen N, Legard Iversen H, Larsen S, Muller-Wohlfeil DI, Svendsen L (2000) Afstromingsforhold i danske vandløb. DMU Report 340, Ministry of the Environment and Energy, Denmark

Blochliger H, Neidhofer F (1998) Impacts des precipitations extremes. La Suisse face au changement climatique. Rapport sur l'e´tat des connaissances. OcCC (Organe consultatif en matiere de recherche sur le climat et les changements climatiques). Berne

Bormann H, Pinter N, Elfert S (2011) Hydrological signatures of flood trends on German rivers: Flood frequencies, flood heights and specific stages. J Hydrol 404:50–66

Brazdil R, Kundzewicz ZW, Benito G (2006) Historical hydrology for studying flood risk in Europe. Hydrolog Sci J 51:739–764

Buisman (1996, 1998, 2000, 2006) Duizend jaar Weer, wind en water in de Lage Landen, deel II–V. Van Wijnen, Franeker

Caspary HJ, Bardossy A (1995) Markieren die Winterhochwasser von 1990 und 1993 das Ende der Stationarität der Hochwasserhydrologie Südwestdeutschlands infolge Klimaänderungen? Wasser Boden 47:18–24

CHR/KHR (1993) Der Rhein unter der Einwirkung des Menschen – Ausbau, Schiffart, Wasserwirtschaft. Bericht I–11 der KHR

Conway D (2013) Securing water in a changing climate. In: Lankford BA, Bakker K, Zeitoun M, Conway D (eds) Water Security: Principles, Perspectives and Practices. Earthscan, London

Cyberski J, Wróblewski A (2000) Riverine water inflows and the Baltic Sea water volume 1901–1990. Hydrol Earth Sys Sci 4:1–11

de Wit MJM, Warmerdam PMM, Torfs PJJF, Uijlenhoet R, Roulin E, Cheymol A, van Deursen, WPA, van Walsum PEV, Ververs M, Kwadijk JCJ, Buiteveld H (2001) Effect of Climate Change on the Hydrology of the River Meuse. Online: www.rivm.nl/bibliotheek/rapporten/410200090.html

Diermanse FLM, Kwadijk JCJ, Beckers JVL, Crebas JI (2010) Statistical trend analysis of annual maximum discharges of the Rhine and Meuse rivers. Bhs Third Int Symp Manag Consequences Chang Glob Env, Newcastle, 1–5

Ebel U, Engel H (1994) The Christmas Floods in Germany 1993/94. Bayerische Rückversicherung, Munich. Special Issue 16

EEA (2008) Impacts of Europe's Changing Climate: 2008 Indicator-based Assessment. European Environment Agency, EEA Rep 4/2008

EEA (2012) Climate Change, Impacts and Vulnerability in Europe 2012: An Indicator-based Assessment. European Environment Agency, EEA Tech Rep 12/2012

Engel H (1997) Die Ursachen der Hochwasser am Rhein – natürlich oder selbstgemacht? In: Immendorf R (ed) Hochwasser, Natur im Überfluss? Verlag CF Müller, Heidelberg

Engel H (1999) Abflusse im Rhein und seine Nebengewässern – Entwicklungen seit Beginn diese Jahrhunderts, Erwartungen für die nahe Zukunft. In: Hydrologische Dynamik im Rheingebiet. IHP/OHP-berichte 13. Koblenz

Forland E, Roald LA, Tveito OE, Hanssen-Bauer I (2000) Past and Future Variations in Climate and Runoff in Norway. DNMI Report 19000/00 KLIMA, Oslo

Glaser R, Stangl H (2003) Floods in Central Europe since 1300. In: Thorndycraft VR, Benito G, Barriendos M, Llasat MC (2003) Palaeofloods, Historical Floods and Climatic Variability: Applications in Flood Risk Assessment. Proceedings of the PHEFRA Workshop, Barcelona, 16–19 October 2002

Hannaford J, Buys G (2012) Trends in seasonal river flow regimes in the UK. J Hydrol 475:158–174

Hannaford J, Marsh T (2006) An assessment of trends in UK runoff and low flows using a network of undisturbed catchments. Int J Climatol 26:1237–1253

Hannaford J, Buys G, Stahl K, Tallaksen LM (2013) The influence of decadal-scale variability on trends in long European streamflow records. Hydrol Earth Sys Sci 17:2717–2733

Hansson D, Eriksson C, Omstedt A, Chen D (2011) Reconstruction of river runoff to the Baltic Sea, AD 1500–1995. Int J Climatol 31:696–703

HELCOM (2013) Climate Change in the Baltic Sea Area: HELCOM thematic assessment in 2013. Balt Sea Env Proc 137

Hellström S, Lindström G (2008) Regional analys av klimat, vattentillgång och höga flöden. SMHI Rapp Hydrol 110

Hennemuth B, Rutgersson A, Bumke K, Clemens M, Omstedt A, Jacob D, Smedman AS (2003) Net precipitation over the Baltic Sea for one year using models and data-based methods. Tellus A 55:352–367

Hyvarinen V (2003) Trends and characteristics of hydrological time series in Finland. Nord Hydrol 34:71–90

IKSE (2012) Hydrologische Niedrigwasserkenngrößen der Elbe und bedeutender Nebenflüsse, Internationale Kommission zum Schutz der Elbe. Online: www.ikse-mkol.org/fileadmin/download/NW-Statistik/IKSE-Niedrigwasserkenngroessen.pdf. Accessed 29 April 2015

Ionita M, Rimbu N, Lohmann G (2011) Decadal variability of the Elbe River streamflow. Int J Climatol 31:22–30

Jones PD, Lister DH, Wilby RL, Kostopoulou E (2006) Extended riverflow reconstructions for England and Wales, 1865–2002. Int J Climatol 26:219–231

Käyhkö J, Apsite E, Bolek A, Filatov N, Kondratyev S, Korhonen J, Kriauciuniene J, Lindström G, Nazarova L, Pyrh A, Sztobryn M (2015) Recent change – river run-off and ice cover. In: The BACC II Author Team, Second Assessment of Climate Change for the Baltic Sea Basin, pp. 99–115. Springer-Verlag

Konnen GP (1999) De toestand van het klimaat in Nederland 1999. Koninklijk Nederlands Meteorologisch Instituut (KNMI)

Lawrence D, Hisdal H (2011) Hydrological projections for floods in Norway under a future climate. Norwegian Water Resources and Energy Directorate (NVE). Report Series 2010–5

Leppäranta M, Myrberg K (2009) Physical Oceanography of the Baltic Sea. Springer-Verlag

Lindström G, Bergström S (2004) Runoff trends in Sweden 1807–2002. Hydrol Sci J 49:69–83

Markovic D, Koch M (2006) Characteristic scales, temporal variability modes and simulation of monthly Elbe River flow time series at ungauged locations. Phys Chem Earth 31:1262–1273

Min T (2006) Assessment of the effects of climate variability and land use change on the hydrology of the Meuse river basin. PhD Thesis, Free University of Amsterdam and UNESCO-IHE Institute for Water Education at Delf

Mudelsee M, Börngen M, Tetzlaff G, Grünewald U (2003) No upward trends in the occurrence of extreme floods in central Europe. Nature 425:166–169

Mudersbach C, Bender J, Kelln V, Jensen J (2013) Analysing flood frequencies at the Elbe River – Do recent extreme events affect design levels? In: Proceeding of the 6th International Conference on Water Resources and Environment Research. Koblenz

O'Connell PE, Ewen J, O'Donell G, Quinn P (2007) Is there a link between agricultural land-use management and flooding? Hydrol Earth Sys Sci 11:96–107

Omstedt A, Gustafsson B, Rohde J, Walin G (2000) Use of Baltic Sea modelling to investigate the watercycle and the heat balance in GCM and regional climate models. Clim Res 15:95–108

Omstedt A, Elken J, Lehmann A, Piechura J (2004) Knowledge of the Baltic Sea physics gained during the BALTEX and related programmes. Progr Oceanogr 63:1–28

Omstedt A, Elken J, Lehmann A, Leppäranta M, Meier HEM, Myrberg K, Rutgersson A (2014) Progress in physical oceanography of the Baltic Sea during the 2003–2014 period. Progr Oceanogr 128:139–171

OSPAR (2000) Quality Status Report 2000, Region II – Greater North Sea. OSPAR Commission, London

Pfister C (1999) Wetternachhersage. 500 Jahre Klimavariationen und Naturkatastrophen 1496–1995. Paul Haupt, Bern

Pfister L, Kwadijk JCJ, Musy A, Bronstert A, Hoffman L (2004) Climate change, land use change and runoff prediction in the Rhine-Meuse basins. River Res Appl 20:229–241

Pinter N, van der Ploeg RR, Schweigert P, Hoefer G (2006) Flood magnification on the River Rhine. Hydrol Proc 20:147–164

Stahl K, Tallaksen LM, Hannaford J, van Lanen HAJ (2012) Filling the white space on maps of European runoff trends: estimates from a multi-model ensemble. Hydrol Earth Sys Sci 16:2035–2047

Störmer O (2011) Climate change impacts on coastal waters of the Baltic Sea. In: Schernewski G, Hofstede J, Neumann T (eds) Global Change and Baltic Coastal Zones. Springer, Dordrecht

Thodsen H (2007) The influence of climate change on stream flow in Danish rivers. J Hydrol 333:226–238

Toonen WHJ (2013) A Holocene Record of the Lower Rhine. Utrecht Stud Earth Sci 141

Uhlenbrook S, Steinbrich A, Tetzlaff D, Leibundgut Ch (2001) Zusammenhang zwischen extremen Hochwassern und ihren Einflussgrößen. In: Arbeitskreis KLIWA (Klimaveränderung und Konsequenzen für die Wasserwirtschaft (ed) KLIWA-Symposium, 29–30 November 2000, Karlsruhe: KLIWA-Berichte 1:187–203

Ward PJ (2009) Simulating discharge and sediment yield characteristics in the Meuse basin during the late Holocene and 21st Century. PhD Thesis, Free University of Amsterdam

Wilby RL (2006) When and where might climate change be detectable in UK river flows? Geophys Res Lett 33:L19407. doi:10.1029/2006GL027552

Wilson D, Hisdal H, Lawrence D (2010) Has streamflow changed in the Nordic countries? Recent trends and comparisons to hydrological projections. J Hydrol 394:334–346

Part II
Future Climate Change

Projected Change—Atmosphere

5

Wilhelm May, Anette Ganske, Gregor C. Leckebusch, Burkhardt Rockel, Birger Tinz and Uwe Ulbrich

Abstract

Several aspects describing the state of the atmosphere in the North Sea region are considered in this chapter. These include large-scale circulation, means and extremes in temperature and precipitation, cyclones and winds, and radiation and clouds. The climate projections reveal several pronounced future changes in the state of the atmosphere in the North Sea region, both in the free atmosphere and near the surface: amplification and an eastward shift in the pattern of NAO variability in autumn and winter; changes in the storm track with increased cyclone density over western Europe in winter and reduced cyclone density on the southern flank in summer; more frequent strong winds from westerly directions and less frequent strong winds from south-easterly directions; marked mean warming of 1.7–3.2 °C for different scenarios, with stronger warming in winter than in summer and a relatively strong warming over southern Norway; more intense extremes in daily maximum temperature and reduced extremes in daily minimum temperature, both in strength and frequency; an increase in mean precipitation during the cold season and a reduction during the warm season; a pronounced increase in the intensity of heavy daily precipitation events, particularly in winter; a considerable increase in the intensity of extreme hourly precipitation in summer; an increase (decrease) in cloud cover in the northern (southern) part of the North Sea region, resulting in a decrease (increase) in net solar radiation at the surface.

W. May (✉)
Research and Development Department, Danish Meteorological Institute (DMI), Copenhagen, Denmark
e-mail: wm@dmi.dk

W. May
Centre for Environmental and Climate Research, Lund University, Lund, Sweden
e-mail: wilhelm.may@cec.lu.se

A. Ganske
Federal Maritime and Hydrographic Agency (BSH), Hamburg, Germany
e-mail: Anette.Ganske@bsh.de

G.C. Leckebusch
School of Geography, Earth and Environmental Sciences, University of Birmingham, Birmingham, UK
e-mail: G.C.Leckebusch@bham.ac.uk

B. Rockel
Institute of Coastal Research, Helmholtz-Zentrum Geesthacht, Geesthacht, Germany
e-mail: burkhardt.rockel@hzg.de

B. Tinz
German Meteorological Service (DWD), Hamburg, Germany
e-mail: Birger.Tinz@dwd.de

U. Ulbrich
Institute of Meteorology, Freie Universität Berlin, Berlin, Germany
e-mail: ulbrich@met.fu-berlin.de

M. Quante and F. Colijn (eds.), *North Sea Region Climate Change Assessment*, Regional Climate Studies, DOI 10.1007/978-3-319-39745-0_5

5.1 Introduction

Wilhelm May

Projections of future climate change are obtained from simulations with global coupled as well as regional climate models (GCMs and RCMs, respectively). In these projections, concentrations or emissions of the well-mixed greenhouse gases and of the anthropogenic aerosol load are prescribed according to different scenarios, which take different possibilities for future developments into account. For the Fourth Assessment Report (AR4) of the Intergovernmental Panel on Climate Change (IPCC), these scenarios were based on the assumptions of the Special Report on Emission Scenarios (SRES; Nakićenović et al. 2000), while for the Fifth Assessment Report (AR5) the newly developed Representative Concentration Pathways (RCPs; Moss et al. 2010; van Vuuren et al. 2011) were applied. The RCP scenarios differ from the SRES scenarios in that they assume different pathways to specific targets of the radiative forcing by the end of the 21st century. The two families of scenarios were also applied in the Coupled Model Intercomparison Project (CMIP), the SRES scenarios in phase 3 (CMIP3; Meehl et al. 2007a) and the RCP scenarios in phase 5 of the project (CMIP5; Taylor et al. 2012).

In the contributions of IPCC Working Group I to AR4, the projections of future climate change based on the SRES scenarios are presented in two different chapters, one addressing the global aspects of climate change (Meehl et al. 2007b) and one covering the regional aspects (Christensen et al. 2007). In the latter, the projected changes in climate are described separately for different continents and/or regions, including Europe. In 2012, the AR4 was complemented by the IPCC Special Report on climate extremes (SREX), where among others the observed and projected future changes in different kinds of extreme climate events were assessed (Seneviratne et al. 2012). In Table 3.3 of the latter report, the projected future changes in temperature and precipitation extremes were summarised for different regions, including northern, central and southern Europe. Also in 2012, the European Environment Agency (EEA) published a report on climate change, impacts and vulnerability in Europe, covering several aspects of climate and climate change (EEA 2012). In particular, the report includes references to several scientific publications based on future climate projections originating from a multi-model ensemble of RCM simulations for Europe performed within the ENSEMBLES project (Van der Linden and Mitchell 2009). The SRES scenarios were applied in these simulations. Recently, a new set of future climate projections for Europe has become available within the World Climate Research Programme (WCRP) Coordinated Regional Downscaling Experiment (CORDEX; Giorgi et al. 2009), with the aim to increase both the number of RCMs and the number of driving coupled climate models compared to the ENSEMBLES project. These scenario simulations are based on the RCP scenarios, with the driving coupled climate model data taken from CMIP5. As for Europe, a specific set of climate scenarios at a horizontal resolution of 12.5 km has become available within the CORDEX initiative, with seven different RCMs to date (Jacob et al. 2014). In ENSEMBLES, the finest horizontal resolution of the climate scenarios was 25 km.

In the contributions of IPCC Working Group I to AR5, the global aspects of the projections of future climate change based on the RCP scenarios are presented in a specific chapter (Collins et al. 2013), while the regional aspects were covered differently to AR4. In AR5, future changes in the characteristics of a number of prominent climate phenomena, i.e., monsoon systems, the El Niño-Southern Oscillation, annular and dipolar modes and large-scale storm systems, and their relevance for regional climate change were assessed (Christensen et al. 2013a), with the regional changes in climate presented in the form of an atlas for as many as 18 different regions distributed over the globe (IPCC 2013). As for Europe, the northern and central parts of the continent and the Mediterranean region were distinguished. The regional aspects of the projections of future climate change were also considered in the contributions of Working Group II to AR5 (Hewitson et al. 2014a, b), again distinguishing between the aforementioned three parts of Europe. A detailed assessment of the impacts of the projected changes in climate for Europe, as for several other regions, is presented in a specific chapter of this part of AR5 (Kovats et al. 2014a, b).

Adaptation strategies are needed in response to the observed as well as to the projected changes in climate (Noble at el. 2014) and these are currently developed at the national and local level in many countries. This is typically done on the basis of national climate scenarios, which are already available for several countries and are likely to become more widespread in the future. Both the Netherlands (KNMI 2014) and Denmark (DMI 2014), for instance, have recently published reports on future climate scenarios for their countries. In Germany, future climate scenarios have even become available at a regional level through so-called regional climate offices, which cover different parts of the country. Despite their high value for the development of adaptation strategies for a particular country or part of a country, these national climate scenarios cannot easily be combined to give a consistent scenario for a larger area, such as the North Sea region. While the climate scenarios for Denmark follow closely the scenarios used in AR4 and AR5 (DMI 2014), the future climate scenarios for the Netherlands were developed by combining numerous climate scenarios originating from different climate models in accordance with the simulated rate of global warming and the simulated change in the large-scale circulation over western Europe (KNMI 2014). This distinction resulted in four categories of climate scenario: one with moderate warming (about 1.5 °C by the end of the 21st century) and a weak influence of circulation change (i.e. a small change in the frequency of the dominant circulation patterns relative to present-day conditions); one with moderate warming and a strong influence of circulation change (i.e. a large change in the frequency of the dominant circulation patterns); one with strong warming (about 3.5 °C by the end of the 21st century) and a weak influence of circulation change; and one with strong warming and a strong influence of circulation change. The dominant circulation patterns are characterised by prevailing westerly winds during winter and prevailing easterly winds in association with high surface pressure during summer, respectively.

In this chapter, the projected changes in the atmosphere in the North Sea region are assessed on the basis of the existing literature, including the recent assessment reports referred to above. Typically, these changes have been projected for the end of the 21st century using conditions at the end of the 20th century as the baseline, but in the last few years several projections have also become available for the middle of the 21st century. Because few studies have focussed specifically on the North Sea region, most of the results described here have been extracted from climate projections for Europe (based on RCM scenario simulations from ENSEMBLES or CORDEX) or even from projections covering the whole globe (based on GCM scenario simulations from CMIP3 or CMIP5). Several aspects describing the state of the atmosphere in the North Sea region have been considered, such as features of the large-scale circulation (Sect. 5.2), the mean and extremes, primarily at daily time scales, in temperature (Sect. 5.3) and precipitation (Sect. 5.4), cyclones and winds (Sect. 5.5), and radiation and clouds (Sect. 5.6).

5.2 Large-Scale Circulation

Uwe Ulbrich, Birger Tinz, Wilhelm May

5.2.1 Prominent Climate Phenomena

Regional climate is affected by various kinds of climate phenomena. Their change under rising greenhouse gas concentrations is thus relevant for future regional climate change (e.g. Christensen et al. 2013a). Prominent climate phenomena include the monsoon systems in different parts of the tropics, the El Niño-Southern Oscillation, different annual or dipolar modes, and blocking and large-scale storm systems. The interannual variability of the climate in the North Atlantic region and specifically the North Sea region is mainly affected by two modes of variability: the North Atlantic Oscillation (NAO) and its hemispheric counterpart, the Northern Annular Mode (NAM) or Arctic Oscillation (e.g. Itoh 2008). Other large-scale factors affecting the climate in the Atlantic-European sector are atmospheric blocking and the strength and position of the Atlantic jet stream. These factors are all related to the strength and location of the Atlantic storm track and in turn to the NAO.

5.2.2 Modes of Interannual Variability

The NAO is a dipolar mode of climate variability, characterised by opposite variations in sea-level pressure between the Atlantic sub-tropical High and the Icelandic Low (e.g. Hurrell et al. 2003). Through its direct effect on westerly air flow into Europe, its link with Atlantic cyclones and atmospheric blocking, it strongly affects the climate over the North Atlantic Ocean and the surrounding continents (e.g. Hurrell and Deser 2009). The NAO can be established throughout the entire year, despite different physical mechanisms initiating and maintaining this mode of variability during winter and summer (e.g. Folland et al. 2009).

The CMIP5 simulations for the intermediate RCP4.5 scenario (i.e. 75 simulations with 37 different global climate models) show an overall amplification of the NAO up to the end of the 21st century in all seasons, with the greatest increase in autumn (Gillett and Fyfe 2013). That is, the pressure difference between the Azores High and the Icelandic Low is projected to increase in these scenario simulations. This is consistent with earlier results from the CMIP3 simulations (Miller et al. 2006). These trends, however, are generally small compared to the natural climate variability (Deser et al. 2012). It should be noted that Gillett and Fyfe's (2013) use of a particular index to define the NAO might have had an effect on the magnitude of the projected change in the NAO, as the respective centres of action over the northern and southern parts of the North Atlantic might have different positions under a changing climate. For instance, Dong et al. (2011) found a poleward and eastward shift in the pattern of NAO variability in response to greenhouse gas forcing, in line with previous findings by Ulbrich and Christoph (1999). Both the future changes in the troposphere and the stratosphere as a direct response to the prescribed greenhouse gas forcing and the associated changes in sea surface temperatures in the North Atlantic contribute to the aforementioned changes in the NAO. In a recent study, Davini and Cagnazzo (2013) pointed at the possibility of misinterpreting the NAO signals in current climate models. This is because some of the models were not able to realistically simulate the physical processes connected to the NAO, namely atmospheric blocking and interaction with the Atlantic jet stream. This is particularly the case for those models that strongly underestimate the frequency of atmospheric blocking in the Greenland area. These shortcomings might affect studies analysing the NAO under different mean climate states, i.e. for future climate scenarios.

The NAO has been interpreted as the manifestation of an annular mode in sea-level pressure, the NAM, over the North Atlantic region (e.g. Thompson and Wallace 2000). Similar to their findings for the NAO, Gillett and Fyfe (2013) also found an overall amplification of the NAM under future climate conditions in all seasons. The increase is greatest in autumn and winter and smallest in summer. Furthermore, none of the climate models simulated a significant decrease in the NAM in any season.

5.2.3 Atmospheric Blocking

Atmospheric blocking is typically associated with persistent stationary or slowly moving high-pressure systems in the extratropics, interrupting the prevailing westerly winds and the usual track of eastward moving cyclones at these latitudes. Blocking occurs most frequently in the exit regions of the storm tracks in both hemispheres and is characterised by marked seasonal variability with high frequencies during winter and spring and low frequencies during summer and autumn (e.g. Wiedenmann et al. 2002; Masato et al. 2013). In the Atlantic-European sector blocking occurs more frequently over the North Atlantic in winter but more frequently over Europe in summer (e.g. Tyrlis and Hoskins 2008). Blocking is a major contributor to intraseasonal variability in the extratropics and can lead to seasonal climate anomalies over large parts of Europe (e.g. Trigo et al. 2004) as well as to climate extremes like cold spells in winter (e.g. Cattiaux et al. 2010) or heat waves in summer (e.g. Matsueda 2011). As previously mentioned, atmospheric blocking in the Atlantic-European sector during winter is strongly related to the NAO (Croci-Maspoli et al. 2007).

The CMIP5 simulations for the high RCP8.5 scenario show an overall decrease in the frequency of atmospheric blocking in the Atlantic-European sector in both winter (Cattiaux et al. 2013; Dunn-Sigouin and Son 2013; Masato et al. 2013) and summer (Dunn-Sigouin and Son 2013; Masato et al. 2013). The decrease in summer is accompanied by an increase on its eastern flank, leading to an eastward shift of the area with high blocking frequencies (Masato et al. 2013). While the decrease in winter is a consistent finding, regardless of how many different simulations from CMIP5 are considered or which method is used to define a blocking event, the situation is less clear in summer. In contrast to the findings of Dunn-Sigouin and Son (2013) and Masato et al. (2013), Cattiaux et al. (2013) found an increase in the frequency of blocking events in the Atlantic-European sector during summer for most of the 19 CMIP5 models considered. The other two studies considered simulations from fewer CMIP5 models and used various indices to define blocking, while Cattiaux et al. (2013) used an approach based on weather regimes, with blocking being one of them. No noticeable changes, however, were found regarding the duration of individual blocking events (Dunn-Sigouin and Son 2013). These results are consistent with findings based on the CMIP3 simulations, which show a significant decrease in blocking frequency, particularly during winter (Barnes and Hartmann 2010; Barnes et al. 2012), but are somewhat less clear. According to Woollings (2010) the effect of greenhouse gas forcing on blocking might to a large extent reflect changes in the mean state of the atmosphere rather than dynamical processes directly associated with blocking. Barnes and Hartmann (2010) demonstrated, for instance, that a poleward shift in the Atlantic jet stream could lead to a decreased frequency of atmospheric blocking in winter due to a reduction in poleward Rossby-wave breaking.

5.2.4 Sea-Level Pressure

The AR5 reported an increase in mean sea-level pressure (MSLP) over western Europe and the adjacent part of the North Atlantic in winter, with a centre over the Mediterranean region, for RCP2.6, RCP4.5 and RCP8.5 (Collins et al. 2013). Further north the MSLP is markedly reduced. In summer, on the other hand, MSLP is reduced over Europe but increased over the North Atlantic, with a centre west of the British Isles. In both cases, the magnitude of the changes in MSLP follows the strength of the radiative forcing with the smallest (largest) changes in MSLP associated with the weakest (strongest) scenario. Van den Hurk et al. (2014) obtained similar results, when regressing changes in MSLP in the Atlantic-European region on the corresponding changes in global mean temperature for a total of 245 climate change simulations from CMIP5, covering 37 different global climate models, four scenarios (including RCP6.0) and ensemble simulations for some of the models. For spring and autumn, the authors found increases in MSLP over much of the North Atlantic and western Europe and decreases further north over the Arctic, but in contrast to winter, the maximum increases are centred over the North Atlantic during the transition seasons. The projected changes in MSLP contribute to the positive trend in the NAO and the NAM mentioned in Sect. 5.2.2, particularly in autumn.

5.2.5 Jet Stream

The CMIP5 simulations show a small (about 1° for the multi-model ensemble means) poleward shift in the position of the Atlantic jet stream for the RCP8.5 scenario, while its speed remains nearly constant (Barnes and Polvani 2013). The poleward shift in the position of the Atlantic jet steam was found to reduce its north-south wobble as well as to enhance the variability of its speed (i.e. more of a pulsing of the jet stream). Woollings and Blackburn (2012) obtained consistent results based on the CMIP3 simulations, both with regard to a poleward shift in the mean position of the Atlantic jet stream and to considerable variations between individual models, particularly in winter. The poleward shifts were often small compared to the errors in the simulation of the jet stream position. Moreover, Woollings and Blackburn (2012) found that the NAO in combination with the East Atlantic pattern (EA) of the large-scale circulation can describe both the climatological changes and the inter-annual variations of both the position and strength of the Atlantic jet stream at the tropopause level. It is largely the NAO that describes shifts in the position of the jet, whereas the NAO and EA are both associated with changes in the strength of the jet.

The mechanisms underlying a poleward shift in the jet stream are still not fully understood. Changes in the activity of large-scale planetary waves or in the characteristics of the synoptic-scale transient wave activity have been suggested to contribute to the poleward shift (e.g. Collins et al. 2013). Haarsma et al. (2013) found an eastward extension to the zonal winds at 500 hPa over the eastern Atlantic Ocean and western Europe, primarily related to changes in the tropospheric temperature profile. The temperature changes in two regions were found to be important for forcing the changes in mean zonal flow: the relatively strong upper-tropospheric warming in the subtropics and the reduced surface warming in the mid-latitudes. Inter-model differences in the projected changes in mean zonal flow over the eastern Atlantic Ocean and western Europe could be partly attributed to uncertainties in the response of the North Atlantic Ocean to the anthropogenic forcing in both the CMIP3 and CMIP5 models.

5.2.6 Summary

Both the CMIP3 and CMIP5 simulations project marked future changes in various aspects of the large-scale circulation over the Atlantic-European region, of which the North Sea region is part. These changes are expected to affect the near-surface climate of the North Sea region, particularly in terms of weather and climate extremes. Examples include the impact of changes in the distribution of the phases of the NAO on the occurrence of climate extremes in Europe (e.g. Scaife at el. 2008), and the role of atmospheric blocking over the North Atlantic on the occurrence of cold winter temperatures in Europe (Sillmann et al. 2011).

5.3 Temperature

Wilhelm May

5.3.1 Global Mean Temperature

The CMIP5 simulations project a global warming with respect to the present day (1986–2005) of between 1.0 (RCP2.6) and 2.0 °C (RCP8.5) by the mid-21st century and between 1.0 (RCP2.6) and 3.7 °C (RCP8.5) by the end of the 21st century for the multi-model ensemble means (see Table 5.1). The projected changes in temperature vary considerably between models, with the uncertainty ranges depending on the magnitude of the projected multi-model changes. For the RCP2.6 scenario 90 % of the projected changes by the middle of the 21st century fall in the range 0.4–1.6 °C (the smallest mean change) and in the range 2.6–4.8 °C by the end of the 21st century for RCP8.5 (the greatest mean change). Assuming a present-day (1986–2005) global warming of 0.61 °C with respect to the pre-industrial period (1850–1900; see Collins et al. 2013), means that under the RCP2.6 scenario global warming is most likely to stay below the internationally agreed target of limiting warming to less than 2 °C with respect to pre-industrial levels throughout the 21st century, while it is most unlikely that global warming will stay below this threshold over the course of the 21st century under the RCP8.5 scenario.

5.3.2 Regional Mean Temperatures

According to Knutti and Sedláček (2012), the CMIP5 multi-model ensemble projects a so-called 'highly robust' mean surface warming in the North Sea region during both winter and summer. Part of this robust warming pattern is

Table 5.1 Projected change in annual global mean surface air temperature (°C) by the mid- and end of the 21st century relative to present day (1986–2005) for RCP2.6 (32 models), RCP4.5 (42 models) and RCP8.5 (39 models) obtained from the CMIP5 multi-model ensemble as well as the 5–95 % ranges from the models' distribution

Period	RCP2.6	RCP4.5	RCP8.5
2046–2065	1.0 (0.4–1.6)	1.4 (0.9–2.0)	2.0 (1.4–2.6)
2081–2100	1.0 (0.3–1.7)	1.8 (1.1–2.6)	3.7 (2.6–4.8)

Adapted from Collins et al. (2013, their Table 12.2)

Table 5.2 Projected changes in mean surface air temperature (°C) by the mid- and end of the 21st century relative to present day (1986–2005) for northern Europe (see Seneviratne et al. 2012, their Fig. 3.1) for RCP4.5 (42 models) and RCP8.5 (39 models) obtained from the CMIP5 simulations, in terms of winter (December through February; DJF), summer (June through August; JJA) and annual means

Period	Season	RCP4.5	RCP8.5
2046–2065	DJF	2.7 (1.8–3.5)	3.4 (2.9–4.7)
	JJA	1.8 (1.2–2.5)	2.5 (1.9–3.2)
	ANN	2.0 (1.6–2.8)	2.9 (2.4–3.5)
2081–2100	DJF	3.4 (2.6–4.4)	6.1 (5.3–7.5)
	JJA	2.2 (1.6–3.0)	4.5 (3.5–5.8)
	ANN	2.7 (2.1–3.5)	5.0 (4.3–6.3)

Data represent the median of the multi-model ensemble results and the 25th and 75th percentiles of the individual model responses. Adapted from Christensen et al. (2013a, their Table 14.1) and Christensen et al. (2013b, their Table 14.SM.1c), respectively

weaker warming over the North Sea than over the adjacent land areas, particularly in winter. This tendency is also evident in the climate change projections for northern and central Europe based on the CMIP5 multi-model ensemble presented in Annex I of AR5 (IPCC 2013). For the RCP4.5 scenario, the ensemble-mean future warming by the end of the 21st century during winter is 1–2 °C over the North Sea and 3–4 °C over eastern Scandinavia. During summer, on the other hand, future warming is 2–3 °C for the entire northern and central European land areas compared to 1–2 °C over the North Sea. The regional patterns of future warming in the North Sea region are characterised by a west-east gradient with the strongest warming in the east during winter and a north-south gradient with the strongest warming in the south during summer. Averaged over northern Europe as a whole, the annual mean warming is between 2.0 °C (RCP4.5) and 3.4 °C (RCP8.5) by the middle of the 21st century and between 2.7 °C (RCP4.5) and 5.0 °C (RCP8.5) by the end of the 21st century (see Table 5.2). The strength of future warming over northern Europe varies between seasons with stronger warming during winter (6.1 °C) than during summer (4.5 °C), for RCP8.5 by the end of the 21st century (see Table 5.2).

The characteristic warming patterns over Europe are also revealed in a multi-model ensemble based on scenario simulations at high horizontal resolution (~12.5 km) with 11 different RCMs for the RCP4.5 and RCP8.5 scenarios (Jacob et al. 2014). In summer (JJA), for instance, projected warming is 1.5–2 °C adjacent to the North Sea except for southern Norway, where the warming exceeds 2 °C (Fig. 5.1). In winter (DJF), on the other hand, warming is 1.5–2 °C in western Europe, 2–2.5 °C in central Europe and over 2.5 °C in northern Europe. In spring (MAM), warming shows a very similar pattern to that for winter, but with slightly (by ~0.5 °C) weaker warming, while in autumn (SON) warming is 2–2.5 °C over the entire area adjacent to the North Sea. Averaged over the Atlantic region, which comprises the North Sea region except for southern Norway but including Ireland, France and the north-eastern part of the Iberian Peninsula (Metzger et al. 2005), the 11 climate scenarios give an annual mean warming of 1.7 °C (RCP4.5) to 3.2 °C (RCP8.5) by the end of the 21st century (see Table 5.3). These estimates of regional warming are somewhat lower than the corresponding estimates for northern Europe (see Table 5.2), which can be explained by northern Europe extending further north than the Atlantic region and not including south-western Europe.

The national climate scenarios also show marked future warming in the respective countries in response to anthropogenic forcing. For Denmark, the CMIP5 multi-model ensemble projects a future annual mean warming of 1.0 (RCP2.6), 1.8 (RCP4.5), and 3.7 °C (RCP8.5) by the end of the 21st century (DMI 2014). These estimates are about 30 % lower than the corresponding estimates for northern Europe (see Table 5.2). For the Netherlands, the projected change in annual mean temperature by the end of the 21st century varies between 1.3 °C for the scenario with moderate warming and a weak influence of circulation change to 3.7 °C for the scenario with strong warming and a strong influence of circulation change (KNMI 2014). The projected annual mean temperature changes for the Netherlands by the mid-21st century are markedly weaker, at 1.0–2.3 °C. Similarly, a multi-model ensemble of climate projections for Germany for the mid-21st century on the basis of seven combinations of RCMs and driving GCMs, gives a warming of 1.0–1.5 °C for northern Germany under the SRES A1B scenario (Wagner et al. 2013).

5.3.3 Temperature Extremes

Changes in long-term averages for variables such as seasonal or annual mean temperature provide insight into relatively slow climatic change. However, in terms of impacts it is changes in the variability of temperature at much shorter time scales that are most relevant. For instance, weather and climate extremes at daily time scales or, in the case of extended warm spells and heat waves, at time scales of

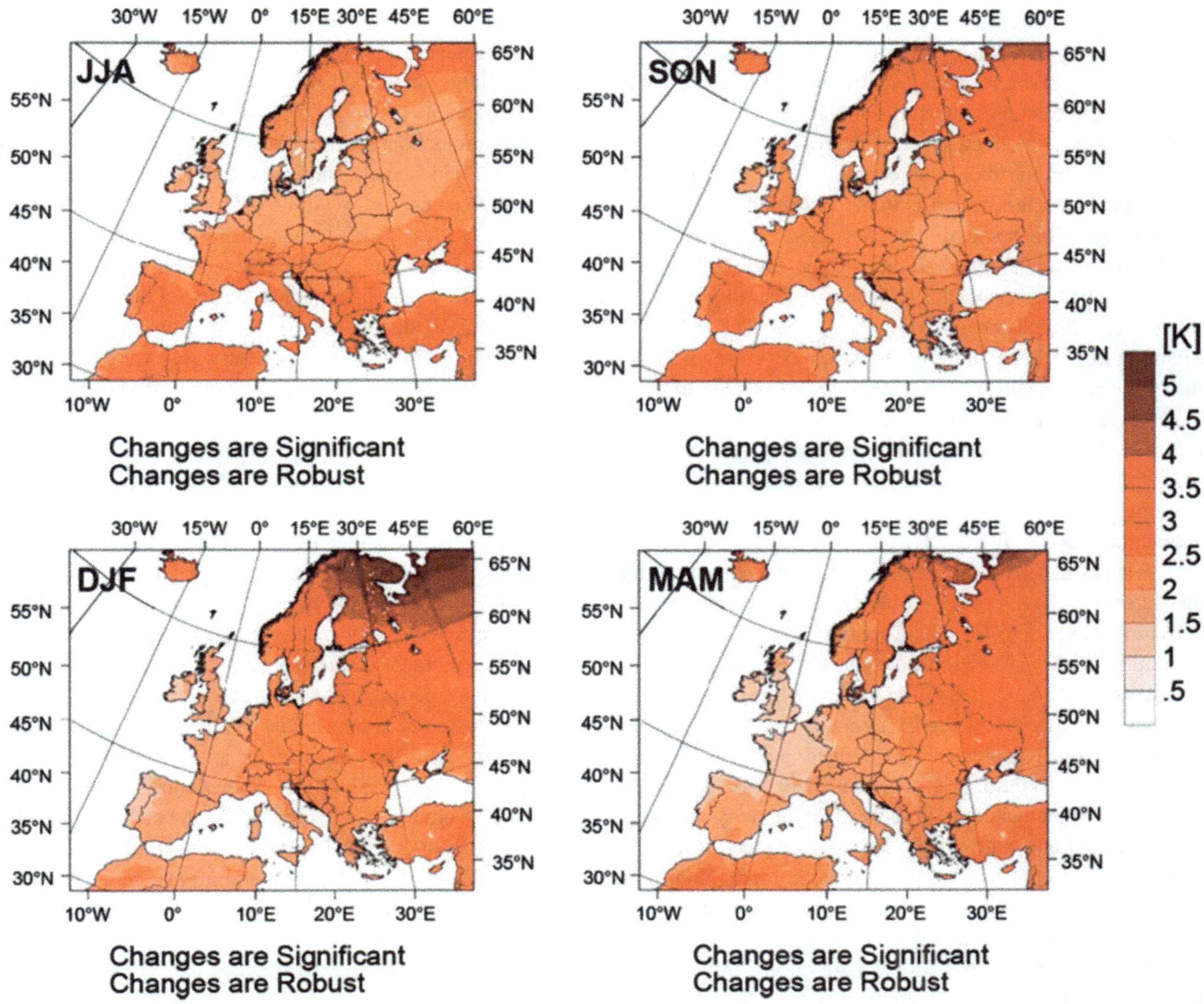

Fig. 5.1 Projected seasonal changes in surface air temperature (K) based on the RCP4.5 scenario for the end of the 21st century (2071–2100) relative to present day (1971–2000). All changes are both robust and statistically significant. From the supplementary material of Jacob et al. (2014)

Table 5.3 Projected changes in selected temperature-related climate variables and indices by the end of the 21st century (2071–2100, with respect to 1971–2000) averaged over the Atlantic region (according to Metzger et al. 2005) for RCP4.5 (eight RCM simulations) and RCP8.5 (nine RCM simulations)

Temperature-related climate indices	RCP4.5	RCP8.5
Annual mean temperature (°C)	1.7 (1.4 to 2.1)	3.2 (2.7 to 3.6)
Frost days per year	−28 (−30 to −15)	−40 (−50 to −26)
Summer days per year	11 (6 to14)	24 (22 to 28)
Tropical nights per year	3 (1 to 5)	7 (3 to 12)
Growing season length (days per growing season)	39 (27 to 43)	58 (47 to 68)
Warm spell duration index (days per year)	21 (19 to 34)	67 (47 to 92)
Cold spell duration index (days per year)	−4 (−5 to −4)	−5 (−6 to −4)

Data represent the median of the multi-model ensemble results and the likely range in these changes, defined to include 66 % of all projected changes around the ensemble median. Adapted from Kovats et al. (2014b, their Table SM23-3) several days to weeks. A number of indices describing climate extremes have been developed based on some of the characteristics of the respective distributions of daily data. In a first attempt to coordinate and standardise the definition of such extremes, Frich et al. (2002) proposed five different indices concerning daily temperature data. Zhang et al. (2011) extended this list of extreme temperature indices to 15, also revising some of the definitions of Frich et al. (2002). In particular, these indices often focus on relative thresholds that describe the tails in the distribution rather than on specific physically-based thresholds. The indices of Zhang et al. (2011) capture both moderately extreme events that typically occur several times per year and extreme events that occur less often (once a year or less). In recognition of the strong impact of weather and climate extremes the IPCC published a special report on managing the risks of extreme events and disasters to advance climate change adaptation (SREX; IPCC 2012).

Kovats et al. (2014b) reported on projected changes in the characteristics of seven different temperature-related extremes based on the multi-model ensemble of high-resolution RCM simulations for the RCP4.5 and RCP8.5 scenarios (Jacob et al. 2014). In Table 5.3 these changes are presented for the end of the 21st century averaged over the Atlantic region. The indices were defined in accordance with Zhang et al. (2011), that is, the number of frost days were defined as the annual count of days when the daily minimum temperature drops below 0 °C, the number of summer days as the annual count of days when the daily maximum temperature exceeds 25 °C, the number of tropical nights as the annual count of days when the daily minimum temperature exceeds 20 °C, growing season length as the annual count between the first span of at least six days with daily mean temperatures above 5 °C and the first span after 1 July of six days with daily mean temperatures below 5 °C, the warm spell duration index as the annual count of days with at least six consecutive days when the daily maximum temperature exceeds the respective 90th percentile, and the cold spell duration index as the annual count of days with at least six consecutive days when the daily minimum temperature drops below the respective 10th percentile.

The projected changes in these indices reveal the overall tendency of a future amplification of the extremes related to daily maximum temperature and a future reduction of the extremes related to daily minimum temperature. The number of summer days by the end of the 21st century, for instance, is increased by 11 (24) for RCP4.5 (RCP8.5), while the number of frost days is reduced by 28 (40) (see Table 5.3). The changes are generally stronger for RCP8.5 than for RCP4.5, and for some indices the likelihood ranges based on individual models for the two scenarios do not show any overlap. This is the case for the number of summer days, growing season length and the warm spell duration index. For the cold spell duration index, on the other hand, the likelihood ranges are similar for the two scenarios. The relatively large likelihood ranges for some indices indicate strong variation between the eight (RCP4.5) and nine (RCP8.5) projections with different RCMs that have been considered, and hence a high degree of uncertainty in the projected changes.

Kovats et al. (2014a) presented the geographical distributions of the projected change in the number of heat waves during May through September at the end of the 21st century on the basis of the same set of RCM simulations for the RCP4.5 and RCP8.5 scenarios. Heat waves were defined as periods of more than five consecutive days with daily maximum temperatures exceeding the mean daily maximum temperature for the reference period (1971–2000) by at least 5 °C. For the North Sea region, the only area with notably more frequent heat waves was in southwestern Norway for RCP8.5, for RCP4.5 the number of heat waves does not change in that region. Jacob et al. (2014) defined heat waves differently, in this case as periods of more than three consecutive days with daily maximum temperatures exceeding the 99th percentile of the daily maximum temperature for the same reference period, and found markedly more heat waves over the North Sea region under RCP8.5 at the end of the 21st century, with increases in the number of heat waves ranging from 10 to 15 for the Netherlands, northern Germany and Denmark, and exceeding 30 in southern Norway.

The CMIP5 simulations have also been used to assess the projected change in various temperature-related extremes in several studies, with some of these assessments being included in AR5 (Collins et al. 2013). Sillmann et al. (2013), for instance, presented global maps of the projected change in annual minimum and maximum temperatures (i.e. the minimum of the daily minimum temperatures and the maximum of the daily maximum temperatures occurring in the course of a year), in the number of frost days and in the number of tropical nights, in the number of cold nights (with daily minimum temperatures below the respective 10th percentile) and in the number of warm nights (with daily maximum temperatures exceeding the respective 90th percentile), as well as in the cold and warm spell duration indices at the end of the 21st century for the RCP2.6, RCP4.5 and RCP8.5 scenarios. Sillmann et al. (2013) also found notable future increases in both the annual minimum and maximum temperatures over western, central and northern Europe. Strong increases in annual minimum temperature of about 9–11 °C for RCP8.5 occur in northern Europe, presumably associated with retreating snow cover in this region. The strongest increases in annual maximum temperature, on the other hand, occur in central and eastern Europe, reaching about 6–8 °C for RCP8.5. Corresponding to these increases in annual temperature extremes, the number of frost days is markedly lower in central and northern Europe, while the number of tropical nights is much higher over southern Europe. In much of the North Sea region the number of tropical nights rises by about 10 days under RCP4.5 and by more than 20 days under RCP8.5, with tropical nights hardly ever occurring in this region under the present-day climate. Similarly, cold spells are projected to become shorter in the North Sea region, by about 3 days for RCP4.5 and about 4–6 days for RCP8.5. Warm spells are projected to become markedly longer in the North Sea region, by about 30 days for RCP4.5 and by 60–120 days for RCP8.5. The national climate change assessment for the Netherlands also considers change in temperature-related extremes (KNMI 2014). In winter, for instance, the number of frost days is projected to decrease by 35–80 % (with respect to 38 days for the reference period 1981–2000) at the end of the 21st century, with the weakest change for the scenario with moderate warming and weak influence of circulation change and the strongest change for the scenario with strong warming and strong influence of circulation change. For the scenarios with strong warming, the differences between a weak and strong influence of circulation change account for 20 % of the projected fall in the number of frost days. The number of summer days, on the other hand, is projected to increase by 30–130 % (relative to 21 days for the reference period 1981–2000), again depending on the overall strength of the scenario.

Kharin et al. (2013) used 20-year return levels to assess future changes in annual extremes of daily temperature at the end of the 21st century for the RCP2.6, RCP4.5 and RCP8.5 scenarios on the basis of the CMIP5 simulations for the entire globe. In the North Sea region, the multi-model ensemble projects increases in the 20-year return levels of the annual minimum temperatures of 4–8 °C for RCP4.5 and 8–12 °C for RCP8.5. The projected increases in the 20-year return levels for annual maximum temperature in the North Sea region are somewhat weaker, at 2–4 °C for RCP4.5 and 6–8 °C for RCP8.5. Nikulin et al. (2011) used 20-year return levels to assess future change in annual extremes of daily temperature in Europe at the end of the 21st century on the basis of an ensemble of six scenario simulations with one particular RCM forced by six different GCMs applying the SRES A1B scenario (Fig. 5.2). According to these scenario simulations, the 20-year return levels for annual minimum temperature increase by about 4–10 °C over most of the North Sea region, while the respective return levels for the annual maximum temperature increase only by about 2–4 °C. Nevertheless, waiting times for a 20-year event of the annual maximum temperature during the reference period (1961–1990) are reduced to 2–5 years in the North Sea region, meaning that at the end of the 21st century such an event is expected to occur every two to five years.

Schoetter et al. (2014) assessed changes in the characteristics of western European heat waves projected in the CMIP5 ensemble at the end of the 21st century. In this case heat waves were defined as periods of three consecutive days, during which at least 30 % of western Europe is affected by extremely high temperatures (exceeding the 98th percentile of the daily maximum temperatures for the period May through October). The study covers the UK, Belgium, the Netherlands and northern Germany as parts of the North Sea region. Heat waves in western Europe become more frequent and of greater duration, increase in extent and

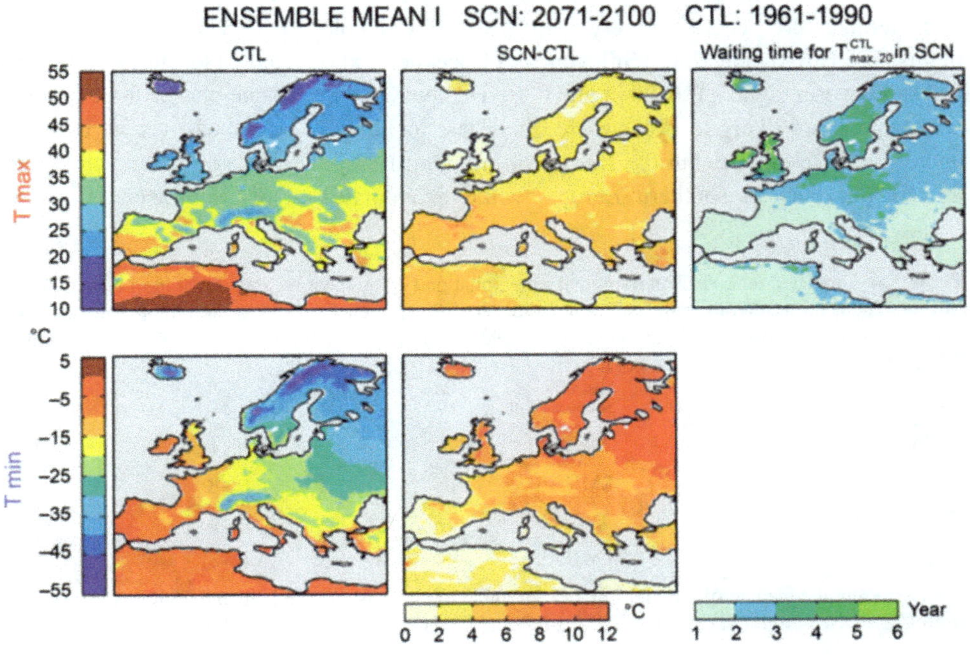

Fig. 5.2 *Left-hand panels* The ensemble mean of (*upper panel*) the 20-year return level of daily maximum temperature ($T_{max,20}$) and (*lower panel*) the 20-year return level of daily minimum temperature ($T_{min,20}$) for 1961–1990 and (*middle panel*) the respective changes of $T_{max,20}$ and $T_{min,20}$ in 2071–2100 relative to 1961–1990 (°C). Only differences significant at the 10 % significance level are shown. *Right-hand Panel* Waiting times (years) of the 1961–1990 $T_{max,20}$ in 2071–2100 (Nikulin et al. 2011)

become more intense. Heat waves that are similar to or stronger than the one observed across Europe in 2003 remain rare under RCP2.6 and RCP4.5, but become the norm under RCP8.5. For the latter, heat waves with five times the severity of the 2003 heat wave were simulated. The severity of heat waves is described by the so-called cumulative heat wave severity, which is defined as the product of the number of heat waves during a 30-year period and the mean severity of the individual heat waves. The latter is defined as the product of the duration, the mean extent and the mean intensity of the respective heat wave. Most of the changes in the temperature-related extremes during summer are partly associated with corresponding changes in the variation in temperature over the course of a day (diurnal cycle) as well as variations in temperature from day to day. According to Cattiaux et al. (2015), both diurnal variability and day-to-day variability in summer temperature increase under the different RCP scenarios, with extremely strong variations over both time scales occurring more frequently. In western Europe, for instance, diurnal and day-to-day variability both increase by about 10 % under the RCP8.5 scenario, with weaker increases over northern Europe of up to 6 %. The increases in variability are primarily linked to a future decrease in surface evapotranspiration as a consequence of drier European summers.

Several extremes related to daily temperature were identified in the SREX report with high confidence for northern Europe (Seneviratne et al. 2012). For instance, the frequency of warm days is very likely to increase, but not as much as in central and southern Europe (Fischer and Schär 2010), there are very likely to be fewer cold days (with daily maximum temperatures below the respective 10th percentile) and a likely increase in the 20-year return levels of annual maximum temperature. There are very likely to be fewer cold nights (Kjellström et al. 2007; Sillmann and Roeckner 2008) and more warm nights (Tebaldi et al. 2006). Heat waves and warm spells are likely to occur more often, last for longer and/or be more intense, but the changes in northern Europe are smaller than in southern Europe, while Scandinavia shows little change at all (Beniston et al. 2007; Koffi and Koffi 2008; Fischer and Schär 2010; Orlowsky and Seneviratne 2012).

5.4 Precipitation

Wilhelm May

5.4.1 Mean Precipitation

At a global scale, the CMIP5 simulations project increases in precipitation in the tropics as well as at mid and high latitudes, and a decrease in the sub-tropics (Knutti and Sedláček 2012). For the North Sea region, the multi-model ensemble projects an increase in winter and a decrease in summer except for Denmark and southern Norway. This tendency is also evident in the projected changes in precipitation for northern and central Europe based on the CMIP5 simulations presented in Annex I of AR5 (IPCC 2013) for the cold (October through March) and warm (April through September) seasons. For the cold season, the RCP4.5 scenario is characterised by increases of up to 10 % in the North Sea region at the end of the 21st century, and the changes projected exceed natural variability over the entire region. For the warm season, on the other hand, precipitation is projected to decrease by up to 10 % in England, Belgium, the Netherlands and northern Germany and to increase by up to 10 % in Denmark and southern Norway. However, the changes projected during the warm season do not exceed natural climate variability anywhere across the region. Averaged over northern Europe, the projected increase in precipitation during the cold season ranges from 8 % (RCP4.5) to 11 % (RCP8.5) for the mid-21st century and from 11 % (RCP4.5) to 20 % (RCP8.5) at the end of the 21st century (see Table 5.4). Precipitation averaged over northern Europe during the warm season is increased, ranging from 3 to 4 % for the mid-21st century and 5–8 % at the end of the century.

Table 5.4 Projected relative changes in mean precipitation (%) by the mid- and end of the 21st century (2046–2065 and 2081–2100, with respect to 1986–2005) for northern Europe (see Seneviratne et al. 2012, their Fig. 3.1) for RCP4.5 (42 models) and RCP8.5 (39 models) obtained from the CMIP5 simulations, distinguishing between the cold season (October through March; ONDJFM) and warm season (April through September; AMJJAS)

Period	Season	RCP4.5	RCP8.5
2046–2065	ONDJFM	8 (3–11)	11 (8–15)
	AMJJAS	3 (2–8)	4 (1–10)
2081–2100	ONDJFM	11 (7–14)	20 (15–29)
	AMJJAS	5 (2–8)	8 (2–12)

Data represent the median of the multi-model ensemble of changes and the 25th and 75th percentiles of the individual model responses. Adapted from Christensen et al. (2013a, their Table 14.1) and Christensen et al. (2013b, their Table 14.SM.1c), respectively

During winter some precipitation in the North Sea region falls as snow. As conditions warm, the fraction falling as snow is expected to decrease. According to Brutel-Vuilmet et al. (2013) the CMIP5 simulations are characterised by several snow-related changes in the mid-latitudes of the northern hemisphere at the end of the 21st century. Between 40° and 60°N the RCP scenarios project a decrease in solid precipitation of about 10 % (RCP2.6) to 30 % (RCP8.5), despite a marked rise in total precipitation at these latitudes. Consistent with this, snow depth declines by about 10 % (RCP2.6) to 40 % (RCP8.5), and the snow season shortens with the decrease ranging from up to a fortnight (RCP2.6) to a month or more (RCP8.5). Räisänen and Eklund (2012) presented consistent results for northern Europe based on an ensemble of regional climate scenarios applying the SRES A1B scenario from the ENSEMBLES project (e.g. van der Linden and Mitchell 2009). They identified future decreases in snowfall and snow depth across all low-altitude parts of northern Europe, including Denmark and southern Norway as part of the North Sea region.

The characteristic changes in precipitation over Europe were also revealed in the multi-model ensemble of high-resolution RCM simulations for Europe used by Jacob et al. (2014). For the RCP4.5 scenario, seasonal mean precipitation in the North Sea region increases in winter and spring by about 10–15 % at the end of the 21st century (Fig. 5.3). In summer and autumn, on the other hand, precipitation increases (exceeding 5 %) in south-western Norway, but there is little change in the rest of the North Sea region, ranging between a slight decrease (of less than 5 %) in the south to a slight increase (of less than 5 %) in the north. Averaged over the Atlantic region, annual mean precipitation increases slightly (1 %) for RCP4.5 and more notably for RCP8.5 (see Table 5.5). For the RCP4.5

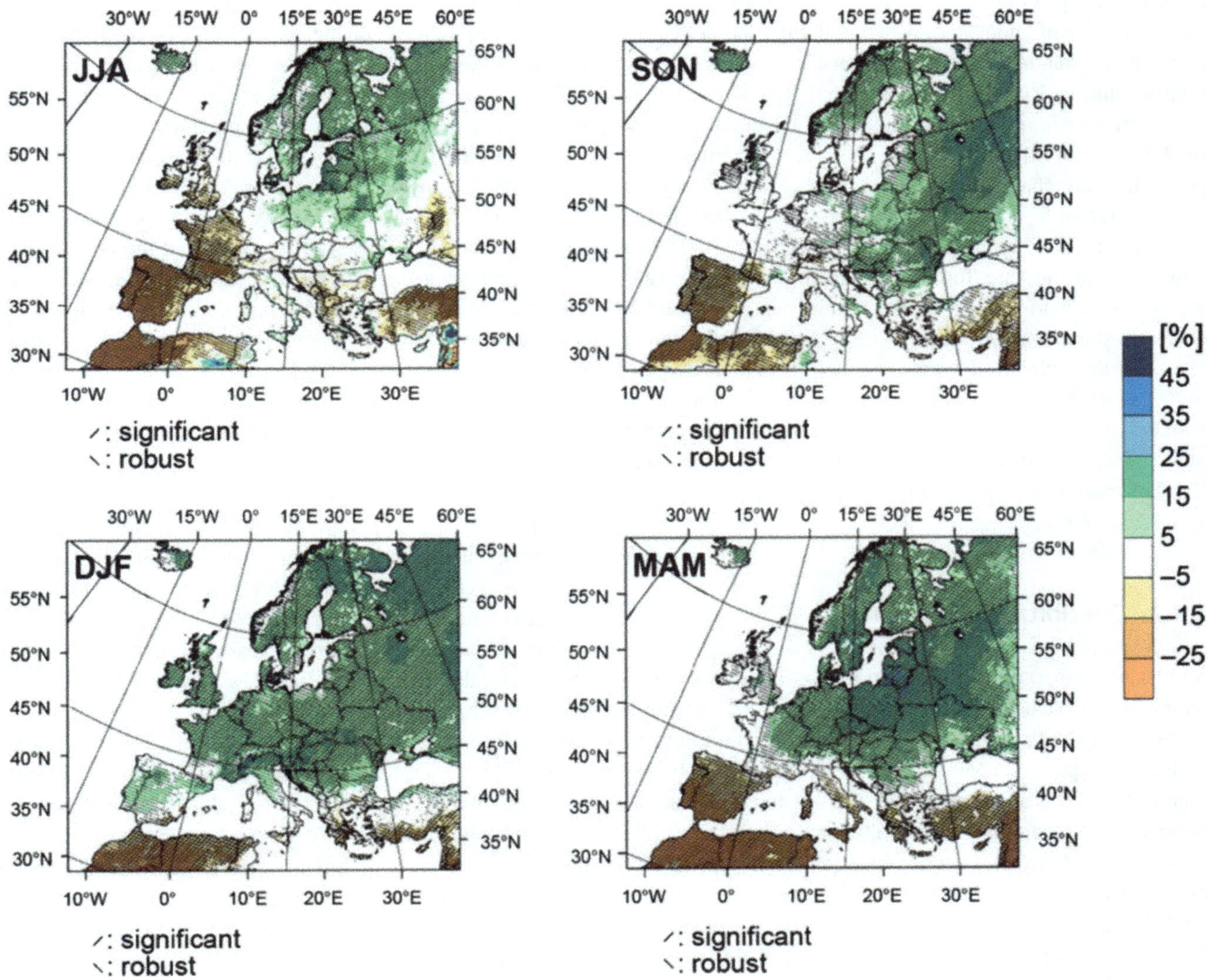

Fig. 5.3 Projected seasonal change in precipitation (%) based on the RCP4.5 scenario for the period 2071–2100 relative to 1971–2000. Hatched areas indicate regions with robust and/or statistically significant change. From the supplementary material of Jacob et al. (2014)

Table 5.5 Projected change in precipitation-related variables and indices for the end of the 21st century (2071–2100 with respect to 1971–2000) averaged over the Atlantic region for the RCP4.5 (eight RCM simulations) and RCP8.5 (nine RCM simulations) scenarios

	RCP4.5	RCP8.5
Annual total precipitation (%)	1 (−1 to 6)	4 (1 to 7)
Annual total precipitation where daily precipitation exceeds the 99th percentile in 1971–2000 (%)	21 (13 to 44)	43 (32 to 68)

The data represent the median of the multi-model ensemble of changes and the likely range of these changes, defined to include 66 % of all projected changes around the ensemble median. Adapted from Kovats et al. (2014b, their table SM23-3)

scenario, however, one sixth of the different RCM simulations are actually characterised by decreasing precipitation across the Atlantic region.

In Denmark the CMIP5 simulations project increases in seasonal mean precipitation at the end of the 21st century in all seasons except summer (DMI 2014). For summer, the RCP8.5 scenario projects a decrease of about 17 % but with an inter-model standard deviation of 21 %. This scenario projects the strongest increase in winter (18 %), and change in the transition seasons are 10 and 11 %, respectively. For annual mean precipitation, the RCP8.5 scenario projects a future increase of about 7 %, which is slightly larger than the inter-model standard deviation. Consistent with this, the high-resolution RCM simulations used by Wagner et al. (2013) project future increases in annual mean precipitation of 2–6 % in northern Germany. For the Netherlands, the projections are characterised by an increase in annual mean precipitation of 5–7 % with little dependence on the strength of impact of the circulation change (KNMI 2014). This is, however, not the case for changes in the seasonal means, where the scenarios with a strong influence of circulation change project stronger changes in precipitation. In winter, the scenarios with strong warming rate project an increase of 30 % by the end of the 21st century in combination with a strong influence of circulation change and 11 % in combination with a weak impact. In summer, on the other hand, the scenarios with strong warming project reductions of 17 and 4.5 %, respectively.

5.4.2 Precipitation Extremes

Similar to temperature, changes in the variability of precipitation at time scales of up to a season are more relevant in terms of impact than changes in seasonal or annual precipitation. Examples are heavy rainfall at sub-daily or daily time scales, wet spells of several days duration and extended dry periods lasting from one to several weeks or months. On the basis of daily time series of precipitation, Frich et al. (2002) proposed five different indices describing climate extremes related to precipitation in order to coordinate and standardise the definition of such extremes. Zhang et al. (2011) extended this list of extreme precipitation indices to 12, also revising some of the definitions of Frich et al. (2002). These indices often focus on relative thresholds that describe the tails of the distribution rather than on physically-based thresholds.

Kovats et al. (2014b) reported on projected changes in the fraction of the annual precipitation originating from extremely wet days (exceeding the 99th percentile of daily precipitation; Zhang et al. 2011). Averaged over the Atlantic region, this is projected to increase by 21 % (RCP4.5) to 43 % (RCP8.5) at the end of the 21st century (see Table 5.5).

Jacob et al. (2014) considered projected change in precipitation on very wet days (exceeding the 95th percentile of daily precipitation; Zhang et al. 2011), distinguishing between seasons. At the end of the 21st century both the RCP4.5 (Fig. 5.4) and RCP8.5 scenarios project significant increases in the intensity of heavy precipitation events over the entire North Sea region and in all seasons. For RCP4.5 the projected increases are typically 5–15 %, while for RCP8.5 the increases are 15–25 % in all seasons except summer. Jacob et al. (2014) also considered future change in very long lasting droughts (defined as the 95th percentile of the length of dry spells) and found no change in the North Sea region for RCP4.5 and a very small increase of 1–2 days in western Europe for RCP8.5.

The CMIP5 simulations have also been used to project change in various precipitation-related extremes, with some referred to in AR5 (Collins et al. 2013). For instance, Sillmann et al. (2013) presented global maps of future change in very high daily precipitation, defined as the 95th percentile of precipitation on wet days. They found pronounced increases in the intensity of heavy precipitation events over western, central and northern Europe at the end of the 21st century for all RCP scenarios considered, with the smallest increases (about 20 %) for RCP2.6 and the largest (40–70 %) for RCP8.5. The magnitude of the relative changes in the intensity of heavy precipitation events is considerably greater than the corresponding changes in the average intensity of daily precipitation on all wet days. In south-western Europe, the intensity of heavy precipitation events is projected to increase despite a projected decrease in average intensity. Consistent with this, Scoccimarro et al. (2013) projected a relatively strong increase in the fraction of precipitation originating from daily precipitation events in the range between the 90th and 99th percentile in western, central and northern Europe. In winter, the contributions of

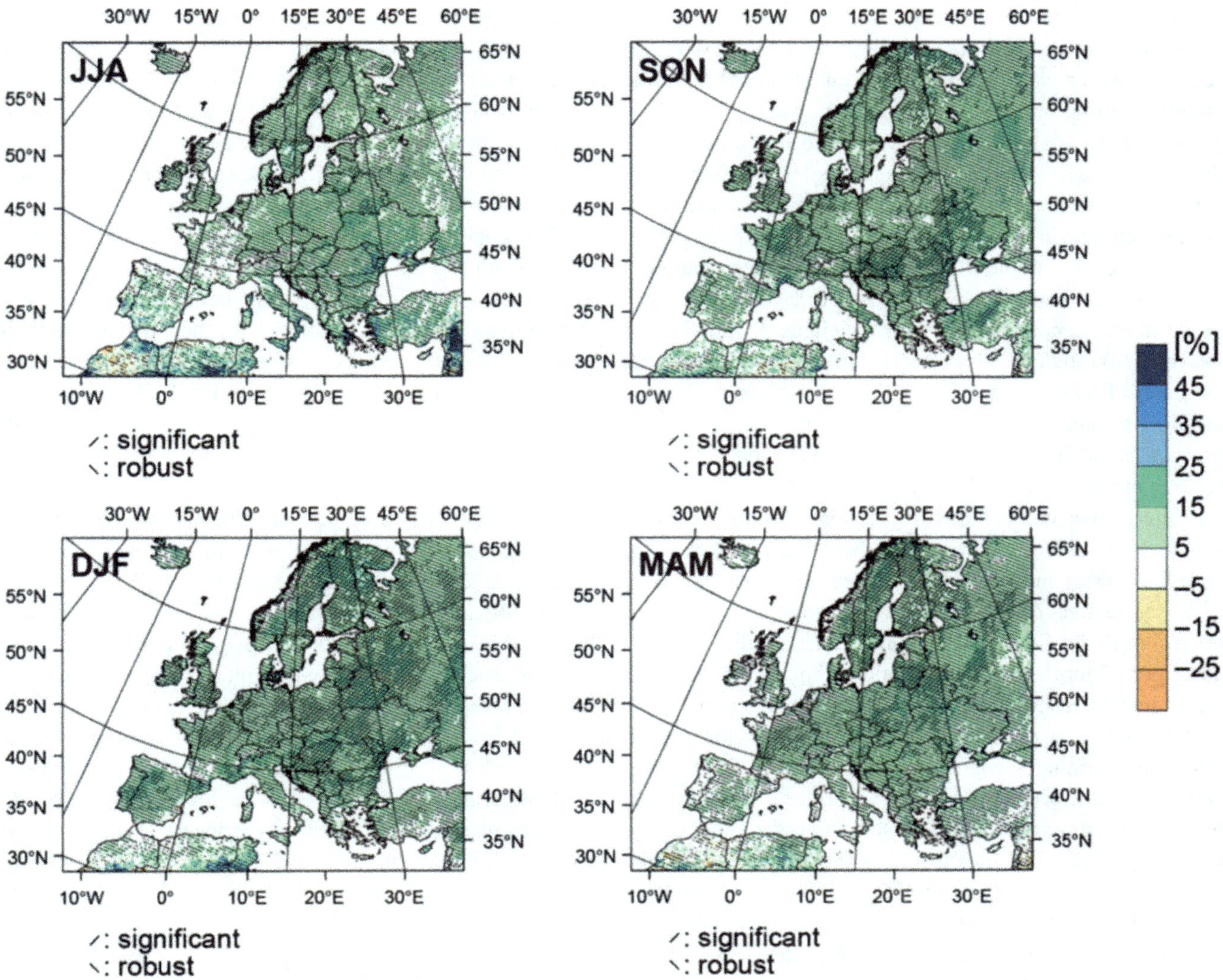

Fig. 5.4 Projected seasonal change in heavy precipitation (%) based on the RCP4.5 scenario for the period 2071–2100 compared to 1971–2000. Hatched areas indicate regions with robust and/or statistically significant change (Jacob et al. 2014)

the heavy daily precipitation events increase by more than 20 % in these areas of Europe, in summer the increases are typically 10–20 % for RCP8.5. It is only in summer that the intensity of heavy precipitation events increases in those parts of western, central and northern Europe, where average intensity decreases. In winter, the intensity of heavy daily precipitation events and the average intensity both increase in western, central and northern Europe.

Another way to depict the projected changes in heavy daily precipitation events is in terms of the number of days for which future daily precipitation exceeds a particular high threshold for the reference period. Applying this approach to an ensemble of RCM simulations, Wagner et al. (2013) found that for more than 5 % of days, the amounts of daily precipitation exceeded the 95th percentile for the reference period in north-western Germany in the mid-21st century. Instead of a variable threshold, another approach is to consider a particular amount of daily precipitation. Sillmann et al. (2013), for instance, analysed future change in the number of days with at least 10 mm precipitation and projected an increase in western, central and northern Europe, ranging from about two additional days (RCP2.6) to about six additional days (RCP8.5) at the end of the 21st century. In contrast to Sillmann et al. (2013), who based their analysis on data covering the entire year, KNMI (2014) distinguished between winter and summer and used different thresholds for the two seasons, 10 mm in winter and 20 mm in summer. In winter, KNMI (2014) found more days with at least 10 mm precipitation in the Netherlands, with increases of 14–24 % for the two scenarios with moderate future warming and 30–60 % for the two scenarios with strong future warming. For each of the two rates of future warming the strongest increases are associated with a strong influence of circulation change (i.e. a more predominantly westerly flow). In summer, however, the situation is different, with more days with at least 20 mm precipitation for the two scenarios with a weak influence of circulation change. In the case of a strong influence of circulation change (i.e. a more predominantly easterly flow), the increase in the number of days with at least 20 mm precipitation is less pronounced in some parts of the Netherlands and the number of such days is even reduced in others.

Over the last couple of years, change in precipitation at sub-daily time scales has also become the subject of scientific study. Lenderink and van Meijgaard (2008), for instance, investigated the potential future change in various extremes of hourly and daily precipitation in central Europe during summer in a scenario simulation with a state-of-the-art RCM. As well as identifying much stronger relative increases in hourly precipitation extremes (19–39 % for different percentiles) than in daily precipitation extremes (9–20 % for different percentiles) they found that the projected increases in hourly precipitation extremes exceeded 7 % per degree of warming, which would be expected according to the Clausius-Clapeyron equation, that is, about 14 % per °C for the 99.9th percentile of hourly precipitation. According to KNMI (2014) the summer maximum hourly precipitation is projected to increase by 8–19 % for the two moderate warming scenarios and by 19–45 % for the two strong warming scenarios by the end of the 21st century. In this case, the difference in the influence of circulation change had little effect. The magnitude of the projected absolute changes in extreme hourly precipitation typically simulated by RCMs, however, is probably smaller than what can actually be expected in the future. Kendon et al. (2014) demonstrated that a numerical model operated at a spatial resolution of 1.5 km, which is typical for numerical weather prediction, gives much stronger changes in hourly precipitation extremes during summer than a model operated at a coarser resolution of 12 km. Nevertheless, the relative increases in extreme hourly precipitation of 45 % for the warm scenario combined with a strong impact of the circulation change are of the same order of magnitude as the relative increases projected over the southern part of the UK by Kendon et al. (2014).

Kharin et al. (2013) depicted future changes in extreme daily precipitation events on the basis of the CMIP5 simulations by means of the 20-year return levels for annual maximum daily precipitation. At the end of the 21st century they found an increase in the 20-year return levels of about 10–20 % in the North Sea region for RCP8.5. This means that the annual maximum daily precipitation amounts with a return period of 20 years under present-day climate conditions are likely to occur about every 10–14 years in the future. Nikulin et al. (2011) analysed future changes in 20-year return levels for maximum daily precipitation in winter and summer, when computing the 20-year return levels combining six RCM simulations for Europe. In summer they found changes in the 20-year return level in the range 10–20 % in the North Sea region at the end of the 21st century, and in winter values of 15–30 %. As a consequence, waiting times for a 20-year event under present-day climate conditions are notably more reduced in winter (about 8–12 years) than summer (about 12–16 years).

The projected intensification of heavy daily precipitation in the North Sea region is accompanied by an increase in the mean duration of periods with consecutive dry days. According to Sillmann et al. (2013), the average length of periods with consecutive dry days increases by 1–5 days for the North Sea region under RCP8.5. For RCP4.5, however, there is little change in the average length of periods with consecutive dry days. This is consistent with the findings of Wagner et al. (2013), who identified only very small changes in the average length of periods with consecutive dry days (in this case lasting more than five days) in northern Germany for the mid-21st century under the SRES A1B scenario.

The SREX report (Seneviratne et al. 2012) identified with high confidence very likely increases in both the intensity and frequency of heavy daily precipitation events in northern Europe, accompanied by increases in the fraction of the days with precipitation, for which the daily precipitation exceeds 10 mm, north of 45°N in winter (Frei et al. 2006; Beniston et al. 2007; Kendon et al. 2008). The report also identified a likely increase in the 20-year return levels of daily precipitation in northern Europe.

5.5 Cyclones and Winds

Anette Ganske, Gregor C. Leckebusch, Wilhelm May

5.5.1 Cyclones

Zappa et al. (2013) analysed future projections of the occurrence of extratropical cyclones in the North Atlantic-European sector on the basis of 19 CMIP5 model simulations for both the RCP4.5 and RCP8.5 scenarios. In this study, cyclones were identified and tracked using the objective feature tracking algorithm developed by Hodges (1999). During winter (December through February) the authors identified a tri-polar pattern over Europe with an increase in storm track density over the eastern North Atlantic centred over the British Isles and the North Sea and decreases centred around Iceland and over the Mediterranean Sea (Fig. 5.5). These changes indicate an extension of the Atlantic storm track to the northeast in combination with a narrowing of the storm track over western Europe. These results are in line with the corresponding changes in storm track density on the

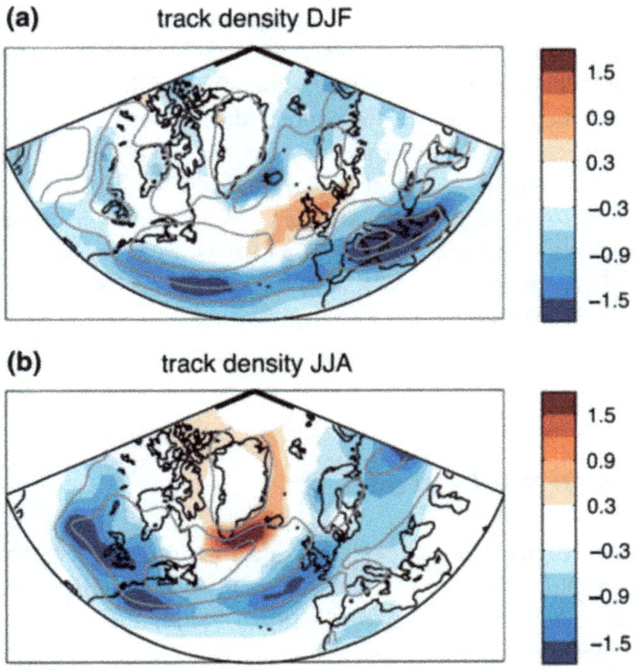

Fig. 5.5 Projected change in mean track density for winter (December through February, DJF; *upper panel*) and summer (June through August, JJA; *lower panel*) based on the RCP8.5 scenario from 19 CMIP5 simulations. Units are number of cyclones per month per unit area. Only responses statistically significant at the 5 % level are shown (Zappa et al. 2013)

basis of the CMIP3 simulations (Ulbrich et al. 2008). The RCP8.5 scenario gives increases in the range 0.6–1.2 cyclones per month in winter at the end of the 21st century over the British Isles, the North Sea and Denmark and only small changes over western Europe. For the RCP4.5 scenario the corresponding changes range between 0.3 and 0.9 cyclones per month. Considering only intense cyclones with pressures below 980 hPa during their lifetimes, Mizuta (2012) found increases of about 0.1 cyclones per month centred over the British Isles for the RCP4.5 scenario on the basis of 11 CMIP5 model simulations. During summer (June through August), on the other hand, Zappa et al. (2013) found an increase in storm track intensity centred between Iceland and southern Greenland and a decrease centred west of the British Isles extending further into the North Sea region (Fig. 5.5). This decrease indicates a marked reduction in the number of cyclones at the southern flank of the storm track over western Europe. For the RCP8.5 scenario the number of cyclones in summer is projected to decrease by 0.6–1.5 cyclones per month over the North Sea and by 0.6–0.9 cyclones per month over western and northern Europe. The RCP4.5 scenario gives increases in the range 0.3–0.6 cyclones per month over the North Sea and about 0.3 cyclones per month over western Europe.

Harvey et al. (2012) assessed the magnitude of projected changes in the Atlantic storm track for both the CMIP3 (SRES A1B scenario) and CMIP5 (RCP4.5 scenario) simulations relative to its typical interannual variations. The storm track was defined via band-pass filtered (2–6 days) variations in the daily surface pressure fields. The authors found that the multi-model ensemble changes in the Atlantic storm track in winter largely agree between the CMIP3 and CMIP5 simulations, when scaling with the respective changes in global mean temperature. The changes simulated by individual models, however, typically have a magnitude similar to the variability at decadal time scales and are locally as strong as the interannual variability. In some parts of the North Atlantic, up to 40 % of the climate models considered were characterised by a positive change in storm track density, exceeding half the magnitude of the interannual variability. With respect to the projected changes in cyclone track density, Ulbrich et al. (2013) noted that part of the uncertainty regarding regional trends in cyclone activity can be related to the choice of a particular method for identification and tracking of cyclones. While different methodologies gave consistent results for intense cyclones, i.e., an increase in the number of cyclones over western Europe in winter, they led to opposing results for weak cyclones with either an increase or decrease in the number of cyclones. According to Chang et al. (2012), the overall tendency of a poleward shift of the Atlantic storm track under future climate conditions is accompanied by an upward extension of the storm track into the upper troposphere and lower stratosphere under the projected global warming, again consistent for the CMIP3 and the CMIP5 simulations.

In a recent review on storminess over the North Atlantic and north-western Europe, Feser et al. (2015) summarised projected changes in both storm frequency and storm intensity on the basis of numerous recent studies that assessed potential future change in these two aspects of storms on the basis of climate scenario simulations with different kinds of models. For the North Sea region, the review considered results from 16 studies published between 1997 and 2013 based on GCMs (either coupled to an ocean model or atmosphere-only) and RCMs with different scenarios for anthropogenic greenhouse gas forcing prescribed. Most of these studies (9 out of 11) showed a future increase in storm frequency, while two found a decrease. Likewise, 10 out of 11 studies showed a future increase in storm intensity; no trend was found in the remaining study. The same trends were also identified over the North Atlantic south of about 60°N, while over northern and central Europe about the same number of studies projected either increases or decreases in storm frequency.

5.5.2 Mean Wind Speeds

The mean winds near the surface (at 10 m height) in the North Sea region are characterised by a clear gradient between the North Sea and the adjacent land areas with considerably higher wind speeds over the ocean than over land, particularly during winter (e.g. Kjellström et al. 2011).

In contrast to other meteorological variables such as precipitation and temperature, very few studies have assessed potential future changes in near-surface winds in response to anthropogenic climate forcing on the basis of scenario simulations originating from GCMs. McInnes et al. (2011) analysed future changes in mean wind speeds at 10 m at a global scale on the basis of the CMIP3 simulations based on the SRES A1B scenario and found an increase in mean wind speeds over both the North Sea and the adjacent land areas in winter at the end of the 21st century, while in summer a notable increase was found over the North Sea only. On an annual basis, mean wind speeds are projected to increase over the entire North Sea region; with the projected changes in mean wind speed typically exceeding 10 %. Despite an overall tendency of increasing mean wind speed in the North Sea region, McInnes et al. (2011) identified marked variations between individual models regarding the sign of the change, particularly in the southern North Sea region. In a recent study, Sterl et al. (2015) analysed projected change in annual mean wind speeds at 10 m over the southern North Sea region for the RCP4.5 and RCP8.5 scenarios using one GCM. In contrast to the overall tendency obtained from the CMIP3 simulations, Sterl et al. (2015) found decreases in annual mean wind speed over the entire region, with little difference between the two scenarios.

More studies exist in which potential future changes in near-surface winds in response to anthropogenic climate forcing for selected regions or continents have been assessed on the basis of scenario simulations with RCMs. The finer spatial resolution not only adds regional detail to the simulations, which is important when looking at the North Sea region, but also affects the magnitude of the projected changes, particularly regarding extreme wind speeds (Winterfeldt and Weisse 2009). This is especially the case when RCMs are applied at very high horizontal resolution. Pryor et al. (2012) showed, for instance, that for the RCA3 RCM an increase in horizontal resolution from 50 to 6.25 km leads to an overall increase in simulated mean near-surface wind speed of 5 % averaged over southern Scandinavia, while the 50-year return level of wind speeds and wind gusts increases by over 10 and 24 %, respectively.

Kjellström et al. (2011) analysed potential future change in mean wind speed on the basis of an ensemble of simulations with the RCA3 RCM driven by six different GCMs for the SRES A1B scenario. These projections are characterised by a small (up to 0.25 ms^{-1}) increase in mean wind speed in the North Sea region in winter but a decrease over land areas and a small increase over the southern part of the North Sea. In particular in winter, the regional distributions of the projected changes vary considerably, both in sign and in strength between the RCA simulations driven by different GCMs. In a similar type of study based on an ensemble of climate projections with the HIRHAM RCM driven by three different GCMs for either the SRES B2 or the SRES A1B scenario, Debernard and Røed (2008) found increases in annual mean wind speed in the North Sea region, reaching up to 2 % over ocean areas.

5.5.3 Wind Extremes

For extremes of near-surface winds, defined via the 99th percentile of daily mean wind speed, the CMIP3 simulations show an overall slight increase (up to 5 %) in the North Sea region during winter and an overall slight decrease (up to 5 %) during summer (McInnes et al. 2011). In this, the projected changes in extreme wind speed are markedly less pronounced than the corresponding changes in mean wind speed when normalised with the climatological values for present-day climate conditions. De Winter et al. (2013) analysed projected changes in annual maximum near-surface wind speed based on scenario simulations with 12 GCMs from CMIP5 for both the RCP4.5 and RCP8.5 scenarios. In contrast to McInnes et al. (2011), they analysed the scenarios from each GCM separately instead of the multi-model ensemble mean. The different GCMs simulated very different changes in the North Sea region, with some models giving either increases or decreases in the intensity of wind extremes over most of the North Sea region and others giving increases over the northern part of the North Sea region and decreases in the southern part. For the RCP8.5 scenario the projected changes typically vary in the range −1.5 to 1.5 ms^{-1}. The individual GCMs simulate not only very different future changes in the intensity of extreme winds, but also very different distributions of the intensity of extreme winds, both with regard to the location of the peak and with regard to the width of the respective probability density functions aggregated over the North Sea.

Donat et al. (2011) presented the projected changes in the intensity of wind extremes (defined via the 98th percentile of daily maximum wind speed) for six different GCMs from CMIP3 for the SRES A1B scenario individually, finding very different changes in the intensity of extreme winds in the North Sea region. The multi-model ensemble mean showed intensified extreme winds in the range 0.25–0.75 ms^{-1} in the North Sea region at the end of the 21st century. Donat et al. (2011) also considered a number of scenario simulations with different RCMs driven by these GCMs, which were part of the ENSEMBLES project (Van der Linden and Mitchell 2009). They found that dynamical downscaling contributed to the uncertainty of the projected changes, as RCMs driven with identical large-scale boundary conditions simulated quite different changes in the intensity of wind extremes.

Nikulin et al. (2011), on the other hand, considered six different scenario simulations with one particular RCM (RCA3) driven by global scenario simulations with six different GCMs for the SRES A1B scenario. Consistent with the studies above, Nikulin et al. (2011) also found very different changes in the 20-year return levels of daily maximum wind speeds in the North Sea region at the end of the 21st century for the individual RCM simulations. The multi-model ensemble means were characterised by a general tendency of more intense wind extremes in the North Sea region. Similarly, Gaslikova et al. (2013) analysed four different scenario simulations with the CCLM RCM driven by four different global scenario simulations with one particular GCM (two realisations of both the SRES B1 and the SRES A1B scenarios). The projected changes in the intensity of extreme winds (defined as the 99th percentile of annual maximum daily wind speeds) over the North Sea were also found to vary considerably between the four scenarios. This was particularly the case for the scenarios driven by the two realisations of the global simulations, where one realisation gave weaker wind extremes over the northern part of the North Sea. The A1B scenario resulted in notably stronger increases in the intensity of extreme winds than the B1 scenario. The multi-model ensemble means are characterised by more intense wind extremes to the south of 58°N, ranging between 0.2 and 0.4 ms^{-1} over most of the area at the end of the 21st century.

The differences between the two realisations over the North Sea are also revealed in the time series of the change in the intensity of wind extremes at different locations in the North Sea for the four different scenario simulations (Fig. 5.6). At the central North Sea location two of the realisations (A1B_2 and B1_2) simulated weaker wind extremes during the entire 21st century, while at the two locations in the German Bight this tendency is only apparent during the first half of the 21st century. The other two realisations (A1B_1 and B1_1), on the other hand, simulated stronger wind extremes during the course of the 21st century. The time series also illustrate the marked internal variability at multi-decadal time scales, making it difficult to identify systematic differences between the SRES A1B and B1 scenario simulations at these locations. For individual 30-year periods, however, marked differences between the A1B and B1 scenarios can occur, i.e., the two realisations A1B_1 and B1_1 at the end of the 21st century.

5.5.4 Wind Direction

McInnes et al. (2011) analysed projected changes in the direction of the mean winds at a global scale on the basis of the CMIP3 simulations. For the North Sea region, they found very small changes in mean wind direction in winter but in summer anticlockwise changes across the entire region, exceeding 15° in the southern areas. The

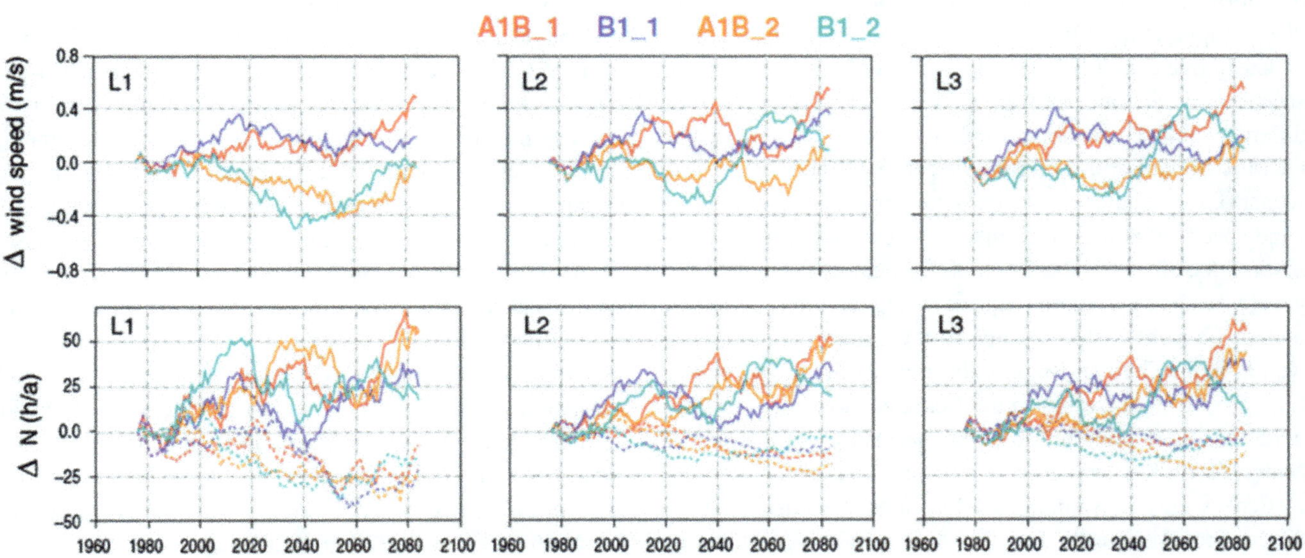

Fig. 5.6 Changes in 30-year running means with respect to 1961–1990 for four different RCM scenario simulations for the annual 99th percentile wind speeds (*upper row*) and (*lower row*) the annual frequencies of strong (≥17.2 ms^{-1}) westerly winds (165–345°; *solid lines*) and strong easterly winds (345–165°; *dashed lines*) at a site in the central North Sea (L1) and two sites in the German Bight (L2 and L3) (adapted from Gaslikova et al. 2013, their Fig. 8)

anticlockwise changes in mean wind direction in the southern North Sea region are consistent between most of the scenario simulations considered.

De Winter et al. (2013), on the other hand, analysed projected changes in the direction of strong winds over the North Sea on the basis of 12 GCMs contributing to CMIP5 for both the RCP4.5 and RCP8.5 scenarios. Strong winds were defined as the annual maxima of daily mean wind speeds and two areas were distinguished, one in the northern part of the North Sea and the other in the southern part. The authors found a common tendency towards less frequent strong winds from south-eastern directions and more frequent strong winds from south-western and western directions in the latter half of the 21st century in both regions for both scenarios. However, it should be noted that due to the rarity of strong wind events the wind direction statistics are characterised by a high degree of variability, which affects the robustness of the projected changes. These changes in the predominant wind directions are consistent with the findings of Donat et al. (2010), who considered storm days (based on daily maximum wind speeds) over western Europe on the basis of six GCMs contributing to CMIP3 for the SRES A1B scenario, and by Sterl et al. (2009) on the basis of a multi-member ensemble of scenario simulations for the SRES A1B scenario with one particular GCM. The projected changes from south-easterly to more south-westerly and westerly winds could indicate a poleward shift in the storm track, because in the North Sea region a storm following a northern track is associated with predominantly westerly winds, while a storm following a more southern track mainly produces south-easterly winds. Both the CMIP3 and CMIP5 simulations are characterised by corresponding changes in the storm track in the North Sea region (e.g. Harvey et al. 2012; Zappa et al. 2013).

Gaslikova et al. (2013) used an ensemble of four different scenario simulations with the CCLM RCM driven by four different global scenario simulations with one particular GCM (two realisations of both the SRES B1 and SRES A1B scenarios) to analyse projected changes in the direction of wind speeds of at least 17.2 ms^{-1} (corresponding to 8 Bft) at several locations in the North Sea region. They found a general tendency of more frequent strong westerly winds and of less frequent easterly winds in the central North Sea as well as in the German Bight in the course of the 21st century (Fig. 5.6). The decreases in the frequency of strong easterly winds are more pronounced in the German Bight than in the central North Sea, while increases in the frequency of strong westerly winds are similar at all locations. The time series of the projected changes for the four scenario simulations reveal both strong temporal variability at multi-decadal time scales and notable differences between the individual scenario simulations, illustrating the important role of internal variability for regional assessments of future change in the characteristics of storms.

5.6 Radiation and Clouds

Burkhardt Rockel, Wilhelm May

Few recent publications describe projected changes in radiation and clouds. Also, the RCMs and GCMs used to derive these projections, the emission scenarios used in the projections, and the time periods analysed are quite diverse. The projected changes are presented separately for solar and terrestrial radiation as well as for cloud cover, with similarities between these changes highlighted. The numbers presented in this section are typically estimated from the geographical distributions that cover a much larger area than the North Sea region, such as the globe or the entire European continent.

5.6.1 Solar Radiation

For annual mean net downward solar radiation at the surface, all studies show a distinct pattern with a decrease in the northern North Sea region and an increase in the south. This tendency is found regardless of which climate model or scenario is used or which time period is considered and so can be considered a robust result. The magnitude of the projected changes in the two areas varies between studies, however.

With increasing numbers of climate scenario simulations available from different coupled climate models, estimates based on a multi-model ensemble are often taken into account. Henschel (2013), for instance, considered results from 39 GCMs from CMIP5 for the RCP8.5 scenario and found a median decrease of about 0.1 Wm^{-2} per year for the southern part of the North Sea region for the multi-model ensemble, corresponding to a decrease of about 4 Wm^{-2} by the middle of the 21st century. In contrast, Henschel (2013) did not find any significant trend for the northern North Sea region (north of about 58°N) until the middle of the 21st century and so did not give any estimate of the change at that point in time.

Trenberth and Fasullo (2009) and Zhou et al. (2009) both considered results from multi-model climate simulations for the SRES A1B scenario to assess projected change until the end of the 21st century. However, the two studies considered different time periods and different parts of the atmosphere. Trenberth and Fasullo (2009) analysed projected change in annual mean net solar radiation at the top of the atmosphere and found an increase in absorbed solar radiation of up to 6 Wm^{-2} in the southern North Sea region and a decrease of up to 1.5 Wm^{-2} in the northern region in the period 2000–2100. Zhou et al. (2009) analysed solar radiation at the surface and found an increase in net surface solar radiation of up to 4 Wm^{-2} in the southern North Sea region and a decrease of up to 6 Wm^{-2} in the north for 2080–2099 relative to 1900–1919.

Ruosteenoja and Räisänen (2013) analysed projected changes in solar radiation from the CMIP3 multi-model ensemble for the SRES A1B scenario by the end of the 21st century, distinguishing between seasons. In their supplemental material, Ruosteenoja and Räisänen (2013) also presented changes by the end of the 21st century for the SRES A2 and B1 scenarios as well as changes by the middle of the 21st century (2020–2049) for the SRES A1B scenario. In contrast to the previously mentioned studies, Ruosteenoja and Räisänen (2013) presented changes relative to the present-day climate (1971–2000) rather than absolute changes. According to their estimates, a relative reduction of 15 % at about 60°N corresponds to a decrease of less than 3 Wm^{-2}. According to their results, the pattern of projected changes at the middle of the 21st century varies little with season except for winter with a decrease in solar radiation over almost the entire North Sea region and a strongest decrease of about 5 %. During summer, on the other hand, solar radiation is increased almost everywhere, with a strongest increase of about 2.5 %. The variations between season are more pronounced at the end of the 21st century than for the mid-century. During both summer and autumn the characteristic north-south structure is evident with a decrease in solar radiation in the northern part (about 5 %) of the North Sea region and an increase in the south (about 10 %; Fig. 5.7). During winter, on the other hand, solar radiation is reduced across the entire North Sea region, particularly in the eastern part with reductions of about 10 %, and over 15 % in the north-eastern part. Consistent with this seasonal variation in the projected changes in solar radiation, KNMI (2014) reported pronounced increases in solar radiation in the Netherlands during summer, of 5.5–9.5 % at the end of the 21st century for the scenarios with a strong influence of circulation change (i.e. scenarios with more frequent high-pressure systems). The projected changes in annual mean solar radiation in the Netherlands are small, ranging from −0.8 to 1.4 % for the different scenarios.

Ruosteenoja and Räisänen (2013) found very similar changes for the SRES A2 scenario, for which in contrast to the A1B scenario forcing by anthropogenic sulphate aerosol is not reduced during the latter half of the 21st century, while the water vapour content of the atmosphere is further enhanced due to the stronger global warming. This led the authors to conclude that the projected changes in solar radiation are mainly caused by changes in meteorological conditions, principally changes in cloudiness.

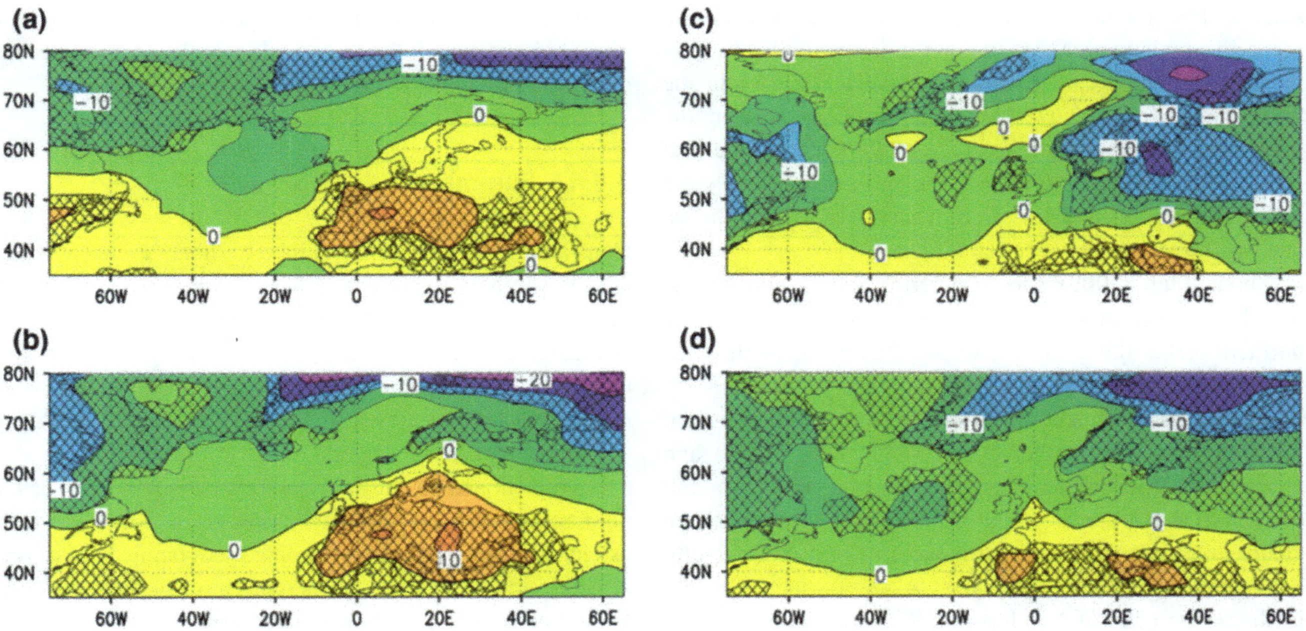

Fig. 5.7 Seasonal change in incident solar radiation (%) from 1971–2000 to 2070–2099 under the SRES A1B scenario as an average of 18 GCMs: **a** summer, **b** autumn, **c** winter, and **d** spring. Areas where more than 85 % of the models (at least 16 of 18 GCMs) agree on the sign of the change are hatched. The contour interval is 5 Wm^{-2}; negative changes are marked by warm colours (*yellow*, *orange* and *red*) and positive changes by cold colours (*green*, *blue* and *purple*) (Ruosteenoja and Räisänen 2013)

5.6.2 Terrestrial Radiation

Compared to solar radiation there are even fewer studies assessing projected changes in terrestrial radiation. This could be considered surprising, since terrestrial radiation plays an important role in the greenhouse effect. Zhou et al. (2009) found, for instance, an increase in annual mean terrestrial radiation across the entire North Sea region, with the increase ranging from 14 Wm^{-2} in the western part to 21 Wm^{-2} in the eastern part by the end of the 21st century. Wild et al. (1997) found a similar pattern with increases of 5–10 Wm^{-2}. Trenberth and Fasullo (2009), who in contrast to other studies considered changes in radiation at the top of the atmosphere, found a decrease in annual mean outgoing terrestrial radiation of about 1.5 Wm^{-2} over the North Sea and about 3 Wm^{-2} over adjacent land areas.

5.6.3 Cloud Cover

Consistent with the projected changes in annual mean net solar radiation at the surface, the aforementioned studies show a distinct pattern with a projected increase in cloud cover over the northern part of the North Sea region and a decrease over the southern part. This can be taken as a robust result, given the different climate models, scenarios and time periods considered. Nevertheless, the magnitude of the projected changes in these two areas does vary between studies. This finding is also supported by the recent RCP scenario simulations. As shown by Collins et al. (2013), both the RCP4.5 and RCP8.5 scenarios give a decrease in the annual mean cloud cover fraction, of up to 5 % in the southern part of the North Sea region by the end of the 21st century for RCP8.5. Moreover, the projected changes are generally weaker (and less significant) over the North Sea itself than over the adjacent land areas.

Zhou et al. (2009) and Trenberth and Fasullo (2009) found a less pronounced effect on cloud cover in the northern part of the North Sea region than in the south. They found a slight increase of up to 0.5 and 0.75 %, respectively, in the northern part, and a considerably stronger decrease of up to 3 % in the southern part. Wild et al. (1997) and Henschel (2013), on the other hand, found a similar amount of change in both areas; about a 2 % increase (decrease) in cloud cover over the northern (southern) part of the North Sea region by the mid-21st century. As these changes are the median from 39 GCMs for the RCP8.5 scenario, these estimates may be considered robust, with two-thirds of the climate models agreeing on a reduction in cloud cover over the southern part of the North Sea region. Consistent with the projected changes in solar radiation, Henschel (2013) did not find any significant trends in cloud cover north of 58°N.

A study by Räisänen et al. (2003) permits a closer look at the North Sea region, as it is based on a set of regional climate simulations for Europe with the RCAO RCM, with lateral boundary conditions originating from two different GCMs for both the SRES A2 and B2 scenarios. By the end of the 21st century they found an increase in annual mean cloud cover of up to 8 % in the northern part of the North Sea region and a decrease of up to 8 % in the southern part. The projected changes in cloud cover are particularly strong during summer, with a typical reduction of 12–20 % in the southern part of the North Sea region, depending on the driving GCM and the scenario used. In the northern part, on the other hand, cloud cover typically increases by 4–12 %. In this, the projected future changes during summer are considerably stronger than during winter. Furthermore, the general structure of the patterns of projected change varies little between the different simulations in summer, emphasising the robustness of these projections. In winter, on the other hand, the patterns of simulated changes in cloud cover are strongly affected by the choice of driving GCM. While the simulations driven by HadAM2H project an increase in cloud cover over all land areas with the exception of the British Isles, the simulations driven by ECHAM4/OPYC project a slight decrease in most of this area. The only exception is the respective simulation for the SRES B2 scenario with enhanced cloud cover over western Europe. According to these results, the projected changes in cloud cover during winter are not as robust as those during summer, presumably owing to the greater uncertainty in the projected changes in the large-scale circulation over Europe due to natural climate variability.

5.6.4 Summary

Considering all the results reported here, a line of zero change can be roughly drawn from the Firth of Forth to the Skagerrak with a tendency for net solar radiation to decrease (increase) in the region to the north (south) of this line. Consistent with this the same zero-line separates areas with an increase (decrease) in cloud cover in the northern (southern) part of the region.

As mentioned in the introductory paragraph, the actual numbers given here for the North Sea region have been estimated from the corresponding geographical part presented in the respective studies, with most of them covering the entire globe. A study on the projected changes in radiation and clouds focusing on the North Sea region is still missing. With the multi-model ensemble of regional climate simulations for Europe, which have become available through CORDEX (Jacob et al. 2014), such a study might be undertaken in the future. Ruosteenoja and Räisänen (2013) took a first step in this direction by considering the changes of the solar radiation for northern and southern Europe separately. Given the importance of the projected changes in cloud cover in the North Sea region, further investigations on the changes in specific cloud properties, i.e. the vertical distribution with low-, mid and high-level clouds or the phase of the clouds (liquid and ice), might help to understand the physical mechanisms behind the projected changes.

5.7 Conclusions

Wilhelm May

The climate projections considered in this chapter reveal changes in the state of the atmosphere in the North Sea region, both in the free atmosphere and near the surface. The changes mostly concern conditions at the end of the 21st century (with the end of the 20th century or the turn of the 20th and the 21st centuries as the baseline), although some relate to the mid-21st century. They comprise:

- Amplification and an eastward shift in the pattern of NAO variability in autumn and winter.
- Changes in the storm track with increased cyclone density over western Europe in winter and reduced cyclone density on the southern flank in summer.
- More frequent strong winds from westerly directions and less frequent strong winds from south-easterly directions.
- A marked mean warming of 1.7–3.2 °C for different scenarios, with stronger warming in winter than in summer and relatively strong warming over southern Norway.
- Intensified extremes related to daily maximum temperature and reduced extremes related to daily minimum temperature, both in terms of strength and frequency.
- An increase in mean precipitation during the cold season and a reduction during the warm season.
- A pronounced increase in the intensity of heavy daily precipitation events, particularly in winter.
- A considerable increase in the intensity of extreme hourly precipitation in summer.
- An increase (decrease) in cloud cover in the northern (southern) part of the North Sea region, resulting in a decrease (increase) in net solar radiation at the surface.

It should be noted that the uncertainty ranges of the future changes projected by the climate scenarios vary between the different meteorological variables. The uncertainty range is particularly large for the projected changes in wind speed and in wind direction, both for mean winds and for wind extremes. Hence, the projected changes in wind characteristics are typically within the range of natural variability and can even have opposite signs for different scenarios either simulated by different climate models or for different future periods.

The projected changes in future climate presented here for the North Sea region have typically been extracted from geographical distributions for either the entire globe, when scenario simulations with GCMs are considered, or for Europe, when scenario simulations with RCMs are used. In some of the respective studies, however, different parts of Europe were considered separately, typically distinguishing between northern and southern Europe. With the multi-model ensemble of regional climate simulations for Europe, which have become available through CORDEX (Jacob et al. 2014), such studies with a special focus on the North Sea region could become available in the near future. The studies considered here vary widely in the choice of underlying scenarios for anthropogenic climate forcing, namely the different SRES scenarios and RCP scenarios. There is, however, a tendency to focus on the SRES A1B scenario in previous studies and the RCP4.5 and RCP8.5 scenarios, respectively, in the most recent studies. Also, the studies vary considerably in the time periods chosen, both for the present-day and future climate conditions, which can make it difficult to directly compare the magnitude of corresponding projected changes between studies. In particular, some studies focus on projections to the middle of the 21st century instead of the end of the 21st century, while some consider projections for both periods. This chapter mostly reports on changes projected at the end of the 21st century. This is mainly because for most of the forcing scenarios the projected changes are stronger at the end of the century, which means there is a higher probability of the projected regional changes exceeding the range of internal variability at that point. Moreover, the differences between RCP scenarios, in particular between the RCP2.6 and RCP4.5 scenarios, develop during the latter half of the century.

Several factors contribute to the uncertainties in the projected changes, that is, the uncertainty in the climate forcing due to different scenarios, the model uncertainty associated with different climate models, and the uncertainty due to the natural variability of the climate system. By coordinating the simulation of future climate scenarios by different research groups in initiatives such as CMIP3, CMIP5 or CORDEX or in the ENSEMBLES project, the importance of some of these sources of uncertainty can be quantified, ultimately leading to estimates of the likelihood at which certain climatic changes can be expected to occur. With the increase in computer power, climate models have been improved in several respects. In particular, components such as vegetation and marine biogeochemical cycles have been added to coupled climate models leading to the development of earth system models (ESMs) and the horizontal and vertical resolutions of both global and regional climate models have been improved, allowing better representation of certain processes in these models. Furthermore, the ongoing development of regionally coupled model systems with an RCM interactively coupled to an ocean model could improve the presentation of climate processes over the North Sea and, hence, the quality of climate simulations for the North Sea region.

References

Barnes E, Hartman DL (2010) Influence of eddy-driven jet latitude on North Atlantic jet persistence and blocking frequency in CMIP3 integrations. Geophys Res Lett 37:L23802. doi:10.1029/2010GL045700

Barnes E, Polvani L (2013) Response of the midlatitude jets, and of their variability, to increased greenhouse gases in the CMIP5 models. J Clim 26:7117–7135

Barnes E, Slingo J, Woollings T (2012) A methodology for the comparison of blocking climatologies across indices, models and climate scenarios. Clim Dyn 38:2467–2481

Beniston M, Stephenson DB, Christensen OB, Ferro CAT, Frei C, Goyette S, Halsnaes K, Holt T, Jylhä K, Koffi B, Palutikof J, Schöll R, Semmler T, Woth K (2007) Future extreme events in European climate: an exploration of regional climate model projections. Clim Change 81:71–95

Brutel-Vuilmet C, Ménégoz M, Krinner G (2013) An analysis of present and future seasonal Northern Hemisphere land snow cover simulated by CMIP5 coupled climate models. Cryosphere 7:67–80

Cattiaux J, Vautard R, Cassou C, Yiou P, Masson-Delmotte V, Codron F (2010) Winter 2010 in Europe: a cold extreme in a warming climate. Geophys Res Lett 37:L20704. doi:10.1029/2010GL044613

Cattiaux J, Douville H, Peings Y (2013) European temperatures in CMIP5: Origins of present-day biases and future uncertainties. Clim Dyn 41:2889–2907

Cattiaux J, Douville H, Schoetter R, Parey S, Yiou P (2015) Projected changes in diurnal and interdiurnal variations of European summer temperatures. Geophys Res Lett 42:899–907

Chang EKM, Guo Y, Xia X (2012) CMIP5 multimodel ensemble projection of storm track change under global warming. GCMs. J Geophys Res 117:D23118. doi:10.1029/2012JD018578

Christensen JH, Hewitson B, Busuioc A, Chen A, Gao X, Held I, Jones R, Kolli RK, Kwon W-T, Laprise R, Magaña Rueda V, Mearns L, Menéndez CG, Räisänen J, Rinke A, Sarr A, Whetton P (2007) Regional climate projections. In: Solomon S, Qin D, Manning M, Chen Z, Marquis M, Averyt KB, Tignor M, Miller HL (eds) Climate Change 2007: The physical science basis. Contribution of Working Group I to the Fourth Assessment Report of the Intergovernmental Panel on Climate Change. Cambridge University Press

Christensen JH, Krishna Kumar K, Aldrian E, An S-I, Cavalcanti IFA, de Castro M, Dong W, Goswami P, Hall A, Kanyanga JK, Kitoh A, Kossin J, Lau N-C, Renwick J, Stephenson DB, Xie S-P, Zhou T (2013a) Climate phenomena and their relevance for future regional climate change. In: Stocker TF, Qin D, Plattner G-K, Tignor M, Allen SK, Boschung J, Nauels A, Xia Y, Bex V, Midgley PM (eds) Climate Change 2013: The Physical Science Basis. Contribution of Working Group I to the Fifth Assessment Report of the Intergovernmental Panel on Climate Change. Cambridge University Press

Christensen JH, Krishna Kumar K, Aldrian E, An S-I, Cavalcanti IFA, de Castro M, Dong W, Goswami P, Hall A, Kanyanga JK, Kitoh A, Kossin J, Lau N-C, Renwick J, Stephenson DB, Xie S-P, Zhou T (2013b) Climate phenomena and their relevance for future regional climate change; Supplementary Material. In: Stocker TF, Qin D, Plattner GK, Tignor M, Allen SK, Boschung J, Nauels A, Xia Y, Bex V, Midgley PM (eds) Climate Change 2013: The Physical Science Basis. Available from www.climatechange2013.org and www.ipcc.ch

Collins M, Knutti R, Arblaster J, Dufresne J-L, Fichefet T, Friedlingstein P, Gao X, Gutowski WJ, Johns T, Krinner G, Shongwe M, Tebaldi C, Weaver AJ, Wehner M (2013) Long-term climate change: Projections, commitments and irreversibility. In: Stocker TF, Qin D, Plattner GK, Tignor M, Allen SK, Boschung J, Nauels A, Xia Y, Bex V, Midgley PM (eds) Climate Change 2013: The Physical Science Basis. Contribution of Working Group I to the Fifth Assessment Report of the Intergovernmental Panel on Climate Change. Cambridge University Press

Croci-Maspoli M, Schwierz C, Davies H (2007) Atmospheric blocking: space-time links to the NAO and PNA. Clim Dyn 29:713–725

Davini P, Cagnazzo C (2013) On the misinterpretation of the North Atlantic Oscillation in CMIP5 models. Clim Dyn 43:1497–1511

De Winter RC, Sterl A, Ruessink BG (2013) Wind extremes in the North Sea basin under climate change: an ensemble study of 12 CMIP5 GCMs. J Geophys Res 118:1601–1612

Debernard JB, Røed LP (2008) Future wind, wave and storm surge climate in the Northern Seas: a revisit. Tellus 60A:427–438

Deser C, Phillips A, Burdette V, Teng H (2012) Uncertainty in climate change projections: The role of internal variability. Clim Dyn 38:527–546

DMI (2014) Fremtidige klimaændringer I Danmark. Dansk Klimacenter rapport 14-06, DMI, Copenhagen, Denmark

Donat MG, Leckebusch GC, Pinto JG, Ulbrich U (2010) European storminess and associated circulation weather types: future changes deduced from a multi-model ensemble of GCM simulations. Clim Res 42:27–43

Donat MG, Leckebusch GC, Wild S, Ulbrich U (2011) Future changes in European winter storm losses and extreme wind speeds inferred from GCM and RCM multi-model simulations. Nat Hazards Earth Syst Sci 11:1351–1370

Dong B, Sutton RT, Woollings T (2011) Changes of interannual NAO variability in response to greenhouse gases forcing. Clim Dyn 37:1621–1641

Dunn-Sigouin E, Son SW (2013) Northern Hemisphere blocking frequency and duration in the CMIP5 models. J Geophys Res 118:1179–1188

EEA (2012) Climate change, impacts and vulnerability in Europe. European Environment Agency (EEA) Report 12/2012

Feser F, Barcikowska M, Krueger O, Schenk F, Weisse R, Xia L C (2015) Storminess over the North Atlantic and northwestern Europe – A review. Q J Roy Met Soc 141:350–382

Fischer EM, Schär C (2010) Consistent geographical patterns of changes in high-impact European heatwaves. Nat Geosci 3:398–403

Folland CK, Knight J, Linderholm HW, Fereday D, Ineson S, Hurrell JW (2009) The summer North Atlantic Oscillation: past, present, and future. J Clim 22:1082–1103

Frei C, Schöll R, Fukutome S, Schmidli J, Vidale PL (2006) Future change of precipitation extremes in Europe: Intercomparison of scenarios from regional climate models. J Geophys Res 111: D06105. doi:10.1029/2005JD005965

Frich P, Alexander LV, Della-Marta P, Gleason B, Haylock M, Klein Tank AMG, Peterson T (2002) Observed coherent changes in climatic extremes during the second half of the twentieth century. Clim Res 19:193–212

Gaslikova L, Grabemann I, Groll N (2013) Changes in North Sea storm surge conditions for four transient future climate realizations. Nat Hazards 66:1501–1518

Gillett NP, Fyfe JC (2013) Annular mode changes in the CMIP5 simulations. Geophys Res Lett 40:1189–1193

Giorgi F, Jones C, Asrar GR (2009) Addressing climate information needs at a regional level: the CORDEX framework. Bull World Met Org 58:175–183

Haarsma RJ, Selten F, van Oldenborgh GJ (2013) Anthropogenic changes of the thermal and zonal flow structure over Western Europe and Eastern North Atlantic in CMIP3 and CMIP5 models. Clim Dyn 41:2577–2588

Harvey BJ, Shaffrey LC, Wollings TJ, Zappa G, Hodges KI (2012) How large are projected 21st century storm track changes? Geophys Res Lett 39:L18707. doi:10.1029/2012GL052873

Henschel F (2013) Projections of insolation changes for solar energy power production. Master Thesis, Swiss Federal Institute for Technology, Zurich

Hewitson B, Janetos AC, Carter TR, Giorgi F, Jones RG, Kwon W-T, Mearns LO, Schipper ELF, van Alst M (2014a) Regional context. In: Barros VR, Field CB, Dokken DJ, Mastrandrea MD, Mach KJ, Bilir TE, Chatterjee K, Ebi KL, Estrada YO, Genova RC, Girma B, Kissel ES, Levy AN, MacCracken S, Mastrandrea PR, White LL (eds) Climate Change 2014: Impacts, Adaptation, and Vulnerability; Part B: Regional Aspects. Contribution of Working Group II to the Fifth Assessment Report of the Intergovernmental Panel on Climate Change. Cambridge University Press

Hewitson B, Janetos AC, Carter TR, Giorgi F, Jones RG, Kwon W-T, Mearns LO, Schipper ELF, van Alst M (2014b) Regional context: Supplementary material. In: Barros VR, Field CB, Dokken DJ, Mastrandrea MD, Mach KJ, Bilir TE, Chatterjee K, Ebi KL, Estrada YO, Genova RC, Girma B, Kissel ES, Levy AN, MacCracken S, Mastrandrea PR, White LL (eds) Climate Change 2014: Impacts, Adaptation, and Vulnerability; Part B: Regional Aspects. Contribution of Working Group II to the Fifth Assessment Report of the Intergovernmental Panel on Climate Change. Available from www.ipcc-wg2.gov/AR5 and www.ipcc.ch

Hodges KI (1999) Adaptive constraints for feature tracking. Mon Wea Rev 127:1362–1373

Hurrell JW, Deser C (2009) North Atlantic climate variability: The role of the North Atlantic Oscillation. J Mar Syst 78:28–41

Hurrell JW, Kushnir Y, Ottersen G, Visbeck M (2003) An overview of the North Atlantic Oscillation. In: Hurrell JW, Kushnir Y, Visbeck M, Ottersen G (eds), The North Atlantic Oscillation: Climate Significance and Environmental Impact. Geophysical Monograph Series, Vol 134, American Geophysical Union, Washington, DC, 1–35, doi:10.1029/134GM01

IPCC (2012) Managing the risks of extreme events and disasters to advance climate change adaptation. In: Field CB, Barros V, Stocker TF, Qin D, Dokken DJ, Ebi KL, Mastrandrea MD, Mach KJ, Plattner G-K, Allen SK, Tignor M, Midgley PM (eds) Special report of the Intergovernmental Panel on Climate Change. Cambridge University Press

IPCC (2013) Annex I: Atlas of global and regional climate projections. In: Stocker TF, Qin D, Plattner GK, Tignor M, Allen SK, Boschung J, Nauels A, Xia Y, Bex V, Midgley PM (eds) Climate Change 2013: The Physical Science Basis. Cambridge University Press

Itoh H (2008) Reconsideration of the true versus apparent Arctic Oscillation. J Clim 21:2047–2062

Jacob D, Petersen J, Eggert B, Alias A, Christensen OB, Bouwer LM, Braun A, Colette A, Déqué M, Georgievski G, Georgopoulou E, Gobiet A, Menut L, Nikulin G, Haensler A, Hempelmann N, Jones C, Keuler K, Kovats S, Kröner N, Kotlarski S, Kriegsmann A, Martin E, van Meijgaard E, Moseley C, Pfeifer S, Preuschmann S, Radermacher C, Radtke K, Rechid D, Rounsevell M, Samuelsson P, Somot S, Soussana J-F, Teichmann C, Valentini R, Vautard R, Weber B, Yiou P (2014) EURO-CORDEX: new high-resolution climate change projections for European impact research. Reg Environ Change 14:563–578

Kendon EJ, Rowell DP, Jones RG, Buonomo E (2008) Robustness of future changes in local precipitation extremes. J Clim 21:4280–4297

Kendon EJ, Roberts NM, Fowler HJ, Roberts MJ, Chan SC, Senior CA (2014) Heavier summer downpours with climate change revealed by weather forecast resolution model. Nature Clim Change 4:570–576

Kharin S, Zwiers FW, Zhang X, Wehner M (2013) Changes in temperature and precipitation extremes in the CMIP5 ensemble. Clim Change 119:345–357

Kjellström E, Bärring L, Jacob D, Jones R, Lenderink G, Schär C (2007) Modelling daily temperature extremes: recent climate and future changes over Europe. Clim Change 81:249–265

Kjellström E, Nikulin G, Hansson U, Strandberg G, Ullerstig A (2011) 21st century changes in the European climate: uncertainties derived from an ensemble of regional climate model simulations. Tellus 63A:24–40

KNMI (2014) KNMI'14 climate scenarios for the Netherlands: a guide for professionals in climate adaptation. KNMI, De Bilt, The Netherlands

Knutti R, Sedláček J (2012) Robustness and uncertainties in the new CMIP5 climate model projections. Nature Clim Change 3:369–373

Koffi B, Koffi E (2008) Heat waves across Europe by the end of the 21st century: multiregional climate simulations. Clim Res 36:153-168

Kovats RS, Valentini R, Bouwer M, Georgopoulou E, Jacob D, Martin E, Rounsevell M, Soussana J-F (2014a) Europe. In: Barros VR, Field CB, Dokken DJ, Mastrandrea MD, Mach KJ, Bilir TE, Chatterjee K, Ebi KL, Estrada YO, Genova RC, Girma B, Kissel ES, Levy AN, MacCracken S, Mastrandrea PR, White LL (eds) Climate Change 2014: Impacts, Adaptation, and Vulnerability; Part B: Regional Aspects. Contribution of Working Group II to the Fifth Assessment Report of the Intergovernmental Panel on Climate Change. Cambridge University Press

Kovats RS, Valentini R, Bouwer M, Georgopoulou E, Jacob D, Martin E, Rounsevell M, Soussana J-F (2014b) Europe; Supplementary material. In: Barros VR, Field CB, Dokken DJ, Mastrandrea MD, Mach KJ, Bilir TE, Chatterjee K, Ebi KL, Estrada YO, Genova RC, Girma B, Kissel ES, Levy AN, MacCracken S, Mastrandrea PR, White LL (eds), Climate Change 2014: Impacts, Adaptation, and Vulnerability; Part B: Regional Aspects. Contribution of Working Group II to the Fifth Assessment Report of the Intergovernmental Panel on Climate Change. Available from www.ipcc-wg2.gov/AR5 and www.ipcc.ch

Lenderink G, van Meijgaard E (2008) Increase in hourly precipitation extremes beyond expectations from temperature changes. Nature Geosci 1:511–514

Masato G, Hoskins BJ, Wollings T (2013) Winter and summer Northern Hemisphere blocking in CMIP5 models. J Clim 26: 7044–7059

Matsueda M (2011) Predictability of Euro-Russian blocking in summer of 2010. Geophys Res Lett 38:L06801. doi:10.1029/2010GL046557

McInnes KL, Erwin TA, Batholds JM (2011) Global climate model projected changes in 10 m wind speed and direction due to anthropogenic climate change. Atmos Sci Let 12:325–333

Meehl GA, Covey C, Taylor KE, Delworth T, Stouffer RJ, Latif M, McAvaney B, Mitchell JFB (2007a) The WCRP CMIP3 multi-model dataset: a new era in climate change research. Bull Am Met Soc 88:1383–1394

Meehl GA, Stocker TF, Collins WD, Friedlingstein P, Gaye AT, Gregory JM, Kitoh A, Knutti R, Murphy JM, Noda A, Raper SCB, Watterson IG, WeaverAJ, Zhao Z-C (2007b) Global climate projections. In: Solomon S, Qin D, Manning M, Chen Z, Marquis M, Averyt KB, Tignor M, Miller HL (eds) Climate Change 2007: The Physical Science Basis. Contribution of Working Group I to the Fourth Assessment Report of the Intergovernmental Panel on Climate Change. Cambridge University Press

Metzger MJ, Bunce RGH, Jongman RHG, Mücher CA, Watkins JW (2005) A climatic stratification of the environment of Europe. Glob Ecol Biogeogr 14:549–563

Miller RL, Schmidt GA, Shindell DT (2006) Forced annular changes in the 20th century Intergovernmental Panel on Climate Change Fourth Assessment Report models. J Geophys Res 111:D18101. doi:10.1029/2005JD006323

Mizuta R (2012) Intensification of extratropical cyclones associated with the polar jet change in the CMIP5 global warming projections. Geophys Res Lett 39:L19707. doi:10.1029/2012GL053032

Moss RH, Edmonds JA, Hibbard KA, Manning MR, Rose SK, van Vuuren DP, Carter TR, Emori S, Kainuma M, Kram T, Meehl GA, Mitchell JFB, Nakicenovic N, Riahi K, Smith SJ, Stouffer RJ, Thomson AM, Weyant JP, Wilbanks TJ (2010) The next generation of scenarios for climate change research and assessment. Nature 463:747–756

Nakićenović NJ, Alcamo J, Davis G, de Vries B, Fenhann J, Gaffin S, Gregory K, Grübler A, Jung TY, Kram T, Lebre La Rovere E, Michaelis L, Mori S, Morita T, Pepper W, Pitcher H, Price L, Riahi K, Roehrl A, Rogner H-H, Sankovski A, Schlesinger M, Shukla P, Smith S, Swart R, van Rooijen S, Victor N, Dadi Z (2000) IPCC Special Report on Emission Scenarios. Cambridge University Press

Nikulin G, Kjellström E, Hansson U, Strandberg G, Ullerstig A (2011) Evaluation and future projections of temperature, precipitation and wind extremes over Europe in an ensemble of regional climate simulations. Tellus 63A:41–55

Noble IR, Huq S, Anokhin YA, Carmin J, Goudou D, Lansigan FP, Osman-Elasha B, Villamizar A (2014) Adaptation needs and options. In: Barros VR, Field CB, Dokken DJ, Mastrandrea MD, Mach KJ, Bilir TE, Chatterjee K, Ebi KL, Estrada YO, Genova RC, Girma B, Kissel ES, Levy AN, MacCracken S, Mastrandrea PR, White LL (eds) Climate Change 2014: Impacts, Adaptation, and Vulnerability; Part A: Global and Sectoral Aspects. Contribution of Working Group II to the Fifth Assessment Report of the Intergovernmental Panel on Climate Change. Cambridge University Press

Orlowsky B, Seneviratne SI (2012) Global changes in extremes events: Regional and seasonal dimension. Clim Change 110:669–696

Pryor SC, Nikulin G, Jones C (2012) Influence of spatial resolution on regional model derived wind climates. J Geopsys Res 117: D03117. doi:10.1029/2011JD016822

Räisänen J, Eklund J (2012) 21st century changes in snow climate in Northern Europe: a high-resolution view from ENSEMBLES regional climate models. Clim Dyn 38:2575–2591

Räisänen J, Hansson U, Ullerstig A, Döscher R, Graham LP, Jones C, Meier M, Samuelsson P, Willén U (2003) GCM driven simulations of recent and future climate with the Rossby Centre coupled atmosphere – Baltic Sea regional climate model. SMHI Rep Met Climatol 101

Ruosteenoja K, Räisänen P (2013) Seasonal changes in solar radiation and relative humidity in Europe in response to global warming. J Clim 26:2467–2481

Scaife A, Folland CK, Alexander LV, Moberg A, Knight JR (2008) European climate extremes and the North Atlantic Oscillation. J Clim 21:72–83

Schoetter R, Cattiaux J, Douville H (2014) Changes of western European heat wave characteristics projected by the CMIP5 ensemble. Clim Dyn. doi:10.1007/s00382-014-2434-8

Scoccimarro E, Gualdi S, Bellucci A, Zampieri M, Navarra A (2013) Heavy precipitation events in a warmer climate: Results from CMIP5 models. J Clim 26:7902–7911

Seneviratne SI, Nicholls N, Easterling D, Goodess CM, Kanae S, Kossin J, Luo Y, Marengo J, McInnes K, Rahimi M, Reichstein M, Sorteberg A, Vera C, Zhan X (2012) Changes in climate extremes and their impacts on the natural physical environment. In: Field CB, Barros V, Stocker TF, Qin D, Dokken DJ, Ebi KL, Mastrandrea MD, Mach KJ, Plattner G-K, Allen SK, Tignor M, Midgley PM (eds), Managing the Risks of Extreme Events and Disasters to Advance Climate Change Adaptation. A Special Report of Working Groups I and II of the Intergovernmental Panel on Climate Change (IPCC). Cambridge University Press

Sillmann J, Roeckner E (2008) Indices for extreme events in projections of anthropogenic climate change. Clim Change 86:83–104

Sillmann J, Croci-Maspoli M, Kallache M, Katz RW (2011) Extreme cold winter temperatures in Europe under the influence of North Atlantic atmospheric blocking. J Clim 24:5899–5913

Sillmann J, Kharin S, Zwiers FW, Zhang X, Bronough D (2013) Climate extreme indices in the CMIP5 multimodel ensemble: Part 2. Future climate projections. J Geophys Res 118:2473–2493

Sterl A, van den Brink H, de Vries H, Haarsma R, van Meijgaard E (2009) An ensemble study of extreme surge related water levels in the North Sea in a changing climate. Ocean Sci 5:369–378

Sterl A, Bakker AMR, van den Brink HW, Haarsma R, Stepek A, Wijnant IL, de Winter RC (2015) Large-scale winds in the southern North Sea region: the wind part of the KNMI'14 climate change scenarios. Env Res Lett 10:035004. doi:10.1088/1748-9326/10/3

Taylor K, Stouffer RJ, Meehl GA (2012) An overview of CMIP5 and the experiment design. Bull Am Met Soc 93:485–498

Tebaldi C, Hayhoe K, Arblaster JM, Meehl GA (2006) Going to the extremes. An intercomparison of model-simulated historical and future changes in extreme events. Clim Change 79:185–211

Thompson DWJ, Wallace JM (2000) Annular modes in the extratropical circulation. Part I: Month-to-month variability. J Clim 13: 1000–1016

Trenberth KE, Fasullo JT (2009) Global warming due to increasing absorbed solar radiation. Geophys Res Lett 36:L07706. doi:10.1019/2009GL037527

Trigo RM, Trigo IF, DaCamara CC, Osborn TJ (2004) Climate impact of the European winter blocking episodes from the NCEP/NCAR Reanalyses. Clim Dyn 23:17–28

Tyrlis E, Hoskins BJ (2008) Aspects of a Northern Hemisphere atmospheric blocking climatology. J Atmos Sci 65:1638–1652

Ulbrich U, Christoph M (1999) A shift of the NAO and increasing storm track activity over Europe due to anthropogenic greenhouse gas forcing. Clim Dyn 15:551–559

Ulbrich U, Pinto JG, Kupfer H, Leckebusch GC, Spangehl T, Reyers M (2008) Changing Northern Hemisphere storm tracks in an ensemble of IPCC climate change simulations. J Clim 21:1669–1679

Ulbrich U, Leckebusch GC, Grieger J, Schuster M, Akperov M, Bardin MY, Feng Y, Gulev S, Inatsu M, Keay K, Kew SF, Liberato MLR, Lionello P, Mokhov II, Neu U, Pinto JG, Raible CC,

Reale M, Rudeva I, Simmonds I, Tilinina ND, Trigo IF, Ulbrich S, Wang XL, Wernli H and The IMILAST TEAM (2013) Are greenhouse gas signals of Northern Hemisphere winter extra-tropical cyclone activity dependent on the identification and tracking algorithm? Meteorol Z 22:61–68

Van den Hurk B, van Oldenborgh GJ, Lenderink G, Hazeleger W, Haarsma R, de Vries H (2014) Drivers of mean climate change around the Netherlands derived from CMIP5. Clim Dyn 42:1683–1697

Van der Linden P, Mitchell JFB (2009) ENSEMBLES: Climate change and its impacts: Summary of research and results from the ENSEMBLES project. Met Office Hadley Centre, Exeter

Van Vuuren DP, Edmonds J, Kainuma M, Riahi K, Thomson A, Hibbard K, Hurtt GC, Kram T, Krey V, Lamarque J-F, Masui T, Meinshausen M, Nakicenovic N, Smith SJ, Rose SK (2011) Representative concentration pathways: an overview. Clim Change 109:5–31

Wagner S, Berg P, Schädler G, Kunstmann H (2013) High-resolution regional climate simulations for Germany: Part II – projected climate changes. Clim Dyn 40:415–427

Wiedenmann JM, Lupo AR, Mokhov II, Tikhonova EA (2002) The climatology of blocking anticyclones for the Northern and Southern Hemispheres: Block intensity as a diagnostic. J Clim 15:3459–3473

Wild M, Ohmura A, Cubasch U (1997) GCM-simulated surface energy fluxes in climate change experiments. J Clim 10:3093–3110

Winterfeldt J, Weisse R (2009) Assessment of value added for surface marine wind speed obtained from two regional climate models. Mon Wea Rev 137:2955–2965

Woollings T (2010) Dynamical influences on European climate: An uncertain future. Phil Trans Roy Soc 368A:3733–3756

Woollings T, Blackburn M (2012) The North Atlantic jet stream under climate change and its relation to the NAO and EA patterns. J Clim 25:886–902

Zappa G, Sheffrey LC, Hodges KI, Sansom PG, Stephenson DB (2013) A multimodel assessment of future projections of North Atlantic and European extratropical cyclones in the CMIP5 climate models. J Clim 26:5846–5862

Zhang X, Alexander L, Hegerl GC, Jones P, Klein Tank A, Peterson TC, Trewin B, Zwiers FW (2011) Indices for monitoring changes in extremes based on daily temperature and precipitation data. WIREs Clim Change 2:851–870

Zhou L, Dickinson RE, Dirmeyer P, Dai A, Min SK (2009) Spatiotemporal patterns of changes in maximum and minimum temperatures in multi-model simulations. Geophys Res Lett 36: L02702. doi:10.1019/2008GL036141

Projected Change—North Sea

6

Corinna Schrum, Jason Lowe, H.E. Markus Meier, Iris Grabemann, Jason Holt, Moritz Mathis, Thomas Pohlmann, Morten D. Skogen, Andreas Sterl and Sarah Wakelin

Abstract

Increasing numbers of regional climate change scenario assessments have become available for the North Sea. A critical review of the regional studies has helped identify robust changes, challenges, uncertainties and specific recommendations for future research. Coherent findings from the climate change impact studies reviewed in this chapter include overall increases in sea level and ocean temperature, a freshening of the North Sea, an increase in ocean acidification and a decrease in primary production. However, findings from multi-model ensembles show the amplitude and spatial pattern of the projected changes in sea level, temperature, salinity and primary production are not consistent among the various regional projections and remain uncertain. Different approaches are used to downscale global climate change impacts, each with advantages and disadvantages. Regardless of the downscaling method employed, the regional studies are ultimately affected by the forcing global climate models. Projecting regional climate change impacts on biogeochemistry and primary production is currently limited by a lack of consistent downscaling approaches for marine and terrestrial impacts. Substantial natural variability in the North Sea region from annual to multi-decadal time scales is a particular challenge for projecting regional climate change impacts. Natural variability dominates long-term trends in wind fields and strongly wind-influenced characteristics like local sea level, storm

C. Schrum (✉)
Geophysical Institute, University of Bergen, Bergen, Norway
e-mail: corinna.schrum@gfi.uib.no

J. Lowe (✉)
Met Office Hadley Centre, Exeter, UK
e-mail: jason.lowe@metoffice.gov.uk

H.E.M. Meier (✉)
Research Department, Swedish Meteorological and Hydrological Institute, Norrköping, Sweden
e-mail: markus.meier@smhi.se

C. Schrum · I. Grabemann
Institute of Coastal Research, Helmholtz-Zentrum Geesthacht, Geesthacht, Germany
e-mail: corinna.schrum@hzg.de

I. Grabemann
e-mail: iris.grabemann@hzg.de

H.E.M. Meier
Department of Physical Oceanography and Instrumentation, Leibniz-Institute for Baltic Sea Research (IOW), Rostock, Germany
e-mail: markus.meier@io-warnemuende.de

J. Holt · S. Wakelin
National Oceanography Centre, Liverpool, UK
e-mail: jholt@noc.ac.uk

S. Wakelin
e-mail: slwa@noc.ac.uk

M. Mathis
Max Planck Institute for Meteorology, Hamburg, Germany
e-mail: moritz.mathis@mpimet.mpg.de

T. Pohlmann
Institute of Oceanography, CEN, University of Hamburg, Hamburg, Germany
e-mail: Thomas.Pohlmann@uni-hamburg.de

M.D. Skogen
Institute of Marine Research (IMR), Bergen, Norway
e-mail: morten.skogen@imr.no

A. Sterl
Royal Netherlands Meteorological Institute (KNMI), De Bilt, The Netherlands
e-mail: sterl@knmi.nl

M. Quante and F. Colijn (eds.), *North Sea Region Climate Change Assessment*, Regional Climate Studies, DOI 10.1007/978-3-319-39745-0_6

surges, surface waves, circulation and local transport pattern. Multi-decadal variations bias changes projected for 20- or 30-year time slices. Disentangling natural variations and regional climate change impacts is a remaining challenge for the North Sea and reliable predictions concerning strongly wind-influenced characteristics are impossible.

6.1 Introduction

This chapter addresses projected future changes in the North Sea marine system focussing on three major aspects, namely changes in sea level, changes in hydrography and circulation, and changes in lower trophic level dynamics, biogeochemistry and ocean acidification. Future changes in the North Sea marine system will be driven by a combination of changes induced by the globally forced oceanic boundary conditions and by regional atmospheric and terrestrial changes. Regional changes in sea level are forced by changes in ocean water mass, spatial changes in the Earth's gravitational field, geological changes, changes in thermal and haline characteristics and the corresponding volume changes, and by the redistribution of water masses. Only the final two are accounted for directly or can be derived from General Circulation Models (GCMs, global climate models that are based on models for atmospheric and oceanic circulation). The first three contributions, which could have substantial impacts on regional sea level, must be estimated by a combination of expert judgement and additional methods and complementary models. In some cases, information from GCMs also plays a role and helps to ensure the development of an internally consistent scenario.

Current GCMs and ESMs (Earth System Models, here used for global models) typically simulate changes in climate at a resolution of 100 km or more, and thus often fail to deliver reliable information on regional-scale circulation such as for the North Sea (e.g. Ådlandsvik and Bentsen 2007). Moreover, GCMs and ESMs are not optimised for shelf sea hydrodynamics and biogeochemistry, and some key processes relevant to North Sea dynamics, such as tides and physical and biogeochemical coupling at the sediment-water interface, are typically neglected. A systematic climate change assessment for the North Sea using GCM and ESM model data is therefore not available, except for climate change impacts on sea level (see Sect. 6.2). Detailed and spatially resolved studies of climate impacts on the North Sea system typically use dynamic downscaling approaches employing regional dynamic models. In a study of water level extremes, such as through storm surges, it is usually possible to make use of computationally inexpensive 2-dimensional barotropic models for water levels. A simplified approach is also possible for sea surface waves and a model of the generation and dissipation of wave energy is typically employed. However, for a detailed and spatially resolved investigation of regional climate change impacts on physical and biogeochemical variables a more complex and computationally expensive approach is needed. This requires high resolution 3-dimensional coupled physical-biogeochemical models with appropriate atmospheric forcing (i.e. air-sea fluxes of momentum, energy and matter, including the atmospheric deposition of nitrogen and carbon), terrestrial forcing (volume, carbon and nutrient flows from the catchment area) and data at North Atlantic and Baltic Sea lateral boundaries. The far-field oceanic changes in hydrography and circulation are almost exclusively projected using GCMs and their results from boundary conditions for regional North Sea studies. Oceanic boundary conditions from ESMs are used to project local changes in North Sea biogeochemistry. Dynamically consistent climate change scenarios for terrestrial drivers are still lacking, both at global and regional scales. Therefore, regional studies typically use a combination of forcing GCMs and ESMs, regional downscaling and impact models (see Annexes 2 and 3 for a general review of methods), and expert judgement based on available evidence for future impact scenarios for freshwater and nutrient fluxes from terrestrial sources. These regional studies typically employ a wide range of different methods to correct the regional bias in forcing GCMs or ESMs, which are necessary to ensure a correct seasonality and coupling of local ecosystem dynamics.

In recent years, a range of regional scenarios have been published for the North Sea, addressing changes in sea level, hydrodynamics, productivity and biogeochemistry. The methods applied and processes considered vary greatly from study to study and could substantially affect the changes projected. Therefore, a classification of the most important methodological aspects used within the different subsections is provided and the projected impacts are discussed in relation to the study configuration where necessary.

6.2 Sea Level, Storm Surges and Surface Waves

The Intergovernmental Panel on Climate Change (IPCC) concluded in its fifth assessment (AR5; IPCC 2013) that it is very likely that the mean rate of global averaged sea-level rise (SLR) was 1.7 mm year^{-1} between 1901 and 2010, and 3.2 mm year^{-1} between 1993 and 2010, with tide-gauge and satellite altimeter data consistent regarding the higher rate during the more recent period. While there has been a statistically significant acceleration in SLR since the start of the 20th century of around 0.009 mm $year^{-2}$ (Church and White 2011), rates similar to that of the 1993–2010 period have been observed previously, for instance between 1920 and 1950. In the North Sea, rates of SLR for the 20th century of around 1.5 mm $year^{-1}$ have been estimated (Wahl et al. 2013). Significant future changes in sea level around the world's coastline are expected over the next century and beyond (IPCC 2013). As a global average, and depending on the choice of future greenhouse gas emission scenarios, SLR to 2081–2100 relative to the 1986–2005 baseline period ranges from 0.26 to 0.82 m. Numerous studies (e.g. Bosello et al. 2012; Hinkel et al. 2013) have highlighted the potential impacts in terms of flooding and loss of coastal wetlands, and the potential damage and adaptation costs. This section reviews recent findings on global and European sea-level changes, including the behaviour of storm surges, tides and waves.

6.2.1 Time-Mean Sea Level Change

This sections addresses changes in the time-average sea level, leaving changes in rapidly varying components such as storm surge, tides and sea surface waves to later sections. The current view based on observations from the recent past and future projections by coupled GCMs is a long-term trend of rising sea level with natural variations superimposed on this general trend on a range of time scales and due to a number of physical drivers including atmospheric pressure and wind, and large-scale steric variations (Dangendorf et al. 2014). This variability obscures the detection of regional climate trends (Haigh et al. 2014) both in observations and scenario simulations.

Variations in the time-average sea level can be driven by a number of processes. First, changes in density due to changing temperature and salinity are important for the sea level on a global and regional basis. Thermal expansion occurs as extra heat is added to the water column. Salinity changes in the water column are also important in some regions. In terms of the global average the thermal expansion effect dominates over the salinity effect on sea level. However, both can be important regionally (Lowe and Gregory 2006; Pardaens et al. 2011a). The other major process driving change in time-average sea level is change in total ocean water mass. Over the next century there is likely to be a transfer of water into the ocean from storage on land in mountain glaciers and the Greenland Ice Sheet, and possibly the West Antarctic Ice Sheet. Smaller contributions to sea level change may come from other terrestrial stores, both natural aquifers and man-made reservoirs—although this input is not expected to exceed the contribution from melting land ice. Geological changes, such as changes in the size of ocean basins can also alter global sea level.

Variations in the spatial distribution of sea level are affected by several factors. From an oceanography perspective, changes in the density structure of the ocean and changes in circulation are likely to be associated with changes in the pattern of sea surface height as the ocean seeks to attain a new dynamic balance (e.g. Gregory et al. 2001; Lowe and Gregory 2006; Landerer et al. 2007; Bouttes et al. 2012). From the perspective of geology and solid earth physics, there are also spatial components associated with change in the Earth's gravity field as water moves from storage in land ice into the ocean and movement of the solid Earth as the mass loading on both the land and ocean basins change (e.g. Milne and Mitrovica 1998; Mitrovica et al. 2001). The local and regional deviations from the global mean change can act in both positive and negative directions—in some cases adding to the global mean change and in others offsetting it. Future projections involving changes in water mass distribution must take account of these effects, typically by scaling the global mean change in water mass terms by an appropriate 'fingerprint' (e.g. Slangen et al. 2014). There is also an ongoing change due to the Glacial Isostatic Adjustment (GIA) associated with the last major deglaciation, although this is typically small in most locations compared to most business-as-usual projections for the 21st century. In the southern North Sea, vertical crust movements are negative and correspond to a future sea level increase. In the northern North Sea and along the Norwegian coastline vertical crust movement is positive and leads to a future decrease in sea level. The rate of GIA is roughly linear, with values between −1.5 and +1.5 mm $year^{-1}$ (Shennan and Horton 2002; Shennan et al. 2009), although some higher values may be found (e.g. Simpson et al. 2014). Taking a wide range of physical effects into account the latest IPCC assessment highlighted that, based on the output of predictive models, around 70 % of the global coastline is expected to experience changes within 20 % of the global mean (IPCC 2013). There may also be land movement changes on a more local scale, for instance associated with subsidence caused by ground water or gas extraction.

6.2.2 Range in Global Time-Mean Sea Level Changes

There are three main approaches to considering future global mean sea level changes in current regular use. The first is the use of complex spatially resolved physically based climate models, which attempt to simulate many of the major processes involved in changing sea level. A typical approach (e.g. Yin 2012) uses a GCM to simulate the large-scale evolution of climate over the next century for a range of alternative pathways of future greenhouse gas forcing. The GCM is able to simulate changes in heat uptake and so thermal expansion can be determined from changes in the in situ simulated ocean temperatures or even simulated directly. The simulated atmospheric temperature and precipitation changes can be used as input to separate physical models of glaciers (e.g. Marzeion et al. 2012; Giesen and Oerlemans 2013; Radić et al. 2014), the Greenland Ice Sheet (e.g. Graversen et al. 2011; Rae et al. 2012; Yoshimori and Abe-Ouchi 2012; Nick et al. 2013) and the Antarctic Ice Sheet (e.g. Vizcaíno et al. 2010; Huybrechts et al. 2011; Bindschadler et al. 2013) to estimate their contributions. The key advantage of this modelling approach is that it can address changes in the relative importance of many different physical processes involved. The disadvantage is that the models may not include all of the important physical processes in the coupled systems or may not represent them with sufficient credibility. This is demonstrated by the latest climate model validation tests (IPCC 2013), which show that although the models clearly have skill at representing many aspects of the real observable climate, other aspects differ sizeably between model and observations. In recent years a significant advance has been to close the global sea level budget (Church et al. 2011). As a result, improved estimates became available for the thermal oceanic contribution, for glaciers and land ice contributions and for terrestrial storage. This credible level of knowledge about the different contributions to SLR in the recent past means it is now possible to model these sufficiently well to make projections of future sea-level change.

The second approach to projecting future global sea level uses climate models with reduced complexity. Here a model that represents the global average climate system is often used. A common approach is to solve the global average heat balance for the upper layer of the ocean, with radiative feedbacks supplying heat upwards from the surface and diffusion of heat downwards into deeper ocean layers (e.g. Raper et al. 2001). In complex models, many key quantities, such as climate sensitivity, are emergent properties. In reduced complexity climate models quantities such as climate sensitivity and mixed-layer depth are set as inputs and provide a means of tuning the simple climate models to emulate the global average behaviour of more complex models. Despite the tuning, there are limitations as to how well the simple model structure is able to achieve this (IPCC 2007). The major advantage of reduced complexity climate models is that they are computationally much less expensive than GCMs and so can be used to explore many more scenarios or to simulate much longer periods. The disadvantage is that they may not capture sufficient physics to be used outside their tuned range. Furthermore, most simple models only simulate long-term trends and do not capture interannual variability. Recent use has also involved combining the simpler models' simulation of global mean values with a scaled spatial pattern of change in sea-surface height from the most complex GCMs (Perrette et al. 2013). This offers the ability to interpolate between the GCM results to generate additional scenarios, although these may be less reliable when addressing stabilised forcing cases. Extra care must be taken if this approach is used for extrapolation.

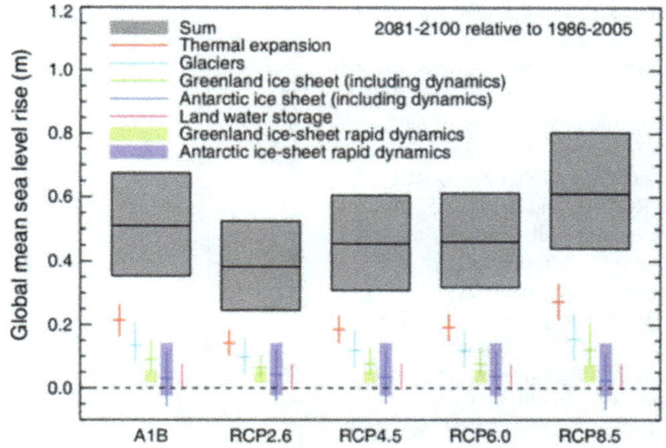

Fig. 6.1 Likely ranges of global mean sea-level rise as reported in the IPCC Fifth Assessment using process based physical models. For comparison, the SRES A1B scenario (the AR4 scenario) has been recalculated using AR5 assessment methods

The IPCC Fifth Assessment (AR5) provides the most comprehensive recent estimates of global SLR from physical models. Figure 6.1 summarises the likely range of 21st century projections. It is important to realise that these ranges are not derived purely from climate models. Expert judgement was used to broaden the range so that model estimates of the 90th percentile range were judged to correspond to the 66th percentile range in the real world. This range is wider than reported in IPCC Fourth Assessment (AR4) although direct comparisons must be undertaken with care, as emission or forcing scenarios, methodologies and even the components of sea level included are different (for emission scenarios see Annex 4). One key difference is that the most recent IPCC assessment (AR5) includes a component from changes in ice dynamics in the likely range of SLR, whereas the previous IPCC assessment (AR4) kept this separate. When this component is included in the AR4 likely range of SLR then for comparable emission or forcing scenarios the two assessments become more similar.

A third class of modelling approach to estimate future global sea level is referred to as semi-empirical and typically uses a relationship derived from observations of sea level and either global temperature (e.g. Rahmstorf 2007) or radiative forcing (e.g. Jevrejeva et al. 2012). By combining the relationship with an estimate of future forcing or surface warming from either a reduced complexity model or a

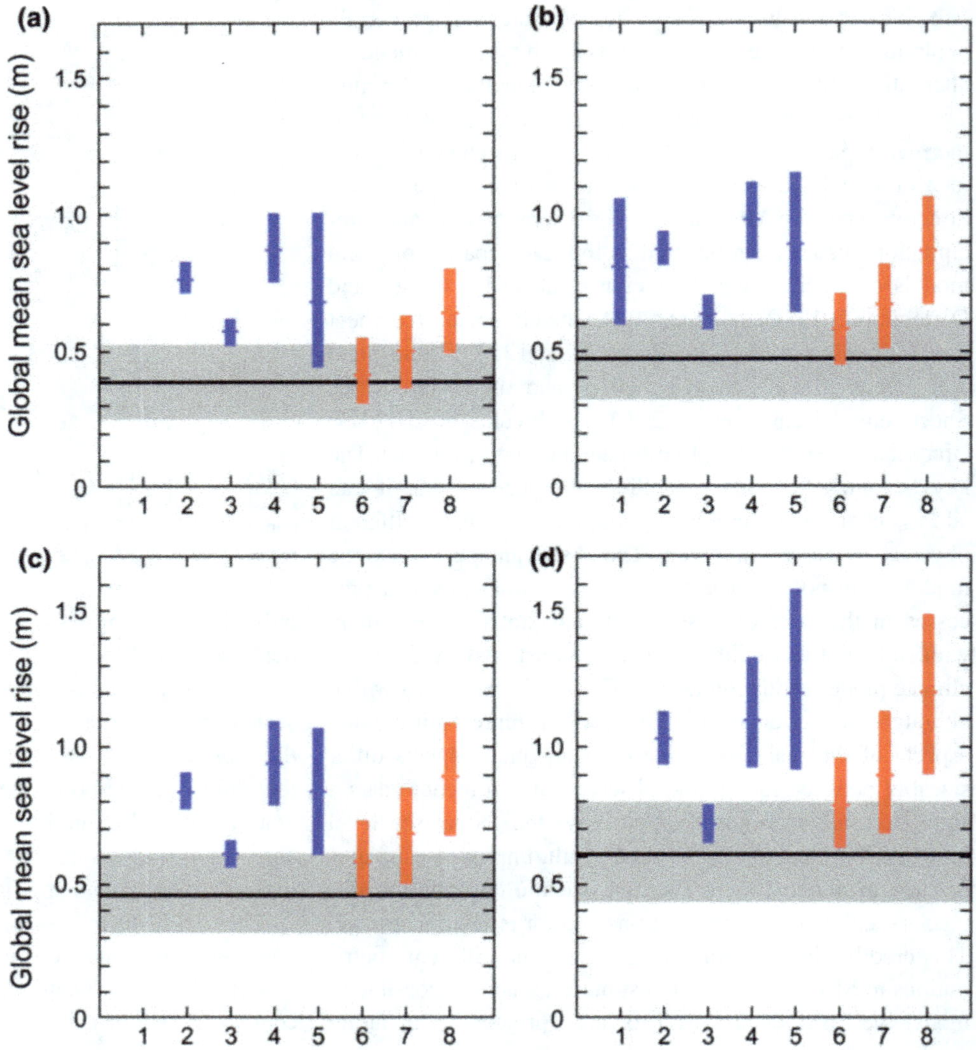

Fig. 6.2 IPCC assessment of the 5–95 % range for projections of global-mean sea level rise (m) at the end of the 21st century (2081–2100) relative to present day (1986–2005) by semi-empirical models for **a** RCP2.6, **b** RCP4.5, **c** RCP6.0, and **d** RCP8.5. *Blue bars* are results from the models using RCP (representative concentration pathway) temperature projections, *red bars* are using RCP radiative forcing. The numbers on the horizontal axis refer to different studies. The likely range (*horizontal grey bar*) from the process-based projections is also shown

complex GCM, an estimate of future sea level can be made. There has been a long debate in the literature (e.g. Lowe and Greogry 2010; Rahmstorf 2010) about the validity of these models. The IPCC AR5 estimate prescribed low confidence in long-term projections from this method (IPCC 2013). However it should be noted that this class of models covers a range of techniques with some likely to be more physically credible than others. Typically semi-empirical methods simulate larger 21st century sea level responses than GCM-based approaches, although there is some recent evidence that ranges estimated from the different approaches are starting to converge (Moore et al. 2013). The range of semi-empirical model estimates in the IPCC AR5 is shown in Fig. 6.2.

It is reasonable to ask if mitigation of emissions will impact significantly on the range of projected future sea level. Recent work has compared the climate response to business-as-usual scenarios, with increasing future emissions and aggressive emission reduction scenarios (Pardaens et al. 2011b; Schaeffer et al. 2012; Koerper et al. 2013). These studies show that mitigation this century (of a size to limit surface warming to no more than 2 °C relative to pre-industrial levels) likely will reduce SLR to 2100 by 25–50 % (Fig. 6.3). Due to the inertia of the climate system larger reductions are expected in the longer term, beyond 2100. However, eventually stabilisation of sea level may not be expected until several hundred years or more after stabilisation of atmospheric greenhouse gas concentrations or radiative forcing (Wigley 2005; Lowe et al. 2006; Levermann et al. 2013). This suggests that to avoid damaging coastal impacts may require both mitigation and adaptation approaches (Nicholls and Lowe 2004). It also raises the question as to whether SLR could be reversed artificially through geo-engineering. Studies such as that by Bouttes et al. (2013) show that the thermal expansion component of SLR can in theory be reversed but that the scenarios of atmospheric greenhouse concentration needed to achieve this are considered unlikely in the next century or so, and possibly even beyond. Land ice melt may be even harder to reverse on a practical time scale because it would take much

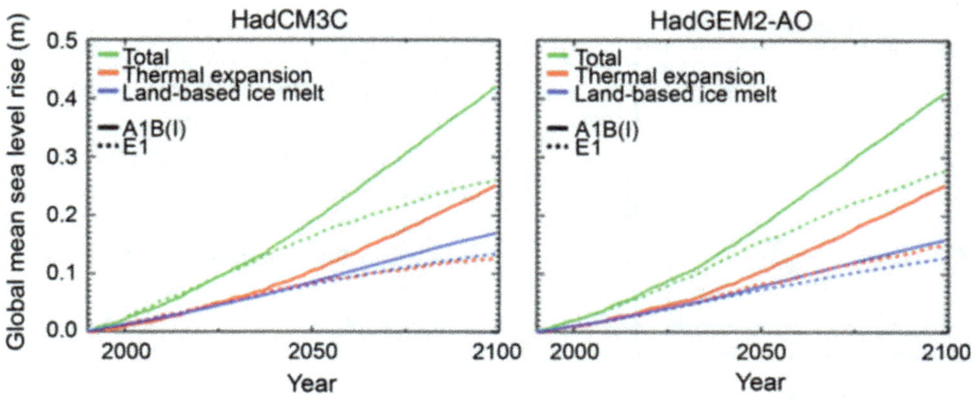

Fig. 6.3 Global mean projections of sea-level rise over the 21st century for the SRES A1B scenario (*solid lines*) and E1 (*dotted lines*) scenarios, together with the thermal expansion and land-based ice melt components. Median projections relative to 1980–1999 are shown for HadCM3C and HadGEM2-AO models (Pardaens et al. 2011b)

longer for the ice sheets to recover, even if greenhouse gas concentrations were significantly reduced (Ridley et al. 2010), than the time needed to reverse thermal expansion.

6.2.3 High-End Estimates of Time-Mean Global Sea Level Change

Another aspect of global mean sea level that has received attention from the adaptation community (e.g. Katsman et al. 2011; Ranger et al. 2013) is the possibility of an increase beyond the likely range projected by physically based climate models. Such a contribution could originate from additional dynamic ice sheet contributions, linked to the movement of fast ice streams and outlet glaciers. Numerous high-end SLR estimates exist (Nicholls et al. 2011) and while the physical processes involved are becoming better understood the global response is still poorly modelled.

Several lines of evidence, such as paleoclimate (Rohling et al. 2008) and consideration of kinematic constraints on ice streams and glaciers (Pfeffer et al. 2008) along with recent consideration of instability of the West Antarctic Ice Sheet suggest it is prudent not to rule out such increases, although the largest increases are considered unlikely. The UK climate assessment in 2009 (UKCP09[1]) (Lowe et al. 2009) concluded that 21st century global sea level increases of up to around 2 m could not be ruled out for design purpose of high risk developments, but clearly stated that rises of under 1 m are much more likely, even in higher emission scenarios. The IPCC AR5 concluded that several tens of centimetres of extra SLR could occur during the 21st century on top of the likely range due to instability of the West Antarctic Ice Sheet, but that other contributions were more unlikely or could not be quantified. When these high-end scenarios are considered, the projected SLR tends to be more similar to that of the semi-empirical method. However, this should not be considered validation of the latter approach because it is unlikely that it is able to capture the physics needed to produce the enhanced rise. Since the publication of the IPCC AR5, evidence has continued to accumulate on the behaviour of the ice sheets and their contribution to future SLR (e.g. Miles et al. 2013; Enderlin et al. 2014; Favier et al. 2014; Khan et al. 2014). This adds further evidence to there being low confidence in the AR5 estimates of the potential contribution of ice sheets to future changes in sea level.

[1]http://ukclimateprojections.defra.gov.uk/.

6.2.4 Time-Mean Sea Level Projections for Europe

Numerous studies report the spatial deviation of regional sea level from the global mean values in GCMs (e.g. Gregory et al. 2001), with a considerable spread between models. Pardaens et al. (2011a) noted the lack of reduction in spread between the third and fourth IPCC assessments. Even the latest IPCC assessment (AR5) shows a wide range in the inter-model spread for regional sea level, although there is some convergence in major features, such as changes across the Antarctic Circumpolar Current and the variations associated with some of the large-scale ocean gyres. Moreover, it is now recognised that this is only part of the total pattern of sea-level response and that locally varying components from changes in land-ice loading must also be included and will further affect the spread (e.g. Simpson et al. 2014).

Two pre-AR5 studies of the North Sea resulted in scenarios of future SLR. Lowe et al. (2009) presented a 5th to 95th percentile range based on IPCC AR4, with a number of regional adjustments. By including scenario uncertainty and model uncertainty they found an increase of 5–70 cm

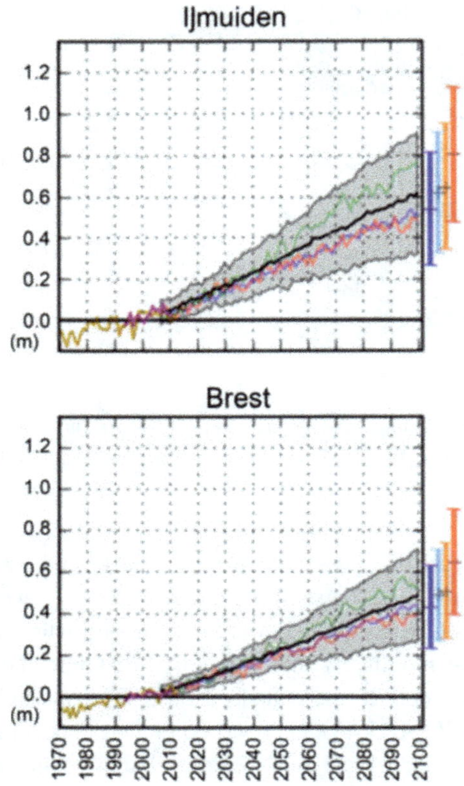

Fig. 6.4 Observed and projected relative net change in sea level for two coastal locations for which long tide-gauge measurements are available. The projected range from 21 RCP4.5 scenario runs (90 % uncertainty) is shown by the shaded region for the period 2006–2100, with the *bold line* showing the ensemble mean. *Coloured lines* represent three individual climate model realisations drawn randomly from three different climate models used in the ensemble. *Vertical bars* at the *right* sides of each panel represent the ensemble mean and ensemble spread (5–95 %) of the likely (medium confidence) change in sea level at each respective location at the year 2100 inferred from RCP2.6 (*dark blue*), RCP4.5 (*light blue*), RCP6.0 (*yellow*), and RCP8.5 (*red*) (IPCC 2013)

relative SLR for Edinburgh and 20–85 cm for the Thames Estuary, both reported for the period 1990–2100 and with the difference between the sites mainly due to different ongoing rates of vertical land movement. Katsman et al. (2008) estimated local increases for the North East Atlantic, for use by planners in the Netherlands. For the moderate climate scenario, they found projected ranges relative to 2005 of 15–25 cm in 2050 and 30–50 cm by 2100. For the warmer climate scenario the corresponding ranges were 20–35 cm in 2050 and 40–80 cm by 2100. In addition to maps of the spatial pattern of change, IPCC AR5 made available some site-specific estimates of future SLR. The time series of the nearest estimates, Ijmuiden in the Netherlands (which is inside the NOSCCA region of interest) and Brest in France (which is outside but near to the NOSCCA region of interest) are shown in Fig. 6.4.

The local time-mean sea-level change values at the end of the 21st century shown in Fig. 6.4 are only slightly different from the global mean estimate for the same scenarios shown in Fig. 6.1. This is not surprising given the IPCC finding that around 70 % of the world's coastline lies within 20 % of the global mean SLR. It also indicates that the global mean estimates for other emission or forcing scenarios can be applied to this European site. Consideration of the spatial patterns also suggests that to a first approximation this value can be applied to the North Sea region.

6.2.5 Future Changes in Extreme Sea Level

Short-lived extreme water levels are often more relevant to many coastal impacts than the time-average changes. A low pressure weather system moving over the North Sea can produce an increase in water level through the inverted barometer effect, and through the winds driving water towards the coastline. The resulting storm surge shows variations on a time scale of a few hours and combines with the tidal water elevations. The highest water levels typically occur with a surge corresponding to the rising limb of the tide rather than the peak of the tide due to non-linear interactions between the tide and surge (Horsburgh and Wilson 2007). The surge is also not a static phenomenon and will move along the coastline as a trapped wave.

Research into future changes in extreme water level uses a range of terminology and sea-surface height metrics, making such estimates difficult to compare. Some studies focus on changes in short-lived extreme water level above present-day mean sea level, while others consider changes in the meteorologically driven surge component only, sometimes expressed as a residual relative to the tidal level but increasingly expressed as changes in the skew surge. Furthermore, some studies refer to return periods while others frame their results as percentiles of the distribution of extreme levels. As the present assessment focuses on identifying the qualitative aspects of past research these complexities should not be a major hindrance.

Changes in extreme coastal water levels can be driven by the time-average sea level changes, which raise the baseline onto which extreme events are added, or by changes in particular atmospheric conditions (e.g. Lowe et al. 2010). There is a strong indication that changes in extreme water levels around the globe during the instrumental record period (about the past 150 years) have been driven predominantly by changes in regional time-mean sea level (Menendez and Woodworth 2010). Similar findings have been published for the English Channel (Haigh et al. 2010). However, there is no way to know a priori whether this will hold in the future, or whether changes in meteorology will alter the characteristics of storm surges. Furthermore, Woodworth et al. (2007) noted a correlation between some aspects of extreme water levels, such as the winter extreme high water level around the UK measured relative to a fixed datum and the winter North Atlantic Oscillation index (NAO index, see Annex 1), a large-scale measure of the atmospheric circulation regime. The pattern of correlation was found to be very similar to that of the correlation of the time-mean water level and the NAO index, although the magnitude was stronger for the winter extreme high water level. As there is sufficient evidence that the changes in extreme water level due to changes in time-mean sea-level rise and changes in storminess combine approximately linearly (e.g. Kauker and Langenberg 2000; Lowe et al. 2001; Howard et al. 2010) over a sizeable range of future sea levels, it is insightful to consider the two components in isolation.

The recent global analysis of Hunter et al. (2013) and extended in the IPCC AR5, examined change in the return period of extreme water level events for a fixed rise in time-mean sea level and a rise following a policy-relevant scenario. Focusing on the European region for a mean SLR of 50 cm, the frequency of extreme events measured relative to a fixed datum in the present day is projected to increase by around a factor of 10 at many sites in the southern North Sea, and by a factor of more than 100 at some points in the northern North Sea. Although the factors can be applied to a range of different return periods of events, this manner of presenting the results must be placed in perspective. The level of protection increase implied by these changes remains less than an 80 cm increase at most locations.

In the EU-funded Ice2sea project (www.ice2sea.eu), Howard et al. (2014) considered how larger regional time-mean sea level increases from enhanced land ice melting might manifest in terms of changes in extreme sea level along the European coastline. Figure 6.5 shows that most of the projected 21st century change in North Sea extreme water levels is likely to come from the time-mean sea-level change. Considering a central ice melt estimate, Howard et al. (2014) found increases in the 50-year return period surge between about 20 and 40 cm. For a high-end scenario, increases in the 50-year return period surge were estimated at around 60 cm and 1 m. The estimated rise was biggest for Esbjerg and smallest for Bergen.

For potential changes in storm surge heights resulting from future changes in meteorology, both modelling approaches (dynamical downscaling and statistical downscaling) are commonly used. It is clear that the large uncertainties about future storm activity in the North Sea (see Chap. 5) are also reflected in future changes in storm surge heights in the North Sea.

A number of early studies looked at the differences between relatively short near present day and future time

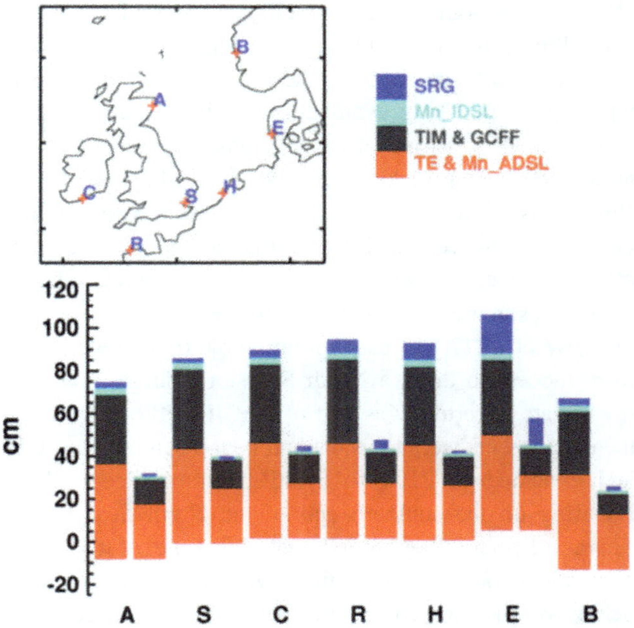

Fig. 6.5 Illustrative addition of high-end and mid-range projections of contributions to changes in the height of the 50-year storm surges in 2100 for seven locations around NW Europe. For each location, the *larger* (*left-hand*) *bar* shows the high-end estimate and the *smaller* (*right-hand*) *bar* shows the mid-range estimate. The projected contribution from Glacial Isostatic Adjustment is shown as an offset to the zero of each bar. The mid-range surge (SRG) projection at Sheerness is negative, and to ensure that this can be seen the mid-range SRG projections are shown as half-width bars. SRG is the change in surge component. Mn_IDSL is the global mean change in dynamic height due to fresh water from ice melt. TIM and GCFF is the ice melt mass component adjusted with a gravitationally consistent fingerprint. TE & Mn_ADSL is the global mean thermal expansion and local dynamic sea surface height pattern. The stations refer to model grid cells, they are close to the following geographical locations: Aberdeen (*A*), Sheerness (*S*), Cork Harbour (*C*), Roscoff (*R*), The Hague (*H*), Esbjerg (*E*) and Bergen (*B*) (Howard et al. 2014)

periods, typically using either barotropic models (Flather and Smith 1998; WASA-Group 1998; Langenberg et al. 1999; Lowe et al. 2001; STOWASUS-Group 2001) or statistical downscaling approaches (Langenberg et al. 1999). Some of these studies suggested significant changes might occur in various measures of extreme water level, although consistency between different studies was not large.

Later studies continued to use the time-slice approach, but focused more on sources of uncertainty. For instance, Lowe and Gregory (2005) attempted to place the results in context by comparing the uncertainty in surge projections with those from other sources, such as uncertainty in mean sea level projections and uncertainty due to emissions scenario choice. Woth (2005) and Woth et al. (2006) analysed simulations for future North Sea storm surge levels for which the forcing data were derived from simulations of the global and regional climate using different global and regional models and the SRES scenarios A2 and B2 (see Annex 4).

However, separating a robust climate signal from natural variability was still problematic. While use of time-slices was a pragmatic approach to the limits of computer power, which prevented long simulations of high resolution atmospheric models, it risks sampling long-period natural variability rather than picking up aspects of a long-term trend. Reanalysis of the 20th century storminess suggested the need for time-slices much longer than a few years or even a couple of decades. Most of the earlier studies also did not credibly estimate uncertainties in the results.

Lowe et al. (2009) used an ensemble of 11 regional climate models to drive a North Sea storm surge model and investigate uncertainty as part of the UKCP09 study. All of the experiments were transient and began before present day and extended to 2100 to avoid the time-slice problem. Focusing on the southern end of the North Sea near the Thames Estuary they found that only one of the model simulations had a statistically significant increase in the height of the 50-year return period storm surge event. However, in physical terms this change of a few centimetres was small compared to the expected time-mean relative change in sea level. This result disagreed with many earlier studies but had the advantage of not needing to use time-slices. A recent reanalysis of the model results for sites outside the United Kingdom (Howard et al. 2014) suggested larger changes in the surge component at some locations, although for sea level extremes the effect of changes in time-mean SLR still typically dominated. Sterl et al. (2009) undertook a similar study using a global model ensemble and found a similar lack of a clear 21st century trend in the storm surge component, adding further weight to the projections from UKCP09. However, an important caveat is that the atmospheric model used for the UKCP09 ensemble was noted to have a particular storm track response; typically showing a southerly movement but with little evidence of an intensification of the storms. While this is one credible future response the possibility of an intensification of storms should not be completely ruled out, because some of the models used in IPCC assessments do show this (Lowe et al. 2009). A simple scaling argument suggested that if the ensemble of driving models had captured the largest increase in storm intensity from additional GCMs available it may have led to a bigger surge increase at some locations, comparable with changes in the future projected time-mean SLR. However, as such large changes in storm intensity were found in only one GCM (using the storm metric applied) the scaled results should be considered a low confidence projection (Lowe et al. 2009).

Gaslikova et al. (2013) investigated a set of four transient regional projections for the North Sea for which the underlying simulations of the global climate includes combinations of one GCM, two initial states and SRES scenarios A1B and B1. Towards the end of the 21st century (2071–2100) they found an increase in extreme surge heights (mean annual 99th percentiles) in the south-eastern North Sea, which are highest in the German Bight by up to about 15 cm. The authors concluded that the increase in the 99th percentile surge height is mainly due to an increase in the frequency of storm events with intensities already occurring in the respective reference climate and that there are relatively few events with greater intensities. 50-year return values calculated from the 100-year long projection period (2001–2100) were compared to 50-year return values calculated from the 40-year long reference period (1961–2000) and resulted in an increase of between about 10 and 80 cm for the two locations examined off the coast of the German Bight (Fig. 6.6). These return values are comparable to those reported by Lowe and Gregory (2005).

Gaslikova et al. (2013) also investigated internal climate variability in North Sea storm surge conditions and found multi-decadal variability within one projection as well as between the four transient projections, which is of the same order of magnitude as the increase towards 2100. Such multi-decadal variability was also found by Weidemann (2009), based on statistical downscaling of 17 projections for the SRES scenario A1B only differing by varying initial conditions. In this study the linear trend over the years 1958–2100 for the five study locations in the German Bight varied between −8 and 18 cm but most of the projections showed an increase in the surge height corrected for time-mean sea-level changes. The trends presented by Gaslikova et al. (2013) for the SRES A1B and B1 projections are within the range presented by Weidemann (2009).

In a recent assessment of the Dutch coastline, KNMI (2014a, b) reported that changes in wind speed are small and that little change is projected over the next century in northerly winds, which are the ones that tend to cause the largest surges along this stretch of coastline. Extremes of water level are expected to continue to rise, however, driven by the rise in time-mean sea level.

Taken together, the more recent studies suggest the possibility of either no significant increase or a relatively small increase in storm surge height in the North Sea. Where an increase is found it is typically largest at the southern end of the North Sea, especially in the south-east, with changes in the western and northern parts of the North Sea being smaller and non-uniform.

It is also useful to consider possible future changes in the propagation of tides due to changes in time-mean sea level. This could be important from both a flood perspective and a consideration of renewable energy generation. The recent study by Pickering et al. (2012) suggests changes in the tides may result in the North Sea due to altered dynamics. They showed that a 2-m SLR would result in a ~ 5 cm increase in M_2 tidal amplitude in the central North Sea and Southern Bight, and a similar decrease in between. An update by

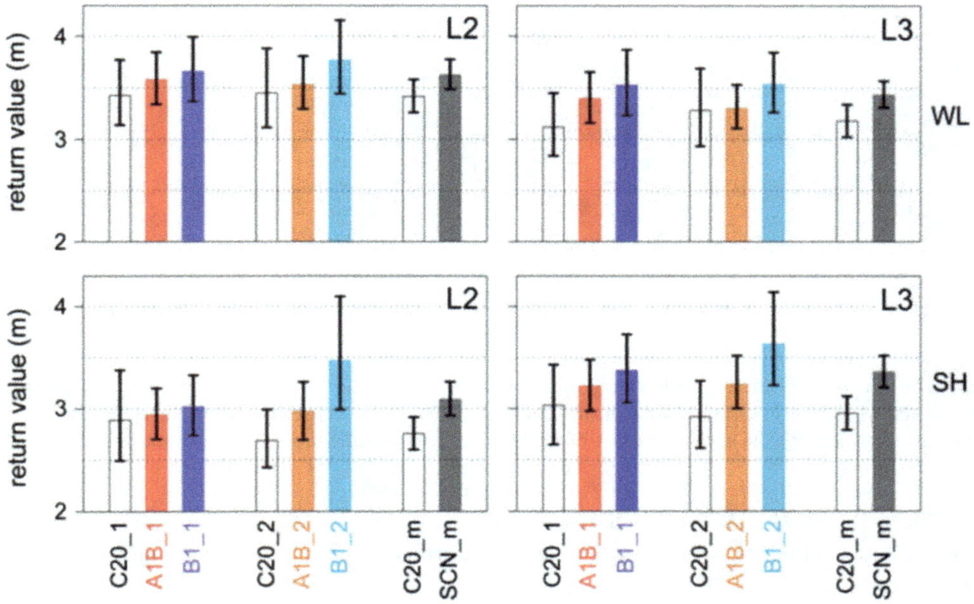

Fig. 6.6 The 50-year return value for water level (WL, *upper panels*) and surge height (SH, *lower panels*) for the control period 1960–2000 from two different ensemble members (C20_1, C20_2) and for four different future scenarios (SRES A1B and B1 scenarios, both simulated using two different GCM ensemble members, for the period 2001–2100). The mean for the control period and the scenario mean are given. The 95 % confidence range for each return value is shown by the *black bars*. L2 and L3 depict locations near the East Frisian and North Frisian coast of the German Bight, respectively (Gaslikova et al. 2013)

Pelling et al. (2013) highlighted a key remaining uncertainty in understanding this response—whether the water is assumed to be contained by a sea wall or allowed to flood the land. Recent work, based on seasonal variations in major tidal constituents (Gräwe et al. 2014; Müller et al. 2014) also suggests the need to consider changes in stratification on the continental shelf in shallow seas, which can alter the eddy viscosity and profile of currents with depth. See Sect. 6.3 for information on how North Sea stratification is projected to change.

6.2.6 Future Changes in Waves

Future changes in waves can be simulated using wind information projected by GCMs and ESMs, sometimes atmospherically-downscaled over the primary region of interest. The studies then typically follow either the statistical approach or the dynamical approach, using models of the generation, transport and dissipation of sea-surface wave energy. Much of the progress in the Northeast Atlantic and the North Sea has used the dynamic wave modelling approach.

Wolf and Woolf (2006) gave a useful overview of how particular aspects of changes in storminess generate changes in the wave climate in the North East Atlantic. The strength of the prevailing westerly winds and the frequency and intensity of storms, the location of storm tracks and the storm propagation speed were all considered. The strength of the westerly winds was found to be most effective at increasing mean and maximum monthly wave height. The frequency, intensity, track and speed of storms have little effect on mean wave height but intensity, track and speed did significantly affect maximum wave height.

The earliest future projection studies, such as those by Rider et al. (1996) of the WASA-Group (1998), used highly idealised climate scenarios, took data from a single or very limited number of climate models and typically used time-slices that were short and did not adequately account for multi-decadal variability. Later studies began to improve their approach, using longer time-slices and modelling policy-relevant future scenarios. The STOWASUS-Group (2001) compared the 30-year time slices 1970–1999 and 2060–2089 for the IPCC scenario IS92a (see Annex 4). For a doubling of carbon dioxide (CO_2) the wave climate responds to projected changes in wind forcing and the mean significant wave height (taken as the mean height of the highest third of waves) increases in the North Sea and north of the British Isles. However, the increase in the mean value throughout the entire year is no more than 15 cm. For extreme waves, expressed as higher percentiles of the distribution of significant wave heights the picture is a more mixed; for the 99th percentile there is an increase of around 0.25–0.5 m in the North Sea, however for the most extreme cases described by the 99th percentile there is little change projected for the North Sea. Debernard et al. (2002) analysed

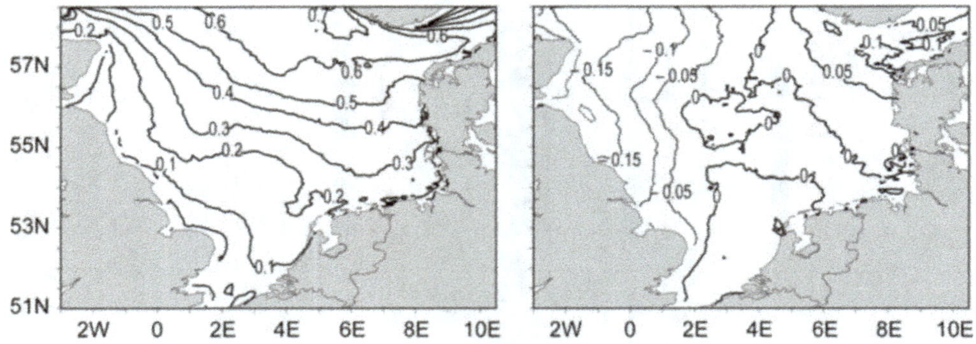

Fig. 6.7 Uncertainties in long-term 99th percentile significant wave height (m) caused by model differences (*left*) and scenario choice (*right*). For the significant wave heights, the uncertainties introduced by different models are generally much larger than those caused by different scenarios. The model uncertainties range from about 0.1 to 0.6 m (adapted from Grabemann and Weisse 2008)

two 20-year time slices for 1980–2000 (reference climate) and 2030–2050 (near future climate). For the global simulations an emission scenario similar to IPCC IS92a was used. The authors reported that the changes in the wave climate for 2030–2050 were mostly small and insignificant.

The next challenge was to try to sample uncertainty in the driving winds by using output from more than one GCM. Debernard and Røed (2008) analysed a set of four climate projections including combinations of SRES scenarios A2, B2, A1B, and three GCMs. The authors compared the results of the 30-year time slices for 1961–1990 (reference climate) and 2071–2100 (future climate) and found changes in the wave conditions for the four climate projections to vary in their spatial pattern and magnitude but that all agree in an increase of severe significant wave heights (99th percentiles) of 6–8 % along the North Sea east coast and in the Skagerrak. Grabemann and Weisse (2008) found comparable changes using a slightly different set of four climate projections, which incorporates two GCMs and SRES scenarios A2 and B2. Comparing the time slices 1961–1990 and 2071–2100 they estimated an increase in extreme wave height (99th percentile) in large parts of the southern and eastern North Sea of about 5–8 % (25–35 cm, average for the four projections). The greatest changes occur in the Skagerrak (an increase of up to 80 cm) in the ECHAM-driven projections. Changes in severe wave height towards the west and north of the North Sea are smaller or even negative. The increase in mean and 99th percentile significant wave height in the eastern North Sea is suggested to result mainly from an increase in the frequency of higher waves. This was also described by Groll et al. (2014). Both Debernard and Røed (2008) and Grabemann and Weisse (2008) reported that model-induced uncertainties and inter-GCM variability are larger than the scenario-related uncertainties (Fig. 6.7). Also Lowe et al. (2009) focused on model uncertainty and used three members of a 17-member GCM ensemble downscaled by a regional model over the North Sea to study future changes in wave heights. Some significant changes in the wave height were noted but further work is needed to understand the patterns. More focus is also needed on how to best select representative ensemble members of the driving GCMs from a larger model ensemble.

Another aspect of uncertainty, the role of natural variability, has been addressed using an 'initial condition ensemble'. De Winter et al. (2012) analysed a 17-member ensemble based on one GCM that was repeatedly started with 17 initial states for the SRES scenario A1B. Again the 30-year time slices 1961–1990 and 2071–2100 were compared. Mean wave heights and wave periods did not change, annual maximum conditions decreased in particular for wave periods and return periods showed no significant change in front of the Dutch coast. Furthermore, the authors found that annual maximum waves propagate more often to easterly directions, which is consistent with an increase in the frequency of extreme westerly winds. The importance of natural variability was investigated by Groll et al. (2014). They used transient projections (1961–2100) to evaluate the internal aspect of climate variability and found strong multi-decadal variability. The changes in median and severe (99th percentile) significant wave heights within a single projection and between projections are of the same order of magnitude as the change (increase in the eastern North Sea) towards the end of the 21st century. Owing to this strong internal variability the largest increase or decrease does not necessarily occur at the end of the 21st century but can occur earlier. Moreover, Groll et al. (2014) noted that the uncertainties from different GCM initial conditions, or arising from the use of different ensemble members are also important.

In a comparative study of ten wave climate projections, including those by Grabemann and Weisse (2008) and Groll et al. (2014) a robust signal was found for the eastern parts of the North Sea where mean and severe wave heights in nine to ten projections tended to increase towards the end of the 21st century (2071–2100). The magnitude of this increase is much more uncertain. For the western parts of the North Sea a decrease is suggested in more than a half of the projections (Grabemann et al. 2015). These findings are in agreement with the results of other studies (e.g. Debernard and Røed 2008). The changes described are also consistent with a projected increase in the frequency of stronger winds from westerly directions.

6.3 Ocean Dynamics and Hydrography

6.3.1 General Aspects and Methodology

North Sea dynamics are controlled by the interplay of the seasonal heating cycle, atmospheric fluxes, tides, river inputs and exchanges with the open ocean. Most physical processes active in the North Sea are to some extent impacted by global change resulting from anthropogenic increases in greenhouse gas emissions. These impacts are, however, highly dependent on time and space scales and the dominant processes under consideration. The impact of climate change in shelf seas is essentially a boundary value problem, due to the shallow depth and short ocean memory relative to the timescale of climate change. Hence it is necessary to consider the external drivers in some detail. These naturally divide into three vectors: atmospheric, oceanic and terrestrial. A fourth important vector is variability in astronomical forcing (top of atmosphere radiation and tidal potential), but this is not a component of anthropogenic change and so not considered in IPCC assessments. However, it should be noted that changes in sea level and stratification will have some effect on local tidal amplitudes and the implications of this require further investigation (e.g. Pickering et al. 2012; Gräwe et al. 2014; see Sect. 6.2). Direct anthropogenic drivers may result as a consequence of climate change adaptation and mitigation measures. These are not specifically considered here, since human effects on the physical marine environment (e.g. arising from the installation of offshore renewable energy structures, mineral extraction, coastal protection measures etc.) tend to be local and/or coastal, and scenarios of anthropogenic drivers not related to climate change have yet to be developed for the North Sea region and integrated into regional future climate change assessments.

To date, the focus of studies to assess potential climate change impacts on the North Sea dynamic system has been on shelf scales (> 10 km from the coast) and seasonal processes. Finer coastal scales and higher frequency processes remain for future work. The downscaling methods and scenarios used are diverse and so this section begins with a short overview of key approaches and methodology. The use of statistical downscaling (von Storch 1995, see Annexes 2 and 3), applied to assess climate change impacts on sea level, storm surges and wave climate (Sect. 6.2) and also frequently marine biota (e.g. Dippner and Ottersen 2001), is unusual for assessing climate change impacts on ocean dynamics and hydrography and all studies reviewed here were undertaken using the more complex and computationally more expensive dynamical downscaling method (see Annexes 2 and 3) using established and validated regional ocean models (ROMs).

The first climate change downscaling studies for the North Sea were performed as research contributions, which focused on method development and provided first quantitative assessments of the potential regional impacts of future climate change (Kauker 1999; Kauker and von Storch 2000; Ådlandsvik 2008; Madsen 2009). These were followed by more comprehensive assessments performed as part of national regional climate change assessments such as the British UKCP09, the German KLIWAS[2] (Auswirkungen des Klimawandels auf Wasserstraßen und Schifffahrt – Entwicklung von Anpassungsoptionen, German Federal Ministry of Transport, Building and Urban Development) and the EMTOX[3] project from the Netherlands (Impacts of climate change effects on natural toxins in plant and seafood production, Dutch Ministry for Economic Affairs, Agriculture and Innovation). In parallel, a few larger European research projects such as the RECLAIM[4] (REsolving CLimAtic IMpacts on fish stocks), ECODRIVE[5] (Ecosystem Change in the North Sea: Processes, Drivers, Future Scenarios) or MEECE[6] (Marine Ecosystem Evolution in a changing Climate) have produced a suite of regional downscaling studies. Most results are published as contributions to peer reviewed literature, but complementary and additional information is available in the form of project reports (e.g. Drinkwater et al. 2008, 2009; Alheit et al. 2012; Wakelin et al. 2012a; Bülow et al. 2014) or made available to the public via the internet (e.g. MEECE via www.meeceatlas.eu).

A wide range of downscaling methods and models (see Table 6.1 for model acronyms) have been applied to assess regional climate change impacts and a best practice on regional marine downscaling is still a matter of research and consensus has so far not been established. The earliest dynamical downscaling exercise using the OPYC model (Kauker 1999; Kauker and von Storch 2000) was carried out well in advance of the IPCC AR4, and utilised GCM forcing from 5-year time slice experiments for a potential 2 × CO_2 world. Most of the more recent regional projections were carried out for the end of the century (2070–2100) and utilise

[2]www.kliwas.de.
[3]www.deltares.nl/en/project/1172392/emtox.
[4]www.climateandfish.eu.
[5]www.io-warnemuende.de/ecodrive.html.
[6]www.meece.eu.

Table 6.1 Model acronyms together with key references

Acronym	Model type	Key publications
BCM Bergen Climate Model	Global climate model, GCM	Furevik et al. (2003)
CCSM3, Community Climate System Model V3	Global climate model, GCM	Public release: www.cesm.ucar.edu/models/ccsm3.0
Delft3D/BLOOM/GEM	Regional model coupled physical-biological	Lesser et al. (2004), Blauw et al. (2008)
DMI-BSHcmod, Danish meteorological institute	Regional ocean model for the North and Baltic seas	Madsen (2009)
DMI HIRHAM RCM	Regional atmospheric model	Christensen et al. (2007)
ECOSMO ECOSystem Model	Regional model coupled physical-biological	Schrum and Backhaus (1999), Schrum et al. (2006), Daewel and Schrum (2013)
ECHAM3/LSG	Global climate model, GCM, first generation coupled model	Roeckner et al. (1992), Maier-Reimer et al. (1993)
ECHAM5-MPIOM, Max-Planck-Institute, Germany	Global climate model, GCM	Marsland et al. (2003); Roeckner et al. (2003, 2006)
ECOHAM ecosystem model Hamburg	Regional ecosystem model	Pätsch and Kühn (2008)
ERSEM	Ecosystem model	Blackford et al. (2004)
GISS, Goddard Institute for Space Studies	Global climate model, GCM	Schmidt et al. (2006)
HadAM3H, Hadley Center Climate Model	Global climate model, GCM	Jones et al. (2001)
HadCM3, Hadley Center Climate Model 3	Global climate model, GCM	Gordon et al. (2000), Pope et al. (2000)
HadRM3 Hadley Center Regional Model 3	Regional model, RCM	Murphy et al. (2009)
HAMOCC, HAMburg Ocean Carbon Cycle model	Ocean carbon cycle model	Maier-Reimer et al. (2005)
HAMSOM HAMburg Shelf Ocean Model	Regional hydrodynamic model	Pohlmann (1996)
IPSL/IPSL-CM4, Institut Pierre-Simon Laplace, France	Earth system model	Marti et al. (2010)
OPYC	Ocean model, isopycnal coordinates	Oberhuber (1993)
MPIOM, Max Planck Institute for Meteorology	Global ocean model	Marsland et al. (2003)
MPIOM-zoom	Global model with Zoom on the North Sea	Gröger et al. (2013)
NORESM, Norwegian Earth System Model	Earth system model	Bentsen et al. (2012)
NORWECOM, NORWegian ECOlogical Model	Ecosystem model	Skogen et al. (1995), Skogen and Søiland (1998)
POLCOMS Proudman Oceanographic Laboratory Coastal Ocean Modelling System	Regional hydrodynamic model	Holt and James (2001)
RACMO	Regional atmospheric model	van Meijgaard et al. (2008)
RCAO	Regional coupled atmosphere-ocean model	Döscher et al. (2002)
RCA4-NEMO	Regional coupled atmosphere-ocean model	Dieterich et al. (2013)
REMO	Regional atmospheric model	Jacob and Podzun (1997)
ROMS	Regional ocean model	Shchepetkin and McWilliams (2005)
WAM	Wave model	Hasselmann et al. (1988)

the SRES scenario A1B (see Annex 4). These experiments were performed either as time slice experiments of 20–30 years for present-day and future (end-of-the-century or middle-of-the-century) climates (Ådlandsvik 2008; Holt et al. 2010, 2012, 2014, 2016; Friocourt et al. 2012; Wakelin et al. 2012a; Pushpadas et al. 2015) or as continuous integrations (e.g. Mathis 2013; Gröger et al. 2013; Bülow et al. 2014; Mathis and Pohlmann 2014). Only one downscaling was performed for the SRES A2 scenario, which considers stronger radiative forcing (Madsen 2009). To date, only the regional ECOSMO model was used to project future changes based on the RCP4.5 scenario (see Annex 4) from IPCC AR5 (Wakelin et al. 2012a; Pushpadas et al. 2015). The downscaling setup and the methods applied for the scenario simulations were different with respect to downscaling chain, coupling of the atmosphere-ocean system, bias correction, consideration of terrestrial climate change impacts, open-ocean climate change impacts, Baltic Sea boundary conditions, and forcing GCM (see Tables 6.2, 6.3 and 6.4 for details). All regional models consider tidal forcing by the M_2 partial tide, which is the major forcing tidal constituent in the North Sea. Most models also consider additional tidal constituents, but the actual tidal setup varies between the different models.

To estimate uncertainties in projections of future climate the multi-model ensemble approach has been introduced in Earth system modelling of the North Sea region following the well-established strategy of IPCC assessments (e.g. Friocourt et al. 2012; Wakelin et al. 2012a; Bülow et al. 2014; Holt et al. 2014, 2016; Pushpadas et al. 2015). Both ensembles using one regional model with different global models (e.g. Wakelin et al. 2012a; Holt et al. 2014, 2016; Pushpadas et al. 2015) and ensemble downscaling from one GCM using different regional model systems are available for the North Sea (Bülow et al. 2014). The ensemble simulations allow for a first estimation of uncertainty arising from different GCMs and RCMs (regional climate models). However, it should be noted that the number of ensemble members is typically only two to three and so too small for a sound final assessment of uncertainty ranges.

Complementary understanding of climate change impacts on the North Sea hydrodynamics and ecosystem dynamics is available from so-called 'what-if' or perturbation experiments that consider hypothetical ranges of forcing parameters. For these numerical experiments, forcing atmospheric boundary conditions (wind speed, air temperature, solar radiation) were separately perturbed by a change roughly of the order of the projected climate change (Schrum 2001; Skogen et al. 2011; Drinkwater et al. 2008) or defined by mixing present day with future forcing GCM variables (Holt et al. 2014, 2016). Such perturbation experiments are not dynamically consistent, but do provide some insight into the sensitivity of the regional system to climate change impacts and so improve process understanding.

6.3.2 Changes in Temperature

Despite huge differences in setup, forcing GCM, bias correction and time slice vs continuous simulations, the future projections for sea-surface temperature (SST) in the North Sea are consistent in sign for the different regional model setups, however there are differences in the magnitude of change. Projected annual mean SST increases for the end of the century are in the range 1–3 °C for the A1B scenario (exact numbers are not given here due to differences in spatial averaging and reference periods from the existing literature). Within the given range, projected temperature changes are consistent for the different regional models used. Projected temperature changes are found to be statistically significant using the Kruskal-Wallis test (Wakelin et al. 2012a) or other measures such as the standard deviation (Ådlandsvik 2008; Mathis 2013) so far investigated. Projected changes in SST are typically more pronounced than changes in depth-averaged (or volume-averaged) temperature, which is the ecologically more relevant parameter since it affects vital rates in organisms that are distributed through the entire water column, and are almost completely driven by changes in atmospheric boundary conditions and air-sea fluxes (e.g. Ådlandsvik 2008; Wakelin et al. 2012a).

A few studies were performed using the same GCM forcing but different regional ocean models and configurations. These use the IPSL-CM4.0 ESM (Wakelin et al. 2012a; Chust et al. 2014; Holt et al. 2014, 2016) and MPIOM (Mathis 2013; Gröger et al. 2013) as global forcing. The resulting changes in SST from these experiments are typically very similar for different regional ocean models and differ only by around a tenth of a degree. On the other hand, ensemble studies performed with one regional ocean model and different forcing GCMs clearly show that the magnitude of the projected changes significantly depend on the forcing GCM (Wakelin et al. 2012a; Holt et al. 2010, 2012, 2014, 2016; Pushpadas et al. 2015; Fig. 6.8). Regional projections using different versions of the Max Planck Institute GCM (ECHAM5/MPIOM) and the Norwegian climate models (BCM and NORESM) are typically at the lower end (Kauker 1999; Wakelin et al. 2012a; Gröger et al. 2013; Mathis 2013; Pushpadas et al. 2015). Stronger warming was projected from simulations using the Hadley-Centre climate model (Holt et al. 2010, 2014, 2016; for the SRES A2 scenario Madsen 2009) and the largest changes were projected when using boundary and initial conditions from the French climate model IPSL-CM4.0 (Wakelin et al. 2012a; Holt et al. 2010, 2012, 2014, 2016; Pushpadas et al. 2015); a GCM that projects stronger warming also on the global scale (e.g. Kharin et al. 2007).

Most of the previously reported downscalings were based on uncoupled ocean downscaling neglecting local atmosphere-ocean feedbacks at the regional scale, which were earlier identified to be potentially important for the North Sea region in a present-day hindcast scenario (Schrum et al. 2003a). To account for these regional air-sea feedbacks, a first multi-model ensemble with coupled atmosphere-ocean regional models (AO regional models) was performed as part of the German climate change impact project KLIWAS (Bülow et al. 2014). Three different

Table 6.2 Uncoupled dynamic downscaling experiments for the North Sea for the end of this century

Model chain: GCM-RAM-ROM	Scenario	Time slice	Bias correction	Runoff/Baltic Sea	Considered forcing: atmosphere-only and atmosphere and ocean change	LTL-model/carbonate chemistry	Related publications
ECHAM3/LSG-no-**OPYG**	2 × CO_2	5-year time slice, 2 × CO_2	Flux correction	Resolved/resolved	A only	No	Kauker (1999), Kauker and von Storch (2000)
BCM-no-**ROMS**	A1B	1972–1997 versus 2072–2097	Sea surface salinity relaxation to BCM	Perturbed by future rainfall/no change	AO change/A-only	No	Ådlandsvik and Bentsen (2007), Ådlandsvik (2008)
HadAM3H-DMI-BSHcmod	A2	1960–1990 versus 2070–2100	Direct, no bias correction	No change/resolved	AO change	No	Madsen (2009)
HadCM3-HadRM3-POLCOMS	A1B	1961–1990 versus 2070–2098	Adjusted fluxes	Future change/no change	AO change	No	Lowe et al. (2009), Holt et al. (2010)
HadCM3-no-**POLCOMS**	A1B	1980–1999 versus 2080–2100	Bias corrected temperature	Runoff perturbed by rainfall/no change	AO change/A only	**ERSEM**/no	Wakelin et al. (2012a), Holt et al. (2014)
IPSLCM4-no-**POLCOMS**	A1B	1980–1999 versus 2080–2100	Bias corrected temperature	Perturbed by future rainfall/no change	AO change	**ERSEM**/yes	Artioli et al. (2012, 2013, 2014), Holt et al. (2012, 2014, 2016); www.meeceatlas.eu
IPSLCM4-no-**POLCOMS**	A1B	1980–1999 versus 2080–2100	Delta change	Perturbed by future rainfall/no change	AO change	**ERSEM**/yes	Wakelin et al. (2012a, b), Holt et al. (2014)
ECHAM5/MPIOM-REMO-HAMSOM	A1B	Transient 1951–2100	Bias corrected, all forcing data	No change/changed by discharge from GCM	AO change, except nutrients	**ECOHAM**/yes	Alheit et al. (2012), Mathis (2013), Mathis et al. (2013), Mathis and Pohlmann (2014)
ECHAM5/MPIOM-zoom	A1B	transient 1860–2100	No	future change/resolved, course resolution	AO change	**HAMOCC**/yes	Gröger et al. 2013
IPSLCM4-no-**ECOSMO**	A1B	1970–1999 versus 2070–2099	Delta change	No change/fully resolved	AO change/A only	**ECOSMO**/yes	Wakelin et al. (2012a), Holt et al. (2014, 2016), Pushpadas et al. (2015); www.meeceatlas.eu

(continued)

Table 6.2 (continued)

Model chain: GCM-RAM-ROM	Scenario	Time slice	Bias correction	Runoff/Baltic Sea	Considered forcing: atmosphere-only and atmosphere and ocean change	LTL-model/carbonate chemistry	Related publications
MPIOM-HAMOCC-no-**ECOSMO**	A1B	1970–1999 vs 2070–2099	delta change	no change/fully resolved	AO change/A only	**ECOSMO**/no	Wakelin et al. (2012a); Pushpadas et al. (2015)
BCM-HAMOCC-no-**ECOSMO**	A1B	1970–1999 versus 2070–2099	Delta change	No change/fully resolved	AO change/A only	**ECOSMO**/no	Wakelin et al. (2012a), Pushpadas et al. (2015)
ECHAM5/MPIOM-RCAO-NORWECOM	A1B	1970–1999 versus 2070–2099	Direct	No/no	AO change, except nutrients	**NORWECOM**/no	Eilola et al. (2013), Skogen et al. (2014)
ECHAM5/MPIOM-zoom-no-no	A1B	Continuous 1860–2100	Bias-corrected restoring to forcing GCM	Resolved, coarse scale	AO	**HAMOCC**/yes	Gröger et al. (2013)
NORESM-no-**ECOSMO**	RCP4.5	1970–1999 versus 2070–2099	Delta change	No change/fully resolved	AO change/A only	**ECOSMO**/no	Wakelin et al. (2012a), Pushpadas et al. (2015)
ECHAM5/MPIOM-HAMOCC-no-**ECOSMO**	RCP4.5	1970–1999 versus 2070–2099	Delta change	No change/fully resolved	AO change	**ECOSMO**/no	Pushpadas et al. (2015)
IPSLCM5-no-**ECOSMO**	RCP4.5	1970–1999 versus 2070–2099	Delta change	No change/fully resolved	AO change	**ECOSMO**/no	Pushpadas et al. (2015)

All SRES A1B and A2 scenarios are forced by IPCC-AR4 generation GCMs and ESMs. The RCP4.5 scenarios are forced by IPCC-AR5 generation ESMs

Table 6.3 Coupled atmosphere-ocean downscaling experiments for the North Sea for the end of this century

GCM-RAM-ROM	Scenario	Time slice	Baltic Sea	Restoring/bias correction	LTL-model/carbonate chemistry	Related publications
ECHAM5/MPIOM-REMO-MPIOM-zoom higher resolution	A1B	Continuous 1860–2100	Resolved	No restoring, bias correction of fresh water fluxes globally, not in the North Sea	No	Bülow et al. (2014), Sein et al. (2015)
ECHAM5/MPIOM-RCA4/NEMO	A1B	Continuous 1961–2100	Resolved	No restoring, bias correction in sea level for North Sea inflow	No	Bülow et al. (2014)
ECHAM5/MPIOM-REMO/HAMSOM	A1B	Continuous 1860–2100	Boundary conditions	No restoring or bias correction in the regional model	No	Bülow et al. (2014), Su et al. (2014)

All coupled downscaling experiments are forced by the IPCC-AR4 generation GCM **ECHAM5/MPIOM**

Table 6.4 Dynamic downscaling experiments for the North Sea for the middle of this century

GCM-RAM-ROM	Scenario	Time slice	Bias correction	Runoff/river load/Baltic Sea	A-only versus AO change	LTL-model/carbonate chemistry	Key publications
GISS-no-**ROMS**	A1B	1986–2000 versus 2051–2065	No	No change/na/no change	AO change	No	Melsom et al. (2009), Alheit et al. (2012)
BCM-no-**ROMS**	A1B	1986–2000 versus 2051–2065	Only ocean boundary conditions	No change/na/no change	AO change	No	Ådlandsvik (2008), Alheit et al. (2012)
CCSM-no-**ROMS**	A1B	1986–2000 versus 2051–2065	Only ocean boundary conditions	No change/na/no change	AO change	No	Melsom et al. (2009), Alheit et al. (2012)
ECHAM3-RACOM-Delft3D/BLOOM/GEM	A1B	1985–2004 versus 2031–2050	No	10 % winter increase and 10 % summer decrease/no change/na	AO change	BLOOM/GEM/no	Friocourt et al. (2012)
ECHAM3-RACOM-NORWECOM	A1B	1985–2004 versus 2031–2050	No	10 % winter increase and 10 % summer decrease/no change/no change	A only	NORWECOM/no	Friocourt et al. (2012)
IPSLCM4-no-**POLCOMS**	A1B	1980–1999 versus 2030–2040	Delta change	Various scenarios	AO change	ERSEM	Zavatarelli et al. (2013a, b)
IPSLCM4-no-**ECOSMO**	A1B	1980–1999 versus 2030–2040	Delta change	Various scenarios	AO change	ECOSMO	Zavatarelli et al. (2013a, b)

All scenario simulations are forced by IPCC-AR4 generation GCMs and ESMs

coupled regional AO models were developed, namely MPIOM-REMO (Sein et al. 2015), HAMSOM-REMO (Su et al. 2014) and RCA-NEMO (Dieterich et al. 2013; Wang et al. 2015). The three models have in common that the atmosphere components are all limited area models while their ocean components differ focusing either on the North Sea (HAMSOM), the North and Baltic seas (NEMO) or the global ocean employing a regional zoom to the North Sea (MPIOM-zoom). An ensemble of transient simulations 1960–2100 from all three models driven by the same GCM (ECHAM5-MPIOM) was performed for the SRES A1B scenario (Bülow et al. 2014). All models show an

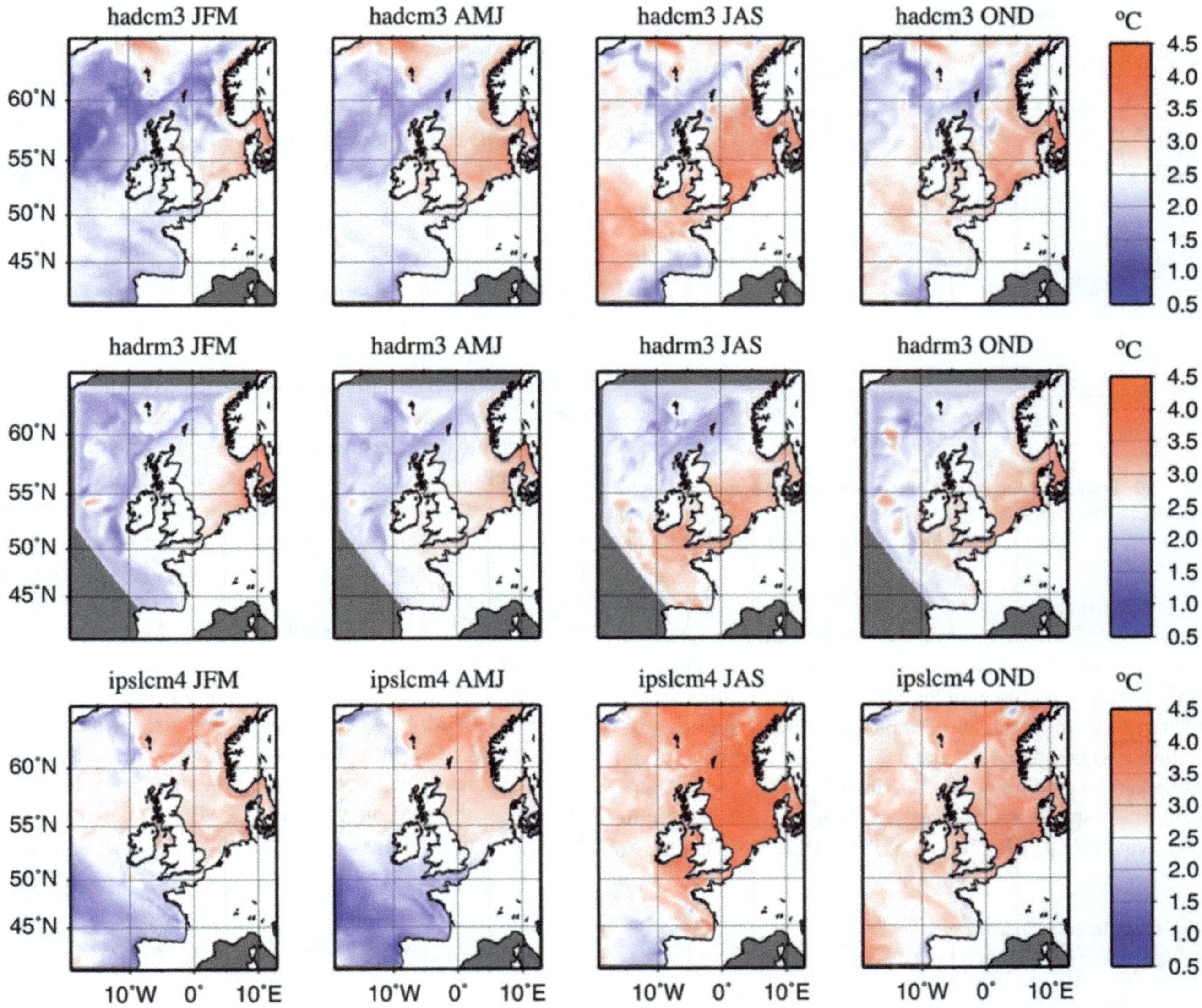

Fig. 6.8 Projected change in seasonal sea-surface temperature from POLCOMS experiments (HADCM3 and IPSL-CM4) and UKCP09 (HADRM3) experiments (redrawn using results from Holt et al. 2010 and Wakelin et al. 2012a)

area-averaged monthly mean SST increase of 1.7–3.0 °C for the end of the century (Fig. 6.9) and annually average SST increases of about 2 °C (Bülow et al. 2014), which is very similar to uncoupled downscalings forced by the ECHAM5/MPIOM model reported by Wakelin et al. (2012a), Mathis (2013) and Mathis and Pohlmann (2014). However, the uncertainty range arising from the different regional models was significantly larger compared to uncoupled model simulations.

An approximately linear trend of about 2 °C per 100 years was projected for SST through continuous simulations (e.g. Mathis 2013: 1.67–1.86 °C). The ensemble projections for the middle of the century were consistent with this trend. When forced by the ECHAM5/MPIOM a change of 0.4–0.8 °C was derived for the near future (2031–2050, Friocourt et al. 2012) and about 0.6–1.3 °C in the coupled simulations (Bülow et al. 2014). The spatial patterns of the projected warming were consistent with time-slice end-of-century projections and increased warming was projected for the coastal zone compared to the northern North Sea. The multi-model ensemble for the near future (+65 years) performed with ROMS, forced by the GISS, BCM, and CCSM GCMs (Alheit et al. 2012; Fig. 6.10), supports the view that the choice of forcing GCM contributes substantial uncertainty to the magnitude of projected warming. The GISS-based downscaling, which has the largest warm bias in the control run, simulates a weak warming. Annual average temperature is rising by 0.3 °C at 25 m in the future scenarios. The BCM downscaling shows an average warming of 0.6 °C. The downscaling based on NCAR CCSM, which has a cold bias in the control simulation, gives the strongest warming at 1.1 °C on average.

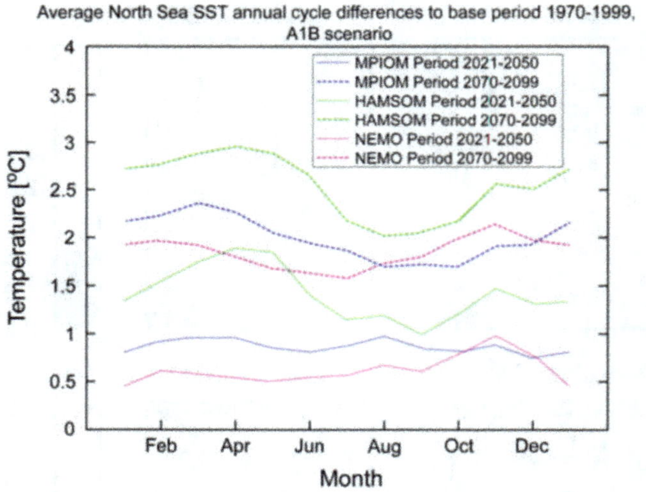

Fig. 6.9 Projected annual cycle of sea surface temperature change for two climate periods and three coupled atmosphere-ocean model downscalings (ocean models: MPIOM, HAMSOM, NEMO) (Bülow et al. 2014)

Simulated future temperature changes are generally seasonally dependent. Many models project larger changes in SST during summer months when a shallow thermocline restricts the incoming heat input to the sea surface (e.g. Figure 6.8) compared to winter, when the North Sea is well mixed and incoming heat is distributed over the entire water column, despite larger changes in heat flux during autumn and winter (e.g. Holt et al. 2010, 2012, 2014). Due to well-mixed conditions during winter, heat flux anomalies result in a larger temperature increase in the shallow south-eastern North Sea than the deeper central and north-western North Sea (e.g. Holt et al. 2010, 2012; Wakelin et al. 2012a; Fig. 6.8). During stratified summer conditions, the southern North Sea also warms more strongly, since mixed-layer thickness is typically shallower than in the northern North Sea (Janssen et al. 1999; Schrum et al. 2003b). However, these regional and seasonal variations in warming are not consistent among the different regional model realisations. Exceptions are those simulated with the HAMSOM model (Mathis 2013), the NORWECOM (Friocourt et al. 2012), and the coupled models (Bülow et al. 2014). These project larger temperature changes in winter, autumn and spring, with significant inter-model differences. From the multi-model ensembles it seems that the strength of the coupling and hence the regional and seasonal pattern of projected changes, are significantly affected by the properties and parameterisations of the regional model. Likely candidates are mixed-layer depth, flux parameterisations and local feedbacks. However, the attribution of a definite cause for the inter-model deviations is not obvious from existing literature.

The ECOSMO model has also been used with forcing from the IPCC AR5 generation models to simulate the RCP4.5 scenario; the first and so far only published attempts to employ the new updated IPCC AR5-scenarios for the North Sea (Wakelin et al. 2012a; Pushpadas et al. 2015).

Fig. 6.10 Projected change in temperature (°C) for the near future (2051–2065 vs. 1986–200) for ROMS simulations forced by BCM (*upper*), GISS (*middle*) and CCSM (*lower*) (Alheit et al. 2012)

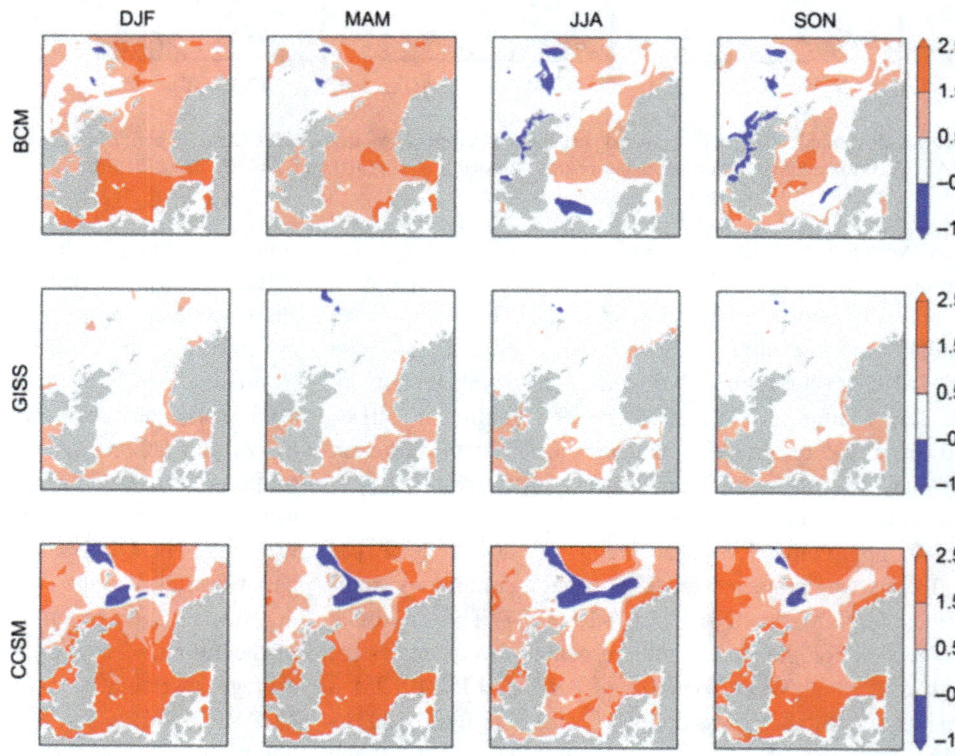

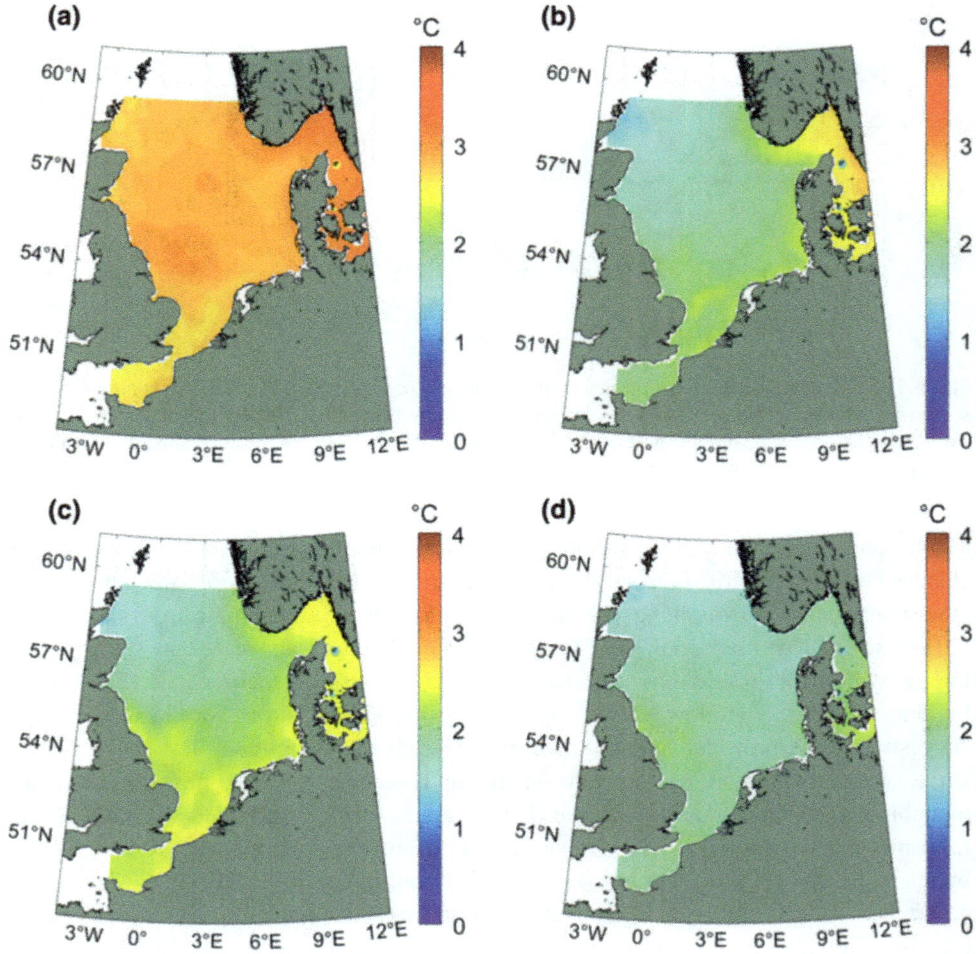

Fig. 6.11 Projected change in temperature (°C) for the end of the century (2070–2100, A1B and RCP4.5 scenario) as projected by the ECOSMO, forced by the IPSL-CM4.0-A1B (**a**), BCM-A1B (**b**), ECHAM5-A1B (**c**) and NORESM-RCP4.5 (**d**) scenarios (Wakelin et al. 2012a)

When comparing the simulation to the older SRES A1B scenario simulations, slightly less warming was found for the simulations forced by the RCP4.5 scenarios and IPCC AR5 generation models (Wakelin et al. 2012a; Pushpadas et al. 2015; Fig. 6.11), which could be largely explained by the fact that both story lines are not fully comparable and the RCP4.5 scenario provides less radiative forcing (see Annex 4). A slight reduction in the ranges of projected SST change was also evident (Pushpadas et al. 2015).

6.3.3 Changes in Salinity

North Sea salinity is influenced by the local balance between precipitation and evaporation, terrestrial runoff and exchange with the North Atlantic and the Baltic Sea. The regional projections of salinity considered in this section utilise full hydrodynamic models. However, their predictive capacity for salt and fresh-water changes is limited and results are biased to an unknown degree by the assumptions and approaches chosen for considering terrestrial fresh-water fluxes (e.g. Wakelin et al. 2012a), Baltic Sea water fluxes (e.g. Mathis 2013), Atlantic boundary conditions (e.g. Friocourt et al. 2012) or the use of a relaxation scheme (e.g. Ådlandsvik 2008), together with the accuracy of cross-shelf circulation and mixing. An attempt to consider all climate change impacts on fresh and salt water sources consistently has only been made for a few studies (e.g. Gröger et al. 2013; Bülow et al. 2014). However, these studies required different global bias- or fresh-water flux corrections in the global forcing model to avoid drift in salinity (see Tables 6.1 and 6.2 for details) and their projections differ despite using the same GCM, possibly due to a regional sensitivity to bias or flux corrections in the GCM.

Gröger et al. (2013) projected substantial freshening of the North Sea manifesting in a reduction in surface salinity of 0.75 (Fig. 6.12), which is coherent with a stronger hydrological cycle and substantial freshening of the North Atlantic under future warming modelled at the global scale. The simulated freshening peaks around 2060–2070, with salinity then increasing towards the end of the century but not to present-day values. A similar freshening of the North Sea is apparent in the HAMSOM regional projections based on the same coarse resolution GCM forcing (Mathis 2013;

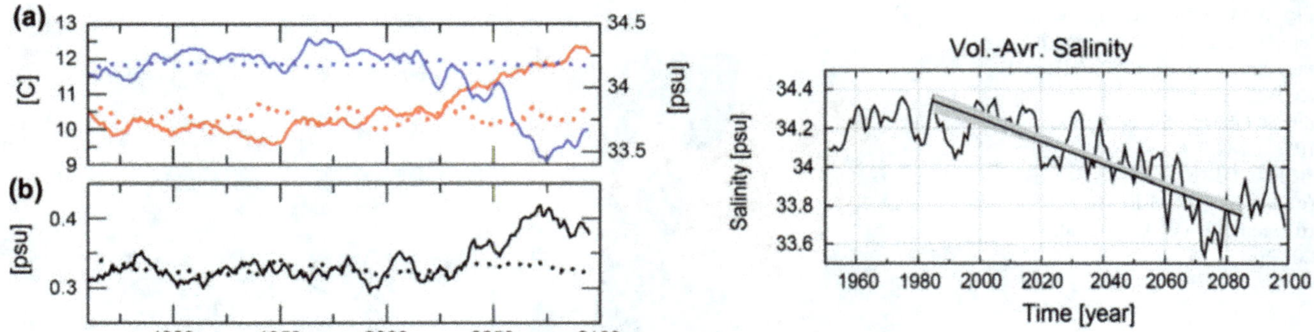

Fig. 6.12 *Left* Time series of projected surface salinity (*a, blue*), sea-surface temperature (*a, red*) and the difference in salinity between the bottom waters and surface waters (*b, black*) in the North Sea simulated by the MPIOM-zoom (Gröger et al. 2013). *Dotted lines* show the results of a control simulation with constant radiative forcing. *Right* Time series of annual (*black*) volume-averaged salinity from HAMSOM uncoupled downscaling (Mathis and Pohlmann 2014)

Mathis and Pohlmann 2014), despite no terrestrial runoff change considered here. In contrast, results from coupled atmosphere-ocean downscaling presented by Bülow et al. (2014), which also used A1B and ECHAM5-MPIOM forcing but with higher resolution, projected salinity to decrease by only about 0.2 (Bülow et al. 2014; Fig. 6.13). This suggests that the projected salinity change is strongly sensitive to the resolution of the atmospheric and oceanic modelling component used and the bias correction or restoring methods used in the global model. Moreover, biases in the flux coupling and internal variability contribute to local deviations and inter-model variability in projected changes. The other regionally coupled AO-projections from NEMO and HAMSOM, which use forcing from the same GCM (but different global realisations, details given by Bülow et al. 2014), project decreases in salinity of the same order of magnitude. However, inter-model differences stemming from the regional models or global runs used are above 0.2 in salinity and significant differences in spatial pattern occur (Fig. 6.14). The differences are particularly strong for the Baltic Sea outflow and the northern boundary inflow.

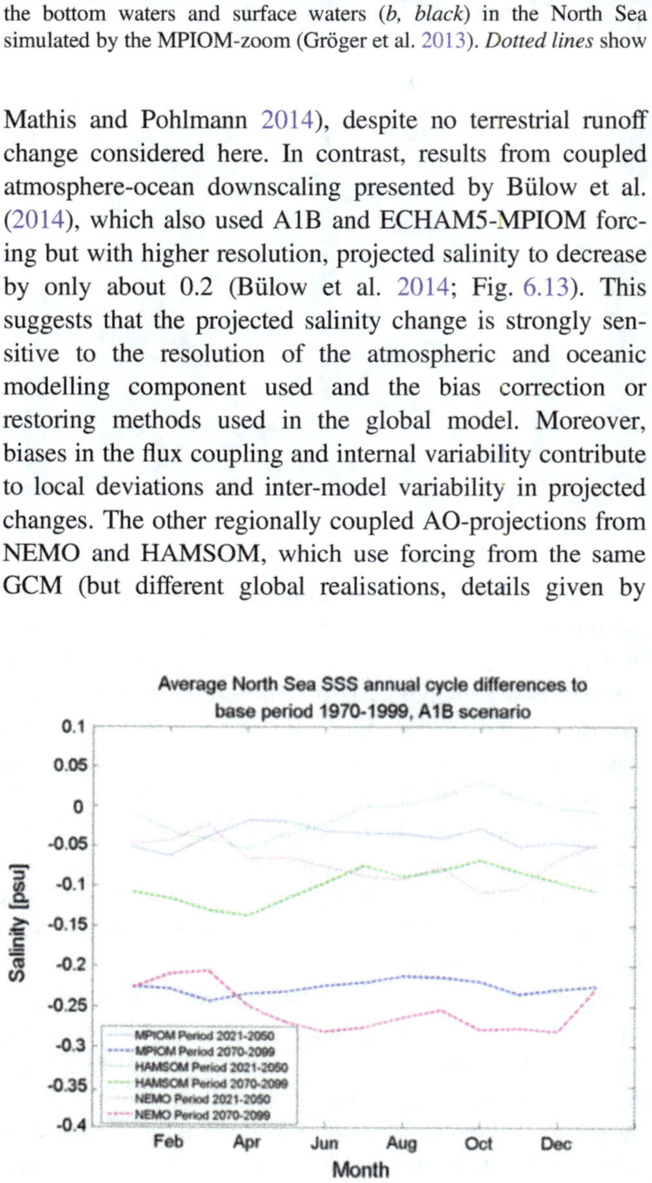

Fig. 6.13 Projected annual cycle of change in sea surface salinity for two periods (2021–2050 vs. 1970–1999, *dotted lines*; and 2070–2099 vs. 1970–1999, *dashed lines*) and three coupled AO-downscalings (ocean models: MPIOM, HAMSOM, NEMO), all with GCM forcing from ECHAM-MPIOM (Bülow et al. 2014)

The projected overall freshening of the North Sea is confirmed by most of the other regional downscaled scenarios (e.g. Kauker 1999; Holt et al. 2010, 2012; Wakelin et al. 2012a; Pushpadas et al. 2015), but the strength of the salinity decrease appeared to be strongly dependent on the choice of GCM (Holt et al. 2014, 2016; Wakelin et al. 2012a; Pushpadas et al. 2015; Fig. 6.15) and inter-GCM related variability in projected surface salinity change is large (≈ O(0.5–1)) and increases from AR4- to AR5-based regional downscaling (Pushpadas et al. 2015). The regional model and assumptions made for runoff and Baltic Sea exchange contribute to inter-model variability. However, these are second order effects compared to GCM-related variability, and projected salinity changes are largely related to North Atlantic salinity changes and the wind-driven inflow to the North Sea. This is confirmed by near future projections: Friocourt et al. (2012) attributed modelled near-future freshening of the North Sea partly to decreasing winter inflow. Potential impacts of circulation changes on salinity were earlier studied by Schrum (2001) in a simple perturbation experiment, which revealed that decreasing westerly wind speed by 25 % would result in a basin-wide freshening of the North Sea of the order of a 0.3–0.4 reduction in average salinity.

6.3.4 Changes in Stratification

During winter the North Sea is generally well mixed due to surface cooling and resulting thermal convection, and winter

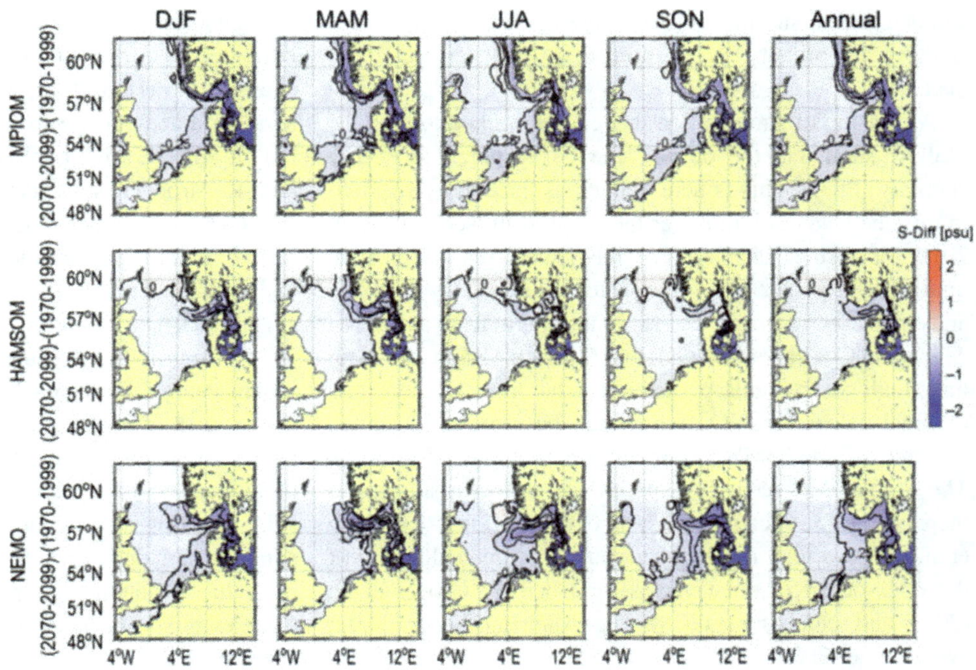

Fig. 6.14 Projected change in sea surface salinity from three regional coupled AO-downscalings (ocean models: MPIOM, HAMSOM, NEMO), all with GCM forcing from ECHAM-MPIOM (Bülow et al. 2014)

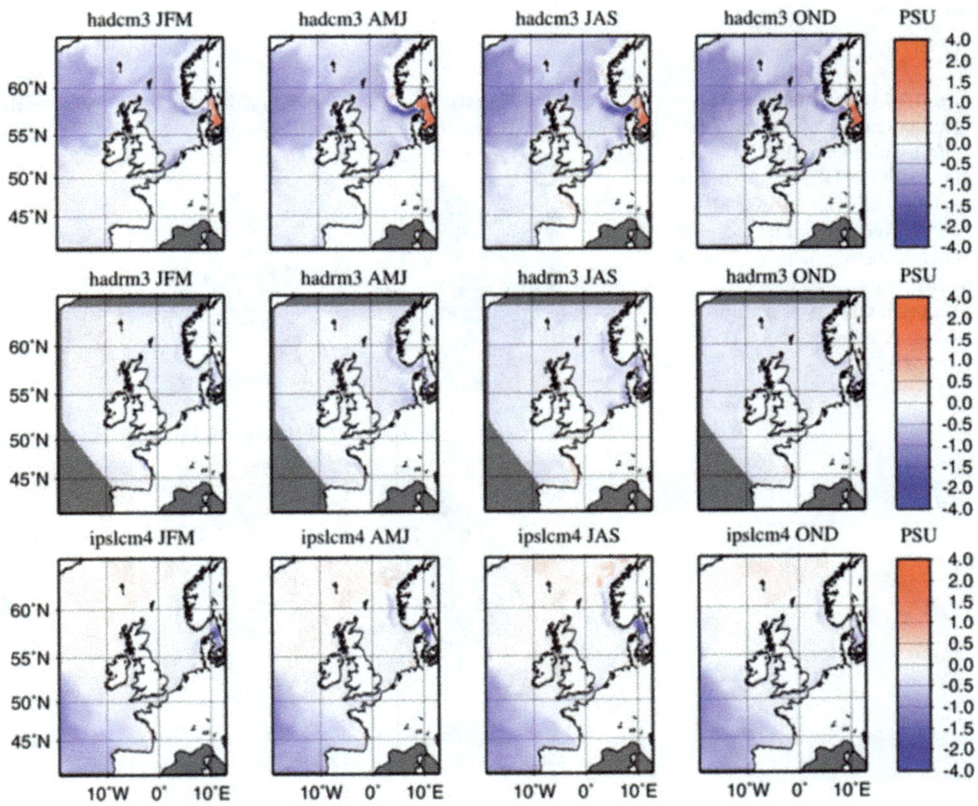

Fig. 6.15 Projected change in sea surface salinity as simulated by the POLCOMS model using forcing from HadCM3, HadRM3 and IPLS-CM4.0 (results combines from Holt et al. 2010, 2012 and Wakelin et al. 2012a, see for time period Table 6.2)

surface temperature generally describes the average temperature of the water column. In late spring, a seasonal thermocline develops (Janssen et al. 1999; Schrum et al. 2003b) and the well-mixed surface layer decouples from the lower layer water. The timing and duration of stratification, thermocline strength and the thickness of the surface mixed layer have implications for air-sea fluxes. Changes in stratification also affect regional ocean characteristics, for example the seasonal variations in tidal constituents through changes in eddy viscosity and current profiles (Sect. 6.2) and sediment transport (Gräwe et al. 2014). Stratification is also an important control of nutrient supply to the euphotic zone and changes in stratification are therefore a major driver for changes in primary production (e.g. Holt et al. 2014; Sect. 6.4).

A measure of stratification that can be applied to both the shallow shelf and the open ocean is the potential energy anomaly (PEA). This is here defined as the energy required to mix the water column over the top 400 m (see Holt et al. 2010 for further details). For the North Sea, PEA is at minimum in winter, when the water column is well mixed, and increases as soon as seasonal thermal stratification develops. Coastal areas and the Southern Bight are well mixed all year round by intense tidal mixing and show minimal values for PEA throughout the year, as illustrated by the POLCOMS control experiment (Fig. 6.16). The POLCOMS scenario experiments project a substantial increase in stratification in open-ocean regions (e.g. Holt et al. 2010), which is consistent with a future shallowing of the open-ocean mixed layer as modelled by Gröger et al. (2013) and seen in most GCMs (e.g. Allen and Ingram 2002; Wentz et al. 2007).

Projections suggest the shelf will remain generally well mixed during winter, but that stratification in spring, summer and autumn will increase significantly, which could be attributed to earlier onset and later breakdown of seasonal stratification (Holt et al. 2010, stratification is here defined as a sustained surface to bottom density difference equivalent to 0.5 °C and a mixed layer shallower than 50 m) and to stronger stratification during summer. Using the POLCOMS-HadRM3-HadCM3 model scenario, Holt et al. (2010) found that stratification would start 5 days earlier and breakdown 5–10 days later by the end of the century (Fig. 6.17). During summer the greatest increase in ocean stratification is to the south of the domain. Ensemble simulations using different GCMs (POLCOMS-based) are shown in Fig. 6.18 (note the graphic shows fractional changes). All ensemble members simulate a positive fractional change almost everywhere throughout the season. The increase in PEA is strongest in winter in the open ocean and lower in summer and on the shelf. The ensemble simulations also indicate substantial inter-model variability in projected changes in stratification during summer on-shelf and at the shelf break and in the open ocean throughout the year. While there is also a significant fractional change in 'well-mixed' regions, absolute values remain low.

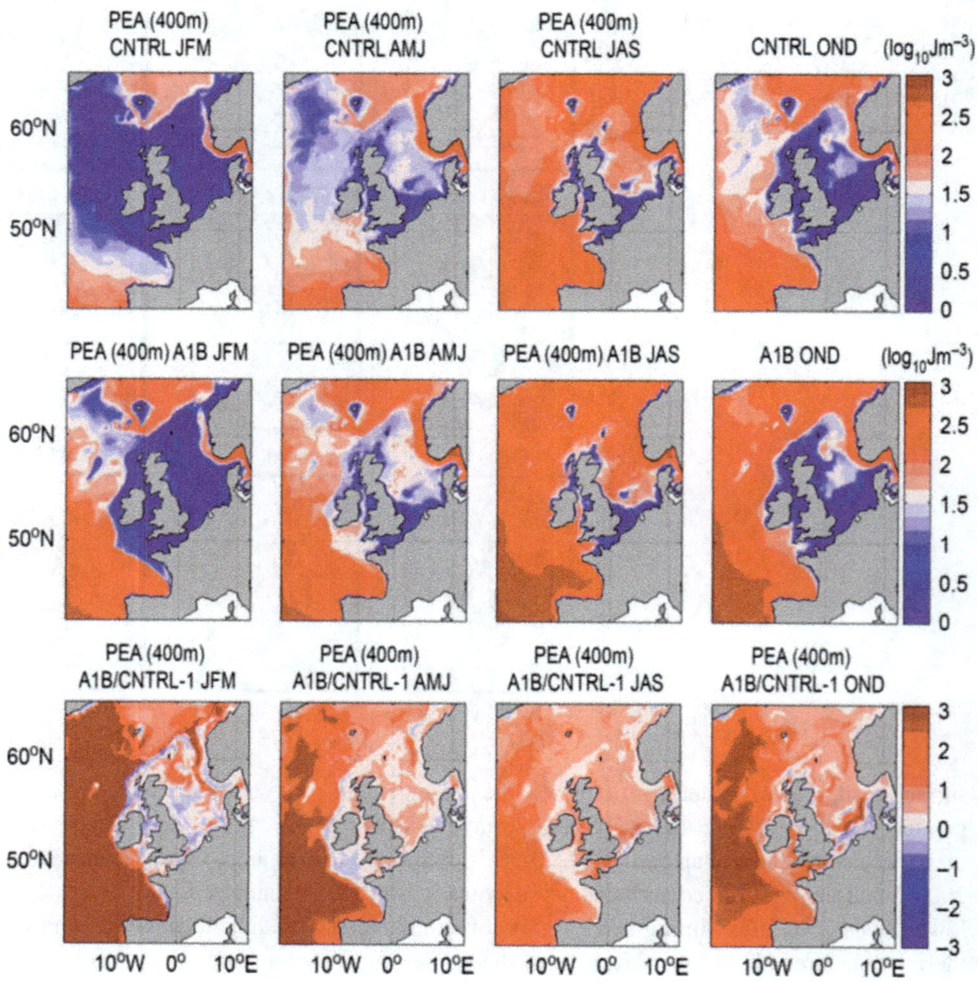

Fig. 6.16 Simulated seasonal mean potential energy anomaly (PEA) with integration limited to 400 m for CNTRL and A1B (note log10 scale, from POLCOMS) and the fractional difference between them. NB this is limited to changes of a factor of 3, maximum change in oceanic regions is a factor of 5.7 (Holt et al. 2012)

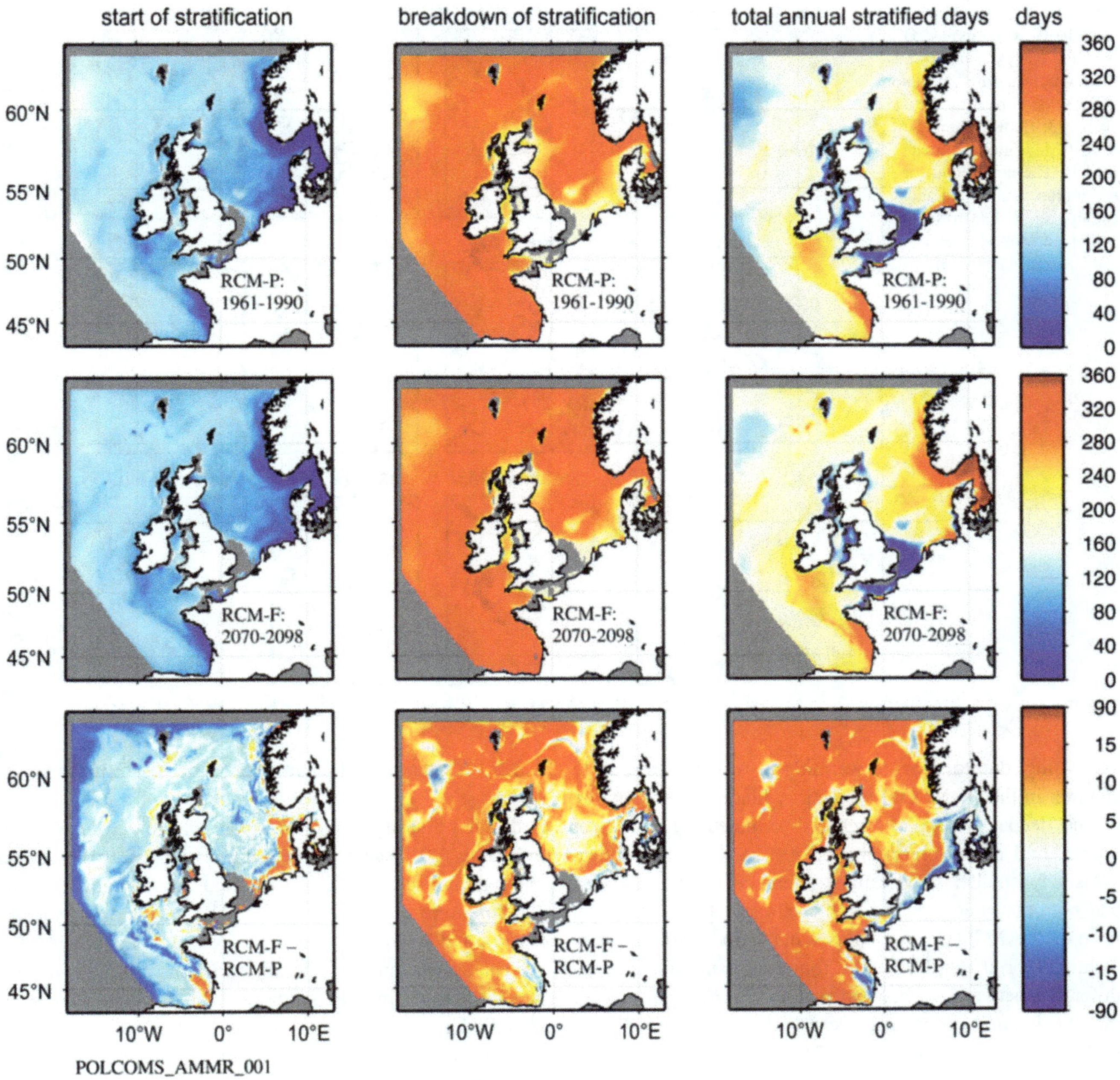

Fig. 6.17 Simulated mean timing of seasonal stratification for present day (RCM-P 1961–1990, *upper*), future climate (RCM-F 2070–2098, *middle*) and the difference between them (i.e. projected change from POLCOMS, *lower*). The graphic shows day of the year (1 January is day 1) when persistent seasonal stratification starts (*left column*) and ends (*middle column*), and the total number of stratified days (*right column*). *Grey shaded* areas are well mixed throughout the year (Holt et al. 2010)

Potential energy anomaly is a metric that does not inform about vertical structure and strength of gradients. Higher PEA could correspond to a deeper thermocline or to a stronger vertical gradient without a change in thermocline depth. Alternative metrics are the depth and strength of the thermocline, which could also provide insight into the nature of the change. Using these metrics Mathis (2013) and Mathis and Pohlmann (2014) identified a weak shallowing of the thermocline in the HAMSOM projection, which they attributed to a weakening of summer wind speeds. In contrast, the seasonal maximum thermocline depth showed a clear and strong deepening trend. Mathis and Pohlmann (2014) attributed this to a delay in thermocline erosion south of the 50 m depth contour in autumn, caused by a decrease in seasonal heat loss and wind speeds in the future (Fig. 6.19).

The spatial extent of stratification is mainly determined by local bathymetry and tidal amplitude and so is not subject to significant change (e.g. Mathis and Pohlmann 2014). Mean and maximum thermocline strength are both

Fig. 6.18 Fractional change (calculated as future/past-1) in potential energy anomaly (PEA) projected by the regional POLCOMS forced by the HadCM3, HadRM3 and IPSL-CM4.0 global climate models. The depth limit for PEA integration is indicated above the colour scale (1400 m) (results combined from Holt et al. 2010, 2012 and Wakelin et al. 2012a)

decreasing, due to stronger warming in winter compared to summer in the HAMSOM projection. Since the temperature of the deeper waters is largely determined by the water temperature of the preceding winter, this results in a progressively smaller temperature difference between surface and bottom waters under a future climate. This conclusion from the HAMSOM simulations is in contradiction to results from Gröger et al. (2013) who found an increasing salinity difference between surface and bottom water and speculated that this is due to enhanced river runoff and a strengthening hydrological cycle through the 21st century. This discrepancy might be explained by different consideration of runoff changes in both downscalings. In contrast to MPIOM simulations, which resolve and consider runoff changes, the HAMSOM-based downscaling experiment is forced by constant climatological river runoff data based on values for the latter half of the 20th century over the entire simulation period. Another reason might be the different downscaling procedures applied, namely bias corrections for HAMSOM downscaling and direct forcing with salinity restoring to the forcing coarse-scale GCM in the coupled atmosphere-ocean model, which have the potential to modulate regional climate change impacts.

6.3.5 Changes in Transport and Circulation

A detailed investigation of transport change was undertaken by Mathis (2013) and Mathis and Pohlmann (2014) based on one regional scenario only. Projected future changes in circulation were analysed for seasonal mean current velocity vectors and trend analyses were applied to depth-averaged current speeds and to volume transports through various lateral sections in the North Sea. Mathis and Pohlmann identified an enhanced general circulation and a stronger northern inflow (Fig. 6.20) in spring, caused by stronger westerly and north westerly winds in the forcing GCM. For the other seasons the slightly decreasing mean wind speeds result in a slightly weaker general circulation. They identified increasing northern inflow in spring (by about +21 %, +0.134 Sv; 1 Sv = 1 million cubic metres per second) as the most significant seasonal 100-year change, which also dominates on annual scales. The other important change is a substantial decreasing inflow through Dover Strait in summer (−38 % or −0.023 Sv). In addition, they found a 12 % (−0.113 Sv) weakening of the Skagerrak recirculation in autumn and a 10 % (−0.055 Sv) reduction of the inflow through the Fair-Isle Passage in winter. A substantial

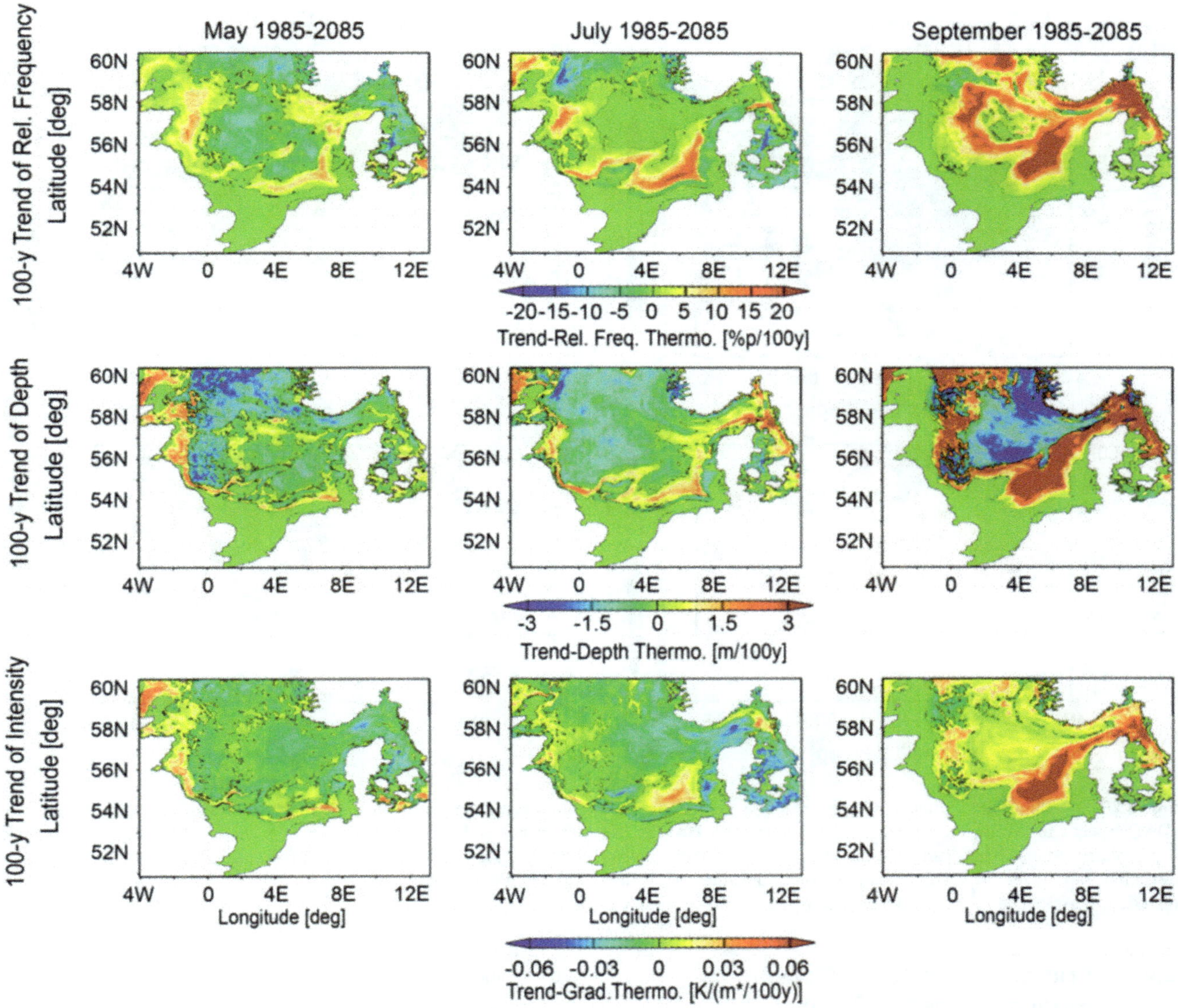

Fig. 6.19 Spatial distribution of modelled representative 100-year trends in the relative frequency of thermocline presence (*upper row*), depth (*middle row*), and intensity (*lower row*) for May (*left column*), July (*middle column*), and September (*right column*) as simulated by the HAMSOM model. The *black contour lines* refer to null trends (Mathis and Pohlmann 2014)

proportion of the northern inflow reverses into the Norwegian Coastal Current shortly after entering the northern North Sea, which consequently increases. Due to a weaker Dooley Current, caused by reduced Fair-Isle inflow, the northern inflow is guided south-eastwards to a lesser extent so that more water of North Atlantic origin is able to enter the central and southern North Sea. A westward strengthening of the northern inflow is indicated through increasing current speed east of the Shetland Islands and in the central North Sea. The changes in depth-averaged current speeds across the entire northern North Sea are statistically significant, as indicated by confidence levels higher than 95 %.

Detailed studies of transport pattern are not available from other regional scenarios and so it remains open how large internal variability and inter-model uncertainty is and whether the projected changes can be considered as robust. However, an overall increasing inflow was also projected by ROMS forced by the global climate model BCM (Ådlandsvik 2008). In contrast to Mathis (2013) and Mathis and Pohlmann (2014), Ådlandsvik projected an increasing inflow for almost the entire seasonal cycle. Only the October and November inflows were projected to decrease slightly. Using a different setup (ECHAM3-RACOM-NORWECOM), Friocourt et al. (2012) simulated for the near future (2031–2050) a decrease in inflow into the North Sea of about 5 %, for almost the entire seasonal cycle (exceptions August and November) with the NORWECOM model.

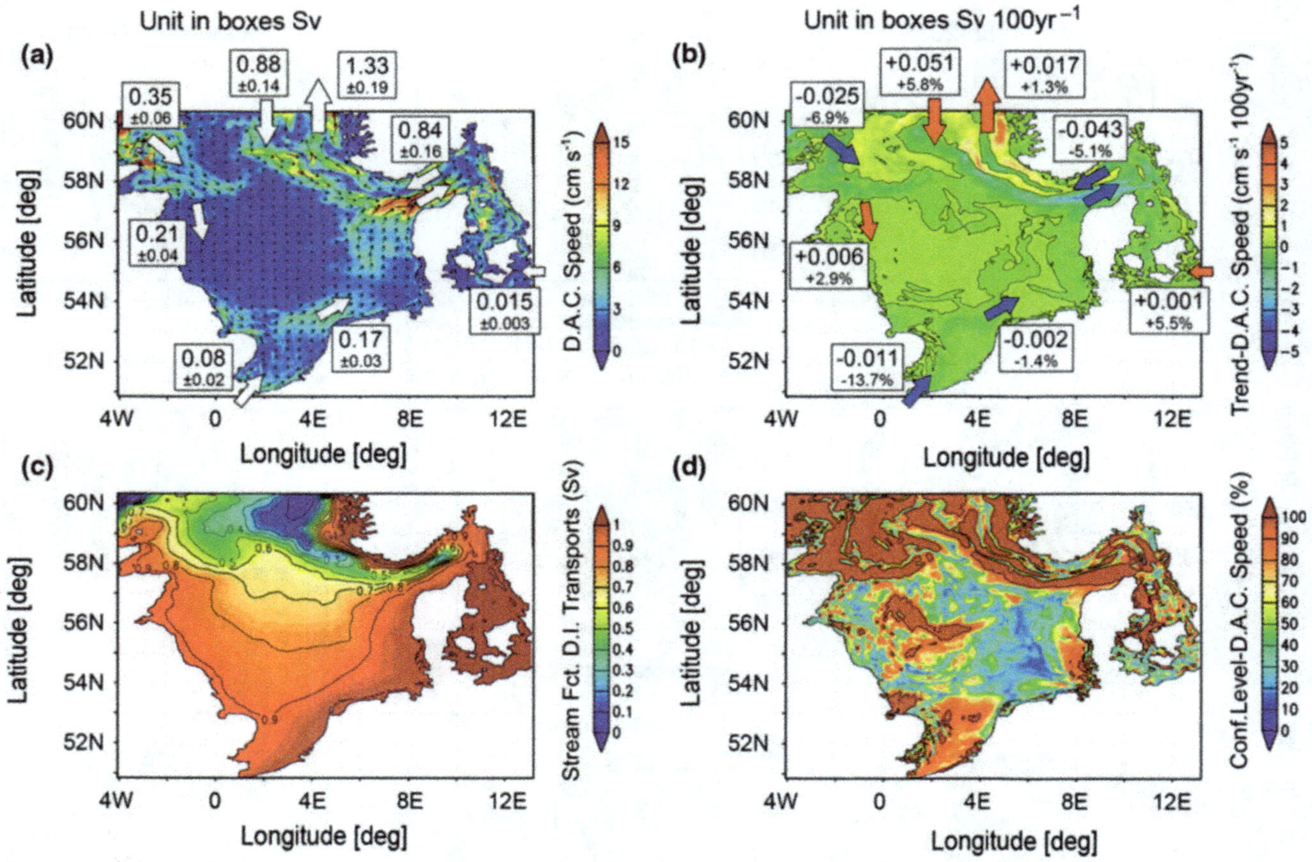

Fig. 6.20 **a** Simulated annual mean depth-averaged current (DAC) speeds and velocity vectors (model results from HAMSOM). Net volume transport at various transverse sections and standard deviations are given in *boxes*. *White arrows* illustrate mean flow direction. **b** Associated linear 100-year trends and relative changes to the magnitudes given in **a**. *Red* (positive) and *blue* (negative) *arrows* indicate trends. **c** Stream function of mean depth-integrated (DI) volume transport. **d** Confidence levels of the linear trends shown in **b** (Mathis and Pohlmann 2014)

6.4 Primary Production, Ocean Biogeochemistry, Ocean Acidification

Climate change impacts on primary production and responsible biogeochemical changes and ocean acidification were studied in a sub-set of the downscaling experiments summarised in Tables 6.2 and 6.4. The POLCOMS, ECOSMO, HAMSOM, Delft3D, NORWECOM and MPIOM simulations were equipped with a lower trophic level model (Alheit et al. 2012; Holt et al. 2012, 2014, 2016; Wakelin et al. 2012a; Gröger et al. 2013; Chust et al. 2014; Skogen et al. 2014; Pushpadas et al. 2015). Some of these downscaling scenarios also considered carbonate chemistry, but published estimates of future ocean acidification are available only from two regional models: POLCOM-ERSEM and ECOSMO (Wakelin et al. 2012a; Artioli et al. 2013, 2014). Although carbonate chemistry was also considered in MPIOM-HAMOCC-zoom (Gröger et al. 2013) and ECOHAM simulations (Alheit et al. 2012), no ocean acidification projections for future climate change have yet been published for these models.

Only the ECOSMO model uses IPCC-AR5 ESM global forcing (Wakelin et al. 2012a; Pushpadas et al. 2015). All other downscaling studies of ocean biogeochemistry and all climate change impact scenarios for ocean acidification are based on global climate change scenarios from the IPCC AR4-generation models (Table 6.2). All regional scenarios lack land-ocean coupling, similar to regional hydrodynamic scenarios (see Annexes 2 and 3 and Sect. 6.3) and climate-driven changes in future river loads are neglected in most regional scenarios in accordance to ESM scenarios (AR4- and AR5-generation models; Regnier et al. 2013). Only for the ECOHAM downscaling scenario was an attempt made to scale river loads by changes in river runoff (Alheit et al. 2012). Different eutrophication scenarios were only considered in a couple of near-future studies undertaken within the MEECE project (Zavatarelli et al. 2013a, b).

6.4.1 Primary Production and Eutrophication

Production of organic carbon through photosynthesis (chemosynthesis is not considered here) is controlled by availability of light and nutrients, and is thus sensitive to climate change. A decrease in annual net primary production (netPP) in the northern North Sea was a consistent impact signal for all scenarios considering North Atlantic impacts and local atmospheric forcing (Holt et al. 2012, 2014, 2016; Wakelin et al. 2012a; Gröger et al. 2013; Pushpadas et al. 2015). The decreasing netPP could be largely attributed to a decrease in cross-shelf nutrient fluxes to the North Sea and a consequent fall in North Sea winter dissolved inorganic nitrogen (DIN; Holt et al. 2012, 2014, 2016; Gröger et al. 2013; Pushpadas et al. 2015). The decrease in nutrient fluxes originates largely from local oceanic stratification changes on the Northwest European Shelf and near the shelf break, but as sensitivity experiments by Holt et al. (2012, 2014) reveal the oceanic far field also contributes. Local stratification changes in the North Sea are less important. Projected decrease in netPP for the end of the century was moderate for the regional models ECOSMO (12 %) and POLCOMS-ERSEM (2 %) when forced by the IPSL-CM4.0 ESM (Wakelin et al. 2012a; Holt et al. 2014, 2016; Fig. 6.21). Projected changes in mean annual netPP from the ECOSMO-IPSL-CM4.0 scenario were significant and could be distinguished from climate-driven variability, although this was not the case for POLCOM-ERSEM results (Wakelin et al. 2012a). Pushpadas et al. (2015) projected greater decreases in netPP for regional model scenarios forced by the ESM MPIOM-HAMOCC (−19 %) and lower primary production decrease for the scenarios forced by BCM-HAMOCC (−2.3 %).

Holt et al. (2012) reported that the North Sea is generally vulnerable to oceanic nutrient changes. However, this is compensated for by on-shelf processes and the actual sensitivity is less than expected. Holt et al. (2014) concluded that, like shelf seas in general, the North Sea is likely to be more robust and less affected by climate change than the global ocean, where the leading process of increasing permanent stratification significantly reduces netPP (e.g. Steinacher et al. 2010). The North Sea is well mixed for almost half the year, and local stratification changes are less important and potentially overridden by other processes (Holt et al. 2014). Strong tidal mixing and a substantial contribution of recycled production, based to a large extent on suspended particulate organic material advected onshore are major controls of ecosystem dynamics in the shallow southern North Sea (Holt et al. 2012). Holt et al. (2014) hypothesised that these properties may shelter the North Sea, similar to other shallow and tidally influenced shelf regions, from some direct impacts of climate change. Consistent with this hypothesis, the regional models ECOSMO and POLCOMS-ERSEM both projected the greatest decreases in production in the deeper northern North Sea and a moderate decrease or even an increase in production in the shallower southern North Sea in most of the scenarios (Holt et al. 2014, 2016; Pushpadas et al. 2015; Figs. 6.21 and 6.22).

Comparing the two regional model scenarios forced by the same ESM (IPSL-CM4.0), shows that the pattern of projected change in netPP and the magnitude of local increases in the southern North Sea are very similar for both regional models, but that the modelled local decreases in other regions are much stronger in the ECOSMO model, which results in a six-fold larger projected decrease in overall netPP for the North Sea with the latter model (Wakelin et al. 2012a; Holt et al. 2014, 2016). A potential cause of this discrepancy is the temperature-dependent metabolic rates in ERSEM, which would speed up growth-and-mortality cycles in a warmer world, and their omission in ECOSMO (Daewel and Schrum 2013). However, according to an assessment by Holt et al. (2014, 2016) this can account for only a small fraction of the discrepancies, mostly along the coast. A more likely candidate is therefore the cross-shelf exchange of nutrients and thus on-shelf production, which is modelled differently by the POLCOMS-ERSEM and IPSL model (Holt et al. 2014). Differences are especially pronounced for the region south and west of Great Britain. These regions appeared to be the most important for nutrient supply to the entire North Sea system (Holt et al. 2012). The different spatial coverage of both regional models is therefore of key importance. ECOSMO, which is forced with boundary nutrients from the global model on the shelf, is more strongly coupled to the global model and its projected changes are more similar to the projected changes by the global model (Chust et al. 2014). In contrast, POLCOMS-ERSEM is forced by boundary conditions from the global model off-shelf and resolves cross-shelf exchange. Hence it can deviate more strongly from the global model and could develop its own regional dynamics. These findings support the hypothesis that the different cross-shelf dynamics in the regional and global model in the shelf break region is a likely cause for deviations in projected climate change impacts between the two regional models. The projected change in the Skagerrak region is also quite different in both regional models. A decrease in netPP is projected by ECOSMO and an increase by POLCOMS-ERSEM. The latter is probably biased by not resolving the Baltic Sea response and artificial boundary assumptions for POLCOMS-ERSEM. Steinacher et al. (2010) reported a drift in the IPSL-CM4.0 ESM A1B simulation. The degree to which the drift affects regional downscaling remains unclear and there has been no attempt

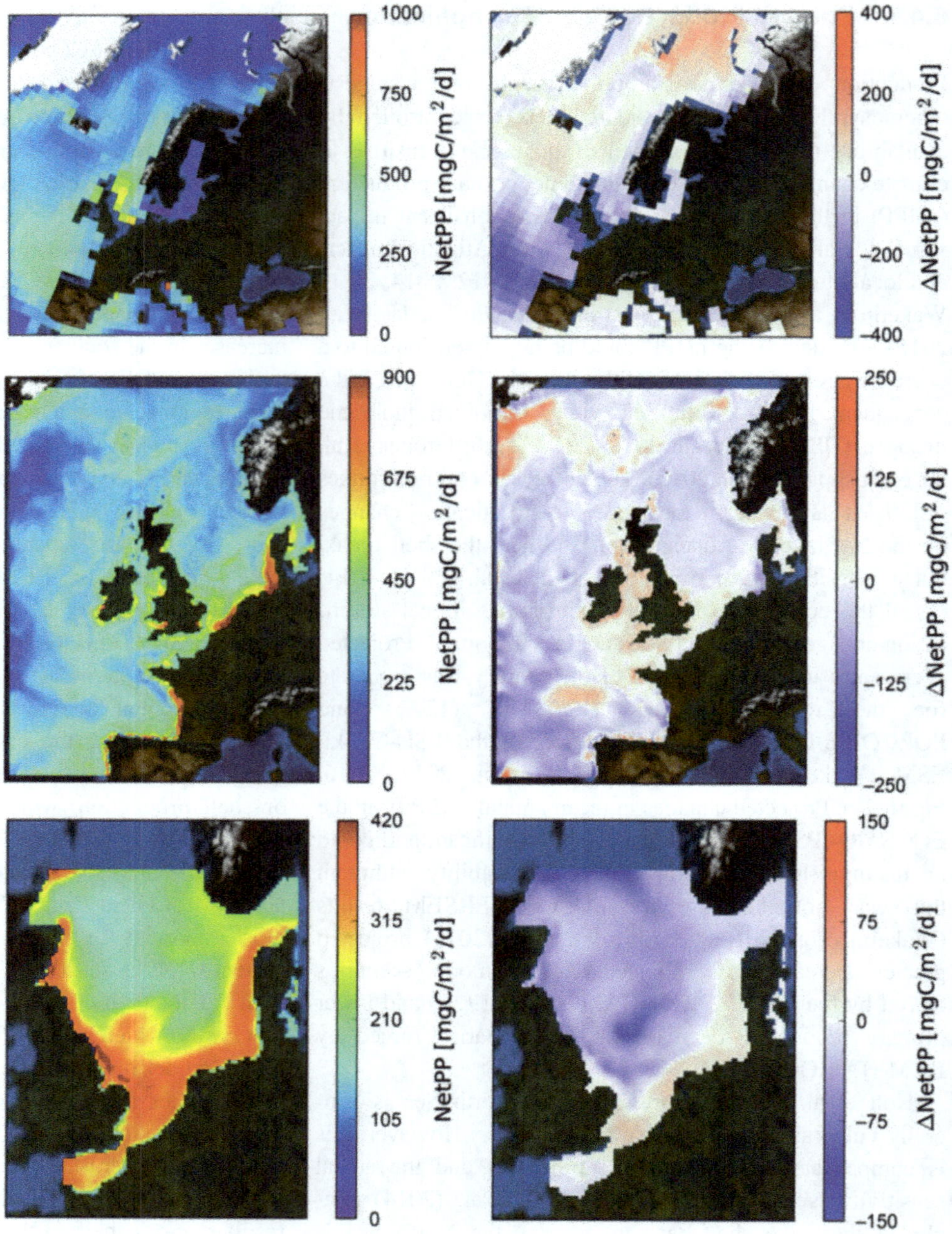

Fig. 6.21 Simulated present-day net primary production (netPP) (*left*, mean 1980–1999) and projected change in daily mean netPP between 1980–1999 and 2080–2099 estimated from the global IPS-CM4 model (*upper*) and from the IPSL-POLCOMS-ERSEM (*middle*) and IPSL-ECOSMO (*lower*) regional downscaling, note the different levels of netPP simulated by both regional models (Holt et al. 2014)

to estimate or remove a potential regional drift in either the POLCOMS-ERSEM or the ECOSMO downscaling.

The projected decrease in netPP from the MPIOM-HAMOCC-zoom model scenarios for the North Sea (Gröger et al. 2013) is 30 % and so substantially larger than estimates from the regional scenarios discussed above. Gröger et al. (2013) concluded that regional impacts on netPP are amplified in the North Sea relative to the global ocean and hypothesised that the shelf is more vulnerable than the open ocean, contradicting the findings and conclusions of Wakelin et al. (2012a), Holt et al. (2014) and Pushpadas et al. (2015). Possible reasons for opposite findings in the regional studies are different sensitivities of the cross-shelf exchange in the global and regional approach caused by different spatial resolution and sensitivity to the GCM bias and bias correction (as shown by Holt et al. 2014), and differences in the regional and global biogeochemical models. That the lower resolution in the MPIOM-HAMOCC zoom configuration (Gröger et al. 2013) compared to the regional models is a major reason for the different sensitivities seems unlikely, since the MPIOM-zoom resolution did not appear to be critical for the representation of hydrodynamics compared to high resolution regional models (not yet published).

More likely factors are differences in the biogeochemical parameterisations and the forcing GCM. The regional multi-ESM ensemble study presented by Pushpadas et al.

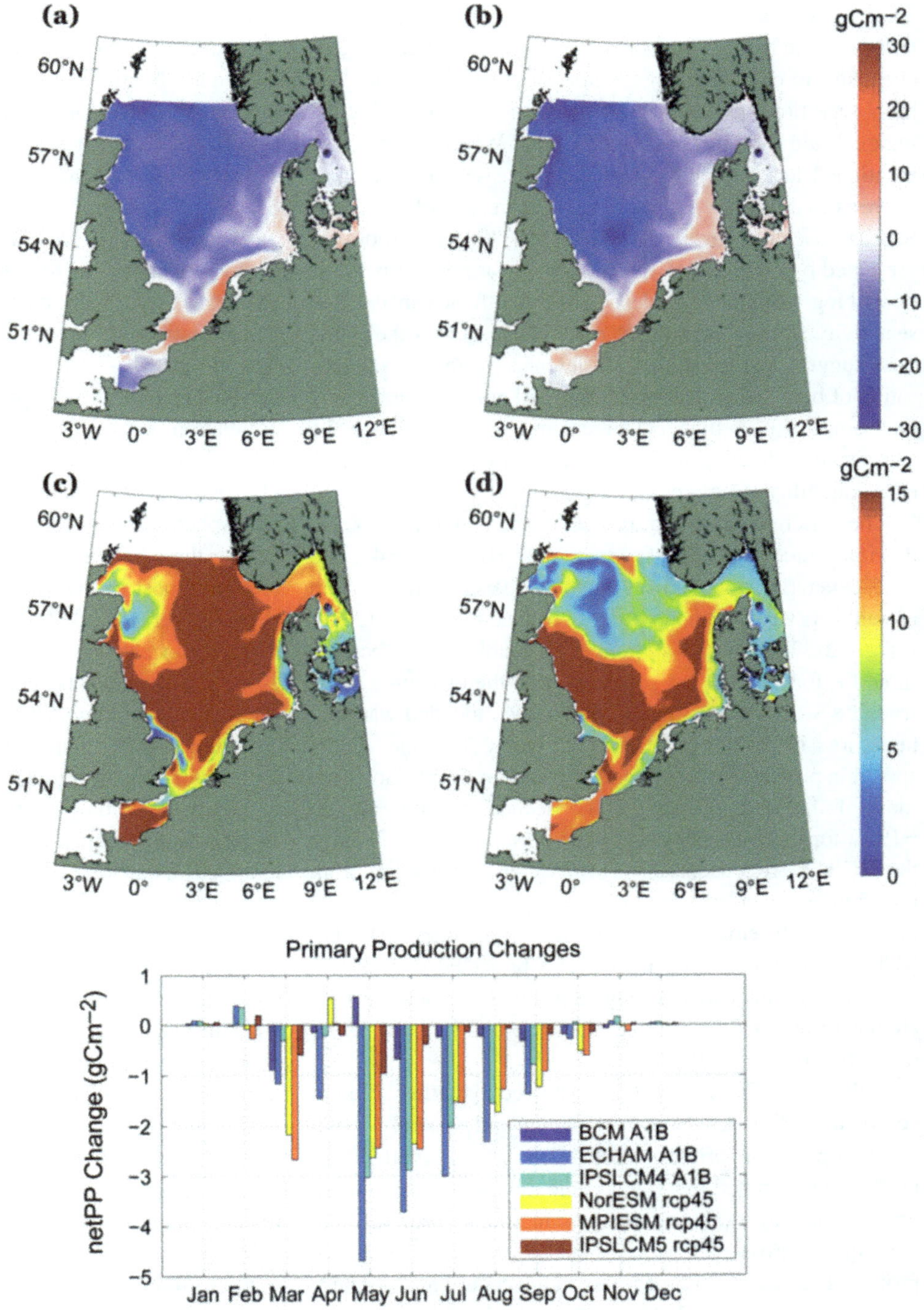

Fig. 6.22 Ensemble mean projected change in net primary production (netPP) from a six-member ensemble simulated with the regional ECOSMO model. *Upper:* **a** ensemble mean for the SRES A1B scenario simulations (forced by BCM-HAMOCC, MPIOM-HAMOCC and IPLS-CM4), **b** ensemble spread for the A1B scenarios, **c** ensemble mean for the RCP4.5 scenarios (forced by NorESM, MPIOM-HAMOCC and IPLS-CM5), and **d** ensemble spread for the RCP4.5 scenarios. *Lower*, projected monthly change in netPP from the different scenarios (redrawn from Pushpadas et al. 2015)

(2015) supported this view, given the large inter-model spread in North Sea projected changes in netPP, caused by the parent ESM (Fig. 6.22). From their six-member ensemble the most pronounced decrease in netPP was modelled by the MPIOM-HAMOCC A1B scenario, however, the projected decrease in the North Sea is weaker (19 %, exact numbers to be used with caution, due to differences in area) when using the regional ECOSMO model compared to the MPIOM-HAMOCC-zoom, which shows that there is also a large sensitivity to the biogeochemical parameterisations on the regional scale. Also, as discussed by Gröger et al. (2013), the global HAMOCC model lacks many processes considered relevant on the shelf, including temperature effects on mineralisation, resolution of the nitrogen cycle and recycled production, realistic benthic remineralisation and re-suspension of organic material. This limits performance of the model in the shallow North Sea and could affect the sensitivity of primary production to climate change. As a result, the North Sea decrease in netPP could be overestimated by the MPIOM-HAMOCC-zoom, which was also considered possible by Gröger et al. (2013). It is important to note that the correct sensitivity of the North Sea biogeochemistry to climate change is not yet clear, due to the lack of observations preventing an assessment of the different regional and global model approaches. Whether the above mentioned lacking biogeochemical processes are critical to the sensitivity of the regional biogeochemistry of the North Sea is also unknown and MPIOM-HAMOCC-zoom as presented by Gröger et al. (2013) was so far not compared to other biogeochemical models of the North Sea in detailed sensitivity studies. Studies for other seas (such as the Baltic Sea) suggest that even small differences in the parameterisation of biogeochemical processes, and not necessarily only the complexity of biogeochemical models may already be significantly affecting model sensitivity to changes in nutrient availability (Eilola et al. 2011).

The ensemble mean projected change in primary production forced by ESMs scenarios (IPCC AR4- and AR5-generation models) and inter-model spread for both ensembles were compared by Pushpadas et al. (2015; Fig. 6.22). They found a stronger ensemble mean decrease in netPP for the RCP4.5 scenarios than the ensemble mean from the SRES A1B scenario, despite the modelled warming being less in the newer RCP4.5 scenarios. The inter-model spread in projected netPP decrease was significantly lower in most of the area for the RCP4.5 scenarios, with −2.3 to −19 % for the A1B-AR4 scenarios versus 2.5 to −13 % for the RCP4.5-AR5 scenarios. However, generalisation from this finding is premature and not supported by the small number of ensemble members (three) for the A1B and RCP4.5 scenarios, respectively. The projected decrease in netPP and its inter-model ranges in the North Sea are very similar to the projected global decrease and its inter-model range (Bopp et al. 2013).

Holt et al. (2014; Fig. 6.23) showed that the different competing processes have contrasting and spatially structured impacts on primary production on the shelf. Their results demonstrated that the oceanic nutrient changes in the upper water layers due to changes in stratification and the consequent cross-shelf fluxes are the primary cause for projected on-shelf netPP decrease in the central and northern North Sea from the SRES A1B scenario forced by the IPSL-CM4.0 ESM. The increase in netPP in the shallow southern North Sea was attributed to changes in wind forcing and thermal forcing. Wind and thermal forcing contribute to faster on-shelf transport of particulate material and faster recycling of organic material due to increased temperature (Holt et al. 2014). In the central and northern North Sea modelled netPP changes were negative due to more stable thermal stratification, a signal that is weak in the central part and stronger towards the northern boundary and off the shelf. In the southern North Sea, netPP increased due to higher air temperature. In this region, the temperature impact on stratification is inconsequential and the temperature effect on biological recycling dominates. Averaged over the entire North Sea, the contributions from temperature increase leading to increases and decreases in netPP tend to cancel out, which is in accordance with results from perturbation experiments by Drinkwater et al. (2009) and Skogen et al. (2011).

Wakelin et al. (2012a) used the ECOSMO model to assess variations in regional projections arising from the forcing GCM for the combined atmospheric drivers. From these simulations no consistent atmospheric driver signal was projected for the North Sea. Both increasing and decreasing netPP were projected and there is low spatial correlation between the different projections (Fig. 6.24), similar to results presented by Holt et al. (2014, 2016). Overall, the local amplitudes of change stemming from the atmospheric drivers remain in the O(10 %). Average changes for the whole North Sea are much lower and wide areas remain almost unchanged. Comparing the atmosphere-only (Wakelin et al. 2012a) and atmosphere-ocean scenarios forced by ocean boundary nutrients from ESMs (Pushpadas et al. 2015) it can be concluded that decrease in oceanic nutrients is the dominant process in these scenarios and that consideration of changes in oceanic nutrient conditions is critical for reliable projections of future climate impact on the North Sea biological system.

Near-future scenarios performed with the NORWECOM and Delft3D-GEM/BLOOM models show minimal changes in near future netPP, using modelled chlorophyll as a proxy (Friocourt et al. 2012). However, potential effects of changes in top-down control on netPP are not addressed when using chlorophyll as a proxy and could be missed. Seasonal variation in average chlorophyll concentrations does not differ much between the control run and the future climate scenario in either model. Whereas the NORWECOM model shows a slight decrease in chlorophyll over most of the year, this is not the case for the Delft3D model. In the latter, the onset of the spring bloom, as indicated by chlorophyll, occurs earlier for the future climate scenario compared to the control run together with a slightly earlier decline in chlorophyll levels. In the NORWECOM model, the onset of the spring bloom is unchanged in the future scenario but, like the Delft3D model, shows a slightly earlier decline in autumn.

Future changes in eutrophication in the North Sea as a consequence of climate change have been investigated through scenario simulations with NORWECOM for the end of the century (Eilola et al. 2013; Skogen et al. 2014). To assess eutrophication impacts in downscaling scenarios the OSPAR Commission Common Procedure was applied (OSPAR 2005), which distinguishes between parameters in four categories (see Almroth and Skogen 2010): the degree of nutrient enrichment (Cat. I); the direct effect of nutrient enrichment, plankton growth (Cat. II); the indirect effect of

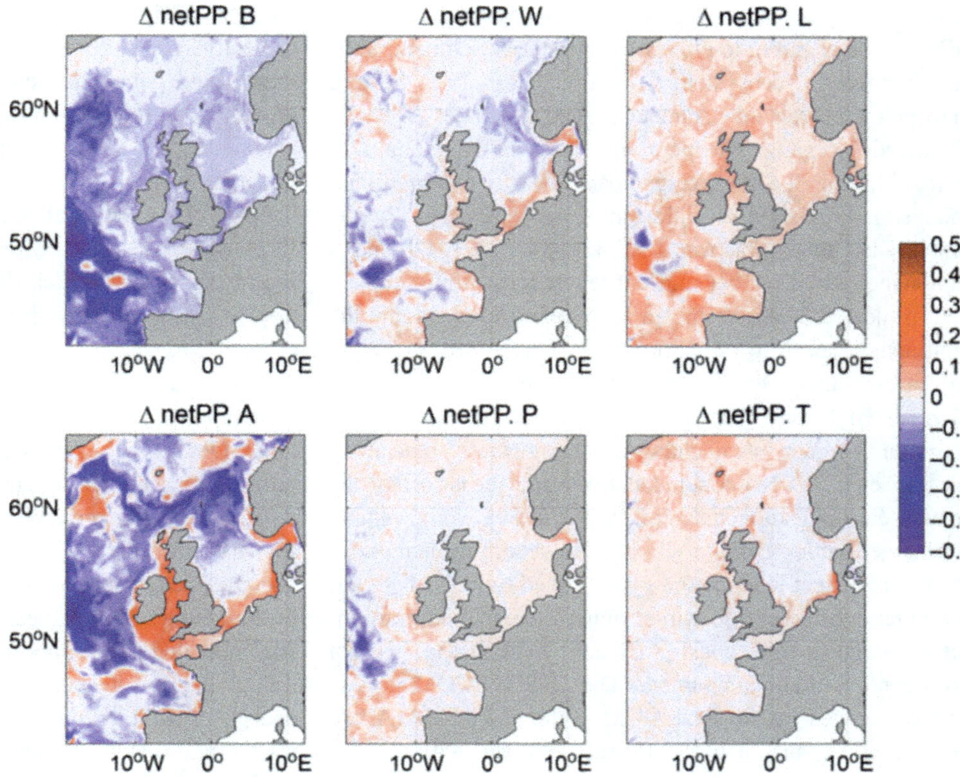

Fig. 6.23 Driver experiments with POLCOMS-ERSEM. Fractional change (estimated as future/past-1) in net primary production (netPP) associated with five external drivers and their non-linear interactions: boundary nutrients (B), wind (W), short-wave radiation (L), air temperature relative humidity (A), precipitation (P), and the direct effects of temperature on growth rates (T) (Holt et al. 2014)

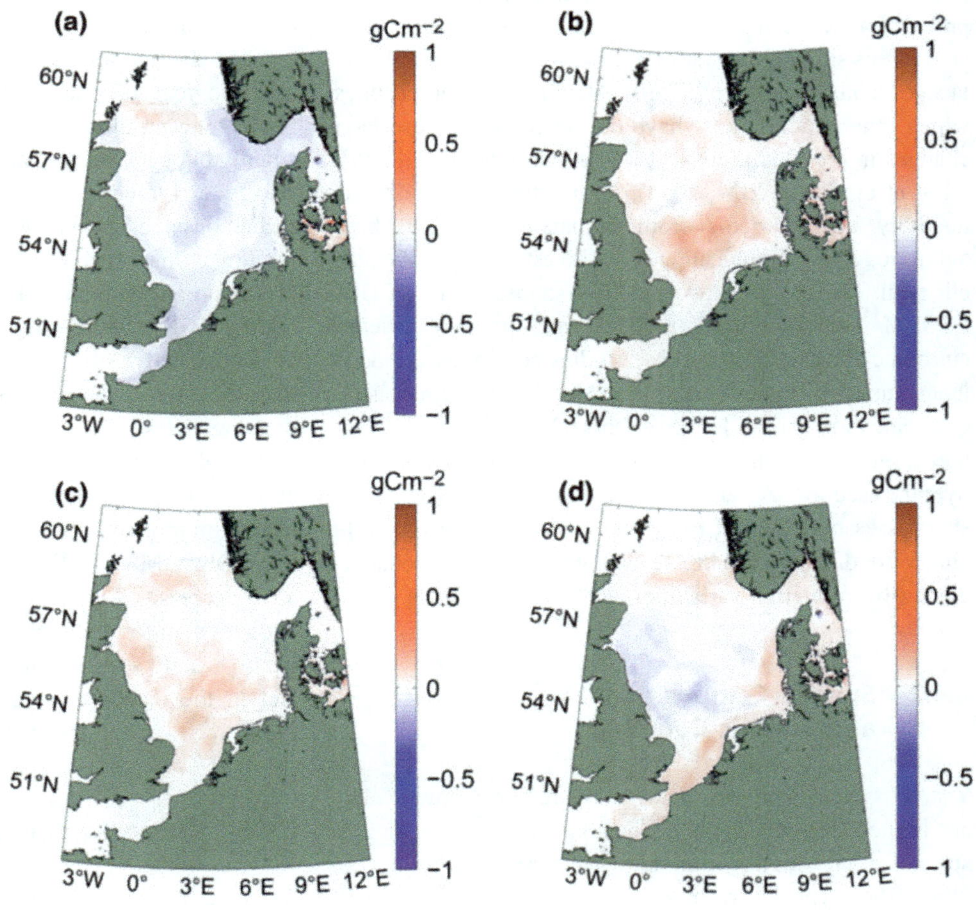

Fig. 6.24 Projected fractional changes (future/past-1) in net primary production (netPP) for the end of the century (1970–1999 vs. 2070–2100). Results from ECOMSO multi-model ensemble, atmosphere-only forced by **a** IPSL-A1B, **b** BCM-A1B, **c** ECHAM5-A1B, **d** NORESM-RCP4.5 (Wakelin et al. 2012a)

nutrient enrichment, increased oxygen consumption (Cat. III); and other possible effects of nutrient enrichment, such as changes in ecosystem structure (Cat. IV). Several eutrophication related parameters, such as winter DIN and dissolved inorganic phosphorous (DIP) and the DIN:DIP ratio, chlorophyll a, and oxygen can be easily explored with models and are used as indicators (in accordance with current management practices). Eilola et al. (2013) found a minor increase in winter nutrient levels for a future climate and projected a slight increase in phosphorus along the continental coast, while nitrogen is unchanged. A slight increase is also seen in summer chlorophyll levels in the German Bight and Kattegat, while the North Sea oxygen minimum is almost unchanged. Using these indicators, Skogen et al. (2014) concluded from one scenario that the overall eutrophication status of the North Sea would remain unchanged under a future climate. However, an increase in the river nutrient load caused by increased runoff, which has the potential to increase winter nutrient levels and eutrophication status near the coast (Zavatarelli et al. 2013a) was not considered in their assessment. The NORWECOM projection was forced by the same GCM as the ECOHAM (Alheit et al. 2012), but using different regional atmospheric, hydrodynamic and biogeochemical models. The projected nutrient levels from ECOHAM are higher near the coast, probably because increasing river loads have been considered in this simulation. A key weakness of both studies is the lack of consideration of North Atlantic nutrient changes, which other studies show have the potential to cause large changes in pre-bloom nutrient levels and thus overall netPP (Holt et al. 2012, 2014, 2016). Near-future change in productivity and nutrients from climate-driven and direct anthropogenic eutrophication were investigated by Zavatarelli et al. (2013b) with two model systems, the POLCOM-ERSEM and ECOSMO models and for two different eutrophication scenarios. These studies confirmed the dominant impact of climate control versus direct anthropogenic eutrophication control for the offshore North Sea system and the larger eutrophication impacts in the coastal zone hypothesised by Zavatarelli et al. (2013a). However, the short assessment period (only ten years) strongly limits the ability to distinguish climate change signals and changes arising from internal variability.

6.4.2 Species Composition and Trophic Coupling

The ecosystem models employed for regional downscaling are limited in terms of their potential to model changes in species composition, community structure and trophic coupling. They typically resolve plankton community structure by addressing a few functional groups such as diatoms, nitrogen-fixing bacteria and flagellates. However, this is a subjective concept and the few groups chosen are arbitrary. Also, fixed parameterisations are used for vital rates (e.g. growth rates, mortality rates), and are often used as a tuning parameter. As a result, these models have very limited potential to resolve changes in species composition, even at the base of the food web and projections are highly uncertain (Follows et al. 2007). Moreover, the models are truncated food-web models, which do not resolve coupling to higher trophic levels and the feedback inherent in this coupling (Fennel 2009).

Holt et al. (2014) investigated the sensitivity of diatoms relative to the rest of the phytoplankton community using the POLCOMS-ERSEM model. They found that both functional groups are sensitive to the projected changes, but that the amplitude of the projected changes is significantly smaller for diatoms than for the other phytoplankton functional groups. The changes are largest during the spring bloom and in the southern North Sea, when an increase in production was modelled to be supported by an accelerated growth rate. The changes were positive for all periods in both groups, except for summer, when production decreased significantly for the non-diatom groups.

Examining changes in the near future with the Delft3D-BLOOM/GEM model (four phytoplankton groups), Friocourt et al. (2012) found substantial differences in the average distribution of the different phytoplankton groups over the year, despite negligible changes in overall chlorophyll concentrations. The spring diatom bloom occurred slightly but consistently earlier in the future climate scenario. The general trend is for an increase in dinoflagellates and an earlier onset of growth for this group. In terms of factors limiting dinoflagellate growth (light-, nitrogen- and phosphorus-limitation), bloom probability and duration are higher for the future climate scenario than for the present day, irrespective of the type of growth limitation. The relative increase is largest for the nitrogen-limited type of dinoflagellates.

Wakelin et al. (2012a), Chust et al. (2014) and Pushpadas et al. (2015) found trophic amplification of the climate impact on productivity in the North Sea, based on ECOSMO and POLCOMS-ERSEM downscaling (Fig. 6.25). The relative decrease in production for the second trophic level is stronger than for the first trophic level, a phenomenon which is also widely seen in downscaling studies for other regions and in the response from the forcing ESM (Chust et al. 2014).

6.4.3 Ocean Acidification

Rising atmospheric CO_2 concentrations result in higher ocean uptake of CO_2. This in turn is driving a decrease in

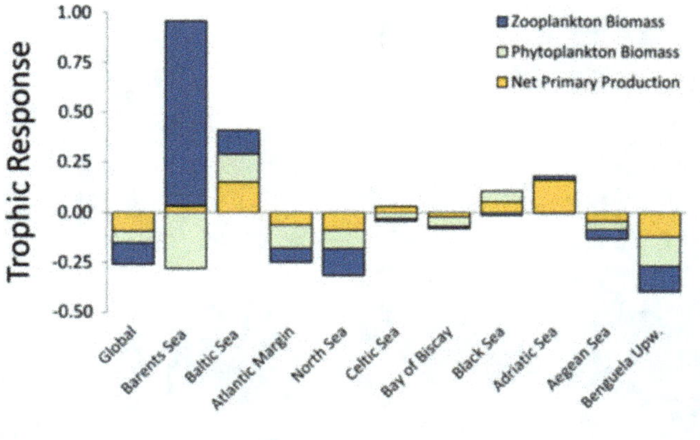

Fig. 6.25 Plankton response to the SRES A1B scenario as projected by different regional models forced by the IPSLCM4 earth system model. The graphic shows fractional change (calculated as future/past-1) for the end of the century (2080–2100) relative to present day (1980–2000) (Chust et al. 2014)

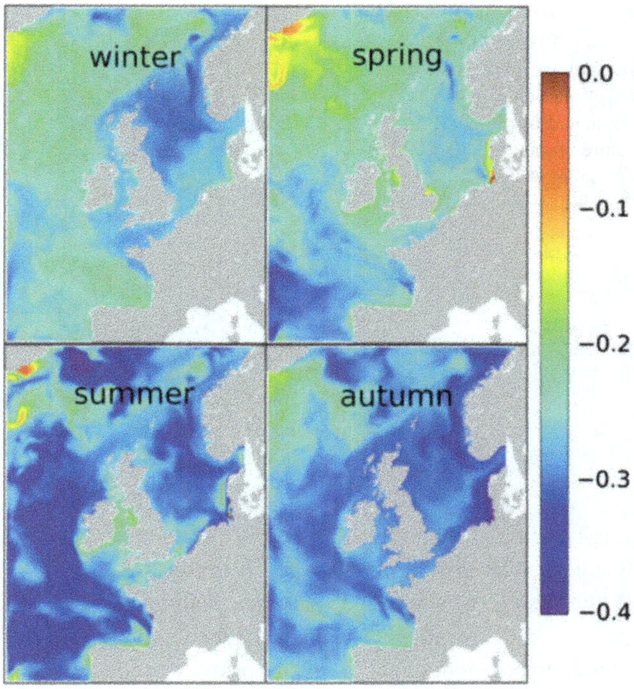

Fig. 6.26 Impacts of climate change and ocean acidification (OA) in the carbonate system as projected by the SRES A1B scenario, the graphic shows absolute difference in surface pH compared to the present-day by season (Artioli et al. 2013, 2014)

ocean pH and thus an increase in ocean acidification (OA) (also known as 'the other CO_2 problem'). However, OA is a complex process and is also influenced by climatic and biogeochemical processes. Artioli et al. (2013, 2014) investigated climate-driven impacts on OA in the North Sea using the POLCOMS-ERSEM model forced by the IPSL-CM4.0 ESM. For the end of the century, they found a significant change in annual mean pH of the order of −0.27, which is consistent in magnitude to the projected change in the annual global mean pH. The major driver for this decrease in pH was clearly the increasing atmospheric CO_2 concentration. The projected temperature rise had contrasting effects on OA in their downscaling experiment; both decreasing the solubility of CO_2, which leads to increased outgassing and lower OA, and increasing dissociation constants, which supports OA. Another but more minor effect is the decrease in total alkalinity due to the projected freshening which reduces the buffering capacity of the system. As Artioli et al. (2013) discussed, this feedback stems from assuming a simple correlation between total alkalinity and salinity (Millero 1995). Uncertainty might therefore be large and the total-alkalinity feedback remains unclear.

Biological processes were identified to be responsible for a strong modulation of the spatial and seasonal patterns of climate-driven impacts on OA, with average local variations of more than 0.4 in pH (Fig. 6.26). In highly productive areas and during the spring bloom less OA was projected. Acidification generally peaks in autumn and aragonite under-saturation was simulated in the present climate for bottom waters in the central North Sea in spring and summer, caused by community respiration and simultaneous stratification, which prohibits ventilation (Artioli et al. 2013, 2014). Artioli et al. (2014) projected an increase in the seasonal aragonite under-saturation in bottom waters in a future climate due to increased respiration in deep waters and benthic systems. The largest seasonal variations in projected pH change occurred in coastal areas; the projected local increase in netPP in this area may also have helped to reduce the projected OA increase by up to 0.1 according to projections by Artioli et al. (2013, 2014). However, uncertainty in the near-coastal projections for OA is large, since river runoff and loads have significant potential to override the OA changes and consistent scenarios and projections for river nutrients and total alkalinity loads are not available. The POLCOMS-ERSEM results are consistent with projections from the ECOSMO model using the same ESM forcing (Wakelin et al. 2012a). The ECOSMO downscaling reveals a decrease in mean North Sea pH from 8.09 to 7.87 (1980–1999 to 2080–2099; Fig. 6.27), indicating slightly weaker OA as projected by Artioli et al. (2013, 2014), and a continuation of the present almost linear trend of OA in the North Sea. The small differences in projected change in pH potentially arise from neglected seasonality in total alkalinity due to biological processes and total alkalinity changes in the ECOSMO study, as well as from different coupling sensitivity to the climate model for both models.

Increasing research efforts in recent years have improved understanding of OA and raised evidence for the broad impacts of OA on the marine ecosystem (e.g. Riebesell et al. 2007; Gattuso et al. 2011). Two such impacts namely the supporting effect of OA on primary production (Riebesell

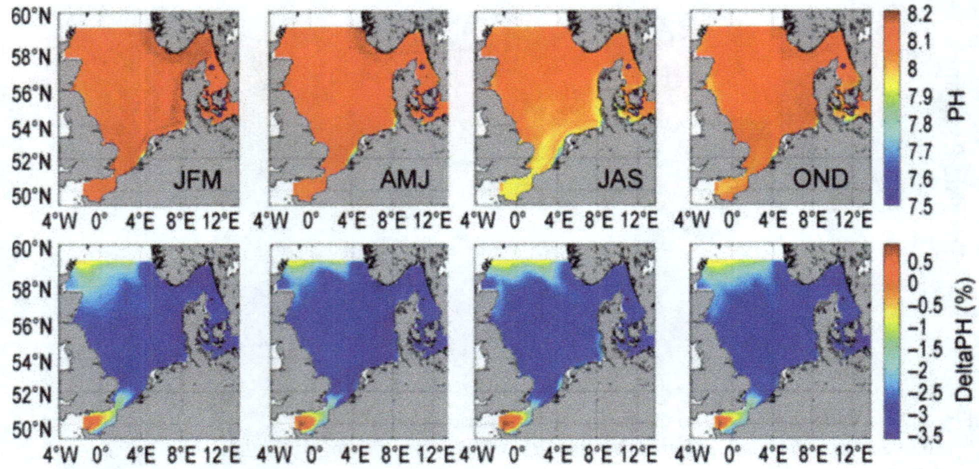

Fig. 6.27 Simulated seasonal mean pH (1980–1999) (*upper panels*) and percentage change (fractional change × 100) simulated by ECOSMO for a future climate (2080–2099) (Wakelin et al. 2012a)

and Tortell 2011) and the effect of OA on the nitrogen cycle (Gehlen et al. 2011) were studied by Artioli et al. (2013). While using a simplified parameterisation of increased growth rates for all phytoplankton functional groups, they found the potential effect of OA on primary production to be similar in magnitude to the climate-driven impact on primary production and to enhance spring production and decrease summer production. This supports a shift from flagellates to diatoms, a signal which was found to move up the trophic chain and support mesozooplankton over microzooplankton. Artioli et al. (2014) concluded that OA and climate change impacts on primary production could cancel out but could also amplify, and that regional hydrodynamics and productivity dynamics need to be taken into consideration. However, it should be mentioned that these are first attempts to model the OA impact on primary production and that a very simple parameterisation was applied to extrapolate available knowledge on a species level to the entire plankton community. This is very likely to be an oversimplification, which could strongly overestimate the potential OA effect on productivity as discussed by Artioli et al. (2013, 2014). The effect of OA on the nitrogen cycle was found to be small compared to the climate-driven impact on the nitrogen cycle via increased mineralisation due to higher temperatures (Artioli et al. 2013).

6.5 Conclusions and Recommendations

Increasing numbers of regional climate change scenario assessments became available for the North Sea and the new developments have contributed important understanding of regional processes mediating climate change impacts in the North Sea. Improved understanding of processes contributing to global sea level rise over the last decade has led to better regional projections of future changes in sea level. Better projections of future storm surges and waves are mainly due to better awareness of the factors driving change in the atmospheric storm track, and a better appreciation of the relative roles of long-term change and natural variability. Assessing climate-driven impacts on hydrography, circulation and biogeochemistry has benefited from new and advanced downscaling methods. Among these are the regional fully-coupled RCMs, initiated by the German KLIWAS project and physical-biological regional downscaling models, which were coupled with ESMs as part of the EU project MEECE. The large number of regional studies now available enables a critical review of current knowledge on climate change impacts in the North Sea region and allows the identification of challenges, robust changes, uncertainties and specific recommendations for future research.

6.5.1 Robustness and Uncertainties

Coherent findings from the climate change impact studies reviewed in this chapter include overall increases in sea level and ocean temperature, a freshening of the North Sea, an increase in ocean acidification and a decrease in primary production. In terms of the drivers of these changes, the impact of natural variability on sea surface temperature and ocean acidification is less dominant compared to projected anthropogenic changes, and their projected future changes appear to be relatively consistent among the different downscaling scenarios. This is also evident when considering GCM simulated time-series of future annual average steric sea level. Unlike atmospheric quantities such as rainfall or temperature, the climate change signal exceeds the simulated natural variability for mean sea level even for future scenarios with a high degree of emissions mitigation. Thus a rise in future global sea level is a robust result, although the precise amount remains uncertain. A projected regional temperature increase towards the end of the century also appears to be a robust result from the multi-model ensemble projections reviewed in this chapter. However, the range in projected future temperature change depends on the choice of GCM and as the range in projected changes is of the order of the amplitude of the projected change itself, the magnitude of the change cannot therefore be considered robust. On smaller spatial scales a lower signal to noise ratio is typically expected. Projected regional patterns and seasonal modulation of temperature increase are variable and their future development is uncertain. The spatial patterns of sea level rise are also more diverse among the different regional projections (Pardaens et al. 2011a) but in the latest IPCC assessment some of the spread across normalised modelled sea level change patterns appears to have been eliminated.

A general decrease in ocean pH was a consistent signal from two regional climate change projections of OA. Offshore inter-model differences in projected future ocean pH appear to be small compared to the magnitude of projected changes, which could be attributed to the strong impact of changes in atmospheric CO_2 levels on ocean pH in comparison to other internal physical and biogeochemical effects. The projected increase in regional OA for the North Sea can thus be considered robust for offshore waters, despite the small number of studies available. In contrast, the importance of terrestrial impacts near the coast is increasing and the projections are adversely affected by the lack of terrestrial coupling and lack of information on river loads and total alkalinity changes.

Wind changes have a strong impact, *inter alia*, on local sea level, storm surges, surface waves, primary production, circulation, advection of salt- and nutrient-rich water from the North Atlantic, mixing, stratification, and offshore transport of river plumes. The North Sea is located in the land-ocean transition zone of the Northwest European shelf, which is characterised by very high variability due to the alternating dominance of the maritime climate of the North Atlantic and the continental climate (e.g. Backhaus 1989; Hawkins and Sutton 2009). There are several modes of variability that are particularly important for the North Sea; the North Atlantic Oscillation (NAO), the Atlantic Multidecadal Oscillation (AMO) and the Atlantic Meridional Mode (AMM, e.g. Grossmann and Klotzbach 2009). The large natural variability has a greater impact on the local North Sea wind field than potential anthropogenic-induced trends, and strong natural climate variability from annual to multi-decadal scales (e.g. Arguez et al. 2009) is a particular challenge when developing projections of climate change in the North Sea. Regional projections for changes in wind in existing scenario simulations are not robust for the North Sea (e.g. Lowe et al. 2009; see also Chap. 5), with many GCMs still unable to accurately capture features such as the placing and timing of atmospheric pressure systems in the UK region (IPCC 2013). The long-term climate trends are superimposed on the natural modes of variability and distinguishing between the two in order to identify the anthropogenic climate change signal is one of the 'grand challenges' of climate change impact studies in marine regions. This is of particular relevance for the North Sea region where reliable predictions concerning strongly wind-influenced characteristics such as local sea level, storm surges, surface waves and thermocline depth are still impossible.

Substantial multi-decadal variability in projected climate change impacts was identified from atmospheric and sea level studies (e.g. Gaslikova et al. 2013). These multi-decadal variations bias projected changes estimated for 20- or 30-year time slices. Whether this is also relevant for ocean hydrodynamics and biogeochemistry has so far not been addressed. However, variability in wind fields appears a strong driver in hydrodynamic and biogeochemical changes in the North Sea (Skogen et al. 2011; Holt et al. 2014, 2016) and substantial multi-decadal variations are also to be expected for hydrodynamics and biogeochemistry (Daewel and Schrum, 2013).

A common regional finding for those scenarios considering future variations in oceanic nutrient conditions is a decrease in future levels of primary production (which are not always statistically significant). However, the projected decrease varies widely (from −2 to −30 %) depending on the driving ESM and the regional model used (Gröger et al. 2013; Holt et al. 2014; Pushpadas et al., 2015). Projections of future regional primary production are therefore less robust than for sea level, temperature and OA, which was also concluded by Bopp et al. (2013) for changes in global primary production projected by recent ESMs. Uncertainties in regional projections from multi-model ensembles are still large for offshore nutrient and salt fluxes and the consequent changes in netPP. Local atmospheric impacts on netPP remain dominated by natural variability and a common response in scenario simulations for lower trophic level dynamics was hardly identified for atmospheric drivers. Moreover, extending the regional models into the Baltic Sea and across the shelf break appears to be critical for the North Sea. The downscaling studies of Holt et al. (2012, 2014, 2016), Wakelin et al. (2012a), Gröger et al. (2013) and Bülow et al. (2014), showed the projected change in cross-shelf exchange probably depends on model resolution and is critically influenced by GCM biases and by the bias correction strategies used in the GCM and regional ocean climate models.

Close to the shelf, the boundary values from ESMs are impaired by a lack of consistent terrestrial coupling for nutrient loads (Regnier et al. 2013) and simplified parameterisation of physics and biogeochemical cycling in some ESMs (such as unconsidered re-suspension and tidal mixing). The reliability of the near shelf and near coastal boundary conditions from ESMs is therefore unclear, and more research is needed to improve understanding of land-ocean coupling and the regional impacts and feedbacks to the global scale. Another source of uncertainty is regional atmosphere-ocean coupling and the advantages and disadvantages of using coupled atmosphere-ocean downscaling versus uncoupled regional atmospheric models to force regional ocean models.

6.5.2 Future Challenges

The lack of consideration given to terrestrial climate change impacts and their coupling to the ocean through runoff and terrestrial carbon and nutrient loads is a major issue. Projecting terrestrial impacts on salinity and especially on nutrients, carbon chemistry and alkalinity is a challenge at both the global and regional scale. Consideration of terrestrial impacts is critical for a shelf sea like the North Sea and improved understanding of the coupled dynamics in the land-ocean transition zone is essential. Many downscaling studies for the North Sea assume that runoff from the catchment area and freshwater outflow from the Baltic Sea will not change under a future climate (e.g. Wakelin et al. 2012a). Only the MPIOM-REMO simulation (see Bülow et al. 2014) closes the water cycle, although a freshwater flux-correction is also used here at the global scale (Sein et al. 2015), which could introduce artificial sources and sinks and bias the projected changes to an unknown degree. To date, no attempt has been made to include changes in terrestrial nutrient loads or alkalinity at the regional scale, nor is this standard for ESMs (Regnier et al. 2013). Although the impact of changes in runoff and river loads and in Baltic Sea outflow properties is probably restricted to the southern coastal North Sea and the Skagerrak, respectively, a more consistent approach is needed to address the water and nutrient budget of the North Sea, one which should consider the entire land-ocean continuum. A new hydrological model, HYPE (HYdrological Predictions for the Environment; Lindström et al. 2010; Arheimer et al. 2012), was recently developed to calculate river flow and river-borne nutrient loads from all European catchment areas; this is known as E-HYPE. In the future, scenario simulations using HYPE should generate more consistent changes in water and nutrient budgets. But despite these recent efforts, the uncertainties in runoff for the end of the 21st century will still be considerable due to precipitation biases in regional atmospheric models, as illustrated by Donnelly et al. (2014) for the Baltic Sea. Projections of nutrient loads are even more uncertain than projections of river flow due to unknown future land use and socio-economic scenarios (Arheimer et al. 2012). Plus, the carbon cycle and carbon loads are still not considered in the present version of HYPE and coupled land-ocean carbon scenarios remain for future work.

Other relevant factors include biogeochemical parameterisations, which have substantial impacts on structuring the ecosystem. For the North Sea system, sediment-water exchange and its parameterisation are particularly important. Further limitations are inherent in present-day regional biogeochemistry models. Changes in plankton community structure (e.g. Follows et al. 2007) and consistent trophic coupling also including higher trophic levels (e.g. Fennel 2009) are not yet incorporated and current models are too simple to provide reliable estimates of changes in community structure or trophic coupling. To date, they are only able to provide first indications of climate change impacts on trophic controls and community structure (e.g. Chust et al. 2014; Holt et al. 2014). Projecting future OA impacts on the regional scale requires better understanding of OA impacts on productivity, which could affect first-order impacts from changes in atmospheric CO_2 levels (Artioli et al. 2013, 2014).

Regardless of the specific methods employed, the downscaled simulations and regional studies are ultimately affected by the driving GCM or ESM. Despite improvements (e.g. Scaife et al. 2010), the latest generation of GCMs and ESMs still has significant biases and the spread in projected global warming among GCMs has not changed from IPCC AR4 to AR5 (e.g. Knutti and Sedláček 2012). Moreover, additional uncertainties arise from the downscaling methods and regional models used (e.g. Holt et al. 2014). From the simulations presented in this chapter it is clear that identifying best practice in climate downscaling is far from trivial and not yet achieved. A range of different approaches is currently used, each with advantages and disadvantages. A review of the literature shows that choice of regional model is not critical for the projected mean change, but is crucial for the projected spatial and seasonal patterns of regional climate change impacts. Regional models are sensitive to climate model biases and bias corrections are necessary for many applications, such as in modelling seasonal cycles of stratification and biological productivity. But bias correction also affects the sensitivity of regional systems to climate change impacts and might shift a critical change in one variable to a non-critical range and vice versa (Holt et al. 2014). Future work is required on the effects of global and regional model bias on regional dynamics and sensitivities to climate change impacts.

While models and observations have long been used together for validation, bias correction and re-analysis there is a movement in general circulation modelling and earth system science towards the use of observations as a constraint on the future change projected by climate models (e.g. Murphy et al. 2004; Stott and Kettleborough 2002; Cox et al.

2013). In some cases (e.g. Stott and Kettleborough 2002), the approach is closely related to optimal detection of past climate change and uses the model's ability to simulate past change in order to derive a distribution of signal weights that are assumed to apply in the future. In other approaches (e.g. Murphy et al. 2004), different model versions, using a range of different parameterisations, are used to simulate the future but the distribution of possible future scenarios is weighted according to each model's ability to simulate recent climatology. While these methods are still evolving they present an exciting opportunity for improving the projection of North Sea climate change. First, the methods could allow a weighting of the boundary conditions to regional simulations. Later they might also be applied directly to regional simulations. Considerable further work is required to understand the relevant constraints and optimum statistical framework for applying them. The result could be a narrowing of the uncertainty range of future projections as the observed signal of climate change becomes ever stronger.

Regional climate change impact assessment in the coastal area requires a greater degree of accuracy and more detailed process consideration than is currently available from GCMs and ESMs. GCM resolution is constantly improving, but so is resolution at the regional scale and the demand for detailed knowledge is expanding. Local planning in relation to climate change impacts might in future require unstructured grids or further local downscaling, as shown by Gräwe and Burchard (2012) for the western Baltic Sea or Zhang et al. (2015) for the North Sea and Baltic Sea. Combined assessment of climatic and direct human impacts such as eutrophication, fisheries, offshore construction, mining and dredging is increasingly required. Downscaling methods are therefore as necessary now as in the future and coupling regional downscaling models to down-stream impact models (such as for inundation or biological production at higher trophic levels; e.g. Daewel et al. 2008) is becoming increasingly important. Regional and seasonal patterns are very important for down-stream impact models and the application of such models requires the identification of robust patterns and estimates of uncertainty, making ensemble simulations, including larger sets of GCMs and regional models mandatory for the North Sea. Comparing results from different models to identify the presence or absence of robust change requires, as a minimum, harmonisation of experiment design, along the lines of the Coupled Model Intercomparison Project (CMIP, http://cmip-pcmdi.llnl.gov), to set standards that endure beyond the length of individual projects. This in turn requires a greater degree of organisation and resources in regional downscaling for the ocean.

References

Ådlandsvik B (2008) Marine downscaling of a future climate scenario for the North Sea. Tellus A 60:451–458

Ådlandsvik B, Bentsen M (2007) Downscaling a twentieth century global climate simulation to the North Sea. Ocean Dyn. 57:453–466

Alheit J, Ådlandsvik B, Boersma M, Daewel U, Diekmann R, Flöter J, Hufnagl M, Johannessen T, Kong S-M, Mathis M, Möllmann C, Peck M, Pohlmann T, Temming A, Shchekinova E, Skogen M, Sundby S, Wagner C (2012) ERANET MarinERA: Verbundprojekt 189592 ECODRIVE. Förderkennzeichen: 03F603, Schlußbericht. Bundesministerium für Bildung und Forschung, Bonn

Allen MR, Ingram WSJ (2002) Constraints on future changes in climate and the hydrologic cycle. Nature 419:224

Almroth E, Skogen MD (2010) A North Sea and Baltic Sea Model Ensemble Eutrophication Assessment. AMBIO 39:59–69

Arguez A, O'Brien JJ, Smith SR (2009) Temperature impacts over Eastern North America and Europe associated with low-frequency North Atlantic SST variability. Int J Climatol 29:1–10

Arheimer B, Dahne J, Donnelly C (2012) Climate change impact on riverine nutrient load and land-based remedial measures of the Baltic Sea Action Plan. AMBIO 41:600–612

Artioli Y, Blackford JC, Butenschön M, Holt J, Wakelin SL, Thomas H, Borges AV, Allen JI (2012) The carbonate system in the North Sea: Sensitivity and model validation. J Mar Sys 102-104:1–13

Artioli Y, Blackford JC, Nondal G, Bellerby RGJ, Wakelin SL, Holt JT, Butenschön M, Allen JI (2013) Heterogeneity of impacts of high CO_2 on the North Western European Shelf. Biogeosci Discuss 10:9389–9413

Artioli Y, Blackford JC, Nondal G, Bellerby RGJ, Wakelin SL, Holt JT, Butenschön M (2014) Heterogeneity of impacts of high CO_2 on the North Western European Shelf. Biogeosciences 11: 601–612

Backhaus J (1989) The North Sea and the climate. Dana 8:69–82

Bentsen M, Bethke I, Debernard JB, Iversen T, Kirkevåg A, Seland Ø, Drange H, Roelandt C, Seierstad IA, Hoose C, Kristjánsson JE (2012) The Norwegian Earth System Model, NorESM1-M – Part 1: Description and basic evaluation. Geosci Model Dev Discuss 5:2843–2931

Bindschadler RA, Nowicki S, Abe-Ouchi A, Aschwanden A, Choi H, Fastook J, Granzow G, Greve R, Gutowski G, Herzfeld U, Jackson C, Johnson J, Khroulev C, Levermann A, Lipscomb WH, Martin MA, Morlighem M, Parizek BR, Pollard D, Price SF, Ren D, Saito F, Sato T, Seddik H, Seroussi H, Takahashi K, Walker R, Wang WL (2013) Ice-sheet model sensitivities to environmental forcing and their use in projecting future sea level (the SeaRISE project). J Glaciol 59:195–224

Blackford JC, Allen JI, Gilbert FJ (2004) Ecosystem dynamics at six contrasting sites: A generic modelling study. J Mar Sys 52:191–215

Blauw AN, Los FJ, Bokhorst M, Erftemeijerm PLA (2008) GEM: a generic ecological model for estuaries and coastal waters. Hydrobiologia 618:175–198

Bopp L, Resplandy L, Orr JC, Doney SC, Dunne JP, Gehlen M, Halloran P, Heinze C, Ilyina T, Séférian R, Tjiputra J, Vichi M (2013) Multiple stressors of ocean ecosystems in the 21st century: projections with CMIP5 models. Biogeosciences 10:6225–6245

Bosello F, Nicholls RJ, Richards J, Roson R, Tol RSJ (2012) Economic impacts of climate change in Europe: sea-level rise. Clim Chan 112:63–81

Bouttes N, Gregory JM, Kuhlbrodt T, Suzuki T (2012) The effect of windstress change on future sea level change in the Southern Ocean. Geophys Res Lett 39: L23602, doi:10.1029/2012GL054207.

Bouttes N, Gregory JM, Lowe JA (2013) The reversibility of sea level rise. J Clim 26:2502–2513

Bülow K, Dieterich C, Heinrich H, Hüttl-Kabus S, Klein B, Mayer B, Meier HEM, Mikolajewicz U, Narayan N, Pohlmann T, Rosenhagen G, Sein D, Su J (2014) Comparison of 3 coupled models in the North Sea region under todays and future climate conditions. KLIWAS Schriftenreihe. www.bsh.de/de/Meeresdaten/Beobachtungen/Klima-Anpassungen/Schriftenreihe/27-2014.pdf

Christensen OB, Drews M, Christensen JH, Dethloff K, Ketelsen K, Hebestadt I, Rinke A, Weidman JP (2007) The HIRHAM Regional Climate Model: Version 5 (β). Tech Rep 06-17, Danish Meteorological Institute

Church JA, White NJ (2011) Sea-level rise from the late 19th to the early 21st century. Surv Geophys 32:585–602

Church JA, White NJ, Konikow LF, Domingues CM, Cogley G, Rignot E, Gregory JM, van den Broeke MR, Monaghan AJ, Velicogna I (2011) Revisiting the Earth's sea-level and energy budgets from 1961 to 2008. Geophys Res Lett 38, L18601, doi:10.1029/2011GL048794

Chust G, Allen I, Bopp L, Schrum C, Holt J, Tsiaras K, Zavatarelli M, Chifflet M, Cannaby H, Dadou I, Daewel U, Wakelin SL, Machu E, Pushpadas D, Butenschon M, Artioli Y, Petihakis G, Smith C, Garçon V, Goubanova K, Le Vu B, Fach BA, Salihoglu B, Clementi E, Irigoien X (2014) Biomass changes and trophic amplification of plankton in a warmer ocean. Glob Chan Biol 20:2124–2139

Cox PM, Pearson D, Booth BB, Friedlingstein P, Huntingford C, Jones CD, Luke CM (2013) Sensitivity of tropical carbon to climate change constrained by carbon dioxide variability. Nature 494:341–344

Daewel U, Schrum C (2013) Simulating long-term dynamics of the coupled North and Baltic Sea ecosystem with ECOSMO II: model description and validation. J Mar Sys 119–120:30–49

Daewel U, Peck MA, Alekseeva I, St. John MA, Kühn W, Schrum C (2008) Coupling ecosystem and individual-based models to simulate the influence of climate variability on potential growth and survival of larval fish in the North Sea. Fish Oceanogr 17-5:333–351

Dangendorf S, Rybski D, Mudersbach C, Müller A, Kaufmann E, Zorita E, Jensen J (2014) Evidence for long-term memory in sea level. Geophys Res Lett 41:5530–5537

de Winter RC, Sterl A, de Vries JW, Weber SL, Ruessink G (2012) The effect of climate change on extreme waves in front of the Dutch coast. Ocean Dyn 62:1139–1152

Debernard J, Røed L (2008) Future wind, wave and storm surge climate in the Northern Seas: a revisit. Tellus 60:427–438

Debernard J, Sætra Ø, Røed L (2002) Future wind, wave and storm surge climate in the Northern Seas. Clim Res 23:39–49

Dieterich C, Schimanke S, Wang S, Väli G, Liu Y, Hordoir R, Axell L, Höglund A, Meier HEM (2013) Evaluation of the SMHI coupled atmosphere-ice-ocean model RCA4-NEMO. SMHI Rep Oceanogr 47

Dippner JW, Ottersen G (2001) Cod and climate variability in the Barents Sea. Clim Res 17:73–82

Donnelly C, Yang W, Dahné J (2014) River discharge to the Baltic Sea in a future climate. Clim Chan 122:157–170

Döscher R, Willén U, Jones C, Rutgersson A, Meier HEM, Hansson U, Graham LP (2002) The development of the regional coupled ocean-atmosphere model RCAO. Boreal Env Res 7:183–192

Drinkwater K, Skogen M, Hjøllo S, Schrum C, Alekseeva I, Huret M, Léger F (2008) The effects of a future climate on the physical oceanography and comparisons of the mean and variability of the physical properties with present day conditions. RECLAIM deliverable 4.1.

Drinkwater K, Skogen M, Hjøllo S, Schrum C, Alekseeva I, Huret M, Woillez M (2009) Report on the effects of future climate change on primary and secondary production as well as ecosystem structure in lower trophic to mid- trophic levels. RECLAIM deliverable 4.2.

Eilola K, Gustafsson BG, Kuznetsov I, Meier HEM, Neumann T, Savchuk OP (2011) Evaluation of biogeochemical cycles in an ensemble of three state-of-the-art numerical models of the Baltic Sea. J Mar Syst 88:267–284

Eilola K, Hansen J, Meier HEM, Molchanov MS, Ryabchenko, Skogen, MD (2013) Eutrophication Status Report of the North Sea, Skagerrak, Kattegat and the Baltic Sea: A model study. Present and future climate. Tech Rep Oceanogr XX/2013, SMHI

Enderlin EM, Howat IM, Jeong S, Noh MJ, van Angelen JH, van den Broeke MR (2014) An improved mass budget for the Greenland ice sheet. Geophys Res Lett 41:866–872

Favier L, Durand G, Cornford SL, Gudmundsson GH, Gagliardini O, Gillet-Chaulet F, Zwinger T, Payne AJ, Le Brocq AM (2014) Retreat of Pine Island Glacier controlled by marine ice-sheet instability. Nat Clim Chan 4:117–121

Fennel W (2009) Parameterizations of truncated food web models from the perspective of an end-to-end model approach. J Mar Sys 76:171–185

Flather RA, Smith JA (1998) First estimates of changes in extreme storm surge elevations due to the doubling of CO_2. Glob Atmos Ocean Sys 6:193–208

Follows MJ, Dutkiewicz S, Grant S, Chisholm SW (2007) Emergent biogeography of microbial communities in a model ocean. Science 315:1843–1846

Friocourt YF, Skogen M, Stolte W, Albretsen J (2012) Marine downscaling of a future climate scenario in the North Sea and possible effects on dinoflagellate harmful algal blooms. Food Addit Contam A 29:1630–1646

Furevik T, Bentsen M, Drange H, Kindem IKT, Kvamstø NG, Sorteberg A (2003) Description and validation of the Bergen Climate Model: ARPEGE coupled with MICOM. Clim Dyn 21:27–51

Gaslikova L, Grabemann I, Groll N (2013) Changes in North Sea storm surge conditions for four transient future climate realizations. Nat Hazards 66:1501–1518

Gattuso JP, Bijma J, Gehlen M, Riebesell U, Turley C (2011) Ocean acidification: knowns, unknowns and perspectives. In: Gattuso JP, Hansson L (eds) Ocean Acidification. Oxford University Press

Gehlen M, Gruber N, Gangstø R, Bopp L, Oschlies A (2011) Biogeochemical consequences of ocean acidification and feedbacks to the earth system. In: Gattuso JP, Hansson L (eds) Ocean Acidification. Oxford University Press

Giesen RH, Oerlemans J (2013) Climate-model induced differences in the 21st century global and regional glacier contributions to sea-level rise. Clim Dyn 41:3283–3300

Gordon C, Cooper C, Senior CA, Banks H, Gregory JM, Johns TC, Mitchell JFB, Wood RA (2000) The simulation of SST, sea ice extents and ocean heat transports in a version of the Hadley Centre coupled model without flux adjustments. Clim Dyn 16:147–168

Grabemann I, Weisse R (2008) Climate change impact on extreme wave conditions in the North Sea: an ensemble study. Ocean Dyn 58:199–212

Grabemann I, Groll N, Möller J, Weisse R (2015) Climate change impact on North Sea wave conditions: a consistent analysis of ten projections. Ocean Dyn 65:255–267

Graversen RG, Drijfhout S, Hazeleger W, van de Wal R, Bintanja R, Helsen M (2011) Greenland's contribution to global sea-level rise by the end of the 21st century. Clim Dyn 37:1427–1442

Gräwe U, Burchard H (2012) Storm surges in the Western Baltic Sea, the present and a possible future. Clim Dyn 39:165–183

Gräwe U, Burchard H, Müller M, Schuttelaars HM (2014) Seasonal variability of M_2 and M_4 tidal constituents and its implications for the coastal residual sediment transport. Geophys Res Lett 41: 5563–5570

Gregory JM, Church JA, Boer GJ, Dixon KW, Flato GM, Jackett DR, Lowe JA, O'Farrell SP, Roeckner E, Russell GL, Stouffe, RJ, Winton M (2001) Comparison of results from several AOGCMs for global and regional sea-level change 1900-2100. Clim Dyn 18: 225–240

Gröger M, Maier-Reimer E, Mikolajewicz U, Moll A, Sein D (2013) NW European shelf under climate warming: implications for open ocean – shelf exchange, primary production, and carbon absorption. Biogeosciences 10:3767–3792

Groll N, Grabemann I, Gaslikova L (2014) North Sea wave conditions: an analysis of four transient future climate realizations. Ocean Dyn 64:1–12

Grossmann I, Klotzbach PJ (2009) A review of North Atlantic modes of natural *variability* and their driving mechanisms. J Geophys Res, Atmos 114: D24107, doi:10.1029/2009JD012728

Haigh I, Nicholls R, Wells N (2010) Assessing changes in extreme sea levels: Application to the English Channel, 1900-2006. Cont Shelf Res 30:1042–1055

Haigh ID, Wahl T, Rohling EJ, Price RM, Pattiaratchi CB, Calafat FM, Dangendorf S (2014) Timescales for detecting a significant acceleration in sea level rise. Nat Commun 5:3635. doi:10.1038/ncomms4635

Hasselmann S, Hasselmann K, Bauer E, Janssen PAEM, Komen GJ, Bertotti L, Lionello P, Guillaume A, Cardone VC, Greenwood JA, Reistad M, Zambresky L, Ewing JA (1988) The WAM model - A third generation ocean wave prediction model. J Phys Oceanog 18:1775–1810

Hawkins E, Sutton R (2009) The potential to narrow uncertainties in regional Climate predictions. Bull Am Met Soc 90:1095–1107

Hinkel J, van Vuuren DP, Nicholls RJ, Klein RJT (2013) The effects of adaptation and mitigation on coastal flood impacts during the 21st century. An application of the DIVA and IMAGE models. Clim Chan 117:783–794

Holt JT, James ID (2001) An s-coordinate density evolving model of the North West European Continental Shelf. Part 1 Model description and density structure. J Geophys Res 106:14015–14034

Holt J, Wakelin S, Lowe J, Tinker J (2010) The potential impacts of climate change on the hydrography of the northwest European Continental shelf. Prog Oceanogr 86:361–379

Holt J, Butenschon M, Wakelin SL, Artioli Y, Allen JI (2012) Oceanic controls on the primary production of the northwest European continental shelf: model experiments under recent past conditions and a potential future scenario. Biogeosciences 9:97–117

Holt J, Schrum C, Cannaby H, Daewel U, Allen I, Artioli Y, Bopp L, Butenschon M, Fach B, Harle J, Pushpadas D, Salihoglu B, Wakelin S (2014) Physical processes mediating climate change impacts on regional sea ecosystems. Biogeosci Discuss 11: 1909–1975

Holt J, Schrum C, Cannaby H, Daewel U, Allen I, Artioli Y, Bopp L, Butenschon M, Fach B, Harle J, Pushpadas D, Salihoglu B, Wakelin S (2016) Potential impacts of climate change on the primary production of regional seas: a comparative analysis of five European seas. Prog Oceanogr 140:91–115

Horsburgh KJ, Wilson C (2007) Tide-surge interaction and its role in the distribution of surge residuals in the North Sea. J Geophys Res, Oceans:112:C08003, doi:10.1029/2006JC004033

Howard T, Lowe J, Horsburgh K (2010) Interpreting century-scale changes in southern North Sea storm surge climate derived from coupled model simulations. J Clim 23:6234–6247

Howard T, Pardaens AK, Lowe JA, Ridley J, Hurkmans R, Bamber J, Spada G, Vaughan D (2014) Sources of 21st century regional sea level rise along the coast of North-West Europe. Ocean Sci Discuss 10:2433–2459

Hunter JR, Church JA, White NJ, Zhang X (2013) Towards a global regionally varying allowance for sea-level rise. Ocean Engin 71: 17–27

Huybrechts P, Goelzer H, Janssens I, Driesschaert E, Fichefet T, Goosse H, Loutre MF (2011) Response of the Greenland and Antarctic ice sheets to multi-millennial greenhouse warming in the Earth System Model of Intermediate Complexity LOVECLIM. Surv Geophys 32:397–416

IPCC (2007) Contribution of Working Group I to the Fourth Assessment Report of the Intergovernmental Panel on Climate Change, 2007. Solomon S, Qin D, Manning M, Chen Z, Marquis M, Averyt KB, Tignor M, Miller HL (eds) Cambridge University Press

IPCC (2013) Climate Change 2013: The Physical Science Basis. Contribution of Working Group I to the Fifth Assessment Report of the Intergovernmental Panel on Climate Change. Stocker TF, Qin D, Plattner GK, Tignor M, Allen SK, Boschung J, Nauels A, Xia Y, Bex V, Midgley PM (eds). Cambridge University Press

Jacob D, Podzun R (1997) Sensitivity studies with the regional climate model REMO. Meteorol Atmos Phys 63:119–129

Janssen F, Schrum C, Backhaus JO (1999) A climatological data set of temperature and salinity for the Baltic Sea and the North Sea. Dt Hydrogr Z 51:5–245

Jevrejeva S, Moore JC, Grinsted A (2012) Sea level projections to AD2500 with a new generation of climate change scenarios. Glob Planet Chan 80-81:14–20

Jones R, Murphy J, Hassell D, Taylor R (2001) Ensemble mean changes in a simulation of the European climate of 2071-2100 using the new Hadley Centre regional modelling system: HadAM3H/HadRM3H. Hadley Centre, Met Office, UK. http://prudence.dmi.dk/public/publications/hadley_200208.pdf

Katsman C, Hazeleger W, Drijfhout S, Oldenborgh G, Burgers G (2008) Climate scenarios of sea level rise for the northeast Atlantic Ocean: a study including the effects of ocean dynamics and gravity changes induced by ice melt. Clim Chan 91:351–374

Katsman CA, Sterl A, Beersma JJ, van den Brink HW, Church JA, Hazeleger W, Kopp RE, Kroon D, Kwadijk J, Lammersen R, Lowe J, Oppenheimer M, Plag H-P, Ridley J, von Storch H, Vaughan DG, Vellinga P, Vermeersen LLA, van de Wal RSW, Weisse R (2011) Exploring high-end scenarios for local sea level rise to develop flood protection strategies for a low-lying delta-the Netherlands as an example. Clim Chan 109:617–645

Kauker F (1999) Regionalization of Climate Model Results for the North Sea. PhD thesis. University of Hamburg

Kauker F, Langenberg H (2000) Two models for the climate change related development of sea levels in the North Sea - a comparison. Clim Res 15:61–67

Kauker F, von Storch H (2000) Statistics of "Synoptic Circulation Weather" in the North Sea as derived from a multiannual OGCM simulation. J Phys Oceanogr 30:3039–3049

Khan SA, Kjaer KH, Bevis M, Bamber JL, Wahr J, Kjeldsen KK, Bjork AA, Korsgaard NJ, Stearns LA, van den Broeke MR, Liu L, Larsen NK, Muresan IS (2014) Sustained mass loss of the northeast

Greenland ice sheet triggered by regional warming. Nat Clim Chan 4:292–299

Kharin V, Zwiers FW, Zhang X, Hegerl GC (2007) Changes in temperature and precipitation extremes in the IPCC ensemble of global coupled model simulations. J Clim 20:1419–1444

KNMI (2014a) Climate Change scenarios for the 21st Century: A Netherlands perspective. Scientific Report WR2014-01, KNMI, DeBilt, The Netherlands

KNMI (2014b) KNMI'14 climate scenarios for the Netherlands: A guide for professionals in climate adaptation. KNMI, De Bilt, The Netherlands

Knutti R, Sedláček J (2012) Robustness and uncertainties in the new CMIP5 climate model projections. Nature Clim change 3:369–373

Koerper J, Höschel I, Lowe JA, Hewitt CD, Salas y Melia D, Roeckner E, Huebener H, Royer J-F, Dufresne J-F, Pardaens A, Giorgetta MA, Sanderson MG, Otterå OH, Tjiputra J, Denvil S (2013) The effects of aggressive mitigation on steric sea level rise and sea ice changes. Clim Dyn 40:531–550

Landerer FW, Jungclaus JH, Marotzke J (2007) Regional dynamic and steric sea level change in response to the IPCC-A1B scenario. J Phys Oceanogr 37:296–312

Langenberg H, Pfizenmayer A, von Storch H, Sundermann J (1999) Storm-related sea level variations along the North Sea coast: natural variability and anthropogenic change. Cont Shelf Res 19:821–842

Lesser GR, Roelvink JA, van Kester JATM, Stelling GS (2004) Development and validation of a three-dimensional morphological model. Coast Eng 51:883–915

Levermann A, Clark PU, Marzeion B, Milne GA, Pollard D, Radic V, Robinson A (2013) The multimillennial sea-level commitment of global warming. Proc Nat Acad Sci USA 110:13745–13750

Lindström G, Pers C, Rosberg J, Strömqvist J, Arheimer B (2010) Development and test of the HYPE (Hydrological Predictions for the Environment) model – A water quality model for different spatial scales. Hydrol Res 41:295–319

Lowe JA, Gregory JM (2005) The effects of climate change on storm surges around the United Kingdom. Phil Trans Roy Soc A 363:1313–1328

Lowe JA, Gregory JM (2006) Understanding projections of sea level rise in a Hadley Centre coupled climate model. J Geophys Res, Oceans 111:C11014, doi:10.1029/2005JC003421

Lowe JA, Greogry JM (2010) A sea of uncertainty. Nature Reports Clim Change doi:10.1038/climate.2010.30

Lowe JA, Gregory JM, Flather RA (2001) Changes in the occurrence of storm surges around the United Kingdom under a future climate scenario using a dynamic storm surge model driven by the Hadley Centre climate models. Clim Dyn 18:179–188

Lowe JA, Gregory JM, Ridley J, Huybrechts P, Nicholls RJ, Collins M (2006) The role of sea-level rise and the Greenland ice sheet in dangerous climate change: implications for the stabilisation of climate. In: Schellnhuber J, Cramer W, Nakicenovic N, Wigley T, Yohe G (eds) Avoiding Dangerous Climate Change. Cambridge University Press

Lowe JA, Howard TP, Pardaens A, Tinker J, Holt J, Wakelin S, Milne G, Leake J, Wolf J, Horsburgh K, Reeder T, Jenkins G, Ridley J, Dye S, Bradley S (2009) UK Climate Projections science report: Marine and coastal projections. Hadley Centre, Met Office, UK

Lowe JA, Woodworth PL, Knutson T, McDonald RE, McInnes KL, Woth K, von Storch H, Wolf J, Swail V, Bernier NB, Gulev S, Horsburgh KJ, Unnikrishnan AS, Hunter JR, Weisse R (2010) Past and Future Changes in Extreme Sea Levels and Waves, in Understanding Sea-Level Rise and Variability. Wiley-Blackwell

Madsen KS (2009) Recent and future climatic changes in temperature, salinity, and sea level of the North Sea and the Baltic Sea. PhD thesis, Niels Bohr Institute, University of Copenhagen

Maier-Reimer E, Mikolajewicz U, Hasselmann K (1993) Mean circulation of the Hamburg LSG OGCM and its sensitivity to the thermohaline surface forcing. J Phys Oceanogr 23:731–757

Maier-Reimer E, Kriest I, Segschneider J, Wetzel P (2005) The Hamburg oceanic carbon cycle circulation model HAMOCC5.1—Technical Description Release 1.1, Tech Rep 14, Reports on Earth System Science, Max Planck Institute for Meteorology, Hamburg

Marsland SJ, Haak H, Jungclaus JH, Latif M, Roeske F (2003) The Max-Planck-Institute global ocean/sea ice model with orthogonal curvilinear coordinates. Ocean Model 5:91–127

Marti O, Braconnot P, Dufresne J-L, Bellier J, Benshila R, Bony S, Brockmann P, Cadule P, Caubel A, Codron F, de Noblet N, Denvil S, Fairhead L, Fichefet T, Foujols M-A, Friedlingstein P, Goosse H, Grandpeix J-Y, Guilyardi E, Hourdin F, Idelkadi A, Kageyama M, Krinner G, Lévy C, Madec G, Mignot J, Musat I, Swingedouw D, Talandier C (2010) Key features of the IPSL ocean atmosphere model and its sensitivity to atmospheric resolution. Clim Dyn 34:1–26

Marzeion B, Jarosch AH, Hofer M (2012) Past and future sea-level change from the surface mass balance of glaciers. Cryosphere 6:1295–1322

Mathis M (2013) Projected Forecast of Hydrodynamic Conditions in the North Sea for the 21st Century. Dissertation, University of Hamburg

Mathis M, Pohlmann T (2014) Projection of physical conditions in the North Sea for the 21st century. Clim Res 61:1–17

Mathis M, Mayer B, Pohlmann T (2013) An uncoupled dynamical downscaling for the North Sea: Method and evaluation. Ocean Model 72:153–166

Melsom A, Lien VS, Budgell WP (2009) Using the Regional Ocean Modeling System (ROMS) to improve the ocean circulation from a GCM 20th century simulation. Ocean Dyn 59:969–981

Menendez M, Woodworth PL (2010) Changes in extreme high water levels based on a quasi-global tide-gauge data set. J Geophys Res, Oceans 115:C10011, doi:10.1029/2009JC005997

Miles BWJ, Stokes CR, Vieli A, Cox NJ (2013) Rapid, climate-driven changes in outlet glaciers on the Pacific coast of East Antarctica. Nature 500:563–566

Millero FJ (1995) Thermodynamics of the carbon dioxide system in the oceans. Geochmic Cosmochim Ac 59:661–677

Milne GA, Mitrovica, JX (1998) Postglacial sea-level change on a rotating Earth. Geophys J Int 133:1-19

Mitrovica JX, Milne GA, Davis JL (2001) Glacial isostatic adjustment on a rotating earth. Geophys J Int 147:562-578

Moore JC, Grinsted A, Zwinger T, Jevrejeva S (2013) Semi-empirical and process-based global sea level projections. Rev Geophys 51:484–522

Müller M, Cherniawsky JY, Foreman MGG, von Storch JS (2014) Seasonal variation of the M_2 tide. Ocean Dyn 64:159–177

Murphy JM, Sexton DMH, Barnett DN, Jones GS, Webb MJ, Collins M, Stainforth DA (2004) Quantification of modelling uncertainties in a large ensemble of climate change simulations. Nature 430:768–772

Murphy JM, Sexton DMH, Jenkins GJ, Boorman PM, Booth BBB, Brown CC, Clark RT, Collins M, Harris GR, Kendon EJ, Betts RA, Brown SJ, Howard TP, Humphrey KA, McCarthy MP, McDonald RE, Stephens A, Wallace C, Warren R, Wilby R, Wood RA (2009) UK Climate Projections Science Report: Climate change projections. Hadley Centre, Met Office, UK

Nicholls RJ, Lowe JA (2004) Benefits of mitigation of climate change for coastal areas. Glob Environ Change 14:229–244

Nicholls RJ, Marinova N, Lowe JA, Brown S, Vellinga P, de Gusmão D, Hinkel J, Tol RSJ (2011) Sea-level rise and its possible impacts given a 'beyond 4 degrees C world' in the twenty-first century. Phil Trans Roy Soc A 369:161–181

Nick FM, Vieli A, Andersen ML, Joughin I, Payne A, Edwards TL, Pattyn F, van de Wal RSW (2013) Future sea-level rise from Greenland's main outlet glaciers in a warming climate. Nature 497:235–238

Oberhuber J (1993) Simulating of the Atlantic circulation with a coupled sea ice-mixed layer-isopycnical General Circulation Model. Part 1: Model Description. J Phys Oceanogr 23:808–829

OSPAR (2005) Common Procedure for the Identification of the Eutrophication Status of the OSPAR Maritime Area. OSPAR, Ref 2005-3. OSPAR Commission, London

Pätsch J, Kühn W (2008) Nitrogen and carbon cycling in the North Sea and exchange with the North Atlantic – a model study. Part I. Cont Shelf Res 28:767–787

Pardaens AK, Gregory JM, Lowe JA (2011a) A model study of factors influencing projected changes in regional sea level over the twenty-first century. Clim Dyn 36:2015–2033

Pardaens AK, Lowe JA, Brown S, Nicholls RJ, de Gusmao D (2011b) Sea-level rise and impacts projections under a future scenario with large greenhouse gas emission reductions. Geophys Res Lett 38: L12604, doi:10.1029/2011GL047678

Pelling HE, Green JAM, Ward SL (2013) Modelling tides and sea-level rise: To flood or not to flood. Ocean Modell 63:21–29

Perrette M, Landerer F, Riva R, Frieler K, Meinshausen M (2013) A scaling approach to project regional sea level rise and its uncertainties. Earth Sys Dyn 4:11–29

Pfeffer WT, Harper JT, O'Neel S (2008) Kinematic constraints on glacier contributions to 21st-century sea-level rise. Science 321:1340–1343

Pickering MD, Wells NC, Horsburgh KJ, Green JAM (2012) The impact of future sea-level rise on the European Shelf tides. Cont Shelf Res 35:1–15

Pohlmann T (1996) Calculating the annual cycle of the vertical eddy viscosity in the North Sea with a three-dimensional baroclinic shelf sea circulation model. Cont Shelf Res 16:147–161

Pope VD, Gallani ML, Rowntree PR, Stratton RA (2000) The impact of new physical parameterizations in the Hadley Centre climate model – HadAM3. Clim Dyn 16:123–146

Pushpadas D, Schrum C, Daewel U (2015) Projected climate change effects on North Sea and Baltic Sea: CMIP3 and CMIP5 model-based scenarios. Biogeosciences Discuss 12:12229–12279

Radić V, Bliss A, Beedlow AC, Hock R, Miles E, Cogley JG (2014) Regional and global projections of twenty-first century glacier mass changes in response to climate scenarios from global climate models. Clim Dyn 42:37–58

Rae JGL, Aðalgeirsdóttir G, Edwards TL, Fettweis X, Gregory JM, Hewitt HT, Lowe JA, Lucas-Picher P, Mottram RH, Payne AJ, Ridley JK, Shannon SR, van de Berg WJ, van de Wal RSW, van den Broeke MR (2012) Greenland ice sheet surface mass balance: evaluating simulations and making projections with regional climate models. Cryosphere 6:1275–1294

Rahmstorf S (2007) A semi-empirical approach to projecting future sea-level rise. Science 315:368–370

Rahmstorf S (2010) A new view on sea-level rise. Nature Rep Clim Change doi:10.1038/climate.2010.29

Ranger N, Reeder T, Lowe JA (2013) Addressing 'deep' uncertainty over long-term climate in major infrastructure projects: four innovations of the Thames Estuary 2100 Project. EURO J Decis Process 1:3–4

Raper SCB, Gregory JM, Osborn TJ (2001) Use of an upwelling-diffusion energy balance climate model to simulate and diagnose A/OGCM results. Clim Dyn 17:601–613

Regnier P, Friedlingstein P, Ciais P, Mackenzie FT, Gruber N, Janssens IA, Laruelle GG, Lauerwald R, Luyssaert S, Andersson AJ, Arndt S, Arnosti C, Borges AV, Dale AW, Gallego-Sala A, Goddéris Y, Goossens N, Hartmann J, Heinze C, Ilyina T, Joos F, LaRowe DE, Leifeld J, Meysman FJR, Munhoven G, Raymond PA, Spahni R, Suntharalingam P, Thullner M (2013) Anthropogenic perturbation of the carbon fluxes from land to ocean. Nat Geosci 6:597–607

Rider KM, Konen GJ, Beersma J (1996) Simulation of the response of the ocean waves in the North Atlantic and North Sea to CO_2 doubling in the atmosphere. KNMI Scientific Rep WR 96-05

Ridley J, Gregory JM, Huybrechts P, Lowe JA (2010) Thresholds for irreversible decline of the Greenland ice sheet. Clim Dyn 35: 1065–1073

Riebesell U, Tortell PD (2011) Effects of ocean acidification on pelagic organisms and ecosystems. In: Gattuso JP, Hansson L (eds) Ocean Acidification. Oxford University Press

Riebesell U, Schulz KG, Bellerby R, Botros M, Fritsche P, Meyerhofer M, Neill C, Nondal G, Oschlies A, Wohlers J, Zollner E (2007) Enhanced biological carbon consumption in a high CO_2 ocean. Nature 450:545–548

Roeckner E, Arpe K, Bengtsson L, Brinkop S, Dümenil L, Esch M, Kirk E, Lunkeit F, Ponater M, Rockel B, Sausen R, Schlese U, Schubert S, Windelband M (1992) Simulation of the present-day climate with the ECHAM model: impact of model physics and resolution. Max-Planck-Institut für Meteorologie, Hamburg. Rep 93

Roeckner E, Bauml G, Bonaventura L, Brokopf R, Esch M, Giorgetta M, Hagemann S, Kirchner I, Kornblueh L, Manzini E, Schlese U, Schulzweida U (2003) The atmospheric general circulation model ECHAM5: Part 1: Model description. Max-Planck-Institut für Meteorologie, Hamburg. Rep 349

Roeckner E, Brokopf R, Esch M, Giorgetta M, Hagemann S, Kornblueh L, Manzini E, Schlese U, Schulzweida U (2006) Sensitivity of simulated climate to horizontal and vertical resolution in the ECHAM5 atmosphere model. J Clim 19:3771–3791

Rohling EJ, Grant K, Hemleben C, Siddall M, Hoogakker BAA, Bolshaw M, Kucera M (2008) High rates of sea-level rise during the last interglacial period. Nat Geosci 1:38–42

Scaife AA, Woollings T, Knight J, Martin G, Hinton T (2010) Atmospheric blocking and mean biases in climate models. J Clim 23:6143–6152

Schaeffer M, Hare W, Rahmstorf S, Vermeer M (2012) Long-term sea-level rise implied by 1.5 degrees C and 2 degrees C warming levels. Nature Clim Chan 2:867–870

Schmidt GA, Ruedy R, Hansen JE, Aleinov I, Bell N, Bauer M, Bauer S, Cairns B, Canuto V, Cheng Y, Del Genio A, Faluvegi D, Friend AD, Hall TM, Hu Y, Kelley M, Kiang NY, Koch D, Lacis AA, Lerner J, Lo KK, Miller RL, Nazarenko L, Oinas V, Perlwitz Ja, Perlwitz Ju, Rind D, Romanou A, Russell GL, Sato M, Shindell DT, Stone PH, Sun S, Tausnev N, Thresher D, Yao M-S (2006) Present day atmospheric simulations using GISS ModelE: Comparison to in-situ, satellite and reanalysis data. J Clim 19: 153–192

Schrum C (2001) Regionalization of climate change for the North Sea and the Baltic Sea. Clim Res 18:31–37

Schrum C, Backhaus JO (1999) Sensitivity of atmosphere-ocean heat exchange and heat content in North Sea and Baltic Sea. A comparative assessment. Tellus 51A:526–549

Schrum C, Hübner U, Jacob D, Podzun R (2003a) A coupled atmosphere/ice/ocean model for the North Sea and the Baltic Sea. Clim Dyn 21:131–151

Schrum C, Siegismund F, St. John M (2003b) Decadal variations in the stratification and circulation patterns of the North Sea. Are the 90's unusual? ICES Symposium of Hydrobiological Variability in the ICES area 1990-1999. ICES J Mar Sci 219:121–131

Schrum C, Alekseeva I, St. John M (2006) Development of a coupled physical–biological ecosystem model ECOSMO Part I: Model description and validation for the North Sea. J Mar Sys 61: 79-99

Sein DV, Mikolajewicz U, Gröger M, Fast I, Cabos W, Pinto JG, Hagemann S, Semmler T, Izquierdo A, Jacob D (2015) Regionally coupled atmosphere ocean sea ice, marine biogeochemistry model ROM: 1. Description and validation. JAMES 7:268–304

Shchepetkin AF, McWilliams JC (2005) The Regional Ocean Modeling System (ROMS): A split-explicit, free-surface, topography-following coordinates ocean model. Ocean Mod 9:347–404

Shennan I, Horton BP (2002) Holocene land- and sea-level changes in Great Britain. J Quat Sci 17:511–526

Shennan I, Milne G, Bradley S (2009) Late Holocene relative land- and sea-level changes: Providing information for stakeholders. GSA Today 19:doi:10.1130/GSATG50GW.1

Simpson MJR, Breili K, Kierulf HP (2014) Estimates of twenty-first century sea-level changes for Norway. Clim Dyn 42:5–6

Skogen MD, Søiland H (1998) A User's Guide to NORWECOM v2.0, a coupled 3-dimensional physical chemical biological ocean model. Tech. Rept. Fisken og Havet 18/98. Institute of Marine Research, Norway. www.imr.no/~morten/norwecom/userguide2_0.ps.gz

Skogen M, Svendsen D, Berntsen E, Aksnes JD, Ulvestad KB (1995) Modelling the primary production in the North Sea using a coupled three-dimensional physical-chemical-biological ocean model. Estuar Coast Shelf Sci 41:545–5658

Skogen MD, Drinkwater K, Hjøllo S, Schrum C (2011) North Sea sensitivity to atmospheric forcing. J Mar Sys 85:106–114

Skogen MD, Eilola K, Hansen JLS, Meier HEM, Molchanov MS, Ryabchenko VA (2014) Eutrophication Status of the North Sea, Skagerrak, Kattegat and the Baltic Sea in present and future climates: A model study. J Mar Sys 132:174–184

Slangen ABA, Carson M, Katsman C, van de Wal RSW, Köhl A, Vermeersen LLA, Stammer D (2014) Projecting twenty-first century regional sea-level changes. Clim Chan 124:317–332

Steinacher M, Joos F, Fröhlicher T, Bopp L, Cadule P, Cocco V, Doney S, Gehlen M, Lindsay K, Moore JK, Schneider B, Segschneider J (2010) Projected 21st century decrease in marine productivity: a multi-model analysis. Biogeosciences 7:979–1005

Sterl A, van den Brink H, de Vries H, Haarsma R, van Meijgaard E (2009) An ensemble study of extreme storm surge related water levels in the North Sea in a changing climate. Ocean Sci 5:369–378

Stott PA, Kettleborough JA (2002) Origins and estimates of uncertainty in predictions of twenty-first century temperature rise. Nature 416:723–726

STOWASUS-Group (2001) Synthesis of the STOWASUS-2100 project: regional storm, wave and surge scenarios for the 2100 century. www.pd.infn.it/AOD/immagini_projects/Stowasus_final.pdf

Su J, Yang H, Pohlmann T, Ganske A, Klein B, Klein H, Narayan N (2014) A regional coupled atmosphere-ocean model system REMO/HAMSOM for the North Sea. KLIWAS Schriftenreihe, KLIWAS-60/2014

Van Meijgaard E, van Ulft LH, van de Berg WJ, Bosveld FG, van den Hurk BJJM, Lenderink G, Siebesma AP (2008) The KNMI regional atmospheric climate model RACMO version 2.1. Tech. Rep TR-302.

Vizcaíno M, Mikolajewicz U, Jungclaus J, Schurgers G (2010) Climate modification by future ice sheet changes and consequences for ice sheet mass balance. Clim Dyn 34:301–324

von Storch H (1995) Inconsistencies at the interface of climate impact studies and global climate research. Meteor Z 4:72–80

Wahl T, Haigh ID, Woodworth PL, Albrecht F, Dillingh D, Jensen J, Nicholls RJ, Weisse R, Wöppelmann G (2013) Observed mean sea level changes around the North Sea coastline from 1800 to present. Earth-Sci Rev 124:51–67

Wakelin S, Daewel U, Schrum C, Holt J, Butenschön M, Artioli Y, Beecham J, Lynam C, Mackinson S (2012a) MEECE deliverable D3.4: Synthesis report for Climate Simulations Part 3: NE Atlantic / North Sea, www.meece.eu/documents/deliverables/WP3/D3%204_Part3_NE%20Atlantic.pdf

Wakelin SL, Holt J, Blackford JC, Allen JI, Butenschön M, Artioli Y (2012b) Modeling the carbon fluxes of the northwest European continental shelf: Validation and budgets. J Geophys Res 117, C05020 doi:10.1029/2011JC007402

Wang S, Dieterich C, Döscher R, Höglund A, Hordoir R, Meier HEM, Samuelsson P, Schimanke S (2015) Development and evaluation of a new regional coupled atmosphere-ocean model in the North Sea and the Baltic Sea. Tellus A 67:24284, http://dx.doi.org/10.3402/tellusa.v67.24284

WASA-Group (1998) Changing waves and storms in the Northeast Atlantic? Bull Am Met Soc 79:741–760

Weidemann H (2009) Statistisch regionalisierte Sturmflutszenarien für Cuxhaven. Thesis. Christian-Albrechts-Universität Kiel

Wentz FJ, Riccaiardulli L, Hilburn K, Mears C (2007) How much more rain will global warning bring? Science 317:233–235

Wigley TML (2005) The climate change commitment. Science 307:1766–1769

Wolf J, Woolf DK (2006) Waves and climate change in the north-east Atlantic. Geophys Res Lett 33 doi:10.1029/2005GL025113

Woodworth PL, Flather RA, Williams JA, Wakelin SL, Jevrejeva S (2007) The dependence of UK extreme sea levels and storm surges on the North Atlantic Oscillation. Cont Shelf Res 27:935–946

Woth K (2005) North Sea storm surge statistics based on projections in a warmer climate: How important are the driving GCM and the chosen emission scenario? Geophys Res Lett 32:122708 doi:10.1029/2005GL023762

Woth K, Weisse R, von Storch H (2006) Climate change and North Sea storm surge extremes: an ensemble study of storm surge extremes expected in a changed climate projected by four different regional climate models. Ocean Dyn 56:3–15

Yin JJ (2012) Century to multi-century sea level rise projections from CMIP5 models. Geophys Res Lett 39: L17709, doi:10.1029/2012GL052947

Yoshimori M, Abe-Ouchi A (2012) Sources of spread in multimodel projections of the Greenland Ice Sheet surface mass balance. J Clim 25:1157–1175

Zavatarelli M, Akoglu E, Artioli Y, Beecham J, Butenschon M, Cannaby H, Chifflet M, Chust G, Clementi E, Daewel U, Fach B, Holt J, Machu E, Petihaks G, Salihoglu B, Schrum C, Shin Y, Smith C, Travers M, Tsiaras K, Wakelin S, Allen JI (2013a) D4.1. Hindcast simulations of isolated direct anthropogenic drivers. MEECE deliverable, www.meece.eu/Deliv.html

Zavatarelli M, Artioli Y, Beecham J, Butenschon M, Chifflet M, Christensen A, Chust G, Daewel U, Holt J, Neuenfeldt S, Schrum C, Skogen M, Wakelin S, Allen JI (2013b) D4.3. Synthesis report on the comparison of WP3 and WP4 simulations: Part 2a Multiple Driver Scenarios, NE Atlantic, North Sea, Baltic Sea and Biscay. MEECE deliverable, www.meece.eu/Deliv.html

Zhang YJ, Stanev E, Grashorn S (2015) Unstructured-grid model for the North Sea and Baltic Sea: Validation against observations. Ocean Model 97:91–108

Projected Change—River Flow and Urban Drainage

7

Patrick Willems and Benjamin Lloyd-Hughes

Abstract

Hydrological extremes, largely driven by precipitation, are projected to become more intense within the North Sea region. Quantifying future changes in hydrology is difficult, mainly due to the high uncertainties in future greenhouse gas emissions and climate model output. Nevertheless, models suggest that peak river flow in many rivers may be up to 30 % higher by 2100, and in some rivers even higher. The greatest increases are projected for the northern basins. Earlier spring floods are projected for snow-dominated catchments but this does not always cause an increase in peak flows; peak flows may decrease if higher spring temperatures lead to reduced snow storage. An increase in rain-fed flow in winter and autumn may change the seasonality of peak flows and floods. The proximity of a river basin to the ocean is also important; the closer the two the greater the potential damping of any climate change effect. In urban catchments, the specific characteristics of the drainage system will dictate whether the net result of the climate change effect, for example the projected increase in short-duration rainfall extremes, is to damp or amplify the impact of this change in precipitation. The response in terms of sewer flood and overflow frequencies and volumes is highly non-linear. The combined impact of climate change and increased urbanisation in some parts of the North Sea region could result in as much as a four-fold increase in sewer overflow volumes.

7.1 Introduction

The hydrological cycle is an intrinsic part of the climate system. Changes within the climate directly and indirectly influence the components of the hydrological cycle. As an illustration, climate change may alter river regimes directly through changes in rainfall, and indirectly through changes in temperature, which may change evaporation and affect snow melt. Differences in rainfall intensity may alter flood hazards through changes in peak discharge and erosion. Additionally, temperature changes, especially during summer, affect the soil water content and groundwater recharge, and thus water input (from ground water and base flow) to rivers. As a result, the risk of low flow alters, which can also impact on water quality, navigation and water availability for agricultural and industrial purposes. In short, climate change affects or controls inputs, losses, storage and transfer into the hydrological system (IPCC 2014). Whether and in what way this is the case for the North Sea region is the focus of this chapter, based on a review of available studies on climate change impacts on river flow in the North Sea region. Impacts in urban catchments are also considered.

To assess the potential impacts of climate change on river flow, methods are applied that make use of both climate models and hydrological models. Climate models simulate

Electronic supplementary material Supplementary material is available in the online version of this chapter at 10.1007/978-3-319-39745-0_7.

P. Willems (✉)
Department of Civil Engineering, KU Leuven, Leuven, Belgium
e-mail: patrick.willems@bwk.kuleuven.be

B. Lloyd-Hughes (✉)
Walker Institute for Climate System Research, University of Reading, Reading, UK
e-mail: B.LloydHughes@reading.ac.uk

M. Quante and F. Colijn (eds.), *North Sea Region Climate Change Assessment*, Regional Climate Studies, DOI 10.1007/978-3-319-39745-0_7

the climate system to determine the response to changes in greenhouse gases in the atmosphere (see Chap. 5) while hydrological models simulate the climate change effect on the water cycle. To assess the potential impact of climate change on river flow, climate change signals are used to alter the input to hydrological models that aim to simulate the climate-driven response in the hydrological environment. The strength of the changes depends on the temporal and spatial scales being examined.

River flow may be affected by changes in land use, ground water abstraction, hydraulic structures (such as reservoirs) along the river course, and urbanisation (see Sect. 7.3.3), among others; none of which are directly linked to climate change. Such features mean not all climate-driven hydrological impacts are easily discernible, and so caution is necessary when attributing hydrological change to 'climate change impact'; see also Chap. 5.

7.2 Methodology

7.2.1 Temporal and Spatial Scales

Uncertainties in climate and hydrological models mean caution must be applied in using the model output to project climate-driven impacts on river flow. This is especially the case for studying local hydrological impacts. Local climates are represented in regional climate models (RCMs) at the spatial resolution of the RCMs, and are less reliable than the coarser resolution climate data obtained from the same RCMs or from global climate models (GCMs). However, the reliability of climate models is improving due to the ongoing research in climate science (see the Supplement S7 to this chapter). The highest resolution RCMs are now in the range of a few tens of kilometres which reduces the mismatch with hydrological models that often operate at resolutions of a few kilometres or less. High resolution models, however, do not completely resolve the physics of the climate system so climate model output still requires further scrutiny before use in regional climate studies.

Although the natural processes addressed in climate and hydrological models are closely linked, because climate science and hydrology are separate disciplines the technical aspects of these different disciplines require an interface linking the respective models. This interface allows a realistic transfer of information between climatic and hydrologic simulations. Methods at the interface range from the direct use of climate model output to correct for bias (systematic over- or underestimations) before use. However, direct use is rarely implemented due to the bias in climate models. Another major interfacing issue is the need for high resolution data in many hydrological applications, both in space and time. Finer-resolution climate models imply a developmental and computational burden which translates to higher resources, in time and money. While efforts to increase the resolution of GCMs and RCMs continue (e.g. HiGEM, Shaffrey et al. 2009; Kendon et al. 2012) the current state-of-the-art is well short of the requirements for local hydrological modelling.

These two main interfacing problems are met by applying statistical downscaling to the climate model output, ultimately in combination with bias correction. The aims of the bias correction and statistical downscaling are to eliminate systematic errors between the climate model output and the corresponding meteorological variables at the finer hydrological impact scales and/or to convert the climate model output to the finer-scale variables using statistical methods (Maraun et al. 2010; Gudmundsson et al. 2012a). More discussion on the mismatch of scales, statistical downscaling and bias correction is available in the Supplement S7. These downscaling and bias correction methods have increased data availability for hydrological assessments. Different approaches have been developed. Several have been applied in the North Sea region, depending on the area, type of hydrological impact, approach and experience of the modeller and available resources, among others.

7.2.2 Analysis

Determining the climate-driven change in river flow typically includes four steps: evaluating the climate models; downscaling/bias correction of the hydrological variables from the climate scenarios; converting climate change signals/perturbations to hydrological parameters; and simulating the hydrological climate change effect.

Different types of hydrological models have been used for studying the impact of climate change, depending on the scale and the processes. Conceptual rainfall-runoff models have been widely applied to individual catchments because of their ease of use and calibration (limited number of model parameters) and because they provide overall runoff estimates at the scale of a catchment or sub-catchment (see Supplement S7 for examples). In order to capture the spatial variability of the hydrological response in larger river basins or regions, spatially-distributed hydrological balance models have been applied. These can be of a conceptual nature or more detailed, depending on the types of impacts studied (e.g. Shabalova et al. 2003; Lenderink et al. 2007; Thompson et al. 2009; Bell et al. 2012; Huang et al. 2013). At the continental and global scale, land surface models and coarse-scale global water balance models are used, such as at the scale of Europe (e.g. Dankers and Feyen 2008; Feyen and Dankers 2009; Prudhomme et al. 2012) or the entire globe (e.g. Arnell and Gosling 2016; Dankers et al. 2014).

Hydrological impact results are typically evaluated for mean annual or seasonal volumes, but also for flow extremes (peak flows and low flows). The latter are of particular relevance for water management, given that they are fundamental to flood and water scarcity risks. Peak and low flow extremes for current and future climatic conditions are typically compared for quantiles, hence for given exceedance probabilities or return periods. Such quantiles, for example the 100-year peak flow, form the basis of water engineering design statistics. They can be obtained empirically from the independent extreme flows extracted from the simulated time series (possible only up to the length of the time series), or after extreme-value analysis (required for extrapolating beyond the length of the time series). Bastola et al. (2011), Arnell and Gosling (2016), Dankers et al. (2014), and Smith et al. (2014), for example, defined peak flows as annual maximum flows and extrapolated these based on the Generalized Extreme Value (GEV) distribution. Lawrence and Hisdal (2011) did the same but used the Gumbel distribution as a special case of the GEV, and Kay and Jones (2012) made use of a generalised logistic distribution. Willems (2013a) selected independent peak flows from a time series by means of hydrological independence criteria to obtain a peak-over-threshold or partial-duration-series. These typically follow the Generalized Pareto Distribution (GPD), or the exponential distribution as a special case. The statistical uncertainty in estimates of large return periods (e.g. 100 years) may be considerable, however, especially when based on relatively short time series (typically 30 years for climate model results) (Brisson et al. 2015). Using information on flood thresholds or hydraulic flood modelling, the flow extremes can be related to flood hazard (e.g. return period of flooding) or even flood risk after considering functions describing the regional or local relationship between flood flow or depth and the flood consequences (Feyen et al. 2012; Ward et al. 2013; Arnell and Gosling 2016).

For impact analysis on urban drainage (sewer floods), because of the quick response of such systems to rainfall, changes in short-duration extremes (hourly to sub-hourly) are considered. These changes are propagated to changes in sewer flow by full hydrodynamic or conceptual sewer models; a recent state-of-the-art review of methods, difficulties/pitfalls, and impact results was made by Willems et al. (2012a, b).

As well as changes in rainfall and evaporation, for impact analysis on water quality in urban drainage systems and along rivers, changes in other variables must also be considered. Impacts on water quality are not only controlled by changes in rainfall, but also by (changes in the length of) dry periods. In the case of longer dry periods in north-western Europe, river pollution will be less diluted and river water quality will deteriorate. Some sources of river pollution might even increase, such as pollution originating from sewer overflows.

Along sewer systems, longer dry periods cause water and wastewater to stay for longer in the sewer pipes. Particularly in the low and flat North Sea region, this will lead to higher sewer solids sedimentation (Bates et al. 2008). An increase in short-duration rainfall extremes will not only increase peak runoff discharges but will also increase wash off from surfaces (impermeable and permeable) in the sewer catchment. An increase in runoff and sewer peak flows, would increase the frequency of sewer overflows or the spilling of storm- and/or waste-water into the receiving river. These effects are studied by integrated urban drainage models comprising the sewer system, wastewater treatment plant and receiving river. Using such a model, Astaraie-Imani et al. (2012) studied the impact of climate change (and urbanisation) on the receiving water quality of an urban river for dissolved oxygen and ammonium using a semi-real case study in the UK. Another application, but for a catchment in Belgium and limited to the flow impacts of sewer systems on receiving rivers was reported by Keupers and Willems (2013). Other types of climatic change effect along sewer systems include changes in temperature, which affect sewer quality processes (Ashley et al. 2008), risk of sulphide production in the sewer pipes, and increased odour problems; as well as increased sewer floods and sewer overflows because of changes in snowmelt patterns in mountainous regions, sea-level rise in low-lying coastal areas, inflow of groundwater during the wet season, and increased leakage of wastewater into the soil during the dry season, among others.

Whatever model type is applied, it is necessary to be aware that parameters calibrated for historical periods may not be valid under a changing climate. For instance, it is known that under dry conditions, soil moisture parameters are likely to change, which may affect the hydrological processes by introducing other complex mechanisms (Diaz-Nieto and Wilby 2005). One way of understanding the changes is to assess longer hydrological and meteorological records with significantly different changes in climate and land use (Refsgaard et al. 2014). However, sufficiently long time series (e.g. over 100 years) are often not available to evaluate this assumption.

It is also necessary to be aware of the limitations of the models in modelling particular types of extremes (e.g. high flow, low flow). For that reason, methods have been developed that explicitly validate model performance for high and low extremes; see Seibert (2003), Willems (2009), and Karlsson et al. (2013). Van Steenbergen and Willems (2012) proposed a data-based method to validate the performance of hydrological models in describing changes in peak flow under changes in rainfall extremes, prior to their use for climate change impact investigations. Vansteenkiste et al. (2013, 2014) compared different hydrological models and concluded that the impact results of climate scenarios might significantly differ depending on the model structure and underlying assumptions, especially for low flow. Gosling et al. (2011) applied two types of distributed hydrological model to different catchments, including the Harper's Brook catchment in the UK, to analyse the impact uncertainty from seven GCM runs. Both models simulated similar climate change signals, but differences occurred in the mean annual runoff, the seasonality of runoff, and the magnitude of changes in extreme monthly runoff. Also, Bastola et al. (2011) emphasised the importance of incorporating hydrological model structure and parameter uncertainty in estimating climate change impacts on flood quantiles. They found that the highest model uncertainty is associated with low frequency flood quantiles and with models that use nonlinear surface storage structures. Lawrence et al. (2009) investigated model parameter calibration uncertainty for the Nordic HBV model calibrated to 115 Norwegian catchments. This was done by selecting 25 parameter sets that lead to almost equal model performance. In general, however, hydrological model related uncertainty is low compared to climate model uncertainty (Minville et al. 2008; Kay et al. 2009). The latter is shown by comparing evaluations of climate and hydrological model performance against observations; however limited to historical (climate) conditions. For drought, Prudhomme et al. (2014) concluded that global hydrological models show a higher uncertainty than global climate models. At the catchment scale, it appears that hydrological model impact uncertainties are greater for low flow than for peak flow (Vansteenkiste et al. 2013, 2014), but are still less than from climate models.

7.2.3 Scenarios

Owing to the high uncertainties involved in the parameterisations of the climate models and the future greenhouse gas scenarios (see Supplement S7), it is better to apply a broad ensemble set of climate model simulations. Uncertainty in the future projections can thus be partly accounted for (Palmer and Räisänen 2002; Tebaldi et al. 2005; Collins 2007; Smith et al. 2009; Semenov and Stratonovitch 2010). Use of ensemble-based probabilistic projections has been proposed but would raises questions and difficulties for impact modellers (New et al. 2007). Linking probabilities to future scenarios is a commendable idea, but it is not clear how the use of probabilities would maintain internal consistency, which is a key requirement for impact analysis. It is pragmatic, therefore, to make use of existing climate change impact methods, albeit with improvements.

Any ensemble of climate model runs best includes a broad set of different climate models and greenhouse gas scenarios (SRES, RCP; see Supplement S7). Note in this respect that hydrological impact analyses of climate change to date have largely ignored the most pessimistic projections for climate change such as the SRES A1FI scenario. It has been argued that emission trends since 2000 are in line with the A1FI projections made in the 1990s (Raupach et al. 2007), which means that the A1FI scenario is becoming more plausible and the most likely projected high flows could be even higher than those reported here. Recent evidence also suggests that GCM projections underestimate the amount of warming that is already being observed in western Europe (van Oldenborgh et al. 2009).

The hydrological impact results reported in this chapter are primarily based on the SRES scenarios. Hydrological impact results for the newer RCP-based climate scenarios were still limited at the time this chapter was drafted (first global results exist: Dankers et al. 2014; Prudhomme et al. 2014), but it would be worth more extensively testing the change and consistency in impact results between the SRES and RCP-based scenarios.

In addition to uncertainties in the climate process modelling and greenhouse gas scenarios, the downscaling method used adds to uncertainty in the climate scenarios (see more discussion in the Supplement S7). Impact modelling based on large ensembles of climate model simulations under different downscaling assumptions remains difficult in practice because of the high computational costs associated with hydrological and hydraulic modelling. A pragmatic approach would be to summarise the different meteorological impact results of climate change in a limited set of (tailored) scenarios. Examples include the UKCIP02 (Hulme et al. 2002), UKWIR06 (Vidal and Wade 2008), and UKCP09 scenarios in the UK (Murphy et al. 2009), the KNMI'06 scenarios in the Netherlands (Van den Hurk et al. 2006; de Wit et al. 2007), and the CCI-HYDR scenarios in Belgium (Willems 2013a; Ntegeka et al. 2014).

Figure 7.1 illustrates how the CCI-HYDR high, mean and low scenarios for one particular season are based on the highest, average and lowest climate factors for the entire set of potential scenarios considered. The changes for different seasons are combined in different versions such that they lead to high, mean, and low impacts for specific (tailored) hydrological applications, for example winter floods, summer flash floods and summer low flows. This is the opposite of the KNMI'06 scenarios that are based on meteorological considerations only (Fig. 7.2, where scenarios W and G refer to higher or lower changes in temperature, and the scenarios W+ and G+ to stronger changes in atmospheric circulation).

7.3 Projections

7.3.1 North Sea Region

Numerous studies indicate that in north-western Europe a warmer climate may lead to an increase in intense rainfall

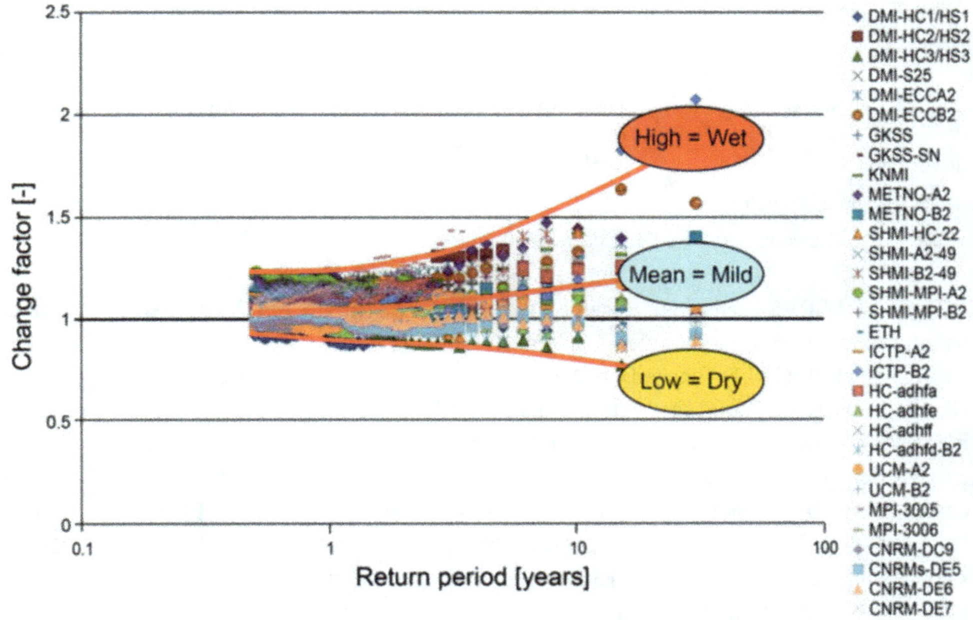

Fig. 7.1 High-mean-low tailored climate scenarios to simplify the flood impact analysis based on an ensemble set of climate model simulations (here: factor change in daily rainfall quantiles from 1961–1990 to 2071–2100 for A2 and B2 SRES scenarios and all RCM runs available for Belgium from the EU PRUDENCE project; Ntegeka et al. 2014)

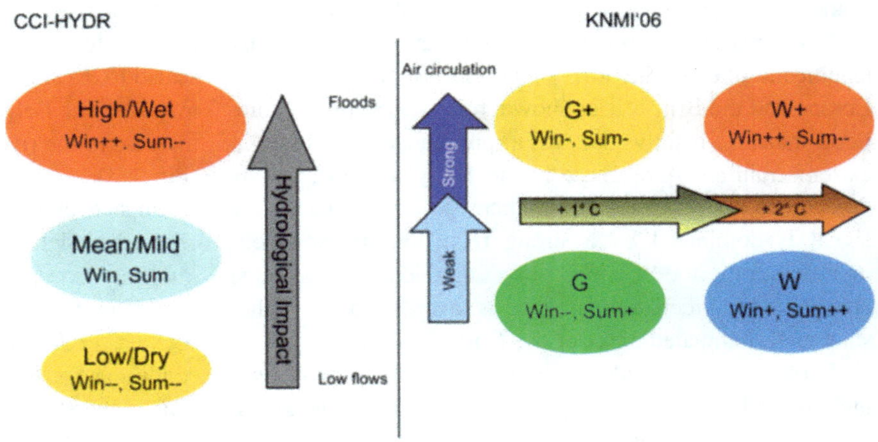

Fig. 7.2 Tailored climate scenarios: hydrological impact based (*left* CCI-HYDR, Ntegeka et al. 2014) versus meteorological based (*right* KNMI'06, van den Hurk et al. 2006)

(e.g. Kundzewicz et al. 2006; Hanson et al. 2007) and to longer dry periods (e.g. Good et al. 2006; May 2008), and consequent changes in river flows, as is shown in Table 7.1 based on a review by the European Environment Agency (EEA 2012) and the Intergovernmental Panel on Climate Change in its Fifth Assessment (IPCC 2014). The projections indicate an intensification of rainfall during both winter and summer, but for summer, although the heavy rainfall events may become more intense the intensity of the light and moderate events will decrease. How these meteorological changes will affect river flow shows strong seasonal and regional differences. For north-western Europe, the intensity and frequency of winter and spring river floods are generally expected to increase (EEA 2012).

Based on climate projections from three GCMs and impact analysis in three relatively coarse resolution global hydrological models, Prudhomme et al. (2012) found that river flow in north-western Europe (e.g. Great Britain) would increase in winter with concurrent increases in regional high flow anomalies, and would decrease in summer. Giving particular attention to daily peak flow and related flood risk, Hirabayashi et al. (2013), Arnell and Gosling (2016) and Dankers et al. (2014) found strong sub-regional variations in Europe with both increases (mostly for the UK, France and Ireland) and decreases in the size of the flood-prone populations. Giving particular attention to hydrological droughts, Prudhomme et al. (2014) found significant increases in the frequency of droughts of more than 20 % in central and western Europe. Also based on a coarse-scale hydrological model, but this time with a focus on the main rivers in Europe, Feyen and Dankers (2009) found stream flow droughts will become more severe and persistent in most parts of Europe by the end of the century, except in the most northern and-north eastern regions.

However, it should be noted that these results are based on only one RCM run (HIRHAM 12-km resolution model;

Table 7.1 Typical change in inland river flows for northern and north-western Europe (EEA 2012; IPCC 2014)

Variable	Northern Europe		North-Western Europe	
	Observed	Projected	Observed	Projected
River flow	+	+	(+)	+
River flood		±	+	+
River low flow (drought)	0	+	0	−

+ increase; − decrease; ± increase and decrease; 0 little change

A2 and A1B SRES scenarios). Based on the same RCM runs and the same hydrological model, Dankers and Feyen (2008) and Rojas et al. (2011) focused on the flood hazard climate change impact and found that extreme discharge levels may increase in magnitude and frequency in parts of western and eastern Europe. In several rivers, the return period of what is currently a 100-year flood may decrease to 50 years or less. Rojas et al. (2012) extended the analysis by applying the same hydrological model to 12 RCM runs, and concluded that results show large discrepancies in the magnitude of change in the 100-year flood for the different RCM runs. Some regions even show an opposite signal of change, but for many regions the projected changes are not statistically significant due to the low signal-to-noise ratio. Western Europe and the British Isles show a robust increase in future flood hazard, mainly due to a pronounced increase in extreme rainfall. A decrease in the 100-year flood, on the other hand, is projected in southern Sweden because the signal is dominated by a strong reduction in snowmelt-induced spring floods, which offsets the increase in average and extreme precipitation. This is also valid for other snowmelt dominated areas of the North Sea region.

Another Europe-wide hydrological impact study was undertaken by Schneider et al. (2013) who applied the global hydrological model WaterGAP3 on a 50′ × 50′ European grid. Climate change impacts were based on three GCMs after bias correction. Looking at their results for the North Sea region, they found that flow magnitude was more affected in the northern parts of the North Sea region, such as Sweden and Norway, with strong increases projected in winter precipitation. The lowest impacts across Europe were found in western Europe (i.e. the UK, Ireland, Benelux, Denmark, Galicia and France). The difference is due to the additional impact of temperature on snow cover in the northern region. The greatest impact on peak flows in Scandinavia occurred in April, rather than May, one month earlier in the future. Earlier snowmelt in spring and sporadic melt events in winter will reduce snow storage. However, in Sweden and Norway, these effects were more than compensated for by higher precipitation. During summer (June to September), increased precipitation is offset by greater evapotranspiration. Scandinavia is the only region in Europe where elevated low flows are projected.

7.3.2 Sub-region or Country-Scale

7.3.2.1 Belgium

Using finer scale hydrological models (c.f. Sect. 7.3.1), more local European climate studies have projected similar climate change impacts. For 67 catchments in the Scheldt river basin in Flanders, Boukhris et al. (2008) found that extreme peak flows in rivers may increase or decrease depending on the climate scenario used. Winter rainfall volumes increase but evapotranspiration volumes also increase. Depending on the balance between rainfall increase versus evapotranspiration increase, the change in net runoff may switch from positive to negative. From a set of 31 statistically downscaled RCM simulations and more than 20 GCM simulations available for Belgium, the most negative change led to an increase in the river peak flows of about 30 % for the 2080s (Fig. 7.3).

Impacts on river low flows were more uniform. All of the climate model simulations projected a decrease in river low flow extremes during summer. For Belgian rivers, the change in low flow extremes projected for the 2080s ranged between −20 and −70 % (Fig. 7.3). The drier summer conditions for Belgium lead to lower groundwater levels, as shown by Brouyére et al. (2004) and Goderniaux et al. (2009) for the Geer catchment, and by Dams et al. (2012) and Vansteenkiste et al. (2013, 2014) for the Nete catchment.

7.3.2.2 Northern France

Within the main river basins in France, Boé et al. (2009) found a decrease in mean discharge for summer and autumn. They also simulated a decrease in soil moisture, and a decrease in snow cover, which was especially pronounced at low and middle altitudes. The low flows in France become more frequent. This was also found by Habets et al. (2013) for the rivers Seine and Somme in northern France, based on seven hydrological models ranging from lumped rainfall-runoff to distributed hydrogeological models, and three downscaling methods. A general decrease in river flow of at least 14 % occurred at the outlets of the Seine and Somme basins by the 2050s and at least 22 % by the 2080s. More than 90 % of projections showed a decrease in summer flow at these outlets. For the winter high flows, about

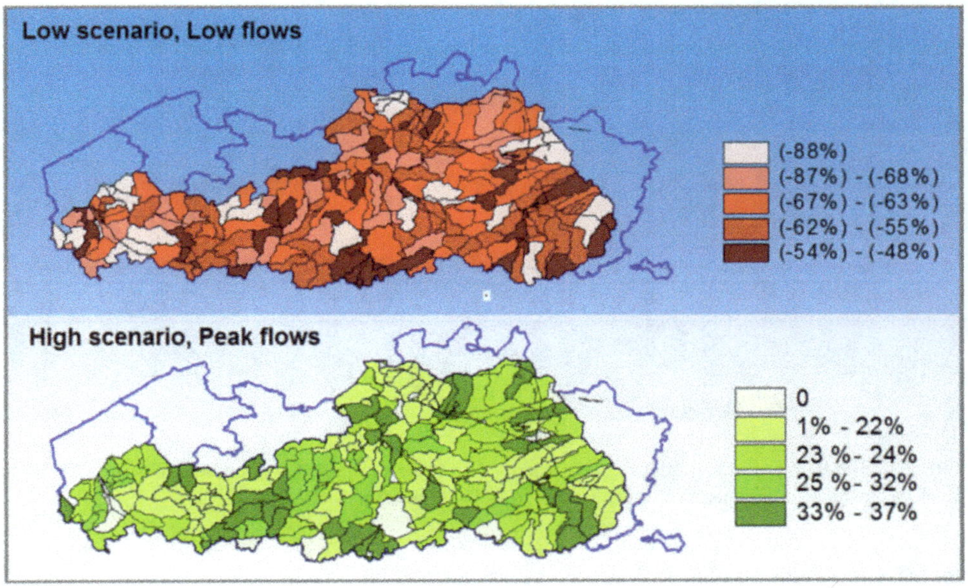

Fig. 7.3 Percentage change in low flows for a low/dry CCI-HYDR climate scenario (*upper*) and peak flows for a high/wet CCI-HYDR climate scenario (*lower*), averaged for return periods of 1–30 years, for 2071–2100 and 67 catchments in Flanders, Belgium (Boukhris et al. 2008)

10 % of projections showed the possibility of increased flow in winter in the River Seine and throughout the year in the River Somme, while 10 % projected a decrease of more than 40 % in river discharge at the basin outlets. For the same basins, Ducharne et al. (2011) found little change in the risk of floods for the 10- and 100-year return periods.

7.3.2.3 Germany

For various river basins in Germany, including the Ems, Weser, Elbe and Rhine (up to the Rees gauge station), based on two RCM simulations Huang et al. (2013) found an increase of about 10–20 % in the 50-year flood levels in the rivers Weser, Rhine, Main, Saale and Elbe. The Ems showed no clear increase and the Neckar a 20 % decrease. In contrast, the Wettreg statistical downscaling method projected a decrease in flood level for the Ems and Weser (10 %), and Saale (20 %) river basins, and no distinct change for the Main and Neckar. For the River Rhine, Shabalova et al. (2003) found future climate scenarios to result in higher mean discharges in winter (about +30 % by the end of the century), but lower mean discharges in summer (about −30 %), particularly in August (about −50 %). Temporal variability in the 10-day discharge increased significantly, even if temporal variability in the climatic inputs remains unchanged. The annual maximum discharge increases in magnitude throughout the Rhine and tends to occur more frequently in winter, suggesting an increasing risk of winter floods. At the Netherlands-German border, the magnitude of the 20-year maximum discharge event increased by 14–29 %; the present-day 20-year event tends to reappear every 3 to 5 years. The frequency of low and very low flows increases, in both scenarios alike. Studying changes in 10-day precipitation sums for return periods in the range 10 to 1000 years in the Rhine basin (within the scope of the RheinBlick2050 project), van Pelt et al. (2012) found changes of up to about +30 %. Pfister et al. (2004) projected increased flooding probably due to higher winter rainfall for the Rhine and Meuse river basins. Most hydrological simulations suggest a progressive shift of the Rhine from a 'rain-fed/meltwater' river to a mainly 'rain-fed' river. Studying projected change in the 1250-year peak flows in the Rhine and Meuse rivers, which are used as the basis for dike design along these rivers, de Wit et al. (2007) found the 1250-year peak flow to increase from 16000 to 18000 $m^3\ s^{-1}$ by 2100 for the Rhine and from 3800 to 4600 $m^3\ s^{-1}$ for the Meuse. For low flow, they found stronger seasonality in the discharge regime of the Meuse: increased low discharge in winter and decreased low discharge in summer. The same findings were obtained by van Huijgevoort et al. (2014).

7.3.2.4 Ireland

For catchments in Ireland, Bastola et al. (2011) simulated monthly changes derived from 17 GCM runs to the input of four hydrological models, and quantified the impact on flood quantiles up to 100-year return periods. They also studied the sensitivity of the impact results within and between hydrological models. The results show a considerable residual risk associated with allowances of +20 % when uncertainties are accounted for and that the risk of exceeding design allowances is greatest for more extreme, low frequency events (Fig. 7.4) with major implications for critical infrastructure such as culverts, bridges, and flood defences.

7.3.2.5 Scandinavia

In the Scandinavian countries, the increase in peak flows is higher than in other North Sea countries due to the higher

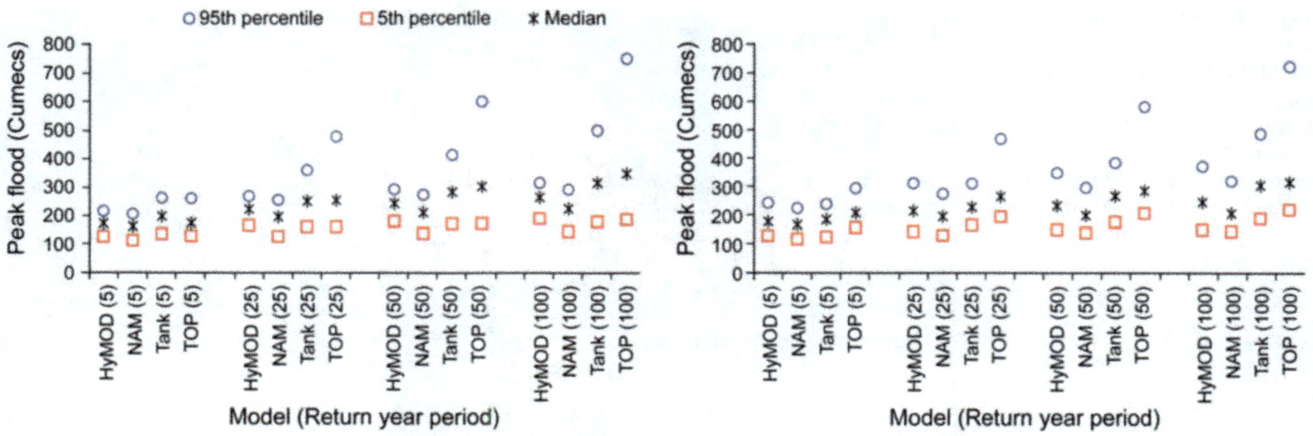

Fig. 7.4 The 95th percentile, 5th percentile and median value for modelled flood quantiles (5-, 25-, 50- and 100-year return periods) for the Moy river basin (*left*) and Boyne river basin (*right*) in Ireland (Bastola et al. 2011)

increase in winter rainfall. In Norway, Lawrence and Hisdal (2011) studied the changes in flood discharges for 115 unregulated catchments. Projected changes in peak flow quantiles for return periods of 200, 500 and 1000 years show strong regional differences (Fig. 7.5). These regional differences are explained by the role of snowmelt versus rainfall and how they increase the peak flows. This is, however, different for catchments where peak flows are mainly due to snow melt in spring. In this case, increased winter temperature will cause reduced snow storage, and thus decreased peak flows. An exception is catchments at higher elevations in areas where winter precipitation continues to fall predominantly as snow and higher spring temperatures produce more rapid snowmelt (SAWA 2012). In addition to changes in snowmelt-induced peak flows, the timing of the peak flows becomes earlier (i.e. spring rather than summer). Changes in the seasonality of peak flows occurs in catchments where flows driven by snowmelt decrease but flows driven by winter and autumn rainfall increase. The median projected change in the ensemble of hydrological projections for Norway at 2071–2100 varied from +10 to +70 % in catchments located in western and south-western regions (Vestlandet), coastal regions of southern and south-eastern Norway (Sørlandet and Østlandet) and in Nordland, and decreased down to −30 % for northernmost areas (Finnmark and parts of Troms) and middle and southern inland areas (Hedmark, Oppland, and parts of Buskerud, Telemark and Trøndelag).

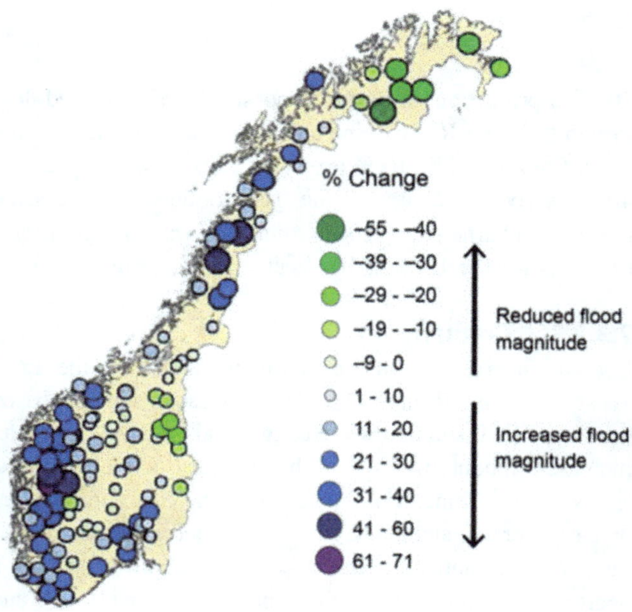

Fig. 7.5 Median projected change in peak flows for 200-year return period and 2071–2100 for 115 unregulated catchments across Norway (Lawrence and Hisdal 2011; SAWA 2012)

Similar results to eastern Norway were obtained by Andréasson et al. (2011) for Sweden; see the regional differences in 100-year peak flows in Fig. 7.6. They are based on spatial interpolation, without taking into account the influence of river regulation effects. The northern catchments in Sweden mostly show decreasing peak flows towards the end of the century, whereas the southern basins show increasing 100-year flows. The changes in peak flows vary from −45 to +45 %. A similar range was found by Teutschbein et al. (2011) and Teutschbein and Seibert (2012) for five catchments in Sweden.

Andersen et al. (2006) studied the climate change impact for six sub-catchments within and for the entire Gjern river basin in Denmark, but only based on one RCM simulation. Mean annual runoff from the river basin increased by 7.5 %, whereas greater changes were found for the extremes. The modelled change in the seasonal hydrological pattern is most pronounced in first- or second-order streams draining loamy catchments, which currently have a low base-flow during summer. Reductions of 40–70 % in summer runoff are projected for this stream type. Similar conclusions were obtained

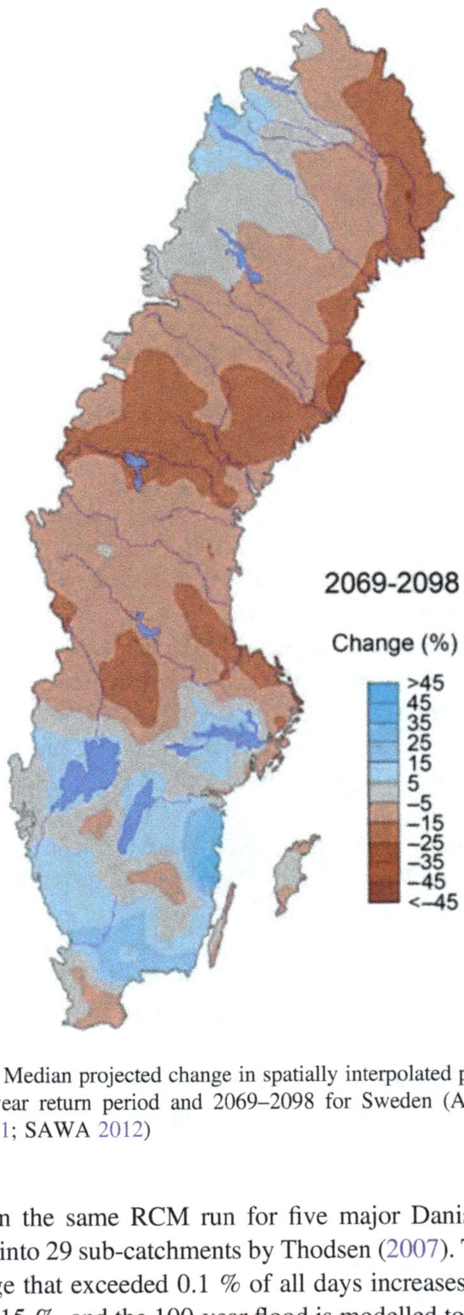

Fig. 7.6 Median projected change in spatially interpolated peak flows for 100-year return period and 2069–2098 for Sweden (Andréasson et al. 2011; SAWA 2012)

based on the same RCM run for five major Danish rivers divided into 29 sub-catchments by Thodsen (2007). The river discharge that exceeded 0.1 % of all days increases approximately 15 %, and the 100-year flood is modelled to increase 11 % on average. Andersen et al. (2006) also studied the climate change impact on diffuse nutrient losses (i.e. losses from land to surface waters). Simulated changes in annual mean total nitrogen load were about +8 %. Even though an increase in nitrogen retention in the river system of about 4 % was simulated in the scenario period, an increased in-stream total nitrogen export occurred due to the simulated increase in diffuse nitrogen transfer from land to surface water.

7.3.2.6 UK

For eight catchments in northwest England, Fowler and Kilsby (2007) used an ensemble set of simulation results with the HadRM3H RCM (UKCIP02 scenarios) and undertook a comprehensive treatment of climate modelling uncertainty. They concluded that annual runoff is projected to increase slightly at high elevation catchments, but to reduce by ~16 % for the 2080s at lower elevations. Impacts on monthly flow distribution are significant, with summer reductions of 40–80 % of mean flow, and winter increases of up to 20 %. The changing seasonality has a large impact on low flows, with 95 %-percentile flows projected to decrease in magnitude by 40–80 % in summer months (Fig. 7.7). In contrast, high flows (>5 %-percentile flows) are projected to increase in magnitude by up to 25 %, particularly at high elevation catchments, providing an increased risk of flooding during winter. Based on the same RCM and with a focus on river flood hazards in winter, Kay et al. (2006) found increased flood hazard particularly in East Anglia and the Upper Thames, with flood peaks in some places increasing by more than 50 % for the 50-year return level. Clear regional differences were also found by Arnell (2011) and Christierson et al. (2012) after analysing many UK catchments and several climate models or scenarios. Based on six catchments across the UK, Arnell (2011) found clear differences between northern and southern catchments, with large climate change effects in winter in the north and summer in the south. After analysing 70 UK catchments, Christierson et al. (2012) found major differences between the western and northern mountainous part of the UK and the rest of the UK, with an increase in winter river flow over the western part but less clear results or a decrease in mean monthly river flows all year round. In summer, most catchments showed negative or very slightly positive changes, with the largest flow decrease in the Thames, Anglian and Severn river basin districts, with decreases of 10 % or more in mean monthly flows all year round and even more in summer.

A specific study for the River Thames by Diaz-Nieto and Wilby (2005) concluded that substantial reductions in summer precipitation accompanied by increased potential evaporation throughout the year, lead to reduced river flow in late summer and autumn. Kay et al. (2006) found the same situation even in winter for some catchments in the south and east of England despite an increase in extreme rainfall events. This was explained by higher soil moisture deficits in summer and autumn that may have an influence up to the start of winter. Also for the Thames basin, but based on the more recent UKCP09 scenarios, Bell et al. (2012) found a 10–15 % increase in winter rainfall by the end of the century. This might potentially lead to higher flows than the River Thames can accommodate. Towards the downstream end, they estimated an average change in modelled 20-year return period flood peaks by the 2080s of 36 % (range −11 to +68 %).

For the River Avon catchment, Smith et al. (2014) obtained changes in the 25-year return period flows of +15,

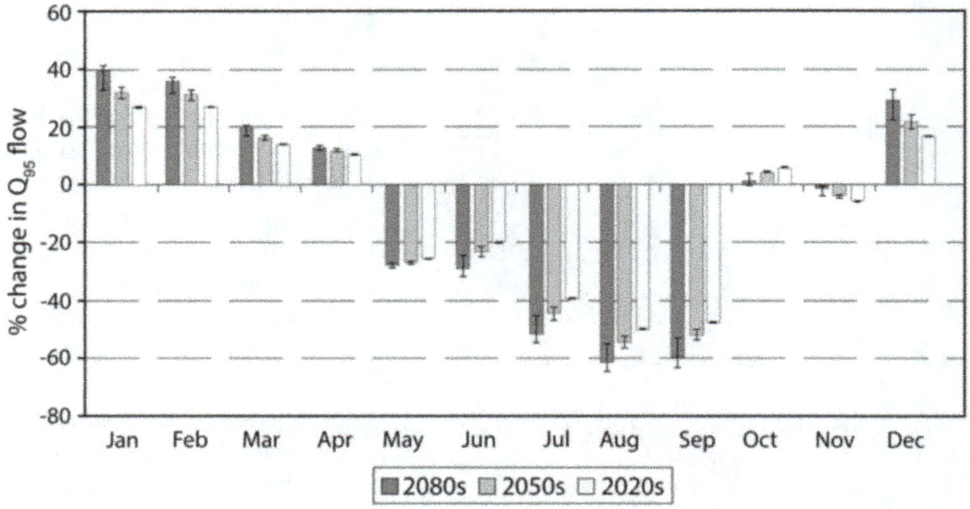

Fig. 7.7 Change in 95 %-percentile flow between the HadRM3H control and future scenarios for 2020s, 2050s and 2080s time-slices. The uncertainty bounds are for the different SRES scenarios (Fowler and Kilsby 2007)

+2 or +7 % based on three different methods for transferring the climate model output to hydrological model input. For 200-year peak flows, these percentages increased to +22 +19 and +6 %. For the River Medway catchment, Cloke et al. (2010) found a significant lowering of summer flow with a more than 50 % reduction for 2050–2080 and up to 70 % in some months. For six other UK catchments, Arnell (2011) simulated changes in summer runoff of between −40 and +20 %.

In terms of groundwater recharge, Herrera-Pantoja and Hiscock (2008) concluded that by the end of the century decreases in recharge of between 7 and 40 % are expected across the UK, leading to increased stress on local and regional groundwater supplies that are already under pressure to maintain both human and ecosystem needs.

The impacts that these hydrological changes may have in terms of flood and water availability risk were assessed by the UK-Government funded initiative AVOID (Warren et al. 2010; MetOffice 2011). Based on an ensemble set of 21 GCMs, it is shown that nearly three-quarters of the models project an increase in flood risk. For the 2030s and averaged over the UK as a whole, the change ranges from −20 to +70 %, with a mean of +4 % (Fig. 7.8). Larger increases are shown for longer time horizons. Overall, the models show a tendency for a large increase in flood risk for the UK as a whole.

The water availability threat in the UK (calculated using the Human Water Security Threat indicator by Vörösmarty et al. 2010) ranges from very high in the south-east to moderate in the south, Midlands, and southern Scotland (Fig. 7.9). For southern England, the loss in deployable water output due to climate change and population growth is estimated to be 3 % by 2035 (Charlton and Arnell 2011). Increased irrigation requirements were also found for the south-east and north-west of England (Henriques et al. 2008).

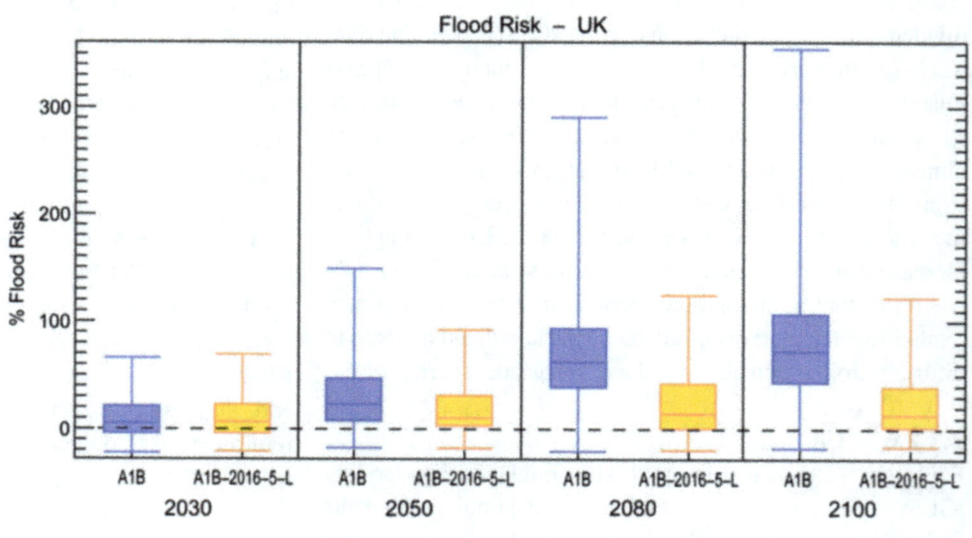

Fig. 7.8 Change in average annual flood risk for the UK, based on 21 GCMs under two emission scenarios (A1B and A1B-2016-5-L), for four time horizons (MetOffice 2011). The plots show the 25th, 50th, and 75th percentiles (represented by the *boxes*), and the maximum and minimum values (shown by the extent of the *whiskers*)

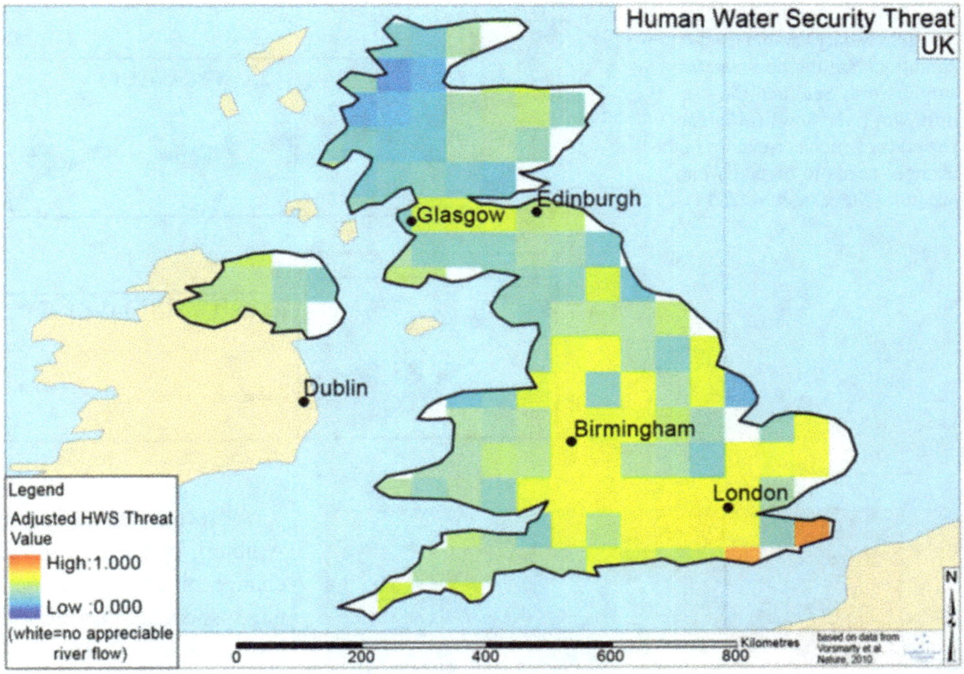

Fig. 7.9 The human water security threat for the UK (MetOffice 2011)

7.3.2.7 Comment on Low Flows

Although models project that climate change will cause a decrease in low flows in north-western European rivers over the coming decades, it should be noted that most models have low accuracy in the simulation of low flow extremes. Evidence for this is provided by Gudmundsson et al. (2012b) based on nine large-scale hydrological models after comparison to observed runoff from 426 small catchments across Europe. Further evidence is provided by Vansteenkiste et al. (2013, 2014) for a catchment in Belgium. Low accuracy for low flows is associated with the representation of hydrological processes, such as the depletion of soil moisture stores (Vansteenkiste et al. 2013, 2014).

7.3.2.8 Estuaries

In addition to changes in inland rainfall, temperature and reference evapotranspiration, which lead to changes at the upstream boundaries of estuaries, it is also important to consider changes in the downstream coastal boundary. In relation to the Scheldt estuary (Fig. 7.10), Ntegeka et al. (2011, 2012) studied projected changes in mean sea level, storm surge levels, wind speed and wind direction, and their correlation with changes in inland rainfall (see also Monbaliu et al. 2014 and Weisse et al. 2014). The changes in storm surge levels were derived from changes in sea-level pressure (SLP) in the Baltic Sea, the Atlantic Ocean area west of France, and the Azores, and a correlation model between SLP and storm surge level. The model was derived after analysing SLP composite maps and SLP-surge correlation maps (Fig. 7.11) for days where the surge exceeds given thresholds (for different return periods). Correlations were identified between the inland (rainfall, runoff) and coastal climatic changes. Based on the ensemble set of change factors, tailored climate scenarios (tailored for the specific application of flood impact analysis along the Scheldt) were developed to the 2080s. After statistical analysis, a reduced set of climate scenarios ('high', 'mean' and 'low') was derived for each boundary condition (runoff upstream, mean sea level, and surge downstream). Smart combinations of the scenarios account for the correlation between boundary changes (Monbaliu et al. 2014; Weisse et al. 2014).

7.3.2.9 Overview

Table 7.2 summarises the hydrological impact studies reviewed in this assessment. Because many of the studies report climate change impacts on peak river flows, the impacts were reported as percentage change by the end of the century. Many other hydrological variables are also of relevance, such as mean or low flows, but fewer studies report percentage change in these variables or the various study results are not directly comparable (e.g. derived at different time scales: annual vs. seasonal or monthly). It should also be noted that in several regions, the sign and order of magnitude of change are not consistent when results from different studies are compared. This reflects differences in methodology (number and type of climate model and greenhouse gas scenario, type of hydrological-hydraulic impact model, and statistical downscaling and analysis approach; see the Supplement S7 for more discussion on such issues), as well as uncertainties in the numerical projections of changes in hydrology. That results on changes in

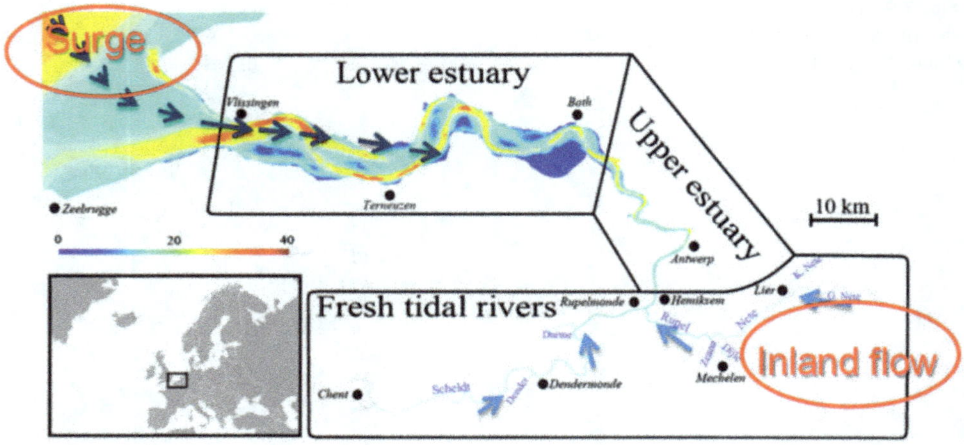

Fig. 7.10 Case study on the Scheldt Estuary: boundary conditions are the downstream surge (North Sea) and the upstream river flows (different rivers) for which correlation in the changes needs to be taken into account (Ntegeka et al. 2012)

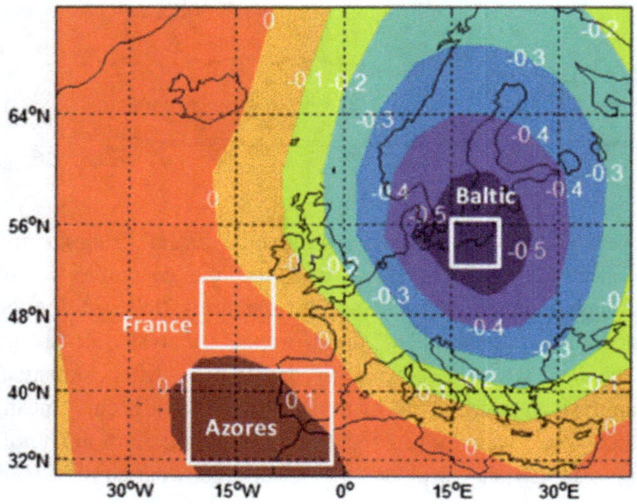

Fig. 7.11 Correlation between storm surges along the Belgian North Sea coast at Ostend and sea-level pressure over the North Atlantic region (mean based on historical events) (Ntegeka et al. 2012)

flood magnitude and frequency resulting from climate change are unclear was also concluded by the international review of Kundzewicz et al. (2013). It makes clear—as stressed in Sect. 7.2—that great care must be taken when conducting model-based impact analyses of climate change and in interpreting the results. Typical issues include consideration of only one or few climate models, greenhouse gas scenarios and/or hydrological models; poor calibration and validation of models; and inaccuracies of the models in extrapolating beyond historical conditions.

7.3.3 Urban Catchments

Hydrological analyses of urban catchments are based on studies with a particular focus on fine-scale meteorological and hydrological processes (as explained in Sect. 7.2.2).

A recent review by Willems et al. (2012a, b) and Arnbjerg-Nielsen et al. (2013) of the impacts of climate change on short-duration rainfall extremes and urban drainage showed that short-duration rainfall extremes were projected to increase by 10–60 % in 2100 relative to the baseline period (1961–1990). An urban drainage system may damp or amplify changes in precipitation, depending on the system characteristics. For the sewer network of Lund, Sweden, Niemczynowicz (1989) found the relative change in urban runoff volume to be higher than for the rainfall input. They found that a 30 % increase in the 40-min rainfall intensity would lead to a 66–78 % increase in sewer overflow volume (depending on a return period of between 1 and 10 years and the type of design storm). In Sweden, Olsson et al. (2009) found an increase in the number of urban drainage system surface floods of 20–45 % for Kalmar in 2100. For Odense in Denmark, Mark et al. (2008) found flood depth and the number of buildings currently affected once in every 50 years would correspond to a return period of 10 years in the future (based on the impacts discussed by Larsen et al. 2009 and Arnbjerg-Nielsen 2012). For Roskilde, also in Denmark, Arnbjerg-Nielsen and Fleischer (2009) found that a 40 % increase in design rainfall intensities would lead to a factor of 10 increase in the current level of damage costs related to sewer flooding. The actual change in cost will depend on catchment characteristics. In a similar study for another location with the same increase in rainfall intensity, Zhou et al. (2012) reported a factor 2.5 increase in annual costs. A common conclusion, however, is that the main impact of an increase in precipitation extremes is not primarily related to the additional cost associated with the most extreme events, but rather with the damage occurring far more frequently.

A higher factor increase in sewer impacts compared to the factor increase in rainfall was also reported by Nie et al. (2009) for Fredrikstad, Norway. They concluded that the total volume of water spilling from overflowing manholes is

Table 7.2 Summary of impact results on river flows available for the North Sea region

Region	Source	GCM-RCM(s) considered	Greenhouse gas scenario(s)	Hydrological-hydraulic impact model(s)	Change in river peak flow by 2100
Belgium	Boukhris et al. (2008), Ntegeka et al. (2014); Vansteenkiste et al. (2013, 2014), Tavakoli et al. (2014)	31 PRUDENCE RCM runs, 18 ENSEMBLES RCM runs	SRES A1, A1B, A2, B1, B2	Lumped conceptual NAM, PDM, VHM spatially distributed MIKE-SHE, WetSpa	Up to +30 %
Denmark	Andersen et al. (2006), Thodsen (2007)	HIRHAM RCM nested in ECHAM4/OPYC GCM	SRES A2	NAM rainfall runoff model /Mike 11–TRANS modelling system	Up to 12.3 %
France	Boé et al. (2009), Habets et al. (2013), Ducharne et al. (2011)	6 IPCC AR4 GCM runs	SRES A1B and A2	Hydrological models MARTHE, MODCOU, SIM, CLSM, EROS, GARDENIA and GR4 for Seine and Somme	No significant change
Germany	Huang et al. (2013)	REMO & CCLM RCMs	SRES A1B, A2, B1	SWIM eco-hydrological model	−20 to +20 %
Germany–Netherlands	van Pelt et al. (2012)	5 RCMs mainly ENSEMBLES + 13 CMIP3 GCMs	SRES A1B	HBV model Rhine basin	
Ireland	Bastola et al. (2011)	17 GCMs AR4	SRES A1B, A2, B1	4 conceptual models (HyMOD, NAM, TANK, TOPMODEL) for 4 catchments	Most up to +20 %
Netherlands	Shabalova et al. (2003), Lenderink et al. 2007	HadRM2 and HadRM3H RCMs	SRES A2 (for Lenderink et al. 2007)	RhineFlow distributed hydrological model	Up to +30 %
	de Wit et al. (2007)	KNMI'06 scenarios		Rhineflow and Meuseflow distributed hydrological models	
	Leander et al. (2008)	3 PRUDENCE RCMs	SRES A2	HBV model Meuse basin	
Norway	Lawrence and Hisdal (2011)	13 RCM runs RegClim & ENSEMBLES	SRES A1B, A2, B2	HBV rainfall runoff model 'Nordic' version	−30 to +70 %
Sweden	Andréasson et al. (2011), Teutschbein et al. (2011), Teutschbein and Seibert (2012)	12 RCM runs SMHI & ENSEMBLES	SRES A1, A2, B1, B2	HBV rainfall runoff model	−45 to +45 %
UK	Cameron (2006)	UKCIP02 climate change scenarios: HadRM3 RCM nested in HadCM3 GCM		TOPMODEL	
	Kay et al. (2006)	1 RCM: HadRM3H (UKCP02)	SRES A2	Simplified PDM lumped conceptual rainfall runoff model	Some up to +50 %
	Fowler and Kilsby (2007)	Ensemble of runs for 1 RCM: HadRM3H (UKCP02)	SRES A2	ADM model	Up to +25 %
	Chun et al. (2009)	7 GCMs & RCMs		pd4-2par conceptual rainfall-runoff model for 6 catchments	
	Cloke et al. (2010)	HadRM3 RCM: subset of UKCP09 scenarios	SRES A1B	CATCHMOD semi-distributed conceptual model for Medway catchment	
	Arnell (2011)	21 CMIP3 GCMs		Cat-PDM conceptual model for 6 catchments	
	Charlton and Arnell (2011)	UKCP09 climate change scenarios		Cat-PDM conceptual model for 6 catchments	
	Christierson et al. (2012)	UKCP09 climate change scenarios	SRES A1B	PDM lumped conceptual rainfall runoff model and CATCHMOD semi-distributed conceptual model for 70 catchments	
	Bell et al. (2012)	UKCP09 climate change scenarios	SRES A1B	G2G model Thames basin	−11 to +68 %
	Kay and Jones (2012)	Perturbed parameter ensemble of 1 RCM		Nationwide grid-based runoff and routing model UK	
	Smith et al. (2014)	18 RCMs from ENSEMBLES and UKCP09	SRES A1B	HBV-light lumped conceptual rainfall runoff model Avon catchment	−1 to +23 %

(continued)

Table 7.2 (continued)

Region	Source	GCM-RCM(s) considered	Greenhouse gas scenario(s)	Hydrological-hydraulic impact model(s)	Change in river peak flow by 2100
UK and NW Europe	Prudhomme et al. (2012)	3 GCMs: ECHAM5, IPSL, CNRM		Global hydrological models JULES, MPI-HM, WaterGAP (WaterMIP project)	
Larger rivers in Europe	Dankers and Feyen (2008), Feyen and Dankers (2009), Rojas et al. (2011, 2012)	HIRHAM5 and 12 ENSEMBLES RCM runs	SRES A2, A1B	Coarse scale spatially distributed model LISFLOOD	Dependent on sub-region
Europe	Schneider et al. (2013)	3 GCMs	SRES A2	Global hydrological model WaterGAP3	Dependent on sub-region
Globe	Arnell and Gosling (2016)	1 GCM: HadCM3	SRES A1B	Water balance model Mac-PDM.09	Dependent on sub-region
	Dankers et al. (2014)	5 GCMs	RCP8.5	9 global water balance models (from WaterMIP)	Dependent on sub-region
	Prudhomme et al. (2014)	5 GCMs	RCP2.6, 8.5	9 global water balance models	
	Hirabayashi et al. (2013)	11 GCMS	RCPs	global river routing model with inundation scheme	Dependent on sub-region

two- to four-fold higher than the increase in precipitation, and that the total sewer overflow volume is 1.5- to three-fold higher. They also found that the number of overflowing manholes and number of surcharging sewers may change dramatically, but the precise magnitude of change in response to the change in precipitation is uncertain.

Willems (2013a) found that for sewer systems in Flanders, Belgium, built for design storms with return periods of two to 20 years, that the present-day design storms would increase for the high-tailored climate scenario by +15 to +50 % depending on the return period (range 1 month to 10 years) (Fig. 7.12). For the mean-tailored scenario, the changes were less: from +4 to +15 %. For the high scenario, the return period of sewer flooding increases by about a factor 2.

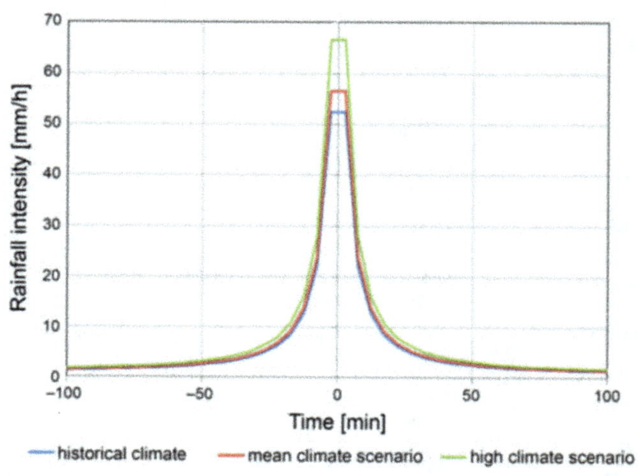

Fig. 7.12 Change in the design storm for sewer systems in Flanders, Belgium, for a 2-year return period for high and mean climate scenarios (Willems 2013a)

For the Windermere drainage area in NW England, Abdellatif et al. (2014) concluded based on the UKCP09 scenarios that an increase in the design storm of as little as 15 % is projected to cause an increase of about 40 % in flood volume due to surface flooding. However, impacts on house basements showed a damping effect (a 35 % increase in design storm leads to 16 % in the number of basements at risk of flooding). This confirms that the precise effects of climate change strongly depend on the type of impacts studies and the specific properties of the sewer system.

The impacts of climate change on sewer flood and overflow frequencies and volumes show wide variation. Studies indicate a range from a four-fold increase to as low as a 5 % increase, depending on the system characteristics (Willems et al. 2012a, b). Floods and overflows occur when runoff or sewer flow thresholds are exceeded. Given that the response of the sewer system to rainfall may be highly non-linear, the changes in the sewer response may be much stronger than the changes in rainfall. And the impact ranges can even by wider when studying the impacts of sewer overflows on receiving rivers. Sewer overflow mainly occurs in summer and as models project the likelihood of lower river flow in summer in north-western Europe, dilution effects in the receiving water might be less, thus increasing impacts on river water quality and aquatic life. Astaraie-Imani et al. (2012) found for a semi-real case study in the UK that changes in rain storm depth and peak rainfall intensity of up to +30 % by the 2080s could cause strong deterioration in river water quality; an increase in rain storm depth of 30 % led to an increase in river ammonium concentration of about 40 % and a decrease in dissolved oxygen concentration of about 80 %. This was found to correspond with a strong increase in the frequency of breaching given concentration thresholds (i.e. immission standards). The frequency of breaching the dissolved oxygen threshold of 4 mg l^{-1} increased from 49 to 99 %; the frequency of exceeding the ammonium threshold of 4 mg l^{-1} increased from 45 to 79 %. The effect of changes in peak rainfall intensity was found to be an order of magnitude lower.

Climate-driven changes in large-scale atmospheric circulation and related wind fields may cause significant changes in the amount and type of sediment on catchment surfaces available for wash-off into urban drainage systems. Higher deposition during prolonged dry periods will increase pollution concentrations in first flushes. This will lead to higher pollution loads in sewer overflows and in inflow to wastewater treatment plants; the latter leading to higher solids loads to clarifiers, different treatment efficiencies and higher pollution loads. Downstream of the treatment plants, receiving rivers during long dry spells in future summers may have reduced capacity to assimilate the more concentrated effluent. Prolonged dilute loading of wastewater treatment plants due to low-intensity long-duration precipitation events can also affect wastewater treatment with potential for major impacts on overall treatment (Plosz et al. 2009).

Changes other than those related to climate may also occur in urban areas and affect or strengthen urban drainage impacts. For example, changes in pavement surfaces, and these should not be seen in isolation but as related to population growth and increase in welfare, and thus partly interrelated with anthropogenic climate change. Semadeni-Davies et al. (2008) analysed the combined impact of climate change and increased urbanisation in Helsingborg, Sweden, and found that this could result in a four-fold increase in sewer overflow volumes. Using a similar approach, Olsson et al. (2010) analysed future loads on the main combined sewer system in Stockholm, Sweden, due to climate change and population increase. They estimated annual total inflow to the treatment plant to increase by 15–20 %, sewer overflow volumes to increase by 5–10 % and critically high water levels to increase by 10–20 % in the first half of the century. For the latter half of the century, they found no further increase in total inflow, but a 20–40 % increase in sewer overflow volumes and a 30–40 % increase in high water levels (within the sewer system). Both studies highlighted the importance of addressing climate change impacts in combination with other key non-stationary drivers of equal importance (e.g. urbanisation trends, sewer system or management changes). In fact, the study by Semadeni-Davies et al. (2008) clearly showed that climate change is not the most important driver of increased pollution levels, and that increases in damage may be effectively counterbalanced by measures not solely related to urban drainage.

Tait et al. (2008) confirmed that increased urbanisation (related to increased population and economic growth) also had a significant impact on urban runoff. For a typical urban area in the UK, in addition to climate change they assumed that paved areas would increase by about 25 % of their current value and roof areas by about 10 %. Model simulations showed that sewer overflow volumes would increase by about 15–20 % when only the increase in paved areas is considered. These changes are comparable to those expected from climate change.

Climatic variability at multi-decadal time scales has been detected by several authors (Stahl et al. 2010, 2012; Hannaford et al. 2012; Boé and Habets 2013; Willems 2013b). This must also be considered, given that it could temporarily limit, reverse or even increase the long-term impacts of climate change (Boé and Habets 2013).

7.4 Conclusion

Hydrological extremes are projected to become more intense. These changes are largely driven by changes in precipitation, which RCM rainfall projections for the North Sea region suggest will become significantly more intense (see Chap. 5; Van der Linden and Mitchell 2009). Future winters are expected to see both an increase in the volume and intensity of precipitation. The intensity of summer extremes may also increase albeit with a reduction in overall volume. These findings are consistent with recent observations at some monitoring stations that show winter extremes in high river flow are already increasing (see Chap. 4).

Quantifying future changes in hydrology is difficult. This reflects the high uncertainties in model output: mainly due to uncertainties in the climate processes, and—to a lesser extent —in knowledge of the hydrological processes and their schematisation in hydrological impact models. The impact uncertainties also reflect the level of uncertainties in future greenhouse gas emissions and concentrations.

Taking the uncertainties into account, the reported overview of impact results for rivers in the North Sea region in Table 7.2, indicates increases in river peak flow by 2100 of up to +30 % for many rivers and even higher for some. An increase in river peak flows is more evident for the northern basins of the North Sea region. The greatest increases are projected for catchments in south-western Norway, up to +70 % for 200-year peak flows. In snow-dominated catchments of Norway and southern Sweden, earlier spring flooding is projected. These spring floods do not always increase, however, peak flows from snowmelt may decrease when higher spring temperatures lead to reduced snow storage. Decreasing snowmelt-induced spring flow, and increased rain-fed flow in winter and autumn may change the seasonality of peak flows and floods. In northern France and Belgium, an increase in river peak flow is less clear in that not all models project an increase. Hence, the spatial differences mainly occur in a north-south direction. The position of a river basin relative to the ocean is also important. Allan et al. (2005) found that the greater the proximity the greater the potential damping of any climate change effect.

The impacts of climate change on sewer flood and overflow frequencies and volumes vary widely. The specific characteristics of an urban drainage system will dictate whether the net result of the projected increase in, for example, short-duration rainfall extremes is to damp or amplify these changes in precipitation. The precise amplitude of response is highly uncertain and non-linear. The combined impact of climate change and increased urbanisation in some parts of the North Sea region could result in as much as a four-fold increase in sewer overflow volumes.

References

Abdellatif M, Atherton W, Alkhaddar R, Osman Y (2014) Application of the UKCP09 WG outputs to assess performance of combined sewers system in a changing climate. J Hyd Eng, doi:10.1061/(ASCE)HE.1943-5584.0001129

Allan JD, Palmer MA, Poff NL (2005) Climate change and freshwater ecosystems. In: Climate Change and Biodiversity. Yale University Press

Andersen HE, Kronvang B, Larsen SE, Hoffmann CC, Jensen TS, Rasmussen EK (2006) Climate-change impacts on hydrology and nutrients in a Danish lowland river basin. Sci Total Environ 65:223–237

Andréasson J, Bergström S, Gardelin M, German J, Gustavsson H, Hallberg K, Rosberg J (2011) Dimensionerande flöden för dammanläggningar för ett klimat i förändring - metodutveckling och scenarier [Design floods for dams in a changing climate – methods and scenarios]. Elforsk rapport 11:25, Stockholm

Arnbjerg-Nielsen K (2012) Quantification of climate change effects on extreme precipitation used for high resolution hydrologic design. Urban Water J 9:57–65

Arnbjerg-Nielsen K, Fleischer HS (2009) Feasible adaptation strategies for increased risk of flooding in cities due to climate change. Water Sci Technol 60:273–281

Arnbjerg-Nielsen K, Willems P, Olsson J, Beecham S, Pathirana A, Bülow Gregersen I, Madsen H, Nguyen VTV (2013) Impacts of climate change on rainfall extremes and urban drainage systems: a review. Water Sci Technol 68:16–28

Arnell NW (2011) Uncertainty in the relationship between climate forcing and hydrological response in UK catchments. Hyd Earth Sys Sci 15:897–912

Arnell NW, Gosling SN (2016) The impacts of climate change on river flood risk at the global scale. Climatic Change 134(3):387–401

Ashley RM, Clemens FHLR, Tait SJ, Schellart A (2008) Climate change and the implications for modeling the quality of flow in combined sewers. Proceedings of the 11th International Conference on Urban Drainage, 31 August - 5 September 2008, Edinburgh, Scotland

Astaraie-Imani M, Kapelan Z, Fu G, Butler D (2012) Assessing the combined effects of urbanisation and climate change on the river water quality in an integrated urban wastewater system in the UK. J Environ Manage 112:1–9

Bastola S, Murphy C, Sweeney J (2011) The sensitivity of fluvial flood risk in Irish catchments to the range of IPCC AR4 climate change scenarios. Sci Total Environ 409:5403–5415.

Bates BC, Kundzewicz ZW, Wu S, Palutikof JP (eds) (2008) Climate change and water. Technical Paper of the Intergovernmental Panel on Climate Change, IPCC Secretariat, Geneva

Bell VA, Kay AL, Cole SJ, Jones RG, Moore RJ, Reynard NS (2012) How might climate change affect river flows across the Thames Basin? An area-wide analysis using the UKCP09 Regional Climate Model ensemble. J Hydrol 442-443:89–104

Boé J, Habets F (2013) Multi-decadal river flows variations in France. Hydrol Earth Syst Sc Disc 10:11861–11900

Boé J, Terray L, Martin E, Habets F (2009) Projected changes in components of the hydrological cycle in French river basins during the 21st century. Water Resour Res 45:W08426 doi:10.1029/2008WR007437

Boukhris O, Willems P, Vanneuville W (2008) The impact of climate change on the hydrology in highly urbanized Belgian areas. Proceedings International Conference on 'Water & Urban Development Paradigms: Towards an integration of engineering, design and management approaches', Leuven, 15-17 September

Brisson E, Demuzere M, Willems P, van Lipzig N (2015) Assessment of natural climate variability using a weather generator. Clim Dynam 44:495–508

Brouyére S, Carabin G, Dassargues A (2004) Climate change impacts on groundwater resources: modelled deficits in a chalky aquifer, Geer basin, Belgium. Hydrogeol J 12:123–134

Cameron (2006) An application of the UKCIP02 climate change scenarios to flood estimation by continuous simulation for a gauged catchment in the northeast of Scotland, UK (with uncertainty). J Hydrol 328:212–226

Charlton MB, Arnell NW (2011) Adapting to climate change impacts on water resources in England - An assessment of draft Water Resources Management Plans. Global Environ Chang 21:238–248

Christierson BV, Vidal JP, Wade SD (2012) Using UKCP09 probabilistic climate information for UK water resource planning. J Hydrol 424-425:48–67

Chun KP, Wheater HS, Onof CJ (2009) Streamflow estimation for six UK catchments under future climate scenarios. Hydrol Res 40:96–112

Cloke HL, Jeffers C, Wetterhall F, Byrne T, Lowe J, Pappenberger F (2010) Climate impacts on river flow: projections for the Medway catchment, UK, with UKCP09 and CATCHMOD. Hydrol Process 24:3476–3489

Collins M (2007) Ensembles and probabilities: a new era in the prediction of climate change. Phil Trans R Soc A 365:1957–1970

Dams J, Salvadore E, Van Daele T, Ntegeka V, Willems P, Batelaan O (2012) Spatio-temporal impact of climate change on the groundwater system. Hydrol Earth Syst Sc 16:1517–1531

Dankers R, Feyen L (2008) Climate change impact on flood hazard in Europe: An assessment based on high-resolution climate simulations. J Geophys Res 113:D19105. doi:10.1029/2007JD009719

Dankers R, Arnell NW, Clark DB, Falloon PD, Fekete BM, Gosling SN, Heinke J, Kim H, Masaki Y, Satoh Y, Stacke T, Wada Y, Wisser D (2014) First look at changes in flood hazard in the Inter-Sectoral Impact Model Intercomparison Project ensemble. PNAS 111:3257–3261

de Wit M, Buiteveld H, van Deursen W (2007) Klimaatverandering en de afvoer van Rijn en Maas [Climate change and the discharge of the Rhine and Meuse]. Memo WRR/2007-006, RIZA, Netherlands

Diaz-Nieto J, Wilby RL (2005) A comparison of statistical downscaling and climate change factor methods: Impacts on low flows in the River Thames, United Kingdom. Climatic Change 69: 245–268

Ducharne A, Sauquet E, Habets F, Deque M, Gascoin S, Hachour A, Martin E, Oudin L, Page C, Terray L, Thiery D, Viennot P (2011) Evolution potentielle du régime des crues de la Seine sous changement climatique. Houille Blanche 1:51–57

EEA (2012) Climate change, impacts and vulnerability in Europe 2012: An indicated based report – Summary. European Environment Agency, Copenhagen

Feyen L, Dankers R (2009) Impact of global warming on streamflow drought in Europe. J Geophys Res 114:D17116. doi:10.1029/2008JD011438

Feyen L, Dankers R, Bódis K, Salamon P, Barredo JI (2012) Fluvial flood risk in Europe in present and future climates. Climatic Change 112:47–62

Fowler HJ, Kilsby CG (2007) Using regional climate model data to simulate historical and future river flows in northwest England. Climatic Change 80:337–367

Goderniaux P, Brouyére S, Fowler HJ, Blenkinsop S, Therrien R, Orban P, Dassargues A (2009) Large scale surface-subsurface hydrological model to assess climate change impacts on groundwater reserves. J Hydrol 373:122–138

Good P, Barring L, Giannakopoulos C, Holt T, Palutikof J (2006) Non-linear regional relationships between climate extremes and annual mean temperatures in model projections for 1961-2099 over Europe. Clim Res 31:19–34

Gosling S, Taylor RG, Arnell N, Todd MC (2011) A comparative analysis of projected impact of climate change on river runoff from global and catchment-scale hydrological models. Hydrol Earth Syst Sc 15:279–294

Gudmundsson L, Bremnes JB, Haugen JE, Engen-Skaugen T (2012a) Technical Note: Downscaling RCM precipitation to the station scale using statistical transformations - a comparison of methods. Hydrol Earth Syst Sc 16:3383–3390

Gudmundsson L, Tallaksen LM, Stahl K, Clark DB, Dumont E, Hagemann S, Bertrand N, Gerten D, Heinke J, Hanasaki N, Voss F, Koirala S (2012b) Comparing large-scale hydrological model simulations to observed runoff percentiles in Europe. J Hydrometeorol 13:604–620

Habets F, Boé J, Déqué M, Ducharne A, Gascoin S, Hachour A, Martin E, Pagé C, Sauquet E, Terray L, Thiéry D, Oudin L, Viennot P (2013) Impact of climate change on the hydrogeology of two basins in Northern France. Climatic Change 121:771–785

Hannaford J, Buys G, Stahl K, Tallaksen LM (2012) The influence of decadal-scale variability on trends in long European streamflow records. Hydrol Earth Syst Sc Disc 10:1859–1896

Hanson CE, Palutikof JP, Livermore MTJ, Barring L, Bindi M, Corte-Real J, Durao R., Giannakopoulos C, Good P, Holt T, Kundzewicz Z, Leckebusch GC, Moriondo M., Radziejewski M, Santos J, Schlyter P, Schwarb M, Stjernquist I, Ulbrich U (2007) Modelling the impact of climate extremes: an overview of the MICE project. Climatic Change 81:163–177

Henriques C, Holman IP, Audsley E, Pearn K (2008) An interactive multiscale integrated assessment of future regional water availability for agricultural irrigation in East Anglia and North West England. Climatic Change 90:89–111

Herrera-Pantoja M, Hiscock KM (2008) The effects of climate change on potential groundwater recharge in Great Britain. Hydrol Process 22:73–86

Hirabayashi Y, Mahendran R, Koirala S, Konoshima L, Yamazaki D, Watanabe S, Kim H, Kanae S (2013) Global flood risk under climate change. Nat Climate Chang 3:816–821

Huang S, Hattermann FF, Krysanova V, Bronstert A (2013) Projections of climate change impacts on river flood conditions in Germany by combining three different RCMs with a regional eco-hydrological model. Climatic Change 116:631–663

Hulme M, Jenkins GJ, Lu X, Turnpenny JR, Mitchell TD, Jones RG, Lowe J, Murphy JM, Hassell D, Boorman P, McDonald R, Hill S (2002) Climate Change Scenarios for the United Kingdom: The UKCIP02 Scientific Report. Tyndall Centre for Climate Change Research, School of Environmental Sciences, University of East Anglia, Norwich, UK

IPCC (2014) Climate Change 2014: Impacts, Adaptation, and Vulnerability. Part A: Global and Sectoral Aspects. Contribution of Working Group II to the Fifth Assessment Report of the Intergovernmental Panel on Climate Change - Chapter 3: Freshwater Resources. Intergovernmental Panel on Climate Change, University Press

Karlsson IB, Sonnenborg TO, Jensen KH, Refsgaard JC (2013) Evaluating the influence of long term historical climate change on catchment hydrology – using drought and flood indices. Hydrol Earth Syst Sc Disc 10:2373–2428

Kay AL, Jones DA (2012) Transient changes in flood frequency and timing in Britain under potential projections of climate change. Int J Climatol 32:489–502

Kay AL, Jones RG, Reynard NS (2006) RCM rainfall for UK flood frequency estimation. II. Climate change results. J Hydrol 318:163–172

Kay AL, Davies HN, Bell VA, Jones RG (2009) Comparison of uncertainty sources for climate change impacts: flood frequency in England. Climatic Change 92:41–63

Kendon EJ, Roberts NM, Senior CA, Roberts MJ (2012) Realism of rainfall in a very high-resolution regional climate model. J Climate 25:5791–5806

Keupers I, Willems P (2013) Urbanization versus climate change: impact analysis on the river hydrology of the Grote Nete catchment in Belgium. Water Sci Technol 67:2670–2676

Kundzewicz ZW, Radziejewski M, Pinskwar I (2006) Precipitation extremes in the changing climate of Europe. Clim Res 31:51–58

Kundzewicz ZW, Kanae S, Seneviratne SI, Handmer J, Nicholls N, Peduzzi P, Mechler R, Bouwer LM, Arnell N, Mach K, Muir-Wood R, Brakenridge GR, Kron W, Benito G, Honda Y, Takahashi K, Sherstyukov B (2013) Flood risk and climate change: global and regional perspectives. Hydrolog Sci J 59:1–28

Larsen AN, Gregersen IB, Christensen OB, Linde JJ, Mikkelsen PS (2009) Potential future increase in extreme one-hour precipitation events over Europe due to climate change. Water Sci Technol 60:2205–2216

Lawrence D, Hisdal H (2011) Hydrological projections for floods in Norway under a future climate. Report no. 5-2011, Norwegian Water Resources and Energy Directorate

Lawrence D, Haddeland I, Langsholt E (2009) Calibration of HBV hydrological models using PEST parameter estimation. Report no. 1-2009, Norwegian Water Resources and Energy Directorate

Leander R, Buishand TA, van den Hurk BJJM, de Wit MJM (2008) Estimated changes in flood quantiles of the river Meuse from resampling of regional climate model output. J Hydrol 351:331–343

Lenderink G, Buishand A, van Deursen W (2007) Estimates of future discharges of the river Rhine using two scenario methodologies: direct versus delta approach. Hydrol Earth Syst Sc 11:1145–1159

Maraun D, Wetterhall F, Ireson AM, Chandler RE, Kendon EJ, Widmann M, Brienen S, Rust HW, Sauter T, Themeßl M, Venema VKC, Chun KP, Goodess CM, Jones RG, Onof C, Vrac M, Thiele-Eich I (2010) Precipitation downscaling under climate change: Recent developments to bridge the gap between dynamical models and the end user. Rev Geophys, AGU, 48: RG3003. doi:10.1029/2009RG000314

Mark O, Svensson G, König A, Linde JJ (2008) Analyses and adaptation of climate change impacts on urban drainage systems. Proceedings of the 11th International Conference on Urban Drainage, 31 August – 5 September 2008, Edinburgh, Scotland

May W (2008) Potential future changes in the characteristics of daily precipitation in Europe simulated by the HIRHAM regional climate model. Clim Dynam 30:581–603

MetOffice (2011) Climate: Observations, projections and impacts, Subreports for United Kingdom, France, Germany. UK MetOffice

Minville M, Brissette F, Leconte R (2008) Uncertainty of the impact of climate change on the hydrology of a Nordic watershed. J Hydrol 358:70–83

Monbaliu J, Chen Z, Felts D, Ge J, Hissel F, Kappenberg J, Narayan S, Nicholls RJ, Ohle N, Schuster D, Sothmann J, Willems P (2014) Risk assessment of estuaries under climate change: lessons from Western Europe. Coast Eng 87:32–49

Murphy JM, Sexton DMH, Jenkins GJ, Booth BBB, Brown CC, Clark RT, Collins M, Harris GR, Kendon EJ, Betts RA, Brown SJ, Humphrey KA, McCarthy MP, McDonald RE, Stephens A, Wallace C, Warren R, Wilby R, Wood RA (2009) Climate change projections. UK Climate Projections Science Report. UK Met Office Hadley Centre

New M, Cuellar M, Lopez A (2007) Probabilistic regional and local climate projections: false dawn for impacts assessment and adaptation? In: Integrating Analysis of Regional Climate Change and Response Options, Nadi, Fiji

Nie L, Lindholm O, Lindholm G, Syversen E (2009) Impacts of climate change on urban drainage systems – a case study in Fredrikstad, Norway. Urban Water J 6:323–332

Niemczynowicz J (1989) Impact of the greenhouse effect on sewerage systems – Lund case study. Hydrolog Sci J 34:651–666

Ntegeka V, Willems P, Monbaliu J (2011) Incorporating the correlation between upstream inland, downstream coastal and surface boundary conditions into climate scenarios for flood impact analysis along the river Scheldt. Geophysical Research Abstracts, EGU2011-6554: EGU General Assembly 2011, Vienna, April 2011

Ntegeka V, Decloedt LC, Willems P, Monbaliu J (2012) Quantifying the impact of climate change from inland, coastal and surface conditions. In: Comprehensive Flood Risk Management – Research for Policy and Practise (Klijn F, Schweckendiek T, eds), CRC Press

Ntegeka V, Baguis P, Roulin E, Willems P (2014) Developing tailored climate change scenarios for hydrological impact assessments. J Hydrol 508:307–321

Olsson J, Berggren K, Olofsson M, Viklander M (2009) Applying climate model precipitation scenarios for urban hydrological assessment: A case study in Kalmar City, Sweden. Atmos Res 92:364–375

Olsson J, Dahné J, German J, Westergren B, von Scherling M, Kjellson L, Ohls F, Olsson A (2010) A study of future discharge load on Stockholm's main sewer system, SMHI Reports Climatology 3 (in Swedish)

Palmer TN, Räisänen J (2002) Quantifying the risk of extreme seasonal precipitation events in a changing climate. Nature 415:512–514

Pfister L, Kwadijk J, Musy A, Bronstert A, Hoffmann L (2004) Climate change, land use change and runoff prediction in the Rhine-Meuse basins. River Res Appl 20:229–241

Plosz BG, Liltved H, Ratnaweera H (2009) Climate change impacts on activated sludge wastewater treatment: a case study from Norway. Water Sci Technol 60:533–541

Prudhomme C, Williamson J, Parry S, Hannaford J (2012) Projections of flood risk in Europe. In: Changes in Flood Risk in Europe, CRC Press

Prudhomme C, Giuntoli I, Robinson EL, Clark DB, Arnell NW, Dankers R, Fekete BM, Franssen W, Gerten D, Gosling SN, Hagemann S, Hannah DM, Kim H, Masaki Y, Satoh Y, Stacke T, Wada Y, Wisser D (2014) Hydrological droughts in the 21st century, hotspots and uncertainties from a global multimodel ensemble experiment. PNAS 111:3262–3267

Raupach MR, Marland G, Ciais P, Le Quéré C, Canadell JG, Klepper G, Field CB (2007) Global and regional drivers of accelerating CO_2 emissions. Proc Nat Acad Sci 104:10288–10293

Refsgaard JC, Madsen H, Andréassian V, Arnbjerg-Nielsen K, Davidson TA, Drews M, Hamilton D, Jeppesen E, Kjellström E, Olesen JE, Sonnenborg TO, Trolle D, Willems P, Christensen JH (2014) A framework for testing the ability of models to project climate change and its impacts. Climatic Change 122:271–282

Rojas R, Feyen L, Dosio A, Bavera D (2011) Improving pan-European hydrological simulation of extreme events through statistical bias correction of RCM-driven climate simulations. Hydrol Earth Syst Sc 15:2599–2620

Rojas R, Feyen L, Bianchi A, Dosio A (2012) Assessment of future flood hazard in Europe using a large ensemble of bias corrected regional climate simulations. J Geophys Res 117:D17109. doi:10.1029/2012JD017461

SAWA (2012) Climate change impacts and uncertainties in flood risk management: Examples from the North Sea Region. Report no. 05 – 2012 of SAWA Interreg IVB Project, Norwegian Water Resources and Energy Directorate

Schneider C, Laizé CLR, Acreman MC, Flörke M (2013) How will climate change modify river flow regimes in Europe? Hydrol Earth Syst Sc 17:325–339

Seibert J (2003) Reliability of model predictions outside calibration conditions. Nordic Hyd 34:477–492

Semadeni-Davies A, Hernebring C, Svensson G, Gustafsson LG (2008) The impacts of climate change and urbanisation on drainage in Helsingborg, Sweden: Combined sewer system. J Hydrol 350:100–113

Semenov MA, Stratonovitch P (2010) Use of multi-model ensembles from global climate models for assessment of climate change impacts. Clim Res 41:1–14

Shabalova MV, van Deursen WP, Buishand TA (2003) Assessing future discharge of the river Rhine using regional climate model integrations and a hydrological model. Clim Res 23:233–246

Shaffrey LC, Stevens I, Norton WA, Roberts MJ, Vidale PL, Harle JD, Jrrar A, Stevens DP, Woodage MJ, Demory ME, Donners J, Clark DB, Clayton A, Cole JW, Wilson SS, Connolley WM, Davies TM, Iwi AM, Johns TC, King JC, New AL, Slingo JM, Slingo A, Steenman-Clark L, Martin GM, (2009) UK HiGEM: The new UK High-resolution Global Environment Model - Model description and basic evaluation. J Climate 22:1861–1896

Smith JB, Schneider SH, Oppenheimer M, Yohe GW, Hare W, Mastrandrea MD, Patwardhan A, Burton I, Corfee-Morlot J, Magadza CHD, Füssel HM, Pittock AB, Rahman A, Suarez A, van Ypersele JP (2009) Assessing dangerous climate change through an update of the Intergovernmental Panel on Climate Change (IPCC) "reasons for concern". PNAS 106:4133–4137

Smith A, Freer J, Bates P, Sampson C (2014) Comparing ensemble projections of flooding against flood estimation by continuous simulation. J Hydrol 511:205–219

Stahl K, Hisdal H, Hannaford J, Tallaksen LM, van Lanen HAJ, Sauquet E, Demuth S, Fendekova M, Jódar J (2010) Streamflow trends in Europe: evidence from a dataset of near-natural catchments. Hydrol Earth Syst Sc 14:2367–2382

Stahl K, Tallaksen LM, Hannaford J, van Lanen HAJ (2012) Filling the white space on maps of European runoff trends: estimates from a multi-model ensemble. Hydrol Earth Syst Sc 16:2035–2047

Tait SJ, Ashley RM, Cashman A, Blanksby J, Saul AJ (2008) Sewer system operation into the 21st century, study of selected responses from a UK perspective. Urban Water J 5:77–86

Tavakoli M, De Smedt F, Vansteenkiste Th, Willems P (2014) Impact of climate change and urban development on extreme flows in the Grote Nete watershed, Belgium. Nat Hazards 71:2127–2142

Tebaldi C, Smith R, Nychka D, Mearns L (2005) Quantifying uncertainty in projections of regional climate change: a Bayesian approach to the analysis of multi-model ensembles. J Climate 18:1524–1540

Teutschbein C, Seibert J (2012) Bias correction of regional climate model simulations for hydrological climate-change impact studies: Review and evaluation of different methods. J Hydrol 456-457:12–29

Teutschbein C, Wetterhall F, Seibert J (2011) Evaluation of different downscaling techniques for hydrological climate-change impact studies at the catchment scale. Clim Dynam 37:2087–2105

Thodsen H (2007) The influence of climate change on stream flow in Danish rivers. J Hydrol 333:226–238

Thompson JR, Gavin H, Refsgaard A, Refstrup Sørenson H, Gowing DJ (2009) Modelling the hydrological impacts of climate change on UK lowland wet grassland. Wetl Ecol Manag 17:503–523

van den Hurk B, Tank AK, Lenderink G, van Ulden A, van Oldenborgh GJ, Katsman C., van den Brink H, Keller F, Bessembinder J, Burgers G, Komen G, Hazeleger W, Drijfhout S (2006) KNMI Climate Change Scenarios 2006 for the Netherlands, KNMI Scientific Report WR 2006-01, KNMI, De Bilt, The Netherlands

Van der Linden P, Mitchell JFB (2009) ENSEMBLES: Climate Change and its Impacts: Summary of research and results from the ENSEMBLES project, UK Met Office Hadley Centre

van Huijgevoort MHJ, van Lanen HAJ, Teuling AJ, Uijlenhoet R (2014) Identification of changes in hydrological drought characteristics from a multi-GCM driven ensemble constrained by observed discharge. J Hydrol 512:421–434

van Oldenborgh GJ, Drijfhout S, Ulden A, Haarsma R, Sterl A, Severijns C, Hazeleger W, Dijkstra H (2009) Western Europe is warming much faster than expected. Clim Past 5:1–12

van Pelt SC, Beersma JJ, Buishand TA, van den Hurk, BJJM, Kabat P (2012) Future changes in extreme precipitation in the Rhine basin based on global and regional climate model simulations. Hydrol Earth Syst Sc 16:4517–4530

Van Steenbergen N, Willems P (2012) Method for testing the accuracy of rainfall-runoff models in predicting peak flow changes due to rainfall changes, in a climate changing context. J Hydrol 414-415:425–434

Vansteenkiste Th, Tavakoli M, Ntegeka V, Willems P, De Smedt F, Batelaan O (2013) Climate change impact on river flows and catchment hydrology: a comparison of two spatially distributed models. Hydrol Process 27:3649–3662

Vansteenkiste Th, Tavakoli M, Ntegeka V, De Smedt F, Batelaan O, Pereira F, Willems P (2014) Intercomparison of hydrological model structures and calibration approaches in climate scenario impact projections. J Hydrol 519:743–755

Vidal JP, Wade SD (2008) Multimodel projections of catchment-scale precipitation regime. J Hydrol 353:143–158

Vörösmarty CJ, McIntyre PB, Gessner MO, Dudgeon D, Prusevich A, Green P, Glidden S, Bunn SE, Sullivan CA, Liermann CR, Davies PM (2010) Global threats to human water security and river biodiversity. Nature 467:555–561

Ward PJ, Jongman B, Weiland FS, Bouwman A, van Beek R, Bierkens PFP, Ligtvoet W, Winsemius HC (2013) Assessing flood risk at the global scale: model setup, results and sensitivity. Env Res Lett 8:044019. doi:10.1088/1748-9326/8/4/044019

Warren R, Arnell N, Berry P, Brown S, Dicks L, Gosling S, Hankin R, Hope C, Lowe J, Matsumoto K, Masui T, Nicholls R, O'Hanley J, Osborn T, Scriecru S (2010) The economics and climate change impacts of various greenhouse gas emissions pathways: A comparison between baseline and policy emissions scenarios. Work stream 1, Deliverable 3, Report 1 of the AVOID programme (AV/WS1/D3/R01)

Weisse R, Bellafiore D, Menendez M, Mendez F, Nicholls R, Umgiesser G, Willems P (2014) Changing extreme sea levels along European coasts. Coast Eng 87:4–14

Willems P (2009) A time series tool to support the multi-criteria performance evaluation of rainfall-runoff models. Environ Modell Softw 24:311–321

Willems P (2013a) Revision of urban drainage design rules after assessment of climate change impacts on precipitation extremes at Uccle, Belgium. J Hydrol 496:166–177

Willems P (2013b) Multidecadal oscillatory behaviour of rainfall extremes in Europe. Climatic Change 120:931–944

Willems P, Arnbjerg-Nielsen K, Olsson J, Nguyen VTV (2012a) Climate change impact assessment on urban rainfall extremes and urban drainage: methods and shortcomings. Atmos Res 103:106–118

Willems P, Olsson J, Arnbjerg-Nielsen K, Beecham S, Pathirana A, Bülow Gregersen I, Madsen H, Nguyen VTV (2012b) Impacts of Climate Change on Rainfall Extremes and Urban Drainage. IWA Publishing

Zhou Q, Mikkelsen PS, Halsnæs K, Arnbjerg-Nielsen K (2012) Framework for economic pluvial flood risk assessment considering climate change effects and adaptation benefits. J Hydrol 414-415:539–549

Part III

Impacts of Recent and Future Climate Change on Ecosystems

8 Environmental Impacts—Marine Ecosystems

Keith M. Brander, Geir Ottersen, Jan P. Bakker, Gregory Beaugrand, Helena Herr, Stefan Garthe, Anita Gilles, Andrew Kenny, Ursula Siebert, Hein Rune Skjoldal and Ingrid Tulp

Abstract

This chapter presents a review of what is known about the impacts of climate change on the biota (plankton, benthos, fish, seabirds and marine mammals) of the North Sea. Examples show how the changing North Sea environment is affecting biological processes and organisation at all scales, including the physiology, reproduction, growth, survival, behaviour and transport of individuals; the distribution, dynamics and evolution of populations; and the trophic structure and coupling of ecosystems. These complex responses can be detected because there are detailed long-term biological and environmental records for the North Sea; written records go back 500 years and archaeological records many thousands of years. The information presented here shows that the composition and productivity of the North Sea marine ecosystem is clearly affected by climate change and that this will have consequences for sustainable levels of harvesting and other ecosystem services in the future. Multi-variate ocean climate indicators that can be used to monitor and warn of changes in composition and productivity are now being developed for the North Sea.

K.M. Brander (✉)
National Institute of Aquatic Resources, Technical University of Denmark, Charlottenlund, Denmark
e-mail: kbr@aqua.dtu.dk

G. Ottersen (✉) · H.R. Skjoldal
Institute of Marine Research (IMR), Bergen, Norway
e-mail: geir.ottersen@imr.no

H.R. Skjoldal
e-mail: hein.rune.skjoldal@imr.no

G. Ottersen
CEES Centre for Ecological and Evolutionary Synthesis, University of Oslo, Oslo, Norway

J.P. Bakker
Groningen Institute for Evolutionary Life Sciences, University of Groningen, Groningen, The Netherlands
e-mail: j.p.bakker@rug.nl

G. Beaugrand
CNRS, Laboratoire d'Océanologie de Géosciences (LOG), Wimereux, France
e-mail: gregory.beaugrand@univ-lille1.fr

H. Herr · A. Gilles · U. Siebert
Institute for Terrestrial and Aquatic Wildlife Research, University of Veterinary Medicine Hannover, Foundation, Büsum, Germany
e-mail: helena.herr@tiho-hannover.de

A. Gilles
e-mail: anita.gilles@tiho-hannover.de

U. Siebert
e-mail: ursula.siebert@tiho-hannover.de

S. Garthe
Research and Technology Centre (FTZ), University of Kiel, Büsum, Germany
e-mail: garthe@ftz-west.uni-kiel.de

A. Kenny
Centre for Environment, Fisheries and Aquaculture Science (Cefas), Lowestoft, UK
e-mail: andrew.kenny@cefas.co.uk

A. Kenny
School of Environmental Sciences, University of East Anglia (UEA), Norwich, UK

I. Tulp
Institute for Marine Resources and Ecosystem Management (IMARES), IJmuiden, The Netherlands
e-mail: ingrid.tulp@wur.nl

M. Quante and F. Colijn (eds.), *North Sea Region Climate Change Assessment*, Regional Climate Studies, DOI 10.1007/978-3-319-39745-0_8

8.1 Introduction

The North Sea is one of the most productive, intensively exploited and well-studied sea areas in the world. It lies just north of the boundary (~50°N) between the warm- and cool-temperate biogeographic regions (Dinter 2001), also referred to as Lusitanian and Boreal. Because of its size, topography, and physical and chemical diversity (described elsewhere), the North Sea encompasses a number of more or less coupled ecosystems with some shared properties. Deep areas of the northern North Sea and Norwegian trench are strongly influenced, both physically and in biota, by inflows from the Norwegian Sea, NW European shelf (Fig. 8.1 is an artist's impression of this ecosystem). The low salinity Baltic Sea outflow affects the Norwegian coastal area and the inflow from the English Channel and several major rivers affects the continental coastal areas of the southern North Sea. Shallower water depth, stronger tidal mixing and diminished ocean influence all contribute to greater seasonal variability in temperature in the southern North Sea, with summer temperature much higher than in the northern North Sea and winter temperatures much lower.

The North Sea has been exploited by humans since they resettled its shifting margins after the last ice age 10,000 years ago. It has also been the subject of conservation concern for many hundreds of years and the focus of many scientific studies of marine life, which show the inexorable decline of easily caught fish and shellfish species with vulnerable life histories, such as sturgeon *Acipenser sturio*, ling *Molva molva*, large elasmobranchs and oysters (see Chap. 12) (Le Masson du Parc 1727; Poulsen et al. 2007).

The long history of exploitation and study of the North Sea means that a great deal of long-term information on fish, shellfish and other biota exists or is currently being reconstructed from archives, archaeological material and museums (Fig. 8.2). Written records go back 500 years in some cases and archaeological records go back many thousands of years (Enghoff et al. 2007), covering a wide range of temperature conditions and providing a basis for establishing the response of the ecosystem to natural climate variability and long-term change, but confounded by the effects of increasing fishing pressure and other anthropogenic drivers. An introductory account of the ecosystems of the North Sea is included in Chap. 1.

Between 1983 and 2007 the sea surface temperature (SST) of the North Sea warmed at rates of up to 0.8 °C decade^{-1} (see Chap. 2), which is an order of magnitude greater than the rate of global warming and among the highest in the world. The high rate of warming in the North Sea is partly due to anthropogenic factors but also to natural multi-decadal regional variability in the North Atlantic. Meyer et al. (2011) used sensitivity experiments to demonstrate that increasing air temperature is the main cause of the warming trend observed in the North Sea, accounting for about 75 % of observed (hindcast) changes in SST. From the record of Central England air temperature (CET, Fig. 8.3), which is the longest instrumented temperature time series in

Fig. 8.1 Artist's view of the ecosystem of the coastal northern North Sea. Artwork by Arild Sæther commissioned by the Institute of Marine Research, Norway

Fig. 8.2 By the time Pieter Breughel the Elder created *Big Fish Eat Little Fish* in 1557, Europeans had been putting pressure on coastal and estuarine parts of the North Sea for centuries (Bolster 2008). Public domain work accessed through Wikimedia Commons

the world, it is known that recent CET is higher than at any time since observations began in 1659, except for a short period in the 1730s. Over the period 1975–2005 the variability in annual mean CET and sea-bed temperature (SBT) during the first quarter of the year match closely (Fig. 8.3).

The rate of warming that the North Sea experienced from 1983 to 2007 is too high to persist, and the component of the warming due to multi-decadal variability is expected to reverse. There are indeed indications in the data since 2008 that temperatures in the North Sea may be returning to lower levels.

The abrupt temperature increase that occurred in the late 1980s (Fig. 8.3) was particularly marked during the period January–March and can be related to a remarkable increase in south-westerly wind strength during the early part of the year (Siegismund and Schrum 2001), which is a useful reminder that in addition to the effects of rising temperature, there are probably several interrelated climatic factors, including wind-driven vertical mixing and changes in inflow to the North Sea, with resultant salinity and nutrient changes, that are also important.

Irrespective of the underlying causes of the changes in sea temperature and other oceanic and atmospheric variables since the early 1980s, it is evident that biota in the North Sea are responding to these strong signals. Changes in advection and mixing may also be driving changes in ocean chemistry that in turn affect biota. Abrupt changes in many components of the biota, sometimes called 'regime shifts', have been observed in the North Sea since the 1980s. Such changes probably have physical as well as biological causes, but the nature of the processes involved is by no means clear and it is notable that the term regime shift is not used in the chapters on physical processes. There is a wide-ranging debate on the extent to which low frequency biological variability reflects external forcing, internal ecological dynamics or a combination of the two (Doney and Sailley 2013). The North Sea is well placed to help resolve the causes and processes behind abrupt and continuous

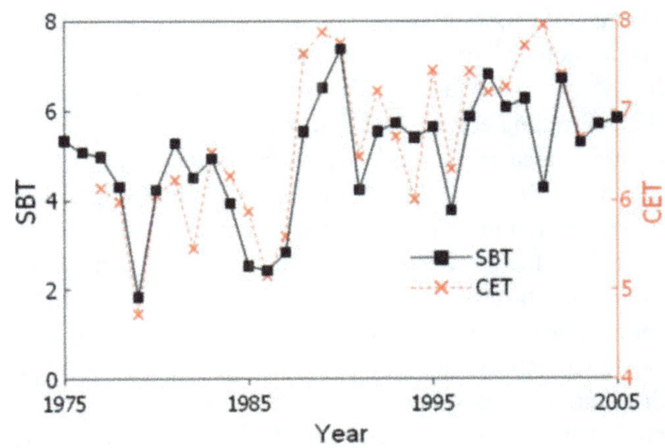

Fig. 8.3 First quarter sea-bed temperature (SBT) and Central England air temperature (CET; R = 0.80; $p < 0.0001$). SBT data from Hiddink and ter Hofstede (2008). CET data from www.metoffice.gov.uk/hadobs/hadcet/cetml1659on.dat

Table 8.1 Changes in North Sea biota in response to climate (from ICES 2008)

	Zooplankton	Benthos	Fish	Seabirds
Total observations	81	85	58	10
No change	8	13	1	0
Change in expected direction	68	47	43	6
Change in opposite direction	5	25	14	4

ecosystem responses to sub-annual-to-decadal physical variability, because spatially resolved, long-term records of physical and biological variables are both available and there have been marked changes in the two over recent decades (Schlüter et al. 2008; Kirby and Beaugrand 2009). The processes may also involve shifts in the biogeographic boundaries between cool- and warm-temperate ecosystems.

Climate change is a recent addition to the human pressures on marine ecosystems, but in this too the North Sea is relatively well studied and described. OSPAR, the intergovernmental body for protecting and conserving the Northeast Atlantic, commissioned a report in 2008 asking whether the impacts of climate change over past decades can be detected on North Sea and Northeast Atlantic biota (ICES 2008). The first step in addressing this question is to decide on the basis of theory and previous observations what the likely change in any feature (distribution, abundance, seasonal pattern) would be under the actual climate of the past decades. The vast majority of long-term data sets assembled for the North Sea (212 out of 234) showed changes in the distribution, abundance or seasonal patterns (maturation, breeding, other seasonal cycles) of zooplankton, benthos, fish and seabirds; 77 % of these changes were in the direction expected due to climate impacts (Table 8.1).

A global study of climate impacts on marine biota (Poloczanska et al. 2013) assembled over 1700 data series and showed a similar proportion (81–83 %) of time series responding consistently with the effects of climate change. The rate of distribution shift for leading edges (i.e. where the distribution is spreading into previously unoccupied areas for the species) is faster ($\sim$72 km decade^{-1}) than the rate of shift of trailing edges ($\sim$15 km decade^{-1}; where a previous occupied area is vacated). The overall global rates of distribution shift ($\sim$30 km decade^{-1} for leading edges, centroids and trailing edges) matched the rates at which ocean surface isotherms had shifted over the same periods and locations (Burrows et al. 2011), but the rates of shift in spring phenology (seasonal timing) were not closely matched with changing seasonality of temperature. Rates of distribution shift varied among taxa and were fastest for phytoplankton and zooplankton. The rates at which both distribution and seasonal timing of marine biota had shifted were comparable to or greater than the rates observed for terrestrial biota. Almost half (45 %) of the data used in the global study of climate impacts on marine biota came from the Northeast Atlantic and a high proportion of these from the North Sea.

This chapter presents a review of what is known about the impacts of climate change on the biota of the North Sea. Plankton and benthos, which are habitat/life history categories, are each considered, as are the taxa fish, birds and marine mammals. Invertebrate taxa are addressed within the sections on plankton and benthos, but viruses, bacteria and the microbial loop are not covered. Other anthropogenic drivers of change such as fishing, habitat disturbance, eutrophication or pollution are dealt with in other chapters.

All North Sea biota are affected by a range of physical and chemical drivers, including inflowing water masses, currents within the North Sea, nutrients, atmospheric warming, winds and other mixing forces (buoyancy flux, tidal mixing) that influence the proximate physical, chemical and biological environment of the biota. Thus climate impacts are by no means limited to temperature effects; the major ocean climate variables are grouped by property type in Table 8.2. This changing environment affects biological processes and organisation at all scales. There are direct effects on the physiology, reproduction, growth, survival, behaviour and transport of individuals and on the distribution, dynamics and evolution of populations. Indirect effects include trophic interactions (predators, prey, competitors), the structure and coupling (e.g. benthic-pelagic coupling) of ecosystems and the effects of pathogens, symbionts and commensals. Life spans of very different length and feedbacks between levels of biological organisation (e.g. ecosystem effects on food supply) can result in cross-scale and lagged effects (Doney and Sailley 2013). Such linkages and ecosystem processes are considered in Sect. 8.7.

Because the North Sea is so well monitored and studied, the effects on marine biota of the very rapid rate of climate change in the region over the past 30 years are more likely to be detected and understood here than in other areas. The following sections review current knowledge of the effects of climate change on functional and taxonomic groups in the North Sea followed by a section on ecosystem effects and a final synthesis that draws together common features and conclusions.

Table 8.2 Ocean climate variables grouped by property type

Property type	Ocean climate variable
Atmospheric and sea surface	Wind
	Cloud cover
	Waves
	Sea level
Chemical and physical	Temperature
	Salinity
	pH
	Oxygen
	Nutrients
Dynamic	Currents
	Stratification
	Turbulence
	Upwelling
	Frontal processes
Seasonal	Storm events (for example)

8.2 Plankton

Many studies have documented the strong influence of both climatic variability and global climate change on plankton ecosystems (Roemmich and McGowan 1995; Edwards and Richardson 2004; Richardson and Schoeman 2004; Mackas et al. 2007), indeed plankton often seem to amplify subtle climatic changes in areas such as the North Sea (Taylor et al. 2002). Some explanations have been proposed to explain the sensitivity of this group to climate (Taylor et al. 2002; Beaugrand et al. 2008), but the precise processes remain to be identified. One contributory factor may be that these organisms are ectotherms and that metabolic rates, growth, reproduction, activity and species interactions are all influenced by temperature (Atkinson 1994; Brown et al. 2004). A second factor could be that they react rapidly to climate change because of their short life cycle. A third that this group is not exploited directly, so that the main drivers are easier to identify. Phenological (Edwards and Richardson 2004) as well as biogeographic shifts (Beaugrand et al. 2009) have been observed in North Sea plankton. Abrupt community or ecosystem shifts (also called regime shifts or critical transitions) (Reid et al. 2001; Scheffer 2009) have been documented (Reid et al. 2001; Weijerman et al. 2005), including changes in phytoplankton and zooplankton (e.g. copepods, euphausiids, gelatinous species) and in holozooplankton (taxa whose whole lifecycle is planktonic) and merozooplankton (taxa with a partly planktonic life history) (Kirby et al. 2008).

8.2.1 Bottom-Up and Top-Down Control

Among environmental factors that may influence population dynamics and individual survival, temperature is probably the major factor. It is often highly correlated with observed changes in biological or ecological systems (Aebischer et al. 1990; Edwards and Richardson 2004; Weijerman et al. 2005; Hatun et al. 2009; Kirby and Beaugrand 2009; Buckley et al. 2012). Temperature modulates predator-prey interaction by influencing locomotion, functioning of sensory organs and activity. It might therefore be difficult to resolve the role of bottom-up (e.g. physics) and top-down (e.g. grazing/predation) effects. This probably also depends on the spatial scale of a study. At the scale of the spatial distribution of a species, the climate variability hypothesis states that the latitudinal range of species is primarily determined by their thermal tolerance (Stevens 1989). Temperature is indeed a key variable in the marine environment because it is affected by many hydro-climatic processes (Beaugrand et al. 2008) and because it exerts an effect on many fundamental biological and ecological processes (Sunday et al. 2012). At smaller scales however, this factor acts in synergy with others and the proportion of all factors acting on a species also varies spatially (i.e. throughout the distributional range of a species) and through time (i.e. seasonal and year-to-year scales) (Kirby and Beaugrand 2009). The level of turbulence in the water column (Rothschild and Osborn 1988), nutrient concentrations and their effect on phytoplankton concentration and composition (Behrenfeld et al. 2009), the amount of photosynthetically active radiation (Asrar et al. 1989) and the length of day (Fiksen 2000) in extratropical regions are key controlling factors. At small scale, the effect of top-down control may start to be detected. On the eastern Scotian Shelf, Frank et al. (2005) suggested a cascading effect of fishing from the top to the bottom level of the ecosystem, although this has been disputed, with changes in stratification being proposed as the driver (Pershing et al. 2015). Nevertheless, despite some evidence of top-down or wasp-waist control in certain marine ecosystems of the world (Cury et al. 2003), bottom-up control seems to be the most frequent type of control in pelagic ecosystems (Richardson and Schoeman 2004). However, data and studies are sparse and some statistical techniques can give ambiguous results. For example, it is statistically difficult to separate bottom-up control from a common response of organisms to climate change (Kirby and Beaugrand 2009). The persistent simplification of marine food webs by overexploitation (Pauly et al. 1998) may have diminished the importance of top-down control in marine ecosystems. Bottom-up control of plankton is likely to be more important than top-down control.

8.2.2 Climate and Changes in Phytoplankton Abundance and Phenology

A substantial literature describes long-term changes in North Sea phytoplankton communities (Reid et al. 1998; Edwards et al. 2009, 2012, 2014; Beaugrand et al. 2010). Total abundance of dinoflagellates has declined since 1960 whereas the total abundance of diatoms has remained virtually unchanged (Hinder et al. 2012). Among the dinoflagellate species, *Ceratium furca*, *Protoperidinium* spp. and to a lesser extent *Prorocentrum* spp., have shown a substantial reduction in summer since the beginning of the 2000s, but the phenology of dinoflagellates has not shifted towards spring, as shown in other studies (Edwards and Richardson 2004). The diatoms *Thalassiosira* spp., *Rhizosolenia imbricata shrubsolei* and *Pseudo-nitzschia seriata* have increased in abundance in spring. The diversity of dinoflagellates has increased in the Northeast Atlantic and in the North Sea (Beaugrand et al. 2010) as both temperature and seasonal stability in temperature have increased, whereas diatoms, which have higher diversity at intermediate and less seasonally stable temperatures, have shown less increase in diversity.

Analysis of daily (work days) sampling of phytoplankton, nutrients and temperature at Helgoland Roads (54° 11′ 3″N, 7° 54′E) from 1962 to 2008 showed that the phenology of three diatom species *Guinardia delicatula, Thalassionema nitzschioides* and *Odontella aurita* did not respond to climate warming in the same way and that overwintering population size, grazing, nutrient levels and water clarity affected their bloom timing (Schlüter et al. 2012). Such species-specific differences in sensitivity to forcing factors could lead to shifts in community structure with potentially far-reaching consequences for ecosystem dynamics. Not only does the marine food web depend on the quantity and timing of phytoplankton production, but also on qualitative features, such as the production of essential fatty acids (Røjbek et al. 2012).

The extratropical North Atlantic Ocean and its adjacent seas may be an important region for carbon export (Sarmiento et al. 2004). The biological pump may be less efficient in a warmer world because of changes in phytoplanktonic types (floristic shifts) but also because upward mixing of nutrients is likely to diminish, due to increased stratification of the oceans (Thomas et al. 2004; Bopp 2005). Deepening of the nutricline, as a result of increased stratification, would shift the phytoplankton community from diatoms (major exporters of carbon to depth) to coccolithophorids (Cermeño et al. 2008) and this latter group has increased in the North Sea (Beaugrand et al. 2013).

8.2.3 Climate Impacts on Biogeographic Boundaries and Biodiversity of Zooplankton

Major biogeographical shifts in zooplankton have been identified in the Northeast Atlantic in response to the warming observed in the region (Beaugrand and Ibañez 2002; Beaugrand et al. 2009) and based on the identification of nine calanoid copepod species assemblages using multivariate analyses (Beaugrand et al. 2002). There was a poleward increase in warm-water species and a reduction in the number of cold-water species in the same areas. All zooplankton assemblages exhibited coherent, long-term shifts but the speed of these biogeographic shifts was surprisingly rapid in comparison to rates of change in terrestrial systems (Parmesan and Yohe 2003). Warm-temperate, pseudo-oceanic species experienced a poleward shift of about 10° of latitude (52–62°N, 10°W) or 23 km y^{-1} for the period 1958–2005 (Beaugrand et al. 2009). The magnitude of the species shifts was however similar to the northward movement of some isotherms (e.g. the 10 °C isotherm moved northwards by about 21.75 km y^{-1}) in the North Sea. The consequence of these shifts has been to increase the diversity of calanoid copepods in the Northeast Atlantic and its adjacent seas (such as the North Sea) (Beaugrand and Ibañez 2002; Beaugrand et al. 2010). Such increases in diversity have also been identified for other taxonomic groups such as dinoflagellates (Beaugrand et al. 2010) and fish (Hiddink and ter Hofstede 2008). The increase in copepod diversity has been paralleled by a concomitant reduction in their mean size (Beaugrand et al. 2010) due to both increased prevalence of species with smaller body size and decrease in body size within species due to increasing temperature. Size reduction may indicate an increase in the metabolism of plankton ecosystems and may have strong consequences for carbon export.

8.2.4 Regime Shift in the North Sea Plankton Community

Marine ecosystems are not all equally sensitive to global climate change and climatic variability (Beaugrand et al. 2008). There are critical thermal boundaries (CTB) where a small increase in temperature triggers abrupt ecosystem shifts (regime shift) and alters the abundance of primary producers, secondary producers and top predators. Such a boundary separates regions where abrupt ecosystem shifts have been reported in the North Atlantic and the North Sea. In these regions, termed vulnerability hotspots, temperature

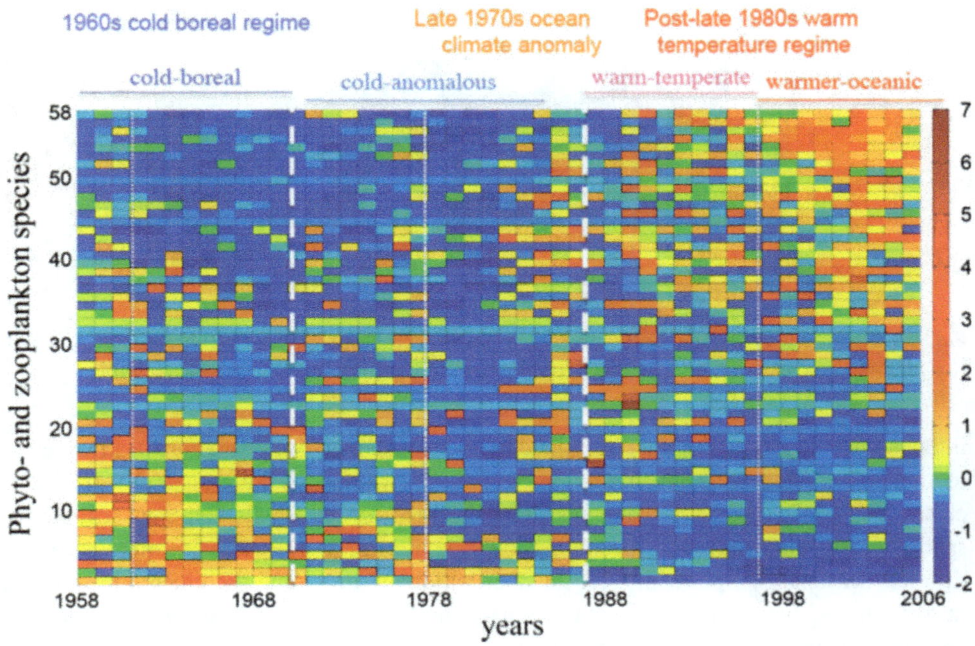

Fig. 8.4 Change in North Sea plankton composition over the past 50 years. Standardised abundance of 83 phytoplankton and zooplankton taxa collected by the Continuous Plankton Recorder (CPR). The taxa are ordered according to the first principal component. Periods characterised by different hydro-climatic conditions are indicated. Adapted from Edwards et al. (2009)

increase has a substantial effect on the community and the ecosystem, modifying their biodiversity and carrying capacity (Beaugrand et al. 2008).

An abrupt ecosystem shift occurred in the North Sea during the mid-1980s (Fig. 8.4) (Reid et al. 2001). The North Sea is one of the most biologically productive ecosystems in the world. This system supports important fisheries leading to the catch of 5 % of the world's total fish and also contributes significantly to biogeochemical cycles (Thomas et al. 2004). The North Sea regime shift has involved an increase in phytoplankton biomass, and changes in plankton community structure, diversity and phenology (Reid et al. 1998; Beaugrand et al. 2003; Beaugrand 2004). The shift was detected in both pelagic and benthic realms (Kröncke et al. 1998; Reid and Edwards 2001; Warwick et al. 2002). Parallel changes occurred in large-scale and regional temperatures, in three trophic levels and in both holozooplanktonic and merozooplanktonic components (Kirby and Beaugrand 2009). The abrupt ecosystem shift that occurred during the 1980s in the North Sea had a detectable effect on about 40 % of species from all taxonomic groups collected by the Continuous Plankton Recorder (CPR) survey (Beaugrand et al. 2014).

The effect of warming on ecosystems is not a gradual process and species and communities are likely to experience a series of sudden and stepwise shifts alternating with periods of greater stability.

8.2.5 Long-Term Changes in Zooplankton

Long-term changes in the zooplankton community have been reviewed from time series at Arendal (northern Skagerrak; 58° 23′N, 8° 49′E), Helgoland Roads (54° 11′ 18″N, 7° 4′E) and Stonehaven (56° 57.80′N, 02° 06.20′W) together with other studies (Hay et al. 2011). Jellyfish abundance has increased (Lynam et al. 2005; Attrill 2007) and there have been reports of incursions of the oceanic scyphozoan *Pelagia noctiluca* into the North Sea, causing mortalities in farmed salmon (Licandro et al. 2010). The ctenophore *Mnemiopsis leidyi* was detected in the Skagerrak in 2006 (Oliveira 2007) and has since occurred in high densities at Arendal Station in late summer and autumn each year, when SSTs are above 20 °C. The Helgoland Roads time series showed abrupt shifts (earlier bloom timing) in the phenology of the ctenophores *Beroe gracilis* and *Pleurobrachia pileus* in 1987/88 (Schlüter et al. 2010) and an inverse relationship between SST anomalies and abundance of small copepods; lowest copepod abundance was observed in the 2000s when sea temperatures were warmest. A comparison of the Helgoland Roads and CPR data indicates a possible time-lagged synchrony (3–5 years) in copepod abundance (Hay et al. 2011): at Stonehaven, total copepod abundance was low in 1997 and 1998. The copepod *Eucalanus crassus*, included in the temperate pseudo-oceanic species assemblage that increased northwards along the European shelf-edge (Beaugrand et al. 2009), has been seen regularly in small numbers at Stonehaven in autumn since 2003.

Since 1958 the copepod *Calanus helgolandicus* has become roughly ten-fold more abundant than *C. finmarchicus* and is now among the most abundant species in the North Sea (Edwards et al. 2014; Fig. 8.5). The underlying climate-related processes have been investigated using life-stage structured models of the two species, combined with a high-resolution 3D circulation model to quantify

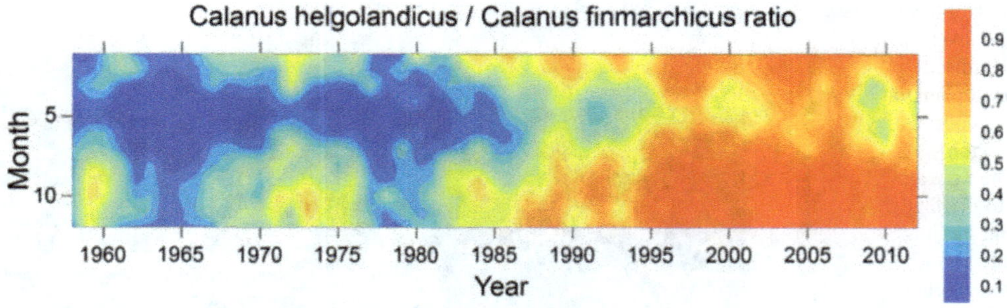

Fig. 8.5 Ratio between the abundance of the temperate-water copepod species *Calanus helgolandicus* and the cold-water species *C. finmarchicus*. *Red* indicates a dominance of *C. helgolandicus* and *blue C. finmarchicus* (Edwards et al. 2014)

inflows into the North Sea and a regional ecosystem model to quantify biogeochemical and foodweb variables (Maar et al. 2013). Model results were tested against the long, spatially-resolved time series from the CPR and against detailed seasonal sampling of vertical distribution of life history stages. Increasing temperature is a major factor in observed changes in *Calanus* phenology, but changes in abundance are also influenced by advection of *C. finmarchicus* through the northern boundary of the North Sea, which is to some degree climate-related. The detailed observational time series available for the North Sea allow testing of quite complex process models on scales encompassing regional physical dynamics, water column processes and species life history. Observed changes in distribution and phenology are consistent with global patterns (Poloczanska et al. 2013), but there are important processes occurring at regional and local scales that modify the simple global pattern. It is probably too early to judge whether adaptive responses by marine zooplankton will keep pace with the current rapid changes in climate (Dam 2013).

The observational time series for North Sea plankton are longer and have better temporal and spatial coverage and resolution than any other in the world. Since many taxa have plankton life stages these time series can be used to analyse long-term change not only in holoplankton, but also in meroplankton, including benthic species and fish, and examples are given in later sections.

8.3 Benthos

This section does not attempt to provide a comprehensive review of all the known processes and mechanisms governing changes in the status of the North Sea benthic ecosystem, rather it attempts to highlight and describe some of the more important and well documented factors which appear to influence the benthos. For example, sediment composition, depth, food availability and water temperature are the main environmental factors governing the large-scale distribution of benthic species in the North Sea (e.g. Glémarec 1973; Duineveld et al. 1991). Small-scale temporal or spatial variability in the benthos, particularly in the shallower areas of the North Sea, may be attributed to temperature, tidal currents, riverine input including nutrient and sediment load, wind-induced swell and sediment resuspension (Rachor and Gerlach 1978; Kröncke et al. 2001) and more rarely to extremely cold winters or anoxia (Duineveld et al. 1991; Kröncke et al. 1998; Armonies et al. 2001).

Macrozoobenthos (relatively large bottom-dwelling animals) form a major component of the North Sea fauna. Most benthic species have pelagic life stages that are likely to be responsive to climate change. For example, the abundance of decapod larvae in the plankton is positively correlated with sea temperature and rising temperatures have resulted in recruitment of large numbers of swimming crabs of the sub-family Polybiinae in the southern North Sea (Luczak et al. 2012). However, once settled on the seabed most species have low mobility and a relatively long lifespan such that individuals reflect integrated effects of climate and other environmental changes over time at their location.

The role of bathymetry and prevailing environmental conditions in structuring the benthic community is examined and long-term temporal patterns are described. Examples of climate effects on macrofauna communities are given from two contrasting intertidal areas of the North Sea, the rocky shores of the British coast and the intertidal flats in the Wadden Sea. Benthic species play an important role in the food web, as a food source for higher trophic levels such as crabs, fishes and migrant birds.

8.3.1 Spatial Patterns

Early studies in Danish waters described the spatial patterns of the benthic fauna (Petersen 1914, 1918) and explained the importance of seabed sediment type as a major structuring force for macro-benthic communities. Later work examined the influence of hydrodynamic mixing and concluded that thermal stability of the water column (i.e. the occurrence and persistence of stratification) was also an important explanatory variable for benthic community structure (Glémarec 1973). Shallow mixed waters in the southern North Sea have benthic species assemblages that are distinct from those in the central North Sea between 50 and 100 m deep, and in the areas deeper than 100 m north of the Dogger Bank, where the water column is stratified for a significant proportion of the year. Benthic animals in the North Sea are generally categorised as northern, southern or cosmopolitan (Glémarec 1973; Rachor et al. 2007).

Wide-scale synoptic benthic surveys of the North Sea in 1986 (Heip et al. 1992; Künitzer et al. 1992) and repeated in 2000 (Rees et al. 2007a, b) showed a clear north-south gradient across a range of habitats in the species of molluscs, annelids, crustaceans and echinoderms present. There were gradients in diversity, abundance, biomass and average individual weight of the soft-bottom infauna. The macro-benthic infauna (animals living within the substratum), epifauna (animals living on or associated with the surface of the substratum) and fish assemblages had significantly correlated spatial patterns (Fig. 8.6).

The correlated distributions suggest that large-scale gradients in bathymetry, temperature and ocean currents were particularly important in structuring the benthos. Benthic community types, and the distribution of biomass and mean individual weights of species have been relatively stable over time (Kröncke and Reiss 2007).

The ability to detect changes in benthic communities in the North Sea is hampered by lack of regular, standardised time series, unlike the spatially and temporally extensive surveys of the North Sea plankton. Consistent long-time series for macrobenthos are limited to a few locations off the north-east coast of England (Frid et al. 2009a, b) and in the southern North Sea, off the Friesian coasts (Kröncke et al. 1998, 2001; Neumann et al. 2009). Thus, much of the understanding of trends in North Sea benthos is based on evidence from relatively few sites. However, because most benthic animals have planktonic life stages the recent advances in molecular analysis mean preserved plankton samples can be now be reanalysed (Kirby and Lindley 2005). North Sea plankton samples, collected monthly by CPR since 1948 from all parts of the North Sea, show increasing abundance of meroplankton (the planktonic life stages of benthic species) and a decline in the abundance of holoplankton (permanent planktonic species) since 1958 (Lindley and Batten 2002). In contrast to deeper water benthic species, rocky shore species are easier to survey due to their greater accessibility and because their community dynamics, biodiversity and ecology can be studied experimentally (Sagarin et al. 1999; Tomanek and Helmuth 2002).

Benthic community structure in the North Sea is affected by hydrographic variables; bottom water temperature has a particularly strong influence, but also bottom water salinity, and tidal stress (for the infauna). Chapter 1 discusses the dynamics of water masses in the North Sea and effects on hydrographic properties which affect the composition and productivity of pelagic (planktonic) communities (see Sect. 8.2) which, in turn, affect the benthic communities. The North Atlantic Oscillation (NAO) drives some of the observed changes in benthic-pelagic coupling and has been shown to affect benthic communities off north-eastern Germany (Kröncke et al. 1998, 2001) and in the Skagerrak off western Sweden (Tunberg and Nelson 1998), by causing variability in nutrient supply, changes in planktonic biomass and so

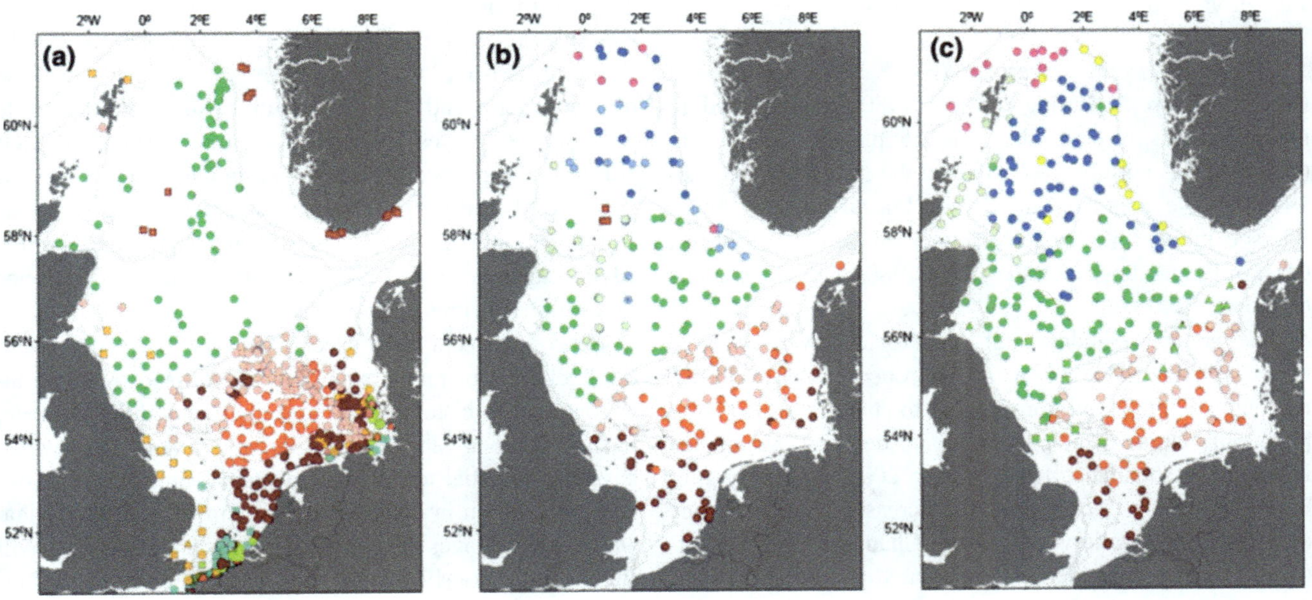

Fig. 8.6 Distribution of **a** infauna, **b** epifauna, and **c** fish assemblages in the North Sea. *Colours* depict different assemblages. The underlying cluster analyses and taxa associated with each assemblage are given in Reiss et al. (2010)

changes in benthos through sedimentation (e.g. Tunberg and Nelson 1998; Reid and Edwards 2001; Kirby et al. 2007).

The occurrence and densities of a wide range of species show distributional depth limits, but also a close association with habitat type, on which biogeographical influences may be superimposed (Künitzer et al. 1992; Zühlke 2001). Water depth, prevalence of fine soft-sediments and community diversity all increase from south to north. Mean annual and maximum temperatures increase along a northwest to southeast gradient, while minimum temperature decreases (Hiddink et al. 2014) and is correlated with a decrease in biomass and individual weight of species (Eggleton et al. 2007; Willems et al. 2007). Coarser substrata in the south-western North Sea and eastern English Channel generally support species-rich communities and hence contrast with the trend for increasing diversity of the fauna of finer sediments to the north, highlighting the importance of sediment heterogeneity and stability in favouring a greater number of species present.

8.3.2 Climate-Driven Temporal Trends

The northern range edge of many benthic invertebrate species in the North Sea has expanded with increased temperature (Hiddink et al. 2014). For example, a southern trochid gastropod that was surveyed in British waters in the 1950s, 1980s and in 2002–2004 showed a range extension of up to 55 km between the 1980s and 2000s (Mieszkowska et al. 2007). Populations sampled over a latitudinal extent of 4° from northern limits towards the centre of the range showed synchronous increases in abundance throughout the years sampled, suggesting that a large-scale factor such as climate was driving the observed changes (Mieszkowska et al. 2007).

The abrupt rise in temperature (Fig. 8.3) and spring wind strength (Chap. 1) during the late 1980s that resulted in major changes (regime shift) in the plankton ecosystem (Fig. 8.4) was not observed to affect the macrobenthos until 1995/96 (Neumann et al. 2009; Luczak et al. 2012). Site-specific species richness increased off the northeast coast of England between the 1970s/1980s and 1990s/2000s (Frid et al. 2009a).

The annual abundance of planktonic larvae of three benthic phyla, Echinodermata, Arthropoda, and Mollusca, respond positively and immediately to changes in SST. The planktonic larvae of echinoderms and decapod crustaceans increased in abundance from 1958 to 2005, especially after the mid-1980s, as North Sea SST increased, but abundance of bivalve mollusc larvae declined. Changes in meroplankton abundance, coincident with increased phytoplankton and declining holoplankton, are probably due to the direct effects of rising SST on the pelagic community and indirect effects of warming on the reproduction and recruitment of many benthic marine invertebrates. The long-term decline in bivalve mollusc larvae may reflect increased predation on the settled larvae and adults by benthic decapods (Kirby et al. 2008). These alterations in the zooplankton may therefore reflect an ecosystem-wide restructuring of North Sea trophic interactions (Kirby et al. 2008).

Mean abundances of macrobenthos at depths of 40 to 120 m at two locations inside Gullmarsfjorden, on the Swedish west coast, and three locations outside the fjord, are negatively correlated with temperature at 600 m in the Skagerrak. This may be due to an NAO-influenced increase in the upwelling of nutrient-rich deep water resulting in increased primary production and food supply to the benthos (Hagberg and Tunberg 2000). Indeed, the impacts of climate change on benthic invertebrates seem to arise from changes in temperature, nutrients and hydrodynamics affecting food supply and hence reproduction (e.g. Kröncke et al. 1998, 2001; Armonies et al. 2001; Clark and Frid 2001).

8.3.3 Climate Impacts on Intertidal Species on Rocky Shores

It has long been known that many intertidal rocky shore species reach their biogeographic limits around the British Isles (Forbes 1858; Hawkins et al. 2009) and changes in the distribution of intertidal species on rocky shores have been related to climate variability and change for decades. For example, the relative abundance of two barnacle species during the early twentieth century—*Balanus balanoides*, a Boreal-Arctic species that reaches its southern limit in the SW British Isles and *Chthamalus stellatus*, a Lusitanian-Tropical species that reaches its northern limit in Scotland—was shown to be related to warm and cold periods (Southward and Crisp 1954). Recent surveys have updated these historic records and show that a number of warm-temperate rocky shore species have extended (or re-extended) their northern limits since the abrupt warming of the late 1980s (Southward et al. 1995; Mieszkowska et al. 2006). It seems that more southern, warm-water species have been recorded advancing polewards than northern, cold-water species retreating (Hawkins et al. 2009).

Responses seem to be species- and habitat-specific, with the likelihood of range extensions determined by a combination of life history traits including reproductive mode, fecundity, larval behaviour and larval duration, all of which have the potential to influence dispersal capability. In contrast to plankton in open pelagic systems, it is unlikely that whole assemblages of intertidal rocky shore species will shift simultaneously (Hawkins et al. 2009) owing to their specific requirements in terms of degree of exposure, vertical zonation and substrate attachment.

The balance between grazers/suspension feeders and fucoids is likely to alter as climate changes. Grazing on algae is likely to increase, and there will be stronger interactions between environmentally-induced stress and increased grazing pressure on early life stages of many species (Coleman et al. 2006; Hawkins et al. 2009).

8.3.4 Changes in Wadden Sea Intertidal Macrofauna Communities and Climate

In recent decades the fauna and flora on intertidal flats in the Wadden Sea have been affected by increasing temperature, accelerated sea-level rise, epidemic diseases, invasion of non-native species, and human pressures from fisheries, habitat alteration (seawall building, harbour construction, dredging), eutrophication and/or pollution (Oost et al. 2009).

A long-term survey (1930–2009) of changes in macrofauna communities in Jade Bay, a shallow sedimentary tidal bay in the German Wadden Sea shows increasing species richness from 65 taxa in the 1930s, to 83 taxa in the 1970s and 114 taxa in 2009 (Schückel and Kröncke 2013). The most striking difference between 1930 and 2009 was the increase in numbers of non-native species, which was attributed to species introduced by shipping. Since many of these species originated from warmer coasts, it is likely that their ability to settle, survive and reproduce in the North Sea is due to increasing temperature (Van der Graaf et al. 2009). Trophic structure in Jade Bay was dominated by surface deposit feeders in the 1930s, but this feeding mode had decreased by the 1970s. Suspension feeders, mainly bivalves, became dominant. Subsurface deposit feeders had increased by 2009 together with deposit and interface feeders, while suspension feeders had again declined (Schückel and Kröncke 2013).

Drivers behind the observed temporal patterns may be the decreasing nutrient levels in Jade Bay and the whole Wadden Sea between 1981 and 2003 (see also Chap. 3), but the decline of bivalve species biomass by 2009 may be due to frequent recruitment failure, related to temperature increase.

One consequence of the increase in winter temperatures of about 1.5 °C since the 1980s is greater body weight loss during winter, with subsequent production of fewer and smaller eggs (Beukema et al. 2002; Beukema and Dekker 2005). An alternative explanation for recruitment failure, also related to changes in sea temperature, might be enhanced shrimp predation of settled bivalve recruits. Juvenile grey (or common) shrimp *Crangon crangon* were more abundant after mild winters than after cold winters caused by earlier arrival of shrimps from the open, colder North Sea (Beukema and Dekker 2005). Grey shrimp abundance in Jade Bay was an order of magnitude higher in spring in the 2000s compared to the 1970s (Schückel and Kröncke 2013). More generally, increasing temperature is expected to favour crustaceans and especially the grey shrimp. The reason may be that increased temperatures are unfavourable for cod, an important predator, thus reducing the predation mortality of epibenthic species including the grey shrimp (Freitas et al. 2007). Predation by grey shrimp on bivalve spat (bottom-settled larvae) may have a knock-on effect on the productivity of the mussel beds as foraging areas for breeding shorebirds and refuelling areas for long-distance migratory birds. Young mussels are also foraged by the common starfish *Asterias rubens*, with little impact during average winter conditions, however this impact may rise with increasing temperature. An increase of 2 °C could double the rate of foraging by common starfish (Agüera et al. 2012). Thus, climate-induced changes on one trophic level can have important consequences for food-web structure and functioning; a change in species composition in favour of some key species may have cascading effects through the food web associated with intertidal areas. On the Dutch tidal flats the effects of increasing temperature have so far been small compared with human impacts, such as mechanical cockle fisheries. However, since dredging for cockles is currently banned, the effects of rising temperature may become more important in the future.

8.4 Fish

Over 200 species of fish have been recorded from the North Sea, including three species of Agnatha (lampreys and hagfish), about 40 species of Chondrichthyes (cartilaginous fishes) and the rest Osteichthyes (bony fishes). The fish fauna includes deepwater species along the northern shelf edge and in the deep Norwegian Trench and Skagerrak, many shelf sea species and also species that occur in shallow water and estuaries. Cold-water species such as Atlantic cod *Gadus morhua* and Atlantic herring *Clupea harengus* occur in the North Sea close to the warm end of their range and southern, warm-water species such as common sole *Solea solea* and sardine *Sardina pilchardus* close to the cold end of their range.

Abundances of each species range from rare to common, but with considerable variability in relative numbers over time, as the North Sea has undergone warmer and cooler periods since the last ice age (Enghoff et al. 2007). The post-glacial inundation of the area south of 55°N occurred about 8000 BP, so much of the North Sea has been invaded by fish fairly recently. Currently about 20 species, most targeted by commercial fisheries, account for 95 % of the total fish biomass.

8.4.1 Long-Term Change in Fish Fauna

The Atlantic Holocene warm period lasted from 7000 to 4000 BP, with temperatures on average 2–2.5 °C above recent annual means. Fish identified from archaeological remains of Mesolithic settlements in Denmark include at least 49 species, most of which are common today, but also several that occur mainly in warm-temperate water, including smoothhound *Mustelus* sp., common stingray *Dasyatis pastinaca*, European anchovy *Engraulis encrasicolus*, European seabass *Dicentrarchus labrax*, black sea bream *Spondyliosoma cantharus* and swordfish *Xiphias gladius* (Enghoff et al. 2007). These warmer water species have all reappeared or increased in abundance over the past thirty years.

Declining catches since the fourteenth century from accessible nearshore areas of the North Sea led to the development of ever more distant fisheries at Iceland, in the NW Atlantic and in the Barents Sea, to feed the growing human population.

8.4.2 Recent Effects of Climate Change on Fish

The increased abundance of warm-temperate fish species has had a remarkable effect on species richness in the North Sea. The ICES-coordinated International Bottom Trawl Survey (IBTS) Programme, samples more than 300 stations throughout the North Sea in the first quarter of each year (January–March). Over the period 1985 to 2006, species richness increased from around 60 species to almost 90 and the increase is positively related to the increase in sea-bottom temperature (SBT) during that quarter, which rose by an average of 0.7 °C decade^{-1} (Fig. 8.7). The increase in species richness is consistent with both the earlier observation that rate of advance of leading edges of distributions is more rapid than the retreat of trailing edges and with the generally higher species richness of warmer areas. The apparent persistence of cool-temperate species during the Mesolithic warm period suggests that the effect is not just a transient one.

Climate (particularly temperature) can affect fish and other biota due to direct and indirect effects. Direct effects include physiological effects on growth and maturation, behavioural effects that alter migration and distribution, and displacement effects brought about by alteration of circulation patterns that transport and disperse eggs and larvae. Indirect effects include changes in the seasonal production of planktonic crustaceans, especially copepods, which form the larval diet of most fish species, and other complex food-web effects that result from changes in prey and predator communities and that can act at all life history stages. Examples of all these types of effect can be found in the North Sea, but because the variety and complexity of the processes (and their interactions) defy a concise, balanced review, the examples presented in the following sections are inevitably selective and partial.

8.4.3 Growth, Phenology and Behaviour

Atlantic cod *Gadus morhua*, being widely distributed, common and harvested both in the wild and in aquaculture is probably the most intensively studied marine fish species. Growth of cod has been shown to depend, among other factors, on food supply and temperature (Bjornsson et al. 2001). When food is not limited, the temperature producing the highest growth rate varies from >12 °C for juvenile fish (body mass <100 g) to <7 °C for adult fish (body mass <5000 g) (Brander 2010). This pattern of change in optimal temperature for growth with body size is due to changing metabolic constraints and means that the same change in temperature can cause both a reduction in growth rate of one life history stage and an increase in another. When food is limited, optimal growth occurs at lower temperatures.

Changes in phenology and growth are linked. Seasonal variations in otolith zone formation have been used to show how changes in temperature in the southern North Sea from 1985 to 2004 affected both phenology and growth of cod. Translucent otolith zones occur up to 22 days earlier in warm than in cold years and appear to be indicative of the onset of metabolic stress that results in slower growth

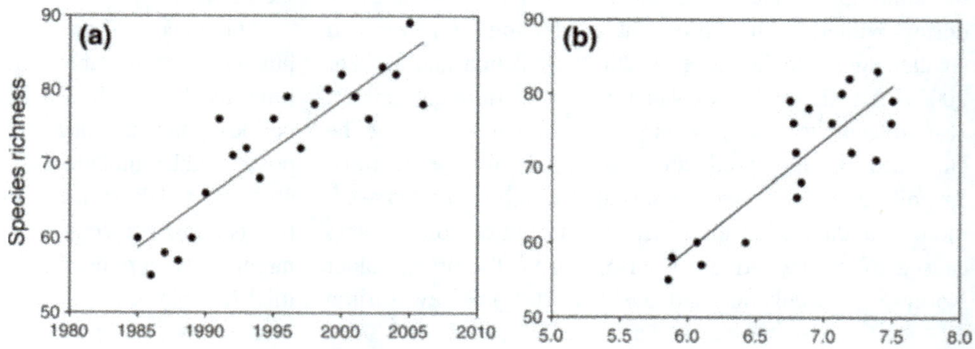

Fig. 8.7 Change in North Sea fish species richness. **a** Total number of species increases with time. **b** Total number of species increases with temperature (Hiddink and ter Hofstede 2008)

(Millner et al. 2011). Although changes in available food (possibly due to seasonal mismatch in production timing) have been suggested as a possible cause of the change in translucent zone formation, experimental evidence indicates that direct temperature effects are more likely (Neat et al. 2008).

The effect of changing temperature on growth of more than 100,000 juvenile cod was investigated using a standardised annual fishing survey in the Skagerrak from 1919 to 2010 (Rogers et al. 2011). Warm springs (SST >4 °C) since 1987 led to increased growth of juvenile cod, but warm summers (SST >16 °C) resulted in reduced growth. Density-dependent effects were detected, but not at the lower population levels of recent years. Fine-scale mapping of fish densities and of local growth dynamics was required to resolve temperature and density effects.

Apart from demonstrating the value of long, detailed, standardised time series, this growth study suggests other important lessons for understanding and predicting the effects of climate. Effects of temperature on growth can be positive at one time of the year and negative at another, particularly where seasonal variability in temperature is high. In the Skagerrak the negative summer effects on cod growth may eventually outweigh the positive spring effects, as temperature rises, but the speed of this will depend on whether the juvenile fish can change their location, particularly their vertical distribution, in order to remain in cooler (deeper) water.

Information collected using data storage tags from eight regions of the North Atlantic shows that cod can tolerate a wide thermal range (typically 12 °C within a stock range) and have sophisticated behavioural thermoregulation. Cod from north of 57° 30′N in the North Sea experienced a range of temperature between 5.5 and 14.5 °C but south of this latitude the range experienced was much wider spanning from 2.3 to 19.5 °C. The temperature range in the southern North Sea is much wider than in the northern North Sea, with lower winter temperatures (Righton et al. 2010). The data storage tags showed that cod in the southern North Sea remained in water above their optimal temperature for growth during the summer, even when there was cooler water nearby, which some fish moved into (Neat and Righton 2007).

The growth rates of co-occurring juveniles of two flatfish species, common sole *Solea solea* (a warm-temperate species) and plaice *Pleuronectes platessa* (a cool-temperate species) responded differently to the effect of rising temperature in the southeast North Sea between 1970 and 2004 (Teal et al. 2008). Warmer winter temperatures significantly lengthened the growing period of juvenile sole but not of plaice and warmer summer temperatures increased the growth rate of sole and, to a lesser extent, plaice. From July to September there was evidence of food-limited growth; thus a reduction in food production in the nursery areas (whether due to increased temperature or other factors such as oxygen limitation) could result in further reduction of growth rates.

Four out of seven sole stocks around the British Isles, including those in the east-central and southern North Sea, showed a significant trend towards earlier spawning over the 40-year period 1970 to 2010, with peak timing of spawning advancing by 1.35 ± 0.19 weeks for every 1 °C rise in winter temperature (Fincham et al. 2013). This shift in phenology is at roughly the same rate as the change observed in cod zone formation.

Growth of haddock *Melanogrammus aeglefinus* in the North Sea seems to have responded to increasing temperature during the period 1970 to 2006, with faster growth rate, a smaller asymptotic size and earlier maturation (Baudron et al. 2011). This may affect the productivity and reproduction of the stock. An extension of this analysis to include seven other North Sea fish species also showed an overall decline in asymptotic size (Baudron et al. 2014), but half of the decline in asymptotic size took place prior to 1988, during a period when temperature declined, suggesting that other factors are at work such as food limitation (as previously mentioned for plaice and sole).

8.4.4 Distribution

Changes in distribution are often thought of as movements of a population brought about by migration, as ecological conditions become less favourable. In the extreme the original stocks or species occupying an area may be imagined moving out and another set moving in. This process clearly does not apply in the case of rooted plants or sessile organisms, which cannot move, and migration probably makes only a limited contribution to observed distribution shifts in fish. For example, genetic and meristic information shows that the European anchovy population expansion in the North Sea since the mid-1990s is due to increasing abundance of a relict North Sea population and not to a northward shift of southern conspecifics from the western English Channel and Bay of Biscay (Petitgas et al. 2012). Changing ecological conditions that affect growth, maturation, survival and reproductive output result in distribution shifts over time, due to population increase or decline within a given area. Nevertheless, there are also many examples of species being detected where they had never occurred before and these obviously require invasion either by passive transport during planktonic stages or by migration of juvenile and adult fish (Quero et al. 1998; Brander et al. 2003). Since most of the North Sea has existed for less than 10,000 years all fish species are relatively recent immigrants.

Forty-five years of annual international standardised scientific trawl surveys provide detailed information on the distribution of fish throughout the whole North Sea.[1] Other national scientific data sources go back to the early 20th century, as do detailed statistics of commercial catch and effort, thus including periods of cooling as well as warming. Most fish species have exhibited northerly shifts in mean latitude and/or movements into deeper water over the past thirty years. Boundary shifts occurred in half of the species with northerly or southerly range margins in the North Sea and all but one shifted northward. Species with northward-shifting distributions had faster life cycles and smaller body sizes than non-shifting species (Perry et al. 2005). The shifts in latitude or into deeper water were correlated with variations in temperature estimated from measurements carried out during the same surveys and implied that shifting species remained within a constant temperature range (Beare et al. 2004; Heath et al. 2012). The landings distributions of cod, saithe *Pollachius virens*, haddock, European hake *Merluccius merluccius*, and European seabass all showed northward shifts of 25–50 km decade^{-1} between the 1970s and 1990s however these are the result of several interacting factors in addition to climate.

A detailed study of the changing distributions of plaice and sole since 1923 using a combination of research survey and commercial catch data shows contrasting patterns. The sole distribution shifted north from the 1920s to 1960 and then south, whereas plaice shifted north from 1947 onwards. Depth distributions also changed in opposite directions (see Fig. 8.8).

The distribution shift in plaice was attributed to climate change rather than fishing, but the sole distribution was influenced by both climate and fishing. However other factors including eutrophication, prey availability and habitat modification probably also need to be considered. There has been a remarkable westward jump in the plaice distribution since the late 1980s, apparently reflecting a collapse of the population in the east-central North Sea and increased abundance off Scotland (Engelhard et al. 2011). A similar analysis of changes in North Sea cod since 1912 (Engelhard et al. 2014) shows that their distribution shifted northward, but only since the late 1990s, and can be related to temperature. A major west to east shift in cod distribution from the early 1980s to 2000 can be related to fisheries-induced reduction in stock biomass rather than climate.

Red mullet *Mullus surmuletus* was not caught in research trawl surveys prior to the late 1980s but has become common in the north-eastern North Sea and also the Skagerrak (Fig. 8.9). This distribution pattern probably indicates that it has migrated into the North Sea from the north. There is some evidence that it migrates northward in winter to avoid the colder water in the southern North Sea (Beare et al. 2005).

The effects of climate and fishing interact to change the structure of fish communities. The scale of industrial (forage fish) fisheries in the North Sea has resulted in a fish community with fewer predatory fish compared to areas such as the Celtic Sea. Fish production in the North Sea is more strongly coupled to zooplankton production than is the case in the Celtic Sea and so it is likely that the effects of climate on North Sea fish are primarily via trophic links to the lower end of the food chain (Heath 2005). In this context the effects of climate change on lesser sandeel *Ammodytes marinus* may be particularly critical, since it is a non-migratory species that is very dependent on the availability of coarse sandy substrate (Heath et al. 2012).

8.4.5 Recruitment

Climate is one of the factors regulating recruitment of fish in the North Sea. It has been shown to influence many species, including cod, sole, plaice and herring. Temperature generally has a positive effect on recruitment of cod stocks at the cold end of their latitudinal range and a negative effect on warm-water stocks, including the North Sea stock (Planque and Fredou 1999). The relation has been shown using different statistical stock-recruitment models (Olsen et al. 2011; Ottersen et al. 2013) and is thought to be responsible for the series of high recruitment during the cold period of the 1960s and early 1970s (O'Brien et al. 2000; Brander and Mohn 2004), contributing also to the 'gadoid outburst' of the late 1960s to mid-1980s, when productivity of cod and other demersal stocks was extraordinarily high (Cushing 1984; Rijnsdorp et al. 2010). Cod recruitment has been low during recent warm years and the stock biomass may remain low unless cooler conditions return (Olsen et al. 2011).

Direct physiological effects of temperature on cod (particularly growth) have already been described. The temperature effect on recruitment is mediated through survival during the planktonic and early life stages and is probably due to a combination of direct and indirect effects. Production of a sufficient supply of the right sizes and quality of zooplankton prey, in particular the cool-temperate copepod *Calanus finmarchicus*, at the right time of year affects cod survival and recruitment (Beaugrand et al. 2003; Mieszkowska et al. 2009). Climate-related changes in spring SST and copepod abundance have consequences for the spatial patterns of recruitment in the North Sea (Nicolas et al. 2014) and probably for adult distribution as well.

Herring in the North Sea have experienced two periods of weak recruitment during recent decades. Poor recruitment during the period 1971–1979 has been ascribed to low

[1] http://ices.dk/marine-data/data-portals/Pages/DATRAS.aspx.

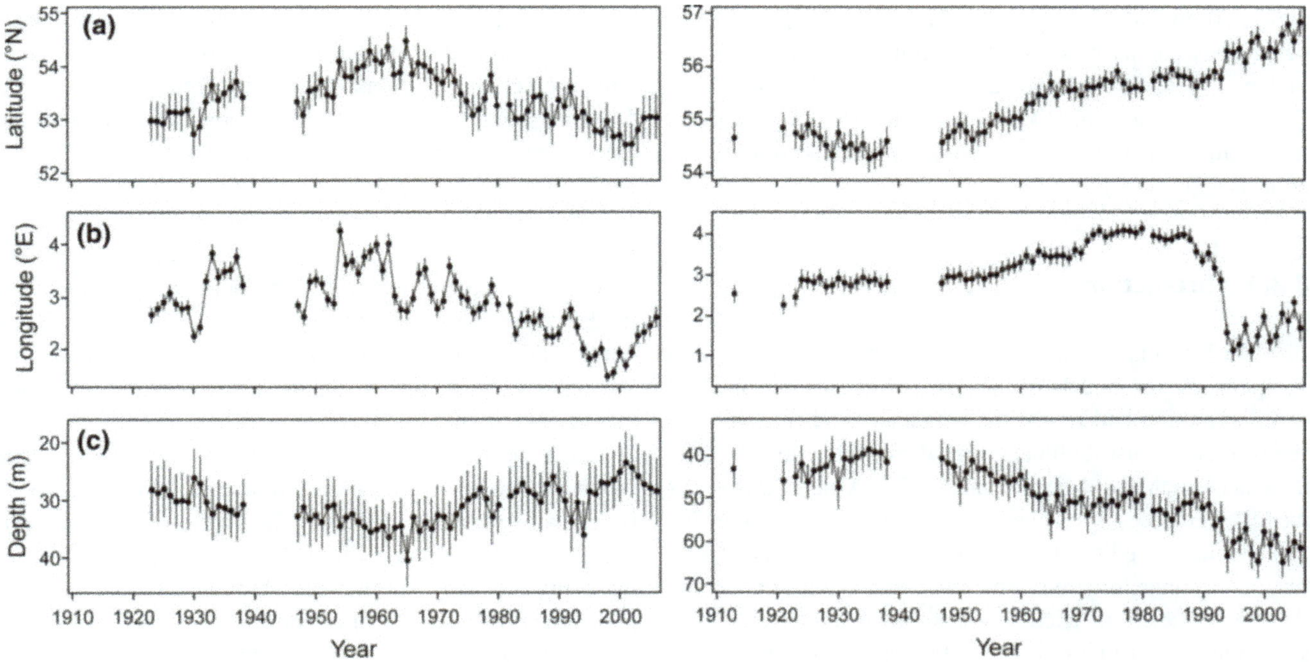

Fig. 8.8 Long-term changes in **a** latitude **b** longitude and **c** depth of North Sea sole (*left panels*) and plaice (*right panels*) using weighted mean catch-per-unit-effort (vertical bars are standard error of means) (Engelhard et al. 2011)

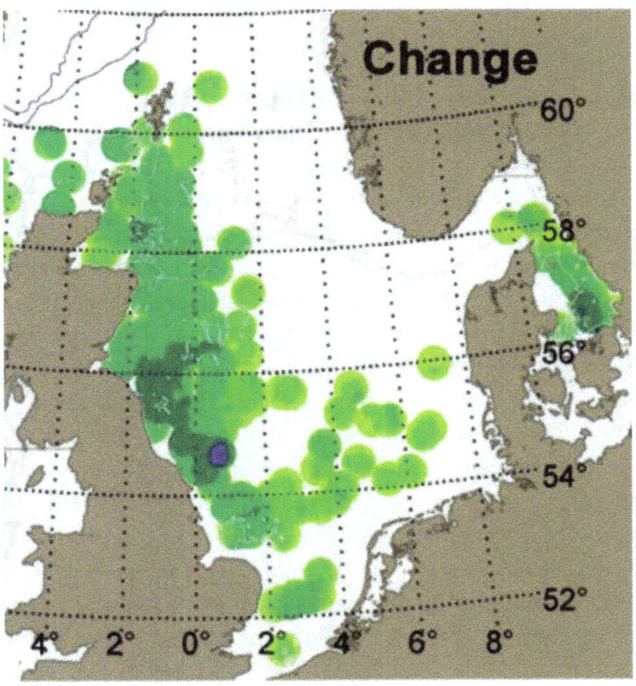

Fig. 8.9 Change in abundance of red mullet in first-quarter research surveys. The *darker the colour*, the greater the increase in abundance in the period 2000–2005 relative to 1977–1989 within each ½° latitude ×1° longitude rectangle (ICES 2008)

spawning biomass and insufficient egg production, however, recruitment to the central and northern populations may also have been affected by reduced Atlantic inflow into the north-western North Sea, resulting in unfavourable environmental conditions (Corten 2013). A second, continuing period of weak recruitment began in 2002, when the adult population was large and exploitation low. This is ascribed to the warming of the North Sea and substantial changes in the zooplankton community described earlier in this chapter (Payne et al. 2009, 2013) which resulted in lower growth rates of larvae and hence probably lower survival. The North Sea herring population consists of several spawning components and recruitment is the sum of the survivors of many spawning events, with different spawning grounds and timing, experiencing different environmental conditions (Hjøllo et al. 2009; Rijnsdorp et al. 2009). It is therefore naïve to search for single environmental drivers; spatial and temporal differences must be taken into account, as must the influence of parental factors (Dickey-Collas et al. 2010).

It is paradoxical that recruitment of both sole and plaice is higher following cold spring conditions (Ottersen et al. 2013), since plaice is close to the warm end of its range in the North Sea and sole is close to the cold end of its range. Cold temperatures in March delay spawning in sole, but it is not known whether recruitment is determined during the pelagic egg and larval stages, or during the early demersal stage (van der Land 1991; Kjesbu et al. 1998; Rijnsdorp and Witthames 2005).

For plaice the higher survival during colder winters is probably related to mortality of their predators during both the pelagic and early demersal stages (Van der Veer et al.

2009). Temperature may also affect the transport of the pelagic egg and larval stages, thus influencing the proportion of larvae reaching coastal nursery grounds (Van der Veer et al. 1998; Bolle et al. 2009). Available habitat for plaice in the North Sea may be reduced with climate change (Petitgas et al. 2013).

8.4.6 Prediction

The North Sea is one of the few areas where it is possible to compare several models of the effects of climate change on fish distribution and to test them against detailed long time series of actual distribution change. Three 'climate envelope' models (AquaMaps, Maxent, and the Sea Around Us project model) were used to project distributions of 14 North Sea fish species, based on data on existing distributions in relation to a range of environmental parameters, with some 'expert guidance' to exclude areas where the species were known not to exist (Jones et al. 2012). The three approaches produced predictions of relative habitat suitability which were reasonable given the occurrence data of each species. However, this analysis does not indicate whether there are differences in the capabilities of each model to expose specific features of the distribution, such as the pattern of relative habitat suitability (Jones et al. 2012).

Uncertainties arise from differences in data-types used, parameterisation and model structure. A multi-model ensemble approach is essential to project distribution ranges. It is evident that there can be an almost limitless number of factors and interactions influencing distribution ranges that are not included in the list of environmental variables in the box (e.g. substrate type, parasites, essential trophic links, oxygen). In the North Sea the dynamics and hydrographic characteristics of inflows from the English Channel, western European Shelf, Norwegian Sea and Baltic Sea have a huge bearing on the potential for species to invade and survive.

The experience of trying to determine the causes of observed changes in fish distributions in the North Sea is valuable because it shows how difficult this can be. The availability of good long-term data tends to show that simple hypotheses (e.g. fish move north when it gets warmer) are incorrect or incomplete. The effects of fish behaviour, genetic adaptation, habitat dependency and the impacts of fishing, result in complex responses that are not explained by simple climate envelope predictions. This point is well illustrated by a recent study that analysed research survey data from 33 years of summer (69 1° × 1° rectangles) and winter (84 1° × 1° rectangles) trawl surveys for the ten most abundant demersal species in the commercial fisheries (Rutterford et al. 2015). General additive models (GAMs) trained on ten-year time periods early in the series can reliably predict later periods for most species using seasonal temperatures, depth and salinity, with co-varying habitat variables also important. The result of coupling these GAMs with projections of North Sea ocean climate for the next 50 years suggests that future distributions of most of the ten current major demersal species will be constrained by the availability of habitat of suitable depth, leading to pronounced changes in community structure, species interactions and fisheries potential for these species. Rutterford et al. (2015) advised caution when applying process-based model projections of distributional shifts, and proposed that interpretations should be informed by data-driven modelling approaches, especially when using predictions for policy and management planning.

Ocean acidification will undoubtedly affect fish in the North Sea, however the nature and time-scale of these effects is difficult to predict. The early life stages of fish are probably more sensitive and vulnerable to acidification, but the main impacts may be indirect, through changes in other more sensitive taxa and in the productivity and structure of the lower trophic levels. Calcifying planktonic organisms are likely to be affected by the end of the 21st century but the direct effect on fish sensory systems may also cause subtle behavioural changes with possible population-level implications (Wittmann and Pörtner 2013).

8.4.7 Climate and Fish Fauna in the Dutch Wadden Sea

The previous sections mainly relate to the open North Sea, and the impacts of climate change on fish may be different in the shallow southern parts such as the Wadden Sea, a nursery area for many fish species, including several commercially fished stocks. The relatively warm water, rich food supply and possibilities to hide from predators provide a safe haven for young fish of species such as plaice, sole, whiting *Merlangius merlangus*, herring, and sprat *Sprattus sprattus*. The adults often spawn further offshore in the North Sea and the eggs and/or larvae drift with the currents towards the coast and into the Wadden Sea (Bolle et al. 2009; Dickey-Collas et al. 2009). Here they can grow rapidly, feeding on invertebrates or plankton. During the first years of life they show seasonal migrations: spending the growing season inside the Wadden Sea and moving to the deeper waters in the North Sea in winter. Besides its role as a nursery area, the Wadden Sea is also home to resident fish species and provides feeding habitat and passage to migrants and seasonal visitors.

Pronounced changes have taken place in the biomass of demersal fish in the Dutch Wadden Sea since monitoring started in 1960–1970. In particular, the marine juvenile guild shows a dome-shaped pattern in abundance, with an increase from the start of the time series, peaking in the 1980s and decreasing towards the present. The role of the Dutch Wadden Sea as a nursery area seems to have changed considerably, a pattern which is most prominent in plaice but also apparent in some other flatfish species. The densities of 0-year old plaice have strongly reduced since the mid-1980s to a stable and low level that has not changed since 2000. The period in which they use the area has also changed: instead of staying from early spring until October, they now tend to disappear in July/August. The 1- and 2-year olds have disappeared completely since the end of the 20th century.

Although several mechanisms may be operating, climate change is a likely cause of these alterations. Using dynamic energy budgets, Teal et al. (2012) showed that the most likely explanation for the recent loss of the nursery function, especially for plaice, is that increased temperatures make coastal areas unsuitable for growth. Growth rate data for 0-year old plaice showed that recent higher summer temperatures result in metabolic activity raised to levels at which food becomes limiting (Teal et al. 2008).

In contrast to the decline observed in overall biomass, dominated by marine juveniles, the resident species show an increase followed by a stable period in the coastal area. However, understanding of the mechanisms acting on the different resident species is still very limited. One exception is the discovery that the decline observed in eelpout *Zoarces vivparus* in the Wadden Sea since 1985 is due to an increase in temperature above the thermal maximum of the species, causing thermally limited oxygen delivery (Pörtner and Knust 2007).

Thus, the mechanisms underlying the large changes observed in the fish fauna of the Wadden Sea are still largely unknown. They are certainly partly climate related, but the impacts of changes in food, predators and abiotic factors acting on the different life stages are still poorly understood (Rijnsdorp et al. 2009).

8.5 Seabirds

Predicting the effects of climate variability on and through the different trophic levels is a major challenge, and one that increases in complexity at successively higher levels of the food web (Myksvoll et al. 2013). Seabirds are typically at the top of the marine food web and are the most numerous and visible of marine top predators. Furthermore, they are considered important indicators of the state of the marine ecosystem (Piatt and Sydeman 2007; Wanless et al. 2007). Worldwide, seabirds have declined faster than terrestrial bird groups with comparable numbers of species (Croxall et al. 2012), with most trends consistent with climate change (Poloczanska et al. 2013).

Seabirds can be affected by changing climate both directly, for example, if extreme weather becomes more frequent, or indirectly, through changes in their food supply. There is a substantial body of evidence suggesting that in most cases indirect effects are the more important of the two, with fluctuations in seabird demography and population dynamics caused in part by climate fluctuations acting through the availability and distribution of food. Effects of climate on life history traits have been documented across many species and populations (Sandvik and Erikstad 2008; Satterthwaite et al. 2012) including black-legged kittiwakes *Rissa tridactyla* (Aebischer et al. 1990; Furness and Tasker 2000) and (northern) fulmars *Fulmarus glacialis* (Thompson and Ollason 2001) in the North Sea.

8.5.1 Trends in Number of Breeding Birds

Seabird populations in the North Sea have shown strong changes in most species over recent decades. While populations may have been at a historic low in the early decades of the 20th century, most species strongly increased in the latter half of the 20th century (e.g. Mitchell et al. 2004; Mendel et al. 2008). The relaxation of persecution, egg collection and exploitation are probably the most important factors underlying these increases at least initially (Camphuysen and Garthe 2000). Commercial fisheries, especially through the vast amounts of discards and offal (Garthe et al. 1996), and overfishing of predatory fish (Furness 2002) are also likely to have been major drivers. Seabird population trends have developed differently since the end of the 20th century and many populations are now in decline, while others are relatively stable and some show increasing abundance (Fig. 8.10). The reasons for these trends are difficult to quantify, but ongoing changes in fisheries practice and climate-related changes are likely to be involved.

8.5.2 Case Studies Highlighting Climate Impacts on North Sea Seabirds

Case Study 1: The role of changes in oceanography and industrial fisheries in the decline of black-legged kittiwakes. In the North Sea, climate is known to affect several seabird populations through their main prey species, lesser sandeel, also called sandlance. Although Arnott and Ruxton (2002) and van Deurs et al. (2009) found this species to be sensitive to changes in sea temperature, this important forage fish is very difficult to study, and little is known about how it is affected by rising sea temperatures.

Studies indicate that in recent warmer years, birds have been struggling to find sufficient food for their chicks, because sandeels have been too few, too small, too lean, or have not been available at the right time. Several species of seabird breed later and less successfully, and survival of adult birds is also lower in warmer years (Daunt and Mitchell 2013).

Breeding success of black-legged kittiwakes in the UK, particularly at colonies along the North Sea coast, has been advocated as a reliable and sensitive indicator of the state of the marine ecosystem for those predators that are reliant on sandeel (Furness and Tasker 2000; Wanless et al. 2007). Breeding success at a given colony of black-legged kittiwakes in the UK is therefore considered to reflect some measure of sandeel availability during the period that birds are associated with the colony, and this assumption is supported by a clear regional clustering of kittiwake breeding success corresponding to the known spatial structuring in sandeel populations (Frederiksen et al. 2005).

Black-legged kittiwake populations have declined by more than 50 % since 1990, a period during which a lesser sandeel fishery was active and profound oceanographic changes occurred. Frederiksen et al. (2004b) studied the role of fisheries and oceanography in kittiwake declines on the Isle of May, southeast Scotland, where sandeels are the main prey. Breeding success and adult survival were low when the sandeel fishery was active (1991–1998) and were also negatively correlated with winter sea temperature, with a 1-year lag for breeding success. An observed improvement in breeding success from 2000 onwards has not been enough to halt the population decline. To stabilise the population, breeding success must increase to unprecedented levels or survival needs to increase substantially. Stochastic modelling indicated that the population was unlikely to increase if the fishery was active or sea temperature increased, and that the population was almost certain to decrease if both occurred. Sandeel recruitment is reduced in warm winters, and Frederiksen et al. (2004a) proposed that this explains the temperature effects on kittiwake survival and breeding success. The sandeel fishery also had a strong effect on kittiwake demographic performance, although the exact mechanism is unclear as kittiwakes and fishermen target different sandeel age groups.

Case Study 2: Breeding success of North Sea seabirds related to copepod abundance and distribution. The copepod *Calanus finmarchicus* is a key species for the trophodynamics of boreal ecosystems of the North Atlantic Ocean (Planque and Batten 2000). The species is a very important prey item for the small fish favoured by seabirds (Beaugrand et al. 2003). In particular, the recruitment of lesser sandeel in the North Sea is strongly positively correlated with *C. finmarchicus* abundance (van Deurs et al. 2009).

It is generally accepted that the distribution of *C. finmarchicus* in the North Atlantic reflects its thermal niche, along with advection from deep-water overwintering areas onto continental shelves such as the North Sea (Speirs et al. 2006; Helaouet and Beaugrand 2007). In accordance with recent warming, large declines in abundance of *C. finmarchicus* have occurred in the North Sea (Beaugrand et al. 2002) and low reproductive success of several forage-fish-dependent seabird species has been linked to these declines (Frederiksen et al. 2006).

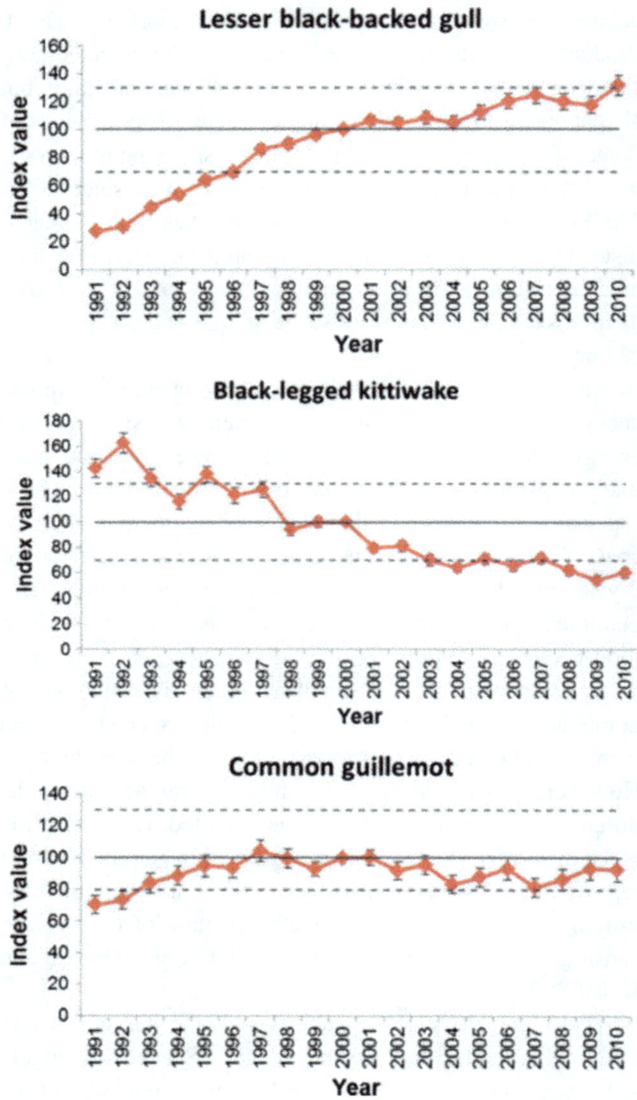

Fig. 8.10 Contrasting population trends for three seabird species breeding in the Greater North Sea from 1991–2010 (OSPAR Region II). Year 2000 was chosen as the baseline with the index value set to 100. *Vertical lines* show standard errors. ICES (2011)

If *C. finmarchicus* is not replaced by other zooplankton suitable as prey for small fish, seabird populations are likely to experience reduced breeding success, leading to further declines in population size (Frederiksen et al. 2013). Indeed, a close relative of *C. finmarchicus*, the warm-temperate *C. helgolandicus* has increased in abundance in the North Sea over recent decades as *C. finmarchicus* abundance has declined (Beaugrand et al. 2002). Nevertheless, *C.*

helgolandicus does not appear to be a full replacement for *C. finmarchicus* in terms of ecosystem functioning, particularly the ability to sustain large stocks of schooling, planktivorous fish (Bonnet et al. 2005). There are several reasons for this: *C. helgolandicus* are smaller, have a lower lipid content, and tend to occur at low densities early in spring when most fish larvae need access to abundant copepod prey (Beaugrand et al. 2003). Frederiksen et al. (2013) anticipated that because of these shifts in the zooplankton community resulting in declines in abundance of fish such as sandeel and the lack of obvious replacements for these as seabird prey, it is likely that breeding populations of piscivorous seabirds in the boreal Northeast Atlantic, including the North Sea will shift northwards. Consequently, the large seabird populations currently present in, for example, eastern Scotland could disappear (Frederiksen et al. 2013).

Case Study 3: Climate impact on breeding phenology in three seabird species. Breeding at the right time of year is essential to ensure that the energy demands of reproduction, particularly the nutritional requirements of growing young, coincide with peak food availability. Global climate change is likely to cause shifts in the timing of peak food availability, which the animals need to be able to adjust the time at which they initiate breeding. Frederiksen et al. (2004a) tested the hypothesis that regulation of breeding onset should reflect the scale at which organisms perceive their environment by comparing phenology of three seabird species at a North Sea colony. As expected, the phenology of two dispersive species, black-legged kittiwake and common guillemot *Uria aalge*, correlated with a large-scale environmental cue, the NAO, whereas a resident species, the European shag *Phalacrocorax aristotelis*, was more affected by local conditions (SST) around the colony. Annual mean breeding success was lower in years in which breeding took place later than normal for European shags, but not for the other two species. Since correlations among climate patterns at different scales are likely to change in the future, these findings have important implications for how migratory animals can respond to future climate change (Frederiksen et al. 2004a).

Case Study 4: Climate effects on the North Sea marine food web may influence coastal ecology through seabirds. Temperature is an important driver of the trophodynamics of the North Sea ecosystem. Recent warming, in combination with overfishing, has caused major changes in trophic interactions within the marine food web (Kirby and Beaugrand 2009). Luczak et al. (2012) studied the relation between lesser black-backed gulls *Larus fuscus graelsii* and swimming crabs (of the Polybiinae sub-family), important food species for the gulls during their breeding season. Luczak and co-workers found a related increase in sea temperature, the abundance of swimming crabs and that of lesser black-backed gulls in 21 major breeding colonies around the North Sea. Interestingly, their cross-correlation analyses suggest the propagation of a climate signal from SST through decapod larvae, adult crabs and lesser black-backed gulls with lags that match the biology of each trophic group. This is indicative of climate-induced changes in the marine fauna extending to the avian fauna, and thus to the terrestrial food web around the seabird colonies (Luczak et al. 2012).

Case Study 5: Vulnerability of the seabird community in the western North Sea to climate change and other anthropogenic impacts. Most seabird studies have tended to consider the impacts of single stressors on single species at specific times of the year, and so may be unrepresentative of the combined effects of pressures experienced by top predator communities over an annual cycle (Burthe et al. 2014). For marine top predators, there is evidence to suggest that interactions between climate and other threats may be additive (Frederiksen et al. 2004b; Burthe et al. 2014). Burthe et al. (2014) studied the cumulative effects of multiple stressors on a community of seabirds in the North Sea. More precisely, they examined vulnerability to climate change and other anthropogenic threats in a seabird community (45 species; 11 families) that used the Forth and Tay region (eastern Scotland) of the North Sea for breeding, overwintering or migration between 1980 and 2011. They found only 13 % of the seabird community in the Forth and Tay region to fall within the categories of low or very low population concern to future warming, whereas in considering multiple anthropogenic threats 73 % of the species in this bird community were considered to be of high or very high population concern for the future (Burthe et al. 2014).

Case Study 6: Effects of extreme climatic events on coastal birds breeding in low-lying saltmarshes (for a more extensive review see Chap. 9). Van de Pol et al. (2010) investigated whether the frequency, magnitude and timing of rare but catastrophic flooding events have changed over time in Europe's largest estuary, the Wadden Sea. They subsequently quantified how this had affected the flooding risk of six saltmarsh nesting bird species (both seabird species and coastal species). Maximum high tide has increased twice as fast as mean high tide over the past four decades, resulting in more frequent and more catastrophic flooding of nests, especially around the time when most eggs have just hatched. By using data on species' nest elevations, on the timing of egg-laying and on the length of time that the eggs and chicks are at risk from flooding, van de Pol et al. (2010) showed that flood risk increased for all six species (even after accounting for compensatory land accretion) and that this could worsen in the near future if the species do not adapt. This study provides the first evidence that increasing flooding risks have reduced the reproductive output below stable population levels in at least one species, the Eurasian oystercatcher *Haematopus ostralegus*. Sensitivity analyses show that birds would benefit most from adapting their nest-site selection to higher areas. However, historically the lower saltmarsh has been favoured for its proximity to the feeding grounds and for its low vegetation, aiding predator detection. Van de Pol et al. (2010) concluded that it is more difficult for birds to infer that habitat quality has decreased from changes in the frequency of rare and unpredictable extreme events than from trends in climatic means. The result is, at present, that the lower parts of the saltmarsh may function as an ecological trap.

8.5.3 Concluding Comments

The Case Studies clearly indicate that climate change influences North Sea seabirds. While this may be true for population developments in some species, it is more obvious for demographic parameters such as the number of chicks hatched and/or fledged and survival rates of adults and young birds. The breeding phenology of several seabird species is also affected (e.g. Frederiksen et al. 2004a). For migrating landbirds air temperature is often the main climate factor (e.g. Cotton 2003). In contrast, temperature changes usually act indirectly on seabirds via changes in the ecosystem, mainly through food supply (Wanless et al. 2007; Frederiksen et al. 2013). However, there may also be direct effects of temperature on seabirds as air and water temperatures may influence energetic costs for birds in maintaining body temperature (e.g. Fort et al. 2009). With generally increasing temperatures this may lead to northward trends for breeding and wintering in some species (Huntley et al. 2007). Northward shifts, probably out of the North Sea, may also result from changes in zooplankton community structure acting through main prey species like sandeel (Frederiksen et al. 2013).

Analyses of possible relationships between climate factors and seabirds are often impeded by difficulties in differentiating between natural variability and anthropogenic factors, thus complicating analyses on direct and indirect effects (Burthe et al. 2014). This is especially true for changes in fisheries practice that include overfishing of predatory fish, production of discards and offal, and direct mortality through fishing gear (reviewed by Tasker et al. 2000).

Some of the many ways in which seabirds respond to climate change were summarised by the International Council for the Exploration of the Sea (ICES 2008) as follows (see also Table 8.3):

- a warming trend may advance the timing of breeding in some species and delay it in others
- seabirds exhibit some flexibility in the timing of breeding, but are ultimately constrained by the often long reproductive period (up to five months from egg-laying to chick-fledging)
- seabirds are long-lived and so often able to 'buffer' short-term (<10 years) environmental variability, especially at the population level
- seabirds are vulnerable to both spatial and temporal mismatches in prey availability, especially when breeding at fixed colony sites with restricted foraging capacities (e.g. foraging distance, diving capacity).

8.6 Marine Mammals

8.6.1 Climate Change Impacts on Marine Mammals

All organisms display tolerance limits that, when exceeded, lead to negative impacts on metabolism, growth, and reproduction, or even death. Endothermic (i.e. 'warm-blooded') organisms such as marine mammals must maintain a relatively constant body temperature, and changes in the ambient temperature outside their preferred range therefore require additional expenditures of energy. If ambient temperatures become too high or too low to maintain body temperature within tolerable limits, adverse effects are likely (Howard et al. 2013). Thus, increasing severity of extreme weather events or changes in average winter or summer temperatures can have negative impacts on endothermic marine species, and repeated mortality events resulting from thermal stress can lead to population decreases (Howard et al. 2013).

In addition to the direct physiological temperature effect, climate change is also expected to affect marine mammals indirectly. This may be through changes in temperature, turbulence and surface salinity inducing productivity shifts at different trophic levels, shifts that can flow up the food web and affect prey availability for top predators. Marine mammals typically exploit patchy prey species that they require in dense concentrations and so their distributions tend to reflect those oceanographic features, both static (e.g. depth and slope) and more mobile (e.g. fronts and upwelling zones), where productivity is high.

Other important indirect pathways by which climate change may affect marine mammals include changes in critical habitats (due to warming) and in nesting and rearing beaches (due to sea-level rise) and increases in diseases and biotoxins (due to rising temperatures and shifts in coastal currents) (Simmonds and Isaac 2007; Howard et al. 2013). Populations may become more vulnerable to climate change owing to interaction with non-climate stressors resulting from human activities, such as pollution and fishing (Howard et al. 2013).

Table 8.3 Examples of links between climate variables and seabird behaviour (including distribution and condition) in the North Sea

Seabird parameter	Species	Region	Climate variable	Sign of correlation with warming
Breeding range	Lesser black-backed gull	UK	Sea temperature	Positive
	Northern gannet	UK	Sea temperature	Positive
Non-breeding range	Lesser black-backed gull	UK	Sea temperature	Positive
Reproductive success	Northern fulmar	Orkney	NAO index	Negative (hatching), positive (fledging)
	Black-legged kittiwake	Isle of May	Sea temperature	Negative
	Black-legged kittiwake	Orkney, Shetland	Sea temperature	Negative
Annual survival	Northern fulmar	Orkney	NAO index	Negative
	Black-legged kittiwake	Isle of May	Sea temperature	Negative
	Atlantic puffin	North Sea	Sea temperature	Negative
Population change	Black-legged kittiwake	Isle of May	Sea temperature	Negative
Nesting date	Black-legged kittiwake	Isle of May	NAO index	Positive
	Common guillemot	Isle of May	NAO index	Positive
	Common guillemot	Isle of May	Sea temperature	Negative
	Razorbill	Isle of May	Sea temperature	Negative
	European shag	Isle of May	Wind	Negative
Foraging cost	Common guillemot	Isle of May	Stormy weather	Positive
	Northern fulmar	Shetland	Wind speed	Negative

ICES (2008)

Macleod (2009) predicted that certain characteristics put some marine mammal species at greater risk from climate-induced changes than others. These include a distribution range that is restricted to non-tropical waters (including temperate species) and a preference for shelf waters (like the North Sea). Conversely, the more mobile (or otherwise flexible) marine mammal species may, to some extent, be able to adapt to climate change (Simmonds and Isaac 2007).

8.6.2 Distributional Shifts in Harbour Porpoise

The harbour porpoise *Phocoena phocoena* inhabits coastal or shelf waters of the northern hemisphere. It is the most abundant cetacean species in the North Sea region (Hammond et al. 2008) and its abundance on the European Atlantic continental shelf was estimated to be around 375,000 in 2005 (Hammond et al. 2013). In the shallow southern North Sea the number of harbour porpoises appears to have increased since the early 1990s (Hammond et al. 2002; Camphuysen 2004; Camphuysen and Peet 2006; SCANS II 2008), however, although still common, numbers in the northern North Sea have declined (SCANS II 2008; Øien 2010; Evans and Bjørge 2014). The reasons for this are not known, but a major distributional shift appears to have taken place from the north-western North Sea in 1994 to the south-western part in 2005 (Hammond et al. 2002, 2013; Fig. 8.11).

The harbour porpoise is a species with high energetic demand, especially as mature females are pregnant and lactating at the same time during most of the year. It is very likely that food availability is a major criterion for habitat selection (Gilles 2009). The shift in distribution shown in Fig. 8.11 may be due to an increase in herring abundance in the southern North Sea (Hammond et al. 2013) but the increase in herring abundance cannot simply be related to higher temperatures, since herring in the southern North Sea are already at the warm boundary of their distribution.

There may, however, be other ways that increasing temperature may affect harbour porpoises. MacLeod et al. (2007) reported that in the Scottish part of the North Sea this species consumed a significantly smaller proportion of sandeels in spring 2002 and 2003 in comparison with their baseline period (1993–2001). Furthermore, in the baseline period only 5 % of the stranded porpoises examined had died of starvation, whereas starvation was the cause of death of 33 % from 2002 and 2003. MacLeod et al. (2007) showed that a lower proportion of sandeels in the diet of porpoises in spring increases the likelihood of starvation. The reduced proportion of sandeels in the porpoise diet is likely to have been because sandeel spawning stock biomass (SSB) and recruitment in the North Sea were substantially lower in 2002 and 2003 than during the baseline period. Fishing is probably the main cause of the decline in sandeel but high winter sea temperatures also tend to reduce their recruitment (Arnott and Ruxton 2002). It follows that climate-induced warming may be the ultimate cause of poor body condition

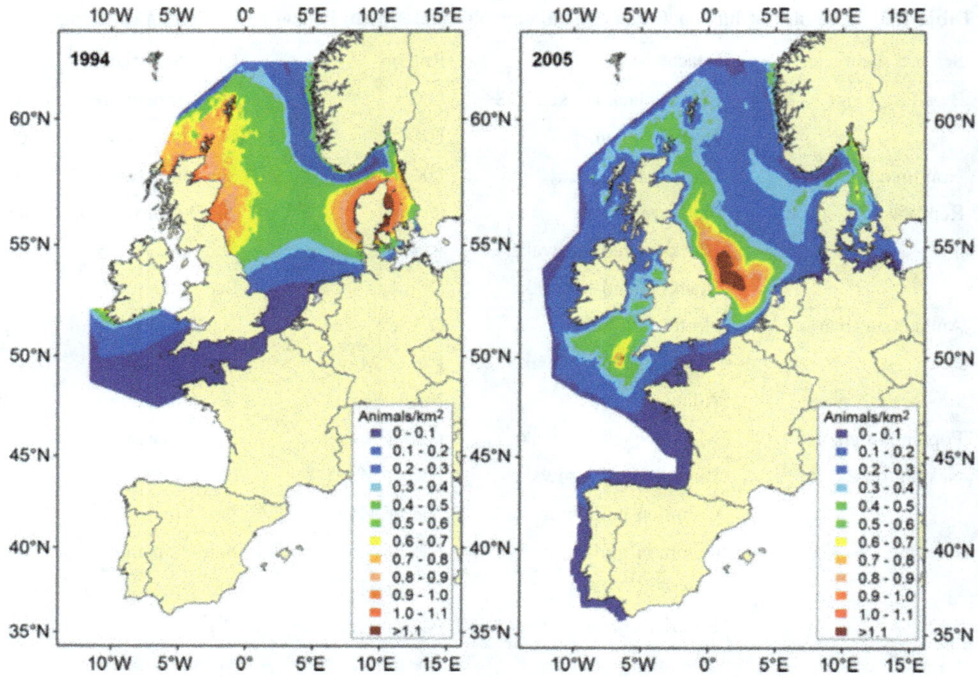

Fig. 8.11 Predicted density surface for harbour porpoises in 1994 and 2005 (SCANS and SCANS II surveys; Hammond et al. 2013)

of harbour porpoises in Scottish North Sea waters, resulting in an increased likelihood of starvation (MacLeod et al. 2007).

8.6.3 Rising Temperatures Favour Warm-Water Dolphins

There is evidence of recent changes in range expansion for several dolphin species in the North Sea region. One such case is the (common) bottlenose dolphin *Tursiops truncatus*, off the northeast coast of Scotland. Here they are at the northern limit of their distribution. The causes behind this increase in distribution are still unknown, but may be related to changes in abundance and/or distribution of prey (Wilson et al. 2004; Learmonth et al. 2006), which may also be linked to climate change.

Another species reported to have exhibited recent range shifts is the white-beaked dolphin *Lagenorhynchus albirostris*; Fig. 8.12). This is a species distributed mainly in cold temperate to Arctic waters. In the North Atlantic they are limited to high latitudes (Evans et al. 2003; MacLeod et al. 2005; Baines et al. 2006; Evans and Smeenk 2008). They are among the most abundant delphinid species in the North Sea in summer; sightings are much rarer during winter. Stranding records have shown a significant increase of the species in the southern North Sea since the 1960s. During recent decades they have regularly been detected in the Southern Bight (Bakker and Smeenk 1987; Kinze et al. 1997; Camphuysen and Peet 2006). However, these changes need not be directly related to changes in sea temperature, but may simply reflect natural or human-induced alterations in particular fish stocks that are favoured prey of the species.

Recent changes in the cetacean community around the British Isles, including the northern North Sea, have been related to increasing local water temperature. (Short-beaked) common dolphin *Delphinus delphis* is a warm-temperate species, commonly found in tropical waters and only sporadically in the North Sea. Over recent years common dolphins have been quite regularly seen in the North Sea even in winter (Sea Watch Foundation, unpubl. data; Evans and Bjørge 2014). This may reflect the expanding range of typically warmer water fish species like anchovy and sardine (ICES 2008; Evans and Bjørge 2014).

However, this northward expansion of common dolphin habitat into the northern North Sea is not necessarily due to global climate change. There were both strandings and sightings in that region during the 1980s and a peak in the number of strandings on the North Sea shores of the UK was reported as far back as the 1930s (Fraser 1946) and along the Dutch coast in the 1940s (Bakker and Smeenk 1987; Camphuysen and Peet 2006). These changes in common dolphin distribution may reflect climatic fluctuations on interdecadal scales, such as caused by the Atlantic Multidecadal Oscillation.

White-beaked and common dolphins have similar habitat and diet preferences. It has been suggested that the two species might partition their otherwise shared niche according to temperature to reduce the potential for competition at this time of year (Macleod et al. 2008). As temperature seems to be important in determining the relative distribution of these species, the range of the white-beaked

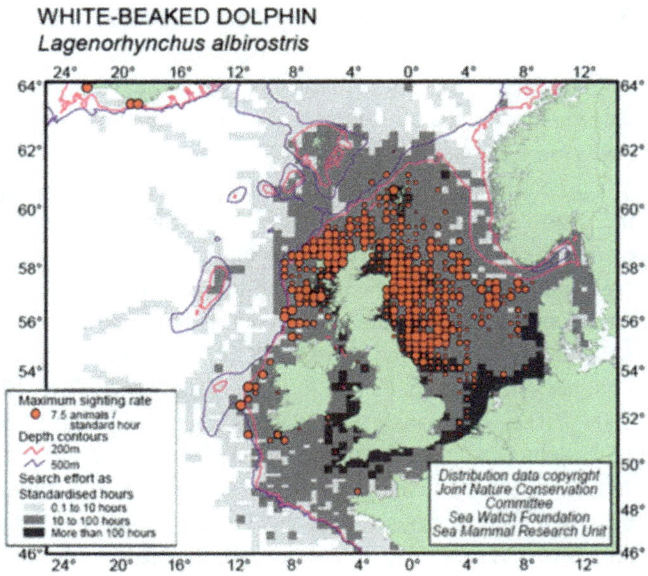

Fig. 8.12 Distribution, relative abundance and associated effort for white-beaked dolphin and Atlantic white-sided dolphin (Reid et al. 2003)

dolphin might be expected to contract in response to rising sea temperature, while that of the common dolphin may expand (Macleod et al. 2008).

Off north-west Scotland (Macleod et al. 2005) the relative occurrence and abundance of white-beaked dolphins has declined and that of common dolphins increased in comparison to previous studies, suggesting a decrease in range of the former and increase of the latter. This may be due to competitive exclusion, as suggested in the previous paragraph, or direct effects of changes in temperature. Independent of the mechanisms, if temperature increase continues some formerly abundant cold-water species, such as white-beaked dolphins and Atlantic white-sided dolphins (*Lagenorhynchus acutus*; Fig. 8.12) may be displaced, in particular from the northern North Sea, by species like the short-beaked common and striped dolphin *Stenella coeruleoalba* (MacLeod et al. 2005; Learmonth et al. 2006; Evans and Bjørge 2014). The white-beaked dolphin, which favours shelf habitats, may be placed under increased pressure if it loses the north-west European continental shelf from within its range (Evans and Bjørge 2014).

8.6.4 Exotic Visitors to the North Sea

A number of warm-water species have in recent decades been recorded for the first time in UK waters, including the North Sea. This includes Blainville's beaked whale *Mesoplodon densirostris* (1993), Fraser's dolphin *Lagenodelphis hosei* (1996), and dwarf sperm whale *Kogia sima* (2011), while ten of eleven strandings of pygmy sperm whale *Kogia breviceps* in Britain and Ireland have occurred since 1980 (Evans et al. 2003; Deaville and Jepson 2011). Between January and April 2008 there were 18 strandings in Wales, Scotland, and Ireland of another typically warm-water species, the Cuvier's beaked whale *Ziphius cavirostris* (Dolman et al. 2010). Although these strandings may not be directly related to climate change, they occurred much further north than would be expected for these species, and generally at times of the year when sea temperatures are at their highest. However, care should be taken in drawing conclusions from such a limited number of records of vagrants (Evans and Bjørke 2014).

If the warming continues, more visits of warm-water vagrants to north-west Europe are to be expected. Likely species include Bryde's whale *Balaenoptera edeni*, pygmy sperm whale, dwarf sperm whale, rough-toothed dolphin *Steno bredanensis*, and Atlantic spotted dolphin *Stenella frontalis*. Baleen whales, like humpbacks *Megaptera novaeangliae* and fin whales *Balaenoptera physalus*, that normally move southwards in winter to warmer waters to breed, may increasingly do so within the waters around the UK, some even in the North Sea (Evans and Bjørge 2014).

8.6.5 Effect of Climate Change on Seals in the North Sea

The harbour seal *Phoca vitulina* and grey seal *Halichoerus grypus* are the most common seal species in the North Sea. Grey seals occur in temperate and subarctic waters on both sides of the North Atlantic Ocean in three distinct populations. The Eastern Atlantic population is found mostly around the coasts of Great Britain and Ireland, as well as on the coasts of the Faroe Islands, Iceland, Norway and north-western Russia as far east as the White Sea. With an estimated number of 415,000 to 475,000 individuals, the species is not threatened as a whole and grey seal numbers are currently increasing at most locations (Thompson and Harkönen 2008). The harbour seal is found throughout the coastal waters of the northern hemisphere. A global population of 350,000 to 500,000 is estimated. Haul-out sites are important for the species, as they are used for resting, moulting, pupping and lactation (Adelung et al. 2004; Reijnders et al. 2005).

There is yet little or no evidence for direct effects of climate change on either of the North Sea seal species, however changes to their physical habitat, through sea-level rise for example may cause haul-out locations in caves or on low-lying coasts to be modified or even lost. More frequent storms and associated storm surges may also have unfavourable effects (Evans and Bjørke 2014).

Harbour seals and grey seals are both opportunistic feeders, but the majority of their diet comprises only a few species, depending on the area. In European waters they are primarily demersal or benthic feeders. Important prey species here include sandeel, Atlantic cod, saithe, herring and some flatfishes (Hall 2002; Santos and Pierce 2003; Hammond and Grellier 2006). Thus, it seems likely that climate may affect seals indirectly, through changes in abundance or distribution of one or more of their most important prey species.

8.7 Ecosystem Effects

Previous sections have presented evidence of substantial changes in plankton, benthos, fish, seabirds and marine mammals in the North Sea over the past century and have related these to climate change. Planktonic and benthic ecosystems are coupled in several ways; many benthic species have planktonic stages and the settled, adult benthic stages are often dependent on planktonic food sources. It is therefore not surprising to find some common patterns of response to climate change, and that the response of plankton generally precedes the benthic response. This section considers whether common patterns, which are found to varying degrees in fish, seabirds and marine mammals, can be described as changes (or regime shifts) in the North Sea ecosystem as a whole. Common features examined include external drivers, timing of changes (both phenological and interannual), common processes (growth, recruitment, survival) and ecosystem characteristics (distribution, diversity, trophic structure).

8.7.1 An Integrated Ecosystem Assessment

An integrated analysis of available North Sea time series data was conducted by Kenny et al. (2009). The analysis was based on 114 variables with long (unbroken) records and broad spatial coverage making them suitable for assessing the North Sea as a whole. The data comprised abiotic environmental variables (including surface and bottom water temperature and nutrient concentrations, and wind speed and direction), plankton, seabirds, fish and fishing pressure. The initial study covered the period 1983 to 2003, but the analysis has since been extended to include data to 2007 (ICES 2009). The later analysis indicates that the North Sea ecosystem as a whole has undergone a series of shifts in state and that the rate of change in the ecosystem has varied over time, with some groups of years having greater similarity than others (Fig. 8.13). The pattern of change reveals three (possibly four) distinct groups of years with a shift in the system occurring between 1990 and 1991 characterised by declines in the dominance of cod SSB, average demersal fish length and *C. finmarchicus* abundance. In contrast, a second shift occurred between 2001/2002 and 2003/2004 which was initially dominated by an increase in average pelagic fish length, sea bottom temperature and *C. helgolandicus* abundance, but was then dominated by an increase or return in state of average demersal fish length and cod SSB.

A number of 'key' signals of environmental change in the North Sea ecosystem, such as bottom temperature, zooplankton and pelagic fish length, demonstrate strong trends over time (Fig. 8.14). It is clear that the period 1989 to 1991 represented a time of rapid change in all three components, with the apparent 1-year lag between them suggesting that the ecosystem shift at this time was driven by a step-change in temperature.

Integrated ecosystem assessments are an essential part of an ecosystem approach to the management of marine resources, that is, "An integrated approach to management that considers the entire ecosystem, including humans, with the goal to maintain an ecosystem in a healthy, productive, and resilient condition so that it can provide the services we want and need" (McLeod et al. 2005). This assessment and the information presented in previous sections of this chapter shows that the composition and productivity of North Sea marine ecosystems are affected by climate change and that this has consequences for sustainable levels of harvesting and for other ecosystem services. It is clearly valuable to have indicators that can be used to monitor and forewarn of changes in composition and productivity and multi-variate ocean climate indicators of this type are now being developed for the North Sea and other well-studied areas such as the California Current (Sydeman et al. 2014).

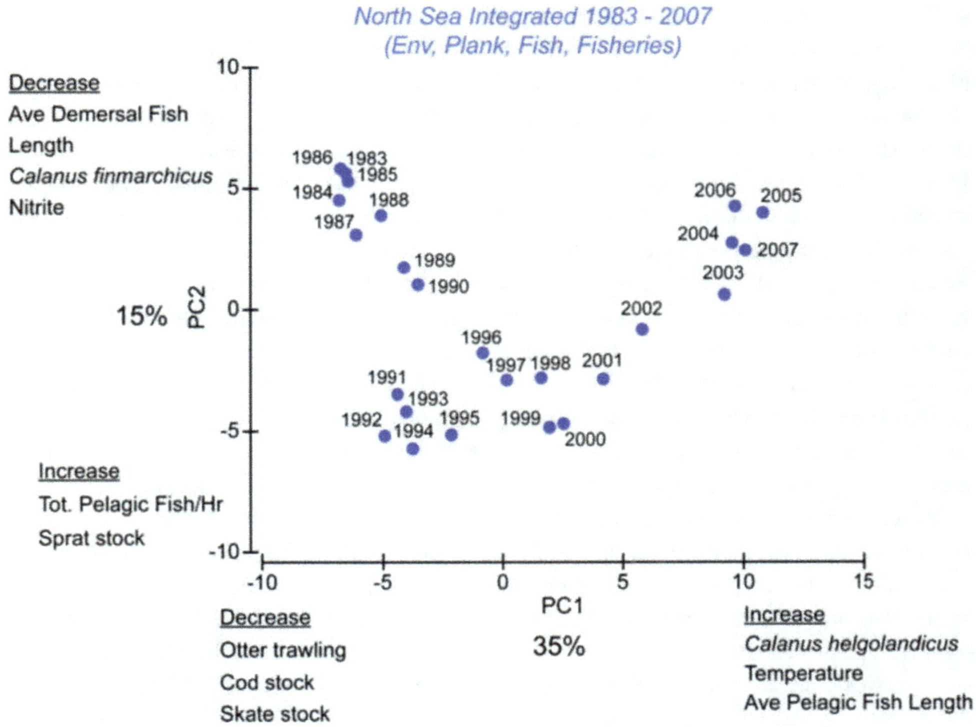

Fig. 8.13 Principal component analysis performed on an integrated data set comprising 106 separate state and pressure variables representing several components of the North Sea ecosystem between 1983 and 2007

8.7.2 Examples of Climate Impacts Across Trophic Levels

Case Study 1: Climate affects cod through zooplankton prey. Survival of North Sea cod larvae is linked to their degree of temporal overlap with zooplankton prey. This is referred to as the 'match–mismatch hypothesis' (Cushing 1990; Durant et al. 2007). Indeed, changes in plankton phenology linked to climate (Sect. 8.2) are seen as a factor contributing to the decline in the North Sea cod stock, although overfishing also plays an important role (Nicolas et al. 2014). Copepod biomass, euphausiid abundance, and prey size have also been shown to influence survival of North Sea cod through early life stages (Beaugrand et al. 2003).

The decline in the quality and quantity of planktonic prey from the 'gadoid outburst' of the 1960s to the periods of low recruitment (after the mid-1980s) was related to an increase in SST. Consequently, high sea temperatures may have had a double negative impact on larval cod survival in the North Sea. Temperature increases metabolic rate and so increases energy demand while at the same time it decreases the quality and quantity of prey available for larvae (the energy supply). The temperature rise may therefore have resulted in an energy imbalance for larval cod, causing increased larval mortality (Beaugrand et al. 2003). The increased rate of development probably resulted in a mismatch with prey (Daewel et al. 2011). On the other hand, high temperatures should also shorten the time from spawning through hatching to metamorphose, which should be favourable for the survival of the cod progeny (Ottersen et al. 2010).

Case Study 2: Climate effects from phytoplankton to seabirds. Many seabirds are at the top of the food chain (see also Sect. 8.5) and it is important to understand which pathways climate signals follow through the food web to influence the different seabird life-history traits (Sandvik et al. 2012; Myksvoll et al. 2013). Aebischer et al. (1990) reported clear similarities between trends in long-term data

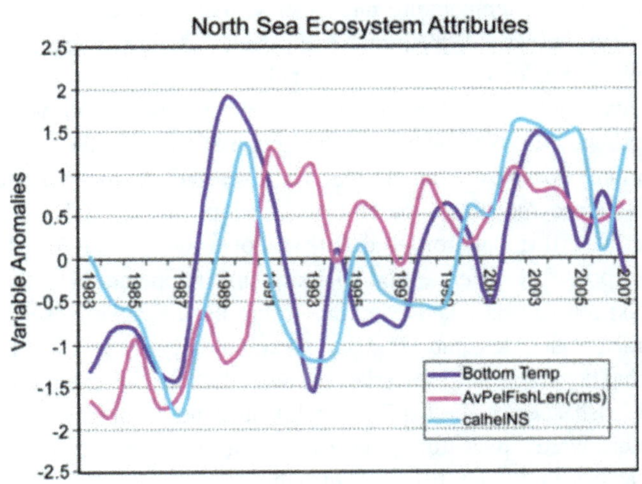

Fig. 8.14 Trends in state of three 'key' components of the North Sea ecosystem; bottom temperature, *Calanus helgolandicus* abundance and average pelagic fish length between 1983 and 2007 (ICES 2009)

series of westerly weather and at four trophic levels in the North Sea: phytoplankton, zooplankton, herring, and black-legged kittiwakes, but the mechanisms behind the similarity were unclear. Frederiksen et al. (2006) also demonstrated consistent trends across four trophic levels, from plankton to seabirds, in the North Sea but again the causal links were undefined. Thompson and Ollason (2001) showed how ocean climate variation had lagged effects on a Scottish pelagic seabird species through cohort differences in recruitment related to temperature changes in summer. Burthe et al. (2012) compared phenological trends for species from four levels of a North Sea food web over the period 1983–2006 when SST increased significantly. The results suggest trophic mismatch between five seabird species breeding in the North Sea and their sandeel prey, but no evidence of an impact on the seabird breeding success or population dynamics (Burthe et al. 2012). Also, the significant increase in the number of lesser black-backed gulls from 1996 onwards has been linked to the earlier mentioned temperature-driven increase in recruitment of swimming crabs in wide areas of the southern North Sea. These crabs are a key prey item for the seagulls (Luczak et al. 2012).

8.8 Brief Synthesis and Reflection on Future Development

This chapter has presented examples of how the changing environment affects biological processes and organisation at all scales, including the physiology, reproduction, growth, survival, behaviour and transport of individuals; the distribution, dynamics and evolution of populations; and the trophic structure and coupling (e.g. benthic-pelagic coupling) of ecosystems. There have been particularly rapid changes in temperature and other climate-related variables since the early 1980s, with many well described effects on North Sea ecosystems. However the examples presented in this chapter also show that biological responses in terms of growth, survival, phenology and population shifts are often more complex than might be expected from thermal response models or bioclimate envelope models (Cheung et al. 2011; Baudron et al. 2014). For example, the growth response of juvenile cod to increasing temperature in the Skagerrak was positive during spring but negative in summer, with a detectable density effect, but only at stock levels that have not been observed for many decades. Distributions of fish species (cod, plaice, sole) have not simply shifted northwards over time in response to temperature; other factors including fishing, eutrophication, prey availability and habitat alteration must also be considered and for some species there are major east-west shifts (see also the change in harbour porpoise distribution shown in Fig. 8.11). The dynamics of the water mass exchanges between the North Sea and the North Atlantic, the English Channel and the Baltic Sea have a major influence on temperature, salinity and nutrient fields within the North Sea and also on invasion routes for biota.

These examples of complex responses can be detected because there are detailed long-term biological and environmental records for the North Sea. The Skagerrak cod records go back to 1929; distributions of commercially important fish species can be inferred from spatially resolved fisheries data going back to the 1920s and before; scientific fishing surveys provide detailed distribution and population structure data on all fish since the early 1970s; the CPR provides spatially resolved monthly records of zooplankton (including larvae of fish and benthic species) and some phytoplankton data back to 1948. In addition to these well maintained observational time-series data are accumulating from historic reconstructions (e.g. Poulsen et al. 2007), archaeology (Enghoff et al. 2007) and other sources. The wealth of sampling and scientific analysis that exists for the North Sea shows the need to look deeper than simple, direct effects and linear responses to one or two variables and to be wary of general conclusions from incomplete models (Heath et al. 2012). It is salutary to find that even in a very well-studied species such as cod in the North Sea, the causes of changes in distribution, abundance and population structure over the past century are still not fully understood, but are undoubtedly complex.

Uncertainty over the causes of observed changes in cod over the past century, despite detailed time series on the physical and chemical environment and on other drivers of change, in particular fishing, must temper our confidence in projections of future changes in cod. The quality and credibility of such biological projections depend on the quality of projections of future changes in environmental variables and on the correct identification and representation of all important processes. A systematic approach that applies basic, mechanistic ecological principles to new situations and that emphasises the testing of hypotheses in experimental frameworks may be useful in identifying and constructing appropriate process models (Kordas et al. 2011). There may also be unknown factors and interactions as the system changes beyond previous limits, including those concerning pH and oxygen.

The changes in biota described in this chapter and in particular the 'regime shift' of the late 1980s might be regarded as evidence of the sensitivity of the North Sea ecosystem to changes in the environment, particularly temperature (Philippart et al. 2011). However given that the environmental changes were very large and rapid and that although the biotic response was evident, it did not include loss of characteristic North Sea species or a complete change in the character of the ecosystems, it could be argued that this demonstrated that the ecosystem response showed great resilience. Increases in species richness seem to have been due to the addition of warm-temperate species without the loss of cool-temperate species. The warm period since the late 1980s is still too short to determine whether the present ecosystem state is transitional and whether the cool-temperate species will gradually disappear, but evidence from archaeological material laid down during the Mesolithic warm period (4000–7500 y BP) indicates that cool-temperate species may remain (Enghoff et al. 2007) at least until temperatures rise by considerably more. The time taken for this to happen depends on the rate of global warming but also on decadal regional variability, which could maintain a cooler state over the next few decades.

Substantial biological changes in the Northeast Atlantic including the North Sea have been associated with shifts in the sub-polar gyre and warming over recent decades (Hatun et al. 2009). This includes large-scale modification of the phenology and distribution of plankton assemblages (Beaugrand et al. 2002; Edwards and Richardson 2004; Richardson and Schoeman 2004), changes in the availability of food resources and species, reproduction of benthic animals, composition of fish assemblages (Attrill and Power 2002; Simpson et al. 2011), and recruitment to the North Sea cod stock (Clark et al. 2003; Olsen et al. 2011).

The diversity of the North Sea ecosystem may lead to contrasting responses to future climate change. For instance, the increase in temperature, light (through improved transparency), and wind in the south-eastern North Sea have probably contributed to the increase in algal biomass in this region during the period 1948–2004 (Llope et al. 2009). While their data suggest that phytoplankton biomass may have reached a maximum in the southernmost parts of the North Sea, Llope et al. (2009) concluded that phytoplankton biomass in the northern North Sea would continue to respond positively to a warmer, brighter, and windier future if current trends are maintained.

Projections of the phenological responses of individual species under climate change have not yet been made, but the empirical evidence suggests that phenological changes will continue as climate warming continues. It is currently uncertain whether genetic adaptations within species populations will be able to cope with these changes, at least partly, or whether the pace of climate change is too fast for genetic adaptations to take place. This uncertainty is further compounded by the difference in phenological responses between species and functional groups. If current patterns and rates of phenological change are indicative of future trends, climate warming may exacerbate trophic mismatching and result in disruption of the functioning, persistence and resilience of North Sea ecosystems.

It is not clear whether general species attributes (e.g. trophic level) are sufficient to predict future outcomes or whether careful study of the individual species is required, however the differences in phenological responses between different diatom species indicates the latter (Schlüter et al. 2012).

References

Adelung D, Liebsch N, Wilson RP (2004) Telemetrische Untersuchungen zur räumlichen und zeitlichen Nutzung des schleswig-holsteinischen Wattenmeeres und des angrenzenden Seegebietes durch Seehunde (*Phoca vitulina vitulina*) in Hinblick auf die Errichtung von Offshore-Windparks. – Teilprojekt 6 im Endbericht zum Verbundvorhaben "Marine Warmblüter in Nord- und Ostsee: Grundlagen zur Bewertung von Windkraftanlagen im Offshore-Bereich". FKZ 0327520

Aebischer NJ, Coulson JC, Colebrookl JM (1990) Parallel long-term trends across four marine trophic levels and weather. Nature 347:753–755

Agüera A, Trommelen M, Burrows F, Jansen JM, Schellekens T, Smaal A (2012) Winter feeding activity of the common starfish (*Asterias rubens* L.): The role of temperature and shading. J Sea Res 72:106–112

Armonies W, Herre E, Sturm M (2001) Effects of the severe winter 1995/96 on the benthic macrofauna of the Wadden Sea and the coastal North Sea near the island of Sylt. Helgoland Mar Res 55:170–175

Arnott SA, Ruxton GD (2002) Sandeel recruitment in the North Sea: demographic, climatic and trophic effects. Mar-Ecol Prog Ser 238:199–210

Asrar G, Myneni R, Kanemasu ET (1989) Estimation of plant canopy attributes from spectral reflectance measurements. In: Asrar G (ed) Theory and Application of Optical Remote Sensing. Wiley

Atkinson D (1994) Temperature and organism size - a biological law for ectotherms. Advances in Ecological Research, vol 25

Attrill MJ (2007) Climate-related increases in jellyfish frequency suggest a more gelatinous future for the North Sea. Limnol Oceanogr 52:480–485

Attrill MJ, Power M (2002) Climatic influence on a marine fish assemblage. Nature 417:275–278

Baines ME, Reichelt M, Anderwald P, Evans PGH (2006) Monitoring a changing world - searching the past for long-term trends in the occurrence of cetaceans around the UK. In: Abstracts, 20th Annual Conference of the European Cetacean Society, Gdynia, Poland, 2-7 April 2006

Bakker J, Smeenk C (1987) Time-series analysis of *Tursiops truncatus*, *Delphinus delphis*, and *Lagenorhynchus albirostris* strandings on the Dutch coast. ECS Newslet 1:14–19

Baudron AR, Needle CL, Marshall CT (2011) Implications of a warming North Sea for the growth of haddock *Melanogrammus aeglefinus*. J Fish Biol 78:1874–89

Baudron AR, Needle CL, Rijnsdorp AD, Marshall CT (2014) Warming temperatures and smaller body sizes: synchronous changes in growth of North Sea fishes. Global Change Biol 20:1023–1031

Beare DJ, Burns F, Greig A, Jones EG, Peach K, Kienzle M, McKenzie E, Reid DG (2004) Long-term increases in prevalence of North Sea fishes having southern biogeographic affinities. Mar-Ecol Prog Ser 284:269–278

Beare DJ, Burns F, Jones E, Peach K, Reid DG (2005) Red mullet migration into the northern North Sea during late winter. J Sea Res 53:205–212

Beaugrand G (2004) The North Sea regime shift: evidence, causes, mechanisms and consequences. Prog Oceanogr 60:245–262

Beaugrand G, Ibañez F (2002) Spatial dependence of calanoid copepod diversity in the North Atlantic Ocean. Mar-Ecol Prog Ser 232:197–211

Beaugrand G, Reid PC, Ibañez F, Lindley JA, Edwards M (2002) Reorganization of North Atlantic marine copepod biodiversity and climate. Science 296:1692–1694

Beaugrand G, Brander KM, Lindley JA, Souissi S, Reid PC (2003) Plankton effect on cod recruitment in the North Sea. Nature 426:661–664

Beaugrand G, Edwards M, Brander K, Luczak C, Ibanez F (2008) Causes and projections of abrupt climate-driven ecosystem shifts in the North Atlantic. Ecol Lett 11:1157–1168

Beaugrand G, Luczak C, Edwards M (2009) Rapid biogeographical plankton shifts in the North Atlantic Ocean. Global Change Biol 15:1790–1803

Beaugrand G, Edwards M, Legendre L (2010) Marine biodiversity, ecosystem functioning, and carbon cycles. P Natl Acad Sci USA 107:10120–10124

Beaugrand G, McQuatters-Gollop A, Edwards M, Goberville E (2013) Long-term responses of North Atlantic calcifying plankton to climate change. Nature Clim Change 3:263–267

Beaugrand G, Harlay X, Edwards M (2014) Detecting plankton shifts in the North Sea: a new abrupt ecosystem shift between 1996 and 2003. Mar Ecol Progr Ser 502:85–104

Behrenfeld MJ, Siegel DA, O'Malley RT, Maritorena S (2009) Global ocean phytoplankton. In: State of the Climate in 2008, pp. 568–573. Peterson TC, Baringer MO (Eds)

Beukema JJ, Dekker R (2005) Decline of recruitment success in cockles and other bivalves in the Wadden Sea: possible role of climate change, predation on postlarvae and fisheries. Mar-Ecol Prog Ser 287:149–167

Beukema JJ, Cadée GC, Dekker R (2002) Zoobenthic biomass limited by phytoplankton abundance: evidence from parallel changes in two long-term data series in the Wadden Sea. J Sea Res 48:111–125

Bjornsson B, Steinarsson A, Oddgeirsson M (2001) Optimal temperature for growth and feed conversion of immature cod (*Gadus morhua* L.). ICES J Mar Sci 58:29–38

Bolle LJ, Dickey-Collas M, van Beek JKL, Erftemeijer PLA, Witte JIJ, van der Veer HW, Rijinsdorp AD (2009) Variability in transport of fish eggs and larvae. III. Effects of hydrodynamics and larval behaviour on recruitment in plaice. Mar-Ecol Prog Ser 390:195–211

Bolster WJ (2008) Putting the ocean in Atlantic history: Maritime communities and marine ecology in the northwest Atlantic, 1500–1800. Am Hist Rev 113:19–47

Bonnet D, Richardson A, Harris R, Hirst A, Beaugrand G, Edwards M, Ceballos S, Diekman R, Lopez-Urrutia A, Valdes L, Carlotti F, Molinero JC, Weikert H, Greve W, Lucic D, Albaina A, Yahia ND, Umani SF, Miranda A, dos Santos A, Cook K, Robinson S, de Puelles MLF (2005) An overview of *Calanus helgolandicus* ecology in European waters. Progr Oceanogr 65:1–53

Bopp L (2005) Response of diatoms distribution to global warming and potential implications: A global model study. Geophys Res Lett 32: L19606. doi:10.1029/2005GL023653

Brander K (2010) Impacts of climate change on fisheries. J Marine Syst 79:389–402

Brander K, Mohn R (2004) Effect of the North Atlantic Oscillation on recruitment of Atlantic cod (*Gadus morhua*). Can J Fish Aquat Sci 61:1558–1564

Brander KM, Blom G, Borges MF, Erzini K, Henderson G, MacKenzie BR, Mendes H, Ribeiro J, Santos AMP, Toresen R (2003) Changes in fish distribution in the eastern North Atlantic: are we seeing a coherent response to changing temperature? ICES Mar Sc 219:261–270

Brown JH, Gillooly JF, Allen AP, Savage VM, West GB (2004) Towards a metabolic theory of ecology. Ecology 85:1771–1789

Buckley LB, Hurlbert AH, Jetz W (2012) Broad-scale ecological implications of ectothermy and endothermy in changing environments. Glob Ecol Biogeogr 21:873–885

Burrows MT, Schoeman DS, Buckley LB, Moore P, Poloczanska ES, Brander KM, Brown C, Bruno JF, Duarte CM, Halpern BS, Holding J, Kappel CV, Kiessling W, O'Connor MI, Pandolfi JM, Parmesan C, Schwing FB, Sydeman WJ, Richardson AJ (2011) The pace of shifting climate in marine and terrestrial ecosystems. Science 334:652–655

Burthe S, Daunt F, Butler A, Elston DA, Frederiksen M, Johns D, Newell M, Thackeray SJ, Wanless S (2012) Phenological trends and trophic mismatch across multiple levels of a North Sea pelagic food web. Mar-Ecol Prog Ser 454:119–133

Burthe SJ, Wanless S, Newell MA, Butler A, Daunt F (2014) Assessing the vulnerability of the marine bird community in the western North Sea to climate change and other anthropogenic impacts. Mar-Ecol Prog Ser 507:277–295

Camphuysen CJ (2004) The return of the harbour porpoise (*Phocoena phocoena*) in Dutch coastal waters. Lutra 47:113–122

Camphuysen CJ, Garthe S (2000) Seabirds and commercial fisheries: population trends of piscivorous seabirds explained? In: Kaiser MJ, de Groot SJ (eds) The effects of fishing on non-target species and habitats: Biological, conservation and socio-economic issues. Blackwell

Camphuysen CJ, Peet G (2006) Whales and Dolphins of the North Sea. Fontaine Uitgewers. pp 420

Cermeño P, Dutkiewicz S, Harris RP, Follows M, Schofield O, Falkowski PG (2008) The role of nutricline depth in regulating the ocean carbon cycle. P Natl Acad Sci USA 105:20344–20349

Cheung WWL, Dunne J, Sarmiento JL, Pauly D (2011) Integrating ecophysiology and plankton dynamics into projected maximum fisheries catch potential under climate change in the Northeast Atlantic. ICES J Mar Sci 68:1008–1018

Clark RA, Frid CLJ (2001) Long-term changes in the North Sea ecosystem. Environ Rev 9:131–187

Clark RA, Fox CJ, Viner D, Livermore M (2003) North Sea cod and climate change – modelling the effects of temperature on population dynamics. Global Change Biol 9:1669–1680

Coleman RA, Underwood AJ, Benedetti-Cecchi L, Aberg P, Arenas F, Arrontes J, Castro J, Hartnoll RG, Jenkins SR, Paula J, Della Santina P, Hawkins SJ (2006) A continental scale evaluation of the role of limpet grazing on rocky shores. Oecologia 147:556–564

Corten A (2013) Recruitment depressions in North Sea herring. ICES J Mar Sci 70:1–15

Cotton PA (2003) Avian migration phenology and global climate change. P Natl Acad Sci USA 100:12219–12222

Croxall JP, Butchart SHM, Lascelles B, Stattersfield AJ, Sullivan B, Symes A, Taylor P (2012) Seabird conservation status, threats and priority actions: a global assessment. Bird Conserv Int 22:1–34

Cury P, Shannon L, Shin YJ (2003) The functioning of marine ecosystems: a fisheries perspective. In: Sinclair M, Valdimarsson G

(eds) Responsible fisheries in the marine ecosystem. FAO and CAB international, Rome
Cushing DH (1984) The gadoid outburst in the North Sea. J Cons Int Explor Mer 41:159–166
Cushing DH (1990) Plankton production and year-class strength in fish populations: An update of the match/mismatch hypothesis. Adv Mar Biol 26:249–294
Daewel U, Peck MA, Schrum C (2011) Life history strategy and impacts of environmental variability on early life stages of two marine fishes in the North Sea: an individual-based modelling approach. Can J Fish Aquat Sci 68:426–443
Dam HG (2013) Evolutionary adaptation of marine zooplankton to global change. Ann Rev Mar Sci 5:349–370
Daunt F, Mitchell I (2013) Impacts of climate change on seabirds. MCCIP Science Review 2013:125–133
Deaville R, Jepson PD (2011) UK Cetacean Strandings Investigation Programme. Final Report to Defra for the period 1 January 2005 – 31 December 2010. Institute of Zoology, London
Dickey-Collas M, Bolle L, van Beek J, Erftemeijer P (2009) Variability in transport of fish eggs and larvae. II. Effects of hydrodynamics on the transport of Downs herring larvae. Mar-Ecol Prog Ser 390:183–194
Dickey-Collas M, Nash RDM, Brunel T, van Damme CJG, Marshall CT, Payne MR, Corten A, Geffen AJ, Peck MA, Hatfield EMC, Hintzen NT, Enberg K, Kell LT, Simmonds EJ (2010) Lessons learned from stock collapse and recovery of North Sea herring: a review. ICES J Mar Sci 67:1875–1886
Dinter WP (2001) Biogeography of the OSPAR Maritime Area. German Federal Agency for Nature Conservation, Bonn, Germany
Dolman SJ, Pinn E, Reid RJ, Barley JP, Deaville R, Jepson PD, O'Connell M, Berrow S, Penrose RS, Stevick PT, Calderan S, Robinson KP, Brownell Jr RL, Simmonds MP (2010) A note on the unprecedented stranding of 56 deep-diving odontocetes along the UK and Irish coast. Mar Biodiv Rec 3:1–8
Doney SC, Sailley SF (2013) When an ecological regime shift is really just stochastic noise. P Natl Acad Sci USA 110:2438–2439
Duineveld GCA, Kuenitzer A, Niermann U, De Wilde PAJ, Gray JS (1991) The macrobenthos of the North Sea. Neth J Sea Res 28:53–56
Durant JM, Hjermann DØ, Ottersen G, Stenseth NC (2007) Climate and the match or mismatch between predator requirements and resource availability. Clim Res 33:271–283
Edwards M, Richardson AJ (2004) Impact of climate change on marine pelagic phenology and trophic mismatch. Nature 430:881–884
Edwards M, Beaugrand G, John AWG, Johns DG, Licandro P, McQuatters-Gollop, Reid PC (2009) Ecological Status Report: results from the CPR survey 2007/2008. SAHFOS Tech Rep 6:1–12
Edwards M, Helaouet P, Johns DG, Batten S, Beaugrand G, Chiba S, Head EJH, Richardson AJ, Verheye HM, Wootton M, Flavell M, Hosie G, Takahashi K, Ward P (2012) Global Marine Ecological Status Report: results from the global CPR survey 2010/2011. SAHFOS Tech Rep 9:1–40
Edwards M, Helaouet P, Johns DG, Batten S, Beaugrand G, Chiba S, Hall J, Head E, Hosie G, Kitchener J, Koubbi P, Kreiner A, Melrose C, Pinkerton M, Richardson AJ, Robinson K, Takahashi K, Verheye HM, Ward P, Wootton M (2014) Global Marine Ecological Status Report: results from the global CPR survey 2012/2013. SAHFOS Tech Rep 10:1–37
Eggleton JD, Smith R, Reiss H, Rachor E, Vanden Berghe E, Rees HL (2007) Species distributions and changes (1986–2000). In: Rees HL, Eggleton JD, Vanden Berghe E (eds) Structure and Dynamics of the North Sea Benthos. ICES Coop Res Rep 288:91–108
Engelhard GH, Pinnegar JK, Kell LT, Rijnsdorp AD (2011) Nine decades of North Sea sole and plaice distribution. ICES J Mar Sci 68:1090–1104
Engelhard GH, Righton DA, Pinnegar JK (2014) Climate change and fishing: a century of shifting distribution in North Sea cod. Global Change Biol 20:2473–2483
Enghoff IB, MacKenzie BR, Nielsen EE (2007) The Danish fish fauna during the warm Atlantic period (ca. 7000–3900 BC): Forerunner of future changes? Fish Res 87:167–180
Evans PGH, Bjørge A (2014) Impacts of climate change on marine mammals. MCCIP Sci Rev 2013:134–148
Evans PGH, Smeenk C (2008) White-beaked dolphin *Lagenorhynchus albirostris*. In: Harris S, Yalden DW (eds) Mammals of the British Isles. Mammal Society, Southampton
Evans PGH, Anderwald P, Baines ME (2003) Status Review of UK Cetaceans. Report to English Nature and Countryside Council for Wales
Fiksen O (2000) The adaptive timing of diapause – a search for evolutionarily robust strategies in *Calanus finmarchicus*. ICES J Mar Sci 57:1825–1833
Fincham JI, Rijnsdorp AD, Engelhard GH (2013) Shifts in the timing of spawning in sole linked to warming sea temperatures. J Sea Res 75:69–76
Forbes E (1858) The distribution of marine life, illustrated chiefly by fishes and molluscs and radiata. In: Johnston AK (ed) AK Johnston's Physical Atlas. Johnston, Edinburgh
Fort J, Porter WP, Gremillet D (2009) Thermodynamic modelling predicts energetic bottleneck for seabirds wintering in the northwest Atlantic. J Exp Biol 212:2483–2490
Frank KT, Petrie B, Choi JS, Leggett WC (2005) Trophic cascades in a formerly cod-dominated ecosystem. Science 308:1621–1623
Fraser FC (1946) Report on Cetacea stranded on the British Coasts from 1933 to 1937. The British Museum (Natural History), London
Frederiksen M, Harris MP, Daunt F, Rothery P, Wanless S (2004a) Scale-dependent climate signals drive breeding phenology of three seabird species. Global Change Biol 10:1214–1221
Frederiksen M, Wanless S, Harris MP, Rothery P, Wilson LJ (2004b) The role of industrial fisheries and oceanographic change in the decline of North Sea black-legged kittiwakes. J Appl Ecol 41:1129–1139
Frederiksen M, Wright PJ, Harris MP, Mavor RA, Heubeck M, Wanless S (2005) Regional patterns of kittiwake *Rissa tridactyla* breeding success are related to variability in sandeel recruitment. Mar-Ecol Prog Ser 300:201–211
Frederiksen M, Edwards M, Richardson AJ, Halliday NC, Wanless S (2006) From plankton to top predators: bottom-up control of a marine food web across four trophic levels. J Anim Ecol 75:1259–1268
Frederiksen M, Anker-Nilssen T, Beaugrand G, Wanless S (2013) Climate, copepods and seabirds in the boreal Northeast Atlantic – current state and future outlook. Global Change Biol 19:364–372
Freitas V, Campos J, Fonds M, Van der Veer HW (2007) Potential impact of temperature change on epibenthic predator-bivalve prey interactions in temperate estuaries. J Therm Biol 32:328–340
Frid CLJ, Garwood PR, Robinson LA (2009a) Observing change in a North Sea benthic system: A 33 year time series. J Marine Syst 77:227–236
Frid CLJ, Garwood PR, Robinson LA (2009b) The North Sea benthic system: a 36 year time-series. J Mar Biol Assoc UK 89:1–10
Furness RW (2002) Management implications of interactions between fisheries and sandeel-dependent seabirds and seals in the North Sea. ICES J Mar Sci 59:261–269
Furness RW, Tasker ML (2000) Seabird-fishery interactions: quantifying the sensitivity of seabirds to reductions in sandeel abundance, and identification of key areas for sensitive seabirds in the North Sea. Mar-Ecol Prog Ser 202:253–264
Garthe S, Camphuysen CJ, Furness RW (1996) Amounts of discards by commercial fisheries and their significance as food for seabirds in the North Sea. Mar-Ecol Prog Ser 136:1–11

Gilles A (2009) Characterisation of harbour porpoise (*Phocoena phocoena*) habitat in German waters. PhD thesis, University of Kiel, Germany

Glémarec M (1973) The benthic communities of the European North Atlantic shelf. Oceanogr Mar Biol 11:263–289

Hagberg J, Tunberg BG (2000) Studies on the covariation between physical factors and the long-term variation of the marine soft bottom macrofauna in Western Sweden. Estuar Coast Shelf S 50:373–385

Hall A (2002) Gray seal *Halichoerus grypus*. In: Perrin WF, Wursig B, Thewissen GM (eds) Encyclopedia of Marine Mammals. Academic Press

Hammond PS, Grellier K (2006) Grey seal diet composition and prey consumption in the North Sea. Final report to Department for Environment Food and Rural Affairs on project MF0319

Hammond PS, Berggren P, Benke H, Borchers DL, Collet A, Heide-Jørgensen MP, Heimlich S, Hiby AR, Leopold MF, Oien N (2002) Abundance of harbour porpoises and other cetaceans in the North Sea and adjacent waters. J Appl Ecol 39:361–376

Hammond PS, Bearzi G, Bjørge A, Forney K, Karczmarski L, Kasuya T, Perrin WF, Scott M, Wang J, Wells R, Wilson B (2008) *Phocoena phocoena*. In: 2008 IUCN Red List of Threatened Species. International Union for Conservation of Nature (IUCN), Gland, Switzerland

Hammond PS, Macleod K, Berggren P, Borchers DL, Burt L, Cañadas A, Desportes G, Donovan GP, Gilles A, Gillespie D, Jonathan Gordon J, Hiby L, Kuklik I, Leaper R, Lehnert K, Leopold M, Lovell P, Øien N, Paxton CGM, Ridoux V, Rogan E, Samarra F, Scheidat M, Sequeira M, Siebert U, Skov H, Swift R, Tasker ML, Teilmann J, Canneyt OV, Vázquez JA (2013) Cetacean abundance and distribution in European Atlantic shelf waters to inform conservation and management. Biol Cons 167:107–122

Hatun H, Payne MR, Beaugrand G, Reid PC, Sando AB, Drange H, Hansen B, Jacobsen JA, Bloch D (2009) Large bio-geographical shifts in the north-eastern Atlantic Ocean: From the subpolar gyre, via plankton, to blue whiting and pilot whales. Prog Oceanogr 80:149–162

Hawkins SJ, Sugden HE, Mieszkowska N, Moore PJ, Poloczanska E, Leaper R, Herbert RJH, Genner MJ, Moschella PS, Thompson RC, Jenkins SR, Southward AJ, Burrows MT (2009) Consequences of climate-driven biodiversity changes for ecosystem functioning of North European rocky shores. Mar-Ecol Prog Ser 396:245–259

Hay S, Falkenhaug T, Omli L, Boersma M, Renz J, Halsband-Lenk C, Smyth T et al (2011) Zooplankton of the North Sea and English Channel. In: O'Brien T, Wiebe PH, Hay S (eds) ICES Coop Res Rep: ICES Zooplankton Stat Rep 2008/2009.

Heath MR (2005) Changes in the structure and function of the North Sea fish foodweb, 1973–2000, and the impacts of fishing and climate. ICES J Mar Sci 62:847-868

Heath MR, Neat FC, Pinnegar JK, Reid DG, Sims DW, Wright PJ (2012) Review of climate change impacts on marine fish and shellfish around the UK and Ireland. Aquat Conserv Freshw Ecosyst 22:337–367

Heip C, Basford D, Craeymeersch JA, Dewarumez JM, Dorjes J, Dewilde P, Duineveld G, Eleftheriou A, Herman PMJ, Niermann U, Kingston P, Kunitzer A, Rachor E, Rumohr H, Soetaert K, Soltwedel T (1992) Trends in biomass, density and diversity of North-Sea macrofauna. ICES J Mar Sci 49:13–22

Helaouet P, Beaugrand G (2007) Macroecology of *Calanus finmarchicus* and *C. helgolandicus* in the North Atlantic Ocean and adjacent seas. Mar-Ecol Prog Ser 345:147–165

Hiddink JG, ter Hofstede R (2008) Climate induced increases in species richness of marine fishes. Global Change Biol 14:453–460

Hiddink JG, Burrows MT, García Molinos J (2014) Temperature tracking by North Sea benthic invertebrates in response to climate change. Global Change Biol 21:doi:10.1111/gcb.12726

Hinder SL, Hays GC, Edwards M, Roberts EC, Walne AW, Gravenor MB (2012) Changes in marine dinoflagellate and diatom abundance under climate change. Nature Clim Change 2:271–275

Hjøllo SS, Skogen MD, Svendsen E (2009) Exploring currents and heat within the North Sea using a numerical model. J Marine Syst 78:180–192

Howard J, Babij E, Griffis R, Helmuth B, Himes-Cornell A, Niemier P et al (2013) Oceans and marine resources in a changing climate. Oceanography and Marine Biology: An Annual Review 51

Huntley B, Green R, Collingham Y, Willis SG (2007) A climatic atlas of European breeding birds. Lynx Edicions, Barcelona

ICES (2008) The effect of climate change on the distribution and abundance of marine species in the OSPAR Maritime Area. ICES Coop Res Rep 293:45. International Council for the Exploration of the Sea (ICES), Copenhagen

ICES (2009) Report of the Working Group on Holistic Assessments of Regional Marine Ecosystems (WGHAME). ICES CM 2009/RMC:13. International Council for the Exploration of the Sea (ICES), Copenhagen

ICES (2011) Report of the Working Group on Seabird Ecology (WGSE). ICES Doc. CM 2011/SSGEF:07. International Council for the Exploration of the Sea (ICES), Copenhagen

Jones MC, Dye SR, Pinnegar JK, Warren R, Cheung WWL (2012) Modelling commercial fish distributions: Prediction and assessment using different approaches. Ecol Model 225:133–145

Kenny AJ, Skjoldal HR, Engelhard GH, Kershaw PJ, Reid JB (2009) An integrated approach for assessing the relative significance of human pressures and environmental forcing on the status of Large Marine Ecosystems. Prog Oceanogr 81:132–148

Kinze CC, Addink M, Smeenk C, García Hartmann M, Richards HW, Sonntag RP, Benke H (1997) The white-beaked dolphin (*Lagenorhynchus albirostris*) and the white-sided dolphin (*Lagenorhynchus acutus*) in the North and Baltic Seas, review of available information. Rep Intern Whaling Commis 47:675–681

Kirby RR, Beaugrand G (2009) Trophic amplification of climate warming. Proc R Soc B 276:4095–4103

Kirby RR, Lindley JA (2005) Molecular analysis of Continuous Plankton Recorder samples, an examination of echinoderm larvae in the North Sea. J Mar Biol Assoc UK 85:451–459

Kirby RR, Beaugrand G, Lindley JA, Richardson AJ, Edwards M, Reid PC (2007) Climate effects and benthic-pelagic coupling in the North Sea. Mar-Ecol Prog Ser 330:31–38

Kirby RR, Beaugrand G, Lindley JA (2008) Climate-induced effects on the meroplankton and the benthic-pelagic ecology of the North Sea. Limnol Oceanogr 53:1805–1815

Kjesbu OS, Witthames PR, Solemdal P, Walker MG (1998) Temporal variations in the fecundity of Arcto-Norwegial cod (*Gadus morhua*) in response to natural changes in food and temperature. J Sea Res 40:303–321

Kordas RL, Harley CDG, O'Connor MI (2011) Community ecology in a warming world: The influence of temperature on interspecific interactions in marine systems. J Exp Mar Biol Ecol 400:218–226

Kröncke I, Reiss H (2007) Changes in community structure (1986–2000) and causal influences. In: Rees HL, Eggleton JD et al (eds) Structure and dynamics of the North Sea benthos. ICES Coop Res Rep 288

Kröncke I, Dippner JW, Heyen H, Zeiss B (1998) Long-term changes in macrofaunal communities off Norderney (East Frisia, Germany) in relation to climate variability. Mar-Ecol Prog Ser 167:25–36

Kröncke I, Zeiss B, Rensing C (2001) Long-term variability in macrofauna species composition off the island of Norderney (East Frisia, Germany) in relation to changes in climatic and environmental conditions. Senck marit 31:65–82

Künitzer A, Basford D, Craeymeersch JA, Dewarumez JM, Dorjes J, Duineveld GC, Eleftheriou A, Heip C, Herman P, Kingston P, Niermann U, Rachor E, Rumohr H, de Wilde PAJ (1992) The benthic infauna of the North Sea: Species distribution and assemblages. ICES J Mar Sci 49:127–143

Le Masson du Parc (1727) Peches et pecheurs du domaine maritime aquitain au XVIII siecle. Proces Verbaux des visites faites par ordre du Roy concernant la peche en mer. Les Editions de l'Entre-deux-Mers

Learmonth JA, MacLeod CD, Santos MB, Pierce GJ, Crick HQP, Robinson RA (2006) Potential effects of climate change on marine mammals. Oceanography and Marine Biology: An Annual Review 44: 431–464

Licandro P, Conway DVP, Daly Yahia MN, Fernandez de Puelles ML, Gasparini S, Hecq JH, Tranter P, Kirby RR (2010) A blooming jellyfish in the northeast Atlantic and Mediterranean. Biol Lett 6:688–691

Lindley JA, Batten SD (2002) Long-term variability in the diversity of North Sea zooplankton. J Mar Biol Assoc UK 82:31–40

Llope M, Chan KS, Ciannelli L, Reid PC, Stige LC, Stenseth NC (2009) Effects of environmental conditions on the seasonal distribution of phytoplankton biomass in the North Sea. Limnol Oceanogr 54:512–524

Luczak C, Beaugrand G, Lindley JA, Dewarumez JM, Dubois PJ, Kirby RR (2012) North Sea ecosystem change from swimming crabs to seagulls. Biol Lett 8:821–824

Lynam CP, Hay SJ, Brierley AS (2005) Jellyfish abundance and climatic variation: contrasting responses in oceanographically distinct regions of the North Sea, and possible implications for fisheries. J Mar Biol Assoc UK 85:435–450

Maar M, Møller EF, Gürkan Z, Jónasdóttir SH, Nielsen TG (2013) Sensitivity of *Calanus* spp. copepods to environmental changes in the North Sea using life-stage structured models. Prog Oceanogr 111:24–37

Mackas DL, Batten S, Trudel M (2007) Effects on zooplankton of a warmer ocean: Recent evidence from the Northeast Pacific. Prog Oceanogr 75:223–252

MacLeod CD (2009) Global climate change, range changes and potential implications for the conservation of marine cetaceans, a review and synthesis. Endang Spec Res 7:125–136

Macleod CD, Bannon SM, Pierce GJ, Schweder C, Learmouth JA, Herman JS, Reid RJ (2005) Climate change and the cetacean community of north-west Scotland. Biol Conserv 124:477–483

MacLeod CD, Begona Santos M, Reid RJ, Scott BE, Pierce GJ (2007) Linking sandeel consumption and the likelihood of starvation in harbour porpoises in the Scottish North Sea: could climate change mean more starving porpoises? Biol Lett 3:185–188

MacLeod CD, Weir CR, Santos MB, Dunn TE (2008) Temperature-based summer habitat partitioning between white-beaked and common dolphins around the United Kingdom and Republic of Ireland. J Mar Biol Assoc UK 88:1193–1198

McLeod KJ, Lubchenco J, Palumbi SR, Rosenberg AA (2005) Scientific consensus statement on marine ecosystem-based management. Signed by 219 academic scientists and policy experts with relevant expertise and published by the Communication Partnership for Science and the Sea (COMPASS) at http://compassonline.org/sites/all/files/document_files/EBM_Consensus_Statement_v12.pdf

Mendel B, Sonntag N, Wahl J, Schwemmer P, Dries H, Guse N, Müller S, Garthe S (2008) Profiles of seabirds and waterbirds of the German North and Baltic Seas. Distribution, ecology and sensitivities to human activities within the marine enviroment. Naturschutz und Biologische Vielfalt 61

Meyer EMI, Pohlmann T, Weisse R (2011) Thermodynamic variability and change in the North Sea (1948–2007) derived from a multidecadal hindcast. J Marine Syst 86:35–44

Mieszkowska N, Kendall MA, Hawkins SJ, Leaper R, Williamson P (2006) Changes in the range of some common rocky shore species in Britain – a response to climate change? Mar Biodiv 183:241–251

Mieszkowska N, Hawkins SJ, Burrows MT, Kendall MA (2007) Long-term changes in the geographic distribution and population structures of *Osilinus lineatus* (Gastropoda : Trochidae) in Britain and Ireland. J Mar Biol Assoc UK 87:537–545

Mieszkowska N, Genner MJ, Hawkins SJ, Sims DW (2009) Effects of climate change and commercial fishing on Atlantic cod *Gadus morhua*. Adv Mar Biol 56:213–273

Millner RS, Pilling GM, Mccully SR (2011) Changes in the timing of otolith zone formation in North Sea cod from otolith records : an early indicator of climate-induced temperature stress? Mar Biol 158:21–30

Mitchell PI, Ratcliffe N, Newton SF, Dunn TE (2004) Seabird populations of Britain and Ireland. Results of the Seabirds 2000 census. Poyser, London

Myksvoll MS, Erikstad KE, Barrett RT, Sandvik H, Vikebo F (2013) Climate-driven Ichthyoplankton drift model predicts growth of top predator young. PLOS One. 8:e79225. doi:10.1371/journal.pone.0079225

Neat F, Righton D (2007) Warm water occupancy by North Sea cod. Proc Biol Sci 274:789–798

Neat FC, Wright PJ, Fryer RJ (2008) Temperature effects on otolith pattern formation in Atlantic cod *Gadus morhua*. J Fish Biol 73:2527–2541

Neumann H, Reiss H, Rakers S, Ehrich S, Kroncke I (2009) Temporal variability in southern North Sea epifauna communities after the cold winter of 1995/1996. ICES J Mar Sci 66:2233–2243

Nicolas D, Rochette S, Llope M, Licandro P (2014) Spatio-temporal variability of the North Sea cod recruitment in relation to temperature and zooplankton. PLOS One 9:e88447. doi:10.1371/journal.pone.0088447

O'Brien CM, Fox C, Planque B, Casey J (2000) Climate variability and North Sea cod. Nature 404:142

Øien N (2010) Report of the Norwegian 2009 survey for minke whales within the Small Management Area EN – the North Sea. IWC SC/62/RMP7

Oliveira OMP (2007) The presence of the ctenophore *Mnemiopsis leidyi* in the Oslofjorden and considerations on the initial invasion pathways to the North and Baltic seas. Aquat Invas 2:185–189

Olsen EM, Ottersen G, Llope M, Chan K, Beaugrand G, Stenseth C, Stenseth NC (2011) Spawning stock and recruitment in North Sea cod shaped by food and climate. Proc R Soc B 278:504–510

Oost AP, Labat P, Wiersma A, Hofstede J (2009) Climate. Thematic Report no. 4.1. In: Marencic H, De Vlas J (eds) Quality Status Report 2009. Wadden Sea Ecosystem 25. Common Wadden Sea Secretariat, Trilateral Monitoring and Assessment Group

Ottersen G, Kim S, Huse G, Polovina JJ, Stenseth NC (2010) Major pathways by which climate may force marine fish populations. J Marine Syst 79:343–360

Ottersen G, Stige LC, Durant JM, Chan K-C, Rouyer TA, Drinkwater KF, Stenseth NC (2013) Temporal shifts in recruitment dynamics of North Atlantic fish stocks: effects of spawning stock and temperature. Mar-Ecol Prog Ser 480:205–225

Parmesan C, Yohe G (2003) A globally coherent fingerprint of climate change impacts across natural systems. Nature 421:37–42

Pauly D, Christensen V, Dalsgaard J, Froese R, Torres F (1998) Fishing down marine food webs. Science 279:860–863

Payne MR, Hatfield EMC, Dickey-Collas M, Falkenhaug T, Gallego A, Groger J, Licandro P, Llope M, Munk P, Rockmann C, Schmidt JO, Nash RDM (2009) Recruitment in a changing environment: the 2000s North Sea herring recruitment failure. ICES J Mar Sci 66:272–277

Payne M, Stine D, Worsøe Clausen L, Munk P, Mosegaard H, Nash R (2013) Recruitment decline in North Sea herring is accompanied by reduced larval growth rates. Mar-Ecol Prog Ser 489:197–211

Perry AL, Low PJ, Ellis JR, Reynolds JD (2005) Climate change and distribution shifts in marine fishes. Science 308:1912–1915

Pershing AJ, Mills KE, Record NR, Stamieszkin K, Wurtzell KV, Byron CJ, Fitzpatrick D, Golet WJ, Koob E (2015) Evaluating trophic cascades as drivers of regime shifts in different ocean ecosystems. Phil Trans R B 370:20130265

Petersen CGJ (1914) Valuation of the sea II. The animal communities of the sea bottom and their importance for marine zoogeography. Rep Dan Biol Sta 21:1–44

Petersen CGJ (1918) The sea bottom and its production of fishfood. Rep Dan Biol Sta 25:1–62

Petitgas P, Alheit J, Peck MA, Raab K, Irigoien X, Huret M, van der Kooij J, Pohlmann T, Wagner C, Zarraonaindia I, Dickey-Collas M (2012) Anchovy population expansion in the North Sea. Mar-Ecol Prog Ser 444:1–13

Petitgas P, Rijnsdorp AD, Dickey-Collas M, Engelhard GH, Peck MA, Pinnegar JK, Drinkwater K, Huret M, Nash RDM (2013) Impacts of climate change on the complex life cycles of fish. Fish Oceanogr 22:121–139

Philippart CJM, Anadon R, Danovaro R, Dippner JW, Drinkwater KF, Hawkins SJ, Reid PC (2011) Impacts of climate change on European marine ecosystems: Observations, expectations and indicators. J Exp Mar Biol Ecol 400:52–69

Piatt J, Sydeman WJ (2007) Seabirds as indicators of marine ecosystems. Theme Section. Mar-Ecol Prog Ser 352:199–309

Planque B, Batten SD (2000) *Calanus finmarchicus* in the North Atlantic: the year of *Calanus* in the context of interdecadal change. ICES J Mar Sci 57:1528–1535

Planque B, Fredou T (1999) Temperature and the recruitment of Atlantic cod (*Gadus morhua*). Can J Fish Aquat Sci 56:2069–2077

Poloczanska ES, Brown CJ, Sydeman WJ, Kiessling W, Schoeman DS, Moore PJ, Brander K, Bruno JF, Buckley LB, Burrows MT, Duarte CM, Halpern BS, Holding J, Kappel C V, O'Connor MI, Pandolfi JM, Parmesan C, Schwing F, Thompson SA, Richardson AJ (2013) Global imprint of climate change on marine life. Nature Clim Change 3:919–925

Pörtner HO, Knust R (2007) Climate change affects marine fishes through the oxygen limitation of thermal tolerance. Science 315: 95–97

Poulsen B, Holm P, MacKenzie BR (2007) A long-term (1667–1860) perspective on impacts of fishing and environmental variability on fisheries for herring, eel, and whitefish in the Limfjord, Denmark. Fish Res 87:181–195

Quero JC, Du Bui MH, Vayne JJ (1998) Les observations de poissons tropicaux et le rechauffement des eaux dans l'Atlantique europeen. Oceanol Acta 21:345–351

Rachor E, Gerlach SA (1978) Changes of macrobenthos in a sublittoral sand area of the German Bight, 1967 to 1975. Rap Proces 172:418–431

Rachor E, Reiss H, Degraer S, Duineveld GCA, Van Hoey G, Lavaleye M, Willems W, Rees HL (2007) Structure, distribution and characterizing species of North Sea macrozoobenthos communities in 2000. In: Rees HL, Eggleton JD et al (eds) Structure and dynamics of the North Sea benthos. ICES Coop Res Rep 288

Rees HL, Eggleton JD, Rachor E, Vanden Berghe E (2007a) Structure and dynamics of the North Sea benthos. ICES Coop Res Rep 288

Rees HL, Eggleton JD, Rachor E, Vanden Berghe E, Aldridge JN, Bergman MJN, Bolam T, Cochrane SJ, Craeymeersch JA, Degraer S, Desroy N, Dewarumez J-M, Duineveld GCA, Essink K, Hillewaert H, Irion G, Kershaw PJ, Kröncke I, Lavaleye MSS, Mason C, Nehring S, Newell R, Oug E, Pohlmann T, Reiss H, Robertson M, Rumohr H, Schratzberger M, Smith R, van Dalfsen J, Van Hoey G, Vincx M, Willems W (2007b) The ICES North Sea Benthos Project 2000: aims, outcomes and recommendations. ICES CM 2007, A:21

Reid PC, Edwards M (2001) Long-term changes in the pelagos, benthos and fisheries of the North Sea. Senck marit 31:107–115

Reid PC, Edwards M, Hunt HG, Warner JA (1998) Phytoplankton change in the North Atlantic. Nature 391:546

Reid PC, de Fatima Borges M, Svendsen E (2001) A regime shift in the North Sea circa 1988 linked to changes in the North Sea horse mackerel fishery. Fish Res 50:163–171

Reid JB, Evans PGH, Northridge SP (eds) (2003) Atlas of Cetacean Distribution in North-West European Waters. Joint Nature Conservation Committee. Peterborough

Reijnders PJH, Abt KF, Brasseur SMJM, Camphuysen K, Reineking B, Scheidat M, Siebert U, Stede M, Tougaard J, Tougaard S (2005) Marine Mammals. In: Essink K, Dettmann C, Farke H, Laursen K, Lüerßen G, Marencic H, Wiersinga W (eds) Wadden Sea Quality Status Report 2004. Wadden Sea Ecosystem No. 19. Trilateral Monitoring and Assessment Group, Common Wadeen Sea Secretariat, Wilhelmshaven

Reiss H, Degraer S, Duineveld GCA, Kroncke I, Aldridge J, Craeymeersch JA, Eggleton JD, Hillewaert H, Lavaleye MSS, Moll A, Pohlmann T, Rachor E, Robertson M, Vanden Berghe E, van Hoey G, Rees HL (2010) Spatial patterns of infauna, epifauna, and demersal fish communities in the North Sea. ICES J Mar Sci 67:278–293

Richardson AJ, Schoeman DS (2004) Climate impact on plankton ecosystems in the Northeast Atlantic. Science 305:1609–1612

Righton D, Andersen K, Neat F, Thorsteinsson V, Steingrund P, Svedäng H, Michalsen K, Hinrichsen H, Bendall V, Neuenfeldt S, Wright P, Jonsson P, Huse G, van der Kooij J, Mosegaard H, Hüssy K, Metcalfe J (2010) Thermal niche of Atlantic cod *Gadus morhua*: limits, tolerance and optima. Mar-Ecol Prog Ser 420:1–13

Rijnsdorp AD, Witthames PR (2005) Ecology of reproduction In: Gibson RN (ed) Flatfish Biology and Exploitation. Academic Press

Rijnsdorp AD, Peck MA, Engelhard GH, Mollmann C, Pinnegar JK (2009) Resolving the effect of climate change on fish populations. ICES J Mar Sci 66:1570–1583

Rijnsdorp AD, Peck MA, Engelhard GE, Möllmann C, Pinnegar JK (2010) Resolving climate impacts on fish stocks. ICES Coop Res Rep 301:371

Roemmich D, McGowan J (1995) Climatic warming and the decline of zooplankton in the California current. Science 267:1324–1326

Rogers LA, Stige LC, Olsen EM, Knutsen H, Chan K-C, Stenseth NC (2011) Climate and population density drive changes in cod body size throughout a century on the Norwegian coast. P Natl Acad Sci USA 108:1961–1966

Røjbek M, Jacobsen C, Tomkiewicz J, Støttrup J (2012) Linking lipid dynamics with the reproductive cycle in Baltic cod *Gadus morhua*. Mar-Ecol Prog Ser 471:215–234

Rothschild BJ, Osborn TR (1988) Small scale turbulence and plankton contact rates. J Plank Res 10:465–474

Rutterford LA, Simpson SD, Jennings S, Johnson MP, Blanchard JL, Schön PJ, Sims DW, Tinker J, Genner MJ (2015) Future fish distributions constrained by depth in warming seas. Nature Clim Change 5:569–573

Sagarin RD, Barry JP, Gilman SE, Baxter CH (1999) Climate-related change in an intertidal community over short and long time scales. Ecol Monogr 69:465–490

Sandvik H, Erikstad KE (2008) Seabird life histories and climatic fluctuations: a phylogenetic-comparative time series analysis of North Atlantic seabirds. Ecography 31:73–83

Sandvik H, Erikstad KE, Saether BE (2012) Climate affects seabird population dynamics both via reproduction and adult survival. Mar-Ecol Prog Ser 454:273–284

Santos MB, Pierce GJ (2003) The diet of harbour porpoise (*Phocoena phocoena*) in the eastern North Atlantic. Oceanogr Mar Biol Ann Rev 41:355–390

Sarmiento JL, Slater R, Barber R, Bopp L, Doney SC, Hirst AC, Kleypas J, Matear R, Mikolajewicz U, Monfray P, Soldatov V, Spall SA, Stouffer R (2004) Response of ocean ecosystems to climate warming. Global Biogeochem Cy 18: doi:10.1029/2003GB002134

Satterthwaite WH, Kitaysky AS, Mangel M (2012) Linking climate variability, productivity and stress to demography in a long-lived seabird. Mar-Ecol Prog Ser 454:221–235

SCANS II (2008) Small Cetaceans in the European Atlantic and North Sea. Final report to the European Commission under project LIFE04NAT/GB/000245. Available from SMRU, Gatty Marine Laboratory, University of St Andrews, Scotland

Scheffer M (2009) Critical Transitions in Nature and Society. Princeton University Press.

Schlüter MH, Merico A, Wiltshire KH, Greve W, von Storch H (2008) A statistical analysis of climate variability and ecosystem response in the German Bight. Ocean Dyn 58:169–186

Schlüter MH, Merico A, Reginatto M, Boersma M, Wiltshire KH, Greve W (2010) Phenological shifts of three interacting zooplankton groups in relation to climate change. Global Change Biol 16:3144–3153.

Schlüter MH, Kraberg A, Wiltshire KH (2012) Long-term changes in the seasonality of selected diatoms related to grazers and environmental conditions. J Sea Res 67:91–97

Schückel U, Kröncke I (2013) Temporal changes in intertidal macrofauna communities over eight decades: A result of eutrophication and climate change. Estuar Coast Shelf S 117:210–218

Siegismund F, Schrum C (2001) Decadal changes in the wind forcing over the North Sea. Clim Res 18:39–45

Simmonds MP, Isaac SJ (2007) The impacts of climate change on marine mammals: early signs of significant problems. Oryx 41:19–26

Simpson SD, Jennings S, Johnson MP, Blanchard JL, Scho P, Sims DW, Genner MJ (2011) Continental shelf-wide response of a fish assemblage to rapid warming of the sea. Curr Biol 21:1565–1570

Southward AJ, Crisp DJ (1954) Recent changes in the distribution of the intertidal barnacles *Chthamalus stellatus* Poli and *Balanus balanoides* L. in the British Isles. J Anim Ecol 23:163–177

Southward AJ, Hawkins SJ, Burrows MT (1995) 70 years observations of changes in distribution and abundance of zooplankton and intertidal organisms in the western English Channel in relation to rising sea temperature. J Therm Biol 20:127–155

Speirs DC, Gurney WSC, Heath MR, Horbelt W, Wood SN, de Cuevas BA (2006) Ocean-scale modelling of the distribution, abundance, and seasonal dynamics of the copepod *Calanus finmarchicus*. Mar-Ecol Prog Ser 313:173–192

Stevens G (1989) The latitudinal gradient in geographic range – how so many species coexist in the tropics. Am Nat 133:240–256

Sunday JM, Bates AE, Dulvy NK (2012) Thermal tolerance and the global redistribution of animals. Nature Clim Change 2:1–5

Sydeman WJ, Thompson SA, García-Reyes M, Kahru M, Peterson WT, Largier JL (2014) Multivariate ocean-climate indicators (MOCI) for the central California Current: Environmental change, 1990–2010. Prog Oceanogr 120:352–369

Tasker ML, Camphuysen CJ, Cooper J, Garthe S, Montevecchi WA, Blaber SJM (2000) The impacts of fishing on marine birds. ICES J Mar Sci 57:531–547

Taylor AH, Icarus Allen J, Clark PA (2002) Extraction of a weak climatic signal by an ecosystem. Nature 416:629–632

Teal LR, Leeuw JJ De, Veer HW Van Der, Rijnsdorp AD (2008) Effects of climate change on growth of 0-group sole and plaice. Mar-Ecol Prog Ser 358:219–230

Teal LR, van Hal R, van Kooten T, Ruardij P, Rijnsdorp AD (2012) Bio-energetics underpins the spatial response of North Sea plaice (*Pleuronectes platessa* L.) and sole (*Solea solea* L.) to climate change. Global Change Biol 18:3291–3305

Thomas H, Bozec Y, Elkalay K, de Baar HJW (2004) Enhanced open ocean storage of CO2 from shelf sea pumping. Science 304:1005–1008. doi:10.1126/science.1095491

Thompson D, Härkönen T (2008) *Halichoerus grypus*. In: IUCN 2009. IUCN Red List of Threatened Species. Version 2009.2. www.iucnredlist.org. Downloaded on 17 December 2009

Thompson PM, Ollason JC (2001) Lagged effects of ocean climate change on fulmar population dynamics. Nature 413:417–420

Tomanek L, Helmuth B (2002) Physiological ecology of rocky intertidal organisms: a synergy of concepts. Integr Comp Biol 42:771–775

Tunberg BG, Nelson WG (1998) Do climatic oscillations influence cyclical patterns of soft bottom macrobenthic communities on the Swedish west coast? Mar-Ecol Prog Ser 170:85–94

Van de Pol M, Ens BJ, Heg D, Brouwer L, Krol J, Maier M, Exo K-M, Oosterbeek K, Lok T, Eising CM, Koffijberg K (2010) Do changes in the frequency, magnitude and timing of extreme climatic events threaten the population viability of coastal birds? J Appl Ecol 47:720–730

Van der Graaf AJ, Jonker S, Herlyn M, Kohlus J, Vinther HF, Reise K, De Jong D, Dolch T, Bruntse C, De Vlas J (2009) Seagrass. Thematic report no. 12. In: Marencic H, De Vlas J (eds) Quality Status report 2009. Wadden Sea Ecosystem No. 25. Common Wadden Sea Secretariat, Trilateral Monitoring and Assessment Group. Wilhelmshaven

Van der Land MA (1991) Distribution of flatfish eggs in the 1989 egg surveys in the southeastern North Sea, and mortality of plaice and sole eggs. Neth J Sea Res 27:277–286

Van der Veer HW, Ruardij P, Van den Berg AJ, Ridderinkhof H (1998) Impact of interannual variability in hydrodynamic circulation on egg and larval transport of plaice *Pleuronectes platessa* L. in the southern North Sea. J Sea Res 39:29–40

Van der Veer HW, Bolle LJ, Geffen AJ, Witte JIJ (2009) Variability in transport of fish eggs and larvae. IV. Interannual variability in larval stage duration of immigrating plaice in the Dutch Wadden Sea. Mar-Ecol Prog Ser 390:213–223

Van Deurs M, van Hal R, Tomczak MT, Jonasdottir SH, Dolmer P (2009) Recruitment of lesser sandeel *Ammodytes marinus* in relation to density dependence and zooplankton composition. Mar-Ecol Prog Ser 381:249–258

Wanless S, Frederiksen M, Daunt F, Scott BE, Harris MP (2007) Black-legged kittiwakes as indicators of environmental change in the North Sea: Evidence from long-term studies. Prog Oceanogr 72:30–38

Warwick R, Ashman C, Brown A, Clarke K, Dowell B, Hart B, Lewis RE Shillabeer N, Somerfield P, Tapp JF (2002) Inter-annual changes in the biodiversity and community structure of the macrobenthos in Tees Bay and the Tees estuary, UK, associated with local and regional environmental events. Mar-Ecol Prog Ser 234:1–13

Weijerman M, Lindeboom H, Zuur AF (2005) Regime shifts in marine ecosystems of the North Sea and Wadden Sea. Mar-Ecol Prog Ser 298:21–39

Willems W, Rees HL, Vincx M, Goethals P, Degraer S (2007) Relations and interactions between environmental factors and biotic properties. In: Rees HL, Eggleton JD et al (eds) Structure and dynamics of the North Sea benthos. ICES Coop Res Rep 288

Wilson B, Reid RJ, Grellier K, Thompson PM, Hammond PS (2004) Considering the temporal when managing the spatial: a population range expansion impacts protected areas-based management for bottlenose dolphins. Anim Conserv 7:331–338

Wittmann AC, Pörtner H (2013) Sensitivities of extant animal taxa to ocean acidification. Nature Clim Change 3:995–1001

Zühlke R (2001) Monitoring biodiversity of epibenthos and demersal fish in the North Sea and Skagerrak. EC Project 98/021. Monitoring Report 2001 to the Commission of the European Community

Environmental Impacts—Coastal Ecosystems

9

Jan P. Bakker, Andreas C.W. Baas, Jesper Bartholdy, Laurence Jones, Gerben Ruessink, Stijn Temmerman and Martijn van de Pol

Abstract

This chapter examines the impacts of climate change on the natural coastal ecosystems in the North Sea region. These comprise sandy shores and dunes and salt marshes in estuaries and along the coast. The chapter starts by describing the characteristic geomorphological features of these systems and the importance of sediment transport. Consideration is then given to the role of bioengineering organisms in feedback relationships with substrate, how changes in physical conditions such as embankments affect coastal systems, and the effects of livestock. The effects of climate change—principally accelerated sea-level rise, and changes in the wind climate, temperature and precipitation—on these factors affecting coastal ecosystems are then discussed. Although the focus of this chapter is on the interaction of abiotic conditions and the vegetation, the potential impacts of climate change on the distribution of plant species and on birds breeding in salt marshes is also addressed. Climate impacts on birds, mammals and fish species are covered in other chapters.

Electronic supplementary material Supplementary material is available in the online version of this chapter at 10.1007/978-3-319-39745-0_9.

J.P. Bakker (✉)
Groningen Institute for Evolutionary Life Sciences, University of Groningen, Groningen, The Netherlands
e-mail: j.p.bakker@rug.nl

A.C.W. Baas
Department of Geography, King's College London, London, UK
e-mail: andreas.baas@kcl.ac.uk

J. Bartholdy
Department of Geosciences and Natural Resource Management, University of Copenhagen, Copenhagen, Denmark
e-mail: jb@ign.ku.dk; jb@geo.ku.dk

L. Jones
Centre for Ecology and Hydrology, Bangor, UK
e-mail: lj@ceh.ac.uk

G. Ruessink
Department of Physical Geography, Faculty of Geosciences, Utrecht University, Utrecht, The Netherlands
e-mail: B.G.Ruessink@uu.nl

S. Temmerman
Ecosystem Management Research Group, Antwerp University, Antwerp, Belgium
e-mail: stijn.temmerman@uantwerpen.be

M. van de Pol
Department of Evolution, Ecology and Genetics, Research School of Biology, Australian National University, Canberra, Australia
e-mail: martijn.vandepol@anu.edu.au

M. van de Pol
Department of Animal Ecology, Netherlands Institute of Ecology (NIOO-KNAW), Wageningen, The Netherlands
e-mail: M.vandePol@nioo.knaw.nl

M. Quante and F. Colijn (eds.), *North Sea Region Climate Change Assessment*, Regional Climate Studies, DOI 10.1007/978-3-319-39745-0_9

9.1 Introduction

This chapter examines the impacts of climate change on the natural coastal ecosystems in the North Sea region. These comprise sandy shores and dunes and salt marshes in estuaries and along the coast. The chapter starts by describing the characteristic geomorphological features of these systems and the importance of sediment transport. Consideration is then given to the role of bioengineering organisms in feedback relationships with substrate, how changes in physical conditions such as embankments affect coastal systems, and the effects of livestock. The effects of climate change—principally accelerated sea-level rise, and changes in the wind climate, temperature and precipitation—on these factors affecting coastal ecosystems are then discussed. Although the focus of this chapter is on the interaction of abiotic conditions and the vegetation, the potential impacts of climate change on the distribution of plant species and on birds breeding in salt marshes is also addressed. Climate impacts on birds, mammals and fish species are covered in other chapters.

9.2 Geomorphology of Sandy Shores and Coastal Dunes

9.2.1 Distribution and Composition

Andreas C.W. Baas, Gerben Ruessink

The North Sea Basin is ringed by sandy shores and coastal dune fields developed from sedimentary deposits on nearly all sides, except the north-east. The sandy shores include most of the Belgian, Dutch, and Danish coasts as well as various parts of the English North Sea coast, where they alternate with muddy and soft rocky coasts, as well as gravel-dominated coastlines. Extensive dune fields are found along the entire coast of Belgium and the Netherlands, along the sequence of Wadden Sea barrier islands to the Elbe Estuary and up along the entire west coast to the northwest coast of Denmark. These sandy shorelines and dune fields constitute a regional complex developed out of sediment delivered mainly from the Rhine-Meuse delta into the coastal zone, transported via longshore currents following the sweep of the semi-diurnal tide anti-clockwise around the south-eastern coastlines of the North Sea Basin, and being accumulated and driven inland from the beaches by the dominant westerly winds. Coastal dunes along the western Danish coastline cover approximately 800 km^2 (Doody and Skarregaard 2007), while 254 of the 350 km of the coastline of the Netherlands is fronted by approximately 450 km^2 of coastal dune fields extending in some places up to 11 km inland (Doing 1995). Sand dunes also fringe most of the 65 km long coastline of Belgium, covering approximately 38 km^2 and with widths of a few kilometres to less than 100 m, although more than 50 % of the coast has been urbanised (Herrier 2008). The western side of the North Sea Basin is also lined by dune fields along the east coast of the United Kingdom, although these are of a less continuous nature (Doody 2013). Significant dune presence can be found along the northeast of East Anglia, along most of the Lincolnshire coast, isolated stretches along the Yorkshire and Northumbrian coast, and more extensive dune fields along the south-eastern coast of Scotland. The North Sea coastal dunes in the UK are smaller and less extensive than their continental counterparts, because sediment delivery to the coastal zone is from smaller regional catchments (the Thames, the Wash, the Humber), and the easterly winds that propel the dune development are generally weaker and less frequent. A tally of dune fields between Dover and Shetland included in the Sand Dune Vegetation Survey of Great Britain (Dargie 1993; Radley 1994) amounts to 93 km^2 (although a significant proportion of dune fields in Scotland are managed as golf courses).

9.2.2 Current Stressors and Management

In general, sandy shores are alongshore elongated sand bodies whose yearly to decadal evolution is primarily driven by waves and wind. Cross-shore and/or alongshore gradients in sand transport cause coastal morphology to change, reflected by erosion or accumulation of sand. According to a recent EU study of coastal geomorphology and erosion (Eurosion 2004), a large proportion (20–26 %) of the North Sea sandy shores is currently experiencing erosion and, as a consequence, is heavily affected by human activities, such as the presence of hard coastal defence measures (seawalls, groynes) or of regular sand nourishments. It is, however, important to realise that coastal erosion or accretion may vary on a wide range of temporal and spatial scales. A study by Taylor et al. (2004), for example, has shown that the bulk of the North Sea coast of England steepened due to erosion and coastal squeeze over the course of the entire 20th century. On a decadal scale, a coast may gently (typically, a few metres per year) accrete or erode, which is often due to small but persistent gradients in alongshore sediment transport. Shorter-term variations around the decadal trend are often significantly larger, and can be due to episodic erosion events (e.g. dune erosion during a storm) or to the alongshore migration of sand bodies (e.g. Ruessink and Jeuken 2002). Coastal changes may be due to long-period variability and oscillations in wave and storm climate unconnected to climate change (Hadley 2009).

The vast majority of coastal dune systems around the North Sea Basin are tightly managed and controlled for the purpose of various socio-economic and ecoservices (see Electronic (E-)Supplement S9). Along the coasts of Belgium and the Netherlands especially, the main purpose of management has been to preserve and, where possible, expand the sand volume of the foredunes to provide coastal flooding protection and more than 40 % of the foredunes have been artificially preserved or established (Arens and Wiersma 1994), usually by dense marram grass *Ammophila arenaria* planting. In the past two decades a more eco-centred concept has slowly been adopted to allow and encourage limited geomorphic re-activation of wind-blown erosion and bare-sand patches to develop a more varied and species-rich coastal dune environment (Arens and Geelen 2006). Other management practices include grazing activities on grey dunes to combat overgrowth of coarse shrubs (e.g. sea-buckthorn *Hippophae rhamnoides*) to preserve biodiversity (Boorman and Boorman 2001) as well as grazing and sod-cutting on species-rich dune grasslands to prevent grass-encroachment due to nitrification from atmospheric nitrogen deposition (Kooijman and Van der Meulen 1996).

9.2.3 Expected Impacts of Climate Change

Climate change could affect both marine and aeolian boundary conditions. In particular, climate change impacts might be felt through accelerated sea-level rise (SLR) and in modifications in the number, severity and location of extra-tropical storms, with associated changes in wind, wave, precipitation and temperature patterns. The potential consequences of climate change on the North Sea Basin sandy shores and coastal dune systems may be separated into direct geomorphic impacts relating to sea forcing (accelerated SLR and changes in wave climate), wind forcing (changes in weather patterns and wind climate), and potential changes in vegetation character and distribution related to overall climatic changes (temperature and precipitation). Consequences for the vegetation are discussed in Sect. 9.3. Some of these direct impacts may be aggravated by adjustment and intensification of human exploitation and management practices.

9.2.3.1 Sea-Level Rise and Wave Characteristics

Historic tide gauge data for the North Sea region from 1900 to 2011 show a mean SLR of 1.5 ± 0.1 mm year^{-1} (Wahl et al. 2013), slightly below the global mean SLR of 1.7 ± 0.3 mm year^{-1} during the latter half of the 20th century (Church and White 2006). Whereas SLR globally appears to have accelerated to 3.3 ± 0.4 mm year^{-1} from 1993 to 2009 (Ablain et al. 2009, see also Chap 3), the Wahl et al. (2013) study suggests that North Sea SLR acceleration in recent decades is not abnormal, but comparable to that of other periods in the last 200 years. Within the North Sea region local differences have been recorded (as indicated in Table 9.1 in Sect. 9.6.1). While the Intergovernmental Panel on Climate Change (IPCC) in its latest assessment (Van Oldenborgh et al. 2013) projects a global mean SLR of 74 cm by 2100 (at a rate of 11 mm year^{-1}), for the 'business-as-usual' Representative Concentration Pathway scenario (RCP8.5), other studies that include semi-empirical forcing models have projected SLR of between 30 and 180 cm by 2100, depending on the model variant (Nicholls and Cazenave 2010). Recent high-end projections suggest that a 1.25 m SLR in the North Sea may be possible by 2100 (Katsman et al. 2011), although the exact magnitude of the rise strongly depends on underlying modelling assumptions.

Studies on the significant wave height H_s in the North Sea (e.g. Grabemann and Weisse 2008; De Winter et al. 2012) project no to small changes along the Dutch-German coast, with magnitudes depending on the type of general circulation model (GCM) or regional climate model (RCM) used and the particular greenhouse-gas emissions scenario adopted, as for example in Fig. 9.1. For example, Grabemann and Weisse (2008), who used the HadAM3H and the ECHAM4/OPYC3 GCM with SRES scenarios A2 and B2, projected a 0.1–0.3 m increase in the 99th percentile of H_s in front of the Dutch coast by the end of the 21st century, while De Winter et al. (2012), using the ECHAM5/MPI-OM and the SRES A1B scenario, found no detectable change in mean wave conditions. This would seem to indicate that model uncertainty in the prediction of H_s is larger than the emission-induced uncertainty, as was found in general for global projections in the latest IPCC assessment (Van Oldenborgh et al. 2013), which reports low confidence in wave projections because of "uncertain storm geography, limited number of model simulations, and the different methodologies used to downscale climate model results to regional scales".

It is important to consider that the wind, and hence wave climate, shows strong natural variability, which poses a difficulty in detecting climate-change induced trends that are smaller than this natural variability. This is especially relevant to wind and wave conditions with high return periods (for example, 1:1–1:10-year return values, or even rarer) that may be most relevant to coastal erosion. Using the 17-member ESSENCE ensemble (Sterl et al. 2008), De Winter et al. (2012) projected a 0.3–0.6 s decrease in the annual wave period T in front of the Dutch coast for the period 2071–2100, as well as a shift in the wave direction of the annual H_s maxima from north-west to south-west. The decrease in T is induced by this wave-angle shift and associated shifts in the fetch; accordingly, the same shift may lead to an increase in T elsewhere in the North Sea Basin. H_s and T with higher return periods, up to 10,000 years, were

Table 9.1 Regional variation in mean sea-level rise (SLR) within the North Sea region (see also Chap. 3)

Area	SLR mm year^{-1}	Period	Source
Europe	1.7	1900–2000	EEA (2012)
	3	1990–2010	EEA (2012)
Wadden Sea	1–2	1900–2000	Oost et al. (2009)
UK east coast	0.5–2.5	1900–2000	Woodworth et al. (2009)
English channel	0.5–2.5	1900–2000	Haigh et al. (2011)
The Netherlands	2.5	1900–2000	Katsman et al. (2008)
German bight	1.2–2.4	1937–2008	Wahl et al. (2011)
	1–2.8	1951–2008	Wahl et al. (2011)
	2.5–4.6	1971–2008	Wahl et al. (2011)
Lower Saxony, Germany	1.7	1936–2008	Albrecht et al. (2011)
Schleswig-Holstein, Germany	2	1936–2008	Albrecht et al. (2011)
Skallingen, Denmark	2.3	1931–1999	Bartholdy et al. (2004)
	5	1980–2000	Bartholdy et al. (2004)

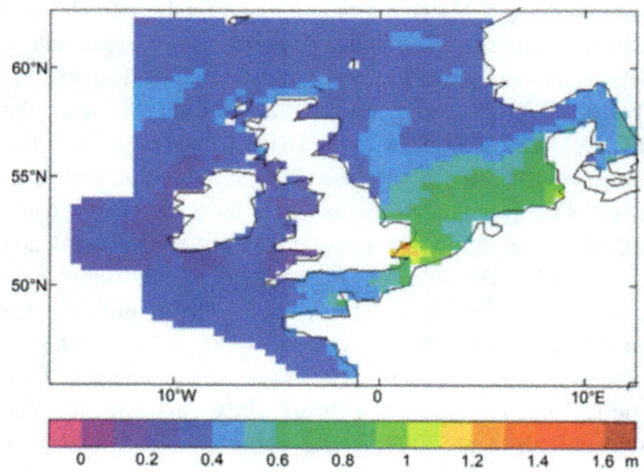

Fig. 9.1 Increase in the height (m) of a 50-year return period extreme water level event, due to combined changes in atmospheric storminess, mean sea-level rise, and vertical land movements, for the SRES A2 scenario, based on the HadRM3H regional atmospheric model and the POL storm-surge model (Lowe and Gregory 2005)

not projected to change significantly at the 95 % confidence level. This also applies to storm surges with similar high return periods (Sterl et al. 2008). In a recent CMIP5 model-comparison study of wind extremes in the North Sea Basin, De Winter et al. (2013) found no changes in annual maximum wind speed or in wind speeds with lower return frequencies for the 2071–2100 period; however, they did find an indication that the annual extreme wind events are coming more often from westerly directions. Thus, these CMIP5 results are not notably different from all earlier CMIP3 (e.g. De Winter et al. 2012) model projections.

The quantitative prediction of the response of sandy shores to climate-change induced effects in the boundary conditions is still under development. The most often used model for the response to accelerated SLR is the Bruun rule (Bruun 1962), a simple two-dimensional mass conservation principle that predicts a landward and upward displacement of the cross-shore profile with SLR (Fig. 9.2). Coastal recession is simply expressed as the product of SLR and the active profile slope, giving typical projected recession distances of 50–200 m depending on profile slope. Although the Bruun rule was routinely used in past decades, its usefulness as a predictive tool is highly controversial (Pilkey and Cooper 2004). For example, the Bruun rule does not include any three-dimensional variability, such as found near engineering structures and tidal inlets, and can thus not be applied in areas with notable gradients in alongshore sediment transport. Also, the Bruun rule predicts sand to be moved offshore during SLR, while overwash and aeolian processes obviously transport sand onshore. Advances have been made in developing more comprehensive conceptual models for understanding and predicting shoreline change in terms of the so-called Coastal Tract (Cowell et al. 2003), but this concept has been aimed at time scales of hundreds to thousands of years. Other more realistic rule-based approaches for predicting shoreline change have been provided by Ranasinghe et al. (2012) and Rosati et al. (2013), but these have not yet been tested on North Sea sandy shores. Beach profile adjustments to accelerated SLR are more likely to involve significant alongshore components and variations as well as greater sensitivities to local changes in sediment budgets (Psuty and Silveira 2010). Even though projections of the North Sea wave climate suggest no to minor change, it is possible that accelerated SLR and changes in wave direction may still aggravate coastal erosion. A modelling study for the UK East Anglian coast, a site with notable alongshore variability in shelf bathymetry, illustrates that

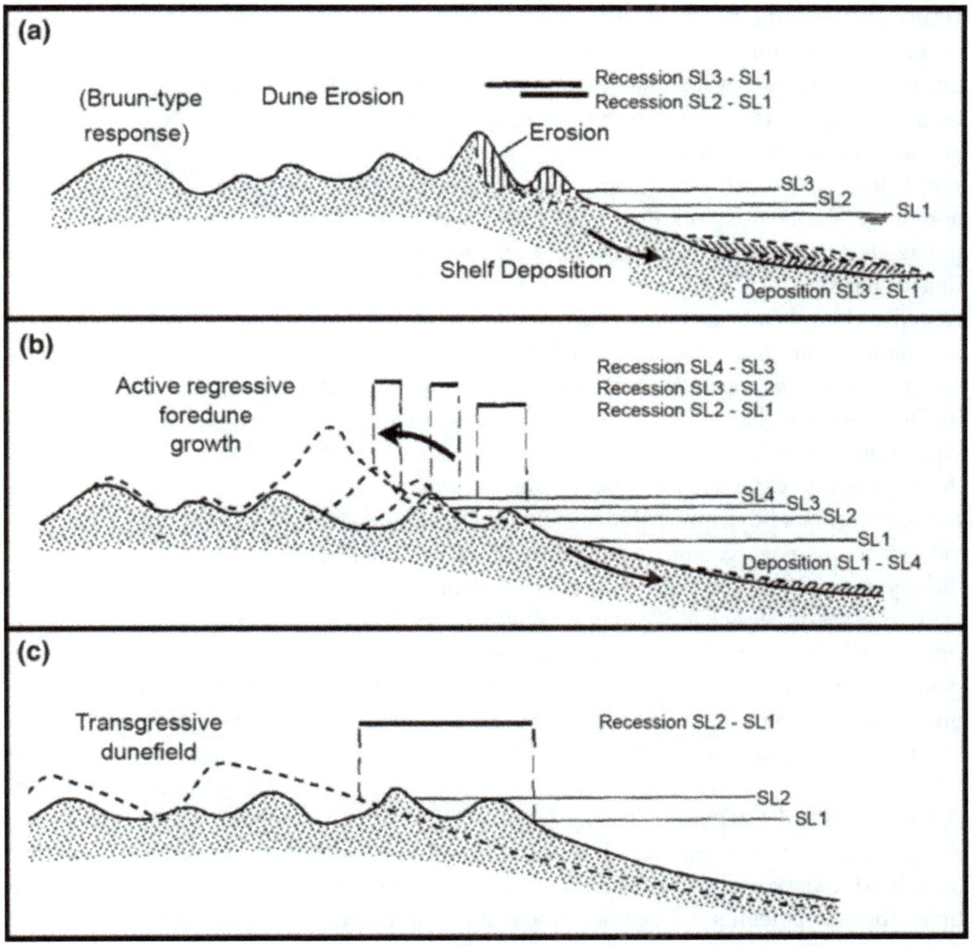

Fig. 9.2 Three potential cross-shore responses to sea-level rise (SLR): **a** Bruun-type response (much disputed) where sediment is redistributed to reach a new equilibrium beach profile; **b** reactivation, growth, and inland migration of foredunes; **c** vegetation cover becomes ineffective leading to a large-scale transgressive dune field (Carter 1991)

inshore wave statistics are sensitive to the trend in SLR and that frequency of occurrence of extreme inshore wave conditions may increase with SLR rates higher than 7 mm year^{-1} (Chini et al. 2010). This, coupled with an increasing occurrence of high water levels by accelerated SLR, may lead to enhanced beach and dune erosion even if wind and wave characteristics remain unaltered under projected climate change; the same effect could also endanger the safety of existing hard coastal defence mechanisms. In a case study for the Dutch coast, De Winter (2014) projected an increase in dune erosion volume by up to 30 % because of a 1-m SLR under unchanged extreme (1:10,000 year) wave conditions. The same case study also illustrated that a change in wave direction by several tens of degrees could lead to a similar increase in dune-erosion volumes as predicted for an approximately 0.4-m rise in sea level, indicating that changes in wind and wave direction are also critically important to coastal response. Such directional changes will also affect the magnitude and, potentially, the sign of gradients in alongshore sand transport, and hence beach width. Current dune-erosion models are primarily based on data collected in laboratory experiments; extensive field validations have, however, not been performed. Furthermore, the models lack descriptions of post-storm coastal recovery by aeolian processes and thus provide an erosion-biased view of coastal evolution. Dunes and beaches are linked in a dynamic and complex sediment exchange system, where losses of sediment from foredunes during storms alternate with inter-storm deposition gains (e.g. Keijsers et al. 2014), potentially leaving the overall shoreline position unchanged in the long term. More frequent storm erosion of the foredune toe may therefore not necessarily result in permanent loss of sediment from the coastal system.

The precise response of the nearshore zone to climatic change and its consequent impact on adjacent coastal dunes is therefore ambiguous and probably subject to substantial regional variation, as in some areas the hydrodynamic boundary conditions may not change very much, while in other areas the coastal sediment budget system will be undergoing complex adjustments. Paradoxically, while sandy beaches in the worst case may get narrower and squeezed between human pressure and accelerated SLR (Carter 1991; Schlacher et al. 2007), the remobilisation and increased dynamics in sediment exchange across the nearshore profile as it is adjusting to SLR may in fact yield an opportunity for reinvigorating coastal dune development

(Psuty and Silveira 2010), as illustrated in Fig. 9.2. Just as during the mid-Holocene, SLR and coastal regression may create opportunities for foredune expansion along some parts of the coastline. This may result from supply of increased amounts of sediment delivered from the adjusting beach profile into the coastal dune system, possibly aided by larger and more frequent storm breach of the dunes, and the potential for greater aeolian sand transport activity in the inland parts of the dune field. This type of landscape response does, however, require the availability of suitable accommodation space for the coastal dune system and in many regions a net landward migration of dunes is arrested by infrastructure and built environment, thus leading to a 'sand dune squeeze' (Doody 2013) and an eventual loss of dune habitat. Coastal squeeze also operates within sites, as the younger, more dynamic habitats including foredunes and yellow dunes may be squeezed against more stable fixed dune grassland, or scrub which occurs in older hind-dune areas. These successionally young habitats are particularly important for many of the dune rare species and, while it is possible for natural remobilisation to occur, it is unlikely under current climatic conditions, since dune mobility is strongly coupled to climate (Clarke and Rendell 2009).

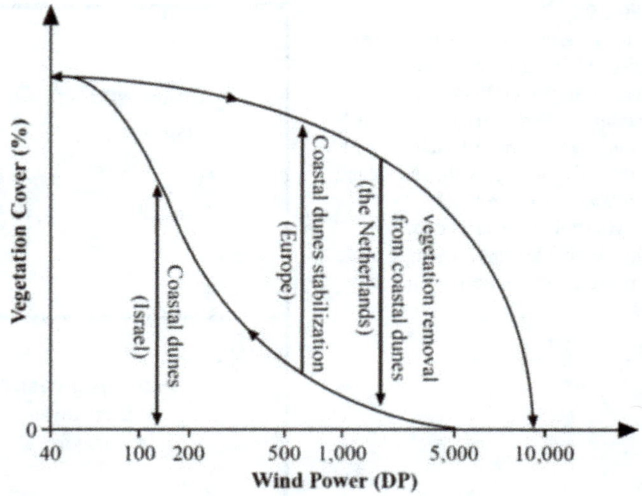

Fig. 9.3 Hysteresis in the relationship between wind power (quantified as a Drift Potential) and the vegetation cover in a coastal dune field (Tsoar 2005)

9.2.3.2 Wind Forcing

As previously mentioned, an increase in storminess and associated changes in wind conditions may have implications for the potential of aeolian reactivation of coastal dunes, as well as the wind-blown transport of the additional sediment mobilised as part of beach profiles that are adjusting to accelerated SLR. Regional climate model projections by Beniston et al. (2007) suggest a 5 % increase in the 90th percentile of daily maximum wind speed over most of the North Sea Basin in winter (Fig. 9.3), with storm winds coming from more north-westerly directions.

These projections conflict with the more recent prediction of more south-westerly storm winds by De Winter et al. (2012, 2013). The IPCC meanwhile indicates a "substantial uncertainty and thus low confidence in projecting changes in NH [northern hemisphere] winter storm tracks, especially for the North Atlantic basin" (Van Oldenborgh et al. 2013). The impact of any potential changes in wind climate on the coastal dune fields in the region has not been considered in detail. A shift in dominant wind direction may have a significant impact on the delivery of sediment from the beach into the foredunes due to changes in sub-aerial fetch distance as a function of the local orientation of the coastline (Bauer et al. 2009). For parts of the Dutch and Danish coastline that are aligned SW–NE, for example, a shift to more north-westerly storm winds may therefore result in comparatively shorter fetch distances and less opportunity for aeolian sand transport into the foredunes, as the north-westerly winds will be blowing more perpendicular to the coast. The contemporary geomorphology of the secondary dunes, meanwhile, is aligned to the typical south-westerly winds of the past and the projected shift in wind direction is likely to make it harder for aeolian activity to reactivate dormant blow-outs and parabolic dunes, as it will be acting perpendicular to their main axes and facing greater topographic roughness. Finally, even though there may be an increase in 'drift potential' (the wind power available for aeolian sand transport), its relationship with vegetation cover and associated reactivation of dormant sand dunes shows a distinct hysteresis (Tsoar 2005; Yizhaq et al. 2007), as illustrated in Fig. 9.3. While the future wind climate conditions may become equivalent to those of bare and actively migrating coastal dunes in present-day Israel, for example (as reported in the studies cited here), most North Sea Basin coastal dune fields are situated on the firmly stabilised limb of the hysteresis curve and a reactivation toward active dunes is likely to require more than just an increase in wind power. Recent attempts in the Netherlands to reactivate blowouts and parabolic dunes by removing existing vegetation have so far met with mixed success, with many opened-up sand areas quickly being recolonised and overgrown with vigorous vegetation within a few years (Arens et al. 2013), accelerated by atmospheric nitrogen deposition (Jones et al. 2004). The future change in wind climate may mean that such reactivation attempts become more successful. If not, remobilisation must be assisted by direct management intervention (Arens and Geelen 2006; Jones et al. 2010).

9.3 Ecology of Sandy Shores and Dunes

Laurence Jones

9.3.1 Climate Change and Dune Ecology

European dune systems have shown constant change over time, with human influences predominating over the last few centuries (Provoost et al. 2011). However, climate change will become an increasingly important influence on dune ecosystems. Climate change is likely to affect coastal dunes in several ways. There will be direct loss of habitat due to accelerated SLR and coastal erosion (see Sect. 9.2), and changes in the climate envelopes affecting the distribution of plant and animal species. There will also be indirect effects through changes in underlying ecosystem processes. These include effects on competition, mediated via plant growth, but also effects on soil development, and groundwater systems which influence the dune wetland communities. This double impact through habitat loss as well as altered climate means that coastal habitats are more sensitive to climate change than the majority of other terrestrial ecosystems. Management of dunes, for example by livestock grazing, may interact with the effects of climate change on the vegetation, but is not discussed here.

9.3.2 Effects of Climate Change on Dry Dune Habitats

General effects of climate change include a lengthening of the growing season, leading to more plant growth where rainfall is not limiting. This is likely to benefit faster growing graminoids and nitrophilous species, and lead to declines in rare species. However, in the drier parts of the North Sea coast such as the Netherlands, increased summer drought may actually reduce the effective growing season, potentially reducing the dominance of some species and allowing other species to flourish. Based on the relationship between drought stress and plant compositional change in Dutch dry dunes (Fig. 9.4), Bartholomeus et al. (2012) suggested that projected increases in the severity of drought may lead to a 15 % increase in the fraction of xerophytes, with potential feedbacks on dune recharge due to reduced evapotranspirative losses (Witte et al. 2012). Levine et al. (2008) predicted an increase in rare dune annuals and Witte et al. (2012) suggested a shift to more xerophytic mosses and lichens and increased cover of bare sand as a consequence of increasing summer drought, with the possible invasion of xeric Ericaceous and broom *Cytisus* species from more southerly European countries. Severe drought has been shown to negatively affect net primary production (NPP) on a European scale (Ciais et al. 2005), and such droughts are predicted to become more frequent.

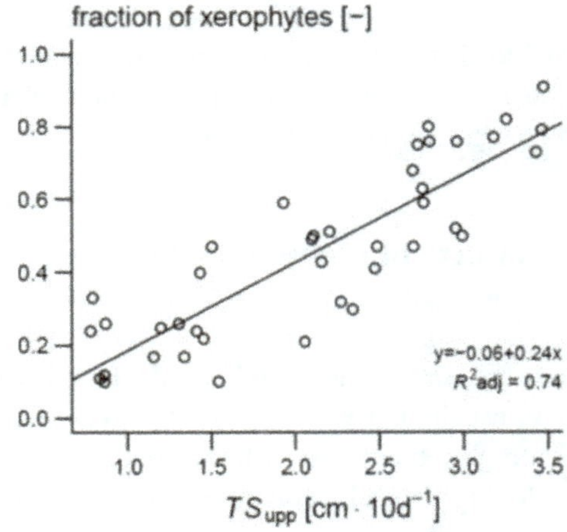

Fig. 9.4 Fraction of xerophytes as a function of the drought stress index TS_{upp} (uppermost transpiration stress) (after Bartholomeus et al. 2012)

The final effects of climate change that may directly impact upon the vegetation that covers most parts of the coastal dune system are changes in temperature and precipitation, and it is here that the great variety of potential climate effects may have a mixture of conflicting positive and negative biological impacts. The Holocene stratigraphic record suggests that warmer and slightly wetter climates in the past have usually resulted in dune stabilisation and full vegetation cover, whereas colder periods with less precipitation, such as the well-documented Little Ice Age, have been associated with more active dune mobility and aeolian sand transport (Pye 2001). The major dune-forming grasses, such as marram grass and sand couch *Elytrigia juncea*, may generally grow better in a CO_2-enriched environment, and may also benefit from the projected 10–15 % increase in precipitation forecast for north-western Europe (Carter 1991). This precipitation is expected to arrive in more extreme and variable events, however, and the potentially drier summers with higher temperatures and greater heatwave risks may also result in greater wildfire incidence in the grasslands covering much of the coastal dune fields, exposing soil to wind erosion and enabling potential dune reactivation. Changes in seasonal temperature and rainfall patterns may also induce changes in species composition. Greater winter precipitation appears to facilitate scrub proliferation and overgrowth, such as sea-buckthorn, while summer droughts have an adverse impact on species diversity in dune slacks through lowering of the water table (Doody 2013). For vegetation on the back beach and foredune toe, meanwhile, simulation studies suggest that accelerated SLR and the narrowing of the beach may constrain plants to such a narrow area that successional processes break down (Feagin et al. 2005). Many of these effects may be difficult to distinguish from more direct anthropogenic impacts, such as grazing, nitrification, groundwater extraction, and changes in land management.

9.3.3 Effects of Climate Change on Wet Dune Habitats

Dune wetlands (slacks) are low-lying depressions between dune ridges, usually in seasonal contact with the water table. They are a highly biodiverse habitat, containing many rare species including plants, invertebrates, and vertebrates (Jones et al. 2011). In addition to the direct influence of temperature, rainfall and length of growing season on plant growth, dune wetlands are highly sensitive to changes in hydrological regime. They are often nitrogen (N) and phosphorus (P) co-limited and dune slack vegetation is dependent on hydrological regime (groundwater level, and seasonal and interannual fluctuations) and groundwater chemistry, particularly buffering capacity (Grootjans et al. 2004). The majority of dune slack plant species of high conservation value are dependent on early successional dune slacks with a high buffering capacity and low nutrient status, and which disappear as slacks decalcify or accumulate nutrients.

9.3.3.1 Groundwater Level

Both accelerated SLR and coastal erosion will lead to a change in dune groundwater tables due to impacts on the hydraulic gradient, with water tables rising (SLR), or falling (steepening of the hydraulic gradient due to coastal erosion). In Denmark it is suggested that slacks which are currently dry will become wetter due to accelerated SLR (Vestergaard 1997). However, some studies suggest that changes in recharge due to altered spatial and seasonal patterns of rainfall and evapotranspiration will have greater effects on groundwater levels than SLR or coastal erosion (Clarke and Sanitwong Na Ayutthaya 2010). In north-west England, dune water tables are predicted to fall over the next 50 years due to negative climate effects on recharge (Clarke and Sanitwong Na Ayutthaya 2010), and this is likely to apply to the majority of English dunes on the North Sea coast. In the Netherlands, recharge is predicted to change relatively little, with either slight decreases or alternatively moderate increases if feedbacks of reduced vegetation cover on the water balance are taken into account (Witte et al. 2012). A modelling study in Belgium assumed there would be increases in recharge to dune groundwater, driven primarily by increased winter precipitation (Vandenbohede et al. 2008). There is as yet little consensus on the likely effects of climate change on dune aquifer recharge and therefore a need for further work on this topic. There have also been very few studies on the impacts of changing water tables on the flora and fauna of dune wetlands. Van Dobben and Slim (2012) using a study of land subsidence due to gas extraction on the barrier island of Ameland in the Netherlands as an analogue for accelerated SLR, suggested that dune vegetation would move towards dune slacks (i.e. become wetter). In the UK, Curreli et al. (2013) showed that differences in mean water table level of only 20 cm differentiate between dune slack vegetation communities, with 40 cm separating the wettest and the driest communities. Based on predictions of falling groundwater levels for a west coast UK dune system (Clarke and Sanitwong Na Ayutthaya 2010), although outside the area of this assessment, Curreli et al. (2013) suggested that climate change may alter what are currently wet slack communities to dry slack or even dry dune grassland by the 2080s (Fig. 9.5). Changes in water table levels are also likely to affect breeding success of the Annex II listed natterjack toad *Epidalea calamita* which requires slacks to dry out in summer to avoid colonisation by fish predators and competitors, but needs water levels to be maintained long enough for the tadpoles to develop into toadlets. April to July is the critical breeding time for this species.

9.3.3.2 Groundwater Chemistry

Dune groundwater composition is sensitive to atmospheric N-deposition and nitrogen and phosphorus inputs from other sources. In dune groundwater, it has been suggested that above total inorganic nitrogen concentrations of 0.2–0.4 mg TIN l^{-1} there may be adverse impacts on dune slack species and soils (Davy et al. 2010), although this threshold has not been tested empirically in dunes.

Although there is relatively little research into the effects of climate change on dune groundwater chemistry, it could be expected to change in the following ways. Rates of mineralisation of soil organic matter are likely to increase due to higher temperatures (Emmett et al. 2004) and an extended growing season (Linderholm 2006) in the majority of North Sea coastal areas, particularly in the north. This will lead to increased production of both nitrates (NO_3) and dissolved organic nitrogen (DON), inevitably increasing N-leaching fluxes to the groundwater. This may not occur in the southern North Sea area however, due to increased soil-moisture deficits in summer acting to reduce mineralisation. In areas with falling groundwater levels due to climate change, or where there is an influence of external groundwater or surface waters on a site, there is potential for nutrient concentrations entering the site to be higher if these source waters become concentrated due to higher evapotranspiration rates and subsequent reduced runoff and lower river flows. Such a mechanism is suggested for altered water composition of the River Meuse under climate change (Van Vliet and Zwolsman 2008). The opposite may occur in the north of the region where runoff is expected to increase (DiPOL 2012).

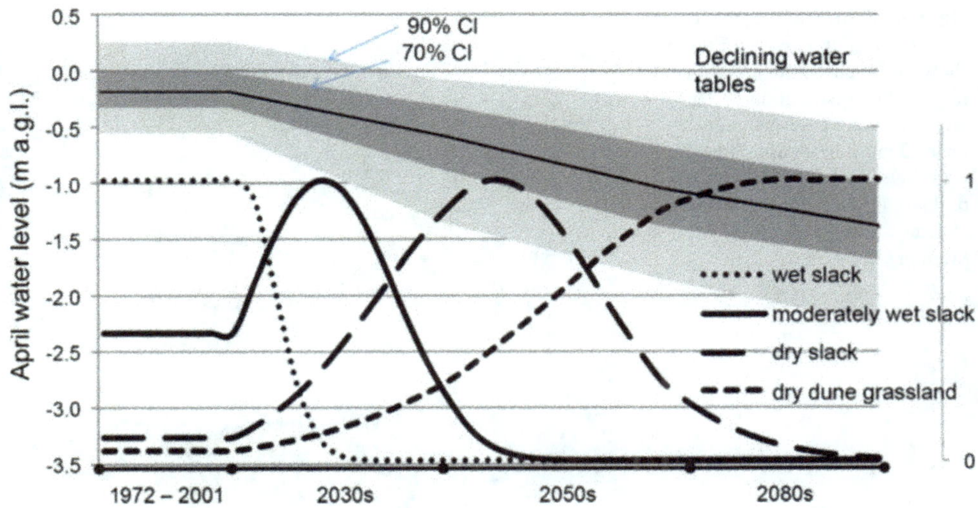

Fig. 9.5 Projections of April groundwater level to the 2080s for a UK site under UKCIP02 medium-high emissions scenario (SRES A2) (Hulme et al. 2002), showing the 70 and 90 % confidence intervals (CI), and corresponding frequency distributions for dune slack communities based on those water levels (after Curreli et al. 2013)

9.3.3.3 Interactions with Hydrological Management

Hydrological management both on- and off-site can affect water composition. A lowering of the water table due to drainage, water abstraction or other causes such as high evapotranspiration from forest, may allow groundwater influx from other sources such as streams or drainage ditches bordering the site or groundwater from inland. The danger is greatest in late summer when dune groundwater tables are at their lowest. These other hydrological inputs to the groundwater may contain high levels of nutrients or other chemicals which affect water composition, particularly if they drain agricultural land. Alterations to the management of hydrological regimes in response to climate change therefore also have the potential to impact dune groundwater chemistry.

9.3.4 Climate Change Effects on Individual Species

Climate change is likely to affect the distributions of individual dune plant species, although there have been relatively few studies on this topic. In Denmark, a study looking at spatial analogues of predicted climate in 2100 suggests that coastal heaths and sand dune vegetation are likely to be vulnerable, particularly the communities "decalcified fixed dunes with *Empetrum nigrum*" (type 2140 according to Annex I of the Habitats Directive; European Commission 1992) and "fixed coastal dunes with herbaceous vegetation" ('grey dunes', type 2130) (Skov et al. 2009). The study lists 17 species which may disappear including crowberry *Empetrum nigrum* and sea brookweed *Samolus valerandi*, and 25 coastal species likely to move north into Denmark, some of which may become problem species. In the UK, the Monarch programme modelled changing climate envelopes for a wide variety of species, including some dune species. Generally species ranges shifted northwards and westwards with marsh helleborine *Epipactis palustris* and the natterjack toad potentially gaining climate space in the UK, while the variegated horsetail *Equisetum variegatum* which has a predominantly northern distribution in the UK would lose climate space (Harrison et al. 2001). Note these are predicted changes in climatic envelopes, and do not take into account the subtleties of changing recharge which might affect dune groundwater levels in a different way (see Sect. 9.3.3.3). The more recent BRANCH project modelled changing climate space of 386 plant and animal species, and predicted that the fen orchid *Liparis loeselii* was likely to lose more than 90 % of its climate space by 2080 (Berry et al. 2007). Metzing (2010) modelled the northward shift of the dune grass *Leymus arenarius* in Europe, which would completely disappear from current parts of its range in the southern North Sea areas of Germany, the Netherlands, Denmark and England under the worst-case scenario by 2050 (Fig. 9.6).

9.3.5 Other Climate Change Impacts

9.3.5.1 Invasive Species

There is much evidence in aquatic systems that invasive species may be favoured by climate change. In terrestrial systems, there is far less evidence for this. However, certain plant species such as bird cherry *Prunus serotina* in the Netherlands and Belgium (Baeyens and Martínez 2004) and Japanese rose *Rosa rugosa* in Germany (Isermann 2008) have become highly invasive in dune systems, despite being present on the systems for many decades. It is unclear why their extent has suddenly increased, but changes in climate (warmer winters and fewer severe frosts are one suggestion), while the combination of changing climate and elevated nutrient levels in dune soils due to N-deposition may be a contributory factor. Although invasive species may initially enhance species richness, once tall species become dominant they can outcompete low-statured species, and decrease species richness.

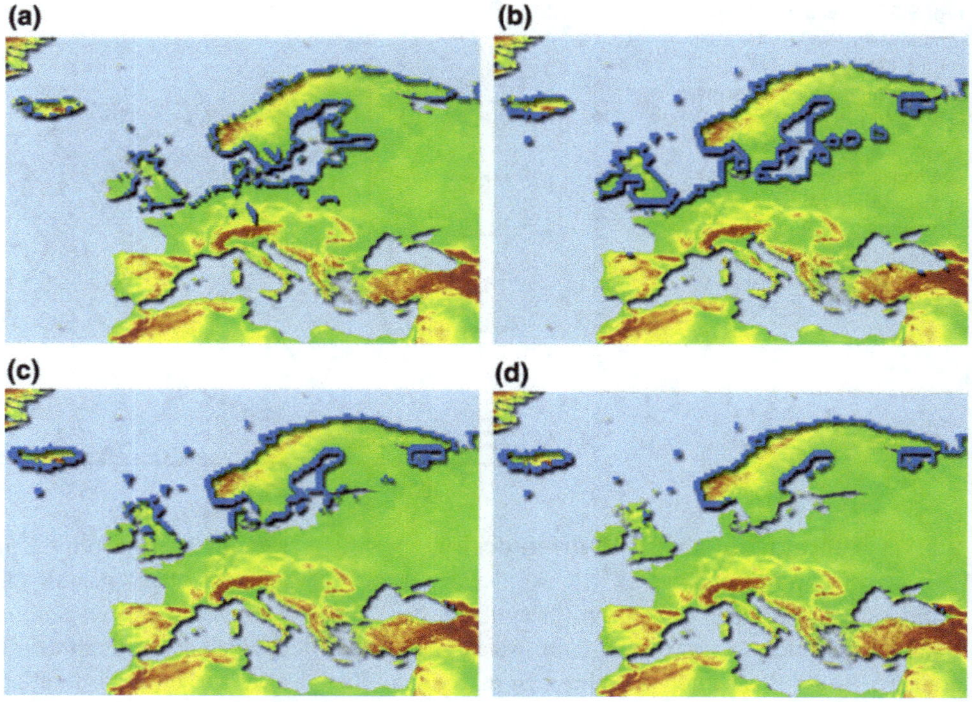

Fig. 9.6 Distribution of the dune grass *Leymus arenarius* in Europe showing: **a**. current distribution, **b**. modelled distribution under recent climate, **c**. distribution under 2050 a best-case climate warming of 1.5 °C and **d**. distribution under a worst-case climate warming of 2.5 °C (Metzing 2010)

9.3.5.2 Impacts on Soils

Where higher temperatures cause greater plant growth, the increased plant production will stimulate soil development through greater litter inputs. Rates of soil organic matter accumulation represent a balance between plant productivity and rates of organic matter decomposition. While the net balance is unknown for many habitats, a dune chronosequence study over the last 60 years linked faster rates of soil development to periods of higher temperatures, and to periods with lower rainfall (Fig. 9.7) (Jones et al. 2008); changes were also correlated with increasing N-deposition. Other changes in soils may include decalcification rates. The rate of leaching of carbonates from dune soils is largely a function of rainfall, therefore changes in rainfall are likely to lead to increases or decreases in decalcification rates of dune soils.

9.3.5.3 Atmospheric Nitrogen Deposition and Interactions with Climate Change

Atmospheric deposition remains a significant source of nitrogen input in many countries around the North Sea, particularly the Netherlands and Denmark. Recent trends in N-deposition in Europe have shown some decline in NO_X emissions and are forecast to decrease further for oxidised nitrogen, but only slightly for reduced nitrogen (Winiwarter et al. 2011). Climate is a strong determining factor in the balance of wet and dry N-deposition, and involves complex interacting processes. Relatively little is known about how climate change may affect N-deposition but is not expected to alter deposition of oxidised nitrogen much (Langner et al. 2005), and it is generally assumed that changes in emissions will have a greater impact on atmospheric deposition than any climate-related effect (Mayerhofer et al. 2002).

Along gradients in atmospheric deposition, studies have shown increased plant production and a decrease in species richness in dunes with higher atmospheric deposition (Fig. 9.8) (Jones et al. 2004; Hall et al. 2011). An increase in biomass occurs above critical levels of 10–20 kg N ha^{-1} $year^{-1}$ in dry dune communities (EUNIS types B1.3, B1.4, B1.5) and above 10–25 kg N ha^{-1} $year^{-1}$ in dune slack

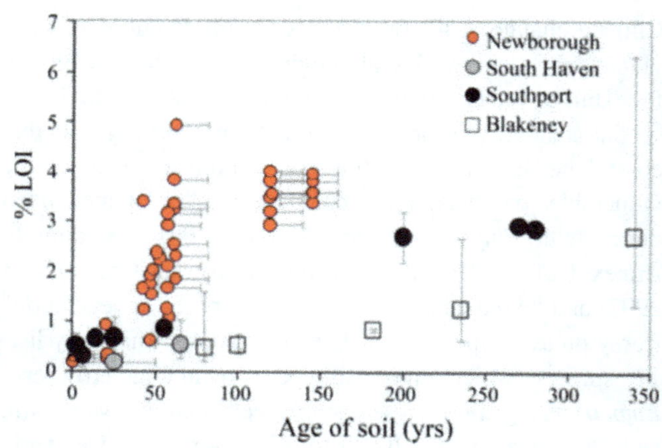

Fig. 9.7 Changes in soil organic matter accumulation in a recent dune soil chronosequence in the UK (Newborough), compared with older UK published chronosequence studies (South Haven 1960; Southport 1925; Blakeney 1922), after Jones et al. (2008)

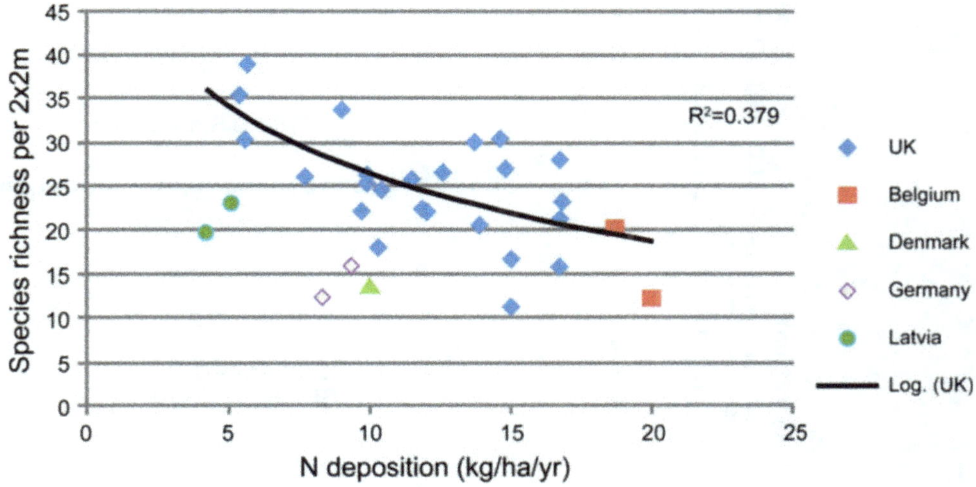

Fig. 9.8 Change in species richness of de-calcified dunes in north-western Europe along a gradient of atmospheric nitrogen (N) deposition (after Hall et al. 2011)

communities (EUNIS type B1.8) (Bobbink et al. 2010). The atmospheric deposition of nitrogen in combination with the release of phosphorous from enhanced mineralisation may result in increased plant production and further spread of grasses (Kooijman et al. 2012).

There is little direct or indirect evidence from dunes of interactions between climate change and N-deposition, although studies in other ecosystems suggest that the combination of higher temperatures and increased rainfall, which are both projected to occur in the northern North Sea dune systems, will lead to faster mineralisation of soil organic matter (Rustad et al. 2001), and therefore faster N-cycling. This may remobilise stored nitrogen in soils and increase leaching of dissolved organic nitrogen (DON) into dune groundwater. There are indications that higher DON concentrations in dune groundwater, which thereby increase the negative impacts of N-deposition are correlated with higher N-deposition (Jones et al. 2002), suggesting that impacts could be worse in areas of high nitrogen inputs. However, in the southern North Sea area where summer soil moisture deficits act to reduce mineralisation rates, there may be less N-leaching into groundwater as a result.

9.4 Fine-Grained Sediment Transport and Deposition in Back-Barrier Areas

Jesper Bartholdy

9.4.1 Physical Conditions

As a shallow semi-enclosed shelf sea, the North Sea imports fine-grained sediment (silt, clay, organic material) from the adjacent North Atlantic. The net import across the transect between Scotland and Norway each year is estimated at 7 million tons and through the Channel at about 14 million tons (Pohlmann and Puls 1993). Due to the generally anti-clockwise circulation in the North Sea, the sediment contribution from the Channel is carried northward where it mixes with sediment of terrestrial origin carried into the North Sea from rivers draining north-western Europe. Part of this flux of fine-grained sediment passes the German Bight close to land and then continues northward as part of the Jutland Current. During this passage, the tidal exchange between the North Sea and the extensive barrier system of the Wadden Sea induces a net landward transport of sediment into the sheltered tidal areas behind the barrier islands.

This import is due to lag effects first described by Postma (1954, 1961, 1967) and Van Straaten and Kuenen (1957, 1958). At high tide, the shift between flood and ebb is slower and involves larger excursions of the water, than the shift between ebb and flood at low tide. Together with the fact that it takes time for a suspended particle to settle out during slack water (*settling lag*) and that erosion/resuspension of a settled fine-grained particle—due to adhesion—demands stronger currents than those in which the particle settled out (*scour lag*), this asymmetry causes a landward shift of sediment for each tidal period. The effect is enhanced by the apparent paradox that the average water depth in a tidal area is smallest during high tide and largest during low tide. A randomly located suspended particle in the exchanged water mass is therefore less prone to settle out during low tide, when only the relatively deep channels are inundated than it is during high tide, where the whole tidal area is inundated with a relatively small mean water depth. A quantitative model of the involved processes during settling and resuspension of sediment particles in tidal currents have shown that *scour lag* is by far the most important mechanism concerning import of fine-grained sediment into the Wadden Sea (Bartholdy 2000). Calibrated on the basis of

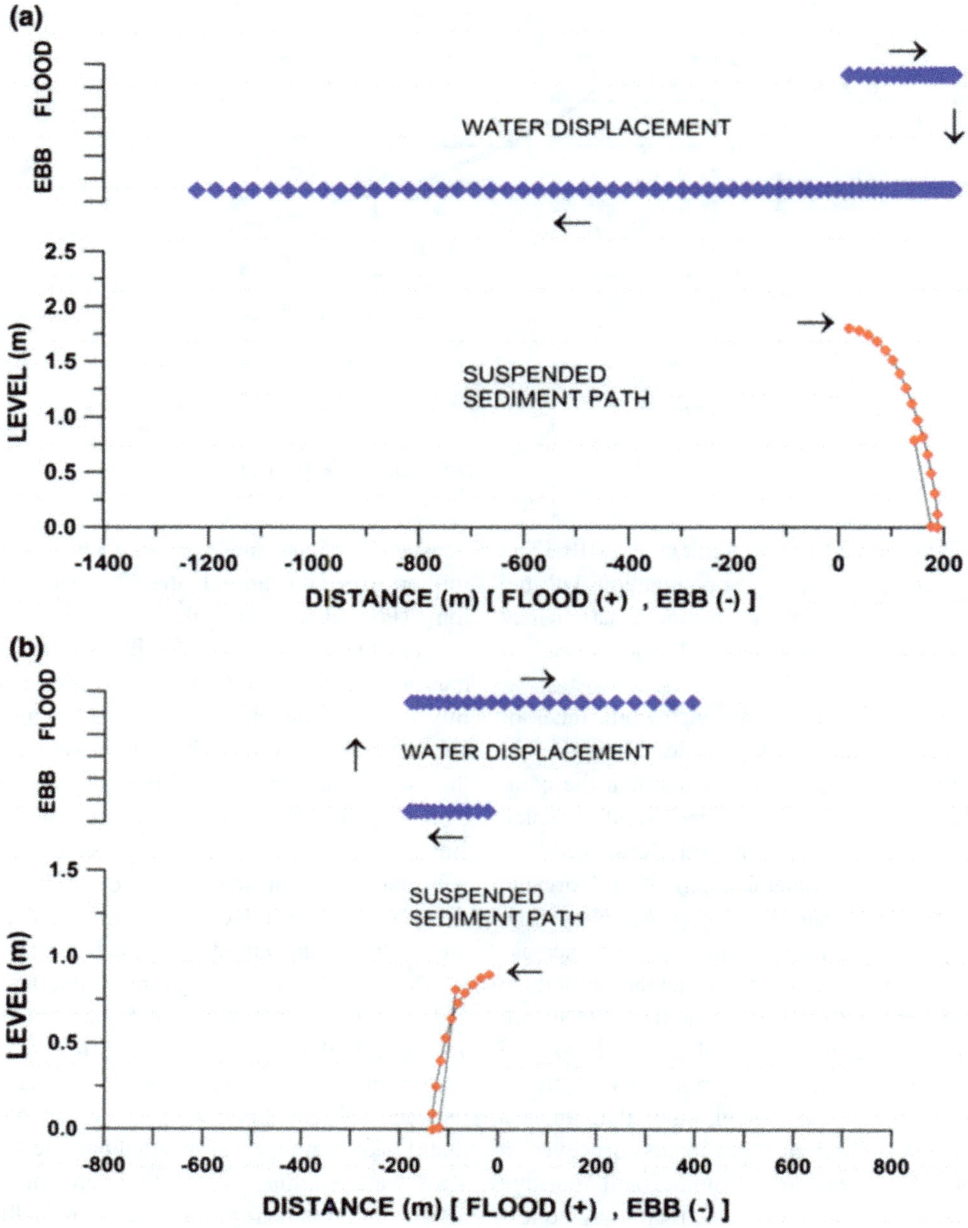

Fig. 9.9 Principals of the import of fine-grained material to tidal areas due to settling lag and scour lag. **a** The location of a water parcel (*blue*) and sediment particle (*red*) is tracked at time intervals of 100 s over a high tide period in the inner part of a tidal area. The tracking takes place from the time the particle starts to settle (at 0 m) to the time it is resuspended (at −1200 m). During the period when the particle settles out, it is transported a distance of 200 m inland (settling lag). From here it is stable on the *bottom* until the water parcel which delivered it has moved 1200 m seawards of the zero point (scour lag). First when the original water parcel has moved to here, current velocity at the location of the settled particle is large enough to resuspend it. Thus, the combined effect of settling lag and scour lag is a shift of the sediment particle to be suspended in a new water parcel located (200 + 1200 m) 1400 m inland. **b** The same dynamics over the low water period in the tidal inlet. Here the settling lag is about 120 m and the scour lag about 400 m. Combined, the seaward shift is therefore (120 + 400 m) 520 m. The joint shift over both flood and ebb is thus (1400–520 m) 880 m in an inland direction (after Bartholdy 2000)

the dynamic conditions in the tidal area Grådyb in the Danish Wadden Sea, the model showed a typical net landward migration on the order of 1 km for each tide due to lag effects (Fig. 9.9).

As a consequence of the lag effects, fine-grained sediments concentrate and form mud flats in inner parts of the Wadden Sea. During storms where wind tide increases the water level and waves resuspend fine-grained sediments from these mud deposits, adjacent salt-marsh areas get inundated by turbid water and sedimentation of fine-grained sediment takes place on the marsh surface during high tide slack water. Because of the sheltering salt-marsh plants, resuspension of

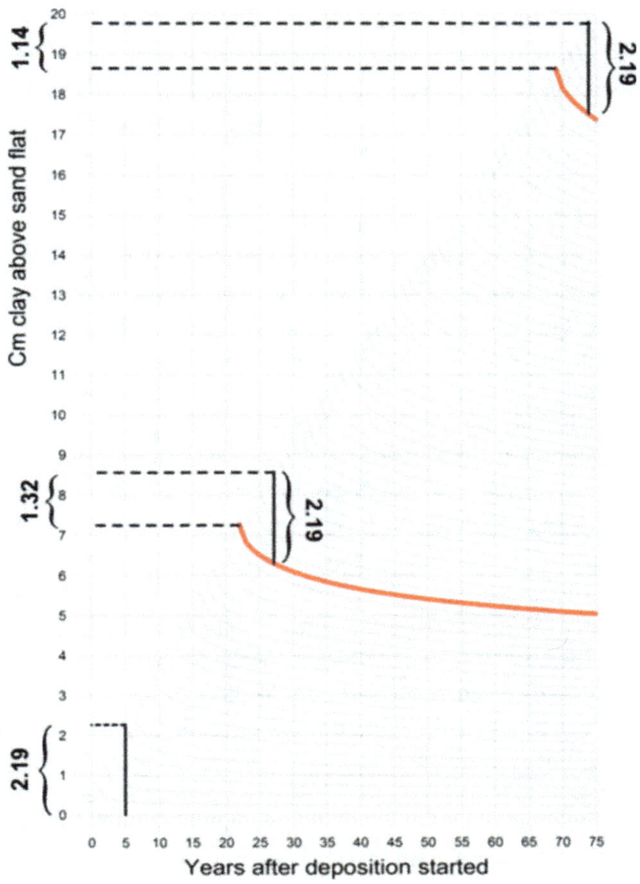

Fig. 9.10 Autocompaction in a salt marsh deposition on top of an incompressible sand flat. The accumulation rate is a constant 1.7 kg m^{-2} $year^{-1}$. The *semi-horizontal lines* represent yearly locations of former salt-marsh surfaces in the sediment column. The *left-hand* end of these *lines* represents the salt-marsh surface location above the sand flat at the year of deposition. After five years of deposition, the salt-marsh thickness above the sand flat is 2.19 cm. The *two red lines* represent years where the salt marsh was marked by a tracer. As it appears five years of deposition also add 2.19 cm on top of the marker horizon; but because of compaction the absolute level increase of the salt-marsh surface, 70 years after deposition, is only about half of this amount, 1.14 cm (after Bartholdy et al. 2010a)

the settled particles is impeded in the succeeding ebb current, and salt marsh therefore forms the end destination of silt, clay and organic material introduced into the Wadden Sea from the North Sea and local sources. The contribution from the North Sea is by far the largest concerning the net deposition. A combined sediment budget based on results from Pejrup et al. (1997) and Pedersen and Bartholdy (2006) for the four northernmost tidal areas in the Wadden Sea (Lister Tief, Juvre Dyb, Knude Dyb and Grådyb) covering an area of 855 km^2 including salt marshes, shows an accumulation of fine-grained material of 230,000 t $year^{-1}$ of which 64 % is derived from the North Sea, 14 % from primary production, 12 % from local rivers, 9 % from coastal erosion and 1 % from atmospheric deposition.

Once deposited on the salt-marsh surface, the fine-grained sediment consolidates due to autocompaction (e.g. Cahoon et al. 1995, 2000). This process causes the bulk dry density to vary from the top layer downwards as a logarithmic increasing function (Bartholdy et al. 2010a). Because of this, it can be misleading to measure salt-marsh accumulation by means of level/thickness change alone. Bartholdy et al. (2010a) showed that on the Skallingen back-barrier marsh in the Danish Wadden Sea, a constant accumulation (weight/area/time) can give half of its initial value of accretion (thickness change/time) after about 70 years of sedimentation (Fig. 9.10).

Sand is also found in salt marshes. Most of it is mobilised from adjacent sandy tidal flats and deposited on the salt marsh during storms. Not many direct measurements have been carried out on the effects of a storm surge on sediment transport. On 3–4 December 1999 during a storm surge at the back-barrier marsh of Skallingen, the suspended sediment concentration (SSC) on the intertidal flats increased from 10 to 200 mg l^{-1}, and the mobile layer of the intertidal flat was removed. The estimated sediment deposition on the salt marsh was 0.15 mm. This is only about 50 % higher than that of a previously monitored storm, and corresponds to <10 % of the annual deposition at the site (Bartholdy and Aagaard 2001). These authors concluded that the effects of storms on deposition depend on the season as well as the sequence of previous import and high-energy events. One extreme periodic storm can be of less importance for annual variations in salt-marsh deposition than more frequent minor surges (Bartholdy and Aagaard 2001; Bartholdy et al. 2004). During deep storms, flooding wave energy may be too strong to be affected by the local vegetation structure and hence prevent settlement of sediment, or may even cause erosion (Silva et al. 2009). Apart from the quantity of material deposited during storm surges, its composition may also be affected. The occurrence of storms can be reflected in the grain-size distribution of the deposited sediment (Allen 2000). The occurrence of dated thin sand layers at the back-barrier marsh of Schiermonnikoog (Netherlands), suggests that storms capable of depositing sand in the marsh occur about every decade (De Groot et al. 2011). Storm-related coarse-grained layers and sand deposits may occur at various locations within a salt marsh. At the back-barrier marsh of Schiermonnikoog, sand layers occur on 20 % of the marsh area and are partly associated with the local sources of the sand (i.e. marsh creeks, the salt-marsh edge and washovers). In total, sand layers contribute less than 10 % of the volume of marsh deposits on Schiermonnikoog (De Groot et al. 2011). The back-barrier marsh of Skallingen contains about 15 % sand (Bartholdy 1997). Deposition of sand layers after storm surges has also been reported from intertidal flats and salt marshes in the Leybucht on the mainland coast of Germany (Reineck 1980).

9.4.2 Effects of Climate Change

9.4.2.1 Sea-Level Rise

Studies on the development of salt marshes in relation to climate change almost exclusively deal with their ability to survive different SLR scenarios. As a rule, the critical level for salt-marsh survival is taken to be close to the mean high tide (MHT) level, usually considered as the level describing the border between the pioneer zone and the lower marsh (see Fig. 9.16 in Sect. 9.6). Assessments of salt-marsh survival are normally based on accretion models built on the basis of the continuity equation for salt-marsh sedimentation (e.g. Allen 1990). French (1993) added different types of semi-empirical equations of deposition, usually based on either a constant characteristic concentration of suspended sediment in the flooding water or concentrations positively correlated with the MHT level (e.g. Temmerman et al. 2003; Bartholdy et al. 2004). The organic component of salt marshes in the Wadden Sea is usually 10–20 %, of which the major part is deposited together with the mineral part of the sediment. An additional source of organic material, the so-called belowground production, is small and either incorporated in the model calibration or added as a constant value.

French (1993) estimated the organic contribution to accretion for a British salt marsh to be about 0.2 mm year^{-1}. Running this type of model for the Skallingen back-barrier marsh, sedimentation was shown to correlate with the North Atlantic Oscillation (NAO) winter index which explained about 63 % of the variation in the period 1970–1999 (Bartholdy et al. 2004). Using an improved version of the model, Bartholdy et al. (2010b) modelled the distribution of salt-marsh sedimentation on the Skallingen back-barrier marsh (Fig. 9.11) and analysed its vulnerability to different SLR scenarios. This type of assessment is either based on a balance between SLR and salt-marsh accretion or on identifying the SLR during which the salt marsh will survive a certain number of years. The latter approach is most common, as the time it takes to reach a genuine balance can be unrealistically long for a constant SLR scenario (e.g. Fig. 9.12). It was found for the Skallingen salt marsh that for the next 100 years, the salt marsh could survive a SLR of about 4 mm year^{-1} while a SLR of 6 mm year^{-1} would drown the inner part, and the outer part—the salt-marsh edge—would just survive. For a salt marsh further south on the German island of Sylt, Schuerch et al. (2013) found a similar although considerable larger mean SLR of close to 20 mm year^{-1}. Both values are considered realistic for Wadden Sea salt marshes.

Predictions of this type depend primarily on the amount of sediment available in the close vicinity of the salt marsh, and this can vary considerably from place to place. This parameter can also vary in time, and represents the most obvious source of error in such assessments. As all models

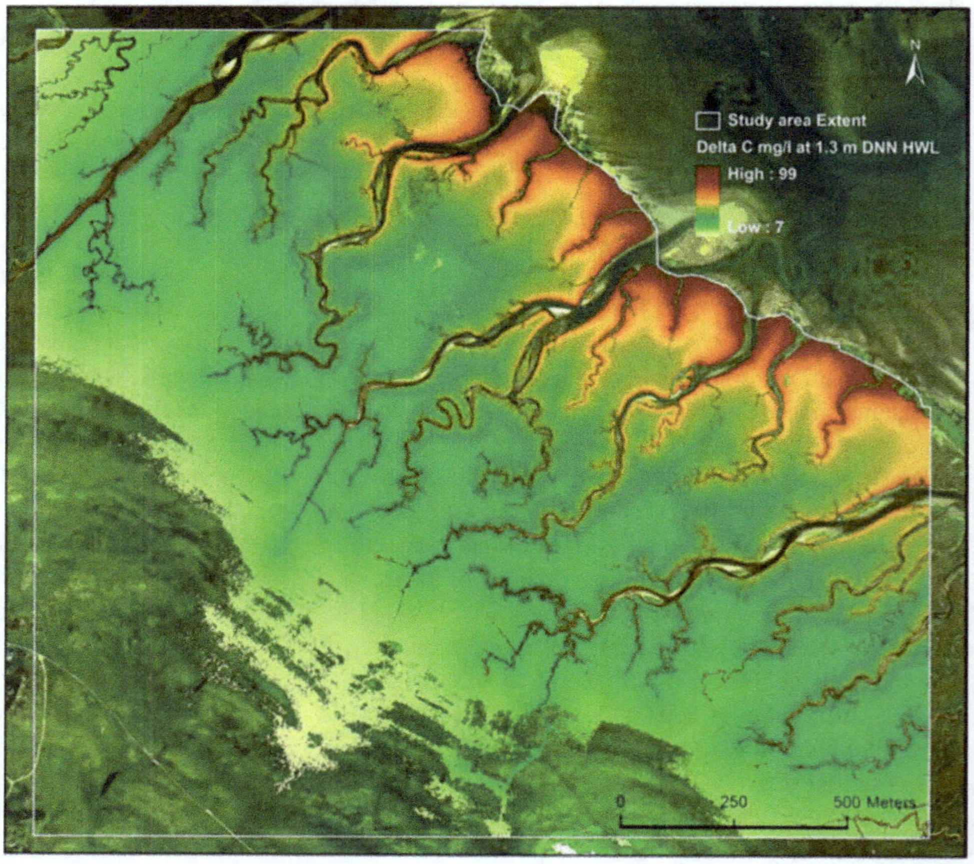

Fig. 9.11 Distribution of the typical deposition of suspended sediment in flooding water over the Skallingen back-barrier marsh at a high water level of 1.3 m Danish Normal Nul DNN (with about 0.5 m of water inundating the salt marsh). This level is the most effective high water level in terms of both frequency and concentration in relation to salt-marsh deposition. The map is superimposed on an aerial photo visible outside the modelled area and in the salt-marsh creeks (after Bartholdy et al. 2010b)

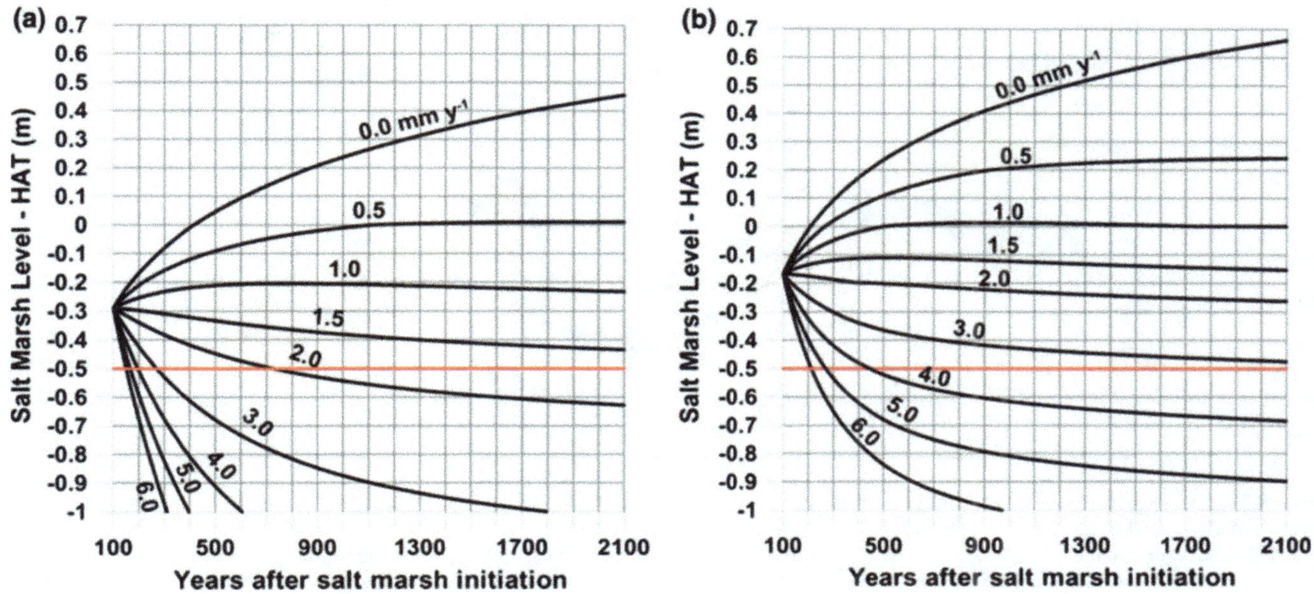

Fig. 9.12 Modelled salt-marsh level at the Skallingen back-barrier marsh minus the highest astronomical high tide level (HAT) for different sea-level rise (SLR) scenarios. A difference between the two of −0.5 m represents the mean high tide level, and is regarded as the lower limit for salt-marsh growth. The start time of 100 years is chosen because the salt marsh is about 100 years old. **a** Conditions characteristic of the central section of the back-barrier marsh. **b** Conditions characteristic for the area close to the salt-marsh edge (Bartholdy et al. 2010b)

are based on a calibration related to actual conditions, they will not make correct predictions of salt-marsh sedimentation if the sediment supply should for whatever reason change over time. The above-mentioned results, however, do not indicate any immediate threat to Wadden Sea salt marshes in general, including back-barrier marshes and mainland marshes. Slow-growing marshes such as that on the Skallingen peninsula, may become threatened in 50 years or so if SLR accelerates to much over 5 mm year^{-1}.

9.4.2.2 Wind Climate

In addition to sediment availability, sediment dynamics (and thus ultimately changes in for example, wind climate) can also play an important role in salt-marsh sedimentation. A major feature related to this is illustrated by the difference between the Wadden Sea and similar areas on the east coast of the USA. Both areas are subject to salt-marsh formation but the salt marshes are very different. In the Wadden Sea, livestock can graze on salt marshes in summer, something which is impossible in the soft mud of the salt marshes in Georgia, for example. The reason for this is the wind-tide effect which is present in the Wadden Sea and absent in Georgia. Because of this, salt marshes in the Wadden Sea quickly grow higher than the highest astronomical tide (HAT), something that never happens in Georgia (Bartholdy 2012). The dry firm summer salt-marsh areas above HAT in the Wadden Sea, therefore appear totally different from the frequently inundated, soft and unsuitable for livestock-grazing salt marshes of Georgia. The two salt-marsh types can therefore be regarded as occurring at opposite ends of a continuum of salt-marsh types directly affected by climate change in terms of changes in wind climate.

9.5 Estuaries: Geomorphology and Sediment Transport

Stijn Temmerman

9.5.1 General Properties

Major estuaries around the North Sea include the Western Scheldt (Belgium and the Netherlands), Eastern Scheldt (the Netherlands), Ems-Dollard (the Netherlands and Germany), Weser (Germany), Elbe (Germany), Firth of Forth (UK), Humber (UK), and Greater Thames (UK). Although historic human modification of these estuaries is very important (see Sect. 9.5.2), their geomorphology is generally characterised by an overall funnel-shaped, landward converging form, including landforms such as subtidal channels, sub- and intertidal sandy shoals or bars, intertidal mudflats and intertidal marshes (ranging from salt, brackish to freshwater tidal marshes) (Fig. 9.13) (Seminara et al. 2001; Prandle 2004; Dronkers 2005; Van Maanen et al. 2013). In the most downstream, wider part of the estuaries the subtidal channel system often comprises multiple channels, including flood and ebb channels, which develop as a consequence of the

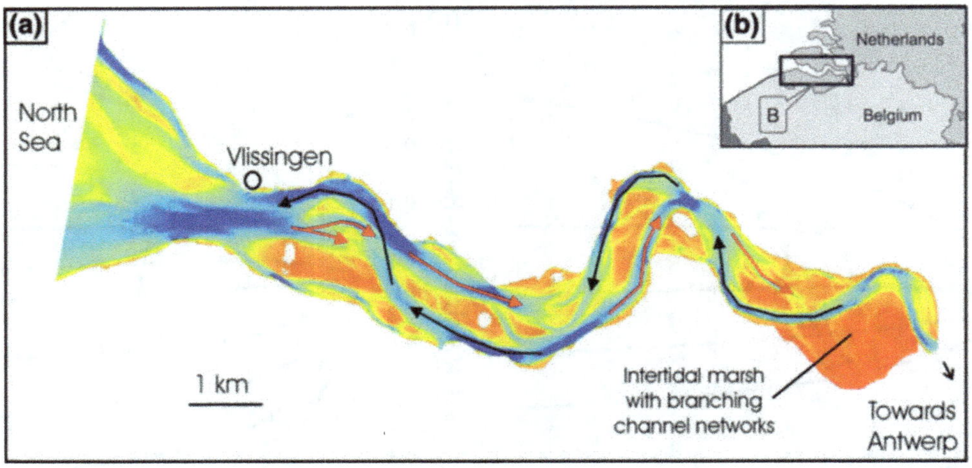

Fig. 9.13 Typical example of a North Sea estuary, the Scheldt estuary, SW Netherlands and Belgium. **a** General location. **b** Bathymetry of the most downstream section of the estuary (up to the Dutch-Belgian border). *Red areas* indicate intertidal flats and marshes; *yellow* and *blue areas* indicate subtidal areas; *darker blue* indicates deeper areas. Flood and ebb channels are indicated with *red* and *black arrows*, respectively. *Note* as for almost all estuaries around the North Sea, human modification of the estuary is important mainly due to large historical embankments of intertidal areas (around 100,000 ha over the last 1000 years) and channel dredging (deepening of sills between the North Sea and port of Antwerp)

semi-diurnal alternating flood and ebb flow directions and inertia in the water movement (Fig. 9.13). More upstream, estuarine channel width and depth decrease and the multiple channel system generally converges towards a single channel system. Branching channel networks typically develop where extensive intertidal flats and marshes are tidally flooded and drained (Fig. 9.13). The most downstream section may transition into a back-barrier tidal lagoon system such as in the Wadden Sea area (e.g. Ems-Dollard, Weser, Elbe) with back-barrier marshes.

The coarse-grained bedload transport is most intense in the subtidal channels and therefore is most determining for the morphodynamic evolution of these channels, while the suspended sediment is more important for the evolution of the intertidal mudflats and marshes. As a consequence of (1) the tidal pumping of sediments, (2) estuarine circulation due to stratification or partial mixing of salt seawater and fresh river water, and (3) settling and scour lag effects as described for the Wadden Sea (see also Sect. 9.4), the more downstream part of estuaries is generally dominated by a net landward transport of sediments of a marine origin; while in the most upstream part of estuaries there is an input of terrestrial sediments through rivers that discharge into the estuary. As a consequence of these converging sediment transport directions, and in combination with flocculation processes, the SSC typically reaches a maximum in the so-called estuarine turbidity maximum (ETM) zone (Fig. 9.14) (e.g. Dyer 1997). The existence of an ETM is well-documented for most estuaries in the North Sea region (e.g. Uncles et al. 2002), and may have important implications for the estuarine ecosystem, because turbidity controls

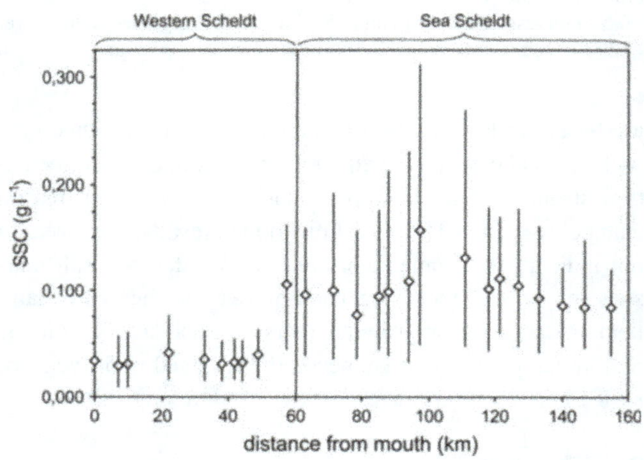

Fig. 9.14 Time-averaged longitudinal variation in suspended sediment concentration (SSC) along the Scheldt estuary, calculated from monthly monitoring data for 1996–2001, showing the presence of an Estuarine Turbidity Maximum (ETM) zone at around 100 km from the estuary mouth in the North Sea. *Error bars* represent the 10th and 90th percentiles of all SSC measurements at a station (after Temmerman et al. 2004)

the depth of light penetration into the water column and thereby the potential limitation of primary production by phytoplankton if turbidity is too high (e.g. Cloern 1987). Furthermore, the SSC determines the capacity of intertidal mudflats and marshes to grow with SLR by sediment accretion (more details in Sect. 9.5.3.1).

The SSC may also depend on sediment surface stability depending on biological controls. These can be divided into biostabilisation and biodestabilisation, with sediment surface stability ultimately dependent on the balance between the two competing sets of processes. This balance varies spatially, both vertically within the tidal frame and horizontally along the estuarine salinity gradient, and temporally, on seasonal, interannual and perhaps longer timescales. These patterns have implications for estuarine morphology; strong biostabilisation will lead to a flatter profile because flood and ebb tidal pulses will be less effective on intertidal flat surfaces. Conversely, destabilisation will lead to lower critical shear stresses for erosion and thus result in steeper profiles. In general, biostabilisation is associated with microorganisms and biodestabilisation with a benthic macrofauna. Thus, for the Humber estuary (UK), Wood and Widdows (2002, 2003) developed a simple (if unvalidated) cross-shore model incorporating biostabilisation (in the form of chlorophyll *a* content) and bioturbation (from the burrowing bivalve Baltic tellin *Macoma balthica*). The model suggests that the erosion or deposition driven by natural fluctuations in biota densities are as large as the changes caused by variations in tidal range and currents over a spring–neap cycle or are equivalent to a doubling of the external sediment supply. Seasonal variations in the density of stabilising diatoms can alter the magnitude of net deposition by a factor of 2 and interannual changes in *Macoma balthica* density change deposition by a factor of 5. In a UK climate change scenario, milder winters result in lower springtime recruitment of *Macoma*, leading to lower rates of bioturbation at mid-intertidal levels and lower sediment supply to the upper intertidal zone and its fringing salt marshes (Wood and Widdows 2003).

Using the same two biotic groups, Paarlberg et al. (2005) extended Wood and Widdows (2002) approach, showing that changes in bioturbation and stabilisation by microphytobenthos can potentially alter the mud content and elevation (by 5–10 cm) of shoal banks in the Western Scheldt estuary (Netherlands). In fact, field observations in the Western Scheldt estuary show that two stable sedimentological states are present at intermediate levels of bed shear stress, either a bare surface with low silt content or a high silt content supporting a high density of diatoms. This bimodal pattern results from the feedback links between the biota and silt content. Diatom growth rates are enhanced by the nutrients present in silt-rich sediments and diatom resuspension falls as diatom density increases, favouring the accumulation of silts. Loss of diatom cover sets this dynamic in reverse (Van de Koppel et al. 2001).

9.5.2 Human Impacts

Human impacts on estuarine geomorphology are particularly significant in the North Sea region. A general phenomenon is the historical reclamation of intertidal flats and marshes in many estuaries, since Medieval and even Roman periods. As a result the present-day area of intertidal flats and marshes is only a tiny fraction of the original area. For example, in the Western Scheldt estuary around 100,000 ha of intertidal areas—mostly marshes—have been reclaimed by seawall building since 1200; nowadays only around 2800 ha of marshes and 8000 ha of intertidal flats (i.e. 10 %) remain (Meire et al. 2005). Apart from the direct impact on the reclaimed land, the geomorphology of the remaining estuary will also react. The large reduction in intertidal areas may considerably reduce the volume of an estuary and hence its tidal prism (the volume of water within the estuary between high and low tides). Due to the reduced tidal prism and tidal currents, the long-term response of the estuary is that channels may fill with sediments and remaining intertidal mudflats in front of the seawalls may silt up promoting the succession towards marshes (e.g. Townend 2005). Extreme cases exist where progressive historical land reclamation resulted in a cascading effect and almost complete silting up and disappearance of the estuary (e.g. Zwin estuary towards Bruges, Belgium). In other cases, estuaries are still responding to historical land reclamation and estuarine sediment infilling may be expected to continue over coming decades. In still more cases, other processes may have compensated for the loss of tidal prism and land reclamation may not induce as much estuarine sediment infilling. As De Swart and Zimmerman (2009) concluded in an extensive review, interactions between intertidal and channel morphodynamics are complex and still not fully understood.

More recent human impacts include the dredging and canalisation of estuarine channels, because many estuaries in the North Sea region provide access to major harbours, including the ports of Rotterdam (Rhine-Meuse), Antwerp (Western Scheldt), and Hamburg (Elbe). Channel deepening reduces hydraulic friction, while land reclamation reduces the water storage capacity of estuaries. As a combined effect tidal penetration has increased in many estuaries (e.g. Friedrichs and Aubrey 1994). An increase in tidal range may increase the stirring of fine-grained sediments and so increase SSC towards highly turbid conditions, such as observed in the Ems estuary (e.g. Van Leussen 2011).

Other human impacts since the 1950s include the building of flood defence structures, such as storm surge barriers (e.g. Eastern Scheldt barrier; Thames barrier) and even the complete closure of estuaries by dams (e.g. Dutch Delta works). In the Eastern Scheldt estuary, the storm surge barrier seriously affects the sediment budget: as the tidal prism is reduced, the channels tend to fill in, but because marine import of sediments is restricted by the storm surge barrier, the intertidal flats inside the estuary experience erosion and sediments are redistributed towards the channels.

9.5.3 Future Expectations

9.5.3.1 Sea-Level Rise

Aspects of climate change that are expected to affect the geomorphology of estuaries include accelerated SLR, increased storminess, and increased fluctuations in freshwater discharge. SLR is of particular concern for the geomorphology and ecology of intertidal flats and marshes. In some parts of the world (e.g. Mississippi Delta; Chesapeake Bay; Venice Lagoon), sediment accretion rates are not enough to compensate for relative SLR (i.e. eustatic SLR combined with local land subsidence), which has resulted over the last century in extensive die-back of marsh vegetation and conversion of marshes and mudflats into open water (e.g. Kearny et al. 2002). A recent study combining five state-of-the-art models of marsh sedimentation in response to SLR, revealed that marshes with a low tidal range and low SSC (<20 mg l^{-1}) are especially at risk from submergence by average SLR projections (Kirwan et al. 2010). Based on Fig. 9.15, and given the large tidal ranges (2–6 m) and generally high SSC values (mostly >20 mg l^{-1} and up to several hundreds of milligrams per litre) for estuaries around the North Sea, the risk of marsh submergence by SLR is low for the coming century. For example, for the Western Scheldt estuary (average SSC ~40 mg l^{-1}; mean tidal range ~4 m), marshes are expected to drown in the long term only if the rate of SLR increases to 50 mm $year^{-1}$, which is far more than expected by 2100 (Temmerman et al. 2004). In theory, SLR may also affect estuarine sediment regimes, as it may increase the landward penetration of tides and hence of the ETM, but only if morphodynamic changes in the estuary are assumed to be zero; however, estuarine morphodynamic changes and human impacts are likely to exert much more control over estuarine sediment regimes than SLR.

9.5.3.2 Storms

Although the precise effects of climate warming on increasing frequency and intensity of storms is subject to debate in the scientific literature, there is growing consensus that climate warming is expected to increase the intensity of extreme tropical and extra-tropical storms for many coastal areas worldwide (e.g. Knutson et al. 2010), including the North Sea (e.g. Knippertz et al. 2000; Leckebusch and Ulbrich 2004; Donat et al. 2010). Increasing storminess may have geomorphological effects as wind waves during storms may affect erosion and sedimentation processes, especially on intertidal flats and along marsh edges (e.g. Callaghan et al. 2010 and see Chap. 18). Wind waves are most important in the seaward, wider parts of the estuaries, where water surface and wind fetch length are longer so that higher

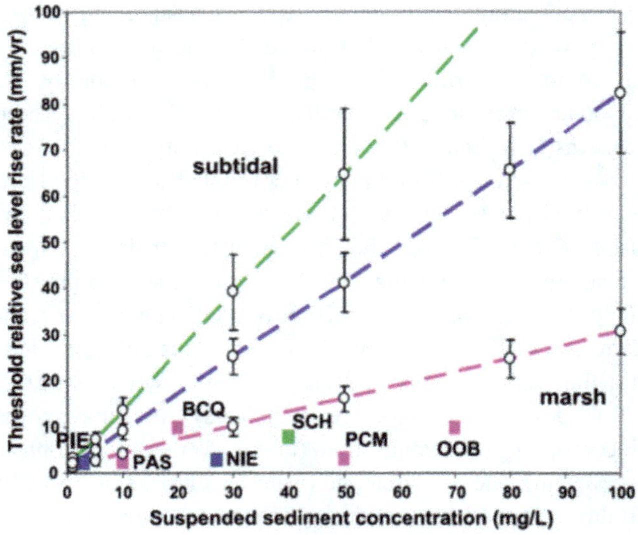

Fig. 9.15 Predicted threshold rates of sea-level rise, above which marshes are replaced by subtidal environments as the stable ecosystem. Each line represents the mean threshold rate (±1 SE) predicted by five models as a function of suspended sediment concentration and spring tidal range. The *hatched line* denotes thresholds for marshes modelled under a 1 m tidal range (*pink*), a 3 m tidal range (*blue*), and a 5 m tidal range (*green*). For reference, examples have been included (denoted with *square markers*) of marshes worldwide in estuaries with different rates of historical sea-level rise, sediment concentration, and tidal range. (*PIE* Plum Island Estuary, Massachusetts; *PAS* Pamlico Sound, North Carolina; *BCQ* Bayou Chitique, Louisiana; *NIE* North Inlet Estuary, South Carolina; *SCH* Scheldt Estuary, Netherlands; *PCM* Phillips Creek Marsh, Virginia; *OOB* Old Oyster Bayou, Louisiana) (after Kirwan et al. 2010)

storm waves can be generated, while in the more landward, smaller parts of the estuaries smaller wind waves are expected. For several North Sea estuaries, including the Greater Thames area (UK) and the Western Scheldt estuary (Netherlands), lateral erosion of the edges of intertidal flats and marshes has been reported and partly attributed to wind wave erosion (e.g. Van der Wal and Pye 2004; Van der Wal et al. 2008), which might increase with increasing storminess due to climate change. However, these studies also highlighted that the wind-wave climate is not the only variable explaining the patterns and rates of lateral marsh erosion, but that other factors also play a major role, including the larger-scale morphodynamic changes in estuarine channel position, which may be affected by human impacts such as dredging and disposal of sediments (e.g. Cox et al. 2003). Several studies also address the role of self-organising mechanisms, driven by feedbacks between marsh vegetation, sediment deposition and erosion, and wave hydrodynamics, leading to cycles of lateral marsh extension and lateral marsh erosion, that are not solely driven by external forcing factors such as storminess (Van de Koppel et al. 2005; Chauhan 2009; Mariotti and Fagherazzi 2010).

9.5.3.3 Precipitation

According to the latest IPCC assessment, winter precipitation (October–March) in central and northern Europe (including the North Sea region) could be up to 40 % higher than present-day by 2100 (this value includes the 5th to 95th percentiles of precipitation projections for all IPCC scenarios) (Van Oldenborgh et al. 2013). For summer (April–September), climate models project a change in precipitation of −10 % to +20 % by 2100. Although the range of projections is quite large, the IPCC projections clearly suggest a future increase in precipitation during winter, with a smaller increase or even a slight decrease, during summer. Larger fluctuations in freshwater river discharge may affect the terrestrial sediment supply to estuaries. More intense rainfall events, especially during winter, may induce larger soil erosion events within the river catchment of estuaries (e.g. Poesen et al. 2003), and so may increase the terrestrial sediment supply to estuaries, potentially contributing to higher SSC values in the ETM zone (see also Chaps. 11 and 13). However, it should be emphasised that in terms of runoff and sediment supply to estuaries and the coast, human impacts such as changes in land use within the river catchment of estuaries are often as important as or even more important than changes in precipitation (Syvitski et al. 2005).

9.5.3.4 Human Impact

It must be stressed that human impacts such as channel deepening for harbour accessibility and flood defence and shoreline protection structures on geomorphology and sediment transport in estuaries are very likely to continue over the coming decades, and may dominate, exacerbate or compensate for the potential impacts of climate change. Channel deepening is likely to exacerbate the increase in tidal range and landward tidal wave penetration over coming decades, increasing risk of sediment resuspension and a potential shift towards hyper-turbid conditions such as happened in the Ems estuary in the Netherlands over past decades (Winterwerp 2011). One particular example of mitigation of climate and human-induced impacts is the conversion and restoration of formerly reclaimed land into intertidal flats and marshes in the UK (so-called managed coastal realignment schemes, such as in the Humber and Blackwater estuaries) (French 2006b) and Belgium (the Sigma plan in the Scheldt estuary) (Maris et al. 2007). Other types of ecosystem-based adaptation to climate change ('soft engineering'), that are starting to be implemented, include the creation of oyster reefs and sand supply on tidal flats such as in the Eastern Scheldt estuary (e.g. Temmerman et al. 2013). Although such schemes provide multiple benefits in the form of ecosystem services (see E-Supplement S9) (e.g. flood storage, erosion protection, water quality regulation, carbon sequestration, fisheries production), societal opposition against conversion of reclaimed land into intertidal flats and marshes may be important and effects on changing tidal conditions in the estuary must be considered (see Temmerman et al. 2013 for a review).

In the Scheldt estuary, up to 3000 ha of historically reclaimed land are designated for conversion into floodplains, of which about 1500 ha are tidal marshes (Broeckx et al. 2011). The new intertidal areas have two objectives: to store extra water and attenuate landward propagating storm surges, thus reducing flood risk in the hinterland; and to provide ecosystem services such as water quality improvement and habitat restoration. Effects on the sediment budget of the estuary may also be expected (Maris et al. 2007). For example, in the Belgian part of the Scheldt estuary, the tidal marsh area is expected to increase from around 420 ha at present to almost 2000 ha by 2030. A pilot project showed that sediment accretion processes are already occurring in the newly created marshes at rates (per surface area) comparable to those on natural marshes (Vandenbruwaene et al. 2011), and so the completion of the whole marsh restoration scheme could reduce the risk of increasing SSC and turbidity within the estuary. Similar tidal marsh creation projects on formerly embanked land have been realised over the last few years and are planned for the near future in other estuaries around the North Sea, in particular in the Humber and Greater Thames estuaries (UK), where this management approach is called 'managed coastal realignment' (e.g. French 2006b; Turner et al. 2007).

The positive outcome of this management approach of tidal marsh creation on formerly reclaimed land should stimulate the wider implementation of this approach in other estuaries around the North Sea as a sustainable and cost-effective manner to mitigate climate- and human-induced impacts.

9.6 Salt Marshes

Jan P. Bakker

9.6.1 Distribution and Dynamics

Salt marshes occur along an elevational gradient from the intertidal flats to middle and high marsh with an associated change in the composition of the vegetation (Fig. 9.16). Vegetation-sedimentation feedbacks are only one of the many potentially important interactions (Fig. 9.17) (Nolte et al. 2013a). The main external controls of sediment deposition are sea level (hydroperiod) and sediment supply. The latter is strongly related to the SSC. Interactions between the physical and biological features of salt marshes

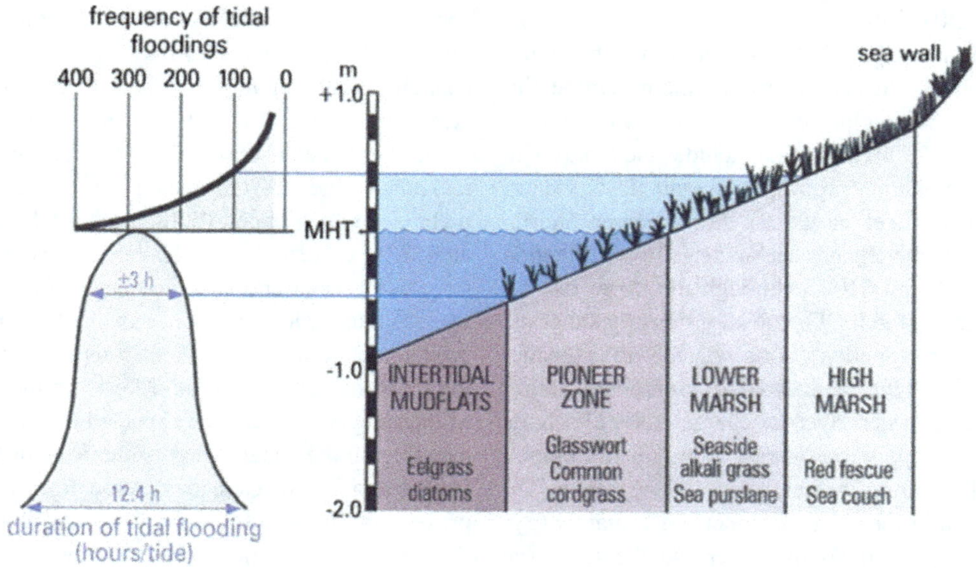

Fig. 9.16 Zonation of salt marshes in relation to the duration and frequency of tidal flooding and marsh elevation for the western German Wadden Sea. On back-barrier marshes the seawall is replaced by dunes (modified after Erchinger 1985). Eelgrass *Zostera* spp., glasswort *Salicornia* spp., common cordgrass *Spartina anglica*, seaside alkali grass *Puccinellia maritima*, sea purslane *Atriplex portulacoides*, *red fescue Festuca rubra*, sea couch *Elytrigia atherica* are the dominant plant species. They may be replaced by other species in the northern North Sea region. Wind flood (Windflut) is defined as 92 cm + MHT, storm flood (Sturmflut) as 198 cm + MHT, hurricane flood (Orkanflut) as 275 cm + MHT. During a period of 20 years, they occur $n \leq 10$ times, $10 > n \leq 0.5$, and $0.5 > n \leq 0.05$, respectively, in the western German Wadden Sea (Niemeyer 2015)

are also important. The accumulation of plant biomass can play an influential role in sediment deposition. Human impacts such as ditching or management practices such as livestock grazing (and then soil compaction) can also affect processes related to hydrodynamics, vegetation composition, and sediment deposition.

Although many salt marshes tend to have a 'natural' appearance, some along the mainland coast result from human interference. Salt marshes emerged about 2500 years BP after the last glacial period. Apart from small-scale embankments from the Roman period onward, most of the coastline became protected by seawalls about 1000 AD, thus reducing brackish marshes further inland and disconnecting them from the salt marshes. Peat reclamation along the coastline and subsidence behind the seawalls combined with sudden falls in the human population due to disease in the Medieval period, made the embankments vulnerable to attack by the sea. This resulted in societal collapse and a subsequent inability to maintain the seawalls that protected the embankments (Wolff 1992). This resulted in embayments such as the Lauwerszee, Dollard (the Netherlands), Leybucht (Germany), and subsequent new salt-marsh development. In past centuries, extensive areas of salt marsh have been embanked for coastal protection and agricultural exploitation (Dijkema 1987). Currently, a decline in the pioneer zone and increase in the high marsh zone in the Wadden Sea has been reported (Esselink et al. 2009). Since the 1960s it has not been economically viable to embank salt marshes for agricultural use (Wolff 1992). Accidental de-embankments after storm surges have occurred since the 19th century, and deliberate de-embankments in north-western Europe since the 1990s (Esselink et al. 2009). In terms of their future survival, it is important to understand the extent to which salt marshes can keep pace with the projected acceleration in SLR. An extreme episodic storm surge may destroy the vegetation at the marsh edge and cause the formation of a cliff (Van de Koppel et al. 2005). Feedbacks between sediment, vegetation and wave hydrodynamics may result in the formation of new marshes (see Sect. 9.5.3.2). Estuarine marshes in south-east England suffer from erosion. An overview of erosion rates revealed a net loss in area between 1973 and 1998 of about 1000 ha, or 33 %. Recent erosion rates (1988–1998) have been slower however (Cooper et al. 2001) and the area covered has recently (2006–2009) increased, resulting in a loss of 'only' 750 ha since 1973 (Phelan et al. 2011).

Lateral erosion can result in a narrowing of salt marshes, or coastal squeeze (Wolters et al. 2005a), particularly in the case of a short foreshore (Bouma et al. 2014) (see also Chap. 18). The potential loss of salt-marsh area through erosion from the seaward edge appears unrelated to the sedimentation processes in the salt marsh itself, but is determined by sedimentation in the pioneer zone, thus allowing dynamic rejuvenation of the lower salt marsh (Boorman et al. 1989; Dijkema et al. 2010), as in the case of a wide foreshore (Bouma et al. 2014). A wider foreshore will

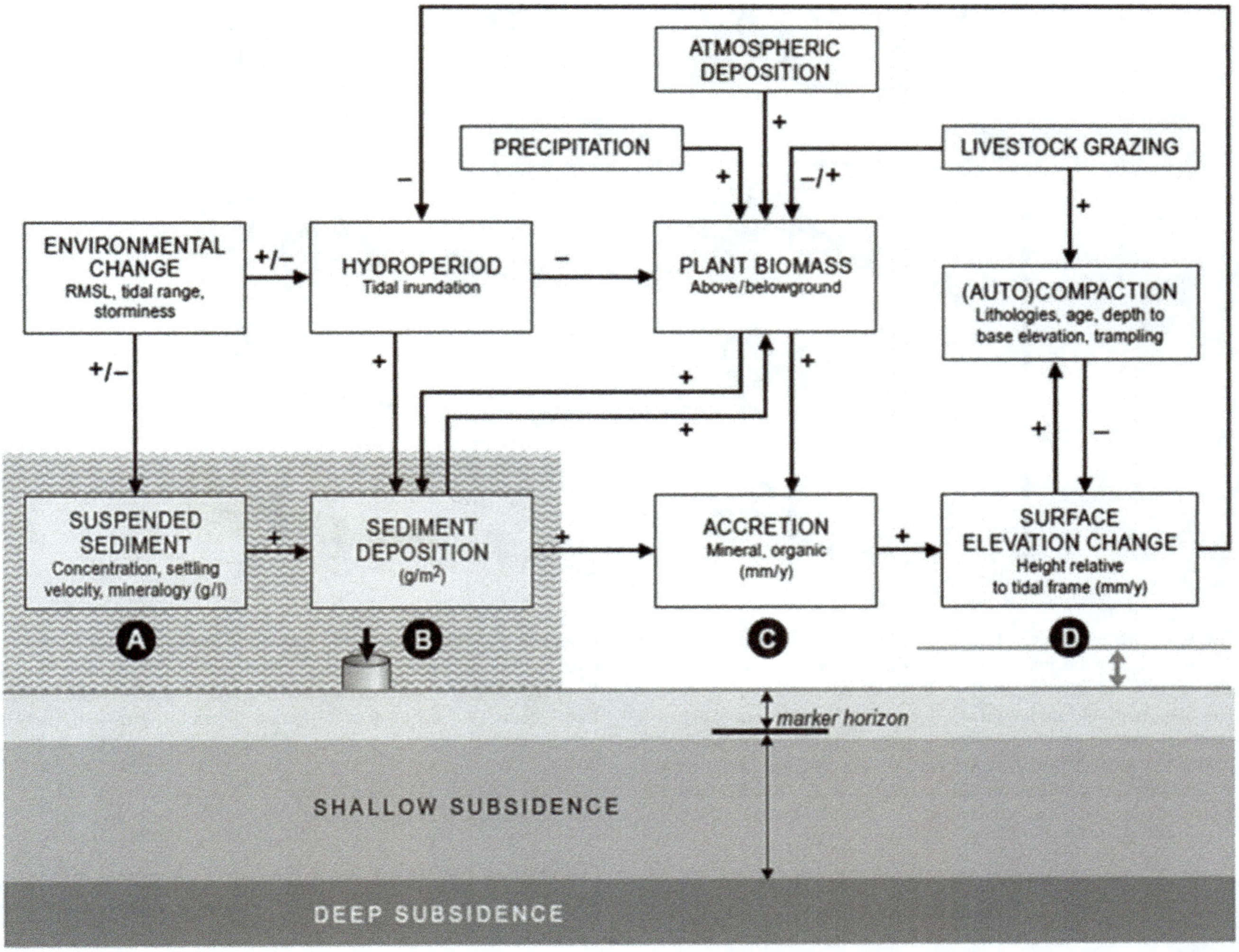

Fig. 9.17 Factors affecting sedimentation and accretion processes in coastal marshes. The letters *A*, *B*, *C*, and *D* indicate the main factors considered in the review (after Nolte et al. 2013a)

inherently offer more space for intertidal ecosystems. Moreover, a wider foreshore will generally have weaker wave energy gradients than a narrower foreshore, thereby making it easier for epibenthic ecosystems to establish. As a result, both the maximum and minimum widths of an intertidal habitat will have a positive relationship with the size of the foreshore (Fig. 9.18). Self-organisation of this type is enhanced by the wave reduction of up to 60 % attributed to vegetation compared to bare soil (Möller et al. 2014). Sea-edge erosion may result in large-scale cliff erosion. If cliff erosion is not prevented by groynes, marshes established from sedimentation fields may disappear in the long term (Dijkema 1994).

The area covered by salt marshes in the North Sea region has increased following accidental de-embankments after breaching of the seawall or summerdike during storm surges, such as occurred in 1953. Accidental de-embankments took place at 35 sites in north-western Europe before 1991. After 1991, deliberate de-embankment took place at 29 sites, and there are plans to increase this number. The de-embanked sites ranged from less than 1 ha to over 500 ha. The total area amounted to more than 5600 ha (Wolters et al. 2005b). Half of the deliberate de-embankments were carried out for habitat restoration and a quarter for flood defence (see E-Supplement S9 for ecosystem services) (Fig. 9.19).

Past SLR is not the same along the entire coastline of the North Sea, nor is it constant over time (Table 9.1 in Sect. 9.2.3.1). Global mean SLR was about 1.7 mm year^{-1} for the 20th century as a whole, but was higher at about 3 mm year^{-1} over the last two decades. These data hold for most of the European coasts, but with variations due to local land movement, either positive or negative (EEA 2012). Mean SLR for the North Sea region over the past 20 years[1] is 1.4 ± 0.4 mm year^{-1}. A recent analysis of data from Dutch

[1]http://ibis.grdl.noaa.gov/SAT/SeaLevelRise/slr/slr_sla_nrs_free_all_66.pdf.

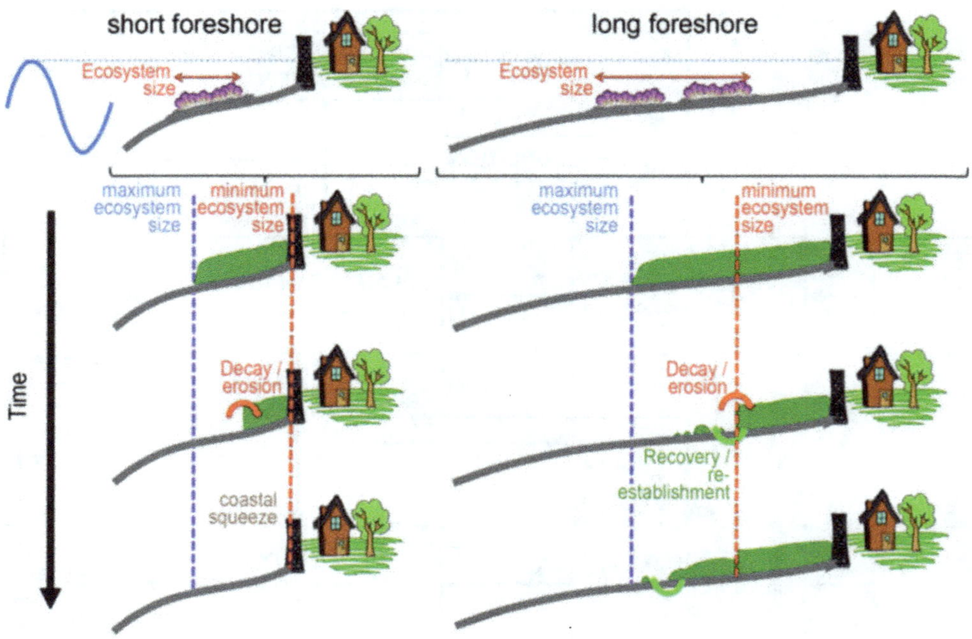

Fig. 9.18 Schematic illustration of the relation between foreshore dimensions and the maximum and minimum widths of an intertidal ecosystem with wave-attenuating aboveground (epibenthic) structures. The maximum and minimum widths relate to the borders reached by intertidal ecosystems with cyclical dynamics: the minimum width is the size of the ecosystem that will maintain, whereas everything between the maximum and minimum will vary over time. A sufficiently wide foreshore is important to enable epibenthic intertidal ecosystems to go through natural cycles of decay and re-establishment (cf. Van de Koppel et al. 2005), without experiencing coastal squeeze. On a narrow foreshore, re-establishment of degrading epibenthic ecosystems may be hampered by gradients in wave energy that are too strong (Bouma et al. 2014)

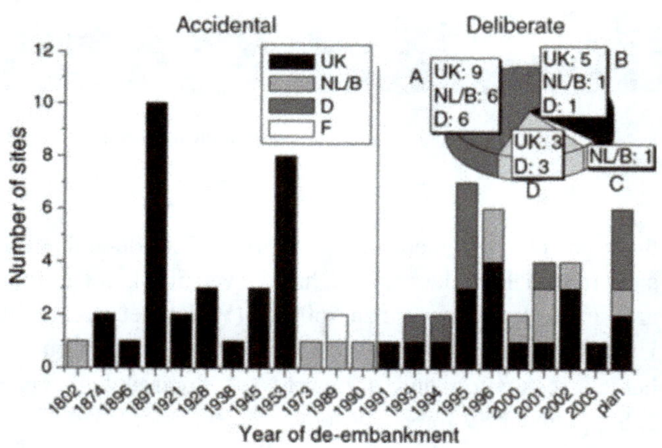

Fig. 9.19 Timing and causes of salt marsh de-embankment by country. **a** habitat creation or restoration; **b** flood defence; **c** gaining experience; **d** unknown (Wolters et al. 2005b)

tidal stations does not show an acceleration in SLR between 1990 and 2010 (Dillingh et al. 2012). SLR can be affected by vertical land movement. This varied from −1.3 to 0.1 mm year^{-1} over the period 1843–2009 in the German Bight (Wahl et al. 2011) and from −1.1 to −0.5 mm year^{-1} in the English Channel through the 20th century (Haigh et al. 2011).

Not only has mean SLR changed over time, but MHT has also varied in different locations and over different periods (Table 9.2). For a given area, the rate of change in MHT was sometimes higher that SLR, sometimes lower, or was even negative.

9.6.2 Sedimentation and Accretion in Intertidal Marshes

Whether coastal marshes can cope with accelerated SLR depends on the change in surface elevation, which in turn depends on availability of sediment (SSC), over marsh tidal events, and autocompaction. These interact with plant biomass and exploitation of salt marshes by livestock grazing (Fig. 9.17). Rates of surface elevation change (SEC) of a broad sample of allochthonous marshes in north-western Europe and North America range from a few millimetres per year to several centimetres per year. There is no obvious trend with tidal range, but the envelope of variability is wider at larger tidal ranges. The sites included no evidence for SEC deficit (i.e. the difference between rates of SEC and SLR) with the exception of a single system (French 2006a).

Accretion and SEC are both often expressed as an average for a given area, including a range on individual measurements. Salt marshes along the mainland coast of the North Sea region reveal higher rates of accretion or SEC than those on barrier islands (Table 9.3).

Table 9.2 Regional variation in change in mean high tide (MHT) within the North Sea region

Area	Change in MHT, mm $year^{-1}$	Period	Source
Wadden Sea	2–2.5	1950–2000	Oost et al. (2009)
Germany–mainland	4.2	1965–2001	Jensen and Mudersbach (2004)
Germany–islands	3.5	1965–2001	Jensen and Mudersbach (2004)
Ameland, Netherlands	6	1963–1983	Dijkema et al. (2011)
	0	1983–2010	Dijkema et al. (2011)
New Statenzijl, Dollard, Netherlands	5.7	1890–1910	Esselink et al. (2011)
	1	1910–1954	Esselink et al. (2011)
	6	1955–1983	Esselink et al. (2011)
	−0.3	1983–2009	Esselink et al. (2011)

Table 9.3 Regional variation in mean rates of accretion (AC), surface elevation change (SEC) or a combination of the two on salt marshes in the North Sea region

Area	AC/SEC mm $year^{-1}$	Period	Source
Mainland			
Friesland, Netherlands	11–29 SEC	1984–2010	Dijkema and Van Duin (2012)
Peazemerlannen, Netherlands	9–14 SEC	1996–2010	Dijkema and Van Duin (2012)
Groningen, Netherlands	8–14 SEC	1984–2010	Dijkema and Van Duin (2012)
Dollard, Netherlands	7–10 SEC	1984–2003	Dijkema and Van Duin (2012)
Schleswig-Holstein south, Germany	2–3 SEC	1998–2009	Suchrow et al. (2012)
Schleswig-Holstein north, Germany	5–9 SEC	1998–2009	Suchrow et al. (2012)
Hamburger Hallig, Germany	4–17 SEC	1996–2009	Stock (2011)
Back-barrier			
Stiffkey, UK	6.5 AC/SEC	1995–1998	Cahoon et al. (2000)
Terschelling, Netherlands	1.3 AC/SEC	1995–1998	Van Wijnen and Bakker (2000)
Schiermonnikoog, Netherlands	0–3 AC/SEC	1995–1998	Van Wijnen and Bakker (2000)
Rømø, Denmark	2.6 AC	1980–2003	Pedersen and Bartholdy (2006)
Fanø, Denmark	2.8 AC	1980–2003	Pedersen and Bartholdy (2006)
Skallingen, Denmark	2.4 AC	1980–2003	Pedersen and Bartholdy (2006)
Skallingen, Denmark	1.4 AC/SEC	1995–1998	Van Wijnen and Bakker (2000)

Data on the age of salt marshes in Table 9.3 are not often available. In contrast, age data are available for back-barrier marshes and these indicate decreasing SEC with increasing age of the marsh (Van Wijnen and Bakker 2001). SEC was 2.5 mm $year^{-1}$ on a marsh of up to 15 years in age, around 1.5 mm $year^{-1}$ on a marsh of 30 years in age, and around 0 mm $year^{-1}$ on a marsh of 100 years in age (Fig. 9.20). Decreasing SEC in older marshes may indicate autocompaction (see Sect. 9.4).

In addition to regional differences in SEC, there are also local differences from high marsh to intertidal flats. At the barrier island of Langli, Denmark, an accretion rate of −1 mm $year^{-1}$ (2001–2009) was recorded on the high marsh and 0.5 mm $year^{-1}$ on the low marsh (Kuijper and Bakker 2012). On the mainland marsh of the Dollard estuary in the Netherlands, the accretion rate was 9–16 mm $year^{-1}$ at 10 cm + MHT and 0–8 mm $year^{-1}$ at 55 cm + MHT over the period 1984–1991 (Esselink et al. 1998). Plots at 20–60 cm + MHT revealed a SEC of 5–9 mm $year^{-1}$, while plots at 70–80 cm + MHT showed only 1 mm $year^{-1}$ for the period 1996–2009 in the mainland marsh of Hamburger Hallig on the German coast. Rates of accretion or SEC were generally higher near the salt-marsh edge in both mainland and back-barrier marshes (Table 9.4). However, this was not found at the southern side of Hamburger Hallig, indicating strong local differences (Stock 2011). Rates were higher on mainland marshes than back-barrier marshes (see also Table 9.3).

The rate of accretion also declines away from creeks and ditches in the mainland marshes of the Dollard; 15 mm $year^{-1}$ next to a creek and 2 mm $year^{-1}$ (over 1984–1991) at 20 m from the creek at a distance of 750 m from the marsh edge (Esselink et al. 1998). Apparently water with very little sediment in suspension inundates the marsh far from the

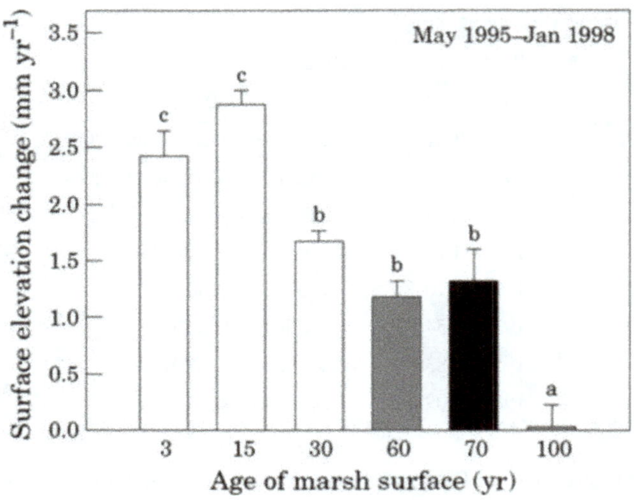

Fig. 9.20 Marsh surface elevation change at 40 cm + MHT in the late 1990s over a three-year period in salt marshes at various successional stages on the ungrazed back-barrier marshes of Schiermonnikoog, the Netherlands (*open bars*), Terschelling, the Netherlands (*grey bar*) and Skallingen, Denmark (*black bar*). Different letters indicate significant differences ($p < 0.05$) (Van Wijnen and Bakker 2001)

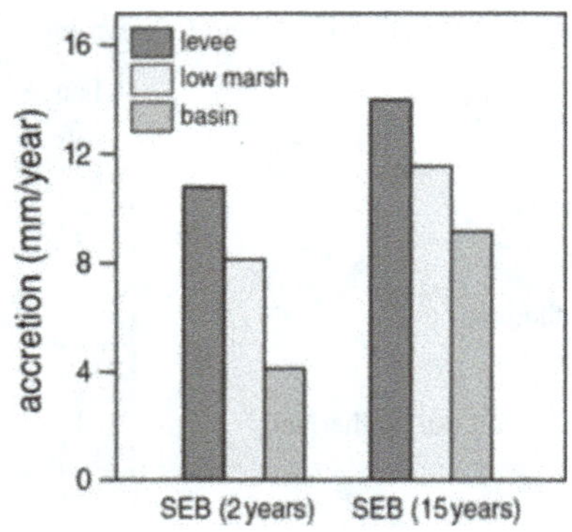

Fig. 9.21 Comparison of accretion rate using Sedimentation Erosion Bar (SEB) measurements for 2 years and 15 years. Measurements took place on levees, in the low marsh, and in basins at the Peazemerlannen mainland salt marsh in the Netherlands. The two-year measurements represent a period without winter storms (Nolte et al. 2013a)

edge. Sediment trapping at the marsh edge is enhanced by plants causing wind wave attenuation (Möller 2006).

Data from a broad sample of allochthonous marshes in north-western Europe and North America, reveal only weak positive linkages between SSC and mean tidal range (French 2006a). Variability in SSC is dependent on the weather conditions. In the eastern Wadden Sea in the Netherlands, SSC of up to 100 mg l^{-1} was recorded during most tides, but could increase to 800 mg l^{-1} during periods of strong western winds (Kamps 1962). Elevated SSC due to wave re-suspension correlates with strong westerly wind events. These local meteorological events seem to overrule sea level and tidal range (French 2006a). An increase in extreme sea level during storm surges in the southern North Sea was found over the period 1850–2000 (Weisse et al. 2012). Wave activities can re-mobilise sediment, especially with onshore winds. However, surges are not always accompanied by strong onshore winds.

Spatial differences in accretion are also found. On the levees, low marsh and depressions in the mainland marsh of Peazemerlannen in the Netherlands, the rate of accretion was higher during a period that included winter storms than periods without (Nolte et al. 2013a) (Fig. 9.21). Most sediment originates offshore (see Sect. 9.4.1), the rest is mobilised from adjacent intertidal flats during storms.

Measurements of SEC over six years along the mainland coast of Friesland in the Netherlands, revealed many floodings of the salt marsh with high tides plus storm surges $\geq$0.90 m + MHT in the period 2002/03–2006/07. A positive relationship was found between the cumulative water column per storm year above the salt marsh and annual SEC for the high marsh (0.57–0.85 m + MHT) and low marsh (0.24–0.34 m + MHT) (Fig. 9.22). Esselink and Chang

Table 9.4 Average rates of accretion (AC) or surface elevation change (SEC) on salt marshes within the North Sea region, at different distances from the salt-marsh edge

Area	AC/SEC (mm year^{-1}) distance to edge		Period	Source
Mainland				
Dengie Peninsula, UK	22 (50 m)	11 (200 m)	1981–1983	Reed (1988)
Hamburger Hallig, Germany	17 (50 m)	4 (650 m)	1996–2009	Stock (2011)
	22 (50 m)	1–2 (1000 m)	1995–1999	Schröder et al. (2002)
Schleswig-Holstein, Germany	8 (100 m)	4 (400 m)	1988–2009	Suchrow et al. (2012)
Dollard, Netherlands	12–16 (50 m)	2–5 (800 m)	1984–1991	Esselink et al. (1998)
Back barrier				
Langli, Denmark	0.5 (50 m)	−1 (150 m)	2001–2009	Kuijper and Bakker (2012)
Norfolk, UK	4.5 (50 m)	3.4 (200 m)	1986–1991	French and Spencer (1993)
Skallingen, Denmark	4.2 (50 m)	1.6 (750 m)	1948–1998	Bartholdy et al. (2004)

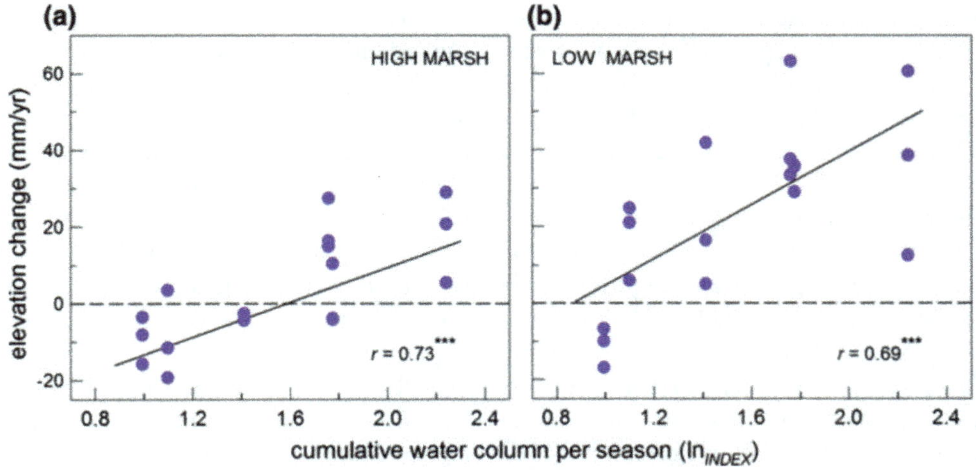

Fig. 9.22 Relation between the annual surface elevation change of the salt marsh and the cumulative water column above 0.9 m + MHT for each storm year (index) along the mainland coast of Friesland in the Netherlands between 2002 and 2007 for the high marsh (**a**) and the low marsh (**b**). Regression lines are based on a joint regression-analysis of the various stations, each station includes three replicates (after Esselink and Chang 2010)

(2010) were able to separate years with various cumulative water columns above the marsh. They found the SEC was 3.8 mm year^{-1} in the high marsh and 31.8 mm year^{-1} in the low marsh in the period 2002–2005, but 4.6 mm year^{-1} in the high marsh and 13.1 mm year^{-1} in the low marsh in the period 2005–2007, which includes the storm surge of 1 November 2006 with a high tide of 2.4 m + MHT compared to average high tides of MHT. Hence, Esselink and Chang (2010) concluded that storm floods contribute strongly to accretion of sediment, but that a hurricane flood does not always result in extra accretion, and may even result in less SEC at the low marsh.

A thick layer of very turbulent water during a hurricane flood might prevent settlement conditions for sediment. Vegetation plays an important role in the accretion of sediment. This was demonstrated in the estuarine Dollard salt marshes in the Netherlands. Both vegetation density and height positively affected accretion rates (Esselink et al. 1998).

9.6.3 Salt-Marsh Ecosystems

9.6.3.1 Plants and Natural Herbivores

Deposited sediment can contribute to SEC, it also contains nitrogen. The nitrogen pool of the rooting zone of 50 cm is positively correlated with the thickness of the clay layer on back-barrier marshes (Olff et al. 1997). In turn, N-mineralisation is positively related to the nitrogen pool (Bakker et al. 2005).

Plant productivity on salt marshes is considered to be limited by nitrogen. N-limitation on lower and higher marshes was demonstrated in west-European salt marshes by Jefferies and Perkins (1977). Nitrogen accumulates during succession in the nitrogen pool of organic matter in the increasing layer of sediment and decaying plants and roots, and N-mineralisation increases with age of the marsh (Van Wijnen et al. 1997). This can be enhanced by the current high rates of atmospheric nitrogen deposition. Although the quantity of plant biomass increases during succession, the quality (more stems that are less palatable) decreases. Hence, the numbers of small herbivores such as winter-staging geese and resident hares and rabbits decrease during succession, after peaking in early successional stages (Van de Koppel et al. 1996). These smaller herbivores need livestock to facilitate for them (Bos et al. 2005).

9.6.3.2 Livestock Grazing

Livestock grazing on European salt marshes can be traced back a couple of millennia (Davy et al. 2009). Older back-barrier marshes are grazed by livestock as mainland marshes. Traditionally many salt marshes were intensively grazed by sheep or cattle (see E-Supplement S9 for ecosystem services). As a result the majority of these marshes were covered by an extremely short homogeneous vegetation of seaside alkali grass *Puccinellia maritima* or red fescue *Festuca rubra* (Kiehl et al. 1996). In the 1980s, grazing by livestock was reduced by up to 60 % on back-barrier marshes and up to 40 % in mainland marshes in the Wadden Sea in 2008 (Esselink et al. 2009). Alongside this change in management regimes there has been an increase in the abundance of the late-successional tall grass species sea couch *Elytrigia atherica* on several salt marshes along the North Sea coast, for example on the Wash, UK (Norris et al. 1997), Schiermonnikoog in the Netherlands (Van Wijnen et al. 1997) and Schleswig-Holstein in Germany (Esselink et al. 2009). There are high numbers of geese and hares on livestock-grazed salt marshes, and low numbers on long-term ungrazed marshes along the coasts of the Wadden Sea (Bos et al. 2005). When grazing ceased, intensive ditching was often discontinued. This resulted in a wetter marsh and reduced spread of sea couch (Veeneklaas et al. 2013).

At the Leybucht, Germany, SEC of 17 mm $year^{-1}$ (1980–1988) at 40 cm + MHT was measured in cattle-grazed marsh, but was higher at 23 mm $year^{-1}$ in the abandoned marsh (Andresen et al. 1990). SEC was about 7 mm $year^{-1}$ over 1996–2009 on ungrazed or low-density grazing on mainland marshes of Hamburger Hallig, Germany. It was however, reduced to 4 mm $year^{-1}$ at sites with high grazing density (Stock 2011). In mainland marshes of the Dollard estuary, intensively grazed sites showed accretion rates of about 8 mm $year^{-1}$ and little-grazed sites about 12 mm $year^{-1}$ over 1984–1991 (Esselink et al. 1998). In Schleswig-Holstein, grazed sites showed an SEC of 3–4 mm $year^{-1}$ and ungrazed sites 8 mm $year^{-1}$ over 1988–2009 (Suchrow et al. 2012). Grazing reduces SEC. However, this is not due to a lower input of sediment in grazed sites, but to an increase in the bulk density of the soil in both a back-barrier marsh (Elschot et al. 2013) and in mainland marshes (Nolte et al. 2013b).

9.6.4 Climate Change and Salt Marshes

9.6.4.1 Sea-Level Rise

Back-barrier marshes with a low vertical accretion might be affected by SLR earlier than mainland marshes with a higher vertical accretion. With little sediment input and continuous SLR, a salt marsh gets wetter, thus preventing the spread of the late successional sea couch *Elytrigia atherica* on Hamburger Hallig even without livestock grazing (Esselink et al. 2009). On the back-barrier marsh of Schiermonnikoog, sea couch has been replaced by the reed *Phragmites australis*, far from the salt-marsh edge (Veeneklaas et al. 2013). This suggest that on broad salt marshes geomorphological changes may occur, resulting in growing differences in elevation, that is, high salt-marsh edge and creek bank levees, and lower depressions between creeks. Especially at the foot of dunes releasing fresh seepage water, brackish conditions can develop.

Accelerated SLR is believed to affect the zonation of salt-marsh plant communities: high salt-marsh communities should turn into lower salt-marsh communities (see Fig. 9.16) (Dijkema et al. 2011). Accelerated SLR can be modelled, but it is not known what will actually happen in reality. An unintended experiment (unique in the North Sea region) taking place as a result of natural gas extraction for energy on the back-barrier salt marsh of Ameland in the Netherlands, may give some idea. Gradual soil subsidence of up to 35 cm by 2050 is expected close to the extraction point (Dijkema 1997). After 25 years, the maximal subsidence is about 25 cm. Near the salt-marsh edge, subsidence is totally counteracted by accretion, with an SEC of zero. About 300 m from the marsh edge, only 4 cm accretion was measured, thus an SEC of −21 cm (Fig. 9.23). Permanent plot data show no decrease in the late successional grass sea couch. At the scale of the salt marsh, however, data from repeated vegetation mapping do show some decrease in the sea couch community (Dijkema et al. 2011). To date, there has been little change in the zonation of plant communities. The future will show whether 25 cm subsidence is sufficient to cause vegetation change, or whether there is a time lag in the response of the vegetation.

Observations of vegetation composition, elevation, soil chemistry, net precipitation, groundwater level, and flooding frequency over the period 1986–2001 were used to predict future changes at the transition between salt marsh and dune due to the combination of ongoing soil subsidence and climate change at the barrier island of Ameland. Climate change was characterised by increases in mean sea level,

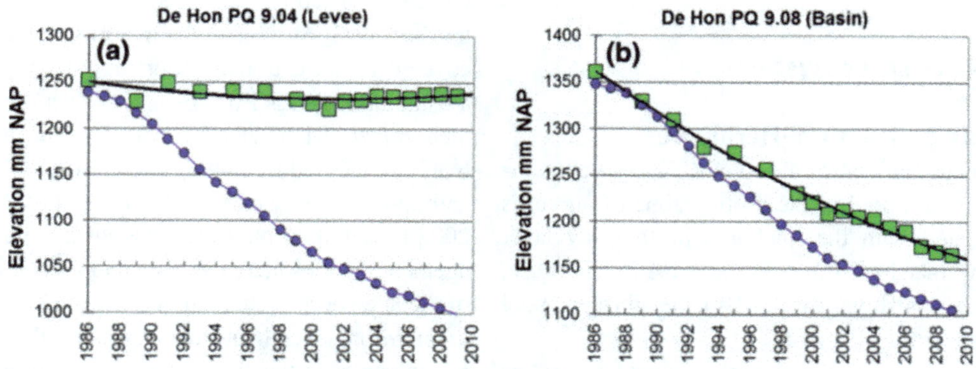

Fig. 9.23 Surface elevation change in the period 1986–2009 (*green*) resulting from soil subsidence due to natural gas extraction (*blue*) and accretion measured near the salt-marsh edge on a creek bank levee (**a**) and about 300 m from the salt-marsh edge in a depression (**b**) (after Dijkema et al. 2011)

storm frequency and net precipitation. Using multiple regression, changes in the vegetation could be subdivided into (1) an oscillatory component due to fluctuations in net precipitation, (2) an oscillatory component due to incidental flooding, (3) a monotonic component due to soil subsidence, and (4) a monotonic component not related to any measured variable but probably due to eutrophication. The changes were generally small during the observation period (1986–2001), but the regression model suggests large changes by 2100 that are almost exclusively due to SLR. Although SLR is expected to cause a loss of plant species, this does not necessarily imply a decrease in nature conservation interest; while common species may be lost rarer species may persist (Van Dobben and Slim 2012).

Comparison of the zonation of beetles and spiders in present mainland salt marshes, and laboratory experiments with enhanced SLR, suggested that species of the lower marsh will move to higher elevation, but species of the higher marsh are unable to escape to higher elevations owing to the barrier of the seawall. These results were interpreted as an example of coastal squeeze (Irmler et al. 2002). Similarly, the lowest-lying vegetation zones will increase at the expense of upper vegetation zones in salt marshes in Denmark with enhanced SLR, according to modelling studies (Moeslund et al. 2011). However, this finding contradicts the previously mentioned results of soil subsidence through gas extraction, which showed no change in the vegetation (Dijkema et al. 2011).

9.6.4.2 Precipitation and Temperature

Although annual deviations from long-term averages are important, there is a general trend of increasing precipitation and rising temperature in the North Sea region. Changes in precipitation may affect plant production (De Leeuw et al. 1990) and plant communities, especially those above MHT (De Leeuw et al. 1991). Long-term trends in precipitation are discussed in Chap. 2 (past) and Chap. 5 (future projections).

Critical loads of atmospheric nitrogen deposition indicate the threshold amount at which ecosystems change dramatically by encroachment of grasses and subsequent loss of species. The empirical range of critical atmospheric nitrogen deposition for 'Pioneer and low-mid salt marshes' has been adjusted in the most recent review (Bobbink et al. 2010) to 30–40 kg N ha^{-1} $year^{-1}$. This range is considered as expert judgment for EUNIS type A2.64 and A2.65. The critical deposition load is now 22 kg N ha^{-1} $year^{-1}$ (1571 mol N ha^{-1} $year^{-1}$) for salt marshes in the Netherlands (Van Dobben and Slim 2012).

Accelerated SLR will have impacts perpendicular to the coastline. In contrast, temperature may have effects parallel to the coastline. The distribution of plant species of salt marshes and dunes may shift along the coast. Based on present patterns of distribution, annual temperature and winter precipitation, Metzing (2010) proposed a model to predict changes in distribution resulting from climate change. An increase of 2.5 °C in annual temperature and 15 % more winter precipitation by 2050 is projected to result in a loss of 16 % of plant species in the Wadden Sea (see also Sect. 9.3.4).

9.7 Coastal Birds

Martijn van de Pol

Coastal birds are typically distinguished from seabirds in that they rely heavily on coastal areas for their food instead of the open sea, although the distinction is not always clear-cut. Coastal and seabirds are phylogenetically closely related and many are part of the same order of Charadriiformes. When discussing the impact of climate change on coastal birds, a rather broad definition of coastal birds is taken here by not only including the large group of waders (sometimes also called shorebirds), but also by including other species that depend heavily on coastal areas for feeding, such as Eurasian spoonbills *Platalea leucorodia*, common eider duck *Somateria mollissima* and various species of terns, geese and gulls. The focus is not on birds that use coastal habitat solely for the purpose of breeding (such as some songbirds).

Since by definition coastal birds rely on coastal areas for their food, they live in the vicinity of coastal shallow waters such as estuaries and intertidal flats and can breed on sandy shores and salt marshes (cliff coasts are generally the domain of seabirds). A substantial proportion of all coastal birds in the North Sea region breed locally, but many others are migratory (e.g. geese) that breed elsewhere (e.g. Arctic), and only use estuaries around the North Sea to overwinter. This distinction between resident and migratory birds is important, as migratory birds may also be impacted by climate change outside the North Sea region.

There is overwhelming evidence from around the globe that the distribution and population size of many bird populations are changing as a direct result of climate change (for an overview see Møller et al. 2010). Around ten species of coastal birds (both breeding and migratory) are declining in parts of the North Sea, such as the Wadden Sea that spans the Netherlands, Germany and Denmark (Fig. 9.24) and estuaries in the United Kingdom (Risely et al. 2012). The functional diversity, however, does not decline (Mendez et al. 2012). But because coastal birds are rarely used as a model system to investigate the effects of climate change, very little is known about the general role that climate change may play in causing changes in numbers of coastal birds in the North Sea region (Ens et al. 2009). Nor is it well understood which species are likely to be adversely affected by climate change and which are likely to benefit.

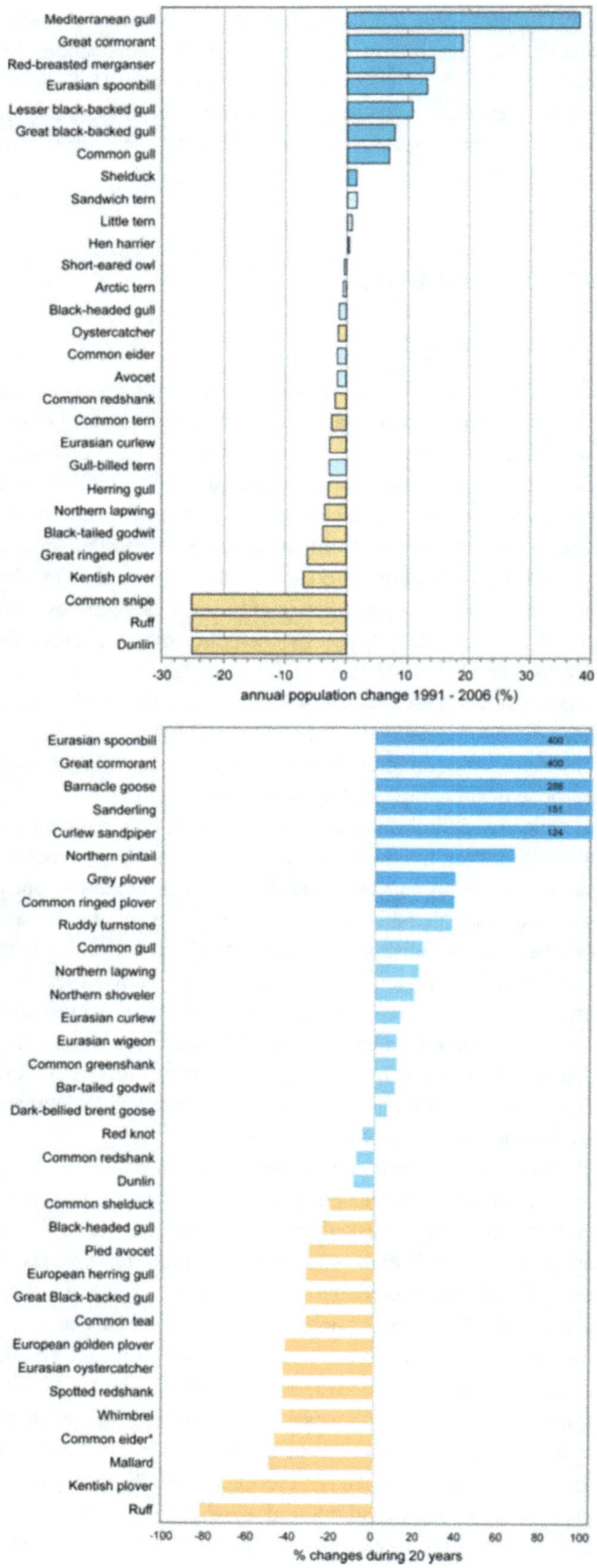

◀ **Fig. 9.24** Recent changes in numbers of breeding (*top*) and migratory (*bottom*) coastal birds in the Danish-German-Dutch Wadden Sea estuary. *Dark blue columns* indicate species with significant, increasing numbers; *light blue columns* indicate species with stable numbers and orange columns indicate species with significant, decreasing numbers (Koffijberg et al. 2009; Laursen et al. 2010)

Nevertheless, one of the aims of this section is to illustrate that although current knowledge about the climate change impact on coastal birds is limited, recent and future climate change are expected to profoundly affect coastal avian communities.

This section focuses on the few available studies on coastal birds from the North Sea region, supplemented by studies from outside the area to help with interpretation. This involves a consideration of those aspects of climate that might be changing and how these may affect coastal birds (Sect. 9.7.1), a review of the ways in which bird populations may respond to climate change, such as the timing of key annual events (migration and egg laying) and changes in reproduction and survival (Sect. 9.7.2), and an examination of how changes in coastal bird population numbers and in distribution range are interpreted in the light of direct and indirect influences of recent climate variability (Sect. 9.7.3).

9.7.1 Effects of Climate Change on Birds

Weather variables such as air and sea temperature, and amounts of precipitation are changing. Intra- and interannual variability in these weather variables, temporal autocorrelation and the frequency of extreme events can also change over time, which may all affect population dynamics of coastal birds (Van de Pol et al. 2010a, b, 2011). However, climate may not change at the same rate throughout the year or in different parts of the world, and this can lead to phenological mismatches in food webs (Visser 2008) and require readjustment of migration schedules (Bauer et al. 2008).

Global warming is causing both air and water temperatures to rise. Air temperatures directly affect the energetic expenditure and thereby food requirements of birds, particularly in small species during winter (Kersten and Piersma 1987). Many benthic and fish species that are prey items of coastal birds are strongly dependent on sea temperatures for growth, reproduction and survival (see also Chap. 8, Sect. 8.4). Severe winters can cause high mortality among many invertebrates, but these same cold winters can also cause shellfish to lose less body mass (Honkoop and Beukema 1997) and result in a good spatfall (Beukema 1992). Thus, it has been suggested that global warming is already

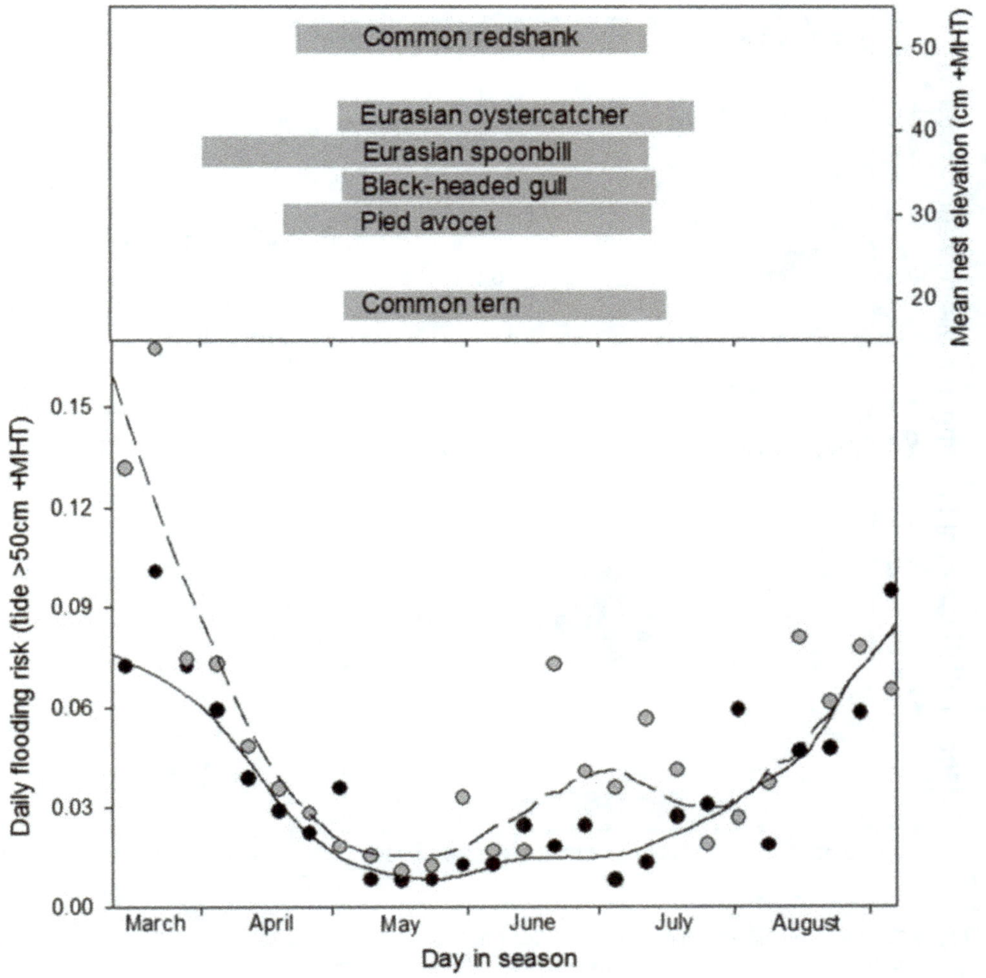

Fig. 9.25 Period during which birds have nests that are sensitive to flooding (eggs or young chicks in the nest) and the mean nest elevation (compared to Mean High Tide) of six salt-marsh breeding species (*upper*). Increase in daily flooding risk of the lower parts of the salt marsh from 1971–1989 (*black circles*) to 1990–2008 (*grey circles*) (*lower*) (Van de Pol et al. 2012)

causing reduced recruitment of shellfish (Beukema and Dekker 2005; Beukema et al. 2009) and a shift in benthic community structure towards less cold-resistant species (e.g. Schückel and Kröncke 2013). Furthermore, invasion of exotic warm-water species, such as Pacific oyster *Crassostrea gigas* which is eaten by few birds, may result in increased competition for existing benthic species (Diederich et al. 2004).

Global warming also causes sea water to expand and land ice to melt thereby directly causing sea levels to rise. Accelerating SLR can have various consequences. The inundation time of intertidal flats may increase or flats used as feeding grounds may become completely inaccessible if sedimentation does not keep up (Flemming and Bartholomä 1997). Accelerated SLR in combination with changing wind patterns (there is currently no definitive evidence in this region for the latter due to the challenge of down-scaling future wind states from coarse resolution climate models) (Van Oldenborgh et al. 2013) can also affect breeding populations on land, as many bird species nest on low salt marshes and beaches that are susceptible to tidal (storm) flooding during the breeding season (Fig. 9.25; Hötker and Segebade 2000; Van de Pol et al. 2010a). Over the long-term, SLR may result in loss of breeding habitat if boundaries such as dunes or seawalls mean that salt marshes cannot move landwards (‘coastal squeeze’; Sect. 9.2.3).

Extreme events may also change in frequency and/or magnitude (Jentsch et al. 2007). In the Netherlands, precipitation and temperature extremes are expected to increase, with more heat waves in summer and higher temperature minima in winter and higher temperature maxima in summer, respectively (KNMI 2014). For example, global warming is expected to decrease the frequency of extremely cold winters in which ice sheets form on intertidal flats. Ice sheets make feeding areas inaccessible and if conditions prevail for long periods, they are known to result in large-scale frost migration and mass mortality of coastal birds (e.g. Camphuysen et al. 1996).

9.7.2 Responses to Climate Change

Individual birds may respond in various ways to a changing environment. Changes in the timing of reproduction (Crick

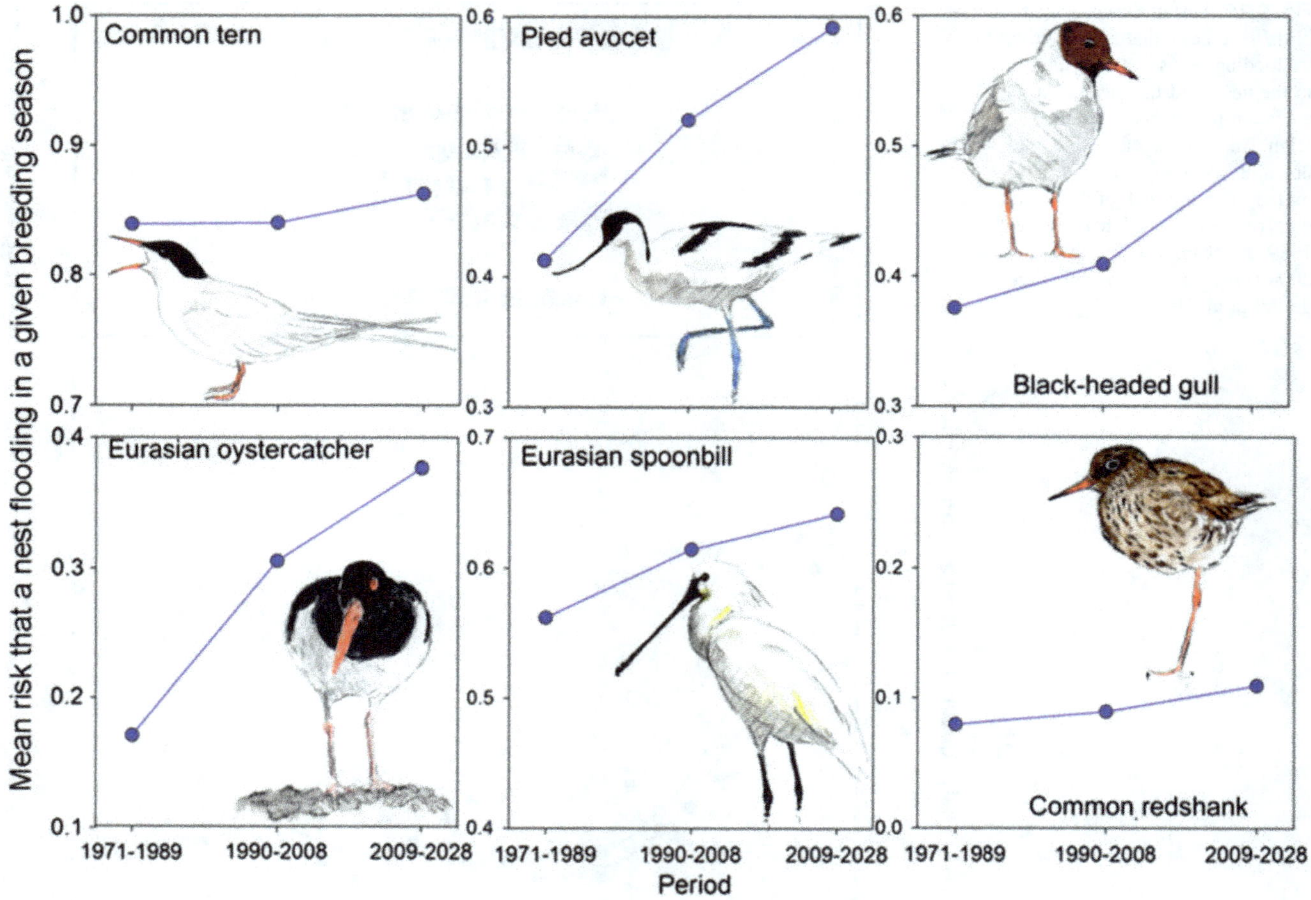

Fig. 9.26 Mean probability that a nest will flood at least once in a given breeding season, calculated for recent years (1990–2008), and projections for earlier (1971–1989) and future periods (2009–2028) based on model projections for sea-level rise and flooding risk that assume birds do not adjust their nest site selection adaptively (Van de Pol et al. 2010a). Please note that the range of the *y*-axis varies between panels

and Sparks 1999) and migration (Cotton 2003) have been most widely reported, since adjusting timing is a flexible way to alter the climatic conditions experienced. If birds cannot alter their timing sufficiently or climate change makes historically adaptive cues maladaptive, then populations may suffer (Visser 2008). Focusing on coastal birds, there appears to be no general pattern in how timing of egg laying has responded, with for example northern lapwings *Vanellus vanellus* showing no response over the past decades, while egg laying in Eurasian oystercatchers *Haematopus ostralegus* advanced in response to increased rainfall, and in ringed plovers *Charadrius hiaticula* advanced in response to rising temperatures (Crick and Sparks 1999).

Studies have also identified various cases where interannual variability in demographic rates was associated with climatic variability. The most common pattern is that annual juvenile and/or adult survival is strongly positively related to winter temperatures in many coastal birds (e.g. Peach et al. 1994; Yalden and Pearce-Higgins 1997; Atkinson et al. 2000). By contrast, a review has suggested that populations of nidifugous species (i.e. most waders) might be more strongly influenced by climatic conditions during the breeding season via effects of summer climate on reproduction (Sæther et al. 2004). A specific example of strong effects of summer climate on the reproduction of a community of coastal birds occurs when salt marshes become flooded during periods of strong wind, which can lead to catastrophic failure of a given breeding season. However, the degree to which species are affected by such events depends strongly on their existing nesting preference and elevation (Fig. 9.26), as well as on their potential to respond to increased frequency of summer flooding events (about which virtually nothing is known).

9.7.3 Changes in Bird Numbers and Distribution

Although there is some evidence that key life-history traits and demographic rates are changing potentially as a direct

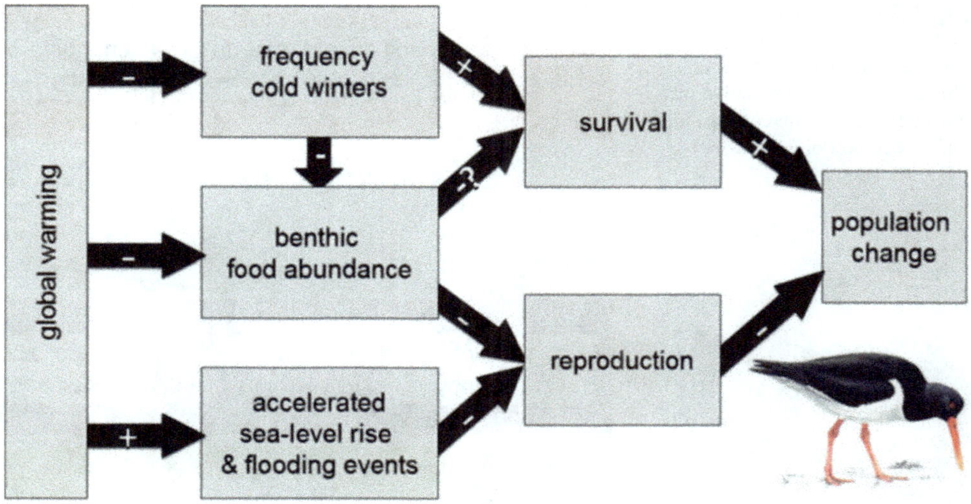

Fig. 9.27 Schematic illustration of how different aspects of a changing climate affect different demographic rates in Eurasian oystercatchers (Van de Pol et al. 2010a)

result of changes in climate variables, it remains very difficult to predict from these patterns what the population consequences of a changing climate might be. The reason for this is two-fold. First, many aspects of the weather (temperature, rainfall) are changing and not necessarily at the same rate throughout the year and in the same area. To assess the impact of climate change as a whole it is necessary to have good knowledge of all major climate drivers affecting bird populations. Second, different parts of the life-cycle may be affected by climate change, and not necessarily in the same direction. This implies that to predict how climate change will affect the population size of a species, requires a thorough understanding of which demographic rates are affected and how this in turn will affect population dynamics. In some cases, even strong climate dependency of demographic rates may not necessarily translate into changes in population size, for example in species with strong density-regulation (Reed et al. 2013).

An example of how complex the situation might be is shown in Fig. 9.27 for Eurasian oystercatchers breeding on a salt marsh on the Wadden Sea island of Schiermonnikoog (Van de Pol et al. 2010a, b, 2011). The frequency of exceptionally cold winters is predicted to decrease in this area, which is expected to result in an increase in both juvenile and adult survival. However, these warmer winters are also expected to reduce the benthic food stocks on which oystercatchers rely for their reproduction. Reproduction is expected to be further negatively affected by the flooding events that have become more frequent and higher in recent decades, especially during the breeding season (Figs. 9.25 and 9.26). Taking all these relationships together it is a priori unclear whether the local population will benefit or suffer, and a population model needs to be built that includes effects on demographic rates across the entire life-cycle in order to establish whether oystercatcher numbers are likely to benefit from climate change (Van de Pol et al. 2010b). This example also illustrates that even though reproduction is likely to be negatively affected by climate change, this does not mean that population numbers will also be reduced. Since reproduction and survival might be affected in opposite directions by different aspects of climate change predicting how population numbers will respond is not straightforward. Translating local population dynamics to meta-population dynamics is even more difficult, as climate does not affect all populations equally (e.g. some salt marshes are more susceptible to flooding than others due to geographic variation in elevation and compensatory sedimentation rates; see Sect. 9.6.2).

Finally, most coastal birds are already affected by many other (anthropogenic) threats, such as changes in land use (see, for example, effects of livestock grazing or abandoning on the vegetation of salt marshes in Sect. 9.6), predation, eutrophication, fisheries and disturbance (Koffijberg et al. 2009), and these other threats may interact with or add to the effects of climate change and thus make it difficult to isolate the contribution of climate change to changes in bird numbers (especially if multiple factors change synchronously over time). For example, as illustrated in Fig. 9.28, when looking at the effects of accelerated SLR on the flooding risk and nesting success of birds, the effects of local soil subsidence due to natural gas extraction may act in a similar cumulative way. Alternatively, the effect of changes in land use may interact indirectly with climate-related SLR via an indirect pathway and feedback loop. Specifically, salt marshes grow vertically in response to more frequent flooding due to sedimentation, but the rate of sedimentation increases with vegetation height, which in turn is affected by the land use in terms of mowing or grazing regimes (see Sect. 9.6).

Direct observations of changes in population numbers are another source of information on how coastal bird populations respond to climate change, although due to the lack of knowledge about the underlying demographic mechanisms, it

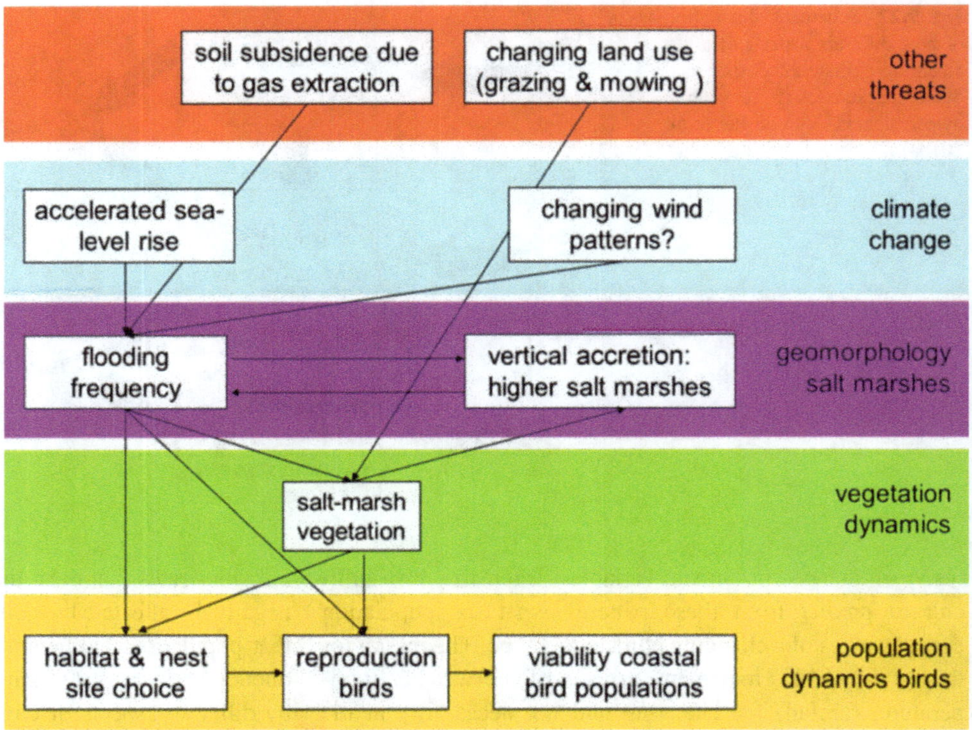

Fig. 9.28 Schematic overview of how specific aspects of climate change and other anthropogenic threats may have cumulative or interactive effects on coastal birds (after Van de Pol et al. 2012)

may be difficult to make reliable predictions for the future. Notwithstanding there is now a strong indication that climate fluctuations are a key driver of population dynamics for coastal birds in Europe's largest estuary, the Wadden Sea. For a striking 30 out of 34 migratory species, changes in meta-population numbers were associated with changes in spring temperatures and/or the NAO index (Laursen et al. 2010). However, the direction of the relationship varied widely among species, and it remains unclear why this is the case.

On a larger scale there have been some clear patterns in observed changes in the distribution of coastal birds, which have been largely attributed to global warming (Ens et al. 2009). It was shown for waders wintering in the United Kingdom that warmer winters led to a shift from wintering on the Atlantic coast (where winters are mild, but food is poor), to wintering on the North Sea coast (where winters are generally more severe, but food supplies are higher) (Austin and Rehfisch 2005). The range shift was especially clear in the smaller species that are affected most from energy stress when temperatures are low (Kersten and Piersma 1987). It has also been suggested that milder winters allow coastal bird species with a greater range of characteristics to overwinter in British estuaries, such that changes in abundance and functional diversity of the community of shorebirds may both change, but in different ways. Another more recent analysis of waders wintering in Europe confirmed a shift in the centre of the distribution to the northeast during the past 30 years, in line with milder temperatures in these areas (Maclean et al. 2008). And also in the Wadden Sea estuary more birds remain to overwinter in the eastern part during mild winters and may also depart at an earlier date to their northern breeding grounds when springs are warm (Bairlein and Exo 2007).

Predictions have also been made for coastal breeding populations on a European-scale based on climate envelope models with mostly northward shifts in the distribution of coastal birds (Huntley et al. 2008). Although these models form a useful starting point, they do not consider the adaptive potential of species. In fact, some coastal species such as common redshanks *Tringa totanus*, black-tailed godwits *Limosa limosa*, northern lapwings and Eurasian oystercatchers have previously shown that they can be very successful in adapting to new environmental conditions, as evidenced by their extremely successful colonisation of non-coastal agricultural areas over the 21st century (Van de Kam et al. 2004).

9.8 Conclusions

9.8.1 Abiotic Conditions

Accelerated SLR, and changes in the wave climate, storms, and local sediment availability all affect the abiotic conditions of coastal systems. The relative importance of these climate change effects and how they interact is poorly understood. It is even more difficult to separate effects of climate change from natural dynamics and the human impacts such as dredging. Human impacts on the geomorphology and sediment transport in estuaries are very likely to continue in the coming decades, and may supersede, exacerbate or compensate for the potential impacts of climate change. Heavy storms may result in coastal squeeze. This is particularly the case for dunes and salt marshes with a short foreshore, that is, a relatively narrow and steep foreshore.

9.8.1.1 Sandy Shores and Dunes

- The general response of sandy shores to climate change in the coming decades will be difficult (if not impossible) to detect and quantitatively predict as it is most likely to be superseded by local natural and/or human-impacted dynamics. Increased wave heights, storm surges and SLR, coupled with observed steepening of beach profiles and a historical decline in sediment availability due to coastal protection mean less sediment is available to replenish erosion of beach sand.
- Future changes in abiotic conditions of coastal dune systems are more likely to be driven by local anthropogenic impacts and natural variability than attributable directly to climate change. Rate of coastal change is strongly affected by local conditions, such as the effect of offshore bathymetry on inshore wave climate, local sand availability, rather than by regional variability and changes in the wave and wind climate.

9.8.1.2 Salt Marshes

- The SSC of fine-grained material is important for salt marshes to cope with sea-level rise. Salt marshes with a low tidal range and low SSC (<20 mg l^{-1}) are at risk of submergence by average SLR projections. The large tidal ranges (2–6 m) and generally large SSCs (mostly > 20 mg l^{-1} and up to several hundreds of mg l^{-1}) in estuaries make the risk low for marsh submergence by SLR. This also holds for most mainland marshes with or without sedimentation fields. The risk for marsh subsidence by SLR may be higher for back-barrier salt marshes with lower SSC values and tidal ranges.
- There is no single figure for surface elevation change for entire marshes. The position on the marsh is important: distance to source of suspended sediment, namely, edge of salt marsh or creek. Depressions away from the salt-marsh edge and creeks on back-barrier marshes are vulnerable. Surface elevation change is reduced in older marshes as a result of autocompaction, and in grazed marshes as a result of increased bulk density.
- Knowledge of vertical accretion rates at the scale of catchment areas on salt marshes (creeks with their drainage area) in salt marshes is poorly developed. Minor storm floods contribute strongly to accretion of sediment, but a heavy storm flood storm does not always result in extra accretion, and may even result in less surface elevation change at the low marsh and pioneer zone.
- More intensive process studies are needed to elucidate the linkages between tidal marshes and adjacent estuarine and coastal systems.
- De-embankment of summer polders can help to enlarge the area of salt marshes.

9.8.2 Plant and Animal Communities

Plant and animal communities can suffer from habitat loss by coastal squeeze in dunes and salt marshes as a result of high wave energy. This is particularly the case for dunes and salt marshes with a short foreshore. Apart from erosion by storm surges in winter, floodings occurred in summer over past decades with subsequent loss of offspring of breeding birds. Plants and animals are also affected by other aspects of climate change, such as changes in temperature and precipitation and atmospheric deposition of nitrogen. Local populations must deal with invasive species that change competitive interactions. Moreover, natural dynamics such as succession, and management practices such as grazing and mowing have a strong impact on plant and animal communities. The key challenge is not only to identify the exact role of climate change, but also to determine the relative importance of climate change compared to other impacts, and how they might interact.

9.8.2.1 Sandy Shores and Dunes

- On the drier southern North Sea coasts, vegetation of dry dunes will increase in xerophytes and bare sand due to moisture limitation of vascular plants.
- Water levels of dune wetlands are highly sensitive to changes in evapotranspiration and therefore recharge. There is currently little consensus on the effects of climate change but implications for dune slack vegetation could be severe. This is a major knowledge gap.
- Dune groundwater chemistry may become more concentrated with solutes due to lower recharge, or altered chemistry of input waters.
- There are relatively few species-specific climate studies, but most suggest a northward shift in species ranges. At a European-scale this may have few consequences, but distributions within individual countries may change markedly. Changes in climate may favour invasive shrub species such as bird cherry *Prunus serotina* or Japanese rose *Rosa rugosa* with subsequent changes in the food web.
- Atmospheric nitrogen deposition is above the critical level for dry and wet dune systems. Nitrogen is available

in excess and causes increased plant production with subsequent loss of slow-growing plant species. There may be interactions between climate change and the effects of nitrogen deposition, linked to faster growth of competitive species in both situations, and enhanced mineralisation of soil organic matter promoting nutrient availability and leaching of nutrients to groundwater.

9.8.2.2 Salt Marshes

- Minor storm floodings in spring negatively affect breeding birds.
- Plant production is significantly positively related to increased precipitation on salt marshes above MHT. Increased plant production could result in outcompeting low-statured species, and hence a decrease in species richness on marshes with subsequent changes in the food web.
- Some southern species will extend northward as a result of higher temperatures. The number of species extinct or emigrating north is smaller than the number of immigrating from the south. The real change in distribution patterns will differ for different species, for example due to migration rates.
- Atmospheric nitrogen deposition is just below the critical level for salt-marsh communities. When the limiting resource nitrogen is available in excess, plant production can easily increase with higher precipitation and temperature.

References

Ablain M, Cazenave A, Valladeau G, Guinehut S (2009) A new assessment of the error budget of global mean sea level rate estimated by satellite altimetry over 1993–2008. Ocean Sci 5:193–201

Albrecht F, Wahl T, Jensen J, Weisse R (2011) Determining sea level changes in the German Bight. Ocean Dyn 61:2037–2050

Allen JRL (1990) Salt-marsh growth and stratification: a numerical model with special reference to the Severn Estuary, southwest Britain. Mar Geol 95:77–96

Allen JRL (2000) Morphodynamics of Holocene salt marshes: a review sketch from the Atlantic and Southern North Sea coasts of Europe. Quat Sci Rev 19:1155–1231

Andresen H, Bakker JP, Brongers M, Heydemann B, Irmler U (1990) Long-term changes of salt marsh communities by cattle grazing. Vegetatio 89:137–148

Arens SM, Geelen L (2006) Dune landscape rejuvenation by intended destabilisation in the Amsterdam water supply dunes. J Coast Res 22:1094–1107

Arens SM, Wiersma J (1994) The Dutch foredunes: inventory and classification. J Coast Res 10:189–202

Arens SM, Mulder JPM, Slings QL, Geelen LHWT, Damsma P (2013) Dynamic dune management, integrating objectives of nature development and coastal safety: Examples from the Netherlands. Geomorphology 199:205–213

Atkinson PW, Clark NA, Clark JA, Bell MC, Dare PJ, Ireland PL (2000) The effects of changes in shellfish stocks and winter weather on shorebird populations: results of a 30-year study on the Wash, England. BTO Res Rep 238

Austin GE, Rehfisch MM (2005) Shifting nonbreeding distributions of migratory fauna in relation to climatic change. Glob Change Biol 11:31–38

Baeyens G, Martínez ML (2004) Animal life on coastal dunes: From exploitation and prosecution to protection and monitoring. In: Martínez ML, Psuty NP (eds) Coastal Dunes: Ecology and Conservation. Ecol Stud 171. Springer

Bairlein F, Exo KM (2007) Climate change and migratory waterbirds in the Wadden Sea. Seriously declining trends in migratory waterbirds: Causes-Concerns-Consequences. Common Wadden Sea Secretariat, Wilhelmshaven

Bakker JP, Bouma TJ, Van Wijnen HJ (2005) Interactions between microorganisms and intertidal plant communities. In: Kristensen K, Kostka JE, Haese RR (eds) Interactions Between Macro- and Microorganisms in Marine Sediments. Coast Estuar Stud 60. American Geophysical Union

Bartholdy J (1997) The backbarrier sediments of the Skallingen peninsula, Denmark. Dan J Geogr 97:11–32

Bartholdy J (2000) Processes controlling import of fine-grained sediment to tidal areas: a simulation model. In: Pye K, Allen JRL (eds) Coastal and Estuarine Environments: Sedimentology, Geomorphology and Geoarcheology. Geol Soc, London, Spec publ 175

Bartholdy J (2012) Salt marsh sedimentation. In: Davis RA, Dalrymple RW (eds) Principles of Tidal Sedimentology. Springer

Bartholdy J, Aagaard T (2001) Storm surge effects on a back-barrier tidal flat of the Danish Wadden Sea. Geo Mar Lett 20:133–141

Bartholdy J, Christiansen C, Kunzendorf H (2004) Long term variations in backbarrier salt marsh deposition on the Skallingen peninsula – the Danish Wadden Sea. Mar Geol 203:1–21

Bartholdy J, Pedersen JBT, Bartholdy AT (2010a) Autocompaction in shallow silty salt marsh clay. Sedim Geol 223:310–319

Bartholdy AT, Bartholdy J, Kroon A (2010b) Salt marsh stability and patterns of sedimentation across a backbarrier platform. Mar Geol 278:31–42

Bartholomeus RP, Witte JPM, Runhaar J (2012) Drought stress and vegetation characteristics on sites with different slopes and orientations. Ecohydrology 5:808–818

Bauer S, van Dinther M, Høgda KA, Klaassen M, Madsen J (2008) The consequences of climate-driven stop-over sites changes on migration schedules and fitness of Arctic geese. J Anim Ecol 77:654–660

Bauer BO, Davidson-Arnott RGD, Hesp PA, Namikas SL, Ollerhead J, Walker IJ (2009) Aeolian sediment transport on a beach: surface moisture, wind fetch, and mean transport. Geomorphology 105:106–116

Beniston M, Stephenson DB, Christensen OB, Ferro CAT, Frei C, Goyette S, Halsnaes K, Holt T, Jylhä K, Koffi B, Palutikof J, Schöll R, Semmler T, Woth K (2007) Future extreme events in European climate: an exploration of regional climate model projections. Clim Change 81:71–95

Berry PM, Jones AP, Nicholls RJ, Vos CC (eds) (2007) Assessment of the vulnerability of terrestrial and coastal habitats and species in Europe to climate change, Annex 2 of Planning for biodiversity in a changing climate – BRANCH project Fin Rep. Natural England

Beukema JJ (1992) Expected changes in the Wadden Sea benthos in a warmer world: Lessons from periods with mild winters. Neth J Sea Res 30:73–79

Beukema JJ, Dekker R (2005) Decline of recruitment success in cockles and other bivalves in the Wadden Sea: possible role of climate change, predation on postlarvae and fisheries. Mar Ecol Prog Ser 287:149–167

Beukema JJ, Dekker R, Jansen J (2009) Some like it cold: populations of the tellinid bivalve *Macoma balthica* (L) suffer in various ways from a warming climate. Mar Ecol Prog Ser 384:135–145

Bobbink R, Hicks K, Galloway J, Spranger T, Alkemade R, Ashmore M, Bustamante M, Cinderby S, Davidson E, Dentener F, Emmett B, Erisman JW, Fenn M, Gilliam F, Norden A, Pardo L, De Vros W (2010) Global assessment of nitrogen deposition effects on terrestrial plant diversity: a synthesis. Ecol Appl 20:30–59

Boorman LA, Boorman MS (2001) The spatial and temporal effects of grazing on the species diversity of sand dunes. In: Houston JA, Edmondson SE, Rooney PJ (eds) Coastal Dune Management: shared experience of European conservation practice. Liverpool Univ Press

Boorman LA, Goss-Custard JD, McGrortyS (1989). Climate change, rising sea level and the British coast. Natural Environment Research Council. ITE Research Publication 1:1-24

Bos D, Loonen M, Stock M, Hofeditz F, Van Der Graaf S, Bakker JP (2005) Utilisation of Wadden Sea salt marshes by geese in relation to livestock grazing. J Nat Conserv 15:1–15

Bouma TJ, Van Belzen J, Balke T, Zhu Z, Airoldi L, Blight AJ, Davies AJ, Galvan C, Hawkins SJ, Hoggart SPG, Lara JL, Losad IJ, Maza M, Ondivila B, Skov MW, Strain EM, Thompson RC, Yang S, Zanuttigh B, Zhang L, Herman PMJ (2014) Identifying knowledge gaps hampering application of intertidal habitats in coastal protection: Opportunities and steps to take. Coast Eng 87:147–157

Broeckx S, Smets S, Liekens I, Bulckaen D, De Nocker L (2011) Designing a long-term flood risk management plan for the Scheldt estuary using a risk-based approach. Natur Haz 57:245–266

Bruun P (1962) Sea level rise as a cause of shore erosion. J Waterw Harb Div, ASCE 88:117–130

Cahoon DR, Reed DJ, Day JW (1995) Estimating shallow subsidence in microtidal salt marshes of the southeastern United States: Kaye and Barghoorn revisited. Mar Geol 128:1–9

Cahoon DR, French JR, Spencer T, Reed D, Möller I (2000) Vertical accretion versus elevational adjustment in UK saltmarshes: an evolution of alternative methodologies. In: Pye K, Allen JRL (eds) Coastal and Estuarine Environments: Sedimentology, Geomorphology and Geoarcheology. Geol Soc Spec Publ 175

Callaghan DP, Bouma TJ, Klaassen PC, Van der Wal D, Stive MJF, Herman PMJ (2010) Hydrodynamic forcing on salt-marsh development: distinguishing the relative importance of waves and tidal flows. Estuar Coast Shelf Sci 89:73–88

Camphuysen CJ, Ens BJ, Heg D, Hulscher JB, van der Meer J, Smit CJ (1996) Oystercatcher *Haematopus ostralegus* winter mortality in the Netherlands: the effect of severe weather and food supply. Ardea 84a:469–492

Carter RWG (1991) Near-future sea level impacts on coastal dune landscapes. Landsc Ecol 6:29–39

Chauhan PPS (2009) Autocyclic erosion in tidal marshes. Geomorphology 61:373–391

Chini N, Stansby P, Leake J, Wolf J, Roberts-Jones J, Lowe J (2010) The impact of sea level rise and climate change on inshore wave climate: a case study for East Anglia (UK). Coast Eng 57:973–984

Church JA, White NJ (2006) A 20th century acceleration in global sea-level rise. Geophys Res Lett 33:L01602

Ciais P, Reichstein M, Viovy N, Granier A, Ogee J, Allard V, Aubinet M, Buchmann N, Bernhofer C, Carrara A, Chevallier F, De Noblet N, Friend AD, Friedlingstein P, Grunwald T, Heinesch B, Keronen P, Knohl A, Krinner G, Loustau D, Manca G, Matteucci G, Miglietta F, Ourcival JM, Papale D, Pilegaard K, Rambal S, Seufert G, Soussana JF, Sanz MJ, Schulze ED, Vesala T, Valentini R (2005) Europe-wide reduction in primary productivity caused by the heat and drought in 2003. Nature 437:529–533

Clarke ML, Rendell HM (2009) The impact of North Atlantic storminess on western European coasts: A review. Quat Int 195:31–41

Clarke D, Sanitwong Na Ayutthaya S (2010) Predicted effects of climate change, vegetation and tree cover on dune slack habitats at Ainsdale on the Sefton Coast, UK. J Coast Conserv 14:115–125

Cloern JE (1987) Turbidity as a control on phytoplankton biomass and productivity in estuaries. Cont Shelf Res 7:1367–1381

Cooper NJ, Cooper T, Burd F (2001) 25 years of salt marsh erosion in Essex: Implications for coastal defence and nature conservation. J Coast Conserv 9:31–40

Cotton PA (2003) Avian migration phenology and global climate change. Proc Nat Acad Sci 100:12219–12222

Cowell PJ, Stive MJE, Niederoda AW, DeVriend HJ, Swift DJP, Kaminsky GM, Capobianco M (2003) The Coastal-Tract (part 1): a conceptual approach to aggregated modeling of low-order coastal change. J Coast Res 19:812–827

Cox R, Wadsworth RA, Thomson AG (2003) Long-term changes in salt marsh extent affected by channel deepening in a modified estuary. Cont Shelf Res 23:1833–1846

Crick HQP, Sparks TH (1999) Climate change related to egg-laying trends. Nature 399:423–424

Curreli A, Wallace H, Freeman C, Hollingham M, Stratford C, Johnson H, Jones L (2013) Eco-hydrological requirements of dune slack vegetation and the implications of climate change. Sci Tot Environ 443:910–919

Dargie TCD (1993) Sand dune vegetation survey of Great Britain. Part 2 - Scotland. Peterborough, Joint Nature Conservation Committee

Davy AJ, Bakker JP, Figueroa ME (2009) Human modification of European salt marshes. In: Silliman BR, Grosholz ED, Bertness MD (eds) Human Impacts on Salt Marshes. University of California Press

Davy AJ, Hiscock KM, Jones MLM, Low R, Robins NS, Stratford C (2010) Ecohydrological guidelines for wet dune habitats, phase 2. Protecting the plant communities and rare species of dune wetland systems. Bristol, Environment Agency

De Groot AV, Veeneklaas RM, Bakker JP (2011) Sand in the salt marsh: contribution of high-energy conditions to salt-marsh accretion. Mar Geol 282:240–254

De Leeuw J, Olff H, Bakker JP (1990) Year-to-year variation in salt marsh production as related to inundation and rainfall deficit. Aquat Bot 36:139–151

De Leeuw J, Van Den Dool A, De Munck W, Nieuwenhuize J, Beeftink WG (1991) Factors influencing the soil salinity regime along an intertidal gradient. Estuar Coast Shelf Sci 32:87–97

De Swart HE, Zimmerman JTF (2009) Morphodynamics of tidal inlet systems. Ann Rev Fluid Mech 41:203–229

De Winter RC (2014) Dune erosion under climate change. PhD Thesis, Utrecht University

De Winter RC, Sterl A, de Vries JW, Weber SL, Ruessink BG (2012) The effect of climate change on extreme waves in front of the Dutch coast. Ocean Dyn 62:1139–1152

De Winter RC, Sterl A, Ruessink BG (2013) Wind extremes in the North Sea Basin under climate change: an ensemble study of 12 CMIP5 GCMs. J Geophys Res: Atmos 118:1601–1612

Diederich S, Nehls G, Beusekom JEE, Reise K (2004) Introduced Pacific oysters (*Crassostrea gigas*) in the northern Wadden Sea: invasion accelerated by warm summers? Helgol Mar Res 59:97–106

Dijkema KS (1987) Changes in salt-marsh area in the Netherlands Wadden Sea after 1600. In: Huiskes AHL, Blom CWPM, Rozema J (eds) Vegetation between Land and Sea. Junk, Dordrecht

Dijkema KS (1994) Auswirkungen des Meeresspieglanstieges auf die Salzwiesen. In: Lozán JJ, Rachor E, Reise K, Von Westernhage H, Lens W (eds) Warnsignale aus dem Wattenmeer. Blackwell

Dijkema KS (1997) Impact prognosis for salt marshes from subsidence by gas extraction in the Wadden Sea. J Coast Res 13:1294–1304

Dijkema KS, Van Duin WE (2012) 50 jaar monitoring van kwelderwerken. De Lev Natuur 113:118–122

Dijkema KS, Kers AS, Van Duin WE (2010) Salt marshes: applied long-term monitoring. In: Marencic H, Eskildsen K, Farke H, Hedtkamp S (eds) Science for Nature Conservation and Management: The Wadden Sea Ecosystem and EU Directives. Proc 12th Int Sci Wadden Sea Symp Wilhelmshaven, Germany, 30 March – 3 April 2009. Wadden Sea Ecosys 26. Common Wadden Sea Secretariat, Wilhelmshaven

Dijkema KS, van Dobben HF, Koppenaal EC, Dijkman EM, van Duin WE (2011) Kweldervegetatie Ameland 1986-2010: effecten van bodemdaling en opslibbing op NeerlandsReid en De Hon. In: Begeleidingscommissie Monitoring Bodemdaling Ameland. Monitoring Bodemdaling Ameland. Deel 2. Evaluatie na 23 jaar gaswinning

Dillingh D, Baart F, Ronde J de (2012) Do we have to take an acceleration of sea level rise into account? In: EGU general assembly 14. Geophys Res Abst. Munich, EGU (TUD)

DiPol (2012) WP "Risk analysis" SCREMOTOX March 2012. Interreg IVb Report

Doing H (1995) Landscape ecology of the Dutch coast. J Coast Conserv 1:145–172

Donat MG, Leckebusch GC, Pinto JG, Ulbrich U (2010) European storminess and associated circulation weather types: future changes deduced from a multi-model ensemble of GCM simulations. Clim Res 42:27–43

Doody JP (2013) Sand Dune Conservation, Management, and Restoration. Springer

Doody JP, Skarregaard P (2007) Sand dune - Country Report, Denmark. Coastal Wiki, Flanders Marine Institute (VLIZ)

Dronkers J (2005) Dynamics of Coastal Systems. World Scientific

Dyer KR (1997) Estuaries: A Physical Introduction. Wiley

EEA (2012) Climate changes impacts and vulnerability in Europe 2012. EEA report 12/2012. European Environment Agency (EEA), Copenhagen

Elschot K, Bouma TJ, Temmerman S, Bakker JP (2013) Effects of long-term grazing on sediment deposition and salt-marsh accretion. Estuar Coast Shelf Sci 133:109–115

Emmett BA, Beier C, Estiarte M, Tietema A, Kristensen HL, Williams D, Penuelas J, Schmidt I, Sowerby A (2004) The response of soil processes to climate change: Results from manipulation studies of shrublands across an environmental gradient. Ecosystems 7:625–637

Ens BJ, Blew J, Roomen MWJ, van Turnhout CAM (2009) Exploring contrasting trends of migratory waterbirds in the Wadden Sea. Common Wadden Sea Secretariat, Trilateral Monitoring and Assessment Group, Joint Monitoring Group of Migratory Birds in the Wadden Sea, Wilhemshaven

Erchinger HF (1985) Dünen, Watt und Salzwiesen. Das Niedersächsische Ministerium für Ernährung, Landwirtschaft und Forsten. Hannover

Esselink P, Chang ER (2010) Kwelderherstel Noard-Fryslân Bûtendyks: invloed van stormactiviteit op zes jaar proefverkweldering. Rapport 01. PUCCIMAR Ecologisch Onderzoek & Advies

Esselink P, Dijkema KS, Reents S, Hageman G (1998) Vertical accretion and profile changes in abandoned man-made tidal marshes in the Dollard estuary, the Netherlands. J Coast Res 14:570–582

Esselink P, Petersen J, Arens S, Bakker JP, Bunje J, Dijkema KS, Hecker N, Hellwig AV, Jensen B, Kers B, Körber P, Lammerts EJ, Lüerssen G, Marencic H, Stock M, Veeneklaas RM, Vreeken M, Wolters M (2009) Salt Marshes. Thematic Report 8. In: Marencic H, De Vlas J (eds) Wadden Sea Quality Status Report 2009.Wadden Sea Ecosystems 25. Common Wadden Sea Secretariat, Wilhelmshaven

Esselink P, Bos D, Oost AP, Dijkema KS, Bakker R, De Jong R (2011) Verkenning afslag Eems-Dollardkwelders. PUCCIMAR Ecologisch Onderzoek en Advies, Altenburg &Wymenga ecologisch onderzoek, Vries

European Commission (1992) COUNCIL DIRECTIVE 92/43/EEC of 21 May 1992, on the conservation of natural habitats and of wild fauna and flora

Eurosion (2004) Living with coastal erosion in Europe: sediment and space for sustainability. Online through www.eurosion.org/report-online/reports.html. Accessed May 2013

Feagin RA, Sherman DJ, Grant WE (2005) Coastal erosion, global sea-level rise, and the loss of sand dune plant habitats. Front Ecol Env 3:359–364

Flemming BW, Bartholomä A (1997) Response of the Wadden Sea to a rising sea level: a predictive empirical model. Deut Hydrogr Z 49:343–353

French JR (1993) Numerical simulation of vertical marsh growth and adjustment to accelerated sea-level rise, North Norfolk. Earth Surf Proc Landf 18:63–81

French J (2006a) Tidal marsh sedimentation and resilience to environmental change: Exploratory modelling of tidal, sea-level and sediment supply forcing in predominantly allochthonous systems. Mar Geol 235:119–136

French PW (2006b) Managed realignment - The developing story of a comparatively new approach to soft engineering. Estuar Coast Shelf Sci 67:409–423

French JR, Spencer T (1993) Dynamics of sedimentation in a tidal-dominated backbarrier salt marsh, Norfolk. Mar Geol 110:315–331

Friedrichs CT, Aubrey DG (1994) Tidal propagation in strongly convergent channels. J Geophys Res 99:3321–3336

Grabemann I, Weisse R (2008) Climate change impact on extreme wave conditions in the North Sea: and ensemble study. Ocean Dyn 58:199–212

Grootjans AP, Adema EB, Bekker RM, Lammerts EJ (2004) Why coastal dune slacks sustain a high biodiversity. In: Martinez ML, Psuty NP (eds) Coastal Dunes: Ecology and Conservation. Ecological Studies. Springer

Hadley D (2009) Land use and coastal zone. Land Use Pol 26S:198–203

Haigh I, Nicholls R, Wells N (2011) Rising sea levels in the English Channel 1990-2100. Proc ICE Mar Eng 164:81–92

Hall J, Emmett B, Garbutt A, Jones L, Rowe E, Sheppard L, Vanguelova E, Pitman R, Britton A, Hester A, Ashmore M, Power S, Caporn S (2011) UK Status Report July 2011: Update to empirical critical loads of nitrogen. Report to Defra under contract AQ801 Critical Loads and Dynamic Modelling Report date: 5th

July 2011. Available online at: http://cldm.defra.gov.uk/PDFs/UK_status_report_2011_finalversion_July2011_v2.pdf

Harrison PA, Berry PM, Dawson TP (2001) Climate Change and Nature Conservation in Britain and Ireland: Modelling natural resource responses to climate change (the MONARCH project). UKCIP Tech Rep, Oxford

Herrier JL (2008) Sand dune - Country Report, Belgium. In: Doody JP (ed) Euromarine Consortium Wiki. www.euromarineconsortium.eu/wiki/Sand_dune_-_Country_Report,_Belgium

Honkoop PJ, Beukema J (1997) Loss of body mass in winter in three intertidal bivalve species: an experimental and observational study of the interacting effects between water temperature, feeding time and feeding behaviour. J Exper Mar Biol Ecol 212:277–297

Hötker H, Segebade A (2000) Effects of predation and weather on the breeding success of Avocets *Recurvirostra avosetta*. Bird Study 47:91–101

Hulme M, Jenkins G, Lu X et al, UKCIP (2002) Climate change scenarios for the United Kingdom: the UKCIP02 scientific report. Norwich, Tyndall Centre for Climate Change Research. School of Environmental Sciences. University of East Anglia

Huntley B, Collingham YC, Willis SG, Green RE (2008) Potential impacts of climatic change on European breeding birds. PLOS ONE 3. doi:10.1371/journal.pone.0001439

Irmler U, Heller K, Meyer H, Reineke HD (2002) Zonation of ground beetles (Coleoptera: Carabidae) and spiders (Araneida) in salt marshes at the North and Baltic Sea and the impact of the predicted sea level increase. Biodiv Conserv 7:1129–1147

Isermann M (2008) Expansion of *Rosa rugosa* and *Hippophae rhamnoides* in coastal grey dunes: effects at different scales. Flora 203:273–280

Jefferies RL, Perkins N (1977) The effects on the vegetation of the additions of inorganic nutrients to salt marsh soils at Stiffkey, Norfolk. J Ecol 65:867–882

Jensen J, Mudersbach C (2004) Zeitliche Änderungen in den Wasserstandzeitreihen an den deutschen Küsten. Klimaänderung und Küstenschutz. Hamburg, Proceedings Universität Hamburg:1–15

Jentsch A, Kreyling J, Beierkuhnlein C (2007) A new generation of climate-change experiments: events, not trends. Front Ecol Env 5:365–374

Jones MLM, Hayes F, Brittain SA, Haria S, Williams PD, Ashenden TW, Norris DA, Reynolds B (2002) Changing nutrient budget of sand dunes: Consequences for the nature conservation interest and dune management. 2. Field Survey. CCW Contract No: FC 73-01-347. CEH Project No: C01919

Jones MLM, Wallace H, Norris DA, Brittain SA, Haria S, Jones RE, Rhind PM, Williams PD, Reynolds B, Emmett BA (2004) Changes in vegetation and soil characteristics in coastal sand dunes along a gradient of atmospheric nitrogen deposition. Plant Biol 6:598–605

Jones MLM, Sowerby A, Williams DL, Jones RE (2008) Factors controlling soil development in sand dunes: evidence from a coastal dune soil chronosequence. Plant and Soil 307:219–234

Jones MLM, Norman K, Rhind PM (2010) Topsoil inversion as a restoration measure in sand dunes, early results from a UK field-trial. J Coast Conserv 14:139–151

Jones MLM, Angus S, Cooper A, Doody P, Everard M, Garbutt A, Gilchrist P, Hansom G, Nicholls R, Pye K, Ravenscroft N, Rees S, Rhind P, Whitehouse A (2011) Coastal margins. In: UK National Ecosystem Assessment. Understanding nature's value to society. Cambridge, Tech Rep. UNEP-WCMC

Kamps LF (1962) Mud distribution and land reclamation in the eastern Wadden Sea shallows. Rijkswaterstaat Communications 4. Den Haag

Katsman C, Hazelager W, Drijfhout S, Van Oldenborgh G, Burgers G (2008) Climate scenarios of sea level rise for the northeast Atlantic ocean: a study including the effects of ocean dynamics and gravity changes induced by ice melt. Clim Change 91:351–374

Katsman CA, Sterl A, Beersma JJ, van den Brink HW, Church JA, hazeleger W, kopp RE, Kroon D, Kwadijk J, et al (2011) Exploring high-end scenarios for local sea level rise to develop flood protection strategies for a low-lying delta – the Netherlands as an example. Clim Change 109:617–645

Kearny MS, Rogers AS, Townshend JRG, Rizzo E, Stutzer D (2002) Landsat imagery shows decline of coastal marshes in Chesapeake and Delaware bays. EOS Trans Am Geophys Union 83:177–178

Keijsers JGS, Poortinga A, Riksen MJPM, Maroulis J (2014) Spatio-temporal variability in accretion and erosion of coastal foredunes in the Netherlands: regional climate and local topography. PLOS One 9,3:e91115. doi:10.1371/journal.pone.0091115

Kersten M, Piersma T (1987) High levels of energy expenditure in shorebirds: metabolic adaptations to an energetically expensive way of life. Ardea 75:175–187

Kiehl K, Eischeid I, Gettner S, Walter J (1996) The impact of different sheep grazing intensities on salt-marsh vegetation in Northern Germany. J Veget Sci 7:99–106

Kirwan ML, Guntenspergen GR, D'Alpaos A, Morris JT, Mudd SM, Temmerman S (2010) Limits on the adaptability of coastal marshes to rising sea level. Geophys Res Lett 37:L23401

Knippertz P, Ulbrich U, Speth P (2000) Changing cyclones and wind speeds over the North Atlantic and Europe in a transient GHG experiment. Clim Res 15:109–122

KNMI (2014) KNMI'14 Climate Scenarios for the Netherlands: A guide for professionals in climate adaptation. KNMI, De Bilt, The Netherlands

Knutson TR, McBride JL, Chan J, Emanuel K, Holland G, Landsea C, Held I, Kossin JP, Srivastava AK, Sugi M (2010) Tropical cyclones and climate change. Nat Geosci 3:157–163

Koffijberg K, Dijksen L, Hälterlein B, Laursen K, Potel P, Schrader S (2009) Thematic Report 8. Breeding birds. In: Marencic H, De Vlas J (eds) Wadden Sea Quality Status Report 2009.Wadden Sea Ecosystems 25. Common Wadden Sea Secretariat, Wilhelmshaven

Kooijman AM, Van der Meulen F (1996) Grazing as a control against 'grass-encroachment' in dry dune grasslands in The Netherlands. Land Urb Plan 34:323–333

Kooijman AM, Van der Hagen HGJM, Noordwijk H (2012) Stikstof depositie in de duinen: alles in beeld? Landschap 29:147–154

Kuijper DPJ, Bakker JP (2012) Vertebrate below- and above-ground herbivory and abiotic factors alternate in shaping salt-marsh plant communities. J Exp Mar Biol Ecol 432-433:17–28

Langner J, Bergström R, Foltescu V (2005) Impact of climate change on surface ozone and deposition of sulphur and nitrogen in Europe. Atmos Env 39 1129–1141

Laursen K, Blew J, Eskildsen K, Günther K, Hälterlein B, Kleefstra R, Lüerßen G, Potel P, Schrader S (2010) Migratory Waterbirds in the Wadden Sea 1987–2008. Common Wadden Sea Secretariat, Wilhelmshaven

Leckebusch GC, Ulbrich U (2004) On the relationship between cyclones and extreme windstorm events over Europe under climate change. Glob Plan Change 44:181–193

Levine JM, McEachern AK, Cowan C (2008) Rainfall effects on rare annual plants. J Ecol 96:795–806

Linderholm HW (2006) Growing season changes in the last century. Agric Forest Meteor 137:1–14

Lowe JA, Gregory JM (2005) The effects of climate change on storm surges around the United Kingdom. Phil Trans R Soc A 363: 1313–1328

Maclean IMD, Austin GE, Rehfisch MM, Blew J, Crowe O, Delany S, Devos K, Deceuninck B, Günther K, Laursen K, van Roomen M, Wahl J (2008) Climate change causes rapid changes in the distribution and site abundance of birds in winter. Glob Change Biol 14:2489–2500

Mariotti G, Fagherazzi S (2010) A numerical model for the coupled long-term evolution of salt marshes and tidal flats. J Geophys Res 115:F01004

Maris T, Cox T, Temmerman S, De Vleeschauwer P, Van Damme S, De Mulder T, Van den Bergh E, Meire P (2007) Tuning the tide: creating ecological conditions for tidal marsh development in a flood control area. Hydrobiologia 588:31–43

Mayerhofer P, de Vries B, den Elzen M, van Vuuren D, Onigkeit J, Posch M, Guardans R (2002) Long-term, consistent scenarios of emissions, deposition, and climate change in Europe. Env Sci Pol 5:273–305

Meire P, Ysebaert T, Van Damme S, Van den Bergh E, Maris T, Struyf E (2005) The Scheldt estuary: a description of a changing ecosystem. Hydrobiologia 540:1–11

Mendez V, Gill JA, Burton NHK, Austin GE, Petchey OL, Davies RG (2012) Functional diversity across space and time: trends in wader communities on British estuaries. Divers Distrib 18:356–365

Metzing D (2010) Global warming changes the terrestrial flora of the Wadden Sea. In: Marencic H, Eskilldsen K, Farke H, Hedtkamp S (eds) Science for Nature conservation and Management: the Wadden Sea Ecosystem and EU Directives. Proc 12th Sci Wadden Sea Symp, 30 March – 3 April 2009. Wilhelmshaven. Wadden Sea Ecosys 26:211–215

Moeslund JE, Arge L, Bøcher PK, Nygaard B, Svenning JC (2011) Geographically comprehensive assessment of salt-meadow vegetation-elevation relation using LIDAR. Wetlands 31:471–482

Möller I (2006) Quantifying saltmarsh vegetation and its effect on wave height dissipation: Results from a UK east coast saltmarsh. Estuar Coast Shelf Sci 69:337–51

Møller AP, Fiedler W, Berthold P (2010) Effects of climate change on birds. Oxford University Press

Möller I, Kudella M, Rupprecht F, Spencer T, Paul M, Van Wesenbeeck B, Wolters G, Jensen K, Bouma TJ, Miranda-Lange M, Schimmels S (2014) Wave attenuation over coastal salt marshes under storm surge conditions. Nat Geosci 7:727–731

Nicholls RJ, Cazenave A (2010) Sea-level rise and its impact on coastal zones. Science 328:1517–1520

Niemeyer HD (2015) Effekte des Klimawandels auf Randbedingungen im Insel- und Küstenschutz -gegenwärtige und zu erwartende Trends. [Effects of climate change on boundary conditions for coastal protection - current and expected trends] Ber Forschungsstelle Küste 44:1–55

Nolte S, Koppenaal EC, Esselink P, Dijkema KS, Schürch M, De Groot AV, Bakker JP, Temmerman S (2013a) Measuring sedimentation in tidal marshes: A review on methods and their applicability in biogeomorphological studies. J Coast Conserv 17:301–325

Nolte S, Müller F, Schuerch M, Wanner A, Esselink P, Bakker JP, Jensen K (2013b) Does livestock grazing influence salt marsh resilience to sea-level rise in the Wadden Sea? Estuar Coast Shelf Sci 135:296–305

Norris K, Cook T, O'Dowd B, Durdin C (1997) The density of Redshank *Tringa tetanus* breeding on the salt marshes of the Wash in relation to habitat and its grazing management. J Appl Ecol 34:999–1013

Olff H, de Leeuw J, Bakker JP, Platerink RJ, van Wijnen HJ, de Munck W (1997) Vegetation succession and herbivory in a salt marsh: changes induced by sea-level rise and silt deposition along an elevational gradient. J Ecol 85:799–814

Oost A, Kabat P, Wiersma A, Hofstede J (2009) Climate Change Thematic Report No.4.1. In: Marencic H, De Vlas J (eds) Wadden Sea Quality Status Report 2009.Wadden Sea Ecosystems 25. Common Wadden Sea Secretariat, Wilhelmshaven

Paarlberg AJ, Knaapen MAF, De Vries MB, Hulscher SJMH, Wang ZB (2005) Biological influences on morphology and bed composition of an intertial flat. Estuar Coast Shelf Sci 64:577–590

Peach WJ, Thompson PS, Coulson JC (1994) Annual and long-term variation in the survival rates of British Lapwings *Vanellus vanellus*. J Anim Ecol 63:60–70

Pedersen JBT, Bartholdy J (2006) Budgets for fine-grained sediment in the Danish Wadden Sea. Mar Geol 235:101–117

Pejrup M, Larsen M, Edelvang K (1997) A fine-grained sediment budget for the Syly-Rømø tidal basin. Helg Meeresunt 51:253–268

Phelan N, Shaw A, Baylis A (2011) The extent of saltmarsh in England and Wales. Environmental Agency, Bristol

Pilkey OH, Cooper JAG (2004) Society and sea level rise. Science 303:1781–1782

Poesen J, Nachtergaele J, Vertstraeten G, Valentin C (2003) Gully erosion and environmental change: importance and research needs. Catena 50:91–133

Pohlmann T, Puls W (1993) Currents and transport in water. In: Sünderman J (ed) Circulation and Contaminant Fluxes in the North Sea. Springer

Postma H (1954) Hydrography of the Dutch Wadden Sea, a study of the relation between water movement, the transport of suspended materials and the production of organic matter. Arch Néerland Zool X:406–511

Postma H (1961) Transport and accumulation of suspended matter in the Dutch Wadden Sea. Netherl J Sea Res 1:148–190

Postma H (1967) Sediment transport and sedimentation in the estuarine environment. In: Lauf H (ed) Estuaries. Am Assoc Adv Sci Publ 83:158–179

Prandle D (2004) How tides and river flows determine estuarine bathymetries. Progr Oceanogr 61:1–26

Provoost S, Jones MLM, Edmondson SE (2011) Changes in landscape and vegetation of coastal dunes in northwest Europe: a review. J Coast Conserv 15:207–26

Psuty NP, Silveira TM (2010) Global climate change: an opportunity for coastal dunes? J Coast Conserv 14:153–160

Pye K (2001) Long-term geomorphological changes and how they may affect the dune coasts of Europe. In: Houston JA, Edmondson SE, Rooney PJ (eds) Coastal Dune Management: shared experience of European conservation practice. Liverpool University Press

Radley GP (1994) Sand dune vegetation survey of Great Britain - Part 1. England. Joint Nature Conservation Committee, Peterborough

Ranasinghe R, Callaghan D, Stive MJF (2012) Estimating coastal recession due to sea level rise: beyond the Bruun rule. Clim Change 110:561–574

Reed DJ (1988) Sediment dynamics and deposition in a retreating coastal salt marsh. Estuar Coast Shelf Sci 26:67–79

Reed TE, Grøtan V, Jenouvrier S, Sæther BE, Visser ME (2013) Population growth in a wild bird is buffered against phenological mismatch. Science 340:488–491

Reineck HE (1980) Sedimentationsbeiträge und Jahresschichtung in einem marinen Einbruchsgebiet/Nordsee. Senckenb Mar 12:380–309

Risely K, Massimo D, Johnston A, Newson SE, Eaton MA, Musgrove AJ, Noble DG, Adams DC, Baillie SR (2012) The Breeding Bird Survey 2011. British Trust for Ornithology

Rosati JD, Dean RG, Walton TL (2013) The modified Bruun Rule extended for landward transport. Mar Geol 340:71-81

Ruessink BG, Jeuken MCJL (2002) Dunefoot dynamics along the Dutch coast. Earth Surf Proc Landf 27:1043–1056

Rustad LE, Campbell JL, Marion GM, Norby RJ, Mitchell MJ, Hartley AE, Cornelissen JHC, Gurevitch J (2001) A meta-analysis of the response of soil respiration, net nitrogen mineralization, and aboveground plant growth to experimental ecosystem warming. Oecologia 126:543–562

Sæther BE, Sutherland WJ, Engen S (2004) Climate influences on avian population dynamics. Adv Ecol Res 35:185–209

Schlacher TA, Dugan J, Schoeman DS, Lastra M, Jones A (2007) Sandy beaches at the brink. Divers Distrib 13:556–560

Schröder HK, Kiehl K, Stock M (2002) Directional and non-directional vegetation changes in a temperate salt marsh in relation to biotic and abiotic factors. Appl Veget Sci 5:33–44

Schückel U, Kröncke I (2013) Temporal changes in intertidal macrofauna communities over eight decades: A result of eutrophication and climate change. Estuar Coast Shelf Sci 117:210–218

Schuerch M, Vafeidis A, Slawig T, Temmerman S (2013) Modeling the influence of changing storm patterns on the ability of a salt marsh to keep pace with sea level rise. J Geophys Res: Earth Surf 118:1–13

Seminara G, Lanzoni S, Bolla Pittaluga M, Solari L (2001) Estuarine patterns: an introduction to their morphology and mechanics. Lect Notes Phys 582:455–499

Silva H, Dias JM, Caçador I (2009). Is the salt marsh vegetation a determining factor in the sedimentation process? Hydrob 621:33-47

Skov F, Nygaard B, Wind P, Borchsenius F, Normand S, Balslev H, Flojgaard C, Svenning JC (2009) Impacts of 21st century climate changes on flora and vegetation in Denmark. Beyond Kyoto: addressing the challenges of climate change. IOP Publishing. IOP Conf Ser: Earth Env Sci 8

Sterl A, Severijns C, Dijkstra H, Hazeleger W, van Oldenborgh GJ, van den Broeke M, Rugers G, van den Hurk B, van Leeuwen PJ, van Velthoven P (2008) When can we expect extremely high surface temperatures. Geophys Res Lett L14703. doi:10.1029/2008GL034071

Stock M (2011) Patterns in surface elevation change across a temperate salt marsh platform in relation to sea-level rise. Coastl Reps 17:33–48

Suchrow S, Pohlman M, Stock M, Jensen K (2012) Long-term surface elevation change in German North Sea salt marshes. Estuar Coast Shelf Sci 98:75–83

Syvitski JPM, Vorosmarty CJ, Kettner AJ, Green P (2005) Impact of humans on the flux of terrestrial sediment to the global ocean. Science 308:376–380

Taylor JA, Murdock AP, Pontee NI (2004) A macroscale analysis of coastal steepening around the coast of England and Wales. Geograph J 170:179–188

Temmerman S, Govers G, Meire P, Wartel S (2003) Modelling long-term tidal marsh growth under changing tidal conditions and suspended sediment concentrations, Scheldt estuary, Belgium. Mar Geol 193:151–169

Temmerman S, Govers G, Wartel S, Meire P (2004) Modelling estuarine variations in tidal marsh sedimentation: response to changing sea level and suspended sediment concentrations. Mar Geol 212:1–19

Temmerman S, Meire P, Bouma TJ, Herman PMJ, Ysebaert T, De Vriend HJ (2013) Ecosystem-based coastal defence in the face of global change. Nature 504:79–83

Townend I (2005) An examination of empirical stability relationships for UK estuaries, J Coast Res 21:1042–1053

Tsoar H (2005) Sand dunes mobility and stability in relation to climate. Physica A 357:50–56

Turner RK, Burgess D, Hadley D, Coombes E, Jackson NA (2007) A cost-benefit appraisal of coastal managed realignment policy. Global Env Change 17:397–407

Uncles RJ, Stephens JA, Smith RE (2002) The dependence of estuarine turbidity on tidal intrusion length, tidal range and residence time. Cont Shelf Res 22:1835–1856

Van de Kam J, Ens BJ, Piersma T, Zwarts L (2004) Shorebirds: An illustrated behavioural ecology. KNNV, Utrecht

Van de Koppel J, Huisman J, Van der Wal R, Olff H (1996) Patterns of herbivory along a productivity gradient: an empirical and theoretical investigation. Ecology 77:736–745

Van de Koppel J, Herman PMJ, Thoolen P, Heip CHR (2001) Do alternate stable states occur in natural ecosystems? Evidence from a tidal flat. Ecology 82:3449–3461

Van de Koppel J, Van der Wal D, Bakker JP, Herman PMJ (2005) Self-organisation and vegetation collapse in salt marsh ecosystems. Amer Natural 165:E1-E12

Van de Pol M, Ens BJ, Heg D, Brouwer L, Krol J, Maier M, Exo KM, Oosterbeek K, Lok T, Eising CM, Koffijberg K (2010a) Do changes in the frequency, magnitude and timing of extreme climatic events threaten the population viability of coastal birds? J Appl Ecol 47:720–730

Van de Pol M, Vindenes Y, Sæther BE, Engen S, Ens BJ, Oosterbeek K, Tinbergen JM (2010b) Effects of climate change and variability on population dynamics in a long-lived shorebird. Ecology 91:1192–1204

Van de Pol M, Vindenes Y, Sæther BE, Engen S, Ens BJ, Oosterbeek K, Tinbergen JM (2011) Poor environmental tracking can make extinction risk insensitive to the colour of environmental noise. Proc Roy Soc B 278:3713–3722

Van de Pol M, Ens BJ, Bakker JP, Esselink P (2012) Sea-level rise and increasing flooding risk of saltmarsh breeding birds. De Lev Nat 113:123–128

Van der Wal D, Pye K (2004) Patterns, rates and possible causes of saltmarsh erosion in the Greater Thames area (UK). Geomorphology 61:373–391

Van der Wal D, Wielemaker-Van den Dool A, Herman PMJ (2008) Spatial patterns, rates and mechanisms of saltmarsh cycles (Westerschelde, The Netherlands). Estuar Coast Shelf Sci 76:357–368

Van Dobben HF, Slim PA (2012) Past and future plant diversity of a coastal wetland driven by soil subsidence and climate change. Clim Change 110:597–618

Van Leussen W (2011) Macroflocs, fine-grained sediment transports, and their longitudinal variations in the Ems estuary. Ocean Dyn 61:387–401

Van Maanen B, Coco G, Bryan KR (2013) Modelling the effects of tidal range and initial bathymetry on the morphological evolution of tidal embayments. Geomorphology 191:23–34

Van Oldenborgh GJ, Collins M, Arblaster JM, Christensen JH, Marotzke J, Power SB, Rummukainen M, Zhou T (2013) IPCC 2013: Annex I: Atlas of global and regional climate projections. In: Stocker TF, Qin D, Plattner GK, Tignor M, Allen SK, Boschung J, Nauels A, Xia Y, Bex V, Midgley PM (eds) Climate Change 2013: The Physical Science Basis. Contribution of Working Group I to the Fifth Assessment Report of the Intergovernmental Panel on Climate Change. Cambridge University Press

Van Straaten LMJU, Kuenen PHH (1957) Accumulation of fine-grained sediments in the Dutch Wadden Sea. Geol Mijnbouw 19:329–354

Van Straaten LMJU, Kuenen PHH (1958) Tidal action as a cause of clay accumulation. J Sedim Petrol 28:406–413

Van Vliet MTH, Zwolsman JJG (2008) Impact of summer droughts on the water quality of the Meuse river J Hydrol 353:1–17

Van Wijnen HJ, Bakker JP (2000) Annual nitrogen budget of a temperate coastal barrier salt-marsh system along a productivity gradient at low and high marsh elevation. Persp Plant Ecol Evol Sys 3:128–141

Van Wijnen HJ, Bakker JP (2001) Long-term surface elevation changes in salt marshes: a prediction of marsh response to future sea-level rise. Estuar Coast Shelf Sci 52:381–390

Van Wijnen HJ, Bakker JP, De VriesY (1997) Twenty years of salt marsh succession on a Dutch coastal barrier island. J Coast Conserv 3:9–18

Vandenbohede A, Luyten K Lebbe L (2008) Effects of global change on heterogeneous coastal aquifers: A case study in Belgium. J Coast Res 24:160–170

Vandenbruwaene W, Maris T, Cox TJS, Cahoon DR, Meire P, Temmerman S (2011) Sedimentation and response to sea-level rise of a restored marsh with reduced tidal exchange: Comparison with a natural tidal marsh. Geomorphology 130:115–126

Veeneklaas RM, Dijkema KS, Hecker N, Bakker JP (2013) Spatio-temporal dynamics of the invasive plant species *Elytrigia atherica* on natural salt marshes. Appl Veget Sci 16:205–216

Vestergaard P (1997) Possible impact of sea-level rise on some habitat types at the Baltic coast of Denmark. J Coast Conserv 3:103–112

Visser ME (2008) Keeping up with a warming world; assessing the rate of adaptation to climate change. Proc Roy Soc B 275:649–659

Wahl T, Jense J, Frank T, Haigh ID (2011) Improved estimates of mean sea level changes in the German Bight over the last 166 years. Ocean Dyn 61:701–715

Wahl T, Haigh ID, Woodworth PL, Albrecht F, Dillingh D, Jensen J, Nicholls RJ, Weisse R, Wöppelmann G (2013) Observed mean sea level changes around the North Sea coastline from 1800 to present. Earth Sci Rev 124:51–67

Weisse R, von Storch H, Niemeyer HD, Knaack H (2012) Changing North Sea storm surge climate: An increasing hazard? Ocean Coast Manage 68:58–68

Winiwarter W, Hettelingh JP, Bouwman AF et al (2011) Future scenarios of nitrogen in Europe. In: Sutton MA, Howard CM, Erisman JW et al (eds) The European Nitrogen Assessment. Cambridge University Press

Winterwerp JC (2011) Fine sediment transport by tidal asymmetry in the high-concentrated Ems River: indications for a regime shift in response to channel deepening. Ocean Dyn 61:203–215

Witte JPM, Runhaar J, van Ek R, van der Hoek DCJ, Bartholomeus RP, Batelaan O, van Bodegom PM, Wassen MJ, van der Zee SEATM (2012) An ecohydrological sketch of climate change impacts on water and natural ecosystems for the Netherlands: bridging the gap between science and society. Hydrol Earth Sys Sci 16:3945–3957

Wolff WJ (1992) The end of a tradition – 1000 years of embankment and reclamation of wetlands in the Netherlands. Ambio 21:287–291

Wolters M, Bakker JP, Bertness M, Jefferies RL, Möller I (2005a) Saltmarsh erosion and restoration in south-east England: squeezing the evidence requires realignment. J Appl Ecol 42:844–851

Wolters M, Garbutt A, Bakker JP (2005b). Salt-marsh restoration: evaluating the success of de-embankments in north-west Europe. Biol Conserv 123:249–268

Wood R, Widdows J (2002) A model of sediment transport over an intertidal transect, comparing the influences of biological and physical factors. Limnol Oceanogr 47:848–855

Wood R, Widdows J (2003) Modelling intertidal sediment transport for nutrient change and climate change scenarios. Sci Tot Environ 314:637–649

Woodworth P, Teferle F, Binley R, Shennan I, Williams S (2009) Trends in UK mean sea level revisited. Geophys J Int 176:19–30

Yalden DW, Pearce-Higgins JW (1997) Density-dependence and winter weather as factors affecting the size of a population of Golden Plovers *Pluvialis apricaria*. Bird Study 44:227–234

Yizhaq H, Ashkenazy Y, Tsoar H (2007) Why do active and stabilized dunes coexist under the same climatic conditions? Phys Rev Lett 98:188001

Environmental Impacts—Lake Ecosystems

10

Rita Adrian, Dag Olav Hessen, Thorsten Blenckner, Helmut Hillebrand, Sabine Hilt, Erik Jeppesen, David M. Livingstone and Dennis Trolle

Abstract

The North Sea region contains a vast number of lakes; from shallow, highly eutrophic water bodies in agricultural areas to deep, oligotrophic systems in pristine high-latitude or high-altitude areas. These freshwaters and the biota they contain are highly vulnerable to climate change. As largely closed systems, lakes are ideally suited to studying climate-induced effects via changes in ice cover, hydrology and temperature, as well as via biological communities (phenology, species and size distribution, food-web dynamics, life-history traits, growth and respiration, nutrient dynamics and ecosystem metabolism). This chapter focuses on change in natural lakes and on parameters for which their climate-driven responses have major impacts on ecosystem properties such as productivity, community composition, metabolism and biodiversity. It also points to the importance of addressing different temporal scales and variability in driving and response variables along with threshold-driven responses to environmental forces. Exceedance of critical thresholds may result in abrupt changes in particular elements of an ecosystem. Modelling climate-driven physical responses like ice-cover duration, stratification periods and thermal profiles in lakes have shown major advances, and the chapter provide recent achievements in this field for northern lakes. Finally, there is a tentative summary of the level of certainty for key climatic impacts on freshwater ecosystems. Wherever possible, data and examples are drawn from the North Sea region.

R. Adrian (✉) · S. Hilt
Leibniz-Institute of Freshwater Ecology and Inland Fisheries (IGB), Berlin, Germany
e-mail: adrian@igb-berlin.de

S. Hilt
e-mail: Hilt@igb-berlin.de

D.O. Hessen (✉)
Department of Biosciences, University of Oslo, Oslo, Norway
e-mail: d.o.hessen@ibv.uio.no

T. Blenckner
Stockholm Resilience Centre, Stockholm University, Stockholm, Sweden
e-mail: thorsten.blenckner@su.se

H. Hillebrand
Institute for Chemistry and Biology of the Marine Environment, Carl von Ossietzky University of Oldenburg, Oldenburg, Germany
e-mail: helmut.hillebrand@icbm.de

E. Jeppesen
Department of Bioscience, Aarhus University, Aarhus, Denmark
e-mail: ej@bios.au.dk

D.M. Livingstone
Department of Water Resources and Drinking Water, EAWAG Swiss Federal Institute of Aquatic Science and Technology, Dübendorf, Switzerland
e-mail: living@eawag.ch

D. Trolle
Department of Bioscience—Lake Ecology, Aarhus University, Silkeborg, Denmark
e-mail: trolle@bios.au.dk

M. Quante and F. Colijn (eds.), *North Sea Region Climate Change Assessment*, Regional Climate Studies, DOI 10.1007/978-3-319-39745-0_10

10.1 Introduction

Freshwaters and freshwater biota are highly vulnerable to climate change (IPCC 2013). The various impacts and threats differ substantially between biomes and among geographical regions, and even within geographical regions hydrological factors and catchment properties will be major determinants of the responses in freshwater ecosystems to climate change. Effects on freshwater resources in agricultural or densely populated catchments for example will differ from those in pristine boreal or alpine catchments. In the North Sea region, water scarcity per se is not seen as the most immediate threat of climate change, with the exception of during summer in some areas. Strong flooding events may pose greater challenges. The North Sea region contains a vast number of aquatic systems, including shallow, highly eutrophic water bodies in agricultural areas; systems highly influenced by the input of terrestrially derived, coloured dissolved organic matter in coniferous areas; and large, deep and oligotrophic systems in pristine, sub-alpine areas. Clearly, the review presented in this chapter cannot cover every aspect related to climate and surface waters in the North Sea region; hence a focus on natural lakes and parameters where there is empirical support for climate change effects, as well as on systems for which long-term data sets exist and for which their responses have major impacts on ecosystem properties such as productivity, community composition and biodiversity. Hydrological effects, wetland effects and past climate effects inferred from paleolimnological surveys are covered elsewhere in this report, although there is no doubt that hydrology will also have strong bearings on water chemistry (e.g. by dilution through increased precipitation) and biota (changed fluxes in key elements like nitrogen, phosphorus, carbon, silicon, iron and calcium), dissolved organic matter and pollutants.

Monitoring the impacts of climate change poses challenges because of the many responses within an ecosystem and the spatial variation within the landscape. A substantial body of research demonstrates the sensitivity of freshwater ecosystems to climate forcing and shows that physical, chemical, and biological lake properties respond rapidly to changes in this forcing (Rosenzweig 2007; Adrian et al. 2009; MacKay et al. 2009; Tranvik et al. 2009; IPCC 2013). Fast turnover times from the scale of organisms to entire lake ecosystems are the prerequisite for these rapid changes. Studies of lake ecosystems have provided some of the earliest indications of the impact of current climate change on ecosystem structure and function (Adrian et al. 1995; Magnuson et al. 2000; Verburg et al. 2003) and the consequences for ecosystem services (O'Reilly et al. 2003). Some climate-related signals are highly visible and easily measured in lakes. For instance climate-driven fluctuations in lake level have been observed on a regional-scale across North America (Williamson et al. 2009), and shifts in the timing of ice formation and melt reflect climate warming at a global scale (Magnuson et al. 2000). Other signals may be more complex and difficult to detect in lakes, but may be equally sensitive to climate or more informative regarding impacts on ecosystem services. Long-term historical records and reconstructions from sediment cores have yielded insight into less visible climate-related changes, thus increasing understanding of the mechanisms driving these changes. In particular, paleolimnological records have been critical for reconstructing the climate record over recent geological periods, making it possible to interpret current climate change and predict its future impacts.

Lake ecosystems are excellent sentinels for current climate change. In this context a sentinel is a lake ecosystem that provides indicators of climate change either directly or indirectly through the influence of climate on the catchment (Fig. 10.1; Carpenter et al. 2007; Adrian et al. 2009; Williamson et al. 2009). The indicators are measureable response variables, such as water temperature, dissolved organic carbon, or phytoplankton composition. Lakes are particularly good sentinels for current climate change for several reasons: they are well-defined ecosystems and studied in a sustained fashion; they respond directly to climate change and incorporate the effects of climate change within the catchment; they integrate responses over time, which can filter out random noise; and they are distributed worldwide and so cover many different geographic locations and climatic regions. However, the large range in lake morphology, geographic location, and catchment characteristics means that broad statements about the ability of lakes to capture the impacts of the current, rapidly changing climate must be made with caution. On the other hand, this also means that there are many different types of sensors in the landscape ranging from small shallow turbid lakes to large, deep, clear lakes that may capture, or provide sentinel information on different aspects of climate change, including temperature and precipitation-related components.

This chapter focuses on both the direct and indirect (e.g. via the catchment) effects of climate change, as well as on internal physical, chemical and biological processes and on the role of temporal scale. Wherever possible, data and examples are drawn from the North Sea region, but for important phenomena where North Sea studies are not available, information is 'borrowed' from other parts of the world.

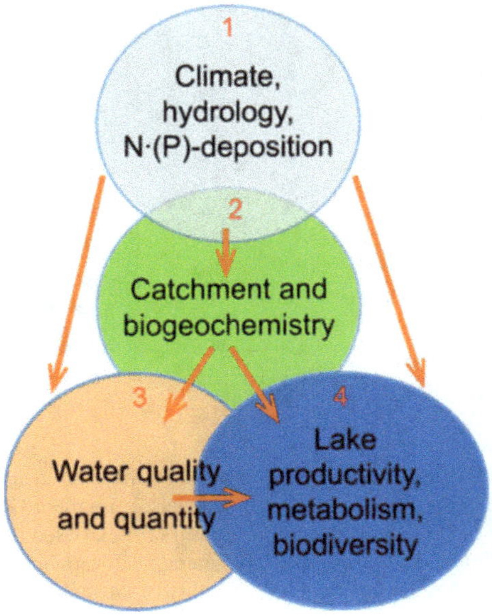

Fig. 10.1 Water and its context: lakes respond to climatic forcing, atmospheric deposition and the properties of their catchments. *1* Atmospheric forcing via temperature, precipitation and deposition of key constituents like nitrogen. *2* The catchment responds to climatic drivers and atmospheric inputs via vegetation and soil processes. *3* This determines the inputs of organic matter, nutrients and key elements determining parameters like retention time, transparency, pH and temperature. *4* The biota respond in terms of phenology, productivity, metabolism, community composition, diversity and food web interactions, to the direct forcing (*1*), catchment properties (*2*) and water properties (*3*)

10.2 Climate Warming and Impacts on Lake Physics

Data-based studies on the impact of climate change on lake temperature began in the early 1990s with the seminal studies of Magnuson et al. (1990) and Schindler et al. (1990). The former demonstrated that near-surface lake temperatures in Wisconsin fluctuate coherently in response to regional climatic forcing, while the latter demonstrated that lake temperatures in the Canadian Experimental Lakes Area were undergoing a long-term increase. Since the 1990s, an increasing number of studies have demonstrated that near-surface lake temperatures can fluctuate coherently over several hundred kilometres (e.g. Benson et al. 2000; Livingstone and Padisák 2007; Livingstone et al. 2010a), and that individual lakes in many parts of the world have been undergoing long-term warming at all depths. Lake warming has been demonstrated in Europe (e.g. Livingstone 2003; Salmaso 2005), North America (e.g. Arhonditsis et al. 2004; Coats et al. 2006; Austin and Colman 2008), East Africa (e.g. O'Reilly et al. 2003; Verburg et al. 2003), Siberia (Hampton et al. 2008) and Antarctica (Quayle et al. 2002). A recent study based on satellite thermal infrared images from 1985 onwards (Schneider and Hook 2010) confirmed that lake surface temperatures have been undergoing a long-term increase over large areas of the northern hemisphere. In an extensive world wide survey of lake's summer surface temperatures O'Reilly et al. (2015) found an average warming trend of 0.34 °C per decade for the period 1985 until 2009.

Modelling studies on strictly dimictic lakes, which are ice-covered in winter and mix twice per year, suggest that surface temperatures will increase faster than deep-water temperatures as a result of climate change (Robertson and Ragotzkie 1990; Hondzo and Stefan 1993). In monomictic lakes, which mix once per year, the divergence between surface and deep-water temperatures is likely to be less strong as a result of heat carry-over in winter (Peeters et al. 2002). However, even in such lakes historical data show that near-surface water temperatures are increasing faster than deep-water temperatures, which implies an increase in thermal stability leading to an increase in the duration of stratification in summer and a corresponding decrease in the duration of homothermy in winter and spring (Livingstone 2003).

At high latitudes or high altitudes, where lakes are generally ice-covered for many months of the year, the more frequent occurrence of mild winters associated with climate warming will imply a general decrease in the duration of ice cover and a corresponding increase in the duration of summer stratification. Since the mid-19th century, there has been a general long-term decrease in the duration of ice cover in northern hemisphere lakes at a mean rate of about 1.2 days per decade (Magnuson et al. 2000), with the rate for individual lakes ranging from 0.9 to 1.7 days per decade (Benson et al. 2012). This decline seems to be accelerating: over the last 30 years, the equivalent is 1.6–4.3 days per decade (Benson et al. 2012). In regions with relatively brief or mild winters, where lakes are ice-covered for a comparatively short period—for instance in southern Sweden, Denmark and northern Germany—increasingly milder winters are likely to result in ice cover becoming intermittent or even disappearing (Livingstone and Adrian 2009; Weyhenmeyer et al. 2011). The disappearance of ice cover from a deep lake, implying a shift in mixing regime from dimixis to monomixis (Boehrer and Schultze 2008; Livingstone 2008), is likely to cause a change in its physical response to further climate warming, as mixing will no longer necessarily occur at the temperature of maximum density (4 °C) before ice-on and after ice-off. However, mixing can still occur at temperatures higher than 4 °C. Deep lakes that are already monomictic will experience individual years in which some form of stratification persists throughout the year, reducing the intensity of mixing and inhibiting deep-water renewal. Thus some deep monomictic lakes may show a tendency towards becoming oligomictic; i.e. will not mix fully every year. Shallow polymictic lakes, which lose their winter ice cover, are likely to undergo permanent mixing during winter.

10.2.1 Lake Water Temperature

Research on lakes within the North Sea region has demonstrated the occurrence of many of the phenomena mentioned in the previous section. In the UK—at the western boundary of the region, where the influence of the North Atlantic is at its greatest—research has focused on the English Lake District, where several decades of data are available from some of the larger lakes. Here, water temperatures both near the surface and in the deep water have been shown to respond coherently to climatic forcing throughout the year (George et al. 2000), with a clear long-term increase recorded in both near-surface temperatures (George et al. 2007a) and deep-water temperatures (Dokulil et al. 2006). Long-term increases in surface water temperatures have also been recorded in lakes in Sweden, at the region's eastern boundary (Adrian et al. 2009), and in northern Germany, at its southern boundary (Adrian et al. 2006, 2009; Wilhelm et al. 2006). It is thus likely that long-term increases in lake temperature are occurring throughout the North Sea region.

It has become evident that the climatic forcing acting on lakes in the North Sea region is extremely large-scale in nature, with the climate prevailing over the North Atlantic playing a major role in determining the physical behaviour of the lakes. Interannual fluctuations in thermal stratification in the lakes of the English Lake District, for instance, are related to north-south displacements of the Gulf Stream, with the early summer thermocline tending to be shallower and more well-defined when the Gulf Stream has its most northern direction (George and Taylor 1995). This effect appears to be related to the effect of the Gulf Stream on wind speed (George et al. 2007b). However, an even more potent determinant of physical lake behaviour is the climate mode known as the North Atlantic Oscillation (NAO), which governs winter weather in western, northern and central Europe to a very large degree (Hurrell 1995; Hurrell et al. 2003) and which is known to play an important role in determining the behaviour of lakes in this region (Straile et al. 2003). This climate mode, which can be considered a regional manifestation of the Arctic Oscillation (AO) (Thompson and Wallace 1998), is associated with interannual fluctuations in the meridional surface air pressure gradient in the north-east Atlantic (between about 35°N and 65°N). A positive NAO index implies a large meridional air pressure gradient, which results in the strong zonal transport of warm, moist maritime air from the North Atlantic towards north-west Europe. In winter, this implies predominantly mild, wet weather in the North Sea region. When the NAO index is negative, however, the eastward transport of warm, moist air from the North Atlantic is much weaker, implying predominantly cold, dry weather in the North Sea region. In lakes throughout northern and central Europe, surface water

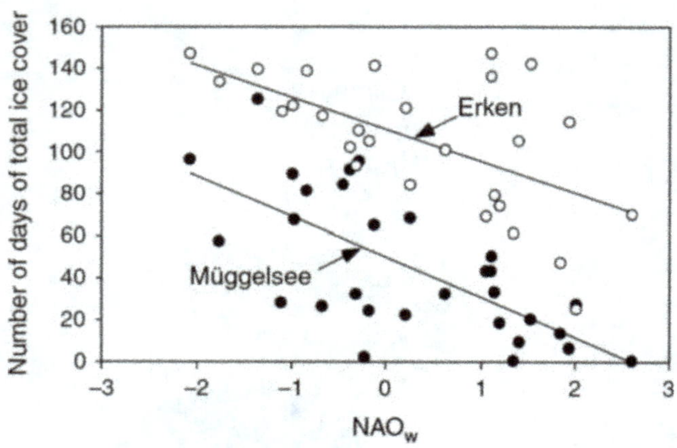

Fig. 10.2 Dependence of the number of days of total ice cover on Lake Erken (southern Sweden) and Müggelsee (northern Germany) on the winter NAO index (NAO$_w$) of Hurrell (1995), based on data from winter 1976/77 to winter 2005/06. Interannual variability in NAO$_w$ explains 35 % of the interannual variability of the duration of ice cover for Lake Erken and 47 % for Müggelsee. The linear regressions (regression lines illustrated) are significant at $p < 0.01$ (from Livingstone et al. 2010b)

temperatures, near-bottom temperatures and the duration of ice cover (Fig. 10.2) are correlated to some extent with the winter NAO index (Blenckner et al. 2007).

In the English Lake District, lake surface temperatures in winter are tightly correlated with the winter NAO index (George et al. 2000, 2004b, 2007a) as, but to a lesser extent, are deep-water temperatures (George et al. 2004b), with the highest correlations being observed in the shallower lakes (George et al. 2004b). Mean annual water temperature of Lake Veluwe, a shallow lake in the Netherlands, is correlated with the winter NAO index (Scheffer et al. 2001), as are the surface water temperatures of Vänern, Vättern, and Mälaren, the three largest lakes in Sweden, in spring (Weyhenmeyer 2004). However, in Mälaren, a morphometrically complex lake with many sub-basins, the significance of the correlation varies substantially among the different sub-basins, suggesting that local lake characteristics can modify the effect of large-scale climatic forcing even on surface water temperature, which apart from the timing of ice-off is probably the lake variable least affected by internal lake processes. In Müggelsee, a well-studied, shallow, polymictic lake in northern Germany, several studies have demonstrated the effect of the winter NAO on lake temperatures (Gerten and Adrian 2000, 2001; Straile and Adrian 2000).

In a comparative study of Müggelsee and two neighbouring lakes, Gerten and Adrian (2001) showed that the winter NAO leaves a signal at all depths in lake temperature, but that the temporal persistence of this signal can differ substantially from lake to lake. Near the surface, the NAO signal is in general confined to late winter and early spring. In the deeper water, however, the persistence of the NAO signal depends on the morphometry and mixing characteristics of the lake. The NAO signal persists only through spring in shallow, polymictic Müggelsee, but throughout much of the following summer in the shallow, dimictic Heiligensee, and throughout the whole of summer and autumn in the much deeper, dimictic Stechlinsee. Thus, although an NAO signal is likely to be present to some extent in the temperature of all lakes within the North Sea region, individual lake characteristics are certain to result in a large degree of variability in the strength and persistence of this signal.

In the context of the NAO, one other phenomenon should be mentioned: the late 1980s climate regime shift. In the late 1980s, an abrupt regime shift occurred in the atmospheric, oceanic, terrestrial, limnological and cryospheric systems in many regions of the world. Evidence suggests that this large-scale regime shift involved abrupt changes in the AO and NAO (Alheit et al. 2005; Rodionov and Overland 2005; Lo and Hsu 2010), had a substantial impact on air temperature in northern Europe (Lo and Hsu 2010), and affected fish populations in the North Sea (Reid et al. 2001; Alheit et al. 2005). It is not surprising therefore that a regime shift in lake temperature in the late 1980s was also detected in Müggelsee (Gerten and Adrian 2000, 2001), and it can be fairly confidently hypothesised that a similar regime shift, at least in lake surface temperatures, is likely to have occurred in many lakes within the North Sea region.

In summer, a more regional approach to determining the effects of climatic forcing on lakes is necessary owing to the smaller spatial scales of the weather systems involved. In the case of the UK, the Lamb synoptic weather classification system (Lamb 1950) has proven useful. For both Windermere, the largest lake in the English Lake District, and Lough Feeagh, a lake located near the west coast of Ireland, the highest lake surface temperatures were recorded during a westerly circulation type in winter (corresponding to a positive NAO index), but a southerly circulation type in summer (George et al. 2007b).

On a multi-annual time scale, large-scale regional coherence is greater in winter than in summer. However, on shorter time scales the opposite appears to be the case. Short-term, high-resolution surface temperature measurements in Scottish Highland lochs (Livingstone and Kernan 2009) show a high degree of regional coherence in daily means from late spring to autumn, but much lower coherence in winter and early spring. Short-term regional coherence is high in summer because the surface mixed layer is thin and surface temperatures respond sensitively to climate forcing, and is high in autumn because surface temperatures are dominated by convective cooling, which is governed by regionally coherent air temperature. Short-term coherence is comparatively low in winter because fluctuations in lake surface temperature are small and may be buffered by partial ice cover, and is low in early spring because the lochs warm up and stratify at different times during the season depending on their altitude and distance from the maritime influence of the Atlantic. This latter effect may be important on the western boundary of the North Sea region, because the ameliorating influence of the Atlantic Ocean acts to increase winter surface temperatures, and hence to reduce the duration of inverse stratification or circulation.

Based on observed air temperature and physical lake characteristics, George et al. (2007b) were able to model well the surface temperatures of the lakes of the English Lake District. Using a regional climate model (RCM) driven by the SRES A2 scenario, they also projected lake surface temperatures in the 2050s. This showed increases of up to 1.1 °C in winter and up to 2.2 °C in summer, with the greatest increase in winter occurring in the shallowest lake, and the greatest increase in summer occurring in the lake with the shallowest thermocline.

10.2.2 Lake Ice Phenology

It is likely that the phenology of ice cover on the lakes of the North Sea region will follow the global trend; with ice-on occurring later, ice-off occurring earlier, and a general reduction apparent in the duration of ice cover (Magnuson et al. 2000; Benson et al. 2012). However, several factors will modify this general trend. Because of the approximately sinusoidal form of the air temperature curve, the dates on which the air temperature falls below and rises above 0 °C, which are crucial for the timing of ice-on and ice-off, respectively, are not linear functions of air temperature. Instead, they are arc cosine functions of air temperature, which implies that the sensitivity of the timing of ice-on, the timing of ice-off, and the duration of ice cover are greater in warmer regions than in colder regions, and so will increase as the climate warms (Weyhenmeyer et al. 2004a, 2011; Jensen et al. 2007; Livingstone and Adrian 2009). Thus, in the North Sea region, the impact of climate warming on lake ice phenology will be disproportionately large in those areas where winters are mild or variable and the duration of ice cover on lakes is already short (i.e. the UK, northern France, Belgium, Netherland, Luxemburg, northern Germany and southern Scandinavia) compared to those areas where winters are consistently cold and the duration of ice cover is much longer (i.e. northern Scandinavia). This is despite the IPCC (Intergovernmental Panel on Climate Change) projections implying that climate warming in winter in northern Scandinavia will be stronger than in the rest of the region (Christensen et al. 2007). In Sweden, several decades of historical data confirm that the timing of ice-off on lakes in the south of the country, where winters are relatively mild, has been responding significantly more sensitively to

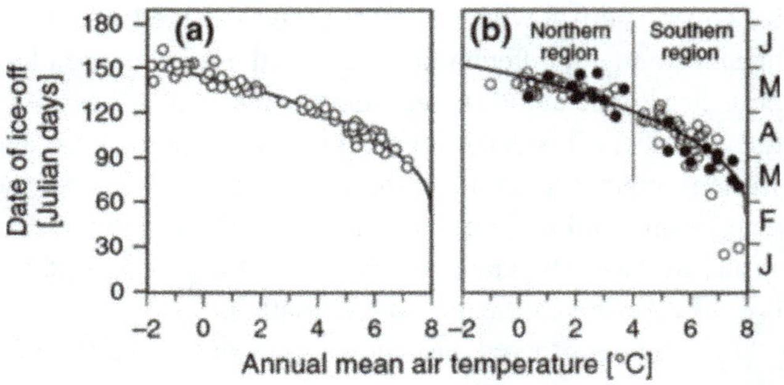

Fig. 10.3 **a** Dependence of the median date of ice-off on annual mean air temperature (1961–1990) for 70 lakes in Sweden. **b** Temporal variation in the relationship between the median date of ice-off and annual mean air temperature for 14 lakes in a northern region of Sweden (61–67°N) and 14 lakes in a southern region of Sweden (56–60°N) for the periods 1961–1990 (*white*) and 1991–2002 (*black*), showing that the form of the relationship is strongly dependent on latitude but not on the time period chosen. Annual mean air temperatures are calculated from July to June, and the curves illustrated are based on the arc cosine model of Weyhenmeyer et al. (2004a) (Livingstone et al. 2010b; after Weyhenmeyer et al. 2004a)

interannual fluctuations in mean winter air temperature than the timing of ice-off on lakes in the north of the country, where winters are longer and more severe (Fig. 10.3; Weyhenmeyer et al. 2004a).

A further study of ice phenology on 54 Swedish lakes during the 30-year IPCC reference period 1961–1990 showed a statistically significant ($p < 0.05$) trend towards earlier ice-off in 47 lakes, with the shift towards earlier ice-off varying between 1 and 29 days (Weyhenmeyer et al. 2005). Again, the shift towards earlier ice-off was stronger in the milder, southern part of Sweden than in the colder, northern part. During the IPCC reference period, the mean air temperature of the northern hemisphere increased by 0.4 °C. This resulted in a shift in the timing of ice-off by ~70 days in southern Sweden, but only ~10 days in northern Sweden. Applying the arc cosine model suggests that interannual variability in the duration of lake ice cover will be far greater in southern Scandinavia, Scotland and northern Germany than in northern Scandinavia (Weyhenmeyer et al. 2011). A more sophisticated probability model applied to Müggelsee, which now shows extremely variable, intermittent ice cover, predicts that the percentage of ice-free winters for this lake will increase from ~2 % now to over 60 % by the end of the 21st century (Livingstone and Adrian 2009).

As for winter lake surface temperatures, the duration of ice cover on lakes in the North Sea region also appears to be strongly related to the NAO. In the case of Müggelsee, 47 % of the interannual variability of the duration of ice cover can be explained statistically ($p < 0.01$) in terms of the interannual variability of the winter NAO index, while for Lake Erken, in east-central Sweden, the equivalent value is 35 % (Fig. 10.2). Even in the UK, where total lake ice cover does not occur frequently, there is evidence that the number of days of partial ice cover is strongly linked to the winter NAO (George et al. 2004a; George 2007).

As well as the duration of ice cover, the timing of both ice-on and ice-off also appear to be determined to some degree by the NAO. For lakes in Sweden, the timing of ice-off is strongly related to the winter NAO while the timing of ice-on shows a weaker relationship to the autumn NAO (Blenckner et al. 2004). These results agree with those of similar studies showing that the winter NAO is an important determinant of ice phenology in the neighbouring Baltic region (Livingstone 2000; Yoo and d'Odorico 2002; Blenckner et al. 2004; George et al. 2004a). As in the case of lake water temperatures, there is evidence to suggest that the late 1980s climate regime shift may have resulted in an abrupt shift in lake ice phenology in the North Sea region: a study of Swedish lakes showed an abrupt shift in 1988/1989 that was substantial in southern Sweden but not in northern Sweden, again emphasising the relative sensitivity of ice phenology in warmer regions to external climatic forcing (Temnerud and Weyhenmeyer 2008).

10.3 Catchment–Lake Interactions

While lakes are commonly seen as closed entities, which is partly true in terms of populations with low or no immigration or emigration, lakes are strongly influenced by catchment properties such as the proportions of forest, bogs, and arable land that serve as major determinants of element fluxes (and water) to lakes. This is however modified by anthropogenic impacts. For example, acidifying elements such as nitrogen (N) and sulphur (S) will modify the catchment export of ions, dissolved organic matter (DOM) and nutrients, and deforestation or afforestation will also have a major impact on the flux of elements to lakes. This is especially striking for nitrogen, where alpine or otherwise sparsely vegetated catchments may have very low N-retention, while forested catchments may retain almost all nitrate (NO_3) and ammonium (NH_4) inputs during the growing season (Hessen 1999).

Different catchment properties have vastly different effects on lakes, even within the North Sea region, and these properties and effects will also be differently modified by climate drivers (Fig. 10.1). Although large, deep and oligotrophic lakes are common in alpine areas of the North Sea drainage basin, the North Sea region is dominated by two major types of lake system and it is these that will be most affected by climate change. These are the boreal lakes generally found in forested, less impacted, catchments with limited input of bioavailable nutrients but high loads of humic substances of terrestrial origin. The other lake type occurs in agricultural areas with higher nutrient loads. Both

Fig. 10.4 Export of coloured dissolved organic carbon (DOC) from wetlands, bogs and forest via rivers (*Photo* D.O. Hessen)

categories of lake/catchment encompass a wide range of size, volume, renewal rates, productivity and community composition.

While catchment responses to climate affect downstream rivers, lakes and ecosystems, hydrology (runoff) also plays a major role both in mobilising and diluting dissolved matter and key elements (such as nitrogen, phosphorus, iron, silicon and calcium). Hydrology also affects the water renewal rate, which has major impacts on the physical, chemical and biological properties of waters. These aspects are less well studied and there is a need for 'bridging the gap' between hydrology and especially aquatic biology.

10.3.1 Boreal Lakes

Rising temperatures and changes in precipitation patterns and amounts will have direct effects on lake ecosystems as well as indirect effects via impacts within the terrestrial catchments (Schindler 2001; Kernan et al. 2010). Atmospheric deposition of nitrogen and sulphur will add to these climate-related impacts on ecosystems (Wright et al. 2010). The well-documented increase in dissolved organic matter in boreal lake ecosystems is one of the most obvious indirect effects (Monteith et al. 2007). This 'browning' is mainly due to coloured dissolved organic carbon (DOC) which absorbs light and so affects lake production at all trophic levels (Karlsson et al. 2009; Xenopoulos et al. 2009) as well as species composition (Watkins et al. 2001). The export of DOC from wetlands, bogs and forest via rivers (Fig. 10.4) to lakes and coastal areas is projected to increase, causing reduced primary and secondary production due to light limitation.

Changes in temperature and precipitation drive seasonal and interannual variations in the export of DOC and nutrients by affecting soil organic matter mineralisation and the production of DOC and mineral nitrogen species (Kalbitz et al. 2000; Hobbie et al. 2002a, b), as well as hydrology within the catchment. On a longer timescale, climate change can allow forests to expand into areas at present under heathland and alpine vegetation (Hofgaard 1997), thereby affecting nutrient retention, nutrient export and soil organic matter pools and quality (Kammer et al. 2009). A major climatic response expected in boreal, coniferous areas is an increase in the concentrations of terrestrially derived DOM and DOC. For a large dataset of boreal lakes (1041 Swedish lakes along a 13° latitudinal gradient), Weyhenmeyer and Karlsson (2009) demonstrated a nonlinear response of DOC to increasing temperature, with the number of days above 0 °C as the major predictor. Analysis of a correspondingly large dataset of boreal Norwegian watersheds within the North Sea region, indicates that even a moderate (2 °C average, downscaled Hadley scenario) increase in temperature with associated increase in precipitation and vegetation density has the potential to increase DOC export substantially (Fig. 10.5; Larsen et al. 2011a, b). For other North Sea regions, especially the UK, reduced deposition of acidifying compounds (notably sulphur) has also been shown to affect DOC concentrations (Evans et al. 2006), although the Nordic studies strongly suggest that climatic drivers like temperature and/or precipitation are the key drivers of rising DOC concentrations in lakes.

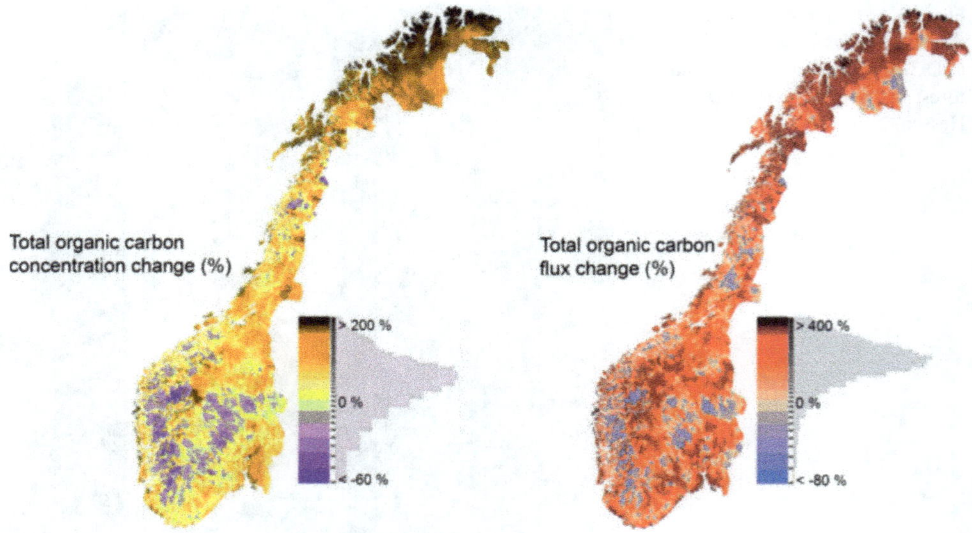

Fig. 10.5 Projected change in total organic carbon concentrations and catchment fluxes at a new steady state based on a downscaled 2 °C increase (Hadley model) (Larsen et al. 2011b)

The ecosystem responses to elevated DOC export may be profound, and may also act in concert with elevated N-deposition and N-export to lakes, which also may increase as a result of elevated N-deposition due to increased precipitation (de Wit et al. 2008; de Wit and Wright 2008; Hessen et al. 2009). This in turn is likely to affect productivity and autotroph community composition; a large survey of US and Scandinavian lakes suggested a transition from N-limitation to P-limitation in many freshwater systems as a result of chronic, elevated N-deposition (Elser et al. 2009).

An increase in allochthonous carbon may fuel microbial production and serve as an alternative food resource for zooplankton (Hessen 1998; Jansson et al. 2007). Changes in water colour and loading of organic carbon may also affect the relative roles of the benthic and pelagic parts of the lake ecosystem, in favour of the pelagic (Ask et al. 2009; Karlsson et al. 2009). However, increased DOC loads are expected to have negative impacts on overall lake productivity. Recent studies from boreal catchments suggest that P-loads especially will decrease in most areas due to increased terrestrial uptake, intensifying P-limitation and in concert with increased light limitation will reduce the overall productivity of boreal freshwaters (Weyhenmeyer et al. 2007; Jones et al. 2012; Thrane et al. 2014). Since DOM may also be a source of phosphorus in pristine catchments, a unimodal response in fish yield over DOC was found for a large number of boreal, Norwegian lakes (Finstad et al. 2014), where an initial stimulus of DOC-associated phosphorus was superseded by light limitation at higher DOC-concentrations.

Hansson et al. (2012) studied how a combination of warming and increased lake colour affects spring plankton phenology and trophic interactions in a mesocosm experiment. Elevated temperature was crossed with increased water colour. Overall, they found temperature to have a stronger effect on phytoplankton and zooplankton abundance than humic substances, but importantly also found synergistic effects between the two stressors. Thermal properties will also be affected by changes in ice cover and basin morphometry (MyLake-model, Saloranta and Andersen 2007), and temperature and reduced nutrients may both induce smaller algal cell size (Daufresne et al. 2009; Hessen et al. 2013) and community changes.

10.3.2 Lakes in Agricultural Areas

While changes in DOC may be the main climatic response in boreal lakes, they will also be affected by changes in nitrogen, phosphorus and silicon loads. The nutrient impact is far more severe in urban or agricultural lakes, however, and such lakes are also more susceptible to catchment erosion promoted by extreme rainfall as well as reduced periods of snow cover or frozen ground in winter. Land use and agricultural practices such as harvesting and fertiliser applications largely determine loads of nutrients and particulate matter to these lakes, but climatic factors, not least precipitation patterns, will add to these effects. Typically the nitrogen load from agricultural areas is expected to increase, but with a seasonal shift to increased N-export in winter, reflecting both land-use practices and climatic change (Jeppesen et al. 2011).

Phosphorus loads will also increase due to higher winter precipitation and erosion, but again the effects of land-use practices will be superimposed on the effects of climate change. The net impact on lake productivity is unclear, not least because increased turbidity has such a profound impact on lake productivity and stability (Mooij et al. 2005; Jeppesen et al. 2011). Specific ecosystem responses in lakes in agricultural areas are addressed in the following sections.

10.4 Ecosystem Dynamics

Climate change is expected to alter community structure and ecosystem functioning within lakes worldwide as well as within the North Sea region. Changes may occur in phenology, species and size distribution, food-web dynamics, life-history traits, growth and respiration, nutrient dynamics and ecosystem metabolism. Temperature-induced changes of this type are expected to interact with the increased nutrient flows resulting from enhanced precipitation and runoff (Blenckner et al. 2007; Jeppesen et al. 2010a; Moss et al. 2011).

10.4.1 Trophic Structure and Function

10.4.1.1 Fish

Several studies indicate that fish community assemblages, size structure and dynamics will change with global warming. A long-term study of 24 European lakes revealed a decline in the abundance of cold-stenothermal fish species, particularly in shallow lakes, and an increase in the abundance of eurythermal fish species, even in deep, stratified lakes (Jeppesen et al. 2012). This occurred despite a reduction in nutrient loading in most of the case studies, supposedly favouring fish in cold-water and low-nutrient lakes. The cold-stenothermic Arctic charr *Salvelinus alpinus* has been particularly affected, showing a clear decline in Lake Elliðavatn in Iceland, Lake Windermere in the UK (Winfield et al. 2010), Lake Vättern in Sweden (Jeppesen et al. 2012), and Scandinavian hydroelectric reservoirs (Milbrink et al. 2011).

Other cold-water-adapted species such as coregonids and smelt *Osmerus eperlanus* are affected at the southern border of their distribution. The harvest of whitefish *Coregonus* spp. has declined substantially in Lake Vättern in Sweden and Lake Peipsi in Estonia (Kangur et al. 2007; Jeppesen et al. 2012). In the UK and Ireland, a decline in the coregonid pollan *Coregonus autumnalis* in recent decades has been attributed to changes in temperature (Harrod et al. 2002). A drastic reduction in the population of smelt has occurred in shallow Lake Peipsi as shown from commercial fishing, with the decline particularly strong in years with heat waves (Kangur et al. 2007; Jeppesen et al. 2012). In contrast, the abundance of eurythermal species, including the thermo-tolerant carp *Cyprinus carpio* (Lehtonen 1996; Jeppesen et al. 2012) is rising in several lakes in the North Sea region (Jeppesen et al. 2012).

It is well-established that high-latitude fish species are not only often larger but also often grow more slowly, mature later, have longer life spans and allocate more energy to reproduction than populations at lower latitudes (Blanck and Lamouroux 2007). Even within species such differences can be seen along a latitudinal gradient (Blanck and Lamouroux 2007) and within North Temperate lakes (Jeppesen et al. 2010b). Thus, changes in life history and size can be expected with warming, and may in fact already have occurred (Daufresne et al. 2009; Jeppesen et al. 2010b, 2012).

10.4.1.2 Plankton

Changes in fish community structure are likely to have cascading effects in lakes, most implying increased predation on larger zooplankton which in turn means less grazing on phytoplankton and so higher algal biomass per unit of available phosphorus (Lehtonen and Lappalainen 1995; Gyllström et al. 2005; Balayla et al. 2010; Ruuhijärvi et al. 2010; Jeppesen et al. 2010a, b; Meerhoff et al. 2012). Decreasing body size has been suggested as a universal biological response to global warming (Gardner et al. 2011; Hessen et al. 2013). However, there is no consensus about the underlying causality.

The predatory effect due to fish that prefer large zooplankton prey could be reversed or partially reversed if the prevailing or additional predators are invertebrates that prefer small prey. In this case, stronger predation at higher temperature would lead to a stronger removal of small species. A shift towards smaller species can also result from stronger resource competition under higher temperatures and competitive advantage for smaller species. Stronger invertebrate predation at higher temperatures has been suggested, particularly for primary producers, because heterotrophic metabolic rates increase faster with rising temperature than photosynthesis (Yvon-Durocher et al. 2011).

Further evidence of warming-induced changes in plankton size structure comes from mesocosm studies that mimic British shallow lakes. In these mesocosms, warming increased the steepness of the plankton community size spectrum by increasing the prevalence of small organisms, primarily within the phytoplankton assemblage. Mean and maximum size of phytoplankton was reduced by about an order of magnitude. The observed shifts in phytoplankton size structure were reflected in changes in phytoplankton community composition, while zooplankton taxonomic composition remained unaffected by warming (Yvon-Durocher et al. 2011). See Sect. 10.6 for more information on responses of lake plankton communities in the context of global warming. Weak changes in the species composition of benthic macroinvertebrates following shifts towards warmer water temperatures were found in Swedish lakes (Burgmer et al. 2007).

10.4.1.3 Cyanobacteria Biomass

Higher phytoplankton biomass, particularly higher biomass of cyanobacteria during summer may be expected as a direct response to enhanced water temperatures and as an indirect response to prolonged stratification. Prolonged stratification causes an increase in internal P-loading (Jensen and Andersen 1992; Søndergaard et al. 2003; Wilhelm and Adrian 2008; Wagner and Adrian 2009b), boosting the decomposition of organic matter and thus oxygen depletion at the water-sediment interface, which further exacerbates the P-release from the sediment (Søndergaard et al. 2003). In polymictic lakes, for example, climate warming extended the periods of stratification, and this lengthening of stratified periods led to more frequent and/or stronger internal nutrient pulses between stratified and mixed periods which again promoted cyanobacteria proliferation during summer (Wilhelm and Adrian 2008; Wagner and Adrian 2009b). In dimictic lakes, on the other hand, longer periods of summer stratification may cause longer periods of nutrient limitation in the epilimnion along with higher water temperature and stronger nutrient pulses during the autumn overturn (Adrian et al. 1995; Huisman et al. 2004; Mooij et al. 2005; Elliott et al. 2006; Jöhnk et al. 2008). Immediate access to the hypolimnetic nutrient pools will be limited to migrating species such as buoyant cyanobacteria species, which are often capable of N-fixation (Reynolds 1984; Paerl 1988). Thus, in addition to causing an increase in algal biomass (particularly for cyanobacteria), climate warming may also lead to a change in ecosystem functionality such as a predominance of species capable of N-fixation (Wagner and Adrian 2009b; Huber et al. 2012).

10.4.1.4 Microbial Loop

How the microbial community and microbial processes are affected by global warming has only been studied in a few large-scale experiments. In mesocosm studies in Denmark the abundance of picoalgae, bacteria and heterotrophic nanoflagellates showed no direct response to experimental warming (Christoffersen et al. 2006). However, experimental warming modified the effects of nutrient addition (Christoffersen et al. 2006; Özen et al. 2013), indicating that interactive effects may be significant in the future given the expected increase in nutrient loading to shallow lakes worldwide (Jeppesen et al. 2009, 2010b; Moss et al. 2011). Increased DOM levels will alter the balance between phytoplankton and heterotrophic bacteria, and thus shift systems (further) towards net heterotrophy (see also Sect. 10.4.4).

10.4.1.5 Macrophytes

Owing to the climate-induced increase in eutrophication there is an increased likelihood of losing submerged macrophytes and thereby shifting shallow lakes from benthic- to pelagic-dominated systems with a consequent reduction in biodiversity. Indications of such developments are based on long-term data of Danish (Jeppesen et al. 2003) and Dutch shallow lakes (van Donk et al. 2003) as well as modelling studies (Mooij et al. 2007, 2009). Moreover, a dominance of filamentous green algae rather than phytoplankton seems possible under elevated temperatures (Trochine et al. 2011). Space-for-time approaches indicate that macrophyte cover will decrease in lakes with fewer days of ice cover, unless nutrient levels also decline (Kosten et al. 2009). Netten et al. (2011) predicted that milder winters may cause submerged macrophytes with an evergreen overwintering strategy as well as free-floating macrophytes, to outcompete submerged macrophytes that die back in winter. Neophytes such as the free-floating species *Salvinia natans* and the submerged species *Vallisneria spiralis* have been shown to be successful under elevated temperatures at the expense of native submerged macrophytes (Netten et al. 2010; Hussner et al. 2014). Mormul et al. (2012) tested the effects of elevated temperature (3 °C) on native and non-native aquatic plant production in mesocosms in combination with 'browning' (increased DOC), a potentially important change in the northern hemisphere and found browning to be more important for species invasion than warming. Climate change is likely to have a direct effect if non-native species respond positively to climate change and an indirect effect through species interactions, for example, because browning impairs the growth of native macrophytes and reduces biotic resistance to invasion.

Native competitors of the invasive *Elodea canadensis* were less successful in browner waters indicating a reduced resistance to invasion. Warming of mesocosms in the UK by 3 °C also significantly altered the proportions of three macrophyte species due to a higher growth rate and higher relative abundance of the neophyte *Lagarosiphon major* (McKee et al. 2002). A subsequent trial with temperatures 4 °C higher and higher nutrient concentrations resulted in a dominance of floating duckweed *Lemna* spp. which severely reduced oxygen availability and resulted in a fish kill (Moss 2010). In general, however, there are few long-term studies on the effect of climate change on macrophyte species distribution and invasions, coverage and subsequent effects on other trophic levels in aquatic ecosystems.

For large areas of northern Europe, mass occurrences ('nuisance growth') of the macrophyte *Juncus bulbosus* have been recorded (Moe et al. 2013). The reasons for this phenomenon are still unclear, but nuisance growth has been linked with hydrology, carbon dioxide and elevated N-deposition (Moe et al. 2013).

10.4.2 Phenology

The most prominent examples of climate-induced changes in lakes are changes in phenology. Coherent changes in ice phenology (see Sect. 10.2.2), and changes in spring and early summer plankton phenology in the North Sea region in recent years have been attributed to climate change (Adrian et al. 1999; Weyhenmeyer et al. 1999; Gerten and Adrian 2000; Straile 2002) as synchronised by large-scale climatic signals such as the NAO (for a review see Blenckner et al. 2007; Gerten and Adrian 2002a; Straile et al. 2003). While indirect temperature effects such as early ice-off, which improves underwater light conditions have brought forward the start of algal bloom development in spring, direct temperature effects caused changes in the timing of rotifer and daphnid spring maxima (Gerten and Adrian 2000; Adrian et al. 2006; Straile et al. 2012) cascading into an earlier clear water phase (Straile 2002). For zooplankton, phenological shifts were most immediate for fast-growing species such as cladocerans and rotifers (but see Seebens et al. 2007), whereas longer-lived plankton such as copepods showed a lag in response. Copepods responded to altered day-length-specific water temperature affecting the timing of the emergence of resting stages in spring (Gerten and Adrian 2002b; Adrian et al. 2006). In addition, warming-induced accelerated ontogenetic development may enable the development of additional generations within a year as has been shown for copepod species (Gerten and Adrian 2002b; Schindler et al. 2005; Adrian et al. 2006; Winder et al. 2009).

10.4.3 Metabolism

Recent research suggests that global warming tends to shift the metabolic regime of entire lakes toward a dominance of respiration (Allen et al. 2005). This fundamental difference in temperature response between autotrophic and heterotrophic processes may have major implications for biological communities and for ecosystems in general. However, different members of a food web react differently to temperature, for example while cell-division rates may increase with temperature, as may grazing rates and metabolic demands of zooplankton. Thus, the net effect of warming on metabolism is not straightforward also because metabolic rates differ at the species level. As a result, how a changing climate interacts with increased nutrient supply to alter ecosystem metabolism is more uncertain than the change in trophic structure. Although evidence suggests that processes such as deoxygenation, decomposition and denitrification are influenced both by nutrients and by warming, interactions are complex and variable and there are discrepancies in study results about the end result for system components as well as for systems as a whole.

Mesocosm studies in the UK indicated that gross primary production and respiration increase with warming, while results for net production and carbon storage differ. Two experiments (Moss 2010; Yvon-Durocher et al. 2010) showed a marked increase (18–35 %) in the ratio of diurnal community respiration rates and gross photosynthesis for a warming of up to 4 °C. If extrapolated to the large number of shallow northern lakes, this could have immense implications for positive feedbacks in the Earth's future carbon cycle (Moss 2010). However, these UK experiments were all of short duration (less than one year) and may only have described the transient state after warming, which may lead to overestimation of the net release of carbon. Long-term mesocosm experiments will provide more reliable indications about the net effect of warming on ecosystem metabolism (Jeppesen et al. 2010a; Liboriussen et al. 2011), complemented by long-term research and modelling of whole lake ecosystems (Trolle et al. 2012).

10.4.4 Greenhouse Gases and Heterotrophy

Ecosystems not only respond as recipients of climatic change, but also provide feedbacks, not least via greenhouse gases such as carbon dioxide (CO_2), methane (CH_4) and nitrous oxide (N_2O). As previously mentioned, the input of organic carbon to lakes through run-off from the catchment has increased in the North Sea region, inducing changes in water colour (Hongve et al. 2004; Erlandsson et al. 2008) (see also Sect. 10.3). Recovery of soils from acidification and changed hydrological conditions are believed to be important factors determining this development (Monteith et al. 2007). On a somewhat longer perspective, climate effects on hydrology and vegetation density may also promote a substantial browning of boreal surface waters (Larsen et al. 2011a, b).

Autotrophs and heterotrophic bacteria compete for the same essential elements, but utilise different energy sources. While DOC is a major energy source for heterotrophic bacteria, it is also an important absorbent in the photosynthetically active part of the spectrum. DOC thus has negative impacts on primary producers both by competing for photons and by stimulating their major nutrient competitors. Increased levels of terrestrially-derived DOC in concert with reduced availability of inorganic phosphorus in lakes may shift systems further towards net heterotrophy and thus a net CO_2 release (Sobek et al. 2003; Larsen et al. 2011c).

Increased loads of DOM in boreal lakes would also promote anoxia in the deeper water layers, which promotes

CH_4 production and its net flux to the atmosphere (Juutinen et al. 2009). Increased N-deposition in concert with elevated export of DOM may promote export of N_2O (Hong et al. 2015). For lakes in agricultural systems, increased productivity due to increased nutrient inputs from the catchment may promote the net efflux of CH_4 and N_2O (Juutinen et al. 2009; Hong et al. 2015), while the net impact of the CO_2-balance will also depend on lake morphometry and stratification and so is harder to predict.

10.5 Biodiversity

Climate conditions are as important for freshwater biodiversity as for terrestrial and marine biodiversity, and consistently explain a major proportion of the geographic variation in species richness of different freshwater taxa such as amphibians, fish, mammals, crayfish and waterbirds (Tisseuil et al. 2013). Even so, other geographic patterns such as the latitudinal gradient in biodiversity is not as strong in freshwater as it is in the marine or terrestrial realms (Hillebrand 2004). However, these geographic trends (often measured at global to regional scales) cannot easily be transferred into predictions on temporal shifts in biodiversity at regional to local scales under climate change. Freshwater systems are particularly vulnerable to climate change for several reasons: owing to the isolated nature of freshwater habitats embedded in a terrestrial matrix; because climate change has direct influences on local temperatures and temperature-associated factors (such as oxygen saturation); and because many freshwater systems already absorb other anthropogenic stressors such as nutrient loading or an altered hydrological regime (Woodward et al. 2010). This section addresses these issues by reviewing existing information on biodiversity shifts in the North Sea region (Sect. 10.5.1) and deriving more general predictions from theoretical and experimental literature identifying research needs in the North Sea region (Sect. 10.5.2).

10.5.1 Shifts in Biodiversity

Predicting shifts in freshwater biodiversity in the North Sea region is difficult, because few studies have been explicitly conducted in lentic or lotic water bodies in this region. Predictions concerning future biodiversity are often derived from bioclimatic envelope models, which project future range shifts based on the current distribution of species (Parmesan and Yohe 2003). These models typically predict a northward (and often eastward) shift in ranges, such that warm-adapted species expand their ranges, and the ranges of cold-adapted species narrow.

Although changes in marine biodiversity in the North Sea region have frequently been predicted using this approach (Beaugrand et al. 2002), studies for freshwater systems in this region are rare. The species richness of macroinvertebrates across running waters in Europe has been predicted to decline in the southern North Sea countries (UK, Germany, Netherlands, Denmark), but to increase in the North (Norway) (Domisch et al. 2013). Changes in distribution and diversity have already been observed in some freshwater groups, such as odonates in the UK (Hickling et al. 2005). Other diversity shifts caused by local, climate-related range retractions have been reported for fishes in Iceland (Jeppesen et al. 2010b) and crustaceans in Norway (Lindholm et al. 2012). Correspondingly, analyses of long-term monitoring data on community composition revealed shifts in macroinvertebrate assemblages associated with ambient temperatures in Greenland, Iceland, Norway, Denmark and Sweden (Burgmer et al. 2007; Friberg et al. 2013). However, few long-term monitoring data sets still exist for freshwater systems within the North Sea region.

Thus it seems clear that specialised communities in colder regions around the North Sea have a high potential for reduced biodiversity. Macroinvertebrates in glacier-fed river systems will be characterised by lower local species richness and lower beta-diversity if warming leads to glacier retreat (Jacobsen et al. 2012). In boreal regions of northern Europe, the riparian zones of running waters are predicted to be affected by additive or interactive combinations of higher temperature, increased annual discharge but less seasonal variation in runoff, changes in groundwater supply, and altered ice regimes (Nilsson et al. 2013). Potential consequences for biodiversity can be negative or positive. Negative consequences are likely if the riparian zone narrows, such as by hydrologic changes, and thus species richness locally declines. In contrast, higher temperatures might allow invasion of exotic species leading to higher local species richness. Other types of change include altered disturbance regimes (e.g. altered freezing and thawing regimes during winter), which could foster a more dynamic and species-rich riparian vegetation, but also a more specific and species-poor assemblage of stress-tolerant species.

Floodplain systems around the North Sea coast have been massively altered by human regulation of flow regimes and inundation (Tockner et al. 2010). At the same time, they harbour a diverse fauna and flora shaped by the interaction of different climatic, hydrological and biological drivers as well as by the interaction between aquatic and terrestrial ecosystems with respect to the exchange of water, nutrients and organisms. Climate-induced shifts in flow regime are thus of primary importance for biodiversity. European-scale modelling scenarios predict that North Sea region floodplains will experience moderately higher flow levels

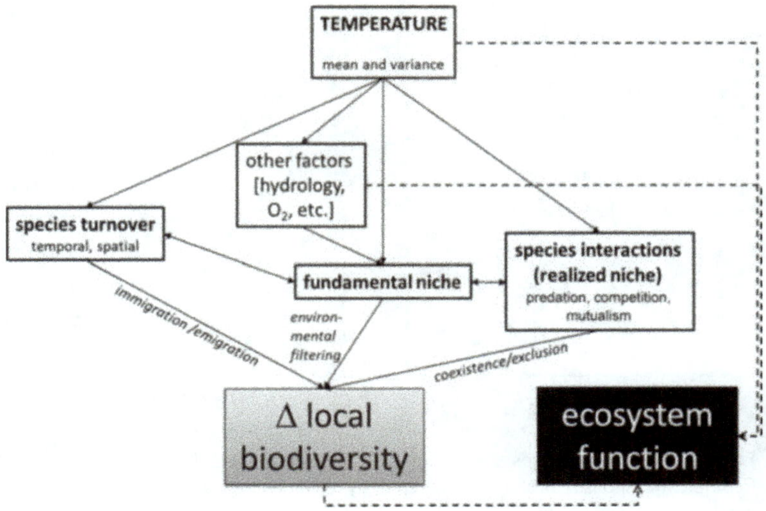

Fig. 10.6 Schematic representation of pathways leading to altered local biodiversity and thus ecosystem function. Main pathways include changes in species pools (species turnover) and on the fundamental as well as realised niches of species

(contrasting with predictions for other regions, such as the Mediterranean floodplains) (Schneider et al. 2013). Consequences for biodiversity remain largely unknown.

Climate-driven changes in biodiversity are very likely to interact with changes associated with other anthropogenic pressures, such as eutrophication (Moss et al. 2009). Some synergistic effects have been found in UK mesocosm studies (Feuchtmayr et al. 2009). For example, warming promoted increased phosphorus concentrations and the frequency of severe benthic anoxia in the mesocosms, with the potential to exacerbate existing eutrophication problems (McKee et al. 2003).

10.5.2 Predictions, Theory and Experimental Studies

Although functional aspects of ecosystem impacts (biomass, productivity, element cycling) under climate change can be predicted with some accuracy (see previous sections), information on shifts in freshwater biodiversity in the North Sea region remains vague (Moss et al. 2009). This is due not only to the scarcity of studies themselves, but also to the focus on larger spatial scales (i.e. regional climate-envelope models on range shifts), which makes predictions for local ecosystems difficult. Information from ecological theory and experiments can help to fill this gap by identifying potential changes in freshwater biodiversity and associated pathways (Fig. 10.6). Local attributes may also be superimposed on large-scale patterns. In an analysis of global lake zooplankton data, Shurin et al. (2010) found increased biodiversity in lakes showing greater temperature variation on different time scales (intra- and interannual). A recent review argued strongly for including temperature variability in climate-change experiments (Thompson et al. 2013).

The survival of a species can be directly impaired if temperatures are shifted beyond the fundamental thermal niche, leading to local extinction (Fig. 10.6). Temperature also affects the fundamental niche of a species with respect to other conditions, which are often directly related to temperature (e.g. oxygen saturation, hydrological regimes, solute concentrations). Indirect consequences of temperature change can be seen in freshwaters from around the North Sea, for example with respect to stratification regimes (Wagner and Adrian 2011) or oxygenation (Wilhelm and Adrian 2008). These changes alter the environmental filtering of colonising species, such that local diversity can increase or decrease depending on the number of species that are precluded or enabled to establish viable populations.

The meta-community context (Leibold et al. 2004) reflects the interplay of dispersal and local processes and will become an important tool to predict future changes in aquatic biodiversity. This is even more likely given that much of the regional biodiversity in aquatic systems is contributed by ponds and shallow lakes as these have high beta-diversity due to their often isolated nature (Scheffer et al. 2006) and provide important ecosystem services such as carbon storage due to their high number (Giller et al. 2004; Downing et al. 2006; Tranvik et al. 2009).

At the same time, there is high potential for small isolated systems to lose species with global climate change (Burgmer and Hillebrand 2011) and spatial dynamics are important for maintaining biodiversity in these systems. Meta-community dynamics can also provide spatial insurance for freshwater ecosystems (Loreau et al. 2003). In a meta-community mesocosm study, colonisation from a regional species pool and higher biodiversity positively affected the recovery of the pond ecosystems from heat stress (Thompson and Shurin 2012).

Changes in species turnover with time reflect changes in biodiversity dynamics, i.e. local immigration and extinction.

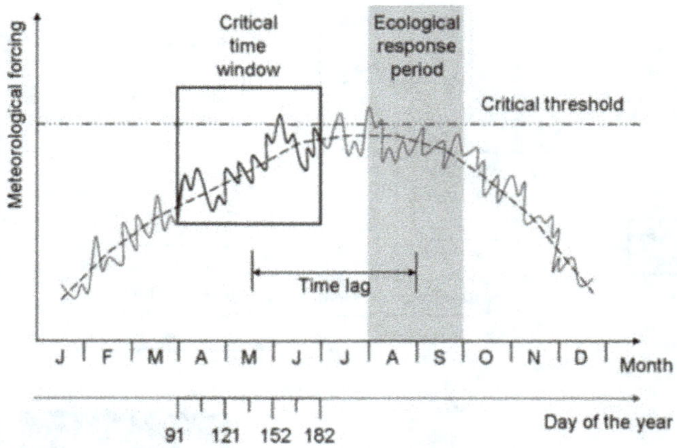

Fig. 10.7 Timing of ecosystem responses to meteorological forcing. Ecological responses are often triggered by changes in critical time windows. The triggering mechanism frequently involves the crossing of critical thresholds in forcing variables (*dashed-dotted line*), and responses tend to occur with a time lag. In this conceptual sketch, analysis at the monthly timescale (*dashed line*) would not be sufficient to detect threshold exceedance, in contrast to analysis at the daily timescale (*solid line*) (Adrian et al. 2012)

Most experiments suggest a more rapid turnover of species at higher temperatures reflecting an acceleration of colonisation–extinction dynamics. In the thermal effluent of a nuclear power plant, higher temperatures resulted in a faster turnover of species composition but without affecting species richness (Hillebrand et al. 2010). Laboratory experiments also indicated faster change in species composition with increasing temperature (Hillebrand et al. 2012). Analysing temporal turnover in community composition could thus be a better means of identifying climate-induced changes in biodiversity than simple univariate measures of biodiversity such as richness or evenness (Angeler and Johnson 2012). Especially because even major changes such as a shift in dominance between functional groups (e.g. from diatoms to cyanobacteria in phytoplankton) might occur without a change in species richness (Wagner and Adrian 2011).

Temperature also modifies species interactions and their consequences for biodiversity (Fig. 10.6). Increased temperatures are often associated with higher rates of consumption (Hillebrand et al. 2009), which can lead to either lower biodiversity (higher mortality) or higher biodiversity (more consumer-mediated coexistence). Competitive interactions are also strengthened by higher temperatures, leading to more rapid exclusion of inferior species—an effect which was shown to depend on consumer presence (Burgmer and Hillebrand 2011). Not only can the strength of interactions be altered by temperature, but also the temporal match of the interacting species through changes in phenology (Berger et al. 2010). Information on temperature-dependent changes in mutualistic interactions in freshwaters is currently missing.

Both experiments and models indicate that warming-induced shifts in biodiversity have functional consequences for ecosystems (Fig. 10.6), among others with respect to primary production, resource use efficiency and temporal stability of ecosystem functions (Hillebrand et al. 2012; Schabhuettl et al. 2013). In a long-term freshwater phytoplankton experiment, temperature-induced reductions in species richness were associated with lower biomass production, and higher extinction rates were associated with higher variability in biomass production (Burgmer and Hillebrand 2011). Similar strong relationships between diversity and resource use efficiency are found in freshwater field data from the UK and Scandinavia (Ptacnik et al. 2008).

10.6 Importance of Temporal Scale

Responses to climate change in lake ecosystems operate on various temporal scales. Physical forces such as variation in temperature and mixing regimes span sub-daily to monthly time scales. Organisms differ in their generation time from daily to yearly time scales (Adrian et al. 2009). Thus, to understand the impacts of single climatic forcing events in the context of longer term dynamics it is necessary to consider not only sufficiently long periods (several decades) but also to consider appropriate small temporal scales within the yearly cycle. For example, ecological variables may respond to meteorological forcing only during short critical time windows, or to short-lived exceedance of ecologically-relevant critical thresholds. Thus, annual, seasonal or monthly climate data may not be enough to capture the thermal dynamics to which organisms actually respond (Fig. 10.7; Adrian et al. 2012).

Members of the grassroots organisation GLEON (Global Lake Observatory Network; www.gleon.org) or the European project NETLAKE (Networking Lake Observatories in Europe; www.cost.eu/domains_ctions/essem/Actions/ES1201) established an international network of automatic stations in lakes to address dynamics of ecosystem properties at sub-hourly scales.

The following sections focus on the role of temporal scale in climate impact research and provide examples of responses at small (sub-daily) to large (decadal) temporal scales, the role of critical time windows, and the significance of exceeding critical thresholds for lakes within the North Sea region (which also applies for lakes throughout the North Temperate Zone).

10.6.1 Critical Temporal Scales

The importance of addressing critical temporal scales in climate impact research has been documented for various lakes in the North Sea region (see Adrian et al. 2012; Sect. 10.2). A closer look at sub-hourly measurements showed that the rate of increase in the daily minima (night-time water temperature) exceeded that of the daily maxima (daytime water temperature) (Wilhelm et al. 2006). The consequences of this day-night asymmetry for the biota are unclear, but may contribute to some of the unexplained changes observed in ecosystem dynamics over time. Day-to-day variation in respiration seems to be common in lakes worldwide, including those in the North Sea region (Solomon et al. 2013). Daily variation in gross primary production explained 5–85 % of the daily variation in respiration. Solomon et al. (2013) found respiration to be closely coupled to gross primary production at a diurnal-scale in oligotrophic and dystrophic lakes, but more weakly coupled in mesotrophic and eutrophic lakes.

Known changes in the thermal regime of lakes in the North Sea region (Sect. 10.2) operate over a broad range of temporal scales, and are closely related to lake morphometry: on sub-daily (Wilhelm and Adrian 2008) to weekly scales in polymictic lakes (Wagner and Adrian 2009b), and on weekly to monthly scales in monomictic or dimictic lakes (Gerten and Adrian 2000; Livingstone 2003). While variation in the timing of spring overturn affects underwater light conditions and thus the start of algal growth (Weyhenmeyer et al. 1999; Gerten and Adrian 2000; Peeters et al. 2007), variation in the timing of summer stratification affects water temperature and internal nutrient loading and subsequent plankton development and species composition in productive lakes (Wilhelm and Adrian 2008; Wagner and Adrian 2011). Differences in water temperature between mixed and stratified periods can be up to 5 °C within days or a just few weeks in summer, favouring thermophilic copepod species for example (Wagner and Adrian 2011).

Changes in phenology in abiotic and biotic variables (see Sect. 10.4.2) operate on time scales of weeks (Weyhenmeyer et al. 1999; Gerten and Adrian 2000; Straile et al. 2003). Thus, in terms of their duration, seasons should be defined by cardinal events within the lake itself, rather than by fixed calendar dates. Important markers successfully used to define phenology-adjusted seasons in lakes include temperature thresholds, ice-off dates, the timing of the clear-water phase, and periods of stable thermal stratification (Rolinski et al. 2007; Wagner and Adrian 2009a; Huber et al. 2010).

Wagner et al. (2012) proposed a seasonal classification scheme tuned to specific hydrographic-sensitive phases for dimictic lakes across a latitudinal gradient: inverse stratification (winter), spring overturn, early stratification and the summer stagnation period. They estimated a mean latitudinal shift of 2.2 days per degree of latitude for the start of these sensitive phases. After accounting for latitudinal time shifts, mean water temperatures during the defined hydrographic cycles were similar in lakes spanning the gradient between 47° and 54°N. Adjusting seasons in this way thus enhances the probability of identifying the major driving forces in climate impacts on lake ecosystems.

10.6.2 Critical Time Windows

Responses to warming trends are often expressed in terms of average changes in temperature on seasonal or annual scales, however species exhibiting short generation times such as planktonic organisms respond only during specific time windows within a season. Shatwell et al. (2008) showed how short time windows can open for cyanobacteria in warm springs in an otherwise diatom-dominated season. They argued that if cyanobacteria attain a critical biomass during that critical time window they can dominate the phytoplankton during summer. In terms of zooplankton, the abundance of cyclopoid (Gerten and Adrian 2002b; Seebens et al. 2009) and calanoid copepods (Seebens et al. 2007) in summer and autumn are determined by conditions in spring—probably related to temperature-induced changes in the emergence of resting stages (Adrian et al. 2006) or short time windows of high food availability which increases offspring survival (Seebens et al. 2009).

Temperature-driven changes in the timing of food availability and of predation by young-of-the-year fish during critical time windows in spring/early summer determined the mid-summer decline in daphnids (Benndorf et al. 2001). More specifically, water temperatures in narrow time windows either before (2.2 weeks) or after the typical clear-water phase (3.2 weeks) affected the start-up populations of summer crustacean zooplankton and explained some of their contrasting success during three hot summers characterised by more or less the same average summer water temperature (Huber et al. 2010).

10.6.3 Critical Thresholds

Threshold-driven responses to environmental forces have gained attention in ecology because of their seeming unpredictability and their potentially large effects at all levels of ecosystems. The crossing of critical thresholds may result in abrupt changes in particular elements of an ecosystem (Andersen et al. 2009; Scharfenberger et al. 2013) or entire ecosystems—the famous example being the alternative stable states of clear versus turbid lakes (Scheffer and

Carpenter 2003). Abrupt changes within ecosystems are already known under warming trends experienced in the recent past for a number of variables spanning abiotic and biotic components, such as nutrients or algal blooms (Wagner and Adrian 2009a; for review see Adrian et al. 2009, 2012). The underlying forces are often unclear, but may involve competition for common resources and the crossing of critical thresholds in the abundance of conspecifics (Scharfenberger et al. 2013) or multiple overlapping environmental forces (Huber et al. 2008).

Critical thresholds are known to have been exceeded within lake ecosystems (Hargeby et al. 2004). For example, Peeters et al. (2007) quantified the exceedance of critical thresholds in spring for several meteorological variables to determine early or late onset of phytoplankton growth in Lake Constance (Germany). In a recent model, Straile et al. (2012) used water temperature phenology as a predictor for *Daphnia* seasonal dynamics in North Temperate lakes. The day of the year when surface water temperatures reached a threshold of 13 °C explained 49 % of the variability of the timing of the spring *Daphnia* maximum in two German lakes (Lake Constance, Müggelsee) and in Lake Washington (USA). The *Daphnia* phenology model also performed well for predicting the timing of the *Daphnia* maxima in 49 lakes within the northern hemisphere—many located in the North Sea region (Straile et al. 2012). Early spawning of zebra mussel *Dreissena polymorpha* in Müggelsee was related to early attainment of the same critical water temperature threshold of 13 °C, known to initiate the first spawning event of the year (Wilhelm and Adrian 2007).

Exceeding direct and indirect temperature thresholds (length of thermal stratification) has been shown to trigger processes such as the onset and magnitude of cyanobacteria blooms (Wagner and Adrian 2009a; Huber et al. 2012). Stratification periods of more than three weeks caused a switch from a dominance of non-N-fixing cyanobacteria to a dominance of N-fixing cyanobacteria species, thus affecting not only biomass of cyanobacteria but also ecosystem functioning (Wagner and Adrian 2009a). In addition to warming-related changes in species composition (Adrian et al. 2009), habitat shifts northward toward temperate-zone lakes over the last few decades have been observed for *Cylindrospermopsis raciborskii*, an invasive freshwater cyanobacterium, originating in the tropics (Padisak 1997). Observations of *Cylindrospermopsis raciborskii* in pelagic populations were found to be temperature-mediated. Filaments emerged in the pelagic habitat when water temperature rose above 15–17 °C in two north German lakes (Wiedner et al. 2007).

10.6.4 Extreme Events

Extreme events, which are expected to become more frequent in the future, are often ecologically more relevant than fluctuations in mean climate. For lakes, extreme events principally refer to exceptionally mild winters, summer heat waves, or extreme storms or heavy rainfall events. For example, an extreme event for a lake that typically freezes in winter would be if the lake did not freeze at all. Based on regional climate model forecasts, Livingstone and Adrian (2009) predicted that the percentage of ice-free winters for a lake in northern Germany would increase from about 2 % at present to over 60 % by the end of the century; see Sect. 10.2.2 for more detail on the frequency of extremely late freeze-up, early break-up and short ice duration. These extremes affect thermal stratification patterns and underwater light conditions, with implications for oxygen conditions and phytoplankton development (see Sect. 10.4.2).

Central Europe has recently experienced extreme heat waves, most notably that of summer 2003. Mean air temperature in summer that year exceeded the long-term average by around 3 °C over much of Europe (Schär et al. 2004). Although heat waves are likely to promote cyanobacteria blooms (Jöhnk et al. 2008), water temperatures above average to the same extent in Müggelsee (Germany) in two recent summers (2003 and 2006) resulted in very different situations: a cyanobacteria bloom in 2006 but a record low cyanobacteria biomass in 2003. This difference was due to the thermal stratification pattern being critically intense only in 2006 (Huber et al. 2012).

Summer fish kills and a change in fish community structure have been attributed to summer temperature extremes in combination with eutrophication in shallow Lake Peipsi (Estonia/Russia) related to a decline in near-bottom oxygen conditions and a decrease in water transparency (Kangur et al. 2013). An extreme rainy period in 2000 caused a strong increase in chemical loading, particularly for organic carbon, in Lake Mälaren (the third largest lake in Sweden) followed by an increase in water colour by a factor of 3.4 and a doubling of spring cryptophyte biomass. This increase in algal mass required changes in the treatment of raw water from Lake Mälaren for the drinking water supply of Stockholm city (Weyhenmeyer et al. 2004b). Extreme summer rain events have also altered CO_2 and CH_4 fluxes in southern Finish lakes, with the systems switching from being a net sink to a net source of CO_2 to the atmosphere (Ojala et al. 2011).

Episodic events of extreme wind speed or rain events exceeding two standard deviations of the seasonal means

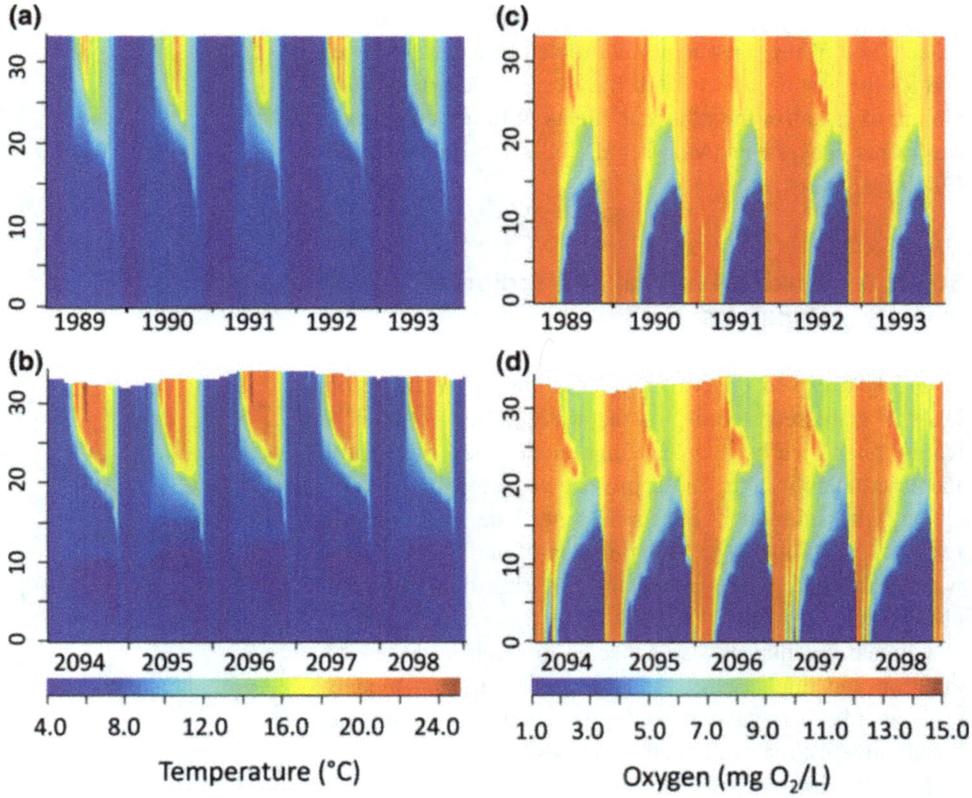

Fig. 10.8 Simulated water temperature for the deep Lake Ravn, Denmark, for current climate (represented by 1989–1993) (**a**) and future climate (represented by 2094–2098) (**b**), and simulated oxygen concentration for current climate (**c**) and future climate (**d**). Simulations were performed using DYRESM-CAEDYM. Future climate forcing was derived by the Danish Meteorological Institute using the regional HIRHAM model and the SRES A2 scenario for Denmark. Y-axis represent water level (m) (D. Trolle, original)

have strong but complex impacts on thermal structure and stability, DOC loading and underwater PAR (photosynthetically active radiation) levels in northern European lakes—the magnitude and direction of change depending on the location of the lake and catchment characteristics (Jennings et al. 2012). A comprehensive summary as to how extreme weather events affect freshwater ecosystems is provided by the British Ecological Society (BES 2013).

10.6.5 Regime Shifts

While ecosystems may be buffered against single short-lived critical threshold exceedance events, gradual changes over longer periods may cause lake ecosystems to switch abruptly from one state to another. The most likely and widespread climate warming-induced shift in lakes will be shifts in thermal regime (see Sect. 10.2). Climate change alters heat redistribution over time (within the annual cycle) and space (vertically within the water column), and eventually leads to transitions in the seasonal mixing regime of a lake much in the sense of scenarios described by regime shift theory (Scheffer and Carpenter 2003).

On the basis of existing climate scenarios, Kirillin (2010) predicted a shift from a dimictic to a monomictic regime in the majority of European dimictic lakes by the end of the 21st century, with the loss of ice cover in the cold season meaning that winter stratification in these lakes would completely disappear. In summer, climate warming has an opposite, stabilising effect that may eventually lead to the mixing regime shifting to a dimictic regime in hitherto polymictic lakes. The ecological consequences of this type of regime shift may be even more far-reaching than for di-/monomictic transitions, because the abrupt detachment of the nutrient-rich hypolimnion from the euphotic layer is likely to trigger stronger competition between autotrophic species resulting in changes in phytoplankton species composition and ecosystem functionality (Wilhelm and Adrian 2008; Wagner and Adrian 2009a, b, 2011).

10.7 Modelling

Predicting the fate of freshwater ecosystem processes under climate change is non-trivial, because many physical, chemical and biological processes interact, and may be affected on different spatial and temporal scales by climate forcing. In attempting to account for these complex interactions, mechanistic numerical models continue to play a greater role in hypothesis testing (system understanding) and for predicting the future state of ecosystems given the projections of future climatic forcing according to climate models (Trolle et al. 2012). Thus, the ability to link—and equally importantly to quantify—complex interactions between physical, chemical and biological processes makes models one of the most important tools of modern science, and for the past decade, models have been used extensively, aiming to establish the potential effects of future climate on freshwater ecosystems (Mooij et al. 2007; Trolle et al. 2011; Elliott 2012).

10.7.1 Linking Physical and Ecological Dynamics

A widely accepted effect of increased climate warming on lakes is increased stability of the water column, which is readily quantified by hydrodynamic models (see Fig. 10.8). Increased stability can cause prolonged periods of stratification, with subsequent effects on biogeochemical cycling, for example by increasing the duration of anoxia in bottom waters and thereby the potential for release of iron-bound phosphorus.

Climate warming is also expected to result in decreased duration and thickness of ice cover on lakes, and models such as MyLake have recently been applied to provide quantitative estimates of such decreases.

Models with a strong focus on food-web dynamics, such as PCLake (Janse 1997) have also been used to quantify the effects of warming in facilitating lake ecosystems to shift from a clear, macrophyte-dominated state to a turbid, phytoplankton-dominated state, under different combinations of warming scenarios and external nutrient load scenarios (Fig. 10.9). Modelling exercises of this type are readily undertaken using a standard desktop computer, in contrast to testing such scenarios in experimental mesocosm studies, for example, which is extremely time consuming and expensive.

One of the key messages from the model study by Mooij et al. (2007) shown in Fig. 10.9 was that the critical loading of phosphorus to a lake, at which a shift from a clear to a turbid state occurs, is likely to decrease as warming continues. This effectively means, that according to the model, external loading will need to be reduced in the future, if lakes are to retain the ecological quality of present day.

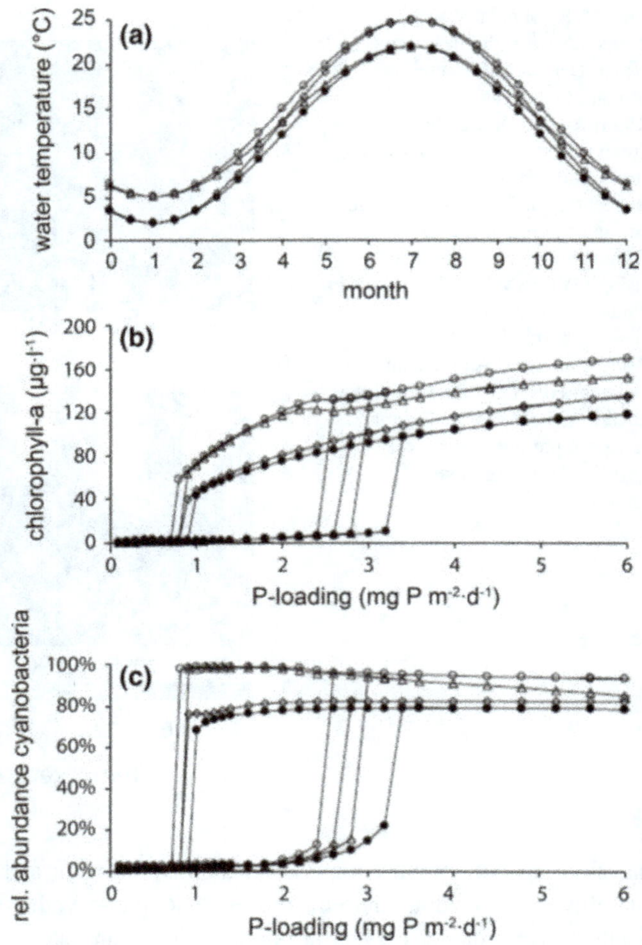

Fig. 10.9 Three scenarios with elevated temperatures (**a**) simulated using PCLake adapted for shallow Dutch lakes, and evaluated for average summer chlorophyll-a concentration (**b**) and the percentage of cyanobacteria relative to total phytoplankton biomass (**c**). The simulations include four scenarios: a control (*closed circles*), an all year round temperature increase of 3 °C (*open circles*), an increase in summer maximum temperature of 3 °C but no change in the winter minimum (*open diamonds*), and an increase in winter minimum temperature of 3 °C, but no change in the summer maximum (*open triangles*) (adapted from Mooij et al. 2007)

10.7.2 Predictions Versus Observations

Modelling studies of the long-term effects of climate change on northern hemisphere temperate lakes (e.g. Elliott et al. 2005; Mooij et al. 2007; Trolle et al. 2011) generally imply that overall phytoplankton biomass is likely to increase, and that cyanobacteria will become a more dominant feature of the phytoplankton species composition. This is in line with empirical observations from time series (Jöhnk et al. 2008; Wagner and Adrian 2009a, b; Posch et al. 2012) and with cross-system analyses (Jeppesen et al. 2009; Kosten et al. 2012). However, the effects may be even more severe than suggested by models, as current models do not fully account for structural changes in the lake ecosystems that could occur due to warming. As stated in Sect. 10.4.1 the composition of fish stocks could change towards smaller and faster reproducing fish, implying more predation on zooplankton and so less grazing on phytoplankton (Jeppesen et al. 2010a; Meerhoff et al. 2012). Such changes have yet to be included in dynamic models.

Table 10.1 Level of certainty for key climatic impacts on freshwater ecosystems

Parameter	Comment
High certainty	
Lake temperature	Several time series, strong modelling tools
Ice development	Several time series, strong modelling tools, satellite data
Phenology	Several time series, modelling tools for autotrophs
Water level, runoff, retention	Strong hydrological modelling tools
Thermal regime	Extension of stable stratification periods, good modelling tools
Oxygen depletion	Several time series, linked with hydrology and trophic state
Medium certainty	
DOC concentration	Long-term data, good modelling tools
Concentrations of key elements	Some data, closely linked with hydrology
Primary production	Contrasting effects of DOC and nutrient inputs. Local differences
Autotroph community composition	Prevalence of cyanobacteria. Local differences
Deterioration of trophic state	Warming acts like eutrophication; linked with hydrology and trophic state
Range expansion of invasive species	Northwards distribution of species
Loss of cold cold-stenothermic fish species	Several time series; results from fish harvest
Low certainty	
Secondary production	Hard to arrive at general conclusions. Taxon- and locality-specific differences
Heterotroph community composition	Taxon- and locality-specific differences. Secondary impacts of autotrophs
Diversity	Limited data availability, taxon- and locality-specific differences; changes in ecosystem function
Food-web responses	Accumulated uncertainty from all other responses
Greenhouse gases (carbon dioxide, methane, nitrous oxide)	Lakes as carbon sink or source, large-scale enclosure experiments
Water clarity	Progress expected with new satellite data

10.7.3 Linking Landscape Activities and Nutrient Losses

A recent trend is to assemble multidisciplinary modelling teams with expertise both in hydrological and ecological modelling. Coupling hydrological and ecological models (Norton et al. 2012; Nielsen et al. 2013) enables a direct and quantitative link between land use activities and surface water quality. For example, Norton et al. (2012) used a farm-scale nutrient budget model (PLANET; Planning Land Applications of Nutrients for Efficiency) combined with a generic nutrient runoff model (GWLF; Generalized Watershed Loading Function) to estimate nutrient losses to Loweswater, a small lake in northwest England. The simulated nutrient losses were subsequently used as input to an ecological model for the lake, PROTECH (Phytoplankton RespOnses To Environmental CHange). This model-chain has been used to forecast the abundance of different phytoplankton types within the lake in response to a range of catchment management measures. Ongoing research projects are currently expanding such model chains to include the effects of future climate on both hydrology and aquatic ecosystems.

10.8 Conclusion

Freshwater systems, and notably lakes, are well suited to trace climate-induced effects both directly via changes in ice cover, hydrology and temperature, but also indirectly via biotic communities since they represent closed bodies (aquatic islands in the landscape). Moreover, they also offer strong tools as 'sentinels' by integrating changes at the catchment scale that elsewhere would be hard to detect (Adrian et al. 2009). The North Sea region has a wide range of freshwaters that reflect changes related directly and indirectly to climate forcing. This includes a number of long-term monitoring sites, and experimental studies offer insights into climate change in freshwaters—and subsequent impacts on downstream coastal recipients. The major impacts on lakes, both those known to date and those predicted to occur, can be addressed with different levels of certainty (tentatively summarised in Table 10.1). The level of certainty falls substantially, once higher trophic levels and the complexity of trophic interactions are included. Nevertheless, this summary may be used to identify systems and parameters that are already well suited for addressing climate responses, those that need further attention, and those that are likely to need addressing at local scales and where general statements and predictions are hard to achieve.

References

Adrian R, Deneke R, Mischke U, Stellmacher R, Lederer P (1995) A long-term study of Heiligensee (1975-1992). Evidence for effects of climate change on the dynamics of eutrophied lake ecosystems. Arch Hydrobiol 133:315–337

Adrian R, Hintze T, Walz N, Hoeg S, Rusche R (1999) Effects of winter conditions on the plankton succession during spring in a shallow polymictic lake. Freshwater Biol 41:621–632

Adrian R, Wilhelm S, Gerten D (2006) Life-history traits of lake plankton species may govern their phenological response to climate warming. Glob Change Biol 12:652–661

Adrian R, O'Reilly CM, Zagarese H, Baines SB, Hessen DO, Keller W, Livingstone DM, Sommaruga R, Straile D, van Donk E, Weyhenmeyer GA, Winder M (2009) Lakes as sentinels of climate change. Limnol Oceanogr 54:2283–2297

Adrian R, Gerten D, Huber V, Wagner C, Schmidt SR (2012) Windows of change: temporal scale of analysis is decisive to detect ecosystem responses to climate change. Mar Biol 159:2533–2542

Alheit J, Möllmann C, Dutz J, Kornilovs G, Loewe P, Mohrholz V, Wasmund N (2005) Synchronous ecological regime shifts in the central Baltic and the North Sea in the late 1980 s. ICES J Mar Sci 62:1205–1215

Allen AP, Gillooly JF, Brown JH (2005) Linking the global carbon cycle to individual metabolism. Funct Ecol 19:202–213

Andersen T, Carstensen J, Hernándes-García E, Duarte CM (2009) Ecological thresholds and regime shifts: Approaches to identification. Trends Ecol Evol 24:49–57

Angeler DG, Johnson RK (2012) Patterns of temporal community turnover are spatially synchronous across boreal lakes. Freshwater Biol 57:1782–1793

Arhonditsis GB, Brett MT, DeGasperi CL, Schindler DE (2004) Effects of climatic variability on the thermal properties of Lake Washington. Limnol Oceanogr 49:256–270

Ask J, Karlsson J, Persson L, Ask P, Bystrom P, Jansson M (2009) Terrestrial organic matter and light penetration: effects on bacterial and primary production in lakes. Limnol Oceanogr 54:2034–2040

Austin J, Colman S (2008) A century of temperature variability in Lake Superior. Limnol Oceanogr 53:2724–2730

Balayla DJ, Lauridsen TL, Søndergaard M., Jeppesen E (2010) Larger zooplankton in Danish lakes after cold winters: are fish kills of importance? Hydrobiologia 646:159–172

Beaugrand G, Reid PC, Ibanez F, Lindley JA, Edwards M (2002) Reorganization of North Atlantic marine copepod biodiversity and climate. Science 296:1692–1694

Benndorf J, Kranich J, Mehner T, Wagner AK (2001) Temperature impact on the midsummer decline of *Daphnia galeata*: An analysis of long-term data from the biomanipulated Bautzen Reservoir (Germany). Freshwater Biol 46:199–211

Benson BJ, Lenters JD, Magnuson JJ, Stubbs M, Kratz TK, Dillon PJ, Hecky RE, Lathrop RC (2000) Regional coherence of climatic and lake thermal variables of four lake districts in the Upper Great Lakes Region of North America. Freshwater Biol 43:517–527

Benson BJ, Magnuson JJ, Jensen OP, Card VM, Hodgkins G, Korhonen J, Livingstone DM, Stewart KM, Weyhenmeyer GA, Granin NG (2012) Extreme events, trends, and variability in Northern Hemisphere lake-ice phenology (1855-2005). Climatic Change 112:299–323

Berger SA, Diehl S, Stibor H, Trommer G, Ruhenstroth M (2010) Water temperature and stratification depth independently shift cardinal events during plankton spring succession. Glob Change Biol 16:1954–1965

BES, 2013. The Impact of Extreme Events on Freshwater Ecosystems. British Ecological Society, www.britishecologicalsociety.org/public-policy/our-position/ecological-issues

Blanck A, Lamouroux N (2007) Large-scale intraspecific variation in life-history traits of 44 European freshwater fish. J Biogeogr 34:862–875

Blenckner T, Järvinen M, Weyhenmeyer GA (2004) Atmospheric circulation and its impact on ice phenology in Scandinavia. Boreal Environ Res 9:371–380

Blenckner T, Adrian R, Livingstone DM, Jennings E, Weyhenmeyer GA, George DG, Jankowski T, Jarvinen M, Aonghusa CN, Nõges T, Straile D, Teubner K (2007) Large-scale climatic signatures in lakes across Europe: a meta-analysis. Glob Change Biol 13:1314–1326

Boehrer B, Schultze M (2008) Stratification of lakes. Rev Geophys 46: RG2005. doi:10.1029/2006RG000210

Burgmer T, Hillebrand H (2011) Temperature mean and variance alter phytoplankton biomass and biodiversity in a long-term microcosm experiment. Oikos 120:922–933

Burgmer T, Hillebrand H, Pfenninger M (2007) Effects of climate-driven temperature changes on the diversity of freshwater macroinvertebrates. Oecologia 151:93–103

Carpenter SR, Benson BJ, Biggs R et al (2007) Understanding regional change: A comparison of two lake districts. Bioscience 57:323–335

Christensen JH, Hewitson B, Busuioc A, Chen A, Gao X, Held I, Jones R, Kolli RK, Kwon W-T, Laprise R, Magaña Rueda V, Mearns L, Menéndez CG, Räisänen J, Rinke A, Sarr A, Whetton P (2007) Regional climate projections. In: Climate Change 2007: The Physical Science Basis. Contribution of Working Group I to the Fourth Assessment Report of the Intergovernmental Panel on Climate Change. Cambridge University Press

Christoffersen K, Andersen N, Søndergaard M, Liboriussen, L, Jeppesen E (2006) Implications of climate-enforced temperature increases on freshwater pico- and nanoplankton populations studied in artificial ponds during 16 months. Hydrobiologia 560:259–266

Coats R, Perez-Losada J, Schladow G, Richards R, Goldman C (2006) The warming of Lake Tahoe. Climatic Change 76:121–148

Daufresne M, Lengfellner K, Sommer U (2009) Global warming benefits the small in aquatic ecosystems. Proc Nat Acad Sci USA 106:12788–12793

de Wit HA, Wright RF (2008) Projected stream water fluxes of NO_3 and total organic carbon from the Storgama headwater catchment, Norway, under climate change and reduced acid deposition. Ambio 37:56–63

de Wit HA, Hindar A, Hole L (2008) Winter climate affects long-term trends in stream water nitrate in acid-sensitive catchments in southern Norway. Hydrol Earth Sys Sci 12:393–403

Dokulil MT, Jagsch A, George GD, Anneville O, Jankowski T, Wahl B, Lenhart B, Blenckner T, Teubner K (2006) Twenty years of spatially coherent deepwater warming in lakes across Europe related to the North Atlantic Oscillation. Limnol Oceanogr 51:2787–2793

Domisch S, Araujo MB, Bonada N, Pauls SU, Jahnig SC, Haase P (2013) Modelling distribution in European stream macroinvertebrates under future climates. Glob Change Biol 19:752–762

Downing JA, Prairie YT, Cole JJ, Duarte CM, Tranvik LJ, Striegl RG, McDowell WH, Kortelainen P, Caraco NF, Melack JM, Middelburg JJ (2006) The global abundance and size distribution of lakes, ponds, and impoundments. Limnol Oceanogr 51:2388–2397

Elliott JA (2012) Is the future blue-green? A review of the current model predictions of how climate change could affect pelagic freshwater cyanobacteria. Water Res 46:1364–1371

Elliott JA, Thackeray SJ, Huntingford C, Jones RG (2005) Combining a Regional Climate Model with a phytoplankton community model to predict future changes in phytoplankton in lakes. Freshwater Biol 50:1404–1411

Elliott JA, Jones ID, Thackeray, SJ (2006) Testing the sensitivity of phytoplankton communities to changes in water temperature and nutrient load, in a temperate lake. Hydrobiologia 559:401–411

Elser JJ, Andersen T, Baron JS, Bergström A-K, Kyle M, Nydick KR, Steger L, Hessen DO (2009) Atmospheric nitrogen deposition distorts lake N:P stoichiometry and alters phytoplankton nutrient limitation. Science 326:835–837

Erlandsson M, Buffam I, Folster J, Laudon H, Temnerud J, Weyhenmeyer GA, Bishop K (2008) Thirty-five years of synchrony in the organic matter concentrations of Swedish rivers explained by variation in flow and sulphate. Glob Change Biol 14:1191–1198

Evans CD, Chapman PJ, Clark JM, Monteith DT, Cresser MS (2006) Alternative explanations for rising dissolved organic carbon export from organic soils. Glob Change Biol 12:2044–2053

Feuchtmayr H, Moran R, Hatton K, Connor L, Heyes T, Moss B, Harvey I, Atkinson D (2009) Global warming and eutrophication: effects on water chemistry and autotrophic communities in experimental hypertrophic shallow lake mesocosms. J Appl Ecol 46:713–723

Finstad AG, Helland IP, Ugedal O, Hesthagen T, Hessen DO (2014) Unimodal response of fish yield to dissolved organic carbon. Ecol Lett 17:36–43

Friberg N, Bergfur J, Rasmussen J, Sandin L (2013) Changing Northern catchments: Is altered hydrology, temperature or both going to shape future stream communities and ecosystem processes? Hydrol Proc 27:734–740

Gardner JL, Peters A, Kearney MR, Joseph L, Heinsohn R (2011) Declining body size: a third universal response to warming? Trends Ecol Evol 26:285–291

George DG (2007) The impact of the North Atlantic Oscillation on the development of ice on Lake Windermere. Climatic Change 81:455–468

George DG, Taylor AH (1995) UK lake plankton and the Gulf Stream. Nature 378:139

George DG, Talling JF, Rigg E (2000) Factors influencing the temporal coherence of five lakes in the English Lake District. Freshwater Biol 43:449–461

George DG, Järvinen M, Arvola L (2004a) The influence of the North Atlantic Oscillation on the winter characteristics of Windermere (UK) and Pääjärvi (Finland). Boreal Environ Res 9:389–399

George DG, Maberly SC, Hewitt DP (2004b) The influence of the North Atlantic Oscillation on the physical, chemical and biological characteristics of four lakes in the English Lake District. Freshwater Biol 49:760–774

George DG, Hewitt DP, Jennings E, Allott N, McGinnity P (2007a) The impact of changes in the weather on the surface temperatures of Windermere (UK) and Lough Feeagh (Ireland). In: Water in Celtic Countries: Quantity, Quality and Climatic Variability. Proceedings of the Fourth Inter-Celtic Colloquium on Hydrology and Management of Water Resources, Guimarães, Portugal, July 2005, IAHS Publications 310

George G, Hurley M, Hewitt D (2007b) The impact of climate change on the physical characteristics of the larger lakes in the English Lake District. Freshwater Biol 52:1647–1666

Gerten D, Adrian R (2000) Climate-driven changes in spring plankton dynamics and the sensitivity of shallow polymictic lakes to the North Atlantic Oscillation. Limnol Oceanogr 45:1058–1066

Gerten D, Adrian R (2001) Differences in the persistency of the North Atlantic Oscillation signal among lakes. Limnol Oceanogr 46:448–455

Gerten D, Adrian R (2002a) Effects of climate warming, North Atlantic Oscillation and El Niño on thermal conditions and plankton dynamics in European and North American lakes. The Scientific World Journal 2: 586–606

Gerten D, Adrian R (2002b) Species specific response of freshwater copepods to recent summer warming. Freshwater Biol 47:2163–2173

Giller PS, Hillebrand H, Berninger UG, Gessner MO, Hawkins S, Inchausti P, Inglis C, Leslie H, Malmqvist B, Monaghan MT, Morin PJ, O'Mullan G (2004) Biodiversity effects on ecosystem functioning: emerging issues and their experimental test in aquatic environments. Oikos 104:423–436

Gyllström M, Hansson LA, Jeppesen E, Garcia-Criado F, Gross E, Irvine K, Kairesalo T, Kornijow R, Miracle MR, Nykanen M, Nõges T, Romo S, Stephen D, Van Donk E, Moss B (2005) The role of climate in shaping zooplankton communities of shallow lakes. Limnol Oceanogr 50:2008–2021

Hampton SE, Izmest'eva LR, Moore MV, Katz SL, Dennis B, Silow EA (2008) Sixty years of environmental change in the world's largest freshwater lake – Lake Baikal, Siberia. Glob Change Biol 14:1947–1958

Hansson LA, Nicolle A, Granéli W, Hallgren P, Kritzberg E, Persson A, Björk J, Nilsson PA, Brönmark C (2012) Food-chain length alters community responses to global change in aquatic systems. Nature Clim Chan 3:228–233

Hargeby A, Blindow I, Hansson LA (2004) Shifts between clear and turbid states in a shallow lake: multi-causal stress from climate, nutrients and biotic interactions. Arch Hydrobiol 161:433–454

Harrod C, Griffiths D, Rosell R, McCarthy TK (2002) Current status of the pollan (*Coregonus autumnalis* Pallas 1776) in Ireland. Arch Hydrobiol 57:627–638

Hessen DO (1998) Food webs and carbon cycling in humic lakes. In: Hessen DO, Tranvik LJ (eds) Aquatic Humic Substances. Springer

Hessen DO (1999) Catchment properties and the transport of major elements to estuaries. In: Nedwell DB, Raffaelli DG (eds) Advances in Ecological Research 29: Estuaries. Elsevier

Hessen DO, Andersen T, Larsen S, Skjelkvåle BL, De Wit HA (2009) Nitrogen deposition, catchment productivity, and climate as determinants of lake stoichiometry. Limnol Oceanogr 54:2520–2528

Hessen DO, Daufresne M, Leinaas HP (2013) Temperature-size relations from the cellular-genomic perspective. Biol Rev 88:476–489

Hickling R, Roy DB, Hill JK, Thomas CD (2005) A northward shift of range margins in British Odonata. Glob Change Biol 11:502–506

Hillebrand H (2004) On the generality of the latitudinal diversity gradient. Amer Natur 163:192–211

Hillebrand H, Borer ET, Bracken MES, Cardinale BJ, Cebrian J, Cleland EE, Elser JJ, Gruner DS, Harpole WS, Ngai JT, Sandin S, Seabloom EW, Shurin JB, Smith JE, Smith MD (2009) Herbivore metabolism and stoichiometry each constrain herbivory at different organizational scales across ecosystems. Ecol Lett 12:516–527

Hillebrand H, Soininen J, Snoeijs P (2010) Warming leads to higher species turnover in a coastal ecosystem. Glob Change Biol 16:1181–1193

Hillebrand H, Burgmer T, Biermann E (2012) Running to stand still: temperature effects on species richness, species turnover, and functional community dynamics. Mar Biol 159:2415–2422

Hobbie SE, Miley TA, Weiss MS (2002a) Carbon and nitrogen cycling in soils from acidic and nonacidic tundra with different glacial histories in Northern Alaska. Ecosystems 5:761–774

Hobbie SE, Nadelhoffer KJ, Hogberg P (2002b) A synthesis: The role of nutrients as constraints on carbon balances in boreal and arctic regions. Plant Soil 242:163–170

Hofgaard A (1997) Inter-relationships between treeline position, species diversity, land use and climate change in the central Scandes Mountains of Norway. Glob Ecol Biogeogr 6:419–429

Hondzo M, Stefan HG (1993) Regional water temperature characteristics of lake subjected to climate change. Climatic Change 24:187–211

Hong Y, Andersen T, Dörch P, Tominaga K, Thrane JT, Hessen DO (2015) Greenhouse gas metabolism in Nordic boreal lakes. Biogeochemistry 126:211–225

Hongve D, Riise G, Kristiansen JF (2004) Increased colour and organic acid concentrations in Norwegian forest lakes and drinking water - a result of increased precipitation? Aqua Sci 66:231–238

Huber V, Adrian R, Gerten D (2008) Phytoplankton response to climate warming modified by trophic state. Limnol Oceanogr 53:1–13

Huber V, Adrian R, Gerten D (2010) A matter of timing: heat wave impact on crustacean zooplankton. Freshwater Biol 55:1769–1779

Huber V, Wagner C, Gerten D, Adrian R (2012) To bloom or not to bloom: contrasting responses of cyanobacteria to different heat waves explained by critical thresholds of abiotic drivers. Oecologia 169:245–256

Huisman J, Sharples J, Stroom JM, Visser PM, Kardinaal WEA, Verspagen JMH, Sommeijer B (2004) Changes In turbulent mixing shift competition for light between phytoplankton species. Ecology 85:2960–2970

Hurrell JW (1995) Decadal trends in the North Atlantic Oscillation: regional temperatures and precipitation. Science 269:676–679

Hurrell JW, Kushnir Y, Ottersen G, Visbeck M (2003) The North Atlantic Oscillation: climatic significance and environmental impact. Geophysical Monograph Series, V 134. American Geophysical Union

Hussner A, Van Dam H, Vermaat JE, Hilt S (2014) Comparison of native and neophytic aquatic macrophyte developments in a geothermally warmed river and thermally normal channels. Fund Appl Limnol 185:155–165

IPCC (2013) Climate Change and Water. IPCC Technical Report VI. Intergovernmental Panel on Climate Change (IPCC)

Jacobsen D, Milner AM, Brown LE, Dangles O (2012) Biodiversity under threat in glacier-fed river systems. Nat Clim Chan 2:361–364

Janse JH (1997) A model of nutrient dynamics in shallow lakes in relation to multiple stable states. Hydrobiologia 342–343:1-8

Jansson M, Persson L, De Roos AM, Jones RI, Tranvik LJ (2007) Terrestrial carbon and intraspecific size-variation shape lake ecosystems. Trends Ecol Evol 22:316–322

Jennings E, Jones S, Arvola L, Steahr PA, Gaiser E, Jones ID, Weathers KC, Weyhenmeyer GA, Chiu CY, DeEyto E (2012) Effects of weather-related episodic events in lakes: an analysis based on high-frequency data. Freshwater Biol 57:589–601

Jensen HS, Andersen FØ (1992) Importance of temperature, nitrate, and pH for phosphate release from aerobic sediments of 4 shallow, eutrophic lakes. Limnol Oceanogr 37:577–589

Jensen OP, Benson BJ, Magnuson JJ, Card VM, Futter MN, Soranno PA, Stewart KM (2007) Spatial analysis of ice phenology trends across the Laurentian Great Lakes region during a recent warming period. Limnol Oceanogr 52:2013–2026

Jeppesen E, Søndergaard M, Jensen JP (2003) Climatic warming and regime shifts in lake food webs - some comments. Limnol Oceanogr 48:1346–1349

Jeppesen E, Kronvang B, Meerhoff M, Søndergaard M, Hansen KM, Andersen HE, Lauridsen TL, Beklioglu M, Ozen A, Olesen JE (2009) Climate change effects on runoff, catchment phosphorus loading and lake ecological state, and potential adaptations. J Env Qual 38:1930–1941

Jeppesen E, Moss B, Bennion H, Carvalho L, DeMeester L, Friberg N, Gessner MO, Lauridsen, TL, May L, Meerhoff M, Olafsson JS, Soons MB, Verhoeven JTA (2010a) Interaction of climate change and eutrophication. In: Kernan M, Battarbee R, Moss Blackwell B (eds). Climate Change Impacts on Freshwater Ecosystems. Publishing Ltd

Jeppesen E, Meerhoff M, Holmgren K, González-Bergonzoni I, Teixeira-de Mello F, Declerk SAJ, De Meester L, Søndergaard M, Lauridsen TL, Bjerring R, Conde-Porcuna JM, Mazzeo N, Iglesias C, Reizenstein M, Malmquist HJ, Balayla D, Lazzaro X (2010b) Impacts of climate warming on lake fish community structure and potential effects on ecosystem function. Hydrobiologia, 646:73–90

Jeppesen E, Kronvang B, Olesen JE, Søndergaard M, Hoffmann CC, Andersen HE, Lauridsen TL, Liboriussen L, Meerhoff M, Beklioglu M, Özen A (2011) Climate change effects on nitrogen loading from cultivated catchments in Europe: implications for nitrogen retention, ecological state of lakes and adaptation. Hydrobiologia 663:1–21

Jeppesen E, Mehner T, Winfield IJ, Kangur K, Sarvala J, Gerdeaux D, Rask M, Malmquist HJ, Holmgren K, Volta P, Romo S, Eckmann R, Sandstrom A, Blanco S, Kangur A, Stabo HR, Tarvainen M, Ventela AM, Sondergaard M, Lauridsen TL, Meerhoff M (2012) Impacts of climate warming on the long-term dynamics of key fish species in 24 European lakes. Hydrobiologia 694:1–39

Jöhnk KD, Huisman J, Sharples J, Sommeijer B, Visser PM, Stroom JM (2008) Summer heatwaves promote blooms of harmful cyanobacteria. Glob Change Biol 14:495–512

Jones SE, Solomon CT, Weidel BC (2012) Subsidy or subtraction: how do terrestrial inputs influence consumer production in lakes? Freshw Rev 5:37–49

Juutinen S, Rantakari M, Kortelainen P, Huttunen JT, Larmola T, Silvola J, Martikainen PJ (2009) Methane dynamics in different boreal lake types. Biogeosciences 6:209–223

Kalbitz K, Solinger S, Park JH, Michalzik B, Matzner E (2000) Controls on the dynamics of dissolved organic matter in soils: A review. Soil Sci 165:277–304

Kammer A, Hagedorn F, Shevchenko I, Leifeld J, Guggenberger G, Goryacheva T, Rigling A, Moiseev P (2009) Treeline shifts in the Ural mountains affect soil organic matter dynamics. Glob Change Biol 15:1570–1583

Kangur A, Kangur P, Kangur K, Möls T (2007) The role of temperature in the population dynamics of smelt *Osmerus eperlanus eperlanus* m. *spirinchus* Pallas in Lake Peipsi (Estonia/Russia). Hydrobiologia 584:433–441

Kangur K, Kangur P, Ginter K, Orru K, Haldna M, Möls T, Kangur A (2013) Long-term effects of extreme weather events and eutrophication on the fish community of shallow lake Peipsi. J Limnol 72:376–387

Karlsson J, Bystrom P, Ask J, Ask P, Persson L, Jansson M (2009) Light limitation of nutrient-poor lake ecosystems. Nature 460:506–580

Kernan M, Battarbee RW, Moss B (eds) (2010) Climate Change Impacts on Freshwater Ecosystems. Wiley-Blackwell

Kirillin G (2010) Modeling the impact of global warming on water temperature and seasonal mixing regimes in small temperate lakes. Boreal Environ Res 15:279–293

Kosten S, Kamarainen A, Jeppesen E, Van Nes EH, Peeters ETHM, Mazzeo N, Sass L, Hauxwell J, Hansel-Welch N, Lauridsen TL, Sondergaard M, Bachmann RW, Lacerot G, Scheffer M (2009) Climate-related differences in the dominance of submerged macrophytes in shallow lakes. Glob Change Biol 15:2503–2517

Kosten S, Huszar VLM, Bécares E, Costa LS, van Donk E, Hansson L-A, Jeppesen E, Kruk C, Lacerot G, Mazzeo N, De Meester L, Moss B, Lürling M, Nõges T, Romo S, Scheffer M (2012) Warmer climate boosts cyanobacterial dominance in Lakes. Glob Change Biol 18:118–126

Lamb HH (1950) Types and spells of weather around the year in the British Isles. Q J Roy Met Soc 76:393–438

Larsen S, Andersen T, Hessen DO (2011a) Predicting organic carbon in lakes from climate drivers and catchment properties. Glob Biogeochem Cyc 25:doi:10.1029/2010GB003908

Larsen S, Andersen T, Hessen DO (2011b) Climate change predicted to cause severe increase of organic carbon in lakes. Glob Change Biol 17:1186–1192

Larsen S, Andersen T, Hessen DO (2011c) The pCO_2 in boreal lakes: Organic carbon as a universal predictor. Glob Biochem Cyc 25:doi:10.1029/2010GB003864

Lehtonen H (1996) Potential effects of global warming on northern European freshwater fish and fisheries. Fish Man Ecol 3:59–71

Lehtonen H, Lappalainen L (1995) The effects of climate on the year-class variations of certain freshwater fish species. In: Beamish RJ (ed). Climate Change and Northern Fish Populations, Vol. 121. Canadian Special Publication of Fisheries and Aquatic Sciences

Leibold MA, Holyoak M, Mouquet N, Amarasekare P, Chase JM, Hoopes MF, Holt RD, Shurin JB, Law R, Tilman D, Loreau M, Gonzalez A (2004) The metacommunity concept: a framework for multi-scale community ecology. Ecol Lett 7:601–613

Liboriussen L, Lauridsen TL, Sondergaard M, Landkildehus F, Sondergaard M, Larsen SE, Jeppesen E (2011) Effects of warming and nutrients on sediment community respiration in shallow lakes: an outdoor mesocosm experiment. Freshwater Biol 56:437–447

Lindholm M, Stordal F, Moe SJ, Hessen DO, Aass P (2012) Climate-driven range retraction on an Arctic freshwater crustacean. Freshwater Biol 57:2591–2601

Livingstone DM (2000) Large-scale climatic forcing detected in historical observations of lake ice break-up. Verhandlungen der Internationalen Vereiningung der theoretischen und angewandten Limnologie 27:2775–2783

Livingstone DM (2003) Impact of secular climate change on the thermal structure of a large temperate central European lake. Climatic Change 57:205–225

Livingstone DM (2008) A change of climate provokes a change of paradigm: taking leave of two tacit assumptions about physical lake forcing. Int Rev Hydrobiol 93:404–414

Livingstone DM, Adrian R (2009) Modeling the duration of intermittent ice cover on a lake for climate-change studies. Limnol Oceanogr 54:1709–1722

Livingstone DM, Kernan M (2009) Regional coherence and geographical variability in the surface water temperatures of Scottish Highland lochs. Fund Appl Limnol 62:367–378

Livingstone DM, Padisák J (2007) Large-scale coherence in the response of lake surface-water temperatures to synoptic-scale climate forcing during summer. Limnol Oceanogr 52:896–902

Livingstone DM, Adrian R, Arvola L, Blenckner T, Dokulil MT, Hari RE, George G, Jankowski T, Järvinen M, Jennings E, Nõges P, Nõges T, Straile D, Weyhenmeyer GA (2010a) Regional and supra-regional coherence in limnological variables. In: George DG (ed) The Impact of Climate Change on European Lakes. Aquatic Ecology Series 4, Springer

Livingstone DM, Adrian R, Blenckner T, George G, Weyhenmeyer GA (2010b) Lake ice phenology. In: George DG (ed) The Impact of Climate Change on European Lakes. Aquatic Ecology Series 4, Springer

Lo TT, Hsu HH (2010) Change in the dominant decadal patterns and the late 1980 s abrupt warming in the extratropical Northern Hemisphere. Atmos Sci Lett 11:210–215

Loreau M, Mouquet N, Gonzalez A (2003) Biodiversity as spatial insurance in heterogeneous landscapes. Proc Nat Acad Sci USA 100:12765–12770

Magnuson JJ, Benson BJ, Kratz TK (1990) Temporal coherence in the limnology of a suite of lakes in Wisconsin, USA. Freshwater Biol 23:145–149

Magnuson JJ, Robertson DM, Benson BJ, Wynne RH, Livingstone DM, Arai T, Assel RA, Barry RG, Card V, Kuusisto E, Granin NG, Prowse TD, Stewart KM, Vuglinski VS (2000) Historical trends in lake and river ice cover in the Northern Hemisphere. Science 289:1743–1746 and Errata 2001. Science 291:254

MacKay MD, Neale PJ, Arp CD, De Senerpont Domis LN, Fang X, Gal G, Jöhnk KD, Kirillin G, Lenters JD, Litchman E, MacIntyre S, Marsh P, Melack J, Mooij WM, Peeters F, Quesada A, Schladow SG, Schmid M, Spence C, Stokes SL (2009) Modeling lakes and reservoirs in the climate system. Limnol Oceanog 54:2315–2329

McKee D, Hatton K, Eaton JW, Atkinson D, Atherton A, Harvey I, Moss B (2002) Effects of simulated climate warming on macrophytes in freshwater microcosm communities. Aquat Bot 74:71–83

McKee D, Atkinson D, Collings SE, Eaton JW, Gill AB, Harvey I, Hatton K, Heyes T, Wilson D, Moss B (2003) Response of freshwater microcosm communities to nutrients, fish, and elevated temperature during winter and summer. Limnol Oceanogr 48:707–722

Meerhoff M, Teixeira-de Mello F, Kruk C, Alonso C, González-Bergonzoni I, Pacheco JP, Lacerot G, Arim M, Beklioğlu M, Brucet S, Goyenola G, Iglesias C, Mazzeo N, Kosten S, Jeppesen E (2012) Environmental warming in shallow lakes: a review of potential changes in community structure as evidenced from space-for-time substitution approaches. Adv Ecol Res 46:259–349

Milbrink G, Vrede T, Tranvik LJ, Rydin E (2011) Large-scale and long-term decrease in fish growth following the construction of hydroelectric reservoirs. Can J Fish Aquat Sci 68:2167–2173

Moe TF, Brysting AK, Andersen T, Schneider SC, Kaste O, Hessen DO (2013) Nuisance growth of *Juncus bulbosus*: the roles of genetics and environmental drivers tested in a large-scale survey. Freshwater Biol 58:114–127

Monteith DT, Stoddard JL, Evans CD, de Wit HA, Forsius M, Høgåsen T, Wilander A, Skjelkvåle BL, Jeffries DS, Vuorenmaa J, Keller B, Kopácek J, Vesely J (2007) Dissolved organic carbon trends resulting from changes in atmospheric deposition chemistry. Nature 450:537–540

Mooij WM, Hülsman S, De Senerpont Domis LN, Nolet BA, Bodelier PLE, Boers PCM, Dionisio Pires LM, Gons HJ, Ibelings BW, Noordhuis R, Portielje R, Wolfstein K, Lammens EHRR (2005) The impact of climate change on lakes in the Netherlands: a review. Aquat Ecol 39:381–400

Mooij WM, Janse JH, De Senerpont Domis, L, Hülsmann S, Ibelings B (2007) Predicting the effect of climate change on temperate shallow lakes with the ecosystem model PCLake. Hydrobiologia 584:443–454

Mooij WM, De Senerpont Domis LN, Janse JH (2009) Linking species- and ecosystem-level impacts of climate change in lakes with a complex and a minimal model. Ecol Modell 220:3011–3020

Mormul RP, Ahlgren J, Ekvall MK, Hansson L-A, Brönmark C (2012) Water brownification may increase the invisibility of a submerged non-native macrophyte. Biol Invas 14:2091–2099

Moss B (2010) Climate change, nutrient pollution and the bargain of Dr Faustus. Freshwater Biol 55:175–187

Moss B, Hering D, Green AJ, Adoud A, Becares E, M. Beklioglu M, Bennion H, Boix D, Brucet S, Carvalho L, Clement B, Davidson TA, Declerck S, Dobson M, van Donk E, Dudley B, Feuchtmayr H, Friberg N, Grenouillet G, Hillebrand H, Hobaek H, Irvine K, Jeppesen E, Johnson R, Jones I, Kernan M, Lauridsen T, Manca M, Meerhoff M, Olafsson J, Ormerod S, Papastergiadou E, Penning W, Ptacnik R, Quintana X, Sandin L, Seferlis M, Simpson G, Trigal C, Verdonschot P, Verschoor A, Weyhenmeyer G (2009) Climate change and the future of freshwater biodiversity in Europe: a primer for policy-makers. Freshw Rev 2:103–130

Moss B, Kosten S, Meerhoff M, Battarbee RW, Jeppesen E, Mazzeo N, Havens K, Lacerot G, Liu Z, De Meester L, Paerl H, Scheffer M (2011) Allied attack: climate change and eutrophication. Inland Waters 1:101–105

Netten JJC, Arts GHP, Gylstra R, van Nes EH, Scheffer M, Roijackers RMM (2010) Effect of temperature and nutrients on the competition between free-floating *Salvinia natans* and submerged *Elodea nuttallii* in mesocosms. Fund Appl Limnol 177:125–132

Netten JJC, Van Zuidam J, Kosten S, Peeters ETHM (2011) Differential response to climatic variation of free-floating and submerged macrophytes in ditches. Freshwater Biol 56:1761–1768

Nielsen A, Trolle D, Me W, Luo L, Han B, Liu Z, Olesen JE, Jeppesen E (2013) Assessing ways to combat eutrophication in a Chinese drinking water reservoir using SWAT. Mar Freshw Res 64:1–18

Nilsson C, Jansson R, Kuglerova L, Lind L, Strom L (2013) Boreal riparian vegetation under climate change. Ecosystems 16:401–410

Norton L, Elliott JA, Maberly SC, May L (2012) Using models to bridge the gap between land use and algal blooms: An example from the Loweswater catchment, UK. Env Modell Softw 36:64–75

Ojala A, Lopez Bellido J, Tulonen T, Kankaala P, Huotari J (2011) Carbon gas fluxes from a brown-water and a clear-water lake in the boreal zone during a summer with extreme rain events. Limnol Oceanogr 56:61–76

O'Reilly CN, Sharma S, Gray DK et al. (2015) Rapid and highly variable warming of lake surface waters around the globe. Geophys Res Lett 42, doi: 10.1002/2015GL066235

O'Reilly CM, Alin SR, Plisnier P-D, Cohen AS, McKee BA (2003) Climate change decreases aquatic ecosystem productivity of Lake Tanganyika, Africa. Nature 424:766–768

Özen A, Šorf M, Trochine C, Liboriussen L, Beklioglu M, Søndergaard M, Lauridsen TL, Johansson LS, Jeppesen E (2013) Long-term effects of warming and nutrients on microbes and other plankton in mesocosms. Freshwater Biol 58:483–493

Padisak J (1997) *Cylindrospermopsis raciborskii* (Woloszynska) *seenayya et Subba Raju*, an expanding, highly adaptive cyanobacterium: worldwide distribution and review of its ecology. Arch. Hydrobiol Suppl 107:563–593

Paerl HW (1988) Nuisance phytoplankton blooms in coastal, estuarine and inland waters. Limnol Oceanogr 33:823–847

Parmesan C, Yohe G (2003) A globally coherent fingerprint of climate change impacts across natural systems. Nature 421:37–42

Peeters F, Livingstone DM, Goudsmit G-H, Kipfer R, Forster R (2002) Modeling 50 years of historical temperature profiles in a large central European lake. Limnol Oceanogr 47:186–197

Peeters F, Straile D, Lorke A, Livingstone DM (2007) Earlier onset of the spring phytoplankton bloom in lakes of the temperate zone in a warmer climate. Glob Change Biol 13:1898–1909

Posch T, Koster O, Salcher MM, Pernthaler J (2012) Harmful filamentous cyanobacteria favoured by reduced water turnover with lake warming. Nature Clim Change 2:809–813

Ptacnik R, Solimini AG, Andersen T, Tamminen T, Brettum P, Lepisto L, Willen E, Rekolainen S (2008) Diversity predicts stability and resource use efficiency in natural phytoplankton communities. Proc Nat Acad Sci USA 105:5134–5138

Quayle WC, Peck LS, Peat H, Ellis-Evans JC, Harrigan PR (2002) Extreme responses to climate change in Antarctic lakes. Science 295:645

Reid PC, de Fatima Borges M, Svendsen E (2001) A regime shift in the North Sea circa 1988 linked to changes in the North Sea horse mackerel fishery. Fish Res 50:163–171

Reynolds C (1984) The Ecology of Freshwater Phytoplankton. Cambridge University Press

Robertson DM, Ragotzkie RA (1990) Changes in the thermal structure of moderate to large sized lakes in response to changes in air temperature. Aquat Sci 52:360–380

Rodionov SN, Overland JE (2005) Application of a sequential regime shift detection method to the Bering Sea ecosystem. ICES J Mar Sci 62:328–333

Rolinski S, Horn H, Petzoldt T, Paul L (2007) Identifying cardinal dates in phytoplankton time series to enable the analysis of long-term trends. Oecologia 153:997–1008

Rosenzweig C (2007) Assessment of observed changes and responses in natural and managed systems. In: Parry ML, Canziani OF, Palutikof JP, van der Linden PJ, Hanson CE (eds) Climate change 2007. Impacts, adaptation and vulnerability. Contribution of Working Group II to the Fourth Assessment Report of the Intergovernmental Panel on Climate Change. Cambridge University Press

Ruuhijärvi J, Rask M, Vesala SA, Westermark A, Olin M, Keskitalo J, Lehtovaara A (2010) Recovery of the fish community and changes in the lower trophic levels in a eutrophic lake after a winter kill of fish. Hydrobiologia 646:145–158

Salmaso N (2005) Effects of climatic fluctuations and vertical mixing on the interannual trophic variability of Lake Garda, Italy. Limno Oceanogr 50:553–565

Saloranta TM, Andersen T (2007) MyLake—A multi-year lake simulation model code suitable for uncertainty and sensitivity analysis simulations. Ecol Model 207:45–60

Schabhuettl S, Hingsamer P, Weigelhofer G, Hein T, Weigert A, Striebel M (2013) Temperature and species richness effects in phytoplankton communities. Oecologia 171:527–536

Schär C, Vidale P, Luthi D, Frei C, Haberli C, Liniger M, Appenzeller C (2004) The role of increasing temperature variability in European summer heatwaves. Nature 427:332–336

Scharfenberger U, Mahdy A, Adrian R (2013) Threshold-driven shifts in two copepod species: testing ecological theory with observational data. Limnol Oceanogr 58:741–752

Scheffer M, Carpenter SR (2003) Catastrophic regime shifts in ecosystems: linking theory to observation. Trends Ecol Evol 18:648–656

Scheffer M, Straile D, van Nes EH, Hosper H (2001) Climatic warming causes regime shifts in lake food webs. Limnol Oceanogr 46:1780–1783

Scheffer M, van Geest GJ, Zimmer K, Jeppesen E, Sondergaard M, Butler MG, Hanson MA, Declerck S, De Meester L (2006) Small habitat size and isolation can promote species richness: second-order effects on biodiversity in shallow lakes and ponds. Oikos 112:227–231

Schindler DW (2001) The cumulative effects of climate warming and other human stresses on Canadian freshwaters in the new millenium. Can J Fish Aquat Sci 58:18–29

Schindler DW, Beaty KG, Fee EJ, Cruikshank DR, DeBruyn ER, Findlay DL, Linsey GA, Shearer JA, Stainton MP, Turner MA (1990) Effects of climatic warming on lakes of the central boreal forest. Science 250:967–970

Schindler DE, Rogers DE, Scheuerell MD, Abrey CA (2005) Effects of changing climate on zooplankton and juvenile sockeye salmon growth in southwestern Alaska. Ecology 86:198–209

Schneider P, Hook SJ (2010) Space observations of inland water bodies show rapid surface warming since 1985. Geophys Res Lett 37: L22405, doi:10.1029/2010GL045059

Schneider C, Laize CLR, Acreman MC, Florke M (2013) How will climate change modify river flow regimes in Europe? Hydrol Earth Sys Sci 17:325–339

Seebens H, Straile D, Hoegg R, Stich HB, Einsle U (2007) Population dynamics of a freshwater calanoid copepod: Complex responses to changes in trophic status and climate variability. Limnol Oceanogr 52:2364–2372

Seebens H, Einsle U, Straile D (2009) Copepod life cycle adaptations and success in response to phytoplankton spring bloom phenology. Glob Change Biol 15:1394–1404

Shatwell T, Köhler J, Nicklisch A (2008) Warming promotes cold-adapted phytoplankton in temperate lakes and opens a loophole for Oscillatoriales in spring. Glob Change Biol 14:1–7

Shurin JB, Winder M, Adrian R, Keller W, Matthews B, Paterson AM, Paterson MJ, Pinel-Alloul B, Rusak JA, Yan ND (2010) Environmental stability and lake zooplankton diversity - contrasting effects of chemical and thermal variability. Ecol Lett 13:453–463

Sobek S, Algesten G, Bergstrom AK, Jansson M, Tranvik LJ (2003) The catchment and climate regulation of pCO_2 in boreal lakes. Glob Change Biol 9:630–641

Solomon CT, Bruesewitz DA, Richardson DC, et al. (2013) Ecosystem respiration: Drivers of daily variability and background respiration in lakes around the globe. Limnol Oceanogr 58:849–866

Søndergaard M, Jensen JP, Jeppesen E (2003) Role of sediment and internal loading of phosphorus in shallow lakes. Hydrobiologia 506:135–145

Straile D (2002) North Atlantic Oscillation synchronizes food web interactions in central European lakes. Proc Roy Soc Lond 269:391–395

Straile D, Adrian R (2000) The North Atlantic Oscillation and plankton dynamics in two European lakes - two variations on a general theme. Glob Change Biol 6:663–670

Straile D, Livingstone DM, Weyhenmeyer GA, George DG (2003) The response of freshwater ecosystems to climate variability associated with the North Atlantic Oscillation. In: Hurrell JW, Kushnir Y, Ottersen G, Visbeck M (eds) The North Atlantic Oscillation: Climatic significance and environmental impact. Geophysical Monograph Series 134. American Geophysical Union

Straile D, Adrian R, Schindler DE (2012) Uniform temperature dependency of the phenology of keystone herbivore in lakes of the Northern Hemisphere. PloSONE 7:345497. doi:10.1371/journal.pone.0045497

Temnerud J, Weyhenmeyer GA (2008) Abrubt changes in air temperature and precipitation: Do they matter for water chemistry? Glob Biochem Cyc 22:GB2008, doi:10.1029/2007GB003023

Thompson PL, Shurin JB (2012) Regional zooplankton biodiversity provides limited buffering of pond ecosystems against climate change. J Anim Ecol 81:251–259

Thompson DWJ, Wallace JM (1998) The Arctic Oscillation signature in the wintertime geopotential height and temperature fields. Geophys Res Lett 25:1297–1300

Thompson RM, Beardall J, Beringer J, Grace M, Sardina P (2013) Means and extremes: building variability into community-level climate change experiments. Ecol Lett 16:799–806

Thrane J-E, Hessen DO, Andersen T (2014) The absorption of light in lakes: negative impacts of dissolved organic carbon on primary productivity. Ecosystems 17:1040–1052

Tisseuil C, Cornu J-F, Beauchard O, Brosse S, Darwall W, Holland R, Hugueny B, Tedesco PA, Oberdorff T (2013) Global diversity patterns and cross-taxa convergence in freshwater systems. J Anim Ecol 82:365–376

Tockner K, Pusch M, Borchardt D, Lorang MS (2010) Multiple stressors in coupled river-floodplain ecosystems. Freshwater Biol 55:135–151

Tranvik LJ, Downing JA, Cotner JB, Loiselle SA, Striegl RG, Ballatore TJ, Dillon P, Finlay K, Fortino K, Knoll LB, Kortelainen PL, Kutser T, Larsen S, Laurion I, Leech DM, McCallister SL, McKnight DM, Melack JM, Overholt E, Porter JA, Prairie Y, Renwick WH, Roland F, Sherman BS, Schindler DW, Sobek S, Tremblay A, Vanni M J, Verschoor AM, von Wachenfeldt E, Weyhenmeyer GA (2009) Lakes and reservoirs as regulators of carbon cycling and climate. Limnol Oceanogr 54:2298–2314

Trochine C, Guerrero M, Liboriussen L, Meerhoff M, Lauridsen TL, Søndergaard M, Jeppesen E (2011) Filamentous green algae inhibit phytoplankton with enhanced effects when lakes get warmer. Freshwater Biol 56:541–553

Trolle D, Hamilton DP, Pilditch CA, Duggan IC, Jeppesen E (2011) Predicting the effects of climate change on trophic status of three morphologically varying lakes: Implications for lake restoration and management. Env Modell Softw 26:354–370

Trolle D, Hamilton DP, Hipsey MR, Bolding K, Bruggeman J, Mooij WM, Janse JH, Nielsen A, Jeppesen E, Elliott A, Makler-Pick V, Petzoldt T, Rinke K, Flindt MR, Arhonditsis GB, Gal G, Bjerring R, Tominaga K, Hoen J, Downing AS, Marques DM, Fragoso CR, Søndergaard M, Hanson PC (2012) A community-based framework for aquatic ecosystem models. Hydrobiologia 683:25–34

van Donk E, Santamaria L, Mooij W (2003) Climatic warming causes regime shifts in lake food webs: a re-assessment. Limnol Oceanogr 48:1350–1353

Verburg P, Hecky RE, Kling H (2003) Ecological consequences of a century of warming in Lake Tanganyika. Science 301:505–507

Wagner C, Adrian R (2009a) Exploring lake ecosystems: hierarchy responses to long-term change? Glob Change Biol 5:1104–1115

Wagner C, Adrian R (2009b) Cyanobacteria dominance: Quantifying the effects of climate change. Limnol Oceanogr 54:2460–2468

Wagner C, Adrian R (2011) Consequences of changes in thermal regime for plankton diversity and trait composition in a polymictic lake: a matter of temporal scale. Freshwater Biol 56:1949–1961

Wagner A, Hülsmann S, Paul L, Paul RJ, Petzoldt T, Sachse R, Schiller T, Zeis B, Benndorf J, Berendonk TU (2012) A phenomenological approach shows a high coherence of warming patterns in dimictic aquatic systems across latitude. Mar Biol 159:2543–2559

Watkins EM, Schindler DW, Turner MA, Findlay D (2001) Effects of solar ultraviolet radiation on epilithic metabolism, and nutrient and community composition in a clear-water boreal lake. Can J Fish Aquat Sci 58:2059–2070

Weyhenmeyer GA (2004) Synchrony in relationships between the North Atlantic Oscillation and water chemistry among Sweden's largest lakes. Limnol Oceanogr 49:1191–1201

Weyhenmeyer GA, Karlsson J (2009) Nonlinear response of dissolved organic carbon concentrations in boreal lakes to increasing temperatures. Limnol Oceanogr 54:2513–2519

Weyhenmeyer GA, Blenckner T, Pettersson K (1999) Changes of the plankton spring outburst related to the North Atlantic Oscillation. Limnol Oceanogr 44:1788–1792

Weyhenmeyer GA, Meili M, Livingstone DM (2004a) Nonlinear temperature response of lake ice breakup. Geophys Res Lett 31: L07203, doi:10.1029/2004GL019530

Weyhenmeyer GA, Willén E, Sonesten L (2004b) Effects of an extreme precipitation event on water chemistry and phytoplankton in the Swedish Lake Mälaren. Bor Env Res 9:409–420

Weyhenmeyer GA, Meili M, Livingstone DM (2005) Systematic differences in the trend towards earlier ice-out on Swedish lakes along a latitudinal temperature gradient. Verhandlungen der Internationalen Vereiningung der theoretischen und angewandten Limnologie 29:257–260

Weyhenmeyer GA, Jeppesen E, Adrian R, Blenckner T, Jankowski T, Jennings E, Jensen JP, Nõges P, Nõges T, Straile D (2007) Nitrate-depleted conditions on the increase in shallow northern European lakes. Limnol Oceanogr 52:1346–1353

Weyhenmeyer GA, Livingstone DM, Meili M, Jensen OP, Benson B, Magnuson JJ (2011) Large geographical differences in the sensitivity of ice-covered lakes and rivers in the Northern Hemisphere to temperature changes. Glob Change Biol 17:268–275

Wiedner C, Rücker J, Brüggemann R, Nixdorf B (2007) Climate change affects timing and size of populations of an invasive cyanobacterium in temperate regions. Oecologia 152:473–484

Wilhelm S, Adrian R (2007) Long-term response of *Dreissena polymorpha* larvae to physical and biological forcing in a shallow lake. Oecologia 151:104–114

Wilhelm S, Adrian R (2008) Impact of summer warming on the thermal characteristics of a polymictic lake and consequences for oxygen, nutrients and phytoplankton. Freshwater Biol 53:226–237

Wilhelm S, Hintze T, Livingstone DM, Adrian R (2006) Long-term response of daily epilimnetic temperature extrema to climate forcing. Can J Fish Aquat Sci 63:2467–2477

Williamson CE, Saros JE, Schindler DW (2009) Sentinels of change. Science 323:887–889

Winder M, Schindler DE, Essington TE, Litt AH (2009) Disrupted seasonal clockwork in the population dynamics of a freshwater copepod by climate warming. Limnol Oceanogr 54:2493–2505

Winfield IJ, Hateley J, Fletcher JM, James JB, Bean CW, Clabburn P (2010) Population trends of Arctic charr (*Salvelinus alpinus*) in the UK: assessing the evidence for a widespread decline in response to climate change. Hydrobiologia 650:55–65

Woodward G, Perkins DM, Brown LE (2010) Climate change and freshwater ecosystems: impacts across multiple levels of organization. Phil Trans Roy Soc 365:2093–2106

Wright RF, Bishop K, Dillon PJ, Erlandsson M, Evans CD, Hardekopf DW, Helliwell R, Hruška J, Hutchins M, Kaste Ø, Kopáček J, Krám P, Laudon H, Moldan F, Rogora M, Sjøeng AMS, de Wit HA (2010) Interaction of climate change and acid deposition. In: Kernan M, Moss B, Battarbee RW (eds) Climate Change Impacts on Freshwater Ecosystems. Wiley-Blackwell

Xenopoulos MA, Leavitt PR, Schindler DW (2009) Ecosystem-level regulation of boreal lake phytoplankton by ultraviolet radiation. Can J Fish Aquat Sci 66:2002–2010

Yoo J, D'Odorico P (2002) Trends and fluctuations in the dates of ice break-up of lakes and rivers in Northern Europe: the effect of the North Atlantic Oscillation. J Hydrol 268:100–112

Yvon-Durocher G, Jones JI, Trimmer M, Woodward G, Montoya JM (2010). Warming alters the metabolic balance of ecosystems. Phil Trans Roy Soc 365:2117–2126

Yvon-Durocher G, Montoya JM, Trimmer M, Woodward G (2011) Warming alters the size spectrum and shifts the distribution of biomass in freshwater ecosystems. Glob Change Biol 17:1681–1694

Environmental Impacts—Terrestrial Ecosystems

11

Norbert Hölzel, Thomas Hickler, Lars Kutzbach, Hans Joosten, Jakobus van Huissteden and Roland Hiederer

Abstract

The chapter starts with a discussion of general patterns and processes in terrestrial ecosystems, including the impacts of climate change in relation to productivity, phenology, trophic matches and mismatches, range shifts and biodiversity. Climate impacts on specific ecosystem types—forests, grasslands, heathlands, and mires and peatlands—are then discussed in detail. The chapter concludes by discussing links between changes in inland ecosystems and the wider North Sea system. Future climate change is likely to increase net primary productivity in the North Sea region due to warmer conditions and longer growing seasons, at least if summer precipitation does not decrease as strongly as projected in some of the more extreme climate scenarios. The effects of total carbon storage in terrestrial ecosystems are highly uncertain, due to the inherent complexity of the processes involved. For moderate climate change, land use effects are often more important drivers of total ecosystem carbon accumulation than climate change. Across a wide range of organism groups, range expansions to higher latitudes and altitudes and changes in phenology have occurred in response to recent climate change. For the range expansions, some studies suggest substantial differences between organism groups. Habitat specialists with restricted ranges have generally responded very little or even shown range contractions. Many of already threatened species could be particularly vulnerable to climate change. Overall, effects of recent climate change on terrestrial ecosystems within the North Sea region are still limited.

N. Hölzel (✉)
Institute of Landscape Ecology, University of Münster, Münster, Germany
e-mail: norbert.hoelzel@uni-muenster.de

T. Hickler (✉)
Senckenberg Biodiversity and Climate Research Centre (BiK-F), and Department of Physical Geography, Goethe University, Frankfurt, Germany
e-mail: thomas.hickler@senckenberg.de

L. Kutzbach (✉)
Institute of Soil Science, CEN, University of Hamburg, Hamburg, Germany
e-mail: lars.kutzbach@uni-hamburg.de

H. Joosten
Institute of Botany and Landscape Ecology, Ernst-Moritz-Arndt University of Greifswald, Greifswald, Germany
e-mail: joosten@uni-greifswald.de

J. van Huissteden
Department of Earth Sciences, University Amsterdam, Amsterdam, The Netherlands
e-mail: j.van.huissteden@vu.nl

R. Hiederer
Institute for Environment and Sustainability, European Commission Joint Research Centre, Ispra, Italy
e-mail: roland.hiederer@jrc.ec.europa.eu

M. Quante and F. Colijn (eds.), *North Sea Region Climate Change Assessment*, Regional Climate Studies, DOI 10.1007/978-3-319-39745-0_11

11.1 Introduction

The chapter starts with a discussion of general patterns and processes (Sect. 11.2), such as impacts of climate change on productivity, phenology and biodiversity. Climate impacts on specific ecosystem types, such as forests, grasslands and mires are discussed in more detail in subsequent sections (Sects. 11.3–11.6). The chapter concludes by discussing links between changes in inland ecosystems and the wider North Sea system (Sect. 11.7) and then summarises the main findings of this assessment in the form of a table (Sect. 11.8). The chapter focuses on the direct impacts of climate change; the potential impacts of indirect drivers are beyond the scope of this chapter.

11.2 General Patterns and Processes

11.2.1 Vegetation Zone Shifts, Productivity and Carbon Cycling

The terrestrial part of the North Sea region lies mainly in the temperate forest zone, with some boreal elements and treeless tundra at higher altitudes in Scandinavia and Scotland (Fig. 11.1). Below the tree line, significant areas of treeless vegetation would naturally occur only in wetlands (marshes, river floodplains and mires), where soil saturation precludes tree growth.

Deforestation and land degradation as a result of grazing and other anthropogenic activities have decreased the natural forest cover over thousands of years (e.g. Simmons 2003; Kaplan et al. 2009; Gaillard et al. 2010). Most of the lowland forests in England, for example, had already been cleared 1000 years ago (Ruddiman 2003). Forest cover over large parts of the UK, the Netherlands, north-western Germany and Denmark is currently less than 15 % (Eurostat 2015; Fig. 11.2).

The current distribution of zonal vegetation types in the North Sea region is influenced climatically mainly by temperature because terrestrial net primary productivity (NPP) is less limited by water supply, which is relatively high during the growing season because this is when most rainfall occurs (see Sect. 1.5). In terms of future changes in climate and weather (see Chap. 5), the warming expected by the end of the century can be expected to lead to a northward shift in zonal vegetation types or up in altitude (Hickler et al. 2012), and an increase in NPP where the warming is not accompanied by substantially drier conditions. Most climate change scenarios project an increase in annual precipitation across the North Sea region by the end of the century, although substantially drier conditions have been projected for summer and in particular for the southern part of the

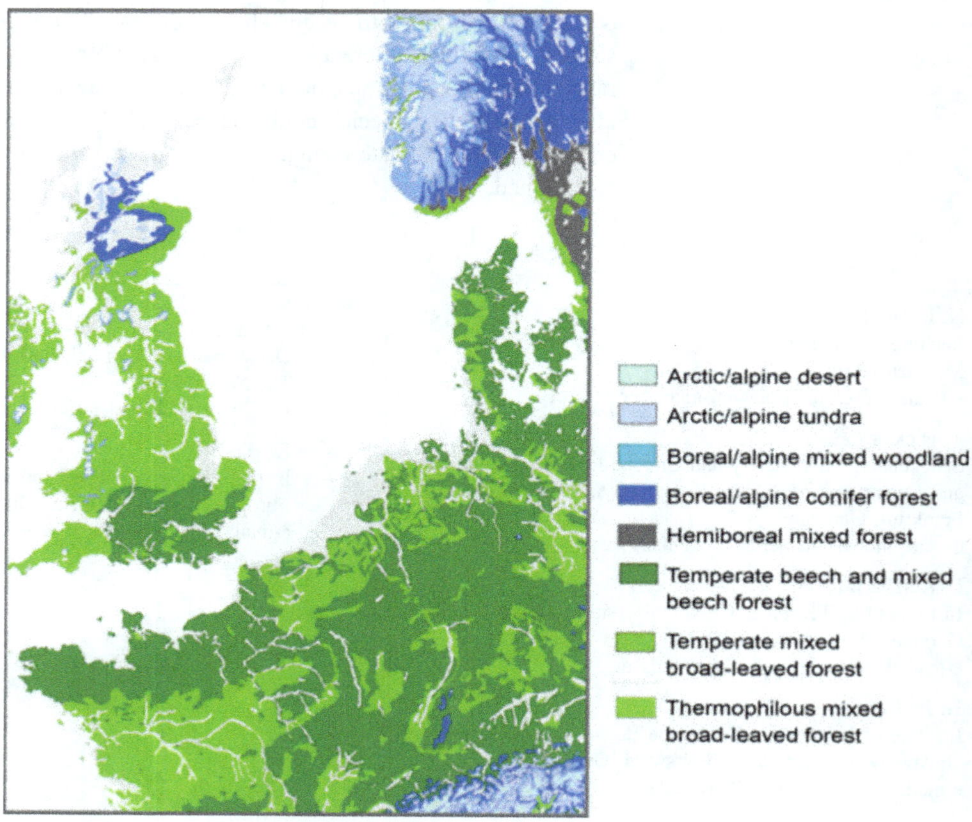

Fig. 11.1 Potential natural zonal (determined by macro-climate) vegetation types in the North Sea region. *Grey areas* were not classified (Bohn et al. 2003; simplified by Hickler et al. 2012)

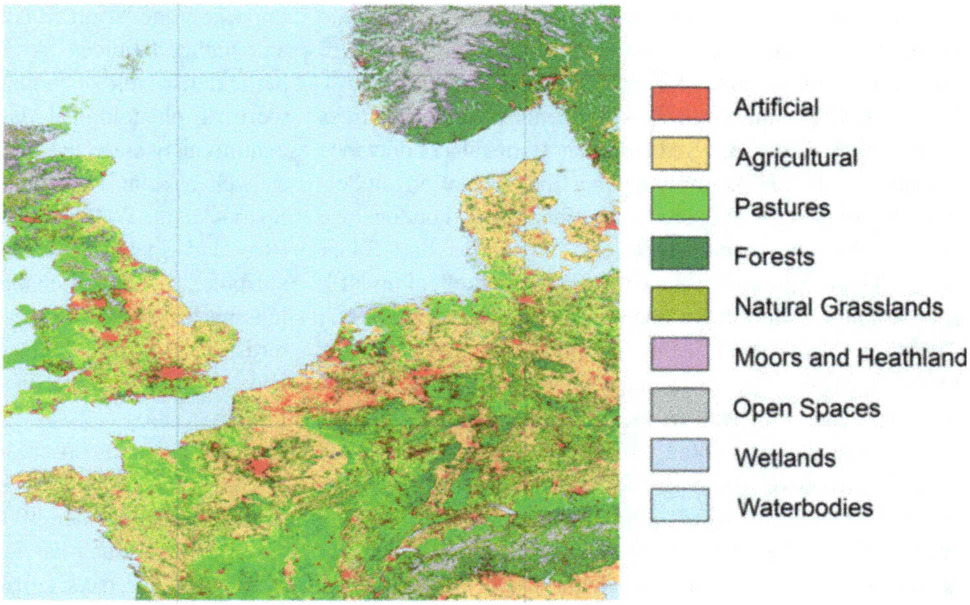

Fig. 11.2 Current land cover in the North Sea region according to CORINE Land Cover (EUROSTAT 2014)

region, where water availability already constrains vegetation productivity (see Chap. 5). Together with the slight projected increase in dry spell length (see Chap. 5 and Jacob et al. 2014), vegetation productivity might, therefore, decrease in the southern North Sea region. However, these projections are based on average results from a number of regional and global climate models (RCMs and GCMs) and because not all models agree in terms of the sign of the change in summer precipitation for different parts of the North Sea region, these projections of future water availability during the main growing season contain uncertainties (see Chap. 5 and Jacob et al. 2014). Furthermore, water availability also controls forest productivity strongly in the south-eastern UK (Broadmeadow et al. 2005), not strictly the southern part of the study region. Here too, increasing drought stress in summer would probably negatively impact NPP. Nevertheless, it should be noted that unchanged precipitation implies less water availability because evapotranspiration will increase with rising temperature. According to the multi-model mean of the CMIP5 models (see Chap. 5), the net outcome of changes in precipitation and evapotranspiration is projected to be an increase in annual run-off in the northern part of the region and a decrease in the south (Collins et al. 2013). These changes in the water balance are particularly important for wetlands (see Sect. 11.6).

The uncertainties in projections of future summer moisture (see Chap. 5) make it difficult to predict the impacts of climate change on terrestrial ecosystems. Morales et al. (2007) simulated the combined effects of climate change and increasing atmospheric carbon dioxide (CO_2) levels on European ecosystems with a dynamic vegetation model, using projections from a variety of combinations of RCMs, bounding GCMs and emission scenarios (Christensen et al. 2007, not accounting for changes in land use). With the exception of north-western France, all simulations indicated increasing NPP in the North Sea region by the end of the century. According to these simulations, the northern part of the study region remains a carbon sink, and the southern part continues to be a small source. However, different climate impact models can yield different results even when driven by the same climate scenario data. Using the SRES high A1Fi scenario (Nakićenović and Swart 2000), a number of dynamic global vegetation models (DGVMs) simulated increasing NPP over most of the North Sea region by the end of the century (Sitch et al. 2008), whereas the Lund-Potsdam-Jena (LPJ) DGVM showed decreased vegetation carbon storage especially in the southern part (Sitch et al. 2008). Most of the models in this study, as well as the model used by Morales et al. (2007), included the potential beneficial plant-physiological effects of increasing atmospheric CO_2 concentrations, but not the constraints on this effect through nutrient limitation.

Increasing levels of atmospheric CO_2 will increase NPP (sometimes referred to as the CO_2 fertilisation effect), and most plants reduce stomatal opening in response to higher CO_2 concentrations (e.g. Ainsworth and Long 2005; Hickler et al. 2015). Reduced stomatal opening leads to lower plant transpiration rates, commonly increasing soil water content and thereby counterbalancing potentially increasing drought stress under climate warming (Arp et al. 1998; Morgan et al. 2004; Körner et al. 2007). Increasing leaf area as a result of higher NPP can counteract this water saving effect (e.g. Gerten et al. 2004), but mostly under conditions of ambient nutrient supply, which enables plants to take advantage of increasing CO_2 and to increase their leaf area (Arp et al. 1998; McCarthy et al. 2006; Norby et al. 2010). According to future simulations with a GCM that includes dynamic vegetation changes, the net outcome of the two effects will be a substantial increase in global run-off (Betts et al. 2007). However, CO_2 enhancement experiments with conifer trees have shown hardly any reduction in stomatal conductance (Körner et al. 2007), implying that the vegetation models probably overestimate the reduction in stomatal conductance and transpiration in conifer forests (Leuzinger and Bader 2012). The magnitude of the CO_2 fertilisation effect on NPP and carbon storage is highly debated (e.g. Körner et al. 2007; Thornton et al. 2007). Although photosynthesis increases under elevated CO_2, this enhancement of carbon assimilation often does not lead to increased biomass as the extra carbon is mainly allocated to below-ground carbon pools with fast turnover (fine roots, root exudates, transfer to mycorrhiza) (Körner et al. 2005; Finzi et al. 2007; Norby et al. 2010; Walker et al. 2014). Nitrogen (N) deposition can also increase NPP, but in the southern North Sea region, N-deposition is already so high that nitrogen is not limiting terrestrial productivity directly (but may decrease productivity through negative side effects such as soil acidification; Bowman et al. 2008; Horswill et al. 2008). N-deposition across the study region is expected to remain at similar levels as today (2014) or to decrease slightly (Tørseth et al. 2012), but N-mineralisation in the soil will probably increase in the northern North Sea region due to warming (Lükewille and Wright 1997; Melillo et al. 2011), which would increase terrestrial productivity particularly in N-limited vegetation on acidic soils (see also Sects. 1.7 and 11.6).

Net primary productivity is an important driver of many ecosystem services, including total carbon storage, but in the North Sea region its dynamics are determined largely by land use, which has not been accounted for in the DGVM study mentioned previously (Sitch et al. 2008). Over most parts of Europe, including the North Sea region, forest carbon stocks, for example, are currently increasing as forests grow older and less timber is harvested than a few decades ago (Janssens et al. 2003; Nabuurs et al. 2003; Ciais et al. 2008).

Total ecosystem carbon storage is further influenced by soil carbon dynamics. Soil respiration, and thereby carbon losses from the soil, is expected to increase under global warming, but the sensitivity of the soil carbon pool remains uncertain (Davidson and Janssens 2006; Luyssaert et al. 2010), and combined effects of potentially increasing NPP (and carbon inputs into the soil) and increasing soil respiration rates (reducing carbon storage) on total ecosystem carbon storage are very difficult to estimate.

11.2.2 Changes in Phenology

Changes in the phenology of biota currently provide the most sensitive and compelling evidence of climate warming impacts in the North Sea region and elsewhere in the middle and higher latitudes. At the same time these changes are particularly well documented due to a pan-European network of phenological data collections that has been run continuously since the mid-20th century (e.g. Menzel 2000) as well as long-term data from bird-ringing stations (e.g. Sparks et al. 2005) and butterfly monitoring programmes (Roy and Sparks 2000). Phenological changes that can be attributed to climate change include leaf unfolding, flowering and leaf colouring as well as the arrival dates of migrant birds, dates of egg laying of birds or the timing of the first appearance of butterflies (Parmesan and Yohe 2003; Parmesan 2006).

An analysis of observational data from the International Phenological Gardens in Europe for the 1959–1996 period (Menzel and Fabian 1999; Menzel 2000) revealed that spring events such as leaf unfolding have advanced on average by 6.3 days (−0.21 days $year^{-1}$), whereas autumn events such as leaf colouring have been delayed on average by 4.5 days (+0.15 days $year^{-1}$). This trend has resulted in an average extension of the annual growing season by 10.8 days since the early 1960s. This trend is of particular significance for regions bordering the North Sea Basin such as Denmark and northern Germany. Similar results were obtained in a more regional study analysing data from the phenological network of the German Weather Service for the period 1951–1996. In this study, Menzel et al. (2001) found the strongest phenological advances in key indicators of earliest and early spring (−0.18 to −0.23 days $year^{-1}$) whereas changes in autumn were less pronounced (delay of +0.03 to +0.10 days $year^{-1}$). Overall, the mean growing season for the period 1974–1996 was up to 5 days longer than for the period 1951–1973. Similar findings were made by van Vliet et al. (2014) in the Netherlands. Using data from the Dutch phenological observation network they found that significant changes in life cycle events started only in the early 1990s. In a large-scale meta-analysis using data from 21 European countries for the period 1971–2000 (Fig. 11.3), Menzel et al. (2006) showed that 78 % of all bud break, flowering and fruiting records advanced, and only 3 % were delayed.

This study clearly demonstrated that phenology is directly linked to the temperature of preceding months with a mean advance of spring/summer by 2.5 days per °C and a mean delay of leaf colouring and leaf fall by 1.0 days per °C. So far, phenological changes of this type are reversible and depend on weather conditions in the year of observation.

In the UK, mean laying dates for the first clutches of 20 bird species advanced on average by 8.8 days between 1971 and 1992 (Crick et al. 1997). Similarly, spawning of two amphibian species (toads) in England advanced by two to three weeks between 1978 and 1994, and the arrival of three newt species in breeding pools advanced by as much as five

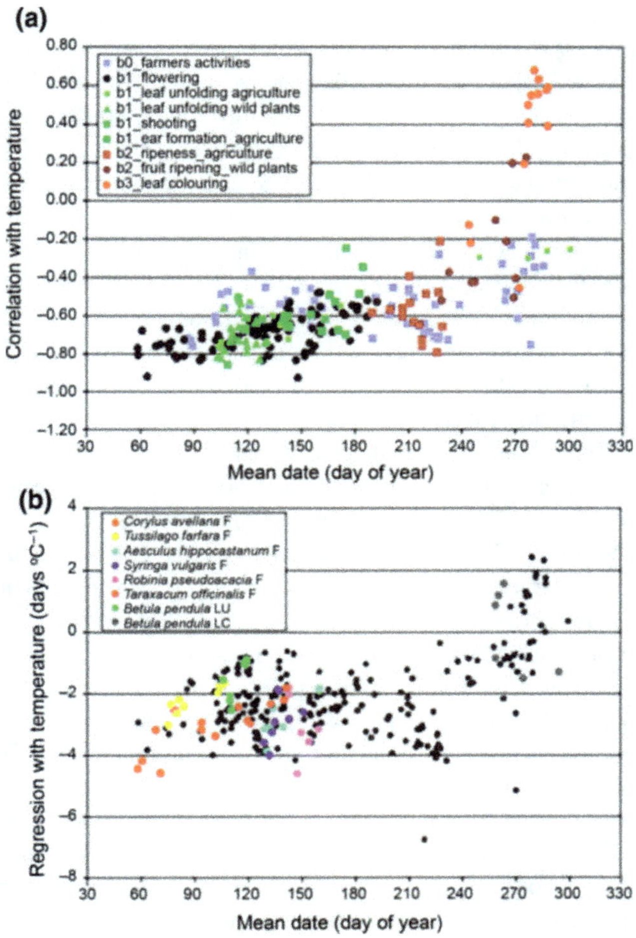

Fig. 11.3 Temperature sensitivity and response across the year. **a** Maximum correlation coefficients for 254 mean national time series of phenophases in nine European countries with mean temperatures of the previous months. **b** Regression coefficients against mean temperature of the previous month. *F* flowering; *LU* leaf unfolding; *LC* leaf colouring. The overall dependence of temperature sensitivity and response on mean date is high **a** $R^2 = 0.59$, $p < 0.001$; **b** $R^2 = 0.47$, $p < 0.001$) (Menzel et al. 2006)

to seven weeks (Beebee 1995). Based on a composite map of 70,000 records for 1998–2007 for the common frog *Rana temporaria*, Carroll et al. (2009) found an average advance of first spawning of about 10 days in the UK compared to map-based data 60 years before.

On the island of Heligoland in the south-eastern corner of the North Sea, mean spring passage times for 24 species of migratory birds advanced by 0.05–0.28 days year^{-1}, which in most species correlated strongly with warmer local temperature during the migration period as well as with the strength of the North Atlantic Oscillation (NAO; Hüppop and Hüppop 2003). Almost identical findings were made at a larger spatial scale from several ringing stations by Sparks et al. (2005). At the continental scale, Both et al. (2004) analysed 23 European populations of pied flycatcher *Ficedula hypoleuca* and found that nine showed an advanced laying date, which were all from those areas with the strongest warming trend and mostly situated at the southern fringe of the North Sea basin. In an area of southern England (Oxfordshire), Cotton (2003) demonstrated that earlier arrival of 20 species of long-distance migratory birds was positively correlated with enhanced air temperatures at wintering grounds in Sub-Saharan Africa.

Climate change also has significant impacts on the winter distribution of migratory birds that fly south to avoid the northern winter. Based on ringing data from the Netherlands, Visser et al. (2009) found that 12 of 24 species studied showed a significant reduction in their migration distance to the south, and that this was strongly correlated with the Dutch winter temperature in the year of recovery. For three common waterfowl species, Lehikoinen et al. (2013) demonstrated that shifts in wintering areas to the northeast correlated with an increase of 3.8 °C in early winter temperature in the north-eastern part of the wintering areas, where bird abundance increased exponentially, corresponding with decreases in abundance at the south-western margin of the wintering ranges. In line with these findings, Maclean et al. (2008) showed that the centres of wintering distribution for five species of wading birds along the north-western European coast flyway shifted 95 km north-eastwards within the period 1981–2000.

For the UK, Roy and Sparks (2000) showed that 26 of 35 species of butterfly exhibited an earlier appearance over the relatively short period 1976–1998 (statistically significant for 13 species). The authors estimated that a warming of 1 °C might advance first and peak appearances of most butterfly species by 2–10 days.

11.2.3 Matches and Mismatches Across Trophic Levels

Shifts in phenology as a response to climate change differ among species and populations. This has been shown for a range of taxonomic groups (Parmesan and Yohe 2003; Parmesan 2006). If climate responses differ between strongly-interacting species, such differences can have immediate impact on key ecological interactions, such as plant–pollinator, herbivore–plant, host–parasite/parasitoid and predator–prey (Visser and Both 2005; Thackeray et al. 2010). This may lead to a phenological mismatch of evolutionary-synchronised species but also to a phenological match of formerly asynchronised species resulting in so far avoided competition, parasitism or predation (Parmesan 2006).

For the Netherlands (Fig. 11.4), it was demonstrated that earlier bud break in sessile oak *Quercus petraea* due to climate warming leads to an earlier appearance of caterpillars, thus disrupting food supply during the main hatching period of the migratory pied flycatcher, which did not keep pace in its arrival at breeding grounds with the advance in peak food supply (Both and Visser 2001; Both et al. 2006). As a consequence of this mismatch in timing, pied flycatcher populations breeding in oak forest declined by 90 % between 1987 and 2003 (Both et al. 2006).

Such phenological mismatches may show effects across four trophic levels leading to deterioration in the timing of food demand and availability for passerines and their avian predators (Both et al. 2009). Biotic mismatches are also considered a major cause of the disproportionate decline in long-distance migratory bird species compared to short-distance migrants that are able to react more flexibly to phenological changes in their breeding areas (Møller et al. 2008; Both et al. 2010; Saino et al. 2011). Evidence for this phenomenon is, however, so far mostly correlative. The genetic basis of mechanisms of adaptation to phenological change is still poorly understood. Although the actual consequences of phenological changes on ecosystem functioning are not clear, several studies highlight the potential risk of desynchronising trophic linkages between primary and secondary consumers (Thackeray et al. 2010). This includes, for example, the distortion of entire food webs, in which top predators moving to cooler regions may trigger trophic cascades that lead to local extinctions and altered ecosystem processes (Montoya and Raffaelli 2010). To date, most of these assumptions are theoretical and more or less unsupported by experiments or empirical data.

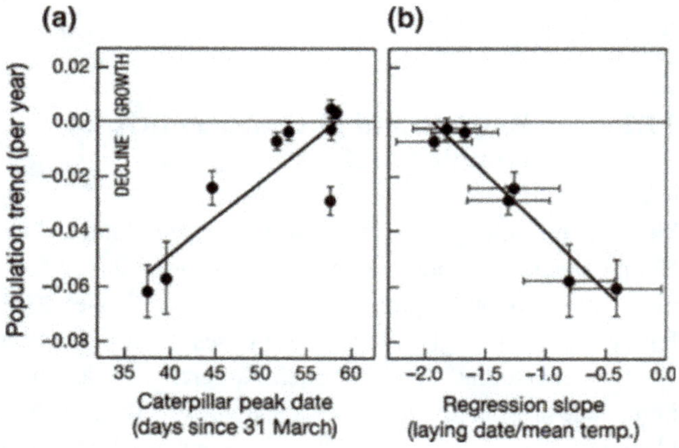

Fig. 11.4 Trends in pied flycatcher populations in response to the local date of peak caterpillar abundance (Spearman rank correlation: $r_s = 0.80$, $n = 9$, $p = 0.013$) (**a**), and the slope of annual median egg laying date on spring (16 April–15 May) temperature ($r_s = -0.86$, $n = 7$, $p = 0.03$) (**b**). Populations of pied flycatchers with an early food peak and a weak response declined most strongly. Population trend is the slope of the regression of the log number of breeding pairs against year. In '**b**', the x axis shows the slope of a linear regression of median laying date against mean temperature from 16 April to 15 May. *Error bars* represent the standard errors of the slopes of the regression lines. All points in '**b**' are also in '**a**', except for one point, for which no data regarding the caterpillar peak were available (Both et al. 2006)

Differences in response to climate change among species may also lead to a matching of originally asynchronous species, also with considerable ecological implications (Visser and Both 2005; Parmesan 2006). Case studies documenting this process are rare. Van Nouhuys and Lei (2004) showed that warmer, early spring-temperatures favoured the parasitoid wasp *Cotesia melitaearum* disproportionally, bringing it into closer synchrony with its host the butterfly *Melitaea cinxia*. Although the authors found no direct effect of the phenological matching on local host population size, the synchrony is likely to be important for overall host meta-population dynamics via variation in the rate of colonisation by the parasitoid.

In addition to phenological mismatches, trophic interactions can also become disrupted if host plants and species feeding on these host plants shift their ranges asynchronously. For the monophagous butterfly *Boloria titania* and its larval host plant *Polygonum bistorta*, Schweiger et al. (2008) showed that climate change may lead to a spatial mismatch of trophically interacting species due to asynchronous range shifts. Schweiger et al. (2012) analysed the potential for such mismatches in the future for 36 European butterfly species by simulating the potential range shifts for butterflies and host plants separately with bioclimatic envelope models, also taking into account land use. They found that those butterflies that are already limited in their distribution by their host plants could suffer most from global climate change, particularly if the host plants have restricted ranges.

11.2.4 Range Shifts and Biodiversity

Recent climate change has already influenced the distribution of species and their abundance (Walther et al. 2002; Parmesan and Yohe 2003; Parmesan 2006; Lenoir et al. 2008; Chen et al. 2011). According to a recent meta-analysis covering a range of taxonomic groups across the globe, species have on average shifted their ranges 16.9 km to higher latitudes and 11 m up in altitude per decade, more than estimates from earlier studies. The average shifts have been larger in those areas that have experienced the strongest warming and have, on average, been sufficient to track temperature changes, but with large variation between species, and most observations are from the temperate zone and tropical mountains (Chen et al. 2011). More than 20 % of species actually shifted in the opposite direction (in latitude and altitude) to the one expected based on temperature changes over the last few decades. Such changes can be explained by drivers other than temperature, such as habitat destruction and water availability, together with biotic interactions such as dependency on certain host plants and special physiological constraints (e.g. minimum or maximum temperatures during crucial phases of the life cycle) (Chen et al. 2011; Crimmins et al. 2011; Tingley et al. 2012). The differences between species reported here were only poorly related to broad taxonomic groups, such as birds and butterflies; rather the differences within such groups were larger than the differences between groups. In all taxonomic groups, habitat specialists and those with a low dispersal and colonisation capability show the lowest or even negative range shifts towards higher latitudes or altitudes (Warren et al. 2001; Chen et al. 2011). However, a recent global analysis of projected rates of temperature shift across landscapes compared to maximum projected speeds at which species can move across landscapes (from observations and modelling studies) showed that many species will probably be unable to track climate change, particularly for the warmer scenarios, which imply faster warming than in the recent past (Settele et al. 2014). Furthermore, this analysis also suggested large differences between organism groups in terms of their dispersal capacity. Herbaceous plants and trees seem to have particularly low dispersal capacity (Settele et al. 2014). An analysis for the British Isles (Fig. 11.5) also showed substantial differences in range expansion for different taxonomic groups (Hickling et al. 2006).

In the North Sea region, substantial average northward shifts have been well-documented for birds, butterflies, moths, dragonflies and damselflies, but mostly with large numbers of species also showing no shift or even retreating northern range boundaries (Parmesan 2006). Among the well-studied groups, plant ranges show the smallest responses to recent climate change, at least in lowland areas, probably because of their limited capacity to disperse and colonise new habitats in highly-fragmented landscapes (Honnay et al. 2002; Bertrand et al. 2011; Doxford and Freckleton 2012). Analyses of community composition, however, show substantial increases in warm-adapted vascular plants and epiphytic lichens across the Netherlands, which have probably been partly driven by climate change (van Herk et al. 2002; Tamis et al. 2005). Also, as these changes have clearly been driven by other factors (such as changes in land use, eutrophication and, in the case of lichens, decreasing sulphur emissions) attributing them to climate change is challenging. Seventy-seven new epiphytic lichen species colonised the area between 1979 and 2001, nearly doubling the total number of species (van Herk et al. 2002) and overall vascular plant richness also increased (Tamis et al. 2005).

Average model projections for the migration rates that would be necessary to track climate change in Europe are substantially larger than those historically observed, but the magnitude of the mismatch depends heavily on the climate change scenario (Skov and Svenning 2004; Huntley et al. 2008; Doswald et al. 2009). Simulations with bioclimate envelope models suggest large local (per grid cell) species losses and turnover rates, assuming that species fully track climate change by migration (Thuiller et al. 2005; Pompe et al. 2008). For the SRES A1 scenario (Nakićenović and Swart 2000) and the HadCM3 climate model, for example, Thuiller et al. (2005) estimated an average turnover of 48 % per grid cell for the European plants considered (1350 for all of Europe) in the European Atlantic region by 2080. However, results from bioclimate envelope models should be

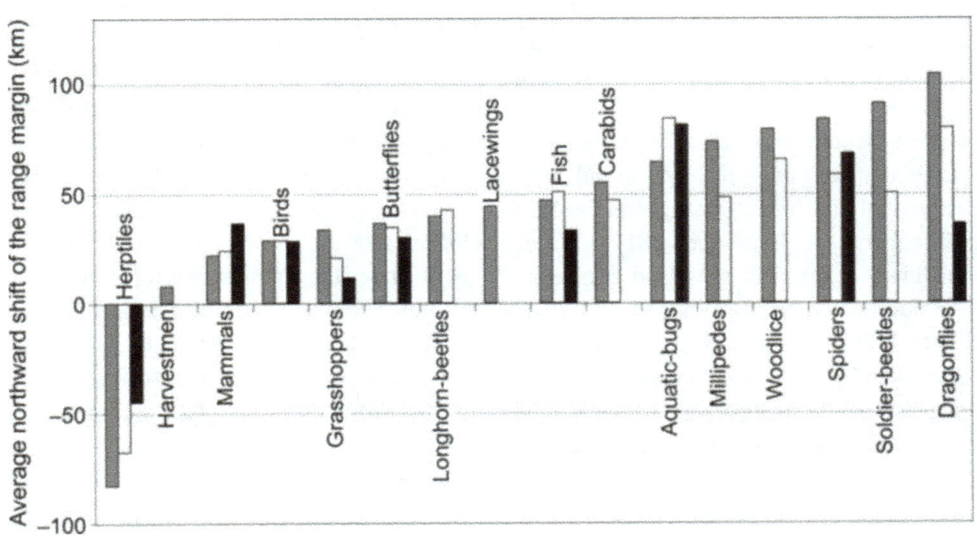

Fig. 11.5 Latitudinal shifts in northern range margins for 16 taxonomic groups in the British Isles during recent climate warming. Results are given for three levels of data subsampling (*grey* recorded; *white* well-recorded; *black* heavily recorded). Only species occupying more than twenty 10 km grid squares across two time periods (between 1960 and 2000, depending on organism group) are included in the analyses; for several of the species-poor groups, these criteria excluded all species from the analysis of 'heavily recorded' squares (Hickling et al. 2006)

interpreted more as potential shifts in the climatic window in which species can thrive rather than projections in range shifts. Furthermore, such models may overestimate change because they are developed based on correlations between species ranges and environmental factors. They do not capture the fundamental niche of species and so underestimate the climatic niche when species have not yet reached their distribution in equilibrium with the climate, which appears to be common, at least for trees (Svenning and Skov 2004; Normand et al. 2011). Furthermore, dispersal is rarely simulated explicitly, and dispersal projections are uncertain, for example, because of large uncertainties in projected wind speeds (Bullock et al. 2012). Nevertheless, it could be expected that many mobile, generalist species will continue to shift their distributions northward and up in altitude in response to climate change, although many habitat specialists (often those that are rare and already endangered) will not, and that many cold-adapted species will probably experience range losses at their southern distribution limit or at lower elevations (Hill et al. 2002; Chen et al. 2011; Sandel et al. 2011; Schweiger et al. 2012). As the area south of the North Sea region is generally more species-rich (e.g. Thuiller et al. 2005), biodiversity in the North Sea region could even increase. Negative impacts on cold-adapted species are expected to be most severe in mountain regions, where species have limited possibilities to migrate upwards or northwards, such as on mountains in the British Isles (Berry et al. 2002; Hill et al. 2002). Recent climate change has also affected the community compositions of birds and butterflies in Europe. Analyses of 9490 bird and 2130 butterfly communities in Europe show large changes, equivalent to a 37 and 114 km northward shift in bird and butterfly communities, respectively. However, these analyses suggest an even larger 'climatic debt', corresponding with a migration lag of 212 and 135 km for birds and butterflies (Devictor et al. 2012).

Intensification of agricultural activities and increasing anthropogenic nitrogen inputs since the 1950s and 1960s have been major drivers of biodiversity changes in the North Sea region (e.g. Ellenberg and Leuschner 2010). Wesche et al. (2012) found large changes in grassland community composition in five floodplain regions in northern Germany between the 1950s and 2008 and a decline in species richness at the plot level of 30–50 %. The decline was particularly strong among nectar-producing herbs, which is likely to have had negative effects on pollinators (Wesche et al. 2012). An analysis of Ellenberg indicator values for nutrient availability and a qualitative comparison with a protected area in the same region suggests that these changes were largely driven by increased nutrient inputs. Also, for a number of insect groups, a decline in species preferring low-productivity habitats and dry grassland specialists has been recorded in northern Germany (Schuch et al. 2012a, b). Pollinators are generally declining in Europe, and this has been particularly well documented for the Netherlands and the UK (Biesmeijer et al. 2006; Potts et al. 2010). However, the reasons for the decline are unclear. Potential drivers include habitat loss and fragmentation, agrochemicals, pathogens, invasion of non-native species, climate change and the interactions between them (Potts et al. 2010). These changes show the significant role of land use practice for biodiversity in north-western Europe. Further intensification of agricultural practices, possibly driven by an increasing demand for biofuels, is likely to have negative effects on biodiversity even if atmospheric N-deposition does not increase.

11.3 Forests

Forests are currently considered the most important carbon sink in Europe (Janssens et al. 2003). Due to the relatively low proportion of forests in the present land cover for countries bordering the North Sea—except Norway—the regional significance of this area as a carbon sink is relatively small or even negative compared to other European regions with higher forest cover (Janssens et al. 2005).

11.3.1 Climate Impacts on Productivity and Carbon Stocks

Although estimates of the mean long-term carbon forest sink (net biome production, NBP) are more reliable than those from grasslands (Janssens et al. 2003), the role of wood harvests, forest fires, losses to lakes and rivers and heterotrophic respiration remains uncertain and difficult to predict. Almost one third of the NBP is sequestered in the forest soil, but large uncertainty remains concerning the drivers and future of the soil organic carbon pool under climate change (Luyssaert et al. 2010). Nevertheless, increasing temperatures, longer growing seasons, higher atmospheric CO_2 concentrations, and in the north, increasing N-mineralisation, are likely to increase the potential forest productivity where summer precipitation does not decline (Lindner et al. 2010). Moreover, it is uncertain to what extent this potential can be realised as forests will increasingly face a climate to which the planted species or provenances are not adapted, which might increase their susceptibility to pests and pathogens, such as bark beetle (Scolytinae) outbreaks, which can lead to major forest die-back events particularly in Norway spruce *Picea abies* stands (Schlyter et al. 2006; Bolte et al. 2010). Furthermore, warmer and longer vegetation periods will accelerate the development of bark beetles, in some regions allowing for additional generations within a growing season (Jönsson et al. 2009). Other insect herbivores will also benefit from warmer conditions (Lindner et al. 2010). In a climate manipulation experiment in a Norwegian boreal forest, raised temperature and CO_2-level stimulated the outbreak of heather beetle *Lochmaea suturalis* and led to a shift in the ground vegetation from common heather *Calluna vulgaris* to blueberry *Vaccinium myrtillus* and cowberry *Vaccinium vitis-idaea* (van Breemen et al. 1998).

The complex interplay between climatic stress, pests and pathogens, and further disturbance such as windfall is hardly captured in the forest models used to project potential future impacts of climate change (e.g. Kirilenko and Sedjo 2007). As a result, it is highly uncertain whether climate change will lead to higher standing biomass in forests.

11.3.2 Shifts in Communities and Species Distribution

Projections of potential climate-driven transient shifts in broadly-defined forest types suggest only moderate changes in the North Sea region by 2100 (Hickler et al. 2012). The most significant changes projected are the spread of broad-leaved and hemi-boreal mixed forests northward in southern Sweden and Norway as well as an upwards shift of the tree-line in the southern Scandes, which is already taking place (Kullman 2002).

Long-term equilibrium of shifts in forest type could be much more substantial, with thermophilous forests dominating in the south-western UK and temperate broadleaved forest along most of the Norwegian coast (Hickler et al. 2012). Recent range shifts northward (Fig. 11.6) have already been observed for cold-hardy, broadleaved, evergreen species such as holly *Ilex aquifolium* at their northern distributional limit in Europe (Walther et al. 2005; Berger et al. 2007).

Many European tree species have not yet filled their potential climatic niche in Europe because of dispersal limitations (Svenning and Skov 2004; Normand et al. 2011). Thus dispersal-limited species may be unable to track future climate change, unless foresters assist migration.

In contrast, there is almost no evidence of range shifts in herbaceous forest plants. Unlike mountain forest with short migration distances, there is some evidence that in lowland forests plant distribution changes will lag behind climate warming (Bertrand et al. 2011). Observational (Honnay et al. 2002) and modelling studies (Skov and Svenning 2004) suggest that this is probably due to dispersal limitation resulting from forest habitat fragmentation in lowlands. Where significant range shifts of forest herbs northward and eastward have been documented, such as for the oceanic annual woodland herb climbing corydalis *Ceratocapnos claviculata*, it is questionable whether this is due to climate change or to other drivers such as eutrophication or assisted migration through the international timber trade (Voss et al. 2012).

Among the major forest tree species, beech *Fagus sylvatica* is expected to extend its range northward in Britain and southern Scandinavia (Kramer et al. 2010; Hickler et al. 2012), whereas environmental conditions for the commercially-important Norway spruce will become less favourable (Pretzsch and Dursky 2002; Schlyter et al. 2006; Hanewinkel et al. 2013). Although beech is often considered to be very sensitive to drought, several studies (Lebourgeois et al. 2005; Meier and Leuschner 2008; Mölder et al. 2011) showed considerable phenotypic plasticity in response to drought stress (e.g. Bolte et al. 2007). The same is true for sessile oak, which proved to be highly resilient even to extreme drought (Leuschner et al. 2001; Lebourgeois et al. 2004; Friedrichs et al. 2009; Merian et al. 2011; Härdtle et al. 2013). Given the generally damp climatic conditions of north-western Europe, major broadleaved forest trees such as beech and sessile oak are probably not constrained by the projected climate change, which is in line with predictions of vegetation models (Kramer et al. 2010; Hickler et al. 2012). However, using older climate projections with lower projected rainfall than the latest average projections (see Chap. 5), simulations with a forest tree suitability model based on climatic and edaphic factors suggested that the majority of native broadleaved species would become unsuitable for commercial timber production in southern England due to increasing drought severity (Broadmeadow et al. 2005).

Forest management includes a wide range of measures to mitigate climate change effects, such as the selection and planting of species and provenances adapted to future climate (Isaac-Renton et al. 2014); a reduction in rotation cycles to accelerate the evolution and establishment of better adapted genotypes (Alberto et al. 2013); and the use of mixtures of high genetic variation across an array of environmental conditions (Hemery 2008; Köhl et al. 2010). Scientifically-sound implementation of such adaptation measures requires a wide range of research and monitoring activities such as testing of the suitability of new tree species and provenances, a regional risk analysis based on retrospective performance as well as the analysis of climate envelope and climate matching under potential future climates (Hulme 2005; Bolte et al. 2009; Hemery et al. 2010).

11.4 Grasslands

After cropland, grasslands are the dominant land use type in the North Sea catchment area. In the UK, grasslands comprise more than 40 % of land cover (EUROSTAT 2015).

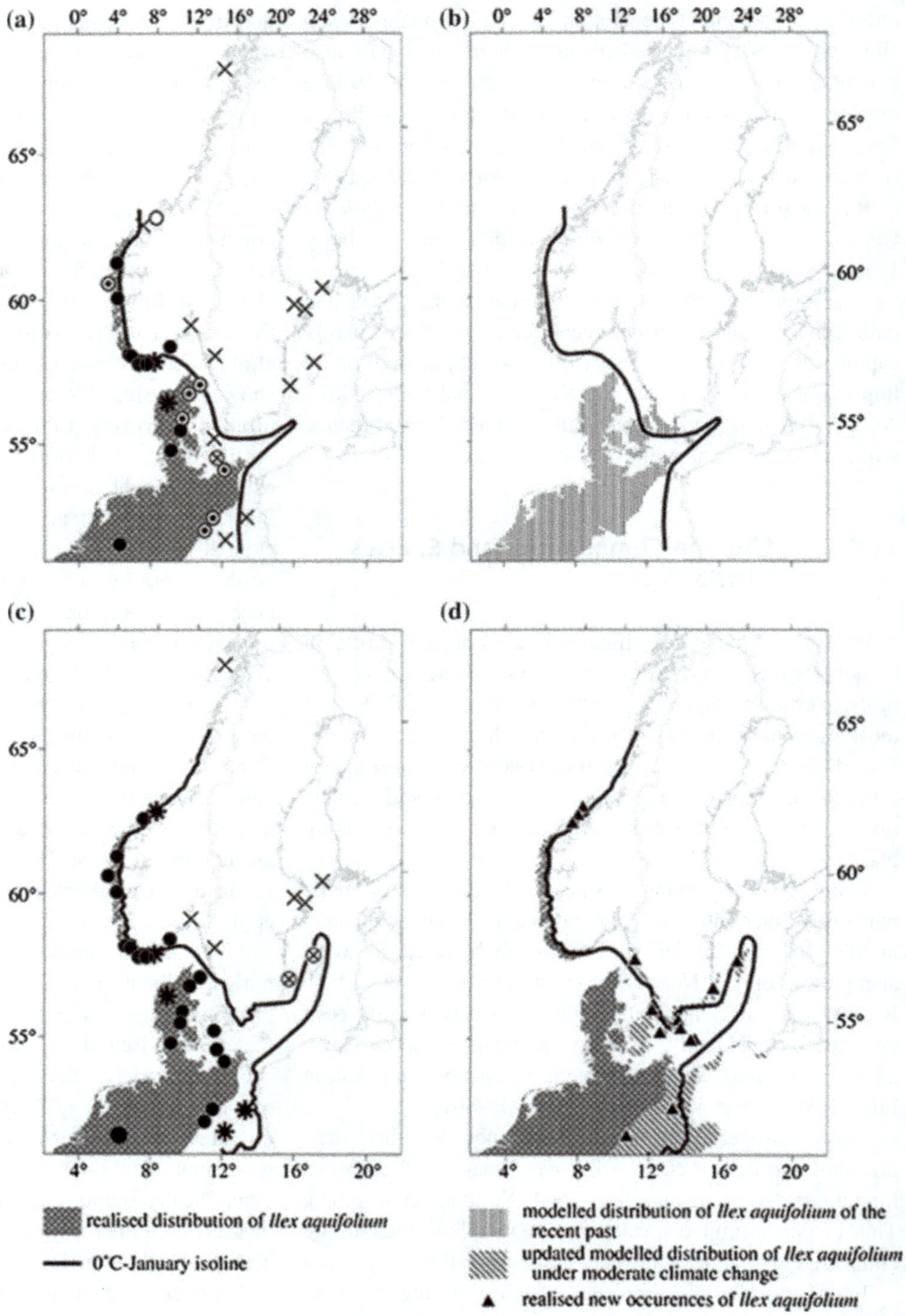

Fig. 11.6 Distribution of holly *Ilex aquifolium* and the 0 °C-January isoline at different times. **a** Former range of *I. aquifolium* based on Enquist (1924) and Meusel et al. (1965), isoline based on Walter and Straka (1970), symbols based on Iversen (1944); *circles*: *I. aquifolium* within or at the border of the station area; *circles*: *with cross I. aquifolium* strayed into woods from gardens; *stars*: *Ilex* area lies immediately outside the station area; *crosses*: *I. aquifolium* missing in the station area. **b** Modelled range of *I. aquifolium* in the recent past (1931–1960), isoline as in '**a**'. **c** Former range of *I. aquifolium* as in '**a**'; isoline updated for 1981–2000 based on Mitchell et al. (2004), symbols as for '**a**'. **d** Former range of *I. aquifolium* complemented by the simulated species distribution under a moderate climate change based on 1981–2000 climate data, isoline as in '**c**'; *triangles* represent locations with new observations of *I. aquifolium* (Walther et al. 2005)

Due to conversion into cropland, and the cessation and intensification of agricultural practices, grasslands underwent fundamental change during the 20th century (Bullock et al. 2011). Changes in management practice and eutrophication are currently the major drivers of ecological change in grasslands. At the same time, grasslands are of major significance for biodiversity and nature conservation in north-western Europe.

11.4.1 Climate Impacts on Carbon Stocks and Cycling

Unlike forests, carbon accumulation in grassland ecosystems occurs mostly below ground. As fluxes of greenhouse gases in grasslands are intimately linked to management and site conditions, grasslands can be either a sink or a source of greenhouse gases. Although many studies consider temperate grassland to be a carbon sink (Soussana et al. 2004), there is still high uncertainty about their current and future net global warming potential (in terms of CO_2 equivalents) at both a regional and continental scale (Janssens et al. 2003, 2005; Smith et al. 2005).

For Britain and Wales, both dominated by grasslands, Bellamy et al. (2005) suggested that significant losses of soil organic carbon (SOC) between 1978 and 2003 must be attributed to climate change because they occurred across all types of land use. However, as shown by Smith et al. (2007), this assumption was precarious and lacked clear empirical evidence. At a global scale (Guo and Giffort 2002), current SOC losses and gains in grasslands can be predominantly attributed to changes in land cover and management, whereas the role of climate change remains uncertain but is predicted to increase over the coming century (Smith et al. 2005). These considerable uncertainties are because grassland ecosystems are particularly complex and difficult to study owing to the wide range in management and environmental conditions to which they are exposed. As a result, studies on the effects of climate change on grasslands are often affected by this variability as well as by other confounding effects such as eutrophication and changes in management practice, which cause difficulties for observational studies and modelling (Soussana et al. 2004).

11.4.2 Climate Impacts on Plant Communities

Few experimental studies attempt to isolate the effects of climate change from other confounding effects. A study simulating warming and extended summer drought in calcareous grasslands at Buxton and Wytham in northern England (Grime et al. 2000) covered two different types of grassland with contrasting effects: after five years of climate manipulation, the more fertile, early successional grassland at Wytham showed significant changes in species composition and aboveground productivity especially for the combination of winter warming and summer drought. In contrast, an oligotrophic and more traditional calcareous grassland at Buxton exhibited almost no response to warming and drought treatments. This was still true even after 13 years of climate modification (Fig. 11.7; Grime et al. 2008). One reason for the high resistance of this infertile grassland may be the small-scale spatial heterogeneity in soil depth allowing the coexistence of

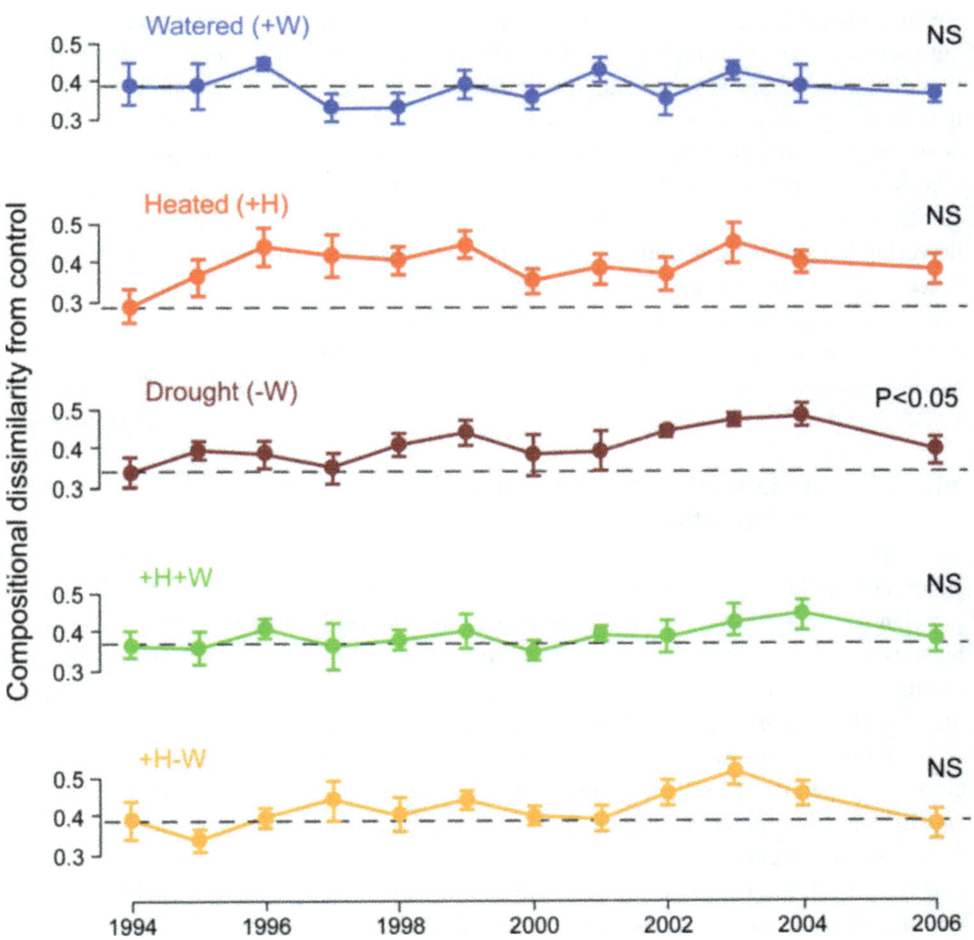

Fig. 11.7 Mean dissimilarity of treatment and control species composition for each year of the Buxton climate change experiment conducted in a nutrient-poor calcareous grassland. Dissimilarity was measured by Sørensen distance estimated separately within each replicate (n = 5 per treatment year). *Dashed lines* indicate mean dissimilarity in year 1 of the experiment. *Error bars* indicate one standard error of the mean. Statistics indicate whether treatment dissimilarity progressively increased over time based on linear autoregressive models (Grime et al. 2008)

drought-tolerant and more mesic species at small spatial scales (Fridley et al. 2011). As indicated by seed addition experiments, the minor changes in species composition, even after long-term climate treatments, are significantly affected by dispersal limitation rather than just biotic resistance (Moser et al. 2011). The prominent role of dispersal limitation as a cause of delayed response to climate effects has also been highlighted in several other studies (Buckland et al. 2001; Stampfli and Zeiter 2004; Zeiter et al. 2006).

Overall, the results of the Wytham and Buxton experiments suggest that more productive grasslands, strongly altered by human activities, might respond more to effects of climate change than infertile and more traditionally managed grasslands with rich species pools that can buffer climate effects (Grime et al. 2000, 2008). However, infertile traditional grasslands show low resilience towards eutrophication and changes in land management, which are currently more important drivers of ecological change in grasslands than climate change. Conversely, future warming potentially in association with increased drought risk could supersede eutrophication as the main driver of change, and in so doing potentially favour the persistence or even spread of dry and infertile grassland types (Buckland et al. 1997).

Observational studies also suggest that changes in the grasslands of north-western Europe can be attributed to recent regional climate change (e.g. Gaudnik et al. 2011), but these are mostly of high uncertainty due to strong confounding effects (McGovern et al. 2011). However, flowering phenology of many typical grassland plants in the UK does reveal significant effects of climate warming. Of 385 plant species, 16 % flowered significantly earlier in the early 1990s compared to previous decades, and earlier onset of flowering was most significant in annual species (Fitter and Fitter 2002). Williams and Abberton (2004) confirmed a significant trend of earlier flowering within different agricultural varieties of the common grassland legume white clover *Trifolium repens*.

11.4.3 Climate Impacts on Animal Communities

More convincing evidence of climate change effects in grasslands comes from animal groups typical of grassland habitats such as butterflies and grasshoppers; several have extended their range northwards significantly over past decades (Parmesan et al. 1999; Hill et al. 2002; Hickling et al. 2006). Unlike most vascular plants, which are often chronically persistent and immobile, highly-mobile animal species can often quickly respond to changing climate by significant range extensions. In north-western Germany, for example, the Roesel's busch-cricket *Metrioptera roeselii*, a typical grassland species, has been rapidly extending its range northward since the early 1990s, which was probably helped by increased rates of macroptery in this normally short-winged species, as a sign of density stress at the range margin (Hochkirch and Damerau 2009; Poniatowski and Fartmann 2011; Poniatowski et al. 2012).

Significant northward range expansions have also been documented for many typical grassland butterflies in the UK (Hill et al. 1999, 2002). However, only particularly mobile habitat generalists can fully exploit the emerging potential offered by climate warming (Fig. 11.8), whereas less mobile species and habitat specialists still suffer from habitat fragmentation and deterioration of habitat quality (Hill et al. 1999; Warren et al. 2001). Thus, range expansion is in many cases significantly lagging behind the current climate and can be reduced or even reversed by non-climatic drivers (Hulme 2005; Oliver et al. 2012).

11.5 Heathlands

In countries bordering the North Sea basin, heathlands dominated by shrubs of the ericaceous family still cover extensive areas especially in highlands of the UK and the southern Scandes (Webb 1998; van der Wal et al. 2011). Due to conversion into cropland and afforestation, heathlands have declined dramatically in the UK-lowlands and in the southern part of the North Sea region (Denmark, Germany, Netherlands). In these regions, they are at the edge of extinction in many sites and have become a major object of biodiversity and nature conservation efforts. Eutrophication and acidification through atmospheric inputs and changes in land management are currently the major drivers of change in these ecosystems (Härdtle et al. 2006), which makes the identification of climate change impacts difficult.

11.5.1 Climate Impacts on Ecosystem Processes

Effects of climate change on ecosystem processes in European heathlands were specifically addressed within the framework of two EU-projects simulating raised temperatures and drought (Wessel et al. 2004). These studies included sites in the UK, Denmark and the Netherlands. Experimental warming of 1 °C induced a significant increase in total above-ground plant biomass growth of 15 % in the most temperature-limited site in the UK, whereas drought treatments led only to a slight decline (Peñuelas et al. 2004). Drought decreased flowering (by up to 24 % in the UK). Warming and drought decreased litterfall in the Netherlands (by 33 and 37 %, respectively). Tissue concentrations of phosphorus (P) generally decreased and the N:P ratio increased with warming and drought except at the UK site,

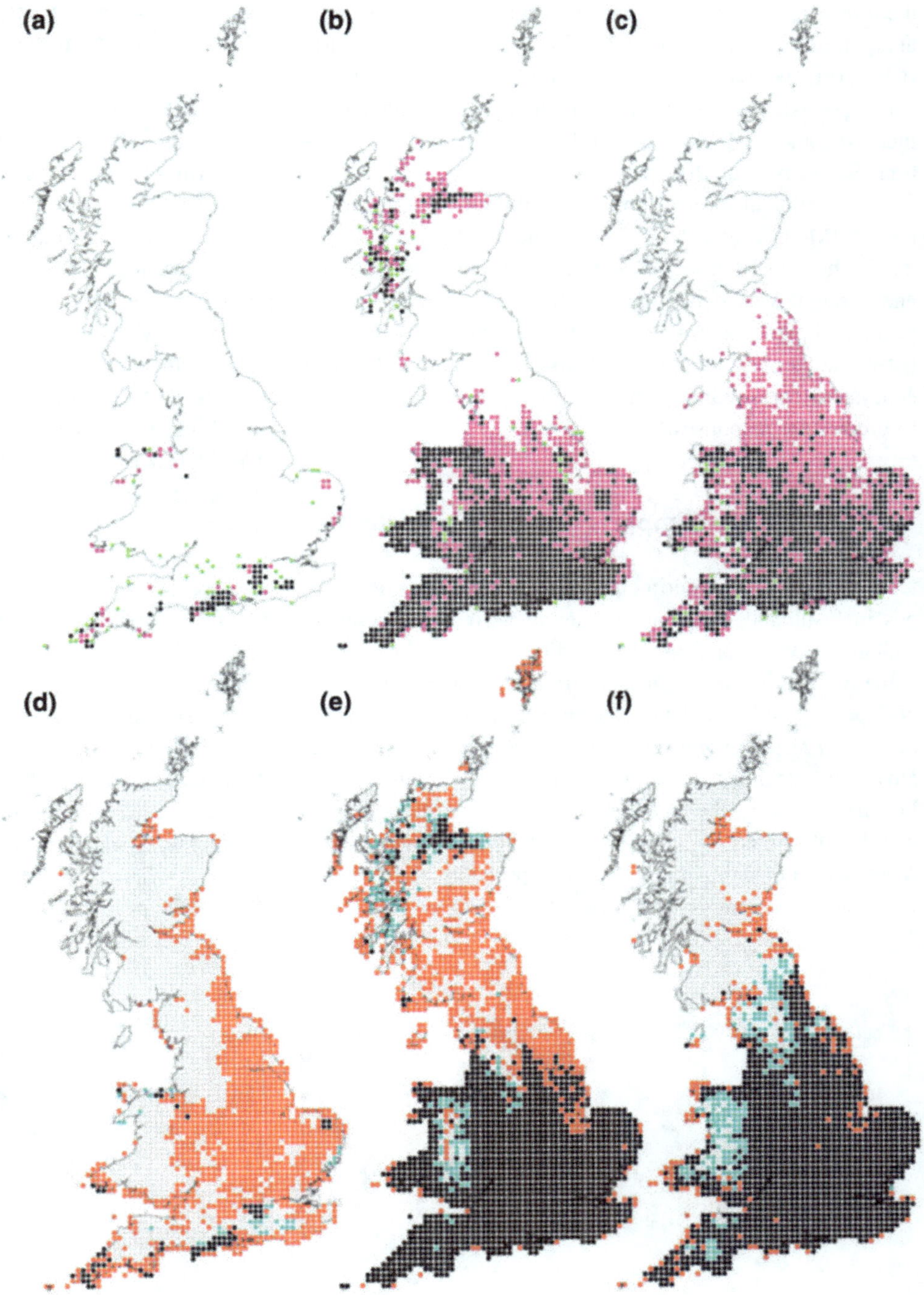

Fig. 11.8 The degree to which three butterfly species have changed their ranges (**a–c**, without subsampling) and are lagging behind current climate in Britain (**d–f**; 10-km grid resolution). **a+d**, silver-studded blue *Plebejus argus*; **b+e**, speckled wood *Pararge aegeria*; **c+f**, comma *Polygonia c-album*. For maps **a–c**, *black circles* show butterfly records for both 1970–1982 and 1995–1999; *green circles* show apparent extinction (recorded 1970–1982; not 1995–1999); *pink circles* show apparent colonisation (no record 1970–1982; record 1995–1999). For maps **d–f**, *black circles* (climate suitable, butterfly recorded) and *grey circles* (climate unsuitable, butterfly not recorded) show where observed 1995–1999 and simulated distributions agree; *red circles* (climate predicted suitable, butterfly not recorded) and *blue circles* (climate deemed unsuitable, butterfly recorded) show mismatches (Warren et al. 2001)

indicating the progressive importance of P-limitation as a consequence of warming and drought.

Owing to their richness in soil organic matter, mature heathlands may become important sources of C and N-release triggered by increasing temperatures and more frequent periods of drought. For the same experiments as above, Schmidt et al. (2004) found mostly weak and insignificant effects of warming and drought treatments on nitrogen and carbon budgets in the soil solution. Only at a strongly N-saturated site with high atmospheric N-deposition in the Netherlands, did warming trigger a significant increase in N-leaching. Similarly, in the same warming and drought experiments, Jensen et al. (2003) and Emmett et al. (2004) found largely weak or inconsistent responses in major soil processes such as decomposition, respiration and N-mineralisation. The latter turned out to be predominantly controlled by soil moisture. The response of soil-related processes to warming and drought treatments was generally found to be strongly dependent on local site conditions (Emmett et al. 2004). At mesic sites in the Netherlands and Denmark, soil respiration decreased in response to drought but recovered quickly to pre-drought levels after re-wetting in the following winter. In contrast, repeated drought treatments at a particularly damp site in the UK, which was particularly rich in organic matter, caused a disturbance in soil structure and a persistent reduction in soil moisture, which induced increased and continuing carbon losses through soil respiration (Sowerby et al. 2008).

The heathland studies conducted within the framework of the CLIMOOR and VULCAN projects (Peñuelas et al. 2007) show that the magnitude of the response to warming and drought was dependent on differences between sites, years, and plant species. Thus there are complex interactions between other environmental factors that condition plant and ecosystem performance, which makes it extremely difficult to predict net responses.

11.5.2 Climate Impacts on Plant Communities

Observational and experimental evidence for floristic changes in heathlands that can be clearly attributed to climate change is weak; for example, Werkman et al. (1996) found some indications that climate warming in combination with N-deposition might enhance the spread of the noxious weed bracken *Pteridium aquilinum* into heathlands in the UK. However, declines in arctic-alpine and boreal-montane lichen species in the heathlands of north-western Europe can be attributed to changes in traditional management practices and acidification, and are probably not directly connected to climate change (Hauck 2009).

11.5.3 Climate Impacts on Animal Communities

A significant decline has been observed over recent decades in artic-alpine bird species inhabiting mountain heathlands in the north of the UK such as ptarmigan *Lagopus mutus*, dotterel *Charadrius morinellus* and snow bunting *Plectrophenax nivalis*. In contrast the thermophilous, submediterranean Dartford warbler *Sylvia undata* has increased its population and spread into southern England, probably due to warmer winters (van der Wal et al. 2011). There is currently almost no empirical evidence of climate change impacts in other heathland-specific animal groups. However, modelling approaches suggest (Thomas et al. 1999; Berry et al. 2002) that eco-thermic animal species in heathlands may benefit from climate warming at their northern range margin.

11.6 Mires and Peatlands

Peatlands store significant quantities of carbon, nitrogen and other elements in their soils (e.g. Limpens et al. 2008; Yu et al. 2010) and are widespread in the North Sea region (Montanarella et al. 2006, Fig. 11.9). According to Joosten and Clarke (2002), peatlands are areas with a naturally accumulated peat layer at the surface, whereas mires are peatlands where peat is currently being formed. While about 80 % of the peatlands in Sweden and Norway are still mires,

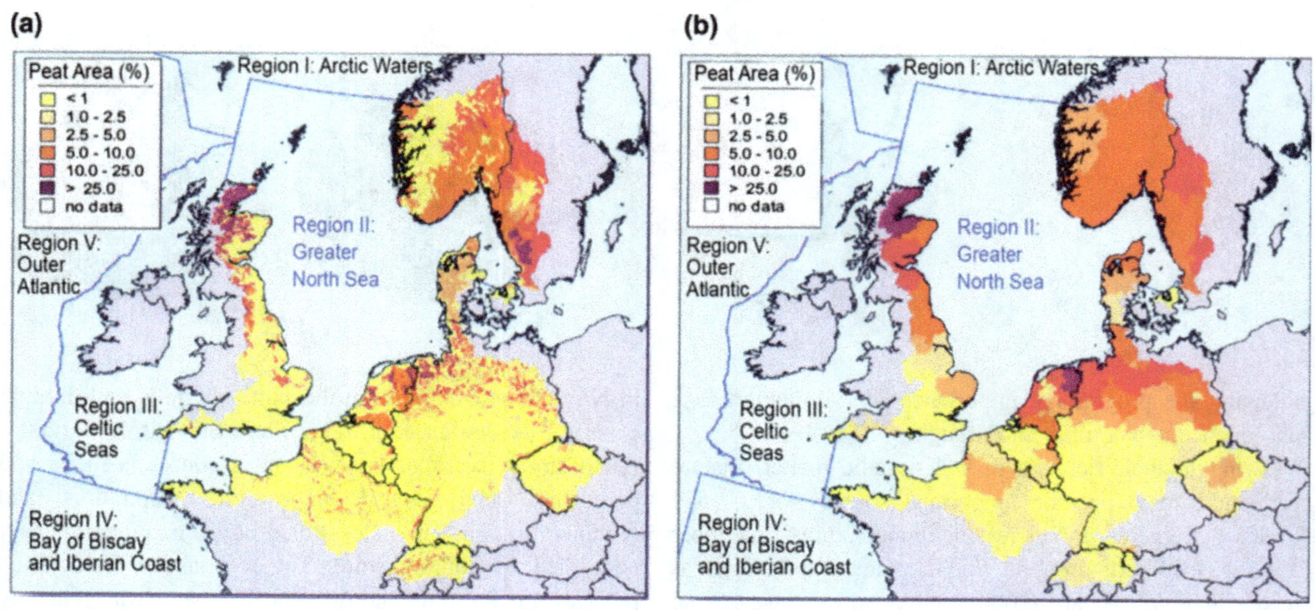

Fig. 11.9 Relative cover of peatlands and peat-topped soils in the North Sea catchment area. Maps based on **a** the soil mapping units of the European Soil Database (King et al. 1994, 1995) and **b** the NUTS (Nomenclature of Territorial Units for Statistics) level 2 administrative regions (R. Hiederer non-published data)

the contribution of mires to the total peatland area in the other, more southern North Sea-bordering states is only 1–7 % due to widespread drainage and intensive land use (Joosten and Clarke 2002).

Peatlands exchange C- and N-containing gases with the atmosphere, particularly the greenhouse gases CO_2, methane (CH_4) and nitrous oxide (N_2O) and thus influence climate (Blodau 2002; Frolking and Roulet 2007; Limpens et al. 2008; Finkelstein and Cowling 2011). Peatlands also exert strong influences on aquatic ecosystems by lateral waterborne export fluxes of elements, especially as particulate and dissolved organic matter (Urban et al. 1989; Freeman et al. 2001; Worrall et al. 2002; Billett et al. 2004; Dinsmore et al. 2010). Importantly, carbon accumulation, and vertical land-atmosphere and lateral waterborne bio-geochemical fluxes of peatlands are affected by climate change, and at the same time by changes in atmospheric chemistry and land use (e.g. Bragg 2002; Belyea and Malmer 2004; Dise 2009; Billett et al. 2010; Charman et al. 2013).

11.6.1 Climatic Impacts on Abiotic Conditions

Climatic change will have direct effects on the energy and water budgets of peatlands. Changing quantities and temporal patterns of precipitation will affect the water table in peatlands; with drought lowering and increased precipitation raising peatland water levels (e.g. Sottocornola and Kiely 2010). Higher temperatures, which are projected for the North Sea region in the future (Chap. 5) will increase evapotranspiration through a larger atmospheric water demand (Kellner 2001; Wu et al. 2010; Peichl et al. 2013). Other important variables influencing the energy and water budget of peatlands are net radiation and incoming short-wave radiation (Moore et al. 2013; Runkle et al. 2014). A continuation of the 'brightening period' through reduced aerosol loading (Wild et al. 2005) in Europe could increase evapotranspiration (Oliveira et al. 2011) leading to lower peatland water tables. However, most atmospheric models simulate a future decrease in shortwave radiation in the northern North Sea region and an increase in short-wave radiation in the south (Chap. 5). A long-term lowering of the water table due to increased evapotranspiration is expected to be modulated by changes in leaf area and the distribution of plant functional groups leading to increased surface resistance, reduced evapotranspiration and an attenuated fall in water tables (Bridgham et al. 1999; Moore et al. 2013). Desiccation of the moss layer during summer droughts can also lead to reduced evapotranspiration (Sottocornola and Kiely 2010). Higher winter precipitation as projected for the North Sea region (Chap. 5) would lead to larger winter discharge from peatlands and probably to a larger lateral export of dissolved organic matter and nutrients (e.g. Tranvik and Jansson 2002; Worrall et al. 2002, 2003; Pastor et al. 2003; Holden 2005). Increasing summer drought and winter rainfall would enhance peatland erosion and export of dissolved organic carbon (DOC) and particulate organic carbon (POC) in susceptible areas, particularly in the upland blanket bogs of the UK and Norway (e.g. Bower 1960, 1961; Francis 1990; Evans et al. 2006b; Evans and Warburton 2010). Drier mire surfaces in summer would enhance the risk of peatland fires, which lead to strong local emissions of CO_2 (Davies et al. 2013) and waterborne DOC (Holden et al. 2007; Clutterbuck and Yallop 2010). At bare peat sites (under peat extraction or crop cultivation), higher wind speeds and rainfall intensities can lead to strong aeolian or water erosion of peat (Warburton 2003).

11.6.2 Climatic Impacts on Biotic Interactions

Climate change is expected to have large impacts on biotic processes in peatlands (e.g. Heijmans et al. 2008; Charman et al. 2013). Higher temperatures will generally increase microbial peat decomposition and carbon mobilisation (Ise et al. 2008), leading to greater concentrations of DOC in peatland surface and soil pore waters (Freeman et al. 2001, Fig. 11.10). The mobilised DOC can be mineralised and released to the atmosphere as CO_2 (e.g. Silvola et al. 1996; Lafleur et al. 2005) or CH_4 (e.g. Dunfield et al. 1993; Moore and Dalva 1993; Daulat and Clymo 1998; Hargreaves and Fowler 1998; Fig. 11.11) depending on the reduction-oxidation status of the organic soils. It can also be laterally exported with the peatland discharge into limnic systems (Tipping et al. 1999; Worrall et al. 2003). Warming-induced increases in soil carbon emissions result partly from the direct temperature effect on soil microbial physiological processes and growth and partly from better substrate availability in the soil pore water caused by higher plant productivity (Mikkelä et al. 1995; Joabsson et al. 1999; Van den Pol-van Dasselaar et al. 1999).

These findings from field and laboratory studies have been incorporated in local, regional and global soil process models (e.g. Walter et al. 2001; van Huissteden and van den Bos 2006; Bohn et al. 2007; Meng et al. 2012). The models predict considerable increases in CH_4 emissions from peatlands over the coming century due to warming as long as wetland area and soil moisture conditions remain unchanged. However, several global models tested within the Wetland and Wetland CH_4 Intercomparison of Models Project (WETCHIMP) predict a decrease in wetland area in the North Sea region in response to higher temperatures, which is likely to lead to lower CH_4 emissions (Melton et al.

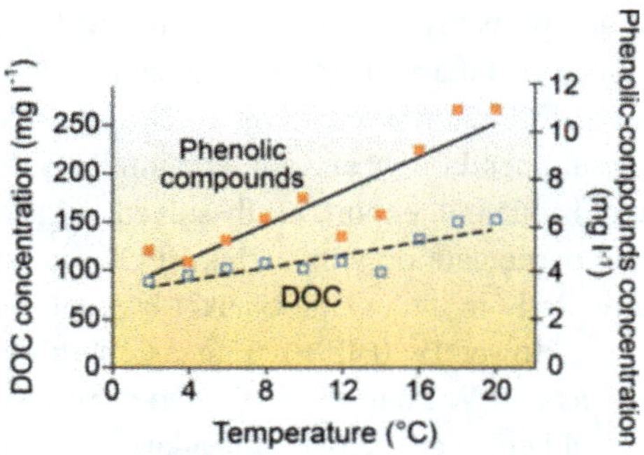

Fig. 11.10 Laboratory observations of increased concentrations of dissolved organic carbon (DOC) and phenolic compounds in peat soil in response to rising temperature (Freeman et al. 2001)

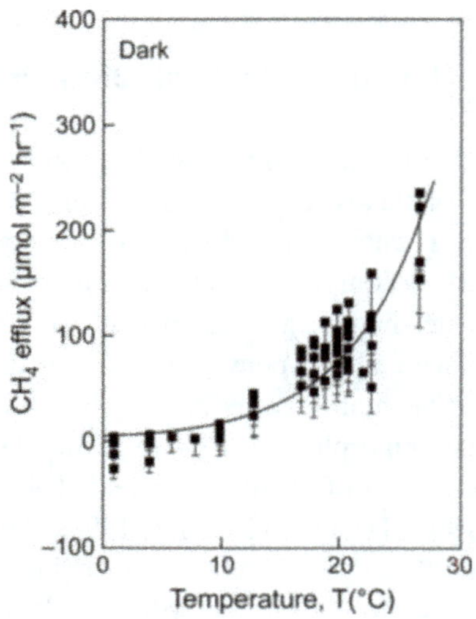

Fig. 11.11 Effects of temperature on methane (CH_4) efflux from 30 cm diameter *Sphagnum papillosum* mire cores. Flux measurements were performed in dark conditions (after Daulat and Clymo 1998)

2013); however, it should be noted that these models show a very wide range of responses.

Whether increased peat decomposition and carbon mobilisation due to higher temperatures leads to lower net ecosystem productivity and to higher net carbon emissions, will depend on the land use of peatlands. For the period 1978–2003, Bellamy et al. (2005) reported carbon losses from all soil types across England and Wales, with particularly strong losses from peat soils. However, it is not clear from such data whether carbon was lost due to climate change or to concomitant changes in atmospheric chemistry and land use (see Sect. 11.4.1 and 11.6.3). Likewise, mapping of peat soils in the Dutch province of Drenthe showed that 42 % of the area of peat soils was converted to mineral soils in the last 30–40 years by carbon loss due to drainage and agricultural management; on average 1 cm peat thickness was lost per year (De Vries et al. 2008). In near-natural peatlands with typical mire vegetation, peat accumulation is expected to increase in response to higher mean annual temperatures because the benefit to primary productivity will be higher than for ecosystem respiration (Loisel et al. 2012; Charman et al. 2013). Longer growing seasons (higher winter temperatures, shorter snow-cover duration) allows the vegetation to take up more photons over the year leading to higher plant productivity and peat accumulation. This increase in net carbon uptake under a warming climate—a negative feedback on climate warming—will be modulated by cloud cover and levels of photosynthetically active radiation (Yu et al. 2010; Loisel et al. 2012; Charman et al. 2013).

If peatland water tables become significantly lower due to increased evapotranspiration and/or decreased summer precipitation, peat decomposition and the release of CO_2 will be enhanced (e.g. Silvola et al. 1996; Laine et al. 2009). Higher N_2O emissions can also be expected with lower peatland water levels (Martikainen et al. 1993; Regina et al. 1996; Goldberg et al. 2010). On the other hand, CH_4 production and emission will decrease, while CH_4 oxidation will be enhanced (e.g. Daulat and Clymo 1998; Hargreaves and Fowler 1998; Nykänen et al. 1998; Le Mer and Roger 2001; Laine et al. 2007, 2009).

The impact of water table drawdown on net ecosystem productivity depends on the response of peatland vegetation, which is difficult to predict and can vary strongly with the micro-topography of mires (Malmer et al. 1994; Strack and Waddington 2007; Lindsay 2010). Since *Sphagnum* mosses have neither roots nor vessels to transport water from deeper soil layers, they rely on high water tables. Water that is lost at the soil surface through evaporation can be replaced by precipitation and capillary rise in *Sphagnum* peat and vegetation. But—depending on the *Sphagnum* species present and the degree of peat decomposition—capillary rise is only efficient in this regard if water tables are not lower than 0.5 m below the land surface (Clymo 1984). Thus, long-lasting drought may damage the vitality of *Sphagnum* mosses (Gerdol et al. 1996; Bragazza 2008; Breeuwer et al. 2009; Robroek et al. 2009).

Medium- to long-term changes in climatic variables such as temperature and precipitation have complex effects on mire vegetation composition that interact with changes in CO_2 concentration and nutrient availability (e.g. Heijmans et al. 2008; Breeuwer et al. 2010, see also Sect. 11.6.3). In a *Sphagnum*-dominated bog in southern Sweden, significant shifts in vegetation composition since the 1950s indicate higher nutrient availability, higher productivity, and drier and shadier conditions due to enhanced tree and shrub growth, which have probably all been triggered by warming (Kapfer et al. 2011). Projected climate change, particularly higher summer temperatures and lower summer precipitation may reduce the bio-climatologically suitable space for blanket

Fig. 11.12 Area covered by the bioclimatic envelope of blanket peatlands as predicted by the model PeatStash using the bioclimatic thresholds associated with the 1961–1990 baseline climate for the UKCIP02 high and low emissions scenarios ('High' and 'Low', respectively) for three time periods: 2020s, 2050s and 2080s. An Ordnance Survey/EDINA supplied service (© Crown Copyright/database 2009) and Met Office/UKCIP gridded climate data (UKCIP02 © Crown Copyright 2002) (Gallego-Sala et al. 2010)

bogs in the UK (Clark et al. 2010; Gallego-Sala et al. 2010; Gallego-Sala and Prentice 2013, Fig. 11.12). Berry and Butt (2002) suggested that the dominant species in lowland raised bogs of Scotland would find suitable conditions, but that some rare species will probably lose suitable climate space under the expected future climatic conditions. Since lower summer precipitation and higher summer temperatures are projected for the entire North Sea region (Chap. 5), mires on the European continent—especially ombrogenous mires—are likely to experience falling summer water levels with negative impacts on typical mire plant communities. However, Lindsay (2010) cautioned that the climate envelope approaches applied might underestimate the importance of local atmospheric humidity, cloud cover and mist frequency for mire occurrence. Other studies even propose—on the basis of biogeographical, bioclimatological and ecophysiological reasoning—that northern oceanic peatlands will expand owing to the increased oceanicity of the future climate (Crawford 2000; Crawford et al. 2003). Lindsay (2010) also stated that mires with a well-developed acrotelm and microtopography might be able to react to climate change without losing their ability to grow and sequester carbon. Mires that are already degraded will have much less resilience to the projected changes in climate (Lindsay 2010).

11.6.3 Competing Effects of Climate Change and Other Influences

In addition to the changes in climate, there have also been changes in the intensity of land use and atmospheric chemistry (such as CO_2 concentration, and atmospheric deposition of nitrogen and sulphur) over recent decades, with strong impacts on near-natural and degraded peatlands. These may intensify, mask or reverse the effects of climate change on peatlands.

11.6.3.1 Atmospheric Chemistry

Research in *Sphagnum* bogs found that elevated atmospheric CO_2 concentrations did not affect NPP due to the strong N-limitation of these ecosystems (Berendse et al. 2001). However, CH_4 emission (Dacey et al. 1994; Hutchin et al. 1995) and DOC export (Freeman et al. 2004; Van Groenigen et al. 2011) from peatlands were enhanced under elevated CO_2 concentrations. Increased N-deposition promotes microbial peat decomposition and thus CO_2 and CH_4 emissions (Aerts et al. 1992; Aerts and de Caluwe 1999) and DOC export fluxes (Bragazza et al. 2006). Large shifts in mire vegetation composition may also occur in response to elevated CO_2 concentrations and increased N-deposition

(Van der Heijden et al. 2000; Berendse et al. 2001; Fenner et al. 2007; Heijmans et al. 2008). Deposition of nitrogen and sulphur leads to acidification of top soils and thus changes the solubility and mobilisation of dissolved organic matter. The rise in DOC concentrations in limnic water bodies observed in the latter half of the 20th century in Great Britain and Sweden seems to have been mainly driven by decreasing S-deposition with the warming effect of minor importance (Freeman et al. 2001; Worrall et al. 2002; Evans et al. 2005, 2006a, 2007; Monteith et al. 2007; Erlandsson et al. 2008). However, S-deposition has also been shown to suppress CH_4 emissions because sulphate reduction is energetically favourable compared to methanogenesis (Dise and Verry 2001). Heavy air pollution can even lead to die-off of *Sphagnum* mosses, triggering for example peat erosion in blanket mires (e.g. Tallis 1985).

11.6.3.2 Land Use

Drained and degraded peatlands are hotspots of greenhouse gas emissions (e.g. Oleszczuk et al. 2008; Couwenberg 2011; Joosten et al. 2012). Mineralisation of peat organic matter and the respective CO_2 emissions are strongly related to drainage depth and management intensity (Aerts and Ludwig 1997; Dirks et al. 2000; Beetz et al. 2013; Leiber-Sauheitl et al. 2013; Schrier-Uijl et al. 2014). A project on peatlands in Germany has shown that the annual greenhouse gas balance of managed peatland areas can be estimated well from two predictor variables—mean annual water level and carbon exported by harvest—which together can be used as a proxy for management intensity (Drösler et al. 2011, 2012). Within the set of managed peatlands, greenhouse gas emissions from deeply-drained peatlands such as cropland or intensively-used pastureland are especially large (Veenendaal et al. 2007; Drösler et al. 2011; Elsgaard et al. 2012; Leiber-Sauheitl et al. 2013). Greenhouse gas emissions can stay high at such sites even when management intensity is moderated by nature conservancy measures (Best and Jacobs 1997; Schrier-Uijl et al. 2010; Hahn-Schöfl et al. 2011). Intensively-used peatlands are more common in lowlands than in uplands due to better accessibility and suitability for high-intensity agriculture. Burning peatland vegetation as a management practice in the UK may strongly affect carbon sequestration and dissolved organic matter export (Garnett et al. 2000; Clutterbuck and Yallop 2010; Yallop et al. 2010) although the evidence is not conclusive (cf. Worrall et al. 2007; Clay et al. 2009; Allen et al. 2013). It should be stressed that change in land use is the primary driver of changes in peatland hydrology and biogeochemistry, and probably has a stronger impact than climate change. However, some climate change effects will exacerbate the impact of human activities such as drainage, grazing, burning and peat mining (e.g. Petrescu et al. 2009). On the other hand, land use-related effects on peatlands often make them more vulnerable to climate change impacts (e.g. Parish et al. 2008).

11.7 Inland Ecosystems and the Wider North Sea System

Inland ecosystems have important functions within the coupled land-ocean-atmosphere system of the North Sea region. Major functions of inland ecosystems are freshwater storage and transmission, carbon storage, carbon sequestration, greenhouse gas emission and the export of dissolved and particulate organic matter to aquatic systems. While forests in the North Sea region currently sequester carbon and act as greenhouse gas sinks (Ciais et al. 2008; Luyssaert et al. 2010), agricultural systems are greenhouse gas sources through CO_2 and N_2O emissions from soils and CH_4 emissions from enteric fermentation of livestock and manure management (Schulze et al. 2009). Greenhouse gas emissions from degraded and agriculturally-used peatlands are significant in several countries of the North Sea region when compared to their total national greenhouse gas emissions, with contributions of about 5 % in Germany and Denmark, 2–3 % in the Netherlands and about 1 % in the UK (Cannell et al. 1999; Van den Bos 2003; Drösler et al. 2008, 2011; Verhagen et al. 2009; Joosten 2010; Worrall et al. 2011; Nielsen et al. 2013). On the other hand, CO_2 uptake by the few remaining near-natural peatlands in the North Sea region is negligible compared to CO_2 release by degraded peatlands or CO_2 uptake by forests. This means that reducing emissions from reclaimed peatlands is more important than the possible contribution of natural peatland to carbon sequestration.

The export of dissolved and particulate organic matter from inland ecosystems has important effects on the biogeochemistry and ecology of the receiving aquatic systems (i.e. lakes, rivers, estuaries and the North Sea) and supplies them with inputs of carbon, nitrogen, phosphorus and other important nutrient elements (e.g. Evans et al. 2005). Because export of DOC and POC is controlled by many interacting factors (e.g. temperature, nutrient supply, precipitation, evapotranspiration, run-off), its future behaviour is difficult to predict. Run-off is projected to increase in the northern part of the North Sea region and to decrease in the south (Alcamo et al. 2007). However, due to the projected warming and higher frequency of heavy rain events in the North Sea region (Chap. 5), enhanced mobilisation of soil organic matter and transport of terrestrial DOC to the limnic ecosystems and the North Sea are likely. Dissolved organic matter affects ecosystem nutrient availability (Carpenter et al. 2005), acidification of limnic systems (Oliver et al.

1983) and solubility, transport and toxicity of heavy metals and organic pollutants (Carter and Suffet 1982; Pokrovsky et al. 2005). It also regulates the photochemistry of natural waters (Zafiriou et al. 1984) and influences aquatic production of algae and bacteria (Wetzel 1992; Carpenter and Pace 1997). The export of organic matter into limnic systems can affect human health adversely since these organic substances support bacterial proliferation and lead to the formation of carcinogens when they react with disinfectants (such as chlorine) during water treatment (Nokes et al. 1999; Sadiq and Rodriguez 2004). The magnitude of DOC fluxes in rivers correlates with organic matter storage in the soils of their catchments (e.g. Hope et al. 1997). Riverine organic matter is modified strongly and largely removed through mineralisation and sedimentation during transport in rivers and estuaries (e.g. Raymond and Bauer 2000; Wiegner and Seitzinger 2001; Abril et al. 2002; Raymond et al. 2013). Thomas et al. (2005) estimated that about one million tons of DOC and POC are transported into the North Sea by rivers each year. Only 10 % of the riverine input of organic carbon is probably buried in the shelf sediments (Hedges et al. 1997; Schlünz and Schneider 2000), with the rest incorporated in the food webs of coastal seas.

Flood risk mitigation is an important issue in coastal and fluvial lowlands bordering the North Sea, especially given the projected acceleration in sea-level rise in the future due to climate change (Chap. 5). Peat soil degradation causes land subsidence by a combination of peat oxidation and compaction after drainage (Schothorst 1977). Historical subsidence—caused by drainage since medieval times—often combined with peat extraction for fuel, in coastal peatlands of the Netherlands, Germany and eastern Britain may have resulted in up to several metres of subsidence (Godwin 1978; Borger 1992; Verhoeven 1992; Hoogland et al. 2012). In the eastern British fenlands, compaction and peat oxidation has resulted in up to 4 m of subsidence in 150 years (Godwin 1978). In Dutch managed peatlands, subsidence is ongoing at up to one centimetre per year (Hoogland et al. 2012, and references therein). Under a warmer climate, peat decomposition would be even faster, particularly in drained peatlands. This would increase flood risk, induce costs for creating and managing flood protection systems and ever deeper drainage, and threaten the economic viability of agriculture. Subsidence also influences peatland hydrology and hydrochemistry. The need for increasingly deeper drainage enhances the upwelling of sulphate-rich brackish or salt water (Hoogland et al. 2012). This in turn may enhance peat decomposition by sulphate reduction, with adverse impacts on water quality by increasing dissolved and particulate organic matter and nutrient mobilisation (Smolders et al. 2006). Replenishing surface water with alkaline river water in agriculturally managed peatlands in dry periods may have a similar effect on peat decomposition.

11.8 Summary

The expected future impacts of climate change on terrestrial ecosystems are summarised in Table 11.1.

Future climate change is likely to increase NPP in the North Sea region due to warmer conditions and longer growing seasons, at least if future climate change is moderate and summer precipitation does not decrease as strongly as projected in some of the more extreme climate scenarios. The physiological effects of increasing atmospheric CO_2 levels and increasing N-mineralisation in the soil may also play a significant role, but to an as yet uncertain extent.

The effects of total carbon storage in terrestrial ecosystems are highly uncertain, due to the inherent complexity of the processes involved. For example, water table effects in mires, large uncertainties in soil carbon modelling, the unknown fate of additional carbon taken up through CO_2 fertilisation, and other important drivers, such as changes in land use (e.g. forest harvest and wetland drainage). For moderate climate change, land use effects are often more important drivers of total ecosystem carbon accumulation than climate change.

Across a wide range of organism groups, range expansions to higher latitudes and altitudes, changes in phenology, and in the case of butterflies and birds, population increases in warm-adapted species and decreases in cold-adapted species have occurred in response to recent climate change. Regarding range expansions, some studies suggest substantial differences between organism groups; for example, herbaceous plants show only small or no responses while variability within other groups is large. Habitat specialists with restricted ranges have generally responded very little or even shown range contractions. Many of these often already threatened species could therefore be particularly vulnerable to climate change. Cold-adapted mountain top species are at particular risk because they have very limited habitat space in which to track climate change.

Overall effects of recent climate change on forest ecosystems within the region are limited, and major impacts on forest type distribution and forest functioning are unlikely if future warming is moderate and summer precipitation does not decrease as much as is projected in some of the more extreme climate scenarios. However, current models simulating potential impacts of climate change on forests rarely include a number of drivers of potentially rapid changes in forest functioning, such as forest pests and diseases (e.g. Kirilenko and Sedjo 2007; Jönsson et al. 2009). As a result, projections of climate-driven changes in future forest productivity, biomass and carbon storage are highly uncertain.

For grasslands, significant range expansion of thermophilous animal species (e.g. Parmesan et al. 1999), changes in flowering phenology (e.g. Fitter and Fitter 2002), and population increases (e.g. Poniatowski et al. 2012) are

Table 11.1 Climate change impacts on terrestrial ecosystems of the North Sea region

Class of impact	Impact of recent climate change	Projected impact of future climate change	Uncertainties
Phenology			
Shift towards earlier spring and summer phases in plants	Spring events advanced on average by 6.3 days ✓✓✓	Further advancement depending on temperature increase ✓✓	Modified responses due to non-linear effects of further increasing temperatures early in the year
Shift towards later autumn phases in plants	Autumn events delayed by on average 4.5 days ✓✓	Further delay depending on temperature increase ✓	Limited data quality
Extension of growing period	Extension of growing season by about 20 days ✓✓	Further extension depending on temperature increase	Limited data quality
Earlier onset of reproduction in animals	Earlier onset of first spawning in amphibians by 10–20 days ✓✓✓	No studies	
	Advances of dates of first clutches in bird species by on average 8.8 days ✓✓✓		
Changed migratory patterns and behaviour	Advances in the arrival of migratory birds ✓✓✓	No studies	
	Shift in winter distribution of waterfowl and waders to the North-East ✓✓✓		
Biogeography and community structure			
Range shifts in vascular plants and cryptogams	Plants: range extensions lagging warming due to dispersal and/or habitat limitation ✓✓	Strongly limited range filling due to dispersal limitation in fragmented landscapes ✓✓	Impact of other abiotic factors (e.g. nutrients)
			Dispersal and recruitment limitation
			Habitat limitation and landscape fragmentation
			Changed biotic interactions (e.g. competition, herbivory, pathogens)
			Changing land-use and disturbance regimes
			Impact of climate extremes
	Lichens: cold-adapted species declining and warm-adapted expanding their ranges northwards	Decline of cold-adapted species at the rear edge and at lower mountain elevations ✓	Poorly known phenotypic plasticity and evolutionary capacity
Range shifts in animals	Substantial range extension to the north in many mobile, generalist animal species ✓✓✓	Continuing range expansion to the north in mobile, generalist species ✓✓	Impact of other abiotic factors (e.g. nutrients)
			Dispersal and recruitment limitation
			Habitat limitation and landscape fragmentation
			Changed biotic interactions (e.g. competition, herbivory, pathogens)
			Changing land-use and disturbance regimes
	No or minor range extension or range contraction in many habitat specialists		Impact of climate extremes
	Decline in some northern species in the south	Decline of cold-adapted species at the rear edge ✓	Poorly known phenotypic plasticity and evolutionary capacity

(continued)

Table 11.1 (continued)

Class of impact	Impact of recent climate change	Projected impact of future climate change	Uncertainties
Changed composition in plant communities	Limited evidence for primarily climate-induced changes so far ✓	Relatively slow shifts in species composition	Impact of other abiotic factors (e.g. nutrients)
			Dispersal and recruitment limitation
			Habitat limitation and landscape fragmentation
			Changed biotic interactions (e.g. competition, herbivory, pathogens)
			Changing land-use and disturbance regimes
			Impact of climate extremes
			Poorly known phenotypic plasticity and evolutionary capacity
			Few studies, 'new' biotic interactions
Changed biome distribution	Upward shift of tree-line in the southern Scandes into arctic-alpine Tundra ecosystems ✓✓	Moderate shifts mostly in the northern part of the region between nemoral and boreal forests (spread of broadleaved trees) and boreal forests and arctic-alpine tundra (spread of shrubs and trees) ✓✓	Impact of other abiotic factors (e.g. nutrients)
			Dispersal and recruitment limitation
			Habitat limitation and landscape fragmentation
			Changed biotic interactions (e.g. competition, herbivory, pathogens)
			Changing land-use and disturbance regimes
			Impact of climate extremes
		Reduction of bioclimatologically suitable space for blanket bogs ✓	Poorly known phenotypic plasticity and evolutionary capacity
Biotic mismatch	Few well documented examples for birds and insects ✓	Increasing spatial mismatches between butterflies and host plants, particularly for those butterflies that are already constrained by specific host plants ✓	
Physiological tolerance and stress			
Tree stress and forest dieback	Limited evidence for drought stress in southern part of the region ✓	Increasing drought risk especially in the southern part of the region ✓	Ecological complexity and lack of process understanding
			Other abiotic factors
	Role of pests and pathogens still rather uncertain. ✓	Increasing risk of pathogens and pests ✓	Species-specific reactions and genetic variability within species

(continued)

Table 11.1 (continued)

Class of impact	Impact of recent climate change	Projected impact of future climate change	Uncertainties
Ecosystem functioning			
Net primary productivity and forest growth	Increased growing season length and NPP ✓✓✓	Further increase of NPP especially in the north of the region ✓✓	Impact of other drivers such as N-deposition and enhanced mineralisation, acidification and potential CO_2 'fertilisation'
		Mixed effects on NPP in the southern part of the region, depending on water supply ✓	Impacts of drought stress and disturbance events poorly captured in vegetation
	Increasing forest growth especially in northern regions and at sites without moisture limitation ✓✓✓	Increased NPP in mires and drained peatlands ✓	Tree species selection and forest management practices
Carbon sequestration capacity	Enhanced vegetation carbon fixation due to increased forest growth ✓✓	Northern areas remain net carbon sinks, southern areas may eventually turn into small to moderate sources ✓	Relative importance and interplay of raised soil respiration and NPP
			Feedbacks of changed hydrology on carbon exchange in wetlands
			Human alterations in land cover and land use
		Increased peat accumulation in mires, accelerated peat decay in drained peatlands✓	Only one regional modelling study included total terrestrial carbon cycle
Greenhouse gas release from mineral and organic soils	Increased soil respiration and CH_4 release from hydrologically intact peatlands. ✓✓	Enhanced carbon release especially from desiccated peat soils and other humus-rich soils ✓✓	Unpredictable hydrological changes associated with climate warming
			Competing effects of climate change and other influences (atmospheric chemistry, land use)
	Increased C-release from desiccated and degraded peat soils, especially in the south ✓✓		Large uncertainties in soil carbon modelling
Lateral waterborne export fluxes of elements	Increased run-off in northern part, decreased run-off in the southern North Sea region ✓✓	Further enhanced mobilisation of DOC, especially from drained peatlands ✓✓	Unpredictable hydrological changes associated with climate warming
			Competing effects of climate change and other influences (atmospheric chemistry, land use)
	Increased export of DOC ✓		Large uncertainties in soil carbon modelling

✓✓✓ strong evidence; ✓✓ moderate evidence; ✓ minor evidence

currently more obvious signs of climate change, than changes in plant community composition and ecosystem processes. Even in experimental studies simulating drought and warming, responses to treatments were modest (Bates et al. 2005; Grime et al. 2008; Kreyling 2010). Evidence from observational and correlative studies is weak and speculative due to many confounding effects such as eutrophication and changes in management practice (e.g. Gaudnik et al. 2011; McGovern et al. 2011).

For heathlands, overall evidence for effects of recent climate change from experimental warming and drought treatments is also weak, variable and inconsistent, suggesting that now and in the near future, climate warming is of low significance compared to other predominant drivers of ecological change in heathland ecosystems such as eutrophication, acidification and altered management practices (e.g. Härdtle et al. 2006). For more extreme climate scenarios, however, substantial effects could be expected in heathlands. Projections of the exact nature of these future effects are highly uncertain.

The projected climatic changes for the North Sea region are likely to have significant impacts on abiotic and biotic processes in mires and drained peatlands. However, the consequences will vary widely between mires and drained peatlands. Higher temperatures and longer growing seasons will increase NPP, but also ecosystem respiration, CH_4 emission and DOC export in mires and drained peatlands. The net effect is expected to result in increased peat accumulation in mires but accelerating peat decay in drained peatlands. In mires, lower water tables due to less summer precipitation and/or higher evapotranspiration will enhance NPP but also—and to a much greater degree—ecosystem respiration, leading to a net loss of peat organic matter and the release of CO_2. On the other hand, CH_4 emission will also be reduced, while effects on DOC export are less clear. In drained peatlands, climatic changes will have less effect on the water budget and biogeochemical fluxes since water tables are regulated.

Low summer precipitation and/or high evapotranspiration can make conditions unsuitable for some mire types. However, well-developed natural mires may have considerable resilience to climate change. The status of peatlands, namely the level of drainage and soil degradation will determine whether peatlands mitigate or exacerbate climate change.

Besides their function as a sink for atmospheric carbon, the export of dissolved and particulate organic carbon and nutrients from terrestrial ecosystems is probably the most significant process directly affecting the North Sea system. Because this export is controlled by many interrelating factors (temperature, precipitation, evapotranspiration, run-off, human impact), its future development is very uncertain and therefore difficult to predict.

Acknowledgements We thank Eva-Maria Gerstner, Senckenberg Biodiversity and Climate Research Centre (BiK-F) for help with the creation of Figs. 11.1 and 11.2.

References

Abril G, Nogueira E, Hetcheber H, Cabec-adas G, Lemaire E, Brogueira MJ (2002) Behaviour of organic carbon in nine contrasting European estuaries. Estuar Coast Shelf Sci 54:241–262

Aerts R, de Caluwe H (1999) Nitrogen deposition effects on carbon dioxide and methane emissions from temperate peatland soils. Oikos 84:44–54

Aerts R, Ludwig F (1997) Water-table changes and nutritional status affect trace gas emissions from laboratory columns of peatland soils. Soil Biol Biochem 29:1691–1698

Aerts R, Wallen B, Malmer N (1992) Growth-limiting nutrients in *Sphagnum*-dominated bogs subject to low and high atmospheric nitrogen supply. J Ecol 80:131–140

Ainsworth EA, Long SP (2005) What have we learned from 15 years of free-air CO_2 enrichment (FACE)? A meta-analytic review of the responses of photosynthesis, canopy properties and plant production to rising CO_2. New Phytol 165:351–372

Alberto FJ, Aitken SN, Alia R, Gonzalez-Martinez SC, Hanninen H, Kremer A, Lefevre F, Lenormand T, Yeaman S, Whetten R, Savolainen O (2013) Potential for evolutionary responses to climate change evidence from tree populations. Glob Change Biol 19:1645–1661

Alcamo J, Flörke M, Märker M (2007) Future long-term changes in global water resources driven by socio-economic and climatic changes. Hydrol Sci J 52:247–275

Allen KA, Harris MPK, Marrs RH (2013) Matrix modelling of prescribed burning in *Calluna vulgaris*-dominated moorland: short burning rotations minimize carbon loss at increased wildfire frequencies. J Appl Ecol 50:614–624

Arp WJ, Van Mierlo JEM, Berendse F, Snijders W (1998) Interactions between elevated CO_2 concentration, nitrogen and water: effects on growth and water use of six perennial plant species. Plant Cell Environ 21:1–11

Bates JW, Thompson K, Grime JP (2005) Effects of simulated long-term climatic change on the bryophytes of a limestone grassland community. Glob Change Biol 11:757–769

Beebee TJC (1995) Amphibian breeding and climate. Nature 374:219–220

Beetz S, Liebersbach H, Glatzel S, Jurasinski G, Buczko U, Höper H (2013) Effects of land use intensity on the full greenhouse gas balance in an Atlantic peat bog. Biogeosciences 10:1067–1082

Bellamy PH, Loveland PJ, Bradley RI, Lark RM, Kirk GJD (2005) Carbon losses from all soils across England and Wales 1978-2003. Nature 437:245–248

Belyea LR, Malmer N (2004) Carbon sequestration in peatland: patterns and mechanisms in response to climate change. Glob Change Biol 10:1043–1052

Berendse F, Van Breemen N, Rydin H, Buttler A, Heijmans M, Hoosbeek MR, Lee JA, Mitchell E, Saarinen T, Vasander H, Wallén B (2001) Raised atmospheric CO_2 levels and increased N deposition cause shifts in plant species composition and production in *Sphagnum* bogs. Glob Change Biol 7:591–598

Berger S, Soehlke G, Walther GR, Pott R (2007) Bioclimatic limits and range shifts of cold-hardy evergreen broad-leaved species at their northern distributional limit in Europe. Phytocoenologia 37:523–539

Berry PM, Butt N (2002) CHIRP – Climate change impacts on raised bogs: a case study of Thorne Crowle, Goole and Hatfield Moors. English Nature Research Report No 457

Berry PM, Dawson TP, Harrison PA, Pearson RG (2002) Modelling potential impacts of climate change on the bioclimatic envelope of species in Britain and Ireland. Glob Ecol Biogeogr 11:453–462

Bertrand R, Lenoir J, Piedallu C, Riofrio-Dillon G, de Ruffray P, Vidal C, Pierrat JC, Gegout JC (2011) Changes in plant community composition lag behind climate warming in lowland forests. Nature 479:517–520

Best EPH, Jacobs FHH (1997) The influence of raised water table levels on carbon dioxide and methane production in ditch-dissected peat grasslands in the Netherlands. Ecol Engineer 8:129–144

Betts RA, Boucher O, Collins M, Cox PM, Falloon PD, Gedney N, Hemming DL, Huntingford C, Jones CD, Sexton DMH, Webb MJ (2007) Projected increase in continental runoff due to plant responses to increasing carbon dioxide. Nature 448:1037–1041

Biesmeijer JC, Roberts SPM, Reemer M, Ohlemuller R, Edwards M, Peeters T, Schaffers AP, Potts SG, Kleukers R, Thomas CD, Settele J, Kunin WE (2006) Parallel declines in pollinators and insect-pollinated plants in Britain and the Netherlands. Science 313:351–354

Billett MF, Palmer SM, Hope D, Deacon C, Storeton-West R, Hargreaves KJ, Flechard C, Fowler D (2004) Linking land-atmosphere-stream carbon fluxes in a lowland peatland system. Glob Biogeochem Cy 18:GB1024

Billett MF, Charman D, Clark JM, Evans CD, Evans MG, Ostle NG, Worrall F, Burden A, Dinsmore KJ, Jones T, McNamara NP, Parry L, Rowson JG, Rose R (2010) Carbon balance of UK peatlands: current state of knowledge and future research challenges. Climate Res 45:13–29

Blodau C (2002) Carbon cycling in peatlands: A review of processes and controls. Environ Rev 10:111–134

Bohn U, Neuhäusle R, Gollub G, Hettwer C, Neuhäuslová Z, Raus T, Schlüter H, Weber H (2003) Map of the natural vegetation of Europe. Landwirtschaftsverlag, Münster

Bohn TJ, Lettenmaier DP, Sathulur K, Bowling LC, Podest E, McDonald KC, Friborg T (2007) Methane emissions from western Siberian wetlands: heterogeneity and sensitivity to climate change. Env Res Lett 2:045015

Bolte A, Czajkowski T, Kompa T (2007) The north-eastern distribution area of European beech - A review. Forestry 80:413–429

Bolte A, Ammer C, Löf M, Madsen P, Nabuurs GJ, Schall P, Spathelf P, Rock J (2009) Adaptive forest management in central Europe: Climate change impacts, strategies and integrative concept. Scand J Forest Res 24:473–482

Bolte A, Hilbrig L, Grundmann B, Kampf F, Brunet J, Roloff A (2010) Climate change impacts on stand structure and competitive interactions in a southern Swedish spruce–beech forest. Eur J For Res 129:261–276

Borger GJ (1992) Draining, digging, dredging; the creation of a new landscape in the peat areas of the low countries. In: Verhoeven JTA (ed) Fens and Bogs in the Netherlands: Vegetation, history, nutrient dynamics and conservation. Kluwer

Both C, Visser ME (2001) Adjustment to climate change is constrained by arrival date in long-distance migrant bird. Nature 411:296–298

Both C, Artemyev AV, Blaauw B et al (2004) Large-scale geographical variation confirms that climate change causes birds to lay earlier. P Roy Soc Lond B 271:1657–1662

Both C, Bouwhuis S, Lessells CM, Visser ME (2006) Climate change and population declines in a long-distance migratory bird. Nature 441:81–83

Both C, Van Asch M, Bijlsma RG, Van den Burg AB, Visser ME (2009) Climate change and unequal phenological changes across four trophic levels: constraints or adaptations. J Anim Ecol 78:73–83

Both C, Van Turnhout CAM, Bijlsma RG, Siepel H, Van Strien AJ, Foppen RPB (2010) Avian population consequences of climate change are most severe for long-distance migrants in seasonal habitats. P Roy Soc B 277:1259–1266

Bower M (1960) The erosion of blanket peat in the southern Pennines. East Midland. Geographer 13:22–33

Bower MM (1961) The distribution of erosion in blanket peat bogs in the Pennines. Trans Inst Brit Geogr 29:17–30

Bowman WD, Cleveland CC, Halada L, Hresko J, Baron JS (2008) Negative impact of nitrogen deposition on soil buffering capacity. Nat Geosci 1:767–770

Bragazza L (2008) A climatic threshold triggers the die-off of peat mosses during an extreme heat wave. Glob Change Biol 14:2688–2695

Bragazza L, Freeman C, Jones T, Rydin H, Limpens J, Fenner N, Ellis T, Gerdol R, Hajek M, Hajek T, Iacumin P, Kutnar L, Tahvanainen T, Toberman H (2006) Atmospheric nitrogen deposition promotes carbon loss from peat bogs. PNAS 103:19386–19389

Bragg OM (2002) Hydrology of peat-forming wetlands in Scotland. Sci Tot Environ 294:111–129

Breeuwer A, Robroek BJ, Limpens J, Heijmans MM, Schouten MG, Berendse F (2009) Decreased summer water table depth affects peatland vegetation. Basic Appl Ecol 10:330–339

Breeuwer A, Heijmans MMPD, Robroek BJM, Berendse F (2010) Field simulation of global change: transplanting northern bog mesocosms southward. Ecosystems 13:712–726

Bridgham SD, Pastor J, Updegraff K, Malterer TJ, Johnson K, Harth C, Chen J (1999) Ecosystem control over temperature and energy flux in northern peatlands. Ecol Appl 9:1345–1358

Broadmeadow MSJ, Ray D, Samuel CJA (2005) Climate change and the future for broadleaved tree species in Britain. Forestry 78:145–161

Buckland SM, Grime JP, Hodgson JG, Thompson K (1997) A comparison of plant responses to the extreme drought of 1995 in Northern England. J Ecol 85:875–882

Buckland SM, Thompson K, Hodgson JG, Grime JP (2001) Grassland invasions: effects of manipulations of climate and management. J Appl Ecol 38:301–309

Bullock JM, Jefferson RG, Blackstock TH, Pakeman RJ, Emmett BA, Pywell RF, Grime JP, Silvertown J (2011) Semi-natural grasslands. The UK National Ecosystem Assessment Technical Report. UNEP-WCMC, Cambridge

Bullock JM, White SM, Prudhomme C, Tansey C, Perea R, Hooftman DAP (2012) Modelling spread of British wind-dispersed plants under future wind speeds in a changing climate. J Ecol 100:104–115

Cannell MGR, Milne R, Hargreaves KJ, Brown TAW, Cruickshank MM, Bradley RI, Spencer T, Hope D, Billett MF, Adger WN, Subak S (1999) National inventories of terrestrial carbon sources and sinks: the UK experience. Climatic Change 42:505–530

Carpenter SR, Pace ML (1997) Dystrophy and eutrophy in lake ecosystems: implications of fluctuating inputs. Oikos 78:3–14

Carpenter SR, Cole JJ, Pace ML, Van de Bogert M, Bade DL, Bastviken D, Gille CM, Hodgson JR, Kitchell JF, Kritzberg ES (2005) Ecosystem subsidies: terrestrial support of aquatic food webs from ^{13}C addition to contrasting lakes. Ecology 86:2737–2750

Carroll EA, Sparks TH, Collinson N, Beebee TJC (2009) Influence of temperature on the spatial distribution of first spawning dates of the common frog (*Rana temporaria*) in the UK. Glob Change Biol 15:467–473

Carter CW, Suffet IH (1982) Binding of DDT to dissolved humic materials. Environ Sci Technol 16:735–740

Charman DJ, Beilman DW, Blaauw M et al (2013) Climate-related changes in peatland carbon accumulation during the last millennium. Biogeosciences 10:929–944

Chen IC, Hill JK, Ohlemüller R, Roy DB, Thomas CD (2011) Rapid range shifts of species associated with high levels of climate warming. Science 333:1024–1026

Christensen JH, Carter TR, Rummukainen M, Amanatidis G (2007) Evaluating the performance and utility of regional climate models: the PRUDENCE project. Climatic Change 81:1–6

Ciais P, Schelhaas MJ, Zaehle S, Piao SL, Cescatti A, Liski J, Luyssaert S, Le-Maire G, Schulze ED, Bouriaud O, Freibauer A, Valentini R, Nabuurs GJ (2008) Carbon accumulation in European forests. Nat Geosci 1:425–429

Clark J, Gallego-Sala AV, Allott TEH, Chapman SJ, Farewell T, Freeman C, House JI, Orr HG, Prentice IC, Smith P (2010) Assessing the vulnerability of blanket peat to climate change using an ensemble of statistical bioclimatic envelope models. Climate Res 45:131–150

Clay GD, Worrall F, Fraser ED (2009) Effects of managed burning upon dissolved organic carbon (DOC) in soil water and runoff water following a managed burn of a UK blanket bog. J Hydrol 367:41–51

Clutterbuck B, Yallop AR (2010) Land management as a factor controlling dissolved organic carbon releases from upland peat soils 2: changes in DOC productivity over four decades. Sci Tot Environ 408:6179–6191

Clymo RS (1984) The limits to peat bog growth. P Trans Roy Soc B 303:605–654

Collins M, Knutti R, Arblaster J, Dufresne J-L, Fichefet T, Friedlingstein P, Gao X, Gutowski WJ, Johns T, Krinner G, Shongwe M, Tebaldi C, Weaver AJ, Wehner M (2013) Long-term climate change: Projections, commitments and irreversibility. In: Stocker TF, Qin D, Plattner G-K, Tignor M, Allen SK, Doschung J, Nauels A, Xia Y, Bex V, Midgley PM (eds), Climate Change 2013: The Physical Science Basis. Contribution of Working Group I to the Fifth Assessment Report of the Intergovernmental Panel on Climate Change. Cambridge University Press, pp 1029-1136

Cotton PA (2003) Avian migration phenology and global climate change. Proc Nat Acad Sci USA 100:12219–12222

Couwenberg J (2011) Greenhouse gas emissions from managed peat soils: is the IPCC reporting guidance realistic? Mires and Peat 8:1–10

Crawford RMM (2000) Tansley Review No. 114: Ecological hazards of oceanic environments. New Phytol 147:257–281

Crawford RMM, Jeffree CE, Rees WG (2003) Paludification and forest retreat in northern oceanic environments. Ann Bot 91:213–226

Crick HQ, Dudley C, Glue DE (1997) UK birds are laying eggs earlier. Nature 388:526

Crimmins SM, Dobrowski SZ, Greenberg JA, Abatzoglou JT, Mynsberge AR (2011) Changes in climatic water balance drive downhill shifts in plant species' optimum elevations. Science 331:324–327

Dacey JWH, Drake BG, Klug MJ (1994) Stimulation of methane emission by carbon dioxide enrichment of marsh vegetation. Nature 370:47–49

Daulat WE, Clymo RS (1998) Effects of temperature and water table on the efflux of methane from peatland surface cores. Atmos Environ 32:3207–3218

Davidson EA, Janssens IA (2006) Temperature sensitivity of soil carbon decomposition and feedbacks to climate change. Nature 440:165–173

Davies GM, Gray A, Rein G, Legg CJ (2013). Peat consumption and carbon loss due to smouldering wildfire in a temperate peatland. Forest Ecol Manage 308:169–177

De Vries F, Hendriks RFA, Kemmers RH, Wolleswinkel R (2008) Het veen verdwijnt uit Drenthe. Omvang, oorzaken en gevolgen. Alterra-rapport 1661. Alterra, Wageningen (in Dutch)

Devictor V, van Swaay C, Brereton T, Brotons L, Chamberlain D, Heliola J, Herrando S, Julliard R, Kuussaari M, Lindstrom A, Reif J, Roy DB, Schweiger O, Settele J, Stefanescu C, Van Strien A, Van Turnhout C, Vermouzek Z, WallisDeVries M, Wynhoff I, Jiguet F (2012) Differences in the climatic debts of birds and butterflies at a continental scale. Nat Clim Change 2:121–124

Dinsmore KJ, Billett MF, Skiba UM, Rees RM, Drewer J, Helfter C (2010) Role of the aquatic pathway in the carbon and greenhouse gas budgets of a peatland catchment. Glob Change Biol 16:2750–2762

Dirks BOM, Hensen A, Goudriaan J (2000) Effect of drainage on CO_2 exchange patterns in an intensively managed peat pasture. Climate Res 14:57–63

Dise NB (2009) Peatland response to global change. Science 326:810–811

Dise NB, Verry ES (2001) Suppression of peatland methane emission by cumulative sulfate deposition in simulated acid rain. Biogeochemistry 53:143–160

Doswald N, Willis SG, Collingham YC, Pain DJ, Green RE, Huntley B (2009) Potential impacts of climatic change on the breeding and non-breeding ranges and migration distance of European Sylvia warblers. J Biogeogr 36:1194–1208

Doxford SW, Freckleton RP (2012) Changes in the large-scale distribution of plants: extinction, colonisation and the effects of climate. J Ecol 100:519–529

Drösler M, Freibauer A, ChristensenT, Friborg T (2008) Observation and status of peatland greenhouse gas emission in Europe. In: Dolman H, Valentini R, Freibauer A (eds) The Continental-Scale Greenhouse Gas Balance of Europe. Ecological Studies 203. Springer

Drösler M, Freibauer A, Adelmann W et al (2011) Klimaschutz durch Moorschutz in der Praxis. Schlussbericht des BMBF-Verbundprojekts „Klimaschutz - Moornutzungsstrategien 2006-2010. Arbeitsberichte aus dem Thünen-Institut für Agrarklimaschutz. Eigenverlag des vTi, Braunschweig

Drösler M, Schaller L, Kantelhardt J, Schweiger M, Fuchs D, Tiemeyer B, Augustin J, Wehrhan M, Förster C, Bergmann L, Kapfer A, Krüger GM (2012) Beitrag von Moorschutz- und Revitalisierungsmaßnahmen zum Klimaschutz am Beispiel von Naturschutzgroßprojekten. Natur Landsch 87:70–76

Dunfield P, Dumont R, Moore TR (1993) Methane production and consumption in temperate and subarctic peat soils: response to temperature and pH. Soil Biol Biochem 25:321–326

Ellenberg H, Leuschner C (2010) Vegetation Mitteleuropas mit den Alpen, 6th edn. Ulmer

Elsgaard L, Görres CM, Hoffmann CC, Blicher-Mathiesen G, Schelde K, Petersen SO (2012) Net ecosystem exchange of CO_2

and carbon balance for eight temperate organic soils under agricultural management. Agr Ecosyst Environ 162:52–67

Emmett BA, Beier C, Estiarte M, Tietema A, Kristensen HL, Williams D, Peñuelas J, Schmidt I, Sowerby A (2004) The response of soil processes to climate change: results from manipulation studies of shrublands across an environmental gradient. Ecosystems 7:625–637

Enquist F (1924) Sambandet mellan klimat och växtgränser. Geol Fören Förhandl 46:202–213

Erlandsson M, Buffam I, Folster J, Laudon H, Temnerud J, Weyhenmeyer GA, Bishop K (2008) Thirty-five years of synchrony in the organic matter concentrations of Swedish rivers explained by variation in flow and sulphate. Glob Change Biol 14:1191–1198

EUROSTAT (2014) CORINE land cover types 2006, www.eea.europa.eu/data-and-maps/figures/corine-land-cover-types-2006. Accessed 24.4.2014

EUROSTAT (2015) Land cover/use statistics (LUCAS), Statistics illustrated, Land over overview2012. http://ec.europa.eu/eurostat/web/lucas/statistics-illustrated. Accessed 10.3.2015

Evans M, Warburton J (2010) Geomorphology of Upland Peat: Erosion, form and landscape change. Wiley and Sons

Evans CD, Monteith DT, Cooper DM (2005) Long-term increases in surface water dissolved organic carbon: Observations, possible causes and environmental impacts. Environ Poll 137:55–71

Evans CD, Chapman PJ, Clark JM, Monteith DT, Cresser MS (2006a) Alternative explanations for rising dissolved organic carbon export from organic soils. Glob Change Biol 12:2044–2053

Evans M, Warburton J, Yang J (2006b) Eroding blanket peat catchments: Global and local implications of upland organic sediment budgets. Geomorphology 79:45–57

Evans CD, Freeman C, Cork LG, Thomas DN, Reynolds B, Billett MF, Garnett MH, Norris D (2007) Evidence against recent climate-induced destabilisation of soil carbon from ^{14}C analysis of riverine dissolved organic matter. Geophys Res Lett 34:L07407

Fenner N, Ostle NJ, McNamara N, Sparks T, Harmens H, Reynolds B, Freeman C (2007) Elevated CO_2 effects on peatland plant community carbon dynamics and DOC production. Ecosystems 10:635–647

Finkelstein SA, Cowling SA (2011) Wetlands, temperature, and atmospheric CO_2 and CH_4 coupling over the past two millennia. Glob Biogeochem Cy 25:GB1002

Finzi AC, Norby RJ, Calfapietra C, Gallet-Budynek A, Gielen B, Holmes WE, Hoosbeek MR, Iversen CM, Jackson RB, Kubiske ME, Ledford J, Liberloo M, Oren R, Polle A, Pritchard S, Zak DR, Schlesinger WH, Ceulemans R (2007) Increases in nitrogen uptake rather than nitrogen-use efficiency support higher rates of temperate forest productivity under elevated CO_2. Proc Nat Acad Sci USA 104:14014–14019

Fitter AH, Fitter RSR (2002) Rapid changes in flowering time in British plants. Science 296:1689–1691

Francis IS (1990) Blanket peat erosion in a mid-Wales catchment during two drought years. Earth Surf Proc Land 15:445–456

Freeman C, Evans CD, Monteith DT, Reynolds B, Fenner N (2001) Export of organic carbon from peat soils. Nature 412:785

Freeman C, Fenner N, Ostle NJ, Kang H, Dowrick DJ, Reynolds B, Lock MA, Sleep D, Hughes S, Hudson J (2004) Dissolved organic carbon export from peatlands under elevated carbon dioxide levels. Nature 430:195–198

Fridley JD, Grime JP, Askew AP, Moser B, Stevens CJ (2011) Soil heterogeneity buffers community response to climate change in species-rich grassland. Glob Change Biol 17:2002–2011

Friedrichs DA, Buntgen U, Frank DC, Esper J, Neuwirth B, Löffler J (2009) Complex climate controls on 20th century oak growth in Central-West Germany. Tree Phys 29:39–51

Frolking S, Roulet NT (2007) Holocene radiative forcing impact of northern peatland carbon accumulation and methane emissions. Glob Change Biol 13:1079–1088

Gaillard MJ, Sugita S, Mazier F, Kaplan JO, Trondman AK, Broström A, Hickler T, Kjellström E, Kuneš P, Lemmen C, Olofsson J, Smith B, Strandberg G (2010) Holocene land-cover reconstructions for studies on land cover-climate feedbacks. Clim Past Discuss 6:307–346

Gallego-Sala AV, Prentice IC (2013) Blanket peat biome endangered by climate change. Nat Clim Change 3:152–155

Gallego-Sala A, Clark J, House J, Orr H, Prentice IC, Smith P, Farewell T, Chapman S (2010) Bioclimatic envelope model of climate change impacts on blanket peatland distribution in Great Britain. Clim Res 45:151-162

Garnett MH, Ineson P, Stevenson AC (2000) Effects of burning and grazing on carbon sequestration in a Pennine blanket bog, UK. Holocene 10:729–736

Gaudnik C, Corcket E, Clément B, Delmas CEL, Gombert-Courvoisier S, Muller S, Stevens CJ, Alard D (2011) Detecting the footprint of changing atmospheric nitrogen deposition loads on acid grasslands in the context of climate change. Glob Change Biol 17:3351–3365

Gerdol R, Bonora A, Gualandri R, Pancaldi S (1996) CO_2 exchange, photosynthetic pigment composition, and cell ultrastructure of *Sphagnum* mosses during dehydration and subsequent rehydration. Can J Botany 74:726–734

Gerten D, Schaphoff S, Haberlandt U, Lucht W, Sitch S (2004) Terrestrial vegetation and water balance– Hydrological evaluation of adynamic global vegetation model. J. Hydrol 286:249– 270

Godwin H (1978) Fenland: Its Ancient Past and Uncertain Future. Cambridge University Press

Goldberg SD, Knorr KH, Blodau C, Lischeid G, Gebauer G (2010) Impact of altering the water table height of an acidic fen on N_2O and NO fluxes and soil concentrations. Glob Change Biol 16:220–233

Grime JP, Brown VK, Thompson K, Masters GJ, Hillier SH, Clarke IP, Askew AP, Corker D, Kielty JP (2000) The response of two contrasting limestone grasslands to simulated climate change. Science 289:762–765

Grime JP, Fridley JD, Askew AP, Thompson K, Hodgson JG, Bennett CR (2008) Long-term resistance to simulated climate change in an infertile grassland. Proc Nat Acad Sci USA 105:10028–10032

Guo LB, Gifford RM (2002) Soil carbon stocks and land use change: A meta analysis. Glob Change Biol 8:345–360

Hahn-Schöfl M, Zak D, Minke M, Gelbrecht J, Augustin J, Freibauer A (2011) Organic sediment formed during inundation of a degraded fen grassland emits large fluxes of CH_4 and CO_2. Biogeosciences 8:1539–1550

Hanewinkel M, Cullmann DA, Schelhaas MJ, Nabuurs GJ, Zimmermann NE (2013). Climate change may cause severe loss in the economic value of European forest land. Nat Clim Change 3:203–207

Härdtle W, Niemeyer M, Niemeyer T, Aßmann T, Fottner S (2006) Can management compensate for atmospheric nutrient deposition in heathland ecosystems? J Appl Ecol 43:759–769

Härdtle W, Niemeyer T, Aßmann T, Aulinger A, Fichtner A, Lang AC, Leuschner C, Neuwirth B, Pfister L, Quante M, Ries C, Schuldt A, Oheimb G (2013) Climatic responses of tree-ring width and $\delta^{13}C$ signatures of sessile oak (*Quercus petraea* Liebl.) on soils with contrasting water supply. Plant Ecol 214:1147–1156

Hargreaves KJ, Fowler D (1998) Quantifying the effects of water table and soil temperature on the emission of methane from peat wetland at the field scale. Atmos Environ 32:3275–3282

Hauck M (2009) Global warming and alternative causes of decline in arctic-alpine and boreal-montane lichens in north-western Central Europe. Glob Change Biol 15:2653–2661

Hedges JI, Keil RG, Benner R (1997) What happens to terrestrial organic matter in the ocean? Org Geochem 27:195–212

Heijmans MMPD, Mauquoy D, van Geel B, Berendse F (2008) Long-term effects of climate change on vegetation and carbon dynamics in peat bogs. J Veg Sci 19:307–320

Hemery GE (2008) Forest management and silvicultural responses to projected climate change impacts on European broadleaved trees and forests. Int For Rev 10:591–607

Hemery GE, Clark JR, Aldinger E, Claessens H, Malvolti ME, O'Connor E, Raftoyannis Y, Savill PS, Brus R (2010) Growing scattered broadleaved tree species in Europe in a changing climate: a review of risks and opportunities. Forestry 83:65–81

Hickling R, Roy DB, Hill JK, Fox R, Thomas CD (2006) The distributions of a wide range of taxonomic groups are expanding polewards. Glob Chang Biol 12:450–455

Hickler T, Vohland K, Feehan J, Miller P, Fronzek S, Giesecke T, Kuehn I, Carter T, Smith B, Sykes M, (2012) Projecting tree species-based climate-driven changes in European potential natural vegetation with a generalized dynamic vegetation model. Glob Ecol Biogeogr 21:50–63

Hickler T, Rammig A, Werner C (2015) Modelling CO_2 impacts on forest productivity. Curr Forestry Rep I:69–80

Hill JK, Thomas CD, Huntley B (1999) Climate and habitat availability determine 20th century changes in a butterfly's range margin. Proc Roy Soc Lond B 266:1197–1206

Hill JK, Thomas CD, Fox R, Telfer MG, Willis SG, Asher J, Huntley B (2002) Responses of butterflies to twentieth century climate warming: implications for future ranges. Proc Roy Soc Lond B 269:2163–2172

Hochkirch A, Damerau M (2009) Rapid range expansion of a wing-dimorphic bush-cricket after the 2003 climatic anomaly. Biol J Linnean Soc 97:118–127

Holden J (2005) Peatland hydrology and carbon release: why small-scale process matters. P Trans Roy Soc B 363:2891–2913

Holden J, Shotbolt L, Bonn A, Burt TP, Chapman PJ, Dougill AJ, Fraser EDG, Hubacek K, Irvine B, Kirkby MJ, Reed MS, Prell C, Stagl S, Stringer LC, Turner A, Worrall F (2007) Environmental change in moorland landscapes. Earth Sci Rev 82:75–100

Honnay O, Verheyen K, Butaye J, Jacquemyn H, Bossuyt B, Hermy M (2002) Possible effects of habitat fragmentation and climate change on the range of forest plant species. Ecol Lett 5:525–530

Hoogland T, van den Akker JJH, Brus DJ (2012) Modeling the subsidence of peat soils in the Dutch coastal area. Geoderma 171-172:92–97

Hope D, Billett MF, Cresser MS (1997) Exports of organic carbon in two river systems in NE Scotland. J Hydrol 193:61–82

Horswill P, O'Sullivan O, Phoenix GK, Lee JA, Leake JR (2008) Base cation depletion, eutrophication and acidification of species-rich grasslands in response to long-term simulated nitrogen deposition. Environ Poll 155:336–349

Hulme PE (2005) Adapting to climate change: is there scope for ecological management in the face of a global threat? J Appl Ecol 42:784–794

Huntley B, Collingham YC, Willis SG, Green RE (2008) Potential impacts of climatic change on European breeding birds. PLOS ONE 3,1:e1439. doi:10.1371/journal.pone.0001439

Hüppop O, Hüppop K (2003) North Atlantic Oscillation and timing of spring migration in birds. P Roy Soc Lond B 270:233–240

Hutchin PR, Press MC, Lee JA Ashenden TW (1995) Elevated concentrations of CO_2 may double methane emissions from mires. Glob Change Biol 1:125–128

Isaac-Renton MG, Roberts DR, Hamann A, Spiecker H (2014) Douglas-fir plantations in Europe: a retrospective test of assisted migration to address climate change. Glob Change Biol 20:2607–2617

Ise T, Dunn AL, Wofsy SC, Moorcroft PR (2008) High sensitivity of peat decomposition to climate change through watertable feedback. Nat Geosci 1:763–766

Iversen J (1944) *Viscum*, *Hedera* and *Ilex* as climatic indicators. A contribution to the study of past-glacial temperature climate. Geol Fören Förhandl 66:463–483

Jacob D, Petersen J, Eggert B, Alias A, Christensen O, Bouwer L, Braun A, Colette A, Déqué M, Georgievski G, Georgopoulou E, Gobiet A, Menut L, Nikulin G, Haensler A, Hempelmann N, Jones C, Keuler K, Kovats S, Kröner N, Kotlarski S, Kriegsmann A, Martin E, Meijgaard E, Moseley C, Pfeifer S, Preuschmann S, Radermacher C, Radtke K, Rechid D, Rounsevell M, Samuelsson P, Somot S, Soussana J-F, Teichmann C, Valentini R, Vautard R, Weber B, Yiou P (2014) EURO-CORDEX: new high-resolution climate change projections for European impact research. Reg Environ Change 14:563–578

Janssens IA, Freibauer A, Ciais P, Smith P, Nabuurs G-J, Folberth G, Schlamadinger B, Hutjes RWA, Ceulemans R, Schulze E-D, Valentini R, Dolman AJ (2003) Europe's terrestrial biosphere absorbs 7 to 12 % of European Anthropogenic CO_2 emissions. Science 300:1538–1542

Janssens IA, Freibauer A, Schlamadinger B, Ceulemans R, Ciais P, Dolman AJ, Heimann M, Nabuurs GJ, Smith P, Valentini R, Schulze E-D (2005) The carbon budget of terrestrial ecosystems at country-scale. A European case study. Biogeosciences 2:15–27

Jensen K, Beier C, Michelsen A, Emmett BA (2003) Effects of experimental drought on microbial processes in two temperate heathlands at contrasting water conditions. Appl Soil Ecol 24:165–176

Joabsson A, Christensen TR, Wallén B (1999) Vascular plant controls on methane emissions from northern peatforming wetlands. Trends Ecol Evol 14:385–388

Jönsson AM, Appelberg G, Harding S, Bärring L (2009) Spatio-temporal impact of climate change on the activity and voltinism of the spruce bark beetle, *Ips typographus*. Glob Change Biol 15:486–499

Joosten H (2010) The Global Peatland CO_2 Picture: Peatland status and drainage related emissions in all countries of the world. Wetlands International

Joosten H, Clarke D (2002) Wise use of peatlands. International Mire Conservation Group, International Peat Society, Jyväskylä, Finland

Joosten H, Tapio-Biström ML, Tol S (2012) Peatlands-guidance for climate change mitigation through conservation, rehabilitation and sustainable use. UN Food and Agriculture Organization, and Wetlands International

Kapfer J, Grytnes J-A, Gunnarson U, Birks JB (2011) Fine-scale changes in vegetation composition in a boreal mire over 50 years. J Ecol 99:1179–1189

Kaplan JO, Krumhardt KM, Zimmermann N (2009) The prehistoric and preindustrial deforestation of Europe. Quat Sci Rev 28:3016–3034

Kellner E (2001) Surface energy fluxes and control of evapotranspiration from a Swedish Sphagnum mire. Agr Forest Meteorol 110:101–123

King D, Daroussin J, Tavernier R (1994) Development of a soil geographical database from the soil map of the European Communities. Catena 21:37–56

King D, Jones RJA, Thomasson AJ (1995) European Land Information Systems for Agroenvironmental Monitoring. EUR 16232 EN, Office for Official Publications of the European Communities

Kirilenko AP, Sedjo RA (2007) Climate change and food security special feature: Climate change impacts on forestry. Proc Nat Acad Sci USA 104:19697–19702

Köhl M, Hildebrandt R, Olschofksy K, Köhler R, Rötzer T, Mette T, Pretzsch H, Köthke M, Dieter M, Abiy M, Makeschin F, Kenter B (2010) Combating the effects of climatic change on forests by mitigation strategies. Carbon Bal Manage 5:doi:10.1186/1750-0680-5-8

Körner C, Asshoff R, Bignucolo O, Hattenschwiler S, Keel SG, Pelaez-Riedl S, Pepin S, Siegwolf RTW, Zotz G (2005) Carbon flux and growth in mature deciduous forest trees exposed to elevated CO_2. Science 309:1360–1362

Körner C, Morgan JA, Norby R (2007) CO_2 fertilisation: when, where, how much? In: Canadell SG, Pataki DE, Pitelka LF (eds) Terrestrial Ecosystems in a Changing World. Springer

Kramer K, Degen B, Buschbom J, Hickler T, Thuiller W, Sykes MT, de Winter W (2010) Modelling exploration of the future of European beech (*Fagus sylvatica* L.) under climate change – range, abundance, genetic diversity and adaptive response. Forest Ecol Manag 259:2213–2222

Kreyling J (2010) Winter climate change: a critical factor for temperate vegetation performance. Ecology 91:1939–1948

Kullman L (2002) Rapid recent range-margin rise of tree and shrub species in the Swedish Scandes. J Ecol 90:68–77

Lafleur PM, Moore TR, Roulet NT, Frolking S (2005) Ecosystem respiration in a cool temperate bog depends on peat temperature but not water table. Ecosystems 8:619–629

Laine A, Wilson D, Kiely G, Byrne KA (2007) Methane flux dynamics in an Irish lowland blanket bog. Plant Soil 299:181–193

Laine AM, Byrne KA, Kiely G, Tuittila ES (2009) The short-term effect of altered water level on carbon dioxide and methane fluxes in a blanket bog. Suo 60:65–83

Le Mer J, Roger P (2001) Production, oxidation, emission and consumption of methane by soils: a review. Eur J Soil Biol 37:25–50

Lebourgeois F, Cousseau G, Ducos Y (2004) Climate–tree growth relationships of *Quercus petraea* Mill. stand in the Forest of Berce ('Futaie des Clos', Sarthe, France). Ann For Sci 61:361–372

Lebourgeois F, Breda N, Ulrich E, Granier A (2005) Climate–tree-growth relationships of European beech (*Fagus sylvatica* L.) in the French permanent plot network (RENECOFOR). Trees Struct Funct 19:385–401

Lehikoinen A, Jaatinen K, Vähätalo AV, Clausen P, Crowe O, Deceuninck B, Hearn R, Holt CA, Hornman M, Keller V, Nilsson L, Langendoen T, Tománková I, Wahl J, Fox AD (2013) Rapid climate driven shifts in wintering distributions of three common waterbird species. Glob Change Biol 19:2071–2081

Leiber-Sauheitl K, Fuß R, Voigt C, Freibauer A (2013) High greenhouse gas fluxes from grassland on histic gleysol along soil carbon and drainage gradients. Biogeosci Discuss 10:11283–11317

Lenoir J, Gégout JC, Marquet PA, de Ruffray P, Brisse H (2008) A significant upward shift in plant species optimum elevation during the 20th century. Science 320:1768–1771

Leuzinger S, Bader M (2012) Experimental versus modelled water use in mature Norway spruce (*Picea abies*) exposed to elevated CO_2. Front Plant Sci 3:229. doi:10.3389/fpls.2012.00229

Leuschner C, Backes K, Hertel D, Schipka F, Schmitt U, Terborg O, Runge M (2001) Drought responses at leaf, stem and fine root levels of competitive *Fagus sylvatica* L. and *Quercus petraea* (Matt.) Liebl. trees in dry and wet years. Forest Ecol Manag 149:33–46

Limpens J, Berendse F, Blodau C, Canadell JG, Freeman C, Holden J, Roulet N, Rydin H, Schaepman-Strub G (2008) Peatlands and the carbon cycle: from local processes to global implications - a synthesis. Biogeosciences 5:1475–1491

Lindner M, Maroschek M, Netherer S, Kremer A, Barbati A, Garcia-Gonzalo J, Seidl R, Delzon S, Corona P, Kolstrom M, Lexer MJ, Marchetti M (2010) Climate change impacts, adaptive capacity, and vulnerability of European forest ecosystems. Forest Ecol Manage 259:698–709

Lindsay R (2010) Peatbogs and Carbon: A critical synthesis. RSPB Scotland

Loisel J, Gallego-Sala AV, Yu Z (2012) Global-scale pattern of peatland *Sphagnum* growth driven by photosynthetically active radiation and growing season length. Biogeosciences 9:2737–2746

Lükewille A, Wright R (1997) Experimentally increased soil temperature causes release of nitrogen at a boreal forest catchment in southern Norway. Glob Change Biol 3:13–21

Luyssaert S, Ciais P, Piao SL, Schulze ED, Jung M, Zaehle S, Janssens IA et al (2010) The European carbon balance. Part 3: forests. Glob Change Biol 16:1429–1450

Maclean IMD, Austin GE, Rehfisch MM, Blew J, Crowe O, Delany S, Devos K, Deceuninck B, Günther K, Laursen K, Van Roomen M, Wahl J (2008) Climate change causes rapid changes in the distribution and site abundance of birds in winter. Glob Change Biol 14:2489–2500

Malmer N, Svensson BM, Wallén B (1994) Interactions between *Sphagnum* mosses and field layer vascular plants in the development of peat-forming systems. Folia Geobot Phytotx 29:483–496

Martikainen PJ, Nykänen H, Crill P, Silvola J (1993) Effect of a lowered water table on nitrous oxide fluxes from northern peatlands. Nature 366:51–53

McCarthy HR, Oren R, Finzi AC, Johnsen KH (2006) Canopy leaf area constrains CO_2-induced enhancement of productivity and partitioning among aboveground carbon pools. Proc Nat Acad Sci USA 103:19356–19361

McGovern S, Evans CD, Dennis P, Walmsley C, McDonald MA (2011) Identifying drivers of species compositional change in a semi-natural upland grassland over a 40-year period. J Veg Sci 22:346–356

Meier IC, Leuschner C (2008) Belowground drought response of European beech: fine root biomass and carbon partitioning in 14 mature stands across a precipitation gradient. Glob Change Biol 14:2081–2095

Melillo JM, Butler S, Johnson J, Mohan J, Steudler P, Lux H, Burrows E, Bowles F, Smith R, Scott L, Vario C, Hill T, Burton A, Zhou YM, Tang J (2011) Soil warming, carbon-nitrogen interactions, and forest carbon budgets. Proc Nat Acad Sci USA 108:9508–9512

Melton JR, Wania R, Hodson EL, Poulter B, Ringeval B, Spahni R, Bohn T, Avis CA, Beerling DJ, Chen G, Eliseev AV, Denisov SN, Hopcroft PO, Lettenmaier DP, Riley WJ, Singarayer JS, Subin ZM, Tian H, Zürcher S, Brovkin V, van Bodegom PM, Kleinen T, Yu ZC, Kaplan JO (2013) Present state of global wetland extent and wetland methane modelling: conclusions from a model inter-comparison project (WETCHIMP). Biogeosciences 10:753–788

Meng L, Hess PGM, Mahowald NM, Yavitt JB, Riley WJ, Subin ZM, Lawrence DM, Swenson SC, Jauhiainen J, Fuka DR (2012) Sensitivity of wetland methane emissions to model assumptions: application and model testing against site observations. Biogeosciences 9:2793–2819

Menzel A (2000) Trends in phenological phases in Europe between 1951 and 1996. Int J Biometeorol 44:76–81

Menzel A, Fabian P (1999) Growing season extended in Europe. Nature 397:659

Menzel A, Estrella N, Fabian P (2001) Spatial and temporal variability of the phenological seasons in Germany from 1951 to 1996. Glob Change Biol 7:657–666

Menzel A, Sparks TH, Estrella N et al (2006) European phenological response to climate change matches the warming pattern. Glob Change Biol 12:1969–1976

Merian P, Bontemps JD, Berge's L, Lebourgeois F (2011) Spatial variation and temporal instability in growth–climate relationships of sessile oak (*Quercus petraea* [Matt.] Liebl.) under temperate conditions. Plant Ecol 212:1855–1871

Meusel H, Jäger E, Weinert E (1965) Vergleichende Chorologie der zentraleuropäischen Flora. Jena: Fischer

Mikkelä C, Sundh I, Svensson BH, Nilsson M (1995) Diurnal variation in methane emission in relation to the water table, soil temperature, climate and vegetation cover in a Swedish acid mire. Biogeochemistry 28:93–114

Mitchell TD, Carter TR, Jones PD, Hulme M, New M (2004) A comprehensive set of high-resolution grids of monthly climate for Europe and the globe: the observed record (1901–2000) and 16 scenarios (2001–2100). Tyndall Centre Working Papers 55, July 2004 www.tyndall.ac.uk/publications/working_papers/wp55.pdf

Mölder I, Leuschner C, Leuschner HH (2011) $\delta^{13}C$ signature of tree rings and radial increment of *Fagus sylvatica* trees as dependent on tree neighbourhood and climate. Trees Struct Funct 25:215–229

Møller AP, Rubolini D, Lehikoinen E (2008) Populations of migratory bird species that did not show a phenological response to climate change are declining. P Natl Acad Sci USA 105:16195–16200

Montanarella L, Jones RJ, Hiederer R (2006) The distribution of peatland in Europe. Mires and Peat vol 1,1

Monteith DT, Stoddard JL, Evans CD, de Wit HA, Forsius M, Høgasen T, Wilander A, Skjelkvale BL, Jeffries DS, Vuorenmaa J, Keller B, Kopacek J, Vesely J (2007) Dissolved organic carbon trends resulting from changes in atmospheric deposition chemistry. Nature 450:537–541

Montoya JM, Raffaelli D (2010) Climate change, biotic interactions and ecosystem services. P Trans Roy Soc B 365:2013–2018

Moore TR, Dalva M (1993) The influence of temperature and water table position on carbon dioxide and methane emissions from laboratory columns of peatland soils. J Soil Sci 44:651–664

Moore PA, Pypker TG, Waddington JM (2013) Effect of long-term water table manipulation on peatland evapotranspiration. Agr Forest Meteorol 178–179:106–119

Morales P, Hickler T, Smith B, Sykes MT, Rowell DP (2007) Changes in European ecosystem productivity and carbon balance driven by regional climate model outputs. Glob Change Biol 13:108–122

Morgan JA, Pataki DE, Körner C, Clark H, Del Grosso SJ, Grünzweig JM, Knapp AK, Mosier AR, Newton PCD, Niklaus PA, Nippert JB, Nowak RS, Parton WJ, Polley HW, Shaw MR (2004) Water relations in grassland and desert ecosystems exposed to elevated atmospheric CO_2. Oecologia 140:11–25

Moser B, Fridley JD, Askew AP, Grime JP (2011) Simulated migration in a long-term climate change experiment: invasions impeded by dispersal limitation, not biotic resistance. J Ecol 99:1229–1236

Nabuurs GJ, Schelhaas MJ, Mohren GMJ, Field CB (2003) Temporal evolution of the European forest sector carbon sink from 1950 to 1999. Glob Change Biol 9:152–160

Nakićenović N, Swart R (eds) (2000) Special report on emission scenarios. A special report of working group III of the Intergovernmental Panel on Climate Change. Cambridge University Press

Nielsen OK, Plejdrup MS, Winther M et al (2013) Denmark's National Inventory Report 2013. Scientific Report from DCE No. 56. Aarhus University, Danish Centre for Environment and Energy, Denmark

Nokes CJ, Fenton E, Randall CJ (1999) Modelling the formation of brominated trihalomethanes in chlorinated drinking waters. Wat Res 33:3557–3568

Norby RJ, Warren JM, Iversen CM, Medlyn BE, McMurtrie RE (2010) CO_2 enhancement of forest productivity constrained by limited nitrogen availability. Proc Nat Acad Sci USA 107:19368–19373

Normand S, Ricklefs RE, Skov F, Bladt J, Tackenbeg O, Svenning JC (2011) Postglacial migration supplements climate in determining plant species ranges in Europe. Proc Roy Soc B 278:3644–3653

Nykänen H, Alm J, Silvola J, Tolonen K, Martikainen PJ (1998) Methane fluxes on boreal peatlands of different fertility and the effect of long-term experimental lowering of the water table on flux rates. Glob Biogeochem Cy 12:53–69

Oleszczuk R, Regina K, Szajdak L, Höper H, Maryganova V (2008) Impacts of agricultural utilization of peat soils on the greenhouse gas balance. Peatlands and climate change. 70. International Peat Society, Jyväskylä, Finland

Oliveira PJC, Davin EL, Levis S, Seneviratne SI (2011) Vegetation-mediated impacts of trends in global radiation on land hydrology: a global sensitivity study. Glob Change Biol 17:453–3467

Oliver B, Thurman E, Malcolm R (1983) The contribution of humic substances to the acidity of colored natural waters. Geochim Cosmochim Ac 47:2031–2035

Oliver TH, Thomas CD, Hill JK, Brereton T, Roy DB (2012) Habitat associations of thermophilous butterflies are reduced despite climatic warming. Glob Change Biol 18:2720–2729

Parish F, Sirin A, Charman D, Joosten H, Minayeva T, Silvius M, Stringer L (eds) (2008) Assessment on Peatlands, Biodiversity and Climate Change: Main Report. Global Environment Centre, Kuala Lumpur and Wetlands International, Wageningen

Parmesan C (2006) Ecological and evolutionary responses to recent climate change. In: Annual Review of Ecology Evolution and Systematics. Annual Reviews, Palo Alto

Parmesan C, Yohe G (2003) A globally coherent fingerprint of climate change impacts across natural systems. Nature 421:37–42

Parmesan C, Ryrholm N, Stefanescu C, Hill JK, Thomas CD, Descimon H, Huntley B, Kaila L, Kullberg J, Tammaru T, Tennent WJ, Thomas JA, Warren M (1999) Poleward shifts in geographical ranges of butterfly species associated with regional warming. Nature 399:579–583

Pastor J, Solin J, Bridgham SD, Updegraff K, Harth C, Weishampel P, Dewey B (2003) Global warming and the export of dissolved organic carbon from boreal peatlands. Oikos 100:380–386

Peichl M, Sagerfors J, Lindroth A, Buffam I, Grelle A, Klemedtsson L, Laudon H, Nilsson MB (2013) Energy exchange and water budget partitioning in a boreal minerogenic mire. J Geophys Res Biogeo 118:1–13

Peñuelas J, Gordon C, Llorens L, Nielsen T, Tietema A, Beier C, Bruna P, Emmett BA, Estiarte M, Gorissen A (2004) Non-intrusive field experiments show different plant responses to warming and drought among sites, seasons and species in a North–South European gradient. Ecosystems 7:598–612

Peñuelas P, Prieto P, Beier C et al (2007) Responses of shrubland species richness and primary productivity to six-years experimental warming and drought in a North–South European gradient: reductions in primary productivity in the hot and dry 2003. Glob Change Biol 13:2563–2581

Petrescu AMR, van Huissteden J, de Vries F, Bregman EPH, Scheper A (2009) Assessing CH_4 and CO_2 emissions from wetlands in the Drenthe Province, the Netherlands: a modelling approach. Neth J Geosci 88:101–116

Pokrovsky O, Dupré B, Schott J (2005) Fe-Al-organic colloids control of trace elements in peat soil solutions: Results of ultrafiltration and dialysis. Aquat Geochem 11:241–278

Pompe S, Hanspach J, Badeck F, Klotz S, Thuiller W, Kühn I (2008) Climate and land use change impacts on plant distributions in Germany. Biol Lett 4:564–567

Poniatowski D, Fartmann T (2011) Weather-driven changes in population density determine wing dimorphism in a bush-cricket species. Agr Ecosyst Environ 145:5–9

Poniatowski D, Heinze S, Fartmann T (2012) The role of macropters during range expansion of a wing-dimorphic insect species. Evol Ecol 26:759–770

Potts SG, Biesmeijer JC, Kremen C, Neumann P, Schweiger O, Kunin WE (2010) Global pollinator declines: trends, impacts and drivers. Trends Ecol Evol 25:345–353

Pretzsch H, Dursky J (2002) Growth reaction of Norway spruce (*Picea abies* (L.) Karst.) and European beech (*Fagus silvatica* L.) to possible climatic changes in Germany: a sensitivity study. Forstwiss Cbl 121:145–154

Raymond PA, Bauer JE (2000) Bacterial consumption of DOC during transport through a temperate estuary. Aquat Microb Ecol 22:1–12

Raymond PA, Hartmann J, Lauerwald R, Sobek S, McDonald C, Hoover M, Guth P et al (2013) Global carbon dioxide emissions from inland waters. Nature 503:355–359

Regina K, Nykänen H, Silvola J, Martikainen PJ (1996) Fluxes of nitrous oxide from boreal peatlands as affected by peatland type, water table level and nitrification capacity. Biogeochemistry 35:401–418

Robroek BJ, Schouten MG, Limpens J, Berendse F, Poorter H (2009) Interactive effects of water table and precipitation on net CO_2 assimilation of three co-occurring *Sphagnum* mosses differing in distribution above the water table. Glob Change Biol 15:680–691

Roy DB, Sparks TH (2000) Phenology of British butterflies and climate change. Glob Change Biol 6:407–416

Ruddiman WF (2003) The anthropogenic greenhouse era began thousands of years ago. Climatic Change 61:261–293

Runkle BRK, Wille C, Gazovic M, Wilmking M, Kutzbach L (2014) The surface energy balance and its drivers in a boreal peatland fen of northwestern Russia. J Hydrol 511:359–373

Sadiq R, Rodriguez MJ (2004) Disinfection by-products (DBPs) in drinking water and predictive models for their occurrence: a review. Sci Tot Environ 321:21–46

Saino N, Ambrosini R, Rubolini D, von Hardenberg J, Provenzale A, Hüppop K, Hüppop O, Lehikoinen A, Lehikoinen E, Rainio K, Romano M, Sokolov L (2011) Climate warming, ecological mismatch at arrival and population decline in migratory birds. Proc Roy Soc B 278:835–842

Sandel B, Arge L, Dalsgaard B, Davies RG, Gaston KJ, Sutherland WJ, Svenning JC (2011) The influence of Late Quaternary climate-change velocity on species endemism. Science 334:660–664

Schlünz B, Schneider RR (2000) Transport of terrestrial organic carbon to the oceans by rivers: re-estimating flux and burial rates. Int J Earth Sci 88:599–606

Schlyter P, Stjernquist I, Barring L, Jonsson AM, Nilsson C (2006) Assessment of the impacts of climate change and weather extremes on boreal forests in northern Europe, focusing on Norway spruce. Climate Res 31:75–84

Schmidt IK, Tietema A, Williams D, Gundersen P, Beier C, Emmett BA, Estiarte M (2004) Soil solution chemistry and element fluxes in three European heathlands and their responses to warming and drought. Ecosystems 7:638–649

Schothorst CJ (1977) Subsidence of low moor peat soils in the western Netherlands. Geoderma 17:265–291

Schrier-Uijl AP, Kroon PS, Leffelaar PA, Van Huissteden JC, Berendse F, Veenendaal EM (2010) Methane emissions in two drained peat agro-ecosystems with high and low agricultural intensity. Plant Soil 329:509–520

Schrier-Uijl AP, Kroon PS, Hendriks DMD, Hensen A, Van Huissteden J, Berendse F, Veenendaal EM (2014) Agricultural peatlands: towards a greenhouse gas sink – a synthesis of a Dutch landscape study. Biogeosciences 11:4559–4576

Schuch S, Bock J, Krause B, Wesche K, Schaefer M (2012a) Long-term population trends in three grassland insect groups: a comparative analysis of 1951 and 2009. J Appl Entomol 136:321–331

Schuch S, Wesche K, Schaefer M (2012b) Long-term decline in the abundance of leafhoppers and planthoppers (Auchenorrhyncha) in Central European protected dry grasslands. Biol Conserv 149:75–83

Schulze ED, Luyssaert S, Ciais P, Freibauer A, Janssens IA (2009) Importance of methane and nitrous oxide for Europe's terrestrial greenhouse-gas balance. Nat Geosci 2:842–850

Schweiger O, Settele J, Kudrna O, Klotz S, Kühn I (2008) Climate change can cause spatial mismatch of trophically interacting species. Ecology 89:3472–3479

Schweiger O, Heikkinen RK, Harpke A, Hickler T, Klotz S, Kudrna O, Kühn I, Pöyry J, Settele J (2012) Increasing range mismatching of interacting species under global change is related to their ecological characteristics. Glob Ecol Biogeogr 21:88–99

Settele J, Scholes R, Betts R, Bunn S, Leadley P, Nepstad D, Overpeck JT, Taboada MA (2014) Terrestrial and inland water systems. In: Field CB, Barros VR, Dokken DJ, Mach KJ, Mastrandrea MD, Bilir TE, Chatterjee M, Ebi KL, Estrada YO, Genova RC, Girma B, Kissel ES, Levy AN, MacCracken S, Mastrandrea PR, White LL (eds), Climate Change 2014: Impacts, Adaptation, and Vulnerability. Part A: Global and Sectoral Aspects. Contribution of Working Group II to the Fifth Assessment Report of the Intergovernmental Panel of Climate Change. Cambridge University Press pp 271-359

Silvola J, Alm J, Ahlholm U, Nykanen H, Martikainen PJ (1996) CO_2 fluxes from peat in boreal mires under varying temperature and moisture conditions. J Ecol 84:219–228

Simmons IG (2003) The moorlands of England and Wales: an environmental history 8000 BC to AD 2000. Edinburgh University Press

Sitch S, Huntingford C, Gedney N, Levy PE, Lomas M, Piao SL, Betts R, Ciais P, Cox P, Friedlingstein P, Jones CD, Prentice IC, Woodward FI (2008) Evaluation of the terrestrial carbon cycle, future plant geography and climate-carbon cycle feedbacks using five Dynamic Global Vegetation Models (DGVMs). Glob Change Biol 14:2015–2039

Skov F, Svenning JC (2004) Potential impact of climatic change on the distribution of forest herbs in Europe. Ecography 27:366–380

Smith JU, Smith P, Wattenbach M, Zaehle S, Hiederer R, Jones RJA, Montanarella L, Rounsevell MDA, Reginster I, Ewert F (2005) Projected changes in mineral soil carbon of European croplands and grasslands, 1990–2080. Glob Change Biol 11:2141–2152

Smith P, Chapman SJ, Scott WA, et al (2007) Climate change cannot be entirely responsible for soil carbon loss observed in England and Wales, 1978–2003. Glob Change Biol 13:2605–2609

Smolders AJP, Lamers LPM, Lucassen ECHET, van der Velde G, Roelofs JGM (2006) Internal eutrophication: How it works and what to do about it – a review. Chem Ecol 22:93–111

Sottocornola M, Kiely G (2010) Energy fluxes and evaporation mechanisms in an Atlantic blanket bog in southwestern Ireland. Wat Resour Res 46:W11524

Soussana JF, Loiseau P, Vuichard N, Ceschia E, Balesdent J, Chevallier T, Arrouays D (2004) Carbon cycling and sequestration opportunities in temperate grasslands. Soil Use Manage 20:219–230

Sowerby A, Emmett BA, Tietema A, Beier C (2008) Contrasting effects of repeated summer drought on soil carbon efflux in hydric and mesic heathland soils. Glob Change Biol 14:2388–2404

Sparks TH, Bairlein F, Bojarinova JG, Hüppop O, Lehikoinen EA, Rainio K, Sokolov LV, Walker D (2005) Examining the total arrival distribution of migratory birds. Glob Change Biol 11:22–30

Stampfli A, Zeiter M (2004) Plant regeneration directs changes in grassland composition after extreme drought: a 13-year study in southern Switzerland. J Ecol 92:568–576

Strack M, Waddington JM (2007) Response of peatland carbon dioxide and methane fluxes to a water table drawdown experiment. Global Biogeochem Cy 21,1:GB1007

Svenning JC, Skov F (2004) Limited filling of the potential range in European tree species. Ecol Lett 7:565–573

Tallis JH (1985) Mass movement and erosion of a Southern Pennine blanket peat. J Ecol 73:283–315

Tamis WLM, Van't Zelfde M, Van Der Meijden R, Udo De Haes HA (2005) Changes in vascular plant biodiversity in the Netherlands in the 20th century explained by their climatic and other environmental characteristics. Climatic Change 72:37–56

Thackeray SJ, Sparks TH, Frederiksen M, Burthe S, Bacon PJ, Bell JR, Botham MS, Brereton TM, Bright, PW, Carvalho L, Clutton-Brock T, Dawson A, Edwards M, Elliott JM, Harrington R, Johns D, Jones ID, Jones JT, Leech DI, Roy DB, Scott WA, Smith M, Smithers RJ, Winfield IJ, Wanless S (2010) Trophic level asynchrony in rates of phenological change for marine, freshwater and terrestrial environments. Glob Change Biol 16:3304–3313

Thomas JA, Rose RJ, Clarke RT, Thomas CD, Webb NR (1999) Intraspecific variation in habitat availability among ectothermic animals near their climatic limits and their centres of range. Func Ecol 13:55–64

Thomas H, Bozec Y, de Baar HJW, Elkalay K, Frankignoulle M, Schiettecatte L-S, Kattner G, Borges AV (2005) The carbon budget of the North Sea. Biogeosciences 2:87–96

Thornton P, Lamarque JF, Rosenbloom NA, Mahowald NM (2007) Influence of carbon-nitrogen cycle coupling on land model response to CO_2 fertilization and climate variability. Glob Biochem Cy 21: GB4018. doi:10.1029/2006GB002868

Thuiller W, Lavorel S, Araújo MB, Sykes MT, Prentice IC (2005) Climate change threats to plant diversity in Europe. Proc Nat Acad Sci USA 102:8245–8250

Tingley MW, Koo MS, Moritz C, Rush AC, Beissinger SR (2012) The push and pull of climate change causes heterogeneous shifts in avian elevational ranges. Glob Change Biol 18:3279–3290

Tipping E, Woof C, Rigg E, Harrison AF, Ineson P, Taylor K, Benham D, Poskitt J, Rowland AP, Bol R, Harkness DD (1999) Climatic influences on the leaching of dissolved organic matter from upland UK moorland soils, investigated by a field manipulation experiment. Environ Int 25:83–95

Tørseth K, Aas W, Breivik K, Fjæraa AM, Fiebig M, Hjellbrekke AG, Lund Myhre C, Solberg S, Yttri KE (2012) Introduction to the European Monitoring and Evaluation Programme (EMEP) and observed atmospheric composition change during 1972-2009. Atmos Chem Phys 12:5447–5481

Tranvik LJ, Jansson M (2002) Terrestrial export of organic carbon. Nature 415:861–862

Urban NR, Bayley SE, Eisenreich SJ (1989) Export of dissolved organic carbon and acidity from peatlands. Wat Resour Res 25:1619–1628

Van Breemen N, Jenkins, A Wright RF, Beerling DJ, Arp WJ, Berendse F, Beier C, Collins R, van Dam D, Rasmussen L, Verburg PSJ, Wills MA (1998) Impacts of elevated carbon dioxide and temperature on a boreal forest ecosystem (CLIMEX project). Ecosystems 1:345–351

Van den Bos RM (2003) Restoration of former wetlands in the Netherlands; effect on the balance between CO_2 sink and CH_4 source. Neth J Geosci 82:325–332

Van den Pol-van Dasselaar A, Van Beusichem ML, Oenema O (1999) Determinants of spatial variability of methane emissions from wet grasslands on peat soil. Biogeochemistry 44:221–237

Van der Heijden E, Jauhiainen J, Silvola J, Vasander H, Kuiper PJC (2000) Effects of elevated atmospheric CO_2 concentration and increased nitrogen deposition on growth and chemical composition of ombrotrophic *Sphagnum balticum* and oligo-mesotrophic *Sphagnum papillosum*. J Bryol 22:175–182

Van der Wal R, Bonn A, Monteith D, Reed M, Blackstock K, Hanley N, Thompson D, Evans D, Alonso I (2011) Mountains, moorlands and heaths. The UK National Ecosystem Assessment Technical Report. UNEP-WCMC, Cambridge

Van Groenigen KJ, Osenberg CW, Hungate BA (2011) Increased soil emissions of potent greenhouse gases under increased atmospheric CO_2. Nature 475:214–216

Van Herk CM, Aptroot A, van Dobben HF (2002) Long-term monitoring in the Netherlands suggests that lichens respond to global warming. Lichenologist 34:141–154

Van Huissteden J, Van den Bos R (2006) Modelling the effect of water-table management on CO_2 and CH_4 fluxes from peat soils. Neth J Geosci 85:3–18

Van Nouhuys S, Lei G (2004) Parasitoid-host metapopulation dynamics: the causes and consequences of phenological asynchrony. J Anim Ecol 73:526–35

van Vliet AJH, Bron WA Mulder S, van der Slikke W, Ode B (2014) Observed climate-induced changes in plant phenology in the Netherlands. Reg Environ Change 14:997–1008

Veenendaal EM, Kolle O, Leffelaar PA, Schrier-Uijl AP, Van Huissteden J, Van Walsem J, Möller F, Berendse F (2007) CO_2 exchange and carbon balance in two grassland sites on eutrophic drained peat soils. Biogeosciences 4:1027–1040

Verhagen A, van den Akker JJH, Blok C, Diemont WH, Joosten JHJ, Schouten MA, Schrijver RAM, den Uyl RM, Verweij PA, Wösten JHM (2009) Climate change scientific assessment and policy analysis. Peatlands and carbon flows. Outlook and importance for the Netherlands. Report 500102 02. Netherlands Environmental Assessment Agency, Bilthoven

Verhoeven JTA (ed) (1992) Fens and Bogs in the Netherlands. Series Geobotany, vol 18. Kluwer

Visser ME, Both C (2005) Shifts in phenology due to global climate change: the need for a yardstick. Proc Roy Soc B 272:2561–2569

Visser ME, Perdeck AC, van Balen JH, Both C (2009) Climate change leads to decreasing bird migration distances. Glob Change Biol 15:1859–1865

Voss N, Welk E, Durka W, Eckstein RL (2012) Biological Flora of Central Europe: *Ceratocapnos claviculata* (L) Lidén. Pers Plant Ecol 14:61–77

Walker AP, Hanson PJ, De Kauwe MG, Medlyn BE, Zaehle S, Asao S, Dietze M, Hickler T, Huntingford C, Iversen CM et al. (2014) Comprehensive ecosystem model-data synthesis using multiple data sets at two temperate forest free-air CO_2 enrichment experiments: model performance at ambient CO_2 concentration. J Geophy Res: Biogeosci 119:937–964

Walter H, Straka H (1970) Arealkunde—Floristischhistorische Geobotanik, 2nd edn. Ulmer

Walter BP, Heimann M, Matthews E (2001) Modeling modern methane emissions from natural wetlands: 2. Interannual variations 1982–1993. J Geophys Res 106:34207–34219

Walther GR, Post E, Convey P, Menzel A, Parmesan C, Beebee TJC, Fromentin JM, Hoegh-Guldberg O, Bairlein F (2002) Ecological responses to recent climate change. Nature 416:389–395

Walther GR, Berger S, Sykes MT (2005) An ecological 'footprint' of climate change. P Roy Soc B 272:1427–1432

Warburton J (2003) Wind-splash erosion of bare peat on UK upland moorlands. Catena 52:191–207

Warren MS, Hill JK, Thomas JA, Asher J, Fos R, Huntley B, Roy DB, Telfer MG, Jeffcoate S, Harding P, Jeffcoate G, Willis SG, Greatorex-Davies JN, Moss D, Thomas CD (2001) Rapid responses of British butterflies to opposing forces of climate and habitat change. Nature 414:65–68

Webb NR (1998) The traditional management of European heathlands. J Appl Ecol 35:987–990

Werkman BR, Callaghan TV, Welker JM (1996) Responses of bracken to increased temperature and nitrogen availability. Glob Change Biol 2:59–66

Wesche K, Krause B, Culmsee H, Leuschner C (2012) Fifty years of change in Central European grassland vegetation: Large losses in species richness and animal-pollinated plants. Biol Conserv 150:76–85

Wessel WW, Tietema A, Beier C, Emmett BA, Peñuelas J, Riis-Nielsen T (2004) A qualitative ecosystem assessment for different shrublands in Western Europe under impact of climate change using the results of the CLIMOOR research project. Ecosystems 7:662–671

Wetzel RG (1992) Gradient-dominated ecosystems: sources and regulatory functions of dissolved organic matter in freshwater ecosystems. Hydrobiologia 229:181–198

Wiegner TN, Seitzinger SP (2001) Photochemical and microbial degradation of external dissolved organic matter inputs to rivers. Aquat Microb Ecol 24:27–40

Wild M, Gilgen H, Roesch A, Ohmura A, Long CN, Dutton EG, Forgan B, Kallis A, Russak V, Tsvetkov A (2005) From dimming to brightening: Decadal changes in solar radiation at Earth's surface. Science 308:847–850

Williams TA, Abberton MT (2004) Earlier flowering between 1962 and 2002 in agricultural varieties of white clover. Oecologia 138:122–126

Worrall F, Burt T, Jaeban RY, Warburton J, Shedden R (2002) Release of dissolved organic carbon from upland peat. Hydrol Process 16:3487–3504

Worrall F, Burt T, Shedden R (2003) Long term records of riverine dissolved organic matter. Biogeochemistry 64:165–178

Worrall F, Armstrong A, Adamson JK (2007) The effects of burning and sheep-grazing on water table depth and soil water quality in a upland peat. J Hydrol 339:1–14

Worrall F, Chapman P, Holden J, Evans C, Artz R, Smith P, Grayson R (2011) A review of current evidence on carbon fluxes and greenhouse gas emissions from UK peatland. JNCC Rep 442, Peterborough, UK

Wu J, Kutzbach L, Jager D, Wille C, Wilmking M (2010) Evapotranspiration dynamics in a boreal peatland and its impact on the water and energy balance. J Geophys Res: Biogeo 115:G04038

Yallop AR, Clutterbuck B, Thacker J (2010) Increases in humic dissolved organic carbon export from upland peat catchments: the role of temperature, declining sulphur deposition and changes in land management. Climate Res 45:43–56

Yu Z, Loisel J, Brosseau DP, Beilman, DW, Unt SJ (2010) Global peatland dynamics since the Last Glacial Maximum. Geophys Res Lett 37:L13402

Zafiriou OC, Joussot-Dubien J, Zepp RG, Zika RG (1984) Photochemistry of natural waters. Environ Sci Technol 18:358–371

Zeiter M, Stampfli A, Newbery DM (2006) Recruitment limitation constrains local species richness and productivity in dry grassland. Ecology 87:942–951

Part IV

Climate Change Impacts on Socio-economic Sectors

12 Socio-economic Impacts—Fisheries

John K. Pinnegar, Georg H. Engelhard, Miranda C. Jones, William W.L. Cheung, Myron A. Peck, Adriaan D. Rijnsdorp and Keith M. Brander

Abstract

Fishers and scientists have known for over 100 years that the status of fish stocks can be greatly influenced by prevailing climatic conditions. Based on historical sea surface temperature data, the North Sea has been identified as one of 20 'hot spots' of climate change globally and projections for the next 100 years suggest that the region will continue to warm. The consequences of this rapid temperature rise are already being seen in shifts in species distribution and variability in stock recruitment. This chapter reviews current evidence for climate change effects on fisheries in the North Sea—one of the most important fishing grounds in the world—as well as available projections for North Sea fisheries in the future. Discussion focuses on biological, operational and wider market concerns, as well as on possible economic consequences. It is clear that fish communities and the fisheries that target them will be very different in 50 or 100 years' time and that management and governance will need to adapt accordingly.

12.1 Introduction

The North Sea remains one of the world's most important fishing grounds. In 2013, around 3.5 million tonnes of fish and shellfish were taken from the region (2.6 million tonnes by EU countries), approximately 55 % of the total for EU countries as a whole. European fisheries are very diverse, ranging from highly industrialised distant-water fisheries to small-scale artisanal fisheries that typically operate near the coast. EU citizens consume large quantities of seafood each year (currently around 23.3 kg on average per person), and rely on fisheries for health and well-being, as well as for supporting more than 120,000 jobs directly and a further 115,000 in fish processing (STECF 2013). There has been

J.K. Pinnegar (✉) · G.H. Engelhard
Centre for Environment, Fisheries and Aquaculture Science (Cefas), Lowestoft, UK
e-mail: john.pinnegar@cefas.co.uk

G.H. Engelhard
e-mail: georg.engelhard@cefas.co.uk

J.K. Pinnegar · G.H. Engelhard
School of Environmental Sciences, University of East Anglia (UEA), Norwich, UK

M.C. Jones
NF-UBC Nereus Program, Fisheries Centre, University of British Columbia, Vancouver, Canada
e-mail: Miranda.c.jones@gmail.com

W.W.L. Cheung
Institute for the Ocean and Fisheries, University of British Columbia, Vancouver, Canada
e-mail: w.cheung@oceans.ubc.ca

M.A. Peck
Institute of Hydrobiology and Fisheries Science, CEN, University of Hamburg, Hamburg, Germany
e-mail: myron.peck@uni-hamburg.de

A.D. Rijnsdorp
Institute for Marine Research and Ecosystem Studies (IMARES), Wageningen University, IJmuiden, The Netherlands
e-mail: adriaan.rijnsdorp@wur.nl

K.M. Brander
National Institute of Aquatic Resources, Technical University of Denmark, Charlottenlund, Denmark
e-mail: kbr@aqua.dtu.dk

M. Quante and F. Colijn (eds.), *North Sea Region Climate Change Assessment*, Regional Climate Studies, DOI 10.1007/978-3-319-39745-0_12

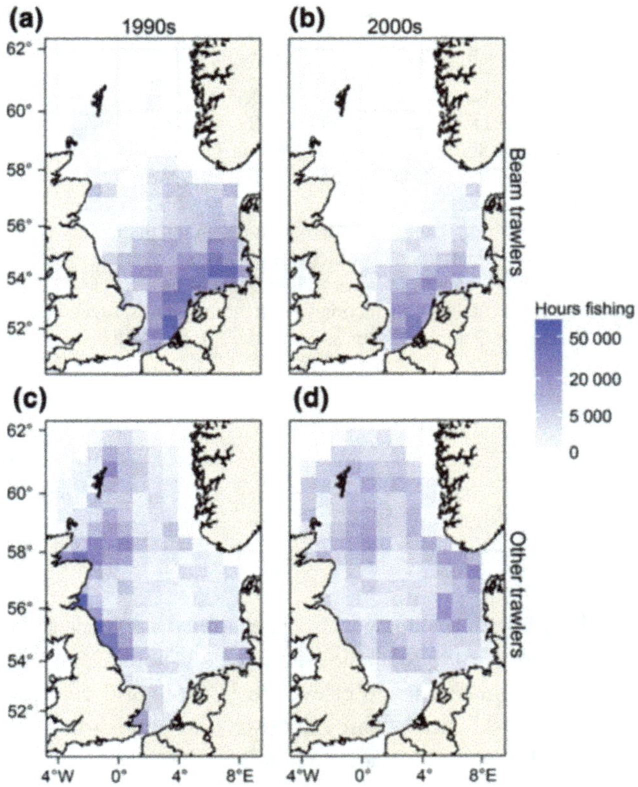

Fig. 12.1 Spatial distribution of international fishing effort in the North Sea by beam trawlers (*upper*) and demersal otter trawlers (*lower*), averaged by year over the periods 1990–1995 (*left*) and 2003–2012 (*right*). *Light to dark shading* indicates the number of hours fishing in each ICES rectangle (redrawn from Engelhard et al. 2015)

much debate in the literature with regard to the extent to which fisheries might be sensitive to climate change and a number of national-scale assessments have been conducted, for example for the United Kingdom (Cheung et al. 2012; Pinnegar et al. 2013). To date, however, no North Sea-wide assessment of climate impacts on the fisheries sector has been carried out and there is limited information for many countries despite the wide-scale and well-documented implications.

12.2 Overview of North Sea Fisheries

Commercial fishing activity in the North Sea is mostly undertaken by fishers from the UK (England and Scotland), Denmark, the Netherlands, France, Germany, Belgium and Norway (Fig. 12.1). Total fish removals are dominated by pelagic species (those that swim in the water column, above the seabed) such as herring *Clupea harengus* (986,471 tonnes), sprat *Sprattus sprattus* (143,581 tonnes), and mackerel *Scomber scombrus* (644,762 tonnes), although demersal fishes are also important. Demersal fish are those that live close to the sea floor and are typically caught by 'otter trawlers'. The most important demersal species include Atlantic cod *Gadus morhua*, haddock *Melanogrammus aeglefinus* and whiting *Merlangius merlangus*, although a wide variety of other species such as saithe *Pollachius virens* and monkfish *Lophius piscatorius* are also caught. Total demersal fishing effort has decreased dramatically over the past 10 years. The estimated overall reduction in effort (kW days at sea) by 2013 amounted to 43 % compared to the average for 2004–2006. Most landings in the demersal otter-trawl fishing sector are taken from the northern North Sea (Fig. 12.1) and the fishery is overwhelmingly dominated by Danish, UK, Norwegian and German vessels.

Major *Nephrops norvegicus* (langoustine) grounds in the North Sea include the Flåden Ground, the Farne Deeps (NE England), Botany Gut (central North Sea) and Horns Reef (west of Denmark). Landings of *Nephrops* have increased in recent years, from 10,613 tonnes in 1990 to a maximum of 90,996 tonnes in 2010, and this reflects restrictions on gear types with larger mesh-size, targeting demersal white-fish.

The North Sea beam trawl fishery mainly targets flatfish (sole *Solea solea* and plaice *Pleuronectes platessa*), but is also known to catch cod, whiting and dab *Limanda limanda*. The average distribution of fishing effort in this sector is illustrated in Fig. 12.1 which suggests that beam trawlers typically operate in the southern North Sea. The Dutch beam trawl fleet is the major player in the mixed flatfish fishery, although Belgian and UK-flagged vessels also operate in this fishery. Total fishing effort by the North Sea beam trawl fleet has reduced by 65 % over the last 15 years and there has also been a shift towards electronic pulse trawls more recently (ICES 2014a).

Fisheries for herring use midwater trawl gears (50–55 mm mesh) and target discrete shoals of fish that are located using echosounding equipment. There is also a purse-seine fishery for herring in the eastern North Sea (Dickey-Collas et al. 2013). The stock is fished throughout the year, with peak catches between October and March. Landings of herring in the autumn are predominantly taken from Orkney and Shetland, off Peterhead, northwest of the Dogger Bank and from coastal waters off eastern England. Landings in the spring are concentrated in the south-western North Sea.

The North Sea is subject to major industrial fisheries targeting sandeel *Ammodytes marinus*, Norway pout *Trisopterus esmarkii*, blue whiting *Micromesistius poutassou*, sprat, and juvenile herring. These fish are mainly caught on offshore sand-banks using fine-meshed (8–32 mm) midwater trawls (Dickey-Collas et al. 2013). The sandeel fishery was the largest single-species fishery in Europe with peak landings in 1997 exceeding 1 million tonnes. The fleet has since declined in size. Total sandeel landings in 2013 were 529,141 tonnes (15 % of total landings), Norway pout

Table 12.1 Factors related to the North Sea fishing industry that could be affected by climate change

Biology—Fish and shellfish	Fishery operations	Fish markets and commodity chains
• Year-class strength (recruitment) • Migration patterns • Distribution/habitat suitability • Growth rate • 'Scope for growth' and energetic balance • Phenology (timing of spawning etc.) • Activity levels • Prey availability (match/mismatch) • Exposure to predators • Pathogen and pest incidence • Calcification and internal carbonate balance (shellfish) • Damage/disturbance of key nursery/spawning habitats	• Catchability (performance of the fishing gear) • Vessel safety and stability (e.g. storminess) • Fuel usage (to follow the shifting fish) • Restrictive TACs and quotas (EU relative stability arrangements) • Effectiveness of spatial closures in protecting spawning/nursery areas • New resource species, requiring new fishing gears • Storm damage to ports, harbours and onshore facilities • Damage to gear and vessels (e.g. storm damage to fixed gears) • Preservation of catch on-board vessels • Fouling of vessel hulls • Unwanted 'choke species' that constrain fishing operations	• New markets for novel species • Demand for fish (nationally and internationally) • Storm damage to processing facilities on land • Storm/flooding disruption to transport routes to market • Availability of alternative resources (nationally and internationally) including imports • Changes in processing requirements • Stability of incomes for fishermen and processors • Quality/robustness of product (e.g. shellfish)

landings were 155,752 tonnes (4 %) and blue whiting 17,645 tonnes (0.5 %). All of these short-lived industrial species are thought to be heavily influenced by climatic variability (e.g. Arnott and Ruxton 2002; Hátún et al. 2009).

12.3 Climate Change and Fisheries

There can be many different manifestations of climate change. The most noticeable effect is an increase in average seawater temperature over time, but the seasonality of warming and cooling is also expected to change. The North Sea has witnessed significant warming over the past century at a rate of around 0.3 °C per decade (Mackenzie and Schiedek 2007). The region has been identified as one of 20 'hot spots' of climate change globally, i.e. discrete marine areas where ocean warming has been fastest, as quantified from historical sea surface temperature data (Hobday and Pecl 2014). Projections suggest that the region will continue to experience warming, by around 2–3 °C over the next 100 years (Lowe et al. 2009). Climate change can also encompass other environmental influences or parameters such as changes in precipitation and run-off (and hence salinity and stratification), and storm frequency and intensity (Woolf and Wolf 2013) that may in-turn greatly impact fishing operations, and changes in chemical conditions such as dissolved oxygen concentrations, carbonate chemistry and seawater pH (Blackford and Gilbert 2007).

In this overview of climate change impacts on North Sea fisheries, all of these climatic influences are considered. Climate change will have consequences not only for the animals supporting fisheries (biological responses—see Table 12.1) but also direct and indirect implications for fishery operations—such as storm damage to gear, vessels and infrastructure, changes in catchability of species and maladaptation of quota allocation, etc. (Table 12.1). Furthermore, climate change elsewhere in the world can have consequences for the fishing industry closer to home, via globalised fish markets and commodity chains.

The following sections outline available evidence for climate change effects on fisheries in the North Sea as well as available projections for North Sea fisheries in the future. This assessment is based on Table 12.1, with a discussion of biological, operational and wider market concerns, including analyses of possible economic implications.

12.3.1 Biological Responses

12.3.1.1 Changes in Fish and Fishery Distribution

Long-term changes in seawater temperature and/or other ocean variables often coincide with observed changes in fish distribution. In an analysis of 50 fish species common in waters of the Northeast Atlantic, 70 % had responded to warming by changing distribution and abundance (Simpson et al. 2011). Specifically, warm-water species with smaller maximum body size had increased in abundance throughout northwest Europe while cold-water, large-bodied species had decreased in abundance.

Distribution and abundance are the traits that are the most readily observed responses. However, many processes interact when considering fisheries and climate change, and these are a manifestation of both biological and human processes. None of these factors act in isolation and many are synergistic. The responses are rarely linear. In fish, it is clear that climate affects physiology and behaviour. These processes interact to influence migration, productivity (growth of populations minus decline in populations), susceptibility to disease and interactions with other organisms. Changes in distribution and abundance are the aggregate responses to these changed processes.

Archaeological evidence can sometimes yield useful insights into historical changes in the distribution and productivity of fish and the response of fisheries. The bones of warm-water species such as red mullet *Mullus surmuletus* have been recovered from archaeological excavations throughout northern Europe. This species has only recently returned to the North Sea in reasonable numbers (Beare et al. 2005), but was apparently widespread during the Roman period (AD 64–400) (Barrett et al. 2004). Enghoff et al. (2007) listed a number of occurrences of warm-water species (e.g. red mullet, seabass *Dicentrarchus labrax*, anchovy *Engraulis encrasicolus*, and seabream *Spondyliosoma cantharus*) among bone assemblages, surrounding the North Sea, from the 1st to the 16th century AD. Alheit and Hagen (1997) identified nine periods, each lasting several decades, during which large quantities of herring were caught close to the shore in the North Sea. Each of these coincided with severe winters in western Europe with extremely cold air and water temperatures and a reduction in westerly winds; physical factors associated with negative anomalies of the North Atlantic Oscillation (NAO) index.

Highly-cited studies using time-series from fishery-independent surveys (Beare et al. 2004a; Perry et al. 2005; Dulvy et al. 2008) have revealed that centres of fish distribution in the North Sea shifted by distances ranging from 48 to 403 km during the period 1977–2001, and that the North Sea demersal fish assemblage has deepened by about 3.6 m per decade over the past 30 years (Dulvy et al. 2008). Species richness increased from 1985 to 2006 which Hiddink and Ter Hofstede (2008) suggested was related to climate change. Eight times as many fish species displayed increased distribution ranges in the North Sea (mainly small-sized species of southerly origin) compared to those whose range decreased (primarily large and northerly species). For a more localised region of the Dutch coast, van Hal et al. (2014) demonstrated latitudinal range shifts and changes in abundance of two non-commercial North Sea fish species, solenette *Buglossidium luteum* and scaldfish *Arnoglossus laterna* that were strongly related to the warming of the coastal waters. For pelagic fish species, a recent paper by Montero-Serra et al. (2015) investigated the patterns of species-level change using records from 57,870 fisheries-independent survey trawls from across the European continental shelf between 1965 and 2012. These authors noted a strong 'subtropicalisation' of the North Sea as well as the Baltic Sea. In both areas, there has been a shift from cold-water assemblages typically characterised by Atlantic herring and sprat from the 1960s to 1980s, to warmer-water assemblages typified by mackerel, horse mackerel *Trachurus trachurus*, sardine *Sardina pilchardus* and anchovy from the 1990s onwards. The primary measure correlated to changes in all species was sea surface temperature (Montero-Serra et al. 2015).

Analyses of Scottish and English commercial catch data in the North Sea spanning the period 1913–2007 have revealed that the locations where peak catches of target species such as cod, haddock, plaice and sole were obtained have all shifted over the past 100 years, albeit not in a consistent way (Engelhard et al. 2011, 2014b). For example, catches of cod seem to have shifted steadily north-eastward and towards deeper water in the North Sea (Engelhard et al. 2014b) and this reflects both climatic influences and intensive fishing. Plaice distribution has shifted north-westwards (Fig. 12.2) towards the central North Sea, again reflecting climatic influences, in particular sea surface temperature as also confirmed by van Keeken et al. (2007). Somewhat confusingly, sole seems to have retreated away from the Dutch coast, southwards towards the eastern Channel although this too is thought to have been a response to warming. Sole is a warm-water species that traditionally moved offshore in winter to avoid excessively low temperatures in the shallows. Cold winters are known to have coincided with mass die-offs of sole (e.g. Woodhead 1964), but in recent years shallower waters surrounding the North Sea have remained habitable all year round (winter conditions are less severe), and hence the apparent southward and shallowing shift (Engelhard et al. 2011). Haddock catches have moved very little in terms of their centre of distribution, but their southern boundary has shifted northwards by approximately 130 km over the past 80–90 years (Skinner 2009).

Theoretically, in the northern hemisphere, warming results in a distributional shift northward, and cooling draws species southward (Burrows et al. 2007). Heath (2007) looked at patterns in international fisheries landings for the whole Northeast Atlantic region. Densities of landings of each species were summed by decade and expressed as a proportion of the total. Both northerly and southerly shifts were observed between decades for individual species, however more species shifted south than north between the 1970s and 1980s (a relatively cool period) and vice versa between the 1980s and 1990s (a relatively warm period). This seems to parallel observed inter-decadal changes in sea and air temperatures.

Distribution shifts will have 'knock on' implications for commercial fisheries catches because changes in migration or spawning location affect the 'availability' of resources to fishing fleets. Populations may move away from or towards the area where particular fishing fleets operate and/or where spatial restrictions on fishing are in place. Furthermore, species distributions may migrate across political boundaries where quotas belong to different nations. A notable example has arisen recently as a result of quota allocations between

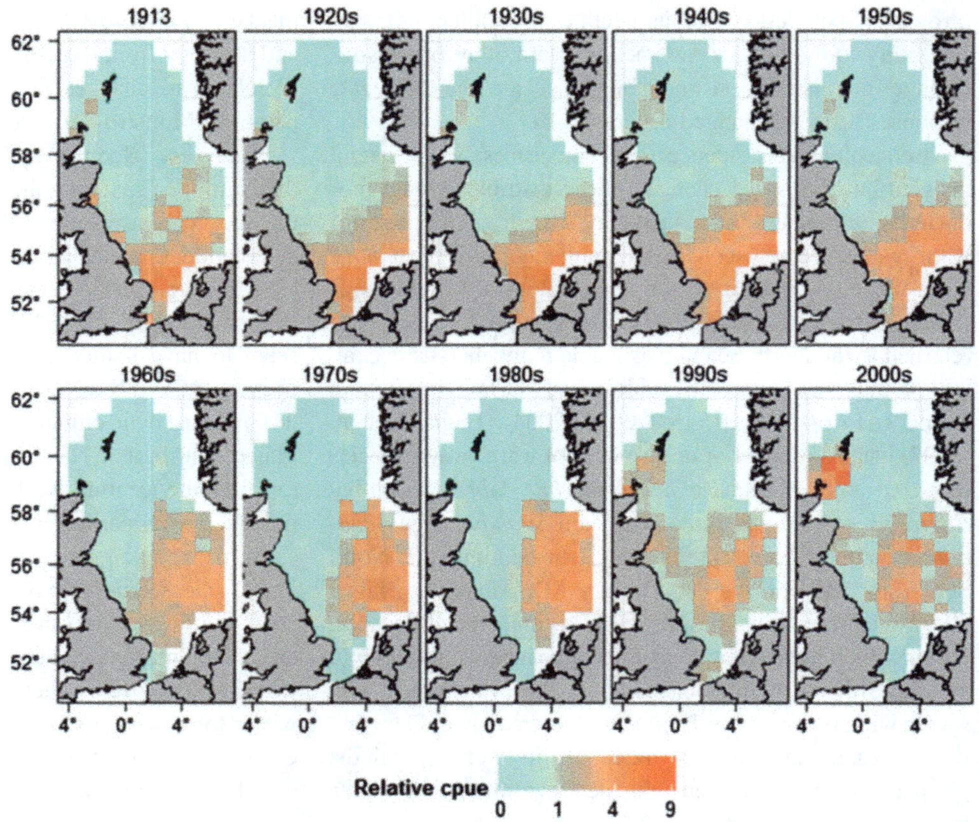

Fig. 12.2 Decadal change in North Sea plaice distribution, 1920s to 2000s, based on fisheries catch-per-unit-effort (CPUE). Shading is proportional to plaice CPUE, normalised by decade and corrected for the average spawning stock biomass (SSB). Adapted from Engelhard et al. (2011)

Norway and the EU, and between Iceland, the Faroe Islands and the EU. In October 2009, North Sea mackerel appeared to have moved away from the Norwegian Sector (possibly as a result of excessively cold conditions near the Norwegian coast), resulting in disagreements over permissible catches by Norwegian boats in EU waters. Norwegian vessels were forcibly evicted by UK fishery patrol vessels, once they had caught their allotted quota (see *Fishing News*, 9 October 2009). At the same time Iceland and the Faroe Islands unilaterally claimed quota for mackerel (146,000 and 150,000 tonnes respectively in 2011 or 46 % of the total allowable catch, TAC), since the species had suddenly attained high abundance in their territorial waters. Whether the apparent changes in mackerel distribution westwards across the northern North Sea were a result of long-term climate change or not remains unclear. Hughes et al. (2014) suggested that sea surface temperature had a significant positive association with the observed northward and westward movement of mackerel, equivalent to a displacement of 37.7 km per °C (based on spring mean sea surface temperature for the region). By contrast, historical appearances of mackerel in the western North Sea and off the coast of Iceland (Beare et al. 2004a) coincided with warming periods linked to the Atlantic Multidecadal Oscillation (AMO) and might not be symptomatic of long-term climate change. Whatever the case—with climate change in the future, more territorial disagreements of this type could be anticipated (Hannesson 2007) and fisheries management will need to adapt accordingly (Link et al. 2011).

A similar phenomenon is now occurring in the English Channel and southern North Sea region with regard to access to European anchovy. Anchovy stocks are currently depleted in the Bay of Biscay where Spanish and French vessels operate, but are increasing further north along southern coasts of the UK and especially along Dutch coasts (Beare et al. 2004b) where they are starting to be targeted by pelagic fishing vessels. Detailed political negotiations are underway to determine whether Spanish and French vessels should be allowed exclusive access in areas where previously they had no quota, and indeed whether the more northerly distributed anchovy represent the same or a genetically different sub-stock to those in the Bay of Biscay. In 2012 a study was published (Petitgas et al. 2012) drawing on four different strands of evidence: genetic studies, larval transport modelling, survey time series and physical oceanographic models. The study concluded that anchovy in the southern North Sea are most likely to be a distinct remnant sub-stock that was previously present (see Aurich 1953), but is now benefiting from greatly improved climatic conditions rather than an invasion of animals from further south. According to

Alheit et al. (2012), the anchovy population from the western Channel (not from the Bay of Biscay) invaded the North Sea and Baltic Sea during positive periods of the AMO. Given this evidence and according to the rules of 'relative stability' within the EU Common Fisheries Policy, Spanish and French vessels would not necessarily be granted exclusive access to this expanding resource, unlike the present situation in the Bay of Biscay.

Under the EU Common Fisheries Policy, a number of closed areas have been implemented as 'technical measures' to conserve particular species and to protect nursery or spawning grounds. In the North Sea, these include closure areas to protect plaice, herring, Norway pout and sandeel. If species shift their distribution in response to climate change then it is possible that such measures will become less effective in the future (van Keeken et al. 2007). Juvenile plaice are typically concentrated in shallow inshore waters of the southeast North Sea and move gradually offshore as they grow. In order to reduce discarding of undersized plaice, thereby decreasing mortality and enhancing recruitment to the fishery, the EU 'Plaice Box' was introduced in 1989, excluding access to beam and otter trawlers larger than 300 hp. However recent surveys in the Wadden Sea have shown that 1-group plaice are now completely absent from the area where they were once very abundant. Consequently, the 'Plaice Box' is now less effective as a management measure for plaice than was the case 10 or 15 years ago. The boundaries of, and expected benefits from marine protected areas (MPAs) may need to be 'adaptive' in the future in the context of climate change. Cheung et al. (2012) looked at other fishery closure areas in the North Sea and noted that they will most likely experience between 2 and 3 °C increases in temperature over the next 80–100 years and consequently it is unlikely that the species they are designed to protect now will occur there in the same numbers in the future given defined temperature tolerances or preferences of specific fishes (Freitas et al. 2007; Pörtner and Peck 2010).

Fishers have witnessed and responded to a number of new opportunities in recent years, as warm-water species have moved into the North Sea and/or their exploitation has become commercially viable for the first time. Notable examples include new or expanding fisheries for seabass, red mullet, John dory *Zeus faber*, anchovy and squid *Loligo forbesi*.

Biomass estimates for seabass in the eastern Channel quadrupled from around 500 tonnes in 1985, to in excess of 2100 tonnes in 2004/2005, with populations also increasing rapidly in the southern North Sea (Pawson et al. 2007). This was attributed to an increase in seawater temperature, especially in the winter and has resulted in a dramatic expansion of seabass fisheries both within the commercial sector and the recreational fishing sector. Seabass are caught by angling on the east coast of Scotland and in Norway, but the northernmost limit of the commercial seabass fishery is around Yorkshire (54°N) in the North Sea. In 2013, 2243 tonnes of seabass were landed by countries surrounding the North Sea and eastern English Channel (Fig. 12.3), compared with only 210 tonnes in 1990. However recent anecdotal evidence (ICES 2012) seems to suggest that the increase in catches may have slowed slightly, as a result of successive cold winters in 2009/10, 2010/11, and 2011/12 likely leading to poor recruitment (Fig. 12.3).

Red mullet is a non-quota species of moderate, but increasing, importance to North Sea fisheries. From 1990 onwards, international landings increased strongly. France is the main country targeting this species although UK and Dutch commercial catches have also increased. Total international landings rose from only 537 tonnes in 1990 to a peak in landings of 4555 tonnes in 2007. Beare et al. (2005) demonstrated that red mullet is one of many species that have become significantly more prevalent in North Sea bottom trawl surveys in recent years, rising from near-absence during surveys between 1925 and 1990, to about 0.1–4 fish per hour of trawling between 1994 and 2004. Red mullet is also among the fish species that have entered the North Sea from both the south and north-west, through the Channel and along the Scottish coast, respectively (Beare et al. 2005).

Although numbers are highly uncertain, there are strong indications that squid are generally becoming more abundant in the North Sea, possibly in response to a change in climate (Hastie et al. 2009a). Cephalopod populations are suggested to be highly responsive to climate change (Sims et al. 2001; Hastie et al. 2009a) and growth in squid availability in the North Sea is generating considerable interest among fishers off the Scottish coast (Hastie et al. 2009b). Off north-east Scotland, where most of the squid are found, more boats are now trawling for squid than for the region's traditional target species, such as haddock and cod (Hastie et al. 2009b). New squid fisheries are also emerging in the Netherlands using bright lamps and hooked lines (*Fish News* September 2007). Total international landings have risen from 2612 tonnes in 1990 (375 tonnes in 1980) to 3417 tonnes in 2013 (see Fig. 12.3). In the English Channel, loliginid squid catches seem to be related to mean sea surface temperature (Robin and Denis 1999). Temperature appears to influence recruitment strength and overall distribution (Hastie et al. 2009a).

The North Sea bottom trawl fleet typically catches many different species in the same haul, thus making it virtually impossible to devise effective management measures that are well suited to the protection or rebuilding of any particular stock without affecting others. In October 2014, the EU introduced reforms to the Common Fisheries Policy that included a ban on discarding and thus a requirement to land all fish caught. To allow fishers to adapt to the change, the landing obligation will be introduced gradually, between

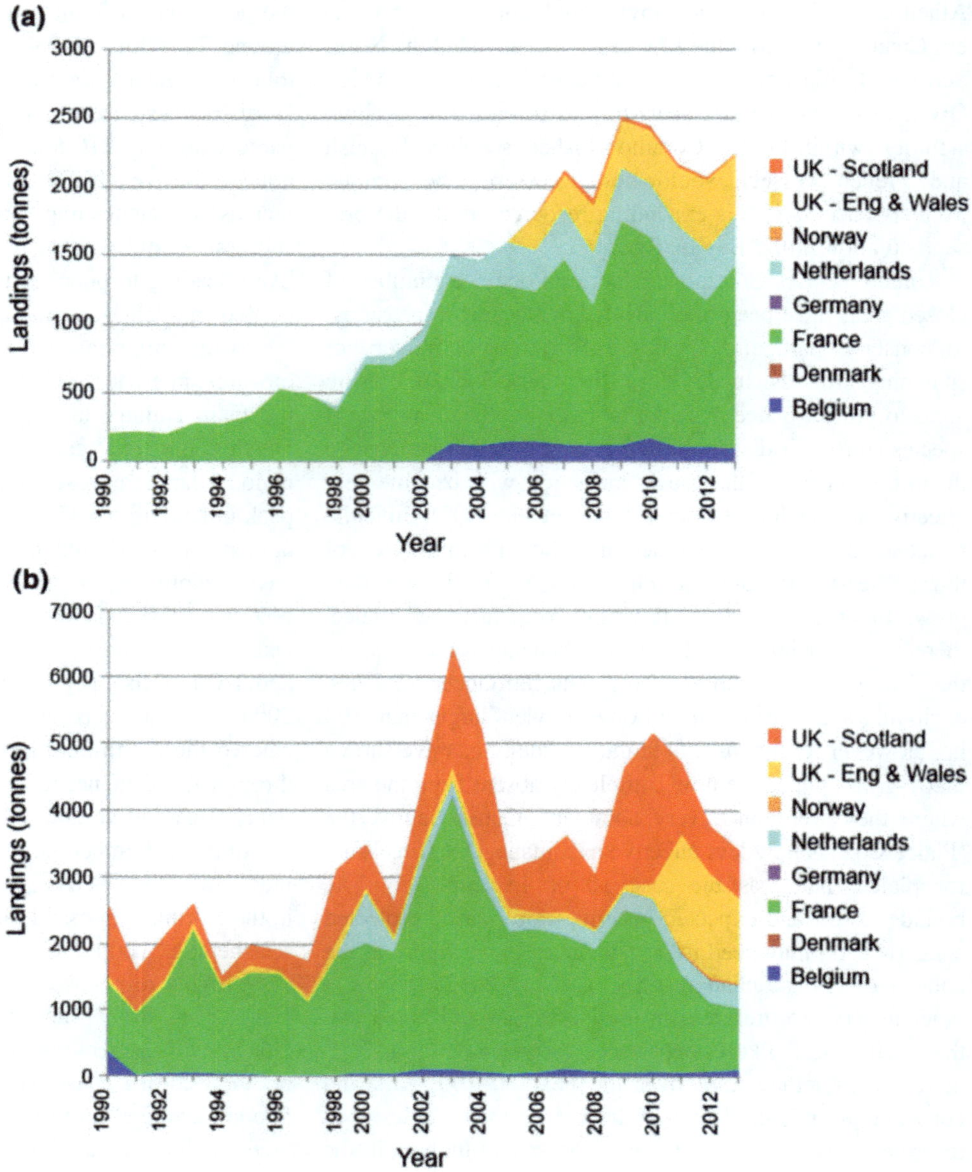

Fig. 12.3 International fishery landings of seabass (*upper*) and squid (*lower*) in the North Sea and eastern English Channel (data for 1999 were excluded as no French data were submitted to ICES in that year)

2015 and 2019 for all commercial fisheries (species under TACs, or under minimum sizes), however this new measure necessitates that once the least plentiful quota species in a mixed fishery—the 'choke species'—is exhausted, the whole fishery must cease operation. Baudron and Fernandez (2015) have argued that many commercial fish stocks are beginning to recover under more sustainable exploitation regimes and, in some cases, as a result of favourable climatic conditions. For example, northern European hake *Merluccius merluccius* a warm-water species, witnessed a dramatic increase in biomass between 2004 and 2011 and has recolonised the northern North Sea where hake had largely been absent for over 50 years. These changes have implications for the management of other stocks. Notably, if discards are banned as part of management revisions, the relatively low quota for hake in the North Sea will be a limiting factor (the so-called 'choke' species) which may result in a premature closure of the entire demersal mixed fishery (Baudron and Fernandez 2015).

Modelling strategies for predicting the potential impacts of climate change on the natural distribution of species and consequently the response of fisheries have often focused on the characterisation of a species' 'bioclimate envelope' (Pearson and Dawson 2003). In other words, by looking at the current range of temperatures inhabited by a species, it is possible to predict future distribution, on the basis that the physical environment in an area is likely to change in the future. Model simulations suggest that distributions of exploited species will continue to shift in the next five decades both globally and in the Northeast Atlantic specifically (Cheung et al. 2009, 2010, 2011; Lindegren et al. 2010).

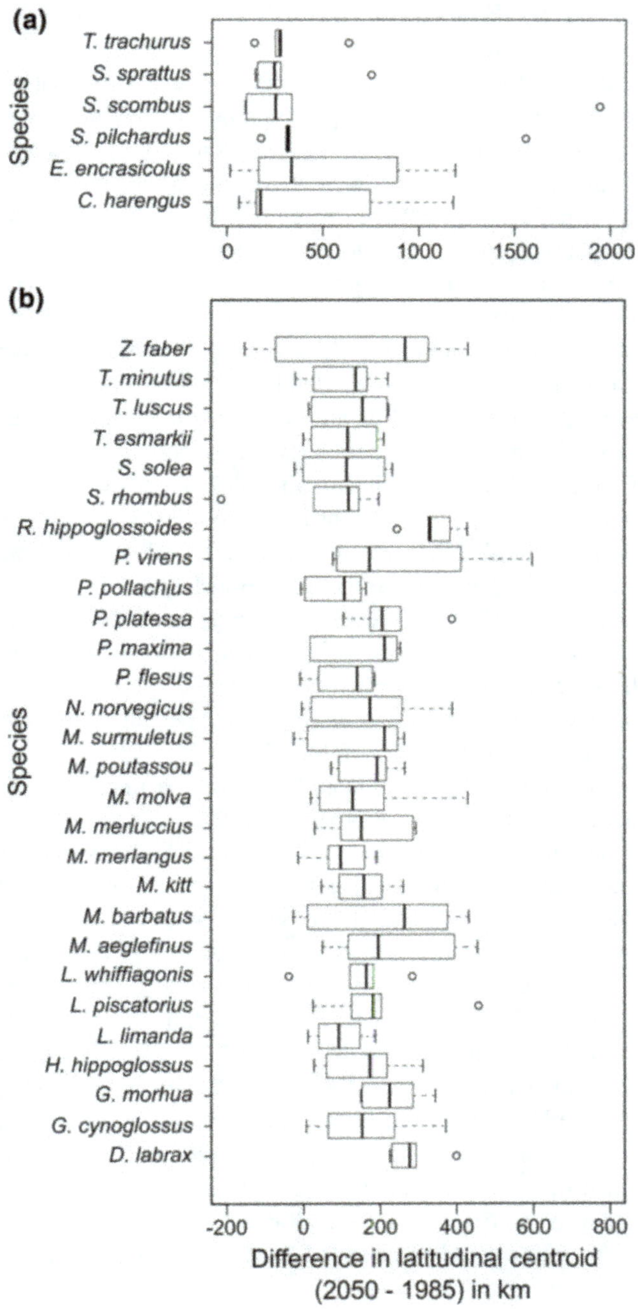

Fig. 12.4 Projected change in latitudinal centroids of habitat suitability surfaces from 1985 to 2050 across species distribution models and climatic datasets for pelagic species (*upper*) and demersal species (*lower*) (Defra 2013). *Thick vertical lines* represent median values, the left and right ends of each *box* show the upper and lower quartiles of the data and the whiskers the most extreme data points no greater than 1.5 times the inter-quartile range. Outliers that were more extreme than whiskers are represented as *circles*

It is important to test the reliability and robustness of tools projecting climate-driven shifts in fisheries resources. Jones et al. (2012) published a localised analysis for the North Sea and Northeast Atlantic whereby three different bioclimate envelope models (AquaMaps, Maxent and DBEM) were applied to the same present distribution datasets and the same environmental input parameters. As indicated by the test statistics, each method produced a plausible present distribution and estimate of habitats suitable for each species (14 commercial fish). When used to make projections into the future, the ensemble of models suggested northward shifts at an average rate of 27 km per decade (the current rate is around 20 km per decade for common fish in the North Sea, Dulvy et al. 2008). This modelling approach was extended to include several additional, commercial species (squid *Loligo vulgaris*, seabass, sardine, sprat, John dory, anchovy, plaice, herring, mackerel, halibut *Hippoglossus hippoglossus*, red mullet etc.) as part of a Defra study (Defra 2013). The species predicted to move the furthest were anchovy, sardine, Greenland halibut, John dory and seabass (i.e. *E. encrasicolus*, *S. pilchardus*, *R. hippoglossoides*, *Z. faber* and *D. labrax* respectively, see Fig. 12.4).

By contrast Rutterford et al. (2015) used the same fish survey datasets for the North Sea, together with generalised additive models (GAMs), to predict trends in the future distribution of species, but came to the conclusion that fish species over the next 50 years will be strongly constrained by the availability of suitable habitat in the North Sea, especially in terms of preferred depths. The authors found no consistent pattern among species in predicted changes in distribution. On the basis of the GAM results the authors suggested that they did not expect or predict substantial further deepening (as previously observed by Dulvy et al. 2008), and that the capacity of fish to remain in cooler water by changing their depth distribution had been largely exhausted by the 1980s, that fish with preferences for cooler water are being increasingly exposed to higher temperatures, with expected physiological, life history and negative population consequences.

Beaugrand et al. (2011) described a model to map the future spatial distribution of Atlantic cod. The model, which they named the non-parametric probabilistic ecological niche model (NPPEN), suggested that cod may eventually disappear as a commercial species from some regions including the North Sea where a sustained decline has already been documented; in contrast, the abundance of cod is likely to increase in the Barents Sea. Lenoir et al. (2011) applied the same NPPEN model with multiple explanatory variables (sea surface temperature, salinity, and bathymetry) to predict the distribution of eight fish species up to the 2090s for the Northeast Atlantic. This study anticipated that by the 2090s horse mackerel and anchovy would show an increased probability of occurrence in northern waters compared with the 1960s, that pollack *Pollachius pollachius*, haddock and saithe would show a decrease in the south, and that turbot *Scophthalmus maximus* and sprat would show no overall change in probability (−0.2 to +0.2) anywhere.

French scientists from IFREMER have used a delta GAM/GLM approach to model future plaice and red mullet distribution in the eastern English Channel and southern North Sea (see Vaz and Loots 2009). Abundance of each species was related to depth, sediment type, bottom salinity and temperature. Results suggested that climate change may strongly affect the future distribution of plaice. For large plaice (>18 cm), distribution will still be centred in the southern part of the North Sea, however for young individuals, the predicted distribution is anticipated to shift north-westwards and to the Dogger Bank area in particular (as has already been observed, see van Keeken et al. 2007; Engelhard et al. 2011). Model outputs indicate that the distribution of red mullet will not change dramatically but that for young individuals (defined as <17.3 cm), the offshore habitat situated on the Dogger Bank may become increasingly favourable. Older individuals seemed little affected by the simulated change in environment, but they may benefit from higher juvenile survival and expand their area of occupation as a result.

There are some concerns about the validity of the bioclimate envelope approach for predicting the future distribution of commercially important fish species (see Jennings and Brander 2010; Heath et al. 2012). First, it may not be possible to assess temperature preferences from current distributions because the observed distributions are modified by abundance, habitat, predator and prey abundance and competition. Second, there may be barriers to dispersal (although this is typically less of an issue in the sea than on land) and species will move at different rates and encounter different local ecologies as temperature changes (Davis et al. 1998). A more detailed, physiologically-based approach has been taken by some authors, whereby the detailed dynamics of individual animals are modelled, often by linking complex biophysical models (forced with the output from Global Climate Models) to sub-routines which replicate the behaviour/characteristics of eggs, larvae, juveniles or adults. Teal et al. (2008) reported a study of plaice and sole distribution in the North Sea, in which they predicted size- and season-specific fish distributions based on the physiology of the species, temperature and food conditions in the sea. This study combined state-of-the-art dynamic energy budget (DEB) models with output from a biogeochemical ecosystem model (ERSEM) forced with observed climatic data for 1989 and 2002, with contrasting temperature and food conditions. The resulting habitat quality maps were in broad agreement with observed ontogenetic and seasonal changes in distribution as well as with long-term observed changes in distribution. The technique has recently been extended to provide future projections up to year 2050, assuming moderate climate warming (L. Teal, pers. comm. IMARES, Netherlands).

12.3.1.2 Year-Class Strength and Implications for Fisheries

Fishers and scientists have known for over 100 years that the status of fish stocks can be greatly influenced by prevailing climatic conditions (Hjort 1914; Cushing 1982). 'Recruitment' variability is a key measure of stock productivity, and is defined as the number of juvenile fish surviving from the annual egg production to be exploited by the fishery. Recruitment is critically dependent on the match or mismatch between the occurrence of the larvae and availability of their zooplankton food (Cushing 1990) as well as other processes that affect early life-history stages (see Petitgas et al. 2013). Empirical data on exploited populations often show strong relationships between recruitment success, fisheries catches and climatic variables. These strong relationships have been demonstrated, for example, for cod (O'Brien et al. 2000; Brander and Mohn 2004; Cook and Heath 2005), plaice (Brunel and Boucher 2007), herring (Nash and Dickey-Collas 2005), mackerel (Jansen and Gislason 2011) and seabass (Pawson 1992). Correlations have been found between fish recruitment and various climate variables, including sea surface temperature, the NAO and even offshore winds (Table 12.2).

A number of publications describing the impact of climate variability (e.g. the NAO and AMO) on small pelagic fishes such as herring, anchovy or sardine in the North Sea have been published in recent years, for example those of Alheit et al. (2012) and Gröger et al. (2010). According to the most recent assessment of the UN Intergovernmental Panel on Climate Change (IPCC), the NAO is one of the climate indices for which it is most difficult to provide accurate future projections (IPCC 2013). Recent multi-model studies (e.g. Karpechko 2010) suggest overall that the NAO is likely to become slightly more positive (on average) in the future due to increased greenhouse gas emissions. Consequently, a slight tendency towards enhanced recruitment and larval abundance of these species in the future could be expected if the relationships observed in the past continue to hold.

In the case of cod, there is a well-established relationship between recruitment and sea temperature (O'Brien et al. 2000; Beaugrand et al. 2003), but this relationship varies with regard to the different cod stocks that inhabit the North Atlantic (Planque and Frédou 1999). For stocks at the northern extremes (e.g. in the Barents Sea) or the western Atlantic (e.g. Labrador), warming leads to enhanced recruitment, while in the North Sea, close to the southern limits of the range, warmer conditions lead to weaker than average year classes (Drinkwater 2005). During the late 1960s and early 1970s, cold conditions were correlated with a sequence of positive recruitment years in North Sea cod

Table 12.2 Demonstrated correlations between recruitment success (year-class strength) and climatic variables for important fish and shellfish stocks in northwest Europe

Species	Source	Sea area	Correlation	Climate parameter
Plaice	van der Veer and Witte (1999) and Brunel and Boucher (2007)	Southern North Sea, English Channel	Negative	SST (Feb)
Herring	Nash and Dickey-Collas (2005)	North Sea	Negative	SST (winter)
Herring	Gröger et al. (2010)	North Sea	Positive	Winter NAO, AMO
Scallop	Shephard et al. (2010)	Isle of Man	Positive	SST (spring)
Cod	Brander and Mohn (2004)	North Sea, Irish Sea	Positive	Winter NAO
Cod	Planque and Frédou (1999) and O'Brien et al. (2000)	North Sea, West of Scotland, Irish Sea	Negative	SST (Feb–Jun)
Sandeel	Arnott and Ruxton (2002)	North Sea	Negative, Positive	Winter NAO, SST (spring)
Mackerel	Jansen and Gislason (2011)	North Sea	Unclear	SST (summer)
Brown shrimp	Siegel et al. (2005) and Henderson et al. (2006)	Bristol Channel, Wadden Sea	Positive, Negative	SST (Jan–Aug), Winter NAO
Sole	Henderson and Seaby (2005)	Bristol Channel	Positive	SST (spring)
Seabass	Pawson (1992)	English Channel	Positive	SST (summer)
Squid	Robin and Denis (1999)	English Channel	Positive	SST (winter)
Turbot	Riley et al. (1981)	Coastal UK	Positive	Offshore wind
Whiting	Cook and Heath (2005)	North Sea	Positive	SST
Saithe	Cook and Heath (2005)	North Sea	Positive	SST

SST Sea surface temperature; *NAO* North Atlantic Oscillation; *AMO* Atlantic Multi-decadal Oscillation

and subsequently high fisheries catches for a number of years thereafter (Heath and Brander 2001). In recent years however, despite several cold winters, cod have suffered very poor recruitment in the North Sea, although it is unclear whether this is a direct consequence of changed climatic conditions, reduced availability of planktonic prey items for larval fish or over-fishing of the parental stock (i.e. some sort of 'Allee effect') (Mieszkowska et al. 2009), or more intensive predation of cod larvae by pelagic fish stocks which have increased, such as herring and sprat (Engelhard et al. 2014a).

A clear seasonal shift to earlier appearance of fish larvae has been described for several species at Helgoland Roads in the southern North Sea (Greve et al. 2005), and this has been linked to marked changes in zooplankton composition and sea surface temperature in this region (Beaugrand et al. 2002). In particular, there has been a decline in the abundance of the copepod *Calanus finmarchicus* but an increase in the closely related but smaller species *C. helgolandicus*. *Calanus finmarchicus* is a key prey item for cod larvae in the northern North Sea, and the loss of this species has been correlated with recent failures in cod recruitment and an apparent increase in flatfish recruitment (Reid et al. 2001, 2003; Beaugrand et al. 2003). *Calanus helgolandicus* occur at the wrong time of the year to be of use to emerging cod larvae. Greve et al. (2005) suggested that in ten cases both the 'start of season' and 'end of season' (Julian date on which 15 and 85 % of all larvae were recorded respectively), were correlated with sea surface temperature. Strongly significant relationships were observed for plaice, sole and horse mackerel as well as for many non-commercial species including scaldfish, and Norway bullhead *Taurulus lilljeborgi*.

Fincham et al. (2013) examined the date of peak spawning for seven sole stocks based on market sampling data in England and the Netherlands. Four out of seven stocks were shown to have exhibited a significant long-term trend towards earlier spawning (including the east-central North Sea, southern North Sea, and eastern English Channel) at a rate of 1.5 weeks per decade since 1970. Sea surface temperature during winter affected the date of peak spawning, although the effect differed between stocks. Recruitment is critically dependent on the match or mismatch between the occurrence of the larvae and the availability of their food (Cushing 1990) and other climate-sensitive processes (Peck and Hufnagl 2012; Llopiz et al. 2014), thus a change in spawning date could have knock-on effects for larval survival and hence future fisheries.

It is important to note that extensive fishing can cause fish populations to become more vulnerable to short-term natural climate variability (e.g. Ottersen et al. 2006) by making such populations less able to 'buffer' against the effects of the occasional poor year classes. Conversely, long-term climate change may make stocks more vulnerable to fishing, by reducing the overall 'carrying capacity' of the stock, such that it might not be sustained at, or expected to recover to, levels observed in the past (Jennings and Blanchard 2004). Cook and Heath (2005) examined the relationship between sea surface temperature and recruitment in a number of North Sea fish species (cod, haddock, whiting, saithe, plaice, sole) and concluded that if the recent warming period were to continue, as suggested by climate models, stocks which express a negative relationship with temperature (including cod) might be expected to support much smaller fisheries in the future. In the case of cod, climate change has been estimated to have been eroding the maximum sustainable yield at a rate of 32,000 tonnes per decade since 1980. Calculations show that the North Sea cod stock, could still support a sustainable fishery under a warmer climate but only at very much lower levels of fishing mortality, and that current 'precautionary reference' limits or targets (such as F_{MSY}), calculated by International Council for the Exploration of the Sea (ICES) on the basis of historic time series, may be unrealistically optimistic in the future.

For Atlantic mackerel, increases in sea surface temperature are known to affect growth, recruitment and migration with subsequent impacts on permissible levels of exploitation (Jansen and Gislason 2011). Jansen et al. (2012) used information on larval fish from the Continuous Plankton Recorder (CPR) to show that abundance has declined dramatically in the North Sea since the 1970s and also that the spatial distribution of mackerel larvae seems to have changed. Whether these trends can be ascribed to changes in climatic conditions remains unclear, although development and/or mortality of mackerel eggs is known to be very sensitive to seawater temperature (Mendiola et al. 2007).

There have been many attempts to include climatic variables in single-species population models (e.g. Hollowed et al. 2009), and thereby to project how the productivity of fish stocks will be affected by climate change in the future. Particular emphasis has been placed on climatic determinants of fish recruitment, and indeed several studies have inserted temperature or other environmental terms within the 'stock-recruit' relationship in order to make medium or long-term forecasts. Clarke et al. (2003) used projections of future North Sea temperatures and estimated the likely impact of climate change on the reproductive capacity of cod, assuming that the high level of mortality inflicted by the fishing industry (in 2003) continued into the future. Outputs from the model suggested that the cod population would decline, even without a significant temperature increase. However, scenarios with higher rates of temperature increase resulted in faster rates of decline. In a re-analysis by Kell et al. (2005), the authors modelled the effect of introducing a 'cod recovery plan' (as being implemented by the European Commission), under which catches were set each year so that stock biomass increased by 30 % annually until the cod stock had recovered to around 150,000 tonnes. The length of time needed for the cod stock to recover was not greatly affected by the particular climate scenario chosen (and was generally around five to six years), although overall productivity was affected and spawning stock biomass (SSB) once 'recovered' was projected to be considerably less than would have been the case assuming no temperature increase (251,035 tonnes compared to 286,689 tonnes in 2015).

12.3.1.3 Ocean Acidification and Low Oxygen

Carbon dioxide (CO_2) concentrations in the atmosphere are rising as a result of human activities and are projected to increase further by the end of the century, as carbon-rich fossil fuels such as coal and oil continue to be burned (Caldeira and Wickett 2003). Uptake of CO_2 from the air is the primary driver of ocean acidification. Modelling and observational studies suggest that the absorption of CO_2 by seawater has already decreased pH levels in the global ocean by 0.1 pH units since 1750 (Orr et al. 2005), which equates to an average increase in surface ocean water acidity worldwide of about 30 % since pre-industrial times, and that the present rate of change is faster than at any time during the previous 55 million years (Pearson and Palmer 2000).

Modelled estimates of future seawater pH in the North Sea are generally consistent with global projections. However, variability and uncertainties are considerable due to riverine inputs, biologically-driven processes and geochemical interactions between the water column and sea-bottom sediments (Blackford and Gilbert 2007; Artioli et al. 2012). Under a high CO_2 emission scenario, model outputs indicate that much of the North Sea seafloor is likely to become undersaturated with regard to aragonite (a form of calcium carbonate used by some marine organisms to build their shells or skeletons) during late winter/early spring by 2100 (Artioli et al. 2012). Ocean acidification may have direct and indirect impacts on the recruitment, growth and survival of exploited species (Fabry et al. 2008; Llopiz et al. 2014) and some species may become more vulnerable to ocean acidification with increases in temperature (Hale et al. 2011). Impacts may be particularly apparent for animals with calcium carbonate shells and skeletons such as molluscs, some crustaceans, and echinoderms (Hendriks et al. 2010; Kroeker et al. 2010), but studies show large variations in responses to ocean acidification between and within taxonomic groups. Several major programmes of research are underway in Europe to determine the possible consequences of future ocean acidification. In laboratory studies, significant effects have been noted for several important commercial shellfish species, notably mussels, oysters, lobster and *Nephrops* (Gazeau et al. 2010; Agnalt et al. 2013; Styf et al. 2013).

A preliminary assessment in 2012 estimated the potential economic losses to the UK shellfish industry under ocean acidification (Pinnegar et al. 2012). Four of the ten most valuable marine fishery species in the UK are calcifying shellfish and the analyses suggested losses in the mollusc fishery (scallops, mussels, cockles, whelks etc.) could amount to GBP 55–379 million per year by 2080 depending on the CO_2 emission scenario. Thus, there is a clear economic reason to improve understanding of physiological and behavioural responses to ocean acidification, and a Europe-wide assessment of economic consequences is currently underway within the German BIOACID programme.

Fin-fish species are thought to be most vulnerable to ocean acidification during their earliest life stages, although experiments on North Sea species (such as cod and herring) have so far shown that they are relatively robust (e.g. Franke and Clemmesen 2011). Indirect food-web effects may be more important for fin-fish, than direct physiological impacts (Le Quesne and Pinnegar 2012). To date, few studies have attempted to investigate the potential bottom-up impacts of ocean acidification on marine food webs, and hence on fisheries (although see Kaplan et al. 2010; Ainsworth et al. 2011; Griffith et al. 2011). In their economic analysis, Cooley and Doney (2009) did account for "fish that prey directly on calcifiers". These authors suggested that indirect economic consequences of ocean acidification could be substantial. Clearly, more work is needed before definitive conclusions can be drawn about the socio-economic implications of ocean acidification for the fishing industry and society as a whole.

Reduced oxygen concentrations in marine waters have been cited as a major cause for concern globally (Diaz and Rosenberg 2008), and there is evidence (Queste et al. 2012) that areas of low oxygen saturation have started to proliferate in the North Sea. Whether these changes are a result of long-term climate change remains unclear and it is also unclear whether such changes will impact on commercial fish species and their fisheries. Unlike parts of the Baltic Sea, which regularly experience complete anoxia (lack of oxygen), regions of the North Sea only experience reduced oxygen conditions (65–70 % saturation, 180–200 μMol dm^{-3}). Therefore it is uncertain whether North Sea fish stocks will suffer major mortality of eggs and larvae due to changes in oxygen levels (as is the case in the Baltic Sea), and it is perhaps more likely that they will experience more subtle, non-lethal, effects in the future. Several authors have highlighted how oxygen concentrations, low pH and elevated temperature interact and determine 'scope for growth' (e.g. Pörtner and Knust 2007). These findings have been used as the basis for models predicting size and distribution in North East Atlantic fishes (Cheung et al. 2013). Laboratory studies concerning low oxygen conditions have been used to predict fish distribution and habitat suitability (Cucco et al. 2012). Some types of organism are more affected than others. Larger fish and spawning individuals can be more affected by low oxygen levels owing to their higher metabolic rates (Pörtner and Farrell 2008). There are certain thresholds below which oxygen levels affect the aerobic performance of marine organisms (Pörtner 2010), although this is very dependent on the species or type of organism, respiration mode, and metabolic and physiological requirements, with highly active species being less tolerant of low oxygen conditions (Stramma et al. 2011). *Nephrops norvegicus* juveniles show sub-lethal effects at oxygen concentrations below 156 μMol dm^{-3}, but adults are more robust, although their ability to tolerate other environmental stresses (for example elevated temperature) is severely compromised (Baden et al. 1990). There are few projections of future oxygen concentrations in the North Sea, although modelling was undertaken for three locations by van der Molen et al. (2013). These authors were able to provide some insight into future conditions, assuming a SRES A1B scenario to 2100. In particular, the model suggested marked declines in oxygen concentration at all sites as a result of simulated changes in the balance between phytoplankton production and consumption, changes in vertical mixing (stratification) and change in oxygen solubility with temperature. A parallel study by Meire et al. (2013) for the 'Oyster Grounds' site (also using the SRES A1B scenario) suggested that bottom water oxygen concentrations in late summer could decrease by 24 μM or 11.5 % by 2100.

12.3.2 Pathogens, Pests and Predators

A key issue for North Sea fish and shellfish is the link between climate change and the prevalence of pathogens or harmful algal bloom (HAB) species. A global review suggested that marine pathogens are increasing in occurrence, and that this increase is linked to rising seawater temperature (Harvell et al. 1999) with possible consequences for commercial fisheries and aquaculture.

The presence of certain pathogens or algal toxins in seawater samples or in tissues harvested from shellfish can result in temporary closure of a fishery. Many pathogens that occur in European shellfish are very sensitive to seawater temperature and salinity, for example the bacteria *Vibrio parahaemolyticus* and *V. vulnificus* that pose a significant threat to human health (Baker-Austin et al. 2013). *Vibrio* species proliferate rapidly at temperatures above 18 °C and incidents of shellfish-associated gastrointestinal illness in Europe have been noted during heat waves (Baker-Austin et al. 2013). In the United States, *Vibrio*-related incidents cost the economy more than any other seafood-acquired pathogen and these incurred costs have increased dramatically in recent years (Ralston et al. 2011). In contrast, *Norovirus*, another major cause of shellfish-acquired gastroenteritis, occurs most frequently in winter and following periods of high precipitation and hence is associated with flash-flooding and runoff from sewers (Campos and Lees 2014). Future projections of precipitation (rain and snowfall) and river run-off for catchments surrounding the North Sea suggest that intense rainfall and hence extreme river flows will occur more often in the future, particularly during winter, and so considerable changes could be anticipated in the epidemiology and proliferation of marine pathogens—and therefore exposure risk for European citizens consuming seafood (Pinnegar et al. 2012).

Reports of increased abundance of jellyfish in the media and in scientific literature over recent decades have raised concerns over the potential role of climate change in influencing outbreaks (Atrill et al. 2007; Purcell 2012) and in possible implications for commercial fisheries and aquaculture. Data obtained from the CPR survey show an increasing occurrence of jellyfish in the central North Sea since 1958 that this may be positively related to the NAO and Atlantic inflow (Lynam et al. 2004; Atrill et al. 2007). High jellyfish numbers are potentially detrimental to fisheries both as competitors with, and predators of, larval fish (Purcell and Arai 2001). In particular, negative impacts of jellyfish on herring larvae have been noted (Lynam et al. 2005) and this is now a major focus of scientific attention.

Climate change can also influence the presence of potential predators. For example Kempf et al. (2014) demonstrated that grey gurnard *Eutriglia gurnardus*, has expanded its high density areas in the central North Sea northward over the last two decades to overlap with that of 0-group cod. Grey gurnard are thought to be important predators of juvenile cod, hence recruitment success of cod was found to be negatively correlated with the degree of spatial overlap between the two species. Similar fears have been voiced by fishers regarding the recent expansion of hake in the northern North Sea (see Sect. 12.3.1.1), since hake is also known to be a voracious predator of smaller fish.

12.3.3 Fishery Operations

Through its effects on seawater temperature as well as its influence on storm conditions climate change can affect the performance of fishing vessels or gears, as well as vessel safety and stability at sea.

Dulvy et al. (2008) explored the year-by-year distributional response of the North Sea demersal fish assemblage to climate change and found that the whole North Sea fish assemblage has deepened by $\sim$3.6 m per decade since 1981. This has important implications since it is known that trawl gear geometry and hence 'catchability' can be greatly influenced by water depth (see Godø and Engås 1989).

In tropical tuna (the main target of Europe's distant-water fleet), strong El Niño events along the west coast of the Americas typically result in a deeper thermocline, and declines in yellowfin tuna *Thunnus albacares* catches (see Miller 2007) because fish are able to spread out in the water column beyond the reach of commercial fisheries. Poor catch rates during the intense 1982–1983 El Niño played a role in the migration of the entire US tuna fleet from the Eastern Pacific to the Western and Central Pacific. Similar processes and mechanisms, but on a smaller spatial scale, appear to influence catch rates in North Sea pelagic fisheries, such as those targeting herring. Maravelias (1997) demonstrated that temperature and depth of the thermocline, appear to be key factors that modulate both the presence and relative abundance of herring within the northern North Sea. Herring appeared to avoid the cold bottom waters of the North Sea during the summer, probably due to the relatively poor food resources there. This greatly affected 'catchability'.

At present, confidence in the wind and storm projections from global climate models (GCMs) and downscaled regional climate models (RCMs) is relatively low, with some models suggesting that northwest Europe might experience fewer storms and others suggesting an increase (Woolf and Wolf 2013). In general, models suggest that climate change could result in a north-eastward shift of storm frequency in the North Atlantic, although the change in storm intensity or frequency that this implies is not clear (Meehl et al. 2000; Ulbrich et al. 2008). The winter of 2013/14 was the stormiest in the last 66 years with regard to the southern North Sea and the wider British Isles (Matthews et al. 2014) and this is known to have coincided with major disruption to the fishing industry throughout the region. Months of high winds and high seas left many fishers unable to work, and caused millions of Euros worth of damage—as well as a lack of fish and higher prices on fish markets.

12.3.3.1 Climatic Influences on 'Catchability'

There is little evidence of significant changes in catchability of demersal trawl gears in the North Sea as a result of climate change or poor weather conditions, although Walden and Schubert (1965) examined wind and catch data, and found that wind direction and force were correlated with catch at a few locations. Similarly Harden Jones and Scholes (1980) investigated the relations between wind and the catch of a Lowestoft trawler. Their analysis showed that over the course of a year, catches of plaice were lowest with northerly winds but that the reverse was true for cod. Long-term projections for winds over the North Sea are highly uncertain, but several authors (notably Wang et al. 2011) have suggested a climate-change-related upward trend in storminess in recent years. Analyses (de Winter et al. 2013) under a range of different climate change scenarios, do not anticipate changes in annual maximum wind speed over the next 50–100 years, however they do suggest that annual extreme wind events will occur more often from western directions. This is particularly relevant for fishing boats in the German

Bight, as such a shift would imply larger extreme waves and surge levels in this region.

Poulard and Trenkel (2007) reported that the impact of wind strength on catchability depends on the habitat preferred by a given species. For a bottom trawl survey in the Bay of Biscay, catches of benthic and demersal species were significantly affected by wind condition whereas no effect was detected for pelagic species. Similarly, Wieland et al. (2011) examined bias in estimates of abundance and distribution of North Sea cod during periods of strong winds. Wind speed had significant effects on catch rates, and specifically catches were reduced during the strongest winds. Strong winds prevailing over a prolonged period lead to poor visibility in shallow coastal waters, caused mainly by resuspension of bottom sediments. North Sea trawlers and especially 'flyshooters' would usually not fish in that area under such conditions due to the expectation of poor catches whereas gillnets may perform well in this case.

Fish are known to behave differently in turbid versus clearer waters. For example, Meager and Batty (2007) examined activity of juvenile cod and found both longer prey-search times and higher activity in turbid conditions, and suggested that such behaviour might increase energetic costs and also make the cod more vulnerable to fishing gears and potential predators. Capuzzo et al. (2015) demonstrated that the southern North Sea has become significantly more turbid over the latter half of the 20th century, and that this may be related to changes in seabed communities, weather patterns, and increased coastal erosion. Gill net catches are typically higher in turbid waters after storms. Ehrich and Stransky (1999) found that catch rates of some groundfish species in the North Sea exhibited significant variability following strong and severe gales (periods of strong winds from 50 to 102 km h^{-1}). Catches of dab, solenette, plaice and sole all changed markedly between the first and second day after the storm.

12.3.3.2 Vessel Stability and Performance

An increase in the frequency or severity of storms could have negative consequences for the ability of fishing boats to access resources in the future or could have consequences for vessel stability and performance. Abernethy et al. (2010) reported that 85 % of fishers interviewed as part of a survey in southern England, elected to stay in port during bad weather due to the risk of gear loss and increased fuel consumption. Fishing remains a dangerous occupation. A research project, published in 2007 showed that the fatal accident rate for UK fishers for the decade 1996–2005 was 115 times higher than that of the general workforce (MAIB 2008). In the United States, severe weather conditions contributed to 61 % of the 148 fatal fishing vessel disasters reported between 2000 and 2009 (Lincoln and Lucas 2010). In Denmark, more than half of fatalities reported were caused by foundering/capsizing due to stability changes in rough weather (Laursen et al. 2008). In the UK, the majority of vessel losses recorded (52 %) were due to flooding/foundering, and most involved small vessels of less than 12 m in length (MAIB 2008). Most flooding/foundering losses occurred in moderate weather. However, this needs to be considered against the likelihood that there would be fewer fishing vessels at sea during extreme weather conditions (MAIB 2008).

Dramatic increases in wave height occurred in the North Sea between 1960 and 1990, but these are now viewed as one feature within a longer history of variability (Woolf and Wolf 2013). Future patterns of storminess are poorly understood, with little consensus between models and highly uncertain model outputs. Changes in storminess and associated consequences for fishing operations is an under-researched topic, with no recent assessments of vessel operating envelopes or the willingness of vessel owners/skippers to put to sea. Laevastu and Hayes (1981) suggested that modern high-sea fishing vessels usually have to stop fishing at wind speeds of 50–78 km h^{-1} (Force 7 to 8 on the Beaufort scale), whereas coastal fishing vessels in the North Sea find difficulty in operating at wind speeds of 39–49 km h^{-1} (Force 6 on the Beaufort scale). A number of modelling approaches have been applied in the North Sea to try to predict the behaviour of fishers and the distribution of fishing vessels (e.g. Hutton et al. 2004) but none have yet included storms or weather disruption in their analyses.

12.3.3.3 Assessing Economic Implications

There has been little research directed towards understanding the future implications of climate change for fishing fleets, fishers, coastal economies and society directly. This is certainly the case with regard to countries surrounding the North Sea. However, a number of studies have set out to investigate the vulnerability and adaptive capacity of the fisheries sector at a global scale (McClanahan et al. 2008; Allison et al. 2009). Vulnerability to climate change depends upon three key elements: exposure to physical effects of climate change; sensitivity of the natural resource system or dependence of the national economy upon economic returns from the fishing sector; and the extent to which adaptive capacity enables these potential impacts to be offset. Allison et al. (2009) ranked North Sea countries very low in terms of overall vulnerability, largely due to low rates of fish consumption in the surrounding countries, highly diversified economies and only moderate exposure to future climate change. Similarly, Barange et al. (2014) categorised all North Sea countries as either low (Norway) or very low in terms of nutritional and economic dependence on fisheries. However fisheries represent an important component of employment in certain North Sea regions (EU 2011), notably in Shetland where 22 % of all jobs are estimated to be in fisheries/fisheries-related industries, and Urk in the Netherlands where 35 % of jobs rely on fisheries.

Cheung et al. (2010) estimated future changes in maximum potential catch (a proxy for maximum sustainable yield) given projected shifts in the distribution of exploited species and changes in marine primary productivity. This study suggested that climate change may lead to large-scale redistribution of global maximum catch potential, with an average of 30–70 % increase in yield of high-latitude regions (north of 50°N in the northern hemisphere), but a drop of up to 40 % in the tropics. North Sea countries are anticipated to gain very slightly in maximum potential catch but not as much as Nordic countries such as Norway and Iceland. This region will witness increases in catches of some commercial species but decreases in others, and thus the gains and losses are expected to broadly balance out.

Working at a national level, Jones et al. (2015) used a similar bioclimate envelope model (see Fig. 12.4) to investigate economic implications for fisheries catch potential in the UK exclusive economic zone (EEZ) specifically. Maximum catch potential was calculated for each species in both the reference and projection periods using an algorithm that takes into account net primary production and range area. Results suggested that the total maximum catch potential will decrease within the UK EEZ by 2050, although this was heavily influenced by an assumed decline in plankton productivity. Extending these projections into a cost benefit analysis resulted in a median decrease in net present value of 10 % by 2050. Net present value over the study period further decreased when trends in fuel price were extrapolated into the future, becoming negative when capacity-enhancing subsidies were removed. This study highlights key factors influencing future profitability of fisheries and the importance of enhancing adaptive capacity in fisheries and resilience to climate change.

Uncertainty is inherent in fisheries management, so there is an expectation of change and a wealth of knowledge and experience of coping with and adapting to this uncertainty. Badjeck et al. (2010) argued that diversification is a primary means by which individuals can reduce risk and cope with future uncertainty. A recent study commissioned by the UK Department for Environment, Food and Rural Affairs included a detailed assessment of whether the fish catching sector might be expected to adapt to the opportunities and threats associated with future climate change over the next 30 years (Defra 2013). This assessment built heavily on the species projections of Jones et al. (2012, 2013—see Sect. 12.3.1.1) and looked for examples of current adaptation by the sector, focusing on species increasing in the UK EEZ (such as anchovy, squid, seabass) and also on past increases in scallops, boarfish *Capros aper*, and hake. The key adaptation actions identified included:

- Travelling further to fish for current species, if stocks move away from existing ports.
- Diversifying the livelihoods of port communities, this may include recreational fishing where popular angling species become locally more abundant (e.g. seabass).
- Increasing vessel capacity if stocks of currently fished species increase.
- Changing gear to fish for different species, if new or more profitable opportunities to fish different species are available.
- Developing routes to export markets to match the changes in catch supplied. These routes may be to locations (such as southern Europe) which currently eat the fish stocks which may move into northern waters.
- Stimulating domestic demand for a broader range of species, through joined-up retailer and media campaigns.

Many of the same adaptation options were also highlighted by McIlgorm et al. (2010) who reviewed how fishery governance may need to change in the light of future climate change, also the ACACIA report (ACACIA 2000) prepared as part of the European impact assessment for the IPCC Third Assessment which included a short chapter on fisheries. Vanderperren et al. (2009) provided a brief overview for Belgian marine fisheries. These authors noted the strong specialisation of the Belgian fleet with regard to a single fishing method (93 % beam trawlers) and target species (mainly flatfish) and that this makes the sector particularly vulnerable to changing circumstances. Possible adaptation measures as well as technological and economic consequences for the fleet were detailed (see Van den Eynde et al. 2011), and the elaboration of scenarios for secondary impacts at different points in time (2040, 2100) is ongoing.

Sumaila and Cheung (2009) attempted to estimate the necessary annual costs of adaptation to climate change in the fisheries sector worldwide. Adaptation to climate change is likely to involve an extension of existing policies to conserve fish stocks and to help communities. In Europe the estimated annual cost of adaptation was USD 0.03–0.15 billion, a small fraction of the costs (USD 1.05–1.70 billion) anticipated for East Asia and the Pacific.

12.4 International Fish Markets and Commodity Chains

Fisheries in the North Sea should not be viewed in isolation given that seafoods are traded globally and many North Sea countries are both exporters and importers of fish and shellfish commodities. It could be expected that prices of a particular commodity would reflect local patterns of availability (supply) and hence that the price of fish might even reflect regional climatic conditions (see Pinnegar et al. 2006). However this is rarely the case, given that supplies can often be secured from elsewhere and thus prices may remain low, even if locally resources become scarce. Cod stock status in the North Sea currently remains very low, possibly as a result of long-term climatic influences on recruitment, but catches are at an all-time high further north in the Barents Sea (ICES 2014b) and thus cod prices in Europe are supressed. A clear adaptation response in the face of ensuing climate change is to obtain fish from sources further north (either by trade or by shifting the location of fishing fleets where this is possible). As other countries around the word also need to secure sufficient food for growing populations, and have considerably higher buying-power (notably China, which in 2030 will account for 38 % of total fish consumption), European countries may find the ability to secure sufficient fish products from Nordic (non-EU) countries, such as Norway, Iceland and Greenland, much more difficult in the future (World Bank 2013).

Another international fisheries topic that has received considerable attention in recent years has been the link between global climate and fishmeal supplies and markets (e.g. Merino et al. 2010a, b). Aquaculture and animal feed production depend on fishmeal and fish oil as their primary source of protein, lipids, minerals and essential Omega 3/6 fatty acids. Every year 30 million tonnes of anchovita *Engraulis ringens, E. mordax* etc., sardines *Sardinops sagax, Sardina pilchardus* and other small pelagic fish are reduced into 6 million tonnes of fishmeal. More than half of this is derived from Peruvian/Chilean anchoveta although Denmark and Norway supply an additional 12 % based on North Sea sandeel, sprat, Norway pout and blue whiting as well as Arctic capelin *Mallotus villosus*. A lack of supply in Peruvian anchoveta (for example during El Niño climatic regimes) raises the price of fishmeal from elsewhere (e.g. the North Sea) and can influence the behaviour of European fishers, with indirect (e.g. predator-prey) consequence for other fish stocks.

12.5 Conclusions

North Sea fisheries may be impacted by climate change in various ways and consequences of rapid temperature rise are already being felt in terms of shifts in species distribution and variability in stock recruitment. While an expanding body of research now exists on this topic, there are still many knowledge gaps, especially with regard to understanding how fishing fleets themselves might be impacted by underlying biological changes and what this might mean for regional economies. Historically, fisheries managers and fishers have had to adapt to the vagaries of weather and climate, however the challenge presented by future climate change should not be underestimated and it is clear that fish communities and the fisheries that target them will almost certainly be very different in 50 or 100 years and that management and governance will need to adapt accordingly.

Acknowledgements This contribution is based heavily on a supporting document (Pinnegar et al. 2013) prepared by the same authors for the UK Marine Climate Change Impacts Partnership (MCCIP). Much of the inspiration for this work came from the EU FP6 project 'RECLAIM' (Resolving the Impacts of Climate Change on Fish and Shellfish Populations) as well as the recent EU FP7 project 'VECTORS' (Vectors of Change in Oceans and Seas Marine Life, Impact on Economic Sectors). Additional funding was provided by the UK Department for Environment, Food and Rural Affairs (Defra) contract MF1228 (Response of ecosystems and fisheries to management in a changing environment).

References

Abernethy KE, Treilcock P, Kebede B, Allison EH, Dulvy NK (2010) Fuelling the decline in UK fishing communities? ICES J Mar Sci 67:1076–1085

ACACIA (2000) Assessment of potential effects and adaptations for climate change in Europe: The Europe ACACIA Project. Jackson Environment Institute, University of East Anglia

Agnalt AL, Grefsrud ES, Farestveit E, Larsen M, Keulder F (2013) Deformities in larvae and juvenile European lobster (*Homarus gammarus*) exposed to lower pH at two different temperatures. Biogeosciences 10:7883–7895

Ainsworth CH, Samhouri JF, Busch DS, Cheung WWL, Dunne J, Okey TA (2011) Potential impacts of climate change on Northeast Pacific marine foodwebs and fisheries. ICES J Mar Sci 68:1217–1229

Alheit J, Hagen E (1997) Long-term climate forcing of European herring and sardine populations. Fish Oceanogr 6:130–139

Alheit J, Pohlmann T, Casini M, Greve W, Hinrichs H., Mathis M., O'Driscoll K, Vorberg R, Wagner C (2012) Climate variability drives anchovies and sardines into the North and Baltic Seas. Prog Oceanogr 96:128–139

Allison EH, Perry AL, Badjeck MC, Adger WN, Brown K, Conway D, Halls AS, Pilling GM, Reynolds JD, Andrew NL, Dulvy NK (2009) Vulnerability of national economies to the impacts of climate change on fisheries. Fish Fisher 10:173–196

Arnott SA, Ruxton GD (2002) Sandeel recruitment in the North Sea: Demographic, climatic and trophic effects. Mar Ecol-Prog Ser 238:199–210

Artioli Y, Blackford JC, Butenschön M, Holt JT, Wakelin SL, Thomas TH, Borges AV, Allen JI (2012) The carbonate system in the North Sea: Sensitivity and model validation. J Mar Syst 102–104:1–13

Atrill MJ, Wright J, Edwards M (2007) Climate-related increases in jellyfish frequency suggest a more gelatinous future for the North Sea. Limnol Oceanogr 52:480–485

Aurich HJ (1953) Verbreitung und Laichverhältnisse von Sardelle und Sardine in der südöstlichen Nordsee und ihre Veränderung als Folge der Klimäanderung. Helgoländer Meeresun 4:175–204

Baden SP, Pihl L, Rosenberg R (1990) Effects of oxygen depletion on the ecology, blood physiology and fishery of the Norway lobster *Nephrops norvegicus*. Mar Ecol-Prog Ser 67:141–155

Badjeck MC, Allison EH, Halls AS, Dulvy NK (2010). Impacts of climate variability and change on fishery-based livelihoods. Mar Policy 34:375–383

Baker-Austin C, Trinanes JA, Taylor NGH, Hartnell R, Siitonen A, Martinez-Urtaza J (2013) Emerging Vibrio risk at high latitudes in response to ocean warming. Nature Clim Change 3:73–77

Barange M, Merino G, Blanchard JL, Scholtens J, Harle J, Allison EH, Allen JI, Holt J, Jennings S (2014) Impacts of climate change on marine ecosystem production in societies dependent on fisheries. Nature Clim Change 4:211–216

Barrett JH, Locker AM, Roberts CM (2004) The origins of intensive marine fishing in medieval Europe: the English evidence. Proc R Soc B 271:2417–2421

Baudron AR, Fernandez PG (2015) Adverse consequences of stock recovery: European hake, a new "choke" species under a discard ban? Fish Fish 16:563–575

Beare DJ, Burns F, Greig A, Jones EG, Peach K, Kienzle M, McKenzie E, Reid DG (2004a) Long-term increases in prevalence of North Sea fishes having southern biogeographic affinities. Mar Ecol-Prog Ser 284:269–278

Beare D, Burns F, Peach K, Portilla E, Greig T, McKenzie E, Reid D (2004b) An increase in the abundance of anchovies and sardines in the north-western North Sea since 1995. Glob Change Biol 10:1–5

Beare D, Burns F, Jones E, Peach K, Reid D (2005) Red mullet migration into the northern North Sea during late winter. J Sea Res 53:205–212

Beaugrand G, Reid PC, Ibańez F (2002) Reorganization of North Atlantic marine copepod biodiversity and climate. Science 296:1692–1694

Beaugrand G, Brander KM, Lindley JA, Souissi S, Reid PC (2003) Plankton effect on cod recruitment in the North Sea. Nature 426:661–664

Beaugrand G, Lenoir S, Ibañez F, Manté C (2011) A new model to assess the probability of occurrence of a species, based on presence-only data. Mar Ecol-Prog Ser 424:175–190

Blackford JC, Gilbert FJ (2007) Ph variability and CO_2-induced acidification in the North Sea. J Mar Syst 64:229–241

Brander KM, Mohn R (2004) Effect of the North Atlantic Oscillation on the recruitment of Atlantic cod (*Gadus morhua*). Can J Fish Aquat Sci 61:1558–1564

Brunel T, Boucher J (2007) Long-term trends in fish recruitment in the north-east Atlantic related to climate change. Fish Oceanogr 16:336–349

Burrows MT, Schoeman DS, Buckley LB, Moore P, Poloczanska ES, Brander KM, Brown C, Bruno JF, Duarte CM, Halpern BS, Holding J, Kappel CV, Kiessling W, O'Connor MI, Pandolfi JM, Parmesan C, Schwing FB, Sydeman WJ, Richardson AJ (2007) The pace of shifting climate in marine and terrestrial ecosystems. Science 334:652–655

Caldeira K, Wickett ME (2003) Anthropogenic carbon and ocean pH. Nature 425:365–365

Campos CJA, Lees DN (2014) Environmental transmission of human noroviruses in shellfish waters. Appl Env Microbiol 80:3552–3561

Capuzzo E, Stephens D, Silva T, Barry J, Forster RM (2015) Decrease in water clarity of the southern and central North Sea during the 20th century. Glob Change Biol 21:2206–2214

Cheung WWL, Lam VWY, Sarmiento JL, Kearney K, Watson R, Pauly D (2009) Projecting global marine biodiversity impacts under climate change scenarios. Fish Fish 10:235–251

Cheung WWL, Lam VWY, Sarmiento JL, Kearney K, Watson R, Zeller D, Pauly D (2010) Large-scale redistribution of maximum fisheries catch potential in the global ocean under climate change. Glob Change Biol 16:24–35

Cheung WWL, Dunne J, Sarmiento JL, Pauly D (2011) Integrating ecophysiology and plankton dynamics into projected maximum fisheries catch potential under climate change in the Northeast Atlantic. ICES J Mar Sci 68:1008–1018

Cheung WWL, Pinnegar JK, Merino G, Jones MC, Barange M (2012) Review of climate change and marine fisheries in the UK and Ireland. Aquat Conserv 22:368–388

Cheung WWL, Sarmiento JL, Dunne J, Frölicher L, Lam VWY, Palomares MLD, Watson R, Pauly D (2013) Shrinking of fishes exacerbates impacts of global ocean changes on marine ecosystems. Nature Clim Change 3:254–258

Clarke RA, Fox CJ, Viner D, Livermore M (2003) North Sea cod and climate change – modelling the effects of temperature on population dynamics. Glob Change Biol 9:1669–1680

Cook RM, Heath MR (2005) The implications of warming climate for the management of North Sea demersal fisheries. ICES J Mar Sci 62:1322–1326

Cooley SR, Doney SC (2009) Anticipating ocean acidification's economic consequences for commercial fisheries. Env Res Lett 4: doi:10.1088/1748-9326/4/2/024007

Cucco A, Sinerchia M, Lefrançois C, Magni P, Ghezzo M, Umgiesser G, Perilli A, Domenici P (2012) A metabolic scope based model of fish response to environmental changes. Ecol Model 237–238:132–141

Cushing DH (1982) Climate and fisheries. Academic Press, London

Cushing DH (1990) Plankton production and year-class strength in fish populations: an update of the match/ mismatch hypothesis. Adv Mar Biol 26:249–293

Davis AJ, Jenkinson LS, Lawton JH, Shorrocks B, Wood S (1998) Making mistakes when predicting shifts in species range in response to global warming. Nature 391:783–786

de Winter RC, Sterl A, Ruessink BG (2013) Wind extremes in the North Sea Basin under climate change: An ensemble study of 12 CMIP5 GCMs. J Geophys Res 118:1601–1612

Defra (2013) Economics of climate resilience: natural environment – fisheries. February 2013. Department for Environment, Food and Rural Affairs (Defra), UK

Diaz RJ, Rosenberg R (2008) Spreading dead zones and consequences for marine ecosystems. Science 321:926–929

Dickey-Collas M, Engelhard GH, Rindorf A, Raab K, Smout S, Aarts G, van Deurs M, Brunel T, Hoff A, Lauerburg RAM, Garthe S, Andersen KH, Scott F, van Kooten T, Beare D, Peck MA (2013) Ecosystem-based management objectives for the North Sea: riding the forage fish rollercoaster. ICES J Mar Sci 71:128–142

Drinkwater KF (2005) The response of Atlantic cod (*Gadus morhua*) to future climate change. ICES J Mar Sci 62:1327–1337

Dulvy NK, Rogers SI, Jennings S, Stelzenmüller V, Dye SR, Skjoldal HR (2008) Climate change and deepening of the North Sea fish assemblage: a biotic indicator of warming seas. J Appl Ecol 45:1029–1039

Ehrich S, Stransky C (1999) Fishing effects in northeast Atlantic shelf seas: patterns in fishing effort, diversity and community structure. VI. Gale effects on vertical distribution and structure of a fish assemblage in the North Sea. Fish Res 40:185–193

Engelhard GH, Pinnegar JK, Kell LT, Rijnsdorp AD (2011) Nine decades of North Sea sole and plaice distribution. ICES J Mar Sci 68:1090–1104

Engelhard GH, Peck MA, Rindorf A, Smout S, van Deurs M, Raab K, Andersen KH, Garthe S, Lauerburg RAM, Scott F, Brunel T, Aarts G, van Kooten T, Dickey-Collas M (2014a) Forage fish, their fisheries, and their predators: who drives whom? ICES J Mar Sci 71:90–104

Engelhard GH, Righton DA, Pinnegar JK (2014b) Climate change and fishing: a century of shifting distribution in North Sea cod. Glob Change Biol 20:2473–2483

Engelhard GH, Lynam CP, García-Carreras B, Dolder PJ, Mackinson S (2015) Effort reduction and the large fish indicator: spatial trends reveal positive impacts of recent European fleet reduction schemes. Environ Conserv 42:227–236

Enghoff IB, MacKenzie BR, Nielsen EE (2007) The Danish fish fauna during the warm Atlantic period (ca. 7000–3900 BC): Forerunner of future changes? Fish Res 87:167–180

EU (2011) Regional social and economic impacts of change in fisheries-dependent communities. Lot 4: Impact Assessment Studies related to the CFP – Final Report. European Commission, March 2011

Fabry VJ, Seibel BA, Feely RA, Orr JC (2008) Impacts of ocean acidification on marine fauna and ecosystem processes. ICES J Mar Sci 65:414–432

Fincham JI, Rijnsdorp AD, Engelhard GH (2013) Shifts in the timing of spawning in sole linked to warming sea temperatures. J Sea Res 75:69–76

Franke A, Clemmesen C (2011) Effect of ocean acidification on early life stages of Atlantic herring (*Clupea harengus* L.). Biogeosciences 8:3697–3707

Freitas V, Campos J, Fonds M, van der Veer HW (2007) Potential impact of temperature change on epibenthic predator bivalve prey interactions in temperate estuaries. J Therm Biol 32:328–340

Gazeau FPH, Gattuso JP, Dawber C, Pronker AE, Peene F, Peene J, Heip CHR, Middelburg JJ (2010) Effect of ocean acidification on the early life stages of the blue mussel *Mytilus edulis*. Biogeosciences 7:2051–2060

Godø OR, Engås A (1989) Swept area variation with depth and its influence on abundance indices of groundfish from trawl surveys. J Northwest Atl Fish Sci 9:133–139

Greve W, Prinage WS, Zidowitz H, Nast J, Reiners F (2005) On the phenology of North Sea ichthyoplankton. ICES J Mar Sci 62:1216–1223

Griffith GP, Fulton EA, Richardson AJ (2011) Effects of fishing and acidification-related benthic mortality on the southeast Australian marine ecosystem. Glob Change Biol 17:3058–3074

Gröger JP, Kruse GH, Rohlf N (2010) Slave to the rhythm: how large-scale climate cycles trigger herring (*Clupea harengus*) regeneration in the North Sea. ICES J Mar Sci 67:454–465

Hale R, Calosi P, Mcneill L, Mieszkowska N, Widdicombe S (2011) Predicted levels of future ocean acidification and temperature rise could alter community structure and biodiversity in marine benthic communities. Oikos 120:661–674

Hannesson R (2007) Global warming and fish migrations. Nat Res Model 20:301–319

Harden Jones FR, Scholes P (1980) Wind and the catch of a Lowestoft trawler. J Cons Int Explor Mer 39:53–69

Harvell CD, Kim K, Burkholder JM, Colwell RR, Epstein PR, Grimes DJ, Hofmann EE, Lipp EK, Osterhaus ADME, Overstreet RM, Porter JW, Smith GW, Vasta GR (1999) Emerging marine diseases-climate links and anthropogenic factors. Science 285:1505–1510

Hastie LC, Pierce GJ, Wang J, Bruno I, Moreno A, Piatkowski U, Robin JP (2009a) Cephalopods in the north-eastern Atlantic: species, biogeography, ecology, exploitation and conservation. Oceanogr Mar Biol, Annu Rev 47:111–190

Hastie L, Pierce G, Pita C, Viana M, Smith J, Wangvoralak S (2009b) Squid fishing in UK waters. A Report to SEAFISH Industry Authority. School of Biological Sciences, University of Aberdeen

Hátún H, Payne MR, Beaugrand G, Reid PC, Sandø AB, Drange H, Hansen B, Jacobsen JA, Bloch D (2009) Large bio-geographical shifts in the north-eastern Atlantic Ocean: From the subpolar gyre, via plankton, to blue whiting and pilot whales. Prog Oceanogr 80:149–162

Heath MR (2007) Responses of fish to climate fluctuations in the Northeast Atlantic. In: Emery LE (ed) The Practicalities of Climate Change: Adaptation and Mitigation. Proceedings of the 24th Conference of the Institute of Ecology and Environmental Management, Cardiff

Heath MR, Brander K (2001) Workshop on Gadoid Stocks in the North Sea during the 1960s and 1970s. The Fourth ICES/GLOBEC Backward-Facing Workshop, 1999. ICES Coop Res Rep 244

Heath MR, Neat FC, Pinnegar JK, Reid DG, Sims DW, Wright PJ (2012) Review of climate change impacts on marine fish and shellfish around the UK and Ireland. Aquat Conserv 22:337–367

Henderson PA, Seaby RM (2005) The role of climate in determining the temporal variation in abundance and growth of sole *Solea* in the Bristol Channel. J Mar Biol Assoc UK 85:197–204

Henderson PA, Seaby RM, Somes JR (2006) A 25-year study of climatic and density-dependent population regulation of common shrimp *Crangon crangon* (Crustacea: Caridea) in the Bristol Channel. J Mar Biol Assoc UK 86:287–298

Hendriks IL, Duarte CM, Alvarez M (2010) Vulnerability of marine biodiversity to ocean acidification: a meta-analysis. Est Coast Shelf Sci 86:157–164

Hiddink JG, Ter Hofstede R (2008) Climate induced increases in species richness of marine fishes. Glob Change Biol 14:453–460

Hjort J (1914) Fluctuations in the great fisheries of Northern Europe viewed in the light of biological research. Rapp p-v Reun (Dan) 20:1–228

Hobday A, Pecl GT (2014) Identification of global marine hotspots: sentinels for change and vanguards for adaptation. Rev Fish Biol Fisher 24:415–425

Hollowed AB, Bond NA, Wilderbuer TK, Stockhausen WT, A'mar ZT, Beamish RJ, Overland JE, Schirripa MJ (2009) A framework for modelling fish and shellfish responses to future climate change. ICES J Mar Sci 66:1584–1594

Hughes KM, Dransfeld L, Johnson MP (2014) Changes in the spatial distribution of spawning activity by north-east Atlantic mackerel in warming seas: 1977–2010. Mar Biol 161:2563–2576

Hutton T, Mardle S, Pascoe S, Clark RA (2004) Modelling fishing location choice within mixed fisheries: English North Sea beam trawlers in 2000 and 2001. ICES J Mar Sci 61:1443–1452

ICES (2012) Report of the Working Group on Assessment of New MoU Species (WGNEW). ICES CM 2012/ACOM:20

ICES (2014a) Report of the Working Group for the Assessment of Demersal Stocks in the North Sea and Skagerrak (WGNSSK), ICES CM 2014/ACOM:13

ICES (2014b) Report of the Arctic Fisheries Working Group (AFWG), ICES CM 2014/ACOM:05

IPCC (2013) Climate Change 2013: The Physical Science Basis. Contribution of Working Group I to the Fifth Assessment Report of the Intergovernmental Panel on Climate Change. Stocker TF, Qin D, Plattner G-K, Tignor M, Allen SK, Boschung J, Nauels A, Xia Y, Bex V, Midgley PM (eds.). Cambridge University Press

Jansen T, Gislason H (2011) Temperature affects the timing of spawning and migration of North Sea mackerel. Cont Shelf Res 31:64–72

Jansen T, Kristensen K, Payne M, Edwards M, Schrum C, Pitois S (2012) Long-term retrospective analysis of mackerel spawning in the North Sea: a new time series and modeling approach to CPR data. PLoS ONE 7:e38758

Jennings S, Blanchard JL (2004) Fish abundance with no fishing: predictions based on macroecological theory. J Anim Ecol 73:632–642

Jennings S, Brander K (2010) Predicting the effects of climate change on marine communities and the consequences for fisheries. J Mar Syst 79:418–426

Jones MC, Dye SR, Pinnegar JK, Warren R, Cheung WWL (2012) Modelling commercial fish distributions: Prediction and assessment using different approaches. Ecol Model 225:133–145

Jones MC, Dye SR, Fernandes JA, Frölicher TL, Pinnegar JK, Warren R, Cheung WWL (2013) Predicting the impact of climate change on threatened species in UK waters. PLoS ONE 8:e54216

Jones MC, Dye SR, Pinnegar JK, Warren R, Cheung WWL (2015) Using scenarios to project the changing profitability of fisheries under climate change. Fish Fish 16:603–622

Kaplan IC, Levin PS, Burden M, Fulton EA (2010) Fishing catch shares in the face of global change: a framework for integrating cumulative impacts and single species management. Can J Fish Aquat Sci 67:1968–1982

Karpechko AY (2010) Uncertainties in future climate attributable to uncertainties in future Northern Annular Mode trend. Geophys Res Lett 37:doi:10.1029/2010gl044717

Kell LT, Pilling GM, O'Brien CM (2005) Implications of climate change for the management of North Sea cod (*Gadus morhua*). ICES J Mar Sci 62:1483–1491

Kempf A, Stelzenmüller V, Akimova A, Floeter J (2014) Spatial assessment of predator–prey relationships in the North Sea: the influence of abiotic habitat properties on the spatial overlap between 0-group cod and grey gurnard. Fish Oceanogr 22:174–192

Kroeker KJ, Kordas RL, Crim RN, Singh GG (2010) Meta-analysis reveals negative yet variable effects of ocean acidification on marine organisms. Ecol Lett 13:1419–1434

Laevastu T, Hayes ML (1981) Fisheries oceanography and ecology. Fishing News Books, Farnham, Surrey

Laursen LH, Hansen HL, Jensen OC (2008) Fatal occupational accidents in Danish fishing vessels 1989–2005. Int J Inj Contr Saf Promot 15:109–117

Le Quesne WJF, Pinnegar JK (2012) The potential impacts of ocean acidification: Scaling from physiology to fisheries. Fish Fisher 13:333–344

Lenoir S, Beaugrand G, Lecuyer E (2011) Modelled spatial distribution of marine fish and projected modifications in the North Atlantic Ocean. Glob Change Biol 17:115–129

Lincoln JM, Lucas DL (2010) Occupational fatalities in the United States commercial fishing industry, 2000–2009. J Agromedicine 15:343–350

Lindegren M, Möllmann C, Nielsen A, Brander K, MacKenzie BR, Stenseth NC (2010) Ecological forecasting under climate change. Proc R Soc B 277:2121–2130

Link JS, Nye JA, Hare JA (2011) Guidelines for incorporating fish distribution shifts into a fisheries management context. Fish Fish 12:461–469

Llopiz JK, Cowen RK, Hauff MJ, Ji R, Munday P, Muhling B, Peck MA, Richardson DE, Sogard S, Sponaugle S (2014) Early life history and fisheries oceanography: new questions in a changing world. Oceanography 27:26–41

Lowe JA, Howard T, Pardaens A, Tinker J, Holt J, Wakelin S, Milne G, Leake J, Wolf J, Horsburgh K, Reeder T, Jenkins G, Ridley J, Dye S, Bradley S (2009) UK Climate Projections science report: Marine and coastal projections. Met Office Hadley Centre, Exeter, UK

Lynam CP, Hay SJ, Brierley AS (2004) Interannual variability in abundance of North Sea jellyfish and links to the North Atlantic Oscillation. Limnol Oceanogr 49:637–643

Lynam CP, Heath MR, Hay SJ, Brierley AS (2005) Evidence for impacts by jellyfish on North Sea herring recruitment. Mar Ecol-Prog Ser 298:157–167

Mackenzie BR, Schiedek D (2007) Daily ocean monitoring since the 1860s shows record warming of northern European seas. Glob Change Biol 13:1335–1347

MAIB (2008) Analysis of UK fishing vessel safety – 1992 to 2006. Marine Accident Investigation Branch, Southampton, UK

Maravelias CD (1997) Trends in abundance and geographic distribution of North Sea herring in relation to environmental factors. Mar Ecol-Prog Ser 159:151–164

Matthews T, Murphy C, Wilby RL, Harrigan S (2014) Stormiest winter on record for Ireland and UK. Nature Clim Change 4:738–740

McClanahan TR, Cinner JE, Maina J, Graham NAJ, Daw TM, Stead SM, Wamukota A, Brown K, Ateweberhan, Venus V, Polunin NVC (2008) Conservation action in a changing climate. Conserv Lett 1:53–59

McIlgorm A, Hanna S, Knapp G, Le Floc' HP, Millerd F, Pan M (2010) How will climate change alter fishery governance Insights from seven international case studies. Mar Policy 34:170–177

Meager JJ, Batty RS (2007) Effects of turbidity on the spontaneous and prey-searching activity of juvenile Atlantic cod (*Gadus morhua*). Phil Trans Roy Soc B 362:2123–2130

Meehl GA, Zwiers F, Evans J, Knutson T, Mearns L, Whetton P (2000) Trends in extreme weather and climate events: issues related to modelling extremes in projections of future climate change. Bull Am Met Soc 81:427–436

Meire L, Soetaert KER, Meysman FJR (2013) Impact of global change on coastal oxygen dynamics and risk of hypoxia. Biogeosciences 10:2633–2653

Mendiola D, Alvarez P, Cotano U, Martínezde Murguía A (2007) Early development and growth of the laboratory reared north-east Atlantic mackerel *Scomber scombrus* L. J Fish Biol 70:911–933

Merino G, Barange M, Mullon C (2010a) Climate change scenarios for a marine commodity: Modelling small pelagic fish, fisheries and fishmeal in a globalized market. J Mar Sys 81:196–205

Merino G, Barange M, Mullon C (2010b) Impacts of aquaculture expansion and environmental change on marine ecosystems. Global Environ Change 20:586–596

Mieszkowska N, Genner MJ, Hawkins SJ, Sims DW (2009) Effects of climate change and commercial fishing on Atlantic cod *Gadus morhua*. Adv Mar Biol 56:213–273

Miller KA (2007) Climate variability and tropical tuna: Management challenges for highly migratory fish stocks. Mar Policy 31:56–70

Montero-Serra I, Edwards M, Genner MJ (2015) Warming shelf seas drive the subtropicalization of European pelagic fish communities. Glob Change Biol 21:144–153

Nash RDM, Dickey-Collas M (2005) The influence of life history dynamics and environment on the determination of year class strength in North Sea herring (*Clupea harengus* L.). Fish Oceanogr 14:279–291

O'Brien CM, Fox CJ, Planque B, Casey J (2000) Climate variability and North Sea cod. Nature 404:142

Orr JC, Fabry VJ, Aumont O, Bopp L et al. (2005) Anthropogenic ocean acidification over the twenty-first century and its impact on calcifying organisms. Nature 437:681–686

Ottersen G, Hjermann DØ, Stenseth NC (2006) Changes in spawning stock structure strengthen the link between climate and recruitment in a heavily fished cod (*Gadus morhua*) stock. Fish Oceanogr 15:230–243

Pawson MG (1992) Climatic influences on the spawning success, growth and recruitment of bass (*Dicentrarchus labrax* L.) in British waters. ICES Mar Sci Symp 195:388–392

Pawson MG, Kupschus S, Pickett GD (2007) The status of sea bass (*Dicentrarchus labrax*) stocks around England and Wales, derived using a separable catch-at-age model, and implications for fisheries management. ICES J Mar Sci 64:346–356

Pearson RG, Dawson TP (2003) Predicting the impacts of climate change on the distribution of species: are bioclimate envelope models useful? Global Ecol Biogeogr 12:361–371

Pearson PN, Palmer MR (2000) Atmospheric carbon dioxide concentrations over the past 60 million years. Nature 406:695–699

Peck MA, Hufnagl M (2012) Can IBMs explain why most larvae die in the sea? Model scenarios and sensitivity analyses reveal research needs. J Mar Syst 93:77–93

Perry AL, Low PJ, Ellis JR, Reynolds JD (2005) Climate change and distribution shifts in marine fishes. Science 308:1912–1915

Petitgas P, Alheit J, Peck MA, Raab K, Irigoien X, Huret M, van der Kooij J, Pohlmann T, Wagner C, Zarraonaindia I, Dickey-Collas M (2012) Anchovy population expansion in the North Sea. Mar Ecol-Prog Ser 444:1–13

Petitgas P, Rijnsdorp AD, Dickey-Collas M. Engelhard GH, Peck MA, Pinnegar JK, Drinkwater K, Huret M, Nash RDM (2013) Impacts of climate change on the complex life cycles of fish. Fish Oceanogr 22:121–139

Pinnegar JK, Hutton TP, Placenti V (2006) What relative seafood prices can tell us about the status of stocks. Fish Fish 7:219–226

Pinnegar J, Watt T, Kennedy K (2012) CCRA Risk Assessment for the Marine and Fisheries Sector. UK 2012 Climate Change Risk Assessment: Department for Environment, Food and Rural Affairs (Defra), London

Pinnegar JK, Cheung WWL, Jones M, Merino G, Turrell W, Reid D (2013) Impacts of climate change on fisheries. MCCIP Sci Rev 2013:302–317

Planque B, Frédou T (1999) Temperature and the recruitment of Atlantic cod (*Gadus morhua*). Can J Fish Aquat Sci 56:2069–2077

Pörtner HO (2010) Oxygen- and capacity-limitation of thermal tolerance: a matrix for integrating climate-related stressor effects in marine ecosystems. J Exp Biol 213:881–893

Pörtner HO, Farrell AP (2008) Physiology and climate change. Science 322:690–692

Pörtner HO, Knust R (2007) Climate change affects marine fishes through the oxygen limitation of thermal tolerance. Science 315:95–97

Pörtner HO, Peck MA (2010) Climate change impacts on fish and fisheries: towards a cause and effect understanding. J Fish Biol 77:1745–1779

Poulard J-C, Trenkel VM (2007) Do survey design and wind conditions influence survey indices? Can J Fish Aquat Sci 64:1551–1562

Purcell JE (2012) Jellyfish and ctenophore blooms coincide with human proliferations and environmental perturbations. Ann Rev Mar Sci 4:209–235

Purcell JE, Arai MN (2001) Interactions of pelagic cnidarians and ctenophores with fish: a review. Hydrobiologia 451:27–44

Queste BY, Fernand L, Jickells TD, Heywood KJ (2012) Spatial extent and historical context of North Sea oxygen depletion in August 2010. Biogeochemistry 113:53–68

Ralston EP, Kite-Powell H, Beet A (2011) An estimate of the cost of acute health effects from food and water-borne marine pathogens and toxins in the USA. J Water Health 9:680–694

Reid PC, Borges MF, Svendsen E (2001) A regime shift in the North Sea circa 1988 linked to changes in the North Sea horse mackerel fishery. Fish Res 50:163–171

Reid CR, Edwards M, Beaugrand G, Skogen M, Stevens D (2003) Periodic changes in the zooplankton of the North Sea during the twentieth century linked to oceanic inflow. Fish Oceanogr 12:260–269

Riley JD, Symonds DJ, Woolner L (1981) On the factors influencing the distribution of 0-group demersal fish in coastal waters. Rap Process 178:223–228

Robin JP, Denis V (1999) Squid stock fluctuations and water temperature: temporal analysis of English Channel Loliginidae. J Appl Ecol 36:101–110

Rutterford LA, Simpson SD, Jennings S, Johnson MP, Blanchard JL, Schön P-J, Sims DW, Tinker JY, Genner MJ (2015) Future fish distributions constrained by depth in warming seas. Nature Clim Change 5:569–573

Shephard S, Beukers-Stewart B, Hiddink JG, Brand AR, Kaiser MJ (2010) Strengthening recruitment of exploited scallops *Pecten mximus* with ocean warming. Mar Biol 157:91–97

Siegel V, Gröger J, Neudecker T, Damm U, Jansen S (2005) Long-term variation in the abundance of the brown shrimp *Crangon crangon* (L.) population of the German Bight and possible causes for its interannual variability. Fish Oceanogr 14:1–16

Simpson SD, Jennings S, Johnson MP, Blanchard JL, Schon PJ, Sims DW, Genner MJ (2011) Continental shelf-wide response of a fish assemblage to rapid warming of the sea. Curr Biol 21:1565–1570

Sims DW, Genner MJ, Southward AJ, Hawkins SJ (2001) Timing of squid migration reflects North Atlantic climate variability. Proc R Soc B 268:2607–2611

Skinner JA (2009) The changing distribution of haddock *Melanogrammus aeglefinus* in the North Sea: effects of fishing effort and environmental variation over the last century. MSc Thesis, University of East Anglia, Norwich

STECF (2013) The 2013 Annual Economic Report on the EU Fishing Fleet (STECF-13-15). 2013. Scientific, Technical and Economic Committee for Fisheries (STECF), JRC 84745

Stramma L, Prince ED, Schmidtko S, Luo J, Hoolihan JP, Visbeck M, Wallace DWR, Brandt P, Körtzinger A (2011) Expansion of oxygen minimum zones may reduce available habitat for tropical pelagic fishes. Nature Clim Change 2:33–37

Styf HK, Nilsson Sköld H, Eriksson SP (2013) Embryonic response to long-term exposure of the marine crustacean *Nephrops norvegicus* to ocean acidification and elevated temperature. Ecol Evol 3:5055–5065

Sumaila R, Cheung WWL (2009) The Costs to Developing Countries of Adapting to Climate Change New Methods and Estimates. In: Global Report of the Economics of Adaptation to Climate Change Study (Consultation Draft). World Bank, Washington

Teal LR, De Leeuw JJ, Van der Veer HW, Rijnsdorp AD (2008) Effects of climate change on growth of 0-group sole and plaice. Mar Ecol-Prog Ser 358:219–230

Ulbrich U, Pinto JG, Kupfer H, Leckebusch GC, Spangehl T, Reyers M (2008) Changing northern hemisphere storm tracks in an ensemble of IPCC climate change simulations. J Clim 21:1669–1670

Van den Eynde D, De Sutter R, De Smet L, Francken F, Haelters J, Maes F, Malfait E, Ozer J, Polet H, Ponsar S, Reyns J, Van der Biest K, Vanderperren E, Verwaest T, Volckaert A, Willekens M (2011) Evaluation of climate change impacts and adaptation responses for marine activities "CLIMAR". Final Report Brussels: Belgian Science Policy Office 2011

van der Molen J, Aldridge JN, Coughlan C, Parker ER, Stephens D, Ruardij P (2013) Modelling marine ecosystem response to climate change and trawling in the North Sea. Biogeochemistry 113:213–236

van der Veer HW, Witte JIL (1999) Year-class strength of plaice *Pleuronectes platessa* in the Southern Bight of the North Sea: a validation and analysis of the inverse relationship with winter seawater temperature. Mar Ecol-Prog Ser 184:245–257

van Hal R, Smits K, Rijnsdorp AD (2014) How climate warming impacts the distribution and abundance of two small flatfish species in the North Sea. J Sea Res 64:76–84

an Keeken OA, van Hoppe M, Grift RE, Rijnsdorp AD (2007) Changes in the spatial distribution of North Sea plaice (*Pleuronectes platessa*) and implications for fisheries management. J Sea Res 57:187–197

Vanderperren E, De Sutter R, Polet H (2009) Climate change: threat or opportunity for Belgian sea fisheries? IOP Conf. Series: Earth Environ Sci 6, 352042. doi:10.1088/1755-1307/6/5/352042

Vaz S, Loots C (2009) Impact of environmental change on plaice and red mullet in the North Sea and eastern English Channel. In: Drinkwater et al. (ed) Report on the effects of future climate change on distribution, migration pattern, growth and reproduction of key fish stocks. RECLAIM – Resolving Climatic Impacts on Fish Stocks. www.climateandfish.eu

Walden H, Schubert K (1965) Untersuchungen über die Beziehungen zwischen Wind und Herings-Fangertrag in der Nordsee. Ber Dtsch Wiss Komm Meeresforsch 28:194–221

Wang XL, Wan H, Zwiers FW, Swail VR, Compo GP, Allan RJ, Vose RS, Jourdain S, Yin X (2011) Trends and low-frequency variability of storminess over western Europe, 1878-2007. Clim Dynam 37:2355–2371

Wieland K, Olesen HJ, Fenger Pedersen EM, Beyer JE (2011) Potential bias in estimates of abundance and distribution of North Sea cod (*Gadus morhua*) due to strong winds prevailing prior or during a survey. Fish Res 110:325–330

Woodhead PMJ (1964) The death of north sea fish during the winter of 1962/63, particularly with reference to the sole, *Solea vulgaris*. Helgoland wiss Meer 10:283–300

Woolf D, Wolf J (2013) Impacts of climate change on storms and waves. MCCIP Science Rev 2013:20–26

World Bank (2013) Fish to 2030: Prospects for Fisheries and Aquaculture. World Bank Report number 83177-GLB. World Bank, Washington

Socio-economic Impacts—Agricultural Systems

13

Jørgen Eivind Olesen

Abstract

Europe is one of the world's largest and most productive suppliers of food and fibre. In the North Sea region, agroecosystems vary from highly productive farming systems such as the arable cropping systems of western Europe to low-input and low-output farming systems with or without livestock. Climate change impacts on agricultural production will vary across the North Sea region, both in terms of crops grown and yields obtained. Given adequate water and nutrient supply, a doubling of atmospheric CO_2 concentration could lead to yield increases of 20–40 % for most crops grown in the North Sea region. The high-input farming systems could also respond favourably to modest warming. Extreme weather events may severely disrupt crop production. Increased temperature and more frequent extreme weather events could affect animal production through changes in feed production, changes in the availability of grazing, direct heat stress, and increased risk of disease. Overall, there seems to be potential for agriculture in the North Sea region to adapt to the changing climate in such a way that productivity and profitability may both increase, particularly over the long term. The challenge will be to ensure sustainable growth in agricultural production without compromising environmental quality and natural resources.

13.1 Introduction

Agriculture is situated at the interface between ecosystems and society with the main aim of ensuring food supply. Located at this interface, agriculture is both affected by and helps drive changes in global environmental conditions, for the latter by contributing to emissions of greenhouse gases, notably methane and nitrous oxide. Management of agricultural ecosystems varies from highly productive farming systems such as the arable cropping systems of western Europe to low-input and low-output farming systems with or without livestock, some of which are also located in Europe.

Europe is one of the world's largest and most productive suppliers of food and fibre (Olesen and Bindi 2002). In 2012, it accounted for 19 % of global meat production and 17 % of global cereal production. About 78 % of the European meat production and 63 % of cereal production occurred within EU countries, with the remaining production primarily in Russia, Belarus and Ukraine. The productivity of European agriculture is generally high, especially in western Europe, and average hectare cereal yields in EU countries are about 40 % higher than the world average (Olesen et al. 2011).

The overall driving force in agriculture is the globally increasing demand for food and fibre. This is primarily caused by a growing world population with a high demand for food production and a wealthier world population with a higher proportion of meat in the diet (Godfray et al. 2010). The result is that agriculture globally exerts increasing pressure on the land and water resources of the earth, which often results in land degradation (such as soil erosion and salinization), and eutrophication. Agriculture is also associated with greenhouse gas emissions (Kirchmann and Thorvaldsson 2000).

Agricultural land use along the Atlantic coast in Europe is dominated by grassland and forage crops, because the wet conditions limit soil trafficability (i.e. capability of supporting

J.E. Olesen (✉)
Department of Agroecology—Climate and Bioenergy, Aarhus University, Aarhus, Denmark
e-mail: jorgene.olesen@agrsci.dk

M. Quante and F. Colijn (eds.), *North Sea Region Climate Change Assessment*, Regional Climate Studies, DOI 10.1007/978-3-319-39745-0_13

agricultural traffic/machinery without degrading the soil) required for cultivating annual crops. In regions with less rainfall such as continental parts of Europe, arable cropping systems often dominate the agricultural landscape. In north-west Europe the arable cropping systems are dominated by cereals, in particular winter wheat and spring barley, and break crops (secondary crops grown to interrupt the repeated sowing of cereals as part of crop rotation) like oilseed rape, grain legumes, and root and tuber crops like sugar beet and potato. Over recent decades the area cultivated with high yielding crops such as winter wheat and silage and grain maize has increased. This increase in area of winter wheat has largely happened at the expense of less productive spring cereal crops.

Increases in winter wheat yield are mostly due to crop breeding and improved crop protection coupled with increased fertilisation; however, wheat yields in Europe have been stagnating over the past 10 to 20 years (Olesen et al. 2011). There also seems to have been greater variability in grain yields for wheat over the past two decades. Stagnating wheat yields in France have been attributed to lower yields under the rising temperature (Brisson et al. 2010), but changes in management may also have played a role in some countries (Finger 2010). In contrast to wheat, yields of grain maize show a continued increase in both France and Germany, such that grain maize yields now exceed those of winter wheat. The area of silage and grain maize is therefore growing in northern Europe (Elsgaard et al. 2012), and this appears to be linked to the warmer climate (Odgaard et al. 2011).

High-input farming systems in western and central Europe generally have a low sensitivity to climate change, because a given change in temperature or rainfall has a modest impact (Chloupek et al. 2004) and because farmers have resources to adapt management. However, there may be considerable difference in adaptive capacity between cropping systems and farms depending on their specialisation (Reidsma et al. 2007). These systems may therefore respond favourably to modest climate warming (Olesen and Bindi 2002). Across the North Sea region there is a large variation in climatic conditions, soils, land use and infrastructure, which greatly influences responsiveness to climatic change.

13.2 Impacts

13.2.1 Crop Responses to Climate Change

Rising greenhouse gas emissions affect agroecosystems directly (primarily by increasing photosynthesis and water use efficiency at higher CO_2 levels) and indirectly via climate change (temperature and rainfall affect several aspects of the functioning of cropping systems). Effects may also be both direct through changes in crop physiology and indirect through impacts on soil fertility, crop protection (weeds, pests or diseases) and the ability to perform field operations in a timely manner. The exact responses depend on the sensitivity of the particular agricultural system to environmental change and on the relative changes in controlling factors.

Increasing atmospheric CO_2 concentration stimulates yield of crops that have the so-called C3-photosynthesis pathway, which constitute almost all crops grown in the North Sea region, with the exception of maize and *Miscanthus* (cultivated for biofuel). A doubling of atmospheric CO_2 concentration is projected to lead to yield increases of 20–40 % in most crops (Ainsworth and Long 2005), provided adequate water and nutrient supply. The response is considerably less for C4-plants, which include tropical grasses such as maize. Higher CO_2 concentration not only increases photosynthesis, but also reduces plant water consumption. This may result in improved tolerance of plants to drought and generally drier conditions.

Higher CO_2 concentrations also affect the quality of plant biomass, because plants accumulate more sugar leading to higher carbon contents of leaves, stems and reproductive organs. This has consequences for the quality of the food and feed, which in some cases are negative. It will thus reduce the protein content of cereal grains and diminish the baking quality of wheat (Högy et al. 2013). The attraction of plants for pests and diseases will also change, which could make the plants more resistant to attack. However, weed growth will also benefit from increased CO_2, which may necessitate intensified or different control measures, for example, due to reduced efficacy of herbicides (Ziska 2001).

Temperature affects crops in different ways, partly through affecting the timing of crop phenological phases (crop development); partly through the efficiency of energy capture, conversion and storage (crop growth); and partly through crop water demands (temperature affects evapotranspiration). With warming, active growth starts earlier, plants develop faster, and the potential growing season is extended. This may have the greatest effect in colder regions (Trnka et al. 2011), and may be most beneficial for perennial crops or crops which remain in their vegetative phase, such as sugar beet and grasslands.

Higher temperature reduces crop duration of determinate species (plants that flower and mature). This concerns all cereals and seed plants such as pulses and oilseed crops. For wheat, a temperature increase of 1 °C during grain fill is estimated to reduce the length of this phase by 5 %, and yield to decline by a similar amount (Olesen et al. 2000). However, in the North Sea region such reductions can often be more than offset by changing to cultivars with longer growth duration (Olesen et al. 2012) and this may even lead

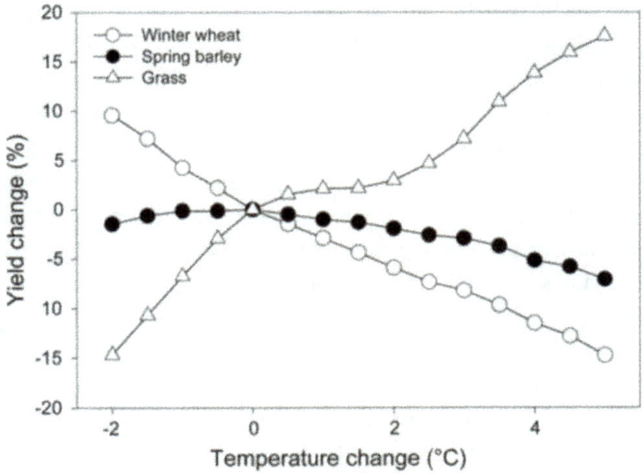

Fig. 13.1 Mean simulated change in yield of winter wheat, spring barley and ryegrass with increasing temperature for a site in Denmark. The simulations were performed with the CLIMCROP model assuming that water is not growth-limiting (Olesen et al. 2000; Olesen 2005)

to improved yields with potential for longer growing seasons at high latitudes (Montesino-San Martin et al. 2014).

These differential responses of crop yield to rising temperature in plants with different responses of crop development are illustrated in Fig. 13.1. The results were produced with a crop simulation model that integrates the biophysical interactions between soil, climate and plants on crop growth and yield over the growing season. Such models are commonly used to assess the effects of climate change on crop yield and quality (Ewert et al. 2015).

The greatest reductions in grain yield in Fig. 13.1 were simulated for winter wheat, where growth duration is reduced, because any changes in sowing date in autumn would have little effect on duration of the vegetative and reproductive phases in the following spring and summer. This response concurs well with observed response of winter wheat yields in Denmark, where the largest reductions were found to be related to high temperatures during the grain filling phase (Kristensen et al. 2011). Figure 13.1 shows a smaller response of spring barley to higher temperatures, because this crop can be sown earlier in spring thus maintaining a productive growing season. In contrast, yields were simulated to increase for a grass crop, which represents crops with a non-determinate growth pattern, where yields depend on the total duration of the growing season with suitable temperatures and rainfall.

Peltonen-Sainio et al. (2010) characterised the coincidence of yield variations with weather variables for major field crops using long-term datasets to reveal whether there are commonalities across the European agricultural regions. Long-term national and/or regional yield datasets were used from 14 European countries for spring and winter barley and wheat, winter oilseed rape, potato and sugar beet. Harmful effects of high precipitation during grain-filling in grain and seed crops and at flowering in oilseed rape were recorded. In potato, reduced precipitation at tuber formation was associated with yield penalties. Elevated temperature had harmful effects for cereals and rapeseed yields. Similar harmful effects of rainfall and high temperature on grain yield of winter wheat were found by Kristensen et al. (2011) in a study using observed winter wheat yields from Denmark.

13.2.2 Impacts of Climatic Variability and Weather Extremes

Extreme weather events, such as periods of high temperature, heavy storms, or droughts, can severely disrupt crop production. Individual extreme events do not usually have lasting effects on the agricultural system. However, if the frequency of such events increases, agriculture will need to respond, either by adapting or by ceasing its activity.

Crops often respond nonlinearly to changes in their growing conditions and have threshold responses, which greatly increases the importance of climatic variability and the frequency of extreme weather events in terms of absolute yield, yield stability and quality (Trnka et al. 2014). This may lead to drastic reductions in yield from short episodes of high temperature during sensitive crop growth phases such as the reproductive period. Temperatures above 35 °C during the flowering period can in most crops severely affect seed and fruit set and thus greatly reduce yield (Porter and Semenov 2005). High temperatures will also greatly increase evapotranspiration leading to higher risk of drought, if rainfall is insufficient to compensate for the water losses (Lobell et al. 2013). Such high temperature stresses may severely impact crop yields, even in the North Sea region (Semenov and Shewry 2011).

An increase in temperature variability will increase yield variability and also result in a reduction in mean yield. Even in the North Sea region there may be a sufficient increase in climatic variability to significantly affect crop yield (Kristensen et al. 2011), although this effect is expected to be more severe in other parts of the world. This risk is likely to be particularly large for high-input production systems (Trnka et al. 2012), where the demand for continued high soil water supply is greater than for low-input systems (having lower rates of evapotranspiration). Also, a given proportional reduction in crop yield will have a greater absolute yield effect on high- rather than low-yielding crops. Therefore increases in climatic extremes will also have greater effects in high-input rainfed systems than in less intensive and diverse systems (Schaap et al. 2011). The high-input rainfed cropping systems may thus be particularly vulnerable to climate change, although some will also benefit in terms of higher average yields from the warming and higher CO_2 concentrations.

13.2.3 Changes in Crop Productivity and Suitability

Climatic warming will in temperate regions result in earlier onset of the growing season in spring and a longer duration in autumn. A longer growing season allows the proliferation of species that have suitable conditions for growth and development and can thus increase their productivity (e.g. crop yield, number of crops per year). This may also allow for the introduction of new species previously unfavourable due to low temperatures or short growing seasons. This is relevant for the introduction of new crops, such as for grain maize or winter wheat in northern Europe (Elsgaard et al. 2012), but will also affect the spread of weeds, pests and diseases that often follow the crops grown (Roos et al. 2011).

Warming has already caused a northward expansion of the area of silage maize in northern Europe into southern parts of Scandinavia, where the system of grass and silage maize for intensive dairy production has largely replaced the traditional fodder production systems (Odgaard et al. 2011; Eckersten et al. 2014; Nkurunziza et al. 2014). Very recently grain maize has started to be grown in southern parts of Denmark, reflecting the warming trends (Elsgaard et al. 2012). Analyses of the effects of observed climate change on yield potential in Europe have shown positive effects for maize and sugar beet, which have benefited from the longer growing season for these crops (Supit et al. 2010). Yield benefits have been greatest in northern Europe. The warming may also have contributed to higher potato yields in northern regions of Europe. In contrast, warmer and more variable climatic conditions with increased occurrence of drought have reduced crop yields in parts of central Europe (Eitzinger et al. 2013).

A further lengthening of the growing season as well as a northward shift for some species are projected to result from further increases in temperature across Europe (Olesen et al. 2011). The date of last frost in spring is projected to reduce by 5–10 days by 2030 and 10–15 days by 2050 throughout most of Europe compared with the period 1961–1990 (Trnka et al. 2011). Since a longer growing season will increase productivity of many crops in northern Europe, this could lead to further intensification of cropping systems.

Projected climate change is expected to result in more favourable conditions for crop production at high latitudes than at low northern European latitudes (Table 13.1). The agroclimatic indices show a substantial lengthening (one month) of the growing season by 2050 in northern regions, but much less in southern parts of the North Sea region. Although the duration of the growing season is projected to increase throughout the North Sea region, in southern and continental parts this increase may be counteracted by drier conditions during summer resulting in reduced crop growth.

Projected impacts of climate change on crop yields depend on crop type, emission scenario and the sensitivity of the underlying climate model used to project climate changes (Olesen et al. 2007). Projections for most crops in the North Sea region show an increase in projected yield during the first half of the 21st century (Supit et al. 2012). However, later in the century yield is projected to decrease due to the effects of temperature rise and reduced summer rainfall that together exceed the benefits achieved from higher atmospheric CO_2 concentration, in particular for cereal crops. For root and tuber crops in Europe (such as sugar beet and potato) yields are projected to continue increasing (Angulo et al. 2013). However, even for potato some regions may become less suitable for production due to drier summer conditions and constraints imposed on the use of irrigation (Daccache et al. 2012).

At high latitudes or at high elevations with wet and cool climates, cropping systems with grasslands and forage production for ruminant livestock currently tend to dominate. Timothy *Phleum pratense* L. and perennial ryegrass *Lolium perenne* L. are the most important forage grasses at high latitudes, and in cold and snow-rich regions, timothy outcompetes perennial ryegrass due to better winter survival (Höglind et al. 2013). Due to the higher productivity and better feed quality of ryegrass compared to timothy, warming leading to less risk of winter kill is expected to shift the patterns in the cultivation of grassland species in Norway and Sweden northwards. Similar shifts may be expected in grazing season duration (Uleberg et al. 2014). In some cases these shifts will be constrained by rainfall, either with conditions too dry during summer or too wet during spring or autumn.

13.2.4 Environmental Impacts

Soils have many functions, of which water and nutrient supply to growing crops are essential for sustained crop production. However, soils are also important in regulating water and nutrient cycles, for carbon storage and greenhouse gas emissions. Soils are habitats for many of the organisms that contribute to the functioning of soils and agroecosystems, having both positive and negative effects on crop yield. Where soil moisture allows, increasing temperatures will enhance decomposition of soil organic matter, which tends to decrease soil organic stocks unless counterbalanced by larger inputs of organic matter in crop residues (Falloon and Betts 2010). A reduction in soil carbon enhances the

Table 13.1 Effects of projected climate change on changes in key agroclimatic indices in northern European agroecological zones by 2050 compared to the period 1961–1990 (Trnka et al. 2011)

Zone	Change in effective solar radiation (%)	Change in effective growing days (days)	Change in date of last frost (days)	Change in dry days in spring (%)	Change in dry days in summer (%)
Alpine North (Norway and north Sweden)	+8	+29	−10	+1	−2
Boreal (Finland, central Sweden, parts of Norway)	+8	+16	−11	−1	+2
Nemoral (south-central Sweden)	+8	+12	−10	+1	+11
Atlantic North (Ireland, British Isles, western Denmark, Netherlands)	−1	+5	−11	−4	+21
Continental (east Denmark, south Sweden, Germany)	−6	−6	−12	−2	+20

contribution of agriculture to global warming through higher net CO_2 emissions. In contrast, the effects of warming on nitrous oxide emissions from agricultural soils are less clear, since effects depend on the balance between the separate effects of temperature, rainfall and CO_2 as well as their seasonal changes, relative to effects of changes in crop growth patterns (Dijkstra et al. 2012).

Any reduction in soil organic matter stocks implies a decrease in fertility and biodiversity, a loss of soil structure, reduced soil water infiltration and retention capacity, and increased risk of erosion and compaction. If these changes are significant, this leads to lower productivity of crops growing on the soils. Changes in rainfall and wind patterns, in particular more intense rainfall, can lead to increased erosion from soils with poor crop cover or with little surface cover of plant residues to protect the soil. Also, increasing frequencies of freeze/thaw cycles during winter, due to reduced snow cover, in combination with stronger rainfall may greatly enhance soil erosion (Ulén et al. 2014). In addition to depleting soil fertility this erosion may also enhance nutrient runoff to sensitive aquatic ecosystems (Jeppesen et al. 2009).

Faster decomposition of soil organic matter at higher temperatures increases mineralisation of soil organic nitrogen. This in turn may increase the risk of nitrate leaching during periods of little or no crop cover with sufficient nitrogen uptake to prevent nitrate being leached in periods of precipitation surplus (Jabloun et al. 2015). This may increase the risk of nitrate leaching to surface and groundwater systems (Stuart et al. 2011; Patil et al. 2012). Current measures to reduce nitrate leaching may not be sufficient to maintain low leaching rates under projected climate change (Doltra et al. 2014) and this could increase the risk of algal blooms and the occurrence of toxic cyanobacteria in lakes (Jeppesen et al. 2011).

13.2.5 Crop Protection

Most pest and disease problems are closely linked with their host crops. Introducing new crops will therefore mean new pest and disease problems. In cool regions, higher temperatures favour the proliferation of insect pests, because many insects can then complete a greater number of reproductive cycles. Higher winter temperatures will also allow pests to overwinter in areas where they are currently limited by cold periods, causing greater and earlier infestation during the following crop season (Roos et al. 2011). Earlier insect spring activity and proliferation of some pest species will favour some of the virus diseases that spread with insects. A similar situation may occur for plant fungal diseases leading to increased need for pesticides.

Unlike pests and diseases, weeds are directly influenced by changes in atmospheric CO_2 concentration. Differential effects of CO_2 and climate change on crops and weeds will alter the weed-crop competitive interactions, sometimes to the benefit of the crop and sometimes to the weeds. Interaction with other biotic factors and with changing temperature and rainfall may also influence weed seed survival and thus weed population development.

Improved climatic suitability will lead to invasion of weeds, pests and diseases adapted to warmer climatic conditions. The speed at which such species invade depends on the rate of climatic change in terms of suitability ranges (e.g. in km per year), the dispersal rate of the species (e.g. in terms of km per year) and on measures taken to combat non-indigenous species. The dispersal rates of pests and diseases are often so high that their geographical extent is determined by the range of climatic suitability. The Colorado beetle *Leptinotarsa decemlineata* L. and the European cornborer *Ostrinia nubilalis* Hubner are examples of pests and diseases that are expected to show a considerable northward expansion in Europe under climatic warming (Olesen et al. 2011).

Studies show projected increases in the occurrence of several crop diseases with projected warming in the currently cooler parts of high-input cropping regions, such as UK (Butterworth et al. 2010; Evans et al. 2010) and Germany (Siebold and von Tiedemann 2012), whereas the risk of some diseases may reduce with warming in regions further south, such as France (Gouache et al. 2013). As well as affecting crop yield, such changes will also affect the quality of the yield, for example through the occurrence of mycotoxins which may increase in northern Europe under the projected climate change (Madgwick et al. 2011; van der Fels-Klerx et al. 2012). This would increase the need for fungicides or alternative strategies such as breeding for resistance.

13.2.6 Livestock Production

Increased temperature and more frequent extreme weather events could affect animal production through changes in feed production, changes in availability of grazing, direct heat effects on animals, and increased risk of disease.

More variable weather and more extreme weather events are projected under climate change (Jacob et al. 2014). This is likely to result in more variable quantities and quality of crops such as cereals, forage crops and protein crops, causing unstable feed prices both globally and locally. This has already occurred in recent years, with large fluctuations in grain price due to heat waves and droughts in wheat-producing regions. This has mostly affected production of monogastric livestock such as pigs and poultry. However, ruminant animals have also been affected through the production of grass and forage, either because conditions are too wet or too dry, which affects grazing. For example, in 2003 a long drought across western and central Europe severely affected not only arable crop production, but also fodder production for ruminants, to the extent that livestock production costs greatly increased (Fink et al. 2004).

Climate change will exacerbate problems with existing animal diseases, which negatively affect animal welfare and livestock production. Global warming and more frequent extreme weather events (droughts and increased rainfall) will provide more favourable climates for some viruses, their vector species, and for fungal or bacterial pathogens. New viral vector-borne diseases may not necessarily originate from nearby regions but may arrive from outside Europe. An example is bluetongue disease, where climate change has allowed the midge *Culicoides imicola* Kiefer that acts as a vector for the disease to spread—causing the virus to expand its distribution northwards in Europe (Purse et al. 2005). The risk of bluetongue and other emerging pathogens and vectors becoming established in the North Sea region will greatly increase under higher temperatures. Blood-sucking midges *Culicoides* spp. are one of the major threats to animal welfare, because they spread viruses that cause serious diseases in animals. Ticks, mosquitoes and lymnaeid snails can also transmit extremely harmful diseases to livestock. Increased annual temperature, milder winters and higher rainfall will favour the propagation of helminth parasites, resulting in disease and pronounced negative effects on the welfare of grazing cattle and sheep (Skuce et al. 2013).

13.3 Adaptation, Vulnerabilities and Opportunities

13.3.1 Adaptation at Farm and Regional Scale

Farmers are already adapting to climate change since farming is very weather dependent. Farmers constantly experiment with new cropping techniques, and the most successful ones spread quickly among the farming community where agricultural advisors and researchers are ready to take up and disseminate new results. This is evident, for example, in the northward spread of silage maize into Denmark and southern Sweden (Odgaard et al. 2011). Such adaptations are autonomous in the sense that they require no external action or planning. In a European context they are also fairly effective due to the high capacity among farmers to incorporate new technologies and management practices.

Adaptation only works when the basic resources for crop growth are still maintained and when the climate allows proper soil and crop management to take place (Table 13.2). In northern areas climate change may have positive effects on agriculture through introducing new crop species and varieties, higher crop production and expansion of areas suitable for cultivation. Negative effects may be an increase in the need for plant protection, risk of increased nutrient leaching and the degradation of soil organic matter. Further south in Europe issues around managing drier summer conditions will dominate adaptation needs.

The responsiveness of agricultural systems to climate change depends on many factors, both how current crops are being affected by climate change, but also on the options available for modifying the systems to reduce negative impacts and take advantage of new opportunities. The capacity for agriculture in the North Sea region to adapt to future changes is expected to be good, since the changes could be largely favourable for production, and because research, educational and advisory capacities are high (Table 13.2). However, there may be barriers to adaptation, not least within the current agricultural and environmental policies that may have to be adjusted to ensure effective adaptation.

Table 13.2 Resource-based policies to support adaptation of agricultural systems to climate change (adapted from Olesen and Bindi 2002)

Resource	Policy
Land	*Reforming agricultural policy to encourage flexible land use.* The great extent of cropland in northern Europe across diverse climates will provide diversity for adaptation
Water	*Reforming water management to ensure balance between maintaining the amount and quality of water resources and the ecosystems that these support, with the needs of agricultural production.* Climate change will affect the demand for irrigation and drainage, which depending on location have consequences for water resources and their ecological quality and may affect needs for revising management and governance schemes
Nutrients	*Improving nutrient use efficiencies through changes in cropping systems and development and adoption of new nutrient management technologies.* Nutrient management needs to be tailored to the changes in crop production as affected by climate change, and utilisation efficiencies must be increased, especially for nitrogen, in order to reduce climate change induced emissions to water and air
Agrochemicals	*Support for integrated pest management systems (IPMS) should be increased through a combination of education, regulation and taxation.* There will be a need to adapt existing IPMS to changing climatic regimes
Energy	*Improving the efficiency of food production and exploring new biofuels and ways to store more carbon in trees and soils.* Reliable and sustainable energy supply is essential for many adaptations to new climate and for mitigation policies
Genetic diversity	*Assembling, preserving and characterising plant and animal genes and conducting research on alternative crops and animals.* Genetic diversity and new genetic material will provide important basic material for adapting crop species to changing climatic conditions, such as by improving tolerance to adverse conditions
Research capacity	*Encouraging research on adaptation, developing new farming systems and developing alternative foods.* Greater investment in agricultural research may provide new sources of knowledge and technology for adaptation to climate change
Information systems	*Enhancing national systems that disseminate information on agricultural research and technology, and encouraging information exchange among farmers.* Fast and efficient information dissemination and exchange to and between farmers using the new technologies (e.g. internet) will increase the rate of adaptation to climatic and market changes
Culture	*Integrating environmental, agricultural and cultural policies to preserve the heritage of rural environments in a new environment.* Integration of policies will be required to maintain and preserve the heritage of rural environments which are dominated by agricultural practices influenced by climate

Some of the adaptation is beyond farm scale, and requires collective action. This is the case for breeding new cultivars and for infrastructure projects that provide water for irrigation or for improving drainage at the catchment scale. Such efforts may have long time perspectives and involve many actors and so require planning and in some cases approval from authorities. Actions for managing water at the catchment scale require a consideration not only of the needs of farmers, but also of the needs of human settlements and nature conservation, including consideration of surface and groundwater quality (Refsgaard et al. 2013).

Plant and livestock breeding is one of the most effective options for adapting to climate change, as well as switching livestock and crop species used. Among the measures required for both plant and livestock breeds is to increase tolerance to heat stress events (Semenov et al. 2014). For plants, there is also a need to enhance tolerance to a wider range of stresses, including drought, extreme heat and flooding. Soil management will need to accommodate the projected increase in frequency and intensity of erosion events associated with more intense rainfall. This may involve trade-offs between various factors that all contribute to crop yield. Therefore there is a risk that higher yield stability may come at the cost of reduced yield in favourable years. Plant breeders will need to deliver cultivars that are more resilient to weather extremes and resistant to new diseases. Plant breeding is a long-term activity, and timely delivery of such cultivars will require good and early predictions of future environmental conditions to allow the development and use of suitable germplasm.

13.3.2 Role of Vulnerability and Uncertainty in Adaptation

Projecting the effects of climate change on agricultural systems involves many uncertainties, some concern the climate change projections themselves while others concern biophysical understanding of how crops and livestock will respond to climate change. However, an even larger uncertainty concerns how well farmers and agricultural systems can and will adapt to climate change in the longer term in order to minimise losses and take advantage of new opportunities (Moore and Lobell 2014). Part of the uncertainty lies in how quickly some of the longer-term adaptations needed to overcome major changes in climate (expanding irrigation or drainage systems, new crops etc.) can be implemented, since short-term adaptations (e.g. changes in varieties or sowing time) are likely to be much less effective (Fig. 13.2). In southern Europe, farming profits are expected to decline

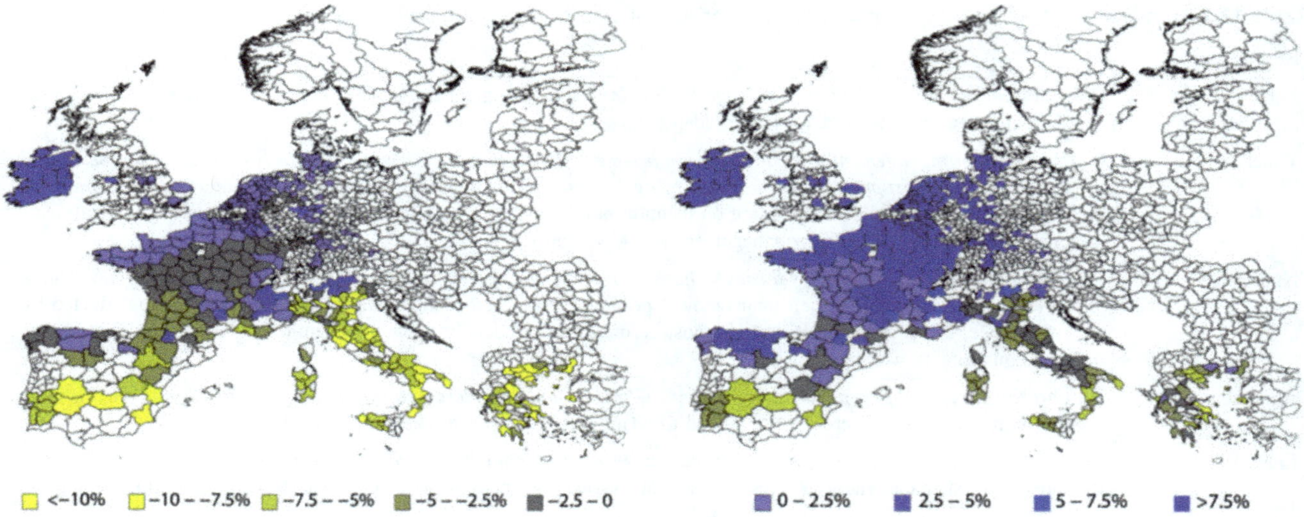

Fig. 13.2 Projected change in farm profit by 2040 under the IPCC A1B scenario for selected growing regions in Europe. Data concern wheat, maize, barley, sugar beet and oilseed. Projections made with short-run response function of crop yield to temperature and precipitation (*left*) and projections made using a long-run response function that includes farm-level adaptations (*right*) (Moore and Lobell 2014). White areas reflect regions with insufficient data

under climate change, while the North Sea region may see a rise in farming profits, particularly over the long term.

Temperature and rainfall regimes in combination with soil properties dictate the potential for agricultural production. Thus climate change, particularly in terms of dryness or wetness will affect agricultural land use, and even moderate changes may have marked effects on land use, especially for soils that are borderline with respect to which crops can be grown (Brown et al. 2011). Such areas are therefore more sensitive to environmental change than areas that are clearly favourable or unfavourable for specific agricultural land uses under both current and future climatic conditions.

Adapting to increased frequency of extreme weather events may be a significant challenge, since extreme events are by nature difficult to predict and so are also difficult to prepare for. Even with statistical evidence that shows extremes are changing in frequency such information may be interpreted differently among decision makers, resulting in over- as well as under-adaptation (Refsgaard et al. 2013).

The European agricultural sector is regulated and financially supported in several ways. Therefore, a major consideration must be how the adaptation responses will interact with regulations on environmental and nature protection, as well as on issues such as food safety and local employment. The vast amount of EU support for farming may be used strategically to support adaptations that maintain the balance between the need for high-production output of healthy and safe foods on the one hand and the need to protect the environment as well as the agricultural resource base on the other.

13.4 Ecosystem Functions and Services

Climate change impacts on agricultural production will vary across the North Sea region, both in terms of crops grown and yields obtained. Overall, there seems to be potential for adapting to the changing climate in such a way that productivity and profitability may both increase. In some parts of the region, a longer growing season would enable a switch to longer season crops such as highly productive grasses or *Miscanthus*, which with the use of biorefinery technologies could increase the output not only of food and feed for livestock, but also of the production of biofuels as a fossil fuel substitute (Smith and Olesen 2010). Because similar increases are not projected for most annual crops, this may facilitate changes in cropping systems, provided the technologies become profitable.

In grasslands, a longer growing season would allow more cuts and higher production, particularly in areas less affected by summer drought. This may facilitate greatly increased production of protein-rich crops by cultivating highly productive grass-clover pastures with little fertiliser and pesticide use. These pastures may be harvested for feed or grazed by ruminant livestock such as dairy and beef cattle and sheep. The pastures may also be a new source of sustainable protein production for monogastric farm animals (pigs and poultry) as well as for farmed fish. This would require the development and implementation of new biorefinery technologies, for which Europe with its ability to combine advanced technologies may be particularly well suited (Parajuli et al. 2015). This would strengthen the role of northwest Europe as a continued supplier of food, but also as a supplier of the technologies for sustainable intensification of production systems that target both adaptation and mitigation to climate change.

Changes in climatic suitability may lead to major changes in land use, which would affect not only the production of goods in agriculture, but also the landscape and ecosystem services (such as the quality of nature, the environment, groundwater and freshwater systems) (Harrison et al. 2013). This would challenge current land use planning, and would call for a strategic, long-term perspective on land-use policy under climate change (van Meijl et al. 2006).

In arable farming systems, higher temperatures will enhance turnover of soil organic matter and this, in combination with increased and more intense rainfall, would enhance the risk of nitrogen and phosphorus losses to the aquatic environment, thereby threatening the quality of these waters for recreational use and fish production. New and revised policy may be needed to manage the environmental impacts of agricultural production. Likewise, an increased need for pesticide use in agricultural production would be problematic in relation to current EU pesticide policies.

Policies will need to promote active resource management and the utilisation of renewable raw materials as substitutes for metal and oil-based products and fossil fuels. This is essential for sustainable resource management, as well as for mitigating climate change. Resource management of this type would need to take multiple needs into consideration, including: provision of biomass for food, feed, bioenergy and biomaterials within the bioeconomy; recycling of nutrients and resilient organic matter to the agricultural systems; maintenance of soil carbon stocks; and provision of other ecosystem goods and services, such as clean water and air and a diverse natural environment.

Cultivation of agricultural crops requires suitable and well-drained soils. The anticipated increase in winter rainfall across large parts of the North Sea region would place additional stresses on current drainage systems. This issue is expected to become increasingly important in areas where agricultural production may expand due to increased suitability. Enhancing drainage of agricultural soils cannot be implemented without ensuring that water can be effectively transported in streams and rivers. Aligning drainage needs with the need to protect parts of the landscape from flooding may cause conflict among actors, and will require new planning at the landscape and catchment level. Similar considerations must be taken into account when preparing for increased risk of summer drought.

13.5 Conclusions

Agricultural systems in northwest Europe are generally characterised by high inputs of fertilisers and pesticides and resulting high crop yields and livestock productivity. Observations over recent decades show consistent changes in crop phenology and geographical shifts towards higher latitudes of intensive crop cultivation in accordance with observed climate change. The observed effects on crop yield range from negative (dominating for cereal and seed crops) to positive (dominating for non-determinate crops such as many forage and grass crops). The combined effects of enhanced CO_2 and changes in temperature and precipitation are expected in many cases to increase productivity. Model-based and empirical studies show an increased risk of higher interannual yield variability with the projected climate change, resulting from changes in interannual temperature variability as well as from nonlinearities in the response of crops to changes in temperature and rainfall, increasing the risk of low yields. Negative effects on crop yield may be further exacerbated by extreme temperature and rainfall events. Climate change will further increase needs to reconsider measures for dealing with soil fertility, crop protection and nutrient retention in intensive cropping systems.

To contribute to global food security and help mitigate agricultural greenhouse gas emissions, there is a need to focus on sustainable intensification of agricultural production (Tilman et al. 2011). The challenges in the North Sea region will be to ensure sustainable growth in agricultural production without compromising environment and natural resources. This is likely to require the development of new production systems with a greater use of perennial crops such as grasses or increased use of cover crops in the rotations to make use of a longer growing season and to protect the soil and wider environment from erosion and nutrient leaching.

References

Ainsworth EA, Long SP (2005) What have we learned from 15 years of free-air CO_2 enrichment (FACE)? A meta-analytic review of the responses of photosynthesis, canopy properties and plant production to rising CO_2. New Phytol 165:351–372

Angulo C, Rötter R, Lock R, Enders A, Fronzek S, Ewert F (2013) Implication of crop model calibration strategies for assessing regional impacts of climate change in Europe. Agric Forest Meteorol 170:32–46

Brisson N, Gate P, Gouache D, Charmet G, Oury FX, Huard F (2010) Why are wheat yields stagnating in Europe? A comprehensive data analysis for France. Field Crops Res 119:201–212

Brown I, Poggio L, Gimona A, Castellazzi M (2011) Climate change, drought risk and land capability for agriculture: implications for land use in Scotland. Reg Env Change 11:503–518

Butterworth MH, Semenov MA, Barnes A, Moran D, West JS, Fitt BDL (2010) North-South divide: contrasting impacts of climate change on crop yield in Scotland and England. J Roy Soc Int 7: 123–130

Chloupek O, Hrstkova P, Schweigert P (2004) Yield and its stability, crop diversity, adaptability and response to climate change, weather and fertilisation over 75 years in the Czech Republic in comparison to some European countries. Field Crops Res 85:167–190

Daccache A, Keay C, Jones RJA, Weatherhead EK, Stalham MA, Knox JW (2012) Climate change and land suitability for potato production in England and Wales: impacts and adaptation. J Agric Sci 150:161–177

Dijkstra FA, Prior SA, Runion GB, Torbert HA, Tian H, Lu C, Venterea RT (2012) Effects of elevated carbon dioxide and increased temperature on methane and nitrous oxide fluxes: evidence from field experiments. Front Ecol Env 10:520–527

Doltra J, Lægdsmand M, Olesen JE (2014) Impacts of projected climate change on productivity and nitrogen leaching of crop rotations in arable and pig farming systems in Denmark. J Agr Sci 152:75–92

Eckersten H, Herrmann A, Kornher A, Halling M, Sindhøj E, Lewan E (2014) Predicting silage maize yield and quality in Sweden as influenced by climate change and variability. Acta Agric Scand Sec B Soil Plant Sci 62:151–165

Eitzinger J, Thaler S, Schmid E, Strauss F, Ferrise R, Moriondo M, Bindi M, Palosuo T, Rötter R, Kersebaum C, Olesen JE, Patil RH, Saylan L, Caldag B, Caylak O (2013) Sensitivities of crop models to extreme weather conditions during flowering period demonstrated for maize and winter wheat in Austria. J Agric Sci 151:813–835

Elsgaard L, Børgesen CD, Olesen JE, Siebert S, Ewert F, Peltonen-Sainio P, Rötter RP, Skjelvåg AO (2012) Shifts in comparative advantages for maize, oat, and wheat cropping under climate change in Europe. Food Add Contam 29:1514–1526

Evans N, Butterworth MH, Baierl A, Semenov MA, West JS, Barnes A, Moran D, Fitt BDL (2010) The impact of climate change on disease constraints on production of oilseed rape. Food Sec 2:143–156

Ewert F, Rötter RP, Bindi M, Webber H, Trnka M, Kersebaum KC, Olesen JE, van Ittersum MK, Janssen S, Rivington M, Semenov M, Wallach D, Porter JR, Stewart D, Verhagen J, Gaiser T, Palosuo T, Tao F, Nendel K, Roggero PP, Bartosova L, Asseng S (2015) Crop modelling for integrated assessment of climate change risk to food production. Environ Modell Softw 72:287–303

Falloon P, Betts R (2010) Climate impacts on European agriculture and water management in the context of adaptation and mitigation - The importance of an integrated approach. Sci Total Env 408: 5667–5687

Finger R (2010) Evidence of slowing yield growth – The example of Swiss cereal yields. Food Pol 35:175–182

Fink AH, Brücher T, Krüger A, Leckebusch GC, Pinto JG, Ulbrich U (2004) The 2003 European summer heat waves and drought - Synoptic diagnosis and impact. Weather 59:209–216

Godfray CJ, Beddington JR, Crute IR, Haddad L, Lawrence D, Muir JF, Pretty J, Robinson S, Thomas SM, Toulmin C (2010) Food security: The challenge of feeding 9 billion people. Science 327:812–818

Gouache D, Bensadoun A, Brun F, Pagé C, Makowski D, Wallach D (2013) Modelling climate change impact on *Septoria tritici blotch* (STB) in France: Accounting for climate model and disease model uncertainty. Agric Forest Met 170:242–252

Harrison PA, Holman IP, Cojocaru G, Kok K, Kontogianni A, Metzger MJ, Gramberger M (2013) Combining qualitative and quantitative understanding for exploring cross-sectoral climate change impacts, adaptation and vulnerability in Europe. Reg Env Change 13:761–780

Höglind M, Thorsen SM, Semenov MA (2013) Assessing uncertainties in impact of climate change on grass production in Northern Europe using ensembles of global climate models. Agric Forest Met 170:103–113

Högy P, Brunnbauer M, Koehler P, Schadorf K, Breuer J, Franzaring J, Zhunusbayerva A, Fangmeier A (2013) Grain quality characteristics of spring wheat (*Triticum aestivum*) as affected by free-air CO_2 enrichment. Env Exp Bot 88:11–18

Jabloun M, Schelde K, Tao F, Olesen JE (2015) Effect of changes in temperature and precipitation in Denmark on nitrate leaching in cereal cropping systems. Eur J Agron 62:55–64

Jacob D, Petersen J, Eggert B, Alias A, Christensen, OB, Bouwer LM, Braun A, Colette A, Déqué M, Georgievski G, Georgopoulou E, Gobiet A, Menut L, Nikulin G, Haensler A, Hempelmann N, Jones C, Keuler K, Kovats S, Kröner N, Kotlarski S, Kriegsmann A, Martin E, van Meijgaard E, Moseley C, Pfeifer S, Preuschmann S, Radermacher C, Radtke K, Rechid D, Rounsevell M, Samuelsson P, Somot S, Soussana J-F, Teichmann C, Valentini R, Vautard R, Weber B, Yiou P (2014) EURO-CORDEX: new high-resolution climate change projections for European impact research. Reg Env Change 14:563–578

Jeppesen E, Kronvang B, Meerhoff M, Søndergaard M, Hansen KM, Andersen HE, Lauridsen TL, Liboriussen L, Bekioglu M, Ozen A, Olesen JE (2009) Climate change effects on runoff, phosphorus loading and lake ecological state, and potential adaptations. J Env Qual 38:1930–1941

Jeppesen E, Kronvang B, Olesen JE, Audet J, Søndergaard M, Hoffmann CC, Andersen HE, Lauridsen TL, Liboriussen L, Larsen SE, Beklioglu M, Meerhoff M, Özen A, Özkan K (2011) Climate change effects on nitrogen loading from catchment: implications for nitrogen retention, ecological state of lakes and adaptation. Hydrobiologia 663:1–21

Kirchmann H, Thorvaldsson G (2000) Challenging targets for future agriculture. Eur J Agron 12:145–161

Kristensen K, Schelde K, Olesen JE (2011) Winter wheat yield response to climate variability in Denmark. J Agr Sci 149:33–47

Lobell DB, Hammer GL, McLean G, Messina C, Roberts MJ, Schlenker W (2013) The critical role of extreme heat for maize production in the United States. Nat Clim Change 3:497–501

Madgwick JW, West JS, White RP, Semenov MA, Townsend JA, Turner JA, Fitt BDL (2011) Impacts of climate change on wheat anthesis and fusarium ear blight in the UK. Eur J Plant Path 130:117–131

Montesino-San Martin M, Olesen JE, Porter JR (2014) A genotype, environment and management (GxExM) analysis of adaptation in winter wheat to climate change in Denmark. Agric Forest Met 187:1–13

Moore FC, Lobell DB (2014) Adaptation potential of European agriculture in response to climate change. Nat Clim Change 4:610–614

Nkurunziza L, Kornher A, Hetta M, Halling M, Weih M, Eckersten H (2014) Crop genotype-environment modelling to evaluate forage maize cultivars under climatic variability. Acta Agric Scand Sec B Soil Plant Sci 64:56–70

Odgaard MV, Bøcher PK, Dalgaard T, Svenning JC (2011) Climatic and non-climatic drivers of spatiotemporal maize-area dynamics across the northern limit for maize production - A case study for Denmark. Agric Ecosyst Env 142:291–302

Olesen JE (2005) Climate change and CO_2 effects on productivity of Danish agricultural systems. J Crop Improv 13:257–274

Olesen JE, Bindi M (2002) Consequences of climate change for European agricultural productivity, land use and policy. Eur J Agron 16:239–262

Olesen JE, Jensen T, Petersen J (2000) Sensitivity of field-scale winter wheat production in Denmark to climate variability and climate change. Clim Res 15:221–238

Olesen JE, Carter TR, Diaz-Ambrona CH, Fronzek S, Heidmann T, Hickler T, Holt T, Minguez MI, Morales P, Palutikof J, Quemada M, Ruiz-Ramos M, Rubæk G, Sau F, Smith B, Sykes M (2007) Uncertainties in projected impacts of climate change on European agriculture and ecosystems based on scenarios from regional climate models. Clim Change 81:123–143

Olesen JE, Trnka M, Kersebaum KC, Skjelvåg AO, Seguin B, Peltonen-Saino P, Rossi F, Kozyra J, Micale F (2011) Impacts and adaptation of European crop production systems to climate change. Eur J Agron 34:96–112

Olesen JE, Børgesen CD, Elsgaard L, Palosuo T, Rötter R, Skjelvåg AO, Peltonen-Sainio P, Börjesson T, Trnka M, Ewert F, Siebert S, Brisson N, Eitzinger J, van der Fels-Klerx HJ, van Asselt E (2012) Changes in flowing and maturity time of cereals in Northern Europe under climate change. Food Add Contam 29:1527–1542

Parajuli R, Dalgaard T, Jørgensen U, Adamsen APS, Knudsen MT, Birkved M, Gylling M, Scjønning JK (2015) Biorefining in the prevailing energy and materials crisis: a review of sustainable pathways for biorefinery value chains and sustainability assessment methodologies. Ren Sust Ener Rev 43:244–263

Patil R, Lægdsmand M, Olesen JE, Porter JR (2012) Sensitivity of crop yield and N losses in winter wheat to changes in mean and variability of temperature and precipitation in Denmark using the FASSET model. Acta Agric Scand Sect B Plant Soil 62: 335–351

Peltonen-Sainio P, Jauhianinen J, Trnka M, Olesen JE, Calanca PL, Eckersten H, Eitzinger J, Gobin A, Kersebaum C, Kozyra J, Kumar S, Marta AD, Micale F, Schaap B, Seguin B, Skjelvåg A, Orlandini S (2010) Coincidence of variation in yield and climate in Europe. Agric Ecosys Env 139:483–489

Porter JR, Semenov MA (2005) Crop responses to climatic variation. Phil Trans Roy Soc B 360:2021–2035

Purse BV, Mellor PS, Rogers DJ, Samuel AR, Mertens PP, Baylis M (2005) Climate change and the recent emergence of bluetongue in Europe. Nat Rev Microbiol 3:171–181

Refsgaard JC, Arnbjerg-Nielsen K, Drews M, Halsnæs K, Jeppesen E, Madsen H, Markandya A, Olesen JE, Porter JR, Christensen JH (2013) The role of uncertainty in climate change adaptation – A Danish water management example. Mitig Adapt Strat Glob Change 18:337–359

Reidsma P, Ewert F, Lansink AO (2007) Analysis of farm performance in Europe under different climatic and management conditions to improve understanding of adaptive capacity. Clim Change 84: 403–422

Roos J, Hopkins R, Kvarnheden A, Dixelius C (2011) The impact of global warming on plant diseases and insect vectors in Sweden. Eur J Plant Path 129:9–19

Schaap BF, Blom-Zandstra M, Hermans CML, Meerburg BG, Verhagen J (2011) Impact of climatic extremes on arable farming in the north of the Netherlands. Reg Env Change 11:731–741

Semenov MA, Shewry PR (2011) Modelling predicts that heat stress, not drought, will increase vulnerability of wheat in Europe. Sci Reps 1:66. doi:10.1038/srep00066

Semenov MA, Stratonovitch P, Alghabari F, Gooding MJ (2014) Adapting wheat in Europe for climate change. J Cereal Sci 59: 245–256

Siebold M., von Tiedemann A (2012) Potential effects of global warming on oilseed rape pathogens in Northern Germany. Fungal Ecol 5:62–72

Skuce PJ, Morgan ER, van Dijk J, Mitchell M (2013) Animal health aspects of adaptation to climate change: beating the heat and parasites in a warming Europe. Animal 7:333–345

Smith P, Olesen JE (2010) Synergies between mitigation of, and adaptation to, climate change in agriculture. J Agric Sci 148, 543–552

Stuart ME, Goody DC, Bloomfield JP, Williams AT (2011) A review of the impact of climate change on future nitrate concentrations in groundwater of the UK. Sci Tot Env 409:2859–2873

Supit I, van Diepen CA, de Wit AJW, Kabat P, Baruth B, Ludwig F (2010) Recent changes in the climatic yield potential of various crops in Europe. Agric Syst 103:683-694

Supit I, van Diepen CA, de Wit AJW, Wolf J, Kabat P, Baruth B, Ludwig F (2012) Assessing climate change effects on European crop yields using the Crop Growth Monitoring System and a weather generator. Agric Forest Met 164:96–111

Tilman D, Balzer C, Hill J, Befort BL (2011) Global food demand and the sustainable intensification of agriculture. Proc Nat Acad Sci USA 108:20260–20264

Trnka M, Olesen JE, Kersebaum KC, Skjelvåg AO, Eitzinger J, Seguin B, Peltonen-Sainio P, Orlandini S, Dubrovsky M, Hlavinka P, Balek J, Eckersten H, Cloppet E, Calanca P, Rötter R, Gobin A, Vucetic V, Nejedlik P, Kumar S, Lalic B, Mestre A, Rossi F, Alexandrov V, Kozyra J, Schaap B, Zalud Z (2011) Agroclimatic conditions in Europe under climate change. Glob Change Biol 17:2298–2318

Trnka M, Brázdil R, Olesen JE, Eitzinger J, Zahradníček P, Kocmánková E, Dobrovolný P, Štěpánek P, Možný M, Bartošová L, Hlavinka P, Semerádová D, Valášek H, Havlíček M, Horáková V, Fischer M, Žalud Z (2012) Could the changes in regional crop yield be a pointer of climatic change? Agric Forest Meteorol 166-167:62–71

Trnka M, Rötter R, Ruiz-Ramos M, Kersebaum KC, Olesen JE, Zalud Z, Semenov MA (2014) Adverse weather conditions for European wheat production will be-come more frequent with climate change. Nature Clim Change 4:637–643

Uleberg E, Hassen-Bauer I, van Oort B, Dalmannsdottir S (2014) Impact of climate change on agriculture in Northern Norway and potential strategies for adaptation. Clim Change 122:27–39

Ulén B, Bechmann M, Øygaarden L, Kyllmar K (2014) Soil erosion in Nordic countries – future challenges and research needs. Acta Agric Scand Sect B Soil Plant Sci 62:176–184

van der Fels-Klerx HJ, Olesen JE, Naustvoll LJ, Friocourt Y, Christensen JH (2012) Climate change impacts on natural toxins in food production systems, exemplified by deoxynivalenol in wheat and diarrhetic shellfish toxins. Food Add Contam 29: 1647–1657

van Meijl H, van Rheenen T, Tabeau A, Eickhout B (2006) The impact of different policy environments on agricultural land use in Europe. Agric Ecosyst Env 114:21–38

Ziska LH (2001) Changes in competitive ability between a C_4 crop and a C_3 weed with elevated carbon dioxide. Weed Sci 49:622–627

14 Socio-economic Impacts—Offshore Activities/Energy

Kirsten Halsnæs, Martin Drews and Niels-Erik Clausen

Abstract

The energy sector has a strong presence in the North Sea and in the surrounding coastal areas. Commercial extraction of offshore oil and gas and related activities (exploration, transportation and distribution; pipelines; oil refining and processing) constitutes the single most important economic sector and renewable electricity generation—mainly from offshore wind—is increasing. Energy and offshore activities in the North Sea are critically vulnerable to climate change along the full supply chain. The major vulnerabilities for offshore installations like rigs, offshore wind energy and pipelines concern wind storms and extreme wave heights, whereas on land coastal installations and transportation may also be adversely affected by flooding. Future renewable energy potentials in the North Sea are also susceptible to climate change. Whereas the hydropower potential is expected to increase, it is highly uncertain how much the future potential of other renewable energy sources such as wind, solar, terrestrial biomass, or emerging technologies like wave, tidal or marine biomass could be positively or negatively affected. Due to the different national energy supply mixes the vulnerability to climate-related impacts will vary among North Sea countries. To ensure safe and reliable future operations comprehensive and systematic risk assessments are therefore needed which account for, for example, the high integration of power systems in the region.

14.1 Introduction

Reliability and security of the energy supply are of critical socio-economic importance and safety at sea is one of the main concerns for offshore industries in general. The offshore energy sector is particularly vulnerable to future changes in climate. This includes changes in metocean conditions (the combined wind, wave and climate conditions as found at a certain location), in relation to the full energy supply chain from resource extraction, to pipelines, refineries, conversion, and transmission (e.g. Ebinger and Vergara 2011). Maintenance and operation as well as energy demand are also likely to be influenced by climate change. This chapter reviews some of the main risks and potential for offshore and energy activities in maritime and coastal areas; with a focus on energy supply and on selected economic sub-sectors within the North Sea region that are considered particularly climate sensitive, including offshore oil rigs and wind farms.

14.2 Climate Vulnerabilities in the North Sea Region

The major climate vulnerabilities in terms of resource extraction in the North Sea region are associated with the operation and maintenance of offshore oil and gas

K. Halsnæs (✉) · M. Drews (✉)
Systems Analysis, DTU Management Engineering, Lyngby, Denmark
e-mail: khal@dtu.dk

M. Drews
e-mail: mard@dtu.dk

N.-E. Clausen
DTU Wind Energy, Roskilde, Denmark
e-mail: necl@dtu.dk

M. Quante and F. Colijn (eds.), *North Sea Region Climate Change Assessment*, Regional Climate Studies, DOI 10.1007/978-3-319-39745-0_14

Table 14.1 Overview of climate change risks on energy conversion

Conversion technology	Gradual climate change	Extreme weather events
Thermal power plant	Rising temperature implies decreased thermal efficiency and cooling efficiency	Damage to plant from storms Lower efficiency of cooling
Oil refinery and gas treatment	Sea-level rise and flooding	Flooding emergency Water scarcity disturbs production
Nuclear power plant	Cooling water scarcity	Damage to plant from flooding or storms
Wind power	Less frequent icing Dust from precipitation Flooding at coastal sites	Structural damage from storms Operation and maintenance
Solar energy	Lower efficiency of photovoltaic systems with higher temperature Higher efficiency of solar thermal heating systems with higher temperatures	Structural damage from storms, hail and heavy precipitation
Hydropower	Decreased potential in some areas and increased potential in others	Damage to dams

Adapted from Troccoli et al. (2014)

infrastructure, principally rigs and pipelines, due to their susceptibilities to wind storms and extreme wave heights (Vanem and Bitner-Gregersen 2012; Bitner-Gregersen et al. 2013; IEA 2013).

Climate change effects are also expected to have a significant impact on renewable energy sources (e.g. EEA 2012; Weisse et al. 2012). Some of the projected impacts include: changes in wind and wave energy potential; changes in hydropower potential (i.e. related to precipitation and temperature); changes in solar energy production (i.e. dependent on solar radiation and temperature); and variations in biomass for energy (i.e. related to the climate-related productivity of dedicated crops, and indirectly influenced by agricultural productivity and food security).

In addition to effects on resource extraction, the energy system is also influenced by vulnerabilities related to energy conversion. Table 14.1 provides an overview of major risks and shows that energy conversion is sensitive both to gradual changes in the mean and variance of climate parameters such as temperature and precipitation and to the projected intensification of extreme weather events in the North Sea region. The efficiency of many existing plants is expected to decline with higher temperatures, for example cooling will be more difficult, and damage from storms and flooding can disrupt energy supply with significant consequences for the economy and for disaster management in the case of extreme weather events. The International Energy Agency (IEA) estimates that for a 1 °C rise in temperature by 2040, 20 % of coal-fired power plants in Europe would need additional cooling capacity, whereas the electricity production capacity could be reduced by up to 19 % during summer (IEA 2013: their Table 3.2). In contrast, Thorsteinsson and Björnsson (2012) concluded that the projected increase in precipitation implied a potential increase in hydropower-based electricity production in the Nordic countries of about 10 % by 2050.

14.3 National Energy Supply Mixes

The vulnerability of energy systems around the North Sea to climate change must be seen in relation to the supply structure of individual countries. Figures 14.1 and 14.2 provide an overview of the present sources of electricity generation in the North Sea region and may be used to highlight some of the key risks.

In 2013, coal and peat accounted for about 32 % of the total electricity generation in the North Sea region, gas for

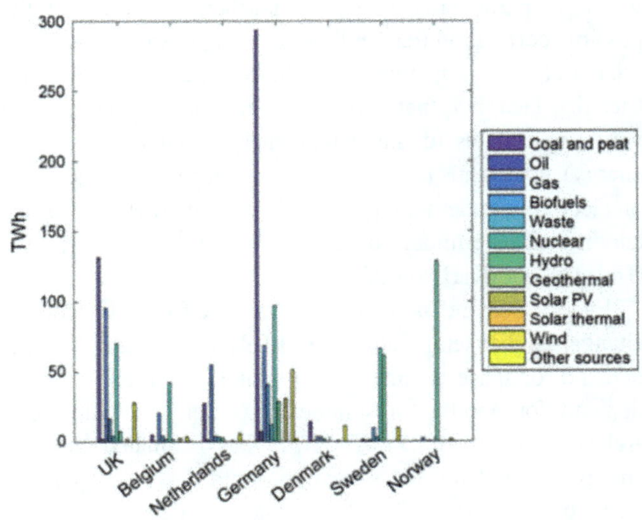

Fig. 14.1 Total electricity generation by source in 2013 (TWh) for North Sea countries (www.iea.org/statistics)

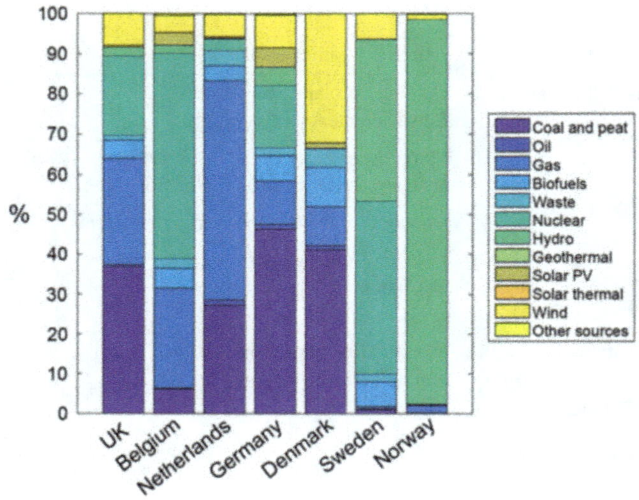

Fig. 14.2 Percentage composition of electricity generation by source in 2013 for North Sea countries (www.iea.org/statistics)

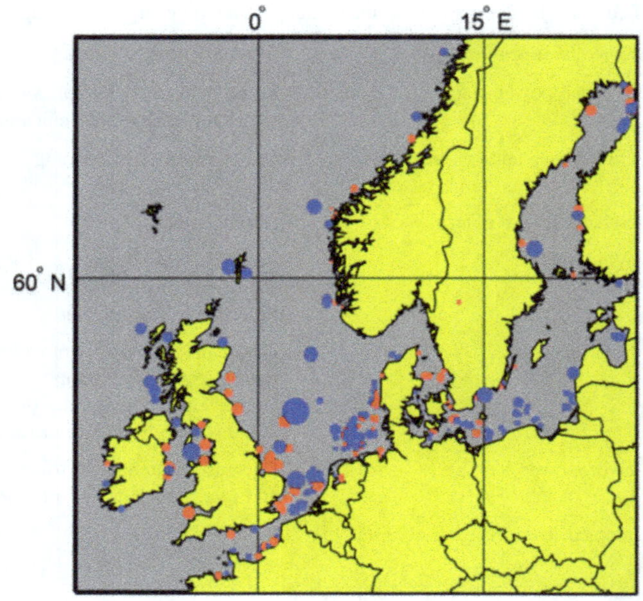

Fig. 14.3 Future offshore wind power development in Northern Europe aligned with the wind energy targets for 2020 (41 GW, *red*) and 2030 (107 GW, *blue*) from the European Wind Energy Association (2014) (Hvidtfeldt Larsen and Sønderberg Petersen 2015)

about 16 %, and nuclear for about 19 % (Fig. 4.1). In terms of renewable energy sources, hydropower accounts for about 15 % of total electricity generation and wind for 7 %. Several countries bordering the North Sea depend largely on coal and gas plants (the UK, Netherlands, Germany and Denmark), while nuclear power is important in others (Belgium and Sweden). Sweden, and especially Norway are highly dependent on hydropower. The different national energy supply mixes (Fig. 4.2) show that the projected climate-related impacts on electricity generation will vary on a country-by-country basis. The combined impacts of climate change on the energy system, whether related to gradual changes in mean climate parameters and their variations or to extreme weather events will also depend on the highly interconnected nature of the electricity markets, which is particularly strong in northern Europe and the possible correlation (e.g. in time) of climate and non-climate related stressors affecting the different fuel sources. Countries like Denmark that aim to base electricity generation on very large shares of fluctuating energy sources (e.g. wind energy), could thus become even more dependent than today on electricity trade with the Scandinavian market, such as for another climate-impacted energy source like hydropower (Halsnæs and Karlsson 2011).

Energy demand is also likely to be affected by climate change. Higher temperatures are likely to lead to decreased demand for space heating during winter, but to an increased demand for cooling in summer, especially in cities (e.g. Aebischer et al. 2007). The IEA has estimated that the energy demand for space heating could decrease by about 12 % by 2050 (IEA 2013) and that this decline could be particularly strong in northern Europe due to the current relatively high demand for space heating.

14.4 Renewable Energy Sources

Energy supply by means of renewable energy sources is expected to increase dramatically as the European Union aims to increase its share of energy consumption from renewable resources to 20 % by 2020. Options include offshore wind, hydropower, bioenergy, solar, wave energy and tidal power. They are all electricity-generating renewable technologies. Two of these technologies—offshore wind energy and hydropower—are well developed and already fully integrated into the energy system, while solar photovoltaics (PV) are in the commercial phase for land-based applications only. Likewise, the production of bioenergy/biofuels, such as from energy crops, has been extensively explored for land-based applications (see Chap. 13), whereas large-scale energy generation from marine biomass is still at an experimental stage.

Wind power is by far the most exploited renewable offshore technology in the North Sea area. Several recent initiatives, including three research and development projects funded by the European Union's Seventh Framework Programme *The Ocean of Tomorrow,*[1] explore the synergies of different renewable technologies in designing new floating offshore platforms powered by a combination of several renewable energy sources. The platforms also include aquaculture, leisure and transport options.

[1] http://ec.europa.eu/research/bioeconomy/fish/research/ocean/index_en.htm.

14.4.1 Offshore Wind Energy

More than 80 % of all offshore wind farms are installed in four countries bordering the North Sea: the UK, Denmark, Germany and Sweden, future development plans are shown in Fig. 14.3. A recent report by the European Wind Energy Association described the current status of the European offshore wind farms (EWEA 2014; Navigant Research 2014):

- 2080 turbines are installed and grid-connected offshore, accounting for a cumulative capacity of 6.5 GW divided onto 69 wind farms in eleven European countries.
- The offshore wind farms generate 24 TWh in an average year. Including land-based wind farms this corresponds to about 204 TWh generated in an average year or about 7.3 % of Europe's total electricity consumption.
- Twelve offshore projects are currently (2014) under construction corresponding to about 3 GW, which will bring the cumulative capacity in Europe to 9.4 GW.
- The average water depth of all wind farms in operation in 2013 is about 16 m and the average distance to shore 29 km.

Offshore wind energy technology is sensitive to changes in average wind speed, extreme wind speed, sea level, atmospheric icing and the extent and duration of sea ice.

The local wind climate represented by the average wind speed is the single most important factor in determining annual electricity generation by a wind farm and thereby the economy of single wind farms and of European offshore wind farms in general. Current climate projections (see Chap. 5) suggest largely unchanged average wind speeds in the North Sea area but as these projections have large uncertainties, considerable changes in generation potential cannot be ruled out.

'Extreme wind', defined as winds with a return period of 50 years (U_{50}) is an important design parameter for offshore wind turbines and used to define the durability of turbines. In general, strong winds are more important than extreme winds for the operation of an offshore wind farm as they occur much more frequently, for example strong winds with wind speeds exceeding 17 m s^{-1} occur as often as for 3–4 days per year at 10 m height in the Fehmarn Belt between Denmark and Germany. Strong winds are commonly described as winds above the 99th percentile (Thorsteinsson and Björnsson 2012) and are important for several reasons. First, during the planning period, the developer must compare potential wind farm sites and different operation and maintenance strategies. Second, in the daily planning of maintenance, strong winds and wind-induced waves have a large influence upon the ability to deploy vessels for installation and maintenance activities and thus on decision-making regarding such operations. Very strong winds (above 25 m s^{-1}) can result in periodic shutdowns due to structural safety. While this leads to a minimal loss of energy generation overall, such events are a challenge for the transmission systems operators who may need to cope with losing a large part of the electricity generating capacity within a few hours and with loads cycling on and off the grid potentially many times within a relatively short period.

Future climate projections for the North Sea (see Chap. 5) indicate that the number of storms towards the end of the century could remain at the same level as in the present-day reference period. The extreme wind speed U_{50} on the other hand could increase by as much as 10 % above present-day extremes in some areas, with the uncertainty of the same order as the estimated increases in extreme wind speed. This is likely to influence slightly the design of the wind turbines and possibly have a (minor) influence on the price of a wind turbine (Tarp-Johansen and Clausen 2006). Changes in strong winds may have a larger impact as they influence the time schedule during construction in general and in particular for crane work during erection of the wind farm. Maintenance activities are also affected by strong winds and boat transport of service crew and spare parts may be periodically difficult or impossible (Fig. 14.4).

Atmospheric icing and sea ice are rare in the North Sea (even though the Baltic Sea partly freezes every year) due to warming from the Gulf Stream. Thus higher temperatures are expected to have only minor influence on offshore wind energy generation.

14.4.2 Hydropower

Hydropower is the most important renewable energy source in Norway, where it currently covers about 99 % of Norway's electricity demand (Chernet et al. 2013). Hydropower also plays an important role in Sweden and makes a significant contribution to energy systems in the UK, Germany and Belgium. Small-scale hydropower plants (<10 MW) generate electricity by converting power from flowing water in rivers, canals or streams, while larger-scale plants often include dams and storage reservoirs to retain water.

Hydropower systems in the North Sea region will be strongly affected by the projected climate change (Thorsteinsson and Björnsson 2012; Chernet et al. 2013). The potential for hydropower is projected to rise by up to 20 % in northern and eastern Europe towards the end of the century. This is due to increased inflow to the hydropower systems from precipitation and snow melt and contrasts with the future decrease in hydropower potential projected for southern Europe.

Fig. 14.4 Installation of wind turbine at Horns Rev II wind farm (Picture provided by DONG Energy)

Hydropower production is sensitive to changes in both total runoff and its timing, and so any increase in climatic variability, even with no change in annual runoff, is likely to affect hydropower performance. Performance also depends on several other factors that are all inherently vulnerable to climate change, including reservoir design, operation strategies, dam safety, and distributions of floods and droughts. Thorsteinsson and Björnsson (2012) showed that climate change may have critical significance for dam safety and flood risk and so is likely to influence the future design and operation of hydropower plants.

14.4.3 Solar Energy

Solar energy is playing an increasingly important role in Europe. Currently, the two main technologies for generating energy from the sun are photovoltaics and solar thermal heating and most applications are land-based. In the North Sea region the largest contributions are found in Belgium and in particular in Germany, where about 7.6 GW of newly connected photovoltaic systems were installed in 2012 alone—the most in the world.

The adverse effects of climate change on solar energy primarily concern damage due to extreme weather events such as storms and heavy precipitation. In addition, some types of photovoltaic systems are sensitive to temperature, that is, their performance declines at higher temperatures (Troccoli et al. 2014). In contrast, solar thermal heating systems generally gain from increasing temperatures. Current climate projections for northern Europe (see Chap. 5) indicate small increases in sun hours (reduced cloud cover) during summer and small decreases in sun hours (increased cloud cover) during winter, but are generally associated with large uncertainties. The performance of both solar thermal heating and photovoltaic systems would thus be expected to improve during summer and decline during winter. Given the dominant uncertainties, however, future technological developments are likely to far outweigh the impacts of climate change.

14.4.4 Wave and Tidal Energy

Wave energy is an emerging technology, which is expected to see future use in the North Sea both in the coastal zone and offshore. Several conceptual designs are being tested or are at a prototype or demonstration stage worldwide. Devices still need to prove their integrity and reliability both during normal operations as well as extreme conditions. If successful some designs, in particular shoreline and near-shoreline devices, could reach commercial status within a decade. Wave energy devices are expected to be highly susceptible to the projected changes in metocean conditions and especially to extreme weather events (see Chap. 5).

The potential of tidal energy generated from either tidal impoundment or tidal streams is very low in the North Sea except near the UK coast, where it is slightly higher (Carbon Trust 2005). Currently, a range of demonstration projects are being implemented, and two commercial-scale power plants are in operation—one in Brittany, France. To date, no studies have highlighted specifically the climate change

Table 14.2 Vulnerability of the oil and gas sector to climate change

<table>
<tr><th>Climate change</th><th>Oil and gas activities</th><th>Potential impacts</th></tr>
<tr><td rowspan="3">Higher temperatures</td><td>Extraction and transportation</td><td>Arctic sea-ice decline could lead to increased exploration and increased access for shipping</td></tr>
<tr><td>(Oil) refining</td><td>Reductions in steam turbine effectiveness might lead to higher energy costs; higher temperatures could affect plant design and operational requirements, materials, and process efficiency</td></tr>
<tr><td>Delivery and distribution</td><td>Low impacts (e.g. extreme temperatures have the potential to cause maintenance problems)</td></tr>
<tr><td rowspan="3">Heavy rain, river floods, sea-level rise</td><td>Extraction and transportation</td><td>Low impacts however onshore transportation could be affected</td></tr>
<tr><td>(Oil) refining</td><td>Flooding of critical infrastructure may cause serious damage and shutdown of operations</td></tr>
<tr><td>Delivery and distribution</td><td>Soil erosion may expose buried pipelines; exposed pipeline sections may suffer damage; transportation by vessel, pipeline, road and rail may suffer flood-induced disruption and damage</td></tr>
<tr><td rowspan="3">Storms and storm surges, extreme wave heights</td><td>Extraction and transportation</td><td rowspan="2">Significant damage to offshore and onshore installations and equipment will disrupt and possibly shut down operations entirely; possible environmental consequences; increased focus on safety; new design standards</td></tr>
<tr><td>(Oil) refining</td></tr>
<tr><td>Delivery and distribution</td><td>Transportation by vessel, pipeline, road and rail may suffer storm or flood-induced disruption and damage</td></tr>
<tr><td rowspan="3">Lightning</td><td>Extraction and transportation</td><td>Oil and gas pipelines may be damaged by lightning strikes, which may lead to increased corrosion, ignition, and operational disruption</td></tr>
<tr><td>(Oil) refining</td><td>Risk of explosions or fires due to hazardous materials</td></tr>
<tr><td>Delivery and distribution</td><td>–</td></tr>
</table>

Based on Cruz and Krausmann (2013 and references therein)

impacts on tidal technologies, but it is clear that tidal turbines like wave energy devices are highly vulnerable to the projected changes in extreme wind and wave conditions in the North Sea and this could influence the operation of such installations and increase the risk of damage.

14.4.5 Marine Biomass

The production of marine biomass like microalgae for bioenergy and/or biofuel has emerged as a promising renewable energy source (e.g. Roberts and Uphamb 2012; Jard et al. 2013). Extensive research and development activities are ongoing; a demonstration case for offshore applications has also been successfully developed by the National University of Ireland, Galway (Edwards and Watson 2011). Results suggest that use of marine biomass if commercially realised could potentially be as large and comparable to existing land-based forestry and agricultural energy crops. The potential of marine biomass as a future energy source would be affected by climate-related changes in the marine ecosystem (see Chap. 8). Likewise, offshore installations would be subject to changes in the frequency and/or intensity of wind and wave extremes.

14.5 Fuel Extraction

Commercial extraction of offshore oil and gas along with related activities such as exploration, transportation and distribution; pipelines; and oil refining and processing at present constitutes the single most important economic sector in the North Sea. Five countries are involved in oil and gas extraction in the North Sea: Norway, Denmark, Germany, Netherlands, and the UK. While oil and gas reserves in the North Sea and thus revenues are expected to decline over the course of the century, industry continues to push the boundaries of oil and gas exploration technology. Even with the expansion in renewable energy sources it is highly likely that the oil and gas sector will continue to be critically important in the North Sea.

A warming climate with stronger and more frequent extreme weather events will pose serious challenges to the oil and gas sector (Bitner-Gregersen et al. 2013; Cruz and Krausmann 2013). Structural failure of offshore structures may result in a loss of lives, severe environmental damage, and large economic consequences. Climate change impacts are likely to affect the entire value-chain of the sector, particularly activities in low-lying areas or areas exposed to extreme weather events. Table 14.2 summarises some of the main vulnerabilities related to oil and gas extraction in the North Sea.

Several researchers, including Cruz and Krausmann (2013) and Bitner-Gregersen et al. (2013), have argued that comprehensive and systematic risk assessment frameworks are needed to manage emerging risks to the offshore oil and gas sector from climate change. This is to ensure that present and future design standards for offshore and onshore infrastructure, maintenance and operations reflect the actual physical threats posed by climate change while remaining acceptable from an economic, societal and environmental perspective. Adaptation options could in some cases require significant investment to upgrade facilities, protect critical infrastructure and build redundancy and robustness into systems.

14.6 Conclusion

Energy systems and offshore activities in the North Sea region of which offshore wind, oil and gas dominate are virtually certain to be affected by climate change. While most studies show that hydropower potential is expected to increase, climate projections are highly uncertain regarding how much the future potential of other renewable energy sources such as wind, solar, terrestrial biomass, or emerging technologies like wave, tidal or marine biomass could be positively or negatively affected. Offshore and onshore activities in the North Sea region are very vulnerable to extreme weather events like extreme wave heights, storms and storm surges. To ensure safe and reliable operations and to mitigate the possible loss of lives and economic assets it is necessary to take action to prevent the potentially negative effects of climate change and to develop comprehensive and systematic risk assessment frameworks, which incorporate climate projections and environmental data.

References

Aebischer B, Henderson G, Jakob M, Catenazzi G (2007) Impact of Climate Change on Thermal Comfort, Heating and Cooling Energy Demand in Europe. In: ECEEE 2007 Summer Study Proceedings: Saving Energy – Just do it! Panel 5 Energy Efficient Buildings: 859-870.

Bitner-Gregersen EM, Eide LI, Hørte T, Skjong R (2013) Ship and Offshore Structure Design in Climate Change Perspective. Springer Briefs in Climate Studies. Springer Verlag.

Carbon Trust (2005). Phase II: UK Tidal Stream Energy Resource Assessment. Report prepared by Black & Veatch Ltd.

Chernet HH, Alfredsen K, Killingtveit Å (2013) The impacts of climate change on a Norwegian high-head hydropower system. J Water Clim Change 4:17–37

Cruz AM, Krausmann E (2013) Vulnerability of the oil and gas sector to climate change and extreme weather events. Clim Change 121:41–53

Ebinger J, Vergara W (2011) Climate Impacts on Energy Systems: Key Issues for Energy Sector Adaptation. Energy Sector Management Assistance Program (ESMAP) and The World Bank

Edwards M, Watson L (2011) Cultivating *Laminaria digitata*. Aquaculture Explained No. 26, National University of Ireland

EEA (2012) Climate Change, Impacts and Vulnerability in Europe 2012 - An Indicator Based Report. European Environment Agency (EEA) Report no. 12/2012

EWEA (2014) The European offshore wind industry – key trends and statistics 2013. European Wind Energy Association (EWEA)

Halsnæs K, Karlsson K (2011) The costs of renewable – past and future. In: Gallaraga I, González-Eguino M, Markandya A (eds), Handbook of Sustainable Energy. Edward Elgar

Hvidtfeldt Larsen H, Sønderberg Petersen L (eds) (2015) DTU International Energy Report 2015. Energy systems integration for the transition to non-fossil energy systems. Technical University of Denmark

IEA (2013) Redrawing the Energy-Climate Map. World Energy Outlook Special Report, International Energy Agency (IEA)

Jard G, Marfaing H, Carrère H, Delgenes JP, Steyer JP, Dumas C (2013) French Brittany macroalgae screening: Composition and methane potential for potential alternative sources of energy and products. Biores Tech 144:492–498

Navigant Research (2014) World Market Update 2013: International Wind Energy Development Forecast 2014-2018.

Roberts T, Uphamb P (2012) Prospects for the use of macro-algae for fuel in Ireland and the UK: An overview of marine management issues. Mar Pol 36:1047–1053

Tarp-Johansen NJ, Clausen NE (2006) Design of Wind Turbines in Typhoon area: A first study of structural safety of wind turbines in typhoon prone areas. EC-ASEAN Energy Facility

Thorsteinsson T, Björnsson H (eds) (2012) Climate Change and Energy Systems. Impacts Risks and Adaptation in the Nordic and Baltic Countries. TemaNord 2011:502, Nordic Council of Ministers

Troccoli A, Dubus L, Haupt SE (eds) (2014) Weather Matters for Energy, Springer-Verlag

Vanem E, Bitner-Gregersen EM (2012) Stochastic modelling of long-term trends in the wave climate and its potential impact on ship structural loads. Appl Ocean Res 37:235–248

Weisse R, von Storch H, Niemeyer HD, Knaack H (2012) Changing North Sea storm surge climate: An increasing Hazard? Ocean Coast Man 68:58–68

Socio-economic Impacts—Urban Climate

15

K. Heinke Schlünzen and Sylvia I. Bohnenstengel

Abstract

About 80 % of the population within the North Sea countries currently lives in an urban area and this percentage is projected to continue to rise. Urban areas are not only impacted by changes in regional climate but are themselves responsible for causing local modifications in regional climate resulting in the so-called 'urban climate'. The urban climate in North Sea cities has several common features: higher temperatures relative to the surrounding regions (especially at night), greater temperature variability, deeper but less stable boundary layers at night, lower average wind speeds but stronger gusts, reduced evapotranspiration, and greater air pollution (local exceedances of limit values for nitrogen oxides, nitrogen dioxide and particulate matter, with ship emissions a relevant contributor in harbour cities). Indications of climate change are now apparent and include hinterland flooding, more intense precipitation, and drier and warmer summers. Cities contribute to greenhouse gas emissions and measures are needed to reduce these. Cities also need to adapt to climate change. Despite broad similarities between urban areas, in terms of mitigation and adaptation to climate change there are large location-specific differences with regard to city planning needs. Hamburg and London are used as examples. Adaptation measures include better insulation of buildings to reduce energy use and anthropogenic heat emissions, higher dykes to protect against increased water levels, and rain water drainage to avoid hinterland flooding. Scenarios are outlined for urban development with greened roofs, higher albedo values and lower sealing of surfaces.

15.1 Introduction

Worldwide every second person lives in a town; in the North Sea region the percentage is even higher. About 80 % of the population within the North Sea countries currently lives in an urban area and this percentage is projected to continue to rise (Fig. 15.1). Nine out of ten citizens are predicted to be living in an urban area by the middle of this century. This level of urbanisation is higher than in Europe as a whole or worldwide, but is similar to that of the United States. The megacity of London (~14 million people) is located on the periphery of the southern North Sea, as are several metropolitan areas with at least 1 million inhabitants (Rotterdam, Hamburg, Amsterdam, Antwerp). Because so many people live in urban areas it is important to understand the interrelations between regional and urban climate and how both will develop over time.

The urban climate is affected by the regional climate and specific local characteristics such as closeness to the ocean or nearby mountains. Most urban areas in the North Sea region (e.g. Amsterdam, Antwerp, Hamburg, London, Rotterdam) experience a warm temperate climate, which is fully

K.H. Schlünzen (✉)
Institute of Meteorology, CEN, University of Hamburg, Hamburg, Germany
e-mail: heinke.schluenzen@uni-hamburg.de

S.I. Bohnenstengel
Department of Meteorology, University of Reading, Reading, UK
e-mail: s.i.l.d.bohnenstengel@reading.ac.uk

Present Address:
S.I. Bohnenstengel
MetOffice@Reading, University of Reading, Reading, UK
e-mail: sylvia.bohnenstengel@metoffice.gov.uk

M. Quante and F. Colijn (eds.), *North Sea Region Climate Change Assessment*, Regional Climate Studies, DOI 10.1007/978-3-319-39745-0_15

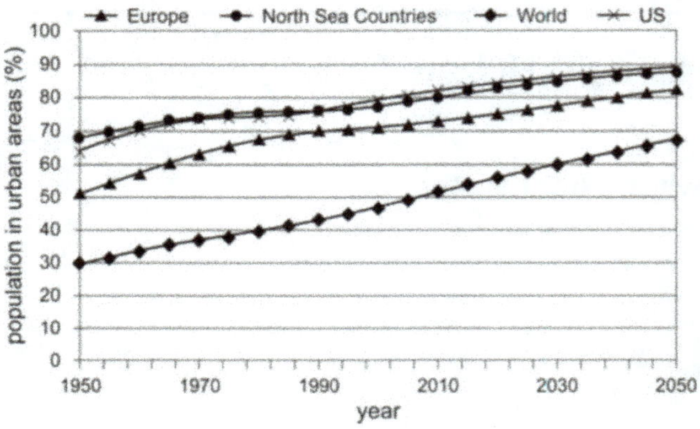

Fig. 15.1 Development of urbanisation in countries bordering the North Sea and other regions of the world (based on UN 2014)

humid with a warm summer (Class Cfb following the Köppen-Geiger climate classification as given by Kottek et al. 2006). Only in the northernmost part of the North Sea region is snow a regular winter feature, which means cities such as Oslo or Bergen are on the margin of the Dfb Köppen-Geiger climate class. More details of the North Sea climate can be found in Chaps. 1 and 2.

The proximity of the large metropolitan areas to the North Sea implies they are generally located in low altitude areas; some parts are even partly situated below sea level (see Annex 5, Fig. A5.1). This is especially true for the urban areas of the Netherlands (Amsterdam, Rotterdam, The Hague and Utrecht), but also for Antwerp, London or Hamburg. Although for the latter at least some parts of the metropolitan area are 10 m or more above sea level. As a result, adaptation to climate change in a coastal urban area means it is important to consider potential changes in sea level, river level, storm surge and connected groundwater level (Schlünzen and Linde 2014). However, while for city planners the rise in sea level (Chaps. 3 and 6) and river level are extremely important, they have little impact on urban climate and so fall outside the scope of this chapter (see Chap. 18 for discussion on this topic). Soil water is relevant, however, since its availability could affect evapotranspiraration and thus temperature and humidity in an urban area (Sect. 15.4.2).

15.2 Urban Climate in the North Sea Region

Any changes in the natural conditions of an area will modify the regional climate such that it is locally altered, resulting in a so-called 'urban climate' in the case of urban areas. Local modifications to regional climate in urban areas largely depend on the urban fabric (e.g. building height, percentage of sealed surfaces, building materials, atmospheric emissions), and for North Sea cities result in several common features:

- *Higher temperatures.* These result from changes in the surface energy budget due to urban fabric having greater heat storage than vegetation in rural areas. In urban areas, heat is stored during the day and then emitted during the evening and at night, supplemented by anthropogenic heat emissions; this increase in air temperature is termed the 'urban heat island' effect (UHI, Sect. 15.4.2). The UHI shows both a diurnal and an annual cycle. The intensity of the night-time warming is even more intense at the surfaces (surface urban heat island).
- *Greater temperature variability.* This results from shading and reflection of short-wave radiation by buildings, radiative trapping, heat storage by buildings, and increased energy use and emission of waste energy (Sect. 15.4.2).
- *Deeper boundary layers and more frequent unstably stratified boundary layers at night.* This is due to the UHI effect and could affect turbulent mixing of pollutants (Chemel and Sokhi 2012) which could in turn increase ozone (O_3) concentrations near the surface at night and reduce nitrogen dioxide (NO_2) concentrations (Zhang and Rao 1999; Sect. 15.4.2).
- *Lower average wind speed and greater gustiness.* The presence of buildings in urban areas causes lower average wind speeds, but local maxima can occur especially within street canyons facing the coastline or river bank. The buildings also trigger an overall increase in gustiness; wind comfort is thus much lower in coastal urban areas of the North Sea region than in inland urban areas (Sect. 15.4.3).
- *Reduced evapotranspiration.* Owing to less vegetation, less water storage capacity and often lower groundwater levels in urban areas, evapotranspiration is smaller. For North Sea cities, even the areas with high groundwater levels have reduced evapotranspiration, if the surfaces are sealed (Sect. 15.4.2).
- *Changed precipitation fields.* The urban fabric and UHI effect lead to convergences and more updrafts in the flow field, often resulting in more downwind precipitation (Shepherd et al. 2002) if anthropogenic pollutant emissions are neglected. In an urban area with high pollutant emissions (e.g. sulphur dioxide, SO_2) the urban area might reduce precipitation; however, aerosol impacts are still uncertain (Pielke et al. 2007). Whether downwind precipitation is higher or lower depends among other things on aerosol composition, meteorological situation, and urban surroundings (Han et al. 2014). Urban precipitation impacts are visible through changes in downwind precipitation (Sect. 15.4.3).
- *More air pollution.* Owing to higher emissions from a range of anthropogenic sources (traffic, households, industry) there are higher levels of primary pollutants. Also, most of the cities mentioned above are harbour cities, with Rotterdam, Hamburg and Antwerp the largest

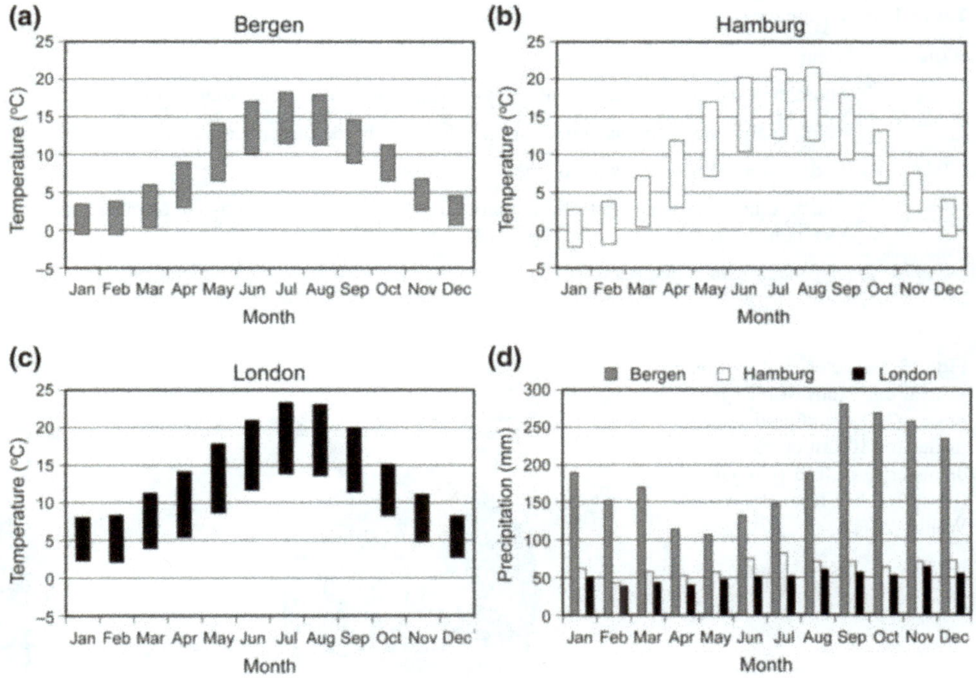

Fig. 15.2 Monthly average minimum and maximum temperatures for Bergen, Hamburg and London, and monthly average precipitation for the three cities. *Data sources* Bergen (http://wetter.welt.de/klimadaten.asp, accessed 16 February 2014), Hamburg (temperature http://wetter.welt.de/klimadaten.asp accessed 16 February 2014; precipitation averaging period 1981–2010, www.dwd.de accessed 3 April 2015), London (averaging period 1981–2010, www.metoffice.gov.uk accessed 3 April 2015)

in Europe. For these cities, emissions from ships add to the air pollution load (Sect. 15.4.1).

The effects of urban areas are referred to collectively in this chapter as the 'urban footprint'.

The following sections examine urban climate in the past (Sect. 15.3) and present (Sect. 15.4), as impacted by climate change (Sect. 15.5) and adaptation measures (Sect. 15.6), using two cities as examples: the megacity of London with an extensive metropolitan area (14.3 million inhabitants[1]) but no international harbour and the comparatively small metropolitan area of Hamburg (2.7 million inhabitants[2]) with one of the largest harbours in Europe. The two cities are about 700 km apart, with Hamburg having a slightly more continental climate, visible in lower winter temperatures, a greater minimum to maximum temperature range and a more pronounced summer precipitation maximum (Fig. 15.2). The busy North Sea harbour cities of Rotterdam and Antwerp have a climate similar to that of London or Hamburg, which implies the external climate drivers interacting with urban-induced changes are similar. In contrast, one of the northernmost North Sea cities, Bergen (Norway), has a lower temperature range in each month and throughout the year, and thus little problem with excessive summer temperatures. However, due to the nearby mountain ranges Bergen experiences much higher precipitation (roughly three-fold higher) and this must be considered in urban planning.

London and Hamburg have experienced urban climate problems, especially regarding heavy air pollution (Sect. 15.3). Only in the past few decades, especially since the very warm summer of 2003, have other parameters characterising the urban climate come into focus (Sects. 15.4–15.6). With a similar climate in both cities, the challenges mainly concern their differences in size and thus urban footprint on regional climate.

15.3 Historical Problems in Urban Climate

Historically, air pollution drove studies on urban climate. A severe pollution event was followed by action to understand and improve air quality (Table 15.1). Elevated sources were found to create widespread pollutant plumes as well as high pollutant concentrations, in urban areas as well as in rural areas. Standards for air quality were initiated by the European Communities Programme for Action on Environment from 1973. This led to the first directive (Council Directive 80/779/EEC, see EC 1980) on levels of SO_2 among EU member states. More EU-wide directives on limit values for pollutant concentrations followed (e.g. Council Directive 96/62/EC, and its later updates given in EC 2008). This initiated national and local strategies to reduce pollutant concentrations. For instance, in London the Clean Air Act of 1993 was followed by the Greater London Authority Act in 1999. With a focus on London, GLA (2002) gives a detailed overview of air pollution control and air quality strategies

[1] www.citypopulation.de/world/Agglomerations.html accessed 11 December 2015.

[2] www.citypopulation.de/world/Agglomerations.html.

Table 15.1 Occurrences related to air pollution

Date	Event
1952	About 4000 people died within five days during a winter smog episode in London (GLA 2002)
1957	Field experiments were performed in the UK to study the dispersion of pollutants (Hay and Pasquill 1957)
1957	Commission on clear air was founded in Germany
1967	First air quality measurement sites established in Germany (financed by the German Research Foundation, DFG); later becoming an operational network
~1970	Dispersion field experiments took place in several countries to better understand dispersion (heavy gases, elevated stack emissions)

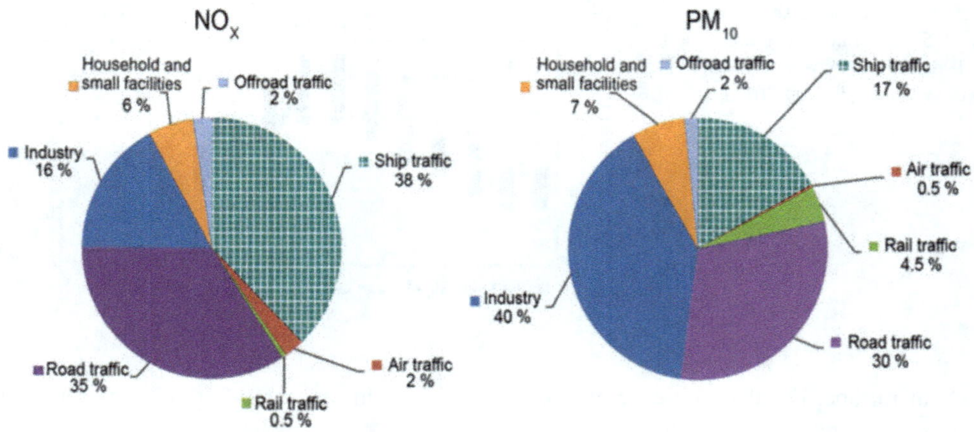

Fig. 15.3 Sector contributions to total emissions of nitrogen oxides (NO_X) and particulate matter (of 10 μm or less in diameter; PM_{10}) in Hamburg (based on data from Böhm and Wahler 2012)

over the last 150 years starting from the control of air pollutants by industry to reduce smoke, to ambient air quality standards.

Heat is an additional health threat. The 2003 heat wave caused around 70,000 excess deaths in Europe, with about 20–38 % attributed to air pollution (Jalkanen 2011). There was an overall 17 % increase in death rates for England and Wales with the excess mortality most pronounced in London, with a 33 % increase in the over 75-year old age group (Kovats et al. 2006). Since regional heat waves and the strongest UHIs are both observed in summer during stationary anti-cyclonic conditions with calm winds, the UHI is even more relevant during heat waves. After summer 2003 it was clear that the additional temperature enhancement in urban areas can lead to unbearable and health-threatening temperatures during a heat wave, even in cities of the North Sea region. This led to the start of several research projects and experimental campaigns to better understand the current urban climate and to develop urban footprint reduction and adaptation measures.

15.4 Current Urban Climate

15.4.1 Air Quality

The contribution of high-stack emissions to the total emissions of primary pollutants and high concentrations recorded in urban areas today is small compared to those of the past (see Sect. 15.3). For example, in 2005 only 25 % of the total nitrogen oxide (NO_X) emissions in Germany were from high stacks. Local traffic and—for harbour cities—ship traffic are now the main sources of several primary pollutants). For Hamburg, 78 % of NO_X emissions and 53 % of PM_{10} (particulate matter of 10 μm or less in diameter) emissions result from traffic, with ship emissions contributing 38 % of the total NO_X emissions (Fig. 15.3). Traffic emissions (except air traffic) are ground-based and so directly increase concentrations within the urban area. Therefore, measurements mainly show exceedances of the NO_2 annual average limit value of 40 μg m^{-3} at traffic-impacted sites, where air masses are confined and so less mixed than in less built-up areas. This is true for Hamburg (Böhm and Wahler 2012) and London (Fuller and Mittal 2012).

However, Fuller and Mittal (2012) found that the limit values are even exceeded at urban background stations in locations such as inner London, close to Heathrow or near the M4 motorway, probably due to the huge commuter belt around London. This is not the case for Hamburg and even in the harbour the NO_2 values are currently below the annual average limit values of 40 μg m^{-3}, but above the values measured at urban background stations (Böhm and Wahler 2012). With the development of new residential areas on the banks of the River Elbe, air masses will be more confined and ship emissions might lead to higher air concentrations that could affect the health of the residents. As a consequence, plans for reducing air pollution concentrations now include ship emissions (Böhm and Wahler 2012).

For NO_X concentrations in London, Fuller and Mittal (2012) found a seasonal cycle with higher concentrations in winter and an overall decline since 1998. The decrease is greatest close to roadsides. Carslaw et al. (2011) reported an increase in the ratio of NO_2/NO_X over the last decade at roadsides and the increase has been more marked in London than at other UK sites. The increase is probably due to higher NO_2 emissions for vehicles conforming to newer emission standards (e.g. through oxidation catalysts and particle filters in light-duty diesel vehicles) (Carslaw et al. 2011). The changes are similar for Hamburg and in Europe as a whole, and so a similar change can be assumed across the whole of the North Sea region.

Annual average PM_{10} concentrations show more or less a decrease for Hamburg between 2001 and 2011, although values are still up to 80 % of the EU annual average limit value of 40 µg m^{-3} (Böhm and Wahler 2012). The interactive map developed by the European Environment Agency[3] gives an annual mean for PM_{10} of the same order (31–40 µg m^{-3}) for London in 2012. This is the highest value in the UK, but comparable to Leiden (Netherlands), Bremen (Germany) and Antwerp (Belgium). According to Fuller and Mittal (2012), monthly mean PM_{10} concentrations vary between 25 and 38 µg m^{-3} depending on location in London (roadside, background, city centre, fringes). They found that several monitoring stations at roadsides in London exceed the 50 µg m^{-3} daily mean limit value on more than 35 days in 2011. However, according to Jones et al. (2012) a large decrease in particle number has occurred in London since 2007 possibly due to the introduction of ultra-low sulphur diesel. Sources of PM_{10} in London depend on the weather pattern and comprise local sources and advection from within the UK and Europe. First results by the ClearfLo campaign measuring the composition of particulate matter in 2011 and 2012 in London at an urban background site suggest that organic aerosol is the most abundant (35 % of the total) followed by secondary inorganic aerosols such as nitrate (18 %), sulphate (11 %) and ammonium (9 %), and smaller contributions from marine aerosol components such as chloride (7 %) and sodium (4 %), and combustion emissions such as elemental carbon (Bohnenstengel et al. 2015). Early analysis indicates that local London emissions have a bigger impact in winter when the lower boundary layer enables a build-up of primary pollutants. See www.londonair.org for a summary of air quality measurements in London from several stations and information on exceedances.

North Sea urban regions have undertaken active measures to reduce pollutant exceedances: The Air Quality Strategy for London (GLA 2010) details some of the measures taken in London to further reduce PM_{10} concentrations. These include low emission zones, cleaner vehicle transport, cycle superhighways, best practice guidance for construction and demolition, and biomass boilers. Measures have also been taken in Hamburg and a reduction in exceedances is expected due to future emission reductions from traffic (including bus-lanes, car-sharing, and land-based energy supply for ships; Böhm and Wahler 2012). However, wood is increasingly used for heating (owing to its CO_2-neutral emissions); without regulatory measures PM_{10} emissions from households and thereby PM load might increase again, especially in winter.

15.4.2 Temperature and Humidity

The UHI is the most well-known feature of urban climate, and describes the temperature difference between urban and rural areas (Oke 1982). It is most pronounced during calm nights with clear skies (e.g. Schlünzen et al. 2010; Richter et al. 2013). This is important because higher night-time temperatures can cause discomfort and increase mortality rates during prolonged hot summer periods, as found for example for London (Armstrong et al. 2011).

In these situations the UHI at night for North Sea cities can be up to 7 K (London: Watkins et al. 2002), 10.5 K (Hamburg: Hoffmann et al. 2012) or 7 K (Rotterdam: Heusinkveld et al. 2014). However, the monthly average values for night-time temperature enhancements are lower. For Hamburg, analyses show monthly average minimum temperature differences between the urban and surrounding rural area of 1 (suburbs) to 2.7 K (inner city) for April through October (Schlünzen et al. 2010). Similar monthly average night-time temperature enhancements were found by Heusinkveld et al. (2014) for Rotterdam (June, July, August: median 0.7–2.26 K depending on location) and Jones and Lister (2009) for London (enhancement of minimum temperatures of 1.6 K for St James Park based on four 30-year averages 1901–1930, 1931–1960, 1951–1980, 1981–2006, and 2.8 K for the central London weather station for 1981–2006). Unpublished long-term simulations with the UK Met Office Unified model at 1-km horizontal resolution show the spatial pattern of positive temperature anomalies in the order of 2–3 K around 1 UTC (Universal Time Coordinated) (Fig. 15.4) and 1–2 K around 4 UTC, averaged for June to August 2006. Using the same model, Bohnenstengel et al. (2011) showed the temperature enhancement to remain constant throughout the night for the London city centre from the evening transition to the morning transition for a case study in May 2008 with moderate winds speeds.

[3] www.eea.europa.eu/themes/air/interactive/pm10.

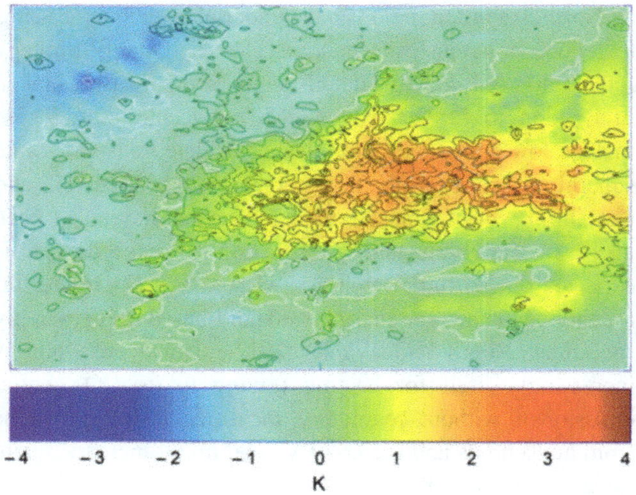

Fig. 15.4 Urban temperature enhancement for London at 1UTC averaged over the period 1 June to 15 August 2006. Values are derived from long-term model simulations with the UK Met Office Unified Model employing the MORUSES urban parameterisation (model setup described by Bohnenstengel et al. 2011). *Black lines* indicate sub-grid scale urban land-use fraction per grid box ranging from 0 (no urban land use) to 1 (grid box entirely covered by urban land use) and *colours* represent the urban temperature anomaly in *K*. A grid box is roughly 1 km^2

As in other regions of the world, urban land-use is the biggest driver of UHI in the North Sea cities (Schlünzen et al. 2010; Bohnenstengel et al. 2011; Hoffmann 2012; Heusinkveld et al. 2014). The effect of greening reduces the enhanced temperatures on a clear night with low wind speeds by 2–3 K (model results by Bohnenstengel et al. 2011 and Grawe et al. 2012, both for London). Similar effects were found for Hamburg and Rotterdam based on measured data, where the heat island is smaller in green areas than in areas with sealed surfaces (Schlünzen et al. 2010; Heusinkveld et al. 2014). Detailed model studies show the significant effect of building height and urban fabric on perceived temperatures (Schoetter et al. 2013). Perceived temperature is a measure of thermal comfort and is based on a heat budget model for the human body; it takes into account temperature, short- and long-wave radiation and wind speed effects on the human body (Kim et al. 2009; Staiger et al. 2011).

Air mass history and the evolution of the urban boundary layer with distance from the rural/urban transition also affect urban air temperature. On a night with moderate wind speeds, air temperature can be around 2 K lower over the upwind fringes of a city such as London than over the city centre and areas downwind of the city centre (Bohnenstengel et al. 2011).

Coastal form and meteorological situation (Crosman and Horel 2010) affect the inland penetration of sea breeze fronts and the front moves further inland the later the afternoon (Simpson et al. 1977). Thus, depending on distance from the coast, sea breezes and marine air intrusions can reduce the intensity of the UHI in the evening or at night for a couple of hours in North Sea cities in spring and early summer. This is especially the case during high pressure situations with calm winds (e.g. Chemel and Sokhi 2012). However, for an inland city like Hamburg (about 100 km inland of the North Sea, 80 km from the Baltic Sea) the impact of sea breezes is rare, since sea breeze fronts typically travel inland by up to 40 km only (Schlünzen 1990), rarely further.

Lane (2014) determined a mean temperature enhancement of 1.9 K for summer (JJA) and 1.6 K for winter (DJF) based on hourly temperature measurements from a roof top site 18 m above ground level in central London and a spatial average of 10 rural stations mostly to the east and west of London. As for other cities, the enhancement of the maximum temperatures is quite small compared to the rural surroundings and most pronounced in winter months (determined for Hamburg; Schlünzen et al. 2010), when anthropogenic heat emissions play a larger role. Schlünzen et al. (2010) found a range of 0.2 (suburb) to 0.7 K (inner city) for Hamburg's monthly average winter maximum temperature enhancements. The maximum temperature enhancement for London is of a similar order at 0.6 K (St James Park; 1901–2006) and 0.9 K (London weather centre; 1981–2006) according to Jones and Lister (2009). They stated that maximum temperature enhancements in St James Park differ marginally between seasons, while minimum temperature enhancements are slightly higher in spring and summer. They found no evidence for climate-related enhanced warming trends in central London compared to the trends found for rural stations around London.

As summarised by Mavrogianni et al. (2011), the excess heat in urban areas affects energy use, comfort and health. Their simulations show that the number of hours with indoor temperatures exceeding 28 °C increases towards the city centre of London for a building without air conditioning. However, building form and urban land-use also play a role in comfort temperatures in London and need to be taken into account when designing strategies that reduce overheating. Iamarino et al. (2011) showed for a resolution of 200 m × 200 m that anthropogenic heat fluxes for the Greater London area are of the order of 10 Wm^{-2}, while the city centre is associated with anthropogenic emissions of the order of 200 Wm^{-2} and, according to Hamilton et al. (2009) and Bohnenstengel et al. (2014), of 400 Wm^{-2} at peak times. Petrik et al. (in prep) determined the anthropogenic heat at 250 m resolution for Hamburg, finding values of 10 Wm^{-2} in suburbs and up to 100 Wm^{-2} at some industrial sites and in harbour areas. These lower values agree well with the findings of Allen et al. (2011) who determined anthropogenic heat fluxes globally on a 2.5 arc minute grid. For North Sea cites, they found higher values for London,

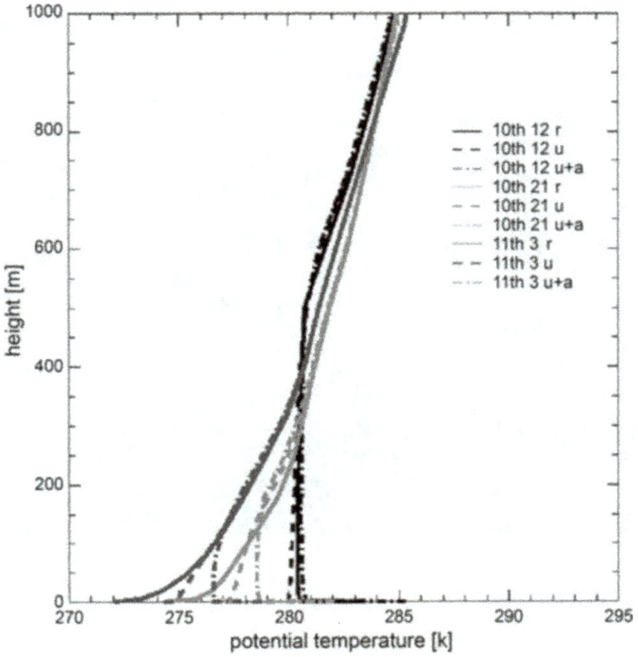

Fig. 15.5 Vertical potential temperature profiles over the London city centre for 9–12 December 2009. *Dark grey lines* depict profiles at noon, *light grey lines* depict profiles at 21UTC and *black lines* depict profiles at 3UTC. *Solid lines* depict the rural simulation, *dashed lines* the urban simulations and *dash-dotted lines* urban simulations with anthropogenic heat fluxes included (Bohnenstengel et al. 2014)

Brussels, Rotterdam and Amsterdam (> 30 Wm^{-2} annual average) and lower values for smaller cities and the Ruhr area, Hamburg or Bremen. Based on their 250-m resolution model studies with the mesoscale model METRAS, Petrik et al. (in prep) found the highest impacts on temperature at night, when the anthropogenic heat is mixed into a shallow boundary layer. Thus, night temperatures are more affected than day temperatures resulting in a summer average night-time temperature increase of up to 0.5 K in those parts of Hamburg with the highest waste heat emissions.

Bohnenstengel et al. (2014) examined the impact of anthropogenic heat emissions on London's UHI using 1-km resolution simulations with the UK Met Office Unified Model for a winter case study with calm winds over the period 9–12 December 2009. They compared three simulations covering London: a 'rural' control run, where London was replaced by grass, and two simulations including the urban surface energy balance—one with and one without high-resolution time-varying anthropogenic heat emissions. During calm and clear winter nights, anthropogenic emissions were found to increase the UHI by up to 1 K. In fact, anthropogenic emissions can tip the balance and maintain a well-mixed boundary layer (Fig. 15.5). This is based on a case study for winter, when the urban boundary layer was shallow and anthropogenic heat emissions affected a very small volume of air. In such cases, anthropogenic heat could affect the mixing properties of the urban boundary layer and thereby pollutant concentrations (Sect. 15.4.1). In spring or summer, when the daytime urban boundary layers are much deeper, the impact of anthropogenic emissions on temperatures (and thus vertical mixing) is within the measurement uncertainty.

Most North Sea cities have a considerable fraction of water surfaces within the urban area. For example, more than 3 % of Hamburg has water surfaces (channels, ponds, small lakes, rivers; Teichert 2013). In summer, the suburbs close to the inner-city water bodies experience advective cooling during the day and warming at night, since the water bodies dampen the diurnal cycle. This results in UHI-like effects at night due to the advection of warm air from the adjacent water bodies (Schlünzen et al. 2010). The water- and urban-fabric-induced reduced night-time cooling are additive and also affect the occurrence of plant species (Bechtel and Schmidt 2011). For large water bodies, such as the River Elbe downstream of Hamburg's harbour the water bodies might cause a river breeze that affects temperatures a few 1000 m off the river, as Teichert (2013) found for a calm meteorological situation in summer simulated with METRAS. It should be noted that daytime cooling by water bodies only occurs if the water temperature is lower than that of the land surfaces. Water temperatures are affected by water use: among others, water is abstracted for drinking water, industrial production or power plant cooling; and discharged in part as waste water, clean but often at higher temperatures than the abstracted water. This can increase river temperature throughout the year, especially if the river is tidal and the same water is used several times. For example, the River Weser regulations aim to prevent river water temperatures of more than 28 °C (www.fgg-weser.de). A river used to discharge the warm waste water might act as an all-year central heating system, especially at night. This can be advantageous in winter, similar to the warm North Atlantic current that acts as a central heating system for all North Sea cities.

To summarise, for industrial cities such as Hamburg or London, waste heat emissions can add to the rise in night-time temperature caused by the urban fabric. In addition, water bodies within built-up areas hinder cooling at night especially in summer when cooling is most needed, for instance during heat waves. Rivers help to cool a city in summer, if their temperature is kept low enough and waste water-related warming is also kept low. Coastal water bodies may cool cities in spring and summer, as observed in Rotterdam or Bergen compared to a city setting more inland and without sea breeze impacts. However, it should be noted that all water bodies reduce urban cooling in autumn and winter and so could help save energy during the cold season. Since urban fabric, heat emissions and water bodies are all very locally structured, a pattern of high temperatures is also

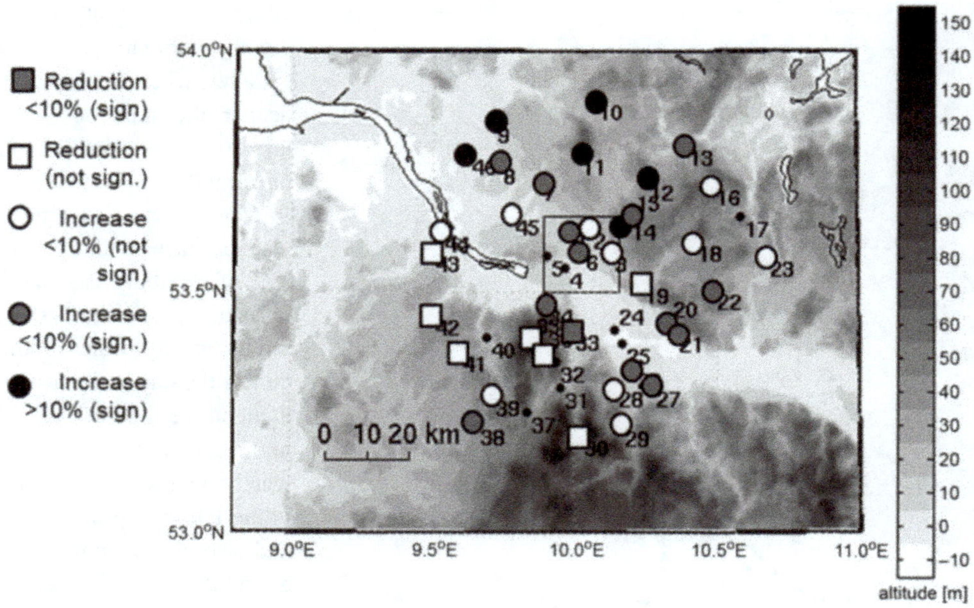

Fig. 15.6 Average percentage increase in precipitation per event, if a site is downwind of the city centre (marked by a *square*) (based on results by Schlünzen et al. 2010). *Black* and *grey filled circles* depict significant increases, the *grey filled rectangle* depicts a significant decrease, and *white filled rectangles* and *circles* depict no significant change

locally structured, with higher night-time temperatures in harbour and industrialised sealed areas, if these are not directly next to a cool river or ocean.

15.4.3 Precipitation

The urban precipitation impact can lead to precipitation enhancement downwind of an urban area. Measured data show this to be the case for Hamburg (Schlünzen et al. 2010), with increases of 5–10 % per precipitation event and found for many (but not all) downwind sites (Fig. 15.6). Assuming only one wind direction throughout the entire year (an extreme and unrealistic assumption), the difference could be 80 mm y^{-1}, which is still less than half of the 200 mm climatological difference with its decrease from the north towards the south-east (Hoffmann and Schlünzen 2010). Thus, despite urban impacts the regional effects might actually be of greater significance.

Schlünzen et al. (2010) also studied long-term changes in precipitation. They found a greater increase in precipitation upwind of Hamburg than downwind of the urban area (trend 1947–2007), which might suggest an overall decline in urban impact. However, these results are speculative, since detailed model studies with METRAS by Schoetter (2013) showed the urban impact of Hamburg is only observable under some meteorological conditions, which agrees with findings by Han et al. (2014). The effects are very local (as can also be inferred from Fig. 15.6) and dependent on the actual meteorological situation. Overall, the impact of Hamburg's urban fabric is not significant for the summer. However, the urban impact might differ in winter or for other urban areas in the North Sea region. Han et al. (2014) pointed out that orography plays an additional role. This was also found for Hamburg, where the highest elevations are only 100–200 m and the urban buildings are low. METRAS model simulations without orography show that orographic effects drive a statistically robust change in the precipitation pattern (Schoetter 2013).

15.5 Scenarios for Future Developments

Adaptation to climate change is of utmost importance for cities to maintain the wellbeing of their inhabitants. Several studies have investigated climate change impacts on urban climate, with some very detailed studies undertaken in Hamburg and London. The ARCC network[4] provides an overview of UK-focussed projects involved with adaptation to 'technological, social and environmental change, including climate change, in the built environment and infrastructure sectors'. Of these the ARCADIA project gives an overview of adaptation and resilience in cities, presenting city-scale climate change scenarios consistent with the UKCP09 scenarios. The Lucid project[5] brought together meteorologists and building engineers to assess the impact of local climate on energy use, comfort and health, while the SCORCHIO project[6] used climate projections to determine adaptation measures focussing on Manchester. Similar multidisciplinary research studies on climate change adaptation were performed for several German cities under the framework of the KLIMZUG program[7] (Climate change in

[4] www.arcc-network.org.uk.

[5] www.homepages.ucl.ac.uk/~ucftiha/index.html.

[6] www.sed.manchester.ac.uk/research/cure/research/scorchio.

[7] www.klimzug.de.

regions): The North Sea region was investigated in northwest 2050[8] (area Bremen/Oldenburg with a focus on the development of roadmaps of climate adaption for three economic sectors: food industry, energy production and distribution, and port management and logistics) and KLIMZUG-Nord[9] (metropolitan area of Hamburg, with a focus on the development of an adaptation master plan that continues until 2050 using the thematic focal points Elbe estuary management, integrated spatial development, nature conservation and governance).

15.5.1 Climate Change Impacts on Urban Climate

Urban areas are a major source of carbon dioxide (CO_2) and anthropogenic heat. While the former drives changes in global climate, the latter has a potentially strong impact on city-scale climate. McCarthy et al. (2011) used an urban land surface scheme (Best et al. 2006) with the Hadley Centre Global Climate Model (HadAM3) and compared the impacts of doubling CO_2 emissions against effects due to urbanisation and anthropogenic heat release in urban areas. They found that urban and rural areas react differently to climate change. While their climate change scenarios (transient SRES A1B scenarios, urbanisation and anthropogenic heat release in urban areas) increased the number of hot days in both areas to the same extent, they found that London has a bigger increase in the number of hot nights (>18.2 °C) than rural areas. The reasons for this difference are local forcing such as anthropogenic heat release and urbanisation leading to the UHI. In fact, local changes such as urbanisation and anthropogenic heat release also increased the frequency of hot days, as did doubling CO_2 emissions. It should be noted that as the UHI is not caused by local CO_2 emissions, it cannot be reduced by lowering them. Oleson (2012) confirmed the results of McCarthy et al. (2011) concerning more frequent hot nights in urban areas compared to rural areas for Europe. Thus, heat risk for the urban population will increase more than for their rural counterparts due to local urban forcing.

Hamdi et al. (2014) studied present (1981–1990) and future climate (2071–2100) for Brussels. On average, the observed nocturnal UHI is of the order of 1.32 K, which agrees well with the simulated average of 1.31 K. Under an A1B scenario, night-time UHIs vary between 0 and 7 K with the frequency of UHIs above 3 K decreasing due to soil dryness in summer. For the city centre the number of heat days will rise by 62.

For Hamburg, Hoffmann et al. (2012) and Grawe et al. (2013) found small changes in the pattern and amplitude of the UHI for climate change scenarios, if the urban fabric remains unchanged. If threshold values are used (such as 18.2 °C for night-time temperatures), these are more frequently exceeded under the future climate due to the higher overall temperatures. The number of exceedances is also higher within urban areas compared to rural areas, because the additional temperature enhancement of urban areas also contributes. However, non-linear effects that contribute to a greater temperature enhancement in urban areas compared to rural areas were not apparent by mid-century under the A1B climate change scenario.

Large precipitation amounts challenge urban infrastructure and could cause streets and houses to flood, or even the total breakdown of some urban infrastructure. Summer precipitation from convective cloud systems may lead to local flooding as was observed in Rostock (Germany) when nearly twice the average monthly rainfall fell within a day (22/23 July 2011; Miegel et al. 2014). A projected increase in winter precipitation of 12–38 % in the climatological mean towards the end of this century (Rechid et al. 2014) poses additional challenges to city planners. Especially in winter, when the already low evapotranspiration in urban areas is even lower and statured soils cannot take up any excess water, this so-called 'hinterland flooding' needs to be addressed in adaptation measures for cities (KLIMZUG-NORD 2014).

15.5.2 City Development Impacts on Urban Climate

As already described (Sect. 15.4), urban areas affect the regional climate through their urban footprint. This relationship provides an opportunity to reduce the regional climate change impact on urban areas—or to enhance it if the wrong mitigation and adaptation measures are applied. Several recent research projects have investigated the impact of planned changes in urban structure on the urban footprint.

The nationally-funded research projects KLIMZUG-NORD (final results in KLIMZUG-NORD 2014) and CliSAP (Schlünzen et al. 2009) investigated different aspects of Hamburg's development on the urban summer climate. In all development scenarios, Hamburg's growth is confined mostly to the current regional area and is restricted vertically. In fact, Hamburg has a ban on high-rise buildings. This means that surface cover changes are relatively small, follow a compact city approach and include aspects of adaptation to the changing climate. More greening (especially of roofs) and higher albedo values on roofs and other sealed surfaces that cannot be greened (e.g. roads) were assumed. Other assumptions include some rebuilding of

[8]www.nordwest2050.de.

[9]www.klimzug-nord.de.

single houses into duplex or terraced houses, replacing terraced houses by blocks, and adding another story to multi-story buildings. All these changes were assumed to be accompanied by a larger greening fraction and more reflective (higher albedo) material. Several simulations were performed with the mesoscale model METRAS to reproduce (at 250-m resolution) a climatological-average summer situation. According to the adaptation measures selected, the average summer temperature could be reduced by 0.2 K with the greatest decreases in those areas where the sealing is very high (KLIMZUG-NORD 2014). Anthropogenic heat emission was prescribed unchanged in these model studies. However, energy use will be more efficient in the future due to retrofitting of houses with better insulation. However, the impacts of better insulation on the surface energy budget are still unclear. Nevertheless, anthropogenic heat emissions will be lower and this will lead to a reduction in the UHI throughout the year.

Impacts on heavy precipitation events were examined for the same scenarios of urban development. As already mentioned (Sect. 15.4.3), the urban fabric has little impact on precipitation, at least for Hamburg, and this was confirmed by the model simulations. Nevertheless, more sealed surfaces pose a challenge for city planners, as these surfaces cannot take up the water and the water must to be drained to avoid hinterland flooding (see Sect. 15.5.1).

Changes in the wind field are expected to be local and possibly very large close to building structures (Schlünzen and Linde 2014). The effects of new buildings on the wind climate of a growing suburb of Hamburg (situated on the large island of Wilhelmsburg) were investigated using the obstacle-resolving model MITRAS (Schlünzen et al. 2003) at a resolution of 5 m. Impacts over a distance of 1000 m from the new buildings were found not just close to the surface but also at higher levels thus affecting ventilation of the buildings in the upper floors (Schlünzen and Linde 2014). Some streets or even balconies on upper stories could become less usable, owing to excessively high wind speeds around the new buildings, while formerly well flushed places could become very calm; this can increase heat stress on sunny days. Furthermore, pollutant dispersion can change due to changes in the wind and temperature fields and this can lead to high concentrations in different sites to before and could change the human exposure pattern.

Studies indicate that changes in temperature and precipitation resulting from urban development scenarios that aim at mitigation and adaptation measures can only slightly reduce the projected climate-driven rise in temperature and change in precipitation patterns. However, although small these local reductions might become more relevant during hot periods by keeping urban temperatures at night at values that reduce health risk. To ensure the cooling effect of urban greening, the watering needs of the vegetation must be ensured (such as by storing water during the wet periods). If the urban vegetation dries out its cooling effect is lost.

15.6 Adaptation and Urban Footprint Reduction Measures

Many North Sea cities are close to the coast or to a river and so must prepare for storm tides. This can be achieved using dykes, as for example in the Netherlands or along the river Elbe for Hamburg, or the Thames Barrier for London. Such measures are expensive, but vital to protect valuable infrastructure and save lives. Hinterland flooding has become an increasing challenge in recent years and similar preparedness needs to be developed here as for storm tides. While upriver dykes help prevent river flooding, and are constantly being improved and strengthened, coastal cities appear to lack focus in terms of rain events that can be equally challenging. Measures are needed to remove rain water following heavy precipitation events. Methods already exist, for example a city like Bergen handles at least twice the precipitation amounts observed in the southern North Sea region every month (Fig. 15.2d). Hamburg has introduced a separate rain water drainage system in recent years and introduced financial penalties, if rain water is not locally drained by home owners. To cope with intense precipitation events (Schlünzen et al. 2010) and increased amounts of winter precipitation, it is essential that urban areas can store water. Storing precipitation in winter would help to cope with future drier summers. Cities will need larger amounts of water in future for two reasons: warmer air can take up more water and so evapotranspiration will be higher, and increased urban greening to help reduce high urban night-time temperatures needs enough water to prevent the vegetation drying out.

Any increase in sealed surfaces should be kept to a minimum (they are sometimes introduced for flood protection, such as new dykes or walls with bitumen or stone cover), since they increase the amount of heat-storing surfaces and thus night-time temperatures. The current replacement of green spaces and gardens in urban and suburban areas by buildings and sealed surfaces should also be limited as this will also cause an increase in urban night-time temperatures. Plus, there is a tendency for urban spread into surrounding rural areas which extends the UHI in space. Hamburg has already begun implementing measures to keep UHI effects within reasonable limits, possibly even causing a reduction. In contrast, London can only grow vertically in the city centre, leading to more heat storage capacity, presumably greater anthropogenic heat release and warmer nights. Any increase in the spatial extent of London, and thus an increase in sealed surface, would also increase the spread of the UHI. Higher temperatures in urban areas would lead to several stress factors. Heat in itself is a recognised health factor. Planners in all North Sea cities should ensure that existing green areas are kept and that new ones are added. In addition, all waste heat emissions (to the atmosphere and to water bodies) should be reduced especially in summer to reduce night-time heat exposure within urban areas. Projections by Iamarino et al. (2011) suggested an increase of 16 % in anthropogenic heat emission due to a larger working population in the city of London by 2025 compared to 2005. Without measures to reduce the urban footprint, this would lead to even higher temperatures within the city of London. This shows a clear synergy between adaptation, mitigation and urban footprint reduction measures: less energy-consuming computers, factories, and vehicles, not only reduce CO_2 emissions (or equivalent) and thus global temperature increase in the long term, but also directly and quickly reduce the amount of waste heat emitted into the urban area and thus night-time temperatures. The same is true for well-insulated buildings: using less energy for heating in winter and cooling in summer means lower CO_2 emission (for energy production). Lower CO_2 emissions implies less global warming, while better insulation reduces UHI at night.

The projected increase in global temperatures could lead to higher biogenic volatile organic compound (VOC) emissions from vegetation, which could in turn increase O_3 levels if NO_X emissions are not considerably reduced (e.g. Meyer and Schlünzen 2011). To avoid additional VOC emissions, new urban vegetation needs to be selected with low VOC emission potential (Kuttler 2013: 281).

Drier summers would mean more particles eroded from dry surfaces and an increase in the already high particle load in urban areas. This supports the argument for increasing the amount of vegetated surface in urban areas and for providing these areas with water during dry periods. An immediate measure used in London to reduce the atmospheric particle load has been to spray adhesives onto roads in some of the most polluted areas, although this does not reduce the source of the particulate matter.

15.7 Conclusions

Urban areas are not only impacted by changes in regional climate, but themselves contribute to climate change through their large greenhouse gas emissions. Urban areas also modify the regional climate through their urban footprint. This is mainly visible in terms of concentration levels above EU limit values for NO_X (daily average value), NO_2 (annual average value) and particulate matter (PM_{10} daily average values). Higher temperatures, especially at night, result from changes in the surface energy budget due to the urban fabric and additional emission of anthropogenic heat. The temperature difference can be in the range of a few degrees in the monthly average. It may be up to ~ 7 K under favourable conditions (clear skies, high radiative impact, low large-scale pressure gradients).

Despite broad similarities between many urban areas, there are also large location-specific differences with regard to city planning needs (such as poorly insulated Victorian housing stock in the UK wasting large amounts of energy). While some cities are growing, others show little change with respect to the number of inhabitants and some are even shrinking. Whatever the future, it will involve change. Changes in climate will become increasingly apparent, especially towards the end of the century, with the first indications of what is to come already apparent (hinterland flooding, more intense precipitation, and drier and warmer summers). Because even in a 'non-changing' city, inhabitants will renovate buildings and young people will adopt new infrastructures and new technologies that will one day be standard for all, there is an opportunity for cities to change for simultaneously adapting to and mitigating climate change such that the worst impacts of climate change can be avoided by mitigation measures, and the unavoidable impacts of climate change can be met by adaptation measures, while the urban footprint becomes ever smaller.

References

Allen L, Lindberg F, Grimmon CSB (2011) Global to city scale urban anthropogenic heat flux: model and variability. Int J Climatol 31:1990–2005

Armstrong BG, Chalabi Z, Fenn B, Hajat S, Kovats S, Milojevic A, Wilkinson P (2011) Association of mortality with high temperatures in a temperature climate: England and Wales. J Epidemiol Commun H 65:340–345

Bechtel B, Schmidt KJ (2011) Floristic mapping data as a proxy for the mean urban heat island. Clim Res 49:45–58

Best MJ, Grimmond CSB, Villani MG (2006) Evaluation of the urban tile in MOSES using surface energy balance observations. Bound-Lay Meteorol 118503–525

Böhm J, Wahler G (2012) Luftreinhalteplan für Hamburg: 1. Fortschreibung 2012. Behörde für Stadtentwicklung und Umwelt. www.hamburg.de/contentblob/3744850/data/fortschreibung-luftreinhalteplan.pdf

Bohnenstengel SI, Evans S, Clark PA, Belcher SE (2011) Simulations of the London urban heat island. Q J Roy Meteor Soc 137: 1625–1640

Bohnenstengel SI, Hamilton I, Davies M, Belcher SE (2014) Impact of anthropogenic heat emissions on London's temperatures. Q J Roy Meteor Soc 140:687–698

Bohnenstengel SI, Belcher SE, Aiken A, Allan JD, Allen G, Bacak A, Bannan TJ, Barlow JF, Beddows DCS, Bloss WJ, Booth AM, Chemel C, Coceal O, Di Marco CF, Dubey MK, Faloon KH, Fleming Z, Furger M, Geitl JK, Graves RR, Green DC, Grimmond CSB, Halios C, Hamilton JF, Harrison RM, Heal MR, Heard DE, Helfter C, Herndon SC, Holmes RE, Hopkins JR, Jones AM, Kelly FJ, Kotthaus S, Langford B, Lee JD, Leigh RJ, Lewis AC, Lidster RT, Lopez-Hilfiker FD, McQuaid JB, Mohr C, Monks PS, Nemitz E, Ng NL, Percival CJ, Prévôt ASH, Ricketts HMA, Sokhi R, Stone D, Thornton JA, Tremper A, Valach AC, Visser S, Whalley LK, Williams LR, Xu L, Young DE, Zotter P (2015) Meteorology, air quality, and health in London: The ClearfLo project. B Am Meteorol Soc 96:779–804

Carslaw DC, Beevers SD, Westmoreland E, Williams ML, Tate JE, Murrells T, Stedman J, Li Y, Grice S, Kent A, Tsagatakis I (2011) Trends in NO_X and NO_2 emissions and ambient measurements in the UK. Version: July 2011

Chemel C, Sokhi RS (2012) Response of London's urban heat island to a marine air intrusion in an easterly wind regime. Bound-Lay Meteorol 144:65–81

Crosman ET, Horel JD (2010) Sea and lake breezes: A review of 527 numerical studies. Bound-Lay Meteorol 137:1–29

EC (1980) Council Directive 80/779/EEC of 15 July 1980 on air quality limit values and guide values for sulphur dioxide and suspended particulates. European Commission (EC)

EC (2008) Directive 2008/50/EC of the European Parliament and of the Council of 21 May 2008 on ambient air quality and cleaner air for Europe (OJL 152, 11.6.2008). European Commission (EC)

Fuller G, Mittal L (2012) Air quality in London – Briefing note to GLA Environment and Health Committee. Environmental Research Group, Kings College London

GLA (2002) 50 years on: The struggle for air quality in London since the great smog of December 1952. Mayor of London, Greater London Authority (GLA)

GLA (2010) Clearing the air: The Mayor's Air Quality Strategy, Mayor of London, Greater London Authority (GLA)

Grawe D, Thompson HL, Salmond JA, Cai X-M, Schlünzen KH (2012) Modelling the impact of urbanisation on regional climate in the Greater London Area. Int J Climatol 32:2388–2401

Grawe D, Flagg DD, Daneke C, Schlünzen KH (2013) The urban climate of Hamburg for a 2K warming scenario considering urban development. Present EMS 2013 Reading, 13.09.2013

Hamdi R, Van de Vyver H, De Troch R, Termonia P (2014) Assessment of three dynamical urban climate downscalling methods: Brussels' future heat island under an A1B emission scenario. Int J Climatol 34:378–399

Hamilton IG, Davies M, Steadman P, Stone A, Ridley I, Evans S (2009) The significance of the anthropogenic heat emissions of London's buildings: a comparison against captured shortwave solar radiation. Build Environ 44:807–817

Han JY, Baik JJ, Lee H (2014) Review: Urban impacts on precipitation. Asia-Pac J Atmos Sci 50:17–30

Hay JS, Pasquill F (1957) Diffusion from a fixed source at a height of a few hundred feet in the atmosphere. J Fluid Mech 2:299–310

Heusinkveld BG, Steeneveld GJ, van Hove LWA, Jacobs CMJ, Holtslag AAM (2014) Spatial variability of the Rotterdam urban heat island as influenced by urban land use. J Geophys Res-Atmos 119:1–14

Hoffmann P (2012) Quantifying the influence of climate change on the urban heat island of Hamburg using different downscaling methods. Dissertation, Univ Hamburg, Germany

Hoffmann P, Schlünzen KH (2010) Das Hamburger Klima. In: Poppendieck HH, Bertram H, Brandt I, Engelschall B, von Prondzinski J (eds) Der Hamburger Pflanzenatlas von a bis z. Dölling und Galitz Verlag GmbH, Hamburg, p28–31

Hoffmann P, Krueger O, Schlünzen KH (2012) A statistical model for the urban heat island and its application to a climate change scenario. Int J Climatol 32:1238–1248

Iamarino M, Beevers S, Grimmond CSB (2011) High-resolution (space, time) anthropogenic heat emissions: London 1970-2025. Int J Climatol 32:1754–1767

Jalkanen L (2011) WMO addressing climate and air quality. Technical Workshop on Science and Policy of Short-lived Climate Forcers, September 9-10, 2011, Mexico City. http://mce2.org/SLCFWorkshop/docs/%28Jalkanen%29%20SLCF%20Mexico%20WMO.pdf

Jones PD, Lister DH (2009) The urban heat island in Central London and urban-related warming trends in Central London since 1900. Weather 64:323–327

Jones AM, Harrison RM, Barratt B, Fuller G (2012) A large reduction in airborne particle number concentrations at the time of the introduction of "sulphur free" diesel and the London Low Emission Zone. Atmos Environ 50:129–138

Kim J, Kim KR, Choi BC, Lee DG, Kim JS (2009) Regional distribution of perceived temperatures estimated by the Human Heat Budget Model (the Klima-Michel Model) in South Korea. Adv Atmos Sci 26:275–282

KLIMZUG-NORD Verbund (ed) (2014) Kursbuch Klimaanpassung. Handlungsoptionen für die Metropolregion Hamburg, TuTech Verlag, Hamburg. www.klimzug-nord.de

Kottek M, Grieser J, Beck C, Rudolf B, Rubel F (2006) World map of the Köppen-Geiger climate classification updated. Meteorol Z 15:259–263

Kovats RS, Johnson H, Griffiths C (2006) Mortality in southern England during the 2003 heat wave by place of death. Health Statist Quart 29:6–8

Kuttler W (2013) Klimatologie. 2. aktualisierte u. ergänzte Ausgabe, Schöningh, Paderborn (UTB 3099)

Lane S (2014) Assessing the validity and impact of urban-scale numerical weather prediction. PhD thesis. University of Reading

Mavrogianni A, Davies M, Batty M, Belcher SE, Bohnenstengel SI, Carruthers D, Chalabi Z, Croxford B, Demanuele C, Evans S, Giridharan R, Hacker JN, Hamilton I, Hogg C, Hunt J, Kolokotroni M, Martin C, Milner J, Rajapaksha I, Ridley I, Steadman JP, Stocker J, Wilkinson P, Ye Z (2011) The comfort, energy and health implications of London's urban heat island. Build Serv Eng Res T 32:35–52

McCarthy MP, Harpham C, Goodess CM, Jones PD (2011) Simulating climate change in UK cities using a regional climate model, HadRM3. Int J Climatol 32:1875–1888

Meyer EMI, Schlünzen KH (2011) The influence of emission changes on ozone concentrations and nitrogen deposition into the southern North Sea. Meteorol Z 20:75–84

Miegel K, Mehl D, Malitz G, Ertel H (2014) Ungewöhnliche Niederschlagsereignisse im Sommer 2011 in Mecklenburg-Vorpommern und ihre hydrologischen Folgen. Teil 1: Hydrometeorologische Bewertung. Hydrologie und Wasserbewirtschaftung 58:18–28

Oke TR (1982) The energetic basis of the urban heat island. Q J Roy Meteor Soc 108:1–24

Oleson K (2012) Contrasts between urban and rural climate in CCSM4 CMIP5 climate change scenarios. J Climate 25:1390–1412

Petrik R, Grawe D, Bungert U, Schlünzen KH (in prep) Quantifying the impact of anthropogenic heat on the urban climate of Hamburg in the summer season. J Urb Clim (in internal review)

Pielke RA, Adegoke J, Beltran-Przekurat A, Hiemstra CA., Lin J, Nair, US, Niyogi D, Nobis TE (2007) An overview of regional land-use and land-cover impacts on rainfall. Tellus B, 59:587–601

Rechid D, Petersen J, Schoetter R, Jacob D (2014) Klimaprojektionen für die Metropolregion Hamburg. Berichte aus den KLIMZUG-NORD Modellgebieten 1. TuTech Verlag, Hamburg. www.klimzug-nord.de

Richter M, Deppisch S, von Storch H (2013) Observed changes in long-term climatic conditions and inner-regional differences in urban regions of the Baltic Sea coast. Atmos Climate Sci 3:165–176

Schlünzen KH (1990) Numerical studies on the inland penetration of sea breeze fronts at a coastline with tidally flooded mudflats. Beitr Phys Atmos 63:243–256

Schlünzen KH, Linde M (2014) Wilhelmsburg im Klimawandel. Ist-Situation und mögliche Veränderungen. Berichte aus den KLIMZUG-NORD Modellgebieten 4. TuTech Verlag, Hamburg. www.klimzug-nord.de

Schlünzen KH, Hinneburg D, Knoth O, Lambrecht M, Leitl B, Lopez S, Lüpkes C, Panskus H, Renner E, Schatzmann M, Schoenemeyer T, Trepte S, Wolke R (2003) Flow and transport in the obstacle layer - First results of the microscale model MITRAS. J Atmos Chem 44:113–130

Schlünzen KH, Ament F, Bechtel B, Böhner J, Eschenbach A, Fock B, Hoffmann P, Kirschner P, Leitl B, Oßenbrügge J, Rosenhagen G, Schatzmann M (2009) Development of mitigation measures for the metropolitan region of Hamburg (Germany). Conference Contribution to Climate Change: Global Risks, Challenges and Decisions. IOP Conf Ser: Earth Env Sci 6 332034, doi:10.1088/1755-1307/6/33/332034

Schlünzen KH, Hoffmann P, Rosenhagen G, Riecke W (2010) Long-term changes and regional differences in temperature and precipitation in the metropolitan area of Hamburg. Int J Climatol, 30:1121–1136

Schoetter R (2013) Can local adaptation measures compensate for regional climate change in Hamburg Metropolitan Region. PhD thesis in Meteorology, Univ Hamburg, Germany

Schoetter R, Grawe D, Hoffmann P, Kirschner P, Grätz A, Schlünzen KH (2013) Impact of local adaptation measures and regional climate change on perceived temperature. Meteorol Z 22:117–130

Shepherd JM, Pierce H, Nergi AJ (2002) Rainfall modification by major urban areas: Observations from spaceborne rain radar on the TRMM satellite. J Appl Meteorol 41:689–701

Simpson JE, Mansfield DA, Milford JR (1977) Inland penetration of sea-breeze fronts. Q J Roy Meteor Soc 103:47–76

Staiger H, Laschewski G, Grätz A (2011) The perceived temperature – a versatile index for the assessment of the human thermal environment. Part A: scientific basics. Int J Biometeorol 56:165–176

Teichert N (2013) Analyse von Oberflächeneinflüssen auf die sommerlichen Temperaturen in der Metropolregion Hamburg unter Anwendung eines GIS. Diploma Thesis Meteorology, Fachber Geowiss, Univ Hamburg, Germany

UN (2014) World Urbanization Prospects: The 2014 Revision. CD-ROM Edition UN, Department of Economic and Social Affairs, Population Division. http://esa.un.org/unpd/wup/CD-ROM Accessed 11 December 2015

Watkins R, Palmer J, Kolokotroni M, Littlefair P (2002) The London heat island: results from summertime monitoring. Build Serv Eng Res T 23:97–106

Zhang J, Rao ST (1999) The role of vertical mixing in the temporal evolution of the ground-level ozone concentrations. J Appl Meteorol 38:1674–1691

Socio-economic Impacts—Air Quality

16

Stig Bjørløw Dalsøren and Jan Eiof Jonson

Abstract

In the North Sea region, poor air quality has serious implications for human health and the related societal costs are considerable. The state of air pollution is often used as a proxy for air quality. This chapter focuses on the two atmospheric pollutants of most significance to human health in Europe—particulate matter and ground-level ozone. These are also important 'climate forcers'. In the North Sea area, the effects on air quality of emission changes since preindustrial times are stronger than the effects of climate change. According to model simulations, this is also the case for future air quality in the North Sea region, but substantial variation in model results implies considerable uncertainty. Short-term events such as heat waves can have substantial impacts on air quality and some regional climate models suggest that heat waves may become more frequent in the coming decades. If the reductions in air pollutant emissions expected through increasingly stringent policy measures are not achieved, any increase in the severity or frequency of heat waves may have severe consequences for air quality. Climate and air quality interact in several ways and mitigation optimised for a climate or air quality target in isolation could have synergistic or antagonistic effects.

16.1 Introduction

The state of air pollution is often expressed as air quality. The concentrations of gaseous pollutants and particulate matter are then used as a measure of air quality. However, it is often not meaningful to discuss air quality without addressing the multiple impacts of air pollution. Major air pollutants may be clustered according to their properties and impacts, and this is shown in Fig. 16.1. After being emitted into the atmosphere pollutants undergo chemical oxidation and form new compounds with different properties and impacts. Pollutants then remain in the atmosphere until they are removed through cloud and precipitation processes or by direct deposition to the earth's surface. Differences in chemical reactivity and removal rates result in atmospheric lifetimes ranging from seconds to months. Air quality and related impacts are therefore influenced by local meteorological features, regional (transboundary) processes, and intercontinental transport. The pathway from emissions to impacts is complex. The focus in this chapter is limited to impacts on human health, climate, and climate-air quality interactions and mainly excludes impacts on ecosystems (acidification, eutrophication, carbon sequestration, crops, and vegetation) and materials. Impacts of climate change on ecosystems are covered in Chaps. 8, 9, 10, and 11.

Air quality and climate interact in several ways. Air pollutants can affect climate both directly and indirectly through their influence on the radiative balance of the atmosphere. Primary particulate matter (primary PM, Fig. 16.1) affects climate directly, while pollutants such as carbon monoxide (CO), non-methane volatile organic compounds (nmVOC), polycyclic aromatic hydrocarbons (PAH), nitrogen oxides (NO_X), sulphur dioxide (SO_2) and ammonia (NH_3) although

S.B. Dalsøren (✉)
CICERO Center for International Climate and Environmental Research-Oslo, Oslo, Norway
e-mail: s.b.dalsoren@cicero.uio.no

J.E. Jonson
Norwegian Meteorological Institute, Oslo, Norway
e-mail: j.e.jonson@met.no

M. Quante and F. Colijn (eds.), *North Sea Region Climate Change Assessment*,
Regional Climate Studies, DOI 10.1007/978-3-319-39745-0_16

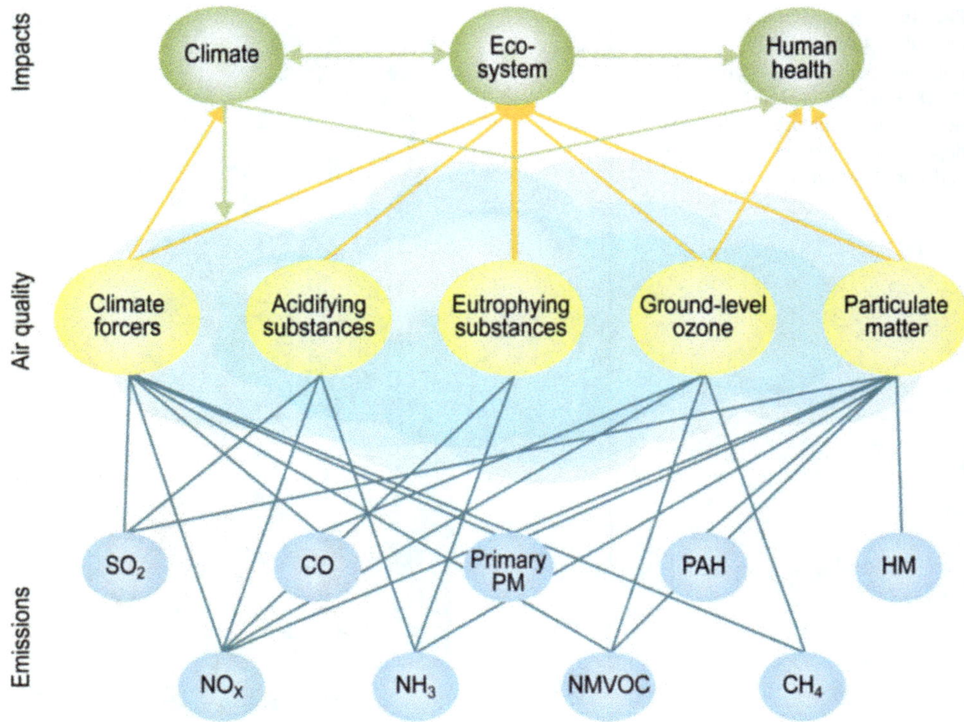

Fig. 16.1 Major air pollutants, clustered according to impacts on climate, ecosystems and human health (EEA 2012). From *left* to *right* the pollutants shown are as follows: sulphur dioxide (SO_2), nitrogen oxides (NO_X), carbon monoxide (CO), ammonia (NH_3), particulate matter (PM), non-methane volatile organic compounds (NMVOC), polycyclic aromatic hydrocarbons (PAH), methane (CH_4), heavy metals (HM)

having a negligible direct radiative ('greenhouse') effect, have an important indirect climate effect by acting as precursors for components that are both harmful pollutants and act as 'climate forcers' (e.g. ozone and particulate matter). On the other hand, air quality is also sensitive to climate change itself, since climate change drives changes in the physical and chemical properties of the earth and atmosphere. Climate policies also imply energy efficiency and technical measures that change emissions of air pollutants. Equally, air quality mitigation measures will impact on greenhouse gas emissions.

The main chemical components addressed in this chapter are PM and ground level ozone (O_3). These are generally recognised as the two pollutants that most significantly affect human health in Europe. The impacts of long-term and peak exposure to these pollutants range in severity from impairing the respiratory system to premature death. Particulate matter in the atmosphere originates from direct emissions (e.g. black and organic carbon, sea salt, dust, pollen) or is derived from chemical reactions involving precursor gases such as SO_2, NO_X, NH_3, PAH and nmVOC. The particles of greatest human health concern are 10 μm in diameter or less (PM_{10}). Of particular concern are those 2.5 μm or less ($PM_{2.5}$) since these could pass from the lungs into the bloodstream. The size and sign of the particulate climate effect (i.e. the particulate-driven temperature change) varies according to particle size, composition, shape, and altitude, and the albedo of the underlying surface. Particles can also change precipitation patterns and surface albedo.

Ozone is a secondary pollutant and greenhouse gas formed by complex chemical reactions involving NO_X, nmVOC, CO and methane (CH_4). In addition to its impact on human health, high O_3 levels may damage plants, affect agriculture and forest growth, and impact on CO_2 uptake. Compared to O_3 and PM, CH_4 is a relatively long-lived and thus well-mixed greenhouse gas and one link with regional air quality is that short-lived air pollutants such as NO_X, nmVOC, and CO may influence its chemical removal in the atmosphere. Changes in CH_4 concentration in turn affect the atmospheric oxidation capacity and thereby the speed of chemical cycles and removal of pollutants. The rise in global CH_4 levels over recent decades has contributed to rising background O_3 concentrations in the northern hemisphere.

16.2 Current Status

Poor air quality in Europe is a serious human health issue and the related external costs (costs imposed by a producer or a consumer on another producer or consumer, outside of any market transaction between them) are considerable. In 2010, annual premature mortalities were over 400,000 in the EU area and the total external costs of the health impacts were estimated at EUR 330–940 billion (EU 2013). Similar estimates are reported in other studies (Watkiss et al. 2005; Anenberg et al. 2010; Amann et al. 2011; Brandt et al. 2013a; Fang et al. 2013; Silva et al. 2013). Particulate matter is the principal pollutant in terms of human health impacts.

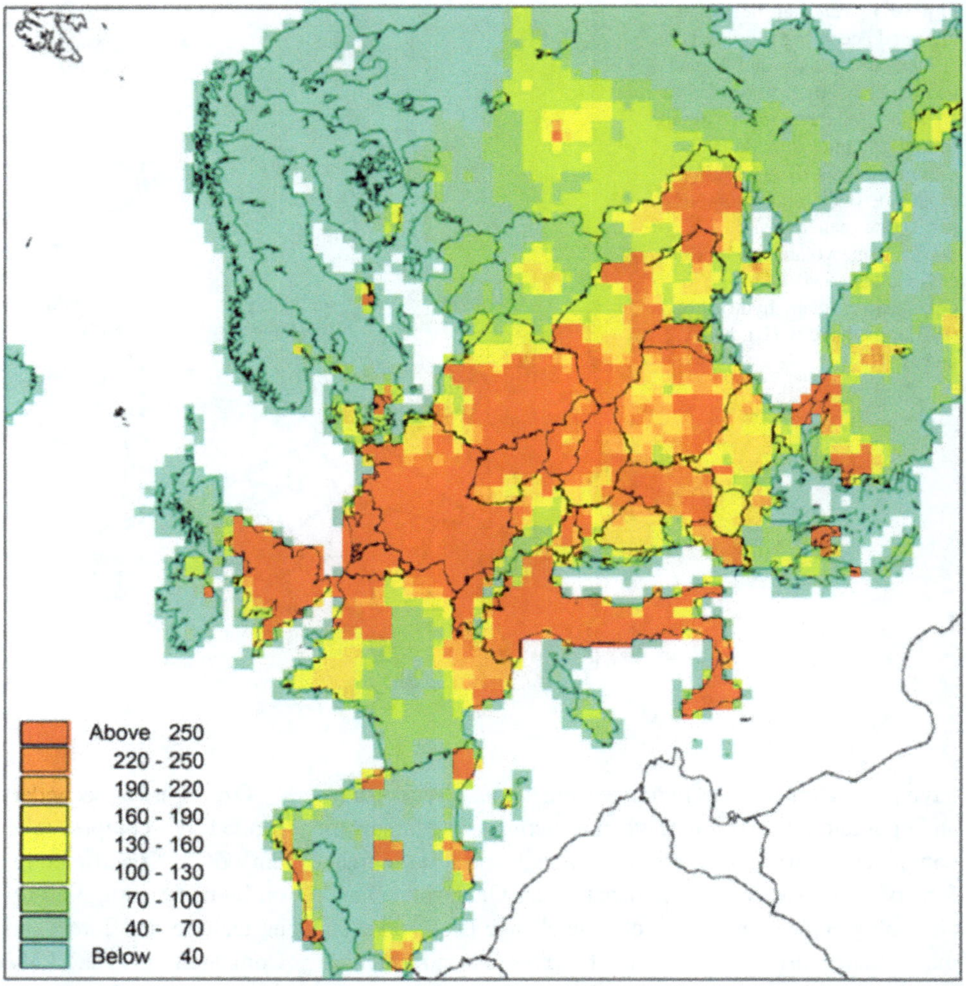

Fig. 16.2 Estimated number of premature deaths due to air pollution per 2500 km^2 grid cell in 2000 (Brandt et al. 2013a). The number of premature deaths is dependent both on pollution level and population density

Figure 16.2 shows an example of a model estimate of the geographical distribution of premature deaths in Europe in 2000 due to PM and O_3 (Brandt et al. 2013a). The number of premature deaths in highly populated regions of the North Sea countries is relatively high. Brandt et al. (2013a) estimated the number of premature deaths in Europe to have declined from 680,000 in 2000 to 570,000 in 2011. This decline reflects measures resulting in lower pollution levels in some regions of Europe such as the North Sea area (see Sect. 16.3).

16.2.1 Current Air Quality

In the European Union, $PM_{2.5}$ resulting from human activity is estimated to have reduced average life expectancy in 2000 by 8.6 months (EEA 2012). Figure 16.3 shows the annual mean measured concentration of surface PM_{10} across Europe in 2010. It should be noted that winter 2010 was colder than normal in the North Sea region (Blunden et al. 2011), resulting in higher PM concentrations than expected from a simple extrapolation of the long-term trend (Tsyro et al. 2012). It is clear that some stations in countries adjacent to the North Sea exceeded the EU daily limit value (orange and red dots). Exceedance at one or more stations occurred in 23 EU Member States in 2010. Of the urban EU population, 21 % was exposed to values above the daily limit value (EEA 2012). The World Health Organization (WHO) has a stricter guideline based on the fact that no threshold is found below which no adverse health effects of PM occur. The WHO guideline value was exceeded in most of the monitoring stations in continental Europe (red, orange and green dots). A similar picture emerges for $PM_{2.5}$ but there are fewer measurement sites.

The EU target value for O_3 of 120 µg m^{-3} (daily maximum of 8-hour running mean values not to be exceeded on more than 25 days per year, averaged over three consecutive years) was exceeded at a large number of stations across Europe in 2010 (dark orange and red dots in Fig. 16.4). However, there are few exceedances of the target value in North Sea countries and none along the North Sea coast. Although winter 2010 was particularly cold in this region, summer was normal or slightly warmer than normal (Blunden et al. 2011). Even so, O_3 levels were low compared to

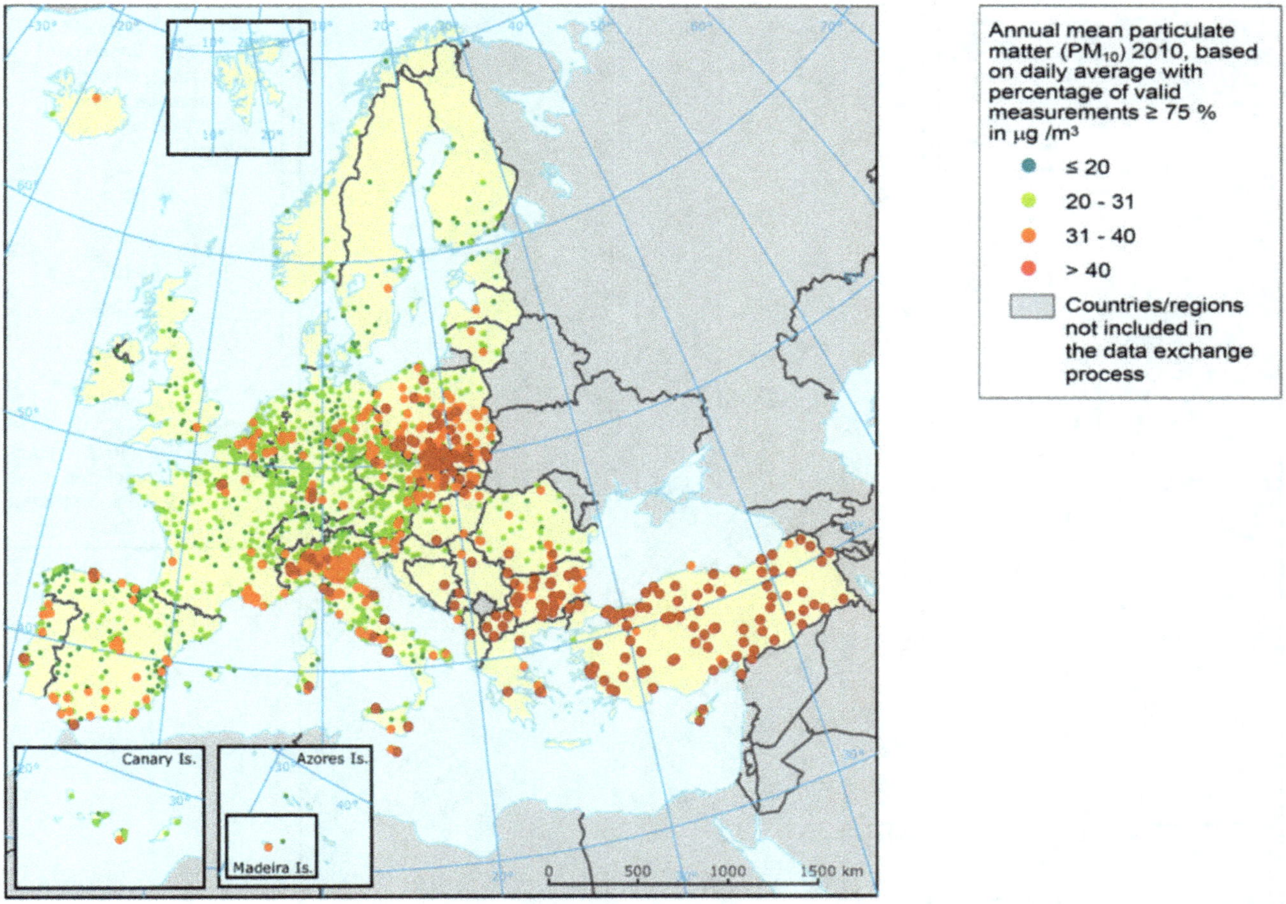

Note: The red dots indicate stations reporting exceedances of the 2005 annual limit value (40 µg/m³), as set out in the Air Quality Directive (EU, 2008c).

The orange dots indicate stations reporting exceedances of a statistically derived level (31 ug/m³) corresponding to the 24-hour limit value, as set out in the Air Quality Directive (EU, 2008c),

The pale green dots indicate stations reporting exceedances of the WHO air quality guideline for PM_{10} of less than 20 µg/m³ but not in exceedance of limit values as set out in the Air Quality Directive (EU, 2008c)

The dark green dots indicate stations reporting concentrations below the WHO air quality guideline for PM_{10} and implicitly below the limit values as set out in the Air Quality Directive (EU, 2008c).

Source: AirBase v, 6.

Fig. 16.3 Annual mean surface concentration of PM_{10} across Europe in 2010 (EEA 2012)

the long-term average (Fagerli et al. 2012). Ozone formation is dependent on sunlight and concentrations increase from north to south across Europe. Ozone is also non-linearly dependent on NO_X and VOC concentrations, and is produced and destroyed in a balance between VOC and NO_X, fuelled by solar radiation. In areas with very high nitrogen oxide (NO) emissions, O_3 will be depleted by the reaction with NO (NO_X titration). As the emission plume moves away from the source region, O_3 may be regenerated. Ozone concentrations are therefore generally higher in rural areas some distance from the main NO_X emission sources. The long-term objective for the protection of human health of 120 µg m^{-3} (daily maximum of 8-hour running mean values), is exceeded at several stations in all North Sea countries (EEA 2011, 2014). For the ambitious WHO guideline (100 µg m^{-3} 8-hour mean) only two of the 510 rural stations, 3 % of urban background stations and 7 % of traffic stations would not exceed this level (EEA 2012). The EU information threshold (180 µg m^{-3} 1-hour mean) is occasionally exceeded in Belgium, the Netherlands and Denmark but is rarely exceeded in other North Sea countries (EEA 2011, 2014).

16.2.2 Contribution from Emission Sectors and Regions

On a country-by-country basis, the ratio between the contribution to air pollution and deposition from domestic versus non-domestic (transboundary) sources varies

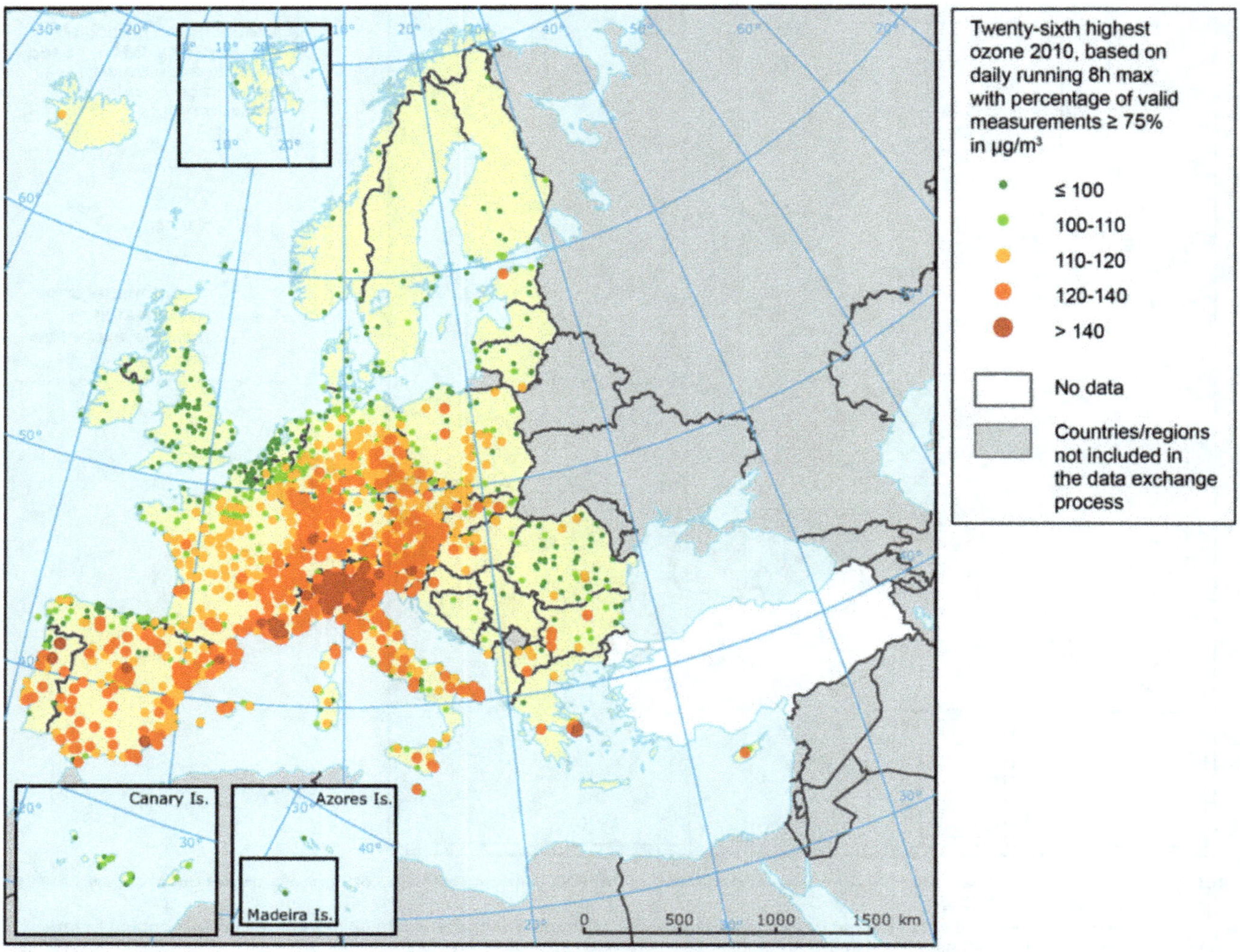

Note: The map shows the proximity of recorded O_3 concentrations to the target value. At sites marked with dark orange and red dots, the twenty-sixth highest daily O_3 concentration exceeded the 120 µg/m³ threshold and the number of allowed exceedances by the target value.

Source: AirBase v. 6.

Fig. 16.4 Twenty-sixth highest daily maximum 8-hour average surface ozone (O_3) concentration recorded at each monitoring station in 2010 (EEA 2012)

substantially, depending on pollutant, local source strength, proximity to major non-domestic sources, and the geographical size of the individual countries (see recent reports by the European Monitoring and Evaluation Programme; EMEP[1]).

For $PM_{2.5}$, contributions to the overall national pollution level are dominated by transport from non-domestic countries (exceptions are Great Britain and Norway, which are heavily influenced by air masses originating over the Atlantic). As an example, the contributions to surface $PM_{2.5}$ and PM_{coarse} ($PM_{coarse} = PM_{10} - PM_{2.5}$) are shown in Fig. 16.5 for the Netherlands (Gauss et al. 2012). The main contributions are clearly from neighbouring countries. The contribution from volcanoes is due to the major volcanic eruption in Iceland in 2010.

The picture is more complicated for O_3 and O_3-derived parameters such as SOMO35.[2] Countries around the North Sea are some of the highest NO_X emitters in Europe. High NO_X emissions in combination with limited solar insolation during winter at these latitudes results in inefficient photochemical O_3 production and substantial NO_X titration. As a result, the present levels of NO_X emissions will, at least as an annual average, decrease the O_3 burden in the North Sea

[1] http://emep.int/mscw/index_mscw.html.

[2] SOMO35, defined as the annual daily sum of 8-hour running average O_3 concentrations over 35 ppb, is a measure of accumulated annual O_3 concentrations and used as an indicator of health hazards, see EMEP (2013) for details.

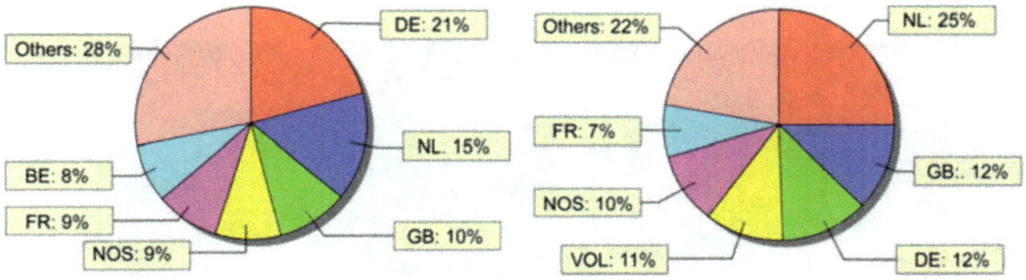

Fig. 16.5 Percentage contribution from individual countries to surface $PM_{2.5}$ (*left*) and PM_{coarse} (*right*) in the Netherlands. *BE* Belgium, *FR* France, *NOS* North Sea, *DE* Germany, *NL* The Netherlands, *GB* Great Britain, *VOL* volcanoes (Gauss et al. 2012)

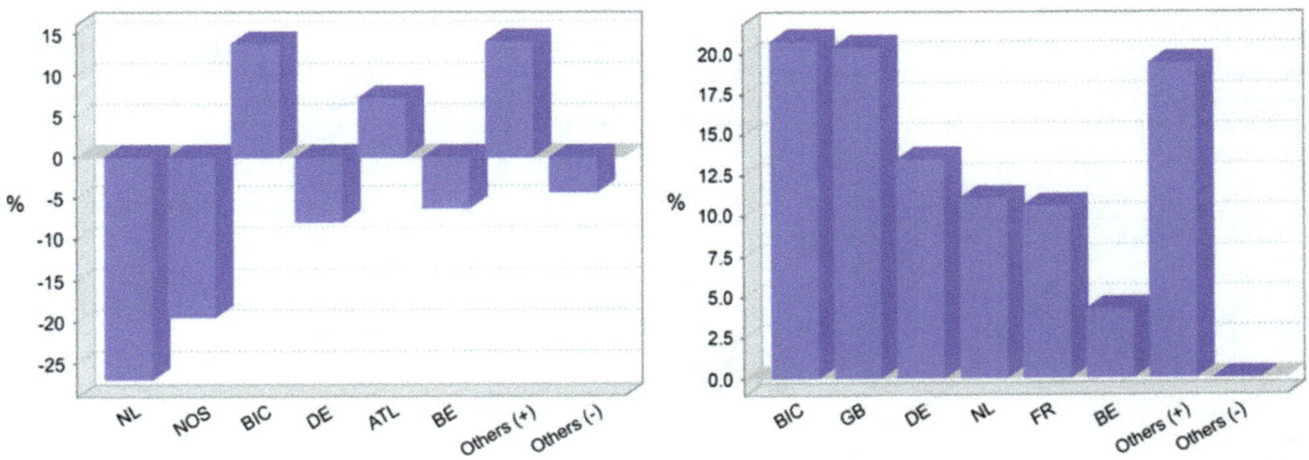

Fig. 16.6 Percentage contribution from individual countries to SOMO35 from NO_X emissions (*left*) and SOMO35 from nmVOC emissions (*right*) in the Netherlands. *NL* Netherlands, *NOS* North Sea, *DE* Germany, *GB* Great Britain, *FR* France, *ATL* remaining Atlantic within model domain, *BE* Belgium. BIC is boundary and initial concentrations (Gauss et al. 2012)

region. On the other hand, regional VOC and CO emissions result in increased O_3 levels. This is illustrated for The Netherlands in Fig. 16.6 (see EMEP country reports for further examples[3]).

The negative contribution from NO_X from several countries is caused by titration. The contribution from BIC (boundary and initial concentration) is calculated by perturbing lateral boundary (and initial) concentrations separately for NO_X and nmVOC in the EMEP model.

The contributions from different domestic emission sectors to model-calculated $PM_{2.5}$ concentrations in European countries are shown in Fig. 16.7. The countries are sorted according to the model-calculated contribution from Sector 1 (combustion in energy and transformation industries). The countries in the North Sea region are all to the right in the figure, with relatively small contributions from industry (Sectors 1–3; see figure caption for definitions).

These countries are characterised by a relatively large share of the emissions from agriculture and transportation (road and shipping). In the North Sea countries, shipping represents a significant share of the transport-related impacts. Studying the external costs from all international ship traffic in relation to the other sources, Brandt et al. (2013a) estimated that ship traffic accounted for 7 % of the total health effects in Europe due to air pollution in 2000. The corresponding value for Denmark, which is surrounded by heavy ship traffic, is 18 %. In Denmark the relative contribution from international ship traffic is comparable to the contributions from road traffic or the domestic use of wood stoves (Brandt et al. 2013b).

The North Sea region is also affected by transport of air pollutants from other continents. Figure 16.8 shows the model-calculated effects on surface O_3 from intercontinental transport due to 20 % reductions in anthropogenic emissions in North America, East Asia and South Asia. As an average for the European continent the calculated contribution to surface O_3 from other continents is about half of the total European contribution (HTAP 2010). The intercontinental contributions show large seasonal (see Fig. 16.8) and geographic variability. Brandt et al. (2012) calculated the effects

[3]http://emep.int/mscw/index_mscw.html.

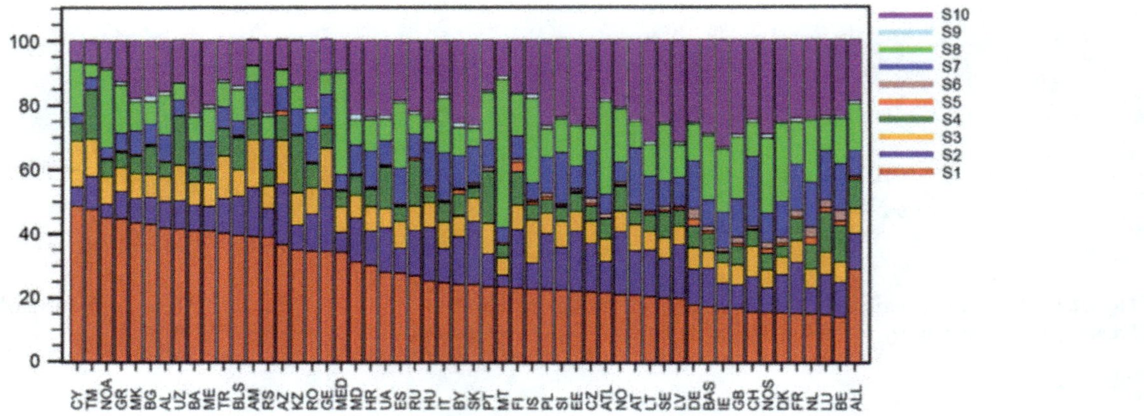

Fig. 16.7 Percentage contribution from individual sectors to $PM_{2.5}$ concentration in European countries: *S1* combustion in energy and transformation industries, *S2* non-industrial combustion plants, *S3* combustion in manufacturing industry, *S4* production processes, *S5* extraction and distribution of fossil fuels and geothermal energy, *S6* solvent and other product use, *S7* road transport, *S8* other mobile sources and machinery (including shipping), *S9* waste treatment and disposal, *S10* agriculture

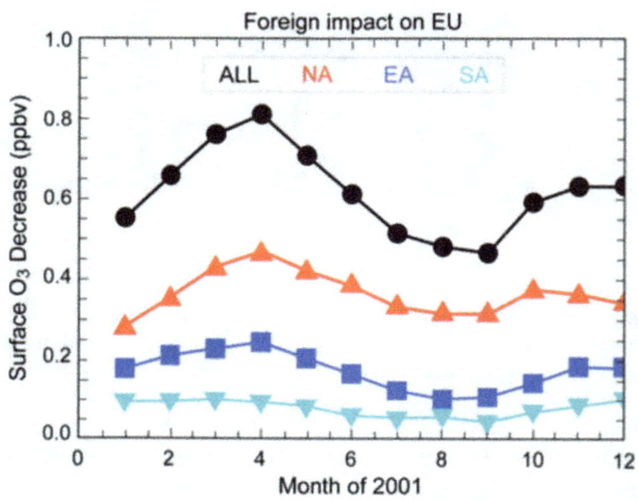

Fig. 16.8 Changes in European surface O_3 levels from 20 % reductions of anthropogenic emissions in North America (NA), East Asia (EA) and South Asia (SA). ALL is the combined effects of the contribution from the three foreign regions (adapted from HTAP 2010, see original report for more details)

of North American emissions on Europe using a tagging method. Their results are at the lower end of the HTAP estimates. The HTAP and Brandt et al. (2012) calculations are for different meteorological years. Located close to the western continental rim, the intercontinental contribution to the North Sea region is higher than the European average as shown by Jonson et al. (2006).

Particulate matter has a short residence time in the atmosphere, and as a result the intercontinental contribution to Europe is in general low. Calculating the ratio of the effect of other (than Europe) source regions to the effect of all source regions (including Europe), indicates that about 5 % of the PM surface concentrations in Europe can be attributed to intercontinental transport (HTAP 2010). However, the temporal variability is large, and the contribution can be significant for specific episodes.

16.3 Recent Change

16.3.1 Emissions

For most air pollutants emission totals reached a maximum in the late 1980s or early 1990s. Since then emissions of air-polluting substances have decreased substantially in most European countries. For parties to the Gothenburg Protocol[4] emission ceilings are set for 2010 for four pollutants: sulphur, NO_X, VOCs and NH_3. For EU Member States these emission ceilings are largely integrated within EU legislation. This is illustrated in Fig. 16.9, showing the evolution of emissions from countries within the EMEP domain (i.e. Europe, large parts of the North Atlantic and the polar basin and parts of North Africa; see www.EMEP.int for definition) from 1990 to 2010. Emissions of PM have only been reported since 2000. There are large differences in trends between individual countries, with some countries even increasing their emissions for one or more species. Parties to the Gothenburg Protocol whose emissions have a more severe impact, and whose emissions are relatively cheap to reduce are obliged to make the largest cuts in emissions. The countries around the North Sea are among those that have had to make the greatest cuts in emissions. Uncertainty in the emission trends is significant. The large drop in PM emissions from 2001 to 2002 may reflect incomplete reporting prior to 2002.

[4] www.unece.org/env/lrtap/multi_h1.html.

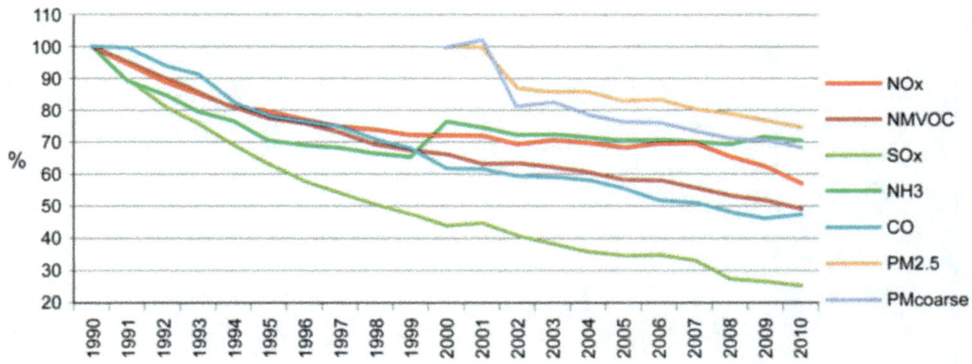

Fig. 16.9 Emission trends 1990–2010 (2000–2010 for $PM_{2.5}$ and PM_{coarse}) (Fagerli et al. 2012)

The North Sea region is strongly influenced by emissions from shipping. Traditionally, emissions from shipping have been largely unregulated. However, recent policy decisions through the International Maritime Organization (IMO MARPOL Annex VI SO_X Emission Control Area SECA requirements) and the EU (Sulphur Directive) affect ship emissions of both SO_X and PM in the region. The former restricts the marine fuel sulphur content in SECAs to 1.0 % as of 1 July 2010 and 0.1 % from 2015 whereas the latter requires ships to use 0.1 % sulphur fuel in harbour areas from 1 January 2010. Prior to 2010 the maximum allowed sulphur content in SECAs was 1.5 % as opposed to the global fleet average of about 2.4 %. Emissions of SO_X, NO_X, VOC, CO and PM from international shipping, from 1990 to present, are listed in appendix B in EMEP (2013).

16.3.2 Air Pollution

There are no PM_{10} or $PM_{2.5}$ measurements extending over decades. It is only for the past decade that enough data exist to derive trends. The monitoring network for $PM_{2.5}$ is sparser than for PM_{10} giving larger uncertainties for the reported $PM_{2.5}$ change. Over the period 2000–2009, Tørseth et al. (2012) found a decrease in PM_{10} and $PM_{2.5}$ concentration at about half of the European rural sites in the EMEP network. Average reductions in the annual means were 18 % (PM_{10}) and 27 % ($PM_{2.5}$). None of the stations in the network showed an increasing trend. Similar results were found by Barmpadimos et al. (2012) using selected EMEP stations corrected for meteorological variability. The trends roughly correspond to reported reductions in emissions of primary PM and precursors for secondary PM (Fig. 16.9). Figure 16.10 shows the change in PM_{10} for the past decade as reported by the European Environment Agency (EEA 2012). These data also include stations in urban surroundings and near roads with heavy traffic. Most of the stations registering a trend showed a decrease, with only 2 % of stations recording an increase. For countries adjacent to the North Sea, moderate decreases are found at most stations although some show no significant trend. There is also a reduction in the number of exceedances of the EU PM_{10} daily limit value for most North Sea countries (EEA 2012).

Devasthale et al. (2006) used satellite measurements complemented by station data to investigate trends in air polluting particles and focused on the English Channel and the top three polluting harbours in Europe. For the period 1997–2002 they found increasing particle concentrations over harbours and coastal areas and decreasing concentrations over land areas. The different evolution is attributed to decreased emissions from land-based sources and increased emissions from shipping. Jonson et al. (2015) and Brandt et al. (2013a) modelled the effects of Baltic Sea and North Sea ship emissions in 2009 and 2011 (before and after the reductions in the sulphur content of marine fuels from 1.5 to 1 % from 1 July 2010). The calculations indicate clear improvements in PM concentration. These are however slightly offset by increasing NO_X emissions, affecting nitrate particle formation. This is particularly the case in and around major North Sea ports owing to partial economic recovery after the financial crisis.

Ozone is strongly coupled to meteorological variability both in terms of regional photochemical production and loss, and the contribution from intercontinental transport. Trends are therefore difficult to detect and long time series are needed. In some regions, lack of long-term data makes trend analysis impossible. Reductions in the highest O_3 values are found (Fig. 16.11) in England, Benelux and Germany (EEA 2009, 2012; Tørseth et al. 2012) for the period 1990–2010, but mainly the 1990s. The frequency of high values has also decreased, especially in the Netherlands, England and Ireland (Tørseth et al. 2012).

Studies report that despite a relatively large decline in anthropogenic emissions in Europe (Fig. 16.9) a corresponding reduction in O_3 concentration is not observed (Jonson et al. 2006; EEA 2009, 2012; Tørseth et al. 2012; Wilson et al. 2012). Models also generally struggle to reproduce some of the observed trends (Solberg et al. 2005; Jonson et al. 2006; Colette et al. 2011; Wilson et al. 2012; Parrish et al. 2014). Likely reasons include increased

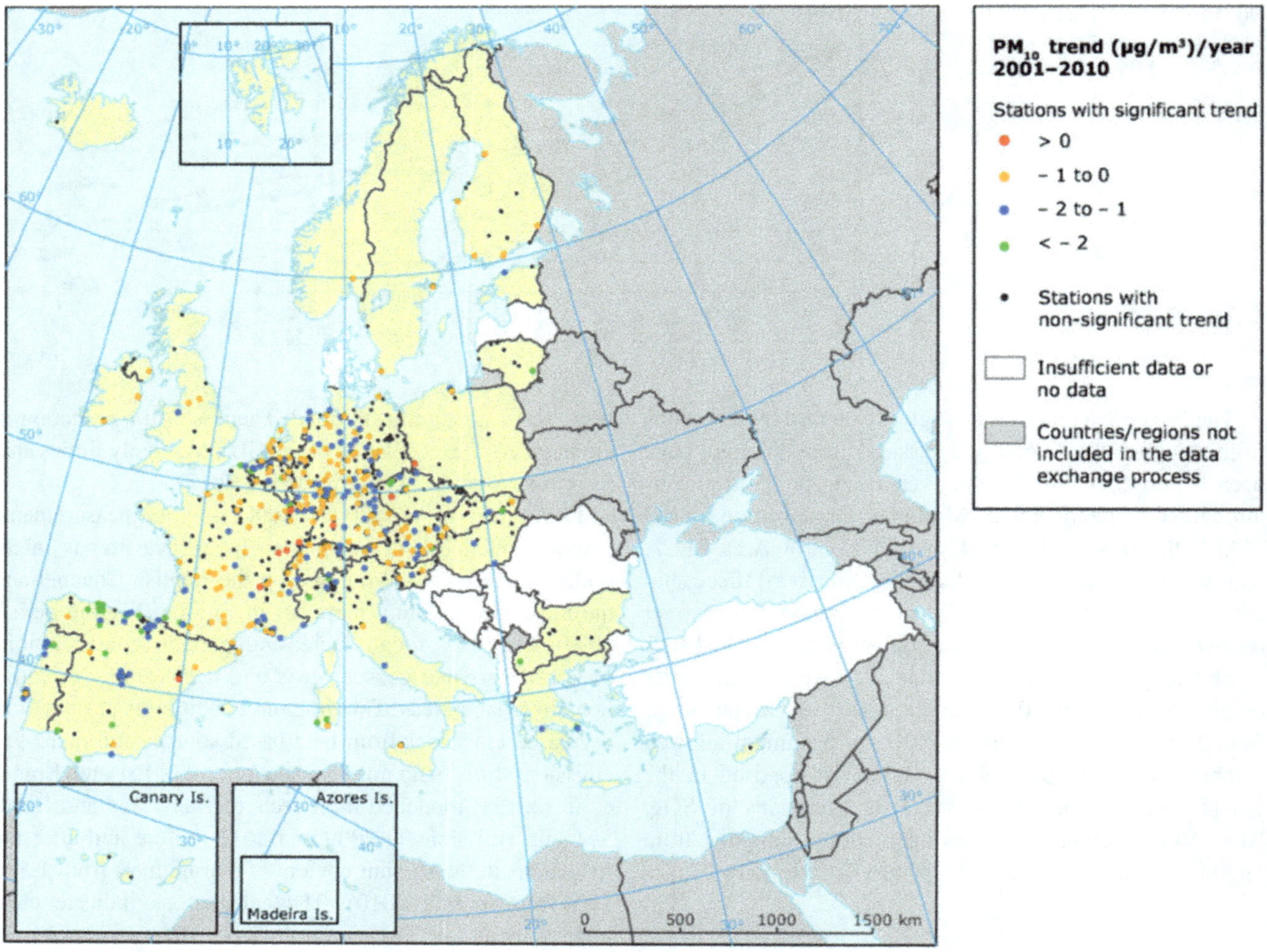

Fig. 16.10 Average annual change in surface PM_{10} concentration for the period 2001–2010 (EEA 2012). Only stations with a statistically significant trend are shown

background (hemispheric) O_3 level—observational evidence suggests that the increase in background O_3 roughly doubled from 1950 to 2000 and then levelled off (Logan et al. 2012; Derwent et al. 2013; Oltmans et al. 2013; Parrish et al. 2013). Other possibilities are limitations in the understanding of photochemistry, coarse model resolution, uncertainties related to anthropogenic emission estimates, variation in poorly constrained natural emissions, and the small number of measurement sites with long-term data sets.

16.3.3 Contribution from Climate Change

Climate change influences air pollution levels through a number of factors (see HTAP 2010), including changes in temperature, solar radiation, humidity, precipitation, atmospheric transport and biogenic emissions. Using a model to compare current conditions (1990–2009) against a baseline period (1961–1990) Orru et al. (2013) found the largest climate-driven increase in O_3-related mortality and hospitalisations to have occurred in Ireland, the UK, the Netherlands and Belgium where increases of up to 5 % are estimated. A decrease is estimated for the northernmost European countries. Hedegaard et al. (2012) compared the 1990s with the 1890s and found climate-driven decreases in surface O_3 in the North Sea region. However, the decrease is not statistically significant over the region as a whole.

Using an ensemble of global models, Silva et al. (2013) found the average number of premature annual deaths attributable to past (1850–2000) climate change in Europe to be 954 for O_3 (respiratory) and 11,900 for $PM_{2.5}$. But the magnitude and even the sign of the values varies between models. In a global study with one model, Fang et al. (2013) found a climate-driven contribution to cardiopulmonary and lung cancer mortality associated with industrial $PM_{2.5}$ since 1860 of up to 14 % over Europe. Ozone was responsible for a small contribution. The calculation does not include the climate-driven effect on emissions of biogenic hydrocarbons.

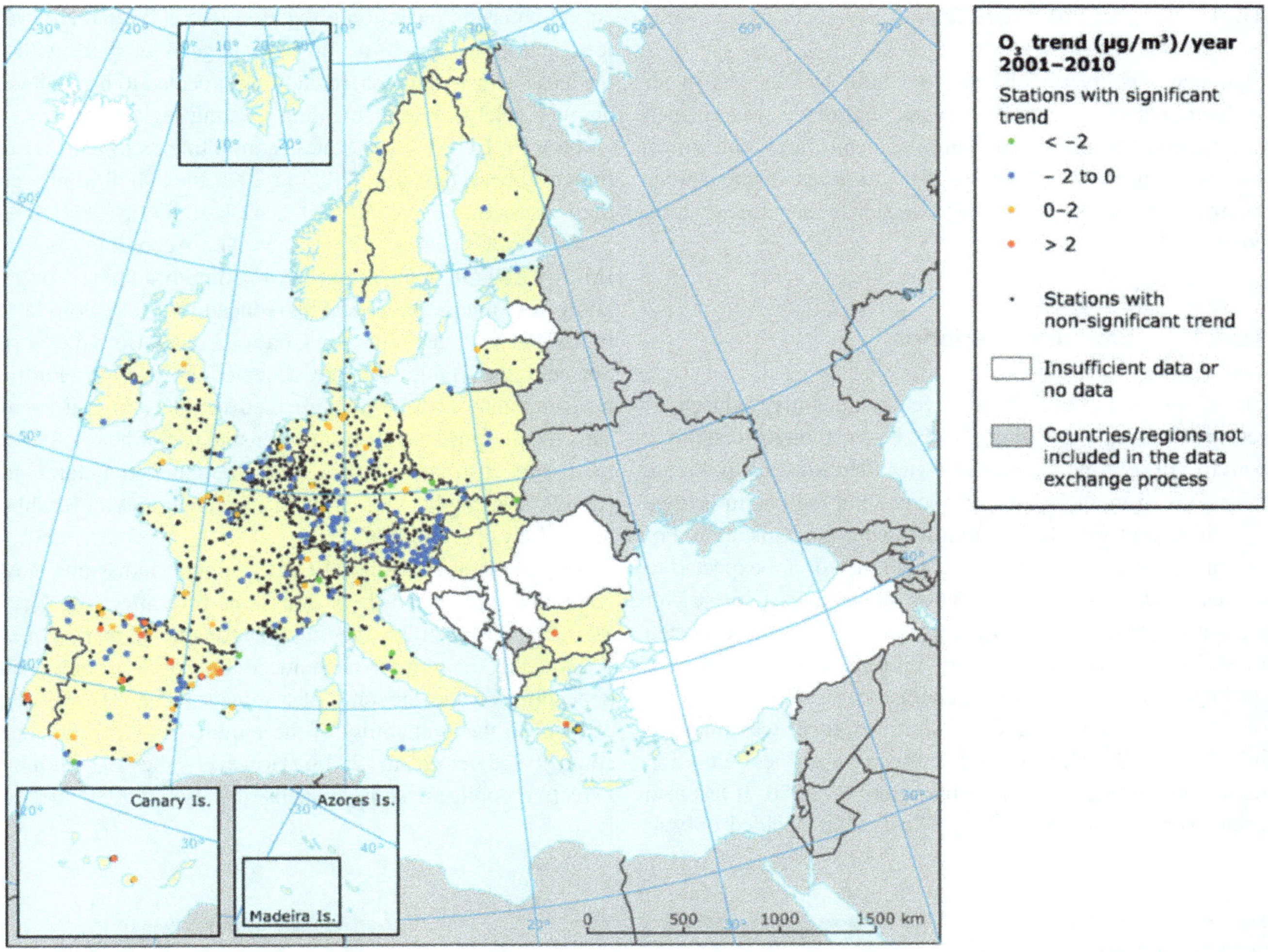

Fig. 16.11 Change in annual mean maximum daily 8-hour ozone (O_3) concentration in the period 2001–2010 (EEA 2012). Only stations (urban/suburban/rural) with a statistically significant trend are shown

Recent heat waves have been linked to climate change (e.g. Stott et al. 2004). Whether these anomalies are exceptional or a signal of changes in climate is still under debate. Summer 2003 was one of the hottest in the history of Western Europe, with surface temperature exceeding the average surface temperature reported for 1901–1995 by 2.4 °C. In fact, this summer was likely to have been warmer than any other back to 1500.

Fischer et al. (2004) estimated that almost half of the excess deaths in the Netherlands during the 2003 heat wave were due to increased air pollution (O_3 and PM_{10}). Stedman (2004) estimated that the same air pollutants were responsible for 21–38 % of excess deaths in the UK. Doherty et al. (2009) found the overall number of deaths attributable to O_3 in England and Wales to be slightly greater than that attributable to heat for the 2003, 2005 and 2006 summers. Several studies have described the high pollution levels observed in 2003 (Vautard et al. 2005; Solberg et al. 2008; Tressol et al. 2008).

The high temperatures during the 2003 heat wave influenced summer O_3 because of its link with high solar radiation, stagnation of the air masses and thermal decomposition of peroxyacetyl nitrate (PAN). Availability of solar radiation favours photolysis yielding radical formation with subsequent involvement in O_3 production. Stagnation of air masses allows the accumulation of pollutants in the planetary boundary layer (PBL) and in the residual layer during the night. Increased temperatures and solar radiation favoured biogenic emissions of isoprene with a potential for enhanced O_3 chemistry in the PBL. High temperature and a spring to summer precipitation deficit reduced dry deposition of O_3. The high temperatures and exceptional drought led to extensive forest fires on the Iberian Peninsula which contributed to the peak in ground level O_3 observed in western central Europe in August (Solberg et al. 2008; Tressol et al. 2008).

16.4 Future Impacts

Future air quality will be affected both by changes in air pollutant emissions and by changes in climate. A large span in emission scenarios and degree of detail in climate simulations are used in different studies. As a result, the studies referred to in the following sections are not directly comparable.

16.4.1 Emission Scenarios

Emissions of air pollutants have significantly reduced in recent decades (see Fig. 16.9). Even though economic activity in Europe is expected to increase, air pollutant emissions from all land-based sectors and regions in Europe are still expected to decline following current legislation. For example, emissions from the transport sector are expected to decrease owing to the penetration of vehicles with stricter emissions standards (Euro 5 and Euro 6), and the expected transition to renewable energy in Europe should drive a decline in emissions from the energy sector. After five years of negotiation, a revised Gothenburg Protocol was finalised in May 2012. The revised protocol specifies emission reduction commitments from base 2005 to 2020. It has also been extended to cover $PM_{2.5}$. Most states decided only to accept emission reduction obligations for 2020 that are even less ambitious than—or at best largely in line with—business-as-usual, that is, reductions expected to be achieved anyway solely by implementing existing legislation.

Overall, EU Member States' commitments to the revised protocol mean that from 2005 to 2020 they shall jointly cut their emissions by 59 % (SO_2), 42 % (NO_X), 6 % (NH_3), 28 % (VOCs) and 22 % ($PM_{2.5}$). According to the IMO MARPOL regulations the maximum sulphur content allowed in marine fuels will be reduced to 0.1 % from 2015 in SECAs (Sulphur Emission Control Areas). Both the North Sea and the Baltic Sea are accepted as SECAs. Further measures may also be implemented for NO_X in these seas. This may impose a shift from extensive use of heavy fuel oil to marine distillates, or a switch to liquefied natural gas (LNG), or using heavy fuel oil in combination with scrubber technology.

In sea areas outside SECAs, sulphur emissions have continued to rise and these emissions also affect the North Sea area. From 2020, the sulphur content in marine fuels outside SECAs should be reduced to 0.5 % globally, but depending on the outcome of a review to be concluded in 2018 as to the availability of the required fuel oil, this date could be deferred to 2025. However, the EU Sulphur Directive obliges ship owners to use 0.5 % fuel in

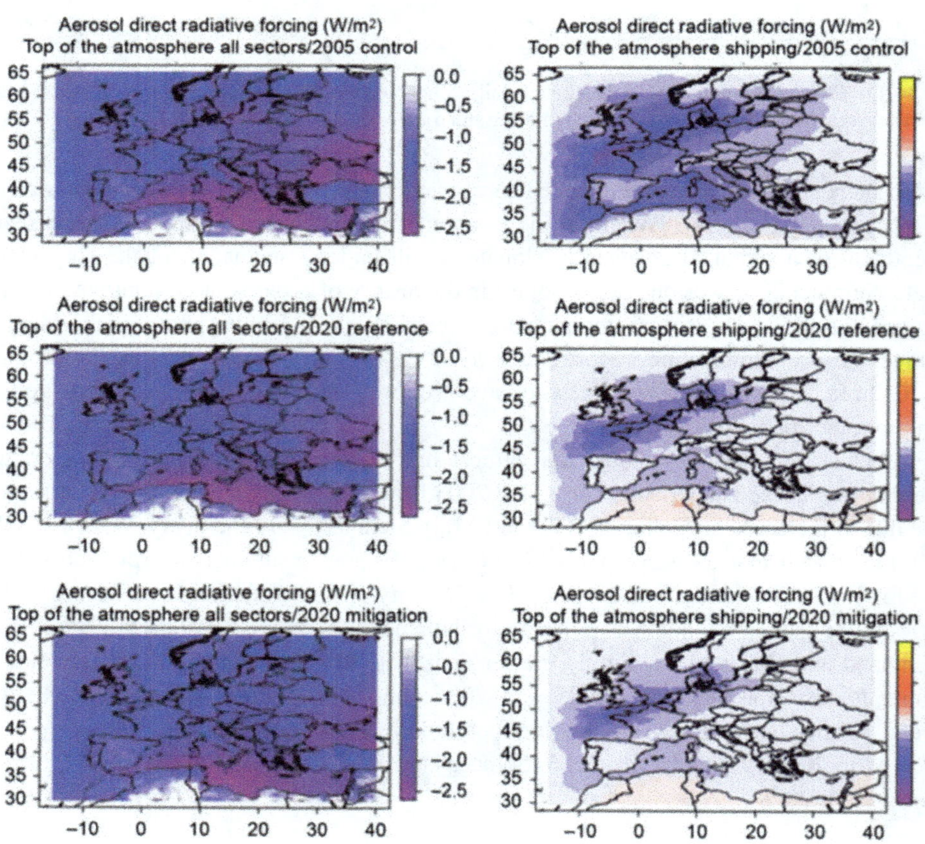

Fig. 16.12 Aerosol-induced direct radiative forcing at the top of the atmosphere in 2005 and for two projections: reference (2020) and mitigation (2020) (*left column*); Contribution attributed to shipping activities (*right column*). The reference scenario includes all current implemented and planned air quality policies. The so-called mitigation scenario in addition includes further climate policies leading to a stabilisation of global warming to not more than 2 °C in 2100. Both scenarios include mitigation measures for shipping that correspond to MARPOL regulations for NO_X and SO_2 (EEA 2013)

non-SECA EU sea areas from 1 January 2020 regardless of the outcome of the IMO review.

Efforts to improve air quality will undoubtedly influence climate. An interesting example is shown in Fig. 16.12. Large direct and indirect aerosol effects lead to a current global net cooling impact from the shipping sector. The aerosol direct and indirect (not shown) effects are likely to be substantially reduced over the North Sea in 2020. Refraining from air pollutant mitigation to favour a (potential) net cooling effect of the shipping sector is however risky from a health and environmental perspective and given large reported uncertainties, especially for the size of the climate impact of shipping aerosols.

To date, there are no NO_X Emission Control Areas (NECAs) in Europe. Hammingh et al. (2012) evaluated the potential impact of establishing a North Sea NECA. This would require new ships to emit 75 % less NO_X, from 2016 onward. A NECA in the North Sea would reduce total premature deaths due to air pollution in the North Sea countries by nearly 1 %, by 2030. This value would approximately double when all ships met the stringent nitrogen standards, a situation expected after 2040. Health benefits would exceed the costs to international shipping on the North Sea by a factor of two.

16.4.2 Impacts on Air Quality

16.4.2.1 Impacts from Emission Changes

Air quality in the North Sea region should improve as a result of expected reductions in emissions. The reductions in emissions should cause a decrease in PM levels. Based on a parameterisation of the HTAP source receptor calculations for the main source regions in the northern hemisphere, Wild et al. (2012) calculated future O_3 trends following the Intergovernmental Panel on Climate Change RCP (representative concentration pathway) emission scenarios. The calculations demonstrated that substantial annual mean surface O_3 reductions can be expected for most RCP scenarios by 2050 over most regions, including Europe. However, as discussed in Sect. 16.2.2 parts of the North Sea region are characterised by extensive NO_X titration of O_3. Unlike most other regions in Europe, reductions in NO_X emissions here are likely to result in increased levels of surface O_3, at least in the short term (Fig. 16.13). Colette et al. (2012) concluded that air pollution mitigation measures (present in both scenarios in Fig. 16.13) are the main factors leading to the net improvement over much of Europe, but an additional co-benefit of at least 40 % (depending on the indicator) is due to the climate policy. However the climate policy has little impact in the North Sea region (Fig. 16.13).

The total health-related external costs in Europe due to the total air pollution levels from all emission sources in the northern hemisphere are calculated to be EUR 803 billion $year^{-1}$ for 2000 decreasing to EUR 537 billion $year^{-1}$ in 2020 (Brandt et al. 2013a). The decrease is due to the general emission reductions in Europe provided that the revised Gothenburg Protocol is implemented and given the regulation of international ship traffic by introducing SECAs in the North Sea and Baltic Sea. For Denmark the external costs are estimated to be EUR 4.54 billion $year^{-1}$ for 2000, decreasing to EUR 2.53 billion $year^{-1}$ in 2020.

Using a baseline emission scenario, Amann et al. (2011) calculated that loss in statistical life expectancy attributed to exposure to $PM_{2.5}$ would decline between 2005 and 2020 from 7.4 to 4.4 months in the EU-27. There are significant improvements for the North Sea region (Fig. 16.14). The improvement in mortality due to ground level O_3 is about 35 % in EU Member States (Fig. 16.15) (Amann et al. 2011), with significant improvements in all North Sea countries. With commercially available emission control technologies, European emissions could be further reduced from baseline by 60 % (SO_2), 30 % (NO_X), 65 % (primary $PM_{2.5}$), and about 35 % (NH_3 and VOC). The measures would cut the loss in statistical life expectancy by 50 % (or another 2.5 months) compared to the baseline case in 2020.

However, the improvements come at a cost. Full implementation of the additional measures would increase costs for air pollution control in Europe in 2020 from EUR_{2005} 110 billion $year^{-1}$ to EUR_{2005} 192 billion $year^{-1}$, i.e. from

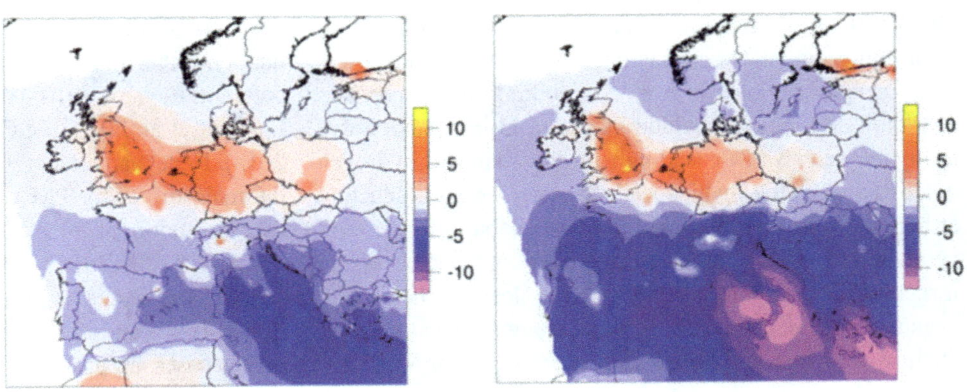

Fig. 16.13 Difference in surface O_3 between 2030 and 2005 for two scenarios: a reference case (*left*) and a sustainable climate policy case (*right*) (Colette et al. 2012)

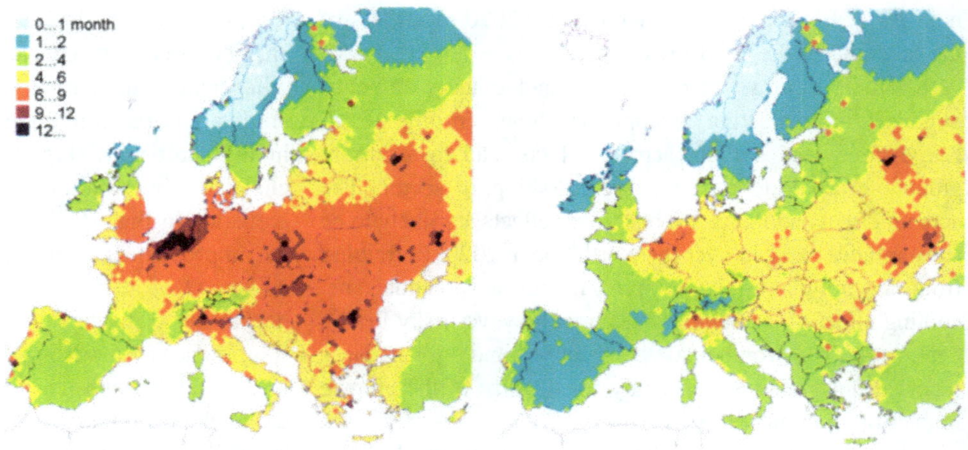

Fig. 16.14 Loss in statistical life expectancy attributable to exposure to $PM_{2.5}$ from anthropogenic sources. 2005 (*left*) and baseline projection for 2020 (*right*) (Amann et al. 2011)

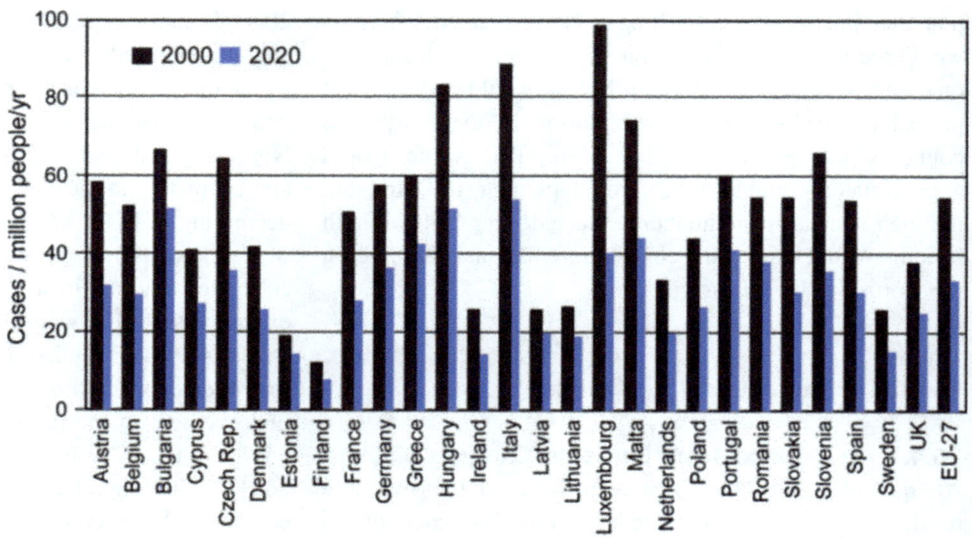

Fig. 16.15 Mortality rates attributable to exposure to ground-level O_3 (Amann et al. 2010)

0.66 % to 1.15 % of the GDP envisaged for 2020. At the same time, some of the measures achieve little environmental improvement. Experience shows that a cost-effectiveness analysis can identify portfolios that realise most of the potential improvements at a fraction of the costs of the total portfolio.

16.4.2.2 Impacts from Climate Change

Climate impacts on air pollution are summarised by HTAP (2010). Future changes in climate are expected to increase local and regional O_3 production and reduce O_3 in downwind receptor regions. Factors contributing to O_3 increases near emission regions include increased O_3 production due to higher water vapour leading to more abundant hydrogen oxide radicals (HO_X) which leads to increased O_3 production at high NO_X concentrations. Increased global average temperature increases photochemistry rates and decreases net formation of reservoir species for NO_X, leaving more NO_X available over source regions. This promotes local O_3 production. In a warmer climate natural emissions of VOC and NO_X (biogenic, lightning) are expected to increase. Such increases will depend on uncertain changes in soil moisture, cloud cover, sunlight, and biogenic responses to a more CO_2-rich atmosphere.

Although there are many factors affecting PM levels, changes in cloud amount and precipitation are the major parameters as wet removal is a major sink for PM. Despite several pathways by which climate change may influence air quality, most model simulations show air pollutant emissions to be the main factor driving change in future air quality, rather than climate or long-range transport (Andersson and Engardt 2010; HTAP 2010; Katragkou et al. 2011; Langner et al. 2012a, b; Coleman et al. 2013; Colette et al. 2013). Hedegaard et al. (2013) found emission changes to be the main driver for PM changes but that climate change is equally important for O_3 in many North Sea countries.

Orru et al. (2013) compared O_3-related mortality and hospitalisation due to climate change for a baseline period (1961–1990) and the future (2021–2050). Increases in O_3-related cases are projected to be greatest in Belgium, France,

Spain and Portugal (10–14 %), whereas in most Nordic and Baltic countries there is a projected decrease in O_3-related mortality of the same magnitude. Overall there is an increase of up to 13.7 % in O_3-related mortality in Europe, which corresponds to 0.2 % in all-cause total mortality and respiratory hospitalisations.

The sensitivity of simulated surface O_3 concentration to changes in climate between 2000–2009 and 2040–2049 differs by a factor of two between the models used in a study by Langner et al. (2012a), but the general pattern of change with an increase in southern Europe is similar across different models. Changes in isoprene emissions from deciduous forests vary substantially across models explaining some of the different climate response. In northern Europe, the ensemble mean for mean and daily maximum O_3 concentration both decrease whereas there are no reductions for the higher percentiles indicating that climate impacts on O_3 could be especially important in connection with extreme summer events such as experienced in summer 2003 (see Sect. 16.2.2). Some regional climate modelling studies suggest that conditions such as those of summer 2003 could become more frequent in coming decades (Beniston 2004; Schär et al. 2004).

Colette et al. (2013) and Hedegaard et al. (2013) found that climate change in the North Sea region would constitute a slight benefit for $PM_{2.5}$ concentrations. Other studies show both small increases and decreases of PM within the region (Nyiri et al. 2010; Manders et al. 2012). The spread of precipitation projections in regional climate models constitutes a major challenge in reducing the uncertainty of the climate impact on PM (Manders et al. 2012). Nevertheless, some conclusions can be drawn from the different climate model projections for the North Sea region (Jacob et al. 2014). Winter precipitation is expected to increase over the coming century, while summer precipitation is expected to decrease over much of the region. Heavy precipitation events are expected to occur more often in all seasons.

16.5 Conclusions

Climate and air quality interact in several ways. Emitted air pollutants could directly impact climate or could act as precursors for components acting both as harmful pollutants and climate forcers. On the other hand, air quality is sensitive to climate change since it perturbs the physical and chemical properties of the environment. Climate policies imply energy efficiency and technical measures that change emissions of air pollutants. Reciprocally, air quality mitigation measures affect greenhouse gas emissions. Mitigation optimised for a climate or air quality target in isolation could have synergistic or antagonistic effects.

In the North Sea area, the effects on air quality of emission changes since pre-industrial times are stronger than the effects of climate change over this period. Despite several pathways by which climate change may influence air quality, model simulations show air pollutant emissions to be the main factor driving change in future air quality in the North Sea region, rather than climate. The variation in climate simulations in different studies results in significant uncertainty in the impacts of climate change on air quality. This is particularly the case for PM where the spread of precipitation projections in regional climate models constitutes a major challenge in narrowing the uncertainty.

Climate impacts on air quality are substantial in connection with heat waves, such as that of summer 2003. Some regional climate models suggest that heat waves could become more frequent in the coming decades. If the anticipated reductions in emissions of air pollutants are not achieved, extreme weather events of this type may cause severe problems in the future.

References

Amann M, Borken J, Bottcher H, Cofala J, Hettelingh J-P, heyes C, Holland M, hunt A, Klimont Z, Mantzos L, Ntziachristos L, Obersteiner M, Posch M, Schneider U, Schopp W, Slootweg J, Witzke P, Wagner A, Winiwarter W (2010) Greenhouse gases and air pollutants in the European Union: Baseline projections up to 2030. EC4MACS Interim Assessment Report, International Institute for Applied Systems Analysis (IIASA)

Amann M, Bertok I, Borken-Kleefeld J, Cofala J, Heyes C, Höglund-Isaksson L, Klimont Z, Nguyen B, Posch M, Rafaj P, Sandler R, Schöpp W, Wagner F, Winiwarter W (2011) Cost-effective control of air quality and greenhouse gases in Europe: Modeling and policy applications. Env Modell Softw 26:1489–1501

Andersson C, Engardt M (2010) European ozone in a future climate: importance of changes in dry deposition and isoprene emissions. J Geophys Res 115, doi:10.1029/2008JD011690

Anenberg SC, Horowitz LW, Tong DQ, West JJ (2010) An estimate of the global burden of anthropogenic ozone and fine particulate matter on premature human mortality using atmospheric modeling. Env Health Persp 118:1189–1195

Barmpadimos I, Keller J, Oderbolz D, Hueglin C, Prévôt ASH (2012) One decade of parallel fine ($PM_{2.5}$) and coarse (PM_{10}–$PM_{2.5}$)

particulate matter measurements in Europe: trends and variability. Atmos Chem Phys 12:3189–3203

Beniston M (2004) The 2003 heat wave in Europe: A shape of things to come? An analysis based on Swiss climatological data and model simulations. Geophys Res Lett 31, doi:10.1029/2003GL018857

Blunden J, Arndt DS, Baringer MO (2011) State of the climate in 2010. Bull Am Met Soc 92:S1–S266

Brandt J, Silver JD, Frohn LM, Geels C, Gross A, Hansen AB, Hansen KM, Hedegaard GB, Skjøth CA, Villadsen H, Zare A, Christensen JH (2012) An integrated model study for Europe and North America using the Danish Eulerian Hemispheric Model with focus on intercontinental transport. Atmos Env 53:156–176

Brandt J, Silver JD, Christensen JH, Andersen MS, Bønløkke JH, Sigsgaard T, Geels C, Gross A, Hansen AB, Hansen KM, Hedegaard GB, Kaas E, Frohn LM (2013a) Assessment of past, present and future health-cost externalities of air pollution in Europe and the contribution from international ship traffic using the EVA model system. Atmos Chem Phys 13:7747–7764

Brandt J, Silver JD, Christensen JH, Andersen MS, Bønløkke JH, Sigsgaard T, Geels C, Gross A, Hansen AB, Hansen KM, Hedegaard GB, Kaas E, Frohn LM (2013b) Contribution from the ten major emission sectors in Europe and Denmark to the health-cost externalities of air pollution using the EVA model system – an integrated modelling approach. Atmos Chem Phys 13:7725–7746

Coleman L, Martin D, Varghese S, Jennings SG, O'Dowd CD (2013) Assessment of changing meteorology and emissions on air quality using a regional climate model: Impact on ozone. Atmos Env 69:198–210

Colette A, Granier C, Hodnebrog Ø, Jakobs H, Maurizi A, Nyiri A, Bessagnet B, D'Angiola A, D'Isidoro M, Gauss M, Meleux F, Memmesheimer M, Mieville A, Rouïl L, Russo F, Solberg S, Stordal F, Tampieri F (2011) Air quality trends in Europe over the past decade: a first multi-model assessment. Atmos Chem Phys 11:11657–11678

Colette A, Granier C, Hodnebrog Ø, Jakobs H, Maurizi A, Nyiri A, Rao S, Amann M, Bessagnet B, D'Angiola A, Gauss M, Heyes M, Klimont Z, Meleux F, Memmesheimer M, Mieville A, Rouïl L, Russo F, Schucht S, Simpson D, Stordal F, Tampieri F, Vrac MC (2012) Future air quality in Europe: a multi-model assessment of projected exposure to ozone. Atmos Chem Phys 12:10613–10630

Colette A, Bessagnet B, Vautard R, Szopa S, Rao S, Schucht S, Klimont Z, Menut L, Clain G, Meleux F, Curci G, Rouïl L (2013) European atmosphere in 2050, a regional air quality and climate perspective under CMIP5 scenarios. Atmos Chem Phys 13:7451–7471

Devasthale A, Krüger O, Graßl, H (2006) Impact of ship emissions on cloud properties over coastal areas. Geophys Res Lett 33, doi:10.1029/2005GL024470

Derwent RG, Manning AJ, Simmonds PG, Gerard Spain T, O'Doherty S (2013) Analysis and interpretation of 25 years of ozone observations at the Mace Head Atmospheric Research Station on the Atlantic Ocean coast of Ireland from 1987 to 2012. Atmos Env 80:361–368

Doherty RM, Heal MR, Wilkinson P, Pattenden S, Vieno M, Armstrong B, Atkinson R, Chalabi Z, Kovats S, Milojevic A, Stevenson DS (2009) Current and future climate- and air pollution mediated impacts on human health. Env Health 8, doi:10.1186/1476-069X-8-S1-S8

EEA (2009) Assessment of Ground-level Ozone in EEA Member Countries, with a Focus on Long-term Trends. EEA Tech Rep 7/2009. European Environment Agency

EEA (2011) Air Pollution by Ozone across Europe during Summer 2010. EEA Tech Rep 6/2011. European Environment Agency

EEA (2012) Air Quality in Europe: 2012 report. EEA Rep 4/2012. European Environment Agency

EEA (2013) The Impact of International Shipping on European Air Quality and Climate Forcing. EEA Tech Rep 4/2013. European Environment Agency

EEA (2014) Air Pollution by Ozone across Europe during Summer 2013. EEA Tech Rep 3/2014. European Environment Agency

EMEP (2013) Transboundary Acidification, Eutrophication and Ground Level Ozone in Europe in 2011. EMEP Rep 1/2013. Oslo

EU (2013) Commission staff working document impact assessment. Accompanying the documents: Communication from the Commission to the Council, the European Parliament, the European Economic and Social Committee and the Committee of the Regions -a Clean Air Programme for Europe. Proposal for a Directive of the European Parliament and of the Council on the limitation of emissions of certain pollutants into the air from medium combustion plants. Proposal for a Directive of the European Parliament and of the Council on the reduction of national emissions of certain atmospheric pollutants and amending Directive 2003/35/EC. Proposal for a Council Decision on the acceptance of the Amendment to the 1999 Protocol to the 1979 Convention on Long-Range Transboundary Air Pollution to Abate Acidification, Eutrophication and Ground-level Ozone, http://ec.europa.eu/environment/archives/air/pdf/Impact_assessment_en.pdf

Fagerli H, Nyiri A, Jonson JE, Gauss M, Benedictow A, Stensen BM, Hjellbrekke A-G, Mareckova K, Wankemüller R (2012) Status of transboundary pollution in 2010. In: Transboundary Acidification, Eutrophication and Ground Level Ozone in Europe in 2010. EMEP Rep 1. Oslo

Fang Y, Naik V, Horowitz LW, Mauzerall DL (2013) Air pollution and associated human mortality: the role of air pollutant emissions, climate change and methane concentration increases from the preindustrial period to present. Atmos Chem Phys 13:1377–1394

Fischer PH, Brunekreef B, Lebret E (2004) Air pollution related deaths during the 2003 heat wave in the Netherlands. Atmos Env 38:1083–1085

Gauss M, Nyiri A, Steensen BM, Klein H (2012) Transboundary air pollution by main pollutants (S, N, O_3) and PM in 2010. EMEP/MSC-W, Data Note

Hammingh P, Holland M, Geilenkirchen G, Jonson J, Maas R (2012) Assessment of the Environmental Impacts and Health Benefits of a Nitrogen Emission Control Area in the North Sea. PBL Netherlands Environmental Assessment Agency, The Hague/Bilthoven 2012. 2194721960

Hedegaard GB, Christensen JH, Geels C, Gross A, Hansen KM, May W, Zare A, Brandt J (2012) Effects of changed climate conditions on tropospheric ozone over three centuries. Atmos Clim Sci 2:546–561

Hedegaard GB, Christensen JH, Brandt J (2013) The relative importance of impacts from climate change vs. emissions change on air pollution levels in the 21st century. Atmos Chem Phys 13:3569–3585

HTAP (2010) Hemispheric transport of air pollution 2010. Part A: Ozone and particulate matter. Task Force on Hemispheric Transport of Air Pollution, UN Economic Commission for Europe, Geneva

Jacob D, Petersen J, Eggert B, Alias A, Christensen OB, Bouwer LM, Braun A, Colette A, Déqué M, Georgievski G, Georgopoulou E, Gobiet A, Nikulin G, Haensler A, Hempelmann N, Jones C, Keuler K, Kovats S, Kröner N, Kotlarski S, Kriegsmann A, Martin E, van Meijgaard E, Moseley C, Pfeifer S, Preuschmann S, Radtke K, Rechid D, Rounsevell M, Samuelsson P, Somot S, Soussana J-F, Teichmann C, Valentini R, Vautard R, Weber B (2014) EURO-CORDEX: New high-resolution climate change projections for European impact research. Reg Env Chan 14:563–578

Jonson JE, Simpson D, Fagerli H, Solberg S (2006) Can we explain the trends in European ozone levels? Atmos Chem Phys 6:51–66

Jonson JE, Jalkanen JP, Johansson L, Gauss M, Denier van der Gon HAC (2015) Model calculations of the effects of present and future emissions of air pollutants from shipping in the Baltic Sea and the North Sea. Atmos Chem Phys 15:783–798

Katragkou E, Zanis P, Kioutsioukis I, Tegoulias I, Melas D, Krüger BC, Coppola E (2011) Future climate change impacts on summer surface ozone from regional climate-air quality simulations over Europe. J Geophys Res Atmos 116 doi:10.1029/2011JD015899

Langner J, Engardt M, Baklanov A, Christensen JH, Gauss M, Geels C, Hedegaard GB, Nuterman R, Simpson D, Soares J, Sofiev M, Wind P, Zakey A (2012a) A multi-model study of impacts of climate change on surface ozone in Europe. Atmos Chem Phys 12:10423–10440

Langner J, Engardt M, Andersson C (2012b) European summer surface ozone 1990–2100. Atmos Chem Phys 12:10097–10105

Logan JA, Staehelin J, Megretskaia IA, Cammas JP, Thouret V, Claude H, De Backer H, Steinbacher M, Scheel H-E, Stubi R, Frohlich M, Derwent R (2012) Changes in ozone over Europe: Analysis of ozone measurements from sondes, regular aircraft (mozaic) and alpine surface sites. J Geophys Res Atmos 117 doi:10.1029/2011JD016952

Manders AMM, van Meijgaard E, Mues AC, Kranenburg R, van Ulft LH, Schaap M (2012) The impact of differences in large-scale circulation output from climate models on the regional modeling of ozone and PM. Atmos Chem Phys 12:9441–9458

Nyiri A, Gauss M, Tsyro S, Wind P, Haugen JE (2010) Future Air Quality, including Climate Change. In: Transboundary Acidification, Eutrophication and Ground Level Ozone in Europe in 2008. EMEP Rep, Oslo

Oltmans S, Lefohn A, Shadwick D, Harris J, Scheel H, Galbally I, Tarasick D, Johnson B, Brunke, E, Claude H, Zeng G, Nichol S, Schmidlin F, Davies J, Cuevas E, Redondas A, Naoe H, Nakano T, Kawasato T (2013) Recent tropospheric ozone changes - A pattern dominated by slow or no growth. Atmos Env 67:331–351

Orru H, Andersson C, Ebi KL, Langner J, Astrom C, Forsberg B (2013) Impact of climate change on ozone-related mortality and morbidity in Europe. Eur Respir J 41:285–94

Parrish DD, Law KS, Staehelin J, Derwent R, Cooper OR, Tanimoto H, Volz-Thomas A, Gilge S, Scheel H-E, Steinbacher M, Chan E (2013) Lower tropospheric ozone at northern mid-latitudes: Changing seasonal cycle. Geophys Res Lett 40:1631–1636

Parrish DD, Lamarque J-F, Naik V, Horowitz L, Shindell DT, Staehelin J, Derwent R, Cooper OR, Tanimoto H, Volz-Thomas A, Gilge S, Scheel H-E, Steinbacher M, Fröhlich M (2014) Long-term changes in lower tropospheric baseline ozone concentrations: Comparing chemistry-climate models and observations at northern mid-latitudes. J Geophys Res Atmos 119:5719–5736

Schär C, Vidale PL, Lüthl D, Frei C, Häberli C, Liniger MA, Appenzeller C (2004) The role of increasing temperature variability in European summer heatwaves. Nature 427:332–336

Silva RA, West JJ, Zhang Y and 27 others (2013) Global premature mortality due to anthropogenic outdoor air pollution and the contribution of past climate change. Environ Res Lett doi:10.1088/1748-9326/8/3/034005

Solberg S, Bergström R, Langner J, Laurila T, Lindskog A (2005) Changes in Nordic surface ozone episodes due to European emission reductions in the 1990 s. Atmos Env 39:179–192

Solberg S, Hov Ø, Søvde A, Isaksen ISA, Coddeville P, De Backer H, Forster C, Orsolini Y, Uhse K (2008) European surface ozone in the extreme summer 2003. J Geophys Res 113 doi:10.1029/2007JD009098

Stedman JR (2004) The predicted number of air pollution related deaths in the UK during the August 2003 heat wave. Atmos Env 38:1087–1090

Stott PA, Stone DA, Allen MR (2004) Human contribution to the European heat wave of 2003. Nature 432:610–614

Tørseth K, Aas W, Breivik K, Fjæraa AM, Fiebig M, Hjellbrekke AG, Lund Myhre C, Solberg S, Yttri KE (2012) Introduction to the European Monitoring and Evaluation Programme (EMEP) and observed atmospheric composition change during 1972–2009. Atmos Chem Phys 12:5447–5481

Tressol M, Ordóñez C, Zbinden R, Brioude J, Thouret V, Mari C, Nedelec P, Cammas J-P, Smit H, Patz H-W, Volz-Thomas A (2008) Air pollution during the 2003 European heat wave as seen by MOZAIC airliners. Atmos Chem Phys 8:2133–2150

Tsyro S, Yttri KE, Aas W (2012) Measurement and model assessment of particulate matter in Europe in 2010. In: Transboundary Particulate Matter in Europe, EMEP Rep 4, Oslo

Vautard R, Honoré C, Beekmann M, Rouil L (2005) Simulation of ozone during the August 2003 heat wave and emission control scenarios. Atmos Env 39:2957–2967

Watkiss P, Pye S, Holland M (2005) CAFE CBA: Baseline Analysis 2000 to 2020. Service Contract for Carrying out Cost-Benefit Analysis of Air Quality Related Issues, in particular in the clean Air for Europe (CAFE) Programme, April 2005. Online at: www.cafe-cba.org/assets/baseline analysis 2000-2020 05-05.pdf

Wild O, Fiore AM, Shindell DT, Doherty RM, Collins WJ, Dentener FJ, Schultz MG, Gong S, Mackenzie IA, Zeng G, Hess P, Duncan BN, Bergmann DJ, Szopa S, Jonson JE, Keating TJ, Zuber A (2012) Modelling future changes in surface ozone: a parameterized approach. Atmos Chem Phys 12:2037–2054

Wilson RC, Fleming ZL, Monks PS, Clain G, Henne S, Konovalov IB, Szopa S, Menut L (2012) Have primary emission reduction measures reduced ozone across Europe? An analysis of European rural background ozone trends 1996–2005. Atmos Chem Phys 12:437–454

17 Socio-economic Impacts—Recreation

Edgar Kreilkamp, Nele Marisa von Bergner and Claudia Mauser

Abstract

Tourism is one of the most highly climate-sensitive economic sectors. Most of its main sub-sectors, including sun-and-beach tourism and nature-based tourism, play a major role in the North Sea region and are especially weather–und climate-dependent. On top of that, most tourist activities in the North Sea region occur in the coastal zones which are highly vulnerable to the impacts of climate change. Climate acts as both a 'push' and 'pull' factor in tourism. Climate-driven changes in tourism demand are hard to determine because the tourist decision-making process is also influenced by factors other than climate. Nevertheless, summer tourism in the North Sea region is expected to benefit from rising temperatures (air and water), decreasing precipitation and longer seasons. Destinations can reduce the negative impacts of climate change on tourism by adapting to the changes. The tourist industry also contributes to climate change. Not only is the tourist industry affected by climate change, it also contributes to climate change itself. Therefore, mitigating the climate effects of tourism is largely the responsibility of politicians, the tourism industry and tourism supply. Despite some negative impacts, the overall consequences of climate change for tourism in the North Sea region are expected to be positive.

17.1 Introduction

The United Nations World Tourism Organization states that tourism is one of the most highly climate-sensitive (and even in some cases climate-dependent) economic sectors. Climate change is therefore a major challenge for global tourism (UNWTO 2009; von Bergner and Lohmann 2014). The main sub-sectors include sun-and-beach tourism, sports tourism, adventure tourism, nature-based tourism, cultural tourism, urban tourism, health and wellness tourism, cruises, theme parks, visiting friends and relatives, and meetings and conferences (Scott and Lemieux 2010). Most of these are weather- and climate-dependent and play a major role in the North Sea region.

17.2 Literature Review

The complex relationship between climate change and tourism has been part of the academic debate since the 1980s (Fig. 17.1). The first papers concerning climate change and tourism were published in 1986 (Harrison et al. 1986; Wall et al. 1986). Since then, the number of publications in this field has increased steadily (Fig. 17.2) up to 83 publications in 2011. From a review of literature between 1986 and 2012, Becken (2013) concluded that half of the studies concerned climate impacts on tourism, 34 % dealt with mitigation, and the remaining 16 % were policy papers or integrative papers.

E. Kreilkamp (✉)
Institute of Corporate Development, Leuphana University of Lüneburg, Lüneburg, Germany
e-mail: edgar.kreilkamp@uni.leuphana.de

N.M. von Bergner
Institute for Tourism and Leisure ITF, HTW Chur, Chur, Switzerland
e-mail: nelemarisa.vonbergner@htwchur.ch

C. Mauser
IFOK GmbH, Bensheim, Germany
e-mail: claudiamauser@gmx.de

M. Quante and F. Colijn (eds.), *North Sea Region Climate Change Assessment*, Regional Climate Studies, DOI 10.1007/978-3-319-39745-0_17

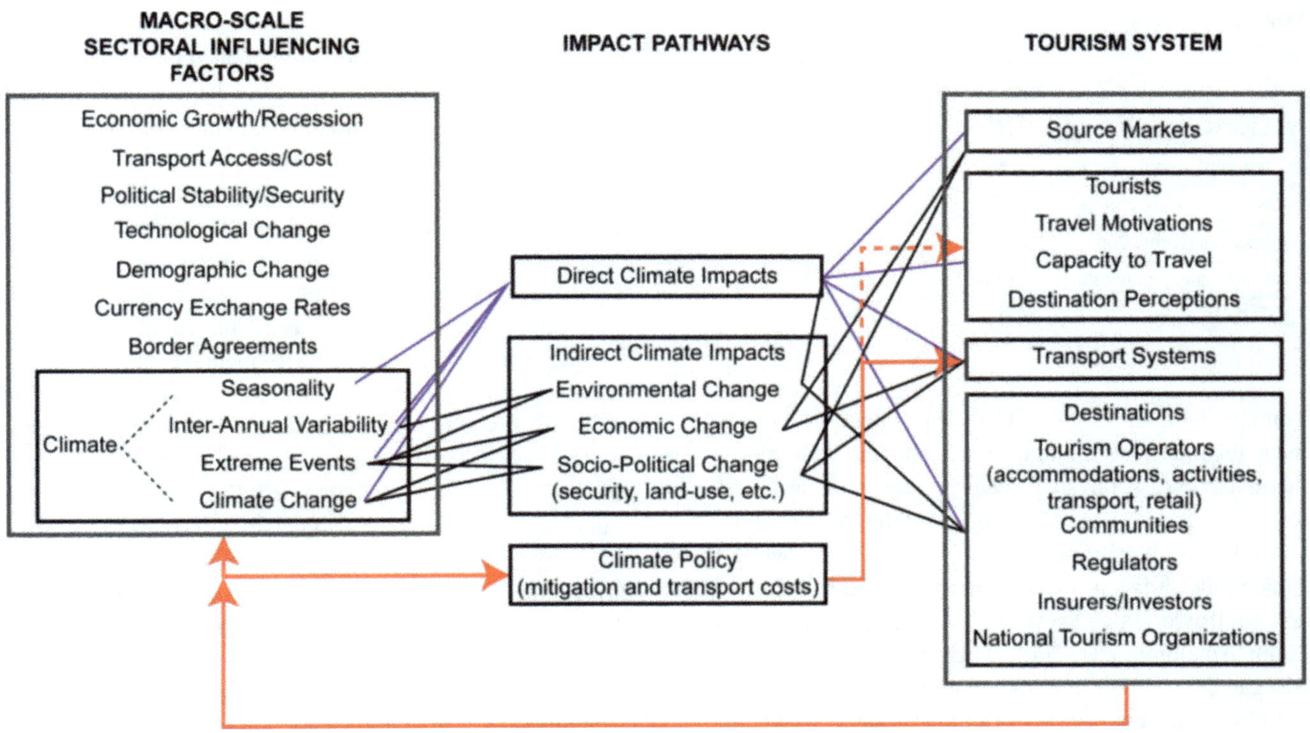

Fig. 17.1 Relationship between climate change and tourism (Scott and Lemieux 2010)

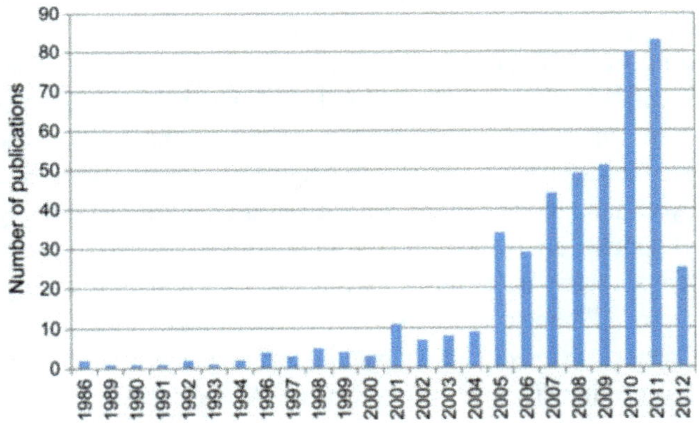

Fig. 17.2 Number of peer-reviewed publications concerning climate change and tourism produced per year between 1986 and 2012 (Becken 2013)

Although climate is not the only determinant of destination choice (Crouch 1995; Witt and Witt 1995; Rosselló et al. 2005; Gössling and Hall 2006; Bigano et al. 2006a), attractiveness is largely determined by thermal environmental assets (Smith 1993; Agnew and Palutikof 2000; Amelung and Viner 2006; for a detailed literature overview on how climate/weather and tourism interact, see Becken 2010). Destinations with better climate resources have a competitive advantage (Perch-Nielsen et al. 2010), especially those for sun-and-sea or winter sports holidays. In tourism, climate acts as both a 'push' and 'pull' factor. A push factor is one where the choice of travel destination is often related to the weather and climate conditions at the point of origin and not just at the holiday region. For example, Hill (2009) found that very rainy weather throughout much of the early summer in the United Kingdom resulted in an increase in foreign holiday bookings abroad compared to the previous year. But climate may also act as a pull factor. In Norway, 84 % of tour operators go to 'sun destinations' (Jorgensen and Solvoll 1996). This is not a recent phenomenon. Even in 1999, an annual survey of German traveller behaviour and tourism-related attitudes showed that 43 % of Germans mentioned weather as the most important factor when choosing a holiday destination (Lohmann and Kaim 1999). Nevertheless, preferences or perceptions of climate differ according to several factors, such as age, cultural and climate contexts, as well as leisure activities or the media, and are therefore hard to predict (Lise and Tol 2002; overview of literature by Gössling et al. 2012, Scott et al. 2012a).

One means of quantifying these preferences is through a 'climate index'; the aim being to provide a measure of the integrated effects of the atmospheric environment on a particular location. This would be useful both for tourists and for the tourism industry to evaluate the potential of tourism in a given area in terms of its perceived climate (de Freitas et al. 2008). The index approach can also be used to analyse the impact of climate change on the climatic attractiveness of tourist destinations (Hamilton 2005). One of the most used indices is that of Mieczkowski (1985), who developed the

Table 17.1 Priority levels for climate aspects (Morgan et al. 2000)

Climate aspect	Relative priority scores (out of 100 for aspects 1–4)
Windiness	26
Absence of rain	29
Sunshine	27
Temperature sensation	18
Bathing water temperature (22–26 °C)	(28)

Tourism Climate Index (TCI). This integrates several climate features into a single index and includes ratings for thermal comfort, physical features (e.g. rain) and aesthetic features (e.g. sunshine duration). Although Mieczkowski's index has been criticised by de Freitas (2003) owing to its subjectivity, it is still being used or adapted by others (Scott and McBoyle 2001; Scott et al. 2004; Amelung and Viner 2006; Amelung et al. 2007; Nicholls and Amelung 2008). Morgan et al. (2000) modified the index to fit beach users in the UK, using five aspects of climate (Table 17.1).

Matzarakis (2007) used the index to develop a Climate Tourism Information Scheme (CTIS) that includes parameters such as cold stress, heat stress and snow fall (i.e. skiing potential). This approach was also used for regional simulations of future conditions along the German North Sea coast (based on two regional climate models—REMO and CLM) that takes into account local-scale differences between the mainland coast and islands (Endler and Matzarakis 2010).

This chapter reviews information on climate change and its impact on recreation and tourism in the North Sea region. Because this is concentrated along the coast, the focus of this chapter is on the impacts of climate change on coastal tourism.

17.3 Impacts of Climate Change

Coastal tourism is the largest component of the tourism industry worldwide (IPCC 2014a). At the same time, coastal zones are especially vulnerable to the impacts of climate change. Even so, until 2012 few studies had addressed the topic of coastal tourism and climate change (Becken 2013). Despite some studies for the Caribbean or Mediterranean Sea and a few other beach or island regions (Nicholls and Hoozemans 1996; Lohmann 2002; Giupponi and Shechter 2003; Nicholls and Klein 2003; Perry 2005, 2006; Amelung and Viner 2006; Hein 2007; Giannakopoulos et al. 2009; Lemelin et al. 2010; Perch-Nielsen 2010; Becken et al. 2011; Jones and Phillips 2011; Scott et al. 2012c) there has been very little published on climate change impacts on tourism in the North Sea region. Those studies that do exist are all discussed in this chapter. Figure 17.2 shows the recent increase in publications on the impact of climate change on coastal tourism. Some of these studies address the response of tourists and tour operators to beach erosion and the tourist's concern about aesthetic appearance (Moreno and Becken 2009; Buzinde et al. 2010), while others examine the vulnerability of coastal tourism infrastructure to sea-level rise (Phillips and Jones 2006; Bigano et al. 2008; Schleupner 2008).

Negative effects of climate change include rising sea levels and extreme weather. Extreme storms and waves together with sea-level rise will increase the extent and frequency of flooding, storm surges and coastal erosion. Not only will this affect natural areas used by tourists, but also cultural assets and tourism infrastructure, especially transportation and accommodation (Phillips and Jones 2006; Amelung and Viner 2007; Scott et al. 2008). From a study of beach tourism in East Anglia, Coombers et al. (2009) showed that although sea-level rise would reduce the width of the beach and cause a possible reduction in visitor numbers, this effect could be outweighed by increased visitation due to better temperatures. However, overall, the economic costs of negative climate change impacts on coastal tourism could become extremely high (IPCC 2014a).

Positive effects of climate change on tourism have also been predicted. The North Sea coastal region has a maritime climate, which means mild winters and relatively warm summers. Climate projections suggest fewer cold stress events in winter, and less significant changes in heat stress events in summer compared to other regions such as the Mediterranean. The higher average temperatures projected by climate models imply a positive effect on the well-being of tourists in the North Sea region. For example, higher temperatures in summer may result in a longer (bathing) season (Pinnegar et al. 2006; Nicholls and Amelung 2008). Changes in precipitation patterns are expected to result in dryer summers and wetter winters. A decrease in summer precipitation may attract more tourists to the North Sea coastal areas. More rain and extreme weather events in winter could reduce the number of visitors in the low season.

Sea-level rise will become a major threat in the North Sea region (see Chap. 6), especially for low-altitude islands with limited tidal range, and coastal areas are particularly vulnerable to extreme weather events (Moreno and Becken 2009). Storm-surge height and the frequency of extreme wave events are expected to increase over large areas—especially in winter (IPCC 2014a). The IPCC cites an increase in future flood losses along the North Sea coast (IPCC 2014b), which may also affect the tourism industry. Even though tourist destinations recover relatively quickly from such disasters, damage to infrastructure and buildings will result in additional costs. Adaptation to climate change will also have economic impacts: Hamilton (2006) showed that protection measures such as longer dikes have a negative impact on accommodation prices along the German sea coast.

Braun et al. (1999) investigated combined scenarios of temperature and precipitation change with sea-level rise and beach loss and their effect on the number of tourists travelling to the Baltic and North Sea coasts of Germany. They concluded that the likelihood of choosing the north German coast for a holiday was slightly higher with increased temperatures. However, for scenarios with potentially negative impacts on the German coasts, such as beach erosion, the likelihood of visiting was substantially lower, even if with adaptations such as greater setback of tourism infrastructure or more diversified outdoor activities. Lohmann (2002) concluded that the effects of climate change in the North Sea area, such as sea-level rise, were likely to destroy infrastructure and that frequent extreme weather events may discourage visitors.

Infrastructure at ports and marinas is a major asset at many destinations. Sea-level rise, coastal erosion and storms might compromise its functionality. There is still a lack of detailed academic research on this topic for the North Sea region.

For the German North Sea coastline, sea-level rise is expected to extend the tides (i.e. lengthen flood duration on mudflats) hampering tidal flat walks for tourists (Regierungskommission Klimaschutz 2012).

Higher average temperatures in the North Sea region will also cause tourism-driven changes in biodiversity. If more tourists visit the North Sea in, before and after the summer season, this might increase pressure on local biodiversity, in particular vegetation cover and habitat for nesting birds (Coombers et al. 2008). Effects on nature-based tourism and activities, like animal watching, still require closer academic study. Travellers to the Baltic Sea coast judged algal blooms negatively, especially for swimming (Nilsson and Gössling 2012). There is also concern that foam algae might pollute the beaches (Regierungskommission Klimaschutz 2012). Further studies are needed to examine the (potentially toxic) effects of new plant species moving into the North Sea region and an increase in harmful algal blooms could impact on bathing water quality and the tourism industry in general over the longer term (Gössling et al. 2012). Broader socio-economic impacts of climate change on destinations, such as those concerning health, security or insurance implications should also be considered. Heat waves in summer might adversely affect health resorts, but it may be that such conditions are still preferable to those of other inland destinations and will therefore have an advantage in the future. Knowledge gaps still remain on health issues, especially the future distribution of vector-borne disease along the North Sea coast.

Overall, tourism in the North Sea area in summer is expected to profit from rising temperatures (air and water), decreasing precipitation and a longer season. But climate-driven changes in tourism demand are hard to determine because the tourist decision-making process is influenced by many other factors in addition to climate (IPCC 2014a). In addition to the direct impacts of climate change on tourism and its infrastructure, the more complex and indirect effects of climate change are also important because climate change affects all economic sectors, politics and society as a whole (Kreilkamp 2011).

17.4 Changing Patterns in Tourism Flow

Climate change may also alter tourism patterns in Europe radically by inducing changes in destination choice and seasonal demand structure (Ciscar et al. 2011: 2680). The scientific literature contains many references to tourists, their preferences and their behaviour, including changes in tourist flows and seasonality (Braun et al. 1999; Maddison 2001; Lise and Tol 2002; Wietze and Tol 2002; Lohmann 2003; Hamilton et al. 2005; Gössling and Hall 2006; Bigano et al. 2006b, 2008; Hamilton and Tol 2007; Moreno and Amelung 2009; Buzinde et al. 2010; Hall 2010; Perch-Nielsen et al. 2010; Denstadli et al. 2011; Rosselló-Nadal et al. 2011; Gössling et al. 2012). One of the major questions these studies raise is whether mass tourism of the type seen today at the Mediterranean Sea coast will shift to destinations in northern Europe, such as the North Sea region. Climate change could also result in a seasonal change in visits.

The climate for tourist activities in the North Sea is expected to improve significantly in summer but less so in autumn and spring for northern continental Europe, Finland, southern Scandinavia, and southern England, especially after 2070 (Amelung et al. 2007; Nicholls and Amelung 2008; Amelung and Moreno 2012). At the same time, the attractiveness of the Mediterranean Sea region is expected to decline as comfort distribution changes from a 'summer peak' to a 'bimodal distribution', with less attractive summers and more attractive springs and autumns (Amelung and Viner 2006; Amelung et al. 2007; Moreno and Amelung 2009; Hein 2009; Perch-Nielsen et al. 2010; Moriondo et al. 2011), see also Table 17.2.

However, studies conclude that by 2030 (or even 2060) the Mediterranean Sea region will not have become too hot for beach tourism (Moreno and Amelung 2009; Rutty and Scott 2010), because surveys show that it is mostly rain that drives beach tourists away (de Freitas et al. 2008; Moreno 2010). Domestic tourism and international visits from southern Europe to locations in northern Europe may increase at the expense of southern locations (Hamilton and Tol 2007; Willms 2007; Hein 2009; Amelung and Moreno 2012; Bujosa and Rosselló 2012). The Intergovernmental Panel on Climate Change stated with medium confidence that tourism activity may increase in northern and

Table 17.2 Qualitative assessment of the impact of climate change (IPCC SRES A1F scenario) on sustainable tourism development in the Balearic Islands in the 21st century (Amelung and Viner 2006)

	Spring	Summer	Autumn	Winter	Net effect
Revenue	↑↑	↓↓↓	↓↓	↑↑	↓↓
Occupancy	↑↑	↓↓	↔/↓	↑↑	↑
Employment	↑↑	↓↓↓	↓↓	↑↑	↓↓
Migration	↑	↓↓↓	↓↓	↔	↓↓
Water use	↑↑	↓↓↓	↓↓	↑↑	↓
Impact on biodiversity	↑↑↑	↓↓↓	↓↓	↑	↓

↑ Increase; ↓ decrease; ↔ little or no change

continental Europe, developing travelling patterns closer to home (IPCC 2014b). Nevertheless, no significant changes in the tourism sector are expected before 2050.

A spatial and temporal redistribution of tourism through climate change could lead to shifts, such as Europeans extending their tourism activities over a longer period, taking trips to the Mediterranean Sea region in spring and autumn, and to northern Europe in summer (Ciscar et al. 2011). However, a key assumption is that the tourism system has full flexibility in responding to climate change (Ciscar et al. 2011: 2681). Studies are also needed to address any new environmental challenges appearing along the North Sea coast, for example if more infrastructure and buildings are needed for the already well visited summer period.

However, there are limitations to those forecasts. Preferred beach temperatures differ among travellers from different countries (Scott et al. 2008; Rutty and Scott 2010). According to Maddison (2001), British tourists are attracted to climates around an average of 30.7 °C, which they are unlikely to find in northern Europe even with climate change. Rutty and Scott (2013) interviewed beach tourists on Caribbean islands and found that travellers from the UK preferred temperatures of 27–30 °C while Germans preferred 30 °C. It was shown that preferred beach temperatures differ among travellers from different countries (Scott et al. 2008; Rutty and Scott 2010). Rutty and Scott (2010) also found that the impact of media news about heat waves on travel decisions varied according to the level of commitment to the trip (planning a holiday or a trip already booked). Hall (2012) listed the major weaknesses of current models in predicting travel flow as follows: validity and structure of statistical databases; temperature assumed to be the most important weather parameter; role of information in decision-making unclear; role of non-climatic parameters unclear (e.g. social unrest, political instability, risk perceptions, destination perception); assumed linearity of change in behaviour unrealistic; and future costs of transport and availability of tourism infrastructure uncertain.

The assumption that rising temperatures will be positive for northern European tourist destinations does not consider the impact of other, potentially negative environmental changes in the region (e.g. Hall 2008). Also, there are tourists that still want to travel to regions where they expect resources other than the weather. For example, Moreno (2010) found that almost three-quarters of visitors from Belgium and the Netherlands questioned would still travel to the Mediterranean Sea region even if their self-defined preferred climatic conditions existed in northern Europe. For some people, certain destinations appeal for reasons largely unaffected by climate change, including uniqueness, travel time, standard and cost of accommodation, perceived safety and security, existing facilities, services, access, and host hospitality (Hall 2005). It becomes clear that climate change is just one out of many factors affecting their attractiveness.

17.5 Mitigation and Adaptation Policies

Destinations can seek to lessen the impact of climate change on tourism by adapting to the changes (Gössling et al. 2012). Several studies address both adaptation and mitigation (see Scott and Becken 2010; Scott et al. 2012b; Becken 2013 and Gössling et al. 2013).

As tourists are flexible in their destination choice and because tourism operators can easily change their portfolio, adaptation measures are of special importance for tourism suppliers on site. The choice of adaptation measures (for example when securing infrastructure, rebuilding accommodation, and changing transport) will depend on the type and magnitude of the climate impacts. They may also affect destination attractiveness. For example, raising seawalls on the North Sea coast could result in a less appealing landscape (Regierungskommission Klimaschutz 2012). More research is needed to understand the role of coastal zone management and tourism activities in climate change adaptation, especially in the North Sea region.

In deciding on long-lasting adaptation measures and investigations, the tourism sector faces two fundamental issues: uncertainty in climate scenarios (Turton et al. 2010) and short investment cycles (Bicknell and McManus 2006).

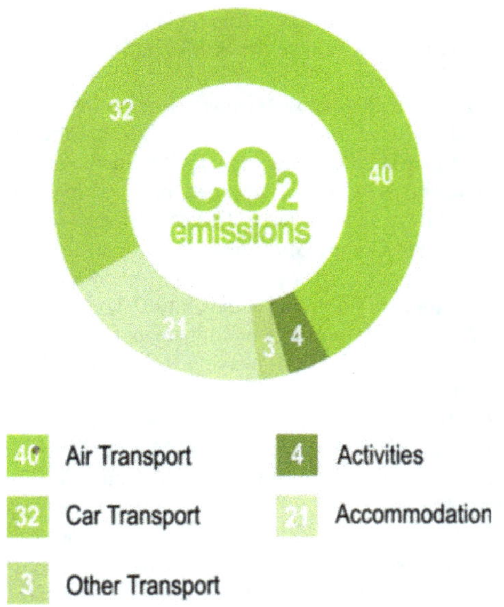

Fig. 17.3 Percentage contribution of tourism sub-sectors to carbon dioxide emissions (UNWTO/UNEP 2008)

Nevertheless, more frequent extreme weather events over the past few years have raised awareness of the need for climate change adaptation and disaster reduction (d'Mello et al. 2009; Becken and Hughey 2013).

Tourism is, in parts, an energy-intensive industry and so itself contributes to climate change. According to UNWTO/UNEP (2008), global tourism accounts for 5 % of global carbon dioxide emissions (Fig. 17.3). A business-as-usual scenario projects emissions from global tourism to grow by 161 % between 2005 and 2035. Emissions from air transport and accommodation are expected to triple. Two alternative emission scenarios show that mitigation solutions using technology only are hard to achieve. Even combined with behavioural changes, no significant reductions in carbon emissions can be gained in 2035 compared to 2005 (IPCC 2014c). In a recent article on tourism's global environmental impact, Gössling and Peeters (2015) predict that tourism-related energy use, emissions, and water, land and food requirements will double within the next 24–45 years. The growth factor for the different components varies from 1.92 (fresh water) to 2.89 (land use) for 2050. An alternative development is possible, but would require a tremendous effort by politics, industry and tourists. But as the demand for tourism is expected to increase (IPCC 2014c), mitigation options are necessary. More research is needed, especially in the transport sector (such as on switching from kerosene to biofuels) and the building sector (such as on retrofitting or energy-efficient new builds) (IPCC 2014c).

A key question is the extent to which tourists will change their travel plans to reduce their impact on global climate. Their apparent unwillingness to adapt their travel behaviour means that the greatest responsibility for mitigation remains with politicians, the tourism industry and tourism supply. According to Kreilkamp (2011), it is a matter of innovativeness: Adaption as well as mitigation actions can be used by companies that aim to differentiate themselves from competitors through innovative approaches and use such actions for effective public relations. Gössling et al. (2013) showed how climate policy may influence travel costs and tourism patterns. Countries with strong climate change policy frameworks (carbon taxes, emissions trading schemes, etc.) also show more interest in tourism-specific policies to address climate change (Becken and Hay 2012). No country has yet adopted a low-carbon tourism strategy (OECD/UNEP 2011) and academic research on tourism policy dealing with climate change is still rare (Becken 2013).

17.6 Conclusions and Future Research

Despite the many papers published up to today, an analysis of the content of four leading tourism journals showed that publications on climate change represented only 1.7 % of all papers published between 2000 and 2009. It demonstrates that 66 % of the 128 papers found were classified as studies of the potential impacts of climate change on destinations or changing visitation patterns, with 40 % on winter ski tourism and less than 10 % on small islands or coastal areas (Scott 2011). For a more detailed analysis of tourism knowledge with respect to climate change adaptation, mitigation and impacts see Hall (2012).

Gössling et al. (2012) summarised in their paper review of the complexity of demand responses and consumer behaviour influenced by climate change that some knowledge gaps remain. It is still difficult to understand the impacts of extreme weather and environmental events on tourist behaviour and this should be considered over both the short and the longer term. There is an assumption that rising temperatures will have positive effects for northern European tourist destinations. However, this does not consider the impact of negative environmental changes in the region or that tourists will still want to travel to climatically disadvantaged regions, since climate change is not the only factor affecting the attractiveness of travel destinations.

Destinations can seek to deal with climate change through adaptation measures and thereby lessen its potential impacts. Further research is needed on the relationship between the impacts of climate change and specific tourist behaviours, activities, or tourism flows to coastal destinations (Moreno and Amelung 2009). Also, as Scott et al. (2012b) pointed out, very few studies address the consequences of mitigation policy in tourism.

Despite some negative impacts, the direct consequences of climate change are expected to be mostly positive for the tourist industry in the North Sea region if supply can keep up with demand. The seasonal distribution of demand will improve substantially in summer, and the region will be able to compete better with other major destinations such as the Mediterranean Sea region due to the warmer, dryer summers expected in the future. As the season lengthens, there will be more days suitable for outdoor recreation. Overall, climate in the North Sea region for tourism will improve. However, other conditions, such as beach width, landscape, and water quality will be affected negatively.

Although the tourism industry has little influence on the behaviour of tourists (IPCC 2014c), it can still take action on tourism supply. Some researchers see a need for drastic changes in the forms of tourism and the uses of leisure time as well as in destinations (Ceron and Dubois 2005; UNWTO/UNEP 2008; Gössling et al. 2010; Dubois et al. 2011; Peeters and Landré 2012).

Further studies in the North Sea region are essential to better understand the role of climate change impacts on the attractiveness of tourist destinations; on a changing Tourism Climate Index on tourism there, on changes in tourism demand and on possible shifts in travelling. As catastrophic events show, such as terrorism or natural disasters, the tourism industry is resilient. Nevertheless, actors in the tourism industry along the North Sea coast need to minimise risks while seeking to take advantage of new opportunities. Multidisciplinary research is needed that considers tourism trends, including climate change, together with social, ecological, economic, technological and cultural developments. More research is needed on how the challenges brought by climate change could be addressed in a proactive and sustainable manner.

References

Agnew MD, Palutikof JP (2000) Impacts of climate on the demand for tourism. In: Falchi MA, Zorini AO (eds) ECAC 2000. Tools for the Environment and Man of the year 2000. Proceedings of the Third European Conference on Applied Climatology, Pisa, Italy, October 2000

Amelung B, Moreno A (2012) Costing the impact of climate change on tourism in Europe: results of the PESETA project. Climatic Change 112:83–100

Amelung B, Viner D (2006) Mediterranean tourism: exploring the future with the tourism climate index. J Sustain Tour 14:349–366

Amelung B, Viner D (2007) The vulnerability to climate change of the Mediterranean as a tourist destination. In: Amelung B, Blazejczyk K, Matzarakis A (eds) Climate Change and Tourism – Assessment and Coping Strategies

Amelung B, Nicholls S, Viner D (2007) Implications of global climate change for tourism flows and seasonality. J Travel Res 45:284–296

Becken S (2010) The Importance of Climate and Weather for Tourism: Literature Review. Lincol University, Research Archive. http://hdl.handle.net/10182/2920

Becken S (2013) A review of tourism and climate change as an evolving knowledge domain. Tourism Manage Persp 6:53–62

Becken S, Hay J (2012) Climate Change and Tourism: From Policy to Practice. Routledge

Becken S, Hughey K (2013) Tourism and natural disaster risk reduction - Opportunities for integration. Tourism Manage 36:77–85

Becken S, Hay J, Espiner S (2011) The risk of climate change for tourism in the Maldives. In: Buter R, Carlsen J (eds) Island Tourism Development: Journeys Towards Sustainability. Cabi

Bicknell S, McManus P (2006) The canary in the coalmine: Australian ski resorts and their response to climate change. Geogr Res 44:386–400

Bigano A, Hamilton JM, Tol RSJ (2006a) The impact of climate on holiday destination choice. Climatic Change 76:175–180

Bigano A, Hamilton JM, Maddison DJ, Tol RSJ (2006b) Predicting tourism flows under climate change – an editorial comment on Gössling and Hall (2006). Climatic Change 79:175–180

Bigano A, Bosello F, Roson R, Tol R (2008) Economy-wide impacts of climate change: a joint analysis for sea level rise and tourism. Mitig Adapt Strat Glob Change 13:765–791

Braun OL, Lohmann M, Maksimovic O, Meyer M, Merkovic A, Messerschmidt E, Riedel A, Turner M (1999) Potential impact of climate change effects on preferences for tourism destinations. A psychological pilot study. Climate Res 11:247–254

Bujosa A, Rosselló J (2012) Climate change and summer mass tourism: the case of Spanish domestic tourism. Climatic Change 117:363–375

Buzinde C, Manuel-Navarrete D, Morais D (2010) Tourists perceptions in a climate of change: eroding destinations. Ann Tourism Res 37:333–354

Ceron JP, Dubois G (2005) The potential impacts of climate change on French tourism. Curr Iss Tourism 8:125–139

Ciscar JC, Iglesias A, Feyen L, Szabó L, Van Regemorter D, Amelung B, Nicholls R, Watkiss P, Christensen OB, Dankers R, Garrote L, Goodess CM, Hunt Al, Moreno A, Richards J, Soria A (2011) Physical and economic consequences of climate change in Europe. Proc Natl Acad Sci U S Am 108:2678–2683

Coombers EG, Jones AP, Sutherland WJ (2008) The biodiversity implications of changes in coastal tourism due to climate change. Environ Conserv 35:319–330

Coombers EG, Jones AP, Sutherland WJ (2009) The implications of climate change on coastal visitor numbers: a regional analysis. J Coast Res 25:981–990

Crouch GI (1995) A meta-analysis of tourism demand. Ann Tourism Res 22:103–118

de Freitas CR (2003) Tourism climatology: evaluating environmental information for decision making and business planning in the recreation and tourism sector. Int J Biomet 48:45–54

de Freitas CR, Scott D, McBoyle G (2008) A second generation climate index for tourism (CIT): specification and verification. Int J Biomet 52:399–407

Denstadli JM, Jacobsen JKS, Lohmann M (2011) Tourist perceptions of summer weather in Scandinavia. Ann Tourism Res 38:920–940

d'Mello C, McKeown J, Minninger S (2009) Disaster Prevention in Tourism: Perspectives on Climate Justice. Ecumenical coalition on tourism. Chiang Mai, Wanida Press

Dubois G, Ceron JP, Peeters P, Gössling S (2011) The future tourism mobility of the world population: Emission growth versus climate policy. Transport Res A 45:1031–1042

Endler C, Matzarakis A (2010) Schlussbericht zum Teilvorhaben "Klima- und Wetteranalyse". In: Mayer H, Matzarakis A (eds) Ber Met Inst Albert-Ludwigs-Univ Freiburg 22

Giannakopoulos C, Le Sager P, Bindi M, Moriondo M, Kostopoulou E, Goodess CM (2009) Climate changes and associated impacts in the Mediterranean resulting from a 2°C global warming. Glob Plan Change 68:209–224

Giupponi C, Shechter M (eds) (2003) Climate Change in the Mediterranean: Socio-economic Perspectives of Impacts, Vulnerability and Adaptation. Edward Elgar

Gössling S, Hall CM (2006) Uncertainties in predicting tourist flows under scenarios of climate change. Climatic Change 79:163–173

Gössling S, Peeters P (2015) Assessing tourism's global environmental impact 1900-2050. J Sustain Tour 23:639–659

Gössling S, Hall M, Peeters P, Scott D (2010) The future of tourism: Can tourism growth and climate policy be reconciled? A mitigation perspective. Tourism Recreation Res 35:119–130

Gössling S, Scott D, Hall CM, Ceron JP, Dubois G (2012) Consumer behaviour and demand response of tourists to climate change. Ann Tourism Res 39:36–58

Gössling S, Scott D, Hall MC (2013) Challenges of tourism in a low-carbon economy. WIREs Climate Change 4:525–538

Hall CM (2005) Tourism: Rethinking the Social Science of Mobility. Pearson

Hall CM (2008) Santa Claus, place branding and competition. Fennia 186:59–67

Hall CM (2010) Climate change and its impacts on tourism: Regional assessments, knowledge, gaps and issues. In: Jones A, Phillips M (eds) Disappearing Destinations: Climate Change and the Future Challenges for Coastal Tourism. Cabi

Hall CM (2012) Tourism and climate change: knowledge gap and issue. In: Singh TV (ed) Critical Debates in Tourism. Short Run Press

Hamilton JM (2005) Tourism, Climate Change and the Coastal Zone. Dissertation Univ Hamburg

Hamilton JM (2006) Coastal landscape and the hedonic price of accommodation. Ecol Econ 62:594–602

Hamilton JM, Tol RSJ (2007) The impact of climate change on tourism in Germany, the UK and Ireland: A simulation study. Reg Env Change 7:161–172

Hamilton JM, Maddison DJ, Tol RSJ (2005) Climate change and international tourism: a simulation study. Glob Env Change A 15:253–266

Harrison R, Kinnaird V, McBoyle G, Quinlan C, Wall G (1986) Climate change and downhill skiing in Ontario. Ont Geogr 28:51–68

Hein L (2007) The impact of climate change on tourism in Spain. CICERO Working Paper 2007:02

Hein L (2009) Potential impacts of climate change on tourism; a case study for Spain. Curr Opin Environ Sustain 1:170–178

Hill A (2009) Holiday deals abroad vanish in rush to flee the rain. The Observer, 9 August 2009

IPCC (2014a) Climate Change 2014: Impacts, Adaptation, and Vulnerability. Part A: Global and sectoral aspects. Working Group II Contribution to the Fifth Assessment Report of the Intergovernmental Panel on Climate Change. Field CB, Barros VR, Dokken DJ, Mach KJ, Mastrandrea MD, Bilir TE, Chatterjee M, Ebi KL, Estrada YO, Genova RC, Girma, Kissel ES, Levy AN, MacCracken S, Mastrandrea PR, White LL (eds). Cambridge University Press

IPCC (2014b) Climate Change 2014: Impacts, Adaptation, and Vulnerability. Part B: Regional Aspects. Contribution of Working Group II to the Fifth Assessment Report of the Intergovernmental Panel on Climate Change. Barros VR, Field CB, Dokken DJ, Mastrandrea MD, Mach KJ, Bilir TE, Chatterjee M, Ebi KL, Estrada YO, Genova RC, Girma B, Kissel ES, Levy AN, MacCracken S, Mastrandrea PR, White LL (eds). Cambridge University Press

IPCC (2014c) Climate Change 2014: Mitigation of Climate Change. Working Group III Contribution to the Fifth Assessment Report of the Intergovernmental Panel on Climate Change. Edenhofer O, Pichs-Madruga R, Sokona Y, Farahani E, Kadner S, Seyboth K, Adler A, Baum I, Brunner S, Eickemeier P, Kriemann B, Savolainen J, Schlömer S, von Stechow C, Zwickel T, Minx JC (eds). Cambridge University Press

Jones A, Phillips M (eds) (2011) Disappearing Destinations: Climate Change and the Future Challenges for Coastal Tourism. Cabi

Jorgensen F, Solvoll G (1996) Demand models for inclusive tour charter: the Norwegian case. Tourism Manage 17:17–24

Kreilkamp E (2011) Klimawandel und Tourismus – Herausforderungen für Destinationen. Z Tourismus 3:203–219

Lemelin H, Dawson J, Stewart EJ, Maher P, Lueck M (2010) Last chance tourism: the boom, doom, and gloom of visiting vanishing destinations. Curr Iss Tourism 13:477–493

Lise W, Tol RSJ (2002) Impact of climate on tourist demand. Climatic Change 55:429–449

Lohmann M (2002) Coastal resorts and climate change. In: Lockwood A, Medlik S (eds) Tourism and Hospitality in the 21st Century. Butterworth-Heinemann

Lohmann M (2003) Über die Rolle des Wetters bei Urlaubsreiseentscheidungen. In: Bieger T, Laesser C (eds) Jahrbuch 2002/2003 der Schweizerischen Tourismuswirtschaft. St. Gallen

Lohmann M, Kaim E (1999) Weather and holiday preference - image, attitude and experience. Rev Tourisme 2:54–64

Maddison D (2001) In search of warmer climates? The impact of climate change on flows of British tourists. Climatic Change 49:193–208

Matzarakis A (2007) Entwicklung einer Bewertungsmethodik zur Integration von Wetter- und Klimabedingungen im Tourismus. Ber Met Inst Univ Freiburg 16

Mieczkowski Z (1985) The tourism climate index: A method of evaluating world climates for tourism. Can Geogr 29:220–233

Moreno A (2010) Mediterranean tourism and climate (change): a survey-based study. Tourism Planning Dev 7:253–265

Moreno A, Amelung B (2009) Climate change and tourist comfort on Europe's beaches in summer: a reassessment. Coast Manage 37:550–568

Moreno A, Becken S (2009) A climate change vulnerability assessment methodology for coastal tourism. J Sustain Tour 17:473–488

Morgan R, Gatell E, Junyent R, Micallef A, Özhan E, Williams A (2000) An improved user-beach climate index. J Coast Conserv 6:41–50

Moriondo M, Giannakopoulos C, Bindi M (2011) Climate change impact assessment: the role of climate extremes in crop yield simulation. Climatic Change 104:679–701

Nicholls S, Amelung B (2008) Climate change and tourism in northwestern Europe: Impacts and adaptation. Tourism Anal 13:21–31

Nicholls RJ, Hoozemans FMJ (1996) The Mediterranean: vulnerability to coastal implications of climate change. Sustainable development at the regional level. Med Ocean Coast Manage 31:105–132

Nicholls RJ, Klein RJT (2003) Climate Change and Coastal Management on Europe's Coast. EVA WP No. 3

Nilsson JH, Gössling S (2012) Tourist responses to extreme environmental events: The case of Baltic Sea algal blooms. Tourism Planning Dev 10:32–44

OECD, United Nations Environment Program (UNEP) (2011) Climate Change and Tourism Policy in OECD Countries. Paris, Organization for Economic Co-operation and Development OECD

Peeters PM, Landré M (2012) The emerging global tourism geography? An environmental sustainability perspective. Sustainability 4:42–71

Perch-Nielsen SL (2010) The vulnerability of beach tourism to climate change – an index approach. Climatic Change 100:579–606

Perch-Nielsen SL, Amelung B, Knutti R (2010) Future climate resources for tourism in Europe based on the daily tourism climatic index. Climatic Change 103:363–381

Perry A (2005) The Mediterranean: how can the world's most popular and successful tourist destination adapt to a changing climate? In: Hall CM (ed) Tourism, Recreation and Climate Change. Channel View Publ

Perry A (2006) Will predicted climate change compromise the sustainability of Mediterranean tourism? J Sustain Tour 14:367–375

Phillips MR, Jones AL (2006) Erosion and tourism infrastructure in the coastal zone: problems, consequences and management. Tourism Manage 27:517–524

Pinnegar JK, Viner D, Hadley D, Sye S, Harris M, Berkhout F, Simpson M (2006) Alternative future scenarios for marine ecosystems. Tech Rep. CEFAS, Lowestoft

Regierungskommission Klimaschutz (2012) Empfehlung für eine niedersächsische Stratgegie zur Anpassung an die Folgen des Klimawandels. Niedersächsisches Ministerium für Umwelt, Energie und Klimaschutz, Hannover

Rosselló J, Aguiló E, Riera A (2005) Modelling tourism demand dynamics. J Travel Res 44:111–116

Rosselló-Nadal J, Riera-Font A, Cárdenas V (2011) The impact of weather variability on British outbound flows. Climatic Change 105:281–292

Rutty M, Scott D (2010). Will the Mediterranean become "too hot" for tourism? A reassessment. Tourism Planning Dev 7:267–281

Rutty M, Scott D (2013) Differential climate preferences of international beach tourists. Climate Res 57:259–269

Schleupner C (2008) Evaluation of coastal squeeze and its consequences for Caribbean island Martinique. Ocean Coast Manage 51:383–390

Scott D (2011) Why sustainable tourism must address climate change. J Sustain Tour 19:17–34

Scott D, Becken S (2010) Editorial introduction. Adapting to climate change and climate policy: progress, problems and potentials. J Sustain Tour 18:283–295

Scott D, Lemieux C (2010) Weather and climate information for tourism. Procedia Environ Sci 1:146–183

Scott D, McBoyle G (2001) Using a 'tourism climate index' to examine the implications of climate change for climate as a natural resource for tourism. In: Matzarakis A, de Freitas CR (eds) Proceedings of the first international workshop on climate, tourism and recreation. International Society of Biometeorology, Commission on Climate, Tourism and Recreation, Halkidi, Greece

Scott D, McBoyle G, Schwartzentruber M (2004) Climate change and the distribution of climatic resources for tourism in North America. Climate Res 27:105–117

Scott D, Gössling S, de Freitas C (2008) Preferred climate for tourism: Case studies from Canada, New Zealand and Sweden. Climate Res 38:61–73

Scott D, Gössling S, Hall CM (2012a) International tourism and climate change. WIREs Climate Change 3:213–232

Scott D, Hall CM, Gössling S (2012b) Tourism and Climate Change. Impacts, Adaptation and Mitigation. Routledge

Scott D, Sim R, Simpson M (2012c) Sea level rise impacts on coastal resorts in the Caribbean. J Sustain Tour 20:883–898

Smith K (1993) The influence of weather and climate on recreation and tourism. Weather 48:398–404

Turton S, Dickson T, Hadwen W, Jorgensen B, Pham T, Simmons D, Tremblay P, Wilson R (2010) Developing an approach for tourism climate change assessment: evidence from four contrasting Australian case studies. J Sustain Tour 18:429–448

UNWTO (2009) From Davos to Copenhagen and Beyond: Advancing tourism's response to climate change. United Nations World Tourism Organization (UNWTO) background paper.

UNWTO/UNEP (2008) Climate Change and Tourism: Responding to global challenges. UN World Tourism Organization (UNWTO), United Nations Environment Programme (UNEP)

von Bergner NM, Lohmann M (2014) Future Challenges for Global Tourism: A Delphi Survey. J Travel R 53(4):420–432

Wall G, Harrison R, Kinnaird V, McBoyle G, Quinlan C (1986) The implications of climate change for camping in Ontario. Recreation Research Review 13:50–60

Wietze L, Tol RSJ (2002) Impact of climate on tourist demand. Climatic Change 55:429–449

Willms J (2007) Climate change = tourism change? The likely impacts of climate change on tourism in Germany's North Sea coast destinations. In: Matzarakis A, de Freitas CR, Scott D (eds) Developments in Tourism Climatology. Commission on Climate, Tourism and Recreation

Witt SF, Witt CA (1995) Forecasting tourism demand: A review of empirical research. Int J Forecasting 11:447–475

Socio-economic Impacts—Coastal Protection

18

Hanz D. Niemeyer, Gé Beaufort, Roberto Mayerle, Jaak Monbaliu, Ian Townend, Holger Toxvig Madsen, Huib de Vriend and Andreas Wurpts

Abstract

All North Sea countries are confronted by climate change impacts such as accelerated sea-level rise, increasing storm intensities resulting in as well higher set-up of storm surges as growing wave energy and a follow-up of morphological changes. Thus it is necessary to question the effectiveness of existing coastal protection strategies and to examine alternative strategies for coastal protection under a range of scenarios considered possible. Scenarios of accelerating sea-level rise leading to changes in sea level of up to 1 m or more by 2100 and higher set-up of storm surges with increasing wave energy have been used for planning purposes. Adaptation strategies for future coastal protection have been established in all North Sea countries with vulnerable coasts, observing two propositions: (1) structures are economic to construct in the short term and their dimensions easily adapted in the future to ensure flexibility in responding to the as yet undeterminable climate change impacts and (2) implementation of soft measures being temporarily effective and preventing counter-action to natural trends. The coastal protection strategies differ widely from country to country, not only in respect of distinct geographical boundary conditions but also in terms of the length of the planning period and the amount of regulations. Their further development is indispensable and emphasis must more and more be laid on strategies considering the effects of long-term development of coastal processes for future coastal protection. Filling gaps in knowledge is essential for developing sustainable adaptation strategies.

H.D. Niemeyer (✉)
Independent Consultant, ret., Coastal Research Station, Norderney, Germany
e-mail: hanz.niemeyer@yahoo.com

G. Beaufort
Hydr. Engineering Cons., De Meern, The Netherlands
e-mail: g.a.beaufort@gmail.com

R. Mayerle
Research and Technology Centre Westcoast, Kiel University, Kiel, Germany
e-mail: rmayerle@corelab.uni-kiel.de

J. Monbaliu
Department of Civil Engineering, KU Leuven, Leuven, Belgium
e-mail: jaak.monbaliu@kuleuven.be

I. Townend
Ocean and Earth Sciences, University of Southampton, Winchester, UK
e-mail: I.Townend@soton.ac.uk

H. Toxvig Madsen
Danish Coastal Authority, Lemvig, Denmark
e-mail: holger.toxvig@kyst.dk

H. de Vriend
Independent Consultant, Oegstgeest, The Netherlands
e-mail: hjdevriend@planet.nl

A. Wurpts
Coastal Research Station, Norderney, Germany
e-mail: andreas.wurpts@nlwkn-ny.niedersachsen.de

M. Quante and F. Colijn (eds.), *North Sea Region Climate Change Assessment*,
Regional Climate Studies, DOI 10.1007/978-3-319-39745-0_18

18.1 Introduction

Climate change will create stronger challenges for coastal protection than experienced in the past. Loads on protection structures are increasing and increased flood risk in the majority of coastal areas has coincided with ongoing growth in population and investment. Since the implementation of measures in coastal protection needs a forerun of decades, the determination of boundary conditions for their design requires an appropriate and sufficiently safe margin for foreseeable developments in the future. This is presently best practice in coastal engineering but becomes more difficult and uncertain the further forward in time considered since there are no reliable forecasts for future climate change impacts, only wide-ranging scenarios. Therefore adaptation strategies for coastal protection must aim to be both economic to construct in the short term and designed such that they can be easily adapted in the future, allowing adequate flexibility in order to respond to the as yet insufficiently determinable effects of future climate change impacts. To meet these requirements, current understanding of climate change effects on coastal protection measures must be used to examine alternative strategies for future coastal protection under a wide range of scenarios for climate change impacts regarded as possible.

18.1.1 Boundary Conditions of Coastal Protection

The aims of coastal protection are first the safety of the hinterland against flooding due to storm surges and second to limit coastal retreat. An essential basis for achieving these objectives is sound knowledge of the governing boundary conditions, such as local hydrodynamic loads or morphological processes. Acceleration of sea-level rise (SLR) due to changing global climate will be a threat in all coastal areas. This threat will be compounded by a number of secondary effects of climate change that will increase loads on coastal protection structures or on dunes and cliffs providing shelter for the hinterland against flooding.

Climate change will also lead to increasing storm intensities which will—particularly in the shallower parts of the North Sea—cause higher set-ups of storm surges (EEA 2012; Woth et al. 2006; Weisse et al. 2012). As a result, water depths at the coastlines will increase for design conditions; the shallower the local coastal waters the greater the increase. Since in areas like the Wadden Sea coasts in the southern North Sea, wave heights and periods on tidal flats are strongly depth-controlled (Niemeyer 1983; Niemeyer and Kaiser 2001), any increase in local water depth would be accompanied by correspondingly higher wave loads on coastal structures or on dunes and cliffs (Niemeyer 2010).

Accelerated SLR will also be accompanied by morphodynamic responses in sedimentary coastal areas which may be unfavourable to coastal protection. For instance, adaption of tidal flat levels may no longer keep pace with SLR, and if rates exceed a certain threshold then tidal flats might even disappear (Müller et al. 2007). Water depth in front of coastal structures would then increase and result in the propagation of higher and longer waves during storm surges and thus stronger wave loads. Adaption of tidal flats to SLR is governed by the hydrodynamics of ordinary tides. In contrast, the vertical growth of saltmarshes depends on hydrodynamics during meteorologically enhanced tides and in particular on storm surges (Townend et al. 2011). In addition to this significant disparity in governing boundary conditions there are indications that salt marshes also have a limited capability to grow with sea level: above a certain threshold in the rate of SLR they will no longer keep pace. The threshold for SLR to limit the vertical growth of saltmarshes will be slightly raised, however, by an increase in the frequency of storm surges (Schuerch et al. 2013).

The response of coastal morphology to accelerated SLR is much more pronounced on wave-exposed sandy coasts and barrier islands than, for example, in front of coastlines on estuaries or tidal basins with tidal flats and salt marshes; areas with a high share of cohesive sediments. Adaption of the shoreface to erosion induced by SLR according to the BRUUN-Rule and its steepening will take place simultaneously (Bruun 1962; Stive and de Vriend 1995). Since shoreface processes affect conditions at adjacent beaches (Mulder and de Vos 1989), erosion and coastal retreat will also occur. At interrupted coasts with estuaries or tidal inlets and basins, SLR will increase basin volume and drive an increasing demand for external sediment supply to enable adaptation towards the moving target of morphodynamic equilibrium (Ranasinghe et al. 2012).

The result is erosion of coastal stretches in the vicinity of the tidal inlets, leading to stronger coastal retreat than would occur through shoreface adaption to SLR alone (Ranasinghe et al. 2012). The volume of ebb-deltas will also decline as they will act as the initial source for meeting the increased sediment demand of the basins (Stive and Eysink 1989). Since the sheltering effects of ebb-deltas depend on their sediment volume (Kaiser and Niemeyer 1999), wave penetration into the basin and onto adjacent beaches will be less restricted causing higher loads on structures, dunes and cliffs and increasing erosion of beaches and dunes and, although to a lesser extent, tidal flats and salt marshes. The impact of all such processes will increase, the more SLR is accompanied by an increase in tidal range.

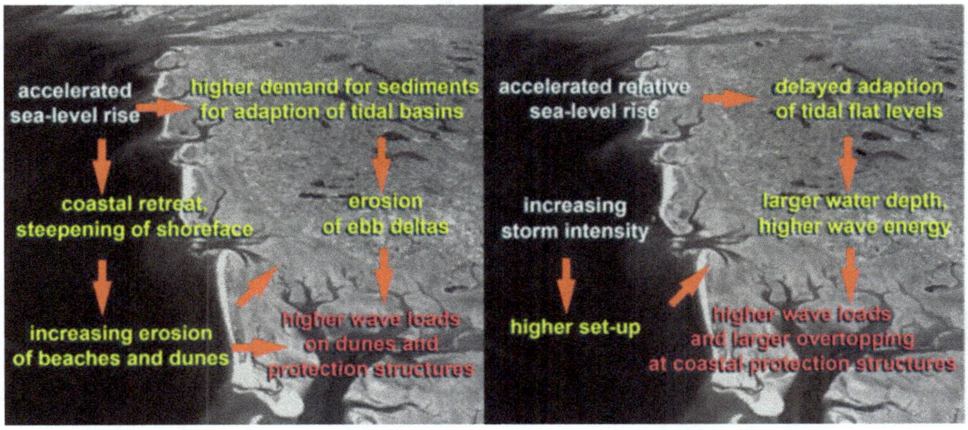

Fig. 18.1 Combined effects of climate change on a North Sea coast: morphodynamics and hydrodynamic loads at a sandy coast (barrier islands) (*left*) and on hydrodynamic loads on coastal structures at a lowland coast along an estuary and tidal basins with cohesive sediments (*right*) (Niemeyer 2015); background image from the Common Wadden Sea Secretariat (www.waddensea-secretariat.org)

These secondary effects of climate change are superimposed on each other, and may even invoke a feedback (Fig. 18.1) which further complicates the prediction of future change (Niemeyer 2015). It will be a major challenge for coastal researchers to develop and apply suitable morphodynamic models that can encompass a sufficiently wide range of scenarios for future climate change effects. Such models are needed to meet the knowledge base required for more detailed planning and development of adaptation measures for coastal protection. This is particularly the case for wave-exposed sandy coasts and barrier islands, where the secondary effects of accelerated SLR on morphology are expected to be stronger, faster and more diverse than those anticipated in front of coastlines with a high degree of cohesive sediments, where morphodynamic adaption is more predictable (Niemeyer 2015).

18.1.2 Coastal Protection Strategies in Response to Climate Change Impacts

Global warming and the resulting acceleration in SLR necessitates a thorough re-evaluation of coastal protection strategies in many parts of the world. This includes the North Sea coasts of Europe, where coastal protection has a history of more than 1000 years. For most of the North Sea coasts, maintaining a protection line through dykes, solid structures or dunes and cliffs was historically the result of human activity. The potential for faster SLR through global warming has alerted coastal managers to question whether this strategy of keeping the line will still be appropriate, or whether alternative strategies should be considered.

The Intergovernmental Panel on Climate Change (IPCC) Coastal Zone Management Subgroup identified alternative adaptation strategies for SLR: retreat, accommodation and protection (IPCC 1990), following an earlier Dutch evaluation, which also included the additional strategy of moving the defence line seaward (Rijkswaterstaat 1989). All four strategies are manifested in historical practice (Niemeyer 2005, 2010). The strategy 'protection' has since been further differentiated by distinguishing between traditional line protection and alternative protection schemes such as set-back or realignment and combined protection (ComCoast 2007). Although moving the defence line seaward is only suitable in very specific situations, and may not always be ethically and politically acceptable, the other strategies are regarded as options for adapting coastal protection in response to possible future climate change effects. Recent investigations have shown that simple conceptual evaluations by graphical schematizations and purely qualitative discussion such as carried out by ComCoast (2007), are unreliable and sometimes even misleading since important boundary conditions such as hydrodynamic loads, topographic features, existing protection structures and necessary resources are ignored or misjudged. Therefore it is essential to evaluate strategies by applying scenarios for design conditions in real-world environments (Niemeyer et al. 2011a, b, 2014).

The same comments also apply to conclusions drawn by Temmerman et al. (2013) concerning the effectiveness of protection strategies that improve the ecological value of coastal and estuarine areas by a set-back of the existing protection line, since the emphasized equivalence of safety against flooding of the hinterland achieved by the protection strategy applied beforehand is only an assumption. Model tests for proving that assumption are to be carried out in the future thus allowing reliable judgements (STW 2013). The expectation of a reduction in hydrodynamic loads on structures by a set-back strategy (Temmerman et al. 2013) contradicts results being achieved for the evaluation of alternative coastal protection strategies by mathematical modelling for design conditions in similar environments (Niemeyer et al. 2011a, b, 2014). Nevertheless, potential improvements in ecological quality by applying alternative strategies should be balanced against the higher capital costs for coastal protection in respect of societal demands.

Although evaluations of protection strategies for coasts with a significant fraction of cohesive sediments yield reliable results, this is not the case for wave-exposed sandy coasts and barrier islands with higher dynamics (Fig. 18.1). The impacts of climate change on coastal processes may require a higher level of adaptation there than at other locations. Taking into account the enormous additional effort this will require, adaptation strategies for wave-exposed sandy coasts and barrier islands will need to accommodate stronger and more variable coastal processes due to future climate change impacts than at present. Such an approach could then serve as a blueprint for the development of flexible coastal protection schemes that are sufficiently adapted to future climate change impacts in order to prevent any incompatibility with future developing trends driven by nature. Such schemes are likely to prove more favourable than some of the traditional coastal protection measures.

18.2 Adaptive Planning and Regulation

Adaptation strategies for future coastal protection have been established in all North Sea countries with vulnerable coasts. These differ widely from country to country, especially in terms of the length of planning period and amount of regulation.

18.2.1 Belgium

The Flemish Government approved a Master Plan Coastal Safety (Afdeling Kust 2011) in June 2011 comprising calculations and safety assessments for the periods 2000–2050 and 2050–2100. A vision for further development of the Flemish coastal zone is on its way aiming at the integration of safety, natural values, attractiveness, sustainability and economic development including navigation and sustainable energy. This concept is referred to as *Vlaamse Baaien* or *Flanders Bays 2100* (Vlaamse Baaien 2015a, b) and includes conceptual plans for responding to climate change effects beyond 2050. This idea was initiated by a concept study launched by a private consortium of different consultant and construction companies under the name *Flanders Bays 2100* (Vlaamse Baaien 2015a). Execution of the Master Plan Coastal Safety, however, is a pre-condition that must be met before implanting the 'Flanders Bays' concept (Vlaamse Baaien 2015a). It is expected that the safety levels incorporating the projected SLR until 2050 will require maintenance nourishments thereafter. For the Belgian part of the Western Scheldt estuary the Sigma Plan was established after the floods of 1976 and was revised in 2005 to include projected SLR until 2050. New understanding of coastal management, which balances safety and environmental protection and also shipping where it plays a key role, have resulted in a vision of multifunctional and sustainable use of the Western Scheldt estuary (Sigmaplan 2016).

18.2.2 Denmark

The Danish Government announced its strategy for adaptation to a changing climate in 2008 (Danish Government 2008). The report provides an overview of the challenges arising from future climate change in terms of 11 sectors, one of which is the coastal zone. The adaptation strategy for coastal protection was developed by the Danish Coastal Authority (2012). The aim is to provide coastal communities with a regionally differentiated basis for adaptation to 2050, and then to 2100. Every five years the Coastal Authority undertakes a safety assessment of the central part of the Danish North Sea as a basis for coastal protection planning as well as financial planning. There is no fixed schedule for safety assessments in the Danish Wadden Sea: two have been undertaken since 1999. In all other parts of the Danish coasts the land owners are themselves responsible for protection.

18.2.3 Germany

For the German North Sea coast, adaptation strategies in the four federal states are regulated differently. In the Free and Hanseatic City of Bremen a sector plan was established by the ministry (SUBV 2012). The building programme matching the safety levels established in 2007 includes a heightened precaution measure for climate change effects and will be finished in 2025. For the Free and Hanseatic City of Hamburg the parliament accepted a proposal made by the state government which is a guideline for planning until 2050 (Senat FFH 2012). A safety assessment will be undertaken every ten years. In Schleswig-Holstein, design boundary conditions were revised with respect to SLR expected by 2050 and 2100 in an update to the Coastal Protection Masterplan (MELUR 2013); safety assessments are planned every ten years. In 2008, the state government of Lower Saxony established a government commission of management experts, scientists and stakeholders to develop an adaptation strategy for climate change effects including coastal protection, supported by expert groups on specific themes. A well-funded research programme was initiated in order to provide the commission with basic information on key issues like coastal protection. A report on the adaptation strategy was delivered in 2012 (MU 2012) and its recommendations for initial actions were approved by the state government in 2013: the optimal strategy for coastal protection at the mainland coast is by keeping the protection line; precautionary observations and investigation programmes are required to address identified knowledge gaps and so enable future substantiation of adaptation measures; and the need to continue the safety assessment programme with a ten-year cycle. Investigations of clay quality in the cover layer of existing dykes as a basis for introduction of increasing overtopping tolerance in future design procedures to balance—at least partly—higher hydrodynamic loads are a major component of this research programme. Of even greater importance is the identification and quantification of morphological effects due to climate change impacts in the dynamic East Frisian barrier islands region to provide the essential basic knowledge for developing a resilient adaptation strategy for the future protection of the area against flooding and effects of structural erosion. An independent commission shall be appointed to provide recommendations on implementing this programme.

18.2.4 Netherlands

In the Netherlands, consideration of climate change effects started earlier than in most other countries (Rijkswaterstaat 1989, 1990). In 2001, a safety assessment procedure was laid down in the Water Act, requiring an assessment every five years, later increased to six. The need for more advanced adaptation to climate change led to the establishment of the second Deltacommissie (2008). Starting from scenarios for SLR and river discharge, this committee produced recommendations which included, among others, the establishment of a Delta Program led by a Delta Commissioner at ministerial level, to recommend how to implement a risk-based flood safety approach and how to establish an effective organisation and legal framework. The Delta Commissie's recommendations were approved by parliament in a Delta Act. A budget of EUR 1 billion per year was initially foreseen for planning and implementing climate adaptation measures, but this has now been revised to EUR 9 billion for the period 2013 to 2028. Adaptation to newly defined safety levels aimed at 2050 is intended to be ready by 2028. The process is accompanied by an annual National Delta Congress. The Delta Program on several strategic decisions regarding future flood safety and freshwater provision is now finished. The new safety norms are currently being laid down in the new Water Act, and are expected to come into effect as of 2017. Future safety assessments will be undertaken every twelve years (Rijkswaterstaat 2015b).

18.2.5 United Kingdom

The Climate Change Act 2008 provides a legally binding and long-term framework to cut carbon emissions in the United Kingdom, but also makes provision for an assessment of the risks of climate change for the United Kingdom to be undertaken on a five-year cycle. The first of these is the 2012 Climate Change Risk Assessment (CCRA) (DEFRA 2012). This was based on climate projections by Lowe et al. (2009) and included an assessment of the economic implications of climate change for different sectors and the potential costs and benefits of different adaptation responses. Building on the outputs of the CCRA, the government and the Devolved Administrations (Northern Ireland, Scotland, and Wales) are developing adaptation programmes that will set out Government objectives for adaptation to climate change as well as proposals and policies to deliver these objectives. The programmes will be subject to regular assessment by the Committee on Climate Change to determine progress towards implementation.

18.3 Safety Margins for Climate Change Effects

18.3.1 Sea-Level Rise Scenarios and Safety Levels

The safety levels of hydrodynamic loads are the criteria used for dimensioning coastal protection structures to ensure their effectiveness in protecting against flooding due to storm surges. Superimposed safety margins ensure that the structures remain effective against flooding over the course of their anticipated lifespan; safety margins for SLR have been in use since the 1950s and are superimposed on the safety levels for hydrodynamic loads on the structures. The safety margins associated with accelerated SLR and other potential climate change effects are considered in distinct rates in the countries along the North Sea coasts. But ultimately, the safety of the protected areas depends on the aggregated safety margins and safety levels; the latter still the more relevant in respect of the order of magnitude. A comparison of safety levels between countries makes little sense. On the one hand, some countries have introduced distinct safety levels on regional scales, while on the other a comparison of exceedance probabilities is sometimes misleading. If distinct extreme value distributions are used to evaluate design parameters, the same exceedance probability might deliver distinctive results; with the larger the difference the lower the probability of occurrence. Moreover methodological differences like choice of used values or data fitting, and length of time series prevent a credible comparison: benchmarking by exceedance probabilities or return periods is only reasonable

Fig. 18.2 North Sea Basin and surrounding countries (base map: http://de.wikipedia.org/wiki/Datei:North_Sea_map-en.png)

and provides reliable results if the methodological basis for their evaluation is compatible. Therefore the following review of current safety margins for SLR and other hydrodynamic effects due to climate change includes only a brief description of safety levels. All locations mentioned in the following texts are shown in Fig. 18.2.

18.3.1.1 Denmark

A SLR of 0.1–0.5 m by 2050 and 0.2–1.4 m by 2100 is assumed in Denmark. This is partly compensated for by a land rise of 0–0.1 m by 2050 and 0–0.2 m by 2100, leading to a relative SLR of 0–0.5 m by 2050 and 0–1.4 m by 2100. An increase in the set-up of severe storm surges of 0–0.1 m by 2050 and 0–0.3 m by 2100 is also assumed due to higher wind velocities resulting also in higher and longer waves. Peak storm surge levels may increase by up to 0.6 m by 2050 and up to 1.7 m by 2100 due to the combined effect of SLR and increasing surge set-up. Information on the changing wave climate is provided by comparing actual conditions with scenarios for the period 2071–2100. For dykes on the Wadden Sea coast, cost estimates for adaptation have been carried out. For a recently strengthened 13-km stretch of the dyke line south of Ribe a safety margin of 40 cm has been considered. The safety level in Denmark is defined for sandy coasts by conditions with a yearly exceedance probability of 10^{-3} for the city of Thyboron and 10^{-2} for the coastal stretch between Agger and Nymindegab. The width of dunes required to meet that safety level was determined empirically from historical data on dune erosion. Safety levels for the dykes at the Wadden Sea coast of Denmark range between 2×10^{-2} and 5×10^{-3}, depending on population density in the protected area. Design is aimed to achieve these safety levels until 2100, and takes into account projections for SLR, increased set-up of storm surges and changes in wave climate. The level of acceptable overtopping tolerance for dykes is 10 %, which is equivalent to approximately 10 [l× (m s)$^{-1}$] for the boundary conditions at the Danish Wadden Sea coast. The other parts of the Danish North Sea coast have no flood risk.

18.3.1.2 United Kingdom

Safety levels in the United Kingdom depend on the degree of development of the protected areas. For London and the developed parts of the Thames estuary a yearly exceedance probability of 10^{-3} is applied, whereas the corresponding safety level for all other urban areas along the North Sea coast is a yearly exceedance probability of 5×10^{-3}. For the other parts, lower safety levels are applied in respect of local circumstances. Since 1999, a SLR of 40 cm is assumed for the North Sea coast north of Flamborough Head for the design of structures with a lifespan of 100 years, and a SLR of 60 cm for the North Sea coast south of Flamborough Head. The flood risk management plan for the Thames estuary takes the following safety margins for SLR into consideration (Environment Agency 2013):

- 4 mm year^{-1} to 2025
- 8.5 mm year^{-1} for 2026–2055
- 12 mm year^{-1} for 2056–2085
- 15 mm year^{-1} for 2086–2115.

National guidance issued in 2011 advises using the UK Climate Projection 09 (DEFRA 2011) for relative SLR based on the medium-emissions 95th percentile projection for the project location. Upper-end (95th percentile) estimates are as follows:

- 4 mm year^{-1} to 2025
- 7 mm year^{-1} for 2026–2050
- 11 mm year^{-1} for 2051–2080
- 15 mm year^{-1} for 2081–2115.

Guidance is also given for storm surges, where an assessment of extremes is recommended and upper-end estimates are provided as follows: 20 cm by the 2020s, 35 cm by the 2050s and 70 cm by the 2080s. Work is underway on developing wave climate projections.

18.3.1.3 Germany

In Germany, the four federal states use three different methods for evaluating design water levels on the North Sea coast and adjacent estuaries. They have been tuned to yield similar values at the Cuxhaven gauge at the mouth of the Elbe estuary between 2010 and 2012. A matching value is achieved for the method practised in Schleswig-Holstein by adding an additional measure for the surge set-up in an estuarine mouth to the value achieved by the commonly used yearly exceedance probability of 5×10^{-3}. Hamburg has developed a new deterministic approach in order to meet the target range. Bremen and Lower Saxony met the anticipated target value beforehand by applying the traditionally used deterministic single-value method by combining the actual mean high water level with the highest values of maximum spring elevation, storm surge set-up and the chosen safety margin for climate change effects for the determination of design water levels. Design water levels in Lower Saxony and in the Netherlands at the Ems-Dollard estuary have similar values, the surge set-up of the design water level has a yearly exceedance probability of 2.5×10^{-4}. Tolerable wave overtopping at dykes is limited to 2 [l× (m s)$^{-1}$] in Schleswig-Holstein and to 3 % in Bremen and Lower Saxony corresponding to an overtopping volume in the range of approximately 0.1–1.5 [l× (m s)$^{-1}$] with a tendency to correspond to the cross-sectional areas of dykes. All four states account for future climate change effects in the evaluation of design water levels by adding a general provision margin of 50 cm for 100 years. This measure would be equivalent to a SLR of about 40–45 cm per 100 years. Since 2012/2013 in Hamburg, 20 cm of the anticipated 50 cm SLR will be taken into account in the design of coastal protection structures with a lifespan to 2050. Whereas in Schleswig-Holstein and Hamburg the provision margin is a comprehensive part of the design water level, in Lower Saxony and Bremen a different approach is used in designing coastal structures: the provision margin is split into a SLR of 25 cm and an additional increase in storm surge set-up of 25 cm. The latter requires higher storm velocities and so also takes into account higher wave energy. Furthermore, for the applied design procedure the—at least partial—adaption of tidal flats to an accelerated SLR is neglected leading to greater water depths and higher and longer waves in front of coastal structures. As a result, the incorporation of dynamic elements in the design procedure generates a higher safety margin than using an additional fixed value for design water levels. Furthermore, in Bremen, Hamburg and Lower Saxony, solid structures are constructed so as to accommodate an increase in water level beyond the anticipated safety margin; this comprises up to an additional 75 cm (Bremen), 80 cm (Hamburg) or 50 cm (Lower Saxony).

18.3.1.4 Belgium

The Flemish authorities are anticipating a SLR of about 6 mm year^{-1} by 2050 and 10 mm year^{-1} between 2050 and 2100 at the Belgian coast, and these values have been considered for planning and construction targeted at safety levels for 2050 and being ready by 2018. The safety level is a yearly exceedance probability of 10^{-3} for both water level and waves, and is based on extreme value distributions for the determinative directions for very high storm surges. The design procedure is based on a storm duration of 45 h, covering three tidal high peaks, for dunes, dykes, sluices, weirs and quay walls in harbours. The threshold of tolerable wave overtopping on dykes is 1 [l× (m s)$^{-1}$] and dune erosion must be limited to a predefined level. Quay levels in harbours, heights of sluices and weirs will be checked with the aim of matching the design water levels. Risk analyses are carried out for four scenarios, including storm surges with higher tidal peaks than considered for the design storm surge up to a yearly exceedance probability of 5.89×10^{-5}. The aim is to derive basic information for the introduction of higher safety levels on the basis of a benefit-cost ratio and risk reduction if events occur for which evacuation is necessary. In the revised Sigma Plan for the Belgian part of the Western Scheldt estuary the design of coastal protection structures was based on a cost-benefit analysis (Broekx et al. 2011; Sigmaplan 2016).

18.3.1.5 Netherlands

To date, safety levels in the Netherlands refer to the recommendations of the first Delta Committee after the 1953 flood: a probabilistic flood safety definition based on the exceedance probabilities of water levels and waves. The safety levels differ between the various parts of the country in respect of population density, economic value and risk of flooding. Two safety levels have been established at the coast: 10^{-4} for the central Holland coast and 2.5×10^{-4} year^{-1} for the southwestern Delta area and the Wadden Sea with the Ems-Dollard estuary in the Northeast. Later overtopping tolerance on dykes has been limited to 0.1–1 [l× (m s)$^{-1}$] depending on the quality of the cover layer. The Second Deltacommissie (2008) recommended raising safety standards ten-fold based on economic and population growth since 1953. Meanwhile, a decision has been made to replace the current procedure by a risk-based approach, incorporating the probability and degree to which a protection structure will fail if its design conditions are exceeded, as well as the loss of life and material damage that would occur in the event of a flood. A basic safety level is introduced, with a yearly probability of 10^{-5} as an upper limit for the loss of life due to flooding as local individual risk. For its evaluation two types of additional study are required: one on the threat to life due to flooding and one based on a societal cost-benefit-analysis. The final operational layout is expected to be introduced in 2016 in order to be ready for the safety assessment in 2017 (MIenM 2013). The Delta Commissioner expects that, to date, the safety levels used in coastal areas have led to protection structures that will meet the requirements of the new safety levels (Helpdesk Water 2015). Explorative studies of some dyke rings, however show that this new approach may lead to very different assessments of flood safety (Rijkswaterstaat 2005, 2015a). A more detailed investigation for the Lake IJssel area (Deltaprogramma IJsselmeergebied 2013) confirms this. The reason is that the failure of different stretches of dyke in a dyke ring may lead to different numbers of individuals being exposed to flooding. Safety margins for accelerated SLR due to the Delta scenarios range from 0.35 to 0.85 m until 2100 (Deltacommissaris 2013).

18.3.2 Coastline Stabilisation and Anticipation of Morphological Changes

Climate change will not only affect the hydrodynamic boundary conditions for coastal protection but will also cause morphological processes unfavourable to coastal protection. Knowledge about such developments and their consequences for coastal protection is much poorer than that available for future hydrodynamic loads. This lack of understanding about future morphological changes not only increases the uncertainties about future hydrodynamic loads but also includes the possibility that parts of the present coastal system could even disappear. A wide range of possible solutions are being considered in the coastal North Sea countries to tackle this problem, although the dimensions of morphological processes due to climate change impacts remain partly unknown. Solutions discussed in the following sections are all based on currently applied means to counter erosion.

18.3.2.1 Germany

In Germany, the Federal States of Bremen and Hamburg are responsible for relatively small sections of the open coast and have left the problem of morphological processes due to climate change impacts untouched in their adaptation scenarios to date. In the 'Masterplan Coastal Protection of the Federal State of Schleswig-Holstein' erosion due the BRUUN-rule is mentioned but only as a term without any consideration in respect of precautionary measures or as a topic for future research (MELUR 2013). Lower Saxony has developed an intensive research programme as part of the adaptation strategy, aiming to provide a robust evidence base for the planning of appropriate measures (MU 2012), but this programme has yet to start. In Schleswig-Holstein and in Lower Saxony structural erosion in sandy environments is typically compensated by artificial nourishments, particularly on barrier islands.

18.3.2.2 Denmark

Some parts of the sandy North Sea coast of Denmark experience structural erosion (Van de Graaff et al. 1991) which is compensated by artificial nourishments of 2–3 million m^3 $year^{-1}$. The total volume required is determined by the sum of:

- the annual average erosion above the 6 m depth contour between 1977 and 1996
- loss of nourished volume between the 6 and 10 m depth contour
- compensation for profile steepening since the middle of the period 1977–1996
- in the future, an extra 15 % of the sum of all three to cover uncertainties.

Since artificial nourishment steepens the shoreface, extra volumes of material are likely to be needed to offset the effects of SLR. The Danish Coastal Authority has carried out intensive empirical studies to determine the volumes required for future nourishments to compensate for erosion due to accelerated SLR, shoreface steepening and increased longshore transport due to anticipated higher wave energy. The additional artificial supply for compensating for anticipated climate change effects under three scenarios averages 17 % in 2050 and 49 % in 2100 relative to the total volume of nourishment in 2008 (Jensen and Sørensen 2008).

18.3.2.3 Belgium

Structural erosion on the Flemish coast is counteracted by shoreface, beach and dune nourishments in order to reduce flood risk. The need for nourishment varies from section to section. Houthuys et al. (2012) noted a long-term general trend along the Flemish coast ranging from slight accretion in the west at the French border shifting to mild erosion east at the Dutch border. For the period 2013–2020, an average yearly volume of 20 m^3 m^{-1} is considered necessary to meet the target safety level and provide a five-year buffer; which gives a total volume of 10 million m^3. To address structural erosion and the projected SLR, an extra annual volume of 7 m^3 m^{-1} corresponding to a total volume of 14 million m^3 is expected to be needed between 2020 and 2050 (Balcaen 2012) of which about half is needed to compensate for SLR. This is based on the assumption of 500 m^3 m^{-1} beach front for an average beach and a foreshore width of 500 m.

18.3.2.4 Netherlands

Since the 1990s, the strategy for the sandy coasts of the Netherlands has been one of dynamic management to stabilise the basal coastline (Rijkswaterstaat 1990). This strategy was extended offshore beyond the shoreface to the 20 m depth contour in 2001, thus including the area known as the coastal foundation (Mulder et al. 2007). On average, 12 million m^3 is used each year for nourishments along the sandy parts of the southwestern Delta, the closed Holland coast and on the West Frisian Barrier islands (Rijkswaterstaat 2011). Following the currently applied procedure (Mulder et al. 2007), increased demand for nourishments due to accelerated SLR and secondary effects will be identified by assessing the annual surveys every four years and then adjusting the amounts compensated within the following four years (Deltacommissaris 2013). However, the presently nourished volume is still insufficient to meet the aims (Mulder and Tonnon 2010): a total volume of 20 million m^3 $year^{-1}$ is needed in relation to current SLR. The reason for this difference is largely due to the demand for sediments from the Western Scheldt estuary and the tidal basins of the Wadden Sea (de Ronde 2008). Although they are excluded from the nourishment programme these coastal areas benefit from sediment import from the coastal foundation. Recent studies on the adaption of the tidal basins of the Wadden Sea to the closure of the Zuider Zee and sand-mining, show that imported sediment volumes have been more than adequate to compensate for current SLR (Elias et al. 2012) which might indicate a sediment transport capacity through the inlets that is large enough to accommodate higher rates of SLR than currently occur. An increase in yearly nourishment volume to 20 million m^3 is anticipated in the National Waterplan (MVenW 2009) but no decision has yet been made. The total amount of material to offset SLR is estimated to be proportional to the rate of SLR; 7 million m^3 per mm $year^{-1}$. The Deltacommissie (2008) suggested that sediment budgets may need to increase to 85 million m^3 $year^{-1}$ by 2050, to compensate for a SLR of 12 mm $year^{-1}$ along the whole Dutch coast including the southwestern Delta and the Wadden Sea, whereas the actually introduced scenarios for SLR assume rates of 3.5 mm $year^{-1}$ until 2050 and 8.5 mm $year^{-1}$ between 2050 and 2100 (Deltacommissaris 2013).

18.3.2.5 United Kingdom

With a coastline of about 18,000 km, the United Kingdom is characterised by a wide range of shoreline types, inlets and estuaries. Historically, responses to coastal stabilisation were piecemeal and highly variable. Solutions included both hard constructions such as seawalls, breakwaters, groynes, and offshore reefs, and soft measures such as shingle recycling, beach nourishment and salt marsh generation. This local response has now been replaced by a more coherent and regional approach, through the adoption of Shoreline Management Plans to balance the requirements for safety against hazards and economic effort. The aim is to determine defence needs at a regional scale before defining the most appropriate form of protection to fulfil the strategic need. Central to this planning is a systematic and risk-based approach, underpinned by regional monitoring. Consideration is given to coastal geomorphology, geology, ecology, exposure, flood and erosion risk, protection type, and management strategy. Programme design focuses on the monitoring requirements needed to deliver new coastal engineering schemes over the next 30 years. Baseline surveys were undertaken for each survey category. Thereafter, a weighted sampling programme was developed according to identified risks, which determines the temporal and spatial frequency of data collection, reflecting factors such as the local geomorphology, exposure to wave climate and management strategy, to determine data requirements. Essentially, those areas that present high risk of erosion or flooding, or are heavily managed have more data collection than stretches of unmanaged coast. Hence, the entire UK coast is monitored at an appropriate level of detail to provide a strategic region-wide overview of coastal change. Consistent observation, specification, quality control, metadata and analysis techniques have been developed for each programme element. Web delivery includes online tools to view data and real-time observations of an extensive network of wave and tidal observations. In addition, a range of end-user products based on annual and cumulative analysis of the data enables coastal managers to develop a region-wide understanding of coastal evolution patterns (Channel Coastal Observatory 2013).

The shoreline management programmes will become more and more effective with an increasing data basis allowing more and more purposeful reactions of regional coastal managers in order to keep coastlines stable following the same basic criteria nationwide.

18.4 Adaptation Strategies

18.4.1 Monitoring Climate Change Effects

All coastal North Sea countries undertake coastal monitoring programmes to support the planning of construction and maintenance of coastal engineering schemes. Such programmes also provide a basis for scientific studies on process analysis, improving design procedures and verifying or driving models. Current monitoring programmes include a wide range of observation techniques including:

- terrestrial surveys by GPS and LIDAR of salt marshes, tidal flats, beaches and dunes or cliffs for moderate conditions, and the upper shoreface, beaches, dunes or cliffs for post-storm conditions
- bathymetric surveys of channels, shoreface and ebb deltas by GPS and sounding
- permanent water level monitoring by gauges
- permanent measurements of currents and salinity
- permanent wave monitoring by buoys or gauges
- monitoring of sediments and habitats.

Measuring campaigns are also undertaken to strengthen the data base for analysing and modelling hydrodynamic and morphodynamic coastal processes. Measurements are supplemented by model results covering hydrodynamic and morphodynamic processes and developments.

Although all North Sea coastal countries regard coastal monitoring as essential the approaches used vary widely, particularly in terms of spatial distribution and sampling frequency. Nevertheless, these data are still useful for detecting climate change impacts and developing coastal protection measures. However, it is important to keep the national monitoring programmes under review in respect of their suitability to deliver basic information for detecting climate change impacts relevant for coastal protection. The layout of monitoring programmes on coastal hydro- and morphodynamics is generally structured according to the knowledge about coastal processes as assembled in currently used coastal classifications like, for example, that of Hayes (1979) which consider tidal range and wave climate as driving forces but no varying SLR (Hayes and Fitzgerald 2013). It is therefore advisable to check whether the existing programmes are already sufficiently structured in respect of data mining and analysis for detecting effects of climate change impacts such as accelerated SLR, increased set-up of storm surges, growing wave energy and morphodynamic adaption.

A promising tool for identifying climate change impacts would be a combination of nationwide knowledge at least at the scale of the countries surrounding the North Sea. International interdisciplinary expert groups could then evaluate which data and information would be helpful in detecting climate change impacts in coastal areas as quickly and accurately as possible. The aim of these efforts should be standardised integrated monitoring around the North Sea supplemented by specific regional programmes addressing specific regional needs. The latter could also generate high quality data sets for driving and verifying mathematical models. Emphasis should also be given to improving and further developing analytical methods for evaluating monitoring data and model results with the aim of early detection of climate change impacts, especially trends. A parallel application of distinct analytical methods and forecast tools could provide comparable results; in case that similar results were found a sounder basis for decision-making could be achieved.

Since the scenarios for climate change impacts are still accompanied by large uncertainties due to the lack of basic knowledge needed for targeted cost-effective planning for coastal protection measures, any reduction in uncertainties by monitoring and the use of models implies a very good benefit-cost ratio.

18.4.2 Belgium

The Flemish authorities aim to keep the protection line at the Belgian North Sea coast. Improvements have taken place in the harbours that are currently considered the weakest links in the protection line and through which 95 % of flooding is expected. In 2007–2008, work was undertaken to ensure a minimum safety level for a storm with a yearly exceedance probability of 10^{-2}. Quay levels must be higher than the water level with an exceedance probability of 10^{-3} and the strength of dykes, sluices and weirs are checked. A storm surge barrier will be constructed in Nieuwpoort at the entrance to the Yser estuary and to the important yacht harbour of Nieuwpoort. Although this barrier will reduce the risk of flooding from the sea, it may also increase the risk of hinterland inundation due to reduced drainage capacity unless additional measures are taken.

Repeated nourishments include a safety margin for climate change effects. In addition, groynes are used to limit longshore transport. Possible positive effects of shoreface nourishments are debated and, for the longer term conceptual ideas of increasing the height of the existing Flemish Banks to reduce wave impact on shores are under consideration. Efforts are being made to limit aeolian transport, so keeping sediments where they can best help reduce hydrodynamic loads. The main design considerations are the use of a broad berm and a mild slope close to the equilibrium beach slope for the sand under consideration, with a preference for relatively coarse sand of about 300 µm in diameter. High sand buffers in front of dykes with a minimum lifespan of five years are suggested.

In the Belgian part of the Western Scheldt an earlier study concluded that the cost of a storm surge barrier near Antwerp would not outweigh the benefits (Berlamont et al. 1982). This study did not include the possible effect of SLR and the Sigma Plan was recently revised: a combination of flood plains and heightening of dykes and quay walls is thought to provide the best solution in terms of costs for investment and maintenance and benefits such as preventing loss of agricultural production, as well as those from ecosystem services and the reduced probability of flooding in high-value areas (Broekx et al. 2011). This also means a change in strategy from a fixed safety level for the basin as a whole to a more flexible approach to safety in different parts of the basin.

18.4.3 Denmark

Protection of the hinterland against flooding at the Danish North Sea coast will continue to be achieved by keeping the protection line in its current position, with the exception of those areas where coastal retreat is regarded as acceptable and no human interference preventing it is deemed necessary. At the Danish Wadden Sea coast existing dykes are strengthened to meet prevailing safety levels and the anticipated safety margins for climate change effects are the measures used.

The protected stretch at the sandy North Sea coast of Denmark comprises those parts where the dunes are being armoured with concrete block revetments and those where the dunes are not. The minimum width for dunes with revetments is 30 m and for dunes without revetments 40 m. These values were determined using erosion data from historical storm surges. Beach and shoreface erosion is currently compensated in front of dunes without revetments and due to a lack of funding is limited to a retreat of 3.2 m year^{-1} in front of dunes with revetments, yielding narrower and lower beaches in front of the revetment. This is acceptable as long as the safety level for the revetments is not reduced beyond the safety threshold.

The adaptation strategy at the Danish North Sea coast has been developed on the basis of experience and understanding and aims less at fixed targets than at a flexible response to changing boundary conditions.

18.4.4 Germany

The Free and Hanseatic City of Hamburg generally employs the strategy of keeping the protection line in its current position. But very recently, some new infrastructure like large public buildings has been erected on dwelling mounds to prevent them flooding if dyke sections fail during a storm surge. The strategy in Schleswig Holstein for dykes at the mainland North Sea coast and on the North Frisian Islands is similar: in the current protection line dykes will be repeatedly strengthened relative to safety levels and safety margins. Since 2010, a new cross-sectional design has been applied enabling dykes to be raised up to 1.5 m for stronger hydrodynamic loads at some future date. The use of older dykes—those no longer in use due to the protection line after embankments moving seaward—is anticipated as a second protection line but is not yet implemented due to budget constraints. Tests on some sections showed the effectiveness of the second dyke line is often very limited. Nevertheless, it is considered worth preserving existing dykes in the second line as a basis for a new dyke line in the future. Information on design, dimensions and costs of strengthening dykes in the second line is lacking (MELUR 2013).

The Free and Hanseatic City of Bremen will keep the protection line in its current position. For its mainland coast and along the tidal estuaries Ems-Dollard, Weser and Elbe, Lower Saxony will do the same. This decision is the result of research on four alternative approaches. The investigations were undertaken in the Ems-Dollard estuary area which is representative of both estuaries and the Wadden Sea coast, with a stepwise increase in design water levels for a SLR of 0.65 and 1.00 m on the one hand, added to by an increase in storm-surge set-up of 0.35 and 0.5 m on the other (Niemeyer et al. 2014). The latter account for higher wind velocities on the one hand and higher and less attenuated wave energy on the other. Tidal flats were assumed not to adapt to accelerated SLR. This pessimistic scenario led to the following results:

- *Retreat* from all areas with flood risk due to storm surges in order to save on the cost of coastal protection. This implies that 1.2 million people in Lower Saxony would need to move to safe areas and about 800,000 people in neighbouring states would be at risk. However, cost savings versus economic losses mean that this strategy is out of the question, even for more pessimistic scenarios than those considered here.
- *Accommodation* by limiting coastal protection to settlements above a certain threshold of inhabitants and economic value. The costs of implementing the new coastal protection schemes are about 25 % of the capital costs of the existing protection line if only the larger cities are safeguarded against storm surges and are of the same order of magnitude if all small villages are also protected. In addition, enormous efforts would be required to keep infrastructure between the protected areas such as railways, streets, energy supply lines operational after storm surge flooding. Even excluding other major disadvantages of this strategy, it is still clear that maintaining and strengthening the existing protection line is a better economic solution.
- *Set-back or realignment* leads to higher hydrodynamic loads than occur at the corresponding outer protection line, in all those areas where it has been moved seaward. Land levels in the areas sheltered by new dyke lines after reclamation have not been subject to sedimentation and are now lower than areas seaward of the dyke, particularly in saltmarshes. The greater water depths in front of the landward-shifted dykes by set-back allow higher wave energy. Without a gain in safety and with extra investment costs exceeding the current yearly budgets for coastal protection 120-fold for new dykes (MU 2012) this alternative is not better than the strategy of keeping the protection line in its current position.
- *Combined protection* with two structures; one for wave attenuation seaward and another to contain storm surge levels landward. Collectively these two structures would require a higher cross-sectional area than a single protection line. The safety achieved by such a scheme is less than that achieved by a conventional dyke and the costs would be significantly higher than for one dyke line.

The results show that strengthening the existing protection line is still the most effective solution both in terms of safety and cost (Niemeyer et al. 2011a, b, 2014). The government commission (MU 2012) and subsequently the State Government decided to follow the strategy of keeping the line in its current position and strengthening the protection structures. This approach was approved by the self-ruling dyke communities and through representative polls of people in the protected areas (MU 2012). To date, further studies have been undertaken for a SLR of 1.0 m, an additional increase in storm surge set-up of 0.5 m and consistently higher and longer waves in the area of the Ems-Dollard estuary (Knaack et al. 2015). These studies led to the same conclusions as the previous studies: keeping the line is the optimal strategy for future protection of lowland coasts at the southern North Sea.

Successful site investigations on wave overtopping of dykes have been undertaken in Denmark (Laustrup et al. 1991), the Netherlands (van der Meer et al. 2009) and

Vietnam (Le et al. 2013). They prove that higher overtopping volumes on dykes than are currently considered tolerable will be acceptable without failure of the structure. Wave overtopping on dykes has been modelled in combination with soil laboratory tests of the covering clay to develop an integrated design that takes into account both hydrodynamics and soil mechanics (Berkenbrink et al. 2010; Richwien et al. 2011). Several tests showed cover layers remained functional for overtopping volumes up to 200 [l× (m s)$^{-1}$]. As overtopping volumes of this magnitude would probably cause severe damage in populated areas, acceptable overtopping volumes should be smaller. Studies were undertaken to quantify the extent to which an enhanced overtopping tolerance could counterbalance the effects of SLR or other climate change effects at three representative cross-sections for coastal and estuarine dykes in Lower Saxony. The results showed that an overtopping tolerance of 10 [l× (m s)$^{-1}$] would allow a reduction in dyke crest heights for presently applied design conditions of 45–60 cm at the Lower Saxony North Sea coast and adjacent estuaries (Niemeyer et al. 2010). This suggests a survey of cover layers for all Lower Saxony coastal and estuarine dykes could help improve estimates of the design parameters for site-specific acceptable wave overtopping volumes, as part of the adaptation strategy for coastal protection (MU 2012). Such a survey would also identify weak points in the existing protection line.

Protection of the East Frisian Islands is currently undertaken using the same guidelines as in the past since the lack of understanding about climate change impacts on morphodynamic processes hampers the development of a resilient adaptation strategy (MU 2012).

18.4.5 Netherlands

The most recent decision on coastal protection strategy in the Netherlands is the adoption of a three-layer safety scheme combined with a new design procedure orientated at the probability of the loss of human life: prevention of flooding by keeping, strengthening and safeguarding the protection line remains the basis, which is extended with supporting measures to reduce the consequential damage of structural failure. The three-layer safety scheme is as follows:

- Layer 1: prevention of flooding by establishing and maintaining an effective flood protection system
- Layer 2: spatial planning such that the impact of flooding after the failure of protection structures is reduced
- Layer 3: disaster control through detailed evacuation plans, making sure that vital infrastructure is still functional in the event of a flood, and the creation of safe havens.

The self-governing waterboards ask for priority to be given to strengthening of the protection structures in order to meet prevailing safety norms, before investing in the second and third layer. The new system of risk-based safety norms differs from the current norm system based on hydrodynamic loads. The Cabinet adopted the new norm system in November 2014; its application is scheduled for 2017 and it is expected to take until 2050 for it to be implemented across the coastal protection system as a whole.

The costs of improving the structure of all existing dykes including those along inland waters, is estimated at about EUR 6.5 billion. An additional EUR 5 billion would be required for adaptation to a SLR of 0.5 m. The sum of both is beyond the likely budget for the Delta Program for 2013 to 2028 (see Sect. 18.2.4). Conceptual studies on very safe dykes (Silva and van Velzen 2008) project an overtopping tolerance of 30 [l× (m s)$^{-1}$] for coastal dykes.

The flood protection scheme for the Netherlands has a unique configuration: dyke rings surrounding protected areas. There are currently 54 dyke rings and the associated protection structures have a total length of 3767 km, about 30 % of these are at open tidal waters. Water Plan Beaufort is currently under development and is aimed at reducing the costs involved in improving protection structures to meet future safety levels, including those associated with climate change effects, as well as increasing options for setting priorities (Beaufort 2010/2013). A major element of the plan is a reduction in the number of dyke rings from 54 to 2 in line with the overall vision of shorter protection lines along sea and rivers and free outflow of rivers to the sea. Stronger dykes and extra locks and sluices are envisaged. Implementation may be phased, and efficient use of budgets should enable an increase in safety levels. Rough estimates indicate a cost saving of 50 % compared to implementing the safety standards according to the Delta program. Water Plan Beaufort includes stronger dykes than at present and more thorough dyke inspections for detecting weak spots (Beaufort 2010/2013). The plan is still under development and not yet included in planning by the responsible Deltacommissaris (2013).

The strategy for keeping coastal dunes and protection structures safe and the coastal foundation stable by nourishments involves significant cost, although considered in terms of an insurance premium for protecting around EUR 1800 billion (Deltacommissie 2008) of invested capital in the protected area the costs seem relatively moderate and more reasonable. Nevertheless, attempts to make artificial nourishments more efficient, particularly by generating and applying knowledge of coastal processes, are still worthwhile.

An impressive example is the ‘Delfland Sand Motor’, a mega-nourishment with a volume of about 21.5 million m^3 (Fig. 18.3) which is almost as big as the currently

Fig. 18.3 Mega-nourishment 'Delfland Sand Motor'. After completion in July 2011 (*left*) and in May 2015 (*right*) after reshaping (https://beeldbank.rws.nl, Rijkswaterstaat/Joop van Houdt)

implemented two-year volume of 24 million m^3. Designing the sand motor required intensive testing by morphodynamic model predictions in order to optimise its shape and to compare its effectiveness with conventional nourishments for maintaining the coastal foundation of Delfland (Fig. 18.3). To keep pace with the present rate of SLR this requires about 5.5 million m^3 within five years (Mulder and Tonnon 2010). Model results and measurements so far indicate that the Delfland Sand Motor will contribute to the maintenance of the coastal foundation of Delfland for around 25 years and that it is more cost-effective than repeated nourishments. But the model results also indicate that five additional nourishments will be necessary to maintain the coastline for this period. The alternatives for the shape of the sand motor have been shoreface nourishment, a bell-shape and a sandy hook (Mulder and Tonnon 2010). Because the coastal processes will rapidly transform any initial shape into a bell-shaped salient, the long-term morphological effects of the alternatives are similar. Combining the aim of the mega-nourishment to create long-term safety conditions as well as extra space for nature and recreation in an innovative manner, the environmental impact assessment showed the hook shape was preferable (Mulder and Tonnon 2010). The reshaping of the mega-nourishment is monitored and analysed. The results will improve the understanding of the effectiveness of this new type of artificial nourishment.

Another option tested recently is a seaward build-out of sandy coasts by over-nourishment, partly combined with supporting solid structures (Stronkhorst et al. 2010): coastal stretches receive excess amounts of sediment, creating beaches and dunes for nature conservation and recreation. Although the Deltacommissaris (2013) stated that this is a viable option, no specific decisions on this have yet been made.

18.4.6 United Kingdom

The approach to coastal protection in the United Kingdom focuses now on 'sedimentary cells' to reflect the adaptation needs of a regionally-varying coastline in terms of landscape, sedimentology and coastal dynamics. A distinction is made between coastal zone management (CZM) and so-called shoreline management. The former is predominantly a planning issue, seeking to reconcile the demands of development with the requirement for adequate protection of the natural environment. In contrast, shoreline management focuses on one aspect of CZM, namely coastal hazards, and concerns efforts to manage flood and erosion risk at the shoreline (Nicholls et al. 2013).

In the early 1990s, the government developed guidance for the preparation of 40 Shoreline Management Plans (SMPs) across England and Wales (MAFF 1995). The main objective was to define management units along the coast and consider the most appropriate Strategic Coastal Defence Options (SCDOs). The SCDOs considered for each management unit comprised four options for the strategy to be applied:

- do nothing
- maintain the existing protection line (while possibly adjusting the protection standard)
- advance the existing protection line
- retreat the existing protection line (subsequently referred to as 'managed realignment').

The management units were then used to initiate a consultation process and the compilation of each SMP, which, in some cases was adopted by the relevant authorities but this was and remains a non-statutory process. Outputs from the first round of SMPs were frequently biased towards the status quo—a fixed shoreline—which was at odds with the desire to move towards a more dynamic and adaptive coast, where appropriate. This led to a careful review of the process (Leafe et al. 1998) and new guidance was developed to promote the preparation of the second round of SMPs (DEFRA 2001). In this new guidance, greater emphasis was placed on:

- ensuring a more consistent evidence base was established
- the engagement of stakeholders throughout the process (but particularly in objective setting and selection of preferred options)
- adoption of the plans by the relevant authorities (DEFRA 2001).

Following a series of trials, this guidance was formally released (DEFRA 2006) and applied to England and Wales (DEFRA 2011). The second generation of SMPs are currently in production and when complete will cover the entire 6000 km shoreline. The intention is that the SMPs provide a 'route map' for local authorities and other decision makers to identify the most sustainable approaches to managing risks to the coast in the short term (0–20 years), medium term (20–50 years) and long term (50–100 years), recognising that changes to the present protection structures may need to be carried out as a staged process. Each SMP will include an action plan that prioritises works needed to manage specific flood and erosion risks, along with details of the coastal erosion monitoring and further research needed to support the plan. The SMPs then inform more detailed strategy studies, which explore the most effective form of delivery, with an increasing focus on adaptation measures that are more likely to be sustainable under a changing climate. For example, the long-term strategy for managing flood risk on the Thames Estuary, termed the Thames Estuary 2100 or TE2100 Project, includes options for managing flood risk to 2100, based on current government projections of climate change. Each option comprises a sequence of interventions to 2100 and beyond and the assessment included consideration of the H++, a low probability, high consequence scenario, which considers the possibility of large contributions to SLR from the Greenland and Antarctica ice sheets. The dates of implementation depend on the rate of climate change and other factors. If change such as rising sea level, or deterioration of the safety status of protection schemes occurs more rapidly than projected in the plan, intervention dates will be brought forward and vice versa. In this way, the timing of interventions on the estuary will be optimised, taking account of actual rates of change and associated updates of scientific knowledge and future projections. While this approach was developed specifically for London and the Thames Estuary, the concepts are now being adopted more widely (HM Treasury/DEFRA 2009).

18.5 Summary and Recommendations

This overview indicates that all countries around the North Sea with coastal areas vulnerable to flooding from storm surges are ready for the challenges that climate change is expected to bring. Scenarios have been developed and investigated as a basis for policy development, regulation and guidance, to provide a structured response that should ensure continued protection with the required level of safety for coastal flood prone areas.

Scenarios of accelerating SLR leading to changes in sea level of up to 1 m or more by 2100 have been used for planning the adaptation of coastal protection schemes. Thus the safety margins considered in all countries around the North Sea are consistent with the upper limit of SLR to 2100 reported in the latest assessment of the Intergovernmental Panel on Climate Change (IPCC 2013). There appears to be a tendency for countries with higher safety levels to consider smaller safety margins for climate change impacts than those with lower safety levels. Increased storm surge set-up and higher wave energy due to higher wind velocities are incorporated in the future design of coastal protection structures in Denmark, Bremen, Lower Saxony and the United Kingdom.

The United Kingdom has established a coastal protection strategy for a flexible response to erosion that reflects the varying conditions around the coast. The resulting strategy ranges from doing nothing or set-back of the protection line by managed realignment, to strengthening of the existing protection line. Denmark allows retreat at some stretches of its North Sea coast and maintains the protection line in the rest. All other countries aim at keeping the current protection line in place to protect the hinterland. In the Netherlands, a decision on implementing additional measures for reducing damage due to the failure of protection structures will be made in 2015. Investigations showed that a reliable basis for evaluating protection strategies is only achievable if real world tests are carried out, since conceptual studies can be misleading.

In all countries, artificial nourishments are traditionally used for combatting structural erosion on sandy coasts and this is expected to increase under future climate change impacts. This approach will thus be used more often and at higher rates for keeping the coastline in position according to the current criteria for intervention. The required increase in nourishment volumes needed to stabilise coastlines has been investigated in Belgium, Denmark and the Netherlands. In the Netherlands, models and large-scale site experiments have been used to gain deeper understanding of the relevant processes with the aim of increasing the efficiency of artificial nourishments or even moving the coastline seaward. In most countries, studies have been carried out to identify borrow areas with appropriate sediments and the volumes available. But there are still knowledge gaps concerning the long-term availability of sediments needed for nourishments to compensate for projected SLR, especially in terms of their being necessary to fulfil the needed volumes for nourishments in order to compensate the impacts of climate change in the long run, in particular in respect of the quality of sediments refilling the borrow pits and their suitability for future nourishments. A good understanding of the availability of suitable sediment reservoirs for nourishment is crucial for a sustainable management strategy to protect sandy coastal environments.

Climate change studies are based on scenarios rather than forecasts and this generates uncertainties, which by the end of a chain of processes may be unquantifiable. As a result, all North Sea countries use ongoing monitoring programmes for coastal management purposes. To help detect the impacts of climate change, some countries will even extend these monitoring programmes. The data provide a sound basis for detecting changes in trends. Testing existing tools and developing new analytical tools would be beneficial. Cooperation at a European scale would not only improve the exchange of knowledge, but would also improve the availability of tools, methodology and resources for problem solving.

Present knowledge already highlights that the effects of climate change at dynamic sandy coasts are stronger than on mainland coasts with cohesive sediments, such as estuaries or tidal basins with large intertidal areas and saltmarshes. Although the morphodynamic processes that are likely to occur due to climate change are reasonably well known, their quantification—if even possible—still involves large uncertainties. Filling the enormous knowledge gaps that still remain will be a challenge for coastal engineering in the future. Mitigating the morphodynamic changes due to climate change impacts will create high budget demands. For efficient measures it is necessary to understand and predict the hydrodynamic and morphodynamic changes that are likely to result from climate change. This justifies much higher budgets for research in this particular field than at present. Advancing process knowledge and improving long-term morphodynamic modelling are indispensable preconditions for providing decision-makers with a sound basis for target-orientated optimised measures. The knowledge potential in this field of expertise is extraordinarily good in Europe. The countries surrounding the North Sea would therefore benefit significantly from a co-ordinated programme aimed at reducing the knowledge gaps highlighted in this chapter.

References

Afdeling Kust (2011) Masterplan Kustveiligheid. [Masterplan Coastal Safety] http://zeeweringenkustbeheer.afdelingkust.be/Userfiles/pdf/110628_RL_Rappor_%20 KustveiligheidWEB4.pdf (In Dutch)

Balcaen N (2012) Masterplan Kustveiligheid [Masterplan Coastal Safety]. www.pianc-aipcn.be/figuren/verslagen%20activiteiten%20Pianc%20Belgi%C3%AB/Powerpoint/Kustveiligheid%20Nathalie%20Balcaen%2015feb2012.pdf (In Dutch)

Beaufort GA (2010/2013) Plan Beaufort, www.bureaubeaufort.com

Berlamont J, Sas M, Van Langenhove G, Thienpont M (1982) Multi-en interdisciplinaire evaluatiestudie betreffende de stormvloedkering te Antwerpen (Oosterweel) [Multi- and interdisciplinary evaluation study on a storm surge barrier at Antwerpen]. Report Hydraulics Laboratory KU Leuven (In Dutch)

Berkenbrink C, Kaiser R, Niemeyer HD (2010) Mathematical modelling of wave overtopping at complex structures: Validation and comparison In: McKee Smith J, Lynett P (eds), Proc 32nd Int Conf on Coastal Engineering, Shanghai, China. http://dx.doi.org/10.9753/icce.v32.structures.4

Broekx S, Smets S, Liekens I, Bulckaen, D, De Nocker L (2011) Designing a long-term flood risk management plan for the Scheldt estuary using a risk-based approach. Nat Hazards 57:245–266

Bruun P (1962) Sea-level rise as a cause of shore erosion. Am Soc Civil Engineers Proc, J Waterw & Harb Div 88:117–130

Channel Coastal Observatory (2013) Strategic regional coastal monitoring programmes of England, www.channelcoast.org

ComCoast (2007) ComCoast flood risk management schemes. www.ec.europa.eu/ourcoast/download.cfm?fileID=773

Danish Coastal Authority (2012) Guidelines for klimattilpasning I kystområder [Guidelines for climate change adaptation in coastal areas], www.masterpiece.dk/UploadetFiles/10852/36/Klimatilpasningikystomr%C3%A5derversion2.pdf (In Danish)

Danish Government (2008) Danish Strategy for Adaptation to a Changing Climate. Danish Energy Agency

De Ronde J (2008) Toekomstige langjarige suppletiebehoefte. [Future long-term need for nourishments] Deltares Report Z4582.24, www.vliz.be/en/imis?module=ref&refid=218721 (In Dutch)

DEFRA (2001) Shoreline Management Plans: A guide for coastal defence authorities. Department for Environment Food & Rural Affairs (DEFRA), London

DEFRA (2006) Shoreline Management Plan Guidance. Vol 1: Aims and requirements, Vol 2: Procedures (plus 10 appendices). Department for Environment Food & Rural Affairs (DEFRA), London

DEFRA (2011) UK Climate Projections. www.ukclimateprojections.defra.gov.uk/09 Department for Environment Food & Rural Affairs (DEFRA), London

DEFRA (2012) UK Climate Change Risk Assessment: Government Report, Department for Environment Food & Rural Affairs (DEFRA), London www.gov.uk/government/publications/uk-climate-change-risk-assessment-government-report

Deltacommissaris (2013) www.deltacommissaris.nl/

Deltacommissie (2008) Working Together with Water: A living land builds for its future. Findings of the Deltacommissie 2008. www.deltacommissie.com/doc/deltareport_summary.pdf

Deltaprogramma Ijsselmeergebied (2013) Bijlage A3 Deltaprogramma 2014 [Supplement A3 Delta Programme 2014]. (In Dutch)

EEA (2012) Climate Change, Impacts and Vulnerability in Europe 2012. An indicator-based report. EEA Report No 12/2012. European Environment Agency (EEA)

Elias EPL, Van Der Spek AJF, Wang ZB, De Ronde J (2012) Morphodynamic development and sediment budget of the Dutch Wadden Sea over the last century. Neth J Geosci 91:293–310

Environment Agency (2013) Thames Estuary 2100 (TE2100) Plan. London

Hayes MO (1979) Barrier island morphology as a function of tidal and wave regime. In: Leatherman S (ed), Barrier Islands: From the Gulf of St. Lawrence to the Gulf of Mexico. Academic, pp 1–27

Hayes MO, Fitzgerald D (2013) Origin, evolution, and classification of tidal inlets. J Coast Res si 69:14–33

Helpdesk Water (2015) Kustveiligheid [Coastal Safety] www.helpdeskwater.nl/onderwerpen/water-en-ruimte/klimaat-0/factsheets/kustveiligheid/ (In Dutch)

HM Treasury/DEFRA (2009) Accounting for the Effects of Climate Change: Supplementary Green Book Guidance. HM Treasury and Department for Environment Food & Rural Affairs (DEFRA), London

Houthuys R, Mertens T, Geldhof A, Trouw K (2012) Long-term autonomous morphological trends of the Belgian shore. In: Belpaeme K, McMeel O, Vanagt T, Mees J (eds) 2012. Book of Abstracts. International Conference Littoral 2012: Coasts of Tomorrow, 27–29 November 2012, Oostende, Belgium. VLIZ Special Publication 61:191–193

IPCC (1990) Strategies for adaption to sea-level rise. Executive Summary of the Coastal Zone Management Subgroup. Intergovernmental Panel on Climate Change (IPCC), Response Strategies Working Group

IPCC (2013) Climate Change 2013: The Physical Science Basis. Contribution of Working Group I to the Fifth Assessment Report of the Intergovernmental Panel on Climate Change. Stocker, T.F., D. Qin, G.-K. Plattner, M. Tignor, S.K. Allen, J. Boschung, A. Nauels, Y. Xia, V. Bex and P.M. Midgley (eds.). Cambridge University Press.

Jensen J, Sørensen C (2008) Fremskrivning af fodringsindsatsen på Vestkysten [Projections of sand nourishment on the Danish west coast] Kystdirektoratet, Denmark (In Danish)

Kaiser R, Niemeyer HD (1999) Changing of local wave climate due to ebb delta migration. In: Edge BL (ed) Proc 26th Int Conf on Coastal Engineering, Copenhagen, Denmark

Knaack H, Berkenbrink C, Wurpts A, Kaiser R, Niemeyer HD (2015). Modellierung von Bemessungssturmfluten und Bemessungsseegang im Ems-Dollard-Ästuar für Klimaänderungsfolgen [Modeling of design storm surges and design waves due to climate change impacts]. Ber Forschungsstelle Küste 44:76–85 (In German)

Laustrup C, Toxvig Madsen H, Jensen J, Poulsen L (1991) Dike failure calculation model based on in situ tests. In: Edge BL (ed) Proc 22nd Int Conf on Coastal Engineering, The Hague, Netherlands. Part V: Coastal Processes and Sediment Transport, 2671–2681

Le HT, Van Der Meer J, Quang Luong N, Verhagen HJ, Schiereck GJ (2013) Wave overtopping resistance of grassed dike slopes in Vietnam. Coastal Structures 2011:191–202

Leafe R, Pethick J, Townend IH (1998) Realizing the benefits of shoreline management. Geogr J 164:282–290

Lowe JA, Howard TP, Pardaens A, Tinker J, Holt J, Wakelin S, Milne G, Leake J, Wolf J, Horsburgh K, Reeder T, Jenkins G, Ridley J, Dye S., Bradley S (2009) UK Climate Projections Science Report: Marine and coastal projections. Met Office Hadley Centre, Exeter

MAFF (1995) Shoreline Management Plans: A guide for coastal defence authorities. Ministry of Agriculture Fisheries and Food (MAFF), London

MELUR (2013) Generalplan Küstenschutz des Landes Schleswig-Holstein [Masterplan Coastal Protection of the Federal State of Schleswig-Holstein]. Ministerium für Energiewende, Landwirtschaft, Umwelt und ländliche Räume des Landes Schleswig-Holstein. www.schleswig-holstein.de/DE/Fachinhalte/K/kuestenschutz/Downloads/Generalplan.pdf;jsessionid=59F186586AAC9A3C825EF3AF0D4754BA?__blob=publicationFile&v=1 (In German)

MIenM (2013) Koersbepaling waterbeleid en toezeggingen WGO van 10 december 2012 [Letter of the Minister for Infrastructure and Environment to the Second Chamber of the Dutch Parliament]. www.rijksoverheid.nl/documenten-en-publicaties/kamerstukken/2013/04/26/koersbepaling-waterbeleid-en-toezeggingen-wgo-van-10-december-2012.html (In Dutch)

MU (2012) Empfehlungen für eine niedersächsische Strategie zur Anpassung an die Folgen des Klimawandels [Recommendations for an Adaption Strategy for Climate Change Effects in Lower Saxony]. www.umwelt.niedersachsen.de/klimaschutz/aktuelles/107128.html (In German)

Mulder JPM, de Vos FJ (1989) Kustverdediging na 1990 - inventarisatie functies onderwateroever [Coastal Protection after 1990: Inventory of shoreface functions]. Rijkswaterstaat. Rep No 9 (In Dutch)

Mulder JPM, Tonnon PK (2010) 'Sand engine': Background and design of a mega-nourishment pilot in the Netherlands. In: McKee Smith J, Lynett P (eds), Proc 32nd Int Conf on Coastal Engineering, Shanghai, China. http://dx.doi.org/10.9753/icce.v32.management.35

Mulder JPM, Nederbragt G, Steetzel HJ, Van Koningsveld M, Wang ZB (2007) Different implementation scenarios for the large scale coastal policy of the Netherlands. In: McKee Smith J (ed) Proc 30th Int Conf on Coastal Engineering, San Diego, California

Müller JM, Zitman T, Stive MJF, Niemeyer HD 2007. Long-term morphological evolution of the tidal inlet 'Norderneyer Seegat', In: McKee Smith J (ed) Proc 30th Int Conf on Coastal Engineering, San Diego, California

MVenW (2009) Nationaal Waterplan [National Waterplan] 2009–2015. www.rijksoverheid.nl/documenten-en-publicaties/rapporten/2009/12/01/nationaal-waterplan-2009-2015.html (In Dutch)

Nicholls RJ, Townend IH, Bradbury AP, Ramsbottom D, Day SA (2013) Planning for long-term coastal change: experiences from England & Wales. Ocean Eng 71:3–16

Niemeyer HD (1983) Über den Seegang an einer inselgeschützten Wattküste [On waves at an island-sheltered Wadden Sea coast]. BMFT-Forschungsber. MF 0203 (In German)

Niemeyer HD (2005) Coastal protection of lowlands: Are alternative strategies purposeful for changing climate? In: Proceedings of the 14th Biennial Coastal Zone Conference, New Orleans, Louisiana

Niemeyer HD (2010) Protection of coastal lowlands: Are alternative strategies a match to effects of climate change? In: Proc 17th IAHR-APD Conference, Auckland, New Zealand

Niemeyer HD (2015) Effekte des Klimawandels auf Randbedingungen im Insel- und Küstenschutz -gegenwärtige und zu erwartende Trends- [Effects of climate change on boundary conditions for coastal protection - current and expected trends]. Ber Forschungsstelle Küste 44:1–55 (In German)

Niemeyer HD, Kaiser R (2001) Design wave evaluation for coastal protection structures in the Wadden Sea. In: Edge BL, Hemsley JM (eds) Ocean Wave Measurement and Analysis, Vol. II
Niemeyer HD, Kaiser R, Berkenbrink C (2010) Increased overtopping security of dykes: a potential for compensating future impacts of climate change. In: McKee Smith J, Lynett P (eds), Proc 32nd Int Conf on Coastal Engineering, Shanghai, China. http://dx.doi.org/10.9753/icce.v32.structures.11
Niemeyer HD, Kaiser R, Knaack H, Dissanayake P, Miani M, Elsebach J, Berkenbrink C, Herrling G, Ritzmann A (2011a) Evaluation of coastal protection strategies for lowlands in respect of climate change. In: Valentine EM, Apelt CJ, Ball J, Chanson H, Cox R, Ettema R, Kuczera G, Lambert M, Melville BW, Sargison JE (eds) Proc 34th World Congress Int Assoc Hydro-Environm Res & Eng, Australia
Niemeyer HD, Berkenbrink C, Miani M, Ritzmann A, Dissanayake P, Knaack H, Wurpts A, Kaiser R (2011b) Coastal protection of lowlands: are alternative strategies a match to effects of climate change. In: Schüttrumpf H, Tomassicchio GR (eds) Proc 5th Int Short Conf Appl Coast Res, Aachen, Germany
Niemeyer HD, Berkenbrink C, Ritzmann A, Knaack H, Wurpts A, Kaiser R (2014) Evaluation of coastal protection strategies in respect of climate change impacts. Die Küste 81:565–578
Ranasinghe R, Duong TM, Uhlenbrook S, Roelvink D, Stive MJF (2012) Climate-change impact assessment for inlet-interrupted coastlines. Nature Clim Change 3:83–87
Richwien W, Pohl C, Vavrina L (2011) Bemessung von Deichen gegen Einwirkungen aus Sturmfluten [Dyke design in dependence of loads during storm surges]. Die Küste 77:1–44 (In German)
Rijkswaterstaat (1989) Kustverdiging na 1990 – discussienota [Coastal Protection after 1990 - discussion]. (In Dutch)
Rijkswaterstaat (1990) A new coastal defence policy for the Netherlands. Public Works Department, Tidal Waters Division
Rijkswaterstaat (2005) Veiligheid Nederland in Kaart. Hoofdrapport onderzoek overstromingsrisico's [The Netherlands on maps, Report on the risk of flooding]. DWW 2005-81 (In Dutch)
Rijkswaterstaat (2011) Suppletieprogramma: 2012-2015 [Nourishment programme 2012-2015] http://slideplayer.nl/slide/2837415/
Rijkswaterstaat (2015a) Dijkringrapporten [Dyke Ring Reports]. www.helpdeskwater.nl/onderwerpen/waterveiligheid/programma'-projecten/veiligheid-nederland/publicaties/dijkringrapporten/ (In Dutch)
Rijkswaterstaat (2015b) Kwaliteit Waterkeringen [Quality of flood protection structures]. http://www.rijkswaterstaat.nl/water/waterbeheer/bescherming-tegen-het-water/kwaliteit-waterkeringen/index.aspx (in Dutch)
Schuerch M, Vafeidis A, Slawig T, Temmerman S (2013) Modeling the influence of changing storm patterns on the ability of a salt marsh to keep pace with sea-level rise. J Geophys Res 118:84–96
Senat FHH (2012) Hochwasserschutz für Hamburg [Flood Protection in the City of Hamburg]. Mitteilung des Senats an die Bürgerschaft-Drucksache 20/5561, www.landtag.nrw.de/portal/WWW/dokumentenarchiv/Dokument/GGD20-5561.pdf (In German)
Sigmaplan (2016) The Sigma Plan www.sigmaplan.be/en
Silva W, van Velzen E (2008) De dijk van de toekomst? [Dyke for the Future?]. Deltares Report No RWS 2008.052, www.rijksoverheid.nl/documenten-en-publicaties/rapporten/2008/10/01/de-dijk-van-de-toekomst-quick-scan-doorbraakvrije-dijken.html (In Dutch)
Stive MJF, de Vriend HJ (1995) Modelling shoreface profile evolution. Mar Geol 126:235–248
Stive MJF, Eysink W (1989) Dynamisch model van het Nederlandse kustsystem [Dynamical model of the Dutch coastal system]. WL Delft Hydraul Rep M 825-IV (In Dutch)
Stronkhorst J, Van Rijn L, De Vroeg H, De Graaff R, Van Der Valk B, Van Der Spek A, Mulder JPM, Oost A, Luijendijk A, Oude Essink G (2010) Technische mogelijkheden voor een dynamische kustuitbreiding. [Technical possibilities of seaward extensions of coastlines] Deltares 1201993-004-VEB-0002, www.rijksoverheid.nl/binaries/rijksoverheid/documenten/rapporten/2010/07/29/technische-mogelijkheden-voor-een-dynamische-kustuitbreiding-een-voorverkenning-t-b-v-het-deltaprogramma/technische-mogelijkheden-voor-een-dynamische-kustuitbreidingbieb.pdf (In Dutch)
STW (2013) Publicatie in Nature voor nieuwe manier van kustverdediging met schorren en kwelders [Publication in 'Nature' on a New Approach in Coastal Protection by Shallows and Salt Marshes]. www.stw.nl/en/node/6502 (In Dutch)
SUBV (2012) Klimawandel in Bremen - Folgen und Anpassung. [Climate change in Bremen – impacts and adaptation] www.bauumwelt.bremen.de/sixcms/media.php/13/BdV_L_S_Klimaanpassung%20Anlage_Endf.pdf (In German)
Temmerman S, Meire P, Bouma TJ, Peter MJ, Herman PMJ, Ysebaert T, De Vriend HJ (2013) Ecosystem-based coastal defence in the face of global change. Nature 504:79–83
Townend IH, Fletcher C, Knappen M, Rossington K (2011) A review of salt marsh dynamics. Water Environ J 25:477–488
Van De Graaff J, Niemeyer HD, Van Overeem, J (1991) Beach nourishment, philosophy and coastal protection policy. Coast Eng 16:3–22
Van Der Meer JW, Steendam GJ, De Raat G, Bernardini P (2009) Further developments on the wave overtopping simulator. In: McKee Smith J (ed) Proc 31st Int Conf on Coastal Engineering, Hamburg, Germany
Vlaamse Baaien (2015a) Flanders Bays 2100. www.vlaamsebaaien.com/flanders-bays-2100
Vlaamse Baaien (2015b) http://www.maritiemetoegang.be/vlaamse-baaien
Weisse R, von Storch H, Niemeyer HD, Knaack H (2012) Changing North Sea storm surge climate: an increasing hazard? Ocean Coast Manage 68:58–68
Woth K, Weisse R, von Storch H (2006) Climate change and North Sea storm surge extremes: ensemble study of storm surge extremes expected in a changed climate projected by four different regional climate models. Ocean Dynam 56:3–15

19 Socio-economic Impacts—Coastal Management and Governance

Job Dronkers and Tim Stojanovic

Abstract

Climate change will have important impacts on the North Sea coastal zones. Major threats include sea-level rise and the associated increase in flood risk, coastal erosion and wetland loss, and hazards arising from more frequent storm surges. The North Sea countries—Belgium, Denmark, France, Germany, the Netherlands, Norway, Sweden and the UK—have developed strategies to deal with these threats. This chapter provides a short introduction to the present adaptation strategies and highlights differences and similarities between them. All the North Sea countries face dilemmas in the implementation of their adaptation strategies. Uncertainty about the extent and timing of climate-driven impacts is a major underlying cause. In view of this, adaptation plans focus on no-regret measures. The most considered measures in the North Sea countries are spatial planning in the coastal zone (set-back lines), wetland restoration, coastal nourishment and reinforcement of existing protection structures. The difficulty of identifying the climate-driven component of observed change in the coastal zone is a critical obstacle to obtaining a widely shared understanding of the urgency of adaptation. A better coordinated and more consistent approach to marine monitoring is crucial for informing policy and the general public and for developing the adaptive capacity of institutions and wider society. A dedicated coastal observation network is not yet in place in the North Sea region.

19.1 Introduction

Climate change will have important impacts in the coastal zones of the eight countries around the North Sea: Belgium, Denmark, France, Germany, the Netherlands, Norway, Sweden and the UK. Major threats include sea-level rise and the associated increase in flood risk, coastal erosion and wetland loss, and hazards arising from more frequent storm surges. The North Sea countries have developed strategies to deal with these threats. For each country a short introduction is given to their present adaptation strategy; differences and similarities are highlighted. All the North Sea countries face dilemmas in the implementation of their adaptation strategies. Uncertainty about the extent and timing of climate-driven impacts is a major underlying cause. Several approaches are available to deal with these dilemmas. The key findings are summarised in a final section.

19.2 Coastal Management in the North Sea Countries

This section briefly reviews coastal management practice in the North Sea countries in relation to climate change. Some working definitions of key terms used within this chapter are given in Box 1.

J. Dronkers (✉)
Deltares, Delft, and Netherlands Centre for Coastal Research, Delft, The Netherlands
e-mail: j.dronkers@hccnet.nl

T. Stojanovic (✉)
Department of Geography and Sustainable Development, University of St Andrews, St Andrews, UK
e-mail: tas21@st-andrews.ac.uk

M. Quante and F. Colijn (eds.), *North Sea Region Climate Change Assessment*, Regional Climate Studies, DOI 10.1007/978-3-319-39745-0_19

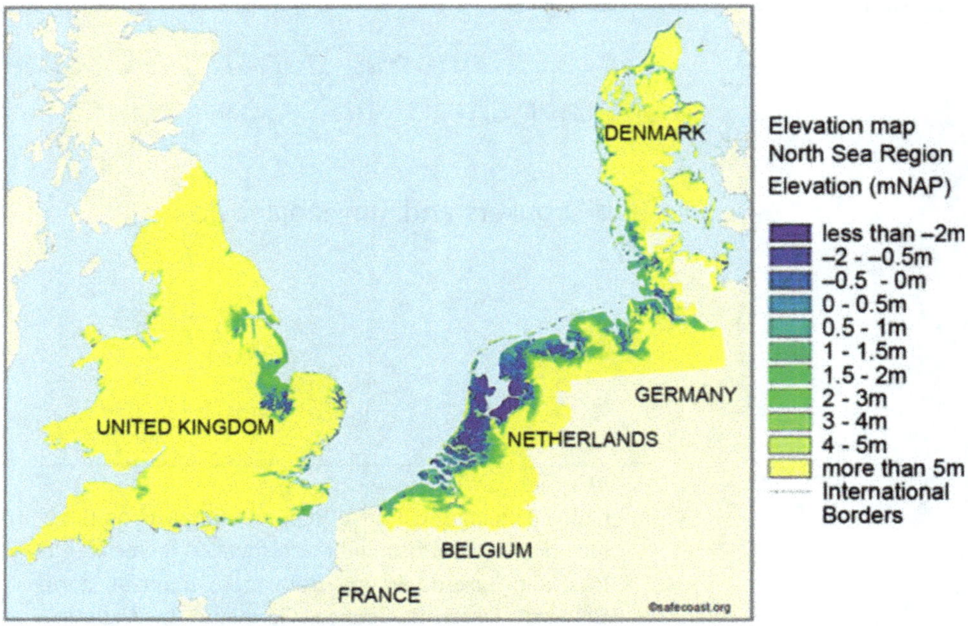

Fig. 19.1 North Sea regions potentially vulnerable to inundation by the sea (Roode et al. 2008)

19.2.1 The Coastal Zone

The shoreline is the most obviously delineated feature of the coastal zone. The North Sea countries have no commonly adopted definition of what else should be considered as the 'coastal zone'. Shoreline management mainly deals with coastal protection; this is the topic of Chap. 18. The present chapter deals mainly with coastal zone governance issues. Whether the societies in North Sea countries effectively adapt to the impacts of climate change in the coastal zone depends on a broad range of factors including continuing drivers for coastal development, and political debate about which measures should be adopted. The framework of 'governance' provides the broadest perspective to consider these issues.

In their climate adaptation strategies, all North Sea countries give particular consideration to marine-related risks. The present chapter therefore equates the coastal zone with the zone of marine-related risks. Figure 19.1 shows North Sea regions subject to marine flooding risk and Fig. 19.2 the North Sea regions with a special protection status under the EU Habitats Directive.

Each North Sea country has its own legal and institutional arrangements for coastal governance. The legal frameworks relating to the coastal zone are complex and diverse, and further complicated by the federal structure or devolution within countries (Gibson 2003). France has specific legislation for the coastal zone (Loi Littoral 1986). The UK has passed the Marine and Coastal Access Act (2009)[1] which has jurisdiction seaward from mean high water. In other countries, the coastal zone is governed through more general legal and institutional frameworks, such as 'Environment', 'Water Management', 'Climate Change Adaptation', 'Territorial Planning', 'Natural Hazards', and 'Fishery', among others. The coordination of national policies rests with the central governments. None of the North Sea countries has an authority dedicated specifically to coastal governance. The implementation of national policies in coastal zone management plans is commonly delegated to regional and/or local authorities.

19.2.2 Coastal Management Issues

The coastal zone is considered a region in its own right because of its dependence on land-ocean interaction. The coastal zone is not only shaped by human interventions, but also by the feedback of natural processes to these interventions. This imposes limitations on the uses of the coastal zone; non-respect of these limitations entails the risk of loss of life and investments. Inappropriate development entails the loss of precious ecosystem values.

Recognition of the particular nature of the coastal zone led to the development of the concept of ICZM (Integrated Coastal Zone Management) in the 1990s. The term 'integrated' points to the need for coordination of the policies of different sectors and different levels of government. The challenges of making disjointed, hierarchical and sector bureaucracies effective, are common to many forms of management and regulation. However, for the coastal zone additional requirements result from the highly dynamic natural land-ocean interaction. Large parts of the European coastal zones received a special protection status through the

[1]www.legislation.gov.uk/ukpga/2009/23/contents.

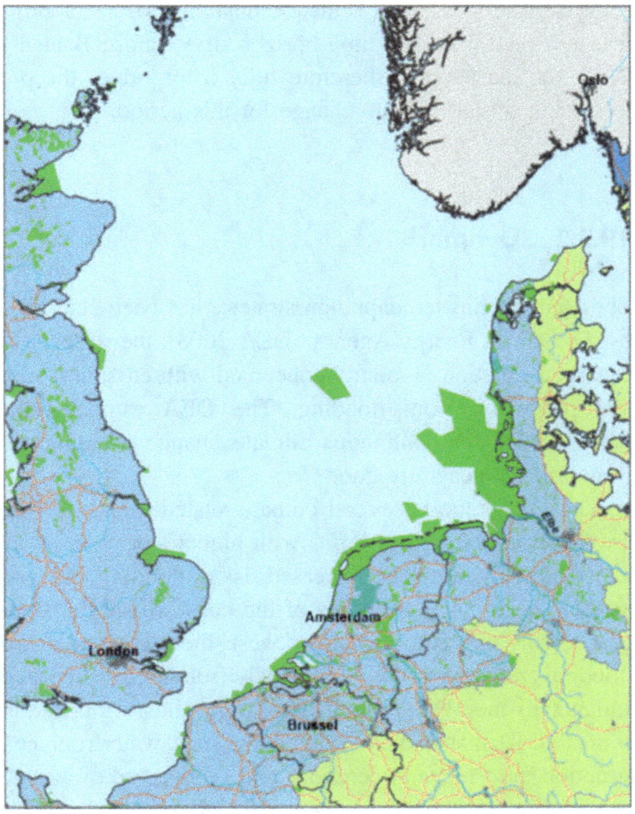

Fig. 19.2 North Sea coastal and marine regions with a special protection status under the EU Habitats Directive (marked in *green*)

EU Habitats Directive and the Natura 2000 network of the European Union. The countries around the Mediterranean Sea agreed on a protocol for ICZM that entered into force in 2011 (Barcelona Protocol 2008). In 2013, the European Commission proposed a directive binding all member states to put into practice the principles of ICZM and to develop spatial marine plans. The directive was adopted in 2014 (EC 2014), but ICZM was excluded following amendments by member states.

According to the evaluation report on IZCM prepared for the European Commission in 2006 (Ruprecht Consult 2006), major coastal issues for the North Sea region include resource management, species and habitat protection, establishment and management of reserves and protected areas, protection of the coast against natural and human induced disasters, and long-term consequences of climate change.

19.2.3 Drivers of Coastal Change

The ELOISE programme (European Land-Ocean Interaction Studies, Vermaat et al. 2005) has collected ample evidence to show that climate change will have serious impacts in the European coastal zones. The effects of climate change will add to the effects of other drivers of change. Other major drivers are related to human population growth and economic expansion. Industrialisation, shipping traffic intensity, fisheries, coastal aquaculture and port development as well as offshore mining for gas and oil have all increased greatly in recent decades, and will probably continue to do so (Stojanovic and Farmer 2013). Together with increased tourism this has led to urbanisation of highly dynamic natural zones. It is expected that climate change will exacerbate most of the adverse impacts of existing drivers of change.

The scale and type of impact that drivers can bring about varies considerably. There are various methods for classifying drivers, for example, PESTLE analysis (Political, Economic, Social, Technological, Legal and Environmental drivers, Ballinger and Rhisiart 2011). Drivers of change in coastal systems are typically external to the coastal zone. Effective coastal zone management therefore requires consideration of policies in many other fields. This implies that coastal adaptation is only a partial response to change.

19.2.4 The Challenge of Adaptation to Climate Change

Development of the coastal zone was accompanied in the past by reclamation and armouring with hard coastal defences, narrowing the active coastal zone (Nicholls and Klein 2005; Vermaat and Gilbert 2006). This process was identified as 'coastal squeeze'. Coastal squeeze is strongly enhanced by sea-level rise and compromises the natural capability of coastal adaptation to climate change. In order to address these problems, new engineering techniques have been developed, following the principle of 'working with nature' (EEA 2006). This practice uses the dynamic response of marine processes, by designing interventions such that the feedback of marine processes is positive (contributes to achieving the objective of the intervention) rather than negative (opposes the intervention). Foreshore nourishment and wetland restoration are typical examples. Further examples of new coastal engineering practices are given in Chap. 18.

Owing to the strong interference of human interventions with natural processes, reversing adverse trends, such as erosion or ecosystem alteration, is not always feasible and is in any case expensive. A long-term perspective is therefore key to coastal governance. Anticipating the effects of climate change is one of the major challenges. Adaptation to climate change may already require a revision of present management strategies in some coastal regions. According to the EEA report *The Changing Faces of Europe's Coastal Areas* (EEA 2006), coastal zones will be subject to many pressures during the 21st century. "These pressures will interact with climate change and exacerbate or ameliorate vulnerability to climate change. Coastal development cannot ignore climate change and development plans should be evaluated with respect to their sustainability under changed climate conditions".

According to Richards and Nicholls (2009), adaptation measures should not be postponed in densely populated and industrial coastal zones. Their calculations indicate that a 'wait and see' strategy generates higher costs in the long run than the costs of protection.

Awareness of the challenges posed by climate change is reflected in coastal policy plans of the North Sea countries. Major features of the coastal policy plans of the North Sea countries are summarised in the following section.

19.3 Adaptation Strategies in the North Sea Countries

19.3.1 Belgium

Most of the effects of climate change at the Belgian coast relate to sea-level rise, resulting in higher storm flood levels, coastal erosion, and deterioration or loss of natural ecosystems, including wetlands. Other impacts associated with higher sea levels are rising groundwater levels and an increase in soil and groundwater salinity in coastal and estuarine areas. Freshwater lenses developed within the dunes are also vulnerable to sea-level rise, leading to threats to drinking water supplies through saltwater intrusion. Climate change will also affect fisheries and coastal tourism (Lebbe et al. 2008; Van den Eynde et al. 2011). One of the most significant social secondary effects is the number of people at risk due to flooding. Economic impacts result not only from direct damage, but also from indirect damage associated with the temporary suspension of production and loss of jobs (Van der Biest et al. 2008, 2009).

The Belgian coastal adaptation strategy for coping with climate change aims at combining flood risk control with the development of ecosystem services (NCC 2010). For controlling flood risks along the Scheldt Estuary, the Sigma-plan has been developed. This provides for the creation of controlled flood zones along the estuary, combining safety against flooding with objectives related to recreation, nature and agriculture.

An ambitious proposal for coastal adaptation has been launched by a group of private investors. The central idea is to combine the need of climate change adaptation with the development of new opportunities for the economy of the Belgian coastal zone. This plan was endorsed by the Flemish government that developed the three-track master plan Vlaamse Baaien (Vlaamse Overheid 2012). This master plan aims at (1) a safe and sustainable coastline with opportunities for economic development, (2) a resilient coastal ecosystem with opportunities for the development of ecosystem services and (3) the establishment of a supportive research platform. The time horizon of Vlaamse Baaien is 2100; the master plan therefore fully incorporates the projected impacts of climate change for this period.

19.3.2 Denmark

The Danish climate adaptation strategy has been elaborated by the Danish Energy Agency (DEA 2008); the strategy for coastal adaptation is mainly concerned with erosion control and protection from flooding. The DEA estimates that opportunities for continuous climate change adaptation in Denmark are generally good.

The DEA reports several climate-related threats. Higher sea levels and stronger storms with higher storm surges are expected. This means an increased risk of flooding and more erosion along many stretches of the coast. Since the strongest storms will come from the west, the increased risk of flooding and erosion will vary widely from the west coast of Jutland, to the Wadden Sea tidal areas and to the interior shores of Danish waters. Moreover, new waterfront construction, port-related operations and sanding up of harbour entrances pose special problems. Cities located at coastal inlets and within fjords may face a very complex set of problems, since they can be under pressure from higher sea levels, increased precipitation and runoff, and changes in groundwater levels.

Increased precipitation, altered precipitation patterns and higher sea levels—with consequent higher water levels in fjords and rivers—will exacerbate problems associated with drainage of low-lying areas, particularly in coastal areas, where about 43 % of Denmark's population occurs. The majority of Denmark's approximately 250,000 summer houses and 73 % of camp sites are within 3 km of the coastal zone. Moreover, increased volumes of water may result in landslides which can affect various types of infrastructure (DEA 2008).

The Danish government considers planning legislation an important means of reducing the negative socio-economic consequences of climate change. Regulations for the coastal zone already restrict new construction areas on open coasts. The Protection of Nature Act 1992 establishes a 300-m protection zone outside urban areas, where most new developments are prohibited, and the Planning Act 1992 defines a coastal planning zone that extends 3 km inland (Gibson 2003). The responsible national authorities continuously evaluate whether there is a need for a follow-up with further restrictions on new building in risk areas. Socio-economic analyses are included as a part of the decision process.

The Danish adaptation strategy allows site owners to raise the beach at their own cost by regular beach nourishment to combat coastal erosion. The same applies to channel dredging, where the amount dredged can be increased as required. Also in the case of reinforcing dikes/dunes or adapting harbour installations and ferry berths, which are relatively simple constructions, it will be possible for owners to adapt to ongoing climate change. Generally speaking, it is a land owner's own choice whether and how to protect themselves from flooding and erosion. Therefore, there are no general laws or regulations stipulating protection, or to what degree owners must or can protect themselves.

An important source of information is municipal planning, which reflects and adapts to the risks and opportunities brought by climate change. Each coastal town must develop an adaptation plan taking into account climate change impacts in the coastal zone. Municipalities are supported in this task by a National Task Force on climate change adaptation. The coastal adaptation plans focus on shoreline management.

However, the general approach of Denmark's climate policy is a stronger focus on mitigation than on adaptation, with no systematic consideration of sea-level rise in present planning policies (Fenger et al. 2008).

19.3.3 France

France has no national coastal management strategy. Coastal management is the responsibility of municipalities. The Loi Littoral imposes restrictions on urban development plans in coastal areas. These restrictions concern mitigation of coastal hazards, assurance of public access to the coast and protection of the environment. In 2013, the Conseil National de la Mer et des Littoraux was installed for the exchange of views and experience among concerned authorities and civil organisations; the Conseil will contribute to the development of a national coastal management strategy. Specific strategies for coastal adaptation in view of climate change are still in a study phase (Idier et al. 2013).

The French macrotidal coasts along the North Sea and the Channel are mostly fairly stable (Anthony 2013; Battiau-Queney et al. 2003). However, at the Pas de Calais a high rate of sea-level rise has been observed over recent decades (Héquette 2010). Some sites (Wissant, in particular) are subject to severe erosion, requiring the construction of seawalls to protect settlements. Climate change will exacerbate erosion and increase the instability of soft cliffs along the French Channel coast (Lissak 2012).

19.3.4 Germany

According to the National Adaptation Strategy on climate change (GFG 2008), coastal regions will be increasingly at risk from sea-level rise and changes in the storm climate. However, there is great uncertainty about the extent of future changes in sea level and the storm climate. One aspect of special importance is the potential danger to wetlands and low-lying areas and to regions with high damage potential, such as the port of Hamburg. There is also concern about saltmarsh ecosystems (Bauer et al. 2010), safety of the estuaries, erosion on coastlines and beaches, safety of shipping traffic and about the future development of the port industry (Reboreda et al. 2007).

The German North Sea coast is part of the Wadden Sea region. The Trilateral Wadden Sea secretariat has developed a climate adaptation strategy for the Wadden Sea, which has been endorsed by the three Wadden Sea countries—Germany, Denmark and the Netherlands (TWS 2014). This strategy comprises seven basic elements: Natural dynamics, Interconnectivity, Integration, Flexibility, Long-term approach, Site specific approach and Participation.

German coastal states are following a strategy mainly based on hard coastal protection measures against flooding, see Chap. 18. This coastline defence policy entails the risk of coastal squeeze on the seaward side, endangering important coastal ecosystems such as tidal flats (Wadden Sea), saltmarshes and dunes when the sea level rises (Sterr 2008).

The German adaptation strategy also attributes importance to 'soft' auxiliary measures such as research, knowledge dissemination, awareness raising and capacity building. Significant organisational and steering measures are also considered necessary. Above all, the National Adaptation Strategy (GFG 2008) places considerable emphasis on the importance of spatial planning, as a means of making a thorough assessment of all relevant adaptation needs within individual regions. Spatial planning provides a formal means through which all concerned parties are able to present their interests and cooperate in the development of a coherent spatial structure and an integrated programme of measures (Swart et al. 2009).

The national adaptation strategy is implemented at state (Länder) level.

19.3.5 Netherlands

As a low-lying country, the Netherlands is particularly vulnerable to sea-level rise and river floods. The damage costs of climate change impacts without adaptation are likely to be substantially higher than for all other North Sea countries combined (Richards and Nicholls 2009). Major impacts expected are increased flood risk in the historic towns of the downstream section of the Rhine-Meuse delta and shortage of fresh water to prevent salinisation of the polders, when river discharges are low. In wet periods, the present capacity of discharge sluices and pumping stations will be insufficient to control inland water levels, in particular in the lake IJssel. There are also concerns related to the loss of ecosystem values in the Wadden Sea and in the heavily modified south-western Delta basins. National study programmes have been launched for assessing other potential climate change impacts and for investigating possible adaptation measures (Oude Essink et al. 2010; Klijn et al. 2012).

The Dutch government has designated a Delta Commissioner, who coordinates a national programme for adapting the Dutch water infrastructure to climate change, in order to secure safety against high water and availability of sufficient fresh water. The Dutch adaptation policy follows a risk-based approach, as in the UK. New adaptation measures are implemented when, as a consequence of climate change and other developments, a tipping point is reached, that is, a point where previous adaptation measures are no longer sufficient to keep damage risks below a certain predefined threshold (Kwadijk et al. 2010).

The Water Test is an important legal instrument that requires regional and local authorities to ensure that water issues, including climate adaptation, are taken into account in spatial and land use planning, such that negative effects on the water system are prevented or compensated for elsewhere.

Sediment management (using sand nourishments) and *Making Space for Water* (realignment of dikes) are the major adaptation strategies for the coastal zone (Aarninkhof et al. 2010) and the lowland fluvial system (Menke and Nijland 2008), respectively.

19.3.6 Norway

Although most of the Norwegian coast is not very sensitive to sea-level rise, there is concern for the low-lying areas in the southwest, which are characterised by soft, erosive coasts. Along the western and northern coastlines, the extensive and well-developed infrastructure of roads, bridges, and ferries linking cities, towns, and villages is likely to be adversely affected by sea-level rise, particularly if this is concurrent with an increased risk and height of storm surges. The potential economic costs of rebuilding and relocating infrastructure and other capital assets in these regions may be considerable (Aunan and Romstad 2008).

The Norwegian Water Resources and Energy Directorate has developed a climate change adaptation strategy that includes monitoring, research and measures to prevent increased damage by floods and landslides in a future climate (NME 2009). Under the Planning and Building Act, municipalities are responsible for ensuring that natural hazards are assessed and taken into account in spatial planning and processing of building applications. Adaptation to climate change, including the implications of sea-level rise and the resulting higher tides, is an integral part of municipal responsibilities. To enable municipalities to ensure resilient and sustainable communities, the central government therefore draws up guidelines for the incorporation of climate change adaptation into the planning activities of municipalities and counties.

The premise of the Norwegian climate adaptation policy is that individuals, private companies, public bodies and local and central government authorities all have a responsibility to take steps to safeguard their own property. If appropriate steps are taken, public and private property are protected from financial risk associated with extreme weather events by adequate national insurance schemes.

19.3.7 Sweden

Rising sea levels are expected to aggravate coastal erosion problems in southern Sweden and increase flood risk along the western and southern coasts. As in the other Scandinavian countries, coastal protection policy in Sweden is mainly focused on spatial planning (EC 2009; OSPAR Commission 2009). The Nature Conservation Act of 1974 states that the first 100–300 m of the coast needs to be free of exploitation. Spatial plans of the different municipalities need to comply with this Act. In addition, new development projects must incorporate a certain safety margin to protect against future erosion or higher water levels. To reduce the vulnerability of Sweden's coasts and to adapt society to long-term climate change and extreme weather events, the Swedish Commission on Climate and Vulnerability made the following recommendations in 2007:

- Spatial planning should be considered the most important tool to protect against marine hazards;
- The risks of coastal erosion in built-up areas should be investigated, bathymetric information should be compiled and evaluated, and extreme weather warning systems should be expanded;
- Compensation and subsidy systems for preventive measures for coastal erosion in built-up areas should be developed;
- Areas of the coastal zones without private or public interests should not be protected but given back to the sea (managed retreat).

19.3.8 UK

Major perceived threats are related to coastal protection. Higher sea level and more intense and frequent storms due to climate change will increase damage to coastal defences.

Approximately one third of existing coastal defences could be destroyed if the level of expenditure on coastal defence does not keep pace with coastal erosion in the coming decades (DEFRA 2010, 2012). Extensive coastal erosion around parts of the UK, in particular along estuaries and the east coast, reduces intertidal area (OST 2004). Loss of intertidal areas (coastal squeeze) occurs mainly where hard defences are present. This in turn causes loss of land, property and coastal habitat, particularly saltmarshes and mud flats, which are also bird feeding grounds.

In the UK, policies for adaptation to sea-level rise are more advanced than in most European coastal countries (De la Vega-Leinert and Nicholls 2008). The UK coastal climate change adaptation policy is based on the appraisal method for dealing with the risks of climate change impacts, as outlined in the DEFRA Policy Statement (DEFRA 2009). This appraisal method is based on a comparison of different options (including the managed adaptive approach, the precautionary approach and the no-regret approach) with respect to costs, benefits and residual risk.

The *no-regret* approach is generally preferred where possible. The *managed adaptive* approach aligns with principles in *Making Space for Water*, which promotes a holistic and long-term approach for flood and coastal management, and reinforces existing climate change policy on 'no-regret' actions and longer term adaptation. This approach promotes flexibility in the appraisal options to respond to future change, during the whole life of a measure, as well as the uncertainties (DEFRA 2009). The *precautionary* approach may be adopted where it is not possible to adapt with multiple interventions on a periodic and flexible basis. Figure 19.3 illustrates the different approaches.

'Managed retreat' as an element of coastal management policy has thus far been applied mainly for ecological reasons and where the retreated area has relatively low value.

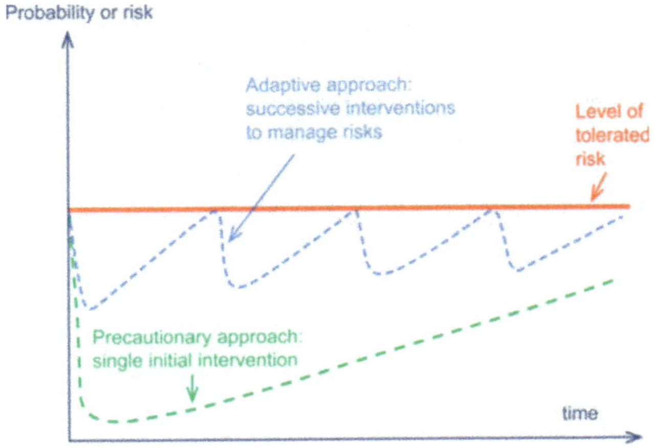

Fig. 19.3 Schematic representation of different adaptation approaches for the UK coastal zones (based on DEFRA 2009)

The Planning Policy Statement (DCLG 2010) obliges local authorities to develop climate adaptation policies and to report on progress. The Marine (Scotland) Act[2] stipulates that forthcoming national and regional marine plans should set objectives relating to the mitigation of, and adaptation to, climate change. An independent UK body, the Adaptation Subcommittee, assesses the preparedness to meet the risks and opportunities of climate change.

In this context adaptation is required to include protecting and restoring marine habitats to increase their resilience to climate change. More than 25 % of English waters is designated as Marine Protected Areas and managed as a network of habitats to aid the movement of species affected by climate change and to decrease threats such as overfishing. The National Heritage Protection Plan sets out how England's landscapes, archaeological sites and historic buildings will be protected from the impacts of climate change. This includes actions such as the continuation of 'Rapid Coastal Zone Assessment Surveys' that record and assess the risk to heritage assets on the coast (DEFRA 2013).

19.4 Governance Issues and Dilemmas for Adaptation

This section compares the various adaptation strategies adopted by the North Sea countries, as well as the dilemmas arising during their implementation and the means by which these dilemmas may be addressed.

19.4.1 Top-Down and Bottom-Up Strategies

The North Sea countries are following different approaches for adapting to change in the coastal zone. In Germany, the Netherlands and Belgium, implementation is steered by national or regional government, whereas in the UK, Sweden, Norway and Denmark, implementation is delegated to local authorities aided by civil organisations and private stakeholders.

Richards and Nicholls (2009) estimated the adaptation costs required for avoiding extra damage related to sea-level rise, and compared them to the costs actually spent on coastal defence measures. They estimated that in Germany, the Netherlands and Belgium more money is presently spent on coastal defence than the avoided damage costs. This can be imputed to a different governance culture, but also to a higher flood-risk awareness and higher standards for acceptable risk. Current adaptation plans in these countries involve large infrastructural projects, with planning

[2]http://www.scotland.gov.uk/Topics/marine/seamanagement/marineact.

procedures similar to other infrastructural projects. National and regional governments bear almost all the costs. In the UK, Sweden and Denmark, governmental steering of adaptation is more indirect, and operates through regulation and guidance. Local and private initiatives play an important role in the implementation plans. In the UK, many local institutions and associations are actively involved in coastal planning and adaptation through the Shoreline Management Planning process.

Several studies (EC 2011; IPCC 2012) have found that national systems play a crucial role in countries' capacity to meet the challenges brought by the observed and projected trends in exposure, vulnerability, and weather and climate extremes. Effective national systems comprise multiple actors from national and regional governments, the private sector, research bodies, and civil society including community-based organisations. Organisations beyond the state are increasingly playing a role in planning and risk management.

Governance theorists highlight different 'modes' of governance, including hierarchies, networks, markets, adaptive management and transition. Coastal management in the North Sea region shares many characteristics with the 'network' mode of governance, focusing on participation, using non-regulatory approaches to achieve progress, and the involvement of multiple actors. However, the evaluation and 'lesson drawing' components have been assessed as somewhat weak (Stojanovic and Ballinger 2009). A key analytical question is which modes of governance have the best 'fit' for the challenges of climate adaptation? (Young et al. 2008).

19.4.2 Public Participation

The recent OURCOAST inventory of coastal management practices in Europe (EC 2011) shows that awareness of coastal and marine issues by the general public and the responsible authorities is strongly stimulated when the public is involved in the development of adaptation strategies. Adaptation strategies are more effective when they are informed by and customised to specific local circumstances and when there is a broadly shared understanding of long-term coastal change. Public participation leads to less conflict between coastal managers or coastal developers and other involved parties. Local populations document their experiences with the changing climate, particularly extreme weather events, in many different ways, and this self-generated knowledge can uncover existing capacity within the community and important current shortcomings. Local participation and community-based adaptation lead to better management of disaster risk and climate extremes. Improvements in the availability of human and financial capital and of disaster risk and climate information customised for local stakeholders can enhance community-based adaptation (IPCC 2012).

Adaptation strategies can widely differ, according to the values to be protected, when, to what extent, how and by whom. The choice between different adaptation strategies is basically a political choice. Valuing coastal assets is intrinsically subjective, even if attempts are made to express some values, such as ecosystem services, in monetary terms. These attempts do not result in generally agreed answers on how to mutually rank different types of damage: loss of human life, loss of economic assets (including ecosystem services), loss of biodiversity and loss of cultural values.

According to the EEA (2006), there is often a fundamental conflict between protecting socio-economic activity and sustaining the ecological functioning of coastal zones in Europe under conditions of rising sea level—a conflict that cannot be resolved by technical or scientific means. Integrated, long-term coastal management should not be exclusively orientated to physical planning and technical solutions, but to combinations of social and physical management mechanisms. The policy and governance strategies for coastal conflict and natural resource management should therefore be improved by developing adaptive, participatory and multi-scale governance (Stepanova and Bruckmeier 2013).

Prerequisites for public participation in coastal adaptation strategies include: political legitimacy through securing broad political support; a process-driven approach in an inclusive, voluntary and culturally sensitive manner; the empowering of historically disadvantaged individuals, groups and communities; building partnerships to provide the basis upon which stakeholders can learn about and appreciate the interest of others; deepening public deliberation through alternative forums and participatory methodologies; and promoting innovation, reflection and feedback in response to changing circumstances and stakeholder interests (Henocque 2013).

Social Impact Assessment (SIA) has been proposed as an instrument to reduce likely future expenditure by the early identification and resolution of potential issues that could otherwise lead to litigation, delays to approval, costs in the form of managing protest actions, and business lost through reputational harm (Vanclay 2012). However, there is little practical experience with SIA to date.

19.4.3 Uncertainty and Awareness

North Sea countries will have to face the implications of climate change and some impacts are already occurring. However, separating the impacts of climate change from change resulting from other natural or human causes is far from obvious. This is illustrated by a study of past ecosystem shifts in the North Sea region. There is evidence, for instance, that these regime shifts are related to decadal-scale fluctuations in the North Atlantic Oscillation index (Kröncke et al. 2013). The full long-term impacts of climate change are still uncertain, especially the question as to when they will occur. For instance, present data do not yet show clear evidence for an increase in the average rate of sea-level rise in the North Sea region (NOAA 2015).

Uncertainty is a serious (perhaps the most serious) obstacle to raising public awareness and to getting climate adaptation high on the political agenda, compared to issues with a more immediate impact (EEA 2014). Uncertainty about the possible impacts of climate change is not the only reason for this. The fact that the greatest impacts are related to exceptional extreme events, plays also a role. According to an enquiry among policymakers, the occurrence of an extreme weather event is presently the most important trigger for progress in climate adaptation (EEA 2014).

While some countries—especially those with low-lying coasts—are traditionally alerted to sea-level rise and flooding, awareness is still low in other countries (Ruprecht Consult 2006). Due to the absence of recent coastal flood disasters in North Sea countries there is a risk of decreasing societal awareness and support for protection measures in specific, flood prone areas. This highlights the need and importance of risk communication and awareness raising to ensure the continuity and support for coastal risk management strategies (Safecoast 2008).

In the Netherlands, risks associated with climate change are made more tangible through tipping-point analysis. This involves testing the robustness of existing policies for addressing anticipated climate-driven changes in environmental conditions, such as temperature, precipitation, and sea level. 'Tipping points' are the thresholds in future environmental conditions at which existing policies fail to keep risk (potential damage) within acceptable limits. Awareness of these tipping points guides policymakers to prepare the necessary adaptation strategies, even if uncertainty remains regarding the timing of required adaptations (Kwadijk et al. 2010).

Greater awareness can also be pursued by internalising costs. Development projects in the coastal zone often increase climate change adaptation costs. According to the EUROSION study (Doody et al. 2004), the costs of reducing coastal risks are mainly supported by national or regional budgets in the North Sea countries and almost never by the developers or the owners of assets at risk. Only in Denmark and Sweden are adaptation costs (partly) supported by owners and the local community. Hence, risk assessment is hardly incorporated in decision-making processes at the local level and risk awareness of the public is poor. The impact, cost and risks associated with coastal development are better controlled through internalising adaptation costs in planning and investment decisions: thus an appropriate part of the risks and risk mitigation costs is transferred to the direct beneficiaries and investors. Risk monitoring and mapping is a prerequisite for incorporating risk into planning and investment policies. The distribution of risks and costs requires due consideration of the interests of all stakeholders in order to guarantee social justice (Safecoast 2008; OST 2004).

19.4.4 Risk-Based Adaptation

The largest climate change impacts in the coastal zone result from extreme events which have a low probability of occurrence within a given time interval. The concept of risk, defined as the product of probability of occurrence and resulting damage, provides an objective measure for the need to adapt to such impacts. By evaluating what damage is avoided at what costs, informed choices can be made among different adaptation strategies. Coastal adaptation strategies of the North Sea countries are increasingly based on risk management considerations. Uncertainty in the probability of occurrence and uncertainty in the extent of damage can be incorporated in risk estimation—for instance, by defining probability distributions for all variables and using a Monte Carlo method. The application of the risk concept in adaptation strategies is limited, however, by the difficulty of quantifying uncertainty in the probability of occurrence and by the more fundamental difficulty of predicting possible damage caused by rare extreme events.

A further complication arises when a choice has to be made among different possible adaptation measures: which temporal and spatial scales must be considered when these measures are evaluated through ranking methods such as cost-benefit, cost-effectiveness or multi-criteria analyses? This choice strongly influences the results. This complication is enhanced by uncertainty about the future in general. How are present values affected by other future global or local change, in addition to climate change? The combination of these different sources of uncertainty is sometimes termed 'deep uncertainty'.

Scenarios provide a way to deal with limitations related to quantifying uncertainty (the probability that some damage will occur) and to quantifying possible damage (loss of certain values). Scenarios describe different futures that can be imagined. These scenarios should be internally consistent, but need not necessarily be expressed in terms of probability and money. Their main function is to open those who are involved in climate adaptation to the wide spectrum of situations and adaptation options that should be considered. Scenarios help in avoiding suboptimal sector approaches and a unilateral focus on certain adaptation options, which are major shortcomings of present coastal adaptation strategies in the North Sea countries (EEA 2005). But scenarios do not of course, in themselves, answer the question as to which adaptation strategy of the options available should be preferred.

The EEA (2007) has provided methodological guidance for quantifying and costing climate change impacts at the global and regional scale. These methods include: treatment of scenarios (both climate and socio-economic projections); issues of valuation (market and non-market effects); indirect effects on the economy; approaches taken to spatial and temporal variation; uncertainty and irreversibility (especially in relation to large-scale irreversible events); and coverage (which climate parameters and which impact categories are included). However, there is limited application of exploratory scenarios at the local level and those applications involving local stakeholders are even rarer. This highlights the need for pilot projects to evaluate, demonstrate and disseminate the effectiveness of scenario approaches to the ICZM community, including predictive, exploratory, and normative scenarios (Ballinger and Rhisiart 2011). To date, few projects have attempted to downscale SRES scenarios to the regional and local level in the North Sea region (Andrews et al. 2005; Holman et al. 2005a, b; Nicholls et al. 2006).

19.4.5 Adaptation Pathways

There is broad agreement that adapting to the impacts of climate change is inevitable and that preparatory actions should already be initiated. But once it becomes clear that a fundamental revision of present coastal policies is needed, questions arise as to which actions are most appropriate to cope with the impacts of climate change at the long term. Revised policies need to deal not only with uncertainty related to the future impacts of climate change, but also with uncertainties related to future social and economic developments. A blueprint plan is inadequate, as the future can unfold differently from what is anticipated. Actions that are appropriate for the foreseeable future could turn out to be inadequate for the long term and could even hinder actions that may become necessary later.

One way of dealing with this problem of 'robust decision making' is the strategy of adaptive pathways (Hallegatte 2009). According to this strategy, adaptation pathways are developed that comprise different sets of successive adaptation actions. Each pathway leads to successful long-term adaptation within a particular scenario of climate change and socio-economic development. Analysis of the different pathways enables the selection of short-term actions that are suitable (no adverse lock-in effects) within different scenarios. The most promising actions are those with the best performance in terms of societal benefits and costs. The exercise of pathway definition and analysis is repeated when new follow-up actions become needed; the lessons of the first actions ('learning-by-doing') as well as the latest knowledge of climate change and socio-economic development serve as input. A sophisticated version of this approach ('strategy of dynamic adaptive policy pathways') was used to underpin the Dutch Delta programme for adaptation to climate change (Haasnoot et al. 2013). A similar method has been developed by Sayers et al. (2013) and applied to the Thames Estuary, UK (McGahey and Sayers 2008).

19.4.6 No-Regret Adaptation Strategy

The measures envisioned in the North Sea countries for adaptation to climate change are similar. Preference for certain measures depends on the nature and seriousness of the climate change threats and on social acceptance. In all North Sea countries there is consensus that adaptation to climate change is inevitable and that some action is already required. Climate change projections for the economic life cycle of coastal infrastructure are currently incorporated in the development of long-term investment plans. This is done, for instance, by adjusting design criteria for the renovation of coastal protection works (see Chap. 18). Spatial planning is recognised as a key instrument for the integration of adaptation measures in a broader coastal management policy and for taking into account developments at larger temporal and spatial scales. Spatial reservations are made for future reinforcement or realignment of coastal defences, and set-back lines for new buildings in the coastal zone are revised. In most North Sea countries, studies are undertaken on how far adaptation should go and whether investment can be postponed. At present, no major public investments are being made with the sole purpose of long-term climate change adaptation.

There is an increasing preference for flexible measures with as much as possible a no-regret character. Potential low-regret measures include early warning systems; risk communication between decision makers and local citizens; sustainable land management, including land use planning; ecosystem management and restoration; improvements to water supply, sanitation, irrigation and drainage systems; climate proofing of infrastructure; development and enforcement of building codes and better education and awareness (IPCC 2012). Such measures deliver additional benefits, such as opportunities for tourism, recreation, nature development and other ecosystem services.

Beach and shoreface nourishment and wetland restoration are examples of no-regret measures already practiced in North Sea countries. They are often part of a broader water management strategy that includes land-use planning in the upstream catchment area. Such measures are implemented step-wise, allowing for adjustment when better knowledge of the impacts of climate change impacts becomes available. They also respond to the insight that natural dynamics generally offer greater long-term resilience (self-regulating capacity) against climate change impacts than hard man-made structures (Dronkers 2005).

An important notion in this context is that present levels of greenhouse gases already imply a commitment to sustained adaptation for several centuries to come (Nicholls et al. 2007; Wong et al. 2014). In some cases, this might lead to more radical strategies, such as the wholesale re-location of coastal settlements, or design of housing infrastructure which can cope with being regularly inundated.

19.4.7 Knowledge and Monitoring

Adaptation efforts benefit from iterative risk management strategies because of the complexity, uncertainties, and long time frame associated with climate change (IPCC 2012). An iterative risk management strategy consists of an iterative process of monitoring, research, evaluation, learning, and innovation. Addressing knowledge gaps through enhanced observation and research reduces uncertainty and helps in designing effective adaptation and risk management strategies.

Because uncertainty is a major obstacle to preparing for climate change adaptation, more reliable predictions of climate change and its impacts are needed (EEA 2014). Many studies address climate change prediction at the global scale. However, there are indications that global-scale projections of climate change may not be representative for the North Sea region, especially in relation to the characteristics of the North Atlantic Gulf Stream (Nicholls et al. 2007). Better understanding of the coupled ocean-atmosphere system for the North Atlantic is therefore a highly relevant and urgent research topic (Vellinga and Wood 2007; Rahmstorf et al. 2015).

Monitoring is also essential for a better understanding of climate change impacts in the North Sea coastal and marine zone. Many data are collected within the different North Sea countries, by public agencies, research institutes and private companies. However, the European Commission (EC 2010) notes that "There are restrictions on access to data, and on use and re-use. Fragmented standards, formats and nomenclature, lack of information on precision and accuracy, the pricing policy of some providers and insufficient temporal or spatial resolution are further barriers." It may be expected that the situation will improve by progress in the implementation of the EU Water Framework Directive, the EU Marine Strategy Framework Directive and the EMODnet marine data network (EC 2012).

A better coordinated and more consistent approach to marine monitoring is essential for a proper analysis of change in the coastal and marine system. This analysis should focus on establishing cause-impact relationships, which make it possible to distinguish climate change impacts from natural variability and other impacts. Monitoring data are often not directly fit for policy evaluation; translating data into indicators pertinent to policy making is a further subject of special attention (Breton 2006; Martí et al. 2007; EEA 2012). This kind of knowledge is crucial for informing policy and the general public and for developing the adaptive capacity of institutions and wider society.

19.5 Summary and Conclusions

1. Strategy

All North Sea countries have developed a climate adaptation strategy. In these strategies special consideration is given to the coastal zone.

2. Perceived Risks

The North Sea countries consider flooding by the sea and coastal erosion as major climate-related coastal risks.

3. Aggravation of Existing Trends

Several studies show that climate change will enhance erosion and habitat loss that occur already, as a result of existing pressures related to use and development of the coastal zone.

4. Governmental Steering

In all North Sea countries, actors at national and regional level have been designated for initiating and coordinating adaptation to climate change. In the Netherlands, the country with the highest number of potentially threatened people, a special governance mechanism, the Delta Commissioner, has been created.

5. Centralised Versus Decentralised Implementation

In Germany, the Netherlands and Belgium coastal adaptation is steered by national and regional programmes and plans. In the UK, Denmark, Sweden and Norway, regional and local governments are responsible for adaptation; coastal communities have the duty to develop adaptation plans and to report (in the UK) on the implementation progress.

6. Public Participation

In all North Sea countries, adaptation plans are subject to public consultation. The UK and the Scandinavian countries pursue active public involvement by accruing adaptation responsibilities to private stakeholders.

7. Risk-Based Adaptation

In all North Sea countries some form of risk assessment (comparison of adaptation costs with costs of avoided risks) is considered for the prioritisation of adaptation measures. However, at present there is no generally accepted methodology.

8. Uncertainty

Uncertainty about the extent and timing of climate-driven impacts is a major obstacle to political and public mobilisation on the issue of climate adaptation. Different methods to deal with uncertainty of climate impacts are being developed, involving scenario development, tipping point analysis and more robust decision-making techniques (such as adaptive pathways).

9. No-Regret Measures

In view of the uncertainties, adaptation plans focus on no-regret measures. The most considered measures in the North Sea countries are spatial planning in the coastal zone (set-back lines), wetland restoration, coastal nourishment and reinforcement of existing protection structures.

10. Monitoring and Research

The climate of the North Sea countries is strongly influenced by the North Atlantic Oscillation (NAO) and the Gulf Stream. Better understanding of ocean-atmosphere dynamics in the North-Atlantic region is important to reduce the uncertainty in climate predictions for the North Sea region. The difficulty of identifying the climate-related component in observed changes of physical and biological parameters in the coastal zone is a critical obstacle to obtaining a widely shared understanding of the urgency of adaptation. A dedicated coastal observation network is not yet in place in the North Sea region.

Box 1
Working definitions of key terms used within this chapter

Governance: The exercise of political, economic and administrative authority in the management of a country's affairs at all levels. Governance comprises the complex mechanisms, processes, and institutions through which citizens and groups articulate their interests, mediate their differences, and exercise their legal rights and obligations (UNDP 1997).

Integrated Coastal (Zone) Management: A continuous process of administration, the general aim of which is to put into practice sustainable development and conservation in coastal zones and to maintain their biodiversity. This involves the coordinated management and synchronised planning of multiple issues and areas of overlapping interest (EC 1999). In Europe this has been characterised by the implementation of the EU Recommendation on Integrated Coastal Zone Management (cf synonyms ICM, ICZM, CZM, ICAM.).

Shoreline Management Planning: Strategic approach to managing the risks of coastal flooding and erosion, especially as they relate to changes in coastal processes (DEFRA 2009).

Coastal Adaptation: Efforts and actions (in the coastal zone) targeted at vulnerable systems to deal with actual or expected problems with the objective of moderating harm (IPPC 2001).

References

Aarninkhof SGJ, Van Dalfsen JA, Mulder JPM, Rijks D (2010) Sustainable development of nourished shorelines. Procs. PIANC MMX Congress Liverpool UK 2010

Andrews J, Beaumont N, Brouwer R, Cave R, Jickells T, Ledoux L, Turner KR (2005) Integrated assessment for catchment and coastal zone management: the case of the Humber. In: Vermaat J, Bouwer L, Turner K, Salomons W (Eds.), Managing European Coasts, Past, Present and Future, pp. 323–354. Springer Verlag

Anthony EJ (2013) Storms, shoreface morphodynamics, sand supply, and the accretion and erosion of coastal dune barriers in the southern North Sea. Geomorphology 199:8–21

Aunan K, Romstad B (2008) Strong coasts and vulnerable communities: potential implications of accelerated sea-level rise for Norway. J Coastal Res 24:403–409

Ballinger R, Rhisiart M (2011) Integrating ICZM and futures approaches in adapting to changing climates. MAST, 10:115–138

Barcelona Protocol (2008) Protocol to the Barcelona Convention on Integrated Coastal Zone Management. http://ec.europa.eu/environment/iczm/barcelona.htm

Battiau-Queney Y, Billet J-F, Chaverot S, Lanoy-Ratel P (2003) Recent shoreline mobility and geomorphologic evolution of macrotidal sandy beaches in the north of France. Mar Geol 194:31–45

Bauer EM, Heuner M, Fuchs E, Schröder U, Sundermeier A (2010) Vegetation shift in German estuaries due to climate change? Procs. Conf. "Deltas in Times of Climate Change", Rotterdam

Breton F (2006) Report on the use of the ICZM indicators from the WG-ID: A contribution to the ICZM evaluation. European Environment Agency (EEA)

DCLG (2010) Planning Policy Statement 25 Supplement: Development and Coastal Change. Department for Communities and Local Government

De la Vega-Leinert AC, Nicholls RJ (2008) Potential implications of sea-level rise for Great Britain. J Coastal Res 24:342–357

DEA (2008) Danish Strategy for Adaptation to a Changing Climate. Danish Energy Agency, Copenhagen

DEFRA (2009) Appraisal of Flood and Coastal Erosion Risk Management. A DEFRA Policy statement, Department for Environment Food and Rural Affairs (DEFRA), London

DEFRA (2010) Charting Progress 2: An Assessment of the State of UK Seas. Department for Environment Food and Rural Affairs (DEFRA), London

DEFRA (2012) UK Climate Change Risk Assessment, London: Evidence Report. Department for Environment Food and Rural Affairs (DEFRA), London

DEFRA (2013) The National Adaptation Programme: Making the Country Resilient to a Changing Climate. Department for Environment Food and Rural Affairs (DEFRA), London

Doody P, Ferreira M, Lombardo S, Lucius I, Misdorp R, Niesing R, Salman A, Smallegange M (2004) Living with Coastal Erosion in Europe: Sediment and Space for Sustainability. EUROSION

Dronkers J (2005) Dynamics of Coastal Systems. World Scientific Pub. Co. Advanced Series on Ocean Engineering, vol 25

EC (1999) Towards a European Integrated Coastal Zone Management (ICZM) Strategy: General Principles and Policy Options. European Commission (EC), Luxembourg

EC (2009) The Economics of Climate Change Adaptation in EU Coastal Areas. European Commission (EC), Directorate-General for Maritime Affairs and Fisheries, Brussels

EC (2010) Marine Knowledge 2020: Marine Data and Observation for Smart and Sustainable Growth. European Commission (EC), Luxembourg

EC (2011) Comparative analyses of the OURCOAST cases. Report No A2213R4v1 European Commission (EC) Luxembourg

EC (2012) Marine Knowledge 2020: From Seabed Mapping to Ocean Forecasting. European Commission (EC), Luxembourg

EC (2014) Directive 2014/89/EU of the European Parliament and of the Council of 23 July 2014 establishing a framework for maritime spatial planning

EEA (2005) Vulnerability and Adaptation to Climate Change in Europe. European Environment Agency (EEA), Technical Report No 7/2005

EEA (2006). The Changing Faces of Europe's Coastal Areas. European Environment Agency (EEA), Report No 6/2006

EEA (2007) Climate Change: The Cost of Inaction and the Cost of Adaptation. European Environment Agency (EEA), Technical Report No 13/2007

EEA (2012) Climate Change, Impacts and Vulnerability in Europe 2012. European Environment Agency (EEA), Report No 12/2012

EEA (2014) National Adaptation Policy Processes in European Countries - 2014. European Environment Agency (EEA), Report No 4/2014

Fenger J, Buch E, Jakobsen PR, Vestergaard P (2008) Danish attitudes and reactions to the threat of sea-level Rise. J Coastal Res 24:394–402

GFG (2008) German Strategy for Adaptation to Climate Change: Climate Change Impact for Coastal Regions. German Federal Government

Gibson J (2003) Integrated coastal zone management law in the European Union. Coast Manage 31:127-136

Haasnoot M, Kwakkel JH, Walker WE, Ter Maat J (2013) Dynamic adaptive policy pathways: A method for crafting robust decisions for a deeply uncertain world. Glob Environ Change 23:485–498

Hallegatte S (2009) Strategies to adapt to an uncertain climate change. Glob Environ Change 19:240–247

Henocque Y (2013) Enhancing social capital for sustainable coastal development: Is satoumi the answer? Estuar Coast Shelf S 116:66–73

Héquette A (2010) Les risques naturels littoraux dans le Nord-Pas-De-Calais, France. VertigO Hors-série 8, Available at: http://vertigo.revues.org/10173

Holman IP, Rounsevell MDA, Shackley S, Harrison PA, Nicholls RJ, Berry PM, Audsley E (2005a) A regional, multi-sectoral and integrated assessment of the impacts of climate change and socio-economic chance in the UK: I methodology. Climate Change 71:9–41

Holman IP, Nicholls RJ, Berry PM, Harrison PA, Audsley E, Shackley S, Rounsevell MDA (2005b) A regional, multi-sectoral and integrated assessment of the impacts of climate change and socio-economic chance in the UK: II results. Climate Change 71:9–41

Idier D, Castelle B, Poumadère M, Balouin Y, Bohn Bertoldo R, Bouchette F, Boulahya F, Brivois O, Calvete D, Capo S, Certain R, Charles E, Chateauminois E, Delvallée E, Falqués A, Fattal P, Garcin M, Garnier M, Héquette A, Larroudé P, Lecacheux S, Le Cozannet G, Maanan M, Mallet C, Maspataud A, Oliveros C, Paillart M, Parisot J-P, Pedreros R, Robin N, Robin M, Romieu E, Ruz M-H, Thiébot J, Vinchon C (2013) Vulnerability of sandy coasts to climate variability. Clim Res 57:19–44

IPCC (2001) Climate Change 2001: Impacts, Adaptation and Vulnerability. Contribution of Working Group II to the Third Assessment Report of the Intergovernmental Panel on Climate Change (IPCC). Cambridge University Press

IPCC (2012) Summary for Policymakers. In: Managing the Risks of Extreme Events and Disasters to Advance Climate Change Adaptation. Field CB, Barros V, Stocker TF, Qin D, Dokken D, Ebi KL, Mastrandrea MD, Mach KJ, Plattner G-K, Allen SK, Tignor M, Midgley PM (eds.). A Special Report of Working Groups I and II of the Intergovernmental Panel on Climate Change (IPCC), Cambridge University Press

Klijn F, De Bruijn KM, Knoop J, Kwadijk J (2012) Assessment of the Netherlands' flood risk management policy under global change. AMBIO 41:180–192

Kröncke I, Reiss H, Dippner JW (2013) Effects of cold winters and regime shifts on macrofauna communities in shallow coastal regions. Estuar Coast Shelf S 119:79–90

Kwadijk JCJ, Haasnoot M, Mulder JPM, Hoogvliet MCM, Jeuken ABM, Van der Krogt RAA, Van Oostrom NGC, Schelfhout HA, Van Velzen EH, Van Waveren H, De Wit MJM (2010) Using adaptation tipping points to prepare for climate change and sea level rise: a case study in the Netherlands. WIREs Climate Change 1:729–740

Lebbe L, Van Meir N, Viaene P (2008) Potential implications of sea-level rise for Belgium. J Coastal Res 24:358–366

Lissak L (2012) Les glissements de terrain des versants côtiers du Pays d'Auge (Calvados): Morphologie, fonctionnement et gestion du risque. Geomorphologie. Thèse Universite de Caen

Loi Littoral (1986) Loi n° 86-2 du 3 janvier 1986 relative à l'aménagement, la protection et la mise en valeur du littoral. Legifrance.gouv.fr

Martí X, Lescrauwaet A-K, Borg M, Valls M (2007) Indicators guidelines: To adopt an indicators-based approach to evaluate coastal sustainable development. Report DEDUCE EU project

McGahey C, Sayers PB (2008) Long term planning – robust strategic decision making in the face of gross uncertainty – tools and application to the Thames. In: Flood Risk Management: Research and Practice, pp. 1543–1553. Proceedings of FLOODrisk 2008. Taylor & Francis

Menke U, Nijland H (2008) Nature development and flood risk management combined along the river Rhine. Procs. 4th ECRR Conference on River Restoration, pp. 329–338. Venice, Italy

NCC (2010) Belgian National Climate Change Adaptation Strategy. National Climate Commission (NCC), Flemish Environment, Nature and Energy Department, Brussels

Nicholls RJ, Klein RJT (2005). Climate change and coastal management on Europe's coast. In: Vermaat JE, Ledoux L, Turner K, Salomons W (eds.): Managing European Coasts: Past, Present and Future, pp. 199–225. Springer Environmental Science Monograph Series

Nicholls RJ, Wong PP, Burkett V, Woodroffe CD, Hay J (2006) Climate change and coastal vulnerability assessment: scenarios for integrated assessment. Sustain Sci 3:89–102

Nicholls RJ, Wong PP, Burkett VR, Codignotto JO, Hay JE, McLean RF, Ragoonaden S, Woodroffe CD (2007) Coastal systems and low-lying areas. In: Climate Change 2007: Impacts, Adaptation and Vulnerability, pp. 315-356. Contribution of Working Group II to the Fourth Assessment Report of the Intergovernmental Panel on Climate Change, M.L. Parry, O.F. Canziani, J.P. Palutikof, P.J. van der Linden and C.E. Hanson (Eds) Cambridge University Press

NME (2009) Norway`s Fifth National Communication under the Framework Convention on Climate Change. Norwegian Ministry of the Environment

NOAA (2015) http://www.star.nesdis.noaa.gov/sod/lsa/SeaLevelRise/slr/slr_sla_nrs_free_all_66.pdf, Accessed 15 July 2015

OSPAR Commission (2009) Assessment of climate change mitigation and adaptation. OSPAR Commission, London

OST (2004) Foresight Future Flooding. Foresight Flood and Coastal Defence Project. Office of Science and Technology (OST), UK

Oude Essink GHP, Van Baaren ES, De Louw PGB (2010) Effects of climate change on coastal groundwater systems: A modeling study in the Netherlands. Water Resour Res 46:doi:10.1029/2009wr008719

Rahmstorf S, Box J, Feulner G, Mann ME, Robinson A, Rutherford S, Schaffernicht E (2015) Evidence for an exceptional twentieth-century slowdown in Atlantic Ocean overturning. Nature Clim Change 5:475–480

Reboreda R, Körfer A, Schernewski G, Pickaver A (eds.) (2007) Coastal management in Germany. EUCC Coastline Special 16(1).

Richards JA, Nicholls RJ (2009) Impacts of climate change in coastal systems in Europe. PESETA-Coastal Systems study. EC Joint Research Centre, Institute for Prospective Technological Studies

Roode N, Baarse G, Ash J, Salado R (2008) Coastal flood risk and trends for the future in the North Sea Region, synthesis report. Safecoast project INTERREG IIIB North Sea Region Programme

Ruprecht Consult (2006) Evaluation of Integrated Coastal Zone Management in Europe. Final Report. Prepared for the European Commission

Safecoast (2008) Coastal Flood Risk and Trends for the Future in the North Sea Region: Results and Recommendations of the Project Safecoast. Synthesis Report. Safecoast project team, The Hague

Sayers P, L.i Y, Galloway G, Penning-Rowsell E, Shen F, Wen K, Chen Y, Le Quesne T (2013) Flood Risk Management: A Strategic Approach. Paris, UNESCO.

Stepanova O, Bruckmeier K (2013) The relevance of environmental conflict research for coastal management. A review of concepts, approaches and methods with a focus on Europe. Ocean Coast Manage 75:20–32

Sterr H (2008) Assessment of vulnerability and adaptation to sea-level rise for the coastal zone of Germany. J Coastal Res 24:380–393

Stojanovic TA, Ballinger RC (2009) Integrated coastal management: A comparative analysis of four UK initiatives. Appl Geogr 29:49–62

Stojanovic TA, Farmer CJQ (2013) The development of world oceans & coasts and concepts of sustainability. Mar Policy 42:157–165

Swart R, Biesbroek R, Binnerup S, Carter TR, Cowan C, Henrichs T, Loquen S, Mela H, Morecroft M, Reese M, Rey D (2009) Europe Adapts to Climate Change: Comparing National Adaptation Strategies. PEER Report No 1.

TWS (2014) TMAP Strategy. Ministerial Council Declaration, Annex 6. 12th Trilateral Governmental Conference on the Protection of the Wadden Sea. 5 February 2014. Trilateral Wadden Secretariat

UNDP (1997) Governance for Sustainable Human Development: A UNDP Policy Document. United Nations Development Programme (UNDP)

Van den Eynde D, De Sutter R, De Smet I, Francken F, Haelters J, Maes F, Malfait E, Ozer J, Polet H, Ponsar S, Reyns J, Van der Biest K, Van der Perren E, Verwaest T, Volckaert A, Willekens M (2011) Evaluation of Climate Change Impacts and Adaptation Responses for Marine Activities "CLIMAR". Final Report Brussels. Belgian Science Policy Office

Van der Biest K, Verwaest T, Reyns J (2008) Assessing climate change impacts on flooding risks in the Belgian coastal zone. In: LITTORAL 2008. A Changing Coast: Challenge for the Environmental Policies. Proc 9th International Conf November 2008, Venice, pp. 1-12

Van der Biest K, Verwaest T, Mostaert F (2009) CLIMAR: Section report 3. Adaptation Measures to Climate Change Impacts along the Belgian Coastline. Version 2_0. WL rapporten, 814_01. Flanders Hydraulics Research, Belgium

Vanclay F (2012) The potential application of social impact assessment in integrated coastal zone management. Ocean Coast Manage 68:149–156

Vellinga M, Wood RA (2007) Impacts of thermohaline circulation shutdown in the twenty-first century. Climatic Change 91:43–63

Vermaat J, Gilbert A (2006) Habitat dynamics at the catchment – coast interface: contributions from ELOISE. J Integr Environ Sci 3:15–37

Vermaat J, Bouwer L, Turner K, Salomons W (eds) (2005) Managing European Coasts: Past, Present and Future. Springer-Verlag

Vlaamse overheid (2012) Vlaamse Baaien. Naar een geïntegreerde visie voor de kust. Available at: www.vlaanderen.be/nl/publicaties/detail/vlaamse-baaien

Wong PP, Losada IJ, Gattuso G-T, Hinkel J, Khattabi A, McInnes KL, Saito Y, Sallenger (2014) Coastal systems and low-lying areas, In: Climate Change 2014: Impacts, Adaptation and Vulnerability, pp. 361-409. Contribution of Working Group II to the Fifth Assessment Report of the Intergovernmental Panel on Climate Change, Field CB, Barros VR, Dokken DJ, Mach KJ, Mastrandrea MD, Bilir TE, Chatterjee M, Ebi KL, Estrada YO, Genova RC, Girma B, Kissel ES, Levy AN, MacCracken S, Masrandrea PR, White LL (Eds) Cambridge University Press

Young OR, King LA, Schroeder H (eds) (2008) Institutions and Environmental Change: Principal Findings, Applications, and Research Frontiers. MIT Press

Annex 1: What is NAO?

Abdel Hannachi and Martin Stendel

The atmospheric circulation in the European/Atlantic sector, which also determines the regional climate of the North Sea region, can be described mainly by the North Atlantic Oscillation (NAO), the zonality or meridionality of the atmospheric flow and the frequency of atmospheric blocking. The NAO is the dominant mode of near-surface pressure variability over the North Atlantic Ocean and Europe, including the 'NOSCCA region', impacting a considerable part of the northern hemisphere (Hurrell et al. 2003). In its positive phase, the pressure difference between the two main centres of action—the Azores High and the Icelandic Low—is enhanced compared to the climatological average, resulting in a stronger than normal westerly air flow (Hurrell 1995). The storm-track extends north-eastward with more storms over the North Sea and northern Europe. These regions have therefore warmer and wetter than average conditions, especially during winter, whereas the Mediterranean region is generally drier and colder than normal. In contrast, during the negative phase of the NAO, the pressure difference between the Azores High and Icelandic Low is reduced, the storm track is more zonal and shifted southward, extending into the western Mediterranean, and the resulting air flow is weaker than normal (Xoplaki 2002; Xoplaki et al. 2004). For strongly negative NAO indices, the flow can even reverse when there is higher pressure over Iceland than over the Azores, with the consequence of harsh winters over large parts of Europe, such as occurred in 2009/2010 (Ouzeau et al. 2011). The strength of the NAO follows an annual cycle with maximum values in January and minimum values in May (Jones et al. 1997; Furevik and Nilsen 2005). Although the largest amplitude and explained variance occur in winter, the impact of the NAO on the North Sea region is present all year round.

Figure A1.1 shows the variability of the NAO over the past 190 years. From a long-term perspective, the behaviour of the NAO appears irregular. However, extended periods of positive or negative NAO indices are apparent. From the mid-1970s to the mid-1990s, positive index values prevailed (e.g. Hurrell et al. 2003). After the mid-1990s, however, there was a tendency towards more negative NAO indices, in other words a more meridional circulation, and it should be noted that the winter of 2010/2011 had the most negative NAO index in the record (Jung et al. 2011; Pinto and Raible 2012).

Fingerprints of the NAO have been known since at least the days of the Scandinavian sailors (Haine 2008), and from the mid-18th century it was noted (Egede 1745; Cranz 1765) that surface air temperatures in Greenland and Scandinavia vary in opposite phase (Stephenson et al. 2003; Pinto and Raible 2012). Depending on the season, the NAO pattern explains between 40 and 60 % of the total variance in sea-level pressure (SLP) over the North Atlantic Ocean (Wanner et al. 2001; Bojariu and Gimeno 2003; Hurrell et al. 2003).

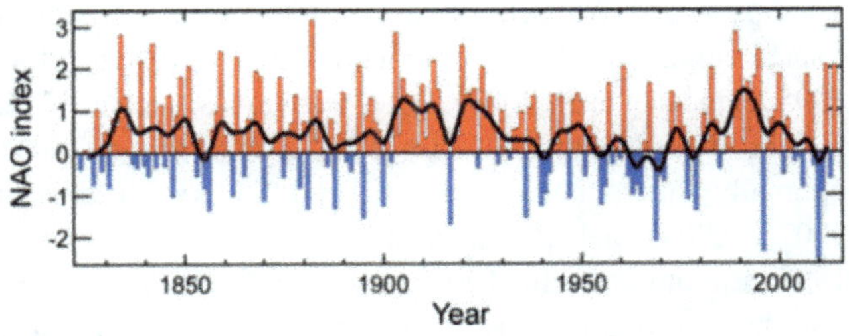

Fig. A1.1 North Atlantic Oscillation (NAO) index for boreal winter (DJFM) 1824/1825 to 2012/2013, calculated as the difference of the normalised station pressures of Iceland and Gibraltar (which is a good measure for the strength of the Azores High) from the monthly means of the period 1951–1980 (Jones et al. 1997, updated at www.cru.uea.ac.uk/~timo/datapages/naoi.htm). The *solid black line* is a 5-year running mean

A. Hannachi (✉)
Department of Meteorology, Stockholm University, Stockholm, Sweden
e-mail: a.hannachi@misu.su.se

M. Stendel
Department for Arctic and Climate, Danish Meteorological Institute (DMI), Copenhagen, Denmark
e-mail: mas@dmi.dk

M. Quante and F. Colijn (eds.), *North Sea Region Climate Change Assessment*, Regional Climate Studies, DOI 10.1007/978-3-319-39745-0

The North Atlantic sea-surface temperature (SST) responds to changes in large-scale atmospheric flow, particularly the NAO. For example, during positive NAO events, there is enhanced cooling of North Atlantic SST north of 45° N. The resulting negative SST anomaly affects air-sea interaction between about 30° and 45°N, leading to positive SST anomalies in this lower latitude band (Marshall et al. 2001). The correlation between the North Atlantic SST anomalies and the NAO index leads to a dipole pattern, known as the Bjerknes' North Atlantic SST dipole (Bjerknes 1962, 1964). The southern lobe of this dipole extends across the Atlantic to the North Sea and thus the NOSCCA region, where the correlation is at a maximum (see Visbeck et al. 2003: their Fig. 2). The NAO affects a whole spectrum of atmospheric and environmental processes, including tropospheric wind (Thompson et al. 2000; see also Fig. 2.2), precipitation (Lamb and Peppler 1987; Zorita et al. 1992; Hurrell and van Loon 1997), ocean surface characteristics (e.g. Moliarini et al. 1997), storminess (Rogers 1997; Serreze et al. 1997), North Atlantic/European atmospheric blocking frequency (Nakamura 1996; Woollings et al. 2010a, b; Häkkinen et al. 2011) and Sverdrup and Ekman transport (Visbeck et al. 2003).

Many approaches have been used to define the spatial structure of the NAO. Historically, (normalised) SLP differences between Iceland and Lisbon (Hurrell 1995), the Azores (Rogers 1997) or Gibraltar (Jones et al. 1997; Vinther et al. 2003) have been used. Several researchers use one-point correlation maps to identify regions of maximal negative correlation near or over Iceland and over the Azores extending to Portugal (e.g. Wallace and Gutzler 1981; Kushnir and Wallace 1989; Portis et al. 2001; Hurrell and Deser 2009). A related approach uses principal components and identifies the NAO by the eigenvectors of the cross-correlation matrix which is computed from the temporal variation of the grid point values of SLP, scaled by the amount of variance they explain (e.g. Barnston and Livezey 1987), or clustering techniques (e.g. Cassou and Terray 2001a,b). Several researchers use unrotated (Horel 1981; Thompson and Wallace 1998; Woollings et al. 2010b) or rotated empirical orthogonal functions (EOFs) (Cheng et al. 1995; Hannachi et al. 2007). Other techniques, such as NAO indices over latitudinal belts (e.g. Li and Wang 2003), optimally interpolated patterns, trend EOFs (Hannachi 2007a, 2008) and cluster analyses (Cheng and Wallace 1993; Kimoto and Ghil 1993; Hannachi 2007b, 2010) have also been proposed. Seasonality can also be taken into account by defining a seasonally and geographically varying NAO index (Portis et al. 2001). All these definitions lead to slightly different NAO indices; but the indices all resemble each other and are in fact highly correlated with each other (Leckebusch et al 2008).

All these definitions have in common that they are based on direct observations or analyses. However, it is also possible to use proxy data to extend the indices back in time. Several reconstructions exist that cover roughly the last millennium. These are based on early instrumental observations (Jones et al 1997; Luterbacher et al. 1999), ship logs (Küttel et al. 2009; Wheeler et al. 2009), other documentary data (Glaser et al. 1999; Luterbacher et al. 2001, 2004), climate field reconstructions (Jones and Mann 2004; Casty et al. 2007), ice cores (Appenzeller et al. 1998), speleothems (Trouet et al. 2009) or strontium/calcium ratios in coral (Goodkin et al. 2008). Multi-proxy reconstructions also exist, based on tree rings and snow accumulation records (Glueck and Stockton 2001) or on tree rings and stable isotope ratios (Cook et al. 2002).

A model-based reconstruction of past atmospheric circulation patterns is in principle possible. While climate models are able to capture the broad spatial and temporal features of the NAO (Gerber et al. 2008), the patterns of variability exhibit substantial differences between models and in comparison to observations (Xin et al. 2008; Casado and Pastor 2012; Handorf and Dethloff 2012). In particular, most models overestimate persistence on time scales from sub-seasonal to seasonal (Gerber et al. 2008). With few exceptions (Selten et al. 2004; Semenov et al. 2008), many climate models are unable to simulate the amplitude of changes in the observed NAO trend since the 1960s (Scaife et al. 2008, 2009; Stoner et al. 2009). This and the apparent underestimation of vertical coupling between troposphere and stratosphere in most models make it difficult to determine the extent to which the underestimation of trends is due to model deficiencies and the extent to which it mirrors anthropogenic forcing (Sigmond and Scinocca 2010; Karpechko and Manzini 2012; Scaife et al. 2012). Further uncertainties arise because there are indications that NAO variability may depend on the mean state of the atmosphere (Branstator and Selten 2009; Barnes and Polvani 2013). It has also been proposed that higher wave numbers could lead to resonance effects and therefore increased persistence of circulation regimes (Coumou et al. 2014), thus corroborating earlier findings, such as those by Kyselý and Huth (2006); see also Rutgersson et al. (2014). It remains an open question how far these drivers of NAO variability are related to changes in the Arctic, such as the decrease in sea ice.

A comparison of the different reconstructions can shed some light on the ability to reconstruct past atmospheric circulation patterns. Pinto and Raible (2012) made such a comparison (after applying a low-pass filter and normalisation) and found reasonable agreement between different reconstructions since the beginning of the 20th century, but also for a few periods in the more distant past (in particular between 1620 and 1720). As these studies rely on different numbers of proxies, different calibration methods and very different types of proxies, including growing-season data to estimate winter NAO, this is not unexpected (e.g. Schmutz et al. 2000). Furthermore, it is also unclear how valid the implicit assumption is that the relation between proxies and the NAO does not change over time.

References

Appenzeller C, Stocker TF, Anklin M (1998) North Atlantic Oscillation dynamics recorded in Greenland ice cores. Science, 282:446–449

Barnes EA, Polvani L (2013) Response of the midlatitude jets and of their variability to increased greenhouse gases in the CMIP5 models. J Climate 26:7117–7135

Barnston AG, Livezey RE (1987) Classification, seasonality and persistence of low frequency atmospheric circulation patterns. Mon Wea Rev 115:1083–1126

Bjerknes J (1962) Synoptic survey of the interaction of the sea and atmosphere in the North Atlantic. Vilhekm Bjerknes Centenary Volume, vol. XXIV, Geofys Publ 3:115–145

Bjerknes J (1964) Atlantic air-sea interactions. Adv Geophys 10:1–82

Bojariu R, Gimeno L (2003) Predictability and numerical modelling of the North Atlantic Oscillation. Earth Sci Rev 63:145–168

Branstator G, Selten F (2009) "Modes of Variability" and climate Change. J Climate 22:2639–2658

Casado MJ, Pastor MA (2012) Use of variability modes to evaluate AR4 climate models over the Euro-Atlantic region. Clim Dyn 38:225–237

Cassou C, Terray L (2001a) Dual influence of Atlantic and Pacific SST anomalies on the North Atlantic/Europe winter climate. Geophys Res Lett 28, 3195–3198

Cassou C, Terray L (2001b) Oceanic forcing of the wintertime low-frequency atmospheric variability in the North Atlantic European sector: A study with the ARPEGE mode. J Climate 14:4266–4291

Casty C, Raible CC, Stocker TF, Wanner H, Luterbacher J (2007) A European pattern climatology 1766-2000. Clim Dyn 29:791–805

Cheng X, Wallace JM (1993) Cluster analysis of the Northern Hemisphere wintertime 500 hPa height field: Spatial patterns. J Atmos Sci 50:2674–2696

Cheng X, Nitsche G, Wallace JM (1995) Robustness of low-frequency circulation patterns derived from EOF and rotated EOF analysis. J Climate 8:1709–1720

Cook ER, D'Arrigo RD, Mann ME (2002) A well-verified, multiproxy reconstruction of the winter North Atlantic Oscillation index since A.D. 400. J Climate 15:1754–1764

Coumou D, Petoukhov V, Rahmstorf S, Petri S, Schellnhuber HJ (2014) Quasi-resonant circulation regimes and hemispheric synchronization of extreme weather in boreal summer. Proc Natl Acad Sci 111:12331–12336

Cranz D (1765) Historie von Grönland enthaltend die Beschreibung des Landes und der Einwohner ec. insbesondere die Geschichte der dortigen Mission der Evangelischen Brüder zu Neu Herrnhut und Lichtenfels. Ebers (Barby)

Egede SH (1745) History of Greenland – A description of Greenland: showing the natural history, situation, boundaries, and face of the country; the nature of the soil; the rise and progress of the old Norwegian colonies; the ancient and modern inhabitants; their genius and way of life, and produce of the soil; their plants, beasts, fishes, etc. (translated from Danish), Pickering Bookseller, London

Furevik T, Nilsen JEØ (2005) Large-scale atmospheric circulation variability and its impacts on the Nordic Seas ocean climate – a review. In Drange H, Dokken T, Furevik T, Gerdes R, Berger W (eds.) The Nordic Seas: An Integrated Perspective. Geophys Mon Ser 158, AGU

Gerber EP, Polvani LM, Ancukiewicz D (2008) Annular mode time scales in the Intergovernmental Panel on Climate Change Fourth Assessment Report models. Geophys Res Lett 35:L22707. doi:10.1029/2008GL035712

Glaser R, Brázdil R, Pfister C, Dobrovolný P, Barriendos Vallvé M, Bokwa A, Camuffo D, Kotyza O, Limanówka D, Rácz, Rodrigo FS (1999) Seasonal temperature and precipitation fluctuation in selected parts of Europe during the sixteenth century. Clim change 43:169–200

Glueck MF, Stockton CW (2001) Reconstruction of the North Atlantic Oscillation, 1429-1983. Int J Climatol 21:1453–1465

Goodkin NF, Hughen KA, Doney SC, Curry WB (2008) Increased multidecadal variability of the North Atlantic Oscillation since 1781. Nature Geoscience 1:844–848

Haine T (2008) What did the Viking discoverers of America know of the North Atlantic Environment? Weather 63:60–65

Häkkinen S, Rhines PB, Worthen DL (2011) Atmospheric blocking and Atlantic multidecadal ocean variability. Science 334:655–659

Handorf D, Dethloff K (2012) How well do state-of-the-art atmosphere-ocean general circulation models reproduce atmospheric teleconnection patterns? Tellus A 64:19777 http://dx.doi.org/10.3402/tellusa.v64i0.19777

Hannachi A (2007a) Pattern hunting in climate: a new method for trends in gridded climate data. Int J Climatol 27:1–15

Hannachi A (2007b) Tropospheric planetary wave dynamics and mixture modeling: Two preferred regimes and a regime shift. J Atmos Sci 64:3521–3541

Hannachi A (2008) A new set of orthogonal patterns in weather and climate: Optimally interpolated patterns. J Climate 21:6724–6738

Hannachi A (2010) On the origin of planetary-scale extratropical circulation regimes. J Atmos Sci 67:1382–1401

Hannachi A, Jolliffe IT, Stephenson DB (2007) Empirical orthogonal functions and related techniques in atmospheric science: A review. Int J Climatol 27:1119–1152

Horel JD (1981) A rotated principal component analysis of the interannual variability of the Northern Hemisphere 500 mb height field. Mon Wea Rev 109:2080–2902

Hurrell JW (1995) Decadal trends in the North Atlantic Oscillation, regional temperatures and precipitation. Science 269:676–679

Hurrell JW, Deser C (2009) North Atlantic climate variability: The role of the North Atlantic Oscillation. J Mar Sys 78:28–41

Hurrell JW, van Loon H (1997) Decadal variations in climate associated with the North Atlantic Oscillation. Clim Change 36:301–326

Hurrell JW, Kushnir Y, Ottersen G, Visbeck M (2003) An overview of the North Atlantic Oscillation. In: The North Atlantic Oscillation: Climatic significance and environmental impact, Geophysical Monograph Series 134:1–36, AGU

Jones PD, Mann ME (2004) Climate over past millennia. Rev Geophys 42:RG2002 doi:10.1029/2003RG000143

Jones PD, Jonsson T, Wheeler D (1997) Extension to the North Atlantic Oscillation using early instrumental pressure observations from Gibraltar and south-west Iceland. Int J Climatol 17:1433–1450

Jung T, Vitart F, Ferranti L, Morcrette JJ (2011) Origin and predictability of the extreme negative NAO winter of 2009/10. Geophys Res Lett 38, L07701, doi:10.1029/2011GL046786

Karpechko AY, Manzini E (2012) Stratospheric influence on tropospheric climate change in the Northern Hemisphere. J Geophys Res 117:D05133 doi:10.1029/2011JD017036

Kimoto M, Ghil M (1993) Multiple flow regimes in the northern hemisphere winter. Part II: Sectorial regimes and preferred transitions. J Atmos Sci 16:2645–2673

Kushnir Y, Wallace JM (1989) Low-frequency variability in the Northern Hemisphere winter: Geographical distribution, structure and time dependence. J Atmos Sci 46:3122–3142

Küttel M, Xoplaki E, Gallego D, Luterbacher J, García-Herrera R, Allan R, Barriendos M, Jones PD, Wheeler D, Wanner H (2009) The importance of ship log data: reconstructing North Atlantic, European and Mediterranean sea level pressure fields back to 1750. Clim Dyn 34:1115–1128

Kyselý J, Huth R (2006) Changes in atmospheric circulation over Europe detected by objective and subjective methods. Theor Appl Climatol 85:19–36

Lamb PJ, Peppler RA (1987) North Atlantic Oscillation: concept and an application. Bull Am Met Soc 68:1218–1225

Leckebusch GC, Weimer A, Pinto JG, Reyers M, Speth P (2008) Extreme wind storms over Europe in present and future climate: a cluster analysis approach. Meteorol Zeit 17:67–82

Li J, Wang JXL (2003) A new North Atlantic Oscillation index and its variability. Adv Atmos Sci 20:661–676

Luterbacher J, Schmutz C, Gyalistras D, Xoplaki E, Waaner H (1999) Reconstruction of monthly NAO and EU indices back to AD 1657. Geophys Res Lett 26:2745–2748

Luterbacher J, Xoplaki E, Dietrich D, Jones PD, Davies TD, Portis D, Gonzalez-Rouco JF, von Storch H, Gyalistras D, Casty C, Wanner H (2001) Extending North Atlantic oscillation reconstructions back to 1500. Atm Sci Lett 2:1-4

Luterbacher J, Dietrich D, Xoplaki E, Grosjean M, Wanner H (2004) European seasonal and annual temperature variability, trends, and extremes since 1500. Science 303:1499–1503

Marshall J, Johnson H, Goodman J (2001) A study of the interaction of the North Atlantic Oscillation with the ocean circulation. J Climate 14:1399–1421

Moliarini RL, Mayer DA, Festa JF, Bezdek HF (1997) Multiyear variability in the near-surface temperature structure of the midlatitude western North Atlantic Oscillation. J Geophys Res 102: 3267–3278

Nakamura H (1996) Year-to-year and interdecadal variability in the activity of intraseasonal fluctuations in the Northern Hemisphere wintertime circulation. Theor Appl Climatol 55:19–32

Ouzeau G, Cattiaux J, Douville H, Ribes A, Saint-Martin D (2011) European cold winter 2009–2010: How unusual in the instrumental record and how reproducible in the ARPEGE-Climate model? Geophys Res Lett 38:L11706. doi:10.1029/2011GL047667

Pinto JG, Raible CC (2012) Past and recent changes in the North Atlantic oscillation. Clim Change 3:79–90

Portis DH, Walsh JE, El-Hamly M, Lamb PJ (2001) Seasonality of the North Atlantic Oscillation. J Climate 14:2069-2078

Rogers JC (1997) North Atlantic storm-track variability and its association to the North Atlantic Oscillation and climate variability of Northern Europe. J Climate 10:1635–1647

Rutgersson A, Jaagus J, Schenk F, Stendel M (2014) Observed changes and variability of atmospheric parameters in the Baltic Sea region during the last 200 years. Clim Res 61:177–190

Scaife AA, Folland CK, Alexander LV, Moberg A, Knight JR (2008) European climate extremes and the North Atlantic Oscillation. J Climate 21:72–83

Scaife AA, Kucharski F, Folland CK, Kinter J, Brönnimann S, Fereday D, Fischer AM, Grainger S, Jin EK, Kang IS, Knight JR, Kusunoki S, Lau NC, Nath MJ, Nakaegawa T, Pegion P, Schubert S, Sporyshev P, Syktus J, Yoon JH, Zeng N, Zhou (2009) The CLIVAR C20C project: selected twentieth century climate events. Clim Dyn 33:603–614

Scaife AA, Spangehl T, Fereday DA, Cubasch U, Langematz U, Akiyoshi H, Bekki S, Braesicke P, Butchard N, Chipperfiels MP, Gettelman A, Hardiman SC, Michou M, Rozanov E, Shepherd TG (2012) Climate change and stratosphere-troposphere interaction. Clim Dyn 38:2089–2097

Schmutz C, Luterbacher J, Gyalistras D, Xoplaki E, Wanner H (2000) Can we trust proxy-based NAO index reconstructions? Geophys Res Lett 27:1135–1138

Selten FM, Branstator GW, Dijkstra HA, Kliphuis M (2004) Tropical origins for recent and future Northern Hemisphere climate change. Geophys Res Lett 31:L21205. doi:10.1029/2004GL020739

Semenov VA, Latif M, Jungclaus JH, Park W (2008) Is the observed NAO variability during the instrumental record unusual? Geophys Res Lett 35:L11701. doi:10.1029/2008GL033273

Serreze MC, Carse F, Barry RG, Rogers JC (1997) Icelandic low cyclone activity: climatological features, linkages with the NAO and relationships with recent changes in the Northern Hemisphere circulation. J Climate 10:453–464

Sigmond M, Scinocca JF (2010) The influence of the basic state on the Northern Hemisphere circulation response to climate change. J Climate 23:1434–1446

Stephenson DB, Wanner H, Brönnimann S, Luterbacher J (2003) The history of scientific research on the North Atlantic Oscillation. In: Hurrell JW, Kushnir Y, Ottersen G, Visbeck M (eds) The North Atlantic Oscillation: Climatic Significance and Environmental Impact. AGU Geophys Monogr Series 134

Stoner AMK, Hayhoe K, Wuebbles DJ (2009) Assessing General Circulation Model simulations of atmospheric teleconnection patterns. J Climate 22:4348–4372

Thompson DWJ, Wallace JM (1998) The Arctic oscillation signature in the wintertime geopotential height and temperature fields. Geophys Res Lett 25:1297–1300

Thompson DWJ, Wallace JM, Hegerl GC (2000) Annual modes in the extratropical circulation: Part II: Trends. J Climate 13:1018–1036

Trouet V, Esper J, Graham NE, Baker A, Scourse JD, Frank DC (2009) Persistent positive North Atlantic Oscillation mode dominated the medieval climate anomaly. Science 324:78–80

Vinther BM, Andersen KK, Hansen AW, Schmith T, Jones PD (2003) Improving the Gibraltar/Reykjavik NAO index. Geophys Res Lett 30:2222, doi:10.1029/2003GL018220

Visbeck M, Chassignet EP, Curry RG, Delworth TL, Dickson RR, Krahmann G (2003) The ocean's response to North Atlantic Oscillation variability In: Hurrell JW, Kushnir Y, Ottersen G, Visbeck M (eds) The North Atlantic Oscillation: Climatic Significance and Environmental Impact. AGU Geophys Monogr Series 134

Wallace JM, Gutzler DS (1981) Teleconnections in the geopotential height field during the Northern Hemisphere winter. Mon Wea Rev 109:784–812

Wanner H, Brönnimann S, Casty C, Gyalistras D, Luterbacher J, Schmutz C, Stephenson DB, Xoplaki E (2001) North Atlantic Oscillation - concepts and studies. Surv Geophys 22:321–381

Wheeler D, Garcia-Herrera R, Wilkinson CW, Ward C (2009) Atmospheric circulation and storminess derived from Royal Navy logbooks: 1685 to 1750. Clim Change 110:257–280

Woollings TJ, Hannachi A, Hoskins B (2010a) Variability of the North Atlantic eddy-driven jet stream. Q J Roy Met Soc 136:856–868

Woollings TJ, Hannachi A, Hoskins B, Turner A (2010b) A regime view of the North Atlantic Oscillation and its response anthropogenic forcing. J Climate 23:1291–1307

Xin XG, Zhou TJ, Yu RC (2008) The Arctic Oscillation in coupled climate models. Chin J Geophys Chinese Edition 51:337–351

Xoplaki E (2002) Climate variability over the Mediterranean. PhD thesis, University of Bern, Switzerland

Xoplaki E, González-Rouco JF, Luterbacher J, Wanner H (2004) Wet season Mediterranean precipitation variability: influence of large-scale dynamics. Clim Dyn 23:63–78

Zorita E, Kharin V, von Storch H (1992) The atmospheric circulation and sea surface temperature in the North Atlantic area in winter: their interaction and relevance for Iberian precipitation. J Climate 5:1097–1108

Annex 2: Climate Model Simulations for the North Sea Region

Diana Rechid, H.E. Markus Meier, Corinna Schrum, Markku Rummukainen, Christopher Moseley, Katharina Bülow, Alberto Elizalde, Jian Su and Thomas Pohlmann

D. Rechid (✉) · K. Bülow
Climate Service Center Germany (GERICS), Helmholtz-Zentrum Geesthacht, Hamburg, Germany
e-mail: diana.rechid@hzg.de

K. Bülow
e-mail: katharina.buelow@hzg.de

H.E. Markus Meier
Research Department, Swedish Meteorological and Hydrological Institute, Norrköping, Sweden
e-mail: markus.meier@smhi.se

H.E. Markus Meier
Department of Physical Oceanography and Instrumentation, Leibniz-Institute for Baltic Sea Research (IOW), Rostock, Germany
e-mail: markus.meier@io-warnemuende.de

C. Schrum
Geophysical Institute, University of Bergen, Bergen, Norway
e-mail: corinna.schrum@gfi.uib.no

C. Schrum
Institute of Coastal Research, Helmholtz-Zentrum Geesthacht, Geesthacht, Germany
e-mail: corinna.schrum@hzg.de

M. Rummukainen
Centre for Environmental and Climate Research, Lund University, Lund, Sweden
e-mail: markku.rummukainen@cec.lu.se

C. Moseley · A. Elizalde
Max Planck Institute for Meteorology, Hamburg, Germany
e-mail: christopher.moseley@mpimet.mpg.de

A. Elizalde
e-mail: alberto.elizalde@mpimet.mpg.de

J. Su · T. Pohlmann
Institute of Oceanography, CEN, University of Hamburg, Hamburg, Germany
e-mail: jian.su@uni-hamburg.de

T. Pohlmann
e-mail: Thomas.Pohlmann@uni-hamburg.de

A2.1 Introduction

Climate models are powerful tools for investigating internal climate variability and the response of the climate system to external forcing, complementing observational studies.

Internal climate variability depicts natural variations due to chaotic processes within the climate system. On annual to multi-decadel time scales internal variability largely arises from the continuous interaction between the atmosphere and the ocean. External forcing involves factors outside the climate system and comprises natural forcing factors (e.g. solar variability, orbital variations or volcanic eruptions) and anthropogenic forcing factors (e.g. emissions of greenhouse gases to the atmosphere, anthropogenic aerosols and changes in land use). Climate variations due to internal processes and external forcing occur at different spatial scales (due to the different spatial extent of the relevant processes) and at different temporal scales (due to the different time scales of the relevant forcing factors and the different response times of the climate system components).

In order to simulate internal and externally driven variability at different temporal and spatial scales with climate models, the relevant components and processes need to be included in the model. To investigate climate system processes, a realistic representation of the coupling between atmosphere and ocean is essential. For this purpose, climate simulations are carried out using coupled Atmosphere–Ocean General Circulation Models (AOGCMs). Such models are able to represent dynamic interactions between atmosphere, ocean and land, and thus also related non-linear feedbacks in the climate system. State-of-the-art Earth System Models (ESMs), which constitute a further development of AOGCMs, also include dynamic land and ocean biosphere models and represent the carbon cycle, and in some

M. Quante and F. Colijn (eds.), *North Sea Region Climate Change Assessment*, Regional Climate Studies, DOI 10.1007/978-3-319-39745-0

cases ice sheet dynamics, aerosol processes and atmospheric chemistry.

A major application of climate models is the simulation of potential future climate changes due to human action within the climate system. Future climate change in the near term (at the scale of several decades) cannot be predicted, due to internal climate variability and unknown external forcings. However, it is possible to examine the impact of some external forcing over the longer term. For example, by using anthropogenic greenhouse gas emission scenarios to project potential future climate evolutions over the coming century and beyond. Each projection is the combined result of the forced climate change signal and a possible course of internal variability under that scenario. Any two projections with one model and for one emission scenario may thus differ with respect to the simulated course of internal variability.

To assess the climate of the North Sea region, regional data from global models are dynamically downscaled using regional climate and ocean models to resolve regional-scale processes in more detail than can be shown at the far coarser resolution of global models. Recent studies for the North Sea region have also applied coupled regional atmosphere–ocean models in order to represent mesoscale feedbacks. One subtask of the German research program KLIWAS is to focus on coupled regional model simulations for the North Sea region.

A2.2 Climate Models

Climate models are models of the climate system based on physical, chemical and biological principles. They can be classified into conceptual models (e.g. one-dimensional energy balance models), earth system models of intermediate complexity (EMICs) and comprehensive global climate models, which are three-dimensional general circulation models (GCMs). Key components of GCMs are atmosphere and ocean general circulation models (AGCMs and OGCMs), which can be dynamically coupled to form atmosphere–ocean general circulation models (AOGCMs). In state-of-the-art Earth System Models (ESMs), further components of the climate system such as ice sheets, vegetation dynamics and biogeochemical cycles may be included. An introduction to climate modelling is given by McGuffie and Henderson-Sellers (2005).

For spatial refinement of GCM simulations, statistical and dynamical downscaling methods are applied. For statistical downscaling, statistical relationships between observed local and large-scale variables are established and then applied to GCM output. According to Wilby and Wigley (1997), statistical downscaling is divided into regression methods, weather pattern-based approaches, and stochastic weather generators. Regression methods are usually applied because they are easy to implement and computationally efficient. Among other things, statistical downscaling has been applied to estimate biological impacts and changes in sea level. For the latter, projected future large-scale meteorology, typically taken from GCMs, is related to local extreme sea level using statistical relationships derived from observations or a limited number of simulations from physically-based models (for a review see Lowe and Gregory 2010). It is unclear how statistical relationships derived from observations or simulations of the past will continue to be applicable under future climate conditions. In the rest of the annex, only dynamical downscaling methods are considered.

Dynamical downscaling involves regional climate models (RCMs). Reviews about RCMs are given, for instance, by Rummukainen (2010) and Rockel (2015). RCMs are local area circulation models for a three-dimensional section of the atmosphere at high spatial resolution, forced by large-scale atmospheric conditions simulated by a GCM. Regional ocean models are circulation models for a three-dimensional section of the ocean, forced by large-scale ocean conditions simulated by a global ocean model, and meteorological forcing from atmospheric models. As in the case of global models, regional models of atmosphere and ocean can be coupled to form regional atmosphere–ocean models, and further complemented by additional components of the climate system, towards regional climate system models.

A2.2.1 Atmosphere–Ocean General Circulation Models

Fluid dynamics and thermodynamics in the atmosphere and ocean are described by fundamental physical laws as the conservation of momentum, mass and energy, and the thermodynamic equation of state. They form a system of non-linear partial differential equations for which no closed analytic solution exists. Rather, they need to be discretised using either the finite difference method or the spectral method and solved numerically. For finite differences, a grid is imposed on the atmosphere and ocean. The grid resolution strongly correlates with available computer power. Typical horizontal resolutions of AGCMs for centennial climate simulations correspond to spatial scales of between 300 and 100 km, in some cases 50 km, with 30–90 vertical levels. Horizontal resolution in OGCMs corresponds to spatial scales of between 160 and 10 km, with 40–80 vertical levels.

Processes which are not resolved at the resolution of the model grid need to be considered by describing their collective effect on the resolved spatial unit. This is done by parameterisations based on theoretical assumptions, process-based modelling or observations and derived empirical relationships. Examples for parameterised subgrid-scale processes in climate models include radiation, convection, processes within the atmospheric and oceanic planetary boundary layers and land surface processes. The fundamental physical understanding behind those parameterisations, together with the numerical methods and model resolutions applied, as well as the treatment of initial and boundary conditions, determine the capabilities of a model. In AOGCMs, the coupling between atmosphere and ocean is of crucial importance. Major difficulties with coupled models arise because the initial states of the ocean and atmosphere are not known precisely and even small inconsistencies in terms of energy, momentum and mass fluxes between atmosphere and ocean can cause a model drift to unrealistic climatic states. In early AOGCM simulations, this problem was addressed by empirical 'flux adjustments' (Manabe and Stouffer 1988). Today, most coupled models no longer need such adjustment owing to improved representation of physical processes, and to finer model resolution.

A2.2.2 Regional Climate Models

Regional climate models are models of a three-dimensional section of the atmosphere and possibly other climate system components. They are based on the same primitive equations for fluid dynamics as global climate models. They are discretised at much finer spatial atmosphere grids (corresponding to spatial scales of 50–2.5 km) for a limited geographical area. At the lateral boundaries of the model domain, meteorological conditions from either global model simulations or observational data are prescribed ('nesting'). Within the model domain, finer-scale processes such as mesoscale convective systems, orographic and land-sea contrast induced circulations are resolved. This method is also called dynamical downscaling. In terms of topography, land-sea distribution and land surface characteristics, regional climate models apply more detailed lower boundary descriptions than global climate models. Compared to global models, the treatment of lateral and lower boundary data in regional models can affect model quality.

The nested regional modelling technique essentially originated from numerical weather prediction. The use of RCMs for climate application was pioneered by Giorgi (1990). The advantages of regional atmosphere models are (1) more detailed orography and improved spatial representation of precipitation, (2) improved representation of the land-sea mask, (3) improved sea surface temperature (SST) boundary conditions if a regional coupled atmosphere–ocean model is used, (4) more accurate modelling of extremes (e.g. low pressure systems), and (5) more detailed representation of vegetation and soil characteristics over land (Rummukainen 2010; Feser et al. 2011 and references therein). Over the sea the added value of the high resolution in the regional atmosphere model is limited spatially to the coastal zone. For the North Sea, added value is found in the Southern Bight and the Skagerrak (Winterfeldt et al. 2010; Feser et al. 2011).

During the last decade, RCMs have been coupled with other climate process models, such as ocean, sea-ice and biosphere models, thus moving towards regional climate system models (RCSMs). RCSMs are able to represent dynamic interactions between the regional climate system components and thus regional-to-local climate feedbacks. RCMs are used in a wide range of applications from paleoclimate to anthropogenic climate change studies. For a comprehensive study of regional climate change in the North Sea region, coupled regional atmosphere–ocean models are appropriate tools. They provide regional to local scale climate information relevant for regional climate and climate change assessments.

A2.2.2.1 Regional Ocean Models

For a detailed and spatially resolved investigation of climate change impacts on physical and biogeochemical variables of the North Sea system a consistent dynamical downscaling approach is needed. Such an approach is usually complex and computationally expensive. It requires coupled physical-biogeochemical models of sufficiently high resolution driven with appropriate atmospheric forcing (i.e. air-sea fluxes of momentum, energy and matter including the atmospheric deposition of nitrogen and carbon), hydrological forcing (water volume, carbon and nutrient flows from the catchment area) and lateral boundary data at locations in the North Atlantic and Baltic Sea depending on the extent of the regional model domain. In addition, consistent initial conditions are needed. For reasons of computational expense, rather than simulating the full transient period from past to distant future, two or more time-slices are often used, with one covering the recent past and the others covering the mid- and/or end of the century. If time slices of present and future climates are calculated instead of the transient evolution under a changing climate, initial conditions are also needed for the future time slice. Due to the relatively short memory of initial conditions in the North Sea the proper choice of initial values for physical variables is not usually a problem. A shorter spin-up period of about 1–3 years guarantees that the state variables are in equilibrium with the model physics. For nutrient and carbon cycling, spin-up periods of 2–5 years are needed, because in the North Sea time scales of the water-sediment fluxes and the biogeochemical system are slightly longer than physical time scales.

For regional North Sea scenario simulations, initial, surface and boundary forcing data can be taken directly from

GCM simulations (e.g. Ådlandsvik 2008). However, due to the coarse resolution of GCMs these data sets suffer from considerable biases at the regional scale, which prevents the realistic modelling of regional hydrodynamic and biogeochemical processes. Either a bias correction method (see Sect. A2.3.2) or a regional atmosphere model and a hydrological model should therefore be used to force the ocean model. As both the ocean and the atmosphere need higher spatial resolution than is usually available from state-of-the-art GCM simulations, the atmospheric forcing of the regional ocean model is often downscaled as well.

A2.2.2.2 Regional Coupled Atmosphere–Ocean Models

While the coarser AOGCMs have been used for some time, a recent major achievement with respect to modelling is the building of high-resolution fully coupled atmosphere–sea-ice–ocean–land-surface models, which allow for consideration and resolution of local feedbacks (Gustafsson et al. 1998; Hagedorn et al. 2000; Rummukainen et al. 2001; Döscher et al. 2002; Schrum et al. 2003; Dieterich et al. 2013, 2014; Ho-Hagemann et al. 2013; Tian et al. 2013; Van Pham et al. 2014; Gröger et al. 2015). The first coupled atmosphere–sea-ice–ocean models were developed to improve short-range weather forecasting (e.g. Gustafsson et al. 1998) or to study processes and the impact of coupling on air-sea exchange (e.g. Hagedorn et al. 2000; Schrum et al. 2003). During the past decade, coupled modelling has become more aligned to perform studies on climate change (e.g. Rummukainen et al. 2001; Räisänen et al. 2004; Meier et al. 2011a) and the first transient centennial climate change simulations became available for the Baltic Sea region (Meier et al. 2011b, 2012a). Transient simulations for the period 1960–2100 using regional coupled atmosphere–ocean models are now available for the North Sea (initialised by the German KLIWAS project; www.kliwas.de) (Bülow et al. 2014; Dieterich et al. 2014; Su et al. 2014b) (see Sect. A2.4).

In a first attempt to model the regional coupled atmosphere–ocean system including the North Sea, Schrum et al. (2003) showed that coupling stabilised the regional model system simulation in a one-year simulation and reduced the drift compared to the uncoupled system. In a decadal simulation, Su et al. (2014b) showed that their coupled model was able to damp the drift seen in an uncoupled regional atmosphere–ocean model system, which had been due to an accumulation of heat caused by heat flux errors. Nevertheless, the impact of air-sea heat fluxes on atmospheric conditions is not the same for different periods. Kjellström et al. (2005) showed that the regional impact of surface fluxes on summer SSTs is greatest during a phase of negative NAO index, when the large-scale atmospheric flow over the North Atlantic is weaker and more northerly, than during a phase of positive NAO index, when the large-scale atmospheric flow is stronger and more westerly. Hence, the impact of the lower boundary condition on near surface atmospheric fields and atmosphere–ocean fluxes is small when horizontal advection is large, for example during years with a positive NAO index.

A2.2.2.3 Towards Regional Climate System Models

In recent years, coupled atmosphere–sea-ice–ocean models have been further elaborated by using a hierarchy of sub-models for the Earth system, combining regional climate models with sub-models for surface waves (e.g. Rutgersson et al. 2012), land vegetation (e.g. Smith et al. 2011), hydrology and land biochemistry (e.g. Arheimer et al. 2012; Meier et al. 2012b), marine biogeochemistry and lower trophic level dynamics (e.g. Allen et al. 2001; Holt et al. 2005; Pätsch and Kühn 2008; Daewel and Schrum 2013), the marine carbon cycle (e.g. Wakelin et al. 2012a, b; Artioli et al. 2013; Gröger et al. 2015, early life stages of fish (e.g. Daewel et al. 2008) and food web modelling (e.g. Niiranen et al. 2013). Hence, there is a tendency to develop Regional Climate System Models (RCSMs), which enables better investigation of the impact of climate change on the entire marine environment. Indeed, RCSMs further enable regional climate simulations which represent dynamical feedback mechanisms such as the ice-albedo feedback (Meier et al. 2011a), by including interactive coupling between the regional climate system components (i.e. atmosphere, ocean, sea ice, land vegetation, marine biogeochemistry).

A2.2.2.4 Regional Coupled Modelling of Land–Sea Processes

Many downscaling studies for the North Sea assume—because more detailed information is lacking—that runoff from the catchment area and the freshwater outflow from the Baltic Sea will not change in a future climate (e.g. Wakelin et al. 2012a). As far as is known, only in the MPIOM-REMO model is the water cycle closed (Sein et al. 2015) and no attempt has so far been made to consider terrestrial changes in nutrient loads or alkalinity at either the global scale in ESMs or for any regional ESM. Although the impact of changing runoff and river load and changing Baltic outflow properties may be restricted to the southern coastal North Sea and the Skagerrak, respectively, a more consistent approach addressing the water and nutrient budget of the North Sea should consider the entire land-sea continuum. Hence, projections of salinity and marine biogeochemical cycles in shelf seas are still uncertain (e.g. Meier et al. 2006; Wakelin et al. 2012a; Artioli et al. 2013). Recently, a new hydrological model, the HYPE model (HYdrological Predictions for the Environment) (Lindström et al. 2010; Arheimer et al. 2012), was developed to calculate river flow and river-borne nutrient loadings from catchment areas.

The HYPE model version developed for Europe is referred to as E-HYPE. In the future, scenario simulations with E-HYPE can be used to calculate changing water and nutrient budgets more consistently. However, a current limitation is that the carbon cycle and carbon loads are not considered in the present version of E-HYPE. Despite these recent efforts, the uncertainties in runoff in scenario simulations for the end of the 21st century are considerable due to biases in precipitation from the regional atmosphere models (Donnelly et al. 2014). Future projections of nutrient loads are perhaps even more uncertain than projections of future river flows, due to unknown future land use and socioeconomic scenarios (Arheimer et al. 2012).

A2.3 Climate Projections

A2.3.1 Methodology

Climate models are applied to project potential future climate evolutions at multi-decadal to centennial time scales. The temporal evolution of future climate will depend on external natural and anthropogenic forcing and on internal climate variability. The following sections explain the methodology of climate model projections, and how external forcings and internal climate variability are considered.

A2.3.1.1 External Forcing

Humans affect climate through emission of substances to the atmosphere and by altering characteristics of the land surface. Future socioeconomic development cannot be foreseen, but it is possible to assume plausible future pathways and derive related emission and land-use scenarios. Potential human pathways are described within global socioeconomic scenarios which assume certain development of demography, policies, technology and economic growth. For each scenario, the related emissions of greenhouse gases and aerosols are quantified, from which the concentrations of the respective substances in the atmosphere are derived. The procedure of defining emission scenarios is described in the Special Report on Emission Scenarios (Nakicenovic and Swart 2000). The latest generation of climate projections for the 21st century build on the more recent Representative Concentration Pathways (RCPs), which are derived from a different scenario process (Moss et al. 2010). RCPs are defined by different levels of radiative forcing at the end of the 21st century. Further information on emission scenarios and RCPs is provided in Annex 4.

The concentrations, in some cases the emissions, are prescribed to climate models, which then simulate the response of the climate system to the forcing. For historical climate simulations, observed concentrations of atmospheric substances are prescribed to the models. The results of climate projections are related to the results of the historical climate simulation in order to derive simulated climate change signals. By prescribing different forcings according to different pathways, a range of potential future climate evolutions can be projected.

Future natural external forcings such as volcanic eruptions and solar variability are not predictable. In the real future of earth, changes in natural factors may occur which could substantially affect future earth climate. This will always be an unknown in climate projections. In most climate projections for the future, natural external forcings are kept constant. For historical climate simulations they are prescribed to the models from available observations. The projected human impact on climate for the 21st century, however, seems significantly larger than the amount of natural external forcing on climate than has occurred over a multi-century and longer historical perspective.

A2.3.1.2 Internal Climate Variability

Assuming one external forcing, a range of climate evolutions are still possible due to the impact of internal climate dynamics. In addition, with external factors changing over time, the internal climate variability itself can also change over time. Internal variability arises from natural processes within the climate system and can lead to stochastic variations in climate parameters at time scales from seconds to centuries. Processes within the atmosphere occur on relatively short time scales, whereas processes within the ocean or ice sheets occur on longer time scales. Interactions and feedbacks between components of the climate system (i.e. atmosphere, biosphere, lithosphere, pedosphere, hydrosphere and cryosphere) lead to natural internal climate variations that are also relevant at the multi-decadal time scales of climate projections. Climate models are able to simulate internal climate variability, but its temporal evolution strongly depends on the initialisation of each model component. To consider different temporal evolutions of natural climate variability, a set of simulations can be performed with the same external forcing but with different initialisation states. The results of such an initial-condition ensemble for a certain time period lie within a range of equally probable climate evolutions.

A2.3.1.3 Regional Climate Change Projections

Global simulations of the historical climate and global projections of the future climate can be dynamically downscaled with RCMs, in order to relate the overall climate change to regional and local consequences in more detail. While RCMs can inherit errors from the GCMs and may also add further uncertainties due to different parameterisations, structures and configurations, they do add value to the modelling results owing to the better representation of local-scale features and processes. Thus, local-to-regional scale climate change patterns simulated by an RCM can decisively differ from the simulation results of a global model.

Models are always simplified images of the earth's climate system. They provide more or less accurate approximations of climate parameters compared to the real system. Many physical processes occur on spatial scales which are not resolved by climate models and thus need parameterisations. Model parameterisations are derived from empirical studies and statistical approaches. Modelling uncertainties arise from an incomplete understanding of processes within the climate system and from the inability to represent all processes and characteristics of the climate system accurately within climate models (see Annex 3). Modelling uncertainties can lead to systematic biases between simulated climate parameters and those based on observations. For some investigations bias correction methods are applied (see Sect. A2.3.2).

Different models apply different physical parameterisations and different numerical approaches. Those structural differences lead to a range of possible climate responses to external forcing, which is addressed with multi-model-ensemble simulations (see Sect. A2.3.3). In the case of regional climate projections, simulations of multi-global model ensembles are downscaled either with a single RCM or with different RCMs. Multi-model ensemble simulations based on a single scenario sample modelling uncertainties, but also different initial conditions of the climate system, as each global model is initialised at a different climate state.

A2.3.2 Bias Correction

To overcome shortcomings in the atmospheric and hydrological forcing and in the lateral boundary data towards the North Atlantic and Baltic Sea, bias correction methods are often applied (e.g. Holt et al. 2012; Wakelin et al. 2012a; Mathis 2013). An advantage of applying bias correction is that the projections become more reliable when the simulated historical climate is closer to the observed climate. The sensitivity of the regional system to projected regional changes is probably also described more realistically. However, a disadvantage is that the projected parameters are among each other no longer dynamically consistent. Furthermore, some bias correction methods assume that internal climate variability is not influenced by external forcing, which can lead to different climate change signals than when they are derived from the original model simulation.

Without loss of generality, the following discussion is restricted to the atmospheric forcing of a regional climate ocean model. Forcing can be handled by three approaches: (1) direct forcing with GCM output (e.g. Ådlandsvik 2008), (2) forcing with regional atmosphere model results driven by GCM data at lateral and surface boundaries (e.g. Holt et al. 2010), and (3) forcing with regional coupled atmosphere–ocean model results driven by GCM data at lateral boundaries (Bülow et al. 2014). In all three cases the atmospheric forcing may be biased compared to observations of historical climate due to the coarse resolution (Case 1), inconsistent SSTs (Case 2) or biases in the large-scale circulation (Cases 1, 2, 3). Furthermore, even in Cases 2 and 3, when a regional climate model is used, the resolution might not be high enough to resolve all the relevant processes with an impact on ocean climate.

Bias correction methods can be applied together with all three approaches. Two main categories of bias correction are the delta approach, and linear or nonlinear bias correction methods. In the delta approach, historical climate forcing is provided by reanalysis data. The climate change signal is derived through perturbing the historical climate forcing with the simulated change from a GCM or an RCM. Both additive and multiplicative perturbations have been used (e.g. Wakelin et al. 2012a; Holt et al. 2014, respectively). The second category methods apply the same, time-independent bias correction to both the historical and climate change forcing to improve agreement between the historical climate and contemporary observations. The correction might either be a linear correction (fractional or additive), for example to correct for a bias of the mean condition (e.g. Mathis 2013), or the correction might be a more complex nonlinear function derived for example from a statistical downscaling approach (e.g. Donnelly et al. 2014).

The overall disadvantage of all bias correction methods is that the simulated changes are affected by the bias correction and are sensitive to the chosen method (e.g. Räisänen and Räty 2013; Donnelly et al. 2014; Holt et al. 2014).

A2.3.3 Ensemble Simulations

Since 1990, the first model intercomparison projects (MIPs) opened a new era in climate modelling. They provide a standard experiment protocol and a worldwide community-based infrastructure in support of model simulations, evaluation, intercomparison, documentation and data access. There are, among others, atmospheric model intercomparison projects (AMIP) for AGCMs and coupled model intercomparison projects (CMIP) for AOGCMs (Meehl et al. 2005), both initiated by the World Climate Research Program (WCRP) and supported by the program for climate model diagnosis and intercomparison (PCMDI).[1] For example, within CMIP phase 3 (Meehl et al. 2007), coordinated climate projections of AOGCMs with interactive sea ice, based on emission scenarios from SRES, were prepared.

[1] www-pcmdi.llnl.gov/projects/model_intercomparison.php.

Within CMIP phase 5 (Taylor et al. 2012), a new set of coordinated experiments of AOGCMs and ESMs, based on RCPs, has been established. The data are available via the earth system grid federation (ESGF) which can be accessed from several nodes world-wide.[2]

The first major effort on Europe-wide coordinated experiments with RCMs was the PRUDENCE project,[3] coordinated by the Danish Meteorological Institute and financed by the EU 5th framework program 2001–2004. This resulted in a series of climate change scenarios for 2071–2100 at a 0.5°–0.22° horizontal resolution for Europe (Christensen and Christensen 2007).

Within the later project ENSEMBLES,[4] coordinated by the Met Office Hadley Centre and financed by the 6th EU framework program 2004–2009, a coordinated matrix of global and regional model simulations, mainly for the SRES A1B scenario, was established for Europe at a 0.22° horizontal resolution (and for Africa at 0.44°) (Hewitt and Griggs 2004). The model data are freely available.[5]

Within the current worldwide initiative on coordinated downscaling experiments (CORDEX), a sample of the global climate simulations of CMIP5 were downscaled for most continental regions of the globe (Giorgi et al. 2009). The CORDEX datasets will be available via the ESGF. Some datasets are already accessible, others will follow successively.

Within the EURO-CORDEX initiative, a unique set of high resolution climate change simulations for Europe on a 0.11° horizontal resolution is currently established (Jacob et al. 2014). Around 26 dynamical downscaling experiments have been or will be conducted, mainly for the scenarios RCP4.5 and RCP8.5. It is possible to track the status of the simulations.[6] Datasets will also be available via the ESGF.

To estimate uncertainties in projections of future climate the multi-model ensemble approach has also been introduced in Earth system modelling of the North Sea region (e.g. Friocourt et al. 2012; Wakelin et al. 2012a; Bülow et al. 2014; Holt et al. 2014). Ensemble simulations sample global and regional model uncertainties, internal variability and potential but unknown greenhouse gas emissions, nutrient and carbon loads, and fishery scenarios (e.g. Meier et al. 2011b, 2012b; Wakelin et al. 2012a). An overview of recent model simulations for the North Sea is provided in Sect. A2.4.

A2.4 Regional Coupled Atmosphere–Ocean Model Simulations for the North Sea

For the assessment of regional climate change in the North Sea region, regional coupled atmosphere–ocean models are essential. They account for local topography and coastline, resolve mesoscale features of oceanic and atmospheric circulation, and are able to simulate small-scale air–sea coupling processes.

Changes in the hydrological system of coastal waters have been investigated within the German Federal Ministry of Transport, Building and Urban Development (BMVBS) research program KLIWAS task 2. The objective of subtask 2.01 'Climate Change Scenarios' is to generate reliable estimates of changes in atmospheric and oceanic conditions, with the help of suitable regional models. To date, simulations for the North Sea are mainly undertaken with regional atmosphere models and regional ocean models separately, which does not account for dynamic atmosphere–ocean interactions. The first coupled regional atmosphere–ocean models have been developed for the North Sea region (BfG 2013) within the activity KLIWAS[7] 'Coast' of the German Federal Maritime and Hydrographic Agency (BSH) in collaboration with the Max-Planck-Institute for Meteorology (MPI-M), the University of Hamburg (UH), the Climate Service Center Germany (GERICS) and the Swedish Meteorological and Hydrological Institute (SMHI).

The final KLIWAS report (Bülow et al. 2014) provides details and results of this activity. A short overview concerning the models and simulations follows. The regional ocean model HAMSOM (Pohlmann 2006) was coupled to the atmospheric model REMO (Su et al. 2014a). The ocean model of MPI, the global MPIOM, had previously been coupled to REMO in a similar way (Sein et al. 2015. A coupled model, comprising the atmospheric regional climate model RCA, and the regional ocean model NEMO, was applied by SMHI (Dieterich et al. 2013, 2014; Wang et al. 2015).

The coupled models were first validated with observed climate data for the past 30–50 years, by performing 'hindcast' simulations driven by reanalysis data. Atmosphere reanalyses data were from the National Center for Environmental Prediction (NCEP) or ERA-40 and ocean reanalysis data from the 'GECCO' data or from a climatology. The historical climate simulations and the climate

[2]http://esgf-data.dkrz.de/esgf-web-fe/.

[3]http://prudence.dmi.dk/.

[4]www.ensembles-eu.org.

[5]http://ensemblesrt3.dmi.dk.

[6]www.euro-cordex.net/EURO-CORDEX-Simulations.1868.0.html.

[7]www.kliwas.de.

Table A2.1 Coupled and uncoupled simulations for KLIWAS 'Küste'. 'MPIOM-NS' denotes the coupling of the global MPIOM on the RCM domain (Mathis et al. 2013; Mathis and Pohlmann 2014)

Simulation	Global ocean	Global atmosphere	Regional ocean model	Regional atmosphere model	Coupling	Period
Hindcast	GECCO	NCEP	HAMSOM	–	No	1950–2000
Hindcast	Levitus climatology	ERA40	NEMO	RCA	No	1961–2002
Hindcast	MPIOM	NCEP	MPIOM-NS	–	No	1948–2007
Hindcast	MPIOM	NCEP	MPIOM-NS	REMO	Yes	1948–2007
Hindcast	MPIOM	NCEP	HAMSOM	REMO	Yes	1985–2000
Hindcast	MPIOM	NCEP	HAMSOM	REMO	No	1985–2000
Hindcast	NCEP/ERA40	NCEP	–	REMO	No	1958–2000
C20+A1B	MPIOM_r3	ECHAM5	NEMO	RCA	Yes	1950–2100
C20+A1B	MPIOM_r2	ECHAM5	NEMO	RCA	Yes	1950–2100
C20+A1B	MPIOM_r3	ECHAM5	HAMSOM	REMO	Yes	1950–2100
C20+A1B	MPIOM	ECHAM5	MPIOM	REMO	Yes	1920–2100
RCP4.5	MPIOM	ECHAM5	MPIOM-NS	REMO	Yes	1950–2100
RCP2.6	MPIOM	REMO	MPIOM-NS	REMO	Yes	1950–2100

projections based on the SRES A1B scenario were driven by global model data from ECHAM5/MPI-OM. A list of regional model simulations (coupled as well as uncoupled) performed within the KLIWAS project is given in Table A2.1.

Detailed information about models and analyses of simulation results are available via the German Federal Maritime and Hydrographic Agency website.[8] The final report of the KLIWAS Coast activity is also available (Bülow 2014).

[8]www.bsh.de/de/Meeresdaten/Beobachtungen/Klima-Anpassungen/index.jsp.

References

Ådlandsvik B (2008) Marine downscaling of a future climate scenario for the North Sea. Tellus A 60:451–458

Allen JI, Blackford J, Holt JT, Proctor R, Asworth M, Siddorn J (2001) A highly spatially resolved ecosystem model for the North west European continental shelf. SARSIA 86:423–440

Arheimer B, Dahne J, Donnelly C (2012) Climate change impact on riverine nutrient load and land-based remedial measures of the Baltic Sea Action Plan. AMBIO 41:600–612

Artioli Y, Blackford JC, Nondal G, Bellerby RGJ, Wakelin SL, Holt JT, Butenschön M, Allen JI (2013) Heterogeneity of impacts of high CO_2 on the North Western European Shelf. Biogeosci Discuss 10:9389–9413

BfG (2013) Auswirkungen des Klimawandels auf Wasserstraßen und Schifffahrt – Entwicklung von Anpassungsoptionen. Ausgewählte, vorläufige Ergebnisse zur 3. Statuskonferenz am 12./13.11. 2013, KLIWAS Kompakt, KLIWAS-22/2013

Bülow K, Dieterich C, Heinrich H, Hüttl-Kabus S, Klein B, Mayer B, Meier HEM, Mikolajewicz U, Narayan N, Pohlmann T, Rosenhagen G, Sein D, Su J (2014) Comparison of 3 coupled models in the North Sea region under todays and future climate conditions. KLIWAS Schriftenreihe, KLIWAS-27/2-14

Christensen JH, Christensen OB (2007) A summary of the PRUDENCE model projections of changes in European climate by the end of this century. Climatic Change 81:7–30

Daewel U, Schrum C (2013) Simulating long-term dynamics of the coupled North and Baltic Sea ecosystem with ECOSMO II: model description and validation. J Mar Sys 119-120:30–49

Daewel U, Peck MA, Kühn W, St. John MA, Alekseeva I, Schrumm C (2008) Coupling ecosystem and individual-based models to simulate the influence of environmental variability on potential growth and survival of larval sprat (*Sprattus sprattus* L.) in the North Sea. Fish Oceanogr 17:333–351

Dieterich C, Schimanke S, Wang S, Väli G, Liu Y, Hordoir R, Axell L, Höglund A, Meier HEM (2013) Evaluation of the SMHI coupled atmosphere-ice-ocean model RCA4-NEMO. SMHI Report Oceanography, 47

Dieterich C, Wang S, Schimanke S, Gröger M, Klein B, Hordoir R, Samuelsson P, Liu Y, Axell L, Höglund A, Meier HEM, Döscher R. (2014) Surface heat budget over the North Sea in climate change simulations. In: International Baltic Earth Secretariat Publication No. 3, June 2014. 3rd International Lund Regional-Scale Climate Modelling Workshop. International Baltic Earth Secretariat

Donnelly C, Yang W, Dahné J (2014) River discharge to the Baltic Sea in a future climate. Climatic Change 122:157–170

Döscher R, Willén U, Jones C, Rutgersson A, Meier HEM, Hansson U, Graham LP (2002) The development of the regional coupled ocean-atmosphere model RCAO. Boreal Environ Res 7:183–192

Feser F, Rockel B, von Storch H, Winterfeldt J, Zahn M (2011) Regional climate models add value to global model data: A review and selected examples. Bull Amer Meteor Soc 92:1181–1192

Friocourt YF, Skogen M, Stolte W, Albretsen J (2012) Marine downscaling of a future climate scenario in the North Sea and possible effects on dinoflagellate harmful algal blooms. Food Addit Contam A 29:1630–1646

Giorgi F (1990) Simulation of regional climate using a limited area model nested in a general circulation model. J Climate 3:941–963

Giorgi F, Jones C, Asrar G (2009) Addressing climate information needs at the regional level: The CORDEX framework. WMO Bull 58:175–183

Gröger M, Dieterich C, Meier HEM, Schimanke S (2015) Thermal air-sea coupling in hindcast simulations for the North and Baltic seas on the NW European shelf. Tellus A 67:26911 http://dx.doi.org/10.3402/tellusa.v67.26911

Gustafsson N, Nyberg L, Omstedt V (1998) Coupling of a high-resolution atmospheric model and an ocean model for the Baltic Sea. Mon Wea Rev 126:2822–2846

Hagedorn R, Lehmann A, Jacob D (2000) A coupled high resolution atmosphere-ocean model for the BALTEX region. Meteorol Zeit 9:7–20

Hewitt CD, Griggs DJ (2004) Ensembles-based predictions of climate changes and their impacts (ENSEMBLES). Eos Trans AGU 85:566 doi:10.1029/2004EO520005

Ho-Hagemann HTM, Rockel B, Kapitza H, Geyer B, Meyer E (2013) COSTRICE – an atmosphere–ocean–sea ice model coupled system using OASIS3. HZG Report 2013-5

Holt J, Allen JI, Proctor R, Gilbert F (2005) Error quantification of a high resolution coupled hydrodynamic- ecosystem-coastal-ocean-model- part 1: Model overview and assessment of the hydrodynamics. J Mar Syst 57:167–188

Holt J, Wakelin S, Lowe J, Tinker J (2010) The potential impacts of climate change on the hydrography of the northwest European Continental shelf. Prog Oceanogr 86:361–379

Holt J, Butenschon M, Wakelin SL, Artioli Y, Allen JI (2012) Oceanic controls on the primary production of the northwest European continental shelf: model experiments under recent past conditions and a potential future scenario. Biogeosciences 9:97–117

Holt J, Schrum C, Cannaby H, Daewel U, Allen I, Artioli Y, Bopp L, Butenschon M, Fach B, Harle J, Pushpadas D, Salihoglu B, Wakelin S (2014) Physical processes mediating climate change impacts on regional sea ecosystems. Biogeosciences Discuss 11:1909–1975

Jacob D, Petersen J, Eggert B, Alias A, Christensen OB, Bouwer LM, Braun A, Colette A, Déqué M, Georgievski G, Georgopoulou E, Gobiet A, Menut L, Nikulin G, Haensler A, Hempelmann N, Jones C, Keuler K, Kovats S, Kröner N, Kotlarski S, Kriegsmann A, Martin E, van Meijgaard E, Moseley C, Pfeifer S, Preuschmann S, Radermacher C, Radtke K, Rechid D, Rounsevell M, Samuelsson P, Somot S, Soussana J-F, Teichmann C, Valentini R, Vautard R, Weber B, Yiou P (2014) EURO-CORDEX: new high-resolution climate change projections for European impact research. Reg Environ Change 14:563–578

Kjellström E, Döscher R, Meier HEM (2005) Atmospheric response to different sea surface temperatures in the Baltic Sea: coupled versus uncoupled regional climate model experiments. Nord Hydrol 6:397–409

Lindström G, Pers C, Rosberg J, Strömqvist J, Arheimer B (2010) Development and test of the HYPE (Hydrological Predictions for the Environment) model – A water quality model for different spatial scales. Hydrol Res 41:295–319

Lowe JA, Gregory JM (2010) A sea of uncertainty. Nat Rep Climate Change doi:10.1038/climate.2010.30

Manabe S, Stouffer RJ (1988) Two stable equilibria of a coupled ocean-atmosphere model. J Climate 1:841–866

Mathis M (2013) Projected Forecast of Hydrodynamic Conditions in the North Sea for the 21st Century. Dissertation, Department of Geosciences, Hamburg University

Mathis M, Pohlmann T (2014) Projection of physical conditions in the North Sea for the 21st Century. Clim Res 61:1–17

Mathis M, Mayer B, Pohlmann T (2013) An uncoupled dynamical downscaling for the North Sea: Method and evaluation. Ocean Model 72:153–166

McGuffie K, Henderson-Sellers A (2005) A Climate Modelling Primer. 3rd Edition. Wiley

Meehl GA, Covey C, McAvaney B, Latif M, Stouffer RJ (2005) Overview of the coupled model intercomparison project. Bull Amer Meteor Soc 86:89–93

Meehl GA, Covey C, Taylor KE, Delworth T, Stouffer RJ, Latif M, McAvaney B, Mitchell JFB (2007) The WCRP CMIP3 multimodel dataset: A new era in climate change research. Bull Amer Meteor Soc 88:1383–1394

Meier HEM, Kjellström E, Graham P (2006) Estimating uncertainties of projected Baltic Sea salinity in the late 21st century. Geophys Res Lett 33:L15705, doi: 10.1029/2006GL026488

Meier HEM, Höglund A, Döscher R, Andersson H, Löptien U, Kjellström E (2011a) Quality assessment of atmospheric surface fields over the Baltic Sea of an ensemble of regional climate model simulations with respect to ocean dynamics. Oceanologia 53:193–227

Meier HEM, Andersson HC, Eilola K, Gustafsson BG, Kuznetsov I, Müller-Karulis B, Neumann T, Savchuk OP (2011b) Hypoxia in future climates: A model ensemble study for the Baltic Sea. Geophys Res Lett 38:L24608 doi:10.1029/2011GL049929

Meier HEM, Hordoir R, Andersson HC, Dieterich C, Eilola K, Gustafsson BG, Höglund A, Schimanke S (2012a) Modeling the combined impact of changing climate and changing nutrient loads on the Baltic Sea environment in an ensemble of transient simulations for 1961-2099. Clim Dyn 39:2421–2441

Meier HEM, Müller-Karulis B, Andersson HC, Dieterich C, Eilola K, Gustafsson BG, Höglund A, Hordoir R, Kuznetsov I, Neumann T, Ranjbar Z, Savchuk OP, Schimanke S (2012b) Impact of climate change on ecological quality indicators and biogeochemical fluxes in the Baltic Sea - a multi-model ensemble study. AMBIO 41: 558–573

Moss RH, Edmonds JA, Hibbard KA, Manning MR, Rose SK, van Vuuren DP, Carter TR, Emori S, Kainuma M, Kram T, Meehl GA, Mitchell JFB, Nacicenovic N, Riahi K, Smith SJ, Stouffer RJ, Thomson AM, Weyant JP, Wilbanks TJ (2010) The next generation of scenarios for climate change research and assessment. Nature 463:747–756

Nakicenovic N, Swart R (eds) (2000) Emission Scenarios. Cambridge University Press

Niiranen S, Yletyinen J, Tomczak MM, Blenckner T, Hjerne O, MacKenzie B, Müller-Karulis B, Neumann T, Meier HEM (2013) Combined effects of global climate change and regional ecosystem drivers on an exploited marine food web. Glob Change Biol 19:3327–3342

Pätsch J, Kühn W (2008) Nitrogen and carbon cycling in the North Sea and exchange with the North Atlantic - a model study. Part I. Cont Shelf Res 28:767–787

Pohlmann T (2006) A meso-scale model of the central and southern North Sea: consequences of an improved resolution. Cont Shelf Res 26:2367–2385

Räisänen J, Räty O (2013) Projections of daily mean temperature variability in the future: cross-validation tests with ENSEMBLES regional climate simulations. Clim Dyn 41:1553–1568

Räisänen J, Hansson U, Ullerstig A, Döscher R, Graham LP, Jones C, Meier HEM, Samuelsson P, Willén U (2004) European climate in the late twenty-first century: regional simulations with two driving global models and two forcing scenarios. Clim Dyn 22:13–31

Rockel B (2015) The regional downscaling approach: a brief history and recent advances. Curr Clim Change Rep 1:22–29

Rummukainen M (2010) State-of-the-art with regional climate models. WIREs Clim Change 1:82–96

Rummukainen M, Räisänen J, Bringfelt B, Ullerstig A, Omstedt A, Willén U, Hansson U, Jones C (2001) A regional climate model for northern Europe: model description and results from the downscaling of two GCM control simulations. Clim Dyn 17:339–359

Rutgersson A, Nilsson EO, Kumar R (2012) Introducing surface waves in a coupled wave-atmosphere regional climate model: Impact on atmospheric mixing length. J Geophys Res – Oceans 117*: C00J15* doi:10.1029/2012JC007940

Schrum C, Hübner U, Jacob D, Podzun R (2003) A coupled atmosphere/ice/ocean model for the North Sea and the Baltic Sea. Clim Dyn 21:131–151

Sein D, Mikolajewicz U, Gröger M, Fast I, Cabos I, Pinto J, Hagemann S, Semmler T, Izquierdo A, Jacob D (2015) Regionally coupled atmosphere-ocean-sea ice-marine biogeochemistry model ROM. Part I: Description and validation. J Adv Model Earth Syst 7:268–304

Smith B, Samuelsson P, Wramneby A, Rummukainen M (2011) A model of the coupled dynamics of climate, vegetation and terrestrial ecosystem biogeochemistry for regional applications. Tellus A 63:87–106

Su J, Sein D, Mathis M, Mayer B, O'Driscoll K, Chen XP, Mikolajewicz U, Pohlmann T (2014a) Assessment of a zoomed global model for the North Sea by comparison with a conventional nested regional model. Tellus A 66:23927 http://dx.doi.org/10.3402/tellusa.v66.23927

Su J, Yang H, Moseley C, Pohlmann T, Ganske A, Klein B, Klein H, Narayan N (2014b) A regional coupled atmosphere-ocean model system REMO/HAMSOM for the North Sea. KLIWAS Schriftenreihe. KLIWAS-60/2014

Taylor KE, Stouffer RJ, Meehl GA (2012) An overview of CMIP5 and the experiment design. Bull Amer Meteor Soc 93:485–498

Tian T, Boberg F, Bøssing Christenssen O, Hesselbjerg Christenssen J, She J, Vihma T (2013) Resolved complex coastlines and land-sea contrasts in a high-resolution regional climate model: a comparative study using prescribed and modelled SSTs. Tellus A 65:19951, http://dx.doi.org/10.3402/tellusa.v65i0.19951

Van Pham T, Brauch J, Dieterich C, Frueh B, Ahrens B (2014) New coupled atmosphere-ocean-ice system COSMO-CLM/NEMO: assessing air temperature sensitivity over the North and Baltic Seas. Oceanologia 56:167–189

Wakelin S, Daewel U, Schrum C, Holt J, Butenschön M, Artioli Y, Beecham J, Lynam C, Mackinson S (2012a) MEECE deliverable D3.4: Synthesis report for Climate Simulations Part 3: NE Atlantic / North Sea

Wakelin SL, Holt J, Blackford JC, Allen JI, Butenschön M, Artioli Y (2012b) Modeling the carbon fluxes of the northwest European continental shelf: Validation and budgets. J Geophys Res 117: C05020 *doi:*10.1029/2011JC007402

Wang S, Dieterich C, Döscher R, Höglund A, Hordoir R, Meier HEM, Samuelsson P, Schimanke S (2015) Development and evaluation of a new regional coupled atmosphere-ocean model in the North Sea and Baltic Sea. Tellus A 67:24284. http://dx.doi.org/10.3402/tellusa.v67.24284

Wilby RL, Wigley TML (1997) Downscaling general circulation model output: a review of methods and limitations. Prog Phys Geogr 21:530–548

Winterfeldt J, Geyer B, Weisse R (2010) Using QuikSCAT in the added value assessment of dynamically downscaled wind speed. Int J Climatol 31:1028–1039

Annex 3: Uncertainties in Climate Change Projections

Markku Rummukainen

A3.1 Introduction

The global emissions of carbon dioxide and other greenhouse gases change the atmospheric composition and enhance the natural greenhouse effect. The climate system responds by warming, sea-level rise, changing precipitation patterns, snow and ice melt, and so on. The overall nature, order of magnitude and many regional characteristics of this response are scientifically well-established. There are also unknowns and uncertainties, but these are not impenetrable. They can be studied in informative ways, which contributes to the utility of climate change projections. This annex provides a pragmatic overview of uncertainties in climate change projections including regional downscaling. The aim is to provide background for the discussion of climate models and climate change projections addressed by different chapters of this book.

A3.2 Climate Models and Climate Projections

Climate models are advanced simulation tools for the climate system, and its characteristics such as temperature, precipitation, clouds, winds, snow, waves, sea ice, ocean salinity, and so on. The basis for climate models is the collected scientific understanding of the fundamental physical, chemical and biological properties and processes of the climate system. The body of climate change projections is made with global climate models (GCM). The latest generation of such projections has been coordinated under CMIP5 (The Coupled Model Intercomparison Project Phase 5; Taylor et al. 2012). Regional climate models (RCMs) are the regional counterpart of GCMs, and are used for downscaling global model projections. For additional information on climate models see Annex 2.

There are three major reasons why climate models are the key scientific tool for making climate change projections. First, the full climate system is complex and its evolution does not lend itself to analytical or statistical representations. Second, the future climate cannot be observed. Third, the present anthropogenic climate forcing combined with the present-day climate baseline, is a unique development of the Earth's climate system.

When run with the present-day atmospheric composition of greenhouse gases, solar variability and land use, climate models simulate the present-day climate. Climate models are also used to model past and alternative future climates under external forcing scenarios, such as anthropogenic greenhouse gas emissions and land use change. It is important to note that all projections are conditional to their underlying assumptions and that specific projections apply for the specific forcing scenarios used, such as the assumed future greenhouse gas emissions.

As we do not know what the 'right' future emissions are, climate simulations are not 'predictions' in the same sense that we tend to view weather forecasts. Thus, the choice of emission scenario can be considered a source of uncertainty in climate projections. Possible major changes in natural climate forcing (solar variability, volcanic eruptions etc.) are another source of uncertainty, but they are not usually considered a climate *projection* uncertainty, since the projections concern climate change due to *anthropogenic* forcing.

A second source of uncertainty in climate projections is related to the different degrees of scientific understanding of climate system processes and to what level of detail they can be modelled with available computing resources. Climate models have different resolutions and differ in terms of how

M. Rummukainen (✉)
Centre for Environmental and Climate Research, Lund University, Lund, Sweden
e-mail: markku.rummukainen@cec.lu.se

M. Quante and F. Colijn (eds.), *North Sea Region Climate Change Assessment*, Regional Climate Studies, DOI 10.1007/978-3-319-39745-0

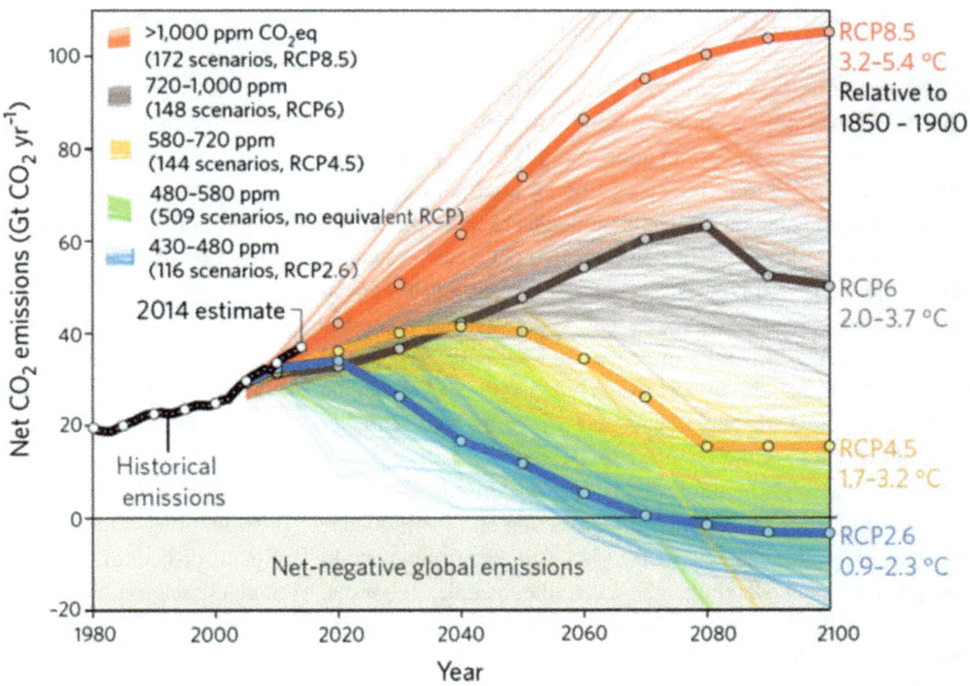

Fig. A3.1 Carbon dioxide emission pathways until 2100: historical emissions from fossil fuel combustion and industry (*black*), and from the early 2000s, possible future pathways based on emissions scenarios also used by the Intergovernmental Panel on Climate Change in its Fifth Assessment (Collins et al. 2013; Cubasch et al. 2013; IPCC 2013a). The Representative Concentration Pathways (RCP) are used in CMIP5 (Fuss et al. 2014). Reprinted by permission from Macmillan Publishers Ltd: Nature Climate Change 4, copyright 2014

climate processes are included and parameterised. This may affect their responses to forcing and lead to different climate models depicting smaller or larger changes compared to other models.

Internal variability is a third source of uncertainty in climate projections. It is created within and inherent to the climate system itself and arises, for example, from large-scale ocean–atmosphere interaction. On regional scales, internal variability is often larger than how it manifests itself in global mean quantities. For example, interannual temperature variability is larger on the scale of, say Europe, than in the global mean.

A3.3 Main Sources of Uncertainty

A3.3.1 Climate Forcing

Climate projections are conditional to their underlying emission scenarios (this is referred to as 'emission uncertainty'). The higher the level of forcing, the greater the response of the climate system. As the 'correct' future emissions are yet unknown, the question of 'how much will climate change' becomes more like 'if the emissions develop this way or that way, how much will climate change?' This collapses the emission uncertainty into specific emission pathway alternatives, with the subsequent projection being specific to the particular emissions. However, as discussed below, such projections are still subject to other sources of uncertainty.

Climate projections are developed for a wide range of emission scenarios—from strong mitigation futures (low emission scenarios) to unabated emissions (high emission scenarios). Figure A3.1 illustrates the greenhouse gas emission pathways for a number of anthropogenic climate forcing scenarios (the four so-called RCP scenarios; Representative Concentration Pathways, see Moss et al. 2010. The 'representative' comes from the fact that they exemplify an even larger body of forcing scenarios from different studies; see the thick and thin lines in the graphic). The global CMIP5 climate projections are driven by these RCP-scenarios. It is not relevant here to describe in detail each of these scenarios, just to stress that each RCP implies a different amount of anthropogenic emissions and that they lead to rather different climate change outcomes for the medium term and even more so on longer time scales beyond the mid-21st century. Over the next couple of decades, anthropogenic emissions and thus atmospheric greenhouse gas levels are more or less already committed due to the existing energy-related infrastructure and investment flows, and land use change, etc. (e.g. Rummukainen 2015). In the longer term, both emission reductions and continued increases are in principle possible, depending on socio-economic developments (for example energy systems, technology, economic growth, policy, …). More information on emission scenarios is provided in Annex 4.

A3.3.2 Model Uncertainty

Climate models employ different resolutions, different numerical techniques and different parameterisations, and these are all sources of some uncertainty. For the purposes of this Annex, this is referred to as 'model uncertainty'.

The basic equations for the atmosphere and the ocean comprise a non-linear system. In climate models, the system of these equations is solved numerically. The solution is thus an approximation. Another issue is that climate system processes occupy a very wide range of spatial and temporal scales, and scale interactions are important. While larger scales can be explicitly simulated, phenomena that occur at scales smaller than the model resolution need to be expressed in terms of resolved large-scale features, that is, 'parameterised'. Examples of such processes are turbulence, convection and the influence of detailed surface characteristics. Also, parameterisations build on physical understanding. However, the complexity of the processes and interactions opens up different ways of describing a certain process. This leads to differences between climate models which may affect their climate sensitivity and subsequently projections.

A summary measure of this is the equilibrium climate sensitivity (ECS) which is defined as the long-term global mean temperature rise due to a doubling of carbon dioxide concentration in the atmosphere. The magnitude of climate sensitivity depends on the net effect of the various changes in the climate system due to warming. For example, a warmer atmosphere can hold more water vapour, which—being a greenhouse gas—enhances the warming (this is an example of a 'positive' feedback). Other key feedback is related, not least, to clouds. How these and other aspects of the climate system respond to emissions in the climate models varies to some extent for different parameterisations. For GCMs, climate sensitivity is not a predetermined parameter but is the combined effect of all processes represented within the models, and varies from about 2 °C to around 5 °C. For emission and atmospheric concentration scenarios other than a doubling of carbon dioxide concentration in the atmosphere, the range in the projected change in temperature will differ from that which corresponds to the equilibrium climate sensitivity.

A3.3.3 Internal Variability

Internal variability is an inherent characteristic of the climate system. Two well-known examples are the El Niño Southern Oscillation (ENSO) and the North Atlantic Oscillation (NAO). These are both intrinsic to the climate system even without external forcing. ENSO, for example, arises from the interplay between the atmosphere and the ocean. Analogous to the real system, climate models generate internal variability in the simulations, which can be compared to observed characteristics. However, climate models simulate *possible* courses of internal variability, whereas the real system follows the *actual* course. When a model is run many times with different initial conditions or other slight changes, the resulting simulations exhibit different courses of internal variability, while still possibly showing comparable climate statistics in terms of averages, trends and so on. This is embodied in the term *projection* (which is used instead of *prediction*).

Addressing internal variability is relevant both when evaluating climate models and when interpreting climate projections. For example, as the courses of internal variability differ in observations and models, the timing of NAO-phases (and their regional imprint on temperature and precipitation) can also differ. When comparing different climate projections, some of the difference in the projected changes may be because the models are in different internal variability states (Räisänen 2001). Internal variability can also mask—or enhance—climate change signals over some specific period. Successive changes relative to some reference period need to become sufficiently large before they become statistically distinguishable from historical climate variability (e.g. Kjellström et al. 2013).

A3.3.4 Relative Importance of Different Sources of Uncertainty

The relative importance of the climate forcing uncertainty, model uncertainty and internal variability varies with the time horizon and the spatial scale (e.g. Hawkins and Sutton 2009).

For the next few decades, the climate forcing uncertainty is small. This is because possible future emission pathways are not likely to diverge significantly over the short term. Also, because the impact of emissions unfolds with a delay, the past and present emissions will continue to affect the climate for some time to come. Towards the end of the 21st century, emission uncertainty typically becomes the largest contributor to climate projection uncertainty if the full range of global emission scenarios is considered. If some subset of emission scenarios is studied instead, for example very ambitious mitigation scenarios, other sources of uncertainty may govern the spread of results.

The relative importance of internal variability diminishes with time, as the climate change signal increases. The relative importance of internal variability is also smaller for global mean values than for regional projections. Thus, near-term regional climate projections may show fairly different results, depending on whether the simulated internal variability enhances or dampens the climate change signals (e.g. Kjellström et al. 2013).

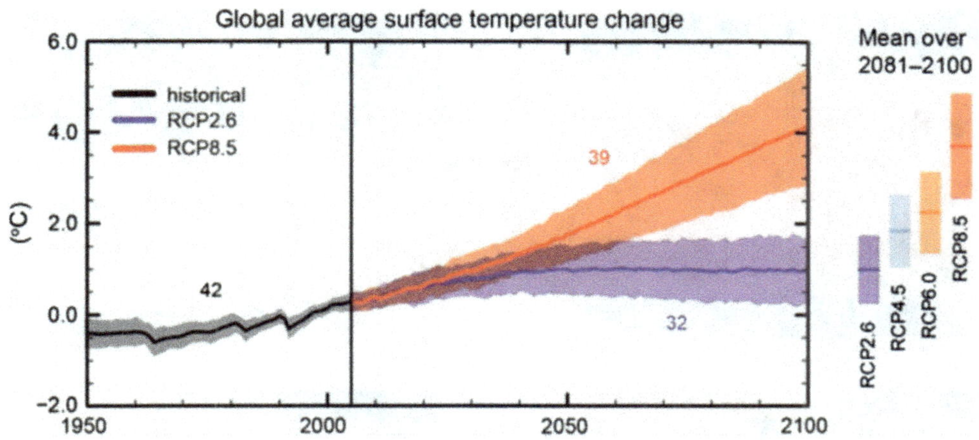

Fig. A3.2 CMIP5 multi-model simulated change in global annual mean surface temperature through the 21st century relative to present-day conditions (1986–2005). Time series of projections (*coloured lines*) and a measure of uncertainty (*shading*) are shown for scenarios corresponding to RCP2.6 (*blue*) and RCP8.5 (*red*). *Black* (*grey shading*) indicates the modelled historical evolution using historical reconstructed forcings. The numbers of CMIP5 models used to calculate the multi-model mean are also shown (IPCC 2013b: Fig. SPM.7, panel (a). Abridged caption)

Climate projections consistently show that anthropogenic greenhouse gas emissions lead to warming, sea level rise, etc., and that smaller (larger) emissions cause less (more) warming. Model uncertainty is nevertheless a factor to consider when assessing the magnitude of the changes and in some cases also the spatial patterns. The emission uncertainty, when considering a wide range of scenarios, catches up with the model uncertainty over time. This is illustrated in Fig. A3.2, where the envelopes show a measure for model uncertainty along high (red) and low (blue) emission scenarios.

A3.4 Quantifying and Qualifying Uncertainties

Climate models undergo continuous evaluation, not least by comparing simulations of the recent past and present-day climate to a range of observations (Räisänen 2006). Model intercomparisons provide additional information.

Overall, climate models simulate well many key aspects of the climate system, but there are also phenomena for which their performance is lower (Flato et al. 2013: e.g. Fig. 9.44). Models' performance also varies to some extent between regions, as is illustrated in Fig. A3.3. The models reproduce the large-scale features of global temperature and precipitation. The latest generation of GCMs have high pattern correlations with observations (0.99 for mean temperature and 0.82 for mean precipitation; Flato et al. 2013). In the case of temperature, relatively large model biases are nevertheless found in some coastal regions, close to sea ice edges, and in regions with major orographic features. In the case of precipitation, bias patterns are more varied. Biases can often be associated with specific physical phenomena (such as coastal temperature bias in upwelling regions) and/or resolution (such as the contrast in characteristics across the sea ice edge, or the lower resolution of orography in climate models than in reality).

Multi-model ensembles are a useful way to provide some quantification of uncertainties. While multi-model mean can be a useful indicator of trends, the spread of model results informs on uncertainty ranges due to internal variability and model uncertainties. A model can also be run a number of times with small variations to parameters in the parameterisations, within reasonable ranges, to gauge the significance of related model uncertainties (Murphy et al. 2004).

A3.5 Downscaling

The resolution of global models is typically lower than is desirable for climate impact studies and regional climate assessments. In regions with homogeneous physiographical features, or for large-scale time-averaged quantities, GCM-data may be sufficient as such or after interpolation. In many regions, however, while being conditioned by the large-scale conditions, local-to-regional climates are also significantly influenced by effects of variable land and ocean basin forms and heterogeneous surface characteristics. High resolution also facilitates simulation of small-scale temporal behaviour, such as extreme precipitation. RCMs are used for downscaling GCM output. This is also coined 'dynamical downscaling'. (Statistical downscaling is another method, but does not concern climate models.)

Dynamical downscaling extends information from global models with additional local-to-regional scale detail

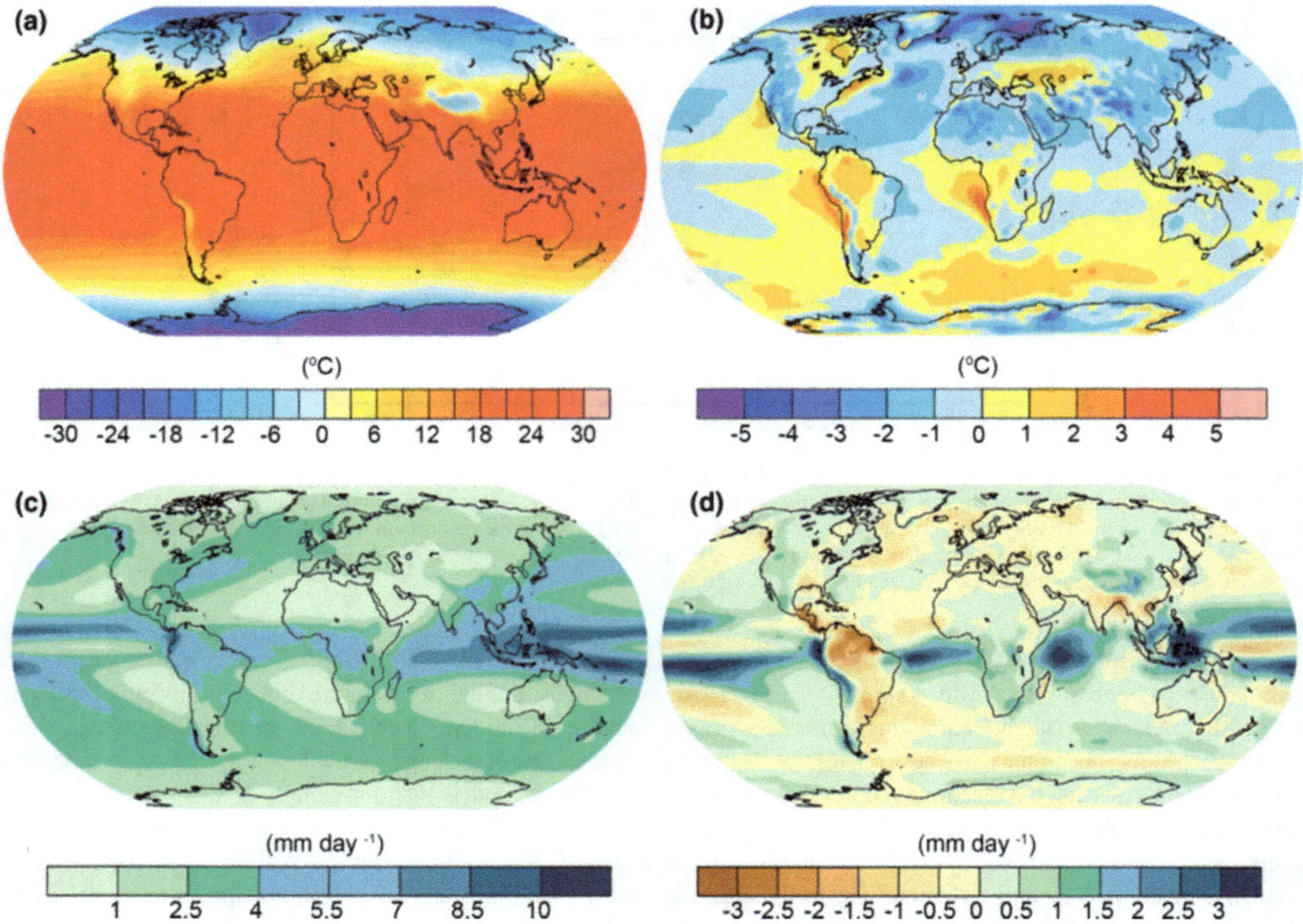

Fig. A3.3 Annual-mean surface (2-m) air temperature (°C) for the period 1980–2005: **a** multi-model mean for the CMIP5 experiment, **b** multi-model-mean bias as the difference between the CMIP5 multi-model mean and the climatology from ECMWF reanalysis of the global atmosphere and surface conditions (ERA)-Interim (Dee et al. 2011). Annual-mean precipitation rate (mm day^{-1}) for the period 1980–2005: **c** multi-model-mean in the CMIP5 experiment, **d** difference between multi-model mean and precipitation analyses from the Global Precipitation Climatology Project (Adler et al. 2003). Note the different scales for the respective mean and bias maps (Flato et al 2013: Figs. 9.2 and 9.4, panels (**a**) and (**b**). Abridged caption)

(Rummukainen 2010; Rockel 2015; Rummukainen 2016). Many RCMs feature an atmospheric and a land surface component, in which case sea-surface temperature and sea ice information is provided from the driving global model, which also provides the other boundary conditions for the regional model (see below). Regional interaction between the atmosphere and the ocean is not dynamic in such RCMs. There are, however, also regional ocean and coupled atmosphere–ocean RCMs (e.g. Döscher et al. 2002; Schrum et al. 2003), for example for the Baltic Sea, the North Sea, the Arctic Ocean and the Mediterranean Sea.

The same overarching sources of uncertainty apply for both global and regional climate models. An RCM covers a specific limited area domain (cf. Fig. A3.4). RCMs feature the same basic equations as GCMs, and are thus subject to emission uncertainty and model uncertainty, and generate internal variability. In terms of projections, RCMs are also affected by their boundary conditions, that is, the GCMs that are being downscaled. In a way, GCM uncertainty could be likened to emission uncertainty in the sense that a particular RCM projection is conditional to the choice of the emission scenario and the boundary conditions. The latter comprise large-scale inflow and outflow (winds, temperature, humidity) into and from the regional domain, from the driving GCM. RCMs can also be provided with boundary conditions from global reanalyses (e.g. Dee et al. 2011), which is often the case in model evaluation studies as comparison with actual observations is more straightforward than in the case of runs with boundary conditions from GCMs.

A key motivation of RCMs is that they facilitate simulations at higher resolution. Today, RCMs are starting to provide climate simulations at resolutions of 1–10 km, compared to around 25 km some 5–10 years ago, and 50 km or more some 10–15 years ago.

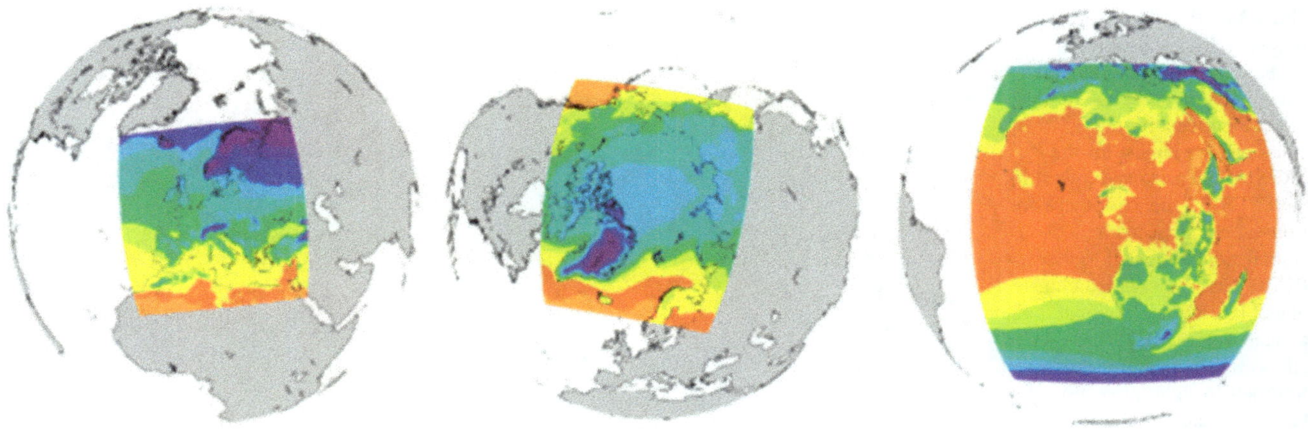

Fig. A3.4 Three examples of a regional climate model domain; for Europe, the Arctic region and Africa. The colours indicate simulated temperature climate. Figure courtesy of the Swedish Meteorological and Hydrological Institute (SMHI)

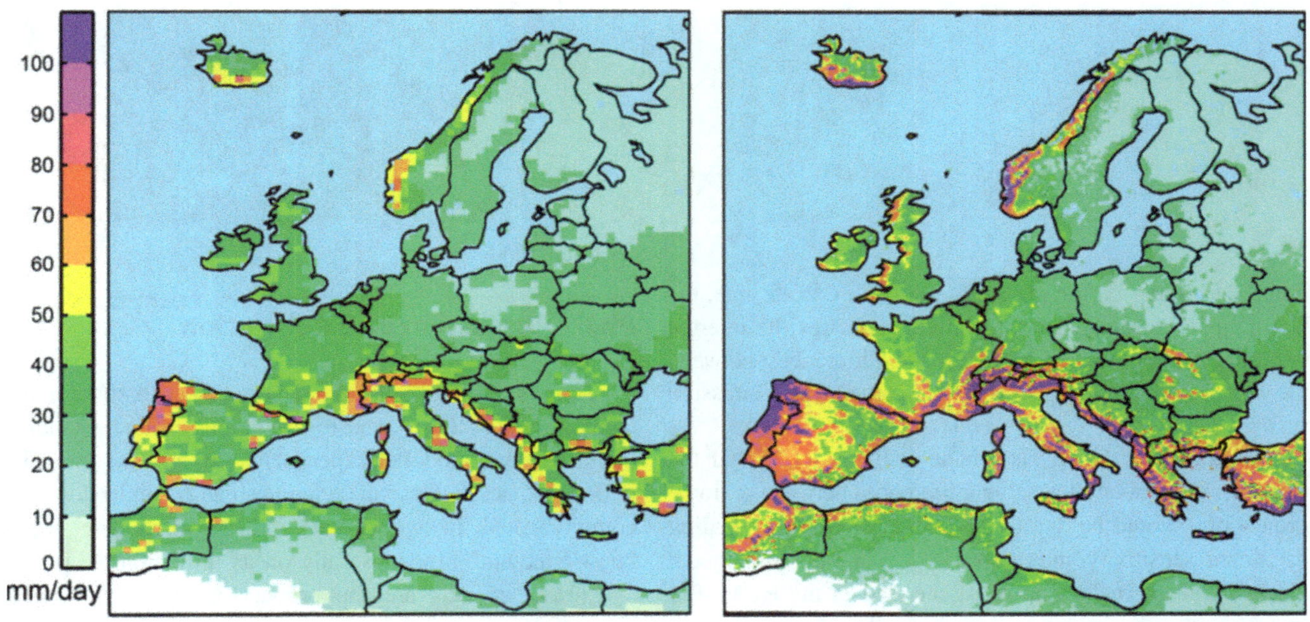

Fig. A3.5 Simulated precipitation intensity with a 20-year return period in winter and for 1971–2000 from a 50-km RCM run (*left*) and a 12-km RCM run (*right*). Differences are evident along many coastlines and in regions of variable orography. Figure courtesy of SMHI

Compared to GCMs, in RCMs extremes can be studied more explicitly, geographical detail resolved better (consider an extreme case of a coarse resolution model for the Nordic region; would it be better to wholly open up the connection between the Baltic Sea and the North Sea by removing Denmark, or totally close off the Baltic Sea?) and suchlike. Figure A3.5 provides an illustrative example of how geographical patterns of extreme precipitation may be simulated in an RCM at two different resolutions. Precipitation patterns and amounts are positively affected, for example, in mountainous regions and along western coastlines.

Figure A3.6 shows an example of RCM projections, for wintertime warming in Europe. Here, a specific RCM has been used to downscale three projections from one GCM which has been run with three different emission scenarios. In all cases, the warming increases with time (compare the panels in each row from left to right), and is greatest towards the north-east. Larger emissions cause greater warming (compare the rows in each column). An indication of internal variability is evident not least in the first two columns. Even though the recent past and near-future emissions are comparable, the regional temperature changes differ. Here, internal variability either enhances or reduces the long-term

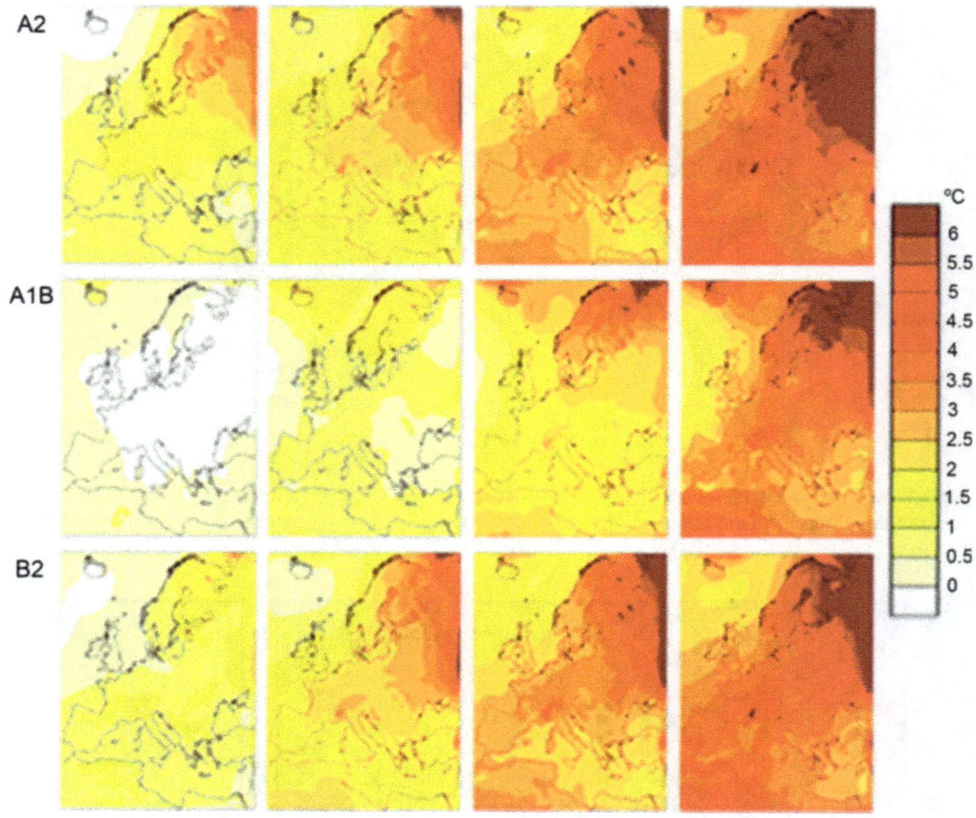

Fig. A3.6 Projected winter season (DJF) temperature increase (°C) for Europe under three emission scenarios (among these, the greenhouse gas emissions are largest for the SRES A2 scenario and lowest for the SRES B2 scenario; these are from an earlier scenario compilation compared to the RCPs). The same global climate model (GCM) and regional climate model (RCM) are used in all cases. The columns depict, from left to right, projections for the thirty-year periods 1981–2010, 2011–2040, 2041–2070 and 2071–2100, compared to 1961–1990. Based on Kjellström et al. (2005)

trend, depending on the particular projection. With time, the warming increases and its magnitude surpasses the internal variability amplitude, after which the differences between the projections are primarily governed by emission scenarios.

The choice of GCM also matters. If the RCM and the emission scenario are the same, differences between regional projections should be due to the choice of GCM (including its climate sensitivity, internal variability and possible model biases), and internal variability generated in the RCM. For large forcing, the latter can be expected to be small especially for temperature change. For other aspects, such as precipitation and wind, it may still be considerable, if the forced change is small and/or the phenomenon is characterised by large variability, such as extreme winds.

Figure A3.7 shows regional temperature and precipitation projections for the Baltic Sea region for the early, mid- and late 21st century, based on data both directly from GCMs and after their downscaling with an RCM. Temperature changes on this scale are comparable between GCMs and RCMs. The same applies for precipitation in winter for this region, but less so in summer. For the latter, the precipitation change in the GCMs varies from decreases to increases, whereas the range after downscaling is from no change to increases. There is also a tendency for larger (smaller) changes after downscaling than the direct GCM results in winter (summer).

A3.6 Discussion and Conclusion

All models and all observations are subject to some uncertainty, whether this is due to limitations in understanding, the instrument, model or experimental design, or some other reason. However, these uncertainties can be understood and communicated in ways that both highlight the robust knowledge and inform usefully on its limitations.

Climate models are the primary means for acquiring scientifically sound information on alternative future climates. Climate projections have utility, but also uncertainties. Uncertainties are, however, bounded and can be studied and characterised in informative ways. Continued climate system observations (such as on the deep ocean heat content) and increasing computing capacity (to allow for increased model resolution, larger model ensembles, incorporation of new model components) can contribute to reducing these uncertainties. Nevertheless, climate projection uncertainty will never be reduced to zero. Even if climate models were perfect depictions of the climate system, uncertainty related to climate forcing, i.e. the future emissions, would still persist. Also, there is no reason to expect that the time evolution of simulated internal variability should match the observed one other than statistically.

In terms of long-term global climate projections, uncertainty on future emissions is of primary importance. Larger

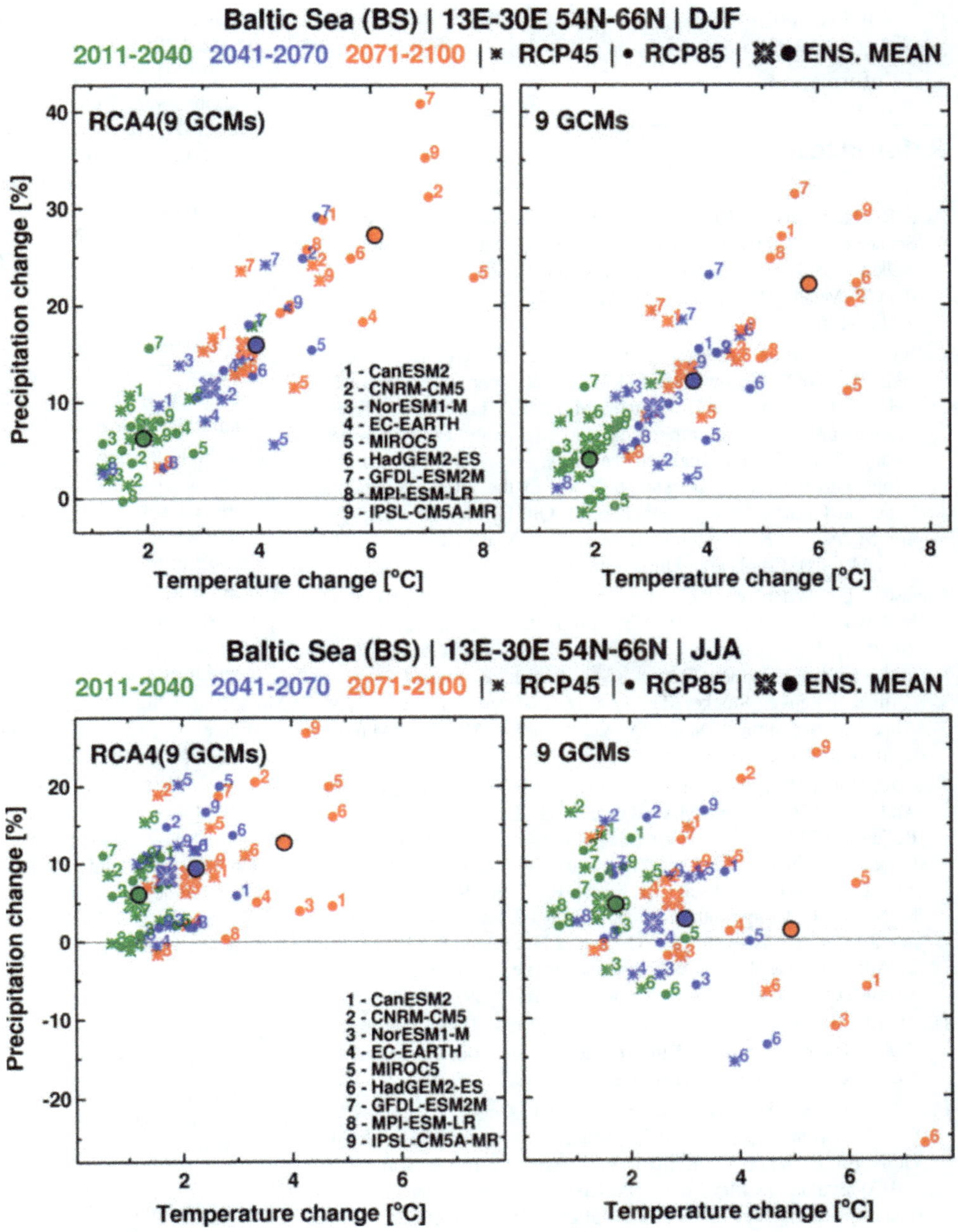

Fig. A3.7 Results from nine GCMs (*right-hand panels*; the numbers identify the GCMs) and after downscaling with the Swedish RCA4 regional climate model (*left-hand panels*). The plots show projected winter (DJF, *upper*) and summer (JJA, *lower*) precipitation and temperature changes for the Baltic Sea region as a whole. Two different climate forcing scenarios (RCP4.5 and RCP8.5) underlie these projections. They are identified by different symbols as depicted at the top of the panel. The colours indicate results for successive 30-year periods during the 21st century. The large symbols correspond to the GCM and RCM ensemble means, in the respective plots. Figure courtesy of SMHI

emissions lead to larger changes and smaller emissions to smaller changes. But how large and, respectively, how small, is subject to model uncertainty, i.e. how well relevant climate processes are represented. For the near-term, uncertainty related to internal variability can be comparable to model uncertainty, whereas emission uncertainty is small. Internal variability becomes less of a concern with increasing projection time horizon (i.e. mounting cumulative emissions), especially at a global scale.

Downscaling inherits uncertainties already present in the driving global model and the underlying emission scenario. Downscaling can, however, improve the projections by taking into account the effect of topography on near-surface climate phenomena, which in many cases is relevant for temporal and spatial information, for example in regions and at scales on which orography and land-sea distribution is important. Downscaling is also useful for studying phenomena with high spatial and/or temporal resolution, such as precipitation extremes.

Uncertainties in climate change projections need to be studied, characterised and managed. Although use of single projections can provide an example of alternative possible future conditions in an application, it is generally advisable to use results from many climate models in climate scenario analysis or impact assessment. This makes it possible to highlight robust outcomes as well as to identify results that should be considered more uncertain. A further alternative to the use of many single scenarios can be the generation of probabilistic projections, such as ensemble means and spreads, for applications which have the possibility to use such information.

References

Adler RF, Huffman GJ, Chang A, Ferraro R, Xie P-P, Janowiak J, Rudolf B, Schneider U, Curtis S, Bolvin D, Gruber A, Susskind J, Arkin P, Nelkin E (2003) The Version 2 Global Precipitation Climatology Project (GPCP) Monthly Precipitation Analysis (1979–Present). J Hydrometeor 4:1147–1167

Collins M, Knutti R, Arblaster J, Dufresne J-L, Fichefet T, Friedlingstein P, Gao X, Gutowski WJ, Johns T, Krinner G, Shongwe M, Tebaldi C, Weaver AJ, Wehner M (2013) Long-term climate change: projections, commitments and irreversibility. In: Climate Change 2013: The Physical Science Basis. Contribution of Working Group I to the Fifth Assessment Report of the Intergovernmental Panel on Climate Change. Stocker TF, Qin D, Plattner G-K, Tignor M, Allen SK, Boschung J, Nauels A, Xia Y, Bex V, Midgley PM (eds). Cambridge University Press

Cubasch U, Wuebbles D, Chen D, Facchini MC, Frame D, Mahowald N, Winther J-G (2013) Introduction. In: Climate Change 2013: The Physical Science Basis. Contribution of Working Group I to the Fifth Assessment Report of the Intergovernmental Panel on Climate Change. Stocker TF, Qin D, Plattner G-K, Tignor M, Allen SK, Boschung J, Nauels A, Xia Y, Bex V, Midgley PM (eds). Cambridge University Press

Dee DP, Uppala SM, Simmons AJ, Berrisford P, Poli P, Kobayashi S, Andrae U, Balmaseda MA, Balsamo G, Bauer P, Bechtold P, Beljaars ACM, Van de Berg L, Bidlot J, Bormann N, Delsol C, Dragani R, Fuentes M, Geer AJ, Haimberger L, Healy SB, Hersbach H, Hólm EV, Isaksen L, Kållberg P, Köhler M, Matricardi M, McNally AP, Monge-Sanz BM, Morcrette J-J, Park B-K, Peubey C, de Rosnay P, Tavoalto C, Thépaut J-N, Vitart F (2011) The ERA-Interim reanalysis: Configuration and performance of the data assimilation system. Q J R Meteorol Soc 137:553–597

Döscher R, Willén U, Jones C, Rutgersson A, Meier HEM, Hansson U (2002) The development of the coupled regional ocean-atmosphere model RCAO. Boreal Environ Res 7:183–192

Flato G, Marotzke J, Abiodun B, Braconnot P, Chou SC, Collins W, Cox P, Driouech F, Emori S, Eyring V, Forest C, Gleckler P, Guilyardi E, Jakob C, Kattsov V, Reason C, Rummukainen M (2013) Evaluation of Climate Models. In: Climate Change 2013: The Physical Science Basis. Contribution of Working Group I to the Fifth Assessment Report of the Intergovernmental Panel on Climate Change. Stocker TF, Qin D, Plattner G-K, Tignor M, Allen SK, Boschung J, Nauels A, Xia Y, Bex V, Midgley PM (eds.). Cambridge University Press, pp. 741–866

Fuss S, Canadell JG, Peters GP, Tavoni M, Andrew RM, Ciais P, Jackson RB, Jones CD, Kraxner F, Nakicenovic N, Le Quéré C, Raupach MR, Sharifi A, Smith P, Yamagata Y (2014) Betting on negative emissions. Nature Climate Change 4:850–853

Hawkins E, Sutton R (2009) The potential to narrow uncertainty in regional climate predictions. Bull Amer Meteor Soc 90:1095–1107

IPCC (2013a) Annex I: Atlas of Global and Regional Climate Projections. van Oldenborgh GJ, Collins M, Arblaster J, Christensen JH, Marotzke J, Power SB, Rummukainen M, Zhou T (eds). In: Climate Change 2013: The Physical Science Basis. Contribution of Working Group I to the Fifth Assessment Report of the Intergovernmental Panel on Climate Change. Stocker TF, Qin D, Plattner G-K, Tignor M, Allen SK, Boschung J, Nauels A, Xia Y, Bex V, Midgley PM (eds). Cambridge University Press

IPCC (2013b) Summary for Policymakers. In: Climate Change 2013: The physical science basis. Contribution of Working Group I to the Fifth Assessment Report of the Intergovernmental Panel on Climate Change. Stocker TF, Qin D, Plattner G-K, Tignor M, Allen SK, Boschung J, Nauels A, Xia Y, Bex V and Midgley PM (eds.). Cambridge University Press, pp. 1-30

Kjellström E, Bärring L, Gollvik S, Hansson U, Jones C, Samuelsson P, Rummukainen M, Ullerstig A, Willén U, Wyser K (2005) A 140-year simulation of European climate with the new version of the Rossby Centre regional atmospheric climate model (RCA3). Report in Meteorology and Climatology 108, SMHI, SE-60176

Kjellström E, Thejll P, Rummukainen M, Christensen JH, Boberg F, Christensen OB, Fox Maule C (2013) Emerging regional climate change signals for Europe under varying large-scale circulation conditions. Clim Res 56:103–119

Moss RH, Edmonds JA, Hibbard Ka, Manning MR, Rose SK, van Vuuren DP, Carter TR, Emori S, Kainuma M, Kram T, Meehl GA, Mitchell JFB, Nakicenovic N, Riahi K, Smith SJ, Stouffer RJ, Thomson AM, Weyant JP, Wilbanks TJ (2010) The next generation of scenarios for climate change research and assessment. Nature 463:747–756

Murphy JM, Sexton DMH, Barnett DN, Jones GS, Webb MJ, Collins M, Stainforth DA (2004) Quantification of modelling uncertainties in a large ensemble of climate change simulations. Nature 430:768–772

Räisänen J (2001) CO_2-induced climate change in CMIP 2 experiments: Quantification of agreement and role of internal variability. J Climate 14:2088–2104

Räisänen J (2006) How reliable are climate models? Tellus A 59:2–29

Rockel B (2015) The regional downscaling approach: a brief history and recent advances. Curr Clim Change Rep 1:22–29

Rummukainen M (2010) State-of-the-art with regional climate models. WIREs Clim Change 1:82–96

Rummukainen M (2015) Our commitment to climate change is dependent on past, present and future emissions and decisions. Climate Res 64:7–14

Rummukainen M (2016) Added value in regional climate modeling. WIREs Clim Change 7:145–159

Schrum C, Hübner U, Jacob D, Podzum R (2003) A coupled atmosphere/ice/ocean model for the North Sea and the Baltic Sea. Clim Dyn 21:131–151

Taylor KE, Stouffer RJ and Meehl GA (2012) An overview of CMIP5 and the experiment design. Bull Am Meteorol Soc 93:485–498

Annex 4: Emission Scenarios for Climate Projections

Markus Quante and Christian Bjørnæs

A4.1 Introduction

Comprehensive climate models are the main tools for projecting future changes in climate (see Annex 2). They are used to develop scenarios for potential climate change impacts which then provide the basis for mitigation and adaptation strategies. Climate projections depend strongly on the underlying assumptions concerning future greenhouse gas (GHG) and particle emissions or their respective precursor gases and are subject to model uncertainties. The latter are addressed in Annex 3. This Annex describes the emission scenarios used by the Intergovernmental Panel on Climate Change (IPCC) in its last three climate change assessments. These scenarios are also relevant to many of the results discussed in this assessment of climate change in the North Sea region.

A scenario is a description of potential future conditions produced to inform decision-making under uncertainty. In addition to a reliable model of the physical climate system, projections of future climate require estimates of the development of forcing agents. Variations in mid-term natural external forcing agents such as incoming solar radiation and volcanic activity are known, at least to some extent, for the past centuries and can be used in model simulations of the past climate. However, their future variability cannot be known because their behaviour is largely unpredictable. Nevertheless, their recent magnitude is less than the present-day human impact on climate. Although future anthropogenic external forcings via GHG emissions and changes in land use are also unknown, their historical growth, present-day magnitude and likely near-future trend are well established, and their longer-term development, such as over the 21st century can be estimated using assumptions concerning global socio-economic developments. As the underlying future GHG emissions will depend on economic, social and political trends that cannot be predicted because they are determined by decisions that have not yet been taken, emission scenarios comprise a wide range of assumptions on the future development of humankind. However, decision-making can narrow the assumptions, if for example, ambitious mitigation developments are chosen.

Thus, scenarios are descriptions of different possible futures, a series of alternative visions of futures (storylines) which are possible, plausible, and internally consistent but none of which is necessarily probable (von Storch 2008). The possibility that any single emission path will occur as described in scenarios is highly uncertain. Because many of the underlying factors are difficult or impossible to predict, a variety of assumptions must necessarily be used in the scenarios. And because emission scenarios for climate change research reflect expert judgements, it is no surprise that some of those expert judgements have been challenged by colleagues (e.g. Pielke et al. 2008).

Early approaches in the assessment of future climate change based on comprehensive general circulation models (GCMs) used a doubling or quadrupling of the pre-industrial carbon dioxide (CO_2) concentration as the driver for so-called equilibrium runs. Simulations using simple time-dependent transient scenarios, such as a steady (for example) 1 or 2 % increase in the atmospheric GHG concentration over the period under consideration, came next. The IPCC-related modelling studies associated with its 1990 assessment started to build on transient emission pathways that played out uncertainties in population and economic

M. Quante (✉)
Institute of Coastal Research, Helmholtz-Zentrum Geesthacht, Geesthacht, Germany
e-mail: markus.quante@hzg.de

C. Bjørnæs
CICERO Center for International Climate and Environmental Research-Oslo, Oslo, Norway
e-mail: christian.bjornas@cicero.oslo.no

M. Quante and F. Colijn (eds.), *North Sea Region Climate Change Assessment*, Regional Climate Studies, DOI 10.1007/978-3-319-39745-0

Table A4.1 Brief description of the SRES 'A' and 'B' storylines (after IPCC 2001)

Scenario	Description
A1	A world of rapid economic growth and rapid introduction of new and more efficient technology. The A1 storyline and scenario family describes a future world of very rapid economic growth, global population that peaks in mid-century and declines thereafter, and the rapid introduction of new and more efficient technologies. Major underlying themes are convergence among regions, capacity building and increased cultural and social interactions, with a substantial reduction in regional differences in per capita income
A2	The A2 storyline and scenario family describes a very heterogeneous world with an emphasis on family values and local traditions (high-CO_2). The underlying theme is self-reliance and preservation of local identities. Fertility patterns across regions converge very slowly, which results in continuously increasing population. Economic development is primarily regionally orientated and per capita economic growth and technological change is more fragmented and slower than other storylines
B1	A world of 'dematerialisation' and introduction of clean technologies (low-CO_2). The B1 storyline and scenario family describes a convergent world with the same global population, that peaks in mid-century and declines thereafter, as in the A1 storyline, but with rapid change in economic structures toward a service and information economy, with reductions in material intensity and the introduction of clean and resource-efficient technologies. The emphasis is on global solutions to economic, social and environmental sustainability, including improved equity, but without additional climate initiatives
B2	A world with an emphasis on local solutions to economic and environmental sustainability. It is a world with continuously increasing global population, at a rate lower than A2, intermediate levels of economic development, and less rapid and more diverse technological change than in the B1 and A1 storylines. While the scenario is also orientated towards environmental protection and social equity, it focuses on local and regional levels

growth as well as different technological futures. See Moss et al. (2010) for a short historical delineation of the development of scenarios for use in climate change research.

The present text focuses on scenarios used by the IPCC in its assessment reports released between 2001 and 2014 (IPCC Third Assessment Report, TAR; IPCC Fourth Assessment Report, AR4; IPCC Fifth Assessment Report, AR5; as well as special reports) as they cover the majority of scenario-driven climate change and impact studies reported in the various chapters of the present assessment. A dedicated activity to build the scenarios used in TAR and AR4 resulted in the so-called *Special Report on Emission Scenarios* (SRES) (IPCC 2000). The latest set of scenarios, used in AR5, followed a new approach for scenario development that uses so-called representative concentration pathways (RCPs) of future forcing and in parallel (or as a follow-up process) examined the range of socio-economic assumptions in model runs consistent with the RCPs, sharing during this step prior experience in the use of narratives and scenarios (Moss et al. 2010; van Vuuren et al. 2011a).

The remainder of this annex draws on material from Quante (2010), Bjørnæs (2013) and WGBU (2014).

A4.2 SRES Scenarios

The IPCC generated three sets of 21st-century GHG emission scenarios, of which the most ambitious and important were produced for the *Special Report on Emissions Scenarios* (IPCC 2000). The SRES Report uses 40 alternative scenarios which differ in terms of their assumptions about the future development of global society. Of these 40 scenarios, which are based on a comprehensive literature review and designed to depict most of the variation in their underlying drivers, the IPCC developed four qualitative storylines for which six 'marker' scenarios were created. One quantification of each storyline was produced plus two technological variants that stressed fossil-intensive and low-carbon energy supply technologies. Related uncertainties in future GHG and short-lived pollutant emissions including sulphur dioxide (SO_2), an important precursor for atmospheric sulphate particles, led to a wide range of driving forces.

The narrative storylines were developed so as to describe consistently the relationships between emission driving forces and their evolution. The scenario groups are known as A1, A2, B1 and B2, each based on diverse assumptions about the factors driving the development of human society through the 21st century. They thus represent different demographic, social, economic, technological, and environmental developments. In general, in the world described by the 'A storylines' people strive for personal wealth rather than environmental quality. In the 'B storylines', by contrast, sustainable development is pursued. However, the SRES scenarios do not include additional climate initiatives, which means that no scenarios are included that explicitly assume implementation of the United Nations Framework Convention on Climate Change or the emissions targets of the Kyoto Protocol. That is, the scenarios do not anticipate any specific mitigation policies for avoiding climate change. The scenario families are characterised in Table A4.1

Illustrative scenarios were chosen for each of the scenario groups A1, A2, B1 and B2, with A1 scenarios split into three distinguishable sub-classes. The A1FI, A1T and A1B illustrative scenarios describe alternative directions of technological change in the energy system, and are therefore quite different in terms of GHG emissions. In A1FI, energy production remains highly dependent on fossil fuels throughout

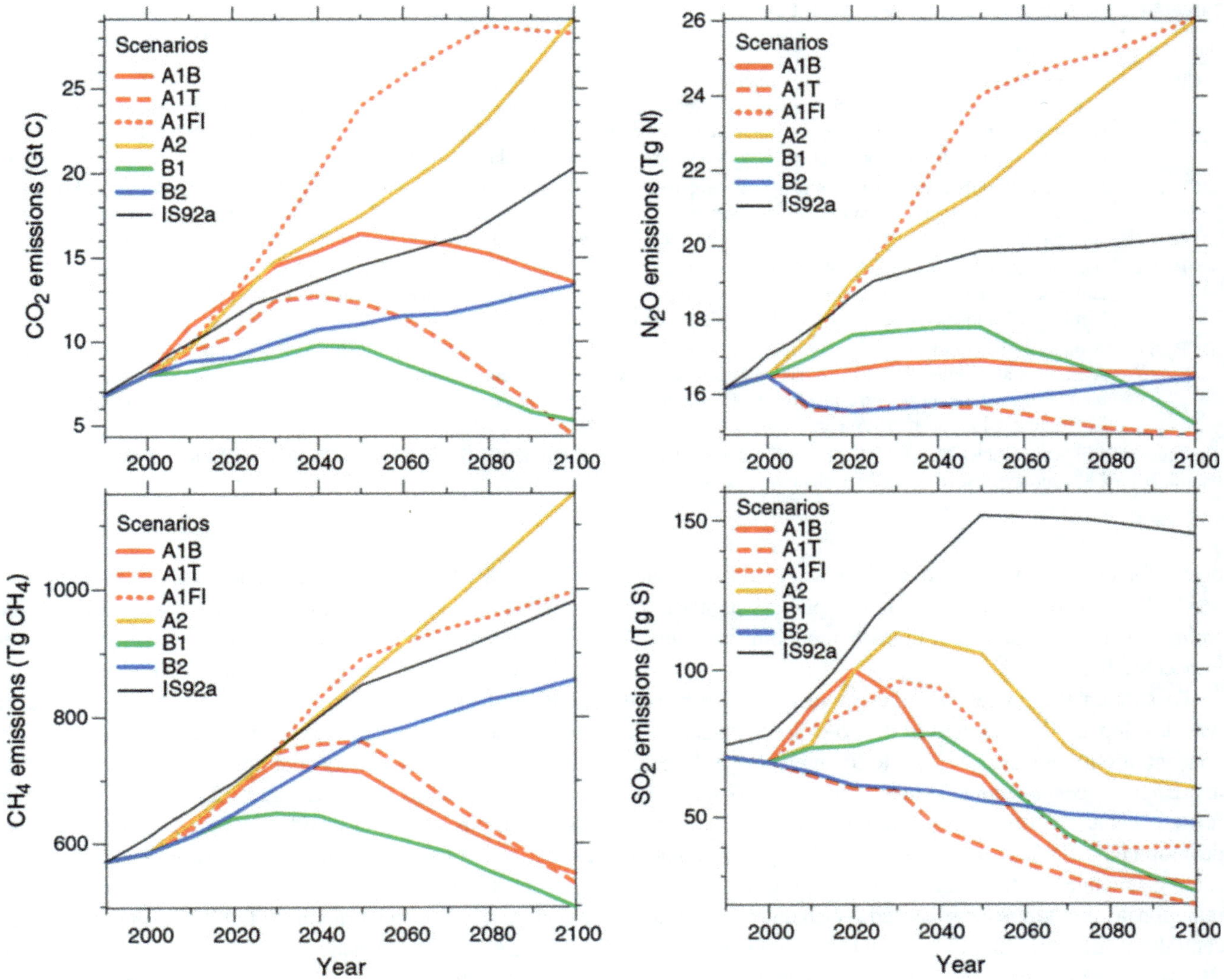

Fig. A4.1 Anthropogenic emissions of carbon dioxide (CO_2), methane (CH_4), nitrous oxide (N_2O) and sulphur dioxide (SO_2) for the six illustrative SRES scenarios, A1B, A2, B1 and B2, A1FI and A1T. One of the scenarios used for projections made in the 1990s (IS92a) is also shown for comparison (IPCC 2000)

the century, whereas A1T represents a rapid migration toward non-fossil energy sources and incorporates the use of advanced technologies. A1B is intermediate between these extreme cases (not relying too heavily on one particular energy source and similar improvement rates for all energy supply and end-use technologies). Of these, A2, A1B and B2 scenarios have been widely used in climate modelling. Figure A4.1 shows the emission time lines of major GHGs and of the sulphate aerosol precursor gas SO_2 aligned with the different SRES scenarios for the 21st century. The increasing spread of the emission curves with time underlines the broadness of the underlying economic and technological developments driving the scenarios. For CO_2 emissions, A2 and B2 show a steady increase throughout the 21st century, A1F1 shows a strong increase until 2080 and then a slight decline, and A1B, A1T and B1 show a decline from around mid-century.

None of the SRES scenarios in the set includes any future policies that explicitly address climate change, an aspect criticised by social scientists. This type of criticism as well as new economic data, new views about emerging technologies and land use and land cover change, called for the development of a new set of scenarios starting just after the release of AR4 in 2007 (Moss et al. 2008). The newest scenarios are the subject of the following section.

A4.3 RCP Scenarios

The SRES scenarios were developed along a sequentially linear chain that started with different socio-economic futures followed by an estimation of related GHG and particle emissions which were converted to concentrations and radiative forcings. Either of the latter could serve as the

Table A4.2 Brief description of the selected RCPs

Pathway	Description
RCP8.5	A high emission pathway for which radiative forcing reaches more than 8.5 Wm^{-2} by 2100 and continues to rise thereafter. This RCP is consistent with a future with no additional policy changes to reduce emissions and is characterised by rising GHG emissions. (The corresponding ECP assuming constant emissions after 2100 and constant concentrations after 2250) (developed by the International Institute for Applied System Analysis in Austria; Riahi et al. 2011)
RCP6.0	Intermediate stabilisation pathway in which radiative forcing is stabilised at approximately 6.0 Wm^{-2} after 2100 through the application of a range of technologies and strategies for reducing GHG emissions. (The corresponding ECP assuming constant concentrations after 2150) (developed by the National Institute for Environmental Studies in Japan; Masui et al. 2011)
RCP4.5	Intermediate stabilisation pathway in which radiative forcing is stabilised at approximately 4.5 Wm^{-2} after 2100 through relatively ambitious emissions reductions. (The corresponding ECP assuming constant concentrations after 2150) (developed by the Pacific Northwest National Laboratory in the USA; Thomson et al. 2011)
RCP2.6	A pathway where radiative forcing peaks at approximately 3 Wm^{-2} before 2030 and then declines to 2.6 Wm^{-2} by 2100. This scenario is also called RCP3-PD (peak and decline). To reach such forcing levels, ambitious GHG emissions reductions would be required over time. (The corresponding ECP is assuming constant emissions after 2100) (developed by PBL Netherlands Environmental Assessment Agency; van Vuuren et al. 2011b)

The references indicate articles that describe the respective RCP-scenario in full detail. The Extended Concentration Pathways (ECPs) cover the period 2100–2300 and are described by Meinshausen et al. (2011)

driver for climate model studies. This sequential approach was seen as a reason for delay in the process as a whole: from scenario generation to climate modelling to climate impact studies.

To shorten the process an alternative parallel approach was developed. This resulted in the so-called representative concentration pathways (RCPs). RCPs represent a different approach to scenario development, one that recognises that many scenarios of socio-economic and technological development can lead to the same pathways of radiative forcing (changes in the balance of incoming and outgoing radiation to the atmosphere caused by changes in the concentrations of atmospheric constituents). Selecting a few RCPs as examples (seen as 'representative') allows researchers to develop scenarios for the different ways the world might achieve those RCPs and to consider the consequences of climate change when those RCPs are achieved via specific scenarios. The word 'pathway' indicates that not only are the values in a reference year (i.e. 2100) of interest but also the trajectory over time. This approach is intended to increase research coordination and simultaneously to reduce the time needed to generate useful scenarios. Climate modelling studies and impact studies can already be conducted before a full set of socio-economic information is available (van Vuuren and Carter 2014).

In a parallel process to climate modelling and impact studies, the scenario community has used Integrated Assessment Models (IAMs) to develop a set of consistent technological, socio-economic and policy scenarios with storylines that could lead to particular concentration pathways (van Vuuren et al. 2011a). These so-called shared socio-economic pathways (SSPs) are intended to guide mitigation, adaptation, and mitigation analysis (O'Neill et al. 2014; van Vuuren and Carter 2014).

The SSPs enable researchers to test various permutations of climate policies and social, technological, and economic circumstances. For example, at a global scale, higher population or increased energy consumption could be compensated by a higher fraction of renewable energy. So rather than prescribing economic development and calculating climate change, researchers could pick an RCP scenario that is compatible with the 2 °C target, for example, and then assess various technology and policy options for achieving the emissions consistent with that pathway and target.

More specifically, RCPs are time and space-dependent trajectories of concentrations of GHGs and pollutants resulting from human activities, including changes in land use. RCPs provide a quantitative description of concentrations of the climate change pollutants in the atmosphere over time, as well as their radiative forcing. One of the goals was to reduce the number of scenarios to a manageable number showing an adequate separation of the radiative forcing pathways at the end of the specified time horizon (Moss et al. 2010). Candidate scenarios were chosen after a thorough selection from the large stock available in the peer-reviewed literature. The eventual selection was four scenarios: RCP2.6, RCP4.5, RCP6.0, and RCP8.5 (see Table A4.2). The RCPs are named to highlight the radiative forcing they achieve in 2100; for example, RCP6.0 achieves 6 Wm^{-2} by 2100.

The GHGs included in the RCPs are CO_2, methane (CH_4), nitrous oxide (N_2O), several groups of fluorocarbons (halogenated) and sulphur hexafluoride. The aerosols and chemically active gases are SO_2, black carbon, organic carbon, carbon monoxide, nitrogen oxides, volatile organic compounds, and ammonia. For the resulting scenarios, Fig. A4.2 shows the development of radiative forcing through the 21st century and attributes the forcing at 2100 among the GHGs.

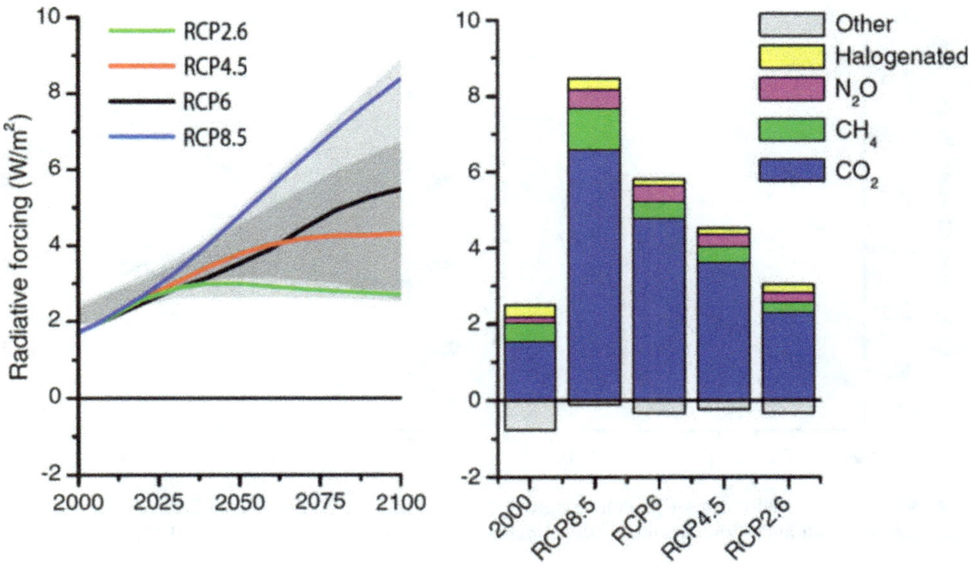

Fig. A4.2 Trends in radiative forcing (*left*) and 2100 forcing level per category (*right*). Forcing is relative to pre-industrial values and does not include land use (albedo), dust, or nitrate aerosol forcing (van Vuuren et al. 2011a)

Table A4.3 Major features of the selected RCP scenarios (after Moss et al. 2010)

RCP	Radiative forcing in 2100 (Wm^{-2})	CO_2 equivalent concentration in 2100 (ppm)	Type of change in radiative forcing
RCP8.5	>8.5	>1370	Rising
RCP6.0	~6.0	~850	Stabilising without overshoot
RCP4.5	~4.5	~650	Stabilising without overshoot
RCP2.6	~2.6 (peak at ~3 Wm^{-2} before 2100)	~450 (peak ~490 ppm before 2100)	Peak and decline

See Moss et al. (2010) and van Vuuren et al. (2011a) for more details of the scenario-building process and resulting scenarios. The main characteristics of the selected RCPs are listed in Table A4.2, while Table A4.3 provides a quick overview of major features.

For the well-mixed GHGs, the emissions and concentrations were harmonised using an IAM (Meinshausen et al. 2011). The emission trends for the four scenarios are given in Fig. A4.3 and the corresponding concentrations are shown in Fig. A4.4. The different developments of the emission and concentration trends are obvious. A striking result is that towards the end of the century for RCP2.6 negative CO_2 emissions occur. RCP2.6 is the only RCP scenario with the potential to meet the so-called 2 °C limit to global warming.

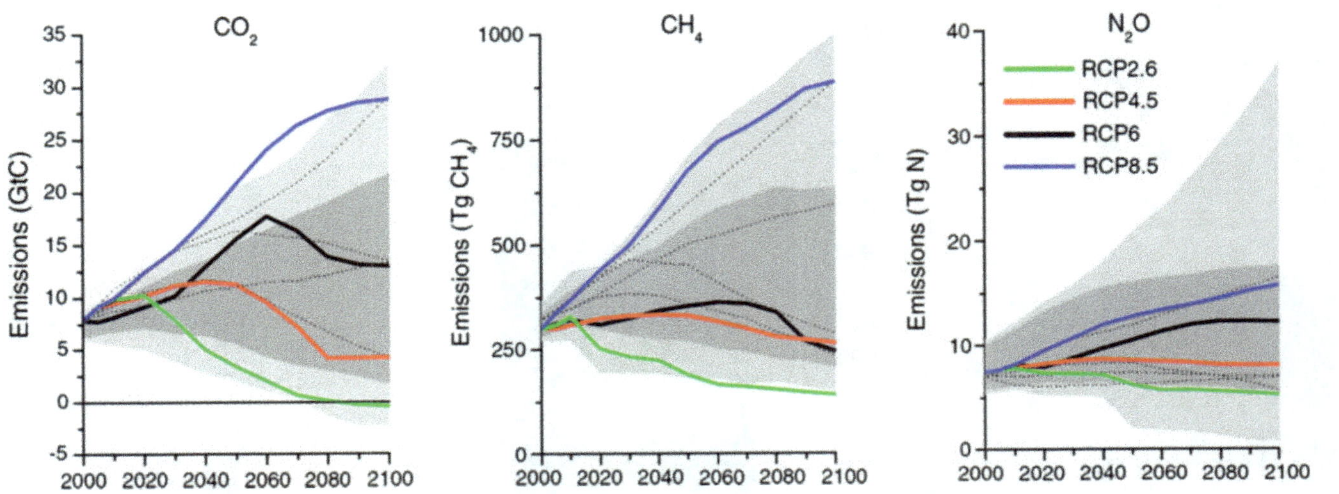

Fig. A4.3 Emissions of the main greenhouse gases carbon dioxide (CO_2), methane (CH_4) and nitrous oxide (N_2O) across the RCPs. The *grey area* indicates the 98th and 90th percentiles (*light/dark grey*) of the underlying scenarios from a literature survey. The *dotted lines* indicate four SRES marker scenarios (van Vuuren et al. 2011a)

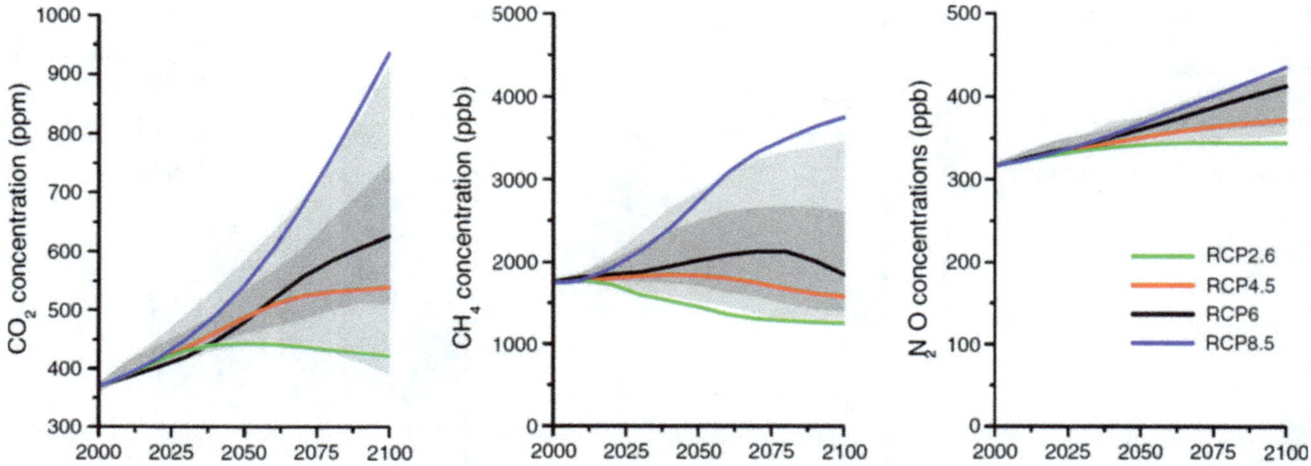

Fig. A4.4 Concentrations of the greenhouse gases carbon dioxide (CO_2), methane (CH_4) and nitrous oxide (N_2O) across the RCPs. The *grey area* indicates the 98th and 90th percentiles (*light/dark grey*) of an earlier emission study (EMF-22) (van Vuuren et al. 2011a)

Fuss et al. (2014) explored the need for negative emissions in more detail using a set of scenarios from IPCC Working Group III AR5 activities. They found that most emission pathways (101 of 116 RCP2.6 pathways) leading to concentrations of 430–480 ppm CO_2-equivalent (CO_2eq; CO_2 plus the other GHGs expressed as CO_2), consistent with limiting warming to below 2 °C, require global net negative emissions in the latter half of this century, as do many scenarios (235 of 653) that reach 480–720 ppm CO_2eq in 2100 (see also Fig. A3.1 in Annex 3).

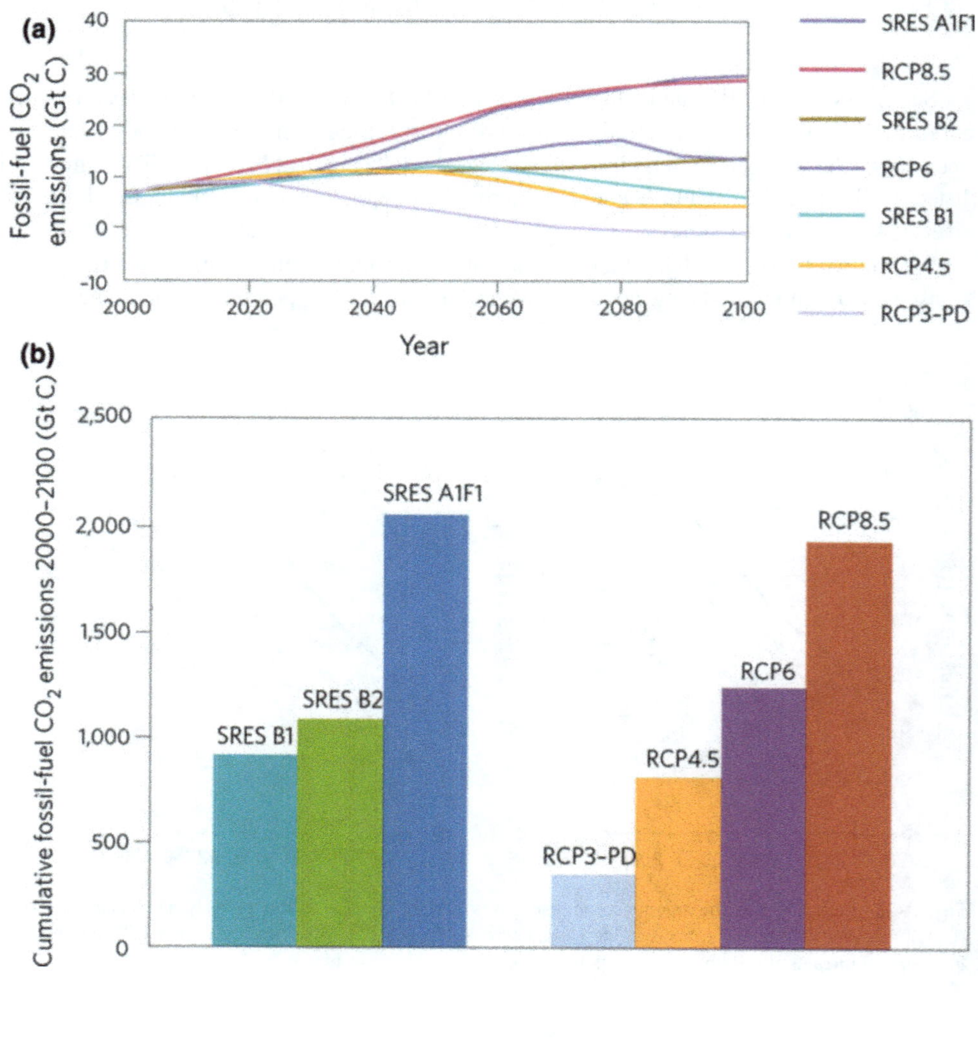

Fig. A4.5 Fossil-fuel carbon dioxide (CO_2) emissions (**a**) and cumulative emissions over the period 2000–2100 (**b**) for the SRES scenarios A1FI, B2 and B1 and as estimated for the four representative RCPs (note RCP3-PD is also known as RCP2.6) (Raper 2012)

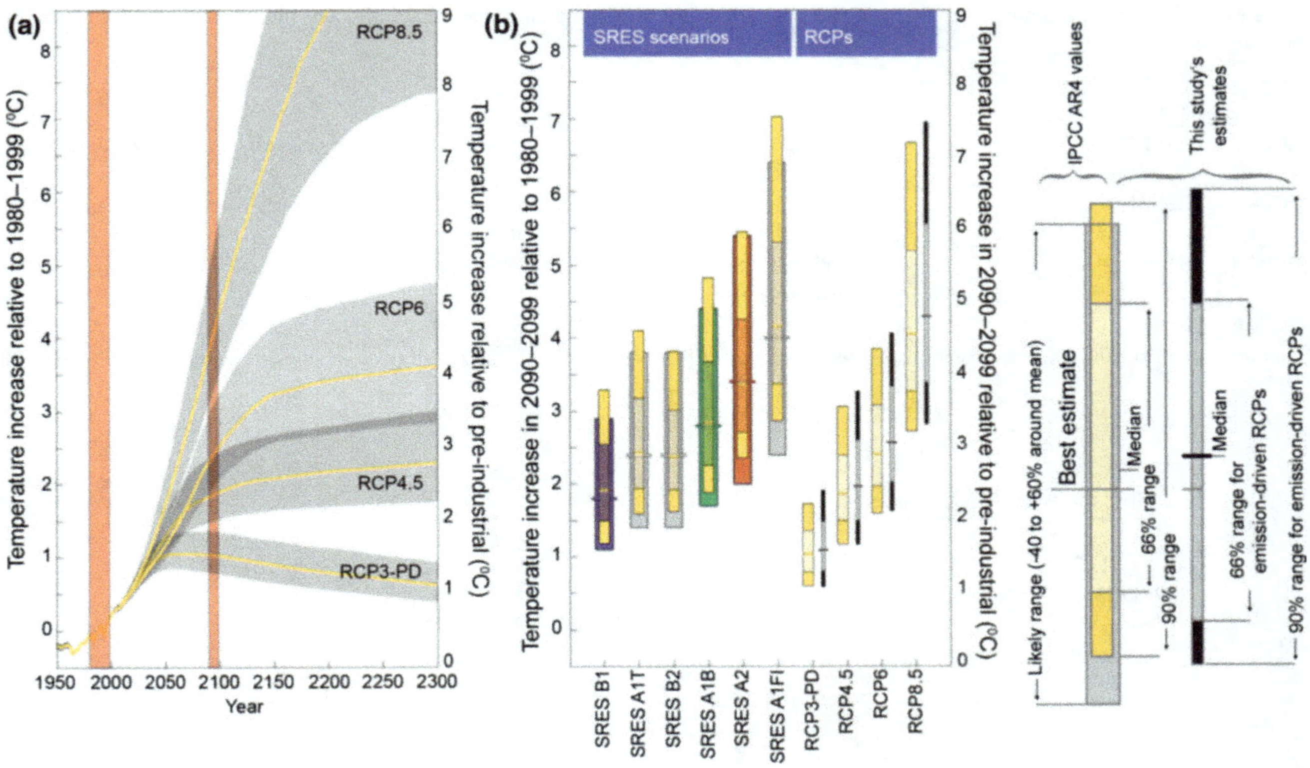

Fig. A4.6 Comparison of temperature projections for SRES scenarios and RCPs. **a** Time-evolving temperature distributions (66 % range) for the four concentration-driven RCPs computed with a representative equilibrium climate sensitivity (ECS) distribution and a model set-up representing closely the climate system uncertainty estimates of IPCC AR4 (*grey areas*). Median paths are shown in *yellow*. *Red shaded areas* indicate time periods referred to in '**b**'. **b** Ranges of estimated average temperature increase between 2090 and 2099 for SRES scenarios and RCPs respectively. Note that results are given both relative to 1980–1999 (*left scale*) and relative to the pre-industrial period (*right scale*). *Yellow* and *thin black ranges* indicate results of the reporting study; other ranges show AR4 estimates (see legend to the *right-hand side*). For RCPs, *yellow ranges* show concentration-driven results, whereas *black ranges* show emission-driven results (Rogelj et al. 2012)

A further key difference to earlier scenarios is that the RCPs are spatially explicit and provide information on a global grid at a resolution of approximately 60 km. This provides a good spatial and temporal distribution of emissions and land use changes. This is an important improvement because the actual location of some short-lived gases and particles has a strong influence on their regional warming potential. The RCPs also include a very wide range of land-use projections, addressing trends in cropland, grassland and other vegetated areas. The final RCP data sets comprise land use data, harmonised GHG emissions and concentrations, gridded reactive gas and aerosol emissions, and ozone and aerosol abundance fields. Global average surface temperature changes based on RCP-driven global projections are presented in Annex 3 (see Fig. A3.2).

A4.4 Relations Between SRES and RCP Scenarios

Reviews of climate change and related impact studies in several chapters of this book reveal that SRES-based as well as RCP-based projections are in use. The question "What is the relation between SRES and RCP scenarios?" is therefore relevant for researchers evaluating different studies to inform, for example, adaptation strategies. A first impression may be gained by comparing CO_2 emissions for the different scenarios (see Fig. A4.5). This indicates that some SRES and RCP scenarios follow a similar path and result in comparable cumulative emissions in 2100.

Rogelj et al. (2012) offered a more detailed comparison that was intended to bridge the gap between the old and new scenarios. They used a common model framework constrained by observations to ensure a low uncertainty link to changes in the past. Rogelj et al. (2012) gave probabilistic climate projections for all SRES and RCP marker scenarios and discussed the associated temperature projections (see also Raper 2012), for the latter see Fig. A4.6. According to this study three pairs of similar scenarios could be identified, they are compared for the 2100 time horizon in Table A4.4. Mapping old and new scenarios was also the focus of a study by van Vuuren and Carter (2014): In principle these authors concluded on the same matching scenario pairs as Rogelj et al. (2014).

A new high-resolution regional climate model (RCM) ensemble has been established for Europe including

Table A4.4 Similarities and differences between RCP and SRES scenarios based on temperature projections (all temperatures in this table are medians)

RCP	SRES with similar temperature increase in 2100	Main differences
RCP8.5	A1FI	Between 2035 and 2080, temperatures with RCP8.5 rise slower than with SRES A1FI, the reverse is true after 2080
RCP6.0	B2	Between 2060 and 2090, temperatures with RCP6.0 rise faster than with SRES B2 and slower during the other periods of the century
RCP4.5	B1	Until mid-century temperatures with RCP4.5 rise faster than with SRES B1, and then slower afterwards
RCP2.6	None	n.a.

The often-used SRES A1B scenario is not listed, because a similar RCP scenario for the 21st century does not exist (modified after Rogelj et al. 2012)

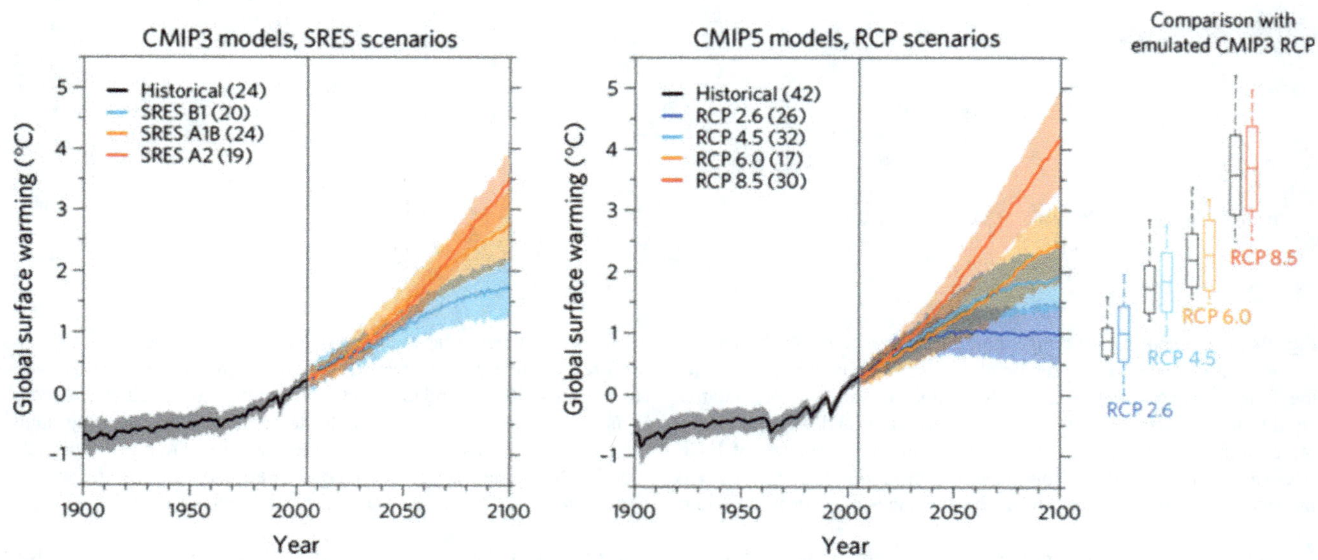

Fig. A4.7 Global surface temperature change (mean and one standard deviation as *shading*) relative to 1986–2005 for the SRES scenarios run by CMIP3 and the RCP scenarios run by CMIP5. The number of models is given in *brackets*. The box plots (mean, one standard deviation, and minimum to maximum range) are given for 2080–2099 for CMIP5 (*colours*) and for an energy balance model (MAGICC) calibrated to 19 CMIP3 models (*black*), both running the RCP scenarios (Knutti and Sedláček 2013)

the entire North Sea region within the World Climate Research Program Coordinated Regional Downscaling Experiment (EURO-CORDEX) initiative. The first set of simulations with a horizontal resolution of 12.5 km was completed for the RCP4.5 and RCP8.5 scenarios. These EURO-CORDEX ensemble results were compared to the SRES A1B results achieved within the ENSEMBLES project by Jacob et al (2014).

An additional point is that most of the SRES-based projections were generated using the older CMIP3 models, whereas the newer CMIP5 models were used for most RCP-based projections (for a review of the coupled model intercomparison projects—CMIP—see Annex 2). A full comparison of available climate change and impact studies needs to address this discrepancy. A study looking at this issue in more depth is that of Knutti and Sedláček (2013), in which emulated CMIP3 models were used for RCP-based projections. The major findings are summarised in Fig. A4.7. The different model generations result in differences in mean, standard deviation and range. The graphic suggests that the CMIP5 models show more warming for a given RCP than the emulated CMIP3 models, while the overall pattern of greater warming with higher forcing is robust.

A4.5 Concluding Remarks

Climate change projections are forced by emission scenarios. This annex describes the SRES and RCP scenarios in order to provide context for the projections discussed in the various chapters of this book. Both scenario sets offer a wide range of emission pathways, although only the RCP scenarios consider ambitious global warming abatement strategies. Of these, the RCP2.6 scenario is the most ambitious and the only one providing an emission pathway towards limiting with a high probability (around 66 %) global warming to below 2 °C above the pre-industrial global temperature.

For the North Sea region, available studies include those presenting results based on the older SRES scenarios as well as those presenting results based on the newer RCP scenarios. Many SRES and RCP scenarios can be paired, which is especially useful for the comparison and continuity of climate-impact studies. The three pairs—SRES A1F1/RCP8.5, SRES B2/RCP6 and SRES B1/RCP4.5—span the range of scenarios considered to date by most impact studies. The often-used SRES A1B scenario has no counterpart among the RCPs, and neither does the strong mitigation scenario RCP2.6 among the SRES scenarios.

In parallel to the construction of the RCPs, so-called shared socio-economic pathways (SSPs) have been developed with the help of integrated assessment models to reveal the driving forces behind the scenarios. An SSP database has been compiled by the research community, which is intended to enhance transparency of the process and to involve a large number of scientists in discussions around newly evolving scenarios (see IIASA 2015 for an update on the database). It is expected that many global and regional scenarios will emerge that are consistent with the new RCPs (Nakićenović et al. 2014).

Finally, it should be mentioned that RCM projections, employed to focus in higher grid resolution on limited areas such as the North Sea region, are linked to the scenarios via the driving GCMs, which provide their meteorological conditions (usually at the lateral boundaries) and sea surface temperatures.

References

Bjørnæs C (2013) A guide to Representative Concentration Pathways. CICERO News. Center for International Climate and Environmental Research, Oslo, Norway (republished online in 2015 at: www.cicero.uio.no/en/posts/news/a-guide-to-representative-concentration-pathways; accessed 28.12.2015)

Fuss S, Canadell JG, Peters GP, Tavoni M, Andrew RM, Ciais P, Jackson RB, Jones CD, Kraxner F, Nakićenović N, Le Quéré C, Raupach MR, Sharifi A, Smith P, Yamagata Y (2014) Betting on negative emissions. Nature Climate Change 4:850–853

IIASA (2015) SSP Database (Shared Socioeconomic Pathways) Version 1.0. online at: https://tntcat.iiasa.ac.at/SspDb/dsd?Action=htmlpage&page=about

IPCC (2000) Special Report on Emissions Scenarios (SRES). Nakićenović N, Swart R (eds). Intergovernmental Panel on Climate Change (IPCC), Cambridge University Press

IPCC (2001) Climate Change 2001: The Scientific Basis. Contribution of Working Group I to the Third Assessment Report of the Intergovernmental Panel on Climate Change. Houghton JT, Ding Y, Griggs DJ, Noguer M, van der Linden PJ, Dai X, Maskell K, Johnson CA (eds). Cambridge University Press

Jacob D, Petersen J, Eggert B, Alias A, Christensen OB, Bouwer LM, Braun A, Colette A, Déqué M, Georgievski G, Georgopoulou E, Gobiet A, Menut L, Nikulin G, Haensler A, Hempelmann N, Jones C, Keuler K, Kovats S, Kröner N, Kotlarski S, Kriegsmann A, Martin E, van Meijgaard E, Moseley C, Pfeifer S, Preuschmann S, Radermacher C, Radtke K, Rechid D, Rounsevell M, Samuelsson P, Somot S, Soussana J-F, Teichmann C, Valentini R, Vautard R, Weber B, Yiou P (2014) EURO-CORDEX: new high-resolution climate change projections for European impact research. Reg Environ Change 14:563–578

Knutti R, Sedláček J (2013) Robustness and uncertainties in the new CMIP5 climate model projections. Nature Climate Change 3: 369–373

Masui T, Matsumoto K, Hijioka Y, Kinoshita T, Nozawa T, Ishiwatari S, Kato E, Shukla PR, Yamagata Y, Kainuma M (2011) An emission pathway for stabilization at 6 Wm^{-2} radiative forcing. Climatic Change 109:59–76

Meinshausen M, Smith SJ, Calvin K, Daniel JS, Kainuma MLT, Lamarque J-F, Matsumoto K, Montzka SA, Raper S, Riahi K, Thonson A, Velders GJM, van Vuuren DPP (2011) The RCP greenhouse gas concentrations and their extensions from 1765 to 2300. Climatic Change 109:213–241

Moss R, Babiker M, Brinkman S, Calvo E, Carter T, Edmonds J, Elgizouli I, Emori S, Erda L, Hibbard KA, Jones R, Kainuma M, Kelleher J, Lamarque JF, Manning M, Matthews B, Meehl J, Meyer L, Mitchell J, Nakićenović N, O'Neill B, Pichs R, Riahi K, Rose S, Runci P, Stouffer R, van Vuuren D, Weyant J, Wilbanks T, van Ypersele JP, Zurek M (2008) Towards New Scenarios for Analysis of Emissions, Climate Change, Impacts, and Response Strategies. Intergovernmental Panel on Climate Change, Geneva.

Moss RH, Edmonds JA, Hibbard KA, Manning MR, Rose SK, van Vuuren DP, Carter TR, Emori S, Kainuma M, Kram T, Meehl GA, Mitchell JFB, Nakićenović N, Riahi K, Smith SJ, Stouffer RJ, Thomson AM, Weyant JP, Wilbanks TJ (2010) The next generation of scenarios for climate change research and assessment. Nature 463:747–756

Nakićenović N, Lempert RJ, Janetos A (2014) A framework for the development of new socio-economic scenarios for climate change research: Introductory essay. Climatic Change 122:351–361

O'Neill BC, Kriegler E, Riahi K, Ebi KL, Hallegatte S, Carter TR, Mathur R, van Vuuren D (2014) A new scenario framework for climate change research: the concept of shared socioeconomic pathways. Climatic Change 122:387–400

Pielke R Jr, Wigley T, Green C (2008) Dangerous assumptions. Nature 452:531–532

Quante M (2010) The Changing Climate - Past, Present, Future. In Habel J, Assmann T (eds), Relict Species – Phylogeography and Conservation Biology, pp 9–56. Springer Verlag

Raper S (2012) Climate modelling: IPCC gazes into the future. Nature Climate Change 2:232–233

Riahi K, Rao S, Krey V, Cho C, Chirkov V, Fischer G, Kindermann G, Nakićenović N, Rafai P (2011) RCP8.5—A scenario of comparatively high greenhouse gas emissions. Climatic Change 109:33–57

Rogelj J, Meinshausen M, Knutti R (2012) Global warming under old and new scenarios using IPCC climate sensitivity range estimates. Nature Climate Change 2:248–253

Thomson AM, Calvin KV, Smith SJ, Kyle GP, Volke A, Patel P, Delgado-Arias S, Bond-Lamberty B, Wise MA, Clarke LE, Edmonds JA (2011) RCP4.5: a pathway for stabilization of radiative forcing by 2100. Climatic Change 109:77–94

Van Vuuren DP, Edmonds J, Kainuma M, Riahi K, Thomson A, Hibbard K, Hurtt GC, Kram T, Key V, Lamarque J-F, Masui T, Meinshausen M, Nakićenović N, Smith S, Rose SK (2011a) The representative concentration pathways: an overview. Climatic Change 109:5–31

Van Vuuren DP, Stehfest E, den Elzen MGJ, Kram T, van Vliet J, Deetman S, Isaac M, Klein Goldewijk K, Hof A, Mendoza Beltran A, Oostenrijk R, van Ruijven B (2011b) RCP2.6: exploring the possibility to keep global mean temperature increase below 2 °C. Climatic Change 109:95–116

Van Vuuren DP, Carter TR (2014) Climate and socio-economic scenarios for climate change research and assessment: reconciling the new with the old. Climatic Change 122:415–429

Von Storch H (2008) Climate change scenarios – purpose and construction. In: von Storch H, Tol RSJ, Flöser G (eds) Environmental Crises – Science and Policy, pp 5–15. Springer

WGBU (2014) Climate Protection as a World Citizen Movement. Special Report, German Advisory Council on Global Change (WBGU), Berlin, Germany

Annex 5: Facts and Maps

Ingeborg Nöhren

Facts concerning the Greater North Sea region are presented in Table A5.1. Figure A5.1 is a physiogeographical map, Fig. A5.2 shows different sea areas of the North Sea. Figure A5.3 gives an overview of existing and prospective uses and nature conservation areas of the North Sea region.

Table A5.1 Facts concerning the Greater North Sea region

Variable	
Length north-south[a]	960 km
Width east-west[a]	580 km
Surface area	750,000 km^2
Volume	94,000 km^3
Average depth[a]	95 m
Maximum depth	700 m (Norwegian Trench)
Annual river input	296–354 km^3
Drainage area	850,000 km^2
Population in drainage area (2000)	184 million
Sea surface temperature amplitude of the yearly cycle	2–7 °C (NW to SE)
Annual mean sea surface temperature	9.5 °C
Mean net inflow from the Atlantic in the north[b]	~2 million m^3 s^{-1} (2.32 Sv)
Mean net outflow to the Atlantic in the north[b, c]	~2 million m^3 s^{-1} (2.33 Sv)
Mean net inflow from the Baltic Sea[b]	15,000 m^3 s^{-1} (0.015 Sv)
Mean net inflow from Dover Strait[b]	160,000 m^3 s^{-1} (0.16 Sv)
Salinity	34–35 psu (central North Sea)
Difference between high and low water	0–8 m

All numbers taken from OSPAR (2000) unless otherwise indicated. OSPAR (2000) Quality Status Report (2000), Region II: Greater North Sea. OSPAR Commission, London

[a]http://www.mumm.ac.be/EN/NorthSea/facts.php

[b]Winter NG, Johannessen JA (2006) North Sea Circulation: Atlantic inflow and its destination. J Geophys Res 111:C12018, doi:10.1029/2005JC003310

[c]Schrum C, Siegismund F (2001) Modellkonfiguration des Nordsee/Ostseemodells. 40 Jahre NCEP-Integration, Ber. Zentr. Meeres- u. Klimaforsch. Univ. Hamburg, Reihe B, Nr. 4

I. Nöhren (✉)
Institute of Coastal Research, Helmholtz-Zentrum Geesthacht, Geesthacht, Germany
e-mail: ingeborg.noehren@hzg.de

M. Quante and F. Colijn (eds.), *North Sea Region Climate Change Assessment*, Regional Climate Studies, DOI 10.1007/978-3-319-39745-0

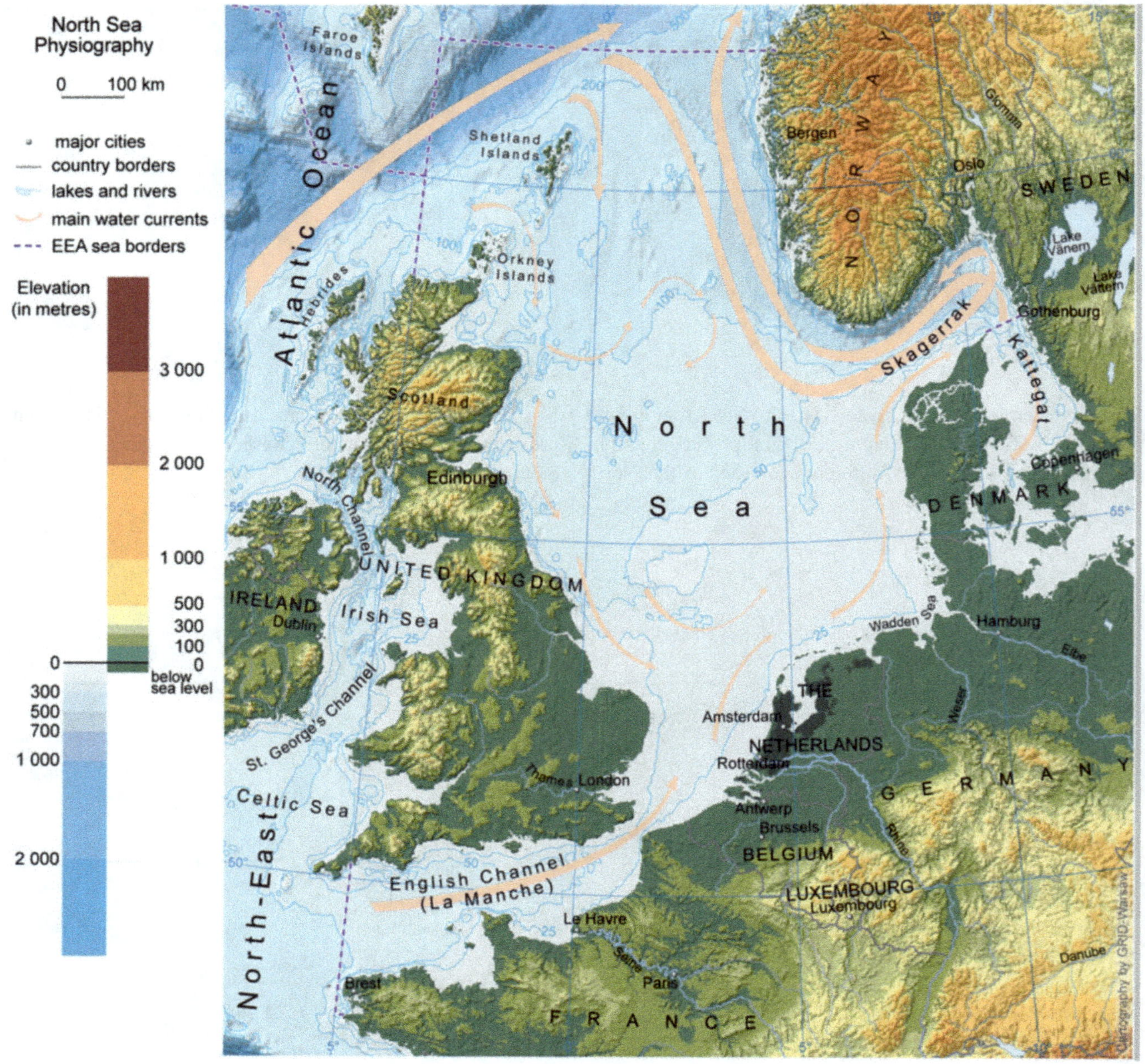

Fig. A5.1 Physiogeographical map of the Greater North Sea region (www.eea.europa.eu/data-and-maps/figures/north-sea-physiography-depth-distribution-and-main-currents)

Fig. A5.2 Different sea areas of the North Sea (Wikimedia Commons, licensed under Creative Commons Attribution-Share Alike 3.0 Unported)

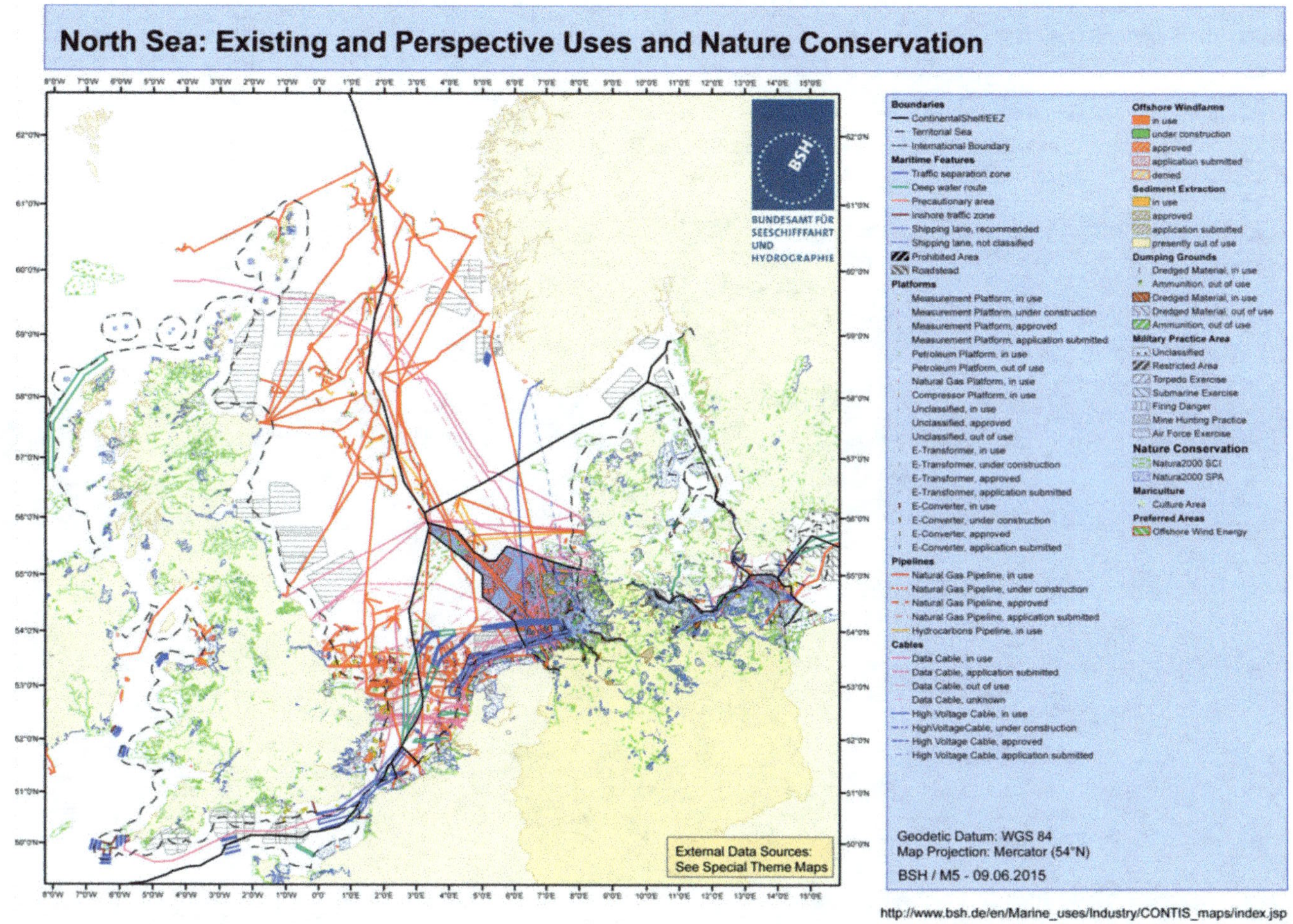

Fig. A5.3 Existing and prospective uses and nature conservation areas of the North Sea region (for higher resolution see: www.bsh.de/en/Marine_uses/Industry/CONTIS_maps/index.jsp, regularly updated)

www.ingramcontent.com/pod-product-compliance
Lightning Source LLC
LaVergne TN
LVHW062154200926
848484LV00007B/1913